Lecture Notes in Civil Engineering

Volume 162

Lecture Notes in Civil Engineering (LNCE) publishes the latest developments in Civil Engineering - quickly, informally and in top quality. Though original research reported in proceedings and post-proceedings represents the core of LNCE, edited volumes of exceptionally high quality and interest may also be considered for publication. Volumes published in LNCE embrace all aspects and subfields of, as well as new challenges in, Civil Engineering. Topics in the series include:

- Construction and Structural Mechanics
- Building Materials
- Concrete, Steel and Timber Structures
- Geotechnical Engineering
- Earthquake Engineering
- Coastal Engineering
- Ocean and Offshore Engineering; Ships and Floating Structures
- Hydraulics, Hydrology and Water Resources Engineering
- Environmental Engineering and Sustainability
- Structural Health and Monitoring
- Surveying and Geographical Information Systems
- Indoor Environments
- Transportation and Traffic
- Risk Analysis
- Safety and Security

To submit a proposal or request further information, please contact the appropriate Springer Editor:

- Pierpaolo Riva at pierpaolo.riva@springer.com (Europe and Americas);
- Swati Meherishi at swati.meherishi@springer.com (Asia - except China, and Australia, New Zealand);
- Wayne Hu at wayne.hu@springer.com (China).

All books in the series now indexed by Scopus and EI Compendex database!

More information about this series at http://www.springer.com/series/15087

Lakshman Nandagiri · M. C. Narasimhan ·
Shriram Marathe · S. V. Dinesh
Editors

Sustainability Trends and Challenges in Civil Engineering

Select Proceedings of CTCS 2020

Springer

Editors
Lakshman Nandagiri
Department of Applied Mechanics
and Hydraulics
National Institute of Technology Karnataka
Surathkal, India

Shriram Marathe
Department of Civil Engineering
N.M.A.M. Institute of Technology
Karkala, India

M. C. Narasimhan
Department of Civil Engineering
National Institute of Technology Karnataka
Surathkal, Karnataka, India

S. V. Dinesh
Department of Civil Engineering
Siddaganga Institute of Technology
Tumkur, India

ISSN 2366-2557 ISSN 2366-2565 (electronic)
Lecture Notes in Civil Engineering
ISBN 978-981-16-2825-2 ISBN 978-981-16-2826-9 (eBook)
https://doi.org/10.1007/978-981-16-2826-9

This Springer imprint is published by the registered company Springer Nature Singapore Pte Ltd.
The registered company address is: 152 Beach Road, #21-01/04 Gateway East, Singapore 189721, Singapore

Organizing Committee

Patron

Sri. N. Vinaya Hegde, Chancellor, Nitte-Deemed to be University; President, Nitte Education Trust, Mangaluru

Steering Committee

Sri. Vishal Hegde, Pro-Chancellor, Nitte-Deemed to be University
Dr. N. R. Shetty, Chancellor, Central University, Karnataka
Dr. Takamoto Itoh, Vice Dean, Mechanical Engineering, Ritsumeikan University, Japan
Dr. Harish Kumar Madhyastha, Professor, University of Miyazaki, Japan
Dr. Omid Ansary, Senior Associate Dean and Professor, Penn State University, USA
Dr. Samson Ojawo, Professor, Ladoke Akintola University of Technology, Nigeria
Dr. Shripad T. Revankar, Professor, Purdue University, USA

General Chair

Dr. Niranjan N. Chiplunkar, Principal

General Co-chairs

Dr. I. R. Mithanthaya, Vice Principal and Dean (Academics)
Dr. Srinivasa Rao B. R., Vice Principal and Controller of Examinations

Chief Conveners

Dr. Sudesh Bekal, Dean (R&D)
Dr. Muralidhara, P. G. Coordinator

Organizing Committee

Dr. G. Srinikethan, Director (R&D—Technical), Nitte-Deemed to be University
Dr. Subrahmanya Bhat, Professor and Dean (Student Welfare, ISO and IQAC Coordinator)
Dr. K. Rajesh Shetty, Dean (Admission and Alumni Affairs)
Dr. Srinivas Pai, Deputy Controller of Examinations
Dr. Srinath K. Shetty, Resident Engineer
Dr. Rashmi Hegde B., Head, Department of Humanities
Dr. Kumudakshi, Associate Professor and Head, Department of Mathematics
Dr. Shivaprasad Shetty M., Assistant Professor and Head, Department of Chemistry
Dr. Shobha R. Prabhu, Assistant Professor and Head, Department of Physics.

Programme Chair

Dr. Arun Kumar Bhat, Professor and Head, Department of Civil Engineering, NMAMIT Nitte

Organizing Secretary

Dr. Bhojaraja B.E., Assistant Professor, Department of Civil Engineering, NMAMIT Nitte
Dr. Shriram P. Marathe, Assistant Professor, Department of Civil Engineering, NMAMIT Nitte

Web Coordinators

Mr. Grynal D'Mello, Department of Mechanical Engineering
Mr. Rajeevan K. V., System Analyst

Conference Advisory Committee

Dr. Mukesh Kashyap, Senior Lecturer (Construction Management), School of Architecture, Design and the Built Environment, Nottingham Trent University, England
Dr. G. L. Sivakumar Babu, Professor, IISc, Bengaluru
Dr. Dharamveer Singh, Associate Professor, IIT Bombay
Dr. Vipul Prakash, Professor, IIT Roorkee
Dr. Amba Shetty, Professor and Head, Applied Mechanics and Hydraulics Department, NITK, Surathkal
Dr. K. Swaminathan, Professor and Head, Civil Engineering Department, NITK, Surathkal
Dr. H. N. Ramesh, Professor and Principal, UVCE, Bengaluru
Dr. Puttaraju, Professor and Principal, SJBIT, Bengaluru
Dr. G. P. Chandradhara, Professor and Head, CT & M, JSS S&T University, Mysuru
Dr. N. Suresh, Professor and Head, Civil Engineering and Director of BFRC, NIE, Mysuru

Technical Advisory Committee

Dr. Annappa, IEEE Mangalore, Sub-ection Chair
Dr. Lakshman Nandagiri, Professor HAG, Applied Mechanics and Hydraulics Department, NITK
Dr. M. C. Narasimhan, Professor, Department of Civil Engineering, NITK
Dr. A. U. Ravishankar, Professor, Department of Civil Engineering, NITK
Dr. R. Shivashankar, Professor, Department of Civil Engineering, NITK
Dr. S. V. Dinesh, Professor and Head, Department of Civil Engineering, SIT Tumakuru
Dr. Narayana Sabhahith, Joint Director, MAHE, Manipal
Dr. Balakrishna Rao, Professor, Department of Civil Engineering, MIT, Manipal
Dr. Purushotham Sarvade, Professor and Head, Department of Civil Engineering. MIT, Manipal
Dr. Kiran Shetty, Professor, Professor, Department of Civil Engineering, MIT, Manipal
Dr. Sandeep J. Nayak, Professor and Head, Department of Civil Engineering, SMVITM, Bantakal, Udupi District
Dr. Chandrashekhar, Professor and Head, Department of Civil Engineering, KVGCE, Sullia
Dr. Ganesh Mogaveer, Professor and Head, Department of Civil Engineering, MITE, Moodabidri
Dr. H. Ajith Hebbar, Professor and Head, Department of Civil Engineering, AIET, Moodabidri, Mangalore

Dr. Yajnheswaran, Professor and HOD, Department of Civil Engineering, SJEC, Mangaluru
Dr. Ramakrishna Hegade, Professor and HOD, Department of Civil Engineering, Srinivas University, Mangalore
Dr. Ananda V. R., Professor and Head, Department of Civil Engineering, VCET, Puttur

List of Reviewers

Dr. A. U. Ravi Shankar, NITK Surathkal
Dr. Anil Kumar, NMAMIT Nitte
Dr. Arun Kumar Bhat, NMAMIT Nitte
Dr. Nitendra Palankar, KLS Gogte Institute of Technology, Belagavi
Dr. N. C. Balaji, The National Institute of Engineering, Mysore
Dr. Ravi Kumar, University B. D. T. College of Engineering, Davangere
Dr. D. S. Prakash, VTU Regional Centre, Mysuru
Dr. Sanjith J., Adhichunchanagiri Institute of Technology
Dr. Ganesh Mogaveera, MITE, Moodabidri
Dr. Santhosh Malkapur, BEC Bagalkot
Dr. Rahavendra Holla, MIT Manipal
Dr. Mithun B. M., NMAMIT Nitte
Dr. P. C. Srinivas, Government Engineering College, Kushal Nagar
Dr. Samson Ojoawo, Ladoke Akintola University of Technology, Ogbomoso
Dr. Sanjeev Sangami, Jain College of Engineering Belagavi
Dr. Akshatha Shetty, Metasite Tandem Pty Ltd., Sydney Australia
Mr. Hemanth Kamplimath, Nirma University, Ahmedabad
Dr. Naveen G. M., Government Engineering College, Chamarajanagara
Dr. Pavan G. S., NITK Surathkal
Dr. Ramakrishna Hegde, Srinivas University, Mangalore
Dr. Durga Prashanth, RVCE Bangalore
Dr. Santhosh L. G., M. S. Ramaiah Institute of Technology, Bangalore
Dr. Chandrashekharappa Agasnalli, BMS Institute of Technology and Management Bangalore
Dr. Vivek S., JSS Academy of Technical Education, Bangalore
Dr. Sridhar Rajagopal, SJB Institute of technology
Dr. S. V. Dinesh, SIT Tumkur
Dr. Jagadish Vengala, Prasad V. Potluri Siddhartha Institute of technology, Vijayawada, Andhra Pradesh
Dr. G. Raghava, NMIT Bangalore
Dr. Naveen B. O., The National Institute of Engineering, Mysuru

Dr. Sangeetha D. M., AJIT Mangalore
Dr. Omar Alobaidi, Middle East College, Muscat-OMAN
Dr. Shivaprasad K. N., JSS Science and Technology University, Mysuru
Dr. Bashir Ahmed Mir, National Institute of Technology Srinagar, J&K
Dr. Shankar Sanni, Basaveshwar Engineering College, Bagalkot
Dr. Saubhagya Kumar Panigrahi, VSSUT BURLA
Dr. Goutham Sarang, VIT Chennai
Mr. Abhishek Mittal, CSIR-Central Road Research Institute
Dr. Santhosh B. S., JSS Academy of Technical Education, Bengaluru
Dr. Raghu Prasad P. S., JSS Science and Technology University
Mr. George Alex, College of Engineering and Management, Punnapra, Alleppey, Kerala
Mr. Sachin Kuckian, Middle East College, Muscat-OMAN
Dr. Radhakrishna, RVCE Bangalore
Mr. Joel Santhosh, RAKNOR
Dr. Narendra Samadhiya, Department of Civil Engineering, IIT Roorkee
Dr. Babloo Chaudhary, NITK Surathkal
Dr. Subrahmanya Bhat, NMAMIT Nitte
Dr. Aarti S. Bhatt, NMAMIT Nitte
Dr. Ayush Mittal, REC Ambedkar Nagar
Dr. Varghese George, NITK Surathkal
Dr. Jayaprakash M. C., MITE, Moodbidri
Dr. Basavaraju Manu, NITK Surathkal
Ms. Aradhana Mishra, IIT Roorkee
Mr. V. S. Sanjay Kumar, NATPAC
Dr. Kundan Meshram, Guru Ghasidas University, Bilaspur (C.G.)
Dr. Harish Kumar S., Jyothy Institute of Technology, Bangalore
Dr. B. J. Akshaya, ATME College of Engineering
Dr. Rahul Agrawal, MIT Aurangabad
Dr. Poornachandra Pandit, MIT Manipal
Mr. Himanshu Gaur, Bauhaus University Weimar
Ms. Shashikala S. Shankar, Moodlakatte Institute of Technology, Kundapur
Dr. Sujatha Unnikrishnan, Christ University, Bengaluru
Mr. Suresh Y. R., Jyothy Institute of Technology, Bangalore
Dr. Paresh Chandra Deka, NITK Surathkal
Dr. Amit Patil, Daulatrao Aher College of Engineering Karad
Dr. Katta Venkataramana, NITK Surathkal
Mr. Subrahmanya V. Bhat P., Srinivas University College of Engineering and Technology
Mr. Ranjith Anand, NMAMIT Nitte
Dr. Chandrashekhar A., K.V.G. College of Engineering, Sullia
Dr. Umeshchandra H. G., Alva's College of Engineering and Technology, Moodbidri
Ms. Shimna Manoharan, BMS Institute of Technology and Management
Dr. G. S. Dwarakish, NITK Surathkal
Dr. Radhakrishnan K., NMAMIT Nitte

Dr. H. Ramesh, NITK Surathkal
Ms. Payal Mehta, Pandit Deendayal Petroleum University
Mr. Mahesh Kumar C. L., NMIT Bangalore
Ms. Shwetha K. G., NMIT Bangalore
Dr. Yogeshwar Navandar, NIT Calicut
Dr. Muneera C. P., MES Engineering College, Kuttippuram
Mr. Sridhar H. N., Dayananda Sagar College of Engineering, Bangalore
Dr. Allamaprabhu Kamatagi, Global Academy of Technology, Bengaluru
Dr. Narasimhan Mattur C., NITK Surathkal
Mr. Likhit M. L., Vemana Institute of Technology
Dr. Namrata Jariwala, S. V. National Institute of Technology, Surat
Dr. Nityanand Singh Maurya, NIT Patna
Dr. Manisha Magdum, Sinhgad College of Engineering, Pune
Dr. Ramesh Bhat, NMAMIT Nitte
Dr. Chandre Gowda, Jyothy Institute of Technology, Bangalore
Dr. Prashanth Janardhan, NIT Silcher
Dr. Bharathi Ganesh, NMIT Bangalore
Dr. Shivakumar Nyamathi, UVCE Bangalore University
Dr. Suman Kundapura, AJIET Mangalore
Dr. Lakshman Nandagiri, NITK Surathkal
Dr. Udayakumar Gaddale, NMAMIT Nitte
Ms. Bhagya Shree, MIT Manipal
Mr. Prasanna P. Rao, Srinivas University College of Engineering and Technology
Mr. Janakaraj Murthy, NMAMIT Nitte
Dr. Harishchander A., Sri Krishna Arts and Science College, Coimbatore
Ms. Deepthishree S. A., Sahyadri College of Engineering and Management, Mangalore
Dr. Shriram Marathe, NMAMIT Nitte
Mr. Adithya Shenoy, MIT Manipal
Mr. Subrahmanya R. M., VCET Putturu
Dr. Durgaprasad J., NMIT Bangalore
Dr. Kumar Raju B. C., Sapthagiri College of Engineering, Bangalore
Dr. Sunandana Reddy Machireddy, Rajeev Gandhi Memorial College of Engineering and Technology, Nandyal, AP
Dr. Vinoth Srinivasan, NITK Surathkal

Preface

N.M.A.M. Institute of Technology, Nitte, Karnataka, India, organized the International Conference on Emerging Trends in Engineering (ICETE 2020) on 22 and 23 December 2020, which is the 10th international conference being organized since 2011. From the year 2019, in an effort to focus on the specific issues associated with various engineering disciplines, the idea of a multi-conference platform has been mooted.

ICETE 2020—a multi-conference platform—was a collection of several international conferences with the themes specific to various engineering streams. Besides, there was an opportunity for the students and research scholars of various branches of engineering and technology, and industrial professionals to present and discuss research papers.

Civil Engineering Trends and Challenges for Sustainability (CTCS 2020) was organized by the Department of Civil Engineering, N.M.A.M. Institute of Technology, Nitte, under the umbrella of ICETE 2020. CTCS 2020 was a platform for the exchange of knowledge from both individual and interdisciplinary researchers pertaining to Civil Engineering Challenges and Sustainability. This international conference aimed to bring together the researchers, scientists, engineers, scholars and students in an international forum for the dissemination of original research results in the domain areas of Civil Engineering.

This proceedings volume contains the full-length research papers, experience reports and empirical study plans. All of these submissions went through a rigorous peer-review process commensurate with their track. In all, 175 research papers were submitted, with each of them being reviewed by a minimum of two experts. Upon peer review, 90 papers were accepted for presentation in the conference. Presentations were reviewed and ranked by the track chairs and discussed with the industry and practice chairs in order to ensure suitable theme sessions were available. Subsequently, after quality and plagiarism checks, 68 papers were accepted (an acceptance rate of 39%) for the publication in the Lecture Notes in Civil Engineering (Springer).

Surathkal, India	Lakshman Nandagiri
Surathkal, India	M. C. Narasimhan
Nitte, India	Shriram Marathe
Tumkur, India	S. V. Dinesh

Contents

About the Editors

Dr. Lakshman Nandagiri is a Professor in the Department of Applied Mechanics & Hydraulics at the National Institute of Technology Karnataka, Surathkal. He has an academic career spanning over 34 years. He possesses a Bachelor's degree in Civil Engineering (University of Mysore, 1983), a Master's degree in Hydraulics (University of Mysore, 1986), and a Doctoral degree in Water Resources Engineering (IISc, 1994). His main areas of research interest are surface water hydrology, hydrological modeling using remote sensing/GIS, vadose zone hydrology, and irrigation & drainage engineering. He has to his credit over 80 technical papers in journals/conference proceedings and has guided 08 doctoral candidates and over 50 M.Tech./B.Tech. dissertations. He has completed 07 R&D projects and several consultancy assignments in the area of hydraulics and water resources engineering and remote sensing/GIS. Prof. Nandagiri has a keen interest in promoting research culture and ethical values amongst students and has delivered a large number of popular lectures on these topics. He is a reviewer for several reputed international and national journals and Associate Editor for the ISH Journal of Hydraulic Engineering (Taylor and Francis).

Dr. M. C. Narasimhan is a proud alumnus of University BDT College of Engineering, Davangere, India [B.E (Civil Engg) - 1982]. He obtained his Masters and a doctoral degree from Indian Institute of Technology Madras, India during the years 1985 and 1993, respectively. He started his teaching career at the National Institute of Technology Karnataka (NITK) [formally, Karnataka Regional Engineering College, KREC] as a lecturer in the Department of Applied Mechanics and Hydraulics in the year 1985 and was promoted as a professor in the Department of Civil Engineering in the year 2002. His research interests are in the areas of concrete technology, special concretes, structural behaviour of RCC, steel and composite constructions, and Numerical methods in structural analysis. He has completed a number of R&D and consultancy projects in his department. He has guided more than fifty postgraduate students for their dissertations and eight students for their doctoral thesis. He has a large number of research publications to his credit--- either published in leading journals or presented and included in proceedings of reputed international conferences. He has been a reviewer of papers submitted to leading international research journals like Construction and Building Materials, Cement and Concrete Composites, Thin-walled structures, Journal of Cleaner Production and Journal of Building Engineering, etc. He has chaired technical sessions at international conferences held at Curtin University, Miri, Malaysia, IIT-Delhi, BITS-Pilani, etc. He is an associated member in the sectional committee on special structures, Bureau of Indian Standards, New Delhi. He is also a life member of leading professional organizations like Institution of Engineers (India), Indian Cement Institute, Association of Consulting Civil Engineers, Indian Society for Technical Education, and Indian Society for Earthquake Technology.

Dr. Shriram Marathe is serving as an Assistant Professor in the Department of Civil Engineering at NMAM Institute of Technology NITTE. He completed his Ph.D. degree at VTU Belagavi in the area of Alkali Activated Concrete Pavements. He completed his M.Tech.[Transportation Engineering] from National Institute of Technology Karnataka, Surathkal, and Master Facilitator Degree (MFLHRD) in Human Resources Development from CLHRD, Mangalore. His areas of research interests include Alkali-activated Concrete applications to pavements, Stabilization of Sub-grade soil, Road Safety, Pavement Material characterization, and Pavement Design. He has guided several projects for B.E. and M.Tech. dissertations. He also successfully executed one funded research project on "Study and to Develop Cost-Effective and Green Masonry Block Using Industrial Waste Materials", funded by NITTE Education Trust. To date, he has published more than 50 research papers out of which 10 articles are published in scopus indexed journals (Three Q1 articles). Out of his research contribution 24 research articles are published in international journals, 06 technical papers in National journals, and also includes conference proceedings papers at 13 National/International conferences, and 5 papers under review. Further, he also filed a patent on his invention entitled, "Sustainable Pervious Alkali Activated Concrete Paver Block Pavement for Ground water Recharge" on 29-07-2020. He also served as conference secretary and successfully completed the CTCS-2020, International Conference held at NMAMIT Nitte during December 2020. Further, he is also serving as a potential reviewer for "International Journal of Pavement Research and Technology", "Construction and Building Materials", "Case Studies in Construction Materials", and "Cogent Environmental Science". Being a resource person, he had delivered several technical talks at various graduate schools. He is a life member of Indian Roads Congress (IRC) New Delhi, Indian Geotechnical Society (IGS) New Delhi, Indian Society for Technical Education (ISTE), Indian Society of Systems for Science and Engineering (ISSE), Technical Institute for Engineers (India), and Kannada Sahithya Parishadh, Karnataka. He worked as Departmental Co-ordinator

(Civil Engineering) for NAAC, NBA, IQAC, Industry Institute Interaction, ISTE, IE(I), and IOV related works.

Dr. S. V. Dinesh is working as Professor of Civil Engineering at Siddaganga Institute of Technology (SIT), Tumkur. He obtained his PhD degree from IISc, Bangalore, in 2003. He completed his postdoc from Yamaguchi University, Japan, under NPEE fellowship from MHRD, Govt. of India, during 2005–06. He is the recipient of the Institution of Engineers fellowship during BE program. He is a life member of Indian Geotechnical Society, Indian Society of Earthquake Technology, Indian Road Congress, Institution of Engineers, and associate Member of ASCE. His areas of interest in research are in geotechnical engineering, geotechnical earthquake engineering, and pavement geo-techniques. He developed research facilities in the area of geotechnical earthquake engineering and pavement geo-techniques. He has guided 7 Ph.D. students and more than 50 M.Tech. theses. He has 30 publications in reputed national and international journals. He has completed several funded projects from various funding bodies. He is the recipient of ISET V H Joshi award for the best journal paper in Soil Dynamics from the Indian Society for Earthquake Technology, Roorkee. Presently, he is an Executive Committee Member, ISET, Roorkee for the period 2019–2021. He has served as a reviewer for several prestigious journals published by Springer and Taylor and Francis. He is a geotechnical consultant and actively involved in geotechnical investigations for private as well as state and central government organizations in India. He also served as a member of BoS at many institutes and involved in the development of many courses at UG and PG levels. He advised many institutions with diploma, UG and PG degrees in the preparation of SAR for NBA accreditation. He is instrumental in developing a skill development centre at the Department of Civil Engineering, SIT to impart hands-on training in construction trades to UG students. He is involved in several outreach activities relevant to societal and technical areas.

Stability Analysis of Embankments on Soft Consolidating Layered Foundation Soil

P. Radhika Bhandary, A. Krishnamoorthy, and Asha U. Rao

Abstract Embankments are being constructed for highways, railways and abutments. Sometimes the foundation soil encountered is soft consolidating soil and can cause settlements due to embankment construction. The method of combining the method of finite elements and the genetic algorithm to evaluate the safety factor of the embankment built on a soft consolidating layered soil is discussed in this paper. The foundation soil has two layers with soil properties weaker than the embankment soil. Two numerical examples are considered for the analysis. In the finite element system, the soil is modeled using the Mohr–Coulomb nonlinear model and the least safety factor is calculated at different time intervals after construction until the consolidation process is completed using the genetic algorithm. The safety factor variant from the study helps to know when the embankment is suitable and safe to use.

Keywords Factor of safety · Consolidation · Embankment · Soft soil

1 Introduction

The stability of embankment is defined by the critical slip surface and the corresponding safety factor. The embankment can be evaluated using the two techniques, namely the method of limiting equilibrium and the method of finite elements. The finite element method has been used to analyze the embankment stability to find realistic stresses when compared to simplified techniques. In this process, non-homogeneous soil properties and nonlinear conditions of stress are easy to implement.

P. R. Bhandary (✉) · A. Krishnamoorthy · A. U. Rao
Civil Engineering Department, Manipal Institute of Technology, MAHE, Manipal, India
e-mail: radhika.st@manipal.edu

A. Krishnamoorthy
e-mail: moorthy.mit@manipal.edu

A. U. Rao
e-mail: asha.prabhu@manipal.edu

L. Nandagiri et al. (eds.), *Sustainability Trends and Challenges in Civil Engineering*,
Lecture Notes in Civil Engineering 162, https://doi.org/10.1007/978-981-16-2826-9_1

The embankment stability depends on the time needed for the complete dissipation of pore pressure when built on soft soils. In the stability analysis of these embankments, consolidation plays an important role. The critical slip surface and the corresponding safety factor define the stability of the embankment. Compared to simplified techniques, the finite element method is used to analyze the stabilization of the embankment to find realistic stresses. The enforcement of non-homogeneous soil properties and nonlinear conditions of stress is simple in this method. During consolidation, the effective stresses change due to dissipation of pore pressure and deformation. This process can be implemented in finite element method by coupling the pore pressure and deformation in soil using the theory proposed by Zienkiewicz [1]. The finite element method has been used to find embankments deformation and behavior has been analyzed and compared to the field measurements [2–4]. As a large number of possible sliding surfaces must be taken into account to determine the critical slip surface, the stability analysis of the embankment is an optimization problem. Compared to the various optimization techniques for finding the global minimum, the genetic algorithm has been found to be more effective in defining the optimum solution for several complex problems [5]. Genetic algorithms can also deliver outcomes more efficiently and can be used to evaluate geotechnical issues involving failures and more decision variables [6]. In the proposed approach the embankment stability study is carried out in terms of safety factor from the construction completion to the end of consolidation. The realistic stresses from the finite element method are obtained for different time intervals and the safety factor is determined by the genetic algorithm. The soil behavior is modified to nonlinear, by adopting the elastic-perfectly plastic satisfying Mohr–Coulomb yield criteria considering the associated flow rule. The present study demonstrates the competence of the finite element to calculate realistic stresses and the combination approach with genetic algorithm in order to find critical slip surface when the foundation soil consolidates softly and has two layers.

2 Methodology

2.1 Steps in Analysis

There are mainly two steps in analysis:

(1) Using finite element procedure for stress determination considering nonlinear soil behavior.
(2) Using the genetic algorithm to define the critical slip surface and the required safety factor.

The discretization is done in order to find the stresses that are realistic using four noded isoparametric plane strain elements with two translational degrees of freedom at each node. The finite element method to obtain stresses at various time intervals developed by Krishnamoorthy and Sanjay [7] is used in the proposed method with modifications in nonlinear model. The soil behavior, taking into account the associated flow law, is modeled as the elastic-perfectly plastic satisfying Mohr–Coulomb yield criteria. In the finite element process, the staged construction of the embankment and the formulation of the time-dependent material behavior are applied.

2.2 Finite Element Method for Stress Determination

The finite element approach is still the most commonly used and potentially the most robust technique in geotechnical engineering. Nonlinear material behavior that can be implemented to analyze the entire structure are the key advantages of using Finite element analysis.

In this technique, even the staged embankment construction and the construction sequences can be easily modeled. In this method, the time-dependent material behavior can also be implemented. Finite element method has been widely used for analysis of consolidating soil for deformation and dissipation of pore pressure.

In the consolidating soil, the displacement u and excess pore pressure p have been implemented in the finite element method by relating them to the displacement vector {u} and the pore pressure vector $\{p\}$ at nodal points using shape functions which defines the displacement and pore pressure distribution of the soil element [7]. Eq. 1 gives the relation between the pore pressure and displacement with the pore pressure vector and displacement vector at each node.

$$p = [Nf]\{p\} \text{ and } u = [Ns]\{u\} \tag{1}$$

Ns is shape function that defines the displacement and *Nf* is shape function defining the pore pressure distribution of the soil element.

The coupled consolidation equation suggested in matrix form by Zienkiewicz [1] is given in Eq. 2, which is used to combine the displacement and pore pressure.

$$\begin{bmatrix} k_s & L \\ 0 & H \end{bmatrix} \begin{Bmatrix} u \\ p \end{Bmatrix} + \begin{bmatrix} 0 & 0 \\ L^T & 0 \end{bmatrix} \begin{Bmatrix} \dot{u} \\ \dot{p} \end{Bmatrix} = \begin{Bmatrix} f \\ 0 \end{Bmatrix} \tag{2}$$

where k_s is the matrix of soil stiffness, *H* is fluid conductivity matrix, *f* is the load vector for externally applied load and the coupling matrix is *L*.

The incremental form is given in Eq. 3 to apply the incremental loads so that the staged construction is incorporated in finite element method.

$$\begin{bmatrix} k_s & L \\ 0 & H \end{bmatrix} \begin{Bmatrix} \Delta u \\ \Delta p \end{Bmatrix} + \begin{bmatrix} 0 & 0 \\ L^T & 0 \end{bmatrix} \begin{Bmatrix} \Delta\dot{u} \\ \Delta\dot{p} \end{Bmatrix} = \begin{Bmatrix} \Delta f \\ 0 \end{Bmatrix} \tag{3}$$

where Δu is the changes in displacement and Δp is excess pore pressure which changes for load increment of Δf.

Using the elastic-perfectly plastic Mohr–Coulomb model, the nonlinear soil behavior is modeled. To model the behavior of the soil, this model requires only two basic parameters, namely cohesion and the angle of internal friction. The stresses are achieved from the completion of the construction of the embankment to the end of the foundation soil consolidation over different time intervals.

2.3 *Genetic Algorithm*

In order to find the safety factor of the slip surfaces created by the genetic algorithm, the stresses obtained from the finite element method are used. John Holland developed Genetic Algorithms in 1975 to study self adaptivity in natural system processes. The genetic algorithm imitates the principles of biological evolution in order to solve optimization problems. Search algorithms are genetic algorithms that emphasizes on mechanisms of natural selection process and natural genetics [5]. Genetic algorithms are perfect for solving complex problems with optimization and are thus suitable for applications requiring techniques for adaptive problem-solving.

A chromosome is known as a solution generated by a genetic algorithm, whereas a population is known as a chromosome array. A chromosome is composed of genes and, depending on the problem, the value of the chromosome may be either symbols, numerical, binary or characters. To evaluate the suitability of solving the problem generated by genetic algorithms, these chromosomes will be subjected to a process called fitness function. By a mechanism called crossover, some population chromosomes can match, thereby producing new chromosomes which are known as offspring. Their gene composition is the combination of their parent's genes. Some of the chromosomes can even undergo mutations in their genes within a generation. The value of the crossover rate and mutation rate governs the number of chromosomes undergoing crossover and mutation. Chromosomes in the population to be preserved for the next generation will be chosen on the basis of Darwinian theory of evolution, and in the next generation, the chromosome with a higher fitness value will be more likely to be selected again. The chromosome value should converge to a certain value after many generations, which is the best solution to the problem.

The problem of slope stability is formulated as a constrained minimization problem to evaluate the critical slip surface and its lowest safety factor. The critical slip surface profile is created as part of the minimization process and is not predefined. The present study uses a computer program implementing genetic algorithms for optimization developed by Deb [8]. The objective function is coded in this program to locate the critical slip surface for the present analysis. The parameters of the genetic algorithm taken for the analysis are: population size 12–20, generation number 2000, 0.4–0.9 crossover rate, and 0.001–0.01 mutation rate. The safety factor is the objective function that needs to be decreased in the stability analysis of the embankment, and the slip surface shape represents the design variables.

2.4 Factor of Safety

The stresses obtained from the finite element method and the slip surface produced by the genetic algorithm are used for the determination of the safety factor. The safety factor is the ratio of the shear strength to the shear stresses developed along the considered slip surface. For each of the slip surfaces generated by genetic algorithms, the factor of safety is determined. The slip surface created is divided into '*n*' number of segments with increasing length ΔL to determine safety factor. Then the aggregate safety factor is determined using the following equation.

$$\text{Safety factor} = \frac{\Sigma \tau_f \Delta L_i}{\Sigma \tau_i \Delta L_i}$$

where τ_i—shear stress that is mobilized and τ_f—shear strength of material. ΔL_i—*i*th segment length.

3 Numerical Examples

When the embankment is constructed on consolidating soil, the effective stresses change with the time. The excess pore pressure dissipates and the soil deforms thus causing a change in the slope geometry. In the present analysis, for two numerical cases, the minimum safety factor for the embankment on consolidating soil is found for different time periods from completion of construction to the end of consolidation. For various time periods, the safety factor is determined from 10 to 6200 days. The numerical examples considered are embankments constructed on foundation soil with two layers.

3.1 Example 1

The embankment details of example 1 is shown in Fig. 1. The embankment is constructed on two-layered foundation soil of thickness 4.5 m each. The foundation soil below the embankment in layer 1 is softer than the soil in layer 2 and the embankment soil. The soil in layer 2 is stiffer than the soil in layer 1. The soil is stiffer as the depth increases. Poisson's ratio and modulus of elasticity of layer 1 considered are 0.2 and 5000 kPa respectively and that of embankment are taken as 0.3 and 30,000 kPa respectively. However, Poisson's ratio and modulus of elasticity of layer 2 considered are 0.3 and 10,000 kPa respectively. The construction of embankment is completed in 10 days. The stability analysis is carried out and, as shown in Fig. 1, the critical slip surface for the slope is calculated using the proposed method. The

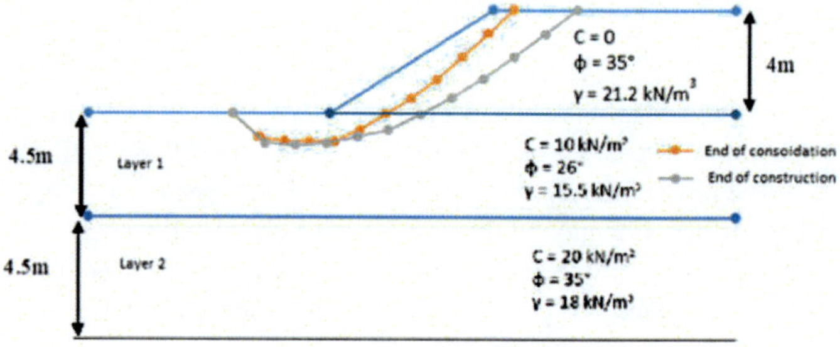

Fig. 1 Numerical example 1 details and critical slip surface for end of construction and end of consolidation

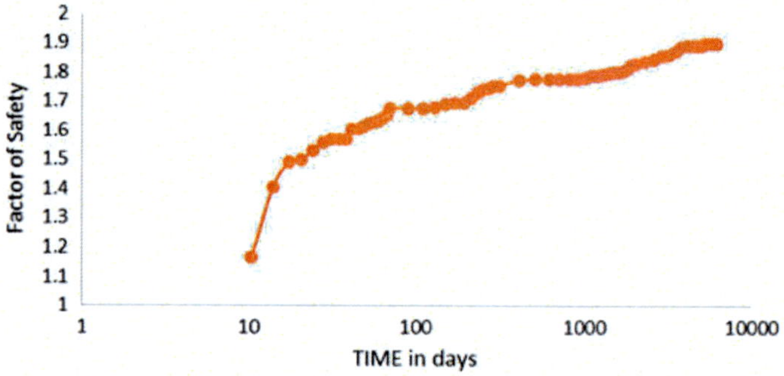

Fig. 2 Variation of factor of safety for example 1 from end of construction to end of consolidation

safety factor is 1.17 for the embankment at the end of construction and 1.89 for the end of consolidation.

The stability of the embankment with time is observed to be increased and the factor of safety variation with respect to the days is as shown in Fig. 2. The safety factor rises up to 100 days at a faster pace and then confirms at around 3810 days to 1.89. After this time interval, the variation in safety factor is insignificant. The variation of safety factor with time provides information about the time when deformation due to consolidation process is negligible.

3.2 *Example 2*

The numerical example is shown in Fig. 3. The embankment is constructed on two-layered foundation soil. The foundation soil below the embankment that is layer 1 is softer than the soil in embankment. The soil in layer 2 is softer than the soil in layer 1. Poisson's ratio and the modulus of elasticity of layer 1 considered are 0.3 and 3000 kPa respectively and that of embankment are taken as 0.3 and 10,000 kPa respectively. However, Poisson's ratio and the modulus of elasticity of layer 2 considered

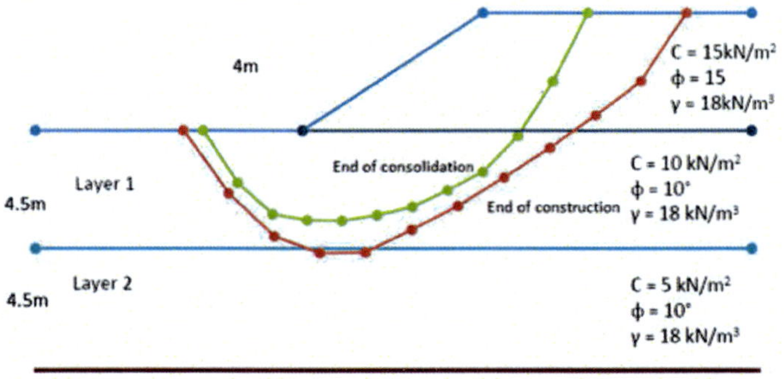

Fig. 3 Example 2 details along with critical slip surface for end of construction and end of consolidation

are 0.35 and 2000 kPa respectively. The construction of embankment is completed in 10 days. Using the proposed approach, the critical slip surface for the slope is calculated and is as shown in Fig. 3.

The critical slip surface passes at the bottommost of layer 1 of foundation. The factor of safety for the slope at the end of construction is obtained as 1.19 and end of consolidation is obtained as 2.14. The embankment stability is improved over time and the safety factor varies with respect to the days are as shown in Fig. 4. The factor of safety increases at faster pace up to 400 days and then confirms to 2.14 at around 4710 days.

When the foundation soil progresses toward the end of consolidation, the safety factor of the embankment tends to be constant.

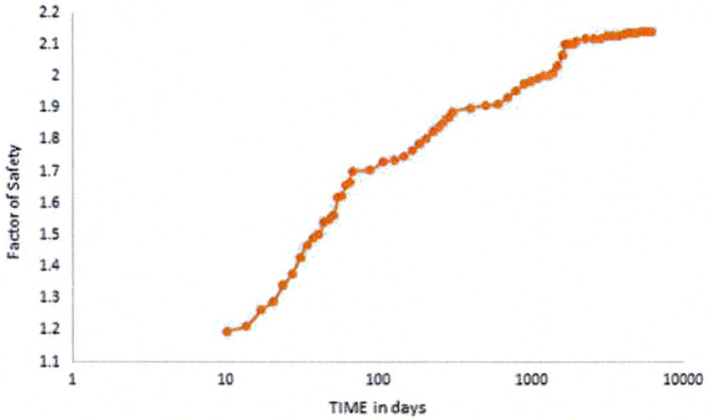

Fig. 4 Variation of factor of safety for example 2 from end of construction to end of consolidation

4 Conclusion

To perform the stability analysis of the embankment built on soft consolidating soil, a method that combines finite element and genetic algorithm is developed. The soil is modeled as nonlinear and the consolidation analysis is implemented in finite element method. The method is appropriate and provides safety factor from completion of construction till the consolidation process ends for different time intervals. The study shows that the method is applicable to derive the time for completion of consolidation and also to check if the embankment is safe during all stages starting from completion of construction till completion of consolidation.

References

1. Zienkiewicz OC (1977) The finite element method, 3rd edn. McGraw–Hill Book Company, UK
2. Muthing N, Zhao C, Hölter R, Schanz T (2018) Settlement prediction for an embankment on soft clay. Comput Geotech 93:87–103
3. Lundström K, Dehlbom B (2019) Soil property changes below existing embankments. In: Proceedings of the XVII ECSMGE-2019, pp 1–8
4. El-Gendy MMS, Fayed AL, El-Mossallamy YM (2020) Behavior of embankments constructed on soft soil deposits reinforced with rigid inclusions. Int J Eng Adv Technol 9(4):2014–2020
5. Goldberg DE (1989) Genetic algorithms in search, optimization and machine learning. Addison-Wesley, Reading, Mass
6. Cui LD, Sheng (2005) Genetic algorithms in probabilistic finite element analysis of geotechnical problems. Comput Geotech 32(8):555–563
7. Krishnamoorthy A, Rao SS (2016) Effects of vertical drains on settlement and load carrying capacity of a footing on soft soil. Int J Geotech Eng 10(1):57–65
8. Deb K (2009) Optimization for engineering design algorithms and examples. PHI Learning Private Limited, New Delhi

Curvature Ductility of Reinforced Masonry Walls and Reinforced Concrete Walls

Jacob Alex Kollerathu

Abstract Research conducted in this work proposes an equation to evaluate and compares the curvature ductility of reinforced masonry (RM) and reinforced concrete (RC) walls. The curvature ductilities are measured at varying levels of axial stresses for walls for aspect ratio (*l*/*h*) of 0.5, 1.0 and 1.5. The percentage of reinforcement is increased from 0.25% (minimum reinforcement for RC walls as per IS-13920) to 1.00%. The curvature ductilities are evaluated by plotting flexural strength (*M*) versus curvature (ϕ) for the walls. The stress–strain curves of masonry, concrete and reinforcing steel are all adopted from existing literature. The compressive strength of masonry and concrete has been chosen as 10 MPa and 25 MPa, respectively. The yield strength of the steel is fixed as 415 MPa. The height and thickness of the wall are 3000 and 230 mm, respectively, and the length of the wall is varied to obtain different aspect ratios. Results obtained from this paper imply due to increase curvature ductility, RM walls provide a better alternative for the construction of structural walls compared to RC walls in regions of significant seismicity.

Keywords Reinforced masonry walls · Reinforced concrete walls · Curvature ductility

1 Introduction

Most household buildings in rural India comprise of masonry walls as the load bearing elements with no reinforcement/structural detailing to enhance their seismic performance. Such structures are termed as unreinforced masonry (URM) structures, and their poor seismic performance has led to it being banned. In an earthquake, depending on their aspect ratios and axial loads levels masonry walls may form any of the following in-plane mechanisms: they develop diagonal cracks along their length, may slide off a plane of weakness and undergo alternating flexural cracking and crushing leading to crushing of masonry.

J. A. Kollerathu (✉)
Civil Engineering Department, CHRIST (Deemed to be University), Bengaluru, Karnataka, India
e-mail: jacob.alex@christuniversity.in

L. Nandagiri et al. (eds.), *Sustainability Trends and Challenges in Civil Engineering*, Lecture Notes in Civil Engineering 162, https://doi.org/10.1007/978-981-16-2826-9_2

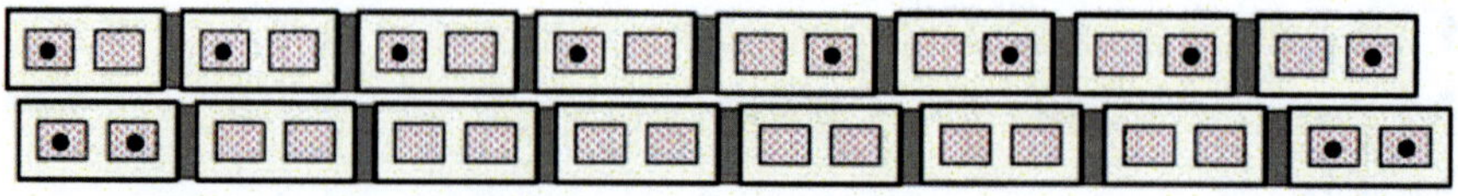

Fig. 1 Detailing of longitudinal steel in reinforced walls

A variant of masonry construction that introduces vertical and horizontal tensile resistant elements in-order to enhance its flexural and shear capacity [1] is termed as reinforced masonry (RM) structures. Bricks used in RM walls should have a *compressive* strength greater than 10 MPa, a minimum density of 19 kN/m^3 and needs to have low absorption rate of moisture [1]. RM structural walls constitute part of the main-load resisting system in structures, located in regions of moderate to high seismicity. Compared to URM walls, RM walls are stronger and efficiently resist earthquake shaking through nonlinear and inelastic actions and energy dissipation. Proper design and detailing of vertical and horizontal reinforcements help in achieving the required strength, ductility and energy dissipation. Figure 1 indicates that the two ways of detailing the vertical reinforcement in reinforced walls are [2]:

(a) Equal distribution of reinforcement along the wall length and
(b) Reinforcement at the two extreme ends of the wall.

Even though the in-plane flexural capacity of walls with the two distributions of steel are not too different, the former often results in marginally greater shear strength and improved shear behaviour in terms of better displacement and ductility, as it resists the propagation of diagonal shear cracks and horizontal shear sliding in the wall. The flexural strength (M) and curvature ductility (μ) of reinforced wall sections can be calculated from their axial load (P)-moment interaction (M) curves and nonlinear moment–curvature (M-Φ) curves. The M-Φ response curve developed should represent the effective (cracked) flexural rigidity, flexural strength and curvature ductility of the section.

In the ensuing sections, the curvature ductilities of reinforced concrete walls and reinforced masonry walls compared for walls with different aspect ratios and subject to various levels of axial loads. Even though RC walls are widely adopted to construct structural members, the lower self-weight of masonry compared to concrete will prove beneficial by providing higher ductilities. Results obtained from this paper may initiate further research in this domain that will see the use of reinforced masonry for structural. The curvature ductilities are measured at varying levels of axial stresses for walls for aspect ratio (l/h) of 0.5, 1.0 and 1.5. The percentage of reinforcement is increased from 0.25% (minimum reinforcement for RC walls as per [3]) to 1.00%.

2 Material Constitutive Laws Used

Due to its more heterogeneous nature, compared to the numerical modelling of reinforced concrete, the numerical modelling of reinforced masonry is complex and challenging. RC structural walls are typically modelled as *midpier frame elements* [4, 5], wherein nonlinear regions are defined using idealised bi-linear or tri-linear M-φ curves.

Masonry is a material characterised by anisotropic nature and heterogeneity, and modelling it is a challenge. Whilst one method of modelling masonry involves, explicitly modelling the brick unit, the mortar and the brick mortar interface, the other extremity involves modelling masonry as a homogenised material. The former modelling approach is called the micro-modelling approach, and the latter is termed as macro-modelling. The nonlinearity in the micro-modelling and macro-modelling approaches is defined at the material level in terms of constitutive relations in compression, tension and shear of the constitutive materials such as units, mortar, vertical and horizontal reinforcement and grout which makes the modelling of RM walls challenging [6].

Micro-modelling approach, in which all the components and their interaction are modelled separately, is useful for more detailed analysis and understanding of all possible behavioural and failure modes [7–10]. In this method, all the constituent elements of masonry, i.e. unit, mortar and unit-to-mortar interface and reinforcement layers (fibre mesh, mortar, fibre-to-mortar interface) are modelled separately. Several material parameters and suitable bond-slip constitutive relationships for the interfaces are necessary for a reliable simulation. This approach is time consuming (even for small scale models) and requires a fine FE mesh size to resolve the mesh dependency of the interface elements, and the analysis is computationally demanding and expensive. On the other hand, more simplified modelling approaches include idealising masonry to behave as a homogenous material with its behaviour being an average representation of the unit, mortar and unit-mortar interfaces. In this approach, also called macro-modelling approach, the reinforcements are modelled as continuum elements, and nonlinearities are homogenised over the elements [11]. Nonlinear regions in the elements are defined based on critical states of damage such as cracking of masonry, yielding of reinforcement bars in tension and compression failure of masonry. Existing research such as [12] indicates that for masonry structures with seismic resistant features, one could adopt the macro-modelling approach to satisfactorily model and analyse masonry structures.

Based on experimental observations [13], the stress–strain behaviour of masonry follows a parabolic variation with its strength (σ'_m) drooping beyond a level of strain. Strain limit of 0.0035 corresponding to f'_m used is from literature [13], for masonry elements having bricks, less deformable than the mortar. The equation for the compressive behaviour of masonry is given by:

$$\sigma_m = \sigma_m^{'}\left(2\left(\frac{\varepsilon}{\varepsilon_{\mathrm{mu}}}\right) - \left(\frac{\varepsilon}{\varepsilon_{\mathrm{mu}}}\right)^2\right) \tag{1}$$

As beyond the value of ε_{mu}, the value of stress and subsequently moment capacity decreases, the stress–strain relation only up to peak stress in considered.

The characteristic stress–strain curve of concrete defined under flexural compressive normal strain is parabolic up to a strain of 0.002 with maximum compressive stress $0.67\sigma_{ck}$, and constant thereafter up to a strain of 0.0035 and is given below.

$$\sigma_{\text{c}} = \left(\begin{cases} 0.67\sigma_{\text{ck}}\left(\left(\frac{\varepsilon}{\varepsilon_{\text{cu}}}\right) - 0.33\left(\frac{\varepsilon}{\varepsilon_{\text{cu}}}\right)^2\right), 0 < \varepsilon < 0.002 \\ 0.67\sigma_{\text{ck}}, 0.002 < \varepsilon < 0.0035 \end{cases} \right)$$

The characteristic steels bars curve of reinforcing bars under flexural normal strain (both tensile and compressive) has three parts: It is linear up to 80% of its yield stress f_y with an elastic modulus of 200 GPa, nonlinear part up to f_y at a non-proportional elongation equal to 0.002 of gauge length, and the third straight part at constant stress of f_y.

3 Flexural Strength and Curvature Ductlity of Sections

Flexural strength and curvature ductility of reinforced wall sections can be determined from their nonlinear moment–curvature (M-φ) curves. Typically, curvature φ of a section is defined as the ratio of the strain at the highly compressed edge to the depth of neutral axis, whilst the ratio between ultimate curvature (φ_{u}) to yield curvature (φ_{y}) is termed as curvature ductility of the section. Curvature ductility is the general measure of ductile response of a structural element, and it significantly depends on the ultimate compressive strain capacity and compressive strength of masonry, yield strength of reinforcement bars, percentage of tension and compression reinforcements, and level of axial load acting on the element.

The axial load and the flexural capacity of RM and RC walls can be estimated by considering the condition of equilibrium of forces, displacement/strain compatibility conditions and constitutive (stress–strain) relations of the respective materials, given in equations below in (3) and (4) for masonry and (5) and (6) for concrete, respectively.

Force-equilibrium equations in masonry walls

$$\Sigma(\sigma_{\text{sci}} - \sigma_{\text{m}})A_{\text{sci}} + \sigma_{\text{c}} t x_{\text{u}} - \Sigma\sigma_{\text{sti}} A_{\text{sti}} = P \tag{3}$$

Compatibility conditions of masonry walls

$$\frac{\varepsilon_{\text{m}}}{x_{\text{u}}} = \frac{\varepsilon_{\text{sti}}}{L' - x_{\text{u}}} \tag{4}$$

Force-equilibrium equations in concrete walls

$$\Sigma(\sigma_{sci} - \sigma_m)A_{sci} + \sigma_c t x_u - \Sigma\sigma_{sti}A_{sti} = P \quad (5)$$

Compatibility conditions of concrete walls

$$\frac{\varepsilon_m}{x_u} = \frac{\varepsilon_{sti}}{L' - x_u} \quad (6)$$

where

P: Axial load demand on walls.
σ_{sci}: Stress in reinforcing bars under compression.
σ_c: Compressive stress of concrete.
σ_m: Compressive stress of masonry.
σ_{sti}: Stress in reinforcing bars under tension.
A_{sci}: Area of reinforcing bars in compression.
t: thickness of the wall.
L': Effective length of the wall (Total length minus cover).
ε_{sti}: Tensile strain in the extreme layer of reinforcement.
x_u: depth of neutral axis.

The curvature φ for any given distribution of strain satisfying equilibrium Eq. (5) of a reinforced masonry and reinforced concrete wall section (representation in Fig. 2) can be written as in Eqs. (7) and (8):

$$\varphi = \frac{\varepsilon_m}{x_u} = \frac{\varepsilon_{st}}{L - x_u} \quad (7)$$

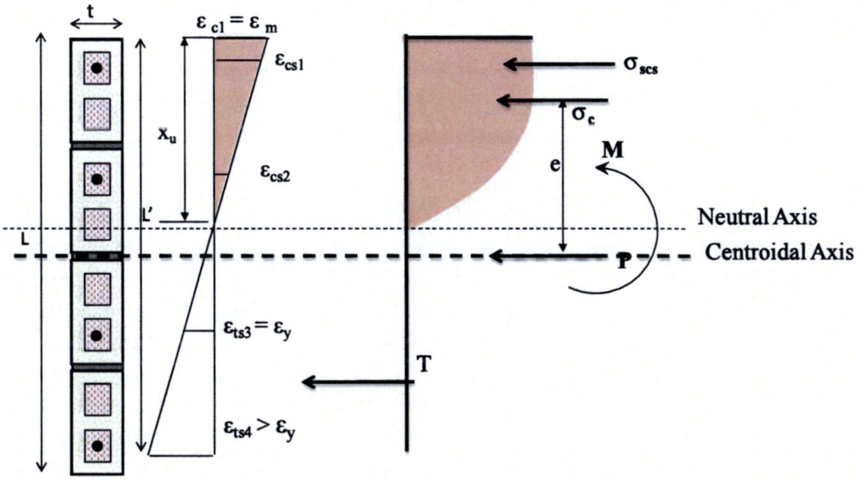

Fig. 2 Typical stress and strain distribution for reinforced wall sections with salient stress, strain distributions and geometric features

$$\varphi = \frac{\varepsilon_m}{x_u} = \frac{\varepsilon_{st}}{L - x_u} \tag{8}$$

Finally, the moment of the section can be determined by considering the moment of tensile and compressive forces about the neutral axis.

4 Idealised Moment–Curvature Relations

Idealised multi-linear M-φ curve of reinforced masonry and concrete wall sections are developed using the following points that correspond to limit states of strain in masonry, concrete or reinforcing steel such as:

- Tensile cracking of masonry/concrete
- Yielding of steel
- Compressive crushing of masonry/concrete.

Additionally, there are two regions in an axial load-flexural strength (P-M) interaction of reinforced wall sections, i.e. compression failure region, above balanced failure point, at which, masonry/concrete and steel reach corresponding limiting ultimate compressive and tensile strains simultaneously), and tensile failure region, below balanced failure point. The limit states of cracking and yielding of tension steel are not mobilised at all levels of axial load, especially above 40 percent of axial load capacity. Hence, simple analytical methods are proposed using the aforementioned limit states to develop idealised multi-linear M-φ curves of reinforced masonry and concrete wall sections subjected of (a) Increasing compressive axial loads of to represent walls in lower stories and (b) Zero axial load (representative of walls in the higher stories).

4.1 Reinforced Wall Sections Subject to Increasing Axial Compressive Load

In the ensuing section, closed form solutions are developed to determine the curvature ductilities of RC and RM walls of aspect ratios (l/h) 0.5, 1.0 and 1.5 for varying axial loads and differing percentages of reinforcements. The axial load ratios are described in terms of the ultimate load carrying capacity of the wall (P_u), and the moment–curvature ductility at axial load ratios of $0.4P_u$, $0.2P_u$, $0.1P_u$ and $0.05P_u$ is examined. Depending on the aspect ratio of the wall and the percentage of reinforcement in the wall, each of the aforementioned axial load levels may be greater than the axial load at balance point ($P > P_{bal.}$), may be equal to axial load at balance point ($P = P_{bal.}$) or may be lesser than axial load at balance point ($P < P_{bal.}$). At very high axial load levels, 0.6–0.8 P_u the only limit state that is activated is crushing of either concrete or masonry. At moderate axial load levels, such as 0.3–0.5 P_u, tensile cracking of the section is also observed. However, none of the reinforcing bars starts to yield. As a result of this at the aforementioned values of axial loads, low levels of ductility (< 2) are observed. At axial loads lesser than the $P_{bal.}$, the limits states that are activated include tensile cracking of the section, yielding of extreme layer and in some cases subsequent layers of reinforcements and compressive crushing of the section. In such cases, the sections exhibit larger ductilities.

The following sections describe as to how can one can arrive at closed form solutions to determine the curvature ductilities of reinforced masonry and concrete walls at axial loads greater than, equal to and lower than the balance axial load and subject to zero axial load. The curvature ductilities of the RM and RC walls are also compared.

4.2 Axial Load P Greater Than P_{bal}

At an axial load greater than P_{bal}, the limits states of strain activated in the walls are crushing of masonry/concrete and in some cases tensile cracking. Hence, three points are required to obtain the idealised *M-Φ* curve that correspond to limit states in masonry in cracking (point 1) and crushing (point 3). If the limit state due to cracking is not activated, then the first point is identified at a small strain increment $\Delta\varepsilon$ of 0.00005 and 0.0001 for masonry and concrete, respectively, at the highly compressed edge. The idealised bi-linear curve obtained from Eq. 9a–c is compared to the nonlinear response of wall and existing bi-linear idealisation based on equivalence of area under curve (see Fig. 3). From the figure, it is observed that the proposed model captures the initial stiffness, flexural strength, ultimate displacement and the curvature of the sections.

Fig. 3 Moment curvature interaction of a wall section with $P > P_{bal}$

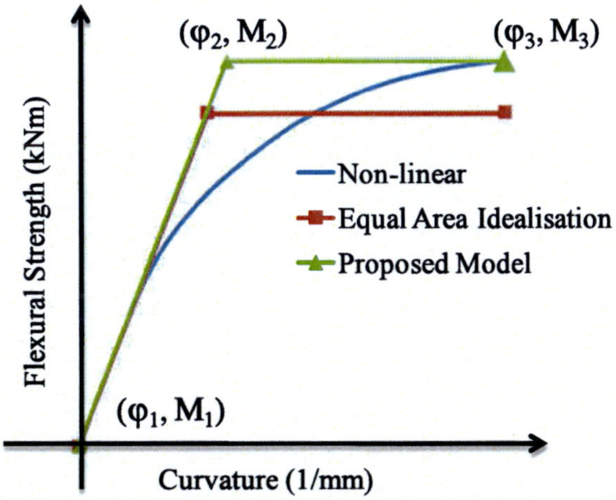

$$\Phi_2 = \frac{M_3}{M_1}\Phi_1 \tag{9a}$$

$$M_2 = M_3 \tag{9b}$$

And curvature ductility is

$$\mu_\phi = \frac{\phi_3}{\phi_2} \tag{9c}$$

4.3 *Axial Load P Lesser Than P_{bal}*

The origin of the idealised curve of reinforced wall section corresponds when there is uniform strain variation across depth *of* the section. Point 1 corresponds to the strain at which masonry/concrete cracks. Since masonry and concrete are materials with negligible tensile strength, a small increment has to be provided at the extreme compressed fibre to determine to cause a strain gradient along the length of the wall. Strain at the opposite edge is then estimated using principle of mechanics. Point 2 and point 3 correspond to yielding of extreme layer of reinforcing steel compression failure limit state of masonry/concrete; whilst point 4 represents yield point of the *M*-φ curve, the coordinates of which are determined as per Eq. (10a–c):

$$\phi_4 = \left[\frac{M_3\phi_2 - M_2\phi_3}{\phi_1(M_3 - M_2) - M_1(\Phi_3 - \phi_2)}\right]\phi_1 \tag{10a}$$

$$M_4 = M_1\frac{\phi_4}{\phi_1} \tag{10b}$$

The curvature ductility is obtained as:

$$\mu_\varphi = \frac{\varphi_3}{\varphi_4} \tag{10c}$$

Figure 4 represents a comparison of the idealised moment curvature relations with the nonlinear response. It is observed from the comparison of different methods represented in Fig. 4 that the equations proposed in Eqs. 10a–c does capture the nonlinear moment curvature interaction.

Fig. 4 Moment curvature interaction of a wall section with $P < P_{bal}$

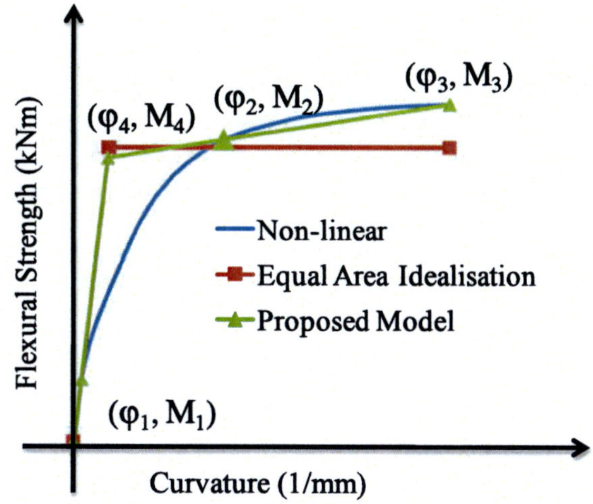

4.4 *Axial Load P Equal to P_{bal}*

At balance axial load, all limit states of strain in masonry/concrete and steel are mobilised. Hence, the equations prescribed in equation hold good in this case as well. The idealised moment curvature response obtained with the proposed equations is compared with the nonlinear response and existing bi-linear idealisation based on equal area idealisation and is represented in Fig. 5, and it is observed that the response in terms of initial stiffness, flexural strength, ultimate displacement and the curvature is well captured.

Fig. 5 Moment curvature interaction of a wall section with $P - P_{bal}$

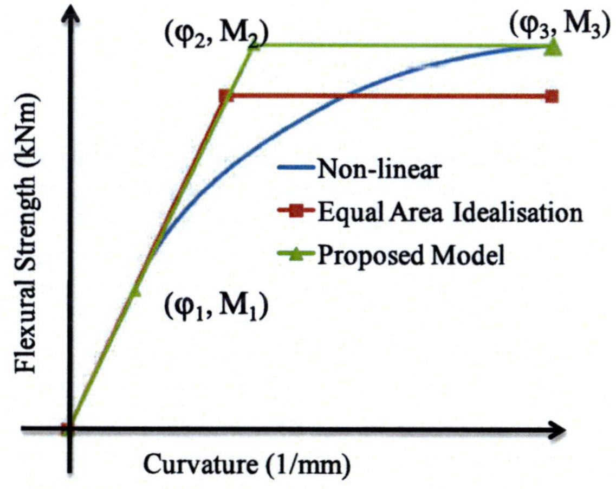

5 Curvature Ductility of RM and RC Walls

Equations prescribed in Sect. 4 indicate that they can be satisfactorily used to determine the moment curvature response of reinforced wall sections. These equations are now used to determine the moment curvature behaviour and curvature ductility of RM and RC walls of three different aspect rations and subject to different axial loads. The aspect ratios of walls studied are 0.5, 1.0 and 1.5, respectively, and the axial loads subjected are 0.4 P_u, 0.2 P_u, 0.1 P_u, 0.0.5 P_u and zero axial loads, respectively. In each of these cases, the percentage of reinforcement is varied from 0.25% (corresponding to minimum percentage of steel in RC walls) to 1.00%. Depending on the aspect ratio of the wall and the amount of reinforcement, the curvature ductility at the level of the aforementioned axial loads is determined according to Eqs. 9 and 10 depending on the values of the axial loads with respect to the balance axial load.

The ductility of RC and RM walls with varying axial loads and reinforced with 0.25% steel is represented in Figs. 6, 7 and 8. For brevity, the M-φ curves of only walls reinforced with 0.25% steel are presented. The curvature ductilities of all walls with varying percentages of reinforcements are given in Tables 1, 2 and 3.

One can observe that due to the lower self-weight of masonry, the curvature ductility of RM walls is greater than RC walls at all axial loads. Whilst this difference is not significant (< 10%) at moderate to high axial loads, at very low axial loads, the ductility offered by RM walls is significantly higher (15–20%). One can also observe that the aspect ratios of the walls have an inverse effect on the ductility. This is expected as an increase in the length of the wall indicates an increase in the in-plane stiffness of the wall. Increase of percentage of reinforcement of steel coincides with a reduction of 12–15% of the ductility. It can also be observed that with the addition of steel results in the adding the dead weight on the wall and subsequently, there is a reduction of the curvature ductility.

6 Results, Discussion and Scope of Future Work

The seismic performance of masonry structures with inadequate seismic resistant features has remained a concern for a very long time. Despite the advantages that it offers, the lack of proper guidelines and adequate technical workforce has limited the use of reinforced masonry for structural purposes.

With the help of analytical equations developed, this research determines the curvature ductility of reinforced masonry and reinforced concrete walls. Results obtained indicate that, at low axial loads, reinforced masonry walls are capable of providing up to 20% higher ductility when compared to RC walls. Hence, preliminary results indicate that RM walls may provide a suitable alternative to RC walls in regions of moderate to high seismicity.

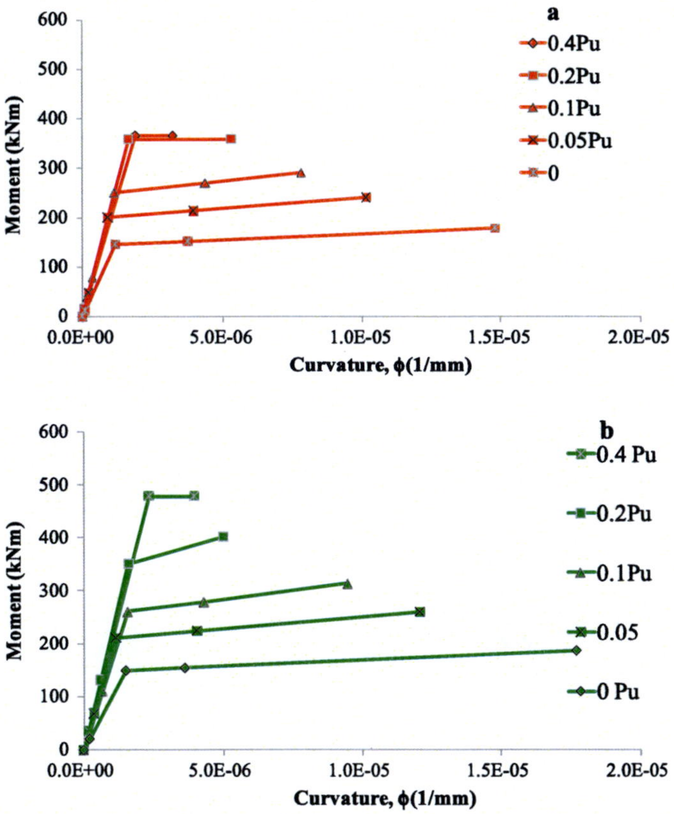

Fig. 6 *M-φ* curves of **a** RM wall and **b** RC wall of aspect ratio 0.5 reinforced with 0.25 percentage steel

Further research includes using the points identified using limiting states of strain as yield points for nonlinear analysis verify the elemental level ductility is also captured at the global level.

7 Conclusions

A numerical model to determine the curvature ductility of reinforced wall sections is proposed in this paper. The curvature ductilities of reinforced concrete and reinforced masonry walls are estimated using the proposed model. The various findings in this paper can be enumerated as follows:

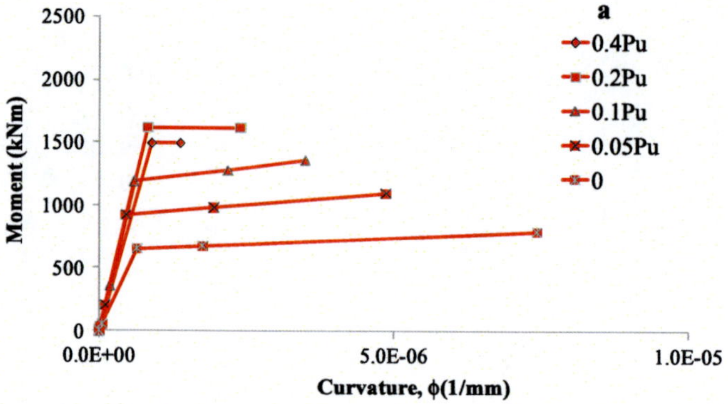

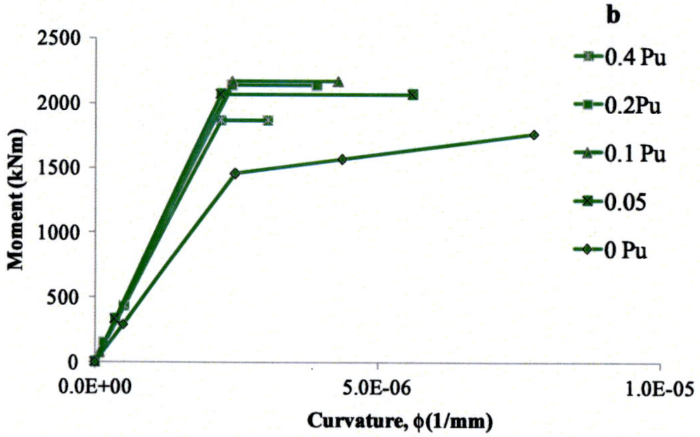

Fig. 7 M-φ curves of **a** RM wall and **b** RC wall of aspect ratio 1.0 reinforced with 0.25 percentage steel

1. Owing to lower self-weight, reinforced masonry walls are observed to provide greater curvature ductilities than reinforced concrete walls. The ductility of RM walls is 15–20% greater than RC walls.
2. Addition of amount of steel is observed to lower the ductility of walls. The reduction is observed to be in the order of 10–12%.
3. Results indicate that due to lower density RM walls provide higher ductilities and can be used as an effective substitute for RC wall in regions of moderate to high seismicity.

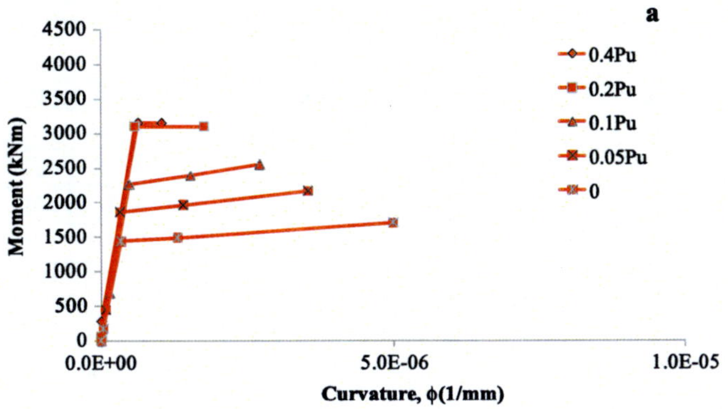

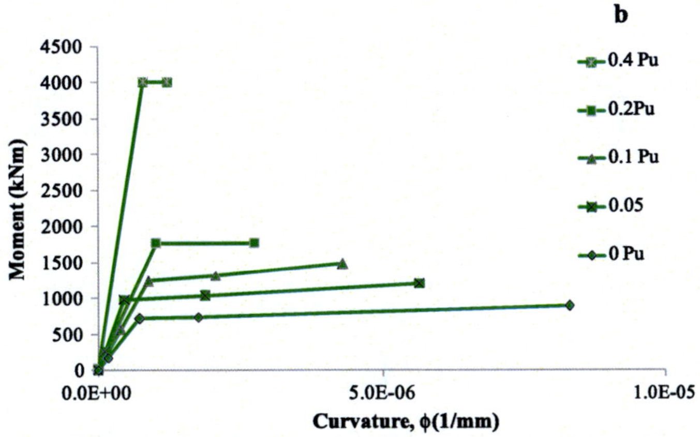

Fig. 8 M-φ curves of **a** RM wall and **b** RC wall of aspect ratio 1.5 reinforced with 0.25 percentage steel

Table 1 Ductility of RC and RM wall of aspect ratio 0.5 with varying axial loads

RM Walls						RC Walls				
% of steel	0.4 P_u	0.2 P_u	0.1 P_u	0.05 P_u	0	0.4 P_u	0.2 P_u	0.1 P_u	0.05 P_u	0
0.25	1.68	3.20	6.86	10.93	12.52	1.67	3.10	5.96	10.15	11.3
0.50	1.65	2.92	4.80	7.29	7.44	1.60	2.78	3.88	5.77	7.25
0.75	1.64	2.33	3.23	4.82	6.76	1.55	2.18	2.83	4.25	5.22
1.00	1.62	2.12	2.81	3.86	5.11	1.5	2.08	2.47	3.56	3.31

Table 2 Ductility of RC and RM wall of aspect ratio 1.0 with varying axial loads

RM walls						RC walls				
% of steel	0.4 P_u	0.2 P_u	0.1 P_u	0.05 P_u	0	0.4 P_u	0.2 P_u	0.1 P_u	0.05 P_u	0
0.25	1.55	2.96	6.46	10.03	10.52	1.50	2.65	4.8	9.91	10.8
0.50	1.52	2.57	4.52	6.59	7.04	1.47	2.27	3.52	4.79	6.25
0.75	1.51	2.28	3.03	4.32	5.66	1.45	2.01	2.54	3.42	4.35
1.00	1.50	1.92	2.61	3.46	4.41	1.38	1.069	2.34	2.92	3.07

Table 3 Ductility of RC and RM wall of aspect ratio 1.5 with varying axial loads

RM walls						RC walls				
% of steel	0.4 P_u	0.2 P_u	0.1 P_u	0.05 P_u	0	0.4 P_u	0.2 P_u	0.1 P_u	0.05 P_u	0
0.25	1.53	2.62	5.38	8.6	9.6	1.49	2.42	4.05	7.75	8.8
0.50	1.50	2.16	4.04	5.52	6.23	1.47	2.02	3.43	4.65	5.51
0.75	1.48	2.12	3.00	3.62	4.26	1.44	1.96	2.28	3.14	3.82
1.00	1.46	1.78	2.36	3.2	3.6	1.36	1.62	2.04	2.72	2.86

References

1. Rushabh G, Menon A (2014) Seismic performance of bed-joint reinforced solid brick masonry walls. In: 9th international masonry conferences, pp 1–12
2. Sunitha P, Goswami R, Murty CVR (2016) Idealised bilinear moment-curvature curves of RC sections for pushover analysis of RC frame buildings. Indian Concr. J 90(4):43–54
3. Bureau of Indian Standards (2014) Bureau of Indian Standards, Ductile detailing of reinforced concrete structures subjected to seismic forces, IS 13920 : 2014, Code of Practice—First Revision of IS 13920, Doc No: CED 39 (7941)
4. Kulkarni R, Goswami K (2015) Comparative study on modelling of RC structural walls for nonlinear static analysis. In: National conference on technological innovations for sustainable infrastructure
5. Kubin MT, Fahjan J, Tan M (2008) Comparison of practical approaches for modelling shear walls in structural analyses of buildings. In: Fourteenth world conference on earthquake engineering
6. Abdellatif A, Shedid M, Okail H, Abdelrahman A (2019) Numerical modeling of reinforced masonry walls under lateral loading at the component level response as opposed to system level response. Ain Shams Eng J 10(2):435–451
7. Penna A, Lagomarsino S, Galasco A (2013) A nonlinear macroelement model for the seismic analysis of masonry buildings. Earthq Eng Struct Dyn 1–21
8. Akhoundi F, Lourenço PB, Vasconcelos G (2014) Numerical modelling of masonry-infilled reinforced concrete frames : model calibration and parametric study, pp 1–13
9. Lourenço PB A review of Qut-of-plane behaviour of masonry 67–73
10. Lourenço PB, Mendes N, Ramos LF, Oliveira DV (2011) Analysis of masonry structures without box behavior. Int J Archit Herit 5(4–5):369–382
11. Wang X, Ghiassi B, Oliveira DV, Lam CC (2017) Modelling the nonlinear behaviour of masonry walls strengthened with textile reinforced mortars. Eng Struct 134:11–24
12. Jacob A, Menon A (2017) Role of diaphragm flexibility modelling in seismic analysis of existing masonry structures. Structures 11:22–39

13. Kaushik HB, Rai DC, Jain SK (2007) Uniaxial compressive stress—strain model for clay brick masonry. Curr Sci 92(4):497–501

Application of Image Analysis in Infrastructure Sector Fields—An Overview

G. Sreelakshmi and M. N. Asha

Abstract Image analysis is one of the recent tools used in the different fields of engineering towards understanding the microscopic behaviour of materials. It extracts information from an image by using automatic or semiautomatic techniques and the result of the image analysis process is a numerical output rather than a picture. Full scale or prototype modelling experimental studies in various Civil Engineering fields involves high-level instrumentation like strain gauges, load cells, data analyzer and data acquisition system. Many researchers in Civil Engineering field have used image analysis along with prototype modelling wherein interface behavioural mechanisms are studied. This avoids complicated experimentation process and permits assessment of discrete behaviours of particles that is impossible otherwise. The current review throws light on various digital image correlation and image processing techniques applied widely in mapping deformation behaviour in different streams of Civil Engineering. It is observed that these techniques offer a greater understanding of crack formation and progression, tracking of subsurface soil movement, monitoring of rail track displacement and visualization of flow in hydraulic channels.

Keywords Behaviour · Deformation · Localized zone · Image processing · Pixel

1 Introduction

In the current world, like the way Internet text search changed our lives, the concept of images has gone beyond the two-dimensional appearance depicted by the same. Image analysis is a technical examination of the complex aspects of an image, by deploying algorithms in artificial intelligence and machine learning. Various methods involved in image analysis processes are image transfer, digitalization, subdivision,

G. Sreelakshmi (✉)
Department of Civil Engineering (VTU RRC), CMR Institute of Technology, Bengaluru, India
e-mail: sreelakshmi.g@cmrit.ac.in

M. N. Asha
Department of Civil Engineering, CMR Institute of Technology, Bengaluru, India
e-mail: asha.n@cmrit.ac.in

L. Nandagiri et al. (eds.), *Sustainability Trends and Challenges in Civil Engineering*,
Lecture Notes in Civil Engineering 162, https://doi.org/10.1007/978-981-16-2826-9_3

image augmentation, organization, recording and evoking. Over the past decades, image analysis is widely used in different fields viz., Biological and Medicinal field for analysing bone fracture pattern and tracking the growth of cancer cells; Material science and Engineering for assessing active mineral participation and interface material behaviour in the field; Aeronautical field for assessing and tracking air traffic volume; Traffic Engineering for real-time road traffic control through the technique of video image processing and so on. Recently, digital image correlation (DIC) technique has been adopted for investigation of surface displacements of concrete specimens, to monitor and measure rail displacements, in subgrades to measure the deformations, to understand the flow analysis in hydraulic structures and investigate soil-structure interaction behaviour of foundations of buildings. From all these, it can be concluded that image processing is considered as one of the most budding areas of digital field, with expanding implementations in all fields of engineering and science. The following sections summarize the significant phases of digital image processing, the principle of digital image processing and application of digital image correlation techniques in different fields of Civil Engineering.

2 Stages of Image Analysis and Steps in Digital Image Processing Methods

2.1 Significant Phases in Digital Image Analysis

Digital image analysis means involves extracting and converting an image into fundamental components through mathematical operation to obtain the spatial variation of position coordinates over a fixed patch size. Figure 1 shows major stages in digital image analysis.

Important steps in digital image analysis can be listed as follows:

- *Image acquisition and enhancement*

It involves the action of retrieving an image from the hardware-based source for processing. It also includes processing, compression, storage, and printing. Figure 2 shows the details of image acquisition and enhancement process.

- *Object/Phase detection*

Generally, object recognition approaches can be:

(i) Edge-based feature
Edge-based feature investigates peripheral details of the object in terms of edges. This feature is subdivided as full-contour (shape) feature and contour fragments feature.

(ii) Patch-based feature

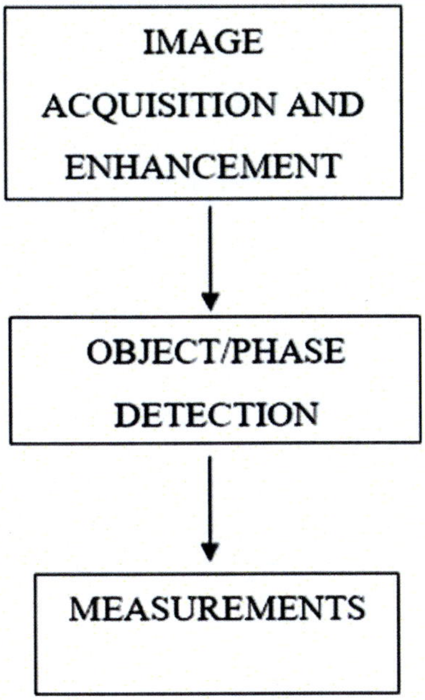

Fig. 1 Major stages in digital image analysis

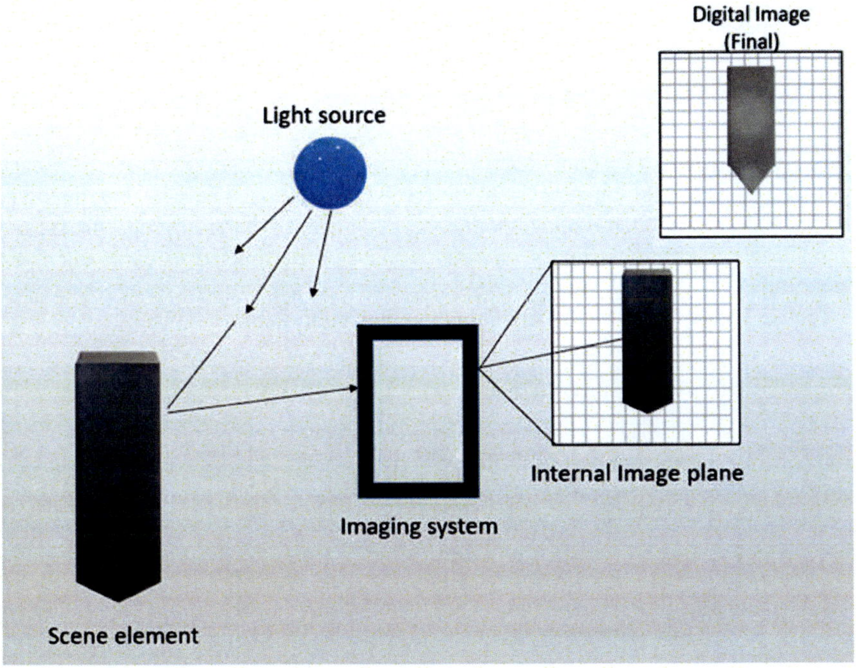

Fig. 2 Image acquisition and enhancement process

Patch-based feature type uses two different variations: Rectangular and Irregular patch type shapes.

(a) Rectangular patch type shapes explain exterior characteristics of the source.
(b) Irregular patch type shapes describe variations in the surface.

The modification in these types is categorized by the edge of the patch size and is named as region-based features. The patch size is an important parameter that affects the quality of details extracted through from any reconstructed image sources [1].

- *Measurements*

There are several types of digital image processing which are available online, as freeware and/or shareware can be listed as Mathtools.net, CVIP Tools, The FFT Home Page, SPRING, SCIKIT-IMAGE, PIV lab, ImageJ, etc. This user-written plugin architecture and built-in development environment has made these platforms versatile and hence facilitating problem-solving in different verticals viz., solve three-dimensional analysis problems, comparisons concerning multiple images, gather displacement data from sequences of digital images captured, measure the profile velocity variation within digital images, exhibit, and transfer numerous features of the flow framework.

2.2 *Steps in Image Analysis Methods*

The first step involves the conversion of unprocessed image source to greyscale image followed by the meshing process to binary image source. The next step involves an analysis of the binary image source to understand the microscopic behaviour of materials. Numerous binary operations must be performed to carry out these operations.

Figure 3 shows the step by step process involved in image analysis

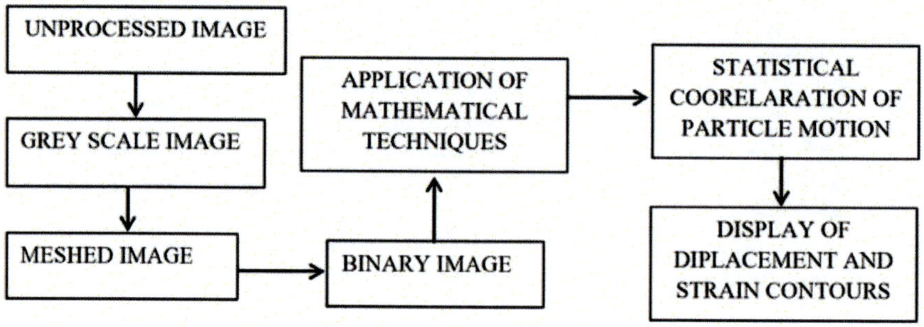

Fig. 3 Steps involved in image analysis methods

3 Preamble

The two-dimensional image correlation algorithm was first developed by Sutton et al. [2] in the year 1983 to estimate the planar strains in a cantilever section with vertical loading. Since then, there is a massive advancement attained in the field of image-based methods and microprocessor-based image processing and analysing techniques. The non-contact measurement method such as X-ray diffraction methods, stereo-photogrammetric techniques, digital image correlation techniques (DIC) is widely used in the Civil Engineering field. These techniques help in visualizing surface displacements, strains at localized zones, estimation of crack size to predict permeability, durability, and strength of structural components and so on. Large-scale DIC techniques are used also for water flow analysis in water resource engineering applications as a diagnostic tool for investigation, forecasting, proposing, risk warning, management, etc. in dynamic flow associated actions. The following sections explain the application of image analysis in measuring planar deformation fields in different fields of Civil Engineering. Figure 4 shows the flowchart of different fields in which image analysis methods are applied.

3.1 Application of Image Processing Technique in the Field of Structural Engineering

Concrete materials undergo fracture process wherein micro-cracks develop due to change in stress–strain behaviour causing material softening finally leading to rupture. A realistic investigation of fracture zones is very important as it helps to understand deterioration problems and predict the surface failure patterns in concrete during the loading process. The destructive and nondestructive tests such as Schmidt hammer, core drilling, ultrasonic pulse velocity and electrical resistivity measurement are used to investigate crushing strength and mechanical properties [3] integrated image analysis technique with BIM software and inspected structural cracks in inaccessible parts of the reinforced concrete tower. Marek Słonski [4] and Skarzyński et al. [5] adopted the DIC technique to measure the size and shape of strain concentrations around a predefined notch in concrete. The objective was to understand deterioration problems in concrete and predict the failure behaviour. Deterministic statistical models beam with 3-point flexural loading conditions were developed and DIC technique was used to track the formation of surface cracks on concrete elements. Figure 5 shows a schematic sketch of a beam section with a notch.

In the studies carried out by Ribeiro et al. [3] image analysis has been used to capture the local deformations above a predetermined failure plane. VIC-2D software has been used for image analysis and search patch sizes used for the study had pixel resolutions of 30, 60, 90, 120, 150–180 pixels. The strain distribution is obtained through two-dimensional Gaussian functions.

Fig. 4 Flow chart showing the applications of image analysis in different fields of civil engineering

Fig. 5 Schematic sketch of beam section with a notch

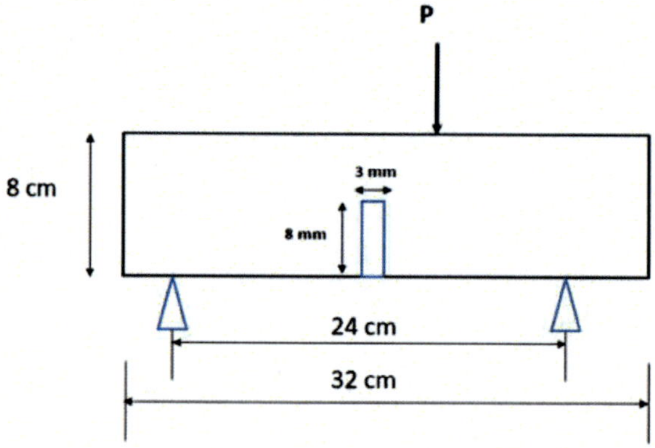

From the test results, it can be concluded that higher image resolution is crucial since surface deformations measured are of smaller magnitudes (in the order of micro-strains). During the deformation process, dimensions of the localized zone near the notch location changed drastically before a macro-crack (2–3 mm) which could be due to a complex nature of concrete. So, the method proposed by Ribeiro et al. [3] precisely determined the deformations of a localized zone and merits than other experiments such as acoustic emission, X-rays, and SEM.

Liu and Yang have used image analysis to identify bug holes in hardened concrete. The image analysis tool of MATLAB was utilized for the estimation of diameter and area ratio of the bug holes. From the analysis, it was also observed that distance at which images are captured also play an important role as far as the accuracy is concerned.

In the distribution of steel fibre influence the performance of reinforced concrete. Ďubek et al. [6] has used the principles of image analysis to determine the size and orientation of the fibres in different cross-sections.

3.2 Application of Image Processing Techniques in the Field of Geotechnical Engineering

Foundation engineering deals with the design of substructures that supports harbour terminals, rail and road bridges, skyscrapers, and offshore structures. But in all these cases it is very difficult to obtain actual ground characteristics about infill medium due to variation in its mechanical behaviour. Physical modelling has been used in recent years to simulate real engineering problems, but these models cannot envisage the failure pattern of infill medium in geotechnical and geo-mechanical processes. Also, experimental studies for these soil-structure interaction problems (full scale or prototype modelling) involve a high level of instrumentation (Strain gauges, load cells, data analyzer and data acquisition system, etc.).

Unlike concrete which is a continuous media, soil comprises of soil grains and voids. Hence the application of DIC is challenging in this stream. Among various approaches, particle image velocimetry (PIV) is widely implemented in geomechanics engineering to study the soil deformation characteristics (idealizing soil as a fluid medium) through velocity and displacement profile. Geo-PIV software developed by Stanier et al. [7] is a MATLAB-based DIC implementation which has been widely in the field of geotechnical engineering. It works by tracking variation in texture captured in a series of images. The source image is processed into numerous patches of desired sizes. The displacement of each patch during a period is determined by associating the patch extracted from the same image at different time frames.

Many researchers like, [7–10] have adopted this tool for laboratory investigation of the shallow and deep foundations in particulate media. PIV adopts fast Fourier transform method wherein cross-correlation between the selected patches of images is calculated by fitting a bicubic interpolation technique. Using this, correlation peaks are traced throughout the selected image that is converted in terms of shear responses. The shear responses estimated through study helps in the assessment of the penetration mechanism at the pile-soil interface.

Offshore piling for naval platforms or piling operations in and around naval harbours is inevitable for berthing of ships. Piling operations on contrasting beds is a real challenge for engineers as the energy required for pile driving is dependent upon the properties of the soil strata. The technology can be extended in offshore pile foundation industry wherein the subsurface soil movements could be visualized that can aid in the optimum selection of hammer load for specific in situ conditions. White and Bolton [9] adopted PIV technique to investigate the phenomenon of friction fatigue by measuring the movement of soil around the outer pile shaft during pile driving progression. In the experimental studies carried out by White and Bolton [9], plane strain chamber with transparent screen has been used and the model piles selected are such that there will be no interference from boundaries. An additional pressure through a rubber-based medium is applied at the top level of embedded soil. Figure 6 shows the line sketch of a plane strain chamber with model pile in the background.

The PIV analysis revealed that there is the greatest displacement of soil particles near the pile shaft (Largest vector of (250 μm, 1.9 pixels) is observed closer to pile shaft. There is greatest horizontal strain observed with a magnitude of 0.6% near the lower pile tip due to the reduction in skin friction over the shaft region. This process is captured effectively through strain measurements data obtained through image analysis. Later [11] adopted Geo-PIV technique to investigate the behaviour of

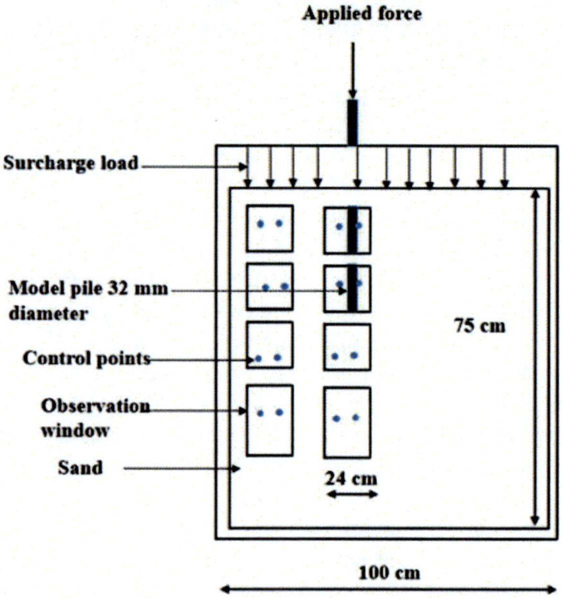

Fig. 6 Representative line sketch of plane strain chamber with a model pile

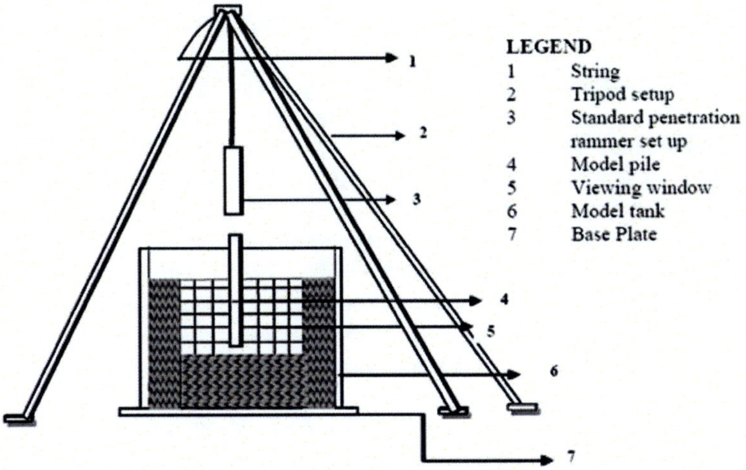

Fig. 7 Graphic sketch of experimental set-up for impact loading

hollow driven piles under impact load conditions and estimated the effect of plugging on bearing capacity using image analysis in-plane strain chamber. Figure 7 shows the graphic sketch of the experimental set-up adopted for the study.

The shear strain contours developed near the pile-soil interface region depicted zones of densified infill formation near the plug zone that contributed towards increased skin friction values in bearing capacity equations. The current approach involves prototype modelling followed by image analysis and this methodology avoids complicated experimentation process and permits assessment of the movement of subsurface soil particles which is impossible otherwise.

Mishra et al. [12] have extended the application of Geo-PIV in analysing the behaviour of five types of geotextiles when subjected to wide width tensile strength test. The DIC technique could capture the non-uniform distribution of strains on the surface of the geotextile. Chen et al. [13] have provided guidelines on using DIC for geotechnical applications. The literature suggests that the accuracy of the analysis is dependent upon the size of the method of approximation and the size of the inspection window. Since the quality of the image plays a crucial role in the analysis, they have also outlined the different methods by which the quality of the image can be improved.

3.3 *Application of Image Processing Techniques in the Area of Pavement Engineering*

For road surfaces, the life span depends on its initial design and construction, environmental factors, drainage conditions, traffic loadings and frequency of preventive maintenance inspections. Among various types of catastrophes, cracks are the most important factor which is used for measuring the state of roadways and its surface conditions. Accordingly, it is necessary to distinguish and diagnose the surface crack precisely before proposing pavement repair techniques. Though there are several manual-based crack detecting methods, they are laborious and offer considerable differences in test results. Hasheminejad et al. [14] and Sun et al. [15] investigated crack development and pattern in geosynthetic pavements using the PIV technique and concluded that this technique efficiently captured changes in crack development path in the case of reinforced pavements.

Hasheminejad et al. [14] adopted DIC technique to examine the deformation developed on the surface of bituminous pavements and found a correlation between crack development and surface weariness through secondary tensile tests. Sun et al. [15] developed laser-based 3-D crack detection technique. This proven technique helps in identifying the surface profile of crack, crack pattern, crack records, etc. Later signal processing techniques are employed to determine crack pattern by using sparse representation method. Martin et al. [16] have tried to estimate the porosity of pervious concrete using image analysis. The results are validated using experimental studies and empirical formulae. With the advent of drone survey, researchers have been trying to correlate the pavement images with the condition of pavements by using the concept of deep learning algorithm [17].

The feasibility of the study is substantiated through wavelet and Gabor dictionary median filtering method using MATLAB image processing module. Wavelet method converts the image file source to a known mathematical model and extracts the transitional behaviour of frequency variation in an image file in the time series domain. Gabor filter method also helps in analysing frequency variation in 2-D images around a specific zone of interest. Table 1 shows the details of the depth of processed crack by various methods.

It is observed that Wavelet method and Gabor median filtering method, misinterpreted cracks development which affected the outcomes of the consequent surface profile. In association with the wavelet and median filtering method, the surface deformations extracted by 3D laser techniques matched the shape of crack very well and showed a noticeable distinction of crack from the surface without disturbing pavement surface.

Table 1 Depths of processed crack by different methods

Original crack	Depth of processed crack		
	Wavelet method	Gabor dictionary	Laser-based method
– 6.2 mm	– 2 mm	– 5 mm	– 6.1 mm

3.4 *Application of Image Processing Techniques in the Field of Railway Engineering*

The railway tracks are subjected to fatigue damages due to continuous movement of trains throughout its service life. This leads to transverse instability of rails causing derailment of tracks. So, it is very important to have an efficient track monitoring mechanism to avoid probable rail accidents. Normally strain gauges are adopted to measure the surface deformation or displacements in the case of structural components which may not be practical in all situations especially for rails. Two-dimensional DIC technique is adopted to measure, detect and record track displacement [18, 19]. Murray et al. [19] monitored "running rail" phenomena using cameras positioned as shown in Fig. 8.

Murray et al. [19] adopted particle image velocimetry (PIV) technique to estimate the magnitude of lateral and vertical movements of rail tracks. The mathematical principles of the PIV method are discussed in previous sections. PIV technique tracks rail, tie, and ballast movement in a defined meshed region and compares the images taken at different instant of time to obtain displacement profile of the region of interest. Numerical modelling of the foundation with subgrade considering the infinite beam on elastic foundation is also undertaken to compare the results. Table 2 shows the evaluation of measured and calculated rail deflection with time captured during motion of the train.

From Table 2, it is evident that measured and calculated rail deflection vs time captured during train movement is following the same trend which indicates the

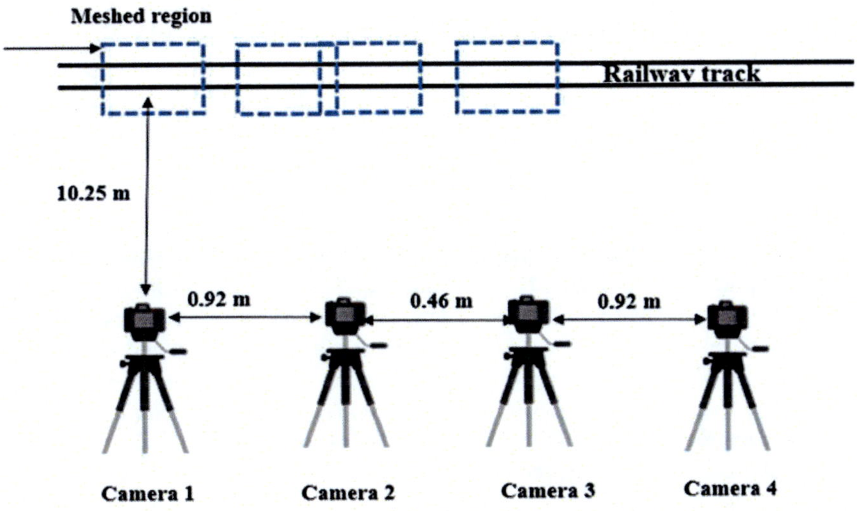

Fig. 8 A basic representation of the ground layout of camera position at Railway track site

Table 2 Evaluation of rail deflection vs time captured during train movement

Time (s)	Rail deflection (mm)	
	Measured value	Calculated value
1	0.25	0
2	− 1.5	− 2.0
3	− 1.45	− 1.40
4	− 1.5	− 1.50
5	− 1.5	− 1.45
6	− 1.35	− 1.25
7	− 1.5	− 1.25
8	0.25	0

feasibility of image analysis. These data's from DIC can be used to develop parameters for designing stiffeners and dampers for an increasing design period of the rail track system.

3.5 *Application of Image Processing Techniques in the Field of Hydraulics and Water Resource Engineering*

For the planning, analysis, and hydraulic design of structures like dams, weirs, and reservoirs, it is very crucial to investigate flow visualization, flow diversion in natural environments and implementation of suitable sediment managing practices. Though there are newer techniques developed to measure velocity and flow discharges, the costs, accuracy and its feasibility to use an inaccessible location is always a matter of concern. Kantoush et al. [20] and Sutarto [21] adopted a Large-Scale Particle Image Velocimetry (LSPIV) (quantitative image-based) method to study the velocity measurements. This technique can classify the flow dynamics and detect water surface velocity by measuring distances between moving particles and time interval between them through consecutive images.

The mathematical approach adopted in LSPIV is same as PIV, but the normalized cross-correlation is calculated between areas of first and second images through Gaussian functions. To plot fluid flow structure in an open canal [21] water is impelled into the channel or canal with a continuous discharge at the upstream end and allowed to flow into a basin at the downstream end. For tracing the motion, tracer particles (Eco-foam) are dispensed on the liquid medium and particle movements are recorded with the camera positioned on top of the area of observations. The video recorded during the experiments is processed into image sequences having a time period of 0.033 s and later analysed using LSPIV method. The sensitivity of the observational area is analysed for longitudinal and lateral velocities by assigning different pixel values of 90, 80, 60 50, 40, 30, and 20 pixels, respectively. The LSPIV method effectively seized the eddying motion and flow streamline pattern in the region of observation. So, the LSPIV measurements employing digital camera and tracing particle for liquid medium provided an appropriate solution for investigating the flow patterns in shallow waters, shallow reservoirs, and rivers.

4 Conclusions

The paper presented collective reviews of the image processing techniques used in different fields of engineering towards understanding the microscopic behaviour of materials. Research papers are selected based on the application of DIC techniques in several small and medium scale experimental studies.

The following conclusions have been listed from the literature reviews:

- DIC technique is an evolving non-surface contact measuring technique used to investigate the strain behaviour in various fields of Civil Engineering.
- DIC reduces complicated instrumentation process and provides a simple, cost-effective, and reliable solutions.
- Crack detection by image-based method for concrete-based structures helped in investigating geometric nature and to forecast the development of the crack formation. Also, enhancement in pixel resolution (pixels) permits in seizing of exhaustive pictures of beam surfaces.
- The open-source Geo-PIV-RG could capture the strain distribution during pile installation stages and proved its suitability for investigating pile-soil interaction.
- The three-dimensional (3D) pavement crack recognition method established on laser technology with sparse representation technique proved successful in crack detection and reconstruction of pavement profile.
- Digital image correlation (DIC) technique served as a simple system to investigate the rail deflection and envisage the fatigue life of railway track.
- The LSPIV technique emerged as flow diagnostic tool for the mapping of complex flow patterns in hydraulic engineering applications.

From the analysis, we can infer that several researchers adopted image processing techniques for the analysis with many segmentation algorithms and extraction technique to investigate complex strain conditions, environmental flow processes and interactions. However, the selection of the tool is dependent upon the application and desired accuracy.

5 Recommendations for Future Work

Rapid urbanization and development in technology have made changes inevitable. However, it is very difficult to predict all changes even if sufficient growth factor or Factor of safety is assumed. In this context, modelling is highly important and the results of a properly scaled prototype can be simulated. The proper understanding of model tests is facilitated by using microscopic analysis. For instance, assessment of pavement performance can be integrated with drone survey wherein remote evaluation of pavement condition can be done and accordingly priority of maintenance can be decided. Diversity in subsoil conditions has made both onshore and offshore piling inevitable. One of the major challenges in pile driving is the selection of energy at which piles can be driven for different conditions of in situ soil. Use of particle image velocimetry to study soil-pile interaction helps in analysing and interpreting the soil movements around the pile as and when it is driven. This also helps in ensuring the verticality of the pile during and after its installation. Performance of high-performance concrete can be studied for compressive strength supported with Scanning Electron Microscopy. However, the results may be validated for crack propagation. Similarly, while designing hydraulic structures, the use of principles of image analysis enables an efficient means of water management. Nowadays, radar images obtained from Remote sensing sources aid in monitoring and comparing land use patterns and helps in estimating damages to the land ecosystem. Hence it is possible to apply the integrate the principles of image analysis to different streams of Civil Engineering.

References

1. Prasad DK (2013) Geometric primitive feature extraction—concepts, algorithms, and applications. Nanyang Technological University
2. Sutton M, Wolters W, Peters W, Ranson W, McNeill S (1983) Determination of displacements using an improved digital correlation method. Image Vis Comput 1(3):133–139. https://doi.org/10.1016/0262-8856(83)90064-1
3. Ribeiro D, Santos R, Shibasaki A, Montenegro P, Carvalho H, Calçada R (2020) Remote inspection of RC structures using unmanned aerial vehicles and heuristic image processing. Eng Fail Anal 117:104813. https://doi.org/10.1016/j.engfailanal.2020.104813
4. Marek Słonski MT (2020) 2D digital image correlation and region-based convolutional neural network in monitoring and evaluation of surface cracks in concrete structural elements. Mater (Basel) 13:1–13. https://doi.org/10.3390/ma13163527
5. Skarzyński JK, Tejchman J (2013) Application of DIC technique to concrete-study on objectivity of measured surface displacements. Exp Mech 53(9):1545–1559. https://doi.org/10.1007/s11340-013-9781-y
6. Ďubek M, Makýš P, Ďubek S, Petro M (2018) The evaluation of the content of fibers in steel fiber reinforced structures and image analysis. J Civ Eng Manag 24(3):183–192. https://doi.org/10.3846/jcem.2018.1642
7. Stanier SA, Blaber J, Take WA, White DJ (2016) Improved image-based deformation measurement for geotechnical applications. Can Geotech J 53(5):727–739. https://doi.org/10.1139/cgj-2015-0253
8. White DJ, Take WA, Bolton MD, Munachen SE (1996) A deformation measurement system for geotechnical testing based on digital imaging, close-range photogrammetry, and PIV image analysis. In: 15th international conference on soil mechanics and geotechnical engineering, vol 1, pp 539–542 [Online]. Available: http://www-civ.eng.cam.ac.uk/geotech_new/people/bolton/mdb_pub/109_ICSMGE_2001_proc_539_542.pdf
9. White DJ, Bolton MD (2002) Observing friction fatigue on a jacked pile. In: Centrifuge and constitutive modelling: two extremes, pp 346–354

10. Dejong JT, White D, Randolph MF (2006) Microscale observation and modeling of soil-structure interface behavior using particle image. Soils Found 46(1):15–28. https://doi.org/10.3208/sandf.46.15
11. Sreelakshmi G, Asha MN, Viswanath D, Kumar YNY, Vinodini S (2020) Construction in geotechnical engineering. In: Image-based measurements to estimate bearing capacity of hollow driven piles under impact loading, pp 285–298. https://doi.org/10.1007/978-981-15-6090-3
12. Mishra VK, Kumar S, Shukla N (2017) Image acquisition and techniques to perform image acquisition. SAMRIDDHI A J Phys Sci Eng Technol 9(01). https://doi.org/10.18090/samriddhi.v9i01.8333
13. Chen ZC, Li K, Omidvar M, Iskander M (2017) Guidelines for DIC in geotechnical engineering research. Int J Phys Model Geotech 17(1):3–22. https://doi.org/10.1680/jphmg.15.00040
14. Hasheminejad N et al (2018) Digital image correlation to investigate crack propagation and healing of asphalt concrete, pp 5381. https://doi.org/10.3390/icem18-05381
15. Sun X, Huang J, Liu W, Xu M (2012) Pavement crack characteristic detection based on sparse representation. EURASIP J Adv Signal Process 2012(1):1–11. https://doi.org/10.1186/1687-6180-2012-191
16. Martin WD, Putman BJ, Kaye NB (2013) Using image analysis to measure the porosity distribution of a porous pavement. Constr Build Mater 48:210–217. https://doi.org/10.1016/j.conbuildmat.2013.06.093
17. Gopalakrishnan K (2018) Deep learning in data-driven pavement image analysis and automated distress detection: a review. Data 3(3):1–19. https://doi.org/10.3390/data3030028
18. Iryani L, Setiawan H, Dirgantara T, Putra IS Development of a railway track displacement monitoring by using pdf
19. Murray CA, Andy Take W, Hoult NA (2014) Measurement of vertical and longitudinal rail displacements using digital image correlation. Can Geotech J 52(2):141–155. https://doi.org/10.1139/cgj-2013-0403
20. Kantoush SA, Schleiss AJ, Sumi T, Murasaki M (2011) LSPIV implementation for environmental flow in various laboratory and field cases. J Hydro-Environ Res 5(4):263–276. https://doi.org/10.1016/j.jher.2011.07.002
21. Sutarto TE (2015) Application of large scale particle image velocimetry (LSPIV) to identify flow pattern in a channel. Procedia Eng 125:213–219. https://doi.org/10.1016/j.proeng.2015.11.031

Microalgae and Sewage Treatment for Developing Countries

Nandini Moondra, R. A. Christian, and N. D. Jariwala

Abstract Wastewater composition can be directly related with lifestyles of the society. Current wastewater effluent has a high nitrogen and phosphorus concentration, which adversely effects ecosystems. Sewage treatment units practiced in undeveloped countries have various limitations. The present study focused on treatment of sewage collected from a sewage treatment plant (STP) of a smart city of a developing nation using microalgae. The study found that the microalgae were efficient in removing nutrients and organic matter. Maximum reduction observed in phosphate, ammonia, and COD was 97.46, 100, and 93.67% at 11 h of detention time. Phycoremediation proved to be an effective method for sewage treatment.

Keywords *C. vulgaris* · Nutrient removal · Photosynthesis · Wastewater treatment

1 Introduction

Population growth and economic development resulted in the increase of water demand worldwide. Globally, over 70–80% of wastewater (industrial and sewage wastewater) is discharged into surface water resources without proper treatment. There is a gap between water demand for current population growth and water supply from the same water resources. The aquatic ecosystem and water quality of the source are adversely affected if wastewater is discharged in it without or with partial treatment. Therefore, the utmost priority is the wastewater treatment to the maximum before discharging and limiting the generation of wastewater [1]. Developing countries such as India (middle and low-income countries) are on the lookout for infrastructures for effective wastewater treatment systems as most of the population resides in rural areas, facing an ever-increasing population and enhanced

N. Moondra (✉) · R. A. Christian · N. D. Jariwala
Civil Engineering Department, SV National Institute of Technology, Surat, India

R. A. Christian
e-mail: rac@ced.svnit.ac.in

N. D. Jariwala
e-mail: ndj@ced.svnit.ac.in

L. Nandagiri et al. (eds.), *Sustainability Trends and Challenges in Civil Engineering*, Lecture Notes in Civil Engineering 162, https://doi.org/10.1007/978-981-16-2826-9_4

lifestyles. The issue is severe because of the fast pace of urban development, and effective effluent measures are mandatory to control and treat.

Secondary treatment units of STPs are inefficient in treating nutrients [2], release of these nutrients changes pH, reduces DO, which affects aquatic life [3] and degrades freshwater ecosystems [4]. Secondary treatments are costly in terms of investment and operations with high sludge generation [5]. These limitations create a setback for secondary treatment in underdeveloped countries [6]. In the activated sludge process, about 50% of electricity consumption is by the blowers/aerators that fulfill aerobic bacteria's oxygen demand to decompose organic matter [7].

Microalgae are favored for the bioremediation process because of its high photosynthetic proficiency, quick take-up of nutrients, and short life expectancy combined with basic growth requirements. Algae consume nitrogen and phosphorus for building biomass, thus contributes to the self-purification of waters [8]. Oxygen is produced in the photosynthesis process during treatment contributes to the waters' oxygen supply. These systems can tolerate extreme environmental conditions. The microalgal biofilm network can hold the biomass while working at a short detention time [9]. Purification capacity of microalgae, which incorporates: increment in pH of wastewater, making adverse conditions for pathogenic organisms, emission of antibacterial substances, generation of lethal extracellular mixes by specific algae, and exhaustion of organic matter and nutrients [10]. Algae absorb carbon dioxide and can create sustainable power sources, energizes, and chemicals, diminishing petroleum product utilization and ozone-depleting substance emissions [3].

The present study aimed to determine microalgae's impact as a treatment approach for the raw sewage collected from a sewage treatment plant (STP) of capacity 200 MLD. The study was conducted at a detention time of 11 h to reduce the physicochemical parameters analyzed to the desired disposable limit without any further treatment.

2 Materials and Methods

The freshwater algal species *Chlorella vulgaris* were obtained from an environmental consultancy firm named Phycolinc Pvt. Ltd., from Ahmedabad, a city in the Gujarat state of India. The culture was kept in a natural environment and provided with nutrients to keep them active. The batch study was performed in a 2000 mL borosilicate beaker with a working volume of 1800 mL. Two glassware was used as an experimental setup, one containing 30% microalgal solution (540 mL) and the remaining were made up of the raw sewage from an STP of a smart city of India. 30% concentration was selected as it was found to be the optimum microalgal concentration for sewage treatment [11]. However, the other beaker contained 1800 mL of raw STP wastewater, which acted as a control. External aeration was provided in both the beakers, which was a mixing mechanism for wastewater and microalgae. In the case of control, it replicated the secondary treatment unit incorporated in STP, shown in Fig. 1.

Fig. 1 Experimental setup for the study

In the study of 11 h of detention time, system was aerated for 9 h to keep algae in suspension. After 2 h of settling, the supernatant was drawn for analysis.

Various parameters analyzed were chemical oxygen demand (COD), dissolved oxygen (DO), nitrate, pH, electrical conductivity (EC), total dissolved solids (TDS), phosphate, ammonia, total solids (TS), and total suspended solids (TSS), as prescribed in the APHA manual [12].

Triplicates were performed for each parameter studied for a sample, and the average was considered for analysis. A study was conducted for both non-filtered and filtered (with a coarse filter of pore size 4–5.5 mm) effluent to get an idea about the presence of filamentous algae and change in characteristics after filtration. But with time, the microalgal settle ability improved, and the clear supernatant was obtained after treatment.

3 Results and Discussion

Removal in various parameters stated was measured at detention time of 11 h for both wastewater with microalgae and the control system. The variation in concentration of raw domestic wastewater and after treatment with *C. vulgaris* and control for various parameters is given in Table 1 except for pH (only mean) for both non-filtered and filtered effluent.

Even at low HRT (11 h), high removal was observed in organics and nutrients, unlike stated in literature [13, 14]. Reductions observed in various parameters during the study for microalgal treatment and control are given in Table 2.

Phosphate concentration in raw wastewater was ranging between 1.87 mg/L and 3.54 mg/L. After treatment, phosphate concentration reduced to 0.08 mg/L (96.33%) and 0.42 mg/L (83.91%) for non-filtered effluents in case of microalgal treatment

Table 1 Variation in concentration of raw wastewater and after treatment with microalgae and control at 11 h HRT

Parameter	Unit	Raw w/w	M (NF)	M (F)	A (NF)	A (F)
pH		7.32	8.30	8.19	7.92	7.84
EC	mS/cm	2.19 ± 0.14	1.86 ± 0.13	1.79 ± 0.13	1.89 ± 0.16	1.85 ± 0.15
TS	mg/L	3014.17 ± 517.25	2292.29 ± 488.84	2034.25 ± 466.38	2377.20 ± 517.38	2098.10 ± 477.20
TDS	mg/L	1086.13 ± 70.88	921.67 ± 62.81	889.58 ± 64.57	936.67 ± 78.61	914.58 ± 75.30
TSS	mg/L	1928.00 ± 533.61	1370.63 ± 483.553	1097.58 ± 446.73	1440.50 ± 500.35	1183.50 ± 462.61
DO	mg/L	0.00 ± 0.00	5.88 ± 0.66	5.83 ± 0.62	4.07 ± 0.65	4.07 ± 0.66
COD	mg/L	236.18 ± 37.00	44.40 ± 27.11	32.93 ± 22.02	120.66 ± 24.28	95.07 ± 25.15
Ammonia	mg/L	15.82 ± 1.32	2.42 ± 2.49	1.76 ± 2.14	7.67 ± 2.95	6.55 ± 2.63
Nitrate	mg/L	0.76 ± 0.50	5.14 ± 3.08	5.03 ± 3.09	2.93 ± 2.58	2.87 ± 2.57
Phosphate	mg/L	2.74 ± 0.47	0.35 ± 0.25	0.27 ± 0.21	0.94 ± 0.32	0.83 ± 0.30

Table 2 Percentage reduction in different parameters after treatment with microalgae and control (only aeration) at 11 h HRT

Parameter	M (NF)	M (F)	A (NF)	A (F)
EC	15.04 ± 3.35	18.05 ± 4.52	13.87 ± 4.22	15.56 ± 3.93
TS	24.18 ± 5.31	32.65 ± 7.04	21.31 ± 6.56	30.48 ± 7.10
TDS	15.10 ± 3.42	18.01 ± 4.74	13.79 ± 4.07	15.82 ± 3.80
TSS	29.54 ± 7.49	43.85 ± 9.43	25.57 ± 8.84	39.00 ± 9.69
COD	80.68 ± 11.83	85.65 ± 9.68	48.11 ± 11.82	59.14 ± 11.99
Ammonia	84.75 ± 15.11	88.99 ± 12.94	51.91 ± 17.63	58.56 ± 15.79
Phosphate	86.68 ± 10.64	89.57 ± 8.94	63.97 ± 15.33	68.46 ± 13.37

and control. However, in the case of filtered effluent, concentration further reduced and reached to 0.05 mg/L (98.08%) and 0.35 mg/L (86.59%) when treated with microalgae and only aeration, respectively, as shown in Fig. 2a, b.

Microalgal cells use phosphate for photosynthesis, DNA and RNA formation, and energy transfer and are stored in excess far beyond the requirement in the form of polyphosphate within algal biomass (luxury uptake) [13, 14] and phosphate precipitation at high pH [15], resulted in higher phosphate removal from the wastewater using microalgae.

Ammonia concentration in the influent varied from 12.6 to 17.72 mg/L. But after treatment, ammonia concentration was reduced to below detectable limit (BDL) (100%) and 4.20 mg/L (73.95%) for non-filtered effluents in case of microalgal treatment and control. However, in the case of filtered effluent, minimum concentration was BDL (100%) and 3.64 mg/L (77.42%) when treated with microalgae and only aeration, respectively, as shown in Fig. 3a, b. Building of genetic material, enzymes, formation of proteins, hormones, vitamins, and energy transfer molecules are all because of nitrogen [13]. Nitrogen removal during the study could be the result of various simultaneous processes, namely (i) biomass uptake of nitrogen (ii) bacterial nitrification–denitrification, (iii) ammonia stripping, and (iv) loss in the atmosphere [16].

In case of COD, raw wastewater characteristics are varied from 176.00 to 297.60 mg/L. However, after treatment, influent concentration reduced to 16.00 mg/L (92.94%) and 89.60 mg/L (70.00%) in case of microalgal treatment and control, respectively, for non-filtered effluents and concentration further reduced to 12.80 mg/L (93.67%) and 54.40 mg/L (75.27%) in case of filtered effluents when treated with microalgae and only aeration, respectively. Variation in COD concentration and removal efficiency during the study is shown in Fig. 4a, b, respectively. The oxygen released as a by-product from microalgae was sufficient to meet bacteria's oxygen requirements while degrading the organics in the treated wastewater, which acted as the sole mechanism for organic reduction [10]. Symbiotic growth of microorganisms (microalgae and bacteria) also contributed to COD removal by disintegrating complex pollutants into simpler ones, hence enhancing the degradation of organic matter [17], which resulted in higher removal of organics in the microalgal system.

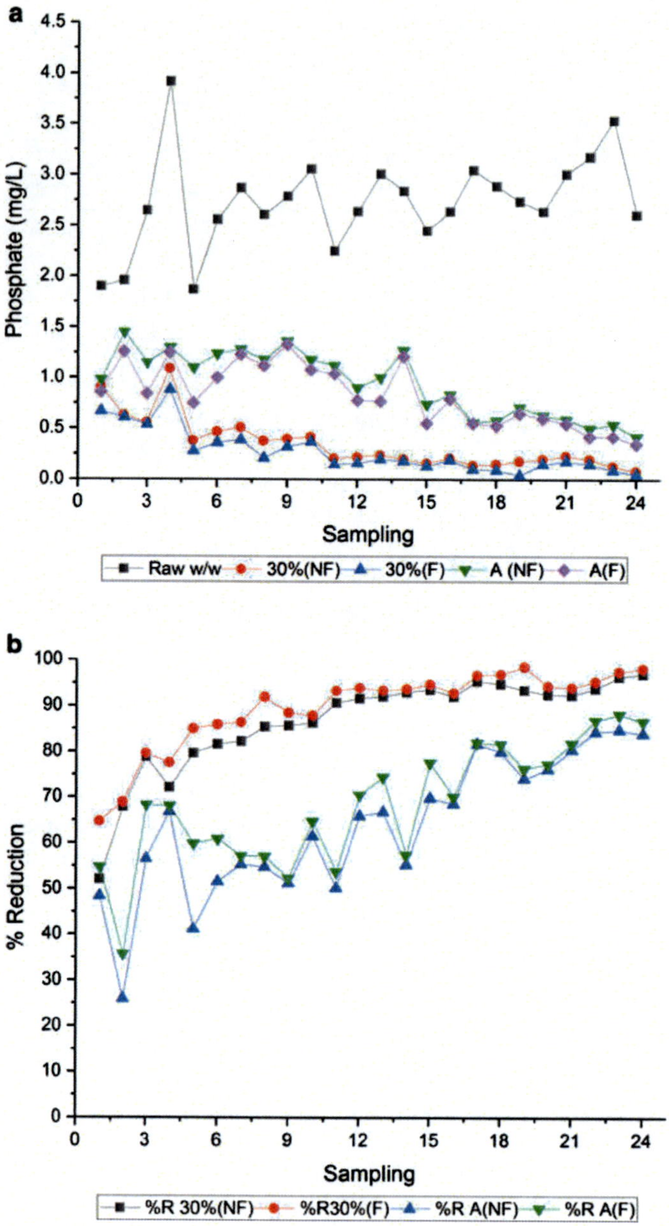

Fig. 2 **a** Variation in concentration of phosphate before and after the treatment at 11 h. **b** Removal efficiency observed in the microalgal system and control for phosphate during the study

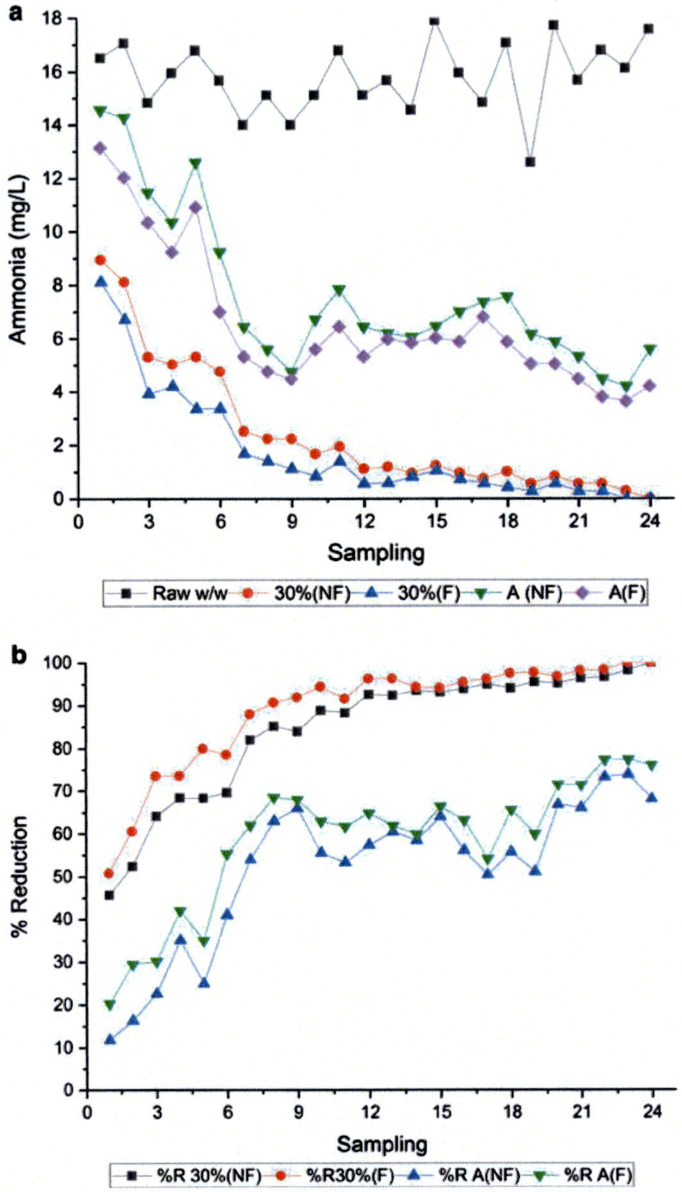

Fig. 3 **a** Variation in concentration of ammonia before and after the treatment at 11 h. **b** Removal efficiency observed in microalgal system and control for ammonia during the study

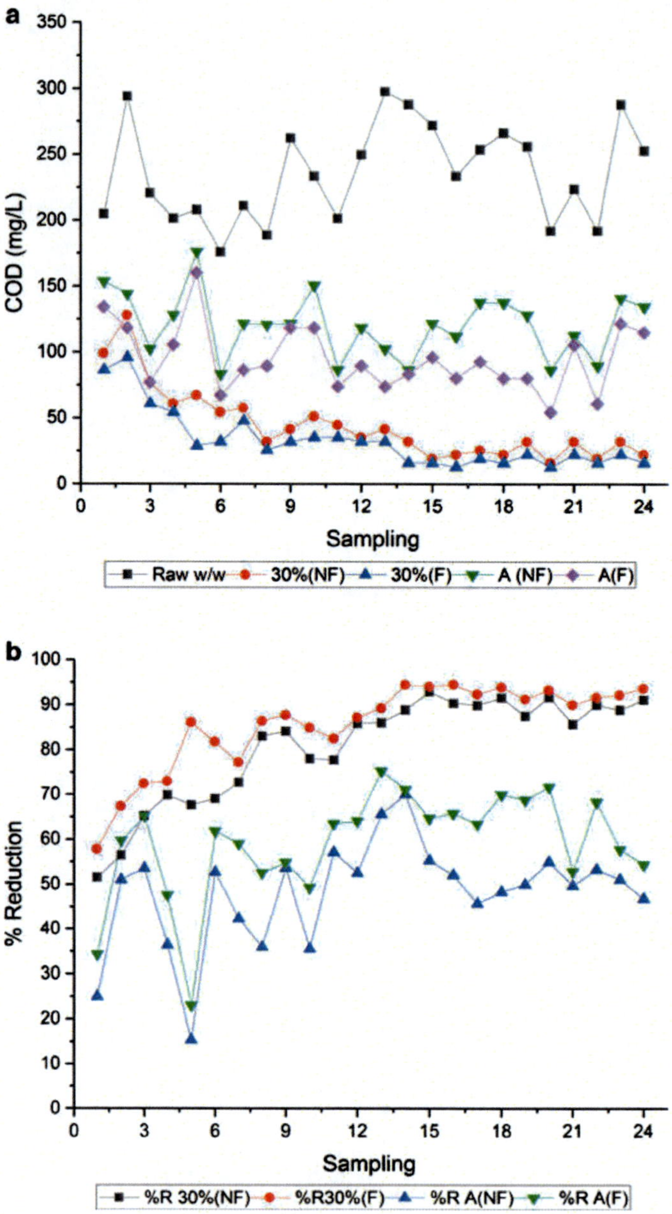

Fig. 4 **a** Variation in concentration of COD before and after the treatment at 11 h. **b** Removal efficiency observed in COD for microalgal treatment and control system

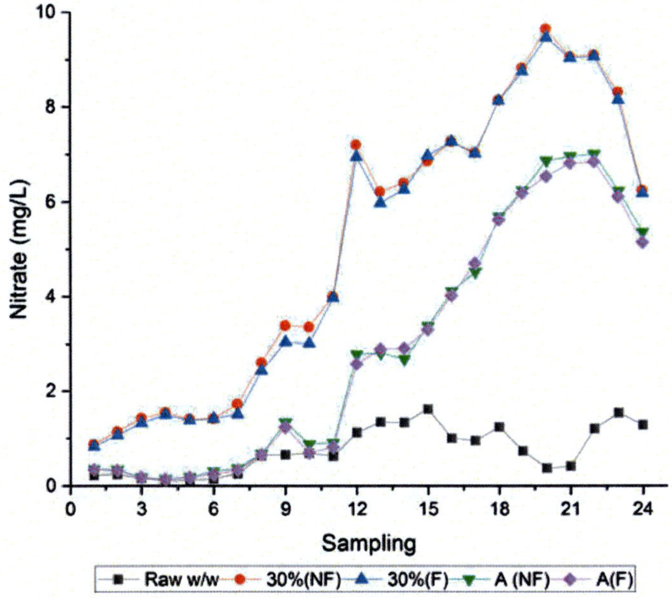

Fig. 5 Variation in nitrate concentration before and after the treatment at 11 h

Nitrate concentration in the raw sewage sample ranged between 0.11 and 1.62 mg/L. An increase in nitrate concentration was observed after treatment in both microalgal and control system due to nitrification [17]. A higher concentration of nitrate was found in the microalgal treatment system in comparison to the control reactor as shown in Fig. 5, as microalgae prefer ammonia the most as the source of energy food in terms of nitrogen source and due to the high DO concentration available in the microalgal reactor [11]. Nitrate concentration in the microalgal treated effluent reached 9.65 mg/L, and in the case of the control reactor, nitrate concentration in the effluent was as high as 7.01 mg/L for non-filtered samples. However, in filtered effluent, nitrate concentration reached 9.47 and 6.85 mg/L with microalgal treatment and aeration only (control), respectively. As from Fig. 5, it can be seen by the end of the experimental phase; nitrate concentration reduces even though ammonia is reduced to nitrate. Reduction in nitrate in the latter phase is because microalgae completely consumed ammonia, and now nitrate is taken as the nitrogen source by algae for its metabolism.

DO concentration increased in both control and microalgal treatment, during the study. However, DO concentration in the microalgal system's effluent was higher because DO was contributed from both microalgae and external aeration. DO in case of microalgal treatment reached 7.0 mg/L. In the case of control, DO was only contributed through external aeration; hence, DO for the effluent of control was lesser than that of microalgal treatment and reached 5.1 mg/L. DO concentration in raw

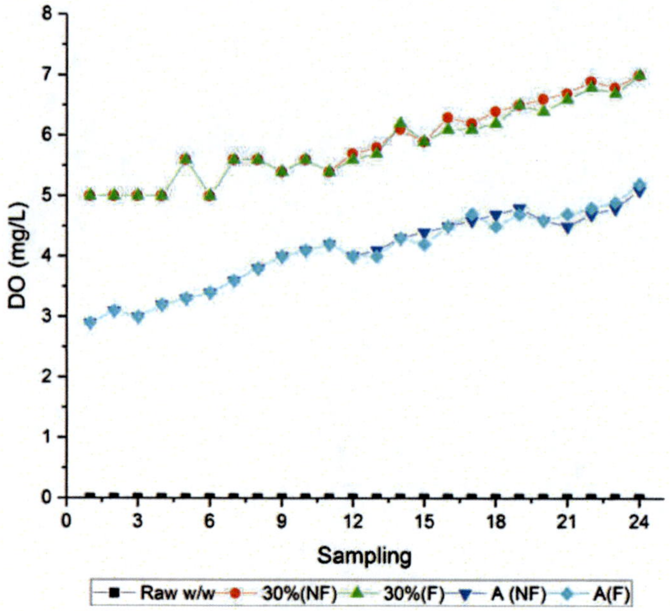

Fig. 6 Variation in DO concentration before and after the treatment at 11 h

wastewater was nil throughout the study. Variation in DO concentration during the study is shown in Fig. 6.

An increase in pH was also observed in the effluents. Large variations in pH were observed in raw wastewater ranging between 6.94 and 7.93. After treatment with microalgae and in the case of control, the maximum pH concentration found was 8.72 and 8.56 in non-filtered effluents, respectively. But in filtered effluents, the pH reached 8.65 and 8.48 for microalgal treated effluent and control-treated effluent, respectively. Variation in the pH concentration during the experimental phase is shown in Fig. 7. An increment in the pH of wastewater during the study also acts as a disinfectant by releasing antibacterial substances making unfavorable conditions for toxic substances. A higher increase in the pH of the effluent generated by microalgal treatment resulted from photosynthetic CO_2 assimilation [18].

Reduction in solids was less when compared to organics and nutrients analyzed during the study. The least reduction observed was in TDS among all the solids (TS, TSS, TDS) analyzed. TDS concentration in raw wastewater is varied from 980 to 1170 mg/L. After treatment, reductions observed in TDS concentration are varied from 9.80 to 23.06% in non-filtered microalgae treated effluent having a minimum concentration of 770 mg/L. When the microalgae treated effluent was filtered with a coarse filter, the reduction improved and ranged between 11.32 and 31.36% with a minimum concentration of 740 mg/L. Reductions were also observed in the control system, but were lesser than that of influent treated with microalgae. Removal efficiency in the case of non-filtered effluent of the control system ranges between 3.92

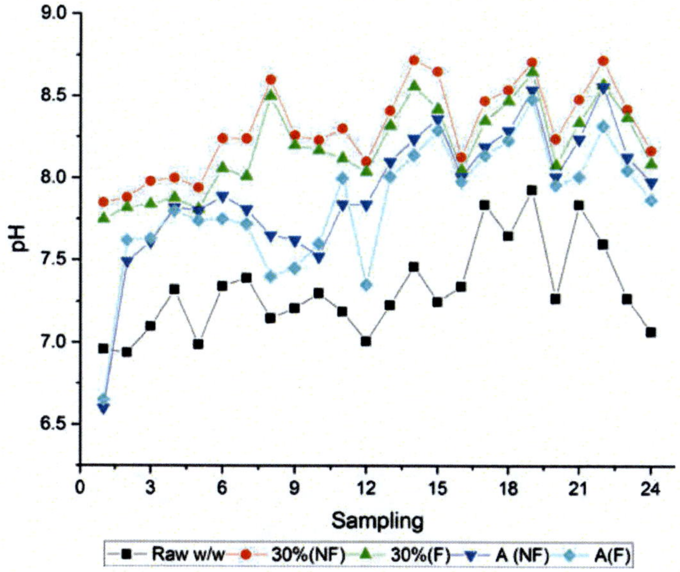

Fig. 7 Variation in pH concentration before and after the treatment at 11 h

and 22.55% with a minimum concentration of 790 mg/L. Removal efficiency in the case of filtered effluent of the control system ranged between 5.88% to 24.51%, with a minimum concentration of 770 mg/L. Variation in TDS concentration in influent and effluent (both treated with *C. vulgaris* and only aeration) and also removal efficiency variation is shown in Fig. 8a, b, respectively.

The highest reduction was observed in TSS among all the solids (TS, TSS, TDS) analyzed. TSS concentration in raw wastewater is varied from 1185 to 3055 mg/L. After treatment, reductions observed in TSS concentration are varied from 7.86 to 44.01% in non-filtered microalgae treated effluent having a minimum concentration of 715 mg/L. After filtration with a coarse filter; reduction improved and ranged between 16.63 and 59.35% with a minimum concentration of 510 mg/L in case of microalgal treatment system. Reductions were also observed in the control system, but were lesser than that of influent treated with microalgae. Removal efficiency in the case of non-filtered effluent of the control system ranges between 2.23 and 40.09%, with a minimum concentration of 809 mg/L. Removal efficiency in the case of filtered effluent of the control system ranged between 8.87 and 55.59%, with a minimum concentration of 623 mg/L. Variation in TSS concentration and removal efficiency in influent and effluent (both treated with *C. vulgaris* and only aeration) is shown in Fig. 9a, b, respectively.

During the study, TS concentration in raw wastewater is varied from 2220 to 4705 mg/L. After treatment, reductions observed in TSS concentration are varied from 8.34 to 33.00% in non-filtered microalgae treated effluent having a minimum concentration of 1655 mg/L. When the microalgae treated effluent was filtered with

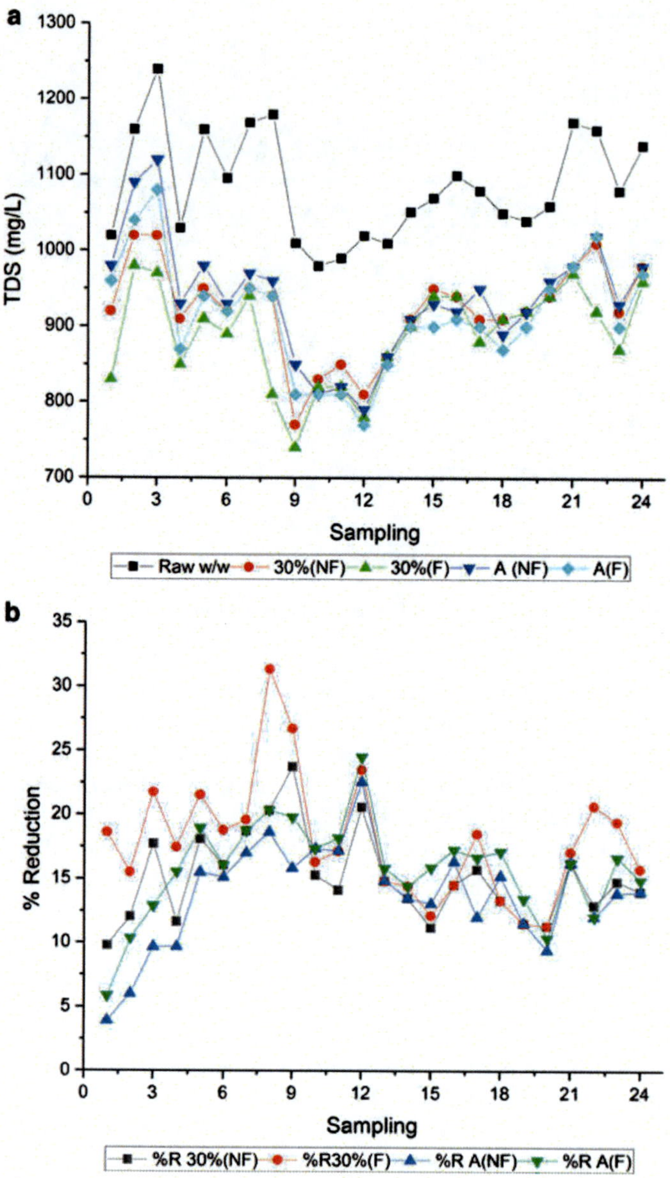

Fig. 8 **a** Variation in TDS concentration before and after the treatment at 11 h. **b** Removal efficiency observed in TDS for microalgal treatment and control system

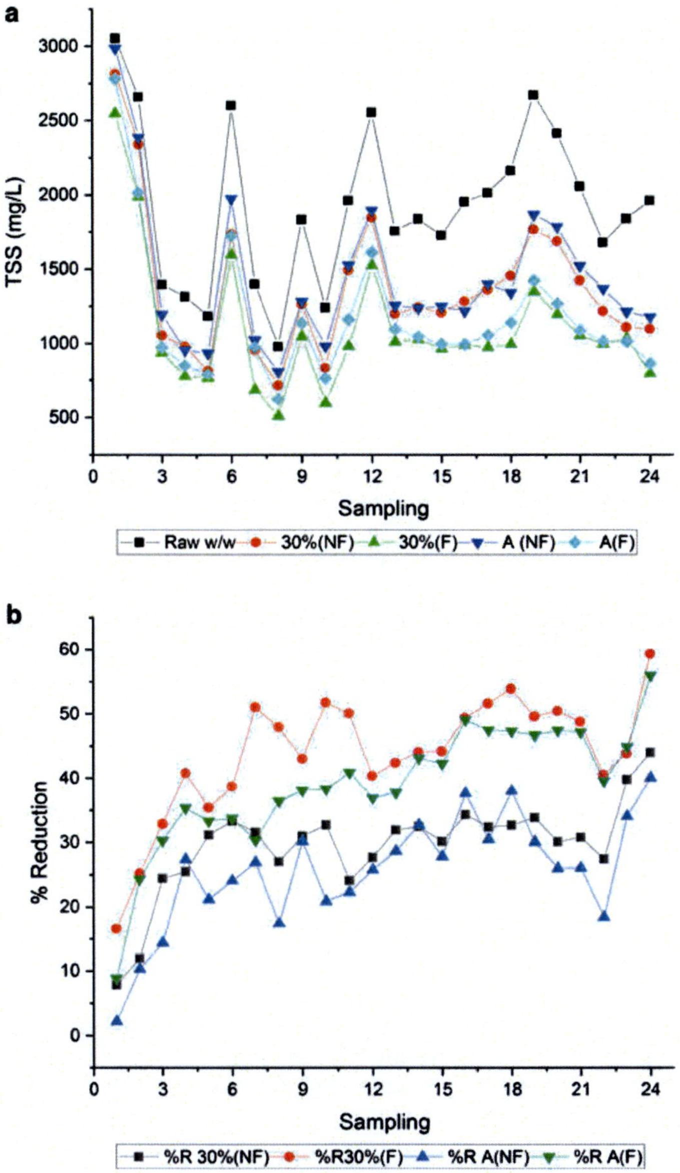

Fig. 9 **a** Variation in TSS concentration before and after the treatment at 11 h. **b** Removal efficiency observed in TSS for microalgal treatment and control system

a coarse filter, the reduction improved and ranged between 13.45 and 42.70% with a minimum concentration of 1408 mg/L. Reductions were also observed in the control system, but were lesser than that of influent treated with microalgae. Removal efficiency in the case of non-filtered effluent of the control system ranges between 2.65 and 30.52%, with a minimum concentration of 1769 mg/L. Removal efficiency in the case of filtered effluent of the control system ranged between 8.12% to 40.90%, with a minimum concentration of 1575 mg/L. Variation in TS concentration in influent and effluent (both treated with *C. vulgaris* and only aeration) and removal efficiency is shown in Fig. 10a, b, respectively.

Variation in EC was similar to that of TDS. During the study, EC concentration in raw wastewater is varied from 1.97 to 2.37 mS/cm. After treatment, reductions observed in EC concentration are varied from 9.26 to 23.30% in case of non-filtered microalgae treated effluent having a minimum concentration of 1.55 mS/cm. When the microalgae treated effluent was filtered, the reduction improved and ranged between 11.50 and 30.94% with a minimum concentration of 1.49 mS/cm. Reductions were also observed in the control system, but were lesser than that of influent treated with microalgae. Removal efficiency in the case of non-filtered effluent of the control system ranged between 3.53 and 23.17% resulting in a minimum concentration of 1.59 mS/cm. Removal efficiency in the case of filtered effluent of the control system ranged between 4.73 and 24.81% resulting in a minimum concentration of 1.55 mS/cm. Variation in EC concentration in influent and effluent (both treated with *C. vulgaris* and only aeration) and also the removal efficiency is shown in Fig. 11a, b.

Mixing acted as an essential factor during the study as it led to homogenized the cells' distribution, warmth, metabolites and encouraged the transfer of gases and prevent settling [19], thus empowers nutrients disintegration also uniform dispersion of light which helped in better performance of both the system.

The study also gave an understanding that microalgae can be an efficient way to treat domestic wastewater. Raw domestic wastewater, when treated with *C. vulgaris,* lead to an excellent removal efficiency for organics and nutrients, and also, a considerable reduction was observed in solids and EC. High photosynthetic effectiveness, fast take-up of nutrients combined with the nutritional necessity of microalgae favors the bioremediation process compared to a conventional system [9]. Microalgae do not rely on organic carbon sources, unlike the heterotrophic bacteria of the secondary treatment systems. Instead, they consume CO_2 for their metabolism and hence also act as a solution to global warming.

4 Conclusion

Microalgae can be an attractive domestic wastewater treatment process. Microalgae, when used for sewage treatment, are alone self-sufficient for the removal of solids, organics, and nutrients. Thus, it can eliminate the need for different sewage treatment units and can be a cost-effective process for the treatment. Also, the oxygen released

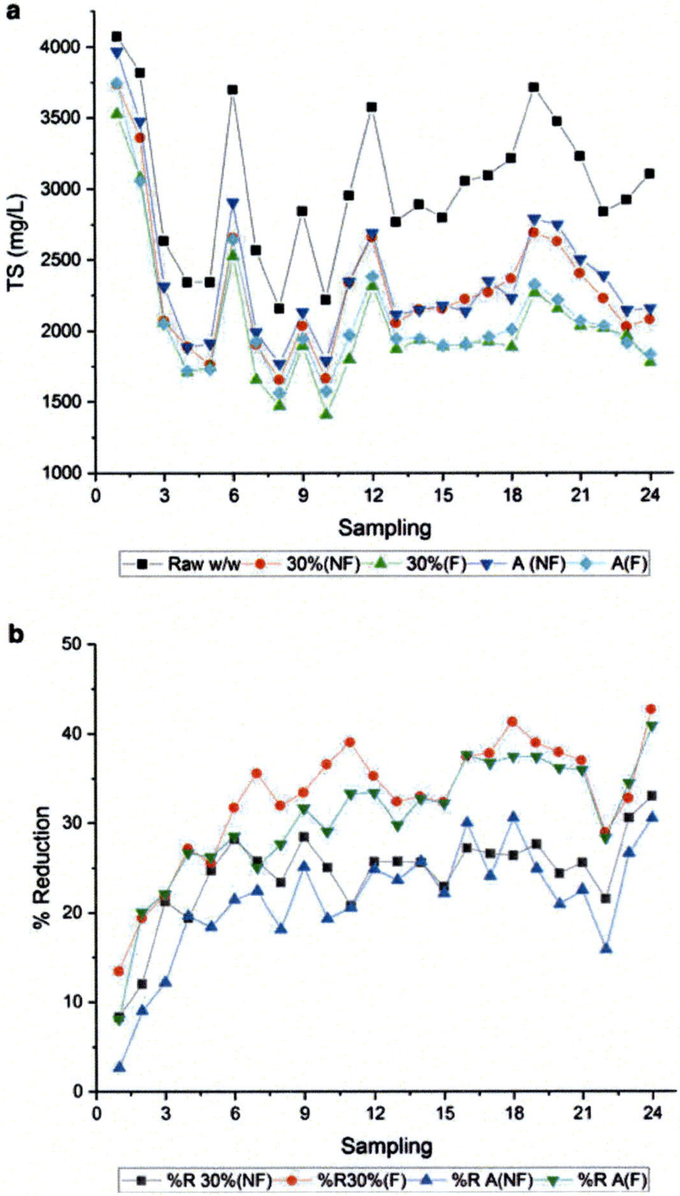

Fig. 10 **a** Variation in TS concentration before and after the treatment at 11 h. **b** Removal efficiency observed in TS for microalgal treatment and control system

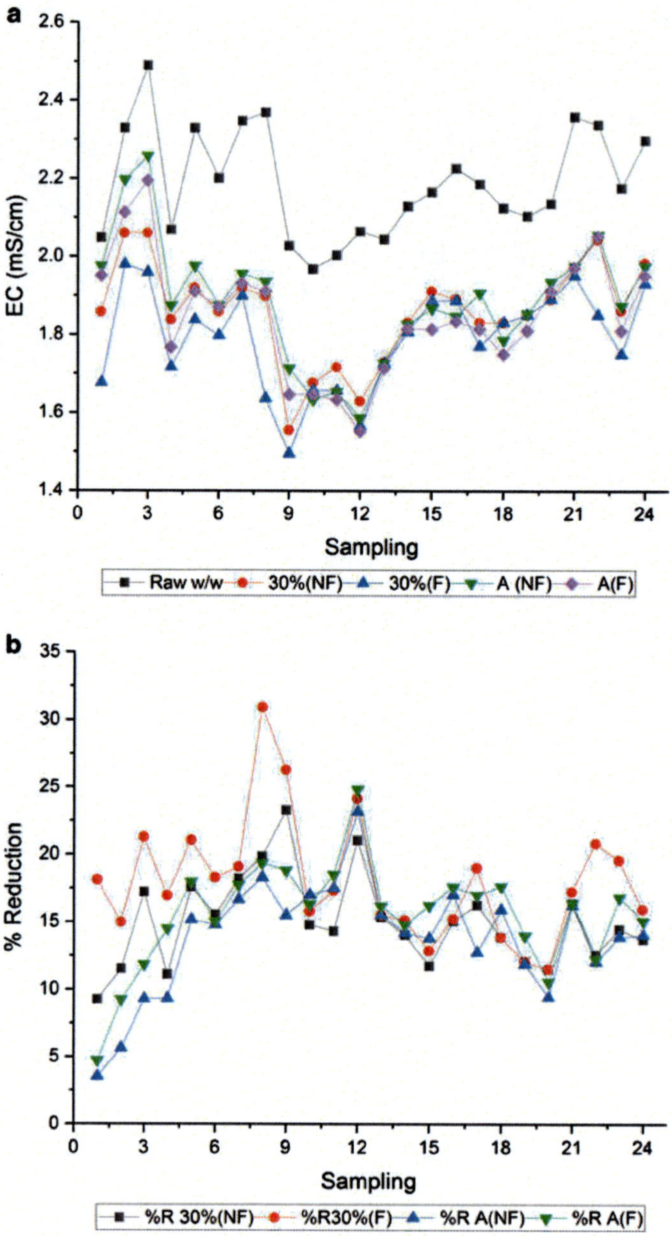

Fig. 11 **a** Variation in EC concentration before and after the treatment at 11 h. **b** Removal efficiency observed in EC for microalgal treatment and control system

as a bio-product can be sufficient for organic and nutrient removal, thus eliminating the necessity of the blowers commonly used in the secondary treatment and alone contributes to 40–50% of the cost for wastewater treatment. In the study, microalgae lead to > 95% reduction in nutrients and > 90% reduction in organics of the raw wastewater, which was far better than the reductions found in the final treated sewage effluent after preliminary, primary, and secondary treatment. Microalgal treatment can be the alluring solution to the eutrophication problem faced by a significant part of the country.

Acknowledgements The authors are thankful to all those people whose technical and non-technical support led to the experimental study's successful conduct.

References

1. Brenner K, You L, Arnold FH (2008) Engineering microbial consortia: a new frontier in synthetic biology. Trend Biotechnol 26:483–489
2. Kaya VM, de la Noue J, Picard G (1995) A comparative study of four systems for tertiary wastewater treatment by *Scenedesmus bicellularis*: new technology for immobilization. J Appl Phycol 7:85–95
3. Abdel-Raouf N, Al-Homaidan AA, Ibraheem IBM (2012) Microalgae and wastewater treatment. Saudi J Biol Sci 19:257–275
4. Renuka N, Sood A, Ratha SK (2013) Evaluation of microalgal consortia for treatment of primary treated sewage effluent and biomass production. J Appl Phycol 25:1529–1537
5. Arbib Z, Ruiz J, Alvarez-Diaz P, Garrido-Perez C, Perales JA (2014) Capability of different microalgae species for phytoremediation processes: Wastewater tertiary treatment, CO2 bio-fixation and low cost biofuels production. Water Res 49:465–474
6. Cabanelas ITD et al (2013) Comparing the use of different domestic wastewaters for coupling microalgal production and nutrient removal. Bioresour Technol 131:429–436
7. Samorı C, Samorı G (2013) Growth and nitrogen removal capacity of *Desmodesmus communis* and of a natural microalgae consortium in a batch culture system in view of urban wastewater treatment. Water Res 47:791–801
8. Mallick N (2002) Biotechnological potential of immobilised algae for wastewater N, P and metal removal: a review. Biometals 15:377–390
9. Boelee NC, Temmink H, Janssen M, Buisman CJN, Wijffels RH (2011) Nitrogen and phosphorus removal from municipal wastewater effluent using microalgal biofilms. Water Res 45:5925–5933
10. Mishra N, Mishra N (2017) Utilization of microalgae for integrated biomass production and phycoremediation of wastewater 8:95–105
11. Moondra N, Jariwala ND, Christian RA (2020) Sustainable treatment of domestic wastewater through microalgae. Int J Phytoremediation 22:1480–1486
12. APHA (2012) Standard methods for the examination of water and wastewater. Am public Heal Assoc
13. Xu M, Li P, Tang T, Hu Z (2015) Roles of SRT and HRT of an algal membrane bioreactor system with a tanks-in-series configuration for secondary wastewater effluent polishing. Ecol Eng 85:257–264
14. Hwang JH, Church J, Lee SJ, Park J, Lee WH (2016) Use of microalgae for advanced wastewater treatment and sustainable bioenergy generation. Environ Eng Sci 33:882–897
15. Cai T, Park SY, Li Y (2013) Nutrient recovery from wastewater streams by microalgae: status and prospects. Renew Sustain Energy Rev 19:360–369

16. Lv J, Feng J, Liu Q, Xie S (2017) Microalgal cultivation in secondary effluent: recent developments and future work. Int J mol Sci 18:1–18
17. Moondra N, Jariwala ND, Christian RA (2020) Microalgal-bacterial consortia: an alluring and novel approach for domestic wastewater treatment. WCM 4:51–56
18. Daliry S, Hallajisani A, Mohammadi Roshandeh J, Nouri H, Golzary A (2017) Investigation of optimal condition for *Chlorella vulgaris* microalgae growth. Glob J Environ Sci Manag 3:217–230
19. Goncalves AL, Pires JCM, Simoes M (2017) A review on the use of microalgal consortia for wastewater treatment. Algal Res 24:403–415

A Review on the Development of Outriggers and Introduction to Hybrid Outrigger System on Tall Buildings

Neethu Elizabeth John and Kiran Kamath

Abstract The concept of outriggers as a lateral load resisting system in high-rise structures has been used since decades. The relevance of outriggers gains prominence due to its efficiency in restraining the rotations in core wall which in turn reduces its bending moment and lateral deflection of the structure. Studies related to the behaviour of outriggers on tall buildings have been undertaken by various researchers, and condensing the concept, mechanism and its behaviour is vital in providing a proper guidance to researchers. Even though researchers have worked on buildings with outriggers, literatures mentioning the analysis of both conventional and virtual outriggers in the same building which acts like a hybrid outrigger system were under-explored. Therefore, this article discusses a review on the background of outriggers, classification of outriggers, parameters influencing the outrigger behaviour and introduces the concept of hybrid outrigger system in high-rise RCC structures. The efficacy of conventional, virtual and hybrid outrigger system is compared by modelling a 30-storey RCC building in a finite element software, ETABS, for both static and dynamic loads. The results from the comparative study showed that hybrid outrigger system may be used as an efficient lateral load resisting system.

Keywords Lateral load resisting system · Conventional outriggers · Virtual outriggers · Hybrid outriggers · Behavioural study

1 Introduction

The development of high-rise structures has been increasing rapidly due to sudden growth in human population and limited land availability. When the design of high-rise structures is concerned, it is necessary to consider wind and seismic loads as a major area of our country is susceptible to these loads. In order to resist the effect of these loads which can cause undesirable damage to the building, new lateral

N. E. John (✉) · K. Kamath
Department of Civil Engineering, Manipal Institute of Technology, MAHE, Manipal, India

K. Kamath
e-mail: kiran.kamath@manipal.edu

L. Nandagiri et al. (eds.), *Sustainability Trends and Challenges in Civil Engineering*, Lecture Notes in Civil Engineering 162, https://doi.org/10.1007/978-981-16-2826-9_5

load resisting structural systems are being discussed and introduced. The structural systems were classified based on their efficacy and building heights in the beginning [1]. After new developments, classifications were based on the capability of the structural systems and can be divided into exterior and interior structural system [2]. Various structural systems include structural wall system, moment frame system, moment frame–structural wall system, core and outrigger system, core, outrigger and belt wall system, framed tube system, multiple tube system and tube-in-tube system [3]. In this article, the concept of outriggers as a lateral load resisting structural system is being discussed. The background, different classifications of outriggers, the force transfer mechanism, structural behaviour of outriggers and concept of hybrid outrigger system which incorporates two different types of outriggers, i.e. conventional and virtual outriggers, in the same buildings and its efficiency in reducing the lateral loads and bending moment is also showcased in this article.

2 Background of Outriggers

Outriggers are stiffened deep beams which connects the exterior columns to the central core [4]. Even though the use of concrete central core was way useful to counter the lateral loads, when increasing the structural height, the core is not able to provide adequate stiffness. So, by providing a deep beam connecting the core to the perimeter columns the load resisting capacity was improved and can be used for structures up to almost 150 storeys [2]. Outrigger usage has been practised in sailing industry since long to withstand the wind load acting on the ship, thus preventing overturning. As a comparison to a structure, the core is related to ship's mast and spreaders as outriggers and stays as perimeter columns [2].

When lateral loads act on the structures, column-restrained outriggers try to restrict the rotation of central core wall, thereby reducing the deflections and moments acting in the core [4]. The ideology is to enhance the effective depth of the building when the core tries to bend as vertical cantilever, generating tension forces in windward columns and compressive force in leeward columns as shown in Fig. 1. The net impact of these forces in the columns is the formation of a couple which in turn reduces the bending moment of the core [5].

Outrigger structure is effective in increasing the flexural stiffness of the building, but it does not improve the resistance to shear; therefore, the shear reduction is taken care by the central core and the bending moment reduction by the outriggers [4].

2.1 *Behaviour of Conventional Outriggers*

The flexibility of the outrigger was considered for analysing the behaviour of the structure with outrigger, and the structure was assumed to be uniform [4]. Multiple regression analysis with up to four outriggers was used to develop formulas to reduce

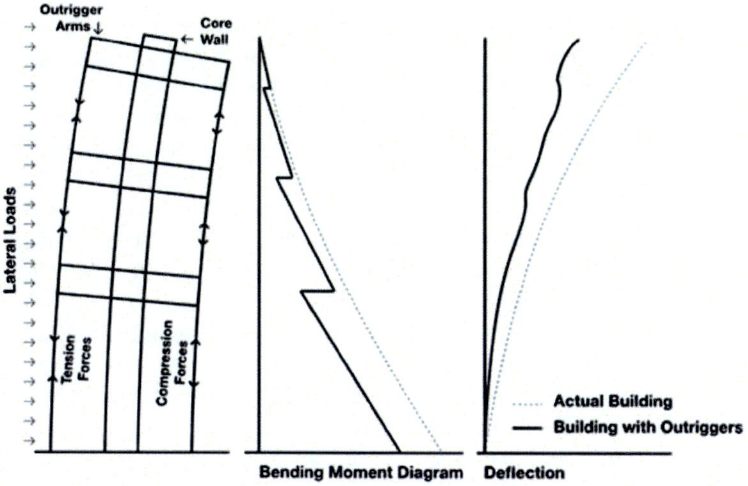

Fig. 1 Outrigger structure under lateral loads

the drift in outrigger braced tall buildings and was solved using compatibility condition analysis of structures [6]. Certain assumptions were considered in the preliminary analysis of outriggers; them being the entire structure to be linearly elastic and uniform, action of axial forces alone in the columns, connection between the core and the outrigger to be rigid and connection between outriggers and columns to be pin jointed [4–6]. A compatibility method of analysis was chosen where the core rotations at the level of outrigger are equated with the rotations formed at the corresponding outriggers [4–8]. For finding the equations for horizontal deflections and the optimum location for n outriggers, the rotations of core and outrigger were equated. The core rotations were written in terms of its bending deformation and that of outriggers in terms of bending of outriggers and axial deformation induced in the columns for uniform horizontal load [4–6]. The optimum positions for n outriggers to resist the lateral deflection and drift were found to be $1/(n + 1)$, $2/(n + 1)$, $3/(n + 1)$, …, $n/(n + 1)$ times height of the structure [5, 6]. Various researchers have worked on finding the optimal location of outriggers, and they have found the positions almost equal to the positions mentioned above.

3 Classification of Outriggers, Their Behaviour and Importance

The main classification of outriggers is based on their connection between the internal core and the external structural systems. The connection between the exterior columns and core can be either direct or indirect. When the core and the exterior columns are connected directly through outrigger beams, it is termed as conventional outrigger system, and when there is no direct connection between core and the perimeter columns but the outrigger action is through a belt wall or truss surrounding the exterior columns, it can be termed as virtual outrigger system [9]. Virtual outriggers consist of a belt wall or a belt truss alone connecting the outer columns as in Fig. 2(a).

Another classification called outrigger and belt truss system was developed to assist the columns which are not connected to outrigger beam to participate in the restraining effect. Here, in addition to the conventional outrigger beams, all the exterior columns in the outrigger level are connected by means of belt truss or belt walls in one- or two-storey depth, which further enhances the stiffness of the building and can reduce the differential axial shortening of the columns and the core [6]. This arrangement termed as outrigger and belt truss/wall system is shown in Fig. 2(b).

The concept of virtual outriggers is based on the use of rigid floor diaphragms which transfer moments from core to the belt wall/truss as shown in Fig. 3. The load transfer in both conventional and virtual is shown below.

The rigid floor diaphragms transfer the moment formed in the core into equivalent horizontal couple and transfer it to vertical wall/truss. Then, the belt truss converts this horizontal couple into vertical couple and transfers it to the exterior columns. So, because of the participation of all the columns for the force transfer, differential shortening of the columns and the core can be avoided as the floor diaphragms are stiff and another advantage of virtual outriggers is that it does not cause space obstruction at the outrigger level as in conventional outrigger system [9, 10]. A new design of distributed belt wall system instead of a continuous belt wall around the columns was developed recently which can minimize the space loss at floor level where the belt walls are installed. The arrangement was found to lessen the drift and displacement values compared to continuous wall and conventional outrigger walls [11]. In distributed belt wall system, pre-stressing strands were introduced to decrease the high shear demand formed in the belt walls and were found that by enhancing the amount of pre-stress and steel reinforcement ratio of the strands, resistance in shear for pre-stressed concrete belt wall has improved [11].

The outriggers can be positioned at an offset distance from the core, and the classification is shown in Fig. 4. In offset outriggers, beams of outrigger elements

(a) (b)

Fig. 2 **(a)** Virtual outrigger and **(b)** conventional outrigger with belt wall

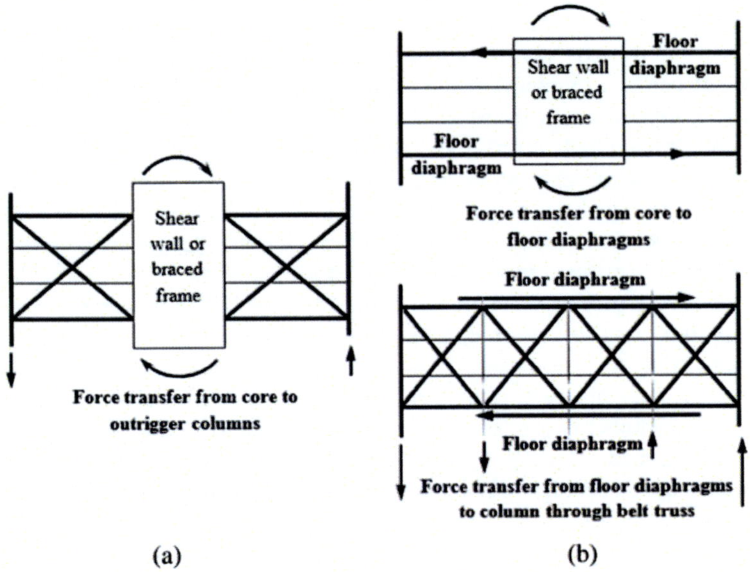

Fig. 3 Force transfer in **(a)** conventional outrigger with belt truss **(b)** virtual outrigger

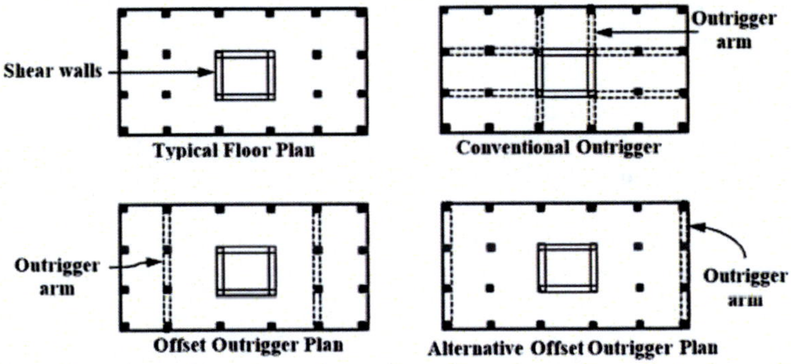

Fig. 4 Different positioning of outriggers

are installed at a definite distance from the central core of the structure and direct connection is eliminated.

The alternative offset outrigger system also termed as facade rigger [12] is another classification in outriggers and is shown in Fig. 4. It is also feasible to use diagonal steel truss members extending over several floors to act as outrigger structures [13].

From the material point of view, outriggers can be classified into steel, concrete and composite outriggers [14]. Steel outrigger can be mainly used for steel building as steel trusses or diagonal truss. The connection between steel and concrete outrigger may be complicated if steel outriggers are installed in concrete frames [9]. Concrete outrigger has comparatively lesser cost and more stiffness when compared to steel outriggers and is used extensively in concrete framed buildings [14]. The brittleness of concrete can be a problem when it comes to ductility of the structure in tall buildings; therefore, combining the benefits of both steel and concrete may be incorporated which can provide high stiffness and ductility which can resist large deformations. This type of arrangement can be termed as composite outrigger arrangement [14]. They can be installed by having a steel truss protected inside a concrete wall or by enclosing a steel core in a concrete casing. When a composite connection is designed, the transfer of horizontal force from the steel braced frame and concrete floor can be practised using shear studs on the concrete slabs. When a concrete shear wall is used, reinforcing steel from concrete wall to concrete floor helps in transferring the forces. Also, the movement of forces between the floor slabs and the belt truss chords can also be practised using shear studs attached to the chords [9]. Vierendeel Outriggers is another type of outrigger which is designed to have high ductility and lower rigidity, thus, allowing them to be used in higher seismic areas and also, this type allows door openings in the outriggers so as to prevent the obstruction of the inner floor area [15].

3.1 Behavioural Study for Various Outrigger Classifications

Studies for finding the optimum location of various outrigger types were undertaken by researchers since long. The following shows a brief review on the behavioural assessment analysis for outriggers in chronological order of their publication year. Outrigger behavioural study had started way back in the 1980s, and the initial analysis on uniform flexible outriggers is mentioned under Sect. 2.1 in Sect. 2 of this article.

Study on drift reduction in rigid outriggers for a non-uniform and uniform belted system subjected to lateral loads was investigated in 1987 [16]. Simple expressions were computed for moment and roof displacement by varying the flexural rigidity of core for assisting in preliminary phase of design and graphical representations showing optimum location and roof displacement for both single and double outriggers were discussed for various loadings [16]. In the 1990s, development on virtual outriggers was studied and is explained in Sect. 3 of this article. Nair in his paper had explained the force transfer mechanism and base restraints for tall structures with basements, analysed a 75-storey steel framed office building with steel belt truss as virtual outrigger and compared the result with conventional outrigger result in GTSTRUDL software [9].

Later in the year 2000, behavioural analysis for facade riggers was performed by Hoenderkamp and Snijder. They developed a simplified method for the preliminary design of façade riggers considering four stiffness parameters; them being bending stiffness of walls, axial stiffness of the columns in the end frames, and bending and racking shear stiffnesses of the facade riggers [12]. The study concluded that the flexibility of the outriggers greatly influenced the exact position of the facade rigger [12]. As an application of Nair's concept of virtual outriggers, an analytical study was performed by Horton on a steel framed office building as analysed by Nair [9] but giving some modifications by removing the chamfers and placing "super-columns" at the corners and continuing the belt truss bracing around the entire outrigger level. He concluded that a twofold increase in outrigger diaphragm thicknesses coupled with a 16% increase in belt truss area enabled the virtual outrigger system to perform within 4% of the lateral deflection capacity achieved by conventional outrigger system [10].

In 2001, performance of diagonal outrigger and belt truss system in tall buildings subjected to lateral loads considering eight models of 40-storey 2-D structures having outrigger belt truss system put through wind load and five numbers of 60-storey 3-D models put through seismic load was analysed to obtain the top displacement [17]. The optimum position was found to be top and at middle of the structure for maximum reduction in lateral displacement and core bending moment [17]. Later in 2003, impact of outrigger locations and member stiffness on the base moment, fundamental vibration period and top drift of tall building structures were analysed using MATLAB program for various loadings and suggested that type of loading also matters the optimum position of outriggers [18]. It was found that, for triangular loads on outrigger structures, the favourable location for outriggers was 4–5% higher compared to horizontal uniform loads, and for maximum reduction in bending moment the position of outriggers was found to be near to the foundation [18].

In the year 2000, facade rigger was analysed taking four stiffness parameters [12] and in 2003, 5 stiffness parameters were considered which includes bending and racking shear stiffness of the facade riggers, bending stiffness of the perimeter frames where riggers were located, and bending and racking shear stiffness of braced wall frames. These when taken into analysis, it was suggested that all the five parameters greatly influenced the location of the facade rigger [19].

When braced outrigger and belt truss were investigated, the racking and bending shear stiffness of the belt truss and outriggers, and stiffness parameter concerning axial shortening and lengthening of the perimeter columns were found to influence the outrigger location for maximum reduction in deflection and bending moment in core [20]. In 2003, behaviour of outriggers for shear walls stiffened by hefty beam and outrigger system was investigated using conventional continuum approach and it was obtained that the favourable outrigger location to reduce the top drift was between 0·4 and 0·6 times the height of building [21].

An analytical study considering the moment in core wall and horizontal top deflection based on the finite rigidities of outriggers was performed in 2007, taking a 50-storey building, and was solved by equating compatibility conditions between the outriggers and the central core wall [8]. It was found that as the outrigger rigidity increased, restraining moment of the core increased rapidly and the horizontal deflection decreased, and when the outrigger rigidity was low, the location of outrigger was higher than the optimal location [8].

In 2008, continuum approach was used for developing equations for outriggers in wall frame buildings, the building was designed as a shear flexural cantilever system with rotational springs as outriggers and impact of flexural and shear deformation of outrigger trusses and the wall-frame was considered for the design equations [22]. It was concluded that the analytical model which considers flexural bending of frame and shear deformation of the wall showed good agreement when compared to the conventional approach [22]. A discussion on the benefits of using virtual outriggers at floor levels and basement in high-rise building as proposed by Nair [9] was made in 2008, and a comparison between conventional and virtual outriggers for an 80-storey building was performed in ETABS software [23]. The results show that there was a huge reduction in bending moment when basement was used as outriggers which in turn can reduce the foundation size, and when same dimensions were used for both conventional and virtual outriggers, the effectiveness of virtual outriggers was found to be less due to indirect force transfer mechanism [23].

In 2008, Hoenderkamp considered outriggers at two levels and framed several equations for finding the optimum outrigger location [24]. From the study, it was found that for the outriggers with unbounded bending stiffness, the optimum position of the 2nd outrigger was at 0·577 from the top and the first one was installed at the top [24]. In 2009, a nonlinear response spectrum analysis for finding the optimum position of outriggers on a 50-storey structure was performed using STRAND7 finite element package for 3 different PGA-to-PGV ratios in each category of seismic records and validation was done by SPACE GASS frame analysis [25]. It was found that the optimal position of the outrigger in that model was between 22 and 24 levels, i.e. is between 0.44 and 0.48 times its height when measured from the bottom of the structure for maximum reduction in lateral deflection [25].

In 2009, a graphical method of evaluation for the preliminary analysis of vertical truss frames and facade riggers was performed considering rigidity of floor diaphragms under horizontal loads [26]. Interconnection between the trussed frames and facade riggers happens with the help of floor diaphragms adjoining the top and bottom frames of the riggers. Therefore, the analysis considered 7 stiffness parameters; them being bending stiffness of the perimeter frame supporting the riggers, bending and shear stiffness of offset riggers, vertical trussed frames and floor structures [26]. It was found that all the seven parameters greatly influenced the mechanism and positioning of facade riggers [26].

Design of tall structures with RC shear walls and belt truss bracing around the perimeter columns put through lateral loading was considered in 2012 using a method similar to the analysis of facade riggers [27]. Flange trusses' racking and bending shear deformations along the windward and leeward facades was considered along with other stiffness parameters mentioned in [26] for accounting the shear lag effect caused due to differential strain in the flange columns and concluded that all the 8 stiffness parameters greatly influenced the working of belt truss connecting the exterior columns [27].

In 2012, a 40-storey building with and without outriggers was analysed using ETABS software for both static and dynamic loads considering PGA of California region using time history analysis [28]. The relative flexural rigidity of core and outrigger was varied from 0.25 to 1.75 in steps of 0.25, and the outrigger position was varied along the height for a relative height of 0.975–0.4 [28]. It was concluded that the performance of the outriggers was most effective when the outrigger was placed in the mid-height for both static and dynamic loadings [28]. Design guidelines for the usage of outriggers were not readily available, so, in 2012, CTBUH had created outrigger working team to form the first guide for outrigger design considering the overview, challengers for outrigger system, applications of outriggers, effects on the behaviour of the building, recommendations for differential axial shortening and construction sequence [29]. Later in 2012, seismic analysis was performed on a 40-storey building using static method of analysis on a multi-storeyed building with multi-outriggers by varying the relative rigidity between core and columns [30]. It was concluded that there was a decrease in horizontal displacement at the top for a multi-outrigger structure having a relative height of 1.5 (storey 20 and 30) and a considerable reduction in bending moment for a relative height of 6.67 (storey 40 and 6) [30].

To determine the ductility characteristics of the structure with outriggers installed, a nonlinear push over analysis was performed for 3 buildings with varying heights and the results were interpreted in terms of internal forces, deformations, capacity and ductility [31]. Three heights of 25, 30 and 35 storeys for 5 different ground accelerations were taken, and three types of strengthening systems included were rigid outrigger, Vierendeel outrigger and braced outrigger [31]. It was concluded that Vierendeel system gave the highest ductility and braced system has the highest lateral resisting ability compared to other models [31]. A 60-storey model having a central core wall was analysed to investigate the effect of differential column shortening (DCS) when outriggers were placed at various levels and were subjected to elastic shortening, shrinkage and creep [32]. It was found from the analysis that the DCS decreased by 34% when a single outrigger was installed and for two outriggers DCS was reduced by 48% when the position of second outrigger was fixed at $H/h1 = 1.33$ [32].

A seismic study for a 2-D steel building using pushover analysis was carried out on a tall structure whose height varied from 20 to 35 storeys for different loading types and outrigger locations [33]. The position of outriggers was measured in terms of base shear, inter-storey drift ratio, storey displacement and performance point in tall structures [33]. It was concluded that addition of an outrigger at the top storey and second outrigger near to the mid-height can effectively increase the performance of the building and the position varied as load type changes [33]. The evolution of outriggers considering concept of outrigger design, favourable topology, installation method, outrigger classification, brief background history of outriggers, issues during construction and axial shortening impact of core and exterior structures was discussed in 2016 by Ho [34].

The optimal position of outrigger on a high-rise building based on inter-storey drift (ISD) factor for multi-outrigger varying from one to three numbers was analysed using MATLAB programming [35]. The distance between the columns and the outrigger sections was the parameters varied for both wind and earthquake loads [35]. It was observed that the optimal location was slightly lower in case of earthquake load compared to wind load [35]. To study the effect of differential axial shortening (DAS) in structural frames with outrigger belt system (OBS), a study was conducted and the DAS effect for outrigger with belt wall and without belt wall was analysed [36]. It was found that the installation of OBS can decrease DAS, but when constructed without belt walls, the DAS effect is increased and also concluded that staged construction or elastic single-step method can decrease the accuracy of DAS results [36].

Using continuum approach, the optimum position of outrigger system in tall buildings was evaluated by maximizing the belt truss outrigger system's strain energy for various lateral loads [37]. The outrigger system was modelled as rotational springs, and when the absorbed energy was maximized, the optimal outrigger position was obtained. It was suggested that the proposed method gave efficient results when the theory was compared with various numerical examples [37]. A relative study of time period based on stiffness was performed on a G + 45 storey RCC building using ETABS software for various plan configurations of outrigger and belt wall system [38]. The results were analysed based on the lateral displacement, time period, storey drift and storey displacement, and from the analysis results it was found that the optimal position of 1st outrigger was at 15th floor and the 2nd outrigger was at 30th storey [38].

A study was conducted to review the existing global formulae which gave the preliminary design values of moments and horizontal deflections for core-supported-with-outrigger (CSOR) system and tube-in-tube-with-outrigger (TTOR) system as the existing formulas neglected the reverse rotation of the outriggers in TTOR [39]. A new methodology which provided better results was proposed based on global and local analysis of RC outrigger beams using nonlinear FE analysis and strut-and-tie modelling [39]. From the results, different challenges obtained were detailing structural members which are heavily reinforced and brittle failure of the members [39].

A study was conducted to minimize the inter-storey drift of a framed-tube structure with outriggers having non-uniform core and perimeter columns using genetic algorithm, and a parameter analysis was performed in MATLAB program to identify the favourable position of the outriggers [40]. The analysis was conducted on a 65-storey building of height 260 m, and the independent parameters varied were the objective functions, thickness of core wall, segments of outriggers, grade of concrete, stiffness of outriggers and type of loading [40]. It was concluded that the value of inter-storey drift (ISD) reduced as the cross section of the peripheral columns, stiffness of outrigger, core and columns was increased making the favourable location of the outrigger go down and as the thickness of core wall was increased, the favourable position of the outrigger was found to go upwards. It was also observed that when outrigger segments were increased from one to five the ISD changed unevenly and remained unchanged for the change in grade of concrete [40]. In 2018, a method to reduce the differential axial shortening (DAS) of structures for various outrigger positions was proposed using piecewise quadratic interpolation [41]. The results demonstrated that the DAS value does not reduce considerably when more than two outriggers were installed, even though the value of DAS reduces with increase in outrigger number [41].

A comparison between various models having shear core, outrigger and belt truss, outrigger and belt wall and its combinations was performed using response spectrum method based on the results from base shear, member shear forces and column axial forces [42]. It was concluded that shear core with outrigger and belt wall system was most effective when compared to other cases [42]. A study on finding optimum construction sequence for outriggers was conducted on a 600 m building, where safety analysis and construction replication of the entire system were well combined [43]. Based on the FEM model, through the outrigger trusses' safety analysis, the lower limits for construction of the outrigger system were calculated and the upper limits were estimated by structural stiffness analysis and construction stage global stability [43]. Further, by analysing the construction simulation models, a criterion was formed for constructing the outriggers in the building and the efficacy of the suggested method was verified with various examples [43].

In 2019, a high-rise structure of steady stiffness, reinforced with framed tube system, shear core and outrigger belt truss (OBT) had been modelled and analysed to obtain the optimum outrigger position using energy method [44]. OBT was modelled as rotational springs and was placed at various positions along the structure's height in such a way to maximize energy dissipation and absorption [44]. The performance of the method was validated with a numerical example using a 50-storey concrete building in SAP 2000, and the results were compared with Stafford Smith's method [44]. It was concluded that the suggested method developed for several OBT systems showed a decrease in the values of axial force and roof displacement when compared with Stafford Smith's method and the building with two OBTs had better efficiency compared to building with three OBTs [44].

Behaviour of a combined special truss moment frame and outrigger system was investigated using nonlinear and linear analysis on a ten-, twenty- and forty-storey building for various lateral load resisting systems [45]. The results concluded that the proposed system had decreased the drift ratios and enhanced the stiffness and energy absorption [45]. Stresses induced due to differential axial shortening (DAS) and lateral displacement (LAT) were studied in terms of the absolute sum of stresses and the decrease ratio of LAT and DAS which affected the strength demand of the outrigger [46]. Analytic equations were suggested to explore the factors concerning the extra outrigger stress, and FEM analysis was performed to explore the decrease in ratio of LAT and DAS quantitatively as well as the stress produced due to location, stiffness, number of outriggers and connection time [46]. It was found the decrease in ratio and the internal stresses of the LAT and DAS enhanced as the area along cross section of the outrigger was enhanced and dual-purpose outriggers were found to be efficiently helpful in delivering the total stress acting in each outrigger more evenly [46].

In 2019, optimum number of outriggers for decreasing the top storey displacement in high-rise buildings was investigated for uniform, triangular and concentrated load followed by employing a MATLAB code to identify the favourable position of outriggers [47]. Effects of variation in rigidity of outriggers, column and core were evaluated, and from the results it was understood that the outrigger number depends on column axial rigidity, flexural rigidity of outriggers, distance of perimeter columns, type of loading and height of structure [47]. The behaviour of high-rise RCC building with outrigger belt truss system was analysed under seismic loads for various arrangements of steel bracings based on the results from lateral deflection, storey drifts and base shear by linear and nonlinear methods in ETABS software [48]. Nine different models were prepared for different outrigger positions for X-type and V-type bracing patterns, and it was found that X-type bracing pattern gave maximum reduction of 18.47 and 22.40% for top storey drift and lateral displacement, respectively [48].

A study on the seismic performance of a damped outrigger system with buckling-restrained brace (BRB-outrigger) based on spectral analysis (SA) procedure was conducted to analyse the nonlinear seismic behaviour of buildings with damped outriggers [49]. The optimal outrigger position, axial stiffness of perimeter column having BRB and core flexural rigidity are the key parameters varied in the study so as to decrease the earthquake response [49]. Models with varying heights of 64, 128, 256 and 384 m were taken for performing nonlinear response history analysis and SA. It was concluded that the BRB-outrigger at top was dominating the structural performance and the BRB-outrigger at bottom improved the seismic behaviour by reducing deflection, rotation and moment in the core [49].

A 64-storey RCC building with outrigger and belt wall was analysed to evaluate its performance along across-wind, along-wind and in direction of torsion using GTSTRUDL software [50]. The along-wind factors were estimated by following the steps from ASCE 7-16, and the across-wind factors were estimated on basis of wind tunnel data as provided in a database of aerodynamic load. Analysis concluded that the incorporation of belt wall can decrease the deflections caused due to wind responses, and also decrease the torsional acceleration of the structure drastically [50]. In 2020, moment resisting frame with shear wall and outrigger braced truss was analysed for both static and dynamic loads and was compared with a system with pendulum-tuned mass damper (PTMD) [51]. Parametric analysis was performed considering drift and top displacement of the building and found that the braced outrigger system provided reduction of 33% for single and 60% for dual outriggers and also the reduction in acceleration of outriggers was estimated as 40% and that of PTMD as 35% which proved the effectiveness of outriggers for controlling the sway in tall buildings [51].

3.2 Summary of Literature Review

Some of the significant literatures for outriggers in the past few years have been briefly explained above. Figure 5 gives an overview of the literatures that has been considered in this article. Various studies were carried out by researchers for identifying the different parameters that influence the optimal positioning and design properties of outriggers. The key parameters that are affecting the design of outriggers are the flexural rigidity of core and outrigger, axial rigidity of perimeter columns, location of outriggers, height of the structure, stiffness of diaphragms and shear rigidity of outriggers and core. The studies were based on certain assumptions such as elastic linearity of the structure, action of only axial forces in columns, rigid connection of outriggers to the core and pinned connection to the columns, uniform sectional property and neglecting shear deformation of the core. In many papers, various equations were formulated using the above parameters and assumptions as a preliminary design

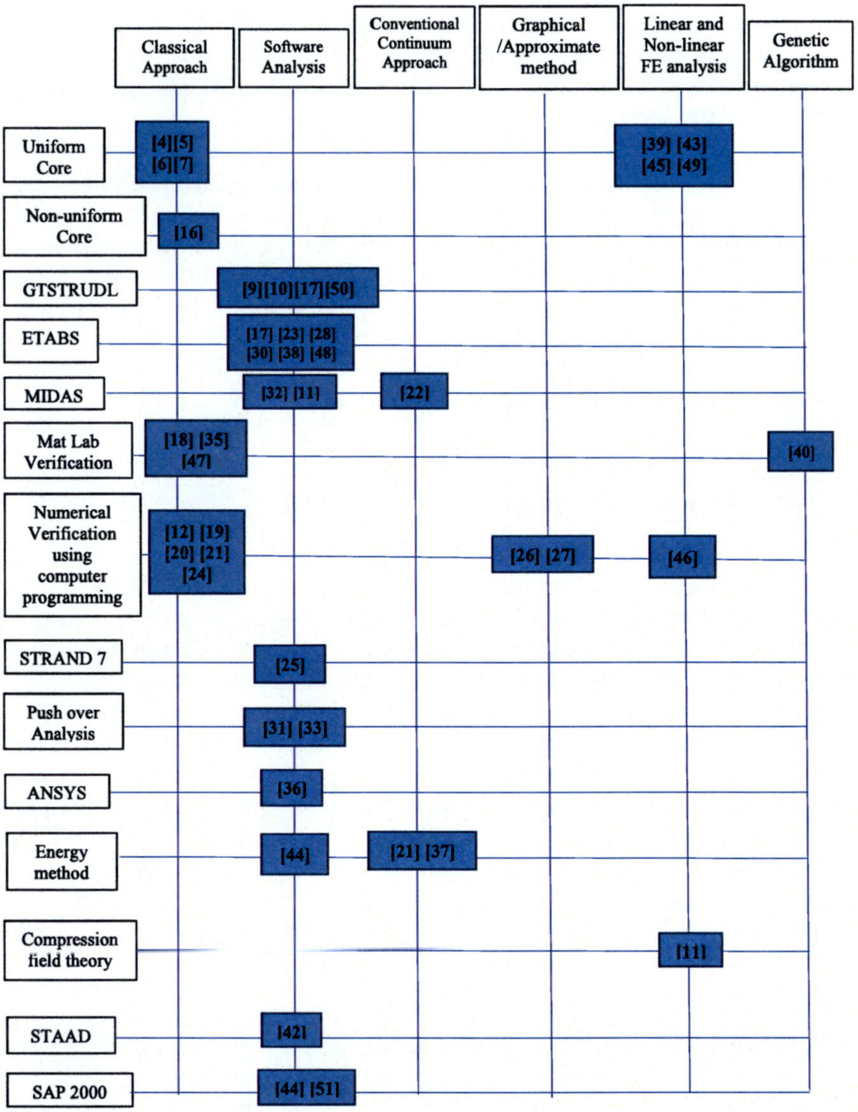

Fig. 5 Overview of selected literatures for outrigger analysis

study for conventional outriggers. Few papers studied the effectiveness of outriggers with belt walls/trusses by analysing lateral displacement, inter-storey drift, bending moment and shear force in core wall. Study on alternative outrigger system like offset and virtual outrigger analysis was taken by few researches, and a very few have done a comparative study between conventional and virtual outriggers to check its effectiveness as a lateral load resisting system. Existing study on virtual outriggers focused mainly on the optimal location of belt truss as virtual outrigger, force transfer mechanism for belt truss and distributed belt walls as virtual outriggers.

From the past literature review, it was found that the main study in outriggers was limited to conventional outriggers, outriggers with belt truss/wall around the outer columns and virtual outriggers. Literature mentioning the analysis of both conventional and virtual outriggers in the same building which acts like a hybrid outrigger system was underexplored. Thus, this study is intended to analyse the efficacy of both conventional and virtual outriggers as a hybrid outrigger system in a high-rise RCC building for various parameters. Plaza Rakyat Tower which is a 77-storey structure in Kuala Lumpur having a concrete shear wall, belt walls as virtual outriggers at two levels and a conventional outrigger system at the roof of the structure can be given as an example for hybrid outriggers [9].

4 Hybrid Outrigger System and Its Significance

As the significance of high-rise structures has been increasing rapidly, an efficient lateral load resisting system needs to be developed. Outriggers are one among the various structural systems available and are gaining importance since the past two decades. A hybrid outrigger system is a lateral load resisting system for high-rise structures comprising both conventional and virtual outriggers at two different positions in the same building. Thus, this research may create a significant contribution to the body of knowledge in terms of the suitability of hybrid outrigger system as a lateral load resisting system in tall buildings. Also, studying the optimal positioning of hybrid outriggers for various structural parameters, loadings and building configuration may contribute significantly to theory building on outriggers. This article gives a brief introduction to hybrid outriggers and a comparative study among various outrigger classifications to evaluate its effectiveness.

5 Comparative Study on Outriggers

The analysis is performed in a finite element software, ETABS, for a 30-storey RCC structure having area 21 m × 21 m. The dimensions and the loads considered for the study are as mentioned in Table 1.

5.1 Modelling Approaches and Descriptions

The structure is symmetrical along both x- and y-axes, fixed at ground level, and the loads acting were calculated with the help of IS codes. The dead load, live

Table 1 Design details for RCC building

Structural elements	Values
No. of storeys	30
Total height	105 m
Storey height	3.5 m
Compressive strength	50 MPa
Beam dimension	400 mm * 600 mm
Column dimension	800 mm* 800 mm
Slab thickness	200 mm
Conventional outrigger dimension	300 mm × 3500 mm
Virtual outrigger dimension	300 mm × 3500 mm
Live load	3.5 kN/m^2
Live load in roof	1.75 kN/m^2
Wind speed	39 m/s
Seismic zone factor	0.24 (zone 4)
Site type	2
Importance factor	1
Response reduction factor	5
Shear wall thickness	300 mm
Super-dead load	1 kN/m^2

load and wind load were assigned based on IS: 875-1987-Part 1, Part 2 and Part 3, respectively. The earthquake loads were assigned as per IS 1893-2002. The masonry load was calculated to 12.67 kN/m and was applied over the beams in the model except on the roof. The slabs were defined as membranes, the outrigger wall and the core wall were designed as shell thin elements, and meshing was done. After loading and assigning all the parameters, the model was checked for errors and analysed for getting the lateral displacement, storey drift, bending moment and shear force in core walls. The concrete design was also checked, and all the concrete members passed the design check. Both static and dynamic loads were taken for the analysis. The dynamic analysis was done by time history method of analysis taking accelerograms from El Centro Array #9, Imperial Valley-02, 5/19/1940, of magnitude 6.95 and PGA of − 0.28 g for x direction and − 0.21 g for y direction.

The case 1 is for control building, and the outriggers are placed at two levels in the other cases to evaluate its effectiveness. Case 2 has two conventional outriggers at 10th and 20th storey, case 3 has two virtual outriggers at 10th and 20th storey, case 4 has conventional outrigger at 20th floor and virtual outrigger at 10th floor, and case 5 has virtual outrigger at 20th floor and conventional outrigger at 10th floor. Case 4 and case 5 are examples of hybrid outriggers. Here, case 4 is having conventional trigger at 20th storey and virtual outrigger at 10th storey and the reverse position is being considered in case 5; this helps to find which case is more effective in reducing the displacement, drift and bending moment. The rendered view for all the models

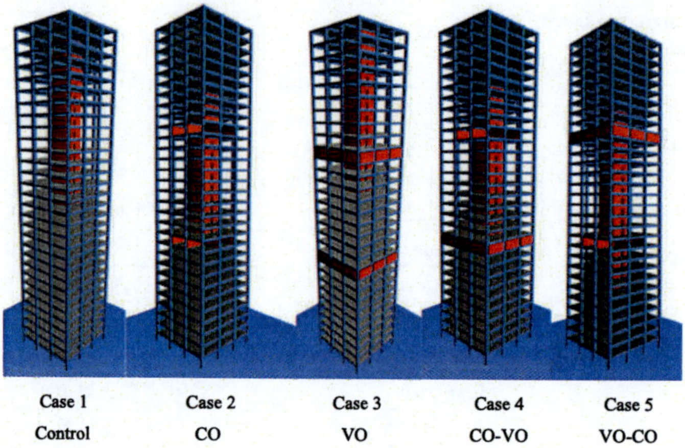

Fig. 6 Models taken for analysis

is given in Fig. 6. CO is the abbreviation for two conventional outriggers, VO for two virtual outriggers, CO–VO for conventional outrigger at 20th floor and virtual outrigger at 10th floor and VO–CO for virtual outrigger at 20th floor and conventional outriggers at 10th floor.

For fixing the position for the outriggers in the above models, the equation given by Smith and Coull is followed. The optimum positions for n outriggers to resist the lateral deflection and drift as per Smith and Coull were found to be $1/(n + 1)$, $2/(n + 1)$, $3/(n + 1)$, …, $n/(n + 1)$ times height of the structure [5, 6]. A 30-storey model with two outriggers is being considered for the study, and therefore, as per the above equation the location of the outriggers was fixed at 10th and 20th storey.

5.2 *Results and Discussion*

After modelling, the building is analysed for both wind and earthquake loads using static and dynamic analysis. The shear force and bending moment values in core walls, storey drift and lateral displacement for all the 5 cases were analysed, compared and plotted in different graphs. Variation of shear force for wind and earthquake loads is given in Fig. 7. Each graph compares the values obtained for all the 5 cases, and loading was taken in both x and y directions.

From the shear force diagrams, it was observed that there is a sudden change in shear force in the storey above the virtual outrigger position and there is a slight change in shear force at place where the conventional outriggers are placed. The variation in bending moment for wind, earthquake and time history case is shown in Fig. 8.

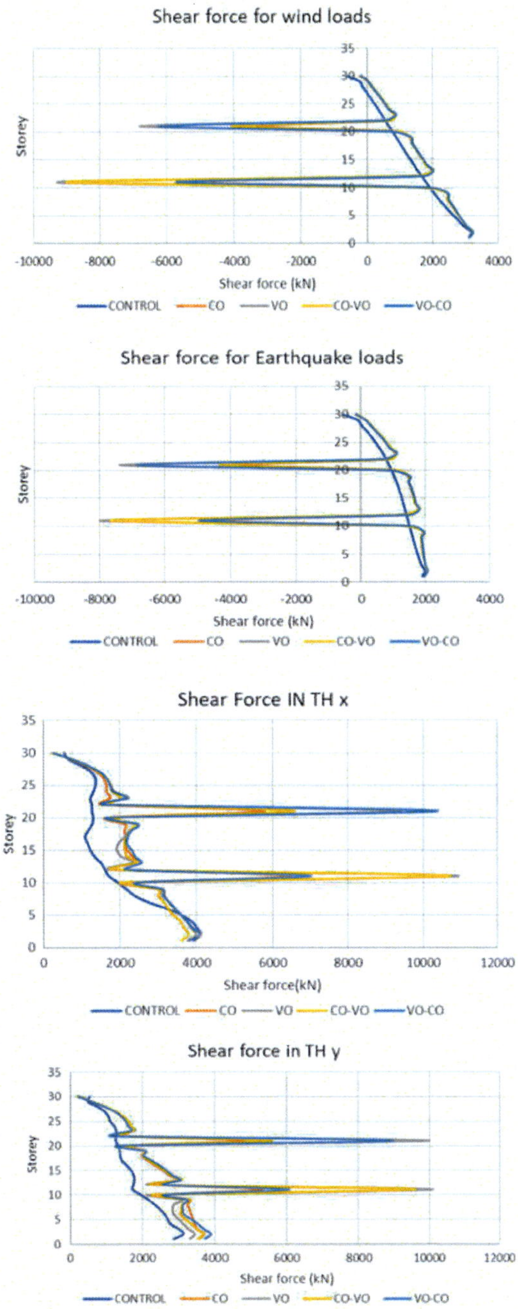

Fig. 7 Shear force diagram for wind and earthquake loads in *y* direction and time history case in *x* and *y* directions

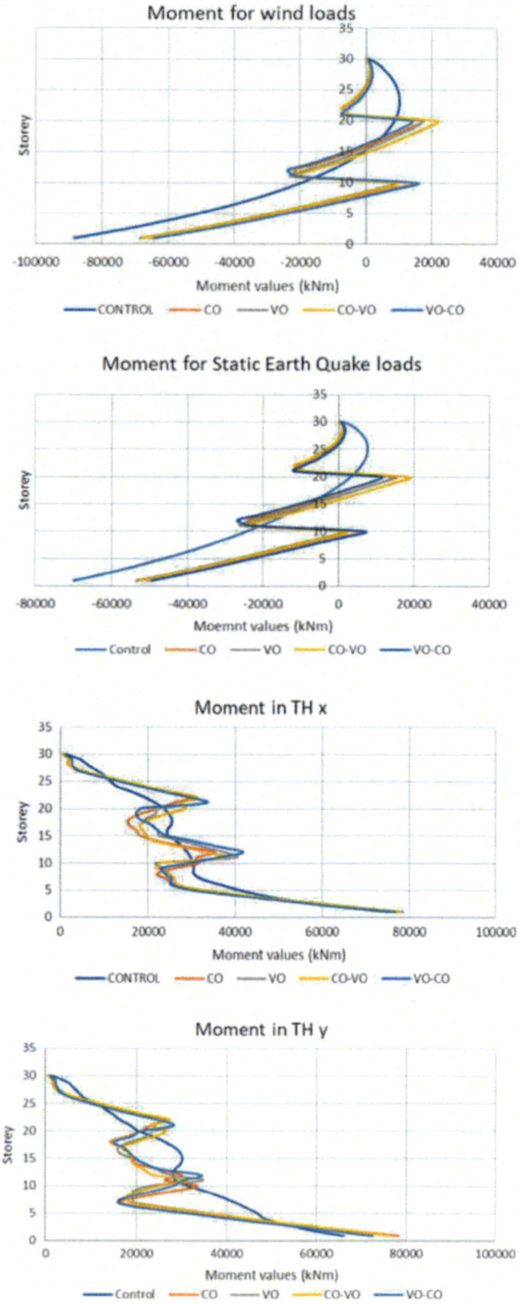

Fig. 8 Bending moment diagram for static wind and earthquake load in *x* direction and time history case in *x* and *y* directions

Table 2 Percentage reduction in shear force at core walls for various loads measured at the base

Loads	Percentage reduction (%)			
	CO	VO	CO–VO	VO–CO
W *X*	− 0.54	− 0.50	− 0.52	− 0.58
W *Y*	− 0.62	− 0.59	− 0.60	− 0.67
EQ *X*	− 2.61	− 3.50	− 3.06	− 3.14
EQ *Y*	− 2.73	− 3.62	− 3.18	− 3.27
TH *X*	− 4.09	0.63	4.77	− 3.63
TH *Y*	− 23.83	− 15.19	− 23.09	− 29.63

Table 3 Percentage reduction in bending moment at core walls for various loads measured at the base

Loads	Percentage reduction (%)			
	CO	VO	CO–VO	VO–CO
W *X*	26.27	22.66	23.39	27.47
W *Y*	26.57	23.34	23.96	27.83
EQ *X*	28.03	23.72	25.03	29.20
EQ *Y*	28.36	24.51	25.67	29.63
TH *X*	− 1.91	− 4.32	− 3.20	− 1.61
TH *Y*	− 18.55	− 8.49	− 15.26	− 9.85

From the graphs, it is observed that there is a sudden change in the bending moment values at the position where the outriggers are placed. The base moment has reduced for all the cases under static loads, and there was an increase in base bending moment for time history case. Tables 2 and 3 show the percentage increase and decrease in the values of shear force and bending moment compared to the values obtained for control model when lateral load resisting systems were installed. From Table 3, it was found that maximum reduction in bending moment at base is for case 5 (VO–CO) followed by case 2 (CO) and in time history there is an increase in bending moment for all the cases when compared to the control model. The increase in shear force at base was found for all the cases, except in TH X for case 3 and case 4. The reduction in base moment is by conversion of bending moment acting in the shear wall into tensile and compressive forces in the perimeter columns, which in turn causes decrease in the value of lateral displacement at the top storey of the structure. The increase in shear force at base may be due to the increase in seismic weight of the building. In case of time history analysis, few variations were observed and may be due to the random nature of ground motions. W *X* and W *Y* are abbreviations for wind loads in x and y directions, EQ *X* and EQ *Y* for earthquake loads in *x* and *y* directions, and TH *X* and TH *Y* for time history loads in *x* and *y* directions.

Figure 9 shows the storey drift variation for wind, earthquake and time history case in *x* and *y* directions.

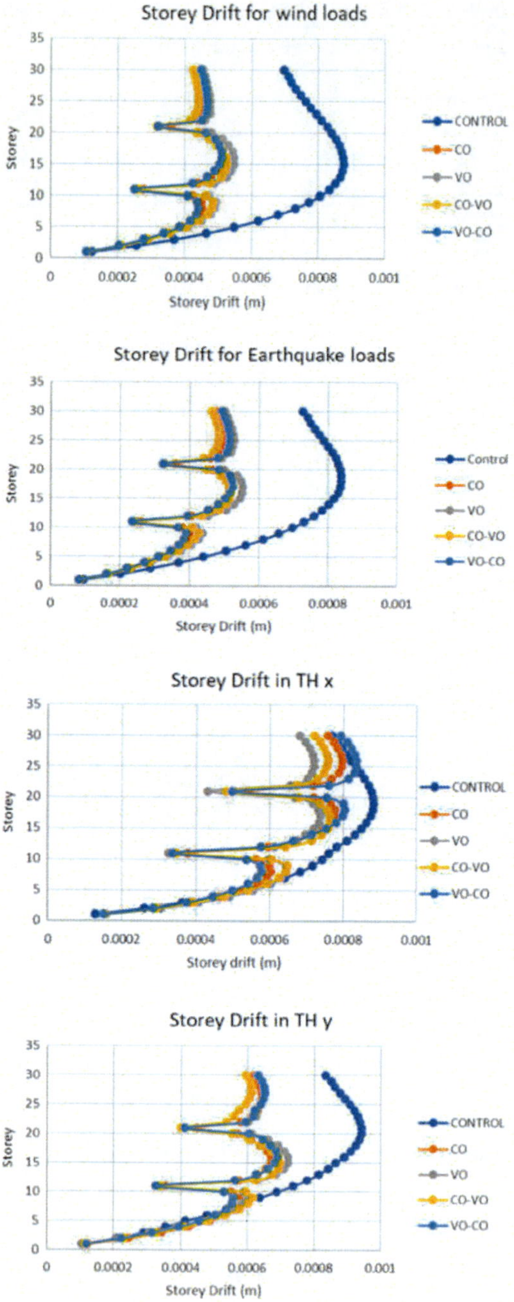

Fig. 9 Storey drift variation for static wind and earthquake loads in *x* direction and time history case in *x* and *y* directions

Table 4 Percentage reduction in storey drift for various loads at 30th storey

Loads	Percentage reduction (%)			
	CO	VO	CO–VO	VO–CO
W *X*	35.80	33.77	38.41	34.93
W *Y*	36.15	34.72	39.02	35.72
EQ *X*	33.75	30.99	35.95	32.23
EQ *Y*	34.06	32.02	36.38	32.97
TH *X*	1.43	11.31	5.98	− 2.99
TH *Y*	25.75	24.79	28.62	24.31

Table 4 gives the percentage reduction in all 5 cases for various loads when top drift was compared and the maximum reduction in drift was found for case 4 (CO–VO) followed by case 2 (CO).

Figure 10 shows the variation in lateral displacement for wind, earthquake and time history case in x and y directions.

Table 5 shows the percentage reduction in maximum lateral displacement values in all 5 cases for different loadings, and the maximum reduction in lateral displacement was almost equal for cases 2, 4 and 5, which again proves the effectiveness of hybrid outriggers in reducing lateral drift and displacement.

From the comparative study done for a 30-storey building, it was found that maximum reduction in bending moment is for case 5 (virtual at top and conventional outrigger at bottom) and lateral displacement and drift for case 4 (conventional at top and virtual outrigger at bottom). Case 4 and case 5 are both cases of hybrid outrigger system. From the analysis results, it can be concluded that hybrid outriggers showed reasonable reduction in storey drift, lateral displacement and bending moment in core walls.

Thus, in this present scenario where the significance of tall buildings is emerging rapidly, the hybrid outrigger system may be used to minimize the adverse effects caused due to wind and earthquake loads. Also, conventional outrigger system can cause space obstruction at the outrigger level; therefore in a building if more number of conventional outrigger is installed, the usage space at those levels reduces. The virtual outrigger can eliminate the problem of space obstruction at the outrigger level. So in a building with hybrid outrigger system, the occupancy area at outrigger level is more compared to a building with conventional outriggers alone.

6 Conclusion

Studies related to the behaviour of outriggers on tall buildings have been undertaken by various researchers, and condensing the concept, mechanism and its behaviour is vital in providing a proper guidance to researchers. Therefore, this article discusses

Fig. 10 Lateral displacement variation for static wind and earthquake loads in *y* direction and time history case in *x* and *y* directions

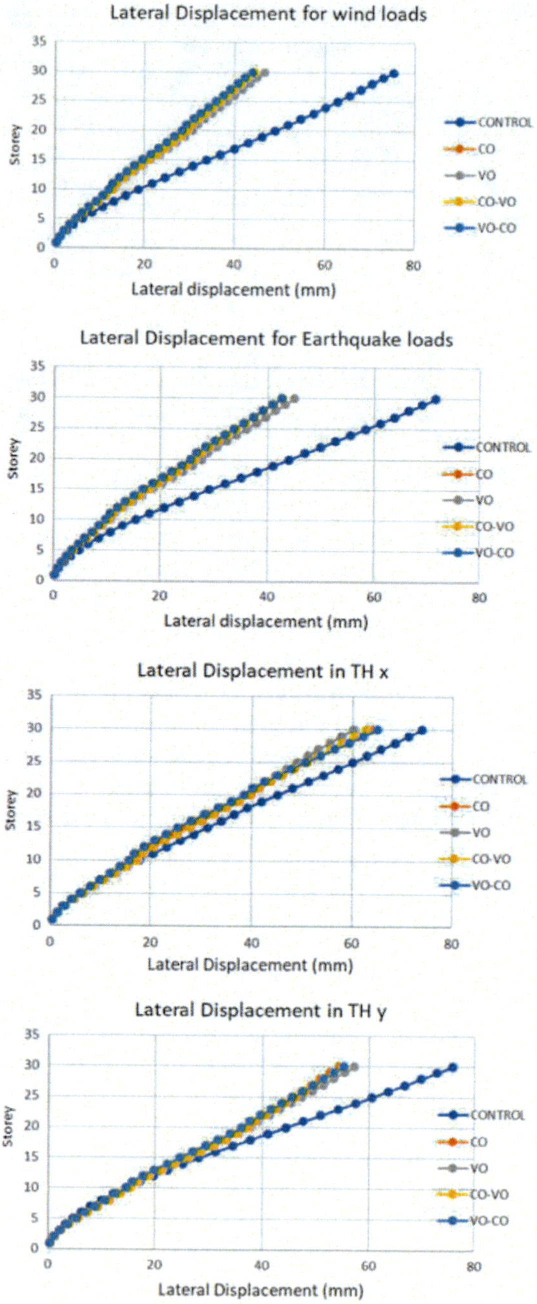

Table 5 Percentage reduction in lateral displacement for various loads at 30th storey

Loads	Percentage reduction (%)			
	CO	VO	CO–VO	VO–CO
W *X*	39.90	36.58	39.68	40.30
W *Y*	41.20	38.35	41.22	41.90
EQ *X*	39.22	35.45	39.16	39.03
EQ *Y*	40.45	37.19	40.63	40.59
TH *X*	13.34	18.65	15.04	12.16
TH *Y*	28.04	24.41	27.43	27.03

a review on the background of outriggers, types of outriggers and parameters influencing the outrigger behaviour and introduces the concept of hybrid outrigger system in high-rise RCC structures. A comparative study was conducted for a 30-storey building, and it was found that hybrid outriggers showed a reasonable reduction in lateral displacement, storey drift and bending moment in core walls. When bending moment was considered, the hybrid outrigger system almost had a percentage reduction of 27% for wind and 29% for earthquake loads when compared to the control model (case 1). When drift was considered, a percentage reduction of 39% for wind loads and 36% for earthquake loads were obtained and for displacement the hybrid outrigger system had a reduction of almost 40% for both lateral loads when compared to the control model. From the above comparative study performed, the hybrid outrigger system was found to be the more effective as a lateral load resisting system when compared to building with two conventional outriggers (case 2) and building with two virtual outriggers (case 3).

Further studies are required to analyse the effectiveness of hybrid outriggers for different loadings and buildings with different configurations. Also, the force transfer mechanism and the parameters that affect its effectiveness need to be studied. These studies are being carried out and will be published in future.

Declaration of Competing Interest

The authors declare that they have no competing personal relationships or financial interests that could have influenced the study presented in this paper.

References

1. Khan FR (1969) Recent structural systems in steel for high-rise buildings. In: Proceedings of the British constructional steelwork association conference on steel in architecture. British Constructional Steelwork Association, London
2. Ali MM, Moon KS (2007) Structural developments in tall buildings: current trends and future prospects. Archit Sci Rev 50(3):205–223. https://doi.org/10.3763/asre.2007.5027
3. IS 16700 (2017) Criteria for structural safety of tall concrete building, BIS, 2017
4. Smith BS, Salim I (1981) Parameter study of outrigger-braced tall building structures. ASCE J Struct Div 107(10):2001–2014

5. Smith BS, Coull A (1991) Tall buildings structures analysis and design. A Wiley Interscience publication, Copyright Wiley Inc.
6. Smith BS, Salim I (1983) Formulae for optimum drift resistance of outrigger braced tall building structures. Comput Struct 17(1):45–50. https://doi.org/10.1016/0045-7949(83)90027-5
7. Taranath BS (2010) Reinforced concrete design of tall buildings. Concrete Reinforcing Steel Institute
8. Zhang J, Zhang ZX, Zhao WG, Zhu HP, Zhou CS (2007) Safety analysis of optimal outriggers location in high-rise building structures. J Zhejiang Univ Sci A 8(2):264–269. https://doi.org/10.1631/jzus.2007.A0264
9. Nair RS (1998) Belt trusses and basements as 'virtual' outriggers for tall buildings. Eng J 35(4):140–146
10. Horton A (2000) Virtual outriggers in tall buildings. https://fdocuments.in/document/virtual-outriggers-in-tall-buildings.html
11. Eom T-S, Kim J-Y, Kim J-Y, Kim J-Y, Kim D-Y (2013) Shear design of concrete belt walls reinforced with post-tensioning bars. J Korean Soc Hazard Mitig 13(1):31–39. https://doi.org/10.9798/kosham.2013.13.1.031
12. Hoenderkamp JCD, Snijder HH (2000) Simplified analysis of facade rigger braced high-rise structures. Struct Des Tall Build 9(4):309–319. https://doi.org/10.1002/1099-1794(200009)9:4%3c309::AID-TAL155%3e3.0.CO;2-Y
13. Nanduri PMBRK, Suresh B, Hussain I (2013) Optimum position of outrigger system for high-rise reinforced concrete buildings under wind and earthquake loadings. Am J Eng Res 02(08):76–89. [Online]. Available: http://www.ajer.org/papers/v2(8)/J0287689.pdf
14. Fu F (2018) Shear wall, core, outrigger, belt truss, and buttress core system for tall buildings. Des Anal Tall Complex Struct 81–107. https://doi.org/10.1016/b978-0-08-101018-1.00003-4
15. Choi HS, Ho G, Joseph L, Mathias N (2017) Outrigger design for high-rise buildings. Outrigger Des High-Rise Build. https://doi.org/10.1201/9781315661971
16. Rutenberg A, Tal D (1987) Lateral load response of belted tall building structures. Eng Struct 9(1):53–67. https://doi.org/10.1016/0141-0296(87)90041-1
17. Kian PS (2001) The use of outrigger and belt truss system for high-rise concrete buildings. Civ Eng Dimens 3(1):36–41. [Online]. Available: http://puslit2.petra.ac.id/ejournal/index.php/civ/article/view/15536
18. Wu JR, Li QS (2003) Structural performance of multi-outrigger-braced tall buildings. Struct Des Tall Spec Build 12(2):155–176. https://doi.org/10.1002/tal.219
19. Hoenderkamp JCD, Snijder HH (2003) Preliminary analysis of high-rise braced frames with facade riggers. J Struct Eng 129(5):640–647. https://doi.org/10.1061/(ASCE)0733-9445(2003)129:5(640)
20. Hoenderkamp JCD, Bakker MCM (2003) Analysis of high-rise braced frames with outriggers. Struct Des Tall Spec Build 12(4):335–350. https://doi.org/10.1002/tal.226
21. Zeidabadi NA, Mirtalae K, Mobasher B (2004) Optimized use of the outrigger system to stiffen the coupled shear walls in tall buildings. Struct Des Tall Spec Build 13(1):9–27. https://doi.org/10.1002/tal.228
22. Lee J, Bang M, Kim JY (2008) An analytical model for high-rise wall-frame structures with outriggers. Struct Des Tall Spec Build 17(4):839–851. https://doi.org/10.1002/tal.406
23. Bayati Z, Mahdikhani M, Rahaei A (2008) Optimized use of multi-outriggers system to stiffen tall buildings. In: Proceedings of 14th world conference earthquake engineering. Beijing, China, Schueller
24. Hoenderkamp JCD (2008) Second outrigger at optimum location on high-rise shear wall. Struct Des Tall Spec Build 17(3):619–634. https://doi.org/10.1002/tal.369
25. Herath N, Haritos N, Ngo T, Mendis P (2009) Behaviour of outrigger beams in high rise buildings under earthquake loads. Aust Earthq Eng Soc. [Online]. Available: http://www.aees.org.au/wp-content/uploads/2013/11/Herath-et-al.pdf
26. Hoenderkamp JCD (2009) The influence of non-rigid floor structures on facade rigger braced high-rise trussed frames. Adv Struct Eng 12(3):385–397. https://doi.org/10.1260/136943309788708428

27. Hoenderkamp JCD, Snijder HH, Hofmeyer H (2012) High-rise structures with belt bracing subject to lateral load. Adv Struct Eng 15(1):65–75. https://doi.org/10.1260/1369-4332.15.1.65
28. Kamath K (2012) A study on static and dynamic behavior of outrigger structural system for tall buildings. Bonfring Int J Ind Eng Manag Sci 2(4):15–20. https://doi.org/10.9756/bijiems.1655
29. Choi HS, Joseph L (2012) Outrigger system design considerations high-rise buildings outrigger system design considerations. Int J High-Rise Build 1(3):237–246 [Online]. Available: www.ctbuh.org
30. Kamath K, Avinash AR, Sandesh UK (2012) A study on static and dynamic behavior of outrigger structural system for tall buildings. Bonfring Int J Eng Manag Sci 2(4):15–20. https://doi.org/10.9756/bijiems.1655
31. Shehu R (2015) Ductility of outrigger typologies for highrise structures. IOSR J Mech Civ Eng 12(2):2320–2334. https://doi.org/10.9790/1684-12263441
32. Upadhyaya SK, Kamath K, Rao S (2015) IJARST20140399 optimum positioning of outriggers to reduce differential column shortening due to long term effects in tall buildings. Int J Adv Res Sci Technol 4(3):353–357 [Online]. Available: www.ijarst.com
33. Patil DM, Sangle KK (2016) Seismic behaviour of outrigger braced systems in high rise 2-D steel buildings. Structures 8:1–16. https://doi.org/10.1016/j.istruc.2016.07.005
34. Ho GWM (2016) The evolution of outrigger system in tall buildings. Int J High-Rise Build 5(1):21–30. https://doi.org/10.21022/ijhrb.2016.5.1.21
35. Zhou Y, Zhang C, Lu X (2016) An inter-story drift-based parameter analysis of the optimal location of outriggers in tall buildings. Struct Des Tall Spec Build 25:215–231
36. Samarakkody DI, Thambiratnam DP, Chan THT, Moragaspitiya PHN (2017) Outrigger-belt and frame interaction in composite tall buildings under differential axial shortening. J Archit Eng 23(3):1–14. https://doi.org/10.1061/(ASCE)AE.1943-5568.0000243
37. Kamgar R, Rahgozar R (2017) Determination of optimum location for flexible outrigger systems in tall buildings with constant cross section consisting of framed tube, shear core, belt truss and outrigger system using energy method. Int J Steel Struct 17(1):1–8. https://doi.org/10.1007/s13296-014-0172-8
38. Mathew M, Mathew M (2017) Optimum position of outrigger with belt wall, pp 10367–10372. https://doi.org/10.15680/IJIRSET.2017.0606033
39. Kafina R, Sagaseta J (2019) Analysis of outrigger-braced reinforced concrete supertall buildings: core-supported and tube-in-tube lateral systems. Struct Des Tall Spec Build 28(1):1–22. https://doi.org/10.1002/tal.1567
40. Chen Y, Cai K, Wang X (2018) Parameter study of framed-tube structures with outriggers using genetic algorithm. Struct Des Tall Spec Build 27(14):1–26. https://doi.org/10.1002/tal.1499
41. Kim HS (2018) Optimum locations of outriggers in a concrete tall building to reduce differential axial shortening. Int J Concr Struct Mater 12(1). https://doi.org/10.1186/s40069-018-0323-y
42. Dangi A, Jamle S (2018) Determination of seismic parameters of R.C.C. building using shear core outrigger, wall belt and truss belt systems. Int J Adv Eng Res Sci 5(9):305–309. https://doi.org/10.22161/ijaers.5.9.36
43. Zhou K, Luo WX, Li S (2018) Decision framework for optimal installation of outriggers in tall buildings. Autom Constr 93:200–213
44. Kamgar R, Rahgozar P (2019) Reducing static roof displacement and axial forces of columns in tall buildings based on obtaining the best locations for multi-rigid belt truss outrigger systems. Asian J Civ Eng 20(6):759–768. https://doi.org/10.1007/s42107-019-00142-0
45. Broujerdian V, Ghamari A, Zolghadr R (2019) Investigating a combinatorial lateral resisting system for tall buildings. Iran J Sci Technol Trans Civ Eng 43(4):697–709. https://doi.org/10.1007/s40996-019-00233-7.
46. Kim HS, Lim YJ, Lee HL (2020) Strength demand of dual-purpose outrigger system for reducing lateral displacement and differential axial shortening in a tall building. Struct Des Tall Spec Build 29(4):1–19. https://doi.org/10.1002/tal.1701

47. Baygi S, Khazaee A (2019) The optimal number of outriggers in a structure under different lateral loadings. J Inst Eng Ser A 100(4):753–761. https://doi.org/10.1007/s40030-019-00379-7
48. Sawant V, Bogar V (2019) Parameters comparison of high rise RCC structure with steel outrigger and belt truss by linear and non-linear analysis. Int J Eng Trends Technol 67(7):15–23. https://doi.org/10.14445/22315381/ijett-v67i7p204
49. Lin PC, Takeuchi T, Matsui R (2019) Optimal design of multiple damped-outrigger system incorporating buckling-restrained braces. Eng Struct 194(May):441–457. https://doi.org/10.1016/j.engstruct.2019.05.078
50. Samat RA, Mohamed Ali N, Kadir Marsono A, Fadzil AB (2019) The role of belt wall in minimizing the response due to wind load. In: MATEC web conference, vol 266, pp 01009. https://doi.org/10.1051/matecconf/201926601009
51. Salman K, Kim D, Maher A, Latif A (2020) Optimal control on structural response using outrigger braced frame system under lateral loads. J Struct Integr Maint 5(1):40–50. https://doi.org/10.1080/24705314.2019.1701799

Efficiency of *Carica papaya* Stem Activated with Phosphoric Acid and Sodium Hydroxide in Mining Wastewater Treatment

Ezekiel A. Adetoro and Samson O. Ojoawo

Abstract Chemically activated carbons prepared from the *Carica papaya* stem using phosphoric acid (CPSAC–H_3PO_4) and sodium hydroxide (CPSAC–NaOH) were used in sequestrating selected toxic metals (Cu^{2+}, Mn^{2+}, Co^{3+}, Fe^{2+}, Zn^{2+}, Cd^{2+}, Pb^{2+} and Cr^{3+}) from mining wastewater. The mining wastewater was characterized for selected toxic metals using atomic absorption spectrophotometer (AAS). Batch adsorption studies were conducted using the CPSACs to remove selected toxic metals from the mining wastewater putting into consideration adsorbent dosage, agitation rate, contact time, pH and temperature parameters. The data obtained were fitted into isotherm (Freundlich, Langmuir, Dubinin–Radushkevich and Temkin) and kinetic (pseudo-first and pseudo-second orders, Elovich and intra-particle diffusion) and thermodynamic (standard enthalpy change, $\Delta H°$; standard entropy change, $\Delta S°$; and standard free energy change, $\Delta G°$) models for its suitability. Three different error functions (sum of absolute errors—EABS, average relative errors—AREs and coefficient of determination—R^2) were used for assessment of the linear and non-linear regression analyses of the selected adsorption isotherms and kinetic models. The removal efficiency of the selected toxic metals in CPSAC samples was between 92.97 and 100.00%. The adsorbent dosage, agitation rate, contact time, pH and temperature for optimum condition were 0.6 g (CPSAC–H_3PO_4) and 0.2 g (CPSAC–NaOH), 150 rpm, 60 min., 7 and 30 °C, respectively. Adsorption data followed the isotherms in the order of CPSAC–NaOH > CPSAC–H_3PO_4 and kinetics (i.e. pseudo-second order and intra-particle diffusion mechanism) in the order of CPSAC-H3PO4 > CPSAC–NaOH. Non-linear regression analyses using EAB, ARE and R^2 error functions indicated to be better in analysing adsorption and kinetic models for the studied toxic metals. The $\Delta H°$, $\Delta S°$ and $\Delta G°$ for the processes were 657.916, 64.288 and 22,012 kJ/mol, respectively. The adsorbents were efficient for removal of the selected toxic metals from the mining wastewater (though CPSAC–H_3PO_4 was most efficient); thus, it can be used for industrial application in wastewater treatment.

Keywords Adsorbent · Batch adsorption · *Carica papaya* stem · Mining wastewater · Toxic metals

E. A. Adetoro · S. O. Ojoawo (✉)
Department of Civil Engineering, Ladoke Akintola University of Technology, Ogbomoso, Nigeria
e-mail: soojoawo@lautech.edu.ng

L. Nandagiri et al. (eds.), *Sustainability Trends and Challenges in Civil Engineering*, Lecture Notes in Civil Engineering 162, https://doi.org/10.1007/978-981-16-2826-9_6

1 Introduction

Industrial wastewater must be treated to get rid of its pollutants, especially toxic metals prior to its ejection into bodies of water for steering clear of ecosystem's devastation and infirmities of people living in the environment [1]. Toxic metals give rise to a noteworthy peril to the human habitat and general fitness due to their noxiousness, cumulation in the food chain and perseverance in character [2]. Toxic metals' sequestrations from contaminated discharges through usual practices are not so successful, very exorbitant particularly for the metals' solubility between 1 and 100 mg/L, frequently limited due to peripheral difficulties with metal-relevance slime that are exceedingly hard to get rid of. The potential of inert and complete biomaterials in remediation of toxic metals has been seriously considered in developing a capable, virtuous and inexpensive method for treating wastewater even at very low concentrations of toxic metals. This method is known as biosorption. It is an ingenious, environmental-friendly, cheap and successful means of sequestrating toxic metals from wastewaters. Moreover, the adsorbent(s) used for the process can be renewed through desorption procedure, thus fostering multiple utilization [3–5].

Activated carbons (ACs) are broadly utilized as adsorbents in treating wastewater due to their energetic functional groups and lofty superficial areas, but their utilization is usually impeded by affordability and availability due to high cost; thus, there is a necessity to substitute low-cost precursor materials (i.e. agricultural biomaterials) with identical function as traditional AC [6, 7]. *Carica papaya* (CP) different parts are broadly utilized amidst biological materials used in biosorption, and their capabilities for biosorption of toxic metals from synthetic wastewater (s) are established by research works of [8] (*C. papaya* leaves), [9] (*C. papaya* seed), [10, 11] (*C. papaya* stem peel), etc. Nevertheless, the biological attributes of CP stem, which is left to spoil, resulting in environmental wastes, are yet to be deeply explored. The reviewed works of [6, 10, 12] showed possibilities of using H_3PO_4 and NaOH activating agents for agricultural biomaterials.

However, no previously known report has been made on the utilization of *C. Papaya* stem activated carbon—CPSAC (using H_3PO_4 and NaOH activating agents) in treating mining wastewater (i.e. real-life wastewater). This study aimed at assessing efficacy (ies) of CPSAC produced with H_3PO_4 and NaOH activating agents in sequestrating eight toxic metal ions (i.e. Cu^{2+}, Mn^{2+}, Co^{3+}, Fe^{2+}, Zn^{2+}, Cd^{2+}, Pb^{2+} and Cr^{3+}) from mining wastewater (MW). These were achieved through batch adsorption experiments and adsorption process model studies (i.e. isotherm, kinetic and thermodynamic studies) of the adsorption data obtained.

2 Materials and Methods

2.1 *Establishment of Toxic Metals' Concentrations Present in the Mining Wastewater (MW)*

MW sample was sourced from Igbeti marble mining site, Nigeria, and its complexity was broken down before being analysed by AAS machine. 10 mL of the MW sample was poured into warming bottle, while 10 mL of blank sample was also poured into another warming bottle. 15 mL of nitric acid (HNO_3) was cautiously added to the contents of the two warming bottles, covered and heated for 5–10 min, and subsequently, the samples were allowed to cool and then filtered into a volumetric flask using Whatman filter papers and funnels. The filtrates were made up to 100 mL with distilled water. The made up filtrates were then poured into transparent tightly corked bottles in readiness for the AAS reading [13].

The AAS machine (PG 990 model) was used to read the toxic metal ions' concentration in the prepared digested and standard samples, and then results recorded. Conversion and calculation of the toxic metal ions' concentrations were then carried out. Reference [13] expressed concentration in part per million (ppm) according to Eq. 1.

$$\text{Concentration (ppm)} = \frac{R \times \text{FV} \times \text{DF}}{W} \quad (1)$$

where R is the reading from the machine, FV is the final volume after digestion, DF is the dilution factor and W is the weight of the sample used. This analysis was used for determination of concentrations of toxic metal ions present in the MW before and after treatment.

2.2 *Preparation of* Carica papaya *Stem Activated Carbon (CPSAC)*

C. papaya stem (CPS) sourced from Ladoke Akintola University of Technology Teaching and Research Farm, Ogbomoso, Nigeria, was sliced and painstakingly rinsed with distilled water in order to remove all feasible contaminants. The cleaned CPS sample was dried at 105 °C in an oven for 24 h before being powdered. The CPS powder was put into cleaned melting pots, placed into muffle furnace with model number 5X1-1008 at 600 °C and left inside it for extra one hour, and the samples are subsequently retrieved from the furnace and allowed to cool. The carbonized sample was rinsed with distilled water till pH of 7.0 was attained, subsequently dried at 105 °C in an oven for 2 h and cooled prior to its activation [12].

50 g of the cooled carbonized CPS sample was poured into two large bottles containing 500 mL of 0.1 M of each of phosphoric acid and sodium hydroxide (i.e.

activating agent), evenly mixed and stayed for 1 day, sieved and rinsed with distilled water till pH of 7.0 was reached, dried at 105 °C, cooled and later kept in a closed-up vessel [12].

2.3 Batch Adsorption Experiments

The adsorption parameters applied in the batch adsorption experiments were adsorbent dosage (0.2, 0.4, 0.6, 0.8 and 1.0 g), agitation rate (50, 100, 150, 200 and 250 rpm), contact time (20, 40, 60, 80, 100 and 120 min), pH (4, 6, 8 and 10) and temperature (30, 40, 50, 60 and 70 °C). 50 mL of the MW sample was put into conical flask and differing dosages of 0.2, 0.4, 0.6, 0.8 and 1.0 g of the prepared CPSAC–H_3PO_4 and CPSAC–NaOH adsorbents being added, set on J.P Selecta orbital shaker with model number 3000974 at differed agitation rate, contact time, pH and temperature. The mixtures were filtered with Whatman filter paper 12.5 cm (100 circles) and its filtrates analysed by AAS and statistically [11].

One factor at a time (OFAT) analyses were performed for all the adsorption parameters (i.e. differing of each parameter while others were fixed during the experiments). The adsorption capacity—q_e (mg/g) and removal efficiency—RE (%) were established through Eqs. 2 and 3, respectively, according to [11, 14].

$$q_e = \frac{(C_o - C_e)V}{W} \tag{2}$$

$$\text{RE} = \frac{(C_o - C_e)}{C_o} \times 100\% \tag{3}$$

where V = solution's volume (mL), W = amount of adsorbent (g), and C_0 and C_e = first and balanced concentration in the solution (mg/L).

2.4 Adsorption Process Models' Studies

These have to do with establishment of suitability of batch adsorption data obtained using adsorption isotherm, kinetic and thermodynamic studies.

(1) *Adsorption isotherm studies*: Equations acquired from plots of batch adsorption data were juxtaposed with four selected adsorption isotherm (i.e. Freundlich, Langmuir, Dubinin–Radushkevich (D–R) and Temkin isotherms) equations for fitness motive. Equations 4 (non-linear) and 5 (linear) were applied for Freundlich isotherm, Langmuir isotherm fitness was determined from Eqs. 6 (non-linear), 7 (linear) and 8 (separation factor), the relevancy of D–R isotherm

was determined using Eqs. 9 (non-linear), 10 (linear), 11 and 12, while Eqs. 13 (non-linear) and 14 (linear) were utilized for Temkin isotherm.

$$q_e = KC_e^{1/n} \tag{4}$$

$$\log q_e = \log K + \frac{1}{n} \log C_e \tag{5}$$

K and n are to be determined from cut-off and gradient of log q_e and log C_e plot. $1/n$, $1/n > 1 = 0$ and $0 < 1/n < 1$ are irreversible, non-optimum adsorption and optimum adsorption process, respectively, according to [11]. The system is more heterogeneous, normal Langmuir isotherm and cooperative adsorption for value close to 0, < 1 and > 1, respectively, as expressed by [14].

$$q_e = \frac{q_m \cdot K \cdot C_e}{1 + K \cdot C_e} \tag{6}$$

$$\frac{C_e}{q_e} = \frac{1}{q_m \cdot K} + \frac{C_e}{q_m} \tag{7}$$

$$R_L = \frac{1}{(1 + K \cdot C_o)} \tag{8}$$

where Langmuir parameters for adsorption capacity and energy are K and q_m, respectively. C_o is the first concentration of toxic metal (mgL^{-1}). $R_L > 1$, 1, 0 and $0 < R_L < 1$ means non-optimum, linear, irreversible and optimum adsorption, respectively, according to [11].

$$q_e = q_m \cdot e^{(-\beta\varepsilon^2)} \tag{9}$$

$$\ln q_e = \ln q_m - \beta\varepsilon^2 \tag{10}$$

$$\varepsilon = RT \ln\left(1 + \frac{1}{C_e}\right) \tag{11}$$

where ε = Polanyi capacity of the required concentration, β = average energy of adsorption E (mol^2/kJ^2), R = appropriate fixed value of gas (8.314 J/mol K) and T = standard temperature (K). q_m and β are determined from the cut-off and gradient of lnq_e and ε^2 plot. E values of 1–8 and 9–16 kJ/mol mean physical and chemical adsorption, respectively, according to [14].

$$E = 1/\left(\sqrt{2\beta}\right) \tag{12}$$

$$q_e = B \cdot \ln(K_t \cdot C_e) \tag{13}$$

$$q_e = B \ln K_t + B \ln C_e \tag{14}$$

where q_e = amount of adsorbed adsorbate at stable condition (mg/g), B (RT/b) = fixed value of heat capacity (L/mg) and C_e = adsorbate's concentration at stable condition (mg/L) [14]. They are usually determined from cut-off and gradient of q_e against lnC_e plot.

(2) *Adsorption kinetic studies*: Equations acquired from plots of batch adsorption data were also juxtaposed with four selected adsorption kinetic models (i.e. pseudo-first order, pseudo-second-order, Elovich and intra-particle diffusion models). Equations 15 (non-linear) and 16 (linear) were used to analyse pseudo-first-order kinetic model with K_1 being determined from gradient of log (q_e–q_t) versus t plot. Pseudo-second-order kinetic study was carried out using Eqs. 17 (non-linear) and 18 (linear) with K_2 being determined from gradient of t/q_t versus t plot. Equation 19 was applied in Elovich kinetic model analysis. The fixed values of α and β were calculated from cut-off and gradient of q_t versus ln t plot, respectively. Intra-particle diffusion analysis was conducted using Eq. 20 and K_{diff} being determined from q_t versus $t^{1/2}$ the plot [11, 14].

$$q_t = q_e\left(1 - e^{-K_1 t}\right) \tag{15}$$

$$\log(q_e - q_t) = \log q_e - \frac{K_1}{2.303}t \tag{16}$$

where q_e and q_t are adsorption capacity at stable condition and time t, respectively, K_1 is the pseudo-first-order fixed rate and t is adsorption process' time of mixing and being determined from gradient of log ($q_e - q_t$) versus t plot according to [11, 14].

$$q_t = \frac{K_2 \cdot t \cdot q_e^2}{K_2 \cdot t \cdot q_e + 1} \tag{17}$$

$$\frac{t}{q_t} = \frac{1}{K_2 \cdot q_e^2} + \frac{t}{q_e} \tag{18}$$

$$q_t = \frac{1}{\alpha}\ln(\alpha\beta) + \frac{1}{\beta}\ln t \tag{19}$$

where K_2 = pseudo-second-order fixed rate, α = first rate of desorption (mg/(g min), β = fixed value of desorption (g/mg) and $1/\beta$ = number of sites obtainable for adsorption as stipulated by [14].

$$q_t = K_{\text{diff}}\left(t^{\frac{1}{2}}\right) + C \tag{20}$$

where K_{diff} = fixed value of intra-particle diffusion (mg/(g min) and C = intercept (cut-off). The plot shows that the cut-off value governs effect of the boundary layer and surface adsorption. Linear plot of intra-particle diffusion with zero as origin portrays only model of the intra-particle diffusion as rate-restricting step. Large cut-off value usually results in great part of the surface adsorption in the rate-restricting step; if not, other mechanisms are involved with intra-particle diffusion [14].

(3) *Adsorption thermodynamic studies*: Three thermodynamic parameters assessed for the adsorption process were standard enthalpy variation ($\Delta H°$), standard entropy variation ($\Delta S°$) and standard free energy variation ($\Delta G°$). $\Delta H°$ and $\Delta S°$ were determined from cut-off and gradient of ln K versus 1/T plot through Eqs. 21 and 22, respectively. $\Delta G°$ was determined from Eq. 23, while Arrhenius equation (i.e. Eq. 24) was utilized in determining if the process is physical or chemical. Ea, the magnitude of activation energy (i.e. the nature of adsorption's determinant), was established from gradient of ln K_2 against 1/T plot. The process has activation energy, and Ea of 5–40 and 40–800 kJ/mol means physisorption and chemisorption process, respectively [14].

$$\ln K = \frac{\Delta S^0}{R} - \frac{\Delta H^0}{\mathrm{RT}} \tag{21}$$

$$K = \frac{C_{\mathrm{ae}}}{C_{\mathrm{e}}} \tag{22}$$

where K = adsorption fixed value of stable condition (L/mg), $\Delta S°$ = standard entropy variation (kJ/mol K), R = appropriate fixed value of gas (8.314 J/mol K), $\Delta H°$ = standard enthalpy variation (kJ/mol), T = standard temperature of solution (Kelvin), C_{ae} = stable concentration adsorbed (mg/L) and C_{e} = stable concentration in solution (mg/L).

$$\Delta G^0 = -\mathrm{RT} \cdot \ln K \tag{23}$$

$$\ln K = \ln A - \frac{\mathrm{Ea}}{\mathrm{RT}} \tag{24}$$

where A = Arrhenius factor and Ea = Arrhenius activation energy of adsorption (kJ/mol). A negative and positive $\Delta H°$ value means that process of adsorption is exothermic and endothermic in nature, respectively. $\Delta S°$ positive value means randomness increment at the solid–solution interface that happens in the adsorption process apart from showing the affinity of the adsorbent towards the adsorbate, while $\Delta G°$ negative value implies that an adsorption process is a natural process at the temperature studied [14].

(4) *Error functions*: Three different error functions (i.e. sum of absolute errors—EABS, average relative errors—AREs and coefficient of determination—R^2) were used for assessment of the linear and non-linear regression analyses of the selected adsorption isotherms and kinetic models. Equations 25–27 were used for EABS, ARE and R^2 error functions, respectively [15, 16].

$$\text{EABS} = \sum_{i=1}^{n} \left(q_e(\exp' t) - q_e(cal.)\right)_i \tag{25}$$

$$\text{ARE} = \left[\left(q_e(\exp' t) - q_e(cal.)\right)^2 / q_e(\exp' t)\right] \tag{26}$$

$$R^2 = 1 - \frac{\sum_{i=1}^{n}\left(q_e(\text{cal}) - q_e^{\text{mean}}(\exp' t)\right)_i^2}{\left(\sum_{i=1}^{n}\left(q_e(\text{cal}) - q_e^{\text{mean}}(\exp' t)\right)_i^2 + \sum_{i=1}^{n}\left(q_e(\text{cal}) - q_e(\exp' t)\right)_i^2\right)} \tag{27}$$

where (q_e (cal.)), (q_e (expt.)) and (q_e^{mean}(expt.)) are the adsorption capacities obtained from the modelled, experimental and experimental mean results.

3 Results and Discussion

3.1 Concentration of Toxic Metal Ions in the MW Sample

The toxic metals' concentrations in the MW sample before and after treatment are shown in Table 1. All the selected toxic metals' concentrations were more than the required standards [17] before treatment, which portrayed possibility of environmental pollution of the study area by the MW.

Iron, lead (in both adsorbents) and cobalt (in CPSAC–NaOH) concentrations were above-required standards [17] after treatment, though drastically reduced, while the remaining toxic metals' concentrations were not detected (ND), which portrayed their removal by CPSAC–H_3PO_4 and NaOH.

3.2 Results of Batch Adsorption Experiments

(1) *Effects of adsorbent dosage*: Figures 1 and 2 present percentage removal trends of toxic metals for CPSAC–H_3PO_4 and NaOH adsorbents' dosages with Pb^{2+} and Co^{3+} having least removal efficiencies, respectively. The observed percentage removal of Cu^{2+}, Mn^{2+}, Zn^{2+}, Cd^{2+} and Cr^{3+} remains 100% with

Table 1 Toxic metals' concentrations present in the MW before and after treatment

Toxic metal	Concentration (ppm)			
	Before treatment	After treatment		[17] standards
	MW	CPSAC–H_3PO_4	CPSAC–NaOH	
Cadmium	37.4	ND	ND	0.003
Chromium	66.2	ND	ND	0.05
Cobalt	9.6	ND	0.675	–
Copper	31.4	ND	ND	1
Iron	756.5	2.31	0.496	0.3
Manganese	5.69	ND	ND	0.2
Lead	45.8	0.38	0.232	0.01
Zinc	94.1	ND	ND	3

ND not detected

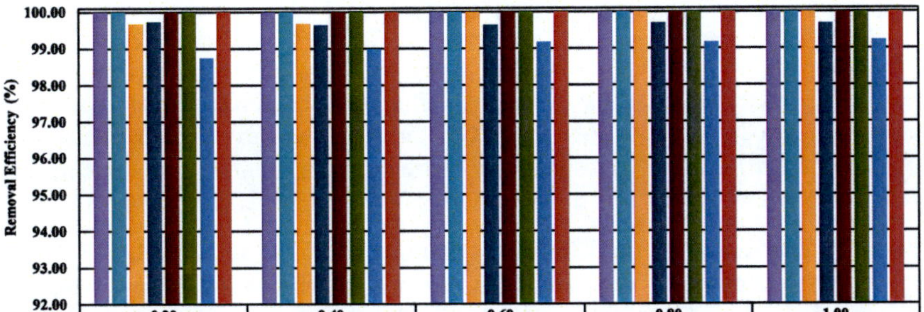

	0.20	0.40	0.60	0.80	1.00
Cu2+	100.00	100.00	100.00	100.00	100.00
Mn2+	100.00	100.00	100.00	100.00	100.00
Co3+	99.68	99.68	100.00	100.00	100.00
Fe2+	99.73	99.64	99.64	99.69	99.68
Zn2+	100.00	100.00	100.00	100.00	100.00
Cd2+	100.00	100.00	100.00	100.00	100.00
Pb2+	98.74	98.99	99.17	99.17	99.22
Cr3+	100.00	100.00	100.00	100.00	100.00

Fig. 1 Percentage removal trend for CPSAC–H_3PO_4 adsorbent dosage

increasing dosage from 0.2 to 1.0 g for both adsorbents. This portrays that increase in the adsorbent dosage has no effect on those toxic metals' removal after 0.2 g adsorbent dosage. The remaining toxic metals displayed varying removal efficiencies with the varying adsorbent dosages.

Optimum % removal efficiencies of Cu^{2+}, Mn^{2+}, Co^{3+}, Fe^{2+}, Zn^{2+}, Cd^{2+}, Pb^{2+} and Cr^{3+} were 100, 100, 100, 99.69, 100, 100, 99.22 and 100, respectively, for the varying CPSAC–H_3PO_4 dosage, while CPSAC–NaOH dosage of 0.2–1.0 g gave maximum % removal efficiencies of 100, 100, 98.34, 99.93, 100, 100, 99.49 and 100, respectively, for Cu^{2+}, Mn^{2+}, Co^{3+}, Fe^{2+}, Zn^{2+}, Cd^{2+}, Pb^{2+} and Cr^{3+}. The optimum CPSAC–H_3PO_4

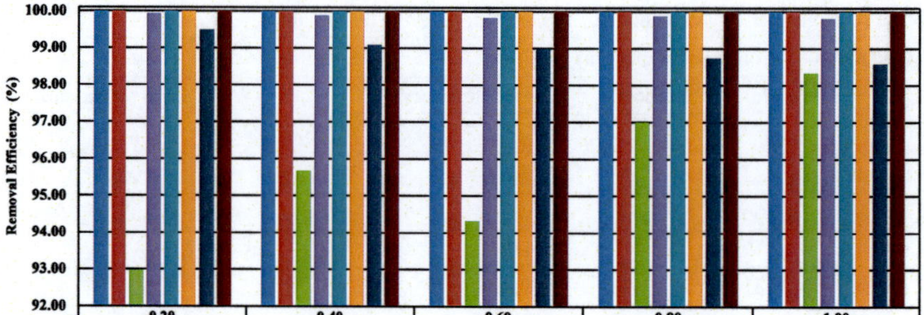

	0.20	0.40	0.60	0.80	1.00
Cu2+	100.00	100.00	100.00	100.00	100.00
Mn2+	100.00	100.00	100.00	100.00	100.00
Co3+	92.97	95.67	94.31	97.00	98.34
Fe2+	99.93	99.89	99.81	99.88	99.81
Zn2+	100.00	100.00	100.00	100.00	100.00
Cd2+	100.00	100.00	100.00	100.00	100.00
Pb2+	99.49	99.09	98.99	98.74	98.59
Cr3+	100.00	100.00	100.00	100.00	100.00

Fig. 2 Percentage removal trend for CPSAC–NaOH adsorbent dosage

and NaOH dosage values for the treatment were 0.6 and 0.2 g, respectively, with total removal of six (Cu^{2+}, Mn^{2+}, Co^{3+}, Zn^{2+}, Cd^{2+} and Cr^{3+}) and five toxic metals (Cu^{2+}, Mn^{2+}, Zn^{2+}, Cd^{2+} and Cr^{3+}), respectively. Adsorption capacities of the toxic metals were of the order $Fe^{2+} > Zn^{2+} > Cr^{3+} > Pb^{2+} > Cd^{2+} > Cu^{2+} > Co^{3+} > Mn^{2+}$.

(2) *Effects of agitation rate*: Effects of agitation rate of CPSAC–H_3PO_4 and NaOH on the toxic metals' removal are shown in Figs. 3 and 4. As the agitation rate increases from 50 to 200 rpm, change in the agitation rate has very little or no

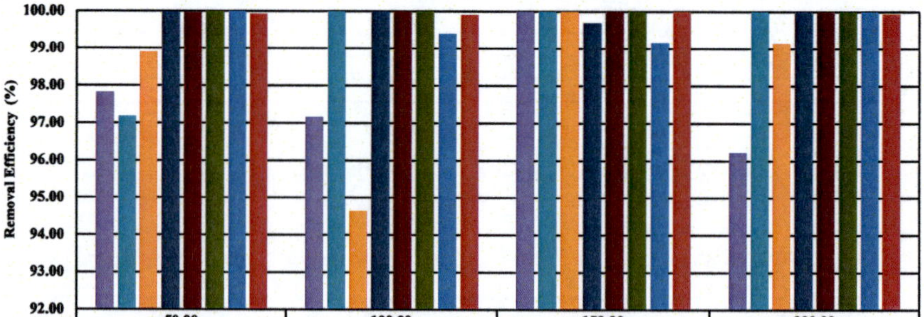

	50.00	100.00	150.00	200.00
Cu2+	97.83	97.16	100.00	96.22
Mn2+	97.19	100.00	100.00	100.00
Co3+	98.91	94.64	100.00	99.16
Fe2+	99.98	100.00	99.69	100.00
Zn2+	99.97	100.00	100.00	100.00
Cd2+	100.00	100.00	100.00	100.00
Pb2+	100.00	99.40	99.17	100.00
Cr3+	99.94	99.90	100.00	99.95

Fig. 3 Percentage removal trend with agitation rate of CPSAC–H_3PO_4 adsorbent

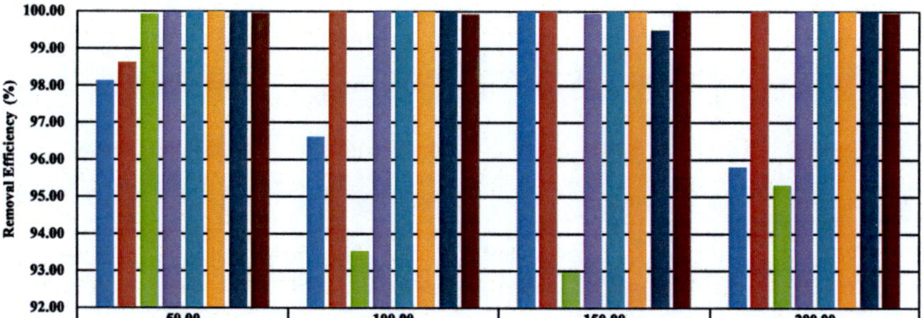

	50.00	100.00	150.00	200.00
Cu^{2+}	98.13	96.61	100.00	95.80
Mn^{2+}	98.63	100.00	100.00	100.00
Co^{3+}	99.92	93.51	92.97	95.31
Fe^{2+}	99.98	100.00	99.93	100.00
Zn^{2+}	100.00	100.00	100.00	100.00
Cd^{2+}	100.00	100.00	100.00	100.00
Pb^{2+}	100.00	100.00	99.49	100.00
Cr^{3+}	99.95	99.92	100.00	99.95

Fig. 4 Percentage removal trend with agitation rate of CPSAC–NaOH adsorbent

noticeable effects on the removal of Zn^{2+} and Cd^{2+} as the percentage removed remained constant all through for both adsorbents. The percentage removal efficiency of Mn^{2+} initially increased from 50 to 100 rpm thereafter remains 100%, while there was little effect around 100–150 rpm for Pb^{2+}. There were, however, irregular patterns noticed in the removal of Cu^{2+}, Co^{3+}, Cr^{3+} and Fe^{2+}. There is optimum % removal efficiencies and agitation rate of 100% and 150 rpm, respectively, for all the toxic metals in both adsorbents. The trend in the removal efficiencies portrays that Co^{3+} was least removed in the treatment for both adsorbents. Six and five toxic metals were removed by CPSAC–H_3PO_4 and NaOH, respectively, at optimum value of 150 rpm.

(3) *Effects of contact time*: Effects of contact time of the CPSAC–H_3PO_4 adsorbent on the toxic metals' removal are shown in Fig. 5. Cu^{2+}, Mn^{2+}, Zn^{2+}, Cd^{2+} and Cr^{3+} have 100% removal efficiencies all through the process, which means that contact time variation has no effect on these toxic metals. The percentage removal of Fe^{2+} decreases with increase in contact time. This was in contrast to Pb^{2+}, which decreases initially, but later increases with increase in contact time. The removal efficiency of Co^{3+} increases from 20 to 60 min and then remains the same from 60 to 120 min. This portrays that contact time variation of 60–120 min has no effect on Co^{3+}. The peak of percentage removal in Fe^{2+} and Pb^{2+} stood at 99.80 and 99.56%, respectively. The optimum contact time for the adsorption process was 60 min. The trend indicated that Pb^{2+} was least removed in the treatment. Figure 6 illustrates the effects of contact time of the CPSAC–NaOH adsorbent on the toxic metals' removal. The observed removal efficiency of Cu^{2+} and Cr^{3+} increases from 20 to 40 min and then remains the same from 40 to 120 min. This portrays that contact time variation of 40–120 min has no effect on Cu^{2+} and Cr^{3+}. Mn^{2+}, Zn^{2+} and Cd^{2+} have 100%

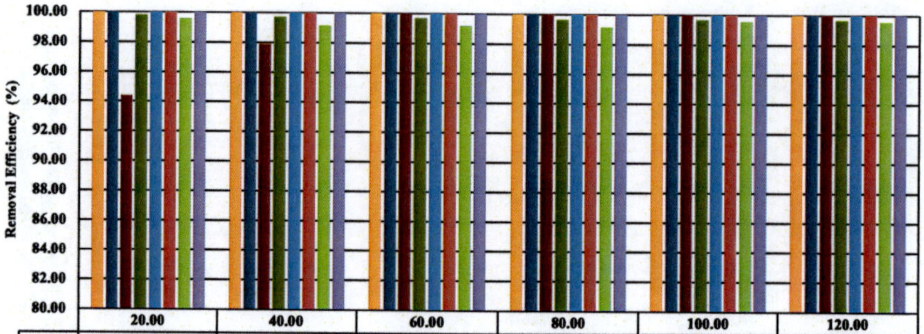

	20.00	40.00	60.00	80.00	100.00	120.00
Cu2+	100.00	100.00	100.00	100.00	100.00	100.00
Mn2+	100.00	100.00	100.00	100.00	100.00	100.00
Co3+	94.40	97.90	100.00	100.00	100.00	100.00
Fe2+	99.80	99.71	99.69	99.65	99.66	99.66
Zn2+	100.00	100.00	100.00	100.00	100.00	100.00
Cd2+	100.00	100.00	100.00	100.00	100.00	100.00
Pb2+	99.56	99.13	99.17	99.13	99.53	99.53
Cr3+	100.00	100.00	100.00	100.00	100.00	100.00

Fig. 5 Percentage removal trend with CPSAC–H_3PO_4 adsorbent contact time

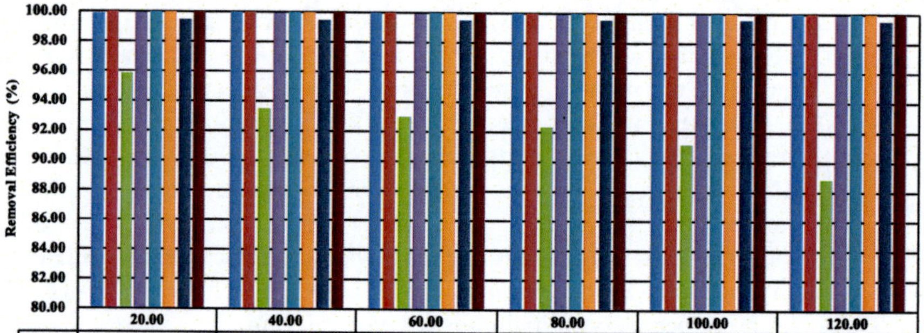

	20.00	40.00	60.00	80.00	100.00	120.00
Cu2+	99.89	100.00	100.00	100.00	100.00	100.00
Mn2+	100.00	100.00	100.00	100.00	100.00	100.00
Co3+	95.81	93.48	92.97	92.31	91.15	88.81
Fe2+	99.91	99.94	99.93	99.93	99.95	99.94
Zn2+	100.00	100.00	100.00	100.00	100.00	100.00
Cd2+	100.00	100.00	100.00	100.00	100.00	100.00
Pb2+	99.45	99.45	99.49	99.53	99.56	99.50
Cr3+	100.00	100.00	100.00	100.00	100.00	100.00

Fig. 6 Percentage removal trend with CPSAC–NaOH adsorbent contact time

removal all through the process, which means that contact time variation has no effect on these heavy metals. The percentage removal of Co^{3+} decreases with increase in contact time, and the trend was in contrast to Fe^{2+} and Pb^{2+} with increasing percentage removal until it reaches 100 min and then decreases from 100 to 120 min. The peak of percentage removal in Co^{3+}, Fe^{2+} and Pb^{2+} stood at 95.81, 99.95 and 99.56%, respectively. The optimum contact time for CPSAC–NaOH was 100 min with Co^{3+} least removed in the treatment.

(4) *Effects of pH*: Effects of pH of CPSAC–H_3PO_4 and NaOH adsorbents on the toxic metals' removal are shown in Figs. 7 and 8. As the pH changes from acidic to being alkaline, the removal efficiencies of most of the toxic metals (i.e. Cu^{2+}, Mn^{2+}, Co^{3+}, Fe^{2+} and Zn^{2+}) increased, Pb^{2+} and Cr^{3+} decreased for CPSAC–H_3PO_4, while the removal efficiencies of Cu^{2+}, Mn^{2+}, Co^{3+} and Fe^{2+} increased, and Zn^{2+}, Pb^{2+} and Cr^{3+} decreased for CPSAC–NaOH. Cd^{2+} remained 100% for both adsorbents. The optimum pH value for the treatment was 8 and 4 for CPSAC–H_3PO_4 and NaOH adsorbents, respectively. The trend in the removal efficiencies portrays that Mn^{2+} was least removed in the treatments. Both adsorbents were negatively influenced at optimum conditions as the number of heavy metals removed was drastically reduced to one (1) and two (2) for CPSAC–H_3PO_4 and NaOH adsorbents, respectively, when compared with that of adsorbent dosage, agitation rate and contact time. Though, CPSAC–H_3PO_4 was more affected.

(5) *Effects of temperature*: Effects of temperature on CPSAC–H_3PO_4 for the toxic metals' removal are shown in Fig. 9. Increase in temperature observed from 30 to 70 °C has no noticeable effect on the removal of Cd^{2+} and Zn^{2+} as the percentage removed remains constant all through. The percentage removal efficiency of Mn^{2+} largely increased, in contrast to that of Co^{3+} and Cr^{3+}, which decreased. Cu^{2+} showed earlier decrease and thereafter remains constant, while Pb^{2+} and Fe^{2+} indicated irregular patterns in their removal efficiencies. There were optimum % removal efficiencies of 99.33, 88.22, 92.65, 99.96, 100, 100, 98.41 and 99.56, respectively, for Cu^{2+}, Mn^{2+}, Co^{3+}, Fe^{2+}, Zn^{2+}, Cd^{2+}, Pb^{2+} and Cr^{3+}. The optimum temperature value for the treatment was 40 °C. The removal efficiency trend showed that Mn^{2+} was least removed. CPSAC–H_3PO_4

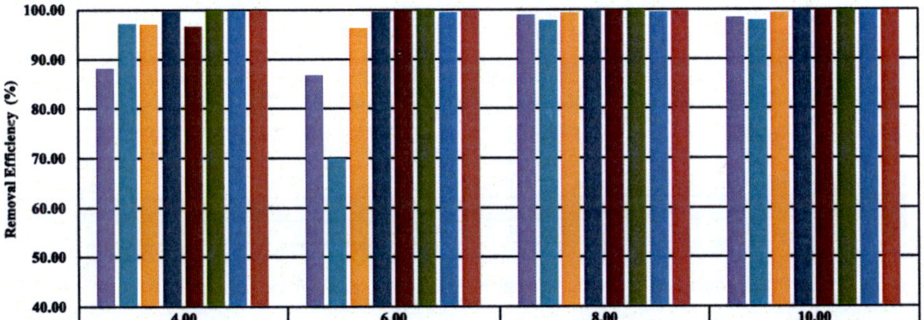

	4.00	6.00	8.00	10.00
Cu2+	88.22	86.78	98.91	98.37
Mn2+	97.19	69.98	97.84	97.84
Co3+	97.04	96.26	99.33	99.33
Fe2+	99.63	99.57	99.98	99.95
Zn2+	96.60	99.85	99.94	99.79
Cd2+	100.00	100.00	100.00	100.00
Pb2+	99.87	99.40	99.50	99.81
Cr3+	99.95	99.94	99.86	99.89

Fig. 7 Percentage removal trend with pH of CPSAC–H_3PO_4 adsorbent

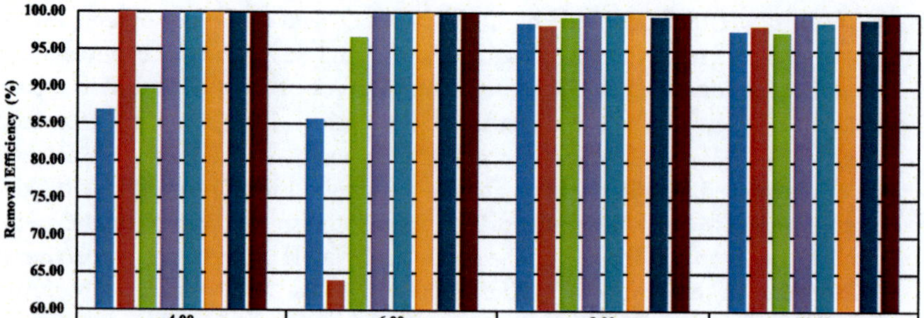

	4.00	6.00	8.00	10.00
Cu2+	86.78	85.65	98.60	97.60
Mn2+	100.00	63.95	98.28	98.28
Co3+	89.68	96.67	99.33	97.41
Fe2+	99.95	99.91	99.96	99.94
Zn2+	99.97	99.81	99.73	98.74
Cd2+	100.00	100.00	100.00	100.00
Pb2+	99.87	99.87	99.50	99.17
Cr3+	99.97	99.95	99.89	99.86

Fig. 8 Percentage removal trend with pH of CPSAC–NaOH adsorbent

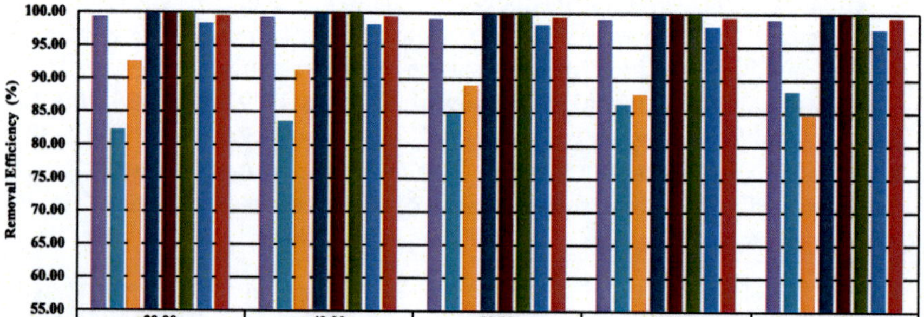

	30.00	40.00	50.00	60.00	70.00
Cu2+	99.33	99.33	99.09	99.09	99.09
Mn2+	82.25	83.59	84.90	86.24	88.22
Co3+	92.65	91.31	89.10	87.77	84.67
Fe2+	99.96	99.95	99.96	99.96	99.96
Zn2+	99.89	100.00	100.00	100.00	100.00
Cd2+	100.00	100.00	100.00	100.00	100.00
Pb2+	98.41	98.24	98.24	98.06	97.66
Cr3+	99.56	99.49	99.45	99.45	99.42

Fig. 9 Percentage removal trend with temperature of CPSAC–H_3PO_4 adsorbent

was negatively influenced at optimum condition as the number of heavy metals removed was drastically reduced to two (2).

The effects of temperature of CPSAC–NaOH on the toxic metals' removal are shown in Fig. 10. Increase in temperature from 30 to 70 °C resulted in increase in percentage removal of Mn^{2+} in contrast to that of Co^3 and Cr^{3+}, which decreased.

Change in the temperature has no noticeable effect on the removal of Cu^{2+}, Zn^{2+} and Cd^{2+} as the percentage removed remains constant all through. Nevertheless,

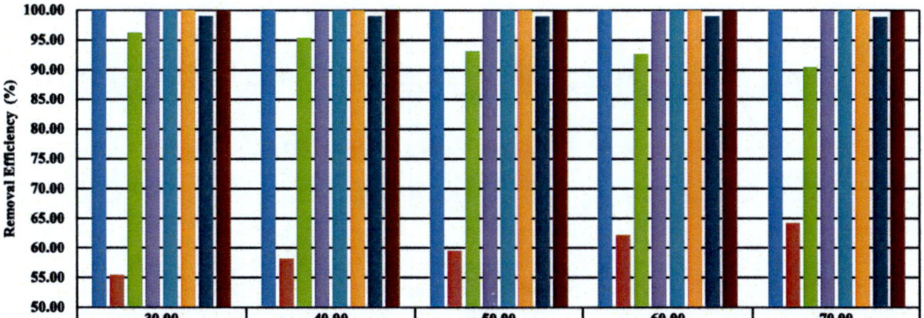

	30.00	40.00	50.00	60.00	70.00
Cu2+	100.00	100.00	100.00	100.00	100.00
Mn2+	55.45	58.15	59.51	62.20	64.22
Co3+	96.20	95.30	93.09	92.65	90.43
Fe2+	99.94	99.89	99.96	99.94	99.94
Zn2+	100.00	100.00	100.00	100.00	100.00
Cd2+	100.00	100.00	100.00	100.00	100.00
Pb2+	98.94	98.94	98.94	98.94	98.83
Cr3+	99.95	99.95	99.91	99.88	99.84

Fig. 10 Percentage removal trend with temperature of CPSAC–NaOH adsorbent

irregular patterns were observed in the removal of Fe^{2+} and Pb^{2+}. Increase in temperature from 30 to 70 °C resulted in optimum % removal efficiencies which were 100, 64.22, 96.20, 99.96, 100, 100, 98.94 and 99.95, respectively, for Cu^{2+}, Mn^{2+}, Co^{3+}, Fe^{2+}, Zn^{2+}, Cd^{2+}, Pb^{2+} and Cr^{3+}. The optimum temperature value for the treatment was 30 °C, and the trend in the removal efficiencies showed that Mn^{2+} was least removed. CPSAC–NaOH was negatively influenced at optimum condition as the number of heavy metals removed was drastically reduced to three (3).

The summary of results of toxic metals removed from mining wastewater treated with CPSAC–H_3PO_4 and NaOH adsorbents is depicted in Table 2. The CPSAC–H_3PO_4 removed six toxic metals (i.e. Cu^{2+}, Mn^{2+}, Co^{3+}, Zn^{2+}, Cd^{2+} and Cr^{3+}) totally at optimum conditions of 0.6 g adsorbent dosage, 150 rpm agitation rate, 60 min contact time, pH of 7 and 30 °C temperature, while CPSAC–NaOH removed five toxic metals (i.e. Cu^{2+}, Mn^{2+}, Zn^{2+}, Cd^{2+} and Cr^{3+}) completely at optimum condition of 0.2 g adsorbent dosage, 150 rpm agitation rate, 40 min contact time, pH of 7 and 30 °C temperature. Both adsorbent efficiencies were positively influenced by adsorbent dosage, agitation rate and contact time, while pH and temperature influenced them negatively, though more pronounced in CPSAC–H_3PO_4.

3.3 Results of Adsorption Isotherm Studies

(1) *Freundlich adsorption isotherm*: The results of Freundlich isotherm for the toxic metals for CPSAC–H_3PO_4 and NaOH are presented in Table 3. Most of the toxic metals' adsorption intensity (1/n) is zero or negative for both adsorbents, which indicated that they have irreversible adsorption, while 1/n

Table 2 Summary of toxic metals removed by CPSAC–H_3PO_4 and NaOH

Adsorbent	Adsorbent dosage (g)					Contact time (min)						pH				Agitation rate (rpm)					Temperature (°C)				
	0.2	0.4	0.6	0.8	1.0	20	40	60	80	100	120	4	6	8	10	50	100	150	200	250	30	40	50	60	70
CPSAC–H_3PO_4	5	5	6	6	6	5	5	6	6	6	6	1	1	1	1	2	4	6	5	5	1	2	2	2	2
CPSAC–NaOH	5	5	5	5	5	4	5	5	5	5	5	2	1	1	1	3	5	5	5	5	3	3	3	3	3

Table 3 Summary of Freundlich isotherm parameters for CPSAC–H_3PO_4 and NaOH

Ads't	Metal	Linear analysis					Non-linear analysis				
		Parameters		Errors			Parameters		Errors		
		$1/n$	K	R^2	EABS	ARE	$1/n$	K	R^2	EABS	ARE
CPSAC–H_3PO_4	Cu^{2+}	0.000	1.000	0.000	3.586	0.717	0.000	0.000	0.500	3.586	0.717
	Mn^{2+}	0.000	1.000	0.000	0.650	0.130	0.000	0.000	0.500	0.650	0.130
	Co^{3+}	− 3.4223	0.613	0.769	1.073	0.234	32.740	0.613	0.991	0.049	0.020
	Fe^{2+}	0.1518	0.101	0.972	87.408	17.713	− 0.870	617.000	0.976	11.008	4.410
	Zn^{2+}	0.000	1.000	0.000	10.743	2.149	0.000	0.000	0.500	10.743	2.149
	Cd^{2+}	0.000	1.000	0.000	4.270	0.854	0.000	0.000	0.500	4.270	0.854
	Pb^{2+}	0.3152	66.085	0.967	4.546	0.798	6.877	0.224	1.000	0.000	0.009
	Cr^{3+}	0.000	1.000	0.000	7.558	1.512	0.000	0.000	0.500	7.558	1.512
CPSAC–NaOH	Cu^{2+}	0.000	1.000	0.000	3.586	0.717	0.000	0.000	0.500	3.586	0.717
	Mn^{2+}	0.000	1.000	0.000	0.650	0.130	0.000	0.000	0.500	0.650	0.130
	Co^{3+}	0.9497	3.173	0.991	0.990	0.204	2.607	0.297	0.993	0.054	0.019
	Fe^{2+}	− 0.6503	65.865	0.995	84.414	16.521	− 1.320	280.200	0.993	5.040	1.669
	Zn^{2+}	0.000	1.000	0.000	10.743	2.149	0.000	0.000	0.500	10.743	2.149
	Cd^{2+}	0.000	1.000	0.000	4.270	0.854	0.000	0.000	0.500	4.270	0.854
	Pb^{2+}	− 0.6394	1.154	0.999	4.567	0.805	-3.960	6.720	0.999	0.373	0.015
	Cr^{3+}	0.000	1.000	0.000	7.558	1.512	0.000	0.000	0.500	7.558	1.512

of few toxic metals is greater than one (i.e. $1/n > 1$), which means that they have non-optimum adsorption. Co^{3+}, Fe^{2+} and Pb^{2+} models best fitted the Freundlich isotherm (i.e. close to unity, $R^2 = 1$) in order of Pb^{2+} (0.9988) > Fe^{2+} (0.995) > Co^{3+} (0.9908) for CPSAC–NaOH, while the order is Fe^{2+} (0.972) > Pb^{2+} (0.967) > Co^{3+} (0.769) for CPSAC–H_3PO_4. Hence, Freundlich isotherm best fitted CPSAC–NaOH adsorption data than that of CPSAC–H_3PO_4. The Co^{3+}, Fe^{2+} and Pb^{2+} adsorption models have low non-linear regression EAB and ARE values with high R^2 in comparison with that of their linear regression, thus best fitted Freundlich isotherm. All the error functions of the remaining toxic metals (which have irreversible adsorption) were the same for both linear and non-linear functions of both adsorbents.

(2) *Langmuir adsorption isotherm*: The results of Langmuir isotherm for the adsorption data of toxic metals for CPSAC–H_3PO_4 and NaOH are presented in Table 4. It depicted that the separation factor, R_L (i.e. the basic and characteristic index of the Langmuir isotherm model) for Cu^{2+}, Mn^{2+}, Zn^{2+} and Cd^{2+} of both adsorbents and Fe^{2+} of CPSAC–NaOH, is one, which means linear adsorption. Co^{3+} and Pb^{2+} of CPSAC–NaOH have their R_L that lies between zero and one, which indicated optimum adsorption. The separation factors of the remaining toxic metals were zero or negative, thus giving irreversible adsorption according to [11]. Langmuir isotherm best fitted models of Co^{3+}, Fe^{2+} and Pb^{2+} of both adsorbents. Though, the isotherm best fitted CPSAC–H_3PO_4 adsorption data (i.e. models) than that of CPSAC–NaOH putting into consideration non-linear regression analyses.

(3) *D–R adsorption isotherm*: The results of D–R isotherm for the adsorption data of toxic metals for CPSAC–H_3PO_4 and NaOH are presented in Table 5. The adsorption energy, *E*, for Cu^{2+}, Mn^{2+}, Zn^{2+}, Cd^{2+} and Cr^{3+} of both adsorbents depicted physical adsorption (i.e. zero). The adsorption energy, *E*, of the remaining heavy metals depicted chemical adsorption (> 16 kJ/Mol) as explained by [14]. D–R isotherm best fitted Co^{3+}, Fe^{2+} and Pb^{2+} models of CPSAC–NaOH, and Co^{3+} and Pb^{2+} model of CPSAC–H_3PO_4. Though, the isotherm best fitted CPSAC–NaOH adsorption data (i.e. models) than that of CPSAC–H_3PO_4 with reference to non-linear regression analyses.

(4) *Temkin adsorption isotherm*: The results of Temkin isotherm for the adsorption data of toxic metals for CPSAC–H_3PO_4 and NaOH are presented in Table 6. The correlation coefficient (R^2) for the isotherm ranges from 0 to 1.0. The adsorption models of Co^{3+}, Fe^{2+} and Pb^{2+} of both adsorbents best fitted the Temkin isotherm, though those of CPSAC–NaOH were greater than that of CPSAC–H_3PO_4 with respect to linear regression analyses as their EAB and ARE error functions were lower than that of non-linear regression analyses.

Table 4 Summary of Langmuir isotherm parameters for CPSAC–H_3PO_4 and NaOH

Ads't	Metal	Linear analysis						Non-linear analysis					
		Parameters			Errors			Parameters			Errors		
		q_{max}	K	R_L	R^2	EABS	ARE	q_{max}	K	R_L	R^2	EABS	ARE
CPSAC–H_3PO_4	Cu^{2+}	0.000	0.000	1.000	0.000	3.586	0.717	0.000	0.000	1.000	0.000	3.586	0.717
	Mn^{2+}	0.000	0.000	1.000	0.000	0.650	0.130	0.000	0.000	1.000	0.000	0.650	0.130
	Co^{3+}	1.595	− 3E + 17	0.000	0.844	1.086	0.216	1.595	− 3E + 17	− 3.5E−19	0.923	0.456	0.091
	Fe^{2+}	8.170	− 0.519	− 0.003	0.999	86.034	17.194	8.170	− 0.519	− 0.003	0.979	23.034	2.727
	Zn^{2+}	0.000	0.000	1.000	0.000	10.743	2.149	0.000	0.000	1.000	0.000	10.743	2.149
	Cd^{2+}	0.000	0.000	1.000	0.000	4.270	0.854	0.000	0.000	1.000	0.000	4.270	0.854
	Pb^{2+}	− 2.150	− 1.470	− 0.015	0.991	5.058	0.990	− 2.150	− 1.470	− 0.015	0.999	0.324	0.022
	Cr^{3+}	0.000	0.000	1.000	0.000	7.558	1.512	0.000	0.000	0.000	0.000	7.558	1.512
CPSAC–NaOH	Cu^{2+}	0.000	0.000	1.000	0.000	3.586	0.717	0.000	0.000	1.000	0.000	3.586	0.717
	Mn^{2+}	0.000	0.000	1.000	0.000	0.650	0.130	0.000	0.000	1.000	0.000	0.650	0.130
	Co^{3+}	1.017	5.098	0.020	0.906	0.436	0.046	0.000	0.000	0.020	0.923	0.380	0.045
	Fe^{2+}	27.322	2.392	0.208	0.717	67.446	10.967	0.000	0.000	1.000	0.992	0.024	2.940
	Zn^{2+}	0.000	0.000	1.000	0.000	10.743	2.149	0.000	0.000	1.000	0.000	10.743	2.149
	Cd^{2+}	0.000	0.000	1.000	0.000	4.270	0.854	0.000	0.000	1.000	0.000	4.270	0.854
	Pb^{2+}	1.573	4.092	0.225	0.727	4.179	0.695	1.573	4.092	0.225	0.998	0.000	0.051
	Cr^{3+}	0.000	0.000	1.000	0.000	7.558	1.512	0.000	0.000	0.000	0.000	7.558	1.512

Table 5 Summary of D–R isotherm parameters for CPSAC–H_3PO_4 and NaOH

Ads't	Metal	Linear analysis					Non-linear analysis		
		Parameters		Errors			Errors		
		q_m	E	R^2	EABS	ARE	R^2	EABS	ARE
CPSAC–H_3PO_4	Cu^{2+}	0.000	0.000	0.000	3.586	0.717	0.000	3.586	0.717
	Mn^{2+}	0.000	0.000	0.000	0.650	0.130	0.000	0.650	0.130
	Co^{3+}	0.617	7070.000	0.718	1.277	0.344	0.769	0.553	0.096
	Fe^{2+}	19.498	500.000	0.267	81.643	15.500	0.581	81.543	15.470
	Zn^{2+}	0.000	0.000	0.000	10.743	2.149	0.000	10.743	2.149
	Cd^{2+}	0.000	0.000	0.000	4.270	0.854	0.000	4.270	0.854
	Pb^{2+}	82.414	1290.000	0.338	4.612	0.826	0.494	3.541	0.494
	Cr^{3+}	0.000	0.000	0.000	7.558	1.512	0.000	7.558	1.512
CPSAC–NaOH	Cu^{2+}	0.000	0.000	0.000	3.586	0.717	0.000	3.586	0.717
	Mn^{2+}	0.000	0.000	0.000	0.650	0.130	0.000	0.650	0.130
	Co^{3+}	3.353	0.000	0.648	0.783	0.197	0.948	0.365	0.056
	Fe^{2+}	47.432	1580.000	0.579	81.682	15.481	0.637	61.567	9.984
	Zn^{2+}	0.000	0.000	0.000	10.743	2.149	0.000	10.743	2.149
	Cd^{2+}	0.000	0.000	0.000	4.270	0.854	0.000	4.270	0.854
	Pb^{2+}	1.051	1580.000	0.356	4.976	0.957	0.565	3.269	0.418
	Cr^{3+}	0.000	0.000	0.000	7.558	1.512	0.000	7.558	1.512

Table 6 Summary of Temkin isotherm parameters for CPSAC–H_3PO_4 and NaOH

Ads't	Metal	Linear analysis						Non-linear analysis		
		Parameters			Errors			Errors		
		β	K_T	b	R^2	EAB	ARE	R^2	EABS	ARE
CPSAC–H_3PO_4	Cu^{2+}	0.000	0.000	0.000	0.000	3.586	0.717	0.000	3.586	0.717
	Mn^{2+}	0.000	0.000	0.000	0.000	0.650	0.130	0.000	0.650	0.130
	Co^{3+}	− 0.336	0.150	− 7371.530	0.992	0.383	0.050	Undefined		
	Fe^{2+}	225.100	0.512	11.007	0.816	36.157	7.460	0.927	264.635	186.476
	Zn^{2+}	0.000	0.000	0.000	0.000	10.743	2.149	0.000	10.743	2.149
	Cd^{2+}	0.000	0.000	0.000	0.000	4.270	0.854	0.000	4.270	0.854
	Pb^{2+}	18.916	3.015	130.978	0.998	0.737	0.035	Undefined		
	Cr^{3+}	0.000	0.000	0.000	0.000	7.558	1.512	0.000	7.558	1.512
CPSAC–NaOH	Cu^{2+}	0.000	0.000	0.000	0.000	3.586	0.717	0.000	3.586	0.717
	Mn^{2+}	0.000	0.000	0.000	0.000	0.650	0.130	0.000	0.650	0.130
	Co^{3+}	1.214	5.027	2040.666	0.913	0.731	0.098	Undefined		
	Fe^{2+}	− 141.330	0.568	− 17.530	0.999	7.907	0.223	0.841	2.866	87.561
	Zn^{2+}	0.000	0.000	0.000	0.000	10.743	2.149	0.000	10.743	2.149
	Cd^{2+}	0.000	0.000	0.000	0.000	4.270	0.854	0.000	4.270	0.854
	Pb^{2+}	− 9.122	0.774	− 271.601	0.996	4.434	0.988	Undefined		
	Cr^{3+}	0.000	0.000	0.000	0.000	7.558	1.512	0.000	7.558	1.512

3.4 Results of Adsorption Kinetic Studies

(1) *Pseudo-first-order kinetic model*: Table 7 presents the results of kinetic model of pseudo-first order for the adsorption data of toxic metals for CPSAC–H_3PO_4 and NaOH. The observed experimental (q_e (expt.)) and modelled (q_e (cal.)) adsorption capacity values were not close or the same. The R^2 of the models of most of the toxic metals for the adsorbent illustrated these as they were low or zero; hence, none of the toxic metals' adsorption data fitted kinetic model of pseudo-first order. However, their non-linear regression EAB, ARE and R^2 values were lower than that of linear regression, and then both regression analyses accepted that most of the toxic metal adsorption models suited kinetic model of pseudo-first order.

(2) *Pseudo-second-order kinetic model*: The results of kinetic model of pseudo-second order for the adsorption data of toxic metals for CPSAC–H_3PO_4 and NaOH are presented in Table 8. The observed adsorption data of the toxic metals absolutely suited the kinetic model of pseudo-second order with R^2 generally tending towards unity (i.e. $R^2 = 1$), EAB and ARE of non-linear regression analyses tending towards zero in all cases. The non-linear regression proved to be better for the analysis due to these low values of the adsorption models of the toxic metals. Moreover, both the q_e (expt.) and q_e (cal.) results were approximately the same in all the cases of the toxic metals for both adsorbents as indicated in Table 8. These were because the rate of adsorption was governed by chemisorption according to [11]. Though, kinetic model of pseudo-second order best fitted CPSAC–H_3PO_4 adsorption data that of CPSAC–NaOH.

(3) *Elovich kinetic model*: Table 9 presents kinetic model of Elovich for the adsorption data of toxic metals for CPSAC–H_3PO_4 and NaOH. The toxic metals' models best suited the Elovich kinetic model depending upon their correlation coefficient values. Most of the toxic metals in both adsorbents did not follow the model, as their correlation coefficients were zero. However, Co^{3+}, Fe^{2+} and Pb^{2+} of CPSAC–NaOH and Fe^{2+} of CPSAC–H_3PO_4 best suited Elovich kinetic model.

(4) *Intra-particle diffusion model*: Table 10 presents kinetic model of intra-particle diffusion for the adsorption data of toxic metals for CPSAC–H_3PO_4 and NaOH. The cut-off value C (i.e. the determinant factor of the model) varies from 0.356 to 47.208 and 0 to 188.94 for CPSAC–H_3PO_4 and CPSAC–NaOH, respectively. Fe^{2+} and Mn^{2+} have the highest and lowest values of cut-off value C, respectively, for both adsorbents. According to [14], the results indicated that there is greater part of the surface adsorption in the rate-restricting step of all the toxic metals for both adsorbents, though, in CPSAC–NaOH than CPSAC–H_3PO_4.

Table 7 Kinetic model of pseudo-first-order parameters for CPSAC–H_3PO_4 and NaOH

Ads't	Metal	Linear analysis						Non-linear analysis		
		Parameters			Errors			Errors		
		q_e (expt.)	q_e (cal.)	K_1	R^2	EAB	ARE	R^2	EABS	ARE
CPSAC–H_3PO_4	Cu^{2+}	1.963	0.394	0.000	0.973	13.082	11.398	0.500	3.586	0.717
	Mn^{2+}	0.356	0.000	0.000	0.965	7.288	18.670	0.500	0.650	0.130
	Co^{3+}	0.600	0.272	0.043	0.628	0.500	0.260	0.987	0.240	0.060
	Fe^{2+}	47.137	20.401	0.051	0.788	38.955	19.118	0.985	21.969	5.748
	Zn^{2+}	5.881	0.000	1.000	0.500	10.742	2.148	1.000	0.000	0.000
	Cd^{2+}	2.338	0.000	1.000	0.500	4.269	0.854	1.000	0.000	0.000
	Pb^{2+}	2.839	0.197	0.019	0.929	8.828	5.808	0.997	0.179	0.026
	Cr^{3+}	4.138	0.070	0.000	0.974	13.711	10.616	0.500	7.558	1.512
CPSAC–NaOH	Cu^{2+}	7.850	0.143	0.340	0.429	7.583	1.472	0.813	4.264	0.595
	Mn^{2+}	1.423	0.000	0.000	0.969	6.515	6.580	0.500	1.423	0.285
	Co^{3+}	2.231	2.481	0.600	0.527	2.220	0.441	0.829	1.198	0.163
	Fe^{2+}	189.001	12.289	0.160	0.570	183.010	35.534	0.802	104.299	14.371
	Zn^{2+}	23.525	0.000	0.000	0.995	23.525	4.705	0.500	3.319	0.518
	Cd^{2+}	9.350	0.000	0.000	0.978	17.494	7.613	0.500	9.350	1.870
	Pb^{2+}	11.392	0.425	0.060	0.686	7.023	0.939	0.671	5.130	1.062
	Cr^{3+}	16.550	0.000	0.000	0.985	16.550	3.310	0.500	8.631	2.072

Table 8 Kinetic model of pseudo-second-order parameters for CPSAC–H_3PO_4 and NaOH

Ads't	Metal	Linear analysis						Non-linear analysis			
		Parameters			Errors			Errors			
		q_e (expt.)	q_e (cal.)	K_2	R^2	EAB	ARE	K_2	R^2	EAB	ARE
CPSAC–H_3PO_4	Cu^{2+}	1.963	1.963	0.000	0.975	1.624	0.797	0.001	1.000	0.000	5.025E−33
	Mn^{2+}	0.356	0.356	0.000	0.976	0.294	0.144	0.004	1.000	0.000	0.000
	Co^{3+}	0.600	0.599	0.000	0.976	0.503	0.264	0.002	1.000	0.009	9.071E−05
	Fe^{2+}	47.137	47.136	0.000	0.975	38.958	19.122	0.000	1.000	0.046	1.160E−05
	Zn^{2+}	5.881	5.882	0.000	0.975	4.862	2.388	0.000	1.000	0.000	1.073E−31
	Cd^{2+}	2.338	2.338	0.000	0.975	1.932	0.949	0.001	1.000	0.000	1.687E−32
	Pb^{2+}	2.839	2.839	0.000	0.975	2.331	1.134	0.001	1.000	0.007	6.238E−06
	Cr^{3+}	4.138	4.137	0.000	0.975	3.420	1.680	0.000	1.000	0.000	3.813E−32
CPSAC–NaOH	Cu^{2+}	7.850	7.849	0.000	0.935	4.262	0.595	0.001	1.000	0.002	3.472E−07
	Mn^{2+}	1.423	1.422	0.000	0.935	0.773	0.108	0.005	1.000	0.000	6.239E−32
	Co^{3+}	2.231	2.116	− 0.192	0.942	1.198	0.163	0.003	1.000	0.064	4.962E−04
	Fe^{2+}	189.001	188.679	0.000	0.934	102.726	14.352	0.000	1.000	0.026	1.127E−06
	Zn^{2+}	23.525	23.529	0.000	0.935	12.782	1.785	0.000	1.000	0.000	2.683E−30
	Cd^{2+}	9.350	9.346	0.000	0.935	5.080	0.709	0.001	1.000	0.000	2.700E−31
	Pb^{2+}	11.392	11.403	0.000	0.933	6.207	0.869	0.001	1.000	0.007	1.359E−06
	Cr^{3+}	16.550	16.549	0.000	0.935	8.992	1.256	0.000	1.000	0.000	1.525E−31

Table 9 Elovich kinetic model parameters for CPSAC–H_3PO_4 and NaOH

Ads't	DESC	Cu^{2+}	Mn^{2+}	Co^{3+}	Fe^{2+}	Zn^{2+}	Cd^{2+}	Pb^{2+}	Cr^{3+}
CPSAC–H_3PO_4	R^2	0	0	0	0.997	0	0	0.000	0
	α	0	0	0	0.000	0	0	3.801	0
	β	0	0	0	− 22.883	0	0	5000.000	0
CPSAC–NaOH	R^2	0	0	0.997	0.997	0	0	0.998	0
	α	0	0	0	4.386	0	0	1.139	0
	β	0	0	− 10.582	27.027	0	0	72.464	0

Table 10 Intra-particle diffusion model parameters for CPSAC–H_3PO_4 and NaOH

Adsorbent	DESC	Cu^{2+}	Mn^{2+}	Co^{3+}	Fe^{2+}	Zn^{2+}	Cd^{2+}	Pb^{2+}	Cr^{3+}
CPSAC–H_3PO_4	R^2	0.000	0.000	0.000	0.993	0.000	0.000	0.020	0.000
	C	1.963	0.356	0.552	47.208	5.881	2.338	2.841	4.138
	K_{diff}	− 7E−16	8E−17	0.005	− 0.010	− 1E−15	− 7E−16	4E−04	0.000
CPSAC–NaOH	R^2	0.000	0.000	0.998	0.990	0.000	0.000	0.997	0.000
	C	7.850	1.423	2.386	188.940	23.525	9.350	11.363	16.550
	K_{diff}	0.000	3E−16	0.001	0.006	− 5E−15	−3.E−15	0.002	0.000

3.5 *Results of Adsorption Thermodynamic Studies*

Table 11 presents the results of thermodynamic studies of the toxic metals' adsorption data for CPSAC–H_3PO_4 and NaOH. The observed activation energy (Ea) values for the adsorption processes range from 0 to 657.916 kJ/mol and 0–168.533 kJ/mol for CPSAC–H_3PO_4 and NaOH, respectively. Increase in the temperature resulted in increase in the Ea values for both adsorbents. The Ea of CPSAC–NaOH (i.e. Cu^{2+}, Fe^{2+}, Zn^{2+}, Cd^{2+} and Pb^{2+}) and CPSAC–H_3PO_4 (i.e. Fe^{2+}, Cd^{2+} and Cr^{3+}) toxic metals depicted physisorption processes, while the remaining toxic metals depicted chemisorption processes. The works of [18, 19] confirmed these. This also indicated that more chemical adsorption processes were involved with CPSAC–H_3PO_4 than CPSAC–NaOH. The negative ΔG and positive ΔH values for the toxic metals' adsorption data portrayed that their adsorption processes were approving, practicable and voluntary and indicate substantial character with feeble attractive forces, while positive ΔS values depict uneven expansion of the changeability at the components of interface of solid solution in the course of the adsorption.

Table 11 Adsorption thermodynamic parameters for CPSAC–H_3PO_4 and NaOH

Temp. (°C)	DESC	CPSAC–H_3PO_4								CPSAC–NaOH							
		Cu^{2+}	Mn^{2+}	Co^{3+}	Fe^{2+}	Zn^{2+}	Cd^{2+}	Pb^{2+}	Cr^{3+}	Cu^{2+}	Mn^{2+}	Co^{3+}	Fe^{2+}	Zn^{2+}	Cd^{2+}	Pb^{2+}	Cr^{3+}
30	R^2	0	0.933	0.878	0.323	0.150	0	0.055	0.824	0	0.933	0.941	0.894	0	0	0.313	0.832
	ΔS	36.456	18.704	10.289	64.288	50.446	0	29.873	41.035	0	6.666	13.489	62.984	0	0	33.335	47.937
	ΔH	49.234	57.887	103.257	4.703	581.191	0	42.8	35.742	0	45.947	126.94	20.588	0	0	15.679	148.879
	ΔG	– 12,586	– 3863	– 6382	– 19,445	– 17,099	0	– 10,390	– 13,661	0	– 551	– 8139	– 18,794	0	0	– 11,430	– 19,006
	Ea	49.234	57.887	103.257	4.703	581.191	0	42.8	35.742	0	45.947	126.94	20.588	0	0	15.679	148.879
40	ΔS	36	18.704	10.289	64.288	50.446	0	29.873	41.035	0	6.666	13.489	62.984	0	0	36.947	47.937
	ΔH	50.859	59.798	106.665	4.858	600.372	0	44.213	36.921	0	47.463	131.129	21.268	0	0	8.688	153.792
	ΔG	– 13,001	– 4236	– 6122	– 19,732	0	0	– 10,467	– 13,712	0	– 856	– 7833	– 17,802	0	0	– 11,807	– 19,633
	Ea	0	59.798	106.665	4.858	600.372	0	44.213	36.921	0	47.463	131.129	21.268	0	0	8.688	153.792
50	ΔS	36.456	18.704	10.289	64.288	50.446	0	29.873	41.035	0	6.666	13.489	62.984	0	0	36.947	47.937
	ΔH	52.484	61.708	110.073	5.013	619.554	0	45.625	38.101	0	48.979	135.318	21.947	0	0	8.966	158.706
	ΔG	– 12,593	– 4638	– 5643	– 20,729	0	0	– 10,801	– 13,966	0	– 1034	– 6985	– 21,143	0	0	– 12,184	– 18,903
	Ea	52.484	61.708	110.073	5.013	619.554	0	45.625	38.101	0	48.979	135.318	21.947	0	0	8.966	158.706
60	ΔS	36.456	18.704	10.289	64.288	50.446	0	29.873	41.035	0	6.666	13.489	62.984	0	0	36.947	47.937
	ΔH	54.109	63.619	113.481	5.168	638.735	0	47.038	39.28	0	50.496	139.508	22.627	0	0	9.243	163.619
	ΔG	– 12,983	– 5081	– 5457	– 21,370	0	0	– 10,862	– 14,398	0	– 1378	– 7014	– 20,655	0	0	– 12,562	– 18,563
	Ea	54.109	63.619	113.481	5.168	638.735	0	47.038	39.28	0	50.496	139.508	22.627	0	0	9.243	163.619
70	ΔS	36.456	18.704	10.289	64.288	50.446	0	29.873	41.035	0	6.666	13.489	62.984	0	0	36.947	47.937
	ΔH	55.734	65.529	116.888	5.324	657.916	0	48.45	40.46	0	52.012	143.697	23.306	0	0	9.521	168.533
	ΔG	– 13,373	– 5743	– 4873	– 22,012	0	0	– 10,637	– 14,647	0	– 1668	– 6404	– 20,959	0	0	– 12,651	– 18,406

(continued)

Table 11 (continued)

Temp. (°C)	DESC	CPSAC–H_3PO_4								CPSAC–NaOH							
		Cu^{2+}	Mn^{2+}	Co^{3+}	Fe^{2+}	Zn^{2+}	Cd^{2+}	Pb^{2+}	Cr^{3+}	Cu^{2+}	Mn^{2+}	Co^{3+}	Fe^{2+}	Zn^{2+}	Cd^{2+}	Pb^{2+}	Cr^{3+}
	Ea	55.734	65.529	116.888	5.324	657.916	0	48.45	40.46	0	52.012	143.697	23.306	0	0	9.521	168.533

4 Conclusions

C. papaya stem activated carbons developed using H_3PO_4 and NaOH activating agents (CPSAC–H_3PO_4 and CPSAC–NaOH) accomplished the aim of sequestrating the selected toxic metals (Cu^{2+}, Mn^{2+}, Co^{3+}, Fe^{2+}, Zn^{2+}, Cd^{2+}, Pb^{2+} and Cr^{3+}) from the mining wastewater. CPSAC–H_3PO_4 and NaOH sequestrated six and five toxic metals, respectively (and others drastically reduced), from the mining wastewater at optimum conditions of 0.6 g (CPSAC–H_3PO_4) and 0.2 g (CPSAC–NaOH) adsorbent dosages, 150 rpm agitation rate, 60 min contact time and pH of 7 at room temperature. The adsorption processes were positively affected by adsorbent dosage, agitation rate and contact time, while pH and temperature negatively affected them. Though, these results were more pronounced in CPSAC–H_3PO_4 than CPSAC–NaOH. Most of the toxic metal adsorption processes were irreversible, while the remaining few ones best fitted Freundlich, Langmuir, D–R and Temkin isotherm models. Though the adsorption data of CPSAC–NaOH best fitted Freundlich, D–R and Temkin isotherm models than that of CPSAC–H_3PO_4. The adsorption data of all the toxic metals for both adsorbents best fitted kinetic model of pseudo-second order with greater part of the surface adsorption in the rate-restricting step, though, in CPSAC–NaOH than CPSAC–H_3PO_4. Non-linear regression analysis (using EAB, ARE and R^2 error functions) indicated to be better in analysing adsorption and kinetic models for the studied toxic metals. The CPSACs (especially the one activated with H_3PO_4) have removal efficiencies of approximately 100% and found to be a better substitute for traditional activated carbon in sequestrating toxic metals from real-life wastewater, particularly mining wastewater.

Credits/Authors' Statement

(1) *Professor Samson O. Ojoawo*: Conceptualization, Methodology, Software, Resources, Supervision, Writing—Reviewing and Editing.
(2) *Ezekiel A. Adetoro*: Data Collation, Methodology, Software, Resources, Fieldwork, Laboratory Analysis, Formal Analysis, Writing—Original Draft Preparation, Visualization, Investigation, Validation.

Funding This research did not receive any specific grant from funding agencies in the public, commercial or not-for-profit sectors.

Conflict of Interest The authors declare that there are no conflicts of interest regarding the publication of this manuscript.

Acknowledgements The authors wish to acknowledge the laboratory assistance rendered by the Department of Civil Engineering, Ladoke Akintola University of Technology, Ogbomoso, Nigeria.

References

1. Alalwan HA, Kadhom MA, Alminshid AH (2020) Removal of heavy metals from wastewater using agricultural byproducts. J Water Supply: Res Technol AQUA 69(2):99–112
2. Al-Fakih AA (2015) Biosorption of Ni^{2+} and Cd^{2+} from aqueous solutions using NaOH-treated biomass of *Eupenicillium ludwigii*: equilibrium and mechanistic studies. Open J Appl Sci 5:376–392. https://doi.org/10.4236/ojapps.2015.57038
3. Abdel-Raouf MS, Abdul-Raheim ARM (2017) Removal of heavy metals from industrial waste water by biomass-based materials: a review. J Pollut Effects Control 5(1):1–13. https://doi.org/10.4172/2375-4397.1000180
4. Czikkely M, Neubauer E, Fekete I, Ymeri P, Fogarassy C (2018) Review of heavy metal adsorption processes by several organic matters from wastewaters. Water (MDPI) 10(1377):1–15
5. Javanbakht V, Ghoreishi SM (2017) Application of response surface methodology for optimization of lead removal from an aqueous solution by a novel superparamagnetic nanocomposite. Adsorpt Sci Technol 35(1–2):241–260
6. Biswas S, Bal M, Behera SK, Tushar KS, Meikap BC (2019) Process optimization study of Zn^{2+} adsorption on biochar-alginate composite adsorbent by response surface methodology (RSM). Water (MDPI) 11(325):1–15
7. Farnane M, Machrouhi A, Elhalil A, Abdennouri S, Qourzal S, Tounsadi H, Barka N (2018) New sustainable biosorbent based on recycled deoiled carob seeds: optimization of heavy metals remediation. Hindawi J Chem 2018:1–16
8. Raju DSR, Rao VN, Prasad PR, Babu NC (2012) Sorption of lead (II) ions from wastewater using *Carica papaya* leaf powder. Int J Eng Sci Adv Technol 2(6):1577–1581
9. Ong S, Yip S, Keng P, Lee S, Hung Y (2012) Papaya (*Carica papaya*) seed as a low-cost sorbent for zinc removal. Afr J Agric Res 7(5):810–819
10. Musa A, Wan Alwi SR, Ngadi N, Abbaszadeh S (2017) Effect of activating agents on the adsorption of ammoniacal nitrogen using activated carbon *papaya* peel. Chem Eng Trans 56:841–846
11. Ojoawo SO, Udayakumar G, Adewale AA, Ogunowo AT (2016) Adsorption potentials of *carica papaya* in remediating Pb^{4+} and Cr^{3+} from metal–galvanizing industrial wastewater. Int J Innov Trends Eng (IJITE) 21(01):52–61
12. Bello OS, Adegoke KA, Fagbenro SO, Lameed OS (2019) Functionalized coconut husks for rhodamine-B dye sequestration. Appl Water Sci 9(189):1–15. https://doi.org/10.1007/s13201-019-1051-4
13. UNIOSUN Laboratory Manual—ULM (2017) Laboratory manual for students. Nigeria, University of Osun, Ejigbo, College of Agriculture: Unpublished Manual
14. Bello OS, Lasisi BM, Adigun OJ, Ephraim V (2017) Scavenging rhodamine B dye using *Moringa oleifera* seed pod. Chem Spec Bioavail 29(1):120–134
15. Yaneva ZL, Koumanova BK, Georgieva NV (2013) Linear and nonlinear regression methods for equilibrium modelling of ρ-nitrophenol biosorption by *Rhizopus oryzae*: Comparison of error analysis criteria. J Chem 517631:1–10. https://doi.org/10.1155/2013/517631
16. Ngakou CS, Anagho GS, Ngomo HM (2019) Non-linear regression analysis for the adsorption kinetics and equilibrium isotherm of phenacetin onto activated carbons. Curr J Appl Sci Technol CJAST 36(4):1–18
17. World Health Organization—WHO (2011) Guidelines for drinking-water quality. Switzerland, Geneva, pp 1–564
18. El-Aila HJ, Elsousy KM, Hartany KA (2016) Kinetic, equilibrium and isotherm of the adsorption of cyanide by MDFSD. Arab J Chem 9:S198–S203
19. Ojedokun AT, Bello OS (2017) Kinetic modeling of liquid-phase adsorption of congo red dye using guava leaf-based activated carbon. Appl Water Sci 7:1965–1977

Ezekiel A. Adetoro possessed postgraduate diploma (FUTA), M.Sc. (Twente, Netherlands) and Ph.D. (Ogbomoso) degrees in civil engineering. He has many publications in reputable journals and conference proceedings and has presented academic papers at some conferences/workshops/seminars both internationally and locally. He is Registered Civil Engineer.

Samson O. Ojoawo possessed B.Tech. (Ogbomoso), M.Sc. and Ph.D. (Ibadan) degrees in civil engineering and a postdoctoral fellowship (India). He is currently Professor of civil engineering (water resources and environmental engineering) in the Faculty of Engineering and Technology, LAUTECH, Ogbomoso, Nigeria. He has many publications in reputable journals and conference proceedings and has presented academic papers at some conferences/workshops/seminars both internationally and locally. He is Registered Civil Engineer and Public Orator.

Response Surface Models for Optimal Concrete Designs

C. Chandre Gowda, B. C. Kumar Raju, and B. E. Bhojaraj

Abstract The merits of response surface models in concrete construction need to be explored. These models have shown enormous use in the field of manufacturing and production. So in the present study, the application of response surface for concrete production is described. Their benefit in determining the results with minimum number of experiments is also discussed. The review summarizes the application of response surface models and shows that the statistical models provide additional support in analyzing the constrained targets. It reduces the test cases making the designs more economical compared to conventional methods.

Keywords Response surface · Concrete design · Statistical model · Construction · Experiments

1 Introduction

Construction of buildings and other essential requirements are evolved through civilization of mankind. Its progressive improvement is displayed at existing ancient temples, monuments, etc. Earlier, the construction practices and its combination for determining the hardened properties were tedious and time-consuming. For instance, the usage of statistical tools in determining the significant variables and its combination in achieving the targets are derived, developed, and practiced [1]. Many statistical models or tools exist and are used, among response surface (RS) models have gained much importance in modeling and determining the optimum values [2, 3]. The importance of statistical models in modern manufacturing and production process has been investigated [4].

C. Chandre Gowda (✉)
Centre of Incubation, Innovation, Research and Consultancy, Jyothy Institue of Technology, VTU, Bengaluru, India

B. C. Kumar Raju
Department of Civil Engineering, Sapthagiri College of Engineering, VTU, Bengaluru, India

B. E. Bhojaraj
Department of Civil Engineering, NMAM Institute of Technology, NITTE VTU, Udupi, India

L. Nandagiri et al. (eds.), *Sustainability Trends and Challenges in Civil Engineering*, Lecture Notes in Civil Engineering 162, https://doi.org/10.1007/978-981-16-2826-9_7

In recent years, many substitutes were used as an alternate for cement (ranging from 1 to 99% in volume) in construction [5–7]. The use of cement and other constituents in the production of concrete is important in infrastructural development such as apartment, complex, hydraulic structure, and pavement design [8]. The particular concrete is designed for specific purpose and target strength, such as quick setting, high flowability, self-compacting, polymer concrete (low thermal conductivity) [9], and high compressive strength. Since the optimum design and combination can be obtained by testing many samples (cubes, cylinder, beam, etc.) [10, 11]. Preparation of samples and conducting tests are time-dependent and non-economical. Some spreadsheet packages have been used in concrete design with constrained usage [12]. In order to overcome these constraints, statistical approaches are established.

The concrete production is bound by its purpose and other numerous materials. Since many polymer wastes are disposed as inland filling, and these materials are major contributors in land and water pollution. In order to reduce the disposal of polymers, some classified polymers (thermoplastic polyurethane, polycarbonate, and polybutylene terephthalate) are used as ingredient for concrete which reduces water absorption [13].

Generally, the optimization problems may have more than one output (conflicting). The properties of the concrete have been determined by using response surface models [14]. Taguchi design model and RS model both have the same origin derived from design of experiments [15, 16]. The Taguchi method is expressed from its design matrix table [17, 18].

Some of the alarming properties in cement are heat of hydration, bleeding, creep, and shrinkage. These properties need to be monitored during the design through RS models. So it is essential to produce concrete with lower heat transfer, minimum creep, and shrinkage. For proportionating of geopolymer concrete, many activator solutions [11] can be considered as design parameters (activator solution to fly ash ratio, w/c ratio, curing conditions, etc.). Cement, plasticizer, curing time, compressing strength, workability, water/cement ratio, fly ash, fine aggregate, and coarse aggregate are the influencing factors in concrete, and these are the varying factors in RS models [5].

The statistical approaches have been used to obtain optimum concrete mixture proportion [19] from experiment data. The utility of these models for designing and optimizing the concrete mixture has been illustrated already [13].

To optimize the concrete design, numerous factors need to be considered among them strength and durability are important. The constituents of concrete classified into two groups, i.e., quality and quantity of concrete. The quality of concrete is governed by its constituents and mode of usage (curing and removal of voids). The quantity of concrete depends on the way mixing and handling it (without segregation, bleeding, etc.). So, an attempt is made to summarize the use of RS models in obtaining the optimum concrete design.

2 Method

The RS model was first introduced in 1951 during the study. It is a statistical model and uses polynomials for the approximation for input–output relationships. They are used in determining the relationship between the input and output parameters [20].

They are efficient approaches to solve a problem by applying all the engineering principles and techniques to ensure group of endurable conclusions. The most important advantage of using them is concurrently considered for collective effects under analysis [20]. In common, the experiments usually involve many variables. On contrary statistically designed test requires a less number of runs compared to the traditional approach.

RS models have 2–5 factors, each factor consists of 2–3 levels. When each factor is applied to two levels, i.e., upper and lower levels, called as two level RS models. When each factor is applied to three levels, i.e., upper, middle, and lower levels, then it is called as three-level RS models. The levels are corresponding to the value of the factor, maximum value is considered as high level, middle value as middle level, and minimum value as low level. Factors are the input or output variables in the model.

Full factorial design (FF), central composite design (CC), and Box–Behnken design (BB) are the three RS models that can be used in optimizing the number of simulations and to obtain the best possible model.

Full factorial design is first order in linear; central composite design and Box–Behnken design are nonlinear in deriving the model. The factorial design is a strategy in which all variables are varied together, rather than one at a time.

In full factorial design, relationship between the response and input parameters is characterized by a regression model, which uses a set of sample experimental data. The polynomial regression model can be developed for any target (strength or other parameters) considered through experiments. It performance may be related to statistical modes. Equation (1) shows linear RS model.

$$Y_i = C_0 + C_1 x_{i1} + C_2 x_{i2} + \cdots + C_n x_{in} + \text{error} \tag{1}$$

where Y = dependent variable, C = the coefficients of the variable x.

Central composite design (CC) is quadratic, with cuboidal shape as region of interest and its graphical representation consists of points in vertices. Additional center and axial points (or star point) are also present to monitor the nonlinearity. It is used to investigate the effects of inputs on performance characteristics of output [21] and is commonly used in second-order design. The design matrix for the nonlinear regression is represented in Eq. (2).

$$y = ax + bx^2 + cx^3 \ldots nx^n \tag{2}$$

where y = dependent variable, a, b, c ... coefficients. The coefficients of the quadratic model can also be determined using method of least square.

The Box–Behnken design (BB) is also a quadratic model. This method is more efficient for higher number of input variables [22]. The designs are formed from combination of 2 k factorials with unfinished block designs, where k is input variables. The CC and BB may be related to nonlinear mathematical model.

After modeling, its significance and adequacy need to be evaluated. The statistical adequacy of the model is tested through analysis of variance (ANOVA) and coefficient of determination (R^2). ANOVA analyzes the variances present. The steps in RS model to determine optimal design are represented in Fig. 1. The models are used very sparsely in prediction.

Fig. 1 Steps in RS model to determine optimal design

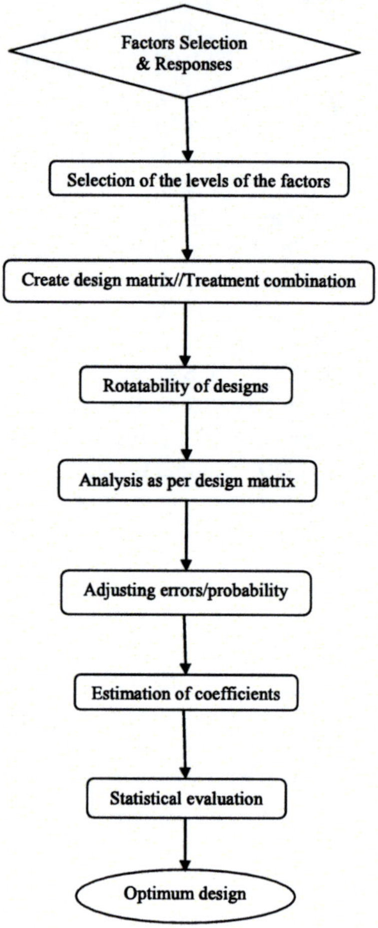

Table 1 Industrial usage of RS models

S. No.	Industry	Application	Usage (%)
1	Automobile	Manufacturing of component and parts	30–75
2	Chemical	Producing chemical products	25–55
3	Construction	Optimal concrete design	< 15
4	Stereolithography	Optimal combination of filament	25–45
5	Plastic	Product best design for the items	40–70

3 Discussion

The statistical models play an important role in chemical, automobile, stereolithography [23], and other industries (Table 1). They are used to conduct test on the expensive sample components. Some of the industries which use the statistical methods (RS and Taguchi) are listed in Table 1. From Table 1, very limited investigations of statistical approaches for concrete and other construction works are observed.

The limited use RS models are may be due to lack of exposure of these models in concreting or due to lesser research activity since the development of a country consumes an enormous amount of components of concrete in construction bound by limits to available strength. So the challenging conditions may occur in designing them. So far in concrete construction and testing, the usage of RS models shows limited applications, as their usages are bound mainly for domestic purposes. Other than the flexural strength, other tests including compressive strength, split tensile strength, slump cone, and water absorption were predicted with very close differences in the earlier study [24].

Other than individual statistical models, the hybrid models including both mathematical and numerical are proven to be more efficient [25]. This work summarizes, that very minimal work has been carried on RS model in the construction field. Contrarily these models have high potential in modeling and can be used optimal design in concrete construction.

4 Conclusion

The investigation on RS models and its application in concrete and construction was brought out. The significance of RS models in construction was discussed and observed keen points are listed below.

1. Though the RS models are explored in automobile and mechanical production, its usage in concreting is still narrow.
2. The RS model show reliable performance with 95% probability, during model evaluation

3. These statistical models are used at lab scales for at limited instances in concrete design
4. These models can be used as an addition to strengthen the physical models in the design factors.
5. The usage of RS models with numerical modeling may result in intangible results since the hybrid models have proved to be better than other models.

Acknowledgements The first author wishes to thank, Director CIIRC for support and the facilities to carry out the research.

References

1. Ahmad S, Alghamdi SA (2014) A statistical approach to optimizing concrete mixture design. Sci World J
2. Bayramov F, Taşdemir C, Taşdemir MA (2014) Optimisation of steel fibre reinforced concretes by means of statistical response surface method. Cement Concr Compos 26:65–75
3. Brown JH, Gomez MJ, Benzo Z, Vaz JE (1996) Application of the response surface methodology for potassium determination in soils by AAS using the slurry technique. Chemom Intell Lab Syst 35:239–247
4. Patel GCM, Krishna P, Parappagoudar MB (2016) Squeeze casting process modeling by a conventional statistical regression analysis approach. Appl Math Model 40:69–88
5. Solikin M, Setiawan B (2017) The effects of design strength, fly ash content and curing method on compressive strength of high volume fly ash concrete: a design of experimental. In: MATEC web of conferences 103:01003 EDP Sciences
6. Akram T, Memon SA, Obaid H (2009) Production of low cost self compacting concrete using bagasse ash. Constr Build Mater 23:703–712
7. Swamy RN, Swamy RN (1986) Cement replacement materials 3. Surrey University Press, Sheffield
8. Joshaghani A, Ramezanianpour AA, Ataei O, Golroo A (2015) Optimizing pervious concrete pavement mixture design by using the Taguchi method. Constr Build Mater 101:317–325
9. Muthukumar M, Mohan D (2004) Optimization of mechanical properties of polymer concrete and mix design recommendation based on design of experiments. J Appl Polym Sci 94:1107–1116
10. Lee BY, Kim JH, Kim JK (2009) Optimum concrete mixture proportion based on a database considering regional characteristics. J Comput Civ Eng 23:258–265
11. Jo M, Soto L, Arocho M, St John J, Hwang S (2015) Optimum mix design of fly ash geopolymer paste and its use in pervious concrete for removal of fecal coliforms and phosphorus in water. Constr Build Mater 93:1097–1104
12. Kasperkiewicz J (1995) Optimization of concrete mix using a spreadsheet package. Mater J 91:551–559
13. Şimşek B, Uygunoğlu T (2018) A full factorial-based desirability function approach to investigate optimal mixture ratio of polymer concrete. Polym Compos 39:3199–3211
14. Ozbay E, Oztas A, Baykasoglu A, Ozbebek H (2009) Investigating mix proportions of high strength self compacting concrete by using Taguchi method. Constr Build Mater 23:694–702
15. Liao HC (2003) Using PCR-TOPSIS to optimise Taguchi's multi-response problem. Int J Adv Manuf Technol 22:649–655
16. Türkmen IIM, Gül R, Çel K, Rboğa DR (2003) Determination by the Taguchi method of optimum conditions for mechanical properties of high strength concrete with admixtures of silica fume and blast furnace slag. Civ Eng Environ Syst 20:105–118

17. Tan O, Zaimoglu AS, Hinislioglu S, Altun S (2005) Taguchi approach for optimization of the bleeding on cement-based grouts. Tunn Undergr Space Technol 20:167–173
18. Hong GB, Su TL (2012) Statistical analysis of experimental parameters in characterization of ultraviolet-resistant polyester fiber using a TOPSIS-Taguchi method. Iran Polym J 21:877–885
19. Myers RH, Montgomery DC, Anderson-Cook CM (2016) Response surface methodology: process and product optimization using designed experiments. Wiley
20. Montgomery DC (2017) Design and analysis of experiments. Wiley
21. Stewart PDS, Baldock PF (1999) Practical experimental design techniques for automatic and manual protein crystallization. J Cryst Growth 196:665–673
22. Özgen S, Yildiz A (2010) Application of box—behnken design to modeling the effect of smectite content on swelling to hydrocyclone processing of bentonites with various geologic properties. Clays Clay Miner 58:431–448
23. Lynn-Charney C, Rosen DW (2000) Usage of accuracy models in stereolithography process planning. Rapid Prototyping J
24. Awolusi TG, Oke OL, Akinkurolere OO, Sojobi AO (2019) Application of response surface methodology: predicting and optimizing the properties of concrete containing steel fibre extracted from waste tires with limestone powder as filler. Case Stud Constr Mater
25. Harish N, Mandal S, Rao S, Patil SG (2015) Particle swarm optimization based support vector machine for damage level prediction of non-reshaped berm breakwater. Appl Soft Comput 27:313–321

Thermal Comfort Studies of Residential Building Models in Vijayawada

Jagadish Vengala, Srinivas Chava, and Premkumar Pydipati

Abstract Brick masonry wall is the more common walling system being used in residential buildings. As the demand for housing is increasing, new construction materials are coming up in the market and being utilized in the construction. Solid concrete blocks (SCB) and autoclave aerated concrete (AAC) blocks are the two materials which are most commonly used in most of the construction sites apart from the clay bricks (CB). The thermal behaviour of the buildings with these materials needs to be studied. It is also necessary to reduce the heat inside the building to improve the human comfort. Effect of coating material over roof also needs to be studied. Living comfort inside any building mainly depends on the "thermal comfort". Thermal comfort mainly involves the interior conditions of the rooms, i.e. mainly temperature and humidity, maintaining and distributing it evenly, and the quality of air (purity, humidity rate, healthiness). By providing a thermal insulation to all the surfaces combined with seasonal adoption of ventilation, thermal comfort can be achieved in different climate conditions. By keeping these points in a view, a thorough study was conducted to know the thermal variations in various house models by varying the wall type. Results indicated the superior thermal performance of the model with AAC block in comparison with the models made with solid concrete block and clay brick.

Keywords AAC block · Thermal comfort · Humidity · Insulation · Temperature

1 Introduction

Anyone who lived in the building or a different climate conditions may have experienced the discomfort of particularly very higher temperatures during the summer

J. Vengala (✉)
Department of Civil Engineering, PVP Siddhartha Institute of Technology, Vijayawada 520007, Andhra Pradesh, India

S. Chava · P. Pydipati
Department of Civil Engineering, VR Siddhartha Engineering College, Vijayawada 520007, Andhra Pradesh, India

L. Nandagiri et al. (eds.), *Sustainability Trends and Challenges in Civil Engineering*, Lecture Notes in Civil Engineering 162, https://doi.org/10.1007/978-981-16-2826-9_8

season compared with the other seasons. In India, most of the newly constructed buildings are constructed of concrete with a flat roof. These surfaces absorb sunlight and transfer the heat inside the building [1]. This hot roof continues to heat up the space in day and night and making the building more hotter especially in the summer season.

Thermal insulation of wall components and cool roof coating of roof is provided to control heat transfer [2]. Temperature and humidity in the building are important factors influencing the human discomfort. The wall components in most of the buildings are constructed either with clay bricks or solid concrete blocks or autoclaved aerated concrete blocks. Out of which, brick is one of the most extensively used construction material which is used in the construction of the walls [3]. Introduction of holes in the clay bricks reduces volume and as well as it increases the insulation.

Autoclave aerated concrete blocks are made by using sand, quick lime, water, gypsum, cement and small amount of chemical used as foaming agent. AAC has good thermal insulation and good absorption properties. AAC blocks are lightweight and offer ultimate workability, flexibility and durability.

A solid cement concrete block (SCB) is extensively used as a building material in the construction of walls. These blocks are available in different sizes. In manufacturing of the solid cement concrete block, concrete mixes use cement, sand, gravel and water in the suitable proportion for the construction. The solid concrete block are water resistant, aesthetic appearance and fire resistant [4].

The main objective of the present work is to study the temperature variations of various building models with different walling systems and also to study the reduction in roof temperature by applying the roof coatings.

2 Literature Review

Vengala et al. [5] conducted limited studies on two models one with AAC blocks and another with solid concrete blocks. The studies indicated that AAC block model showed better thermal comfort in comparison with the solid concrete block model, i.e. during day time cooler by 1–2 °C. Between these two models, the maximum temperature difference observed was 3 °C.

Patil et al. [6] carried out a theoretical analysis on thermal analysis of RC Roofing with various composite materials. Composite materials comprised of RC Roofing with polyurethane foam, RC Roofing with polystyrene, RC Roofing with polyurethane foam and GI sheet, RC Roofing with polystyrene. Based on the analysis, the combination of RC Roofing with polystyrene and GI sheet was found to be feasible in improving the thermal comfort. Techno-economical analysis carried out showed the least cost for a combination of RC Roofing with polystyrene and GI sheet.

Perez et al. [7] studied the thermal performance of a solid roof top with and without insulation by varying two colours using the finite volume method. The insulating material comprised of polystyrene, and colour variation considered was grey and white. The roofing with white coloured coating significantly reduced the interior surface temperature when compared with grey coloured coating.

Dharek et al. [8] investigated the efficacy of cool roof coatings w.r.t thermal analysis for the Bengaluru region. Aluminium-based coating and acrylic-based coating were considered as part of the study. The temperature variations were recorded with sensor-based digital temperature data loggers on the scaled models with these coatings. The models coated with aluminium-based coatings showed better performance in bringing down the temperature levels inside the scaled building models. Parametric analysis carried out as part of this study indicated total heat loss being more for models without coating.

3 Experimental Programme

The house models were set up on the roof portion of a building in the college. The models are directly exposed to the sun to simulate the real-time conditions. The models were constructed in the month of May. Based on the temperature data obtained [9] in Fig. 1, the temperature values in Vijayawada used to reach peak in the month of May.

For temperature measurements, three model houses were built. The size of the house model was 1.5 m × 1.5 m × 0.87 m. Each model building has same size and

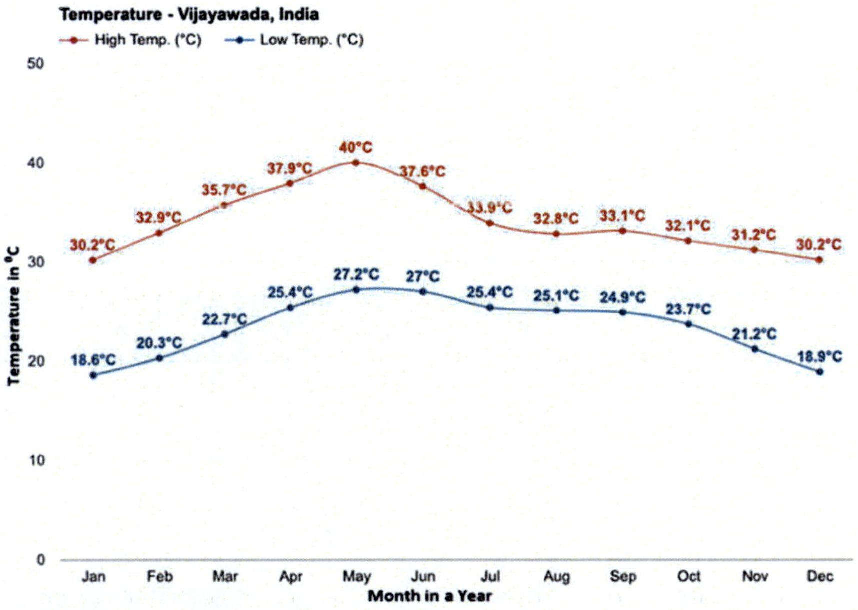

Fig. 1 Vijayawada temperature map [9]

shape, but each one was built with different building materials. One was constructed using solid concrete blocks (SCB) and clay brick (CB), and the other was constructed using autoclave aerated concrete blocks (AAC). Table 1 gives the details of the materials used in both the model rooms used for thermal comfort studies.

The model buildings were equipped with DH11 sensors to measure the internal temperature and the humidity. Raspberry Pi module was used to record the temperature measurements. Figure 2 shows the testing of the sensor set-up before installing on to the models.

The model buildings are constructed in the Vijayawada region. The models have the two windows on either side with dimension of 0.3 m × 0.6 m one on the southern wall and another on northern wall. The height between the inner slab surface and floor level is 0.87 m, i.e. inside model. Figure 3 shows the view of building models.

To understand the effect of coating on the roof surface, three more building models constructed using AAC Block, CB and SCB, and white acrylic cool roof coating was applied on these models as shown in Fig. 4. Before applying the coating, surface

Table 1 Details of the materials used in all the model rooms

Model room	AAC	SCB	CB
Blocks used	Autoclaved aerated concrete blocks	Conventional solid cement concrete blocks	Clay brick
Material used for joints	Cement mortar as 1 in 3	Cement mortar as 1 in 3	Cement mortar as 1 in 3
Material used for plastering	Cement mortar as 1 in 6	Cement mortar as 1 in 6	Cement mortar as 1 in 6
Roofing material	Reinforced cement concrete	Reinforced cement concrete	Reinforced cement concrete
Curing period	14 days	14 days	14 days
Size of the block	600 mm × 200 mm × 100 mm	400 × 200 × 100mm	228 × 107 × 69 mm

Fig. 2 Testing of the sensor and Raspberry Pi set-up

Fig. 3 View of the building models used in the study

Fig. 4 Application of acrylic cool roof coating over the roof portion

preparation was carried out. Two coats was applied on the surface. Temperature values were measured for the building models with and without coating.

4 Results and Discussions

This section presents the temperature and humidity results of block specimens for 30 days under varying climate conditions. The temperature readings were obtained after continuous monitoring. Data loggers were placed on the top surface of roof and inside the model house. DH11 sensors were placed in each type of building model. Temperature and humidity data was recorded for all the models, and average temperatures were calculated. Table 2 gives the temperature and humidity data of a typical day in the month of May for model house with AAC Block. The table gives the average hourly temperature and humidity data between 10 AM to 4 PM. The peak temperature values were observed between 1 and 2 PM. From Table 2, it can be seen that the difference in temperature is almost 9–10 °C during the peak hours.

Table 2 A typical day [15.05.2019] temperature and humidity measured for model house with AAC block

Time during the day	Avg. int. temp in °C	Avg. ext. temp in °C	Diff. in temp in °C	Avg. int. humidity in %	Avg. ext. humidity in %	Diff. in humidity in %
10–11 AM	27.0	32.0	5.0	55.0	65.5	− 16.8
11–12 AM	29.0	37.0	8.0	67.0	42.5	− 20.1
12 AM–1 PM	32.0	42.0	10.0	65.5	59.0	− 32.5
1–2 PM	38.0	47.0	9.0	45.0	39.8	− 26.5
2–3 PM	38.1	44.5	6.4	45.2	19.5	− 24.5
3–4 PM	38.1	40.0	1.9	40.4	25.5	− 25.5

Tables 3 and 4 give the temperature and humidity data for the house model with CB and SCB, respectively. Even in these house models, significant difference in temperature was observed during the peak hours.

Table 3 A typical day [15.05.2019] temperature and humidity measured for model house with CB

Time during the day	Avg. int. temp in °C	Avg. ext. temp in °C	Diff. in temp in °C	Avg. int. humidity in %	Avg. ext. humidity in %	Diff. in humidity in %
10–11 AM	29.0	32.0	3.0	67.0	65.5	− 1.5
11–12 AM	30.0	33.0	3.0	67.0	62.5	− 4.5
12 AM–1 PM	38.3	42.0	3.7	46.3	59.0	12.7
1–2 PM	40.0	47.0	7.0	46.0	39.8	− 6.3
2–3 PM	39.0	44.5	5.5	44.0	19.5	− 24.5
3–4 PM	37.0	40.0	3.0	42.0	25.5	− 16.5

Table 4 A typical day [15.05.2019] temperature and humidity measured for model house with SCB

Time during the day	Avg. int. temp in °C	Avg. ext. temp in °C	Diff. in temp in °C	Avg. int. humidity in %	Avg. ext. humidity in %	Diff. in humidity in %
10–11 AM	29.0	32.0	3.0	69.0	65.5	− 3.5
11–12 AM	31.5	37.0	5.5	62.0	62.5	0.5
12 AM–1 PM	38.0	42.0	4.0	39.4	59.0	19.6
1–2 PM	39.0	47.0	8.0	38.0	39.8	1.8
2–3 PM	38.2	44.5	6.3	38.5	19.5	− 19.0
3–4 PM	38.0	42.0	4.0	38.2	25.5	− 12.7

Table 5 presents the average hourly temperature during the entire month, i.e. May for the model house with AAC blocks.

Figure 5 represents the average hourly temperature values for all the three houses during the peak hour. From the figure, it is evident that AAC block house model has the less value followed by SCB model and CB model. From Fig. 6, it can be seen that

Table 5 Details of avg. difference in temperature in °C during May 2019 for model house with AAC block

Day	10–11 AM	11–12 AM	12 AM–1 PM	1–2 PM	2–3 PM	3–4 PM
	Difference in temperature in °C					
1	5	9	10	10	9.4	5.9
2	4.8	8.3	11	10.9	12.1	5
3	8.7	9.5	8.5	11.5	9.5	7
4	7.3	9.2	10.2	8.5	8.3	8.5
5	6.7	9.8	10.6	9.5	5.7	7
6	4.5	5.3	10	12	8.4	8
7	2	8.8	11	10.9	12.9	5
8	7.1	10.5	9.5	11.5	7.5	9
9	8.1	9.2	10.2	7.5	7.3	9.5
10	6.7	10.8	12.3	10.5	5.7	7
11	2	3	8.6	10	8.5	5
12	2.6	9.7	12	11.9	13.4	7
13	7.1	8.7	7.2	7.5	8	8
14	10.1	9.4	9.7	6.2	5	5.5
15	5	8	10	9	6.4	1.9
16	1.8	8.3	11	10.9	12.1	5
17	6.7	9.5	7.5	7.5	6.5	7
18	7.3	8.2	9.2	7.5	6.3	7.5
19	6.7	9.8	10.6	9.5	5.7	7
20	3.5	5.3	10	9	6.4	4
21	2	8.8	11	10.9	12.9	5
22	7.1	9.5	7.5	7.5	6.5	7
23	7.1	8.2	9.2	7.5	6.3	7.5
24	6.7	9.8	11.3	9.5	5.7	7
25	1	2.4	8	11	8.5	5
26	2.6	9.7	12	11.9	13.4	7
27	7.1	8.7	7.2	7.5	8	8
28	7.1	8.4	9.7	7.2	5	5.5
29	7.1	9.5	8.5	9.5	9.5	7
30	7.1	8.2	9.2	7.5	6.3	7.5

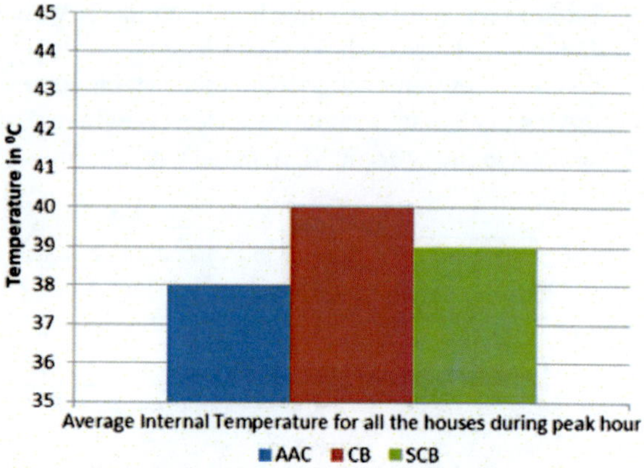

Fig. 5 Average hourly temperature for all the houses during peak hour

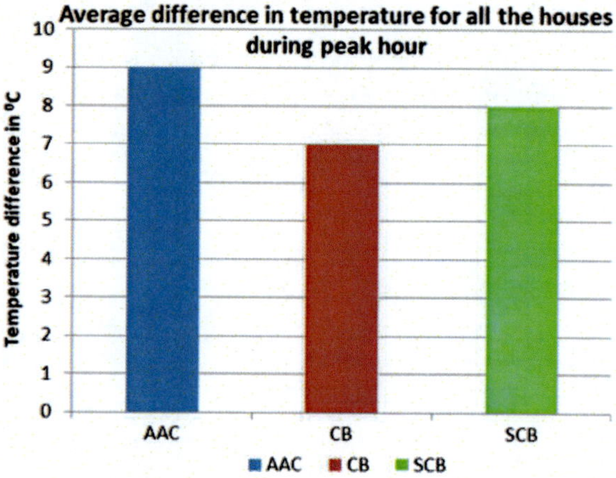

Fig. 6 Average difference in temperature for all the houses during peak hour

the temperature difference with ambient is high in case of AAC block house model followed by SCB model and CB model. This clearly shows the better performance of AAC block house model in comparison with other two house models. Table 6 presents the average hourly humidity values during the entire month, i.e. May for the model house with AAC blocks. Table 7 gives the average temperature difference (Fig. 7) and humidity values for all the three houses during the month of May.

Relative humidity affects both thermal comfort and indoor air quality. As air temperature increases, the RH decreases. High RH make people hot and sticky in warm weather. Low RH (very dry air) can cause dryness and discomfort in the nose and make skin feel dry and itchy.

From Fig. 8, it can be seen that the humidity difference with external value is high in case of AAC block house model; hence, the humidity inside the model is less when compared with other two house models. This indicates the good thermal comfort inside the AAC house model.

Table 6 Details of avg. difference in humidity values in % during May 2019 for model house with AAC block

Day	10–11 AM	11–12 AM	12 AM–1 PM	1–2 PM	2–3 PM	3–4 PM
	Difference in humidity in %					
1	10.5	− 24.5	− 6.5	− 5.2	− 25.7	− 14.9
2	0.3	− 13.3	− 5.8	− 31.2	− 34.9	− 28.9
3	− 15.7	− 24.7	− 30.4	− 34.5	− 29.5	− 17
4	− 15.6	− 18.4	− 30.5	− 25.3	− 34.5	− 42
5	− 17.1	− 17.4	− 31.2	− 29	− 24.2	− 30
6	10	− 4.5	− 6.5	− 5.2	− 25.7	− 14.9
7	0.7	− 13.2	− 5.8	− 31.2	− 34.9	− 28.9
8	− 15.3	− 24.7	− 30.4	− 34.5	− 29.5	− 17
9	− 15.6	− 18.4	− 30.5	− 25.3	− 34.5	− 42
10	− 17.1	− 17.4	− 31.2	− 29	− 24.2	− 30
11	7.5	9.4	14	− 4.2	− 20.2	− 28.9
12	− 0.7	− 17.7	− 15.2	− 40.2	− 42.7	− 16.5
13	− 9.3	− 12.2	− 14.8	− 25.3	− 32	− 40.5
14	− 16.8	− 20.1	− 32.5	− 26.5	− 24.5	− 25.5
15	10.5	− 24.5	− 6.5	− 5.2	− 25.7	− 14.9
16	0.3	− 13.3	− 5.8	− 31.2	− 34.9	− 28.9
17	− 15.7	− 24.7	− 30.4	− 34.5	− 29.5	− 17
18	− 15.6	− 18.4	− 30.5	− 25.3	− 34.5	− 42
19	− 17.1	− 17.4	− 31.2	− 29	− 24.2	− 30
20	10	− 4.5	− 6.5	− 5.2	− 25.7	− 14.9
21	0.7	− 13.2	− 5.8	− 31.2	− 34.9	− 28.9
22	− 15.3	− 24.7	− 30.4	− 34.5	− 29.5	− 17
23	− 15.6	− 18.4	− 30.5	− 25.3	− 34.5	− 42
24	− 17.1	− 17.4	− 31.2	− 29	− 24.2	− 30
25	7.5	9.4	14	− 4.2	− 20.2	− 28.9
26	− 0.7	− 17.7	− 15.2	− 40.2	− 42.7	− 16.5
27	− 9.3	− 12.2	− 14.8	− 25.3	− 32	− 40.5
28	− 16.8	− 20.1	− 32.5	− 26.5	− 24.5	− 25.5
29	− 15.3	− 24.7	− 30.4	− 34.5	− 29.5	− 17
30	− 15.6	− 18.4	− 30.5	− 25.3	− 34.5	− 42

Table 8 gives the details of the average difference in temperature values for the models with and without white acrylic coating. The effect of coating over the roof surfaces reduces an internal temperature of 2 °C for all the house models.

Based on the studies conducted on house models, AAC model shown better thermal comfort in comparison with the model with SCB and clay brick (CB). The

Table 7 Average difference in temperature and humidity measured for all the model house for the entire month of May 2019

House model	10–11 AM	11–12 AM	12 AM–1 PM	1–2 PM	2–3 PM	3–4 PM
	Average difference in temperature in °C					
AAC block	5	9	10	11	9.4	5.9
Brick	3.3	4.8	5.0	5.6	5.4	5.0
solid Concrete block	4.4	7.3	7.6	7.7	8.0	5.4
	Difference in humidity in %					
AAC block	− 7.3	− 15.9	− 19.2	− 25.1	− 29.9	− 27.1
Brick	− 5.9	− 3.6	− 5.2	− 19.0	− 26.0	− 29.9
solid Concrete block	− 5.7	− 8.7	− 6.9	− 18.9	− 26.1	− 25.8

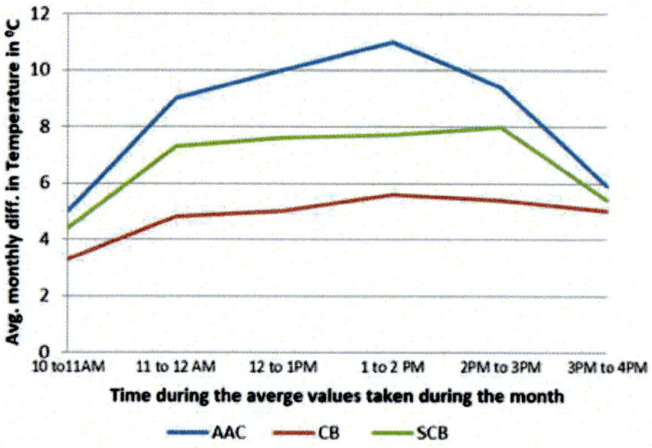

Fig. 7 Average monthly difference in temperature for all the houses during day time

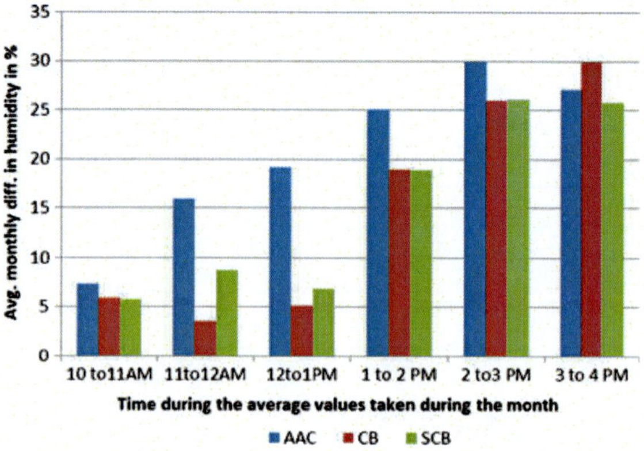

Fig. 8 Average monthly difference in humidity for all the houses during day time

Table 8 Details of avg. difference in temperature values in °C for the models with and without coating during peak hour (1–2 PM)

Model Room	AAC (°C)	SCB (°C)	CB (°C)
Without coating	8	5	6
With coating	10	7	8

thermal conductivity of clay bricks and solid concrete blocks is comparably high in comparison with AAC blocks, and hence, heat transfer from solid block/clay brick is more than AAC block.

5 Conclusions of the Present Study

The following conclusions have been drawn from the above study.

- Average difference of outside ambient temperature and internal temperature variation of nearly 4 °C in the brick model, 6 °C in a solid cement block and 8 °C in an autoclaved aerated concrete block in models without coating.
- With a white acrylic coating (cool roof based coating), the difference in temperature with and without coating was 10 °C, 8 °C and 6 °C, respectively for AAC block model, SCB model and clay brick building model.
- Among the three models, AAC block house model showed the more reduction in humidity inside the model when compared to external humidity. This resulted in better comfort inside the AAC model.
- Among the solid concrete block model and clay brick model, SCB showed better performance and this may be due to the thickness of the walling material when compared with the clay brick wall.

References

1. Vengala J, Nath SK, Pandey CN (2011) Energy consumption of wood based panel products: a review. J Indian Acad Wood Sci 8:80–83. https://doi.org/10.1007/s13196-012-0052-9
2. Vengala J, Dharek MS, Sachin D, Ghanashyam TB (2021) Thermal analysis of building model with acrylic and aluminium based roof coating materials. Materials today: Proceedings. https://doi.org/10.1016/j.matpr.2021.03.008
3. Vengala J (2020) Comparison of embodied energy in different bamboo-based houses. In: Pancharathi R, Sangoju B, Chaudhary S (eds) Advances in sustainable construction materials. Lecture notes in civil engineering, vol 68. Springer, Singapore. https://doi.org/10.1007/978-981-15-3361-7_15
4. Vengala J, Raju S, Shiva B, Manjunatha LR, Yogananda MV (2020) Use of GGBS in manufacturing of solid concrete blocks. In: Babu K, Rao H, Amarnath Y (eds) Emerging trends in civil engineering. Lecture notes in civil engineering, vol 61. Springer, Singapore. https://doi.org/10.1007/978-981-15-1404-3_16

5. Vengala J, Mangloor S, Goud TKC (2019) Performance of autoclaved aerated concrete blocks under varying temperatures. Int J Recent Technol Eng (IJRTE) 7(6C2). ISSN: 2277-3878
6. Patil SE, Shinde Nn (2014) Theoretical analysis of composite roof with respect to comfort in building envelope. Curr Trends Technol Sci 3(3):168–172
7. Hernández-Pérez I, Álvarez G, Gilbert H, Xamán J, Chávez Y, Shah B (2014) Thermal performance of a concrete cool roof under different climatic conditions of Mexico. Energy Procedia 57:1753–1762. https://doi.org/10.1016/j.egypro.2014.10.164
8. Dharek MS, Vengala J, Sachin D, Ghanashyam TB (2020) Model based comparative thermal analysis on cool roof coatings in Bengaluru. Int J Adv Sci Technol 29(7):4272–4280. Retrieved from http://sersc.org/journals/index.php/IJAST/article/view/23219
9. https://www.weather-ind.com/en/india/vijayawada-climate

Investigations on Compression Behaviour of Short Reinforced NSC Columns

J. Sanjith, R. Prabhakara, M. S. Sudarshan, and Jayachandra

Abstract The analysis is carried out for studying subsequent parameters to know structural performance of NSC columns and aim is to predict the values of deformation, yielding load and ultimate load of short reinforced Normal Strength Concrete of three different grades M20, M30 and M40 subjected to axially compressive load in loading frame. ANN tool was trained with proper inputs like fresh properties of materials, spacing of stirrups and percentage of longitudinal reinforcement and keeping target values obtained from experiment and results are compared with the ANN values accompanied by marginal errors which are around 0.5%.

Keywords NSC · Short columns · ANN

Nomenclature

P_{cr}	Initial crack load
Δ_{cr}	Deformation at initial crack
P_y	Load at yielding
Δy	Deformation at yielding load
P_u	Ultimate load
Δu	Deformation at ultimate load
MP_{cr}	Predicted initial crack load
$M\Delta_c$	Predicted deformation at initial crack

J. Sanjith
Department of Civil Engineering, Adichunchanagiri Institute of Technology, Chikkamagalur, India

R. Prabhakara
Brindavan College of Engineering, Bengaluru, India

M. S. Sudarshan
Stedrant Technoclinic Pvt. Ltd., Bengaluru, India

Jayachandra (✉)
School of Civil Engineering, REVA University, Bengaluru, India

L. Nandagiri et al. (eds.), *Sustainability Trends and Challenges in Civil Engineering*,
Lecture Notes in Civil Engineering 162, https://doi.org/10.1007/978-981-16-2826-9_10

MP_y	Predicted load at yielding
$M\Delta_y$	Predicted deformation at yielding load
Mp_u	Predicted ultimate load
$M\Delta_u$	Predicted deformation at ultimate load

1 Introduction

Concrete is a building material composed of cement, fine and coarse aggregates combined with water for hydration. Portland cement is typically used for the manufacture of concrete. In architecture, concrete is used for columns, beams, slabs and for the construction of foundations. Load bearing walls are also made of concrete. Other than cement, the binding agent used is lime for lime concrete and bitumen for asphalt concrete used for road construction.

Concrete production is the consistency of w.r.t. cement. The two key types of concrete mix are the nominal mix and the design mix. Nominal mix is used for residential buildings and the construction mix consists of a combination of proportions and is finalised on the basis of a cube or cylinder lab test for its compressive power. This is called the Mix Design Process.

Columns are vertical elements that are designed to withstand compressive loads. Columns have a bending moment on all the axes or around the single axis of the cross-section. This bending creates tensile strength and tensile stress at the cross-section. The RC column will have longitudinal reinforcement and lateral relations. The RC column includes a steel frame embedded in the concrete.

Columns are categorized as long or short, braced or unbraced on different structural and dimensional factors. Generally, the columns are square, circular and rectangular. The larger dimension should not be more than 4 times the smaller dimension. If $h > 4b$, it is called a wall. Columns can fail due to failure to compress concrete or steel reinforcement, buckling and combination of compression and buckling failure. Compression failures occur for short column and buckling failure occur for longer columns.

ANN is the system of neurons that are connected to each other. This system is called Connectionist system. ANN in new trend are used to study the systems which are complex and are difficult for modelling by using mathematical and conventional modelling. The researcher named Ashour et al. [1] researcher used a form of ANN that is, "*GP*" where *G* stands for Genetic and *P* stands for Programming i.e., Genetic-Programming, which performs the empirical modelling of shear strength of RC deep beam. Researcher achieved in experiment successfully by making good arrangement between the model predictions and experiment [1]. Another, researchers named Saka and Sanad [2], the predicted the ultimate shear strength of RC beams using ANN. It provides the better result when compared with strut-and-tie method, Mau-Hsu method, and ACI code method. In ANN shear strength ratio was 0.97 and the ACI method shows the ratio of 2.08 [2]. Structural Engineering involves the understanding of properties of building components and modelling of natural phenomenon, material behaviour and laws of mechanics, intuition, past experience or ability and investigation strategies [4]. The modern computer can bring speed, aptitude and meticulousness in investigation of structures. But to computerize the areas such as intangible design, reasonable structure, representing of normal wonder and material conduct and harm appraisals is very troublesome as it requires human skill. Structural design is an iterative process. The fundamental plan is the initial phase in configurational design process. Programming assumes a focal job in practically all parts of everyday life. Softwares may be characterized as the guidelines, which give usefulness of the program that requires playing out a particular kind of information handling in an expert way to achieve an assignment. The number, size, and application areas of PC programs have developed significantly. In RC Frame Structures, Column sections are the most significant primary structural components and have the vital job both under vertical and seismic burdens [5, 6].

1.1 Artificial Neural Network

Interconnected neurons form the ANN the connection is always in the one direction hence it is called unidirectional communication. Weight is the term used to show the neurons strength connection which is a numerical value. Weights are combined together to storage the knowledge or the data required. The function of sum of inputs received from adjacent node by means of weights is the activation value which neuron is having. To form interconnected neurons ANN have been described as the second-best way to connect and work on it. These neurons are used to model brains and also to perform definite computational tasks. Successful ANN application will have the ability of determining the perfect character recognition [3]. This present study is an effort to apply artificial neural network-based system identification of methods to predict the shear behaviour based on the NSC, SCC, and GPC mix proportions. For this phenomenon, a computer program is developed using ANN design toolbox in the software called MATLAB (v. R2018a). Using this platform (Programs), neural networking model with different hidden layers will be constructed, trained, and then tested using the available test data sets, respectively [7, 8]. The original moto of ANN methodology is to solve problems in the same way that human brain would. But as the time passed, attention has been stimulated to performing specific tasks, and leading deviation from the biology. ANN's have been used on multiplicity of tasks, as these tasks include computer visionary, machine understanding and translation, social network filtering, Playing video games, speech recognition, medical diagnosis, and also been used in the activities which are reserved to humans like painting. ANN began as an attempt to feat the planning of human brain to perform tasks that conventional algorithms had little success with. Warren McCulloch and Walter Pitts opened this topic by creating a computational model for neural networks.

Neural networks are well-trained and adjusted to specified inputs to obtain the required target output. The network is so formed that itself makes the comparison between the target and output, till the target is matched with the network output.

2 Objectives of Study

The following are the Objectives of the research work: 1. To Prepare an Mathematical model using an ANN tool with MATLAB(v. R2018a) to predict the values of deformation, yielding load and ultimate load of short reinforced Normal Strength Concrete with varying reinforcement percentage 2. To validate the results obtained by MATLAB trained with ANN with Experimental results.

3 Materials and Methodology

Here the cement is used for the binding of fine and coarse aggregates. 53 grade OPC as per IS 12269-1987 with specific gravity 3.15. Fly ash confirming to class-F IS-3812:2003 [9, 10]. Coarse aggregate of 20 mm downsize aggregate of specific gravity 2.65 for NSC. The coarse aggregate must be confirming to IS 383:2016. Fine aggregates are M-Sand of specific gravity of 2.57 and fineness modulus 3.05 confirming to zone II and for IS 383:2016 (Table 1).

3.1 Test Specimen

For the investigation purpose, 36 [12 for each grade] columns of NSC specimens were casted for each M-20, M-30, and M-40. The cross-section of 125 × 125 × 1000 mm, respectively and the slenderness ratio was 8 which shows the behaviour

Table 1 Mix proportions, fresh properties, and cube compression strength of NSC [20] NSC [30], and NSC [40]

Materials	Proportion		
	M-20	M-30	M-40
Cement (kg/m^3)	358.47	383.16	465
Fine aggregates (kg/m^3)	690.68	670.72	650.80
Coarse aggregates (kg/m^3)	1113.93	1128.41	1094.89
Water (l/m^3)	197.16	191.58	186
Slump cone (mm)	115	80	45
Cube compressive strength (MPa)	27.23	36.28	49.61

of short column [11]. Lateral ties of 8 mm dia at a spacing of 100 mm, 200 and 300 mmC/C is provided. The percentage of longitudinal steel was varied from 1.29, 2.01, 2.89, and 5.15% (Table 2).

Table 2 Specimen details of columns

Specimen	Longitudinal. reinforcement		Transverse reinforcement (mm)
	Dia. of bar (mm)	Reinforcement (%)	
NSC/M20/1.29/100	8	1.29	100
NSC/M20/2.01/100	10	2.01	100
NSC/M20/2.89/100	12	2.89	100
NSC/M20/5.15/100	16	5.15	100
NSC/M20/1.29/200	8	1.29	200
NSC/M20/2.01/200	10	2.01	200
NSC/M20/2.89/200	12	2.89	200
NSC/M20/5.15/200	16	5.15	200
NSC/M20/1.29/300	8	1.29	300
NSC/M20/2.01/300	10	2.01	300
NSC/M20/2.89/300	12	2.89	300
NSC/M20/5.15/300	16	5.15	300
NSC/M30/1.29/100	8	1.29	100
NSC/M30/2.01/100	10	2.01	100
NSC/M30/2.89/100	12	2.89	100
NSC/M30/5.15/100	16	5.15	100
NSC/M30/1.29/200	8	1.29	200
NSC/M30/2.01/200	10	2.01	200
NSC/M30/2.89/200	12	2.89	200
NSC/M30/5.15/200	16	5.15	200
NSC/M30/1.29/300	8	1.29	300
NSC/M30/2.01/300	10	2.01	300
NSC/M30/2.89/300	12	2.89	300
NSC/M30/5.15/300	16	5.15	300
NSC/M40/1.29/100	8	1.29	100
NSC/M40/2.01/100	10	2.01	100
NSC/M40/2.89/100	12	2.89	100
NSC/M40/5.15/100	16	5.15	100
NSC/M40/1.29/200	8	1.29	200
NSC/M40/2.01/200	10	2.01	200
NSC/M40/2.89/200	12	2.89	200
NSC/M40/5.15/200	16	5.15	200

(continued)

Table 2 (continued)

Specimen	Longitudinal. reinforcement		Transverse reinforcement (mm)
	Dia. of bar (mm)	Reinforcement (%)	
NSC/M40/1.29/300	8	1.29	300
NSC/M40/2.01/300	10	2.01	300
NSC/M40/2.89/300	12	2.89	300
NSC/M40/5.15/300	16	5.15	300

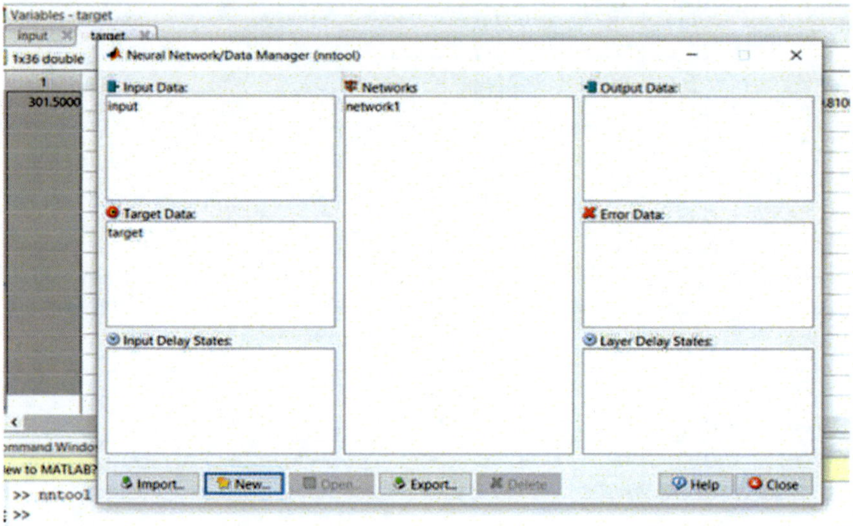

Fig. 1 Providing input and target values of NSC columns

3.2 *Inputs for ANN*

See Figs. 1, 2, 3, 4 and 5 and Tables 3 and 4.

4 Results

A comprehensive analysis was carried out to study subsequent parameters to know structural performance of NSC Columns (Table 5).

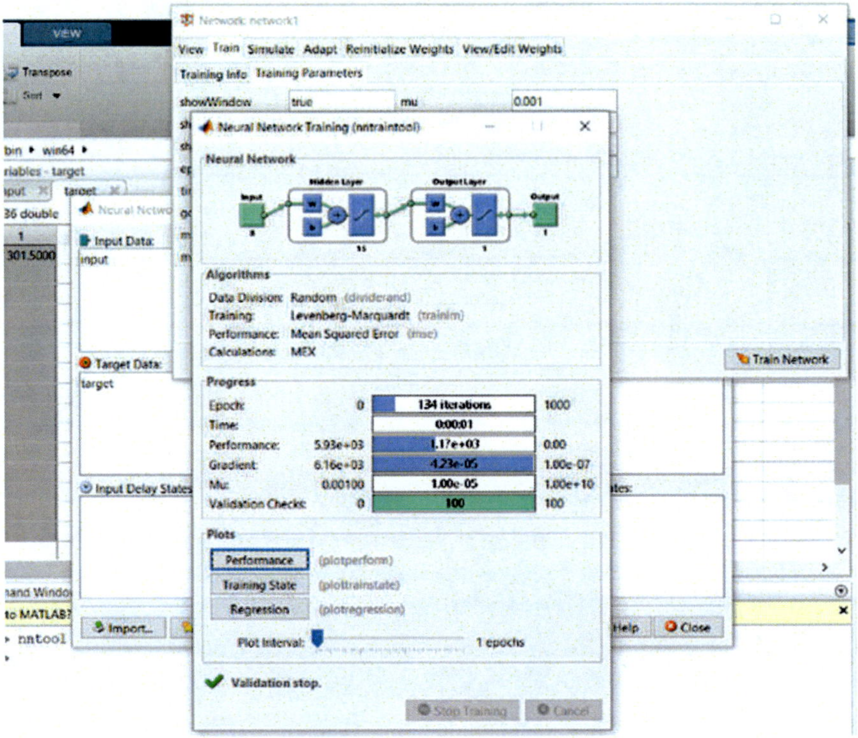

Fig. 2 Neural network training

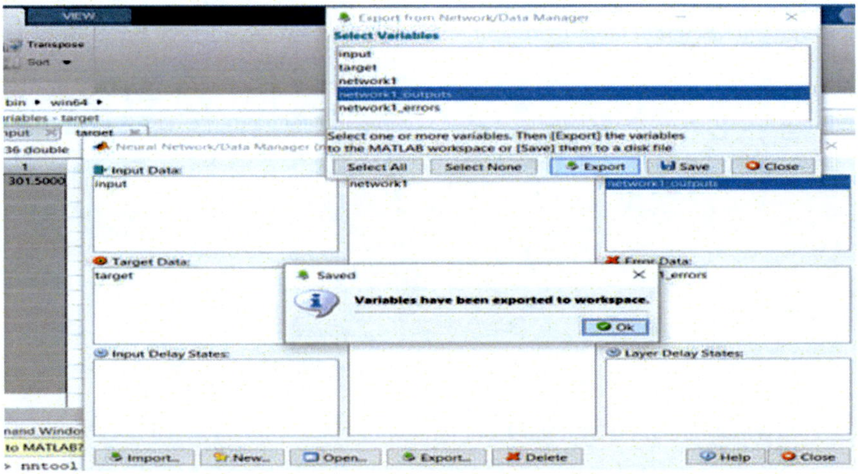

Fig. 3 Neural network training

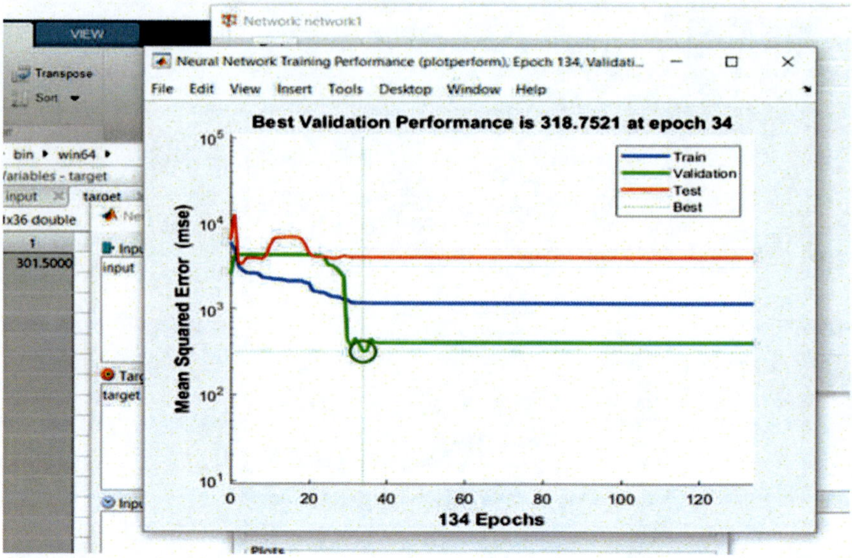

Fig. 4 Best validation performance

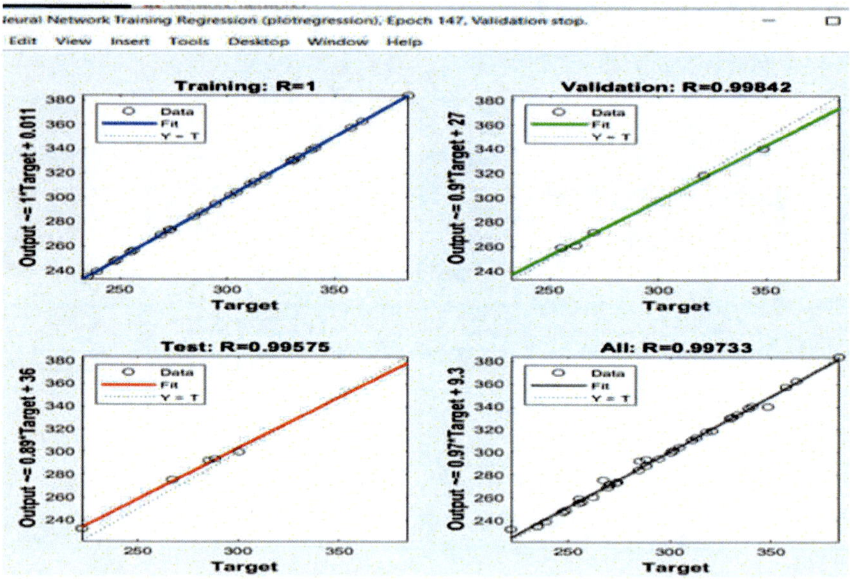

Fig. 5 Regression plots for NSC columns

Table 3 Input data

Cement (kg/m^3)	FA (kg/m^3)	CA (kg/m^3)	W (l/m^3)
310	612	1084	155
348.33	681.66	1146.8	191.58
450	873.36	1224.3	206.5
310	612	1084	155
348.33	681.66	1146.8	191.58
450	873.36	1224.3	206.5
310	612	1084	155
348.33	681.66	1146.8	191.58
450	873.36	1224.3	206.5
310	612	1084	155
348.33	681.66	1146.8	191.58
450	873.36	1224.3	206.5
310	612	1084	155
348.33	681.66	1146.8	191.58
450	873.36	1224.3	206.5
310	612	1084	155
348.33	681.66	1146.8	191.58
450	873.36	1224.3	206.5
310	612	1084	155
348.33	681.66	1146.8	191.58
450	873.36	1224.3	206.5
310	612	1084	155
348.33	681.66	1146.8	191.58
450	873.36	1224.3	206.5
310	612	1084	155
348.33	681.66	1146.8	191.58
450	873.36	1224.3	206.5
310	612	1084	155
348.33	681.66	1146.8	191.58
450	873.36	1224.3	206.5
310	612	1084	155
348.33	681.66	1146.8	191.58
450	873.36	1224.3	206.5
310	612	1084	155
348.33	681.66	1146.8	191.58
450	873.36	1224.3	206.5

(continued)

Table 3 (continued)

Slump (mm)	Fck (MPa)	Spacing (mm)	Ast (%)
115	27.23	100	1.29
80	36.28	100	1.29
45	49.61	100	1.29
115	27.23	200	1.29
80	36.28	200	1.29
45	49.61	200	1.29
115	27.23	300	1.29
80	36.28	300	1.29
45	49.61	300	1.29
115	27.23	100	2.01
80	36.28	100	2.01
45	49.61	100	2.01
115	27.23	200	2.01
80	36.28	200	2.01
45	49.61	200	2.01
115	27.23	300	2.01
80	36.28	300	2.01
45	49.61	300	2.01
115	27.23	100	2.89
80	36.28	100	2.89
45	49.61	100	2.89
115	27.23	200	2.89
80	36.28	200	2.89
45	49.61	200	2.89
115	27.23	300	2.89
80	36.28	300	2.89
45	49.61	300	2.89
115	27.23	100	5.15
80	36.28	100	5.15
45	49.61	100	5.15
115	27.23	200	5.15
80	36.28	200	5.15
45	49.61	200	5.15
115	27.23	300	5.15
80	36.28	300	5.15
45	49.61	300	5.15

Table 4 Target data

P_{cr} (kN)	Δ_{cr} (mm)	P_y (kN)	Δ_y (mm)	P_u (kN)	Δ_u (mm)
301.50	4.50	351.75	6.05	502.50	10.68
317.92	4.88	370.90	6.34	529.86	10.36
322.27	5.32	388.82	6.4	555.45	9.72
255.42	3.97	297.99	5.4	425.70	11.32
270.27	4.42	315.32	6.19	450.45	10.60
284.85	4.96	332.33	6.4	474.75	9.91
221.88	6.06	258.86	7.2	369.80	12.54
234.96	5.77	274.12	6.6	391.60	11.00
247.43	6.12	288.67	7.1	412.39	10.54
310.76	4.40	362.56	5.29	517.94	9.20
330.81	4.96	385.95	5.4	551.35	8.80
348.45	5.14	406.58	5.9	580.75	8.21
269.28	4.21	314.16	6.1	448.80	9.68
284.52	4.62	331.94	5.1	474.20	9.24
300.30	4.76	350.35	5.6	500.50	8.68
234.90	4.26	274.05	7.7	391.50	10.20
248.70	4.71	290.15	5.6	414.50	9.46
262.02	5.17	305.69	6.18	436.70	8.90
320.76	3.80	374.22	4.4	534.60	6.84
338.88	4.30	395.36	4.51	564.80	6.54
357.42	4.22	416.99	4.62	595.70	6.04
273.24	4.21	318.78	4.5	455.40	7.20
304.50	3.91	355.25	4.5	507.50	6.23
239.10	3.58	278.95	4.17	398.50	7.45
255.30	3.89	297.85	4.32	425.50	7.30
267.12	4.05	311.64	4.5	445.20	6.78
340.68	3.06	397.46	3.38	567.80	5.20
362.70	3.30	423.15	3.55	604.50	4.80
383.88	2.98	447.86	3.21	639.80	4.56
294.48	2.76	343.56	3.42	490.80	5.64
312.90	3.27	365.05	3.48	521.5	5.10
329.76	3.30	384.72	3.56	549.60	4.69
256.50	3.29	299.25	4	427.50	6.30
273.75	3.00	319.38	3.48	456.25	5.65
288.12	2.91	336.14	3.65	480.20	5.20

Table 5 ANN results of column specimen

MP_{cr} (kN)	$M\Delta_c$ (mm)	MP_y (kN)	$M\Delta_y$ (mm)	Mp_u (kN)	$M\Delta u$ (mm)
303.19	4.78	353.63	6.12	505.24	10.33
319.90	5.05	373.11	5.96	533.07	9.81
335.61	5.21	391.42	6.14	559.24	9.29
260.80	4.86	6.19	6.35	434.60	10.76
277.51	5.13	323.66	6.19	462.43	10.24
293.22	5.29	341.97	6.37	488.60	9.72
218.41	4.94	254.73	6.58	363.96	11.19
235.12	5.21	274.21	6.42	391.79	10.67
250.83	5.37	292.52	6.60	417.96	10.15
310.57	4.41	362.24	5.58	517.54	9.33
327.28	4.68	381.72	5.41	545.37	8.82
342.99	4.83	400.03	5.60	571.54	8.30
268.18	4.49	312.79	5.81	446.90	9.76
284.89	4.76	332.27	5.64	474.73	9.25
300.60	4.91	350.58	5.83	500.90	8.73
225.79	4.57	263.34	6.04	376.26	10.19
242.50	4.84	282.82	5.87	404.09	9.68
258.21	4.99	301.13	6.06	430.26	9.16
319.59	3.95	372.76	4.92	532.58	8.12
336.29	4.22	392.24	4.75	560.41	7.60
352.01	4.37	410.55	4.94	586.58	7.08
277.20	4.03	323.31	5.15	461.94	8.55
293.90	4.30	342.79	4.98	489.77	8.03
309.62	4.45	361.10	5.17	515.94	7.51
234.81	4.11	273.86	5.38	391.30	8.98
251.51	4.38	293.34	5.21	419.13	8.46
267.23	4.53	311.65	5.40	445.30	7.94
342.75	2.78	399.78	3.22	571.18	4.99
359.46	3.05	419.27	3.05	599.01	4.48
375.17	3.20	437.58	3.24	625.18	3.96
300.36	2.86	350.33	3.45	500.54	5.42
317.07	3.13	369.82	3.28	528.37	4.91
332.78	3.28	388.13	3.47	554.54	4.39
257.97	2.94	300.88	3.68	429.90	5.85
274.68	3.21	320.37	3.51	457.73	5.34
290.39	3.36	338.68	3.70	483.90	4.82

5 Summary and Conclusions

To predict the ultimate Compressive strength on Normal strength concrete columns. ANN model was constructed, trained and tested by conducting experiment of columns casted.

The formed models are used to perform the parametric study to evaluate the results of variables of column on ultimate compressive strength. Percentage of error for the predicted value is less than 1.

ANN model was utilized to predict the shear behaviour of RC column. The models developed by the ANN showed the goodness in fit in the results.

The *R*-square value of the model of flow diameter of slump by use of ANN is greater than 0.9 and the regression model values is 0.792. The normal strength concrete was applied with feedforward back-propagation neural network.

ANN includes 4-layers: the input layer will be containing 7 neurons, 1st and 2nd hidden layer will be having 5 neurons each and the output layer will be having only one neuron.

Acknowledgements We acknowledge the support extracted by Management of AIT, Principal, HOD, Faculty and Staff of Civil Dept. of AIT, Affiliated to VTU Belagavi.

References

1. Ashour AFA, Alvarexand LF, Toropov VV (2003) Empherical modelling of shear strength of RS beam by genetic programming. Comput Struct 81:331–338
2. Saka MP, Sanad A (2001) Prediction of ultimate shear strength of reinforced concrete deep beam using neural network. J Struct Eng 127:818–828
3. Kirkegaard PH, Rytter A (1993) Use of neural networks for damage detection and location in steel member. In: Proceedings of 3rd international conference on the application of AI in the field of structural engineering. Heriot-Watt University, Edinburgh, UK, 17–19 Aug 1993, pp 1–9
4. Arafa M, Alogedra M, Najjar HA, Neural models for predicting shear strength of reinforced and normal and high-strength concrete deep beams. https://doi.org/10.3923/jas.2011.266.274
5. Jayaramappa N, Krishna A, Annpurna BP, Kiran T (2014) Prediction of base shear for three dimensional RC frame subjected to lateral load using artificial neural network. Indian J Sci Technol 7(6):729–733
6. Aga AAA, Adam FM (2015) Design optimization of reinforced concrete frames. Open J Civ Eng 5:74–83
7. Kirkegaard PH, Rytter A (1993) The use of neural networks for damage detection and location in a steel member. In: Proceedings of the 3rd international conference on the application of artificial intelligence to civil and structural engineering. Heriot-Watt University, Edinburgh, UK, 17–19 Aug 1993, pp 1–9
8. Nawy EG (2008) Reinforced concrete: a fundamental approach, 6th edn. Prentice Hall, New York. ISBN-13: 9780132417037
9. Sumajouw DMJ, Hardjito D, Wallah SE, Rangan BV (2007) Fly ash-based geopolymer concrete: study of slender reinforced columns. J Mater Sci 42(9):3124–3130

10. Barbosa VF, MacKenzie KJ, Thaumaturgo C (2000) Synthesis and characterization of materials based on inorganic polymers of alumina and silica: sodium polysialate polymers. Int J Inorg Mater 2(4):309–317
11. Kaarthikeyana K, Shanth RM (2016) Optimization of RC columns using artificial neural Network. Int J Sci Eng Res 7(4). ISSN 2229-5518

Seismic Analysis of Multi-storey Building on Sloping Ground with Ground, Middle and Top Soft Storey

Ratnakala S. Bidreddy and Shankar H. Sanni

Abstract Nowadays, fast development is occurring in uneven zones because of lack of plain ground. As a result, the bumpy territories have checked impact on the structures in terms of style, material and technique of construction leading to popularity of multi-storied structures in hilly regions. Because of inclining profile, the different degrees of such structures step back towards the slope slant and may likewise have set back also at the simultaneity. In the present study, G+12 storey building is considered, and this building is assumed to be present in flat ground and on sloping ground 20°. The models are also equipped with infill walls and two different shear wall arrangements. The models are also a soft storey structure. By this, total ten models are created for analysis purpose. Here, equivalent static analysis is carried out for the study.

Keywords ETABS-2017 · Shear wall · Equivalent static method · Displacement · Drift · Time period · Base shear and ground middle and top (GMT) soft storey

1 Introduction

In hilly areas, the building configuration will be different and may have vertical irregular, variation in column heights, irregular mass distribution, mass and stiffness unequal distributions. These parameters alter the building responses.

Due to functional requirement, many of the tall structures are kept open at the ground first as open, i.e. termed as soft storey. This will make the structure weak and less stiff. However, in many of the column beam framed structure instead of glazing, they are undergoing masonry structure. This will reduce the extent of damages to the structure and life.

A. Soft Storey: These are multiple floors in which one or more floors have openings for windows, large doors and vehicle parking. The rigidity of this floor is less than that of the normal floor.

R. S. Bidreddy (✉) · S. H. Sanni
Department of Civil Engineering, Basaveshwar Engineering College, Bagalkot 587102, India

L. Nandagiri et al. (eds.), *Sustainability Trends and Challenges in Civil Engineering*, Lecture Notes in Civil Engineering 162, https://doi.org/10.1007/978-981-16-2826-9_11

B. Brick Masonry: Brick masonry is very durable in construction. It is built by placing brick and mortar. In this analysis, brick masonry is very helpful in reducing displacement. Currently, 230 mm thick brick masonry wall is used for analysis.

Properties of masonry wall.

- Masonry is considered as an anisotropic composite material because of different geometries and material properties of masonry, including shape of units (bricks/blocks), dimensions of units, perforations, slenderness ratio, strengths of materials, modulus of elasticity of materials, water absorption, etc.
- Elastic modulus (or Young's modulus) is one of the most important parameters in determination of the stiffness of structural elements prior to cracking.

C. Shear Wall: Shear walls are vertical RCC members that resist lateral loads. The shear wall reduces the displacement of floors that meet the earthquake.

The main objective of this work is to improve the structure which is on a sloping ground. The structure is analysed for equivalent static method using the E-TAB software.

Properties of Shear Wall

- As shear force in earth quake prone zone increases, hence there is necessity of placing shear walls in those areas.
- Shear walls should have more strength and stiffness. For slender walls where the bending deformation is more, shear wall resists the loads.

2 Literature Review

Siddiqui and Vidyadhara [1], investigated 12 storey structure having 5 m width in each direction with 4 bays. They observed that the earthquake behaviour is found out for infill-framed structure. The hinge formation is first found at soft storey location than shear wall, and base shear is also found maximum in case of model with shear wall. The infill walls provided between frames give more stability and drastically reduces the displacements than bare frame models.

Vaidya [2] studied the structure located on sloping ground. The shear wall is provided in the structure, and its positions are varied to check the performance of the structure. Main ideas are to understand the optimum location of the shear wall by increasing the effectiveness in the seismic behaviour. The results are interpreted in terms of displacement, storey drift and maximum forces in the columns. Time period of the structure is found to be with the model having shear wall and shorter columns. It is also found that the short columns are more critical in sloping ground and shear wall provided in plan symmetry is having better results and having control over displacements than other models.

Kalsulkar and Rathod [3] carried out step back and set back configuration. The analysis was carried out by using response spectrum method. From the results, it is found that the step back building is having highest base shear value than set back configuration. And time period is also found to be maximum in case of step back building. It is also observed from the study that the number of storey and number of bays reduces top storey displacement. And hence, it is concluded that the building spread area reduces the displacement drastically than building with more height.

Thombre and Makarande [4] studied the structures located on sloppy ground and plain ground as well. The structure is low-rise RCC structure with G + 5 storeys with storey height of 3 m. The structure is considered to be located in zone 5. The analysis is carried out, and it is found that the building having shorter slope having less displacement than other structure.

Birajdar and Nalawade [5] carried out high rise structure with 24 storeys with set back, step back and step back–set back structure for investigation. Here, 3D analyses are carried out using ETABS software. The lateral loads such as seismic loads are considered for both structures resting on ground and sloppy ground. The structures are assumed to have moderate slope at ground of around 27°. The response spectrum analysis is carried out and its torsional behaviour is also captured.

Khadiranaikar and Masali [6] their study focused the response of structure located on hilly slope. It is found that the behaviour of structure resting on slope area is different from resting on plane ground. The dynamic analysis is carried out for the study. Study is also carried out for step back and set back buildings. It is noticed that, the short column is attracting higher forces than other column and which is most affected than other columns. The presence of shear wall and infill wall is benefitting. The noticeable reduction in displacement and drift is found out. It is better to increase the number of ways to reduce the displacement, drift and time period.

Halkude et al. [7] used response spectrum method for both set back and step back buildings. The dynamic nature of the structure is different from simple static analysis. From the study, it is noticed that both step back and set back buildings are more favourable than set back building alone. The results such as displacement, fundamental time period. The base shear value is more for step back building.

3 Objectives of the Project

A. To study the effect of infill in frame when subjected to seismic loading on sloping ground.
B. To study the effect of soft storey in multi-storey buildings when subjected to seismic loading on sloping ground.
C. To study how shear wall can enhance the performance of soft storey RC buildings on sloping ground.
D. To find displacement, drift, base shear, time period for 0° and 20°, using equivalent static method.

4 Methodology

This paper attempts to investigate the seismic effect on the RCC multi-storey building model G + 12 with masonry infill, GMT with soft storey, L type and C type shear wall. The 13-storey RCC building models are created and analysed by ETAB software [8]. After successfully completing the models, the best position of the different sloping terrains is found by changing the different degree to minimize the seismic effect. Different models have been created and the results are compared to additional models.

The height of each storey is maintained at 3 m. The seismic zone considered is V and the soil is medium (Type II). In this work, the structure includes live load, seismic load and dead load, and the codes are used in accordance with IS-875 part 1 and 2 [9, 10], IS 1893-2002 [11] and IS 456-2000 [12]. The structure is analysed using the equivalent static method and response spectrum method. Such as displacement, storey drift, time period, and base shear are calculated. After analysing the structure, the obtained values are used to form a table, graphs and finally, the conclusion.

Equivalent Static Method
In the linear static procedure, building modelled is an equivalent single degree freedom system. The equivalent static analysis procedure includes the following steps.

- Estimate the first mode response period of the building from the design response spectra.
- Use the specific design response spectra to determine that the lateral base shear of the complete building is consistent with the level of post-elastic (ductility) response assumed.

5 Modelling Description

The proposed models are RCC flat slab structure with RCC columns and composite columns. Plan layout is kept the same for all configurations of building frame.

5.1 Properties of Members

Poisson's ratio: 0.2
Density: 25 KN/m^3
Thermal coefficient: 0.0000055/°C
Grade of concrete: M40
Yield strength of steel: Fe_{500}.

5.2 *Seismic Parameter*

Zone value (Z): 0.36
Response reduction factor (R): 5 (S.M.R.F)
Importance factor: 1
Damping ratio: 0.05
Soil Type: Medium.

5.3 *Size of Members*

Height of building: 39 m (G+12 storeys)
Storey to storey height: 3 m
Span length: 5 m
Column sizes: 1200 mm × 1200 mm, 750 mm × 1000 mm, 550 mm × 750 mm
Beam size: 230 mm × 600 mm
Slab thickness: 150 mm
Wall thickness: 230 mm.

5.4 *Load Intensity*

Live load on each floor: 3 kN/m^2
Floor finish: 1.5 kN/m^2.

5.5 *Load Combinations*

The load combination is itself calculated by the E-TAB software and the models are analysed as the calculated load combination.

6 Plan, Elevation and 3D View of Different Models

See Figs. 1, 2, 3, 4, 5, 6, 7 and 8.

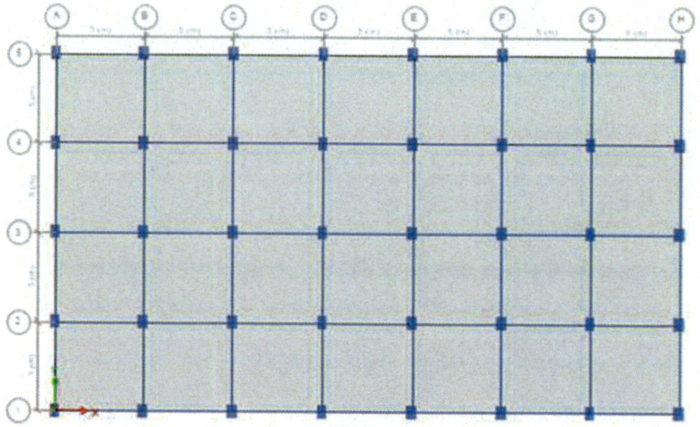

Fig. 1 Plan

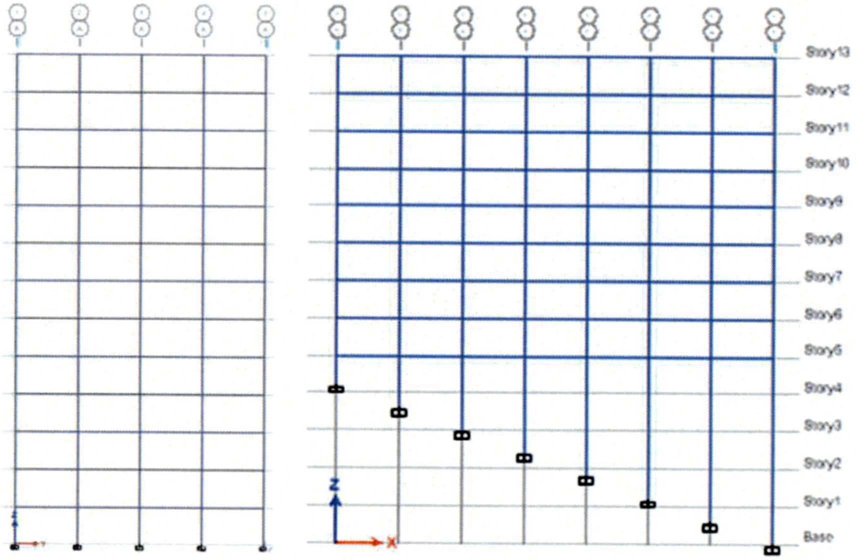

Fig. 2 Elevation of 0° and 20°

7 Results and Discussion

The results of normal building and sloping ground building is compared by considering masonry brick infill, GMT soft storey, GMT with L type shear wall and GMT with C type Shear wall. The lateral responses like displacement, storey drift, time period and base shear is evaluated and compared.

Fig. 3 3D view

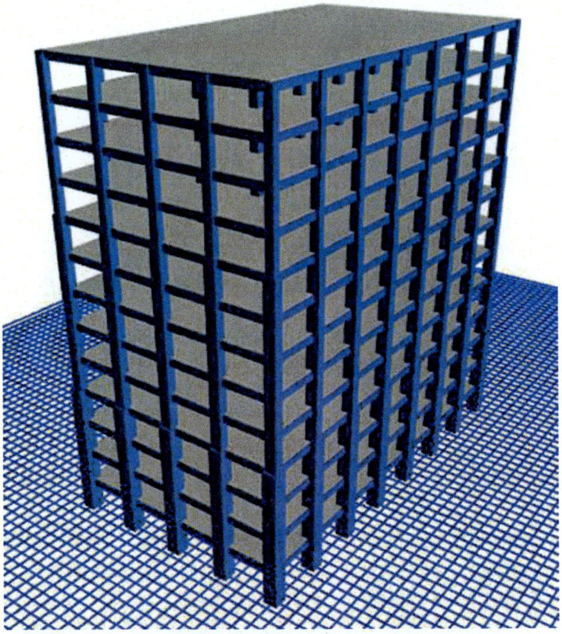

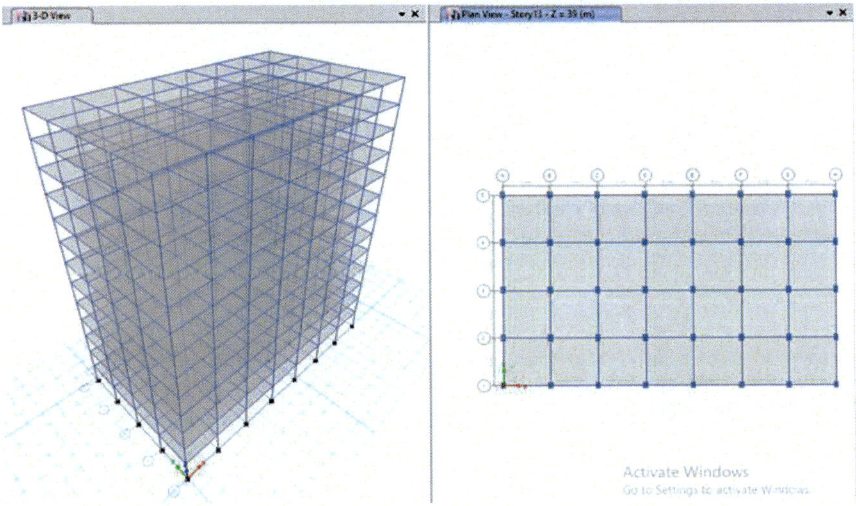

Fig. 4 Bare frame

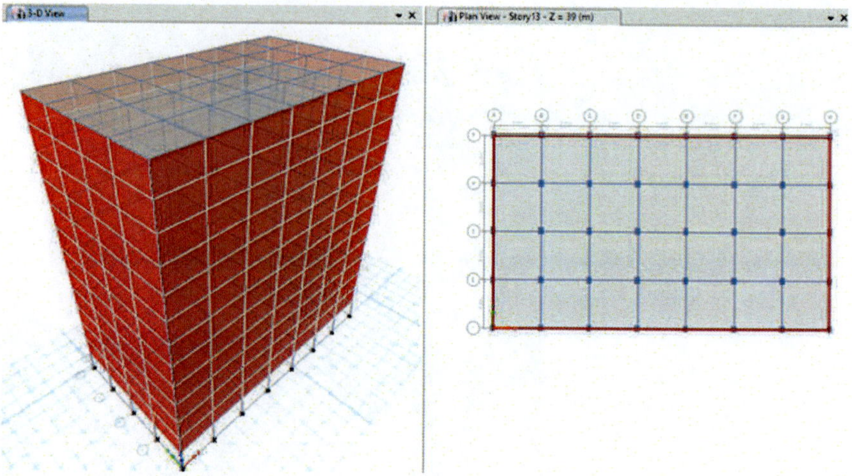

Fig. 5 Masonry brick infill

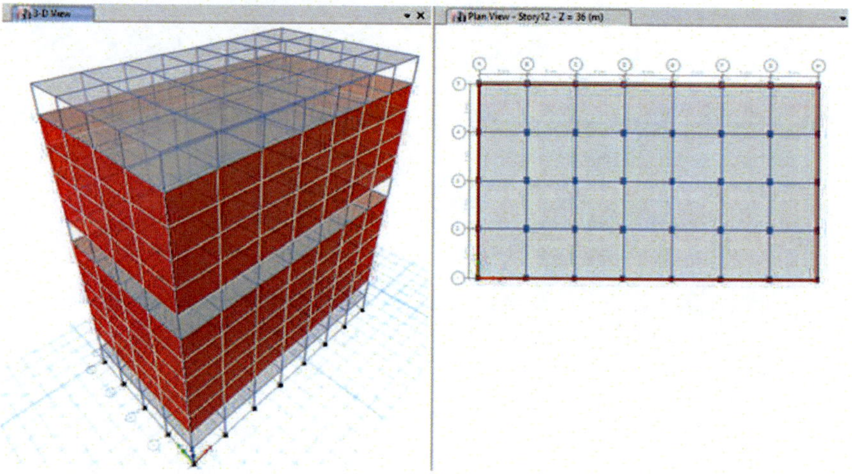

Fig. 6 GMT with soft storey

MODELS A1, A2, A3, A4 and A5 for 0° plane ground.
MODELS C1, C2, C3, C4 and C5 for 20° plane ground.

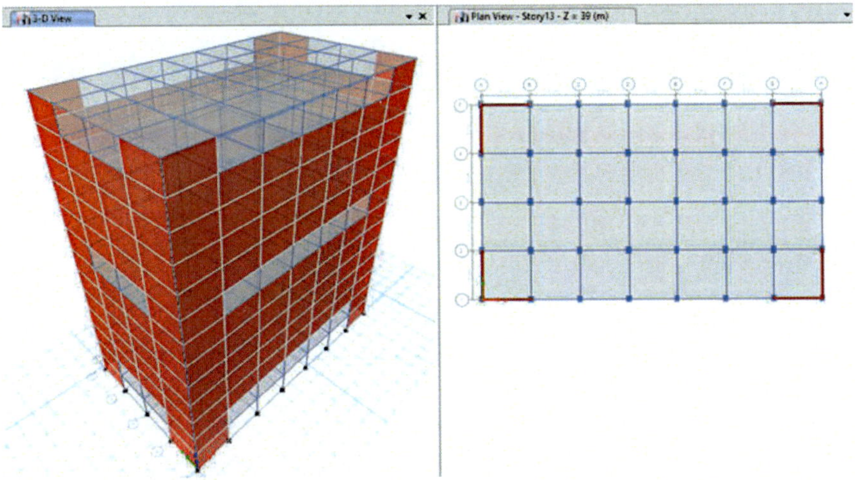

Fig. 7 GMT with L type shear wall

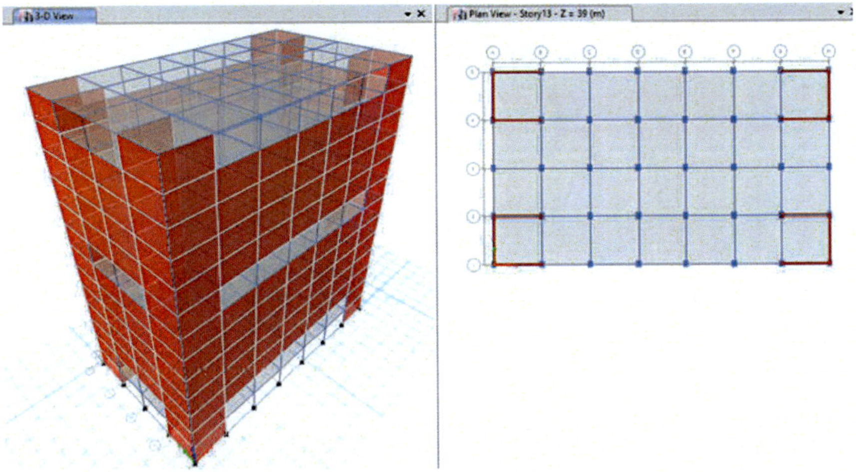

Fig. 8 GMT with C type shear wall

7.1 Displacement (mm) due to Equivalent Static Method

From Table 1 and Fig. 9, it can be observed that the model A1 is having maximum displacement compared to other models. The reduction of displacement is found maximum in model A5. The displacement is increased nominally for A3 model compared to A2.

From Table 2 and Fig. 10, it can be observed that the model C1 is having maximum displacement compared to other models. In this method of analysis, it can be noticed

Table 1 Displacement EQX 0° slope

Storey	Model A1	Model A2	Model A3	Model A4	Model A5
13	36	11	13	7	6
12	35	11	12	7	6
11	33	10	11	6	5
10	30	10	11	6	5
9	27	9	10	5	4
8	23	8	9	4	3
7	19	7	7	3	3
6	16	6	6	3	2
5	12	4	5	2	2
4	8	3	4	2	1
3	5	2	3	1	1
2	3	1	1	1	0
1	1	0	0	0	0
0	0	0	0	0	0

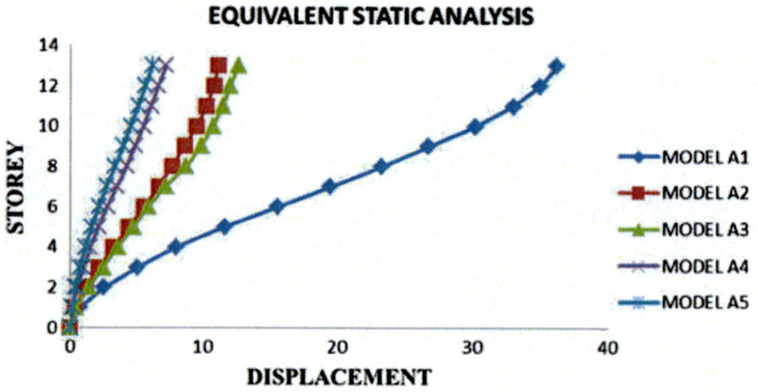

Fig. 9 Displacement versus storey EQX 0° slope

that there is reduction in displacement for models C1 to C5, which shows that C1 is more critical, the model C5 with shear wall has lesser displacement.

7.2 Drift (mm) Due to Equivalent Static Method

From Table 3 and Fig. 11, it can be observed that the model A1 is having maximum drift compared to other models. The reduction of displacement is found maximum in model A5. The maximum drift found is critical in storeys 5–9 in bare frame.

Table 2 Displacement EQX 20° slope

Storey	Model C1	Model C2	Model C3	Model C4	Model C5
13	19	6	7	4	4
12	18	6	7	4	3
11	16	5	6	3	3
10	14	5	6	3	3
9	12	4	5	3	2
8	9	3	4	2	2
7	6	3	3	2	1
6	4	2	2	1	1
5	2	1	1	1	1
4	0	0	0	0	0
3	0	0	0	0	0
2	0	0	0	0	0
1	0	0	0	0	0
0	0	0	0	0	0

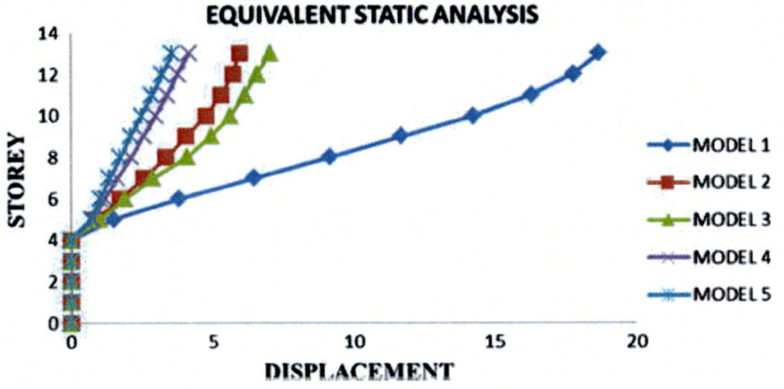

Fig. 10 Displacement versus storey EQX 20° slope

From Table 4 and Fig. 12, it can be observed that the drift values are highest for model 1 as comparing with other models. However, the remaining models are having almost similar drift values. This shows that the slope of the ground is not the governing parameter.

Table 3 Drift EQX 0° slope

Storey	Model A1	Model A2	Model A3	Model A4	Model A5
13	1	0	1	1	1
12	2	1	1	1	1
11	3	1	1	1	1
10	3	1	1	1	1
9	4	1	1	1	1
8	4	1	2	1	1
7	4	1	1	1	1
6	4	1	1	1	1
5	4	1	1	1	0
4	3	1	1	1	0
3	3	1	1	0	0
2	2	1	1	0	0
1	1	0	0	0	0
0	0	0	0	0	0

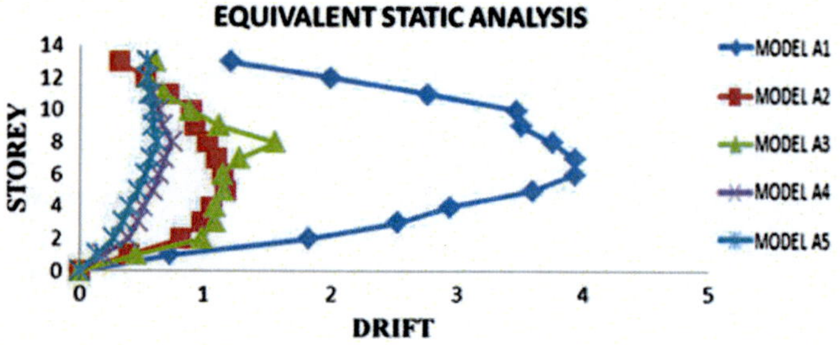

Fig. 11 Drift versus storey EQX 0° slope

7.3 *Base Shear (kN)*

From Table 5 and Fig. 13, it can be observed that the base shear is higher for model A5 and model A2. The increase in base shear depends on seismic weight of the structure. The model A3 is having lesser base shear compared to all other models.

From Table 6 and Fig. 14, it can be observed that the base shear is higher for model 5 and model 2. The increase in base shear depends on seismic weight of the structure. The model 3 is having lesser base shear compared to all.

Table 4 Drift EQX 20° slope

Storey	Model C1	Model C2	Model C3	Model C4	Model C5
13	1	0	0	0	0
12	1	0	0	0	0
11	2	1	1	0	0
10	3	1	1	0	0
9	3	1	1	0	0
8	3	1	1	0	0
7	3	1	1	0	0
6	2	1	1	0	0
5	2	1	1	1	1
4	0	0	0	0	0
3	0	0	0	0	0
2	0	0	0	0	0
1	0	0	0	0	0
0	0	0	0	0	0

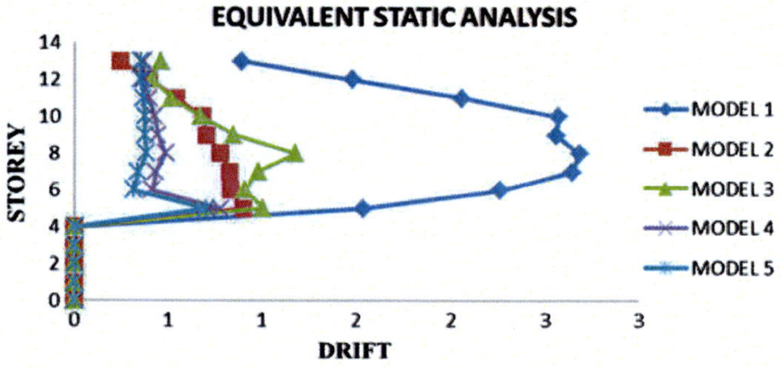

Fig. 12 Drift versus storey EQX 20° slope

Table 5 Base shear EQX 0° slope

Base shear				
Model A1	Model A2	Model A3	Model A4	Model A5
5185	5253	5062	5168	5325

7.4 *Time Period (s)*

From Table 7 and Fig. 15, it can be observed that the time period of model A1 is more and found to be flexible among other models. However, models with shear wall shows rigid and having less time period.

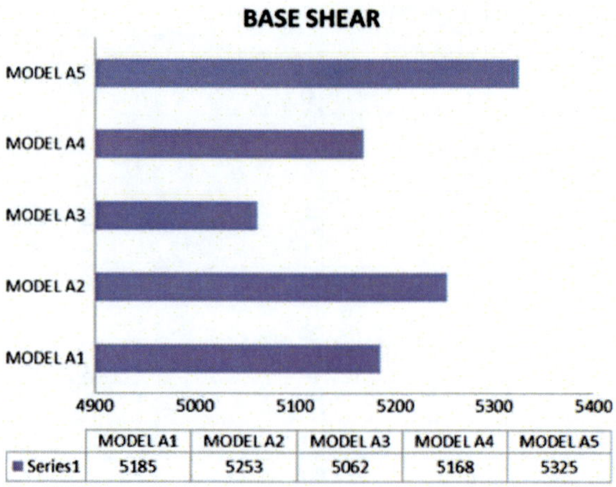

Fig. 13 Base shear EQX 0° slope

Table 6 Base shear EQX 20° slope

Base shear				
Model 1	Model C1	Model C2	Model C3	Model C4
3726	3825	3628	3714	3831

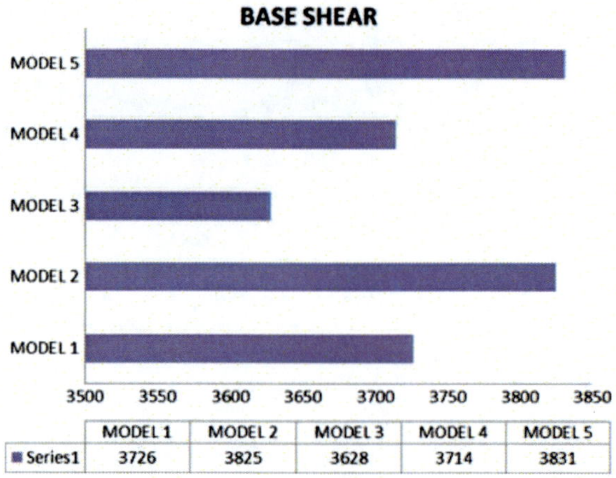

Fig. 14 Base shear EQX 0° slope

The similar kind of observation was true here in sloping ground also as seen from Table 8 and Fig. 16. The time period of a model C1 is more and compared to other models. However, model C5 with shear wall shows rigid and having less time period.

Table 7 Time period EQX 0° slope

Time period					
Mode	Model A1	Model A2	Model A3	Model A4	Model A5
1	1.28	0.87	0.89	0.59	0.57
2	1.25	0.72	0.76	0.54	0.49
3	1.17	0.59	0.62	0.35	0.35
4	0.42	0.29	0.30	0.18	0.17
5	0.41	0.25	0.26	0.17	0.14
6	0.39	0.20	0.22	0.11	0.11
7	0.24	0.16	0.17	0.09	0.09
8	0.22	0.14	0.16	0.09	0.07
9	0.21	0.12	0.13	0.06	0.06
10	0.15	0.11	0.11	0.06	0.06
11	0.14	0.10	0.11	0.06	0.05
12	0.13	0.08	0.09	0.04	0.04

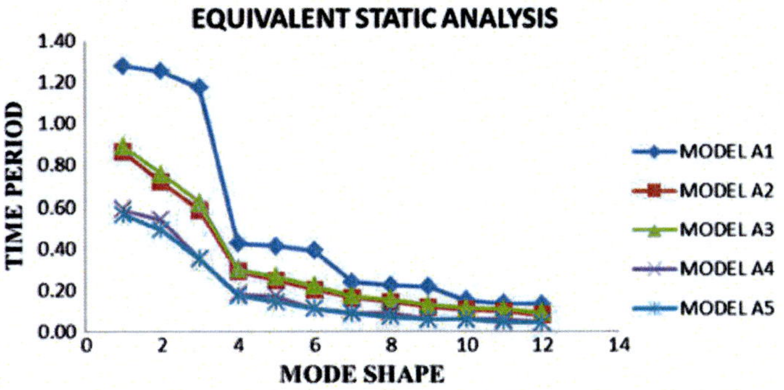

Fig. 15 Time period EQX 0° slope

8 Conclusion

The following conclusions can be drawn with the present investigation.

- The observed parameters like displacement, drift, base shear and time period are found to be maximum in bare frame model (A1 and C1)
- The displacement of models can be reduced by introducing masonry and shear wall in the models. The reduction is found to be 69, 63, 80 and 83% for models 2, 3, 4 and 5 when compared with model 1.
- The drift values of all the models are within the limits. i.e., $h/250 = 3000/250 = 12$. And hence, all models in 20° sloping angle are safe and within the permissible limits.

Table 8 Time period EQX 20° slope

Time period					
Mode	Model C1	Model C2	Model C3	Model C4	Model C5
1	1.04	0.72	0.78	0.57	0.57
2	0.96	0.55	0.60	0.45	0.41
3	0.87	0.45	0.50	0.31	0.30
4	0.33	0.23	0.26	0.16	0.16
5	0.32	0.19	0.24	0.15	0.13
6	0.28	0.18	0.19	0.11	0.10
7	0.18	0.16	0.17	0.10	0.10
8	0.17	0.12	0.16	0.08	0.08
9	0.14	0.11	0.14	0.08	0.07
10	0.11	0.10	0.13	0.08	0.06
11	0.11	0.09	0.11	0.07	0.06
12	0.09	0.09	0.09	0.07	0.06

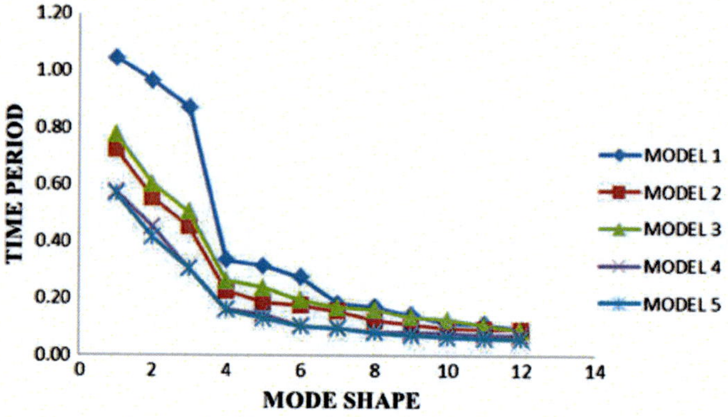

Fig. 16 Time period EQX 20° slope

- The base shear is higher for model 5 and model 2. The increase in base shear depends on seismic weight of the structure. The model 3 is having lesser base shear compared to all other models.
- From the model analysis, it is found that, the model 1 is very much flexible and having highest time period and time period will decreases by 32, 30, 54 and 55% for model 2, 3, 4 and 5 when compared with model 1.
- From the overall analysis, it is proved that the models with shear wall, significantly improves the performance of the structure in terms of displacement, drift and time period.
- In case of providing the shear wall is difficult due to some constraint, the masonry wall will work efficiently in case of normal or slope ground.

References

1. Siddiqui R-U-H, Vidyadhara HS (2013) Seismic analysis of earthquake resistant multi bay multi storied 3D—RC frame. Int J Eng Res Technol 2(10). ISSN: 2278-0181
2. Vaidya PR (2015) Seismic analysis of building with shear wall on sloping ground. Int J Civ Struct Eng Res 2(2):53–60. ISSN 2348-7607
3. Kalsulkar N, Rathod S (2015) Seismic analysis of RCC building resting on sloping ground with varying number of bays and hill slopes. Int J Curr Eng Technol 5(3). E-ISSN 2277-4106. P-ISSN 2347-5161
4. Thombre P, Makarande SG (2016) Seismic analysis of building resting on sloping ground. J Emerg Technol Innov Res (JETIR) 3(6). ISSN-2349-5162
5. Birajdar BG, Nalawade SS (2004) Seismic analysis of buildings resting on sloping ground. In: 13th world conference on earthquake engineering. Paper no. 1472. Vancouver, BC, Canada
6. Khadiranaikar RB, Masali A (2015) Seismic performance of buildings resting on sloping ground. IOSR J Mech Civ Eng (IOSR-JMCE) 11(3):12–19
7. Halkude SA, Kalyanshetti MG, Ingle VD (2013) Seismic analysis of buildings resting on sloping ground with varying number of bays and hill slopes. Int J Eng Res Technol (IJERT) 2(12). ISSN: 2278-0181
8. ETABS (2017) Integrated building design software
9. IS 875 (part 1) (1987) Code of practice for dead loads. Bureau of Indian Standards, New Delhi
10. IS 875 (part 2) (1987) Code of practice for imposed loads. Bureau of Indian Standards, New Delhi
11. IS 1893(part 1) (2002) Criteria for earthquake resistant design of structure. Bureau of Indian standards, New Delhi
12. IS 456 (2000) Plain and reinforced concrete-code of practice. Bureau of Indian standards, New Delhi

Investigations on Compression Behaviour of Short Reinforced SCC Columns

J. Sanjith, R. Prabhakara, M. S. Sudarshan, and H. K. Thejas

Abstract The objective of this work is to predict the values of deformation and load at cracking point, yielding point and ultimate point of short reinforced self-Compacting Concrete columns which was subjected to axially compression in loading frame. An ANN tool by giving proper inputs like fresh properties of materials, spacing of stirrups and percentage of longitudinal reinforcement and keeping target values obtained from experiments, it is compared with the experimental values accompanied by marginal errors.

Keywords SCC · Short columns · ANN

1 Introduction

"A concrete that is able to flow and fill every part and corner of the formwork, even in the presence of dense reinforcement, purely by means of its own weight and without the need for any vibration or other type of compaction" [1].

SCC was developed in Japan to offset a growing shortage of skilled labour. In the present-day, self-compacting concrete can be classified as an advanced construction material. Self-compacting concrete has been described as "the most revolutionary development in concrete construction for several decades" [1].

An important improvement of health and safety is also achieved through elimination of handling of vibrators and a substantial reduction of environmental noise loading on and around a site [2]. The composition of SCC mixes includes substantial

J. Sanjith
Department of Civil Engineering, Adichunchanagiri Institute of Technology, Chikmagalur, India

R. Prabhakara
Brindavan College of Engineering, Bengaluru, India

M. S. Sudarshan
Stedrant Technoclinic PVT. LTD., Bengaluru, India

H. K. Thejas (✉)
School of Engineering and Technology, CHRIST (Deemed to be University), Bengaluru, India
e-mail: thejas.hk@christuniversity.in

L. Nandagiri et al. (eds.), *Sustainability Trends and Challenges in Civil Engineering*,
Lecture Notes in Civil Engineering 162, https://doi.org/10.1007/978-981-16-2826-9_12

proportions of finely-grained inorganic materials. This offers possibilities for utilization of waste products which have no practical applications and are costly to dispose [3]. The SCC, as the name suggests, does not require to be vibrated to achieve full compaction. This offers many benefits and advantages over conventional concrete. These include an improved quality of concrete and reduction of on-site repairs, faster construction times, lower overall cost and facilitation of introduction of automation into concrete construction [4].

It is a new kind of high-performance concrete with distinctive deformability. This concrete has three abilities like segregation resistance flowing, passing ability and filling ability. Hence, it flows without segregation and bleeding and has ability to cover the voids in reinforcement without any blockage. The important thing is to understand the fresh properties of cement paste in order to evaluate the flow behaviour of SCC. It is the advanced technique where any mechanical compaction is not necessary as like conventional concrete. SCC in fresh state is stable to have homogeneity in the concrete property [5]. It also avoids some site properties like blockage, settlement or segregation and bleeding.

ANN is computer model which minimizes the learning necessity of human brain. It helps to learn the software from generalizing of previous example and from experience. It provides answers even if there is error and incomplete inputs when we are trying to analyse the complex problems [6]. It contains developing of a network for the required material behaviour and training them to get the results. If there is an exact information about the material behaviour, the neural network will provide the sufficient information about the material [7].

2 Artificial Neural Network

An artificial neural network (ANN) is demonstrated in the brain where neurons are associated in complex patterns to process data from the senses, create memories and control the body. It is based entirely on the operational fundamentals of biological neural networks. The artificial neural network is defined as the system of interconnected neurons. Artificial neural network is also called as connectionist systems. We can store information in computer as well. Both brain and computer can store information, but their mechanisms are very different. These are the computing system ambiguously inspired by the biological neural networks that institute animal brains [7]. If we memorize something, our brain stores that information. In computer, the information is stored at specific location, whereas brain alters the association of neurons to store information. Researchers are designing ANN to solve the problems of pattern recognition, prediction, optimization, association memory and control. The neuron itself has no storage capability. So, it transmits signals from one neuron to other neurons [8]. Warren McCulloch and Walter Pitts opened this topic by creating a computational model for neural networks. For this phenomenon, a computer programme is developed using ANN design toolbox in the software called MATLAB (v. R2018a). To form interconnected neurons, ANN has been described as the second bestway to connect and work on it. When we describe about our brains, the neural network imitates the mechanism of the brain. Brain is composed of connections of neurons, while neural network is composed of connections of nodes. These neurons are used to model brains and also to perform definite computational tasks [9]. We can call the brain a gigantic network of neurons and association of neurons forms specific information. Successful ANN application will have the ability of determining the perfect character recognition. This present study is an effort to apply artificial neural network-based system identification of methods to predict the shear behaviour based on the SCC mix proportions.

3 Materials and Methodology

The cement is used for the binding of fine and coarse aggregates. 53 grade OPC cement as per IS 12269-1987 with specific gravity 3.15. Fly ash confirming to class-F IS-3812:2003. Coarse aggregate of 20 mm downsizes aggregate of specific gravity 2.62 are used for SCC. The coarse aggregate must be confirming to IS 383:2016. A fine aggregate is M-Sand of specific gravity of 2.57 and fineness modulus 3.05 confirming to zone II and for IS383:2016 are used (Table 1).

Table 1 Mix proportions, fresh properties and cube compression strength of SCC [20], SCC [30] and SCC [40]

Materials	Proportion		
	M-20	M-30	M-40
Cement (kg/m^3)	257	314.28	365
Fly ash (kg/m^3)	173	188.43	198
Fine aggregates (kg/m^3)	970	925.63	964
Coarse aggregates (kg/m^3)	815	743.69	781
Water (l/m^3)	197	148.07	182
Super plasticizer (%)	0.7	1.3	1.9
VMA (%)	0.15	0.18	0.21
Slump cone (mm)	820	680	520
T_{50} sec	2.8	3.2	3.6
J-ring	3.3	4	4.7
V- funnel sec	8	11	14
L-box	0.62	0.85	1.13
U-box (mm)	7	9	11
Cube compressive strength (MPa)	29.93	38.93	47.64

Test Specimen

For the investigation purpose, 36 [12 for each grade] columns of SCC specimens were casted for each M-20, M-30 and M-40.The cross-section of 125 × 125 × 1000 mm, respectively was selected, and the slenderness ratio was kept 8, which shows the behaviour of short column.

Confinement reinforcement was kept same for all columns. Lateral ties provided are 8 mm at a spacing of 100, 200 and 300 mm C/C respectively. The percentage of longitudinal steel was varied from 1.29, 2.01, 2.89 and 5.15% (Table 2).

Table 2 Specimen details of columns

Specimen	Longitudinal reinforcement			Transverse.rmt
	No. of bars	Dia. of bar (mm)	RMT (%)	
SCC/M20/1.29/100	4	8	1.29	100
SCC/M20/2.01/100		10	2.01	100
SCC/M20/2.89/100		12	2.89	100
SCC/M20/5.15/100		16	5.15	100
SCC/M20/1.29/200		8	1.29	200
SCC/M20/2.01/200		10	2.01	200
SCC/M20/2.89/200		12	2.89	200
SCC/M20/5.15/200		16	5.15	200
SCC/M20/1.29/300		8	1.29	300
SCC/M20/2.01/300		10	2.01	300
SCC/M20/2.89/300		12	2.89	300
SCC/M20/5.15/300		16	5.15	300
SCC/M30/1.29/100		8	1.29	100
SCC/M30/2.01/100		10	2.01	100
SCC/M30/2.89/100		12	2.89	100
SCC/M30/5.15/100		16	5.15	100
SCC/M30/1.29/200		8	1.29	200
SCC/M30/2.01/200		10	2.01	200
SCC/M30/2.89/200		12	2.89	200
SCC/M30/5.15/200		16	5.15	200
SCC/M30/1.29/300		8	1.29	300
SCC/M30/2.01/300		10	2.01	300
SCC/M30/2.89/300		12	2.89	300
SCC/M30/5.15/300		16	5.15	300
SCC/M40/1.29/100		8	1.29	100
SCC/M40/2.01/100		10	2.01	100
SCC/M40/2.89/100		12	2.89	100

(continued)

Table 2 (continued)

Specimen	Longitudinal reinforcement			Transverse.rmt
	No. of bars	Dia. of bar (mm)	RMT (%)	
SCC/M40/5.15/100		16	5.15	100
SCC/M40/1.29/200		8	1.29	200
SCC/M40/2.01/200		10	2.01	200
SCC/M40/2.89/200		12	2.89	200
SCC/M40/5.15/200		16	5.15	200
SCC/M40/1.29/300		8	1.29	300
SCC/M40/2.01/300		10	2.01	300
SCC/M40/2.89/300		12	2.89	300
SCC/M40/5.15/300		16	5.15	300

Inputs for ANN

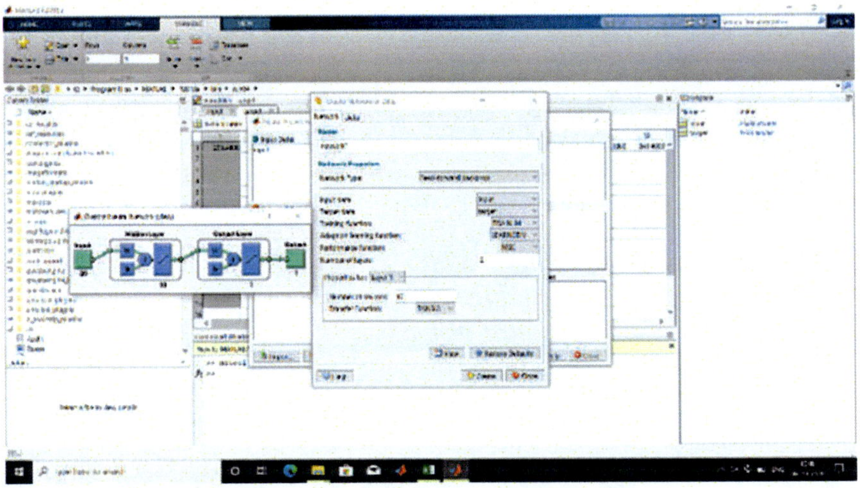

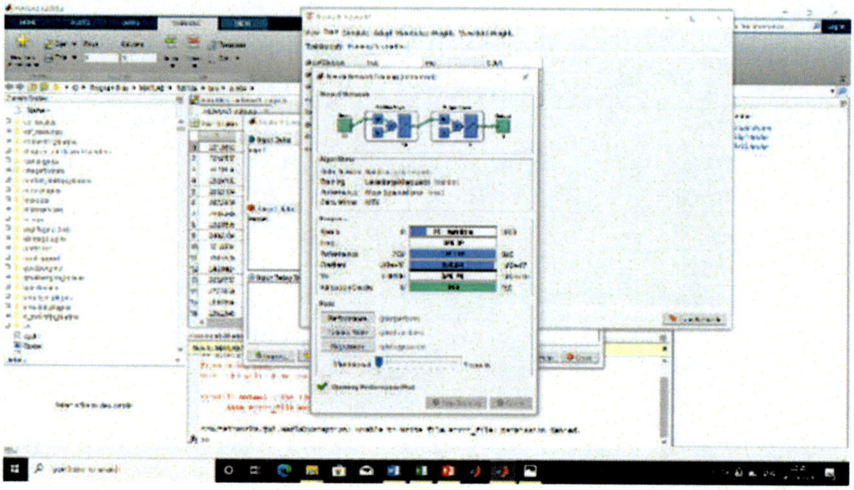

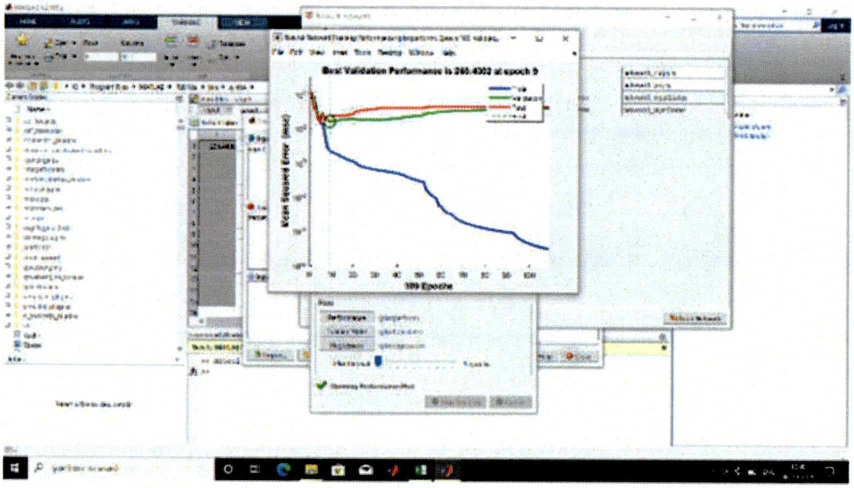
Mean Squared Error (mse)

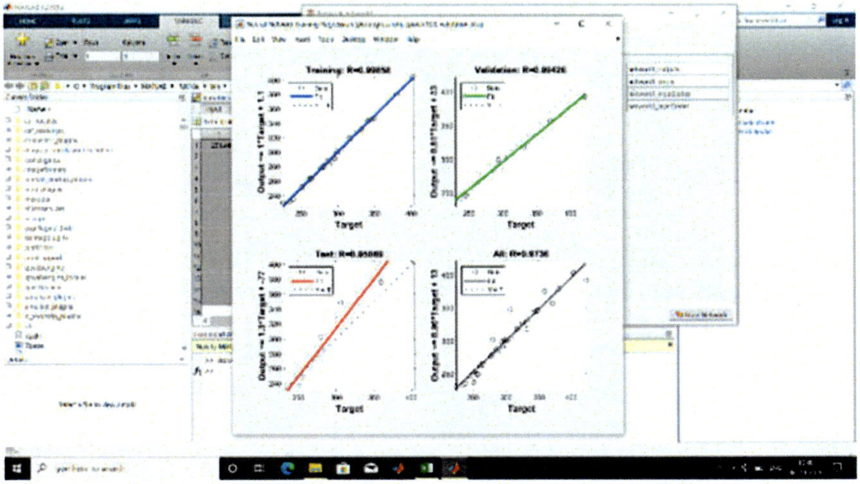

See Tables 3, 4 and 5.

Table 3 Input data

Cement	FA	CA	W	Slump
257	970	815	197	820
314.28	925.63	743.69	148.07	680
365	964	751	182	520
257	970	815	197	820
314.28	925.63	743.69	148.07	680
365	964	751	182	520
257	970	815	197	820
314.28	925.63	743.69	148.07	680
365	964	751	182	520
257	970	815	197	820
314.28	925.63	743.69	148.07	680
365	964	751	182	520
257	970	815	197	820
314.28	925.63	743.69	148.07	680
365	964	751	182	520
257	970	815	197	820
314.28	925.63	743.69	148.07	680
365	964	751	182	520
257	970	815	197	820
314.28	925.63	743.69	148.07	680

(continued)

Table 3 (continued)

Cement	FA	CA	W	Slump
365	964	751	182	520
257	970	815	197	820
314.28	925.63	743.69	148.07	680
365	964	751	182	520
257	970	815	197	820
314.28	925.63	743.69	148.07	680
365	964	751	182	520
257	970	815	197	820
314.28	925.63	743.69	148.07	680
365	964	751	182	520
257	970	815	197	820
314.28	925.63	743.69	148.07	680
365	964	751	182	520
257	970	815	197	820
314.28	925.63	743.69	148.07	680
365	964	751	182	520

Fck	Spacing	Ast	Flyash	Sup.plr
29.93	100	1.29	173	0.7
38.93	100	1.29	188.43	1.3
47.64	100	1.29	198	1.9
29.93	200	1.29	173	0.7
38.93	200	1.29	188.43	1.3
47.64	200	1.29	198	1.9
29.93	300	1.29	173	0.7
38.93	300	1.29	188.43	1.3
47.64	300	1.29	198	1.9
29.93	100	2.01	173	0.7
38.93	100	2.01	188.43	1.3
47.64	100	2.01	198	1.9
29.93	200	2.01	173	0.7
38.93	200	2.01	188.43	1.3
47.64	200	2.01	198	1.9
29.93	300	2.01	173	0.7
38.93	300	2.01	188.43	1.3
47.64	300	2.01	198	1.9
29.93	100	2.89	173	0.7
38.93	100	2.89	188.43	1.3

(continued)

Table 3 (continued)

Fck	Spacing	Ast	Flyash	Sup.plr
47.64	100	2.89	198	1.9
29.93	200	2.89	173	0.7
38.93	200	2.89	188.43	1.3
47.64	200	2.89	198	1.9
29.93	300	2.89	173	0.7
38.93	300	2.89	188.43	1.3
47.64	300	2.89	198	1.9
29.93	100	5.15	173	0.7
38.93	100	5.15	188.43	1.3
47.64	100	5.15	198	1.9
29.93	200	5.15	173	0.7
38.93	200	5.15	188.43	1.3
47.64	200	5.15	198	1.9
29.93	300	5.15	173	0.7
38.93	300	5.15	188.43	1.3
47.64	300	5.15	198	1.9

VMA	T50.se	J.Ring	V.Funn	L.bo	U.Box
0.15	2.8	3.3	8	0.62	7
0.18	3.2	4	11	0.85	9
0.21	3.6	4.7	14	1.13	11
0.15	2.8	3.3	8	0.62	7
0.18	3.2	4	11	0.85	9
0.21	3.6	4.7	14	1.13	11
0.15	2.8	3.3	8	0.62	7
0.18	3.2	4	11	0.85	9
0.21	3.6	4.7	14	1.13	11
0.15	2.8	3.3	8	0.62	7
0.18	3.2	4	11	0.85	9
0.21	3.6	4.7	14	1.13	11
0.15	2.8	3.3	8	0.62	7
0.18	3.2	4	11	0.85	9
0.21	3.6	4.7	14	1.13	11
0.15	2.8	3.3	8	0.62	7
0.15	2.8	3.3	8	0.62	7
0.18	3.2	4	11	0.85	9
0.21	3.6	4.7	14	1.13	11
0.15	2.8	3.3	8	0.62	7
0.18	3.2	4	11	0.85	9

(continued)

Table 3 (continued)

VMA	T50.se	J.Ring	V.Funn	L.bo	U.Box
0.21	3.6	4.7	14	1.13	11
0.15	2.8	3.3	8	0.62	7
0.18	3.2	4	11	0.85	9
0.21	3.6	4.7	14	1.13	11
0.15	2.8	3.3	8	0.62	7
0.18	3.2	4	11	0.85	9
0.21	3.6	4.7	14	1.13	11
0.15	2.8	3.3	8	0.62	7
0.18	3.2	4	11	0.85	9
0.21	3.6	4.7	14	1.13	11
0.15	2.8	3.3	8	0.62	7
0.18	3.2	4	11	0.85	9
0.21	3.6	4.7	14	1.13	11
0.15	2.8	3.3	8	0.62	7

Table 4 Target data

Pcr	Δ_{cr}	P_y	d_y	P_u	Δu
295.44	4.38	492.40	10.45	344.68	5.7
307.22	4.64	512.04	9.81	358.43	6.1
332.64	5.41	554.40	9.42	388.08	6.4
251.22	3.84	418.70	10.70	293.09	5.2
263.85	4.21	439.75	10.10	307.83	5.8
285.48	4.98	475.80	9.67	333.06	6.4
224.76	6.13	374.60	11.54	262.22	7.3
233.70	5.64	389.50	10.86	272.65	6.56
251.10	6.18	418.50	10.40	292.95	7.08
297.84	4.15	496.40	10.21	347.48	5.1
315.18	4.66	525.30	9.12	367.71	5.3
341.40	5.08	569.00	8.70	398.30	5.58
264.07	4.11	440.12	10.50	308.08	4.2
276.84	4.55	461.40	9.42	322.98	4.9
292.75	4.67	487.92	8.94	341.54	4.7
239.04	5.34	398.40	11.12	278.88	6.58
251.10	4.91	418.50	10.50	292.95	5.85
260.70	5.16	434.50	9.22	304.15	6.12

(continued)

Table 4 (continued)

Pcr	Δ_{cr}	P_y	d_y	P_u	Δu
331.29	3.94	552.15	9.42	386.51	4.54
348.84	5.17	581.40	8.70	406.98	5.72
370.74	4.00	617.90	6.60	432.53	4.54
288.18	5.01	480.30	9.79	336.21	5.8
301.74	5.40	502.90	9.10	352.03	5.9
318.24	4.08	530.40	6.89	371.28	4.66
245.34	5.55	408.90	10.80	286.23	6.46
256.83	5.89	428.05	10.20	299.64	6.6
280.26	5.70	467.10	8.80	326.97	6.62
380.76	4.50	634.60	7.60	444.22	4.8
403.31	4.55	672.18	6.90	470.53	5.12
422.44	3.00	704.07	5.74	492.85	3.39
326.16	4.34	543.60	8.40	380.52	5.11
345.42	4.38	575.70	7.30	402.99	4.86
359.25	4.30	598.75	6.20	419.13	4.77
279.65	4.65	466.08	9.10	326.26	5.93
299.22	4.12	498.70	7.90	349.09	4.99

Table 5 Results

MP_{cr}	$M\Delta_{cr}$	MP_y	MΔy	Mp_u	$M\Delta u$
297.0968	4.6876	495.27	10.6283	346.8203	5.5899
312.0551	4.8667	520.2036	9.8274	364.2677	5.6775
330.2234	4.7493	550.4926	8.8048	385.4674	5.4731
254.633	5.0976	424.5	11.2383	297.2803	6.1199
269.5913	5.2767	449.4336	10.4374	314.7277	6.2075
287.7597	5.1593	479.7226	9.4148	335.9274	6.0031
212.1693	5.5076	353.73	11.8483	247.7403	6.6499
227.1276	5.6867	378.6636	11.0474	265.1877	6.7375
245.2959	5.5693	408.9526	10.0248	286.3874	6.5331
311.394	4.5458	519.0985	10.0714	363.5006	5.3724
326.3523	4.7248	544.0322	9.2705	380.948	5.46
344.5207	4.6074	574.3212	8.2479	402.1477	5.2556
268.9302	4.9558	448.3285	10.6814	313.9606	5.9024

(continued)

Table 5 (continued)

MP_{cr}	$M\Delta_{cr}$	MP_y	$M\Delta y$	Mp_u	$M\Delta u$
283.8886	5.1348	473.2622	9.8805	331.408	5.99
302.0569	5.0174	503.5512	8.8579	352.6077	5.7856
226.4665	5.3658	377.5585	11.2914	264.4206	6.4324
241.4248	5.5448	402.4922	10.4905	281.868	6.52
259.5932	5.4274	432.7812	9.4679	303.0677	6.3156
328.8684	4.3724	548.2223	9.3907	383.8877	5.1066
343.8267	4.5515	573.156	8.5898	401.335	5.1941
361.9951	4.4341	603.4449	7.5672	422.5347	4.9897
286.4047	4.7824	477.4523	10.0007	334.3477	5.6366
301.363	4.9615	502.386	9.1998	351.795	5.7241
319.5313	4.8441	532.6749	8.1772	372.9947	5.5197
243.9409	5.1924	406.6823	10.6107	284.8077	6.1666
258.8992	5.3715	431.616	9.8098	302.255	6.2541
277.0676	5.2541	461.9049	8.7872	323.4547	6.0497
373.7459	3.9272	623.0174	7.6426	436.2453	4.4238
388.7042	4.1063	647.9511	6.8417	453.6927	4.5114
406.8725	3.9889	678.2401	5.8191	474.8924	4.307
331.2821	4.3372	552.2474	8.2526	386.7053	4.9538
346.2404	4.5163	577.1811	7.4517	404.1527	5.0414
364.4088	4.3989	607.4701	6.4291	425.3524	4.837
288.8184	4.7472	481.4774	8.8626	337.1653	5.4838
303.7767	4.9263	506.4111	8.0617	354.6127	5.5714
321.945	4.8089	536.7001	7.0391	375.8124	5.367

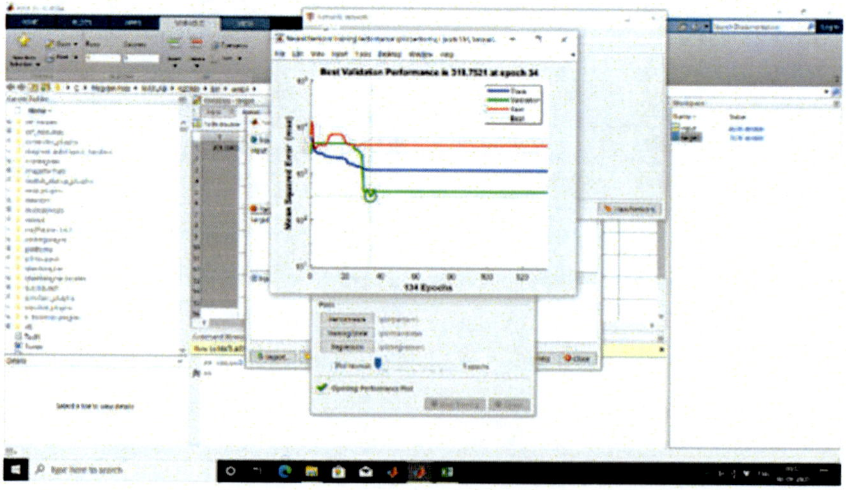

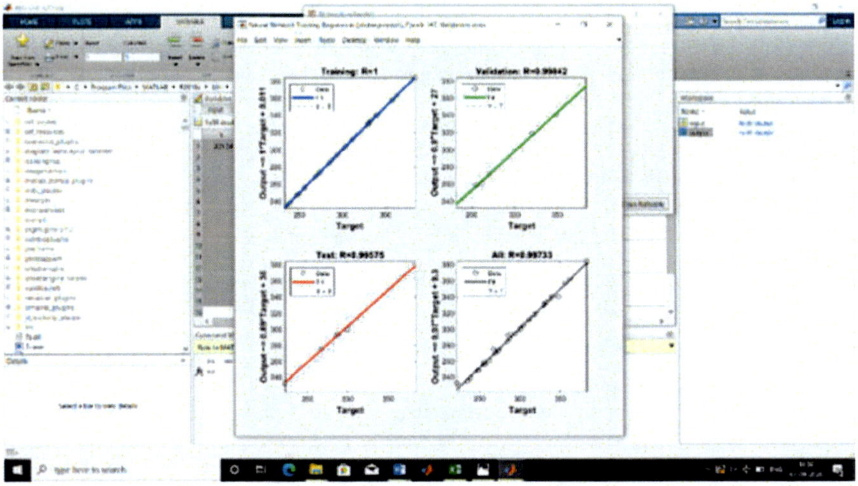

4 Summary and Conclusions

Finally, we conclude that ANN is helpful to predict the fresh properties of SCC model. Input parameters were mix composition, cement content, water, fine aggregates, coarse aggregate, super plasticizer and fresh properties were taken for input parameters, and load deformation values were considered as a target value.

Percentage of error for the predicted value is than 1. ANN model was utilized to predict the shear behaviour of RC column. The models developed by the ANN showed the goodness in fit in the results.

The *R*-square value of the model of flow diameter of slump by use of ANN is greater than 0.9, and the regression model values is 0.792.

The self-compacting concrete was applied with feed forward back-propagation neural network. ANN includes four layers: the input layer will be containing seven neurons, 1st and 2nd hidden layer will be having five neurons each and the output layer will be having only one neuron.

Acknowledgements We acknowledge the support extracted by Management of AIT, Principal, HOD, Faculty and Staff of Civil Dept. of AIT, affiliated to VTU Belagavi.

References

1. Yeh IC (2008) Prediction of workability of concrete using design of experiments for mixtures. Comput Concr 5(1):1–20

2. Yeh IC (2008) Modeling slump of concrete with fly ash and superplasticizer. Comput Concr 5(6):559–572
3. Okamura H, Ouchi M (2003) Self-compacting concrete. J Adv Concr Tech 1(1):5–15
4. Pala M, Ozbay E, Oztas A, Yuce M (2007) Appraisal of long-term effects of fly ash and silica fume on compressive strength of concrete by neural networks. Constr Build Mater 21:384–394
5. Saka MP, Sanad A (2001) Prediction of ultimate shear strength of reinforced concrete deep beam using neural network. J Struct Eng 127:818–828
6. Sonebi M, Cevik A (2009) Genetic programming based formulation for fresh and hardened properties of self-compacting concrete containing pulverized fuel ash. Constr Build Mater 23:2614–2622
7. Cussigh F, Sonebi M, De Schutter G (2003) Project testing SCC segregation test methods. In: Wallevik O, Nielson I (eds) Self compacting concrete, third international RILEM symposium. RILEM Publications, pp 311–322
8. Yeh IC (1998) Modeling of strength of high-performance concrete using artificial neural networks. Cem Concr Res 28(12):1797–1808
9. Lippman RP (1988) An introduction to computing with neural nets. In: Artificial neural networks. Computer Society Theoretical Concepts, Washington, pp 36–54

Performance Evaluation of Bituminous Pavement with High Recycled Asphalt Pavement Material content—A Case Study

P. B. Gnanamurthy and B. V. Kiran Kumar

Abstract In India to meet the requirement of growing travel demand and to help rapid growth of economic activity, infrastructure development is taken up by central and state governments on large scale. Majority of the road network in India has flexible pavement constructed using quarry aggregates and bitumen. Because of this process, there has been faster depletion of road aggregates and also an increase in emission of carbon gases causing environmental pollution. The use of eco-friendly bitumen, reuse of aggregates obtained from milling, and material reclaimed from old pavement to construct or rehabilitate roads will be the most promising alternative to the current practice toward achieving sustainability in road construction. In this regard, present study aims at formulating design methodology through extensive field and laboratory studies for the construction of surface course containing high recycled asphalt pavement content material and waste plastic. A test track of 25 mm thick close graded premix carpet (CGPC) containing virgin aggregates by 20%, recycle asphalt pavement aggregates (RAP) 80%, and waste plastic 8% by weight of bitumen was laid and subjected to performance evaluation under existing road and traffic condition. Based on the laboratory and field test results, it was found that bituminous mix containing high recycled asphalt pavement material content (80% recycled aggregates) with 8% of waste plastic by weight of bitumen as recycling agent performs better and found economical than the stretch with conventional hot mix asphalt (HMA) mix. Performance evaluation was carried out in two cycles one before the monsoon in May and the second post-monsoon season in December. From the evaluation study, it was found that there was no considerable surface distress found on the test track both during cycle-1 and cycle-2. Roughness test results imply that bituminous pavement surface with RAP and waste plastic provides comparatively better riding quality and comfort than the conventional bituminous mix. Based on the findings of the study, it may be inferred that it is attainable to delineate acceptable quality bituminous mix with high RAP content that meets needed properties and performance criteria as per standards and specifications. From the present study, it is also clear that recycling technology is an effective and economical alternative for

P. B. Gnanamurthy (✉) · B. V. Kiran Kumar
Department of Civil Engineering, Government S.K.S.J.Technological Institute, Bengaluru, India

L. Nandagiri et al. (eds.), *Sustainability Trends and Challenges in Civil Engineering*,
Lecture Notes in Civil Engineering 162, https://doi.org/10.1007/978-981-16-2826-9_13

the construction of low or medium-traffic intensity roads in India which constitute the major portion of the Indian road network.

Keywords RAP · Performance evaluation · Waste plastic · Low volume roads · Close graded premix surfacing

Abbreviations and Acronyms

BBD	Benkelman Beam Deflection
CGPC	Close Graded Premix Carpet
HMA	Hot Mix Asphalt
IRC-SP	Indian Road Congress-Special Publication
KSHIP	Karnataka State Highways Improvement Project
MDR	Major District Road
MERLIN	Machine for Evaluating Roughness using Low-cost Instrumentation
MORTH	Ministry of Road Transport and Highways
NH	National Highway
PCI	Pavement Condition Index
PIU	Project Implementation Unit
PRE	Panchayat Raj Engineering
RAP	Reclaimed Asphalt Pavement
SH	State Highway
SR	Schedule of Rates
ToR	Terms of Reference.
TRRL	Transport and Road Research Laboratory
VDF	Vehicle Damaging Factor
VG	Viscosity Grade
WEP	World Bank Assisted Engineering Procurement and Construction Project

1 Introduction

The development of road infrastructure is currently being given high priority by the central and state governments in India to meet the requirement for increased travel demand and to help the faster growth of economic activity. Most of the roads in India were constructed using quarry aggregates and bitumen, a petroleum product mixed at high temperature to produce hot mix asphalt (HMA). Due to the use of virgin aggregates in the construction of roads, it is observed that there is a faster depletion of road aggregates and also an increase in emission of carbon gases into the atmosphere causing environmental pollution [1]. In the construction industry today, the two main factors affecting the use of recycled, locally available, marginal materials, and other low-cost technologies are economic savings and environmental benefits in asphalt pavement [2]. Low emission bitumen (green bitumen materials), aggregates obtained from milling, and reuse of material collected from old existing pavement to construct or rehabilitate highways will be the most promising alternative to achieve sustainability in highway construction. Bituminous mixes with RAP have proven not only economical and environmentally friendly but also conserving the natural resources and exhibiting structural performance similar to or even better than conventional HMA [3]. Comparatively, the use of recycled material in different layers of flexible pavements proven economical, as there was a reduction in thickness compared with the conventional HMA mix [4]. Transportation professionals and practicing engineers are looking for methods that optimize the value of in-place materials which minimizes traffic congestion and the environmental impact of paving operations [1]. Also, the increase in the volume of heavy goods vehicles annually to cater to the needs of various sectors has led to numerous distresses within the pavement. The aging of bitumen binder is another crucial drawback inflicting deterioration of pavements. A common maintenance measure practiced across the globe for flexible pavements is overlaying the distressed pavements with virgin courses. However, this results in the thickening of pavement layers, exhaustion of natural resources, and discharge of harmful gases. A better alternative for the rehabilitation of bituminous pavement which also replaces overlaying of the bituminous surface is by recycling of existing surface course and reconstruction [3]. In the recycling process, the material from deteriorated pavement known as reclaimed asphalt pavement (RAP) is used to ensure desirable pavement performance of recycled mixtures. To compensate for the effect of aged, and stiff binder on recycled aggregates use of rejuvenating or softening additives such as plastic, polymer rubber, softer virgin binder grade are to be considered [5]. Several studies carried out so far to investigate the performance of HMA which incorporates RAP and rejuvenators in suitable proportion have shown constructive results compared to conventional HMA mix [6]. Significant factors governing the construction cost of the pavements are the cost of aggregates and the cost of binder materials. Replacing a part of the virgin mix with recycled aggregate has proven as an optimal solution especially for low-volume roads [7].

2 Study Outline

2.1 Objective and Research Approach

The study aims at evaluating the performance of newly overlaid close graded premix carpet (type-B) surface course containing virgin aggregate (20% by weight of mix), recycled asphalt pavement material (80% by weight of mix), and waste plastic (8% by weight of bitumen) on a distressed Major District Road (MDR) in the state of

Karnataka under low to medium traffic flow condition. The objectives of the study are summarized below:

- Characterization of construction materials through laboratory investigation.
- Construction of performance track to validate the suitability of high RAP content bituminous mix under prevailing road and traffic conditions.
- Performance evaluation of test track through various field studies.

The present study is a research work sponsored by the Government of Karnataka to promote the use of locally available material and recycled material in the construction of low and medium-traffic intensity roads. In this connection, a test track is laid on the stretch identified by the Karnataka State Highway Improvement Project (KSHIP) authority for a length of 500 m. Since the present traffic was 200–250 vehicles per day, close graded premix carpet (CGPC) of 25 millimeter (mm) thick was considered for overlay. The test track constructed using RAP and waste plastic material is evaluated to ascertain the performance of the mix in two cycles one before monsoon and the other post-monsoon season as per the contract agreement. A brief outline of the performance studies conducted at the field is road inventory and pavement condition survey, pavement roughness evaluation by Machine for Evaluating Roughness using Low-cost Instrumentation (MERLIN), structural evaluation by Benkelman Beam Deflection (BBD) study, and axle load survey.

2.2 *Materials and Methods*

(1) *Aggregate*: Virgin aggregates used in the present study were procured from the quarry near Hosakote town, Bangalore. Assessment of physical properties of the aggregates done through laboratory investigations and test results is presented in Table 1.

Table 1 Physical properties of virgin aggregate

Property	Test method	Virgin aggregates test results	Specification as per MoRTH table-500-8
Combined flakiness and elongation index (%)	IS-2386 part-1	22.50	35% maximum
Water absorption (%)	IS-2386 part-3	0.18	1% maximum
Aggregate impact value (%)	IS-2386 part-4	13.14	27% maximum
Aggregate crushing value (%)	IS-2386 part-4	21.50	30% maximum (as per IS:383-1970)
Aggregate abrasion value (%)	IS-2386 part-4	14.15	35% maximum
Specific gravity	IS-2386 part-3	2.59	2.5–2.9

Fig. 1 Equipment for processing RAP

(2) *RAP Material*: RAP material was obtained from the reclaimed material dump yard near Hosakote–Gownipalli via Chintamani road (State Highway-82), Karnataka. Obtained RAP material (Figs. 2 and 3) processed mechanically with the help of equipment (Fig. 1) developed by M/s KK Waste Plastic Management Pvt. Ltd. Bengaluru before used in the construction process. Processing of reclaimed aggregates includes removal of dust and deleterious material from the surface of RAP aggregates. Reclaimed material obtained from the field is to be placed in the pour bin from where it will be taken to the brush chamber, where aggregates are cleaned with the help of a mechanical brush by churning the surface of the aggregates. Properties of RAP aggregate play a crucial role in the design of HMA mix, hence, properties such as particle size distribution and bitumen content of RAP material were assessed to check the suitability of physical and engineering properties of the RAP material. The bitumen extraction test was conducted as per IRC: SP 11-1988 (Appendix-5), and the results obtained are tabulated in Table 2. Aggregate gradation and physical properties of the RAP material are demonstrated in Tables 3 and 4.

(3) *Virgin binder*: VG-30 Bitumen was used to produce the desired bituminous mix. The properties of the virgin binder and its requirements are shown in Table 5.

(4) *Waste plastic*: waste plastic for the present study was provided by the M/s KK Waste Plastic Management Pvt. Ltd. Bengaluru in shredded form, conforming to IRC SP-98:2013. 8% of shredded plastic by weight of the bitumen was blended with the VG-30 virgin bitumen during mix design at the lab and the same quantum was added in situ during the construction of the test track with the binder.

(5) *Mix design*: The main aim of the mix design is to optimize material requirements and to bring out a mix that provides better performance, and the designed mix must meet the required volumetric properties [8]. The material requirement of the CGPC mix was reckoned following the Ministry of Road Transport and Highways (MoRTH) section:508 specification. Laboratory mix designs

Fig. 2 RAP Material obtained from the site

Fig. 3 Sieved RAP material

Table 2 Bitumen extraction test results

Description	Sample 1	Sample 2	Sample 3
Weight of RAP material in grams (A)	500	500	500
Weight of RAP material after extraction in grams (B)	484	479	480
Percentage of Bitumen contents in RAP material = $\frac{(A-B)}{B} \times 100$	3.3	4.3	4.1
Average Bitumen contents in RAP material	3.9%		

provide different proportions of mineral aggregate combinations in terms of individual sieve sizes. For practical on-site operational purposes, blend two or more sizes of aggregate (each size contains a range of individual sieve sizes). This mix ratio is obtained on a weight basis and indicates the weight percent of coarse aggregate, fine aggregate, and filler needed to give the final aggregate gradation (MoRTH Section:511.2.3). The combined gradation of materials

Table 3 RAP aggregate gradation

Sieve size in millimeters	% Passing by weight	
	Source 1 (10–6 mm)	Source 2 (less than 6 mm/dust)
13.2	100	100
11.2	96.5	100
5.6	1	82.5
2.8	0.2	42.3
0.09	0.2	5.2

Table 4 Physical properties of RAP

Property	Test method	RAP aggregates test results	Specification as per MoRTH table-500-8
Water absorption (%)	IS-2386 part-3	0.15	1% maximum
Aggregate impact value (%)	IS-2386 part-4	12.10	27% maximum
Aggregate crushing value (%)	IS-2386 part-4	24.50	30% maximum
Specific gravity source 1 (10–6 mm) source 2 (less than 6 mm/dust)	–	2.48	
2.30	2.5–3.0		

Table 5 Properties of virgin binder

Property	Test method	Virgin binder (VG-30)	Permissible limit as per Table 1 of IS 73:2013 (min)
Specific gravity	IS-1202	0.995	0.99
Penetration at 25 °C, 5 s	IS-1203	64	45
Softening point (°C)	IS-1205	48.00	47
Viscosity at 135 °C in centistokes (cSt)	IS-1206 (Part 3)	380	350
Ductility at 25 °C	IS-1208	96.00	40
Flashpoint (°C)	IS: 1448 [P:69]	276.00	220

considered for the design is demonstrated in Table 6. In the present study, the road considered for the laying of test track connects two higher category roads, i.e., State Highway (SH) 82 on the one end and National Highway (NH) 75 on the other end. Because of this major road connectivity, this stretch was experiencing a traffic volume flow of 200 to 250 vehicles per day. Since the existing traffic falls in the range of low to medium traffic category, a surface course which is economical and suitable to cater to the present traffic needs is to be

Table 6 Job mix formula

Sieve size in millimeter	RAP				Virgin aggregate		Combined % passing ($A+B+C$)	Required gradation for as per MoRTH
	Source 1 (10–6 mm)		Source 2 (less than 6 mm/dust)		Less than 6 mm/dust			
	% passing by weight	60% (A)	% passing by weight	20% (B)	% passing by weight	20% (C)		
13.2	100	60	100	20	100	20	100.00	100
11.2	96.5	57.9	100	20	97.5	19.5	97.40	88–100
5.6	1	0.6	82.5	16.5	95.5	19.1	36.20	31–52
2.8	0.2	0.12	42.3	8.46	60	12	20.58	5–25
0.09	0.2	0.12	5.2	1.04	10.6	2.12	3.28	0–5

provided. As the annual rainfall of the study area is less than 1600 mm, CGPC type-B surfacing overlay of 25 mm (mm) thick was considered for the design of the surface course following MoRTH specifications.

(6) *Binder content calculation*: As per MoRTH specification,

- Total quantity of binder required per 10 m^2 area = 19 kilograms (kg).
- Total quantity of aggregates used for 10 m^2 area = 0.27 m^3.
- Bulk density of combined aggregate from the test = 1350 kg/m^3.
- Weight of aggregates in kilograms (kg) = 0.27 × 1350 = 364.5 kg.
- Weight of total mix (Bitumen + Aggregate) in kg = 364.5 + 19 = 383.5 kg.
- Required binder content in percentage by weight of mix = $\frac{\text{weight of bitumen}}{\text{weight of mix}} \times 100$

$$= \frac{19}{383.5} \times 100 = 4.95\%$$

(7) *Test track*: The Government of Karnataka through KSHIP an initiative of Karnataka Public Works, Port and Inland Water Transport Department is implementing up-gradation of State Highways in the State *of Karnataka.* The up-gradation includes widening and strengthening of existing highways from single/intermittent lane to two-lane carriageway with the paved and unpaved shoulder. The KSHIP under its environmental policy is committed to protecting and conserving natural resources by innovative technologies. In this regard, reclaimed asphalt from the World Bank assisted Engineering procurement construction Project (WEP1) contract package has been recycled and used on Chikkanahalli road at Shivanapura cross for a length of 500 m in Hoskote taluk Bangalore rural district, Karnataka. Figure 4 displays the location of the test track (Fig. 4). Performance evaluation of the above-mentioned newly paved road (Fig. 6) using the reclaimed asphaltic material was carried out under prevailing traffic and climatic conditions to validate the suitability of reclaimed material in the construction of surface course for medium and low-volume roads. The road selected for laying of test track is a Major District Road

Fig. 4 Location map of test track

(MDR) that comes under the supervision of Karnataka Public Works, Port and Inland Water Transport Department and designated as chikkanahalli road. The road connects two major roads on either end. It provides connectivity to SH-82 (Hosakote-Gownipalli Road via Chintamani) at shivanapura cross (chainage: 16 + 00 km) in the west and NH-75 (Bangalore-Kolar road) at yelachahalli (chainage:0 + 00 km) in the east. The total length of the road is 16 km out of which 500 m, i.e., from chainage: 14 + 500 km to chainage:1 5 + 000 km is considered for laying test track to carry out performance evaluation.

(8) *Methods*: In the present study, four field tests (Fig. 7) were chosen to evaluate the performance of the test track. Pavement roughness evaluation was carried

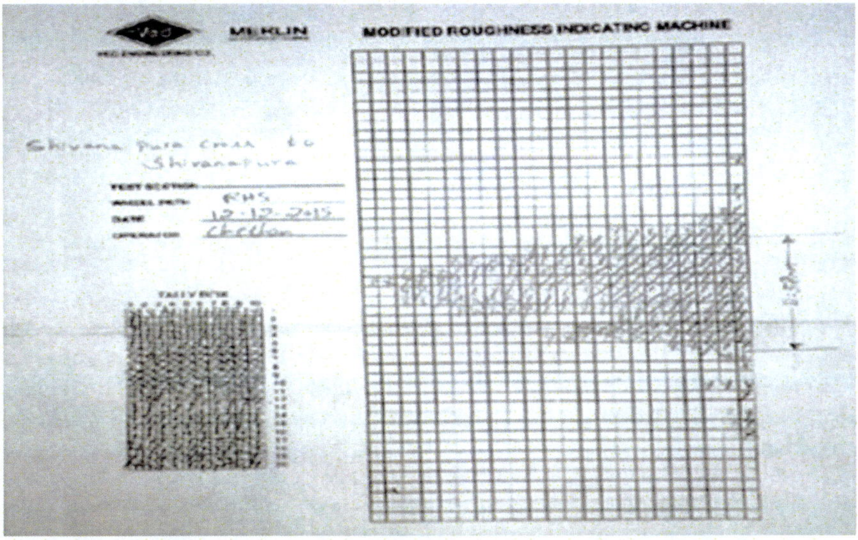

Fig. 5 MERLIN chart

Fig. 6 Various stages of test track overlay

out using MERLIN (TRRL Research Report No: 301), road inventory, and pavement condition survey per IRC:82-2015 guidelines. Structural evaluation of pavement was done using BBD (IRC:81-1997) and axle load survey as per IRC publication (Indian Highways, vol 28, 2000). The evaluation was taken up in two cycles one before the monsoon in May and the second cycle post-monsoon in December. The interval between the two cycles was six months.

(a) *Road inventory and pavement condition survey*: Pavement conditions represent the general appearance of the pavement surface. An absolute pavement is leveled and has a continuous, smooth, and good riding quality surface, on the contrary, a distressed pavement may be distorted, fractured, or disintegrated. To assess the surface condition

Fig. 7 Field studies

of the pavements, unbiased and repeatable survey procedures must be used like walking along the pavement section or from a moving vehicle called windscreen survey. Transect walk survey along the project road provides the most precise data about the condition of the pavement. Pavement conditions may deteriorate with time due to the following factors, i.e., wrong design procedures, overloading of pavement, material aging, construction deficiencies, and prevailing environmental and traffic factors. To plan for maintenance and rehabilitation strategies, one

Table 7 Pavement distress rating for highways

Defects	Range of distress				
Cracking (%)	>30	21–30	11–20	5–10	<5
Raveling (%)	>30	11–30	6–10	1–5	0
Pothole (%)	>1	0.6–1.0	0.1–0.5	0.10	0
Shoving (%)	>1	0.6–1.0	0.1–0.5	0.10	0
Patch (%)	>30	16–30	6–15	2–5	<2
Settlement and depression (%)	>5	3–5	Upto 2	Upto 1	0
Rutting (mm)	>50	21–50	11–20	5–10	<5
Rating	1	2	3	4	5
Condition	Very poor	Poor	Fair	Good	Very good

should aware of pavement conditions at various stages of their life. Pavement condition studies, also known as distress studies, are conducted as part of the Pavement Management System (PMS) to assess current pavement conditions and the need for maintenance and rehabilitation treatments. The main objective of these studies is to determine the condition of the pavement surface. These studies usually help to understand the causes of the breakdowns present in the pavement structure. The necessary data such as the type of failure, severity, extent, and location are collected from the inventory survey, based on the intended use of data. Condition surveys are conducted periodically using a visual inspection while walking on the pavement to identify and determine the type of distresses present. For conducting the pavement condition survey, the stretch is divided into subsections of some specified length. Table 7 shows guidelines as per IRC: 82-2015 for deciding pavement condition from various pavement distress aspects. As per IRC recommendations, pavement distresses recorded through visual observations in the prescribed format are as follows: cracking, raveling, potholes, shoving, patching, settlement, edge failure, and rut depth.

(b) *Pavement roughness evaluation by MERLIN*: The Machine for Evaluating Roughness using Low-cost Instrumentation (MERLIN) can be used for the direct measurement of road roughness and to calibrate instruments such as the vehicle-mounted bump integrator. Merlin Study gives the unevenness/roughness index of the pavement surface which is used to rate riding quality or comfort. The undulations present on the surface is recorded at regular interval on a chart mounted on the machine. Data is collected by repeated measurements of undulations along the wheel track and is presented in the form of a histogram. The dimension in the horizontal direction of this histogram (Fig. 5) can be used to estimate road roughness on the international roughness index (IRI) scale. The format of the histogram is shown in Fig. 5. The spacing between the two marks, D on the histogram as shown in Fig. 5, is then measured in millimeters and is used to obtain roughness using Eq. (1).

$$\mathrm{IRI} = 0.593 + 0.0471(D) \qquad (1)$$

where

IRI = International roughness index in meter (m) per kilometer (km).
D = Roughness in millimeter (mm) obtained from the merlin chart corresponding to 90% of the data points range.

Generally, IRI values below three-meter (m) per kilometer (km) are generally not important as it nominally affects the vehicle operating cost.

(c) *Benkelman beam deflection (BBD) method*: The structural adequacy of the pavement is evaluated by the Benkelman beam deflection methodology as per IRC:81-1997 guidelines. The magnitude of rebound deflection of the pavement is measured under a standard wheel load. The deflection was measured after the removal of the standard load away from the selected marked point on the pavement surface. Overall characteristic deflection (D_c) for the section is reckoned by statistical approach considering factors governing the pavement deflection value and after application of suitable correction factors. The characteristic deflection so obtained is used to obtain overlay thickness from the design curves given in IRC:81-1997.

(d) *Axle load survey*: The axle load survey is carried out to ascertain the complete spectrum of axle loads and to differentiate the damaging effect of overloaded vehicles on a given stretch of road. In practice, the laden weights of the vehicles can be measured using static or dynamic methods. In a static method, vehicles are stopped at the roadside in order to measure their axle loads. Weighing pads designed in accordance with Transport Research Laboratory (TRL) guidelines are widely used to measure axle load. A dynamic method employing weigh-in-motion (WIM) technology records axle loads to be measured without stopping the vehicle. In the present study, the static method was adopted to measure axle loads. A pair of portable weigh pads of capacity 20 tons each were employed to record axle loads. The axle loads measured for various categories of commercial vehicles are used to calculate vehicle damaging factors (VDF). The damaging effect of overloading vehicles concerning standard axle loads is established following the AASHTO Road Test which states that the destructive effect of axles on both rigid and flexible pavements is approximately proportional to the fourth power of the axle load.

3 Results and Discussion

3.1 *Mix Properties*

From the job mix formula, aggregates blend comprising 80% RAP material and 20% virgin aggregates were found to be ideal as the obtained gradation meets required limits as per the specification. 20% of virgin aggregates by weight of the mix passing 5.6 mm sieve is considered to compensate the gap grading of the RAP aggregates. Required binder content by weight of mix for the obtained aggregate gradation was reckoned as per the MoRTH guidelines and was works out to be 4.95%. Eight percent of waste plastic by weight of the bitumen was added along with the virgin bitumen as a recycling agent.

3.2 *Road Inventory and Pavement Condition Survey*

Performance of the test track was assessed through four important field studies in two cycles as per contract agreement specifications. Results of the same are presented in Table 8.

Pavement condition index (PCI) for cycle-1 was performed as per the IRC:82-2015 guidelines in May 2015. Inventory data indicates that there is no considerable surface distress in any of the form of failures under prevailing traffic and environmental conditions. The surface condition and riding quality were found to be good. It was also observed that the adjacent shoulder material is worn-out leading to a change in the profile of shoulder cross slope which has led to the loss of support and confinement at the edge of the pavement.

The cycle-2 evaluation was taken up in December 2015 from the inventory, it was observed that even after the monsoon season there was no considerable surface distress found on the test track except at few locations, where the adjacent shoulder material is worn-out leading to a change in the profile of shoulder cross slope which has led in loss of support and confinement at the edge of the pavement, and however, the surface condition and riding quality is found to be good.

3.3 *Axle Load Survey*

Since the project road considered for the study was experiencing a considerable volume of commercial vehicle traffic due to major road connectivity, an axle load survey was carried out to ascertain the spectrum of vehicular load and damaging effect due to overloading. Axle load survey data show that movement of single axle goods trucks was more than the multi-axle trucks, and the equivalency factor or vehicle damage factor (VDF) was found to be 1.4 for cycle-1 and 2.32 for cycle-2.

Table 8 Summary of performance evaluation test

State	Karnataka	
District	Bangalore rural	
Block	Hosakote	
Package number	PIU/KSHIP-II/ENV./Scarified Bitumen/2015–16/167)	
Name of road	Chikkanahalli road (classification/category: major district road)	
Chainage, and length of the test track	Chainage: 14 + 500 km to 15 + 000 km and length = 500 m	
Type of surface course	Close graded premix carpet (CGPC) Type-B	
Particulars	**Measurement/observation**	
	Cycle-1	**Cycle-2**
Date of survey	28-05-2015	12-12-2015
Observations on-road inventory and pavement condition survey	The pavement surface is in good condition, shoulder material is worn-out leading to change in the profile of shoulder cross slope which has led to the loss of support and confinement at the edge of the pavement	No considerable surface distress found on the test track even after monsoon except at few locations where the adjacent shoulder material is worn-out leading to change in the profile of shoulder cross slope which has led to the loss of support and confinement at the edge of the pavement, however, the surface condition and riding quality is found to be good

(continued)

Table 8 (continued)

MERLIN Roughness Value (IRI): *Note* If the length of stretch is 500 m and above conduct MERLIN survey with 200 observations in each direction. For stretches less than 500 m in length, a survey with 100 (instead of 200) observations in each direction. In both cases, for finding the roughness from the datasheet, eliminate 5% observations on each end (tail) of the datasheet	Left	LHS-3.41 m/km	LHS-3.18 m/km
	Right	RHS-2.49 m/km	RHS-3.32 m/km
	The average for both direction	Average-2.95 m/km	Average-3.25 m/km
Pavement evaluation by Benkelman beam deflection test as per IRC 81-1997		Characteristic deflection (D_c) = 0.2 mm	Characteristic deflection (D_c) = 0.22 mm
Axle load survey		Average VDF 1.4	Average VDF 2.32

Indicating an increase in the number of commercial vehicles between cycle-1 and cycle-2 as a result damaging effect on the pavement was also increased by 40%.

3.4 Pavement Roughness Evaluation by MERLIN

To determine the unevenness of the pavement as per the terms of reference (ToR) of the contract agreement, the MERLIN survey was carried out following the procedure laid down in TRRL Research Report No: 301. The international roughness index (IRI) value calculated on the test track was found to be 2.95 m/km for cycle-1 and 3.25 m/km for cycle-2. The road stretches having IRI value well within the permissible limit of 3.0 m/km indicate that the effect of vehicle operating cost (VOC) and discomfort to the passengers is less. IRI value of 2.95 m/km for cycle-1 indicates good surface condition without any distress in the form of rutting, settlement, raveling, etc., similarly for cycle-2 IRI value is slightly higher than the permissible limits which indicate minor distress at few locations along wheel path, nevertheless riding quality and comfort will be satisfactory.

3.5 Benkelmen Beam Deflection(BBD)study

BBD study was carried out as per IRC:81-1997 guidelines. BBD study gives the structural behavior of the pavement in terms of deflection bowl analysis, to compare structural behavior of flexible pavement with and without RAP content in surface course. BBD study was carried out on both, test stretch (RAP overlay) as well as on the road stretch beyond the test track with conventional hot mix asphalt (HMA) overlay. Summary of test results is tabulated in Table 9. It was observed that characteristic deflection (Dc) found to be less for stretch containing RAP and waste plastic material than the regular HMA stretch, where the other parameters such as environmental, moisture, traffic remain similar for both the cases. The existence of aged bitumen in RAP material makes the mix stiffer which deflects less under the application of load and also use of waste plastic with RAP aggregates has helped in improving the mechanical properties of the mix.

From the above field study results, it can be observed that regular bituminous mix deforms more under the application of wheel load than bituminous mix containing RAP and waste plastic. Comparatively, un-rebound deflection is minimal in the case of a bituminous mix containing RAP and waste plastic. Deflection of pavement is more after post-monsoon than that of pre-monsoon because pavement experiences moisture ingress which causes more deflection in the pavement. Test track with RAP and waste plastic has shown lesser deflection and performed better compared to regular conventional bituminous stretch. Merlin test results imply that bituminous pavement surface with RAP provides comparatively better riding quality and comfort than the conventional HMA mix.

Table 9 Summary of BBD and MERLIN test

	Characteristic deflection, D_c (in mm)	IRI (in m/km)
	Cycle-1 (pre-monsoon, tests conducted in the month of May 2015)	
Test stretch constructed with a bituminous mix containing RAP + waste plastic	0.2	2.95
Pavement beyond the control stretch (without overlay)	1.07	3.20
	Cycle-2 (post-monsoon, tests conducted in the month of December 2015)	
Test Stretch constructed with bituminous mix containing RAP + waste plastic	0.22	3.25
Pavement beyond the control stretch with overlay using VG-30 grade	0.39	3.60

3.6 Cost Analysis

To assess the cost-effectiveness of the CGPC containing RAP aggregates, virgin aggregates, and waste plastic over conventional HMA mix, a cost analysis was carried out. Rate analysis of the CGPC and conventional HMA mix was obtained as per the material cost and specifications provided in Panchyat Raj Engineering (PRE) Bangalore circle Schedule of Rates (SR). From the analysis, cost of CGPC mix with RAP and waste plastic per square meter works out to be Indian Rupees 14 (fourteen), whereas conventional HMA mix with virgin aggregate and waste plastic works out to be Indian rupees 43 (forty-three). This implies recycling and reuse of RAP aggregates in low- or medium-traffic intensity roads with suitable recycling agents and technology is economical and a saving of 70% in the construction cost of the surface course can be achieved.

4 Conclusions

The principal aim of the present study was to validate the use of a bituminous mix containing high RAP content in the construction of surface course for low- and medium-traffic intensity roads in India. In this regard, a test track with RAP and waste plastic was laid and subjected to performance evaluation under existing road and traffic conditions. Based on the laboratory and field tests, it was found that bituminous mix (CGPC type-B) containing high RAP content (80% recycled aggregates) with 8% of waste plastic by weight of bitumen as a rejuvenator exhibits better results in terms of physical properties, performance properties and meets all the requirements of close graded premix type-B specifications as per MoRTH specifications. From the field performance evaluation study, it was found that there was no considerable surface distress found on the test track both during cycle-1 and cycle-2 evaluation except at few locations, where the edge drop was observed due to adjacent shoulder material worn-out leading to loss of support and confinement at the edge of the pavement, however, the surface condition and riding quality are found to be good for both the cycle. Axle load survey results show that the average vehicle damage factor (VDF) is 1.4 for cycle-1 and 2.32 for cycle-2; this implies the damaging effect on the pavement surface for the standard vehicle is more by 40% during cycle-1 evaluation and 132% during cycle-2 evaluation. Despite overloading and damaging effect more than the permissible limits, bituminous overlay with RAP and waste plastic was performing well with good riding quality and comfort. From the BBD study results, it can be observed that regular conventional bituminous mix deforms more under the application of wheel load than bituminous mix containing RAP and waste plastic. Comparatively, un-rebound deflection is minimal in the case of a bituminous mix containing RAP and waste plastic. The existence of aged bitumen in RAP material and the use of waste plastic makes the mix stiffer which deflects less under the application of load. Merlin test results imply that the bituminous pavement surface with RAP provides comparatively better riding quality and comfort than the conventional bituminous mix even after the monsoon season.

Based on the findings of the study, it can be concluded that it is achievable to outline an acceptable quality bituminous mixture with a high RAP content that meets the required volumetric properties and performance criteria as per standards and specification for the construction of low or medium-traffic intensity roads. It is clear from the performance evaluation study that the use of recycling agents such as waste plastic is essential to enhance the performance of bituminous mix with high RAP content. Promoting recycle technologies contribute to conserving natural resources by reducing the use of virgin construction materials, reduces waste sent to landfills, saves energy, and reduces greenhouse emissions. It is also evident from the present study that since the use of RAP aggregates in higher percent reduces the consumption of virgin aggregates and the use of waste plastic replaces the quantity of virgin binder to some extent, and the overall cost for the production of a bituminous mix containing RAP and waste plastic is less compared to the production of conventional bituminous mix with virgin aggregates. However, it would be very accurate if an energy/carbon audit is conducted to understand the reduction of the carbon footprint.

Acknowledgments The present study has received funding from "Project Implementation Unit (PIU)," Karnataka State Highways Improvement Project (KSHIP) Bangalore under grant agreement "Performance Evaluation of the Pavement using Reclaimed Asphalt Mix from WEP-1".

The authors would like to acknowledge M/s KK Waste Plastic Management Pvt. Ltd. Bengaluru for providing RAP material, waste plastic, and sponsoring the construction of the test stretch.

The authors acknowledge assistance extended by post-graduation degree M.Tech. Highway Technology students of Government S K S J Technological Institute, K R Circle, Bangalore in carrying out field studies.

References

1. Aravind K, Das A (2007) Pavement design with central plant hot-mix recycled asphalt mixes. ASCE J Transp Eng 21:928–936
2. Izaksa R, Haritonovsb V, Klasac I, Zaumanisd M (2015) Hot mix asphalt with high RAP content. Proc Eng 114:676–684. https://doi.org/10.1016/j.proeng.2015.08.009
3. Hussain A, Yanjun Q (2012) Evaluation of asphalt mixes containing reclaimed asphalt pavement for wearing courses. In: IPCSIT, vol 26. IACSIT Press, Singapore
4. Saride S, Avirneni D, Javvadi SCP (2016) Utilization of reclaimed asphalt pavements in Indian low-volume roads. ASCE J Mater Civ Eng 28(2)
5. Pradyumna TA, Mittal A, Jain PK (2013) Characterization of reclaimed asphalt pavement (RAP) for use in bituminous road construction. Soc Behav Sci 104:1149–1157. https://doi.org/10.1016/j.sbspro.2013.11.211
6. Pradyumna TA, Jain PK (2016) Use of RAP stabilized by hot mix recycling agents in bituminous road construction. Transp Res Proc 17:460–467. https://doi.org/10.1016/j.trpro.2016.11.090
7. Anil R, Jayakumar J (2018) Use of recycled premix chipping carpet (RPCC) for rural road construction. Int J Eng Adv Technol (IJEAT) 8(2C) (2018). ISSN: 2249-8958
8. Ministry of Road Transport and Highways (2003) Ministry of road transport and highways specifications for road and bridgeworks, 5th rev. Indian Road Congress, Apr 2013
9. IRC: 81 (1997) Guidelines for strengthening of flexible road pavements using Benkelman beam deflection technique, 1st rev. Indian Road Congress, July 1997
10. IRC 14 (2004) Recommended practice for open graded premix carpets, 3rd rev. Indian Road Congress, June 2004
11. IS 73 (2013) Paving bitumen—specification, 4th rev. Bureau of Indian Standards, Apr 2013
12. IRC: 82 (2015) Code of practice for maintenance of bituminous road surfaces, 1st rev. Indian Road Congress, June 2015
13. NCHRP SYNTHESIS: 421 (2011) Recycling and reclamation of asphalt pavements using in-place methods. Project 20-05 (topic 40–13). Library of Congress
14. Jahren C, Yu JJ, Williams RC (2016) Technical report on alternate design methods to renew lightly traveled paved roads. Minnesota Department of Transportation Research Services and Library, MN/RC 2016-14, 2016
15. Research Report 301 (1991) The merlin low-cost road roughness measuring machine. Transport and Road Research Laboratory

Life Cycle Costing on a Building, An Approach to Make Building Energy Efficient

Shashwath M. Nanjannavar, Samreen Hullur, Darshan Baddi, Shadab Mulla, and A. G. Ramyashree

Abstract The research paper presents implementation of life cycle costing (L.C.C.) approach to a building on five parameters which makes the existing hostel building energy efficient and sustainable. According to GRIHA (Green Rating for Integrated Habitat Assessment) manual, five criteria (parameters) are chosen viz. solar panels, low VOC paints, double-glazed windows, fly ash bricks and LED bulbs. To choose the best alternative, each parameter is analyzed using L.C.C with present worth method as per IS: 13174 part 1 and 2. The L.C.C results show that implementation of the energy efficient parameters, makes the building closer in gaining one-star rating according to GRIHA evaluation system, so if the chosen five parameters are implemented in the existing building then the building's green rating points will be increased from 11 to 46. Using Revit software (insight 360), conventional building and energy efficient building models are developed to generate energy consumption per annum. The analysis report obtained from the Insight 360 software shows lower peaks in energy demand suggesting that building with the selected parameters is more efficient in terms of cost and energy which helps the building to gain one-star rating according to GRIHA.

Keywords Life cycle costing (L.C.C) · GRIHA (green rating for integrated habitat assessment) · Present worth method · Revit software (insight 360) · Energy efficient

1 Introduction

In the forefront of sustainability, construction industry holds a primary role in achieving it compared to other industries. With an ever increase in Indian economy and to cater for future energy demand, a distribution which is equal to 3–4 times higher than the total energy utilized today will be required [1]. Life cycle cost is a

S. M. Nanjannavar (✉) · S. Hullur · D. Baddi · S. Mulla · A. G. Ramyashree
School of Civil Engineering, K.L.E. Technological University, Hubballi, India
e-mail: shashwath.nanjannavar@kletech.ac.in

L. Nandagiri et al. (eds.), *Sustainability Trends and Challenges in Civil Engineering*, Lecture Notes in Civil Engineering 162, https://doi.org/10.1007/978-981-16-2826-9_14

technique of estimating monetary costs of the investment, design, execution, operation, maintenance and repair, replacement and sometimes demolition of a building [2]. Within the context of the built environment L.C.C is a method used to access the anticipated economic performance of the building throughout its life cycle [3]. L.C.C is a tool that determines how much money is spent on an asset over the course of its useful time. L.C.C includes four cost parameters:

Initial cost, maintenance cost, operational and salvage cost. The total cost of a building during the life span can be known at the early stage of a project.
Initial costs: All costs which include, market research, initial development (production cost), testing, material cost, labor cost involved in developing the product, trials, construction and commissioning of an asset.
Maintenance costs: All the costs associated with maintenance of equipment or any asset. In the view of a building maintenance cost will generally be painting of building/MEP works.
Operational costs: All the costs pertaining to the operation of a instrument, parts, piece of equipment or facility. Electrical use/consumption is the operational cost of a building.
Salvage or Resale value: The financial sum expected from the disposal of a property at the end of its economic life, or at the end of the study period [4, 5].

GRIHA, Green Rating for Integrated Habitat Assessment has a hundred-point evaluation system consisting of some important points which are mandatory and remaining points can be earned by satisfying the criteria. GRIHA has 36 criteria in the evaluation system. Based on the percentage of the points earned, the star rating for the building is decided. It has one to five-star ratings, which is assigned based on the criteria evaluation system [6].

The L.C.C is a mathematical technique that makes use of basic monetary evaluation approaches, like the annual worth method, the net present value method, and the savings/investments ratio (SIR) method to ascertain the various cash flows of project lifecycle [5].

Through the use of L.C.C, it is expected that one can make better decisions in selecting the type of building system only pertaining to energy usage like, thermal shielding, passive solar features, HVAC, domestic hot water, smart building lighting and automation systems. However, L.C.C can also be used on building parts or involve costs related to resident productivity, system monitoring, environmental impact and any other concerns that impact costs over the course of time.

A life cycle costing technique is useful in collecting factors that can be monetized, but there are often non-financial organization or society goals that also need to be inculcated in the design.

A tool is essential to accumulate the existing data, determine the life cycle costs in a methodical and transparent way, and present the results in a way that ease comparison of alternative designs [5, 7].

2 Summary of Literature Review

- L.C.C (life cycle costing) is a method of financial analysis of all costs related to construction project over a defined period of time.
- Use of L.C.C in the construction industry will allow us to predict the present costs and the future costs which will be incurred during the life cycle of the building.
- L.C.C is one good possibility to meet 3 Ecriteria, i.e., economy, effectiveness and efficiency.
- L.C.C helps to reduce the expenditure which may be incurred in future due to maintenance and operations
- Life cycle costing has been for a long time but literature review shows that very less research is carried out on buildings whole cycle assessment.

3 Problem Statement

This research focuses on determining the cost effectiveness of implementing different energy efficient parameters like solar panels, low VOC paints, glazed windows, LED bulbs and masonry blocks on a building using life cycle costing approach to upgrade the building to a higher green rating which supports the cause of sustainable development.

4 Objectives

- To carry out L.C.C calculations using present worth method and compare the alternatives chosen for each parameter (Solar panels, LED and CFL bulbs, Low VOC paints and VOC paints, double-glazed windows and normal windows, Fly ash bricks and concrete blocks)
- To create two models: Conventional building and energy efficient building and analyze the buildings for energy efficiency using insight 360 software tools.

5 Methodology

The research enables to predict the L.C.C of residential infrastructure through developing deterministic forecasting models. This research is focused on a hostel building studied modeled as two building types: Conventional and sustainable one. In the first phase of the research five parameters of a building are selected based on GRIHA manual, namely solar panels, window panels, paints, bulbs and bricks. Then for each of the parameters selected different alternatives are analyzed using L.C.C method. Four main criteria are investigated in L.C.C model, which include: initial costs (cost of construction), cost of operations (energy, operational and maintenance, and major repairs expenses), environmental and residual value. The life cycle costing method and calculations is done as per IS: 13174 part 1 and part 2 [5, 7]. Based on the L.C.C results, the second phase of research involves the implementation of the best alternatives determined on the conventional hostel building. Therefore, 2 models of same building are prepared, one is conventional hostel building and other is sustainable hostel building. The sustainability assessment model includes three major groups, such as energy and aerosphere (energy utilization), material and resources (recyclability and material re-usage), and creation and design process (life cycle assessment).

- Building considered for study: Hostel Building
- Area of the building understudy

S. No.	Floors	Gross area (m^2)
1	Cellar floor	883.25
2	Ground floor	1255.94
3	First floor	1255.94
4	Second floor	1255.94
5	Third floor	1255.94
6	Fourth floor	1255.94
7	Fifth Floor	469.40
	Total	7632.35

Factors considered for overall electricity usage and maintenance.

(1) Solar panels
(2) Paints
(3) Window panels
(4) Bulbs
(5) Bricks.

5.1 *Life Cycle Costing Approach and Its Components*

i. Initial cost is the cost incurred during the purchase of the material, its installation cost, transportation cost. The sum of all these costs refers to the Initial cost [4].
ii. Operating (Operational) costs are the expenditures which are corresponding to the operation of an instrument part, piece of equipment or the facility. Electrical use/consumption is the operational cost of a building [4].
iii. Maintenance costs are the expenses pertaining to the maintenance of any device or property. It includes the costs of secondary staff as-well who is involved in

problem identification and repair. In the view of a building, maintenance cost will generally be painting of building components/MEP works [4].

iv. Residual value or salvage value is the estimated value that an owner gains when the product is sold at the end of its useful life [4].

5.2 *Present Worth Method*

The L.C.C method includes various methods as mentioned earlier in the paper; this research utilizes the technique of present worth method (NPV technique) for performing life cycle costing on various green alternatives.

The net present value (NPV) technique is used to transform all cash flows occurring at different times to a single sum equivalent at the beginning stage of the study period. With the help of this technique, all expenses and profits, regardless of the occurrence time, will be compared over a certain standard year, defined as a baseline year [8].

Money spent and future money inflow will be suitably discounted to compensate their time value. When these future money in-flow are discounted, they will be comparable to those expended on the present day, or throughout the "baseline year". When this discounting takes place, all expenses and profits are considered on a common basis and summed up together to calculate the total net present value. As most of the initial costs occur almost at the same time, they are considered to occur during the base year of the period under study. Hence, there is no need to calculate the present worth of these initial costs because their present worth will be equivalent to their actual cost. The technique for life cycle costing for various materials is conducted as per IS-13174(Part 2):1994. This IS code provides two methods present worth method (PW) and annualized cost method (AC) [7]. The research is conducted on L.C.C using present worth method. This method provides all expenditure over the life cycle of a item or project. PW and AC methods are mathematically similar. The choicc of thc mcthod is complctcly dcpcndcnt on thc rcscarchcr or thc managcmcnt personnel [8]. The components considered for L.C.C are initial costs, maintenance costs, operational costs, salvage or residual value.

$$\begin{aligned} \text{L.C.C}_{\text{alternative}} &= \text{PW(initial cost)} + \text{PW(maintenance costs)} \\ &\quad + \text{PW(operational cost)} - \text{PW(salvage value)} \end{aligned}$$

where

$b = 1/(1 + 0.01 \times i)$;

i = rate of discount per annum;

t = numbers of years

PW = Present worth

Present worth Factor (pwf) without inflation = $(1 + 0.01 \times i) - t$

$$\text{upwf} = \frac{b(1 - b^t)}{1 - b}$$

where

$b = 1 + 0.01 * r/1 + 0.01 * i.$

$i =$ discount rate.

$r =$ inflation rate.

Rate of inflation = (Present GDP − Old GDP)/(Old GDP).

There can be repetitive annual expense but not for complete building under study. In such circumstances, the expenditure for each year has to be reduced to present worth by multiplying the respective pwf. Alternatively, one can opt for uniform present worth factor (upwf)

$$\text{upwf} = \frac{b(1 - b^t)}{1 - b}$$

where

$b = 1/(1 + 0.01 \times i).$

$i =$ rate of discount per annum = 8%

$t =$ numbers of years = 20 years.

The discount rate value is selected as the average rate of interest of loans in Indian banks for construction industry, i.e., decided by reserve bank of India, pwf (present worth factor) can be obtained by Referring Annex A of IS Code 13174 part 2. The inflation rate is decided as per IS code 13174 part 2 [7].

6 Results

The results obtained are categorized as:

(a) The initial cost, maintenance cost, operational cost and salvage values are collected from the local material dealers and through online help to calculate Life cycle costing of solar panels, LED and CFL bulbs, fly ash and concrete blocks, double-glazed and normal window, low VOC and VOC paints and Rain water harvesting.
(b) The present worth values are determined on all the alternatives chosen under study.
(c) Use of Revit insight 360 software to generate and compare energy usage between conventional and energy efficient building (Table 1).

Table 1 Data of initial, operational, maintenance, salvage values for energy efficient alternatives

Costs (Rs.) parameters	Initial (Rs.)	Operational (Rs.)	Maintenance (Rs.)	Salvage (Rs.)
Solar panel poly crystalline mono crystalline	4,410,100	0	28,019	851,370
	4,776,400	0	28,019	851,370
LED bulb	9,259,660	346,598	0	2086
CFL bulb	2,329,128	346,598	0	2086
Low VOC paint VOC paint	1,654,284	0	5,323,719	0
	1,168,260	0	6,842,647	0
Double glazed window plane window	2,148,432	0	0	2074
	573,663	0	0	2074
Fly ash brick	3,201,197	0	8,367,306	28,594
Concrete blocks	5,514,780	0	8,367,306	142,971

6.1 Life Cycle Costing Values of Various Parameters

(1) L. C. C. of Solar Panels
According to the result, polycrystalline solar panels are cost advantageous and more effective as compared to mono crystalline solar panels as in 20 years life span it can save Rs. 372,495/– if polycrystalline solar panels are used. The operational cost is zero for solar panels, as no cost is incurred after installing the solar panels only maintenance is required (Table 2).

(2) L.C.C of Low VOC and VOC paints
As per LCC results by adopting low VOC paints in building one can save Rs. 1,032,901.79/– in the complete life span of the paints as compared to VOC paints (Table 3).

(3) L.C.C of Double glazed and plane window

Table 2 L.C.C components for alternative solar panels

Item (costs in Rs.)	Alternative 1 polycrystalline	Alternative 2 mono crystalline
Initial	4,410,100	4,776,400
Operational	0	0
Maintenance	28,019	28,019
Salvage (solar panels, batteries, mounting stand)	851,370	851,370
Discount rate (%)	8	8
Inflation rate (%)	4	4
Life cycle cost	3,558,730	3,925,030

Table 3 L.C.C components for paint alternatives

Item (costs in Rs.)	Alternative 1 low VOC	Alternative 2 VOC
Initial	1,654,284	1,168,260
Operational	0	0
Maintenance	5,323,719	6,842,648
Salvage value	0	0
Discount rate (%)	8	8
Inflation rate (%)	4	4
Life cycle cost	6,978,003	8,010,908

Table 4 L.C.C components for window alternatives

Item (costs in Rs.)	Alternative 1 double gazed	Alternative 2 plain window
Initial	2,148,432	573,663
Operational	0	0
Maintenance	0	0
Salvage value	2074	2074
Discount rate (%)	8	8
Inflation rate (%)	4	4
Life cycle cost	214,658	571,589

According to the result, alternative 1 is energy efficient when compared to alternative 2. Considering these glass windows to be manually operated the operational costs is taken as zero. Double-glazed window has many advantages as compared to normal plane window, few of them are, it avoids thermal flow through the window as the barrier insulates the indoor area from varying air temperatures at either side of the glazing panel and also double glazing also provides a better sound isolation (Table 4).

(4) L.C.C of LED and CFL bulbs
According to the result, LED bulbs are more cost advantageous because in 20 years life span it can save Rs. 1,403,162/– if LED bulbs are adopted (Table 5).

(5) L.C.C of fly ash bricks and concrete blocks
Alternative 1 is cost advantageous and energy efficient as compared to alternative 2 and is more durable, a sum of rupees 2,199,207/– can be saved if fly ash bricks is used for construction during its initial time. Since operational and maintenance cost for bricks are assumed to be zero (Table 6).

Table 5 L.C.C components lighting alternatives

Item (costs in Rs.)	Alternative 1 LED	Alternative 2 CFL
Initial	925,966	2,329,128
Operational	346,598	346,598
Maintenance	0	0
Salvage value	2086	2086
Discount rate (%)	8	8
Inflation rate (%)	4	4
Life cycle cost	1,270,479	2,673,641

Table 6 L.C.C components for brick alternatives

Item (costs in Rs.)	Alternative 1 flash ash bricks	Alternative 2 concrete blocks
Initial	3,201,197	5,514,780
Operational	0	0
Maintenance	8,367,306	8,367,306
Salvage value	28,594	142,971
Discount rate (%)	12	12
Inflation rate (%)	4	4
Life cycle cost	11,539,909	13,739,116

6.2 Revit, Insight 360 Energy Usage Analysis

Revit, a 3D modeling design software, is used to make a 3D model of the selected building. Then, the energy analysis is carried out with the help of Insight360 by adding the chosen parameters (Solar panels, LED bulbs, low VOC paint, double-glazed window, Fly ash bricks) and their properties to the model (Fig. 1).

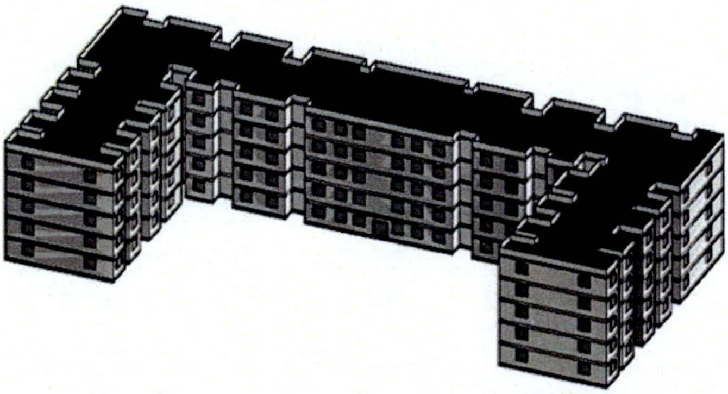

Fig. 1 Revit model of hostel building

(1) Energy Use Intensity of conventional and Energy efficient building is shown.

EUI	Conventional	Energy efficient
Electricity (kWh/sm/year)	149	140
Fuel (MJ/sm/year)	16	28
Total (MJ/sm/year)	551	532

(2) Life cycle Energy Use/cost of conventional and energy efficient building is shown

Life cycle	Conventional	Energy efficient
Electricity use (kWh)	28,239,915	23,259,447
Fuel use (MJ)	3,085,944	4,689,603
Energy cost (dollars)	1,036,081/–	860,598/–

Annual energy use/cost of conventional versus energy efficient building. This energy consumption is throughout the life cycle of building of 20 years (assuming all the electrical appliances are turned on throughout) (Figs. 2 and 3).

Annual energy use/cost	Energy efficient	Conventional
Electricity use (%)	97	95

(1) Simulated Electricity Peak Conventional versus energy efficient building (Figs. 4 and 5).

X-axis represents months.
Y-axis represents electricity peak demand.

The analysis report obtained from the Revit software shows lower peak demands in energy demand suggesting that building with the selected parameters is more efficient in terms of cost and energy which helps the building to gain one- star rating according to GRIHA [6].

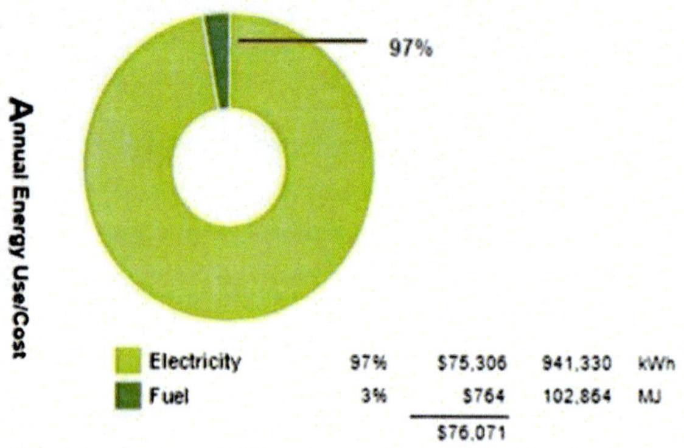

Fig. 2 Annual energy use/cost of conventional building

Fig. 3 Annual energy use/cost of energy efficient building

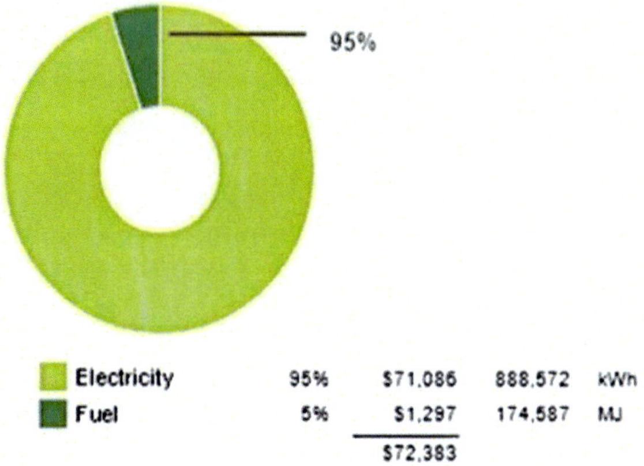

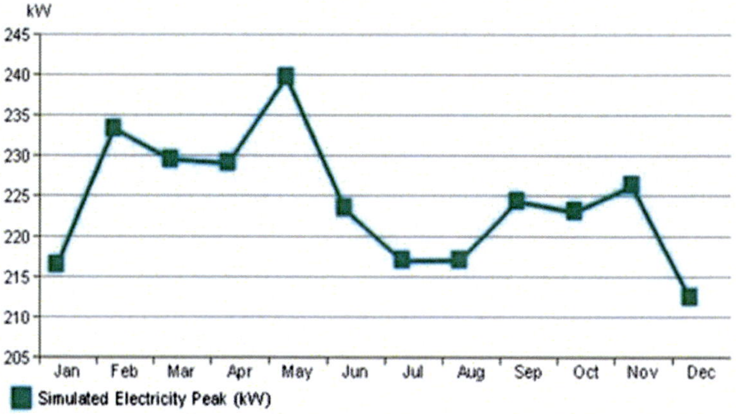

Fig. 4 Monthly electricity demand of conventional building

7 Conclusions

1. The analysis of parameters using L.C.C technique with present worth method is carried out, the results shows that the parameters chosen to implement, make the building cost effective and energy efficient.
2. With the assumption that all the electrical appliances are turned on throughout, the electricity bill of the building is Rs. 876,000/– per year, approximately Rs. 170,00,000 for 20 years. By installing the solar panel, we can save up to Rs. 9,500,000/– in 20 years, i.e., 56% of the electricity bill can be saved.
3. LED bulbs use 70–75% less energy, Rs. 5040/– can be saved per month and in a year Rs. 60,480/– can be saved and in 20 years approximately Rs. 1,209,600/- can be saved.

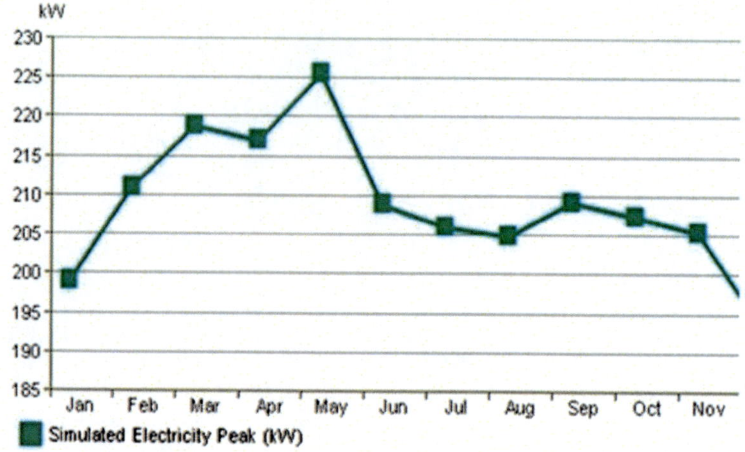

Fig. 5 Monthly electricity demand of energy efficient building

4. Double glazed window has many advantages, it can reduce the outer noise up to 60% and reduces interior fading. In 20 years, one can save Rs. 350,000/– of electricity bill, i.e., Rs. 17,500/– every year, if a Double glazed window is installed.
5. Low VOC paints are durable and do not have any bad impact on human health and it provides better thermal comfort. Finally, by using Low VOC paints Rs. 1,032,902/– can be saved.
6. Fly ash bricks are cost effective as compared to concrete bricks, more durable and environmentally friendly, by using fly ash bricks Rs. 2,199,207/– can be saved compared to concrete blocks.
7. Two models are created conventional building and an energy efficient building using Revit software and analyzed for energy efficiency using In-sight 360 tool. Through the results obtained, the annual energy consumption of conventional building is more than the energy efficient building.
8. The identified parameters can be implemented in the existing building. All these parameters show that they have a significant impact on making the building cost effective and energy efficient, which contributes to a sustainable built environment.

8 Future Scope

- A database for LCA and sustainability evaluation based on various material characteristics, need to be developed, which may facilitate the product designers and manufacturers in the design and delivery of products from the stand point of life cycle costing.

- Environmental impact costs can be included in L.C.C as a cost driver to have a more reliable analysis by conducting environmental impact assessment.
- Life cycle costing technique can be used to evaluate the best bidder for a contract.

9 Limitations

- The study is restricted to six criterions (parameters) out of 36 criterions due to time and resource availability as constraints.
- The accuracy of outcomes of life cycle costing depends on the duration of the study period adopted. The accuracy decreases with the increase in study period. This is due to the large variation in the value of discount rates and the inflation rates over a period of time.
- There is a dearth of reliable data regarding the various cost drivers in India, hence the construction industry fails to apply the L.C.C technique.

Acknowledgements The contentment that accompanies the successful completion of any work would be undone without mentioning of the number of individuals whose professional guidance and encouragement helped us in the successful completion of this project.

In view of the above we express our gratitude toward the whole management of K.L.E technological university for providing us with the necessary data and facility to conduct research.

Also we like to extend our gratitude to the open source software who support in conducting valuable research in the field of civil engineering.

References

1. Sawant Priyadarshi H, Sayed SSM (2015) Life-cycle cost and financial analysis of energy components in mass housing projects—a case project in sub-urban India. Elsevier
2. Clift M (2003) Life-cycle costing in the construction sector. Ind Environ 26(2):37–40
3. Dwaikat LN, Ali KN (2018) Green buildings life cycle cost analysis and life cycle budget development: practical applications. J Build Eng 18:303–311
4. Indian standard codes-13174 (1991) Part 1, life cycle costing
5. Sugandhini HK, Nanjannavar SM (2019) Life cycle costing for the analysis of cost-effectiveness of alternative concretes and masonry blocks. In: Sustainable construction and building materials. Springer, Singapore, pp 409–421.
6. Ministry of New and Renewable Energy Government of India, The Energy and Resource Institute New Delhi (2010) GRIHA manual, introduction to national rating system—GRIHA
7. Indian Standard Codes-13174 (1994) Part 2, life cycle costing
8. Gluch P, Baumann H (2004) The life cycle costing (LCC) approach: a conceptual discussion of its usefulness for environmental decision-making. Build Environ 39(5):571–580

An Experimental Study on Usage of Treated Waste Water (Domestic) on the Fresh and Hardened Properties of Conventional Vibrated Concrete for Sustainable Construction

N. M. Rakshit Jain, N. Ajay, and P. U. Vinyas Gowda

Abstract Concrete is the most greatly employed unnatural heterogenous mixture in the world. Also, concrete is one among the industries that expends huge amount of water. Approximately, 130–150 liters of water is prerequisite per cu-m of concrete mixture, beyond the considerations of other exercises of water in the construction industry. Water is a pivotal environmental concern; and supply and quality of water are becoming more bounded worldwide. Thus, there is a demand to conserve the water and on the other side, there are million liters of wastewater produced and wasted every day from various sources. The usage of treated waste water in production of concrete is very important for the present scenario. Taking this into consideration, an experimental study was carried out to assess the fresh and strength properties of conventional vibrated concrete (CVC) by using domestic treated wastewater. In the present work, the concrete mixes were produced with water-to-cement ratio about 0.45 along with different proportions of treated waste water which ranges between 25 to 100% of potable water and the specimens were casted. For each concrete mixtures, the fresh properties and strength properties were assessed as per IS codal provisions. In addition to these, three of the each specimens were cured separately in potable water and treated wastewater to assess the influence of treated wastewater as curing water. Results signify that the concrete mixtures produced with treated wastewater showed similar strength properties with concrete produced from potable water. Treated wastewater, while employed as curing water made very less impact on the strength parameters of concrete.

Keywords Compressive strength · IS-10262 2019 · Treated wastewater

N. M. Rakshit Jain (✉) · N. Ajay · P. U. Vinyas Gowda
RASTA-Center for Road Technology, Bengaluru, India

L. Nandagiri et al. (eds.), *Sustainability Trends and Challenges in Civil Engineering*, Lecture Notes in Civil Engineering 162, https://doi.org/10.1007/978-981-16-2826-9_15

1 Introduction

The groundwork of existence is water, but still very less importance is given toward conservation of potable water [1]. Of the total global water, world's freshwater accounts for only 2.5% and larger portion of it are frosted in glaciers and ice caps [1]. Rest of the unfrozen freshwater is primarily presented as groundwater, with less amount present in air. Also, population has been increasing in a burdensome manner. From 1959 to 1999, the world's population has been doubled [2], from three billion to six billion. By 2043, the world's population is envisaged to reach 9 billion (USCB 2009) [2]. Increasing population increases per capita water consumption which intensifies water demand, finally straining the water supply. Growing water demand means growing water scarcity.

Concrete is the most greatly employed construction material. The concrete is majorly made of four ingredients namely, cement, fine aggregate, coarse aggregate and water. The water is not only required for blending of concrete, but also to render strength to concrete by the phenomenon called hydration of cement and also, it maintains the cohesiveness of concrete in the fresh state [3]. IS 456-2000 [4] recommends potable water to be used for mixing, placing and curing of concrete. Potable water is habitually acquired from various resources, but mainly through lakes and ground in the form of groundwater. In recent decades, encroachment and destruction of lakes and canals have minimized the potential of groundwater recharge. Due to various economic activities and population growth, there is an increasing demand for enormous quantity of water. So, there is a need to control the usage of freshwater and reduce the pressure on potable water resources. As a result, substitute resources must be explored [3]. One of the efficient ways is to reuse the recycled treated wastewater for construction purposes, which may mitigate the severity of depletion rate of potable water. However, some harmful impurities in treated waste water need to be skimmed off by sedimentation process [3].

Kosmatka and Panarese reports that "almost any natural water fit for drinking and has no pronounced taste or odor can be used as mixing water for making concrete [5]. However, some water which is not fit for drinking, but may be suitable for concrete production". As per ASTM C 94-98 [6], the compressive strength of standard mortar cubes at 7 days made with the untried water to be not less than 90% of cubes made with distilled water or tap water. British note [7] states that the strength reduction upto 20% is acceptable when untried water is utilized, but the proportions should be harmonized as appropriate.

Cebeci and Saatci [8] studied the usage of treated sewage as well as raw sewage in the production of concrete as mixing water. Properties of the concrete, i.e., initial setting time, compressive strength of mortar and concrete was not affected by treated wastewater. These tests proved that treated sewage water is equivalent to distilled water while employed as mixing water. However, compressive strength was reduced by 9% with the usage of raw sewage. They also proclaimed the relaxation in the setting time when raw sewage was employed for mixing the concrete.

Swami et al. [9] studied on the suitability of utilizing treated sewage as mixing water for ordinary concrete. The tests show that there was an improvement in compressive strength by 7% and a gain of 10% in pull strength.

Al-Jabri et al. [10] inspected the impact of using wastewater on parameters of high-strength concrete. The chemical composition of the effluent was found to be more than tap water, however not beyond the limits specified in ASTM. Various tests were performed over concrete with varying percentages (0, 25, 50, and 100%) of wastewater. After 28 days of curing, no representative differences were observed in the compressive strength among the concrete mixes. This paper concludes that the utilization of effluent obtained from car washing station has insignificant impact on the strength of concrete.

At present, around 1600 million liters per day (MLD) of sewage is generated in Bengaluru, but only 1060 MLD is treated [11]. In that 1060 MLD, only around 310 MLD is reused and utilized for agriculture, environmental rejuvenation, flushing of toilets and industrial uses. As stated by NITI Aayog's composite water management index report, 21 cities in India will run short of water by 2020 and Bengaluru is one among them [12]. Therefore, there is need to study on the utilization of treated wastewater which minimizes the consumption of freshwater and thus, cutting down the burden on potable water resources which also establishes the balance between serviceable resources and requirements.

In present work, an experimental study was carried out to assess the feasibility of usage of domestic treated wastewater to prepare the conventional vibrated concrete (CVC) and to assess the fresh and strength parameters of concrete.

2 Experimental Investigation

2.1 Materials

The characteristics of the materials used are shown in Table 1.

Table 1 Material properties

Materials	Specific gravity	Specific surface (m^2/kg)	Water absorption (%)	Remarks
Cement—OPC 53 grade	3.10	265	–	Conforming to IS: 12269-2013 [13]
Fine aggregate—M sand	2.6	–	2.2	Conforming to Zone-II as per IS: 383-2016 [14]
Coarse aggregate (20 mm)	2.77	–	0.46	Crushed angular aggregates

Table 2 Chemical composition of water samples

Parameters	Result of PW	Result of TW	Requirement as per IS 456: 2000	Test method
Organic solids	73 mg/l	91 mg/l	200 mg/l	IS:3025/Part 16
Inorganic solids	603 mg/l	1059 mg/l	3000 mg/l	IS:3025/Part 16
Sulfates (as SO_4)	18.6 mg/l	34.7 mg/l	400 mg/l	IS:3025/Part 24
Chlorides (as Cl)	160 mg/l	310 mg/l	500 mg/l	IS:3025/Part 32
Suspended matter	<1.0 mg/l	18 mg/l	2000 mg/l	IS:3025/Part 17
pH	7.7	7.1	≥ 6	IS:3025/Part 24

Legend PW—potable water, *TW*—treated wastewater

Both potable water and treated wastewater were chemically tested for parameters. The Treated wastewater sample was procured from domestic sewage treatment plant (STP) of Volvo Group India Pvt. Ltd. The chemical results of both potable and treated wastewater given in Table 2.

2.2 *Mix Design*

Mix design of concrete (M-25 grade) was developed in accordance with IS 10262: 2019 [15]. Figure 1 depicts the steps involved in concrete mix design. Concrete mix

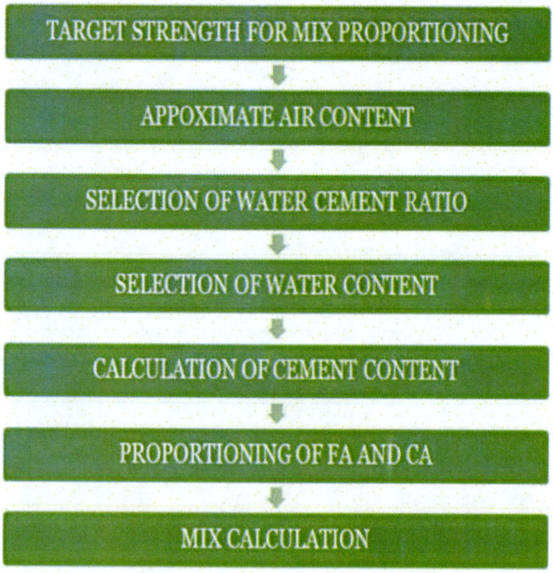

Fig. 1 Flow chart for mix design of M25 grade of concrete as per IS 10262-2019

Table 3 Particulars of mix proportions

Mixes	% of TW	Water (l/m^3)		w/c ratio	Ce (kg/m^2)	Fa (kg/m^2)	Ca (kg/m^2)
		PW	TW				
Mix 1	0	149	0	0.45	329	722	1307
Mix 2	25	112	37	0.45	329	722	1307
Mix 3	50	75	75	0.45	329	722	1307
Mix 4	75	37	112	0.45	329	722	1307
Mix 5	100	0	149	0.45	329	722	1307

Legend PW—potable water, *TW*—treated wastewater, *Ce*—cement, *fa*—fine aggregate, *Ca*—coarse aggregate

proportioning was developed with w/c ratio of 0.45. The concrete mix was designed for a medium slump range of 50–75 mm which usually does not require chemical admixtures. Henceforth, chemical admixtures have not been employed in the study. Totally, five mixes were developed with varying % of treated wastewater, i.e., 0, 25, 50, 75 and 100%. The particulars of each mix are tabulated in Table 3.

2.3 *Testing Procedures*

Figure 2 shows the tests that were conducted to reckon the fresh and strength parameters of concrete.

(1) *Slump Test*

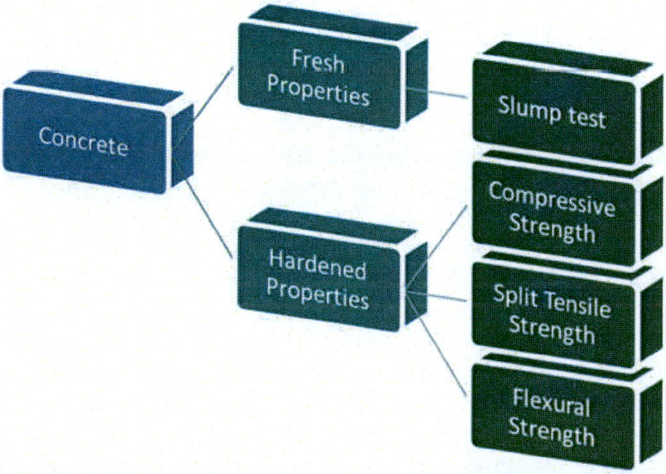

Fig. 2 Tests on concrete

Fig. 3 Experimental setup for slump test

Workability of concrete was evaluated by performing slump test and was conducted according to IS 1199: 1959 [16]. Slump test apparatus is depicted in Fig. 3. Then, specimens were casted and de-molded after 24 hrs, cured separately in potable water and treated wastewater to analyze the effect of treated wastewater as curing water; and then, the specimens are tested at room temperature at the prescribed ages.

(2) *Compression Test*
The compression test was performed in accordance with IS 516: 1959 [17]. Twelve cubes of 150 mm × 150 mm were casted for each mix and tests were performed over the specimens at recognized period of 7 and 28 days. The compression testing machine is as shown in Fig. 4.

(3) *Split Tensile Test*
The split tensile test was executed in accordance with IS 5816: 1999 [18]. Six cylinders of 150 mm × 300 mm were casted for each mix. Tests were conducted at definite interval of 28 days. Figure 5 shows the experimental setup of split tensile test.

(4) *Flexural Strength Test*
The flexural strength test was performed according to IS 516: 1959 [17]. Six beams of 100 mm × 100 mm × 500 mm were casted for each mix. Tests were conducted at recognized age of 28 days. The beam specimens and flexural strength testing machine are shown in Fig. 6.

Fig. 4 Compression testing machine

3 Results and Discussions

3.1 Slump Test

Table 4 shows the results of slump test of each mixes. Figure 7 shows the variation of slump against percentage of treated wastewater.

3.2 Compression Test

Tables 5 and 6 shows results of compressive strength test on concrete specimen with respect to curing of specimens in potable water and treated wastewater, respectively.

Figure 8 shows the improvement of compressive strength with hike in proportion of treated wastewater when potable water is employed as curing water. The same kind of trend was reported in a technical note [8, 9]. Presumable explanation for the development was provided by Swami et al. [9] which states that pore filling effect of deposition of suspended and dissolved solids present in treated wastewater was considered accountable for evolution of strength. And when treated wastewater is considered as curing water, the compressive strength increases till 50% and then

Fig. 5 Experimental setup of split tensile test

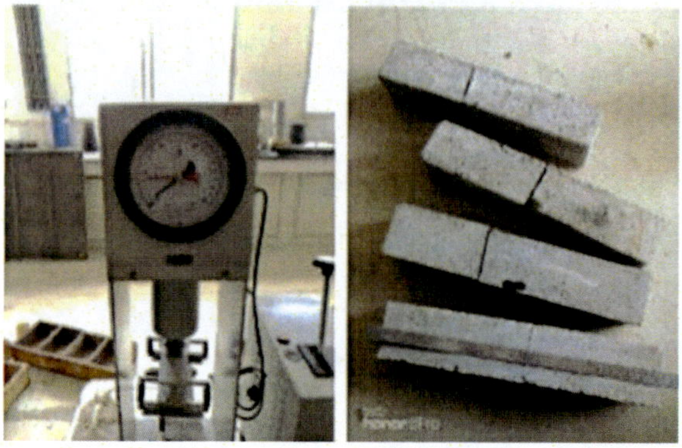

Fig. 6 Flexural strength testing machine and beam specimens

drops down. Cube with 100% treated wastewater (cured in potable water) gives the maximum strength value of 27.23 MPa, about 18.8% of increase in compressive strength can be noticed when treated sewage is employed as mixing water and potable water as curing water which gives positive sign of hope toward utilization of treated wastewater as mixing water. The influence of treated wastewater being curing water decreases the compressive strength by 11.8% with 100% treated wastewater.

Table 4 Results of slump test of concrete mixes

Mixes	% of TW	Water content (l/m^3)		w/c ratio	Slump (mm)
		PW	TW		
Mix 1	0	149	0	0.45	20
Mix 2	25	112	37	0.45	22
Mix 3	50	75	75	0.45	35
Mix 4	75	37	112	0.45	46
Mix 5	100	0	149	0.45	50

Legend PW—potable water, *TW*—treated wastewater

Fig. 7 Comparison of slump against percentage of treated wastewater

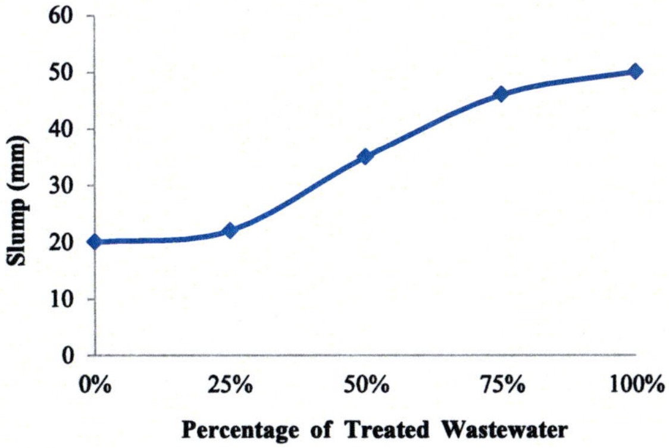

Table 5 Results of compressive strength test (cured in potable water)

Mixes	% of TW	Water content (l/m^3)		w/c ratio	Compressive strength (MPa)	
		PW	TW		7 days	28 days
Mix 1	0	149	0	0.45	14	22
Mix 2	25	112	37	0.45	16	22
Mix 3	50	75	75	0.45	19	24
Mix 4	75	37	112	0.45	21	25
Mix 5	100	0	149	0.45	23	27

Legend PW—potable water, *TW*—treated wastewater

3.3 Split Tensile Test

Table 7 displays the 28 days split tensile strength values of specimens cured in both potable water and treated waste water.

From Fig. 9, it is evident that the 28 days split tensile strength improves with addition in treated wastewater content when curing water is potable water. When treated water is curing water, split tensile strength increases till 50% of treated water

Table 6 Compressive strength test (cured in treated wastewater)

Mixes	% of TW	Water Content (l/m^3)		w/c ratio	Compressive Strength (MPa)	
		PW	TW		7 days	28 days
Mix 1	0	149	0	0.45	17	24
Mix 2	25	112	37	0.45	18	25
Mix 3	50	75	75	0.45	22	26
Mix 4	75	37	112	0.45	21	25
Mix 5	100	0	149	0.45	19	24

Legend PW—potable water, *TW*—treated wastewater

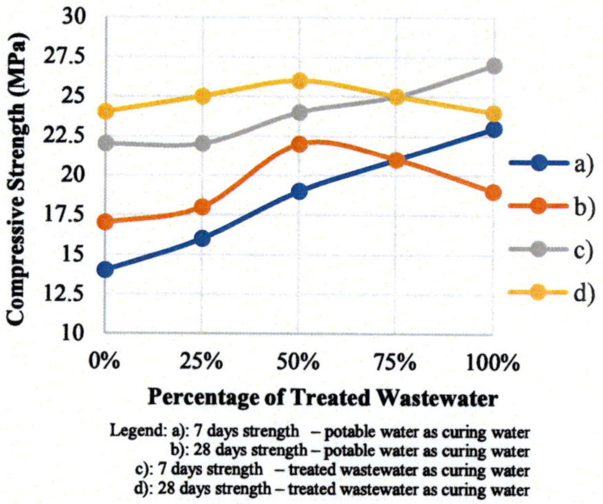

Fig. 8 Comparison of compressive strength of concrete against percentage of treated wastewater for both potable and treated wastewater curing methods

Table 7 Split tensile strength test results

Mixes	% of TW	Water content (kg/m^3)		28 days split tensile strength (MPa)	
		PW	TW	Potable water as curing water	Treated water as curing water
Mix 1	0	149	0	2.38	2.75
Mix 2	25	112	37	2.62	2.79
Mix 3	50	75	75	2.83	2.84
Mix 4	75	37	112	2.90	2.81
Mix 5	100	0	149	2.92	2.79

Legend PW—potable water, *TW*—treated wastewater

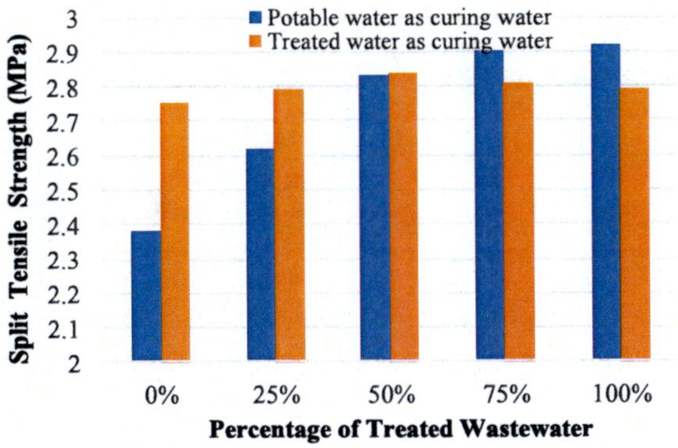

Fig. 9 Comparison of split tensile strength against proportion of treated wastewater

and then decreases. Cylinder with 100% substitution of treated sewage water over potable water cured in potable water shows the maximum value of 2.92 MPa. When the potable water is completely replaced with treated waste water, 18.5% increase in strength can be noticed. 4.4% of decrease in split tensile strength can be noticed when specimens were cured in treated wastewater.

3.4 *Flexural Strength Test*

Table 8 shows the 28 days flexural strength values of specimens cured in both potable water and treated waste water.

The variation of 28 days flexural strength against the proportion of treated wastewater with difference in curing water materials can be seen from Fig. 10. Flexural strength also follows the identical shift in variation as compressive strength test. The topmost flexural strength of 3.65 MPa can be noticed in beam specimen made with 100% treated water and cured in potable water. 14.7% increase in flexural strength can be observed when beam specimens are produced using treated wastewater and

Table 8 Flexural strength test results

Mixes	% of TW	Water content (kg/m^3)		28 days flexural strength (MPa)	
		PW	TW	Potable water as curing water	Treated water as curing water
Mix 1	0	149	0	3.12	3.45
Mix 2	25	112	37	3.27	3.49
Mix 3	50	75	75	3.54	3.56
Mix 4	75	37	112	3.63	3.51
Mix 5	100	0	149	3.66	3.47

Legend PW—potable water, *TW*—treated wastewater

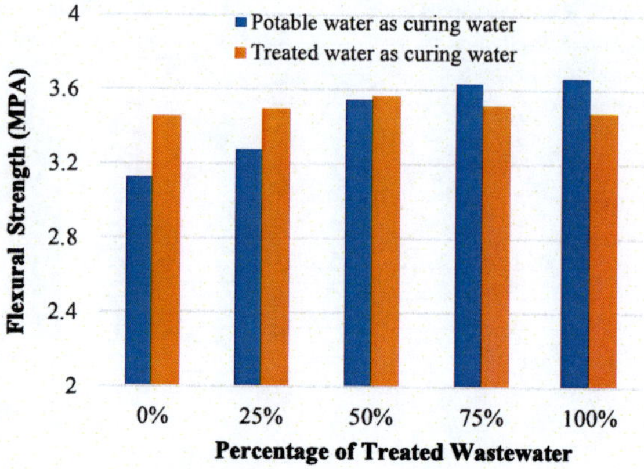

Fig. 10 Comparison of flexural strength against proportion of treated wastewater

cured in potable water. Maximum decrease in 5.2% in flexural strength can be noticed between different mixes and different curing options.

4 Conclusion

The following are the concluding remarks in accordance with the experiments conducted.

- The chemical composition of treated sewage water is more than potable water; however, they are not beyond the IS limits. And the health risk of its utilization is minimal because of limits of impurities and pH is as same as normal water.
- As the volume of treated sewage water hikes, the workability of the mixes increases with which it can be concluded that less effort is required for handling and placement of concrete with hike in portion of treated waste water. The shift in workability portrays the impact of treated sewage water on fresh parameters of concrete.
- The compressive strength of the mixes increases with increasing in curing period regardless of treated wastewater percentage used. The average compressive strength accelerates with hikes in percentage of treated sewage water (as mixing water).
- 18.8% development in cube compressive strength can be perceived when potable water is completely replaced with treated water, i.e., mix 5 showed higher compressive strength when correlated with mix 1.
- The compressive strength increases till 50% of replacement and then declines down when treated waste water is employed as curing water. The average compressive strength decreases by 11.8% when treated wastewater is utilized as curing water.

- The split tensile strength rose with addition of treated wastewater. 18.5% raise in split tensile strength can be identified, while treated waste water is employed as mixing water.
- The graph of split tensile strength follows a hill trend with maximum value at 50% replacement when curing water is treated waste water. 4.4% decrease in split tensile strength can be recognized when cured in treated water.
- As the quantity of treated wastewater increases, the flexural strength increases. 14.7% increase in flexural strength can be observed when blended using treated sewage.
- When treated waste water is employed as curing water, flexural strength increases till 50% replacement and then declines down. 5.2% decrease in flexural strength can be observed when curing water is treated waste water.
- From these above remarks, the influence of treated sewage water as both mixing water and curing water on mechanical properties of concrete can be understood. Although there are decrease in strengths of concrete specimens, but the reduction rates are not beyond the limits as recommended by American (ASTM), British code and German guidelines.
- Henceforth, the conclusion can be made as treated waste water can be utilized as both mixing water and treated waste water.

5 Future Scope of Work

- The study is bounded to M25 grade of concrete, whereas the scope can be extended for different grade of concrete mixes.
- Examining cement paste under scanning electron microscope (SEM) to know the chemical compositions and to estimate how each element would affect the strength parameters of concrete.
- Utilization of chemical admixtures in concrete production.
- Conduction of durability tests over concrete specimens.

Acknowledgements The authors gratefully acknowledge the assistance rendered by RASTA—Center For Road Technology, Bengaluru for their technical support in conduction of this experimental study. The authors also express their gratitude to the authority of Volvo Group India Pvt. Ltd. for facilitating the use of treated wastewater in the study.

References

1. https://www.usgs.gov/special-topic/water-science-school/science/where-earths-water
2. https://www.census.gov/
3. Kanwal H, Arif S, Javed MA, Farooq A, Khan MA (2018) Effect on compressive strength of concrete using treated wastewater for mixing and curing of concrete. Mehran Univ Res J Eng Technol 37(2):445–452

4. IS 456: 2000, Plain and reinforced concrete—code of practice. Bureau of Indian Standards, New Delhi, India
5. Kosmatka SH, Panarese WC (1995) Mixing water for concrete. In: Design and control of concrete mixtures, 6th Canadian edn. Portland Cement Association, pp 32–35
6. ASTM C 94-98, Standard specification for ready-mixed concrete. ASTM International
7. British Standard BS 3148: 1980 (1980) Methods of test for water for making concrete (including notes on the suitability of the water). British Standards Institution, London
8. Cebeci OZ, Saatci AM (1989) Domestic sewage as mixing water in concrete. ACI Mater J 86(5):503–506
9. Swami D, Sarkar K, Bhattacharjee B (2015) Use of treated domestic effluent as mixing water for concrete: effect on strength and water penetration at 28 days. Indian Concr J 89
10. Al-Jabri KS, Al-Saidy AH, Taha R, Al-Kemyani AJ (2011) Effect of using wastewater on the properties of high strength concrete. Proc Eng 14:370–376
11. https://www.thehindubusinessline.com/specials
12. https://niti.gov.in/
13. IS 12269: 2013, Ordinary Portland cement, 53 grade-specifications. Bureau of Indian Standards, New Delhi
14. IS 383: 2016, Coarse and fine aggregate for concrete—specification. Bureau of Indian Standards, New Delhi
15. IS 10262:2019, Concrete mix proportioning—guidelines. Bureau of Indian Standards, New Delhi, India
16. IS 1199: 1959, Methods of sampling and analysis of concrete. Bureau of Indian Standards, New Delhi
17. IS 516: 1959, Methods of tests for strength of concrete. Bureau of Indian Standards, New Delhi
18. IS 5816: 1999, Splitting tensile strength of concrete: method of test. Bureau of Indian Standards, New Delhi

Parametric Study of Shear Behavior of Compacted Kaolinite-Bentonite–Sand Mix Proportions

D. N. Jyothi, H. S. Prasanna, C. V. Amrutha Lakshmi, and D. K. Nageshwar

Abstract Fine-grained soils mainly consist of kaolinite, montmorillonite, and illite clay minerals in varying proportions along with sand and silts. The studies related to the shear behavior of such soils are being limited to the geographic locations and is having limited applications rather than the generic one. To have a thorough understanding of the general trend in variation of shear behavior of fine-grained soils having different clay mineralogical compositions around the world is tedious and time-consuming. The behavior of kaolinitic soils predominantly controlled by flocculant fabric whereas the behavior of montmorillonite soils is governed by dispersive fabric effects which also makes the study more complex. In the present experimental study, an attempt has been made, to investigate the behavior of natural fine-grained soils through a mixture of kaolinite-bentonite–sand in varying proportions. Physical properties and compaction characteristics such as MDD and OMC for Indian standard Light compaction for the mix proportions investigated were determined according to BIS specifications. The kaolinite-bentonite–sand mixture compacted at MDD and OMC was subjected to various load applied using the Box shear test and shear failure behavior was critically analyzed. Compaction characteristics of the mix proportions can be effectively correlated. The percent fines of the mix proportions have a significant bearing on the magnitude of cohesion and angle of internal friction. Useful regression models were also developed correlating shear strength behavior of the mix proportions with percent fines and sand which exhibits behavior akin to natural fine-grained soils when subjected to compactive energy.

Keywords Clay mineralogy · Compaction energy levels · Fine-grained soils shear behavior

D. N. Jyothi · H. S. Prasanna · C. V. Amrutha Lakshmi · D. K. Nageshwar (✉)
Department of Civil Engineering, National Institute of Engineering (Affiliated to VTU), Mysuru, India

H. S. Prasanna
e-mail: prasanna@nie.ac.in

L. Nandagiri et al. (eds.), *Sustainability Trends and Challenges in Civil Engineering*, Lecture Notes in Civil Engineering 162, https://doi.org/10.1007/978-981-16-2826-9_16

1 Introduction

The understanding of the mechanical properties of sand with silt and clay content is essential to do a realistic analysis of the geotechnical problems involving these materials. It has been recognized that the existence of clay content affects the shear strength and compression response of sands. The shear strength plays a major role in the construction of footing because of an increase in the stress in the soil which causes shear failure. In many of the engineering constructions, such as the construction of foundation, embankment, sheet piling, retaining wall, the value of cohesion and angle of internal friction is needed for the design purpose. The major factors affecting the shear behavior of the soil are the type and amount of clay minerals present, water content, gradation, dry density, soil structure, thixotropy, and the normal effective stress acting on the failure plane.

To have a thorough understanding of the general trend in variation of shear behavior of fine-grained soils having different clay mineralogical compositions around the world is tedious and time-consuming. Hence, any laboratory study of shear behavior of artificial mix proportion depicting the natural fine-grained soil is a valuable addition to the existing literature. Fine-grained soils predominantly consist of fine size clay particles which impart plasticity characteristics like liquid limit and plastic limit. Clay minerals are essentially hydrous aluminum silicates lesser than 2 μm in size. These soil types are made of two building blocks namely silica tetrahedron and alumina octahedron (Gibbsite), See details in Fig. 1. The degree of swelling and shrinking of a clay mineral depends on the bond between these two building blocks.

1.1 Literature Review

Many researchers have attempted to study the shear behavior of natural fine-grained soils and artificial mix proportions mainly consisting of kaolinite and bentonite mixed in limited proportions.

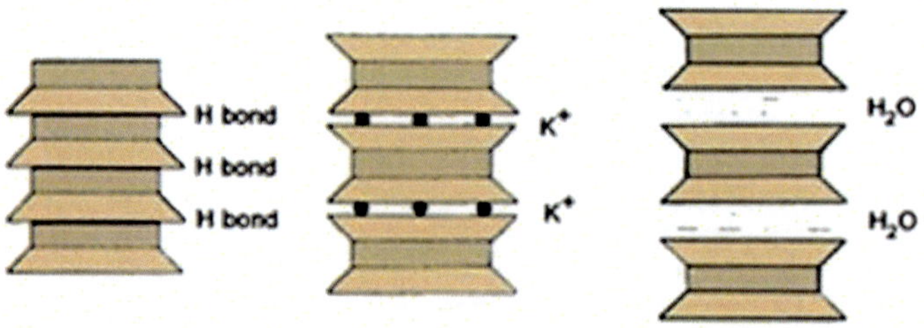

Fig. 1 Types of clay minerals (Kaolinite, Montmorillonite and Illite)

Wasti and Alyanak [1] have studied the mixture of sand and clay proportions mentioned that, the role of voids to be filled with the different percentages of clay at a higher value of porosity, which brings a change in particle orientation of a mixture. Also, the relationship between the liquid limit (W_L) and the plastic limit was found to follow a straight line. The various mixes of sand and clay indicated that the soil samples tested possessed characteristics of sandy soils and clayey soils. At W_L, they observed that the threshold value exists about 25% of kaolin content, for the kaolin clay involved mixture also.

Novais-Ferreira [2] based on their study on consolidated drained (CD) box shear tests on mixtures of artificial soils with different percentages of clay, fine sand, coarse sand, and clayey soil which is montmorillonitic dominant found that with an increase in the percentage of clay there is a tendency of decrease in shear stresses (maximum and limited).

Georgiannou [3] conducted a study on the characteristics of clayey sandy soils with two different types of loadings, i.e., constant and cyclic. He found that the stress–strain behavior of soil was influenced by the percentage of fines. It was also seen that there existed a suppression of the dilatant nature of soil samples tested with an increase in the percentage of fines including the samples with 40% fines.

Georgiannou et al. [4] have made an experimental investigation on load-deformation characteristics of anisotropically consolidated clayey sands using digitally maintained triaxial specimens. By immersing the Ham river sand into the kaolin solution, the samples were prepared. By this, the effect of change in percentage of clay and void ratio of granular sized particles was observed. The chosen method was creating a material which is having less stability, the maximum void ratio in granular level, maximum brittleness character, and ductility character.

Pitman et al. [5] have studied the effect of fines content and arrangement of soil particles on characteristics of loosely arranged sandy soils. These samples were formed by changing the percentage of plastic and non-plastic fines with moist compaction procedures and consolidated by maintaining the same effective stress level. The samples were subjected to constant undrained triaxial loading and consolidated isotropically. They have observed that the behavior of undrained brittleness was decreased with the percentage of fines and increased with plastic and non-plastic fines content.

Bayoğlu et al. [6] have attempted to study the effects of the fine-sized particles (particle size <75 μm) concerning shear strength and compressibility. Different proportions of soil mixtures were considered from sand to silt–clay. The tests conducted on the soil mixtures were drained box shear and consolidated undrained triaxial compression with different percentages of fines, i.e., 5, 15, 35, 50, 75, and 100%. They concluded that until 50% fines (optimum) the internal friction angles varied between 30° and 38°, thereafter a slight decrease in angle of internal friction with an increase in the percentage of fines and variation is about 10°.

Thevanayagam [7] conducted an experimental investigation on large strain undrained shear strength (s_{us}) in triaxial compression for particular host sand mixed with various percentages of non-plastic fines. The test results indicated that the intergranular void ratio (e_s) was playing a very crucial role in large strain undrained shear strength (s_{us}) of silty sands. Silty sand having less magnitude of large strain undrained shear strength (s_{us}) compared with host sand at the same void ratio level. For less than the value of maximum void ratio level of host sand ($e_{max, HS}$), the two grains of sand are showing the same s_{us} and these are independent of initial confining pressure. For the e_s (silty sand) > $E_{max, HS}$ condition, the s_{us} depends on the initial confining pressure. The value of s_{us} was decreased with an increase in e_s when the consolidation stress is very low (loose state).

Naser Al Shayea [8] based on their study reported that the characteristics of soil mass as small fractions having clay mineralogical dominance and the amount of clay percentage was playing an important role in finding the shear strength and compressibility. By increasing the percentage of fines, the plasticity characteristics and coefficient of secondary compression (C_R) values were increased and the angle of internal friction and hydraulic conductivity values were decreased.

Olmez [9] conducted three different series of undrained and drained triaxial tests and box shear tests to know the shear strength characteristics of soil mixtures with the effect of kaolin in sand-clay mixtures. At an optimum percentage of kaolin content of 20%, a significant change in shear strength and stress–strain response were observed.

Pakbaz and Moqaddam [10] attempted to study the effect of sand particles arrangement on the behavior of sand-clay mixtures. They conducted the box shear tests on existing over consolidated expansive clays and densified soil mixtures of sand and bentonite. The drained shear strength and angle of internal friction of the soil mixtures were decreased with an increase in the percentage of fines. They also observed that the optimum percentage of the fine is 30%, for significant change in shear strength and stress–strain behavior.

Li et al. [11] made a study to understand the effect of test conditions on the shear behavior of composite soil. They observed that a homogeneous soil specimen becomes heterogeneous after shearing. The shear zone is the most affected part of the specimen in terms of structure and void ratio, whereas these properties remain unchanged in the outer zones.

Ali Shafiee et al. [12] have attempted to study the undrained behavior of compacted sand-clay mixtures under monotonic loading paths and reported that, the undrained shear strength in compression increases with sand content, particularly when sand content reaches as high as 60% and the initial confining stress is low. On the other hand, undrained shear strength in extension gradually increases with sand content.

Jyothi and Prasanna [13] conducted an experimental study on the Shear behavior of compacted clay and sand mixtures using a triaxial shear test experiment and concluded that pure-clay have maximum cohesion and with the addition of sand in increasing proportions, the value of cohesion decreases gradually since sand is a cohesion-less soil, whereas the angle of internal friction is less for pure-clay and increases as the percentage of sand increases because the voids in the clay get occupied by sand and friction between them increases.

Based on a detailed literature review, the study related to shear behavior of compacted kaolinite-bentonite–sand mix proportions with particular reference to clay mineralogy is very scanty.

2 Materials and Methodology

2.1 Preparation of Materials

Materials used for the present experimental study are

- River sand passing 425 μm.
- Commercially available pure kaolinite and bentonite clay minerals.

Commercially available clay minerals (kaolinite and bentonite) were procured and stored in airtight containers.

The river sand used in the present study was retrieved from a riverbed. This is composed of subrounded particles with a specific gravity of 2.65. The sand was wet washed on a 75 μm sieve to take away the organic content.

The sand was kept in an oven for a period of 24 h for oven drying, at a temperature of 100 °C ± 5 °C and brought to room temperature.

Figure 2 shows the grain-size distributions of riverbed sand used in this present study. The sand particles were found to be subrounded with a uniformity coefficient (c_u) and coefficient 'c_c' of 2.33 and 1.4, respectively. Also, the median particle sizes, D_{10}, D_{30}, and D_{60} were found to be 0.3, 0.54, and 0.7, respectively.

The sand was then mixed with different percentages of clay minerals like kaolinite and montmorillonite to represent various naturally occurring soil mixes. These were kept in a double-barreled airtight container with a plastic cover to prevent moisture losses. The percentage of bentonite in each sample was maintained the same. The

Fig. 2 This represents the failure sample

Table 1 Mix proportions of the samples used in the study

Bentonite (%)	Kaolinite (%)	Sand (%)
10	10	80
10	20	70
10	30	60
10	40	50
10	50	40
10	60	30
10	70	20
10	80	10

details on the mix proportions of samples used in the present study were shown in Table 1.

2.2 *Tests Conducted*

a. Grain-size analysis test (IS: 2720-part 4-1985)—to classify the soil.
b. Specific gravity test (IS: 2386-part 3-1963)—to find the specific gravity of fine aggregate.
c. Atterberg limits tests (IS: 2720-part 5-1985)—to determine plastic limit liquid limit and plasticity index.
d. Standard and modified compacted test (IS: 2720-part 7-1980)—for obtaining optimum moisture content (OMC) and maximum dry density (MDD).
e. Consolidated drained direct shear test (IS: 2720(Part-13)-1986)—to find shear parameters, i.e., cohesion (*C*) and friction angle (□).

An intermediated strain-controlled digital direct shear apparatus, with a specimen dimension 60 mm × 60 mm× 20 mm was used in the experimental study. The apparatus was equipped with two digital dial gauges to measure the vertical and horizontal (shear) displacements. Normal stress was applied to the specimen through a lever loading device. Shear force was measured using two proving rings with measuring capacities of 2.0 and 4.5 KN, and accuracy within ±1% of the indicated load. Horizontal and vertical displacements and shear forces applied were recorded by a data logger at the minimum shear displacement intervals of 2 μm.

3 Results and Discussions

Table 2 shows compaction characteristics of artificial mix proportions keeping bentonite constant.

Figure 3 shows the relation between compaction characteristics.

Table 2 Compaction characteristics of mix proportions (Bentonite constant)

Combination	Γdmax (KN/m^3)	OMC (%)
10B + 10K + 80S	19.51	9.8
10B + 20K + 70S	18.98	11.6
10B + 30K + 60S	18.36	12.78
10B + 40K + 50S	17.75	13.9
10B + 50K + 40S	17.12	16.1
10B + 60K + 30S	16.35	17.4
10B + 70K + 20S	15.68	19.8
10B + 80K + 10S	14.56	23.5
20B + 10K + 70S	19.46	13.62
20B + 20K + 60S	18.61	16.11
20B + 30K + 50S	17.71	18.42
20B + 40K + 40S	17.11	20.72
20B + 50K + 30S	16.62	22.61
20B + 60K + 20S	16.18	24.68
20B + 70K + 10S	15.58	27.83
30B + 10K + 60S	18.14	15.76
30B + 20K + 50S	17.31	17.32
30B + 30K + 40S	16.96	19.5
30B + 40K + 30S	16.04	21.4
30B + 50K + 20S	15.21	24.3
30B + 60K + 10S	14.57	26.4
40B + 10K + 50S	18.06	17.3
40B + 20K + 40S	17.12	20.1
40B + 30K + 30S	16.24	23.3
40B + 40K + 20S	15.05	22.8
40B + 50K + 10S	14.51	27.1
50B + 10K + 40S	17.15	18.11
50B + 20K + 30S	15.8	21.8
50B + 30K + 20S	15.15	22.75
50B + 40K + 10S	14.55	26.21
60B + 10K + 30S	16.17	19.41
60B + 20K + 20S	15.92	22.73
60B + 30K + 10S	15.2	26.5
70B + 10K + 20S	15.74	23
70B + 20K + 10S	14.43	26.4
80B + 10K + 10S	14.71	25

Fig. 3 Relationship between OMC and MDD

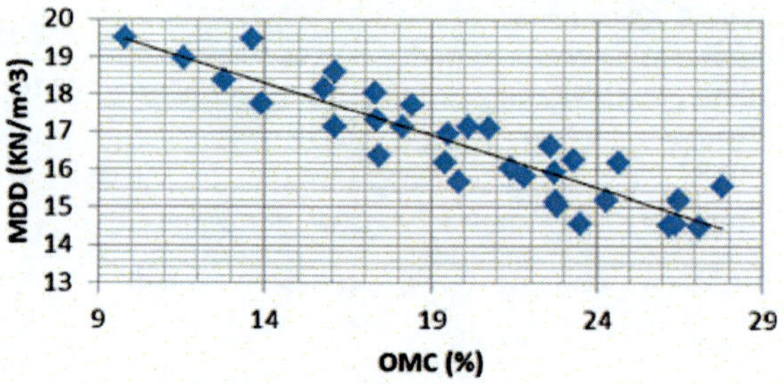

From Fig. 3, it can be observed that there is a very good correlation between compaction characteristics such as between MDD and OMC as shown in Eq. (1) with a coefficient of correlation $r = 0.90$, and a coefficient determination of R-square = 0.81.

$$\text{MDD} = -0.2771 * \text{OMC} + 22.184 \tag{1}$$

It can be also observed that as OMC increases, the MDD decreases, which is a characteristic feature of natural fine-grained soils.

Figure 4 shows the relation between shear strength and MDD of mix proportions.

Table 3 shows details on the regression model developed correlating shear strengths to the MDD for various mix proportions.

From Fig. 4, it can be observed that as the percent bentonite in the mix proportion increases, shear strength decreases, which very well agrees with the observation made in document literature [2]. Further MDD can be effectively correlated with shear strength of mix proportion (Table 3).

Figure 5 shows the relationship between shear strength and OMC of the mix proportions.

Table 4 shows details on the regression model developed correlating shear strengths to the OMC for various mix proportions.

From Fig. 6 it can be observed that as percent fines of the mix proportion increase, the water holding capacity also increases which leads to a higher magnitude of OMC.

Fig. 4 Relationship between shear strength and MDD

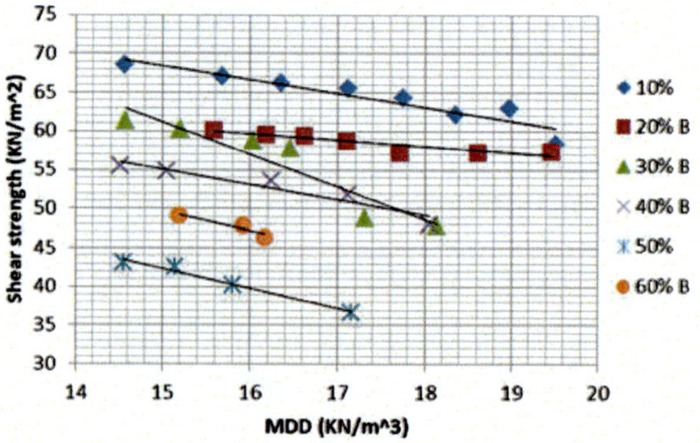

Table 3 Regression model developed correlating shear strength with the MDD

Combination	Equation	R-square value
10% bentonite	$\tau = -1.772(\text{MDD}) + 95.009$	0.86
20% bentonite	$\tau = -0.8134(\text{MDD}) + 72.612$	0.83
30% bentonite	$\tau = -4.1921(\text{MDD}) + 124.06$	0.88
40% bentonite	$\tau = -1.9502(\text{MDD}) + 84.351$	0.90
50% bentonite	$\tau = -2.5745(\text{MDD}) + 80.951$	0.98
60% bentonite	$\tau = -2.484(\text{MDD}) + 86.919$	0.90

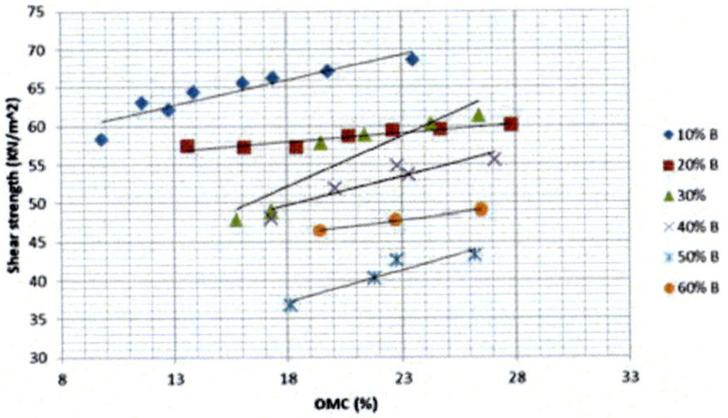

Fig. 5 Relationship between shear strength and OMC

Table 4 Regression equations showing the correlations of shear strength with OMC

Combination	Equation	R-square value
Figure 10a, b	$\tau = 0.6565(\text{MDD}) + 54.126$	0.86
Figure 11a, b	$\tau = 0.2308(\text{MDD}) + 53.773$	0.86
Figure 12a, b	$\tau = 1.3178(\text{MDD}) + 28.393$	0.83
Figure 13a, b	$\tau = 0.7512(\text{MDD}) + 36.15$	0.86
Figure 14a, b	$\tau = 0.8172(\text{MDD}) + 22.47$	0.88
Figure 15a, b	$\tau = 0.3711(\text{MDD}) + 39.273$	0.98

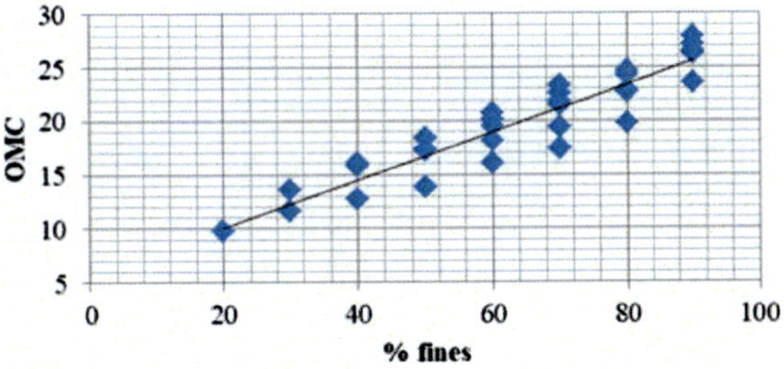

Fig. 6 Relationship between OMC and % fines

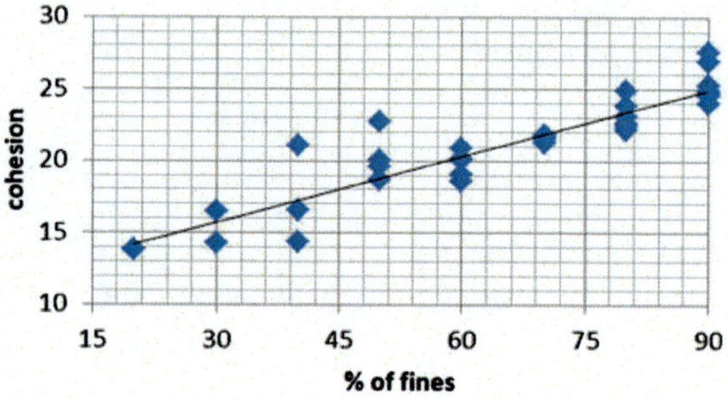

Fig. 7 Relationship of percentage fines with cohesion

The relationship between OMC and percent fines is given by Eq. (2) with a correlation coefficient of $R = 0.94$.

$$\text{OMC} = 0.22 * (\text{FINES}) + 5.63 \tag{2}$$

From Fig. 7 it can be observed that as percent fines increase, the magnitude of specific surface increases which leads to a higher degree of interparticle attraction (cohesion). The relationship between percent fines (PF) and cohesion is given by Eq. (3) with a correlation coefficient of $R = 0.90$.

$$C = 0.1532 * (\text{FINES}) + 11.163 \tag{3}$$

From Fig. 8 it can be seen that the magnitude of the angle of internal friction of the mix proportion decreases as percent fines increase, this is since the intermolecular granular friction reduces because of the higher percentage of fines. The relationship between friction angle and cohesion is given by Eq. (4) with a correlation coefficient of $R = 0.93$.

$$\text{FA} = -0.1603 * (\text{FINES}) + 24.913 \tag{4}$$

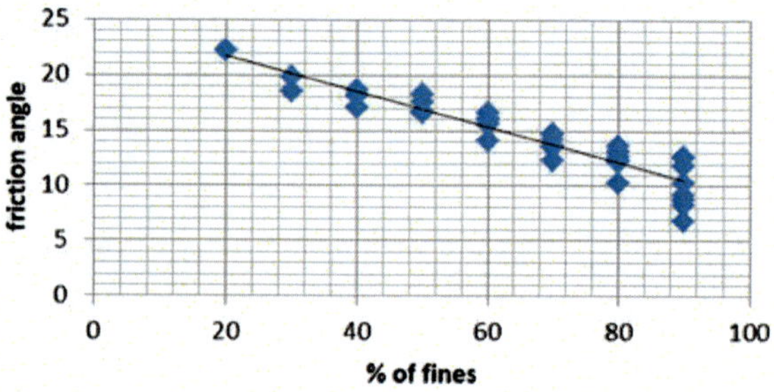

Fig. 8 Relationship between the angle of internal friction and percentage fines

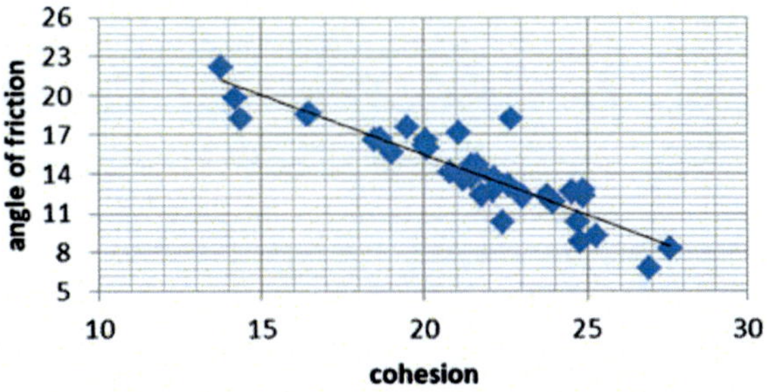

Fig. 9 Relationship between the angle of cohesion and angle of friction

From Fig. 9, it is seldom observed that as the magnitude of cohesion of mixed proportion increases the magnitude of the angle of internal friction decreases which is akin to natural fine-grained soil behavior. The relationship between friction angle and cohesion is given by Eq. (5) with a correlation coefficient of $R = 0.89$.

$$\text{FA} = -0.9181 * (C) + 33.853 \tag{5}$$

Table 5 shows details on the regression model developed correlating shear strength (SS) to percentage of sand (PS) and percentage of fine (PF), respectively (keeping 10% bentonite constant) mix proportions.

Table 6 shows details on the regression model developed correlating shear strength to percentage of sand (PS) and percentage of fine, respectively (keeping 20% bentonite constant) mix proportions.

Table 7 shows details on the regression model developed correlating shear strength to percentage of sand and percentage of fine, respectively (keeping 30% bentonite constant) mix proportions.

Table 8 shows details on the regression model developed correlating shear strength to percentage of sand and percentage of fine, respectively (keeping 40% bentonite constant) mix proportions.

Table 9 shows details on the regression model developed correlating shear strength to a percentage of sand and percentage of fine, respectively (keeping 50% bentonite constant) for the mix proportions.

Table 5 Regression equations showing the correlations of shear strength with percentage sand (10% Bentonite constant)

Combination	Equation	R^2 value
Figure 10a	SS = 0.1035PF + 59.034	0.98
Figure 10b	SS = −0.1118PS + 69.631	0.98

Table 6 Regression model developed correlating shear strength with percent sand (20% Bentonite constant)

Combination	Equation	R^2 value
Figure 11a	SS = 0.0486PF + 55.777	0.96
Figure 11b	SS = −0.0486PS + 60.634	0.96

Table 7 Regression model developed correlating shear strength with percent sand (30% Bentonite constant)

Combination	Equation	R^2 value
Figure 12a	SS = 0.3009PF + 36.478	0.96
Figure 12b	SS = −0.3009PS + 66.563	0.96

Table 8 Regression model developed correlating shear strength with percent sand (40% Bentonite constant)

Combination	Equation	R^2 value
Figure 13a	SS = 0.1375PF + 43.551	0.96
Figure 13b	SS = −0.1375PS + 57.301	0.96

Table 9 Regression model developed correlating shear strength with percent sand (50% Bentonite constant)

Combination	Equation	R^2 value
Figure 14a	SS = 0.2337PF + 23.12	0.98
Figure 14b	SS = −0.2337PS + 46.49	0.98

Table 10 Regression model developed correlating shear strength with percent sand (60% Bentonite constant)

Combination	Equation	R^2 value
Figure 15a	SS = 0.132PF + 37.203	0.98
Figure 15b	SS = −0.132PS + 50.403	0.98

Table 10 shows details on the regression model developed correlating shear strength to percentage of sand and percentage of fine, respectively (keeping 60% bentonite constant) mix proportions.

The shearing strength of fine-grained soil is primarily affected by cohesion, water content, gradation, dry density, soil structure and thixotropy.

From Figs. 10a through 15b, it can be observed that percent fines and sand of the mix proportion can be very well correlated with shear strength of the mix proportion which is very useful for the estimation of shear strength, i.e., by knowing the percent fines or sand of the mix proportion, the shear strength can be estimated with a fair degree of accuracy.

From Figs. 10a, 11a, 12a, 13a, 14a, 15a it can be observed that the shear strength increases with the increase in percent fine content of the mix proportions which imparts magnitude of the cohesion of the mix proportions mainly because of the change in the orientation of the particles from flocculated to disperse state due to the application of the normal stress thereby increasing the shear strength. The magnitude of shear strength of each group increases from 13.96% for Group-1 to 5.7% Group-5 respectively for percent fines.

The variation of the values here is mainly attributed due to the formation of kaolinite ($Al_2Si_2O_5(OH)_4$) and bentonite ($Al_2H_2Na_2O_{13}Si_4$) itself, i.e., kaolinite forms as a result of weathering of aluminum silicate minerals such as feldspar whereas bentonite forms from volcanic ash in the presence of water. The naturally available

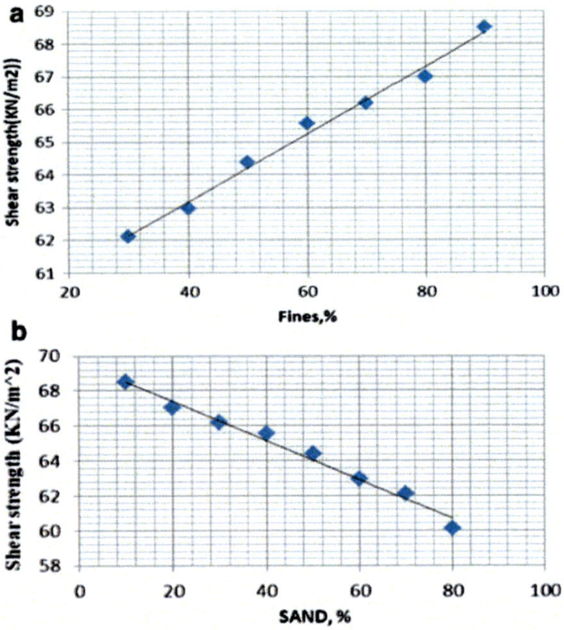

Fig. 10 **a** Relationship of percentage fines with shear strength (10% Bentonite Constant). **b** Relationship between the percentage of sand and shear strength (10% Bentonite Constant)

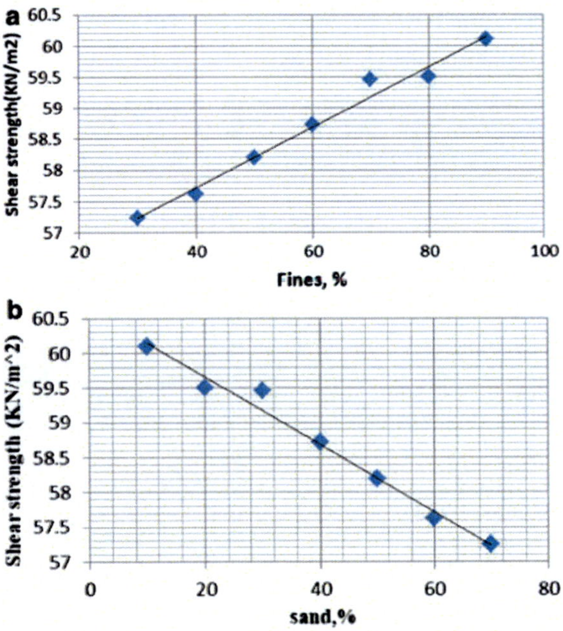

Fig. 11 **a** Relationship between percentage fines and shear strength (20% Bentonite Constant). **b** Relationship between the percentage of sand and shear strength (20% Bentonite Constant)

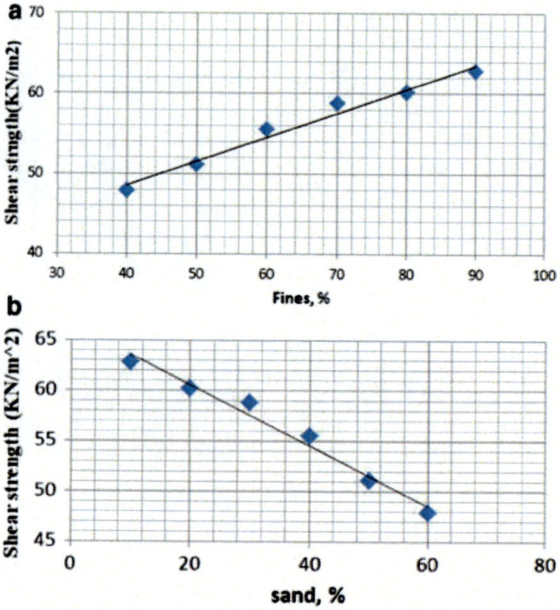

Fig. 12 **a** Relationship between the percentage of fines and shear strength (30% Bentonite Constant). **b** Relationship between the percentage of sand and shear strength (30% Bentonite Constant)

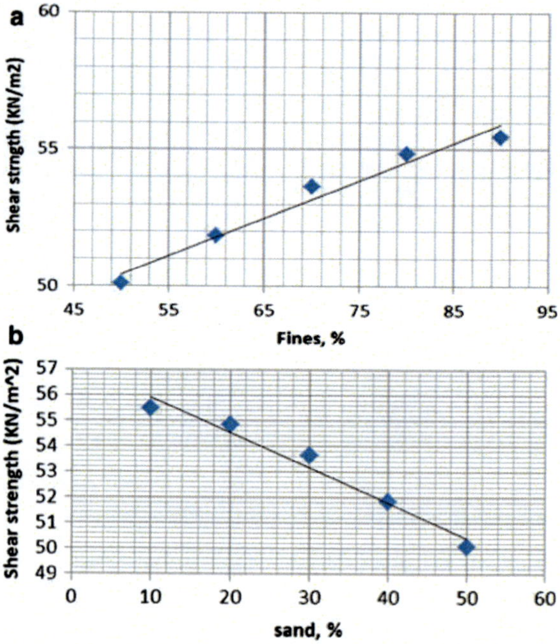

Fig. 13 **a** Relationship between the percentage of fines and shear strength (40% Bentonite Constant). **b** Relationship between the percentage of sand and shear strength (40% Bentonite Constant)

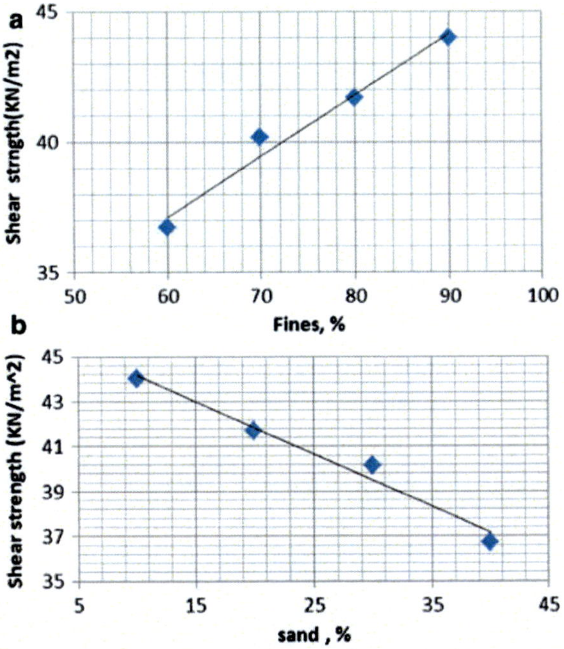

Fig. 14 **a** Relationship between the percentage of fines and shear strength (50% Bentonite Constant). **b** Relationship between the percentage of sand and shear strength (50% Bentonite Constant)

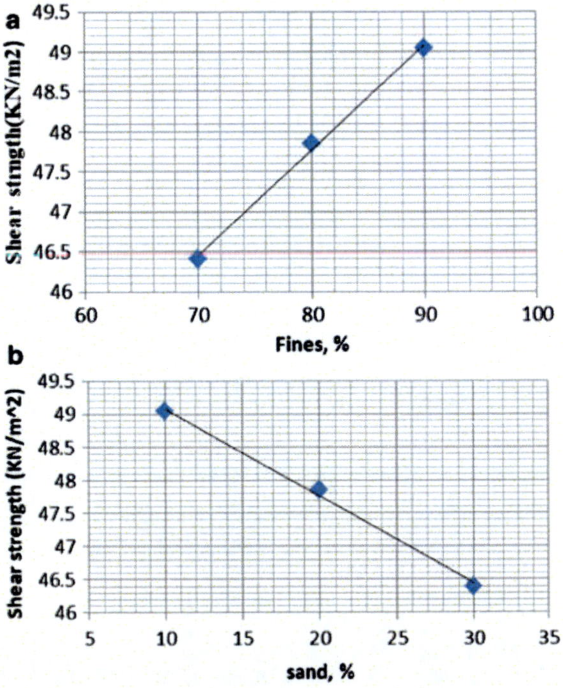

Fig. 15 **a** Relationship between the percentage of fines and shear strength (60% Bentonite Constant). **b** Relationship between the percentage of sand and shear strength (60% Bentonite Constant)

clay soils with lower cation exchange capacity have more remediation potential and are more susceptible to dispersion.

The kaolinite-bentonite mixtures with higher amounts of montmorillonite (or higher amounts of bentonite) experience prolonged swelling compared with the compacted kaolinite-bentonite samples with lower amounts of montmorillonite, due to the higher water retention capacity of the samples rich in montmorillonite, leading to a lower rate of water absorption.

It is noteworthy to observe here, that in the present experimental study; water content that influences the shear strength is not only controlled by the molding water content but also includes any changes in moisture content after placement.

4 Conclusion

Based on the above-mentioned experimental study, the following conclusions can be made

1. Compaction characteristics, i.e., MDD, and OMC of the mix proportions can be correlated with a fair degree of accuracy.
2. The percent fines and sand of the mix proportions can be correlated with shear parameters C and φ.
3. The shear parameters C and φ of the mix proportion can be correlated with the shear strength of the mix proportion.
4. The correlation between compaction characteristics, percent fines with shear parameters and shear parameters with shear strength very well agrees with the behavior of natural fine-grained soils having kaolinite and montmorillonitic clay dominance.

Acknowledgements The authors would like to acknowledge the support given by the Research Center Department of Civil Engineering, The National Institute of Engineering Mysuru in carrying out the experimental study.

References

1. Wasti Y, Alyanak I (1968) Kil Muhtevasinin Zeminin Davramsina Tesiri, Insaat Miihendisleri Odasi, Tiirkiye insaat Miihendisligi 4, Teknik Kongresi. Ankara
2. Novais-Ferreira H (1971) The clay content and the shear strength in sand-clay mixtures. In: Proceedings of the 5th African regional conference on soil mechanics and foundation engineering. Luanda
3. Georgiannou VN (1988) Behavior of clayey sands under monotonic and cyclic loading. Ph.D. thesis. Department of Civil Engineering, Imperial College of Science, Technology and Medicine, London, England
4. Georgiannou VN, Burland JB, Hight DW (1990) The Undrained behavior of clayey sands in triaxial compression and Extension. Geotechnique 40(30):41–449

5. Pitman TD, Robertson PK, Sego DC (1994) Influence of fines on the collapse of loose sands. Can Geotech J 31:728–739
6. Bayoğlu E (1995) Shear strength and compressibility of sand-clay mixtures, Middle East Technical University
7. Thevanayagam S (1998) Effect of fines and confining stress on undrained shear strength of silty sands. J Geotech Geoenviron Eng Div ASCE 124(6):479–491
8. Al-Shayea NA (2001) The combined effect of clay and moisture content on the behavior of remolded unsaturated soils. Eng Geol 62 (4):319–342
9. Olmez MS (2008) Shear strength behavior of sand-clay mixtures. Middle East Technical University
10. Pakbaz MS, Moqaddam AS (2012) Effect of sand gradation on the behavior of sand-clay mixtures, University of Shahid Chamran Ahwaz, Iran
11. Li Y, Chan LS, Yeung AT, Xiang X (2013) Effects of test conditions on shear behavior of composite soil. Proc Inst Civ Eng Geotech Eng 166(3):310–320
12. Afacan KB, Yniesta S, Shafiee A, Stewart JP, Brandenberg SJ (2019) Total stress analysis of soft clay ground response in centrifuge models. J Geotech Geoenvironmental Eng 145(10):04019061
13. Jyothi DN, Prasanna HS (2020) Shear behavior of compacted kaolinite-sand mixtures. IRJET 07(special issue)

A Study on Volumetric Shrinkage of Compacted Fine-Grained Soils Subjected to Various Energy Levels

H. S. Prasanna, C. Thrupthi, and B. Varshini

Abstract Plasticity characteristics of the soils (fine-grained) assume greater importance in view of engineering behaviour. It mainly depends on the clay mineralogical composition, predominantly controlled by physico-chemical phenomenon. This is due to the fact that the contradictory behaviour of the presence of the least active clay mineral kaolinite to most active montmorillonite in addition to Illite and other clay minerals, in different proportion. The contradictory behaviour of kaolinite and montmorillonite makes the study more complex. Shrinkage limit, among the Atterberg limits (plasticity characteristic) of the soils (fine-grained), associates with degree of packing, while the liquid and plastic limits classify the soils more qualitatively. The shrinkage and expansion of fine-grained soils lead to distress of foundation. The study related to volumetric behaviour of montmorillonite soils assumes a greater practical significance. Role of compaction energy on shrinkage behaviour of soils with particular reference to soil fabric also plays a vital role in the design of earthen-related structures. The potential shrinkage of soil is quantitatively described by the newly proposed method for testing expansive clayey soils, known as the ring shrinkage test. The measurements of lateral, vertical and linear shrinkages are provided by the test. Volumetric shrinkage and shrinkage limit along with compaction energy are determined alternatively by this method. The present experimental study mainly focuses on the effects of placement conditions, role of clay minerology and compaction effort on the shrinkage (linear, lateral and volumetric) behaviour of compacted fine-grained soils using ring shrinkage test. Correlations were developed between characteristics of compaction, i.e. optimum moisture content (OMC), maximum dry density (MDD), volumetric (VS) and lateral shrinkage (LS) with compaction energy (CE).

Keywords Compaction energy · Lateral shrinkage · Volumetric shrinkage

---

H. S. Prasanna · C. Thrupthi (✉) · B. Varshini
Department of Civil Engineering, The National Institute of Engineering, Mysuru, India

H. S. Prasanna
e-mail: prasanna@nie.ac.in

L. Nandagiri et al. (eds.), *Sustainability Trends and Challenges in Civil Engineering*, Lecture Notes in Civil Engineering 162, https://doi.org/10.1007/978-981-16-2826-9_17

1 Introduction

For the sites with expansive clayey soils to be investigated in detail, the effects of changes in soil wetness conditions on shrinkage and swelling phenomenon (soil volume change potential) are needed in general. According to the site conditions various techniques are employed to modify the soil, in order to upgrade its engineering soil properties. Generally, the shrinkage effect of soils is correlated with the different compaction energies and its behaviour. The linear shrinkage can be defined as the ratio of reduction in the vertical and lateral dimensions to its original dimensions expressed as the percentage, when the water content of the specimen is reduced from its in-situ value to its shrinkage limit. The basic ring shrinkage test method allows monitoring of the volume change with a gradual lowering in water content due to desiccation.

Knowledge of shrinkage effect of fine-soils at various compaction energy levels assumes a huge importance regarding the significance for practical purposes. An attempt to understand shrinkage effect on compacted soils of fine-grain with respect to clay mineralogy is made. It is observed that compaction control and clay minerology are the significant factors controlling the behaviour of earthen constructions with fine-grained soils which are compacted.

2 Literature Review

Sridharan and Prakash [1] have established the independency of the shrinkage limit over the plasticity characteristics, which rather depends upon the grain-size distribution relation of soil. Daniel with Wu [2], Benson along with Albrecht [3] and Osinubi et al. [4, 5] have all suggested more shrinkage in compacted specimen with greater moulding water content. While, Benson and Albrecht [3] further report, that the compaction with higher efforts and the use of nearly optimum content of water in the specimen have reduced the strains of volumetric shrinkage and the amount of water per unit volume during saturation. Anand et al. [6] have made an effort to compare volumetric shrinkage strains via digital and conventional manual approaches. [7] explicitly attempted in developing an equation explaining the relationship of swell per cent and swell pressure. [8] made an effort to obtain correlation between optimum content of moisture and dry density for different soils compacted at various energy levels of compaction. George et al. [9] state that, in relation to optimum content of moisture (OMC) in soil, by increasing the content of water used for moulding results in the increase in volumetric shrinkage strain. Kundiri et al. [10] have shown that the lower magnitudes of Volumetric shrinkage strain were observed for soils compacted at higher compactive efforts. Mahvash et al. [11] have observed that the fly ash percentage added to the soil (as an admixture) increases the OMC values while reducing the magnitudes of MDD. Yamusa et al. [12, 13] report that the effects of gradation, content of fines at different gradations and water content used for moulding on the VS of laterite soil which is compacted. Julina and Thyagaraj [14] have tried to determine the volume of soil material under shrinkage (other than the crack volume) with digital camera images. Jiangtao with Jiang [15] proved experimentally that through the volume adjustable process (repeated cycles of contraction and expansion), the magnitude of the volumetric shrinkage can be reduced to about 0.1%. Udukumburage et al. [16] emphasized the applicability of the conventional shrinkage test using oedometer, for an approximate estimation of shrinkage vertically for expansive soils. Mishra et al. [17] have reported a systematic investigation and analysis on the evolution of geometrical samples during evaporative de-watering. After the literature review in detail, it is observed that the studies related to ring shrinkage test to evaluate the expansive behaviour of natural soils (fine-grained) having no common clay minerology and subjected to different compactive energy is forcing the construction industry to use such soils mainly because of non-availability of gravelly soils at large. The study related to volumetric behaviour of fine-grained soils is of paramount importance from societal needs and eco-friendly construction methods.

3 Materials and Methodology

For the present experimental study, six soils from Mysore district and Chamarajanagar district were selected based on the clay mineralogical composition. The soils were carefully brought from the field to the laboratory and subjected to air drying. After ensuring air dried process completion, the samples were kept in hot air oven for 24 h at 105 ± 5 °C. The samples were brought to room temperature and powdered, sieved through 425 μm IS sieve. The samples were thoroughly mixed to ensure the uniform degree of mix. Then they were stored in airtight containers. These soil samples were selected on the basis of Free Swell Ratio (IS;2911, Part-3-1980), free swell index (IS;1498-1970), modified free swell index (Sridharan et al. [18]) to know the dominant clay mineral composition. All the samples of soil were tested for their index property by the tests such as free swell (IS: 2720, Part 40-1977), specific gravity (IS:2720, Part-3 (section-10)-1980), and analysis of grain size (IS:2720, Part-4-1985). The compaction tests like reduced standard proctor (60% of energy as that of standard proctor), standard proctor (IS:2720, Part-7-1974), reduced modified proctor (60% of energy as that of modified proctor) and modified proctor (IS:2720, Part-7-1980/1987) tests were performed on all the samples of soil under study. For each of these tests, about 6–8 samples weighing 2.5 kg each were kept in airtight double-walled polythene covers, after a thorough mix with varying moisture contents for moulding. The different soil samples depend upon their types. Depending upon the soil type, the samples were undisturbed for about 5–10 days to ensure the equilibration of moisture in them. The soil samples were subjected to compaction tests, after the period of equilibration. The tabulation that follows (Table 2) contains the maximum density of dry soil and moisture content that is optimum for every soil under study relating to the compaction curves obtained. The soil samples were statically compacted into the shrinkage rings (to the respective MDD and OMC of various energy levels) which is made up of mild steel of diameters varying from 30 to 120 mm.

Volumetric shrinkage (VS), lateral shrinkage (LS) and compaction energy (CE) of the specimen were calculated using the following Eqs. (1), (2) and (3):

$$\mathrm{VS} = \left(V_i - V_f\right)/V_i \tag{1}$$

$$\mathrm{LS} = \left(h_i - h_f\right)/h_i \tag{2}$$

$$\mathrm{CE} = (N \times B \times R_W \times H)/M_V \tag{3}$$

where

V_i initial volume of moulded soil sample
V_f final volume of the shrunk soil sample
h_i initial height of moulded soil sample
h_f final volume of the shrunk soil sample
N Number of soil layers
B Number of blows
H Height of fall
R_W Weight of the ram
M_V Volume of mould.

Figure 1 shows the ring shrinkage testing equipment and method.

Tables 1 and 2 shows the index property and characteristics of compaction of the soils under study.

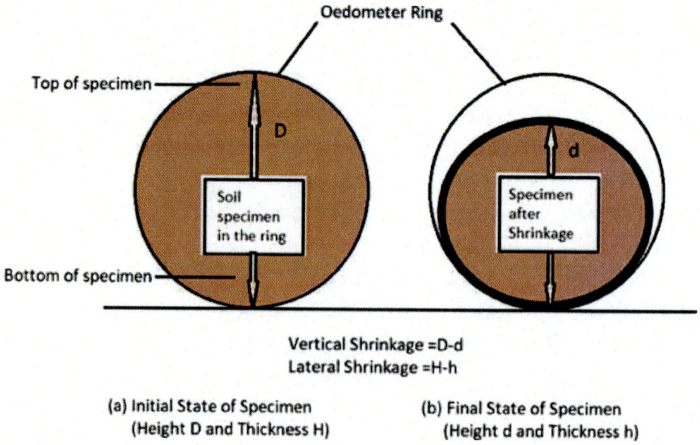

Fig. 1 Ring shrinkage testing equipment and method

Table 1 Classification of soils and physical soil properties under study

Soil sample	Specific gravity	Grain size			W_L (%)	W_P (%)	Plasticity index $I_P = W_L - W_P$ (%)	Classification based on FSR	Classification based on FSI and degree of compaction	Classification based on MFSI
		sand (%)	Silt (%)	Clay (%)						
Soil-1	2.59	20.0	31.0	49.0	38.0	14.0	24.0	KAOLINITIC	Low	Slight
Soil-2	2.66	30.0	53.0	17.0	38.0	14.0	24.0	MONTMORILLONITIC	Medium	Slight
Soil-3	2.80	10.0	54.0	36.0	42.0	22.0	20.0	MONTMORILLONITIC	Medium	Slight
Soil-4	2.54	47.0	28.0	25.0	54.0	23.0	31.0	MONTMORILLONITIC	Medium	Slight
Soil-5	2.79	45.0	38.0	17.0	54.0	29.0	25.0	KAOLINITIC and MONTMORILLONITIC	Low	Negligible
Soil-6	2.60	32.0	48.0	20.0	32.0	19.0	13.0	KAOLINITIC and MONTMORILLONITIC	Low	Slight

Table 2 Maximum density of dry soil (MDD) and optimum content of moisture (OMC) of the soils under study

Type of compaction energy	Soil-1		Soil-2		Soil-3		Soil-4		Soil-5		Soil-6	
	OMC (%)	ρ_d max (g/cc)	OMC (%)	ρ_d max (g/cc)	OMC (%)	ρ_d max (g/cc)	OMC (%)	ρ_d max (g/cc)	OMC (%)	ρ_d max (g/cc)	OMC (%)	ρ_d max (g/cc)
RSP	14.0	1.56	14.0	1.61	18.3	1.46	32.7	1.45	15.4	1.56	34.4	1.58
SP	12.8	1.62	12.8	1.63	17.0	1.60	29.7	1.48	14.4	1.66	30.0	1.63
RMP	13.5	1.60	18.0	1.58	16.1	1.57	23.4	1.51	13.9	1.75	26.8	1.68
MP	9.8	1.84	9.8	1.65	21.5	0.75	20.7	1.54	12.9	1.84	24.0	1.72

4 Discussion of the Results

Figures 2, 3, 4, 5, 6 and 7, represents the correlation between maximum density at dryness (MDD) and optimum content of moisture (OMC) of soils under study.

Table 3 shows the correlation equations and correlation coefficients of the soils under study.

From these figures, it can be observed that very good correlations were observed between maximum density of dry soil and optimum content of moisture of the soils under study. From Figs. 2, 3, 4, 5, 6 and 7 representing the correlation between

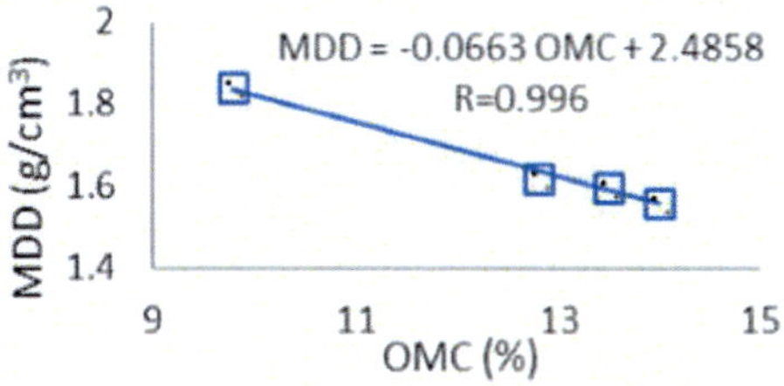

Fig. 2 Correlation of optimum content of moisture and maximum density of dry soil (soil 1)

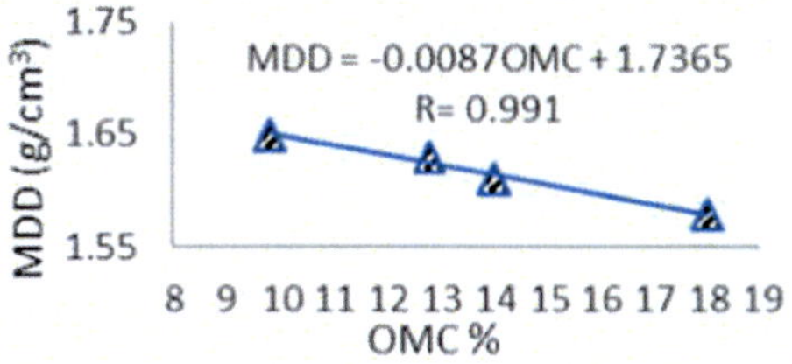

Fig. 3 Correlation of optimum content of moisture and maximum density of dry soil (soil 2)

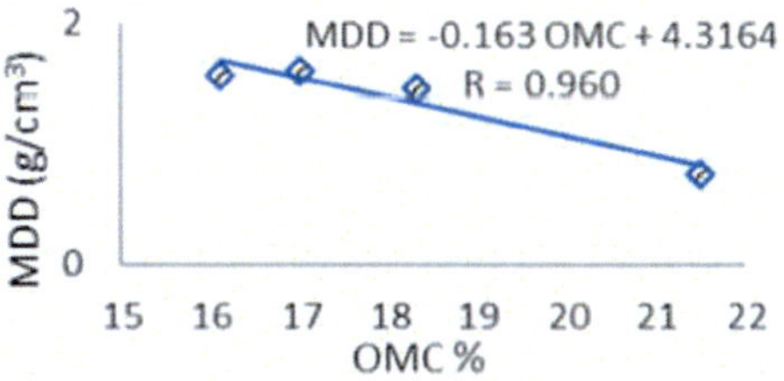

Fig. 4 Correlation of optimum content of moisture and maximum density of dry soil (soil 3)

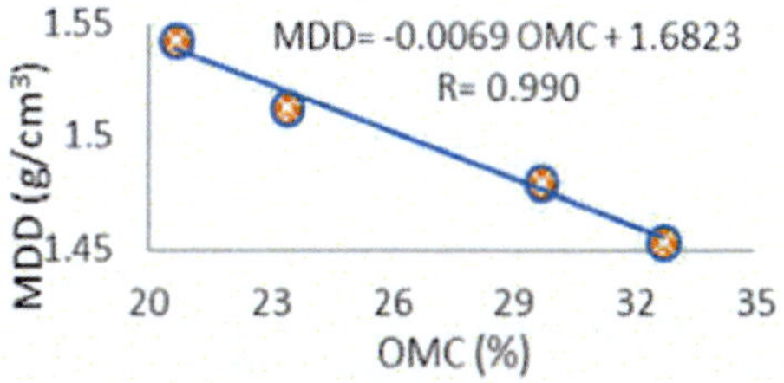

Fig. 5 Correlation of optimum content of moisture and maximum density of dry soil (soil 4)

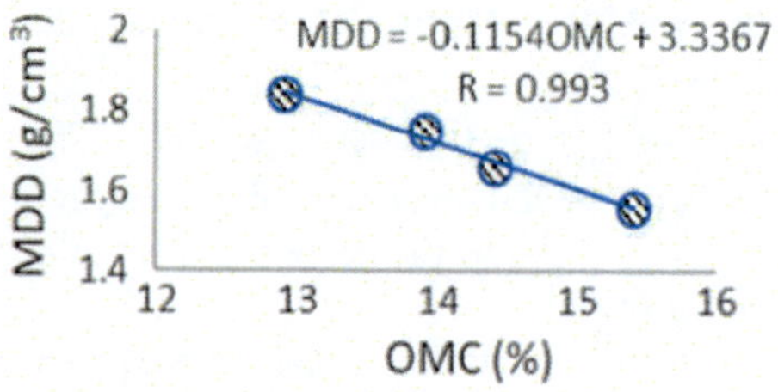

Fig. 6 Correlation of optimum content of moisture and maximum density of dry soil (soil 5)

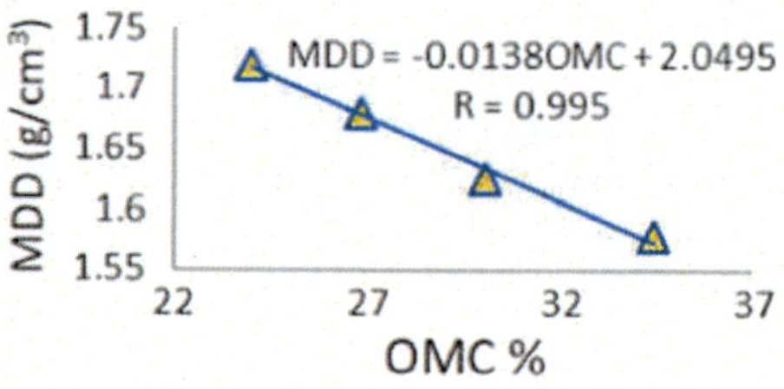

Fig. 7 Correlation of optimum content of moisture and maximum density of dry soil (soil 6)

Table 3 Correlation equations and correlation coefficients of the soils under study for Figs. 2, 3, 4, 5, 6 and 7

Figures	Correlation equation	Correlation coefficient
Figure 2	MDD = −0.0663 OMC + 2.4858	0.99
Figure 3	MDD = −0.0087 OMC + 1.7364	0.99
Figure 4	MDD = −0.1630 OMC + 4.3164	0.96
Figure 5	MDD = −0.0069 OMC + 1.6830	0.99
Figure 6	MDD = −0.1154 OMC + 3.3367	0.99
Figure 7	MDD = −0.0138 OMC + 2.0495	0.99

optimum content of moisture and maximum density of dry soil, it can be observed that as the moisture content in the soil increases, the value of maximum density of the dry soil decreases. As the value of OMC increases from 9.8 to 34.4%, the value of MDD reduces from 1.84–1.45 g/cc. The montmorillonite soil (soil 3) has maximum value of MDD and OMC on the other hand the minimum value is observed in kaolinitic soil. Almost all the soils under study have a common value of regression coefficient, i.e. $R = 0.99$. The optimum moisture content would permit the soil to reach the lowest permeability and allows the soil to attain a higher degree of compaction with application of least effort. The optimum moisture content, the type and amount of clay present would influence the soil in any kind of changes in its volume. Basically, kaolinite is 1:1 clay mineral while montmorillonite is 2:1 clay mineral. Montmorillonite soil has a tendency to swell because of the dispersive effect and also due to the formation of diffused double layer. Kaolinite has a basal-spacing smaller (0.72 nm) due to hydrogen bonding than Montmorillonite soil (4 nm) depending on the dominant exchangeable cation. The results of the present study are value addition to the existing literature.

Figures 8, 9, 10, 11, 12 and 13 represent the correlation between compaction energy (CE) and volumetric shrinkage (VS) of soils under study. Tables 4, 5, 6, 7, 8 and 9 show the corresponding correlation equations along with correlation coefficients.

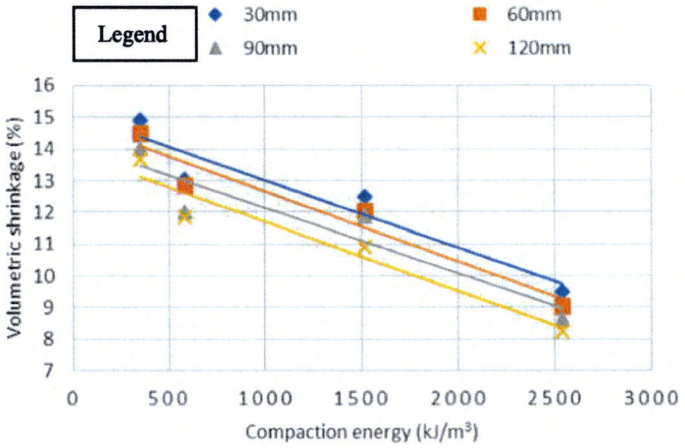

Fig. 8 Variation of compaction energy and volumetric shrinkage (soil 1)

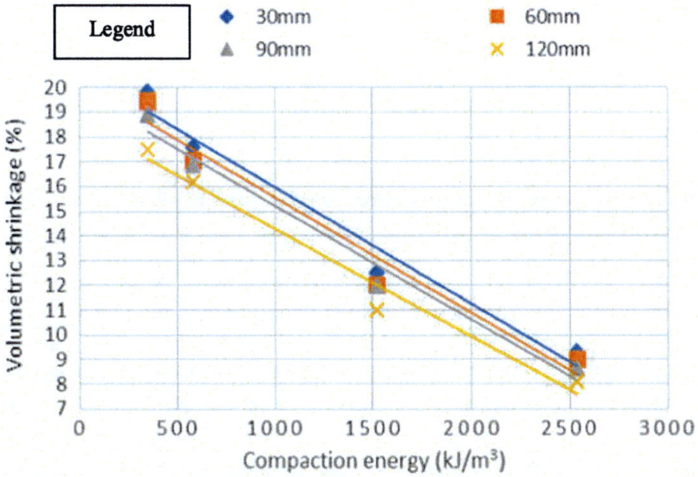

Fig. 9 Variation of compaction energy and volumetric shrinkage (soil 2)

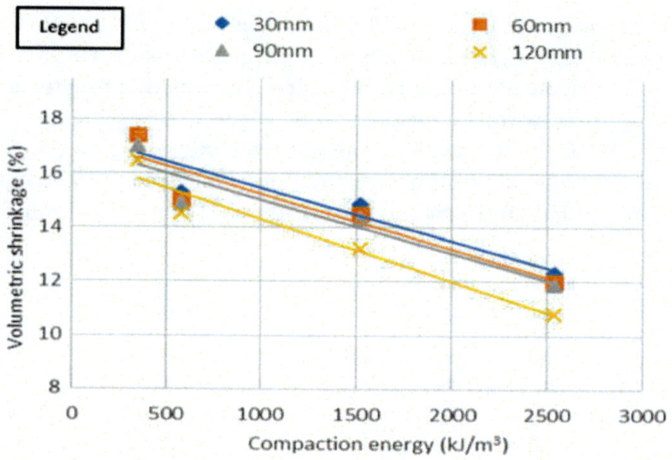

Fig. 10 Variation of compaction energy and volumetric shrinkage (soil 3)

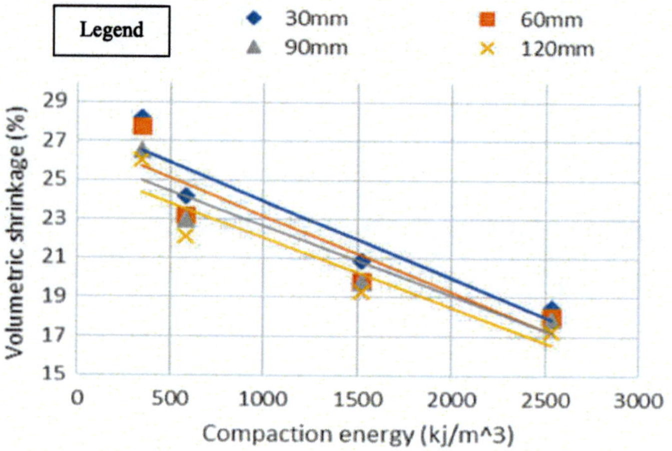

Fig.11 Variation of compaction energy and volumetric shrinkage (soil 4)

From Figs. 8, 9, 10, 11, 12 and 13, it can be observed that very good correlations were observed between volumetric shrinkage and compaction energy for the soils under study. As the diameter of the mould goes on increasing the tendency of the soil to shrink goes on decreasing. In turn, as the diameter of the mould increases keeping the depth of the mould as constant the volume of the soil also increases, which means higher the volume of the soil lesser the shrinkage observed. This phenomenon has taken place due to the close packing of soil particles, as the volume of the soil increases the particles become tightly packed due to which void ratio reduces, simultaneously reducing the shrinkage.

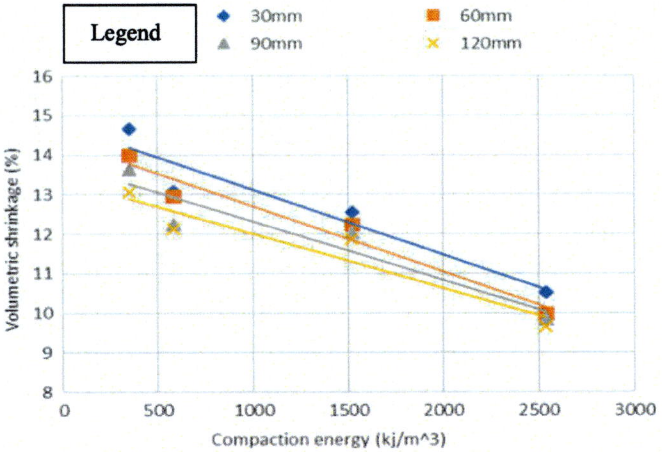

Fig. 12 Variation of compaction energy and shrinkage of volume (soil 5)

Fig. 13 Variation of compaction energy and shrinkage of volume (soil 6)

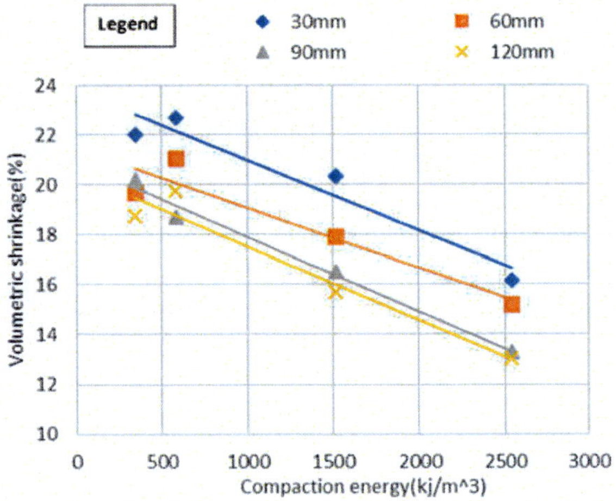

Table 4 Correlation equations with correlation coefficients (Fig. 8)

Diameter (mm)	Correlation equation	Correlation coefficient
30	VS = −0.0021 CE + 15.141	0.95
60	VS = −0.0022 CE + 14.867	0.96
90	VS = −0.0021 CE + 14.214	0.93
120	VS = −0.0022 CE + 13.878	0.96

Table 5 Correlation equations with correlation coefficients (Fig. 9)

Diameter (mm)	Correlation equation	Correlation coefficient
30	VS = −0.0047 CE + 20.680	0.98
60	VS = −0.0047 CE + 20.226	0.98
90	VS = −0.0046 CE + 19.847	0.99
120	VS = −0.0044 CE + 18.643	0.99

Table 6 Correlation equations with correlation coefficients (Fig. 10)

Diameter (mm)	Correlation equation	Correlation coefficient
30	VS = −0.0020 CE + 17.434	0.93
60	VS = −0.0021 CE + 17.303	0.93
90	VS = −0.0020 CE + 16.986	0.94
120	VS = −0.0023 CE + 16.616	0.97

Table 7 Correlation equations with correlation coefficients (Fig. 11)

Diameter (mm)	Correlation equation	Correlation coefficient
30	VS = −0.0040 CE + 27.899	0.94
60	VS = −0.0039 CE + 27.083	0.91
90	VS = −0.0036 CE + 26.222	0.94
120	VS = −0.0035 CE + 25.584	0.93

Table 8 Correlation equations with correlation coefficients (Fig. 12)

Diameter (mm)	Correlation equation	Correlation coefficient
30	VS = −0.0016 CE + 14.740	0.95
60	VS = −0.0017 CE + 14.361	0.97
90	VS = −0.0015 CE + 13.782	0.94
120	VS = −0.0014 CE + 13.378	0.95

Table 9 Correlation equations with correlation coefficients (Fig. 13)

Diameter (mm)	Correlation equation	Correlation coefficient
30	VS = −0.0028 CE + 23.835	0.96
60	VS = −0.0024 CE + 21.469	0.95
90	VS = −0.0030 CE + 20.193	0.99
120	VS = −0.0030 CE + 20.503	0.97

From these correlation equations by knowing the compaction energy imparted to the soil, it is possible to estimate the volumetric shrinkage for different diameter with fair degree of accuracy. It can be observed that, as compactive energy increases the volumetric shrinkage decreases. For the soils having different clay minerology and compactive energy, the maximum value of volumetric shrinkage is observed in the soil moulded in 30 mm diameter ring, whereas, minimum values of volumetric shrinkage were observed in the soil samples in the moulds of 120 mm diameter. This tendency highlights the role of volume of soil associated with the shrinkage of the soil volume. The maximum and minimum values of volumetric shrinkage observed in kaolinitic soil are 15% and 8%, respectively. In case of montmorillonite soil, the values of volumetric shrinkage vary from the maximum per cent of 29 to the minimum per cent of 7.5. As for kaolinitic and montmorillonite soil, the magnitude of volumetric shrinkage varies from minimum to maximum values of 9–23%. The minimum value of volumetric shrinkage is observed in kaolinitic soil (soil 1) whereas the maximum value of volumetric shrinkage is observed in montmorillonite soil (soil 4). Soil 2, 3 and 4 have the same clay minerology (montmorillonite) but they differ with respect to the percentage of silt and clay present. Soil 2 has 17% of clay, whereas, soil 3 and 4 has 36% and 25% of clay, respectively. The presence of clay plays a vital role in deciding the amount of volumetric shrinkage. Soil 4 has the maximum value of volumetric shrinkage even if the percentage clay present in the soil is low when compared to that of soil 3. For higher values of OMC, higher volumetric shrinkage is observed, i.e. as the presence of moisture increases the tendency of soil to shrink also increases, due to inter-particle orientation in presence of water. Figures 14, 15, 16, 17, 18 and 19, represents the correlation between compaction energy (CE) and lateral shrinkage (LS) of soils under study. Tables 10, 11, 12, 13, 14 and 15 show the corresponding correlation equations along with correlation coefficients.

From these figures, it can be observed that very good correlations were observed between lateral shrinkage and compaction energy for the soils under study. From these correlation equation by knowing the compaction energy imparted to the soil it is possible to estimate the lateral shrinkage for different diameter with fir degree of accuracy. It can be observed that maximum lateral shrinkage is found in the soil of mould 30 mm diameter and the minimum in the soil of mould 120 mm diameter, this

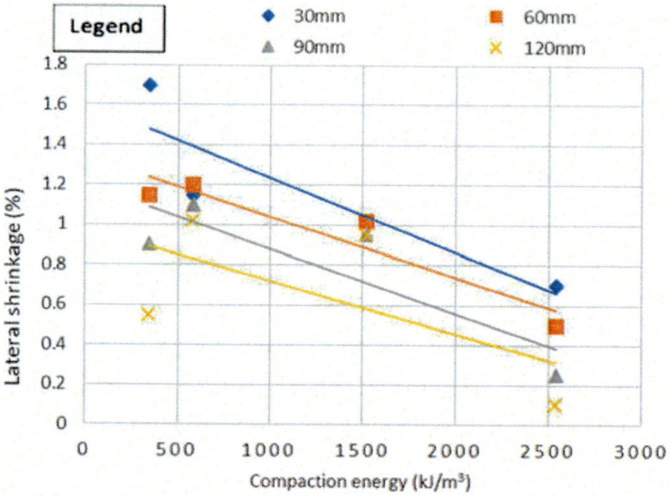

Fig.14 Variation of compaction energy and lateral shrinkage (soil 1)

Fig. 15 Variation of compaction energy and lateral shrinkage (soil 2)

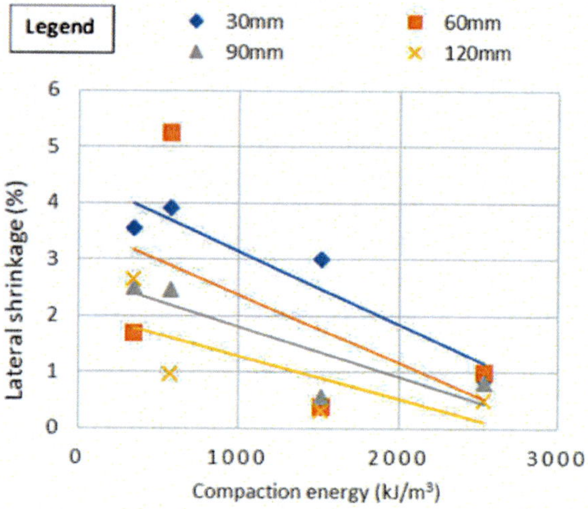

trend is followed by all the soils under study having different clay minerology. As the magnitude of compaction energy increases the values of lateral shrinkage decreases.

The maximum value of lateral shrinkage for the kaolinitic soil is observed to be 1.7% and the minimum value is observed to be 0.1%. In case of montmorillonite soil, the maximum and minimum values of lateral shrinkage vary from 10% to 0.1%, respectively. As for the kaolinitic and montmorillonite soil, the value of lateral shrinkage changes from the maximum of 5.9% to the minimum of 0.7%.

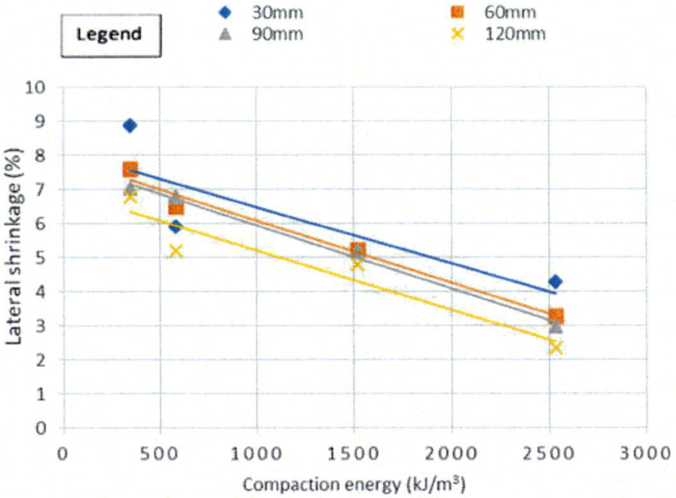

Fig.16 Variation of compaction energy and lateral shrinkage (soil 3)

Fig.17 Variation of compaction energy and lateral shrinkage (soil 4)

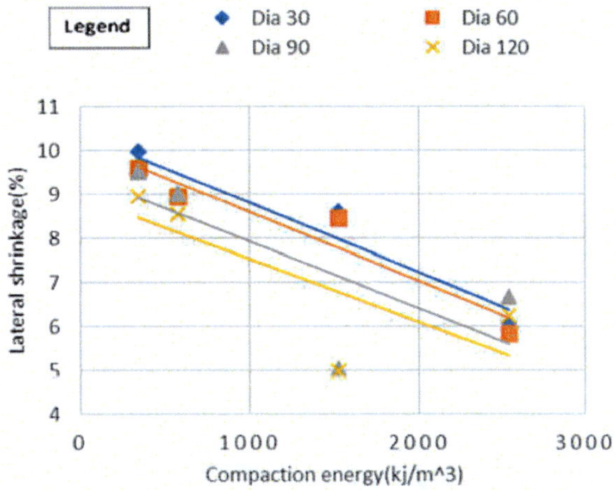

The value of lateral shrinkage is observed to be minimum in kaolinitic soil, whereas, the montmorillonite soil has the maximum value of lateral shrinkage. Soil-2, 3 and 4 have the same clay minerology but vary with the percentage of clay present. Among the three montmorillonite soils, soil-4 has the maximum value of lateral shrinkage, due to the higher values of OMC. The presence of higher moisture content increases the tendency of the soil to shrink. The percentage of OMC and the percentage of clay present plays a major role in deciding the volumetric shrinkage along with lateral shrinkage soil properties, highlighting the role of clay minerology.

Fig. 18 Variation of compaction energy and lateral shrinkage (soil 5)

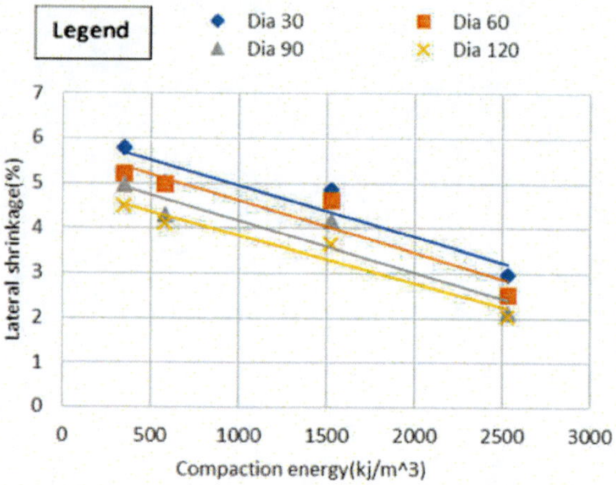

Fig. 19 Variation of compaction energy and lateral shrinkage (soil 6)

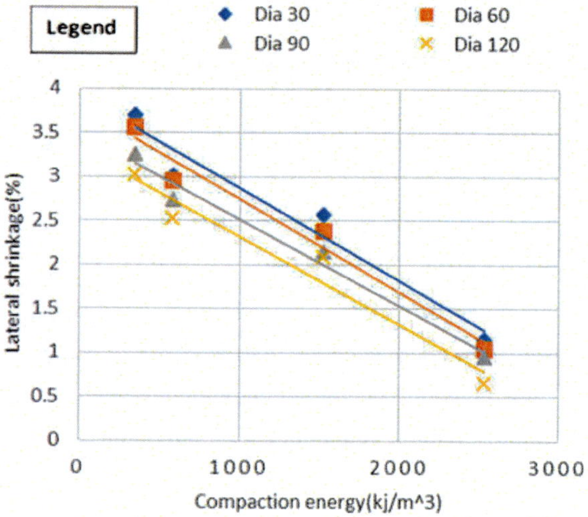

Table 10 Correlation equations with correlation coefficients (Fig. 14)

Diameter (mm)	Correlation equation	Correlation coefficient
30	LS = −0.0004 CE + 1.6061	0.90
60	LS = −0.0003 CE + 1.3463	0.94
90	LS = −0.0003 CE + 1.1999	0.95
120	LS = −0.0003 CE + 0.9821	0.90

Table 11 Correlation equations with correlation coefficients (Fig. 15)

Diameter (mm)	Correlation equation	Correlation coefficient
30	LS = −0.0013 CE + 4.4393	0.94
60	LS = −0.0012 CE + 3.6002	0.90
90	LS = −0.0009 CE + 2.6971	0.90
120	LS = −0.0008 CE + 2.0547	0.90

Table 12 Correlation equations with correlation coefficients (Fig. 16)

Diameter (mm)	Correlation equation	Correlation coefficient
30	LS = −0.0017 CE + 8.1426	0.90
60	LS = −0.0018 CE + 7.9391	0.99
90	LS = −0.0019 CE + 7.8293	0.99
120	LS = −0.0017 CE + 6.9685	0.95

Table 13 Correlation equations with correlation coefficients (Fig. 17)

Diameter (mm)	Correlation equation	Correlation coefficient
30	LS = −0.0016 CE + 10.401	0.96
60	LS = −0.0016 CE + 10.184	0.96
90	LS = −0.0015 CE + 9.4603	0.90
120	LS = −0.0014 CE + 8.9867	0.90

Table 14 Correlation equations with correlation coefficients (Fig. 18)

Diameter (mm)	Correlation equation	Correlation coefficient
30	LS = −0.0011 CE + 6.0732	0.94
60	LS = −0.0012 CE + 5.7804	0.94
90	LS = −0.0012 CE + 5.3213	0.94
120	LS = −0.0011 CE + 4.8864	0.97

Table 15 Correlation equations with correlation coefficients (Fig. 19)

Diameter (mm)	Correlation equation	Correlation coefficient
30	LS = −0.0011 CE + 3.9171	0.97
60	LS = −0.0011 CE + 3.8090	0.98
90	LS = −0.0010 CE + 3.4954	0.99
120	LS = −0.0010 CE + 3.3192	0.98

Figures 20, 21 and 23, represents the correlation between compaction energy (CE) and volumetric shrinkage (VS) of soils under study for diameters 30 mm, 60 mm, 90 mm and 120 mm, respectively. Tables 16, 17, 18 and 19 show the corresponding correlation equations along with correlation coefficients.

From these figures, it can be observed that very good correlations were observed between volumetric shrinkage and compactive energy for the soils under study tested with different diameters of the rings used. From the figures, Figs. 20, 21, 22 and 23 represent the correlation of the compactive energy and volumetric shrinkage, keeping the diameter of the mould as constant. Here, it can be observed that for all the soil moulds of varying diameter (30, 60, 90 and 120 mm) the maximum volumetric shrinkage is obtained for the montmorillonite soil (soil-4), and the minimum value is obtained for the kaolinitic and montmorillonite soil (soil-5). This implies that the clay minerology is crucial in deciding the shrinkage properties of any soil under investigation.

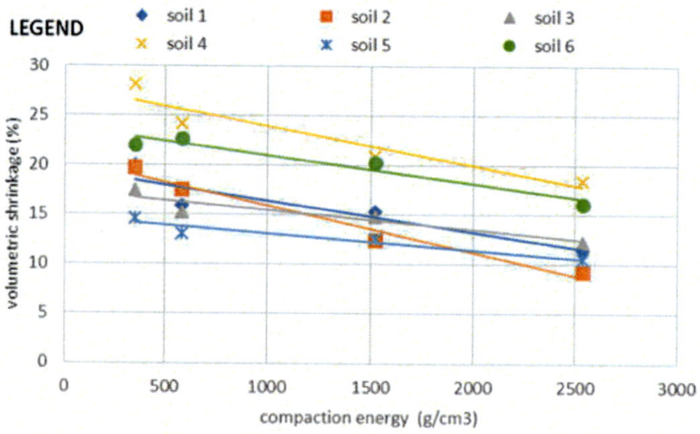

Fig. 20 Variation of volumetric shrinkage with compaction energy keeping diameter 30 mm of the ring as constant

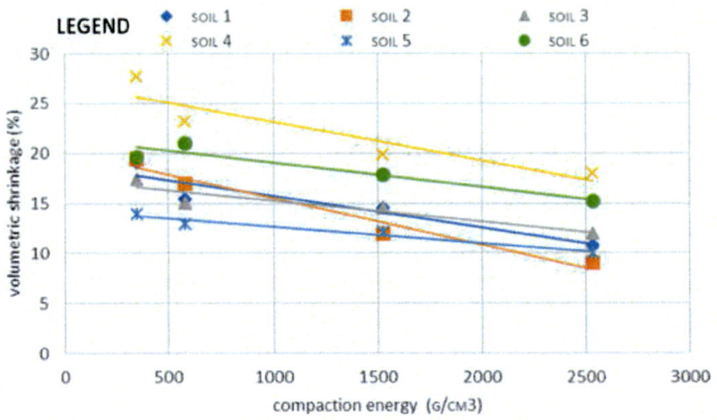

Fig. 21 Variation of volumetric shrinkage with compaction energy keeping diameter 60 mm of the ring as constant

Table 16 Correlation equations with correlation coefficients (Fig. 20)

Soil sample	Correlation equation	Correlation coefficient
Soil 1	VS = −0.0032 CE + 19.652	0.91
Soil 2	VS = −0.0047 CE + 20.680	0.98
Soil 3	VS = −0.0020 CE + 17.434	0.93
Soil 4	VS = −0.0040 CE + 27.899	0.93
Soil 5	VS = −0.0016 CE + 14.750	0.95
Soil 6	VS = −0.0028 CE + 23.835	0.96

Table 17 Correlation equations with correlation coefficients (Fig. 21)

Soil sample	Correlation equation	Correlation coefficient
Soil 1	VS = −0.0032 CE + 18.932	0.92
Soil 2	VS = −0.0047 CE + 20.226	0.99
Soil 3	VS = −0.0021 CE + 17.303	0.92
Soil 4	VS = −0.0039 CE + 27.083	0.90
Soil 5	VS = −0.0017 CE + 14.361	0.95
Soil 6	VS = −0.0024 CE + 21.469	0.95

Table 18 Correlation equations with correlation coefficients (Fig. 22)

Soil sample	Correlation equation	Correlation coefficient
Soil 1	VS = −0.0036 CE + 18.092	0.91
Soil 2	VS = −0.0044 CE + 18.643	0.98
Soil 3	VS = −0.0020 CE + 16.986	0.94
Soil 4	VS = −0.0035 CE + 25.584	0.90
Soil 5	VS = −0.0014 CE + 13.378	0.91
Soil 6	VS = −0.0030 CE + 20.503	0.98

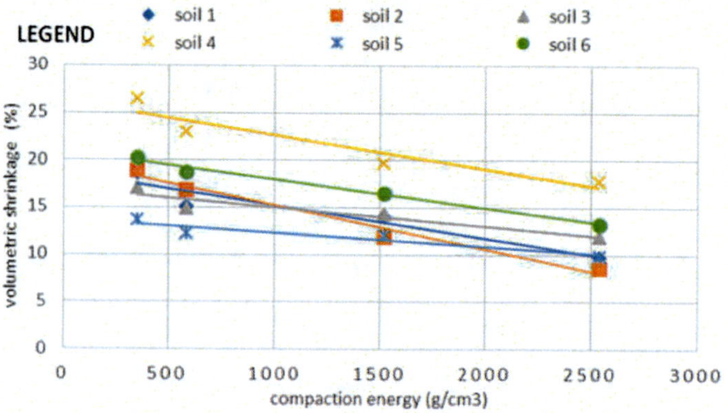

Fig. 22 Variation of volumetric shrinkage with compaction energy keeping diameter 90 mm of the ring as constant

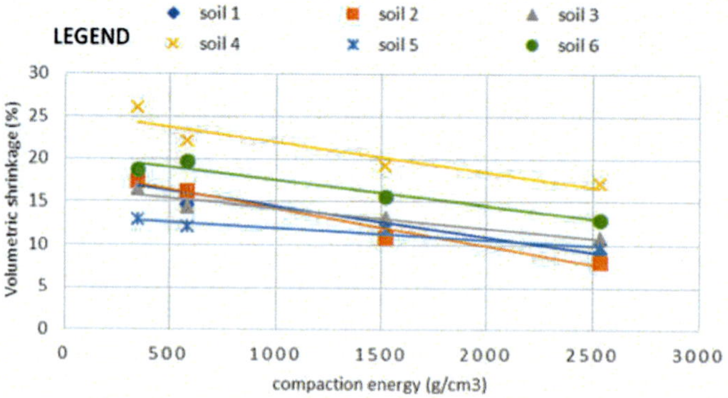

Fig. 23 Variation of volumetric shrinkage with compaction energy keeping diameter 120 mm of the ring as constant

Table 19 Correlation equations with correlation coefficients (Fig. 23)

Soil sample	Correlation equation	Correlation coefficient
Soil 1	VS = −0.0035 CE + 18.682	0.92
Soil 2	VS = −0.0046 CE + 19.847	0.98
Soil 3	VS = −0.0020 CE + 16.986	0.92
Soil 4	VS = −0.0036 CE + 26.222	0.91
Soil 5	VS = −0.0015 CE + 13.782	0.93
Soil 6	VS = −0.0030 CE + 20.913	0.97

5 Potential Field Applications

Due to alternate swelling and shrinkage of expansive soils of structures founded on compacted expansive soils would be severely damaged which occurs in the region of active zone. The volume of soil associated below any foundation level within the influential zone relies on any.

- lateral pressure such as earthquake, tectonic movements
- deep cuttings like mining
- overstressing indicating heavy loads
- groundwater fluctuations.

Any experimental study related to volumetric and lateral shrinkage of expansive soils is of paramount importance from practical applications. The magnitude of volumetric and lateral shrinkage mainly depends upon the clay mineralogical composition, i.e. presence of active clay mineral, soil fabric and the magnitude of compactive energy imparted to the soil. The present experimental study highlights, the role of clay minerology and compactive energy in controlling the volumetric and lateral shrinkage of expansive soils which can be assessed by knowing the compactive energies imparted to the soil like reduced standard Proctor, standard Proctor, reduced modified Proctor and modified Proctor, respectively, while designing the shallow and deep foundations in addition to clay liner applications.

6 Conclusions

Conclusions that follow are derived from the present experimental study:

1. The kaolinitic soil has lower magnitude of optimum content of moisture (OMC) and higher magnitude of maximum density of dry soil (MDD) in comparison

with the montmorillonite soil, which has a higher magnitude of OMC and a lower magnitude of MDD.
2. Volumetric shrinkage of the soils under study vary with the magnitudes of compactive energy imparted, i.e. as the compactive energy increases, the magnitude of volumetric shrinkage decreases.
3. The maximum volumetric shrinkage is observed for the soil in the mould of diameter 30 mm, which is nearly onefold greater than the maximum volumetric shrinkage observed in the soil in the mould of 120 mm diameter.
4. For all the soils under study lower the values of compactive energy, higher values of the lateral shrinkage are observed.
5. The maximum lateral shrinkage is observed for the soils in mould of diameter 30 mm which is nearly 3-folds more than the lateral shrinkage observed in the soils in the mould of 120 mm diameter.

Acknowledgements The authors would like to acknowledge the laboratory staff of Geotechnical Engineering, Department of Civil engineering, The National Institute of Engineering, Mysuru for the assistance provided in carrying out the present experimental work.

References

1. Sridharan A, Prakash K (2000) Shrinkage limit of soil mixtures. Geotech Test J 23(1):3–8
2. Daniel DE, Wu YK (1993) Compacted clay liners and cones for arid site. J Geotech Eng 119(2):223–237
3. Benson CH, Albrecht BA (2001) Effects of Desiccation on Compacted Natural Clays. J Geotech Geo-environ Eng 127(1):67–75
4. Osinubi KJ, Nwaiwu CMO (2005) Hydraulic conductivity of compacted lateritic soils. J Geotech Geo-environ Eng Am Soc Civ Eng 131(8):1034–1041
5. Osinubi KJ, Nwaiwu CMO (2008) Desiccation-induced shrinkage in compacted lateritic soil. J Geotech Geol Eng 26(5):603–611
6. Puppala AJ, Katha B, Hoyos LR (2004) Volumetric shrinkage strain measurements in expansive soils using digital imaging technology. Geotech Test J 27(6):547–556
7. Erzin Y, Gunes N (2013) The unique relationship between swell percent and swell pressure of compacted clays. Bull Eng Geol Environ 72(1):71–80
8. Gurtug Y, Sridharan A (2015) Prediction of compaction behaviour of soils at different energy levels. In: All in one conferences, vol 47
9. George M, Oriola FOP, Osinubi KJ (2016) Desiccation-induced volumetric shrinkage of compacted metakaolin treated black cotton soil for a hydraulic barriers system. Solvak J Civ Eng 24(1):1–5
10. Kundiri AM, Muhammed AS, Abah G (2016) Effect of compactive efforts on desiccation-induced volumetric shrinkage strain of some compacted tropical soils. Malays J Civ Eng 28(2)
11. Mahvash S, Lopez-Querol S, Bahadori-Jahromi A (2017) Effect of class-F Flyash on fine sand compaction through soil stabilization. Heliyon 3(3):C00274
12. Yamusa YB, Ahmad K, Rahman NA (2017) Hydraulic conductivity and volumetric shrinkage properties review on gradation effect of compacted laterite soil liner. Malays J Civ Eng 29
13. Yamusa YB, Ahmad K, Rahman NA, Yunus NZ (2018) Volumetric shrinkage of compacted soil linear for sustainable waste landfill. Chem Eng Trans 63:613–618

14. M. Julina, T. Thyagaraj (2018) Determination of volumetric shrinkage of an expansive soil using digital camera images. Int J Geotech Eng 1–9
15. Chen J, Jiang S, Goa Y, Sun F (2018) Reducing volumetric shrinkage of photopolymerizable materials using reversible disulfide bond reactions. J Mater Sci 53(23):16169–16181
16. Udukumburage RS, Gallage C, Dawes L (2019) Oedometer based estimation of vertical shrinkage of expansive soil in a large instrumental soil column. Heliyon 5(9):C02380
17. Mishra PN, Zhang Y, Bhuyan MH, Scheuermann A (2020) Anisotropy in volume change behaviour of soils during shrinkage. Acta Geotech 1–16
18. Sridharan A, Rao Sudhakar M, Murthy NS (1985) Free swell index of soils; a need for redefinition. Ind Geotech J 15(2):94–99

A Study on Secondary Compression of Compacted Fine-Grained Soils

H. S. Prasanna, B. V. Rachana, R. Anusha, and Ganesh Basavaraj Badaradinni

Abstract A manifestation of the effects of structural mechanics is the relationship linking the structure of soil and mechanical behavior. Compaction is a mechanical process modification of soil structure by the expulsion of pore air from the voids, whereas consolidation is a process of expulsion of pore water pressure in which settlement takes place. Secondary compression generally happens after the end of primary consolidation may demonstrate a steady decrease or increase over a substantial period. In order to calculate long-term foundation deformation, understanding the impact of secondary compression is fundamental. Foundation failures can happen by either shear failure or settlement failure, whereas shear failure is instantaneous and settlement failure occurs over a while. Depending on the magnitude and demographic location of the structure, the permissible settlement of the structure should not settle more than 25 and 40 mm allowable as per IS guideline to minimize distress to the foundation. The present experimental study is made on five field soils, and two soils having a liquid limit of 55% and another three soils of 68% of a liquid limit have been also considered. These soils were subjected to IS light and heavy compaction energy levels and static compaction to attend the desired compaction characteristics and achieve the required OMC and MDD; additionally, soils were subjected to a one-dimensional consolidation test from a seating load of 6.25–1600 kPa through a load increment ratio of 1 for different placement conditions. The variation of void ratio with pressure for the soil having a different clay mineralogy for different placement conditions was analyzed. Correlations were developed between pressure and the magnitude of secondary compression, for soils having different clay mineralogy.

Keywords Clay mineralogy · Energy levels · Secondary compression

---

H. S. Prasanna · B. V. Rachana (✉) · R. Anusha · G. B. Badaradinni
Department of Civil Engineering, The National Institute of Engineering (Affiliated to VTU), Mysuru, India

H. S. Prasanna
e-mail: prasanna@nie.ac.in

L. Nandagiri et al. (eds.), *Sustainability Trends and Challenges in Civil Engineering*, Lecture Notes in Civil Engineering 162, https://doi.org/10.1007/978-981-16-2826-9_18

1 Introduction

As known from various case studies, generically the land available for the construction activities is reducing day by day. Invariably, we are supposed to use the marginal lands and landfills for the construction purpose. It is only made out if the landfills and marginal lands are composed of coarse-grained soil. For this purpose, fine-grained soils are used in the process. The coarse-grained soil is a majority preferred as the engineering behavior of this soil is described as a physical phenomenon, but as for that of fine-grained soil contains different clay minerals, and hence the engineering behavior is described as a physicochemical process. The major problem faced with that of fine-grained soil is that of the poor average weighing capacity and settlement, which associates compaction and consolidation. As far as compaction of fine-grained soil is considered, the effect of primary and secondary settlement characteristics are to be considered while designing the sub-structures on such soils. Furthermore, if the consolidation of the fine-grained soil is considered, the clay mineralogy has to be considered since the fine-grained soil contains different clay minerals. For example, kaolinitic soil has a flocculating behavior whereas montmorillonitic soil has high water-absorbing properties and it is subjected to shrinkage and swelling and hence leads to distress of the complete structure. To obtain a better perceptive of the soil, the characteristic settlements must be considered with the realistic application.

The present experimental research aims to investigate the condition of the ability of consolidation characteristics of fine-grained soils and secondary compression characteristics of fine-grained compacted soils at different levels of energy.

2 Literature Review

Sridharan and Rao [1] based on an experimental study mechanism to regulate secondary clay compression concluded that secondary compression decreased co-efficiently with a decrease in the void ratio and an increase in pressure.

Mesri and Castro [2] observed that C_α/C_c of any individual soil exhibits secondary compression behavior. They conclude that in order to achieve the behavior of the coefficient of earth pressure at rest K$_\circ$ throughout secondary compression K_0, the C_α/C_c definition was used.

Navarro and Alonso [3] observed that secondary compression of local clay dehydration processes can be described in terms of water conversion from clay aggregates to bulk water.

Bui et al. [4] based on their experimental study on the secondary compression of clay through 1-D consolidation tests showed that the relationship of time settlement does not follow Terzaghi theory. Solutions for the calculation of creep settlements of a porous viscoelastic base material have been presented. Direct numerical inversion of the solution is used to obtain the time-dependent solution.

Mesri and Vardhanabhuti [5] made an experimental study on the large volume of accurate 1-D settlement measurements observed in the laboratory. He noted that the $\Delta s/\Delta \log t$ can remain steady increases or decreases for a considerable amount of time.

Sompie et al. [6] conducted a study on secondary behavior in three types of consolidation tests. The time-dependent existence of the standard consolidation (SC), constant strain rate consolidation (CSRC), and constant loading rate consolidation (CLRC) analyses is systematically defined by a clear assumption that time-dependent volume changes are due to shearing.

Takeda et al. [7] made an experimental study taking place 1-D consolidation analysis described for predicting the consolidation time curve of clay exhibiting secondary compression during primary consolidation.

El Amin Bourouis et al. [8] conducted a study on two approaches to faster prediction. One is through the machines by learning a multi-gene genetic program and the other with the algorithm and artificial neural networks.

Raheena and Robinson [9] carried out an experimental analysis on a technique using the $\sqrt{t}$ method to measure the secondary compression coefficient from the rapid cumulative loading consolidation test protocol. Each of the pressure increments applied in this experimental study is to achieve a linear secondary compression in the log t plots.

Jiang et al. [10] observed that $C_{\alpha\max}$ is associated with the soft soil-bound water content and represents the impact of soil composition on soft soil secondary compression characteristics.

Studies related to the variation of C_α have compacted fine-grained having dissimilar clay mineralogy, and placement conditions are very scanty.

3 Materials and Method

3.1 Selection of Natural Soils

It has been decided to conduct the experimental investigation on two field soils—one kaolinitic and the other montmorillonitic, having the same liquid limit in the lower liquid limit range—and on two field soils—one kaolinitic and the other montmorillonitic, having the same liquid limit in the higher liquid limit range along with one pure clay mineral (China clay).

Nearly, thirty soils from different locations in Mysore and Chamarajanagar districts were subjected to preliminary laboratory investigation involving the liquid limit and free swell tests. Their liquid limits were determined using the Casagrande percussion method according to IS: 2720-Part 5, 1985, and the nature of their clay mineralogical composition was judged by the free swell ratio technique (Prakash et al. [11]).

3.2 Preparations of Natural Soils for Investigation

To extract the coarser fraction, the field soils from Kollegala and Kuderu were wet-sieved through 425 μm IS sieve. To have soil fractions finer than 425 μm scale, they were then oven-dried and powdered. The resulting soil was placed in separate airtight plastic containers.

The field soils from the CFTRI layout and Bannur were wet-sieved through 75 μm IS sieve to remove the sand content. They were oven-dried, powdered, and placed in separate airtight plastic containers.

The following experiments were carried out on the soil sample prepared by the Bureau of Indian Standard specifications.

1. Specific gravity test: Using the density bottle method with kerosene as the test liquid, the standards of the specific gravity of the soil have been collected (IS:2720-Part-3/Sec 1, 1980).
2. Grain size analysis: The particle size distributions of the Bannur, CFTRI soil, china clay, Kollegala soil, and Kuderu soil were obtained by combined wet sieving and hydrometer analysis (IS: 2720, Part 4, 1985).
3. Free swell test: Bureau of Indian Standards (IS:2720, Part 40, 1977) suggests that the free swell index of the soil be calculated as

$$\text{FSI, Percent} = \left(\frac{V_d - V_k}{V_k}\right)100 \tag{1}$$

4. where V_d is a balanced sediment volume of 10 g of oven-dried soil passing through a 425 μm sieve, completely mixed with distilled water to create a 100 ml preliminary volume soil–water suspension in a 100 ml measuring container, and V_k is a balanced sediment volume of 10 g oven-dried soil passing through a 425 μm kerosene sieve with an initial volume of 100 ml of soil–kerosene suspension in 10 g oven-dried soil.
5. Liquid limit test: The Casagrande percussion method was adopted to determine the liquid limits of the soils (IS: 2720-Part 5, 1985). Also, liquid limit tests by the fall cone method (IS: 2720-part 5, 1985) were also conducted on soils from Kollegala, with water and kerosene as test liquids.
6. Plastic limit test: The plastic limit of the soils were determined using the conventional 3 mm thread rolling process (IS: 2720-Part 5, 1985).
7. Shrinkage limit test: By the mercury displacement process, the shrinkage limits of the soils were obtained (IS: 2720-Part 6, 1972).

Sridharan et al. [12], furthermore, observed that unconstructive free swell indices for kaolinite-rich soils were provided by the IS process. They proposed the use of the modified free swell index (MFSI) to indicate the degree of expansivity of soils.

$$\text{i.e., MFSI} = V_d/10\left(\text{cm}^3/\text{g}\right) \tag{2}$$

In the present experimental work, both FSI and MFSI of soils were determined. Since Sridharan et al. [13] showed with the aim the balanced sediment volume of soil in kerosene and that in CCl_4 are essentially the same, kerosene was used in the present investigation instead of carbon tetrachloride.

3.3 Classification of Soils

Soils used in this experimental investigation have been classified according to the Unified Soil Classification System as specified by IS: 1498-1970.

Table 1 represents the index properties of all the soils obtained as per procedures indicated above, including the IS soil classification.

3.4 Clay Mineralogy of Soils and Soil Expansivity

A plasticity chart is used in differentiating inorganic clays from inorganic silts. Figure 1 represents the plasticity chart indicating the position of all the soils of the present investigation on it. However, it has been noted that the plasticity chart cannot be used to identify soils based on their degree of expansivity.

Table 2 gives the free swell ratios of all the soils under study and their dominant clay mineralogical composition based on FSR. Table 2 also shows that the field soil from Kuderu is montmorillonitic. However, it indicates that the field soils from Kollegala contain both *K* and *M* clay minerals.

All geotechnical laboratories cannot meet the expense to have sophisticated instruments such as X-ray diffractometer and the like for the qualitative identification of clay minerals present. In such cases, identification of clay mineral type by a simple method, which can serve the required purpose with a fair degree of accuracy, is a welcome move. In this context, FSR has been shown to serve this purpose admirably well. Hence, following the procedure is suggested to classify the soils depending upon their degree of expansivity.

- Determine the stabilized sediment volumes of fine-grained oven-dried soils passing through 425 μm sieve using two 100 ml measuring jars of distilled water and carbon tetrachloride as test liquids and the actual volumes of soil–liquid suspensions in the two jars being 100 ml.
- Calculate the free swell ratio as the ratio of V_d to V_k (i.e., Eq. 3).
- Identify the soil as kaolinitic if FSR is less than unity or as montmorillonitic if FSR is more than 1.5.
- If FSR is between 1.0 and 1.5, the soil under consideration is of mixed clay mineral type.

To identify the dominant clay mineral present in soils containing more than one clay mineral, Prakash et al. [11] proposed that the liquid limit of the soils should

Table 1 Physical properties of studied soils

S. No.	Soil	Specific gravity	Casagrande method					FSR	Clay mineralogy	Grain size distribution			IS classification
			Liquid limit (w_L) (%)	Plastic limit (w_P) (%)	Plasticity index (I_p) (%)	Shrinkage limit (w_S) (%)	Shrinkage index (I_s) (%)			Clay size (%)	Silt size (%)	Sand size (%)	
1	Kollegala soil	2.74	55	26	29	15.9	39.1	1.11	Kaolinitic	37.0	34.5	28.5	CH
2	Kuderu soil	2.85	54	26	28	11.5	42.5	1.42	Montmorillonitic	39.0	21.0	40.0	CH
3	Bannur soil	2.69	67	30	37	16.1	13.9	1.11	Kaolinitic–montmorillonitic	45.0	55.0	–	CH
4	CFTRI	2.72	68	33	35	13.9	19.1	1.42	Montmorillonitic	51.0	49.0	–	MH
5	China clay	2.61	68	30	38	24.8	5.2	0.58	Kaolinitic	63.0	37.0	–	CH

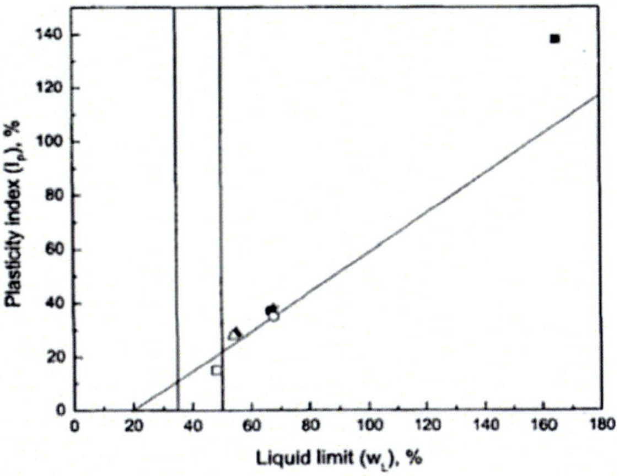

Fig. 1 Position of the soil understudy on the plasticity chart

Table 2 Free swell ratio of the soils under study and corresponding dominant clay mineral type

S. No.	Soil	V_d (cm^3)	V_k (cm^3)	FSR	Dominant clay mineral types
01	Kollegala soil	14.0	11.0	1.27	Kaolinitic
02	Kuderu soil	16.0	8.0	2.00	Montmorillonitic
03	Bannur soil	19.5	17.5	1.11	Kaolinitic–montmorillonitic
04	CFTRI layout soil	17.0	12.0	1.42	Montmorillonitic
05	China clay	20.0	34.5	0.58	Kaolinitic

be calculated by means of the fall cone procedure with distilled water and CCl_4 as pore liquids. If the fall cone liquid limit of the soil in distilled water is more than that of CCl_4, it implies that the prevalent clay mineral in the soil is montmorillonitic. On the other hand, if the fall cone liquid limit of soil in CCl_4 is more than that in distilled water, it indicates the dominance of kaolinite clay mineral in the soil. Table 3 presents the details of the fall cone liquid limit test results on the soils. Table 4 represents the re-designation of soils under study.

1. Compaction Tests

Table 3 Results from fall cone liquid limit tests

S. No.	Soil	Liquid limit (%)		Dominant clay mineral(s) type
		In distilled water	In CCl_4	
1	Kollegala soil	43	54	Kaolinitic
2	Bannur soil	61	63	Kaolinitic–montmorillonitic
3	CFTRI layout soil	61	46	Montmorillonitic

Table 4 Re-designation of soil understudy

S. No.	Soil source	Re-designation of soils	
S1	Kollegala soil	Soils of the same group of low liquid limit	*K*-soil
S2	Kuderu soil		*M*-soil
S3	Bannur soil	Soils of the same group of high liquid limit	*KM*-soil
S4	CFTRI layout soil		*M*-soil
S5	China clay		*K*-soil

Light and heavy compaction tests were performed on the soils under study (IS: 2720, Part-7, 1980; IS: 2720, Part-8, 1983). Approximately 6 to 8 samples, each with a mass of 2.5 kg, were thoroughly mixed with dissimilar molding water contents for every one of these tests, including light and heavy compaction tests, and kept within separate polythene covers. Depending on the soil quality, these samples were preserved for moisture equilibrium of between five and ten days. After this equilibrium duration, the compaction tests are performed on these soil samples. The compaction characteristics of the soil under the study i.e., MDD and OMC for both IS light and heavy compaction test were determined and tabulated.

2. Consolidation Tests on Compacted Soils

- *Sample preparation for consolidation testing*

Consolidation tests were conducted on compacted soils of the same group of low liquid limit (i.e., *K*-soil and *M*-soil) and compacted soils of the same group of high liquid limit (i.e., *KM*-soil, *M*-soil, and *K*-soil). As it was noticed that the consolidation was taking a prohibitively long duration of time, it was not considered in this experimental program. Since both kaolin and China clay are pure kaolinite clay minerals, only China clay was considered for a consolidation test. The consolidation tests were done at three levels of initial molding water contents—water contents corresponding to ρ_{dmax} (i.e., OMC), 0.95 ρ_{dmax} on the dry side of optimum, and 0.95 ρ_{dmax} on the wet side of optimum.

Every soil under analysis was mixed by the requisite water content and acceptable for moisture equilibration later than ensuring the steadiness of the appropriate water content of the soil sample, and the intended capacity of matured soil was compacted into the consolidation ring to accomplish the requisite dry density.

The consolidation ring was then assembled with the compacted soil sample having the appropriate molding water content and dry density in its location on the consolidation cell. The consolidation cell was used for the investigational work is of a double drainage fixed ring type and has the facility to perform the variable head permeability test on the soil sample as well. The inner diameter of the consolidation ring is 60 mm, and the height is 20 mm. To decrease the friction, the inside surface of the consolidation ring was soiled with silicone lubricant before compacting the soil into it.

• *Load–deformation–time measurements for compacted soils*

Consolidation tests were conducted according to IS: 2720, Part 15 (1986) (28). To gauge the vertical deformation of the soil sample, a seating consolidation stress of 6.25 kPa was functional and diluted water was supplementary to the consolidation cell after the consolidation cell was installed in its spot on a consolidation loading frame designed with a sensitive dial gauge with a least count of 0.002 mm.

Beneath the seating stress, the soil samples were allowed to balance. Some soil samples exhibited swelling on the addition of water into the consolidation cell. In such cases, time-swelling readings were recorded till the equilibrium was reached. The samples were loaded from 6.25 to 1600 kPa through a stress increment ratio of the accord. Beneath each consolidation stress increment, time compression readings were recorded till the near-equilibrium state was reached.

Beneath consolidation stress of 1600 kPa, the samples were unloaded in stages to seating stress of 6.25 kPa after hitting the near-equilibrium state (IS: 2720, Part 15, 1986). Then, the samples were dismantled and weighed. Their final heights were measured.

4 Experimental Results and Discussion

4.1 e-logσ' Behavior

i. Effect of placement conditions.

Figure 2a, b presents the e-log σ' curves of K-soil and M-soil belonging to the same group of low liquid limit, respectively.

One of these two figures comprises three curves—one corresponding to the soil at an optimum compacted condition, one corresponding to the soil compacted at an optimum dry side, and the other to the soil at the wet side at an optimum compacted at a limit of 0.95 ρ_{dmax}. The following observations can be made from these illustrations.

- The soils compacted to the same placement density (i.e., 0.95 ρ_{dmax}) show dissimilar e-log σ' curves under different preliminary water contents adopted during compaction.
- The e-log σ' curve of the wet-side compacted soil is above that of the soil trampled on the dry side of the optimum. This means that the equilibrium void ratio of wet-side trampled soil at any successful consolidation stress level is often greater than that of dry-side compacted soil. While this similar behavior is demonstrated by both K-soil and M-soil, the regulation mechanisms in both cases may be different.

In general, soil fabric changes from a more flocculent type on the dry side of the optimum to a comparatively more oriented/less flocculent type on the wet side of the optimum. In K-soil, flocculent fabric on the dry side of the optimum is susceptible to

Fig. 2 **a**: e-log σ' curves for *K*-soil of the same group of the low liquid limit under different placement conditions. e-log σ' curves for *M*-soil of the same group of the low liquid limit under different placement conditions

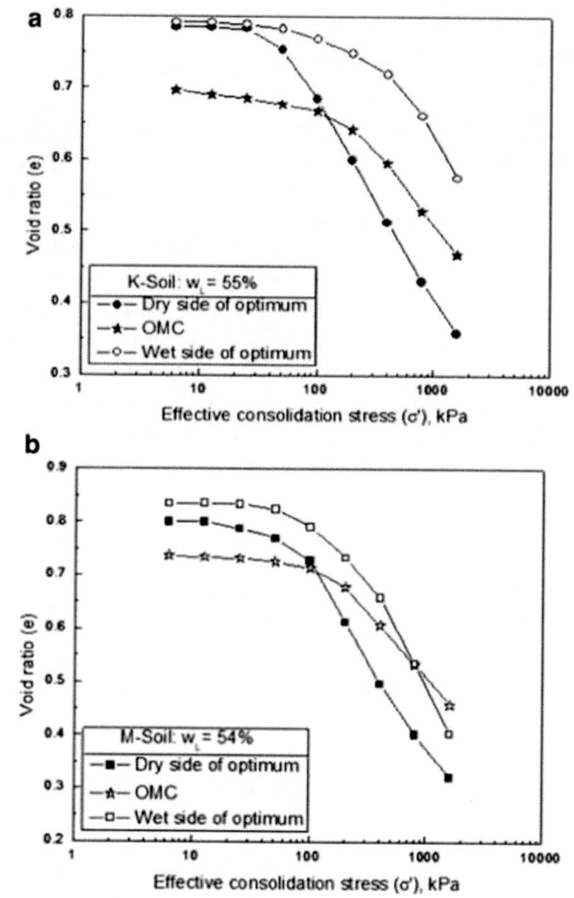

break down even at lower effective consolidation stresses resulting in more compression, which is evident from Fig. 2a. In *M*-soil, due to the presence of relatively more oriented fabric, which is favorable for the full development of the diffused double layer, the inter-particle repulsive forces develop. This leads to a more equilibrium void ratio on the wet side of optimum under several effective consolidation stresses.

Figures 3a–c presents the e-log σ' curves of soils of the same group of high liquid limit, namely *KM*-soil, *M*-soil, and *K*-soil, respectively.

All these soils often have their e-log-like curves compacted on the wet side of optimum above those of soil samples firmed on the dry side of optimum. In Fig. 3a, the e-log σ' curves of *KM*-soil corresponding to be firmed on dry and wet sides of optimum states appear to be almost identical. Due to the almost equal dominance of both kaolinitic and montmorillonitic clay minerals, which leads to counteracting effects, no distinct difference can be seen between this e-log σ' curves. The dominance of inter-particle repulsive forces due to the relatively more oriented fabric on

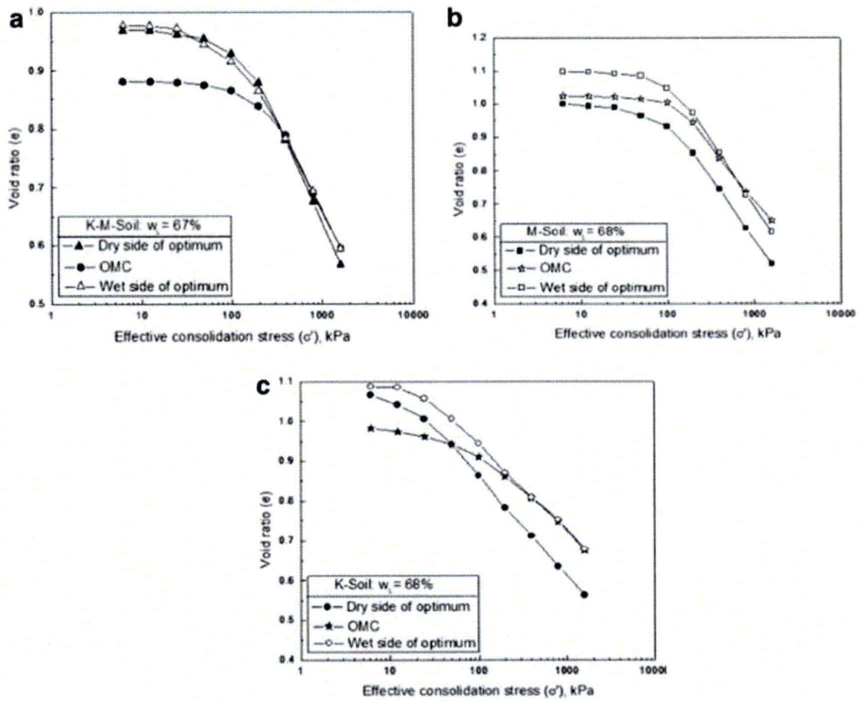

Fig. 3 **a**: e-log σ' curves for *KM*-soil of the same group of the high liquid limit under different placement conditions. **b** e-log σ' curves for *M*-soil of the same group of the high liquid limit under different placement conditions. **c**: e-log σ' curves for *K*-soil of the same group of the high liquid limit under different placement conditions

the wet side of optimum, which results in the higher equilibrium void ratio under any effective consolidation stress level, can be seen from the e-log σ' curves of *M*-soil (Fig. 3b). Similarly, Fig. 3c illustrates the early breakdown of the flocculent fabric of *K*-soil on the dry side of optimum leading to the lower equilibrium void ratio of the dry side compacted soil samples.

For more clarity, the void ratios of soil samples with different placement conditions are normalized using the void ratio at seating effective consolidation stress (i.e., at 6.25 kPa), and these normalized void ratios are plotted against log σ'.

4.2 δ-log *t* Curves

The conventional δ-log t curve following Terzaghi's one-dimensional consolidation theory has a reverse *s*-shape with an initial well-defined parabolic region followed by a linear portion leading to a reverse curve and a linear portion representing the secondary consolidation zone.

The δ-log σ' curves of compacted soils can be considered to have three regions, namely distinctly over the consolidated or pre-yield zone, zone concerning the transition from over-consolidated state to normally consolidated state, and distinctly normally consolidated zone or post-yield zone.

The secondary compression is computed for the δ-log t plots in the post-yield region (after the end of primary consolidation that is the slope of the δ-log t plots after the soil attaining primary consolidation) because the soil exhibits a typical reverse *s*-shaped curve as it is a characteristic feature of normally consolidated soil as proposed by Terzaghi for the soil under study containing a different clay mineralogy and liquid limit.

Figure 4a through c represents typical δ-log t curves (*K*-soil) belonging to a group of high liquid limits compacted on dry, OMC, and wet placement conditions, respectively.

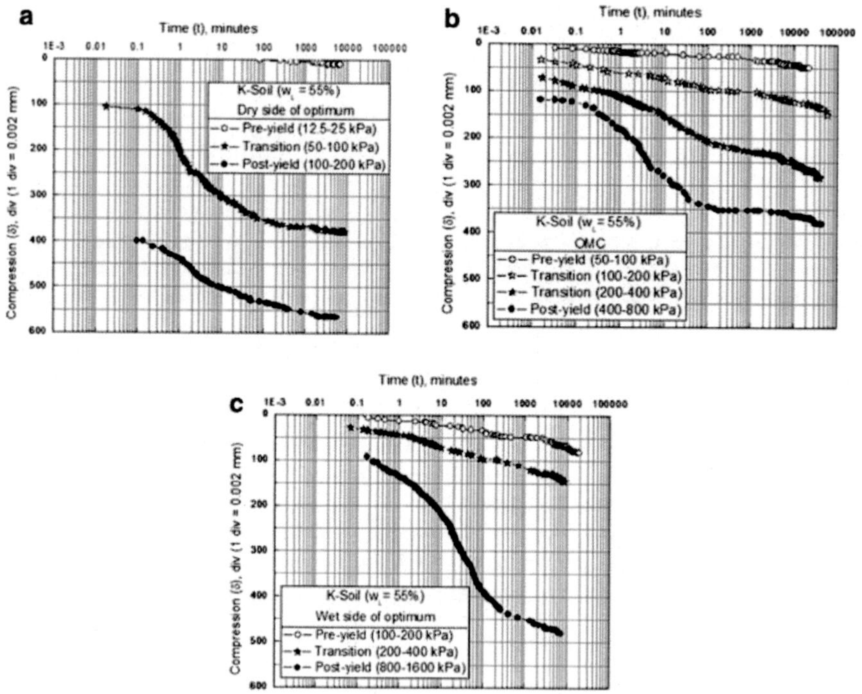

Fig. 4 **a** Typical δ-log t curves for *K*-soil ($w_L = 55\%$) compacted on dry side of optimum. **b** Typical δ-log t curves for *K*-soil ($w_L = 55\%$) compacted at OMC. **c**: Typical δ-log t curves for *K*-soil ($w_L = 55\%$) compacted on wet side of optimum

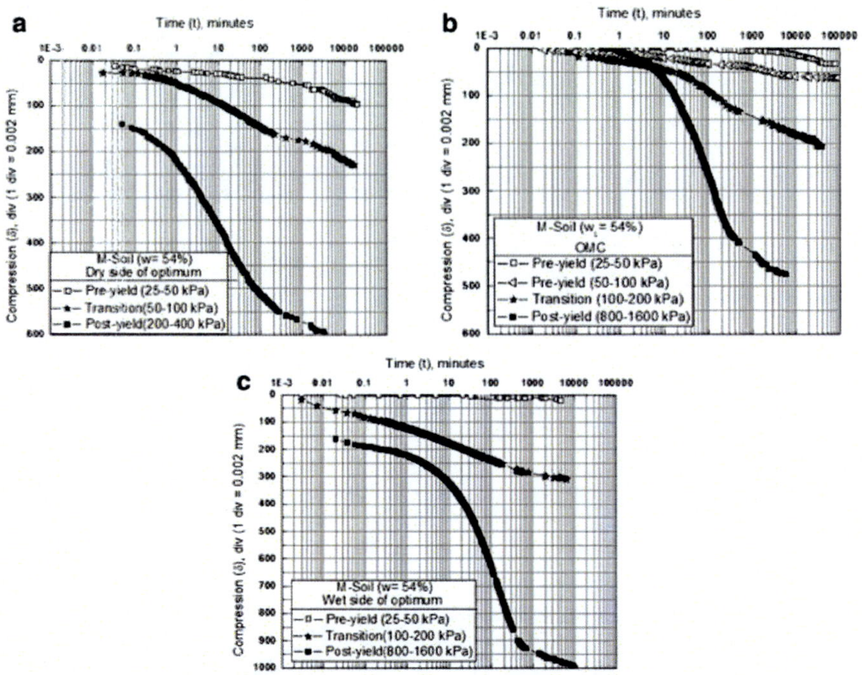

Fig. 5 **a** Typical δ-log t curves for M-soil ($w_L = 54\%$) compacted on dry side of optimum. **b** Typical δ-log t curves for M-soil ($w_L = 54\%$) compacted at OMC. **c** Typical δ-log t curves for M-soil ($w_L = 54\%$) compacted on wet side of optimum

Figure 5a through c represents typical δ-log t curves (M-soil) belonging to high liquid limit group compacted on dry, OMC, and wet placement conditions, respectively.

Figure 6a through c represents typical δ-log t curves (KM-soil) belonging to high liquid limit group compacted on dry, OMC, and wet placement conditions, respectively.

Figure 7a through c represents typical δ-log t curves (M-soil) belonging to high liquid limit group compacted on dry, OMC, and wet placement conditions, respectively.

Figure 8a through c represents typical δ-log t curves (K-soil) belonging to high liquid limit group compacted on dry, OMC, and wet placement conditions, respectively.

Each of Figs. 4a through 8c shows the validity of the δ-log t curves of consolidation loading in the pre-yield zone, the transition zone, and the post-yield zone. The analysis of these figures may be used to make subsequent observations.

The delayed compression describes the δ-log t curves of the pre-yield region. This is expressed through the preliminary linear portion of the δ-log t curve over a while of consideration, afterward the concave down the curve. Most of the δ-log t curves in

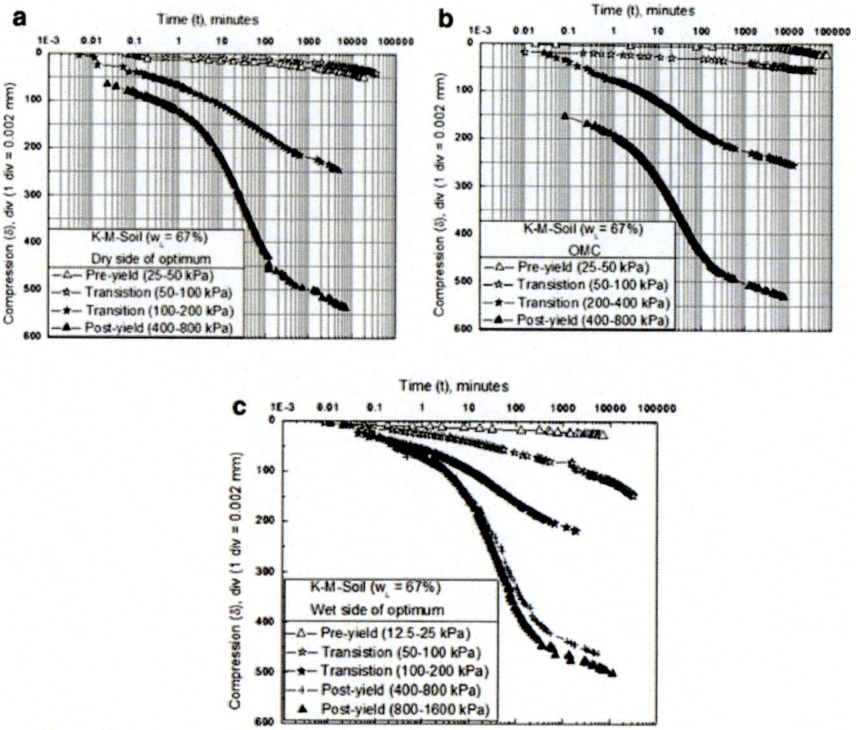

Fig. 6 **a** Typical δ-log t curves for *KM*-soil ($w_L = 67\%$) compacted on dry side of optimum. **b** Typical δ-log t curves for *KM*-soil ($w_L = 67\%$) compacted at OMC. **c** Typical δ-log t curves for *KM*-soil ($w_L = 67\%$) compacted on wet side of optimum

the pre-yield zone do not demonstrate a predictable reverse s-shaped curve, which is distinctive of over-consolidated soils, despite the fact that stress increased the ratio being one. Because of this, it is not achievable to be relevant to the time fitting the logarithm method to measure the consolidation coefficient in such situations.

Over the transition region, the tendency of delayed compression is reduced and distinct reverse *s*-shaped curves result from consolidation loading in the post-yield zone. Figures 4b, 5a, 6a–c, 7a–c, and 8a, b show the plodding shift in the form of the δ-log t curve from that of over-consolidated soil to that of the normally consolidated soil is quite evident. It should be noted that a stress increase ratio of unity corresponds to all the δ-log t curves shown above bearing different shapes.

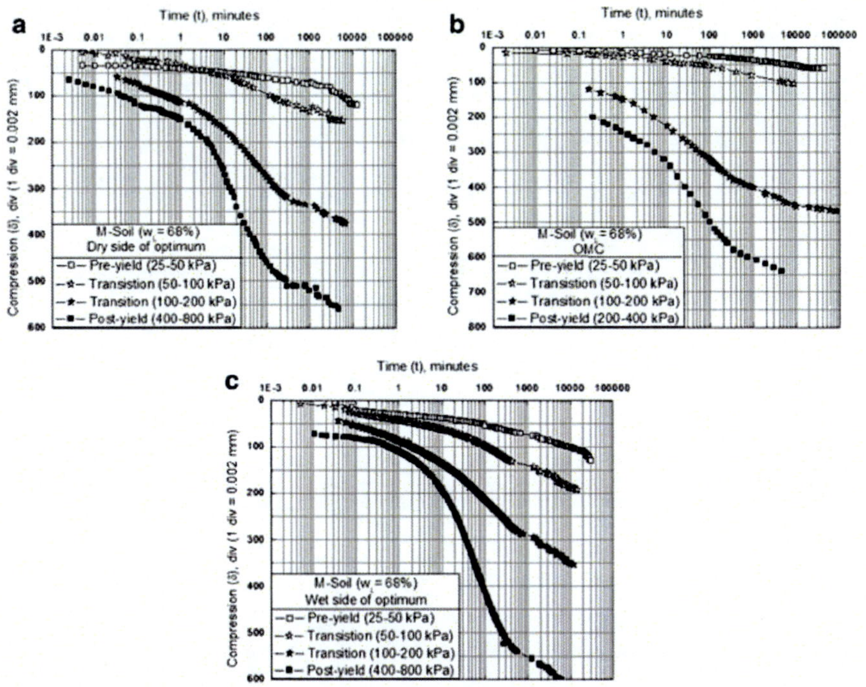

Fig. 7 **a** Typical δ-log t curves for M-soil ($w_L = 68\%$) compacted on dry side of optimum. **b** Typical δ-log t curves for M-soil ($w_L = 68\%$) compacted at OMC. **c** Typical δ-log t curves for M-soil ($w_L = 68\%$) compacted on wet side of optimum

4.3 Variation of C_α with Pressure (log α)

Instant and delayed compressions for normally consolidated natural soils are components of a continuous volume change mechanism. Soil compressibility is determined by the composition and physicochemical environment. The magnitude of the coefficient of secondary compression (C_α) of compacted fine-grained soils having poles apart clay mineralogy subjected to IS light and heavy compaction energy levels was computed by using Eq. (3) with the help of δ-log t curves.

$$C_\alpha = \Delta e / \Delta \log t \tag{3}$$

(where e = void ratio and t = time).

Figure 9a through e shows the variation of secondary compression (C_α) with respect to pressure (log α).

The secondary compression (C_α) value of the group of the low liquid limit and high liquid limit soils is compared with different placement conditions. The following observations are made from the above figures.

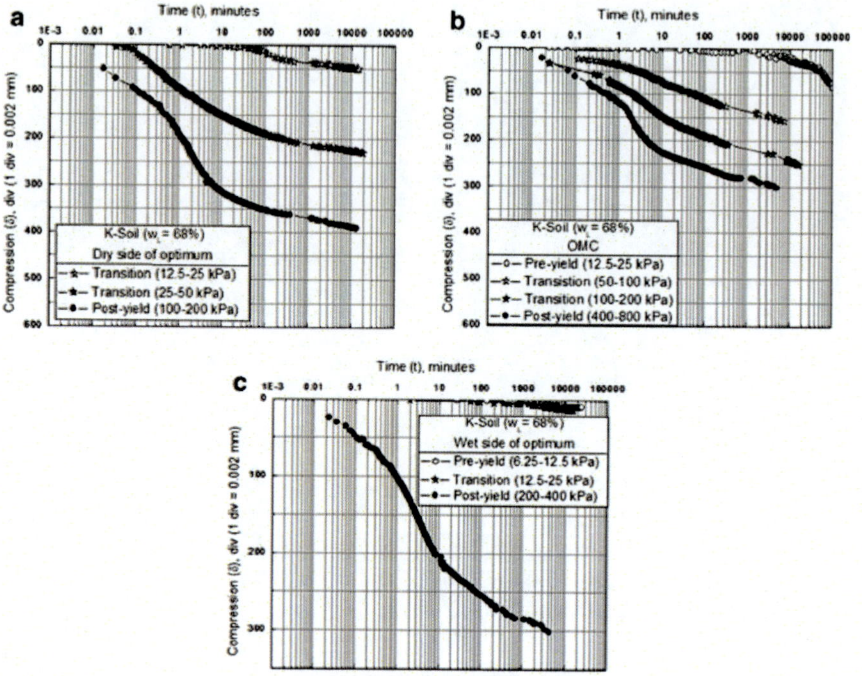

Fig. 8 **a** Typical δ-log *t* curves for *K*-soil ($w_L = 68\%$) compacted on dry side of optimum. **b** Typical δ-log *t* curves for *K*-soil ($w_L = 68\%$) compacted at OMC. **c** Typical δ-log *t* curves for *K*-soil ($w_L = 68\%$) compacted on wet side of optimum

K-soil of both high and low liquid limit groups has lower values of C_α at OMC when compared to dry and wet sides of optimum conditions, due to the cause of flocculation. *M*-soil of the lower liquid limit group has higher values of C_α on dry side of optimum and lower values on wet side of optimum. This tendency is reversed in a higher liquid limit group owing to the effect of diffuse double layer formation which leads to a higher magnitude of compression which was observed during the one-dimensional consolidation test. It is also observed that *KM*-soil of the group of the high liquid limit also has higher values of C_α on dry side of optimum and lower value of C_α OMC at optimum due to the balancing of attractive and repulsive forces of kaolinitic and montmorillonitic soils. The magnitude of C_α for *KM*-soil increases with respect to pressure up to pre-consolidation stress value beyond which the variation becomes insignificant. (Shrinkage index of *KM*-soil is slightly lower than *M*-soil and higher than *K*-soil of the same group of liquid limit.)

Figure 9f through h presents the variation of secondary compression (C_α) with pressure (log α) for the soils having the same range of liquid limit (*K*-soil and *M*-soil) for dry, OMC, and wet side of optimum placement conditions, respectively.

From Figs. 9f through h, it can be observed that as pressure increases the magnitude of C_α decreases for montmorillonitic soils, whereas, for kaolinitic soil, C_α values

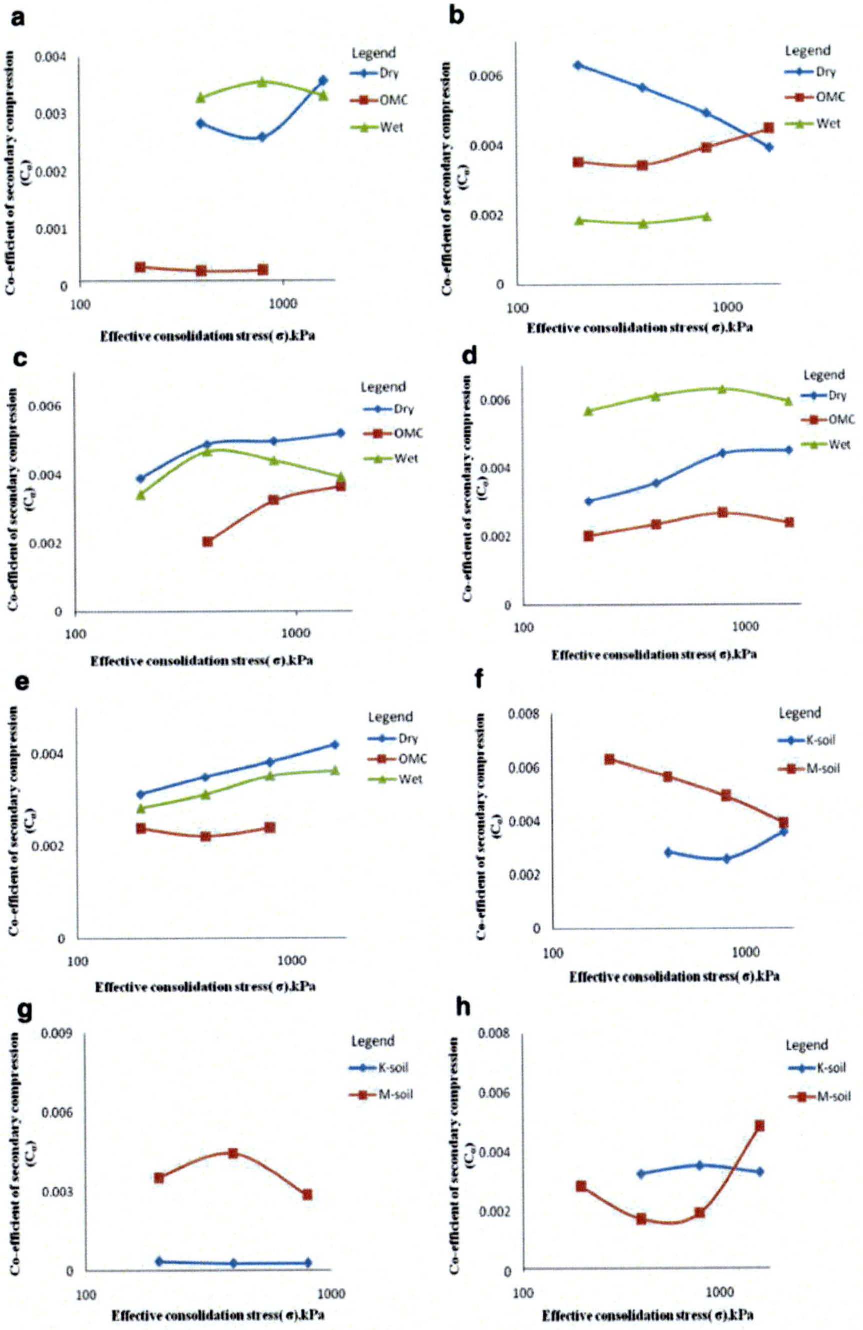
a
Legend
Dry
OMC
Wet
Co-efficient of secondary compression ($C_α$)
0.004
0.003
0.002
0.001
0
100
1000
Effective consolidation stress(σ).kPa
b
Legend
Dry
OMC
Wet
Co-efficient of secondary compression ($C_α$)
0.006
0.004
0.002
0
100
1000
Effective consolidation stress(σ).kPa
c
Legend
Dry
OMC
Wet
Co-efficient of secondary compression ($C_α$)
0.006
0.004
0.002
0
100
1000
Effective consolidation stress(σ).kPa
d
Legend
Dry
OMC
Wet
Co-efficient of secondary compression ($C_α$)
0.006
0.004
0.002
0
100
1000
Effective consolidation stress(σ).kPa
e
Legend
Dry
OMC
Wet
Co-efficient of secondary compression ($C_α$)
0.004
0.002
0
100
1000
Effective consolidation stress(σ).kPa
f
Legend
K-soil
M-soil
Co-efficient of secondary compression ($C_α$)
0.008
0.006
0.004
0.002
0
100
1000
Effective consolidation stress(σ).kPa
g
Legend
K-soil
M-soil
Co-efficient of secondary compression ($C_α$)
0.009
0.006
0.003
0
100
1000
Effective consolidation stress(σ).kPa
h
Legend
K-soil
M-soil
Co-efficient of secondary compression ($C_α$)
0.008
0.006
0.004
0.002
0
100
1000
Effective consolidation stress(σ).kPa

◂**Fig. 9** **a** Variation of C_α with pressure (log α) for *K*-soil ($w_L = 55\%$) for different placement conditions. **b** Variation of C_α with pressure (log α) for *M*-soil ($w_L = 54\%$) for different placement conditions. **c** Variation of C_α with pressure (log α) for *KM*-soil ($w_L = 67\%$) for different placement conditions. **d** Variation of C_α with pressure (log α) for *M*-soil ($w_L = 68\%$) for different placement conditions. **e** Variation of C_α with pressure (log α) for *K*-soil ($w_L = 68\%$) for different placement conditions. **f** Variation of C_α with pressure (log α) for *K*-soil and *M*-soil of the same liquid limit ($w_L = 55\%$) for dry side of optimum. **g** Variation of C_α with pressure (log α) for *K*-soil and *M*-soil of the same liquid limit ($w_L = 55\%$) for OMC at optimum. **h** Variation of C_α with pressure (log α) for *K*-soil and *M*-soil of the same liquid limit ($w_L = 55\%$) for wet side of optimum. **i** Variation of C_α with pressure (log α) for *KM*-, *K*-, and *M*-soils of the same liquid limit ($w_L = 68\%$) for dry side of optimum. **j** Variation of C_α with pressure (log α) for *KM*-, *K*-, and *M*-soils of the same liquid limit ($w_L = 68\%$) for OMC at optimum. **k** Variation of C_α with pressure (log α) for *KM*-, *K*-, and *M*-soils of the same liquid limit ($w_L = 68\%$) for wet side of optimum

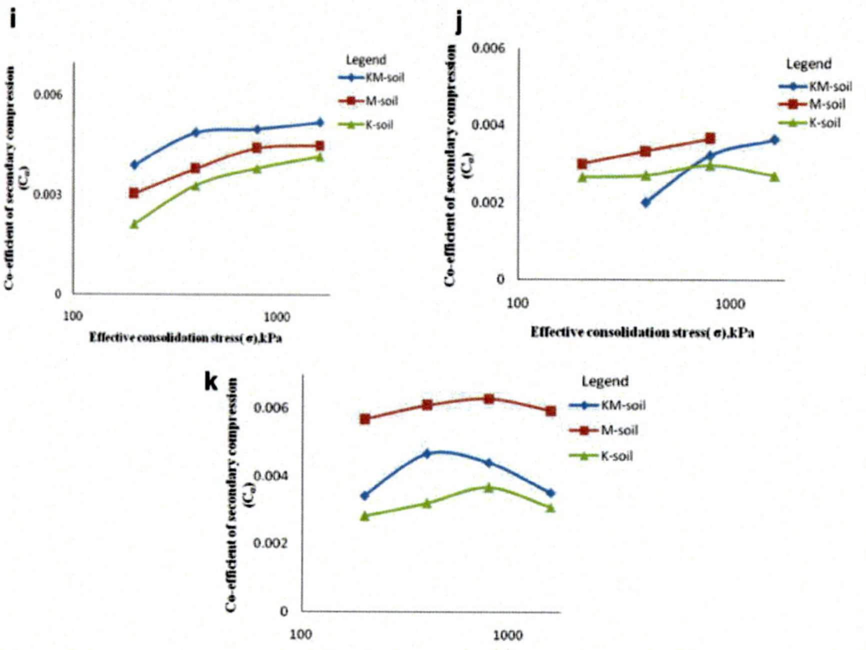

Fig. 9 (continued)

slightly increase on the dry side of optimum. Similarly, at OMC, as pressure increases the value of C_α increases up to the pre-consolidation stress value, beyond which it decreases for *M*-soil. But no significant variation of C_α values was observed for *K*-soil. The magnitude of C_α increases for *M*-soil when pressure increases, whereas for *K*-soil, C_α values do not vary much with respect to pressure on wet side of optimum. Kaolinitic soil and montmorillonitic soil of the same low liquid limit group, C_α values of montmorillonitic soil is higher than kaolinitic soil for all placement conditions.

The primary reason for this behavior can be attributed due to the high dispersive effect of montmorillonitic soils, which affects the particles to move away from each other in the presence of water. As the percentage of soil water in soil mixture starts reducing, causing the particles to orient themselves due to the effect of compaction energy imparted which in turn leads to the lower value of shrinkage index.

Figure 9i through k presents the variation of secondary compression (C_α) with pressure (log α) for the soils having the same range of liquid limit (*KM*-soil, *K*-soil, and *M*-soil) for dry, OMC, and wet side of optimum placement conditions, respectively.

The following observations were made from these figures.

For *KM*-, *M*-, and *K*-soils of a high group liquid limit, the magnitude of C_α tends to increase as pressure increases on the dry side of optimum. Similarly, at OMC, the variation of C_α for *K*-soil increases and decreases marginally by 10% as pressure increases, while for *M*-soil, there was an 18% increase, and for *KM*-soil, a 45% increase with respect to pressure. The values of C_α increase up to pre-consolidation stress value beyond which the variation becomes insignificant on the dry side and OMC at optimum. On wet side of optimum, for *K*- and *M*-soils, C_α values increase and then decrease as the pressure increases, whereas for *KM*-soil, the variation of C_α values is insignificant with respect to pressure. The magnitude of C_α increases up to pre-consolidation stress value, beyond which it decreases on the wet side of optimum. Further, *K*-soil of the high group of liquid limits shows lower values of C_α when compared with *M*- and *KM*-soils of the same liquid limit group for all placement conditions, due to the flocculating effect, whereas for *KM*-soil, higher values of C_α were observed in relative comparison to *K*- and *M*-soils of the same group of high liquid limit. These observations highlight the role of placement condition and clay mineralogy of fine-grained soils with respect to the coefficient of secondary compression.

4.4 Field Applications

Generally, during the construction phase itself, elastic settlement takes place, beyond which primary consolidation progress consumes a few months to decades in order to reach equilibration. The secondary compression of the soil takes place under sustained static loading which occurs over a very long period of time after the end of primary consolidation and generally leads to creep behavior. Any prediction of the magnitude of secondary compression through the experimental study will help us to find the resistance offered by the foundation under the effect of creep. It is also observed that creep behavior mainly depends upon the profile of the soil and the extent of soil in the influence zone. The spacing between the columns of the frame structures should be large enough to avoid over-stressing in the influence zone. The pressure distribution zone of the adjoining foundation should not overlap which may lead to over-stressing leading to excessive settlement beyond the permissible limits of settlements as per codal provisions. Soils in the stressed zone will also influence the magnitude of the secondary compression due to fluctuation in the groundwater table leading to wetting and drying alternatively, which further induces cyclic stress. The present experimental study helps in understanding the variation in the magnitude of secondary compression with effective consolidation stress for soils having a different clay mineralogy and liquid limit which is of immense practical applications.

5 Conclusions

The conclusions can be drawn based on a detailed experimental study.

- The compaction process induces stress history effect in the compacted fine-grained soil.
- Soils having lower shrinkage index compress less than soils having a higher degree of shrinkage index, thus imparting that the shrinkage limit is a degree of packing phenomenon rather than plasticity characteristic.
- The fabric, clay mineralogy, and placement conditions play a leading role as controlling mechanisms in the e-log α behavior of compacted fine-grained soils having unusual clay mineralogy.
- There exists a definite and opposite trend of variation of coefficient of secondary compression with effective consolidated stress for kaolinitic, montmorillonitic, and kaolinitic–montmorillonitic soils.

Acknowledgements The authors would like to acknowledge the support given by the Research Centre at the Department of Civil Engineering, The National Institute of Engineering, Mysuru, in carrying out the experimental study.

References

1. Sridharan A, Rao AS (1982) Mechanisms controlling the secondary compression of clay. Géotechnique 32(3):249–260
2. Mesri G, Castro A (1987) C_α/C_c Concept and K_0 during secondary compression. J Geotech Eng 113(3):230–247
3. Navarro V, Alonso EE (2001) Secondary compression of local dehydration process of clay. Géotechnique 51(10):859–869
4. Bui et al (2001) Secondary compression of clays. In: 14th Southeast Asian geotechnical conference, geotechnical engineering, vol 2, pp 1081–1084
5. Mesri G, Vardhanabhuti B (2005) Secondary compression. J Geotech Geoenviron Eng 131(3):398–401
6. Sompie OBA, Arai K, Kita A (2008) Secondary behavior in three types of consolidation tests. J Southeast Asian Geotechn Soc 39(3):145–158
7. Takeda T et al (2017) Secondary compression behavior in one-dimensional consolidation tests. J Geotechn Eng 7(2):053–058

8. El Amin Bourouis M et al (2020) Contribution of two artificial intelligence techniques in predicting the secondary compression index of fine-grained soils. Innov Infrastruct Solut 5. Article number 96
9. Raheena R (2020) Determination of coefficient of secondary compression in accelerated incremental loading consolidation test. Geotech Charact Modell 85:1113–1125
10. Jiang N et al (2020) Influence of structure and liquid limit on the secondary compressibility of soft soils. J Mar Sci Eng 8(9):627–632
11. Prakash K, Sridharan A, Prasanna HS (2016) Dominant parameters controlling the permeability of compacted fine-grained soils. Indian Geotech J 46(4):408–414
12. Sridharan A, Rao SM, Murthy NS (1985) Free swell index of soils: a need for redefinition. Indian Geotech J 50(2):94–99
13. Sridharan A, Rao SM, Joshi S (1990) Classification of expansive soils by sediment volume method. Geotech Test J 13(4):375

Slag–Fly Ash–Glass Powder-Based Alkali-Activated Concrete—A Critical Review

Shriram Marathe, I. R. Mithanthaya, and Siddhivinayaka Hegde

Abstract This paper presents some key research findings in the field of "alkali-activated concretes" (AAC) incorporating ingredients which are industrial by products like slag, fly-ash, etc., as a binding ingredient. The paper gives the brief outlook of the current development in the field alkali-activated materials for concrete applications starting from the historical developments. The literature review reveals that, the alkali-activated concretes proven to be feasible (sustainable) and satisfactory substitute for conventional Portland cement-based cementitious substances. In addition to the conventional ingredients, the use of powdered waste glass has shown an improved engineering performance in concrete production as per the research outputs from many potential researchers. The paper also presents the possibility of a wide scope for the research as a research gap in the field of AAC and their applications.

Keywords Alkali activation · Fly-ash · Glass powder · Slag · Na_2O dosage · Strength

1 Introduction

The environmental aspects involved in the manufacturing and use of "ordinary Portland cement" ("OPC") is a challenge and lot of research is going on in this regard. OPC has been a suitable binder for structural applications for more than 150 years, but not without drawbacks. Cement production consumes massive quantities of natural elements. Again it is energy intensive with high energy consumption at 100–150 kW per ton of cement produced. On the other hand, sulfur dioxide emission is also in a high rate, depending upon the kind of the fuel used for its manufacture. It is becoming increasingly capital intensive to erect new cement plants to meet a huge demands for housing and infrastructural needs. Finally, many structures of concrete have showed early distress and problems, which have an unfavorable consequence

S. Marathe · I. R. Mithanthaya · S. Hegde (✉)
Department of Civil Engineering, NMAM Institute of Technology, Nitte, Karkala, India

I. R. Mithanthaya
e-mail: irmithanthaya@nitte.edu.in

L. Nandagiri et al. (eds.), *Sustainability Trends and Challenges in Civil Engineering*, Lecture Notes in Civil Engineering 162, https://doi.org/10.1007/978-981-16-2826-9_19

on the resource output of the industry. The call for the day is, hence "Sustainable Development" which demands the new concrete technology which uses minimum natural resource, energy and generates "carbon dioxide," without compromising on strengths and robustness behaviors [1–3].

Alkali-activated cements are considered as new generation binders, which have a potential to give a sustainable replacement to OPC. There is an urge to limit the usage of the OPC in infrastructural development due to its adverse environmental issues and its structural limitations [4]. Because of low internal energy and low carbon dioxide emission associated with production of alkali-activated binder (AAB)-based mixtures compared to conventional Portland Cement Concretes, this novel material may be considered to be more eco-friendly [3, 5].

These AABs have been extensively addressed and encouraged as a constituent of the recent and forthcoming toolkit of "sustainable cementing binder systems." "Alkali activation" is the process of reaction of a solid aluminosilicate ("precursor") under alkaline conditions (induced by the "alkali activator"), to generate a toughened binder which is based on an amalgamation of "hydrous alkali–aluminosilicate" and/or "alkali–alkali-earth–aluminosilicate" phases [6].

This chapter gives an overview of the various key literatures available in the field of alkali activation and geopolymerization techniques of concrete production. Also, few essential background studies for this research work—including few standard IS codal provisions, literatures related to paver and masonry blocks, IRC codal provisions, and also few key literatures on FE-based software utilized in the present research exploration.

2 Brief Historical Overview of Alkali-Activated Materials

The initial works in alkali-activated systems started from the year 1930. Way back in 1930, for the first time Kuhl has invented the hardening behavior of slag in the incidence of an alkali (caustic potash, i.e., potassium hydroxide). Then in 1937, Chassevent had measured the rate of reactivity of slag using "caustic potash and soda" solution. In 1940, Purdon carried out laboratory study on clinker-free cements which consisted slag as the binder and caustic alkalis formed by a base and an alkaline salt. In 1959, a researcher from Russia, Glukhovsky discovered a new group of binders which included both of low calcium or free from calcium, "aluminosilicate" and alkali metal solutions which were termed "soil cements" and the respective concretes termed as "soil silicates." Further, Glukhovsky was the first author to investigate the cements used in prehistoric Egyptian and Roman constructions. He stated these structural creations were made of "aluminosilicate calcium hydrates," which are alike to the constituents of "Portland cement" and also of "crystalline phases of analcite," a rock which is natural that would elaborate the binders durability [4]. He (Glukhovsky) also divided the binders into two groups based on the mineral constituents of preliminary materials; the first group of binders were called as "alkaline binding systems" and the second group as "alkaline-earth–alkali binding systems." The study by Glukhovsky was considered as a significant step stone for the further researches in the field of "alkali-activated cements," following which the development of "alkali-activated concretes" [7]. In the year 1979, Davidovits developed an innovative faction of binders produced by combining sintering kaolinite products, "limestone" or "dolomite" as the "aluminosilicate" ingredients, and called those binders as "geopolymers." The term "geopolymer" was given to such binders because of the presence of a polymeric structure. He proposed that any source, either from natural geological source or an industrial waste material such as rice husk ash and fly-ash, that are rich in alumina and silica can be effectively activated by means of highly alkaline liquid solutions [8]. The classification of alkali-activated binders based on their chemical configuration and characterization of hydration products was done by Krivenko in 1994 [7]. The "alkaline aluminosilicate systems," "(R–A–S–H, where R=Na or K)" were called "geocements," whose formation process is similar to process of natural zeolites and "alkaline–alkaline-earth systems" (R–C–A–S–H) in which the products of hydration included the formation of "calcium silicate hydrates," i.e., "C–S–H" with low "Ca/Si" ratio. Further, it has been suggested by few researchers that the use of the term "geopolymer" in alkali-activated cementitious materials only if there is a presence of a zeolite-like phase with amorphous to semi-crystalline characteristics [9]. The chief historic developments of alkali-activated binders are abridged by Roy [10] as shown in Fig. 1. The field of "alkali-activated cements" was witnessed further contributions from many researchers [1, 2, 4, 11–13].

3 Source Materials, Alkaline Activator, and Mechanical Properties

The raw constituents used for the making of AABs may be classified as pozzolanic or latent hydraulic materials [7]. Pozzalans are the materials rich in silica and alumina, possess diminutive or no cementitious behavior by themselves, but in finely divided state and in the occurrence of moistness, experience chemical reaction with calcium hydroxides at ambient temperature to form composites having cementing behavior [14, 15]. Low calcium "(class F) fly-ash" and "silica fume" are the commonly used "pozzolanic materials." Latent hydraulic materials are finely divided materials similar to pozzolans, which contain sufficient amount of calcium to form complexes with binding property after reacting with water. In general, the latent hydraulic materials cannot undergo direct setting and hardening after reacting with water in normal conditions [14]. The hardening energy is quiescent and requires an "activator" such as "calcium hydroxide" or other alkaline compound that is strong to release the cementitious properties. The "latent hydraulic materials" (LHM) when combined with OPC and water get activated under the influence of "calcium hydroxide" which is generated during the hydration reaction of cement. GGBS is one of the examples of LHM, which have a good potential to be used in cement manufacturing. Both the pozzolans and latent hydraulic materials may be naturally occurring in nature or may

Author(s)	Year	Significance
Feret	1939	Slags used for cement
Purdon	1940	Alkali-slag combinations
Glukhovsky	1959	Theoretical basis and development of alkaline cements
Glukhovsky	1965	First called "alkaline cements" because natural substances used as components
Davidovits	1979	"Geopolymer" term—emphasizes greater polymerization
Malinowski	1979	Ancient aqueducts characterized
Forss	1983	F-cement (slag-alkali-superplasticizer)
Langton and Roy	1984	Ancient building materials characterized (Roman, Greek, Cyprus)
Davidovits and Sawyer	1985	Patent leading to "Pyrament"
Krivenko	1986	D.Sc. Thesis, R_2O-RO-R_2O_3-SiO_2-H_2O
Malolepsy and Petri	1986	Activation of synthetic melilite slags
Malek et al.	1986	Slag cement-low level radioactive waste forms
Davidovits	1987	Ancient and modern concretes compared
Deja and Malolepsy	1989	Resistance to chlorides shown
Kaushal et al.	1989	Adiabatic cured nuclear waste forms from alkaline mixtures including zeolite formation
Roy and Langton	1989	Ancient concrete analogs
Majumdar et al.	1989	$C_{12}A_7$ - slag activation
Talling and Brandstetr	1989	Alkali-activated slag
Wu et al.	1990	Activation of slag cement
Roy et al.	1991	Rapid setting alkali-activated cements
Roy and Silsbee	1992	Alkali-activated cements: overview
Palomo and Glasser	1992	CBC with metakaolin
Roy and Malek	1993	Slag cement
Glukhovsky	1994	Ancient, modern and future concretes
Krivenko	1994	Alkaline cements
Wang and Scrivener	1985	Slag and alkali-activated slag microstructure

Fig. 1 Historical development of AAC systems and alkaline cements

be produced artificially from industrial processes. Davidovits discovered that during the production of geopolymeric binders, the addition of GGBS, which is a "latent hydraulic" cementitious end product, that quickens the setting time and improves compressive and flexural strengths [8].

Many researchers reported [16] the use of mineral admixtures in the GGBS-based alkali-activated concrete to obtain the enhanced concrete performance. The silica-rich materials such as powdered waste glass, silica fume, alumina-rich red mud, natural pozzolan, calcium-rich powdered lime were utilized to enhance the engineering performance. The study by Rostami and Behfarnia revealed the enhanced 90 days compressive strength by 32% by the use of 15% substitution of GGBS by silica fume. It also enhanced the resistance to chloride penetration, reduced voids, and reduced water captivation potential of the concrete [17].

Numerous researchers have used "fly-ash" as source substance in the making of "geopolymer" concrete. Since the fly-ash is available as an abundant waste material obtained mainly from coal based on the thermal power plants; it is considered as most used binding material in production of geopolymer concrete. The yearly production of "fly-ash" is assessed to be approximately 780 million tonnes. It mainly consists of amorphous alumina and silica along with a favorable particle size and shape which has proved to improve the workability property of geopolymer mix production [1, 18]. Many studies were also carried out by considering other waste materials which have pozzolanic properties in the production of geopolymer concrete. However, except GGBS and fly-ash, other industrial left-over materials such as "palm oil fuel ash," "rise husk ash (RHA)," "biomass ash," red mud (i.e., bauxite residue), heavy metal containing wastes were also used in geopolymer binder production in concrete applications [18].

The source materials for alkali-activated binders need to be activated using strong alkalis in order to form the resulting binding material. "caustic alkalis" or "alkaline salts" are the most widely used "alkaline activators." The alkaline activators generally consist of mixtures of silicates and hydroxides of alkali. "sodium silicate" (Na_2SiO_3), "sodium hydroxide" (NaOH), "sodium carbonate" (Na_2CO_3) or mixture of "sodium–potassium hydroxide" (NaOH, KOH) with "sodium silicate–potassium silicate," or any other combinations are the most widely used "alkaline activators." A combination of "sodium hydroxide" with liquid sodium silicate has been agreed to provide the best strength performance for activation of alkali-activated binders [19] The strength of "AABs" is governed by the alkaline activator type, activator modulus, and dosage of alkaline activator [20]. The "activator modulus" (Ms) is defined as the ratio of the mass ratio of "SiO_2" to "Na_2O" components present in the alkaline activator, while the dosage (usually referred as %Na_2O) is the total sum of the mass of "Na_2O" present in the alkaline activator (Mass of "Na_2O" present in liquid sodium silicate + mass of Na_2O equivalent in sodium hydroxide if mixture of "sodium silicate" and "sodium hydroxide" used as alkaline activator).

Wang et al. [21, 22] stated that mechanical strength and other properties of AAS mortars were influenced by the nature of the alkaline activators. The dosage and the activator modulus have significant effects on the properties of AABs. They provided a range of activator modulus within maximum compressive strength may be obtained based on the type of GGBS. Wang et al. suggested that, optimum "modulus of alkaline activator" solution is ranging from 0.75 to 1.25 for acid slag, 1.0 to 1.5 for basic slag, and 0.9 to 1.3 for neutral slag [22]. It was observed that, for all types of GGBS in AAS-based mixes of concrete, as the dosage of activator modulus increases, there is a rise in the compressive strength value up to certain optimum value of dosage and further increment of the modulus leads to an decrease of strength value.

Few researchers [23] studied the activation of GGBS using diverse activators such as "sodium carbonate," "sodium silicate," "sodium phosphate," "sodium hydroxide," and combinations of mentioned activators. They recommended that sodium silicate solution was the best giving higher value of compressive strength with 8% Na_2O dosage with an activator modulus of 1.25 for better results (shown in Fig. 2). They also observed that, at a higher modulus dosage, there is a significant decrease in setting time and a reduced early strength gain was seen. At a very higher dosage of modulus, there was a sign of high shrinkage in concrete—which further performed like a fast-setting cement (Fig. 3).

Fernandez-Jimenez et al. [20] reported that the strength of "AAS" mortars is mostly influenced by the type and nature of alkaline activator, dosage of alkaline activator (AA) and observed that the optimal dosage of "alkaline activator" varies

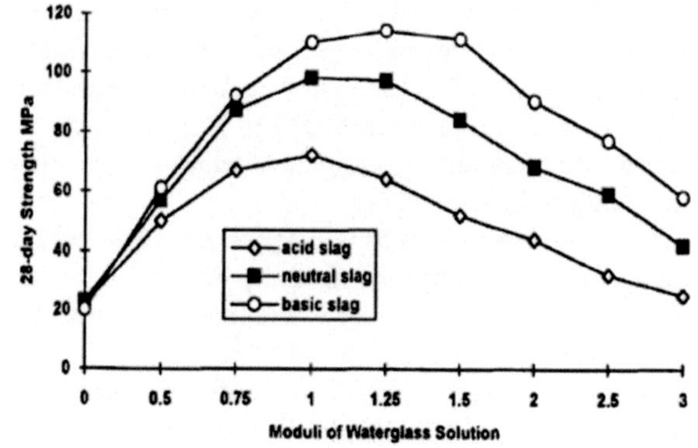

Fig. 2 Variation of "moduli of sodium silicate solution" with 28-day compressive strength value for different types of GGBS

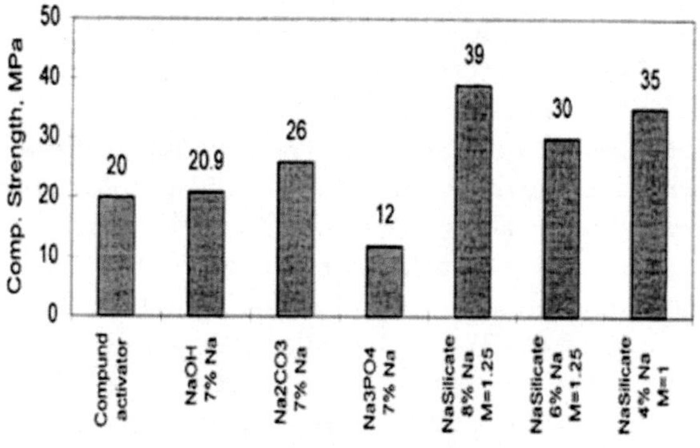

Fig. 3 Compressive strengths of slag-based alkali-activated cements

from 3 and 5.5% of "Na_2O" by the mass of GGBS. "Na_2O" dosage above this limit may cause efflorescence problems along with inefficient uneconomical mixtures.

Study by Krizan and Zivanovic [24] shown that an higher ultimate strength than OPC can be achieved with GGBS-based alkali-activated cements activated using sodium silicate with "activator modulus" ranging in between 0.6 and 1.5 with the suitable Na_2O dosage. They also shown from the study of hydration properties of AAC mixes that the liquid sodium silicate-activated GGBS displays two heat evolution peaks during preinduction period, and by increasing the modulus of liquid sodium silicate, there is a rise in the value of additional initial peak and lasting induction period. However, the rise in the dosage of activator leads to a decrease in the induction period.

Fernandez and Palomo [25] studied the activation of FA using different types of alkaline activators and by varying the Na_2O dosage between 5 and 15% (mass of binder). They reported that the activator modulus along with a "water/binder" ratio (w/b), affects the mechanical strength. Also observed that the Na_2O dosage of 5.5% (mass of fly-ash) led to very less pH, that affected the reaction development negatively, while the increase in Na_2O dosage led to higher strength with 14% Na_2O dosage (mass of FA) providing the maximum compression strength.

Wardhono et al. [26] studied the activation of AASFA mixes with different GGBS:FA mixes of 100:0, 90:10, 80:20, 70:30, 60:40, and 50:50 activated using sodium silicate and sodium hydroxide solutions. The "activator modulus" dosages were varied from 4.06 to 4.72. The results shown that the maximum compression strength of about 62 MPa was obtained for 50:50 mix for an alkaline activator modulus of 4.72.

Cengiz et al. examined the AAS mortars using different types of activators such as "sodium hydroxide," "sodium carbonate," and "sodium silicate," with varying activator modulus and with Na_2O dosages in the range of 4–8% (mass of GGBS). It was detected that the compressive and tensile strengths of "AAS" mortars increased with Na_2O dosage of alkaline activators and also suggested that there exists an optimal alkaline modulus for which the highest compressive and tensile strengths can be obtained [27].

Palankar et al. deliberated the use of slag–fly-ash-based AAC for the application of concrete pavements. In this study, "alkali-activated slag concrete" (AASC) and "alkali-activated slag–fly-ash concrete" (AASFAC) were developed which was then related with OPC concrete of alike design grade. GGBS and fly-ash were used in the mix, in ratio 100:0, 75:25, 50:50, and 25:75 as the binder ingredient. The modulus of alkaline activator was varied from 0.75 to 1.75, and the dosage of Na_2O was varied from 4.0 to 5.5. All the AAC mixes were subjected to air curing in laboratory conditions. The results indicated that, all OPC, AASC, and AASFC mixes have obtained the target slump value. The density of AAFC was found to decrease as there is a rise in the dosage quantity of fly-ash. The compressive strength test for the mixes was carried out at 3, 7, 28, and 90 days curing; results indicated that the performance of alkali-activated mixes had shown better outcome than OPC-based mixes. Effect of activator modulus on compressive strength of AASC and AASFC is shown in Fig. 4.

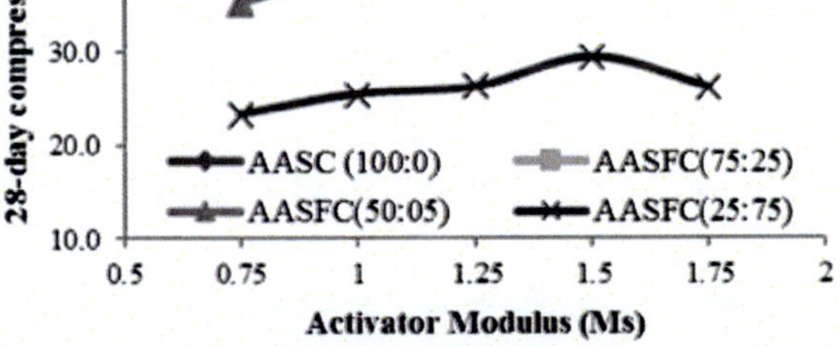

Fig. 4 Effect of "activator modulus" on strength of "alkali-activated mixes"

The early strength gain was also better in the mixes where the dosage of GGBS was more. However, the 28 days and 90 days strength of all mixes were similar. AASFAC displayed a max. compressive strength values of 67.4 MPa at 90 days for Ms 1.25 and for the mix of 75:25. The "flexural strength" and the "modulus of elasticity" were also investigated. "OPCC" and "AAC" mixes displayed similar "modulus of elasticity," but the flexural strength of AAC combinations were higher than OPCC. It was a peculiar observation that, as there is a rise in "fly-ash" content, the workability of the mix increased, but there was a little reduction in the strength and "modulus of elasticity" values. Absorption of water and total porosity tests were conducted on the concrete mixes according to "ASTM C 642–06" at curing of 28 days. The AASC displayed the less porosity and absorption of water, among all the concrete mixes. The AASFC with 25:75 displayed the maximum porosity and water absorption [28].

Karim et al. studied consequence of various alkaline activators, i.e., "NaOH," "KOH," and "$Ca(OH)_2$" on a concrete mix using triple blend of binders ("slag," "fly-ash," and "rice-husk ash") and comparing the results with conventional Portland-based concrete. They called the cement as zero-cement, which was abbreviated as Z-Cem. The alkaline activators were used in varied weights. They have stated that the setting time and consistency of the Z-Cem raised with the increase in the rice-husk ash dosage. The maximum compressive strength of about 42–44 N/mm^2 was achieved at 28 days of curing with 5% of NaOH dosage. They studied the microstructure and FTIR analysis which revealed that the development of the strength is due to the advancement of hydration products called silica-hydrates which are similar to the CSH gel in Portland-based concretes [29].

A extensive variety of "Si, Al, and Ca" wealthy sources of mineral has been explored as solo forerunners and as admixtures in dual and ternary alkali-activated concretes based on fly-ash, GGBS, MK, and powdered waste glass. Some studies have also shown that the usage of OPC as an admixture to produce alkali-activated cementitious systems. Aliabdo et al. reported the use of Portland cement in "fly-ash"-based alkali-activated concrete decreased the setting time of the concrete from 24 h to half an hour. This also leads to an improved mechanical performance of the hardened concrete (i.e., increased tensile and compressive strength, reduced porosity, and absorption of water) [30]. Assi et al. have deliberated the effect of addition of Portland cement (15% by weight) to the fly-ash and silica fume-based AAC, which lead to a very superior mechanical performance and the reduction in cost of construction. The study suggested that good mechanical performance is achieved without the use of external heat which makes this type of concrete mix more eco-friendly [31].

4 Durability and Flexural Fatigue Studies on AAC Mixes

The durability property of any material is one of the key parameters to use the material in practice. In the conventional OPC concrete, studies have exposed that the usage of pozzolanic "mineral admixtures" such as fly-ash, GGBS will enhance the durability performance [32], whereas in the case of AAC mixes, the durability is ruled by many factors such as type of mix, composition of binders, Na_2O dosage, "water/binder ratio" (w/b), and the modulus of activator [33, 34]. Several researchers have made the effort to study the performance of durability of various types of alkali-activated cementitious systems and several research outcomes reported that the alkali-activated concretes performed far well than that of the "OPC"-based concrete systems [7, 35–38]. The better durability performance of "alkali-activated slag and fly-ash"-based concrete mixes are mainly because of the nonexistence of calcium hydroxide (Portlandite) and little calcium content in the reaction products [39]. Few of the works were presented in this section.

Karim investigated the durability performance of two types of "slag–fly-ash–rice husk ash"-based "alkali-activated concrete" mixes. He studied the durability of these two types of mixes which gave the satisfactory performance (in strength) were done by testing it for porosity, absorption of water, sulfate resistance, chloride penetration, sulfate resistance, thermal resistance, and corrosion. Further, the durability results were related with the OPC mix. All the results shown that the alkali-activated binder-based concrete shown better performance in all the durability tests when compared it with the OPC-based concrete. He also developed the regression equations showing the connection between "porosity" and "compressive strength" of the AAC and "OPC"-based concrete mixes. To investigate the sulfate attack, the prepared mortar mixes were immersed in 5% $MgSO_4$ solution, and reduction in strength is tested at age of 30, 60, and 90 days [29].

Palankar et al. studied the long-term aging performances of GGBS–fly-ash-based alkali-activated cementitious systems by studying their long-term exposure to sulfate environment ($MgSO_4$), and acid environment (sulfuric acid) is carried out. Along with this, the absorption of water and their volume of permeable voids are also determined. In the study, the incorporation of "steel slag aggregates" is done. Results were compared with that of the Portland cement-based cementitious systems. The economical and ecological analysis was also performed on both type of mixes. For the acid and sulfate resistance test, the 100 mm cube specimens were immersed in the prepared solution after curing for 28 days (air curing for AAC and water curing for OPC were adopted), and testing at regular intervals till 360 days. The solution for sulfate attack test was the aqueous solution comprising 10% "$MgSO_4$" by maintaining a pH value of 6.5–7.5 using nitric acid. The solution for acid attack test was the aqueous solution containing 1% H_2SO_4 of pH of 1.0. The alkali-activated binder mixes have shown better performance than the OPC-based concrete mixes [40]. Similar results were reported by Mithun and Narasimhan where a type of AAC performed better than OPC-based concrete mixes subjected to sulfate attack and acid test [5].

El-Didamony et al. studied the durability of "alkali-activated slag" (AAS) concretes subjected to extreme saline exposure (sea water). The performance of the pastes developed by utilization of sea water and tap water was done by using two types of cements, i.e., GGBS-based alkali-activated cement and the sulfate-resistant cement (SRC) were compared. The study involved IR spectroscopy, thermogravimetric analysis, scanning electron microscopy, bulk density determination, and compressive strength determination. The studies on durability were carried out by immersing the concrete specimens which were cured, (both AAS and SRC) in sea water up to 12 months. The results show that the for SRC mixes, a drastic reduction in compressive strength were reported after 6 months of immersion, whereas for AAS mixes, there is no reduction strength were reported till 12 months; which was mainly due to the absence of chloro-aluminate and ettringite in the hydration product of AAS systems [37].

"Fatigue failure" is one of the vital forms of failure in concrete pavement structures which subjected to frequent application of loads. This type of failure in pavements arises under the manipulation of recurring loads or cyclic loads, whose ultimate values are considerably lesser than safe loads guesstimated through static tests. In the case of fatigue breakdown, the material fails by recurring application of traffic loading, which is not of a sufficient amount to cause failure due to single load application. In any concrete constructions, the failure due to fatigue causes localized, progressive, and permanent harm due to dynamic loads. Typically, these changes escort the development of cracks, further this leads to the growth of this crack which may lead to failure. This kind of failure is most common in the case of concrete pavements which will happen due to the cracks with progressive growth under the working of cyclic traffic loading, mostly when the stresses induced are greater than the strength of flexure of the concrete [41, 42].

Mithun et al. studied, the "flexural fatigue behavior" of AAS-based concrete mixes prepared at alkaline modulus of 1.25 by utilizing "copper slag" as fine aggregate, replacement to natural sand. Results were compared with OPC-based mixes designed for same target strength as that of alkali-activated concrete mixes. The studies were conducted on six types of concrete mixes, one being OPC mix by incorporating OPC as binder with the utilization of natural sand; other five are "AASC" mixes by varied dosage of copper slag with the replacement to the natural sand (replacement level varied from 0 to 100%, at an interval of 25%). The fatigue tests, done at different values of stress ratios varied from 0.70 to 0.85, and the fatigue life were determined. The fatigue lives of all the mixes were represented using "S–N curves." Further, the probabilistic analyses were carried out using Weibull distribution. The results indicate that both the static and fatigue flexural strength of the AAC mix was better than that of the OPC mix. This was due to the presence of dense and uniform ITZ in the case of AAC mixes when compared it with that of the OPC concrete mix [36, 43]. Comparable results were reported by Palankar et al., where a type of alkali-activated concrete developed using steel slag aggregate replaced in place of conventional granite aggregates performed better under flexural fatigue loading while compared with that of the similarly target designed conventional OPC-based concrete mixes [33, 38].

5 Usage of Glass Powder in Making Concrete

There is a continuous search for new supplementary materials which can be used as a complete or partial replacement as cementitious systems in place of conventional OPC-based concretes. Mineral concrete admixtures such as "GGBS, Silica fume, fly-ash, rice husk-ash" and limestone fines are few famous examples for such supplementary materials, use of which as a fractional standby in OPC/AAC-based concrete production is well known. Powdered glass is a new addition to such supplementary materials, and the several research has shown that the powdered glass which can be used as accompanying material in concrete production [44]. Another reason to test out the utilization of glass in some fruitful manner is to solve the disposal problems associated with waste glass. It is a renowned fact that there is a generation of more than million tonnes of waste glass yearly all over the world, and this will create lot of disposal problems as this waste material can be considered as a non-biodegradable waste. The reuse of waste glasses from customer utilization and industrialized progression pretense a major difficulty for metropolis worldwide [12]. Thus, making an attempt to utilize such a waste in producing sustainable concrete is really appreciable [45].

Zidol et al. have worked on the use of glass powder in producing OPC-based concrete. The studies were done by 0%, 20%, and 30% replacement of conventional cement with glass powder; and the result was compared with the 30% fly-ash + 70% OPC concrete mix and 30% GGBS + 70% OPC concrete mix. Four types of mixes were made with four different w/c ratios (i.e., 0.35, 0.40, 0.55, and 0.65) were investigated. The results show that the conventional control OPC mix have shown better performance in case of compressive strength and chloride ion permeability. They also reported that glass powder-based concrete behaved in a similar manner as that of the class F fly-ash-based concrete mixes. Use of GP reflects the slow and continuous pozzolanic behavior which may lead to better durability of glass powder-based concrete systems. Thus, from the results obtained, Zidol et al. [44] concluded that the performance of glass powder can be considered to be similar to those of other major supplementary materials (such as "fly-ash" and "GGBS"), hence where the conventional supplementary materials are not available, and where powdered glass is easily available, this can be effectively used.

Schwarz et al. deliberated, that the long-term compressive strength of 10% glass powder-based OPC mixes will obtain the strength which is just 5% different than that of the 10% fly-ash-based OPC mixes. They also experimentally established that there is a "potential" of glass powder which will help to minimize the expansion in concrete due to "alkali–silica" reaction. Further, 10% glass powder–10% fly-ash blended with 80% OPC mix have shown that there is a similarity between the mix prepared using 20% fly-ash with 80% OPC binder. The long-term performance with regard to the strength, sorptivity, and coefficient of moisture diffusion values indicates that the "glass powder-modified concretes" perform equal or better than concretes with improved fly-ash at beginning ages [46].

Many studies were also reported with the usage of glass powder in the manufacture of "alkali-activated cementitious systems." The results reveal that there is a huge potential to use powdered waste glass in production of quality alkali-activated binder-based concretes. Some of the key results were presented in the following subsections.

Zhang and Yue stated that use of 14.57% of powdered glass had led to 28 day compressive strength and flexural strength of 66.4 MPa and 7.1 MPa with the optimum use of 8.31% Na_2O dosage. Further, this GGBS–glass powder-based AAC had shown a better resistance to sulfate attack; but the mix had limitation that it shown a higher shrinkage than the Portland cement-based concrete mix. From this research, it was revealed that the use of powdered "waste-glass" is a potential admixture for preparing "alkali-activated slag-based concrete" when the mixture of "sodium silicate" and "sodium hydroxide" were used as alkaline activators. Further, it was recommended that the optimization between glass powder content and Na_2O equivalent should be regarded when waste glass powder is used in alkali-activated cementitious systems [47].

Redden and Neithalath studied the strength, moisture stability, and "microstructure" of alkali-activated concretes formed using fly-ash and powdered glass as binders. The study revealed that a high value of compressive strength achieved with glass powder-activated mixes when compared with that of the "fly-ash"-activated mixes at lower heat curing temperatures. They found that the hydration product of glass powder-activated mixes was sodium silicate gel, whereas for "glass powder-fly-ash"-blended-activated mixes gave the combination of "sodium silicate and sodium aluminosilicate (N–A–S–H) gels form" as a hydration product. Further, they observed that there was a drastic loss in strength when the concrete mixes containing more % of glass powder, when it is bare to the moisture or vulnerability to an alkaline solution. To eradicate this problem, the use of aluminum containing mineral such as metakaolin or slag was suggested to be used in the alkali-activated mixes containing glass powder, which may lead to the development of moisture-stable hydration products of reaction. Also, the changes in structure upon the contact to moisture are explained using microstructural and "FTIR spectroscopical" observations. Micrograph showing the hydration reaction products of alkali-activated concrete mix upon heat curing for 48 h at 75 °C containing 50% powdered glass and 50% fly-ash blend activated using 8 M NaOH is shown in Fig. 5, where the structure of NASH and silica-rich gel can be seen [48].

Similar studies were conducted by Tashima et al. who investigated that strength and microstructure characteristics of "alkali-activated mortars" by using "glass fiber waste" product. The alkaline activator solutions used for the study are "NaOH and KOH" solution. The compressive strength of the mortar samples obtained was in the order of 77 N/mm^2 just after 3 days of heat curing done at 65 °C when "10 mol/L" sodium hydroxide solutions was used as activator. Whereas the similar mixes produced using KOH as activator have shown that the development of the strength around 70 N/mm^2. This was because of a higher degree of reaction when NaOH is used as activator when compared it with that of KOH activator-based mixes. This fact was proved even by the microstructure studies on the two types of mixes,

Fig. 5 SEM Image showing hydration products of fly-ash glass powder-blended AAC

where the NaOH-based mixes shown a dense and compact microstructure while compared with that of KOH-based systems [49].

Pascual et al. studied the combination of "powdered glass with metakaolin." The use of metakaolin will lead to the induction of aluminum content and also to stabilize alkali ions in the glass powder-based alkali-activated systems. There was a rise in the compressive strength value of the mortars when the dosage of metakaolin was up to 8%. When the metakaolin content was less than 3%, there was decrease in the strength values were observed, which significantly shows that there must be aluminum content in the mix when glass powder is used to produce alkali-activated cementitious mixes, otherwise it will be adversely affect the performance [50].

Banjare et al. studied the use of an amalgamation of "calcined kaolin clay," dross from "aluminum recycling," and "lead–silica glass" from recycled fluorescent lamps to produce a bubbled "alkali-activated binder-based cementitious systems." They could produce a very lightweight concrete weighing 460–560 kg/m^3 with around 82% porosity and a maximum compressive strength up to 2.3 N/mm^2 [51].

Puertas and Torres Carrasco explored the utilization of powdered waste glass for the potential activation of slag in producing alkali-activated concrete. The findings showed that the possibility of using waste glass in producing alkali-activated slag. Treating waste glass with "$NaOH/Na_2CO_3$" (pH = 13.6) favors the partial liberation of the reactive silica from the powdered glass. The resulting solution from the treatment of glass waste acts as "alkaline activators", partially dissolving the "vitreous blast furnace slag." The compressive strength was over 60 N/mm^2 at the age of 28 days when the "$NaOH/Na_2CO_3$" mixture and glass waste mixed solution were used as activators, indicating that the glass was potentially useful as an additional silica source in place of commercial silicate solutions. The strength, composition, and microstructure of the products of reaction recognized in conventional activator and activated powdered glass waste as activator were comparable.

6 Key Findings from Literature Review

From the comprehensive "literature review," it is noticed that the durability and strength of AAC mixes are influenced by several factors, namely the nature of binder, chemical characteristics of the binder, % of sodium oxide dosage (Na_2O), water content and water-to-binder ratio, nature of alkaline activator, activator modulus of the alkali, curing regime, etc. Properly designed AAC mixes have an ability to perform far better (with respect to the durability and strength) than that of the similarly designed conventional OPC-based concrete mixes. AAC mixes containing higher dosages of slag (i.e., GGBS dosage more than 50% by weight of binding ingredient) may arrive at adequate amount of strength even at room temperatures, i.e., when subjecting it to air curing, without need for any kind of heat curing or other methods. The addition of higher quantities of fly-ash in AAC mixes may lead to a reduction in the strength performance, but this will lead to a rise in workability of the concrete mixes. Further, the "activation energy" required for AAC mixes containing massive proportions of fly-ash is higher (i.e., in the mixes where fly-ash dosage greater than 50% by weight of binding ingredient); thus, the replacement of GGBS with FA ahead of 50% in AAC mixes requires a higher amount of dosage of the "alkaline activator" solution; also these mixes may require heat curing for attaining the required target strength. Among many alkaline activators, the incorporation of NaOH flakes and sodium silicate (Na_2SO_3) solution will provide the most excellent activation solution in producing alkali-activated concretes. The strength durability performances of AAC mixes are appreciably influenced by the modulus of activation (Ms) of alkaline activators and the dosage of sodium oxide (Na_2O). It is essential to find the "Na_2O" dosage and the "optimal activator modulus" (Ms) for which the target strength of AAC can be conquered. The optimal value of Na_2O dosage for any AAC mixes may differ somewhere between 3 and 6% by weight of binder; whereas higher dosages might lead to uneconomical concrete assortment and efflorescence problems. Whereas, the optimal modulus of activation (Ms) for the AAC mixes may fluctuate between 0.50 and 2.0. The Na_2O dosage, Ms Value of alkaline activator, water-to-binder ratio, etc., required to be controlled in order to acquire mixes of the required workability and desired strength. Waste glass is an industrial end product and can be looked upon as a potential substitution for traditional ingredients in concrete production. A satisfactory and an improved engineering performance of this powdered glass in concrete production have been reported by many researchers which were used in producing both conventional OPC and AAC-based concrete as a cementing ingredient. However, there is limited research available on strength, durability, and fatigue performance of AAC mixes incorporating powdered waste glass as binding ingredient along with conventional slag and fly-ash. The further research works can be focused to investigate the properties of sustainable air-cured AAC mixes with powdered waste glass as binding ingredient with an object of conserving environment by utilizing waste product.

References

1. Singh B, Ishwarya G, Gupta M, Bhattacharyya SK (2015) Geopolymer concrete: a review of some recent developments. Constr Build Mater 85:78–90
2. Pacheco-Torgal F, Castro-Gomes J, Jalali S (2008) Alkali-activated binders: A review. Constr Build Mater 22(7):1305–1314
3. Yang KH, Song JK, Il Song K (2013) Assessment of CO_2 reduction of alkali-activated concrete. J Clean Prod 39:265–272
4. Pacheco-Torgal F, Castro-Gomes J, Jalali S (2008) Alkali-activated binders: a review. Part 1. Historical background, terminology, reaction mechanisms and hydration products. Constr Build Mater 22(7):1305–1314
5. Mithun BM, Narasimhan MC (2015) Performance of alkali activated slag concrete mixes incorporating copper slag as fine aggregate. J Clean Prod 112:837–844
6. Provis JL (2017) Alkali-activated materials. Cem Concr Res 114:40–48
7. Shi C, Roy D, Krivenko P (2006) Alkali-activated cements and concretes, 1st edn. Taylor & Francis, London
8. Davidovits J (2002) 30 years of successes and failures in geopolymer applications. Market trends and potential breakthroughs. In: Geopolymer 2002 conference, pp 1–16
9. Pacheco-Torgal F, Abdollahnejad Z, Camões AF, Jamshidi M, Ding Y (2011) Durability of alkali-activated binders: a clear advantage over Portland cement or an unproven issue? Constr Build Mater 30:400–405
10. Roy DM (1999) Alkali activated cements, opportunities and challenges. Cem Concr Res 29:249–254
11. Pacheco-Torgal F, Castro-Gomes J, Jalali S (2008) Alkali-activated binders: a review. Part 2. About materials and binders manufacture. Constr Build Mater 22(7):1315–1322
12. Provis JL, Palomo A, Shi C (2015) Advances in understanding alkali-activated materials. Cem Concr Res 78:110–125
13. Ding Y, Dai J, Shi C (2016) Mechanical properties of alkali-activated concrete: a state-of-the-art review. Constr Build Mater 127:68–79
14. Zhang MH, Malhotra VM (1995) Characteristics of a thermally activated alumino-silicate pozzolanic material and its use in concrete. Cem Concr Res 25(8):1713–1725
15. Keerio VMM, Malhotra MA (1996) rice husk ash as a potential supplementary cementing. ACI Mater J 1(1):629–636
16. Rakhimova NR, Rakhimov RZ (2019) literature review of advances in materials used in development of alkali-activated mortars, concretes, and composites. J Mater Civ Eng 31(11):03119002
17. Rostami M, Behfarnia K (2017) The effect of silica fume on durability of alkali activated slag concrete. Constr Build Mater 134:262–268
18. Toniolo N, Boccaccini AR (2017) Fly ash-based geopolymers containing added silicate waste. A review. Ceram Int 43(17):14545–14551
19. Rashad AM (2013) Properties of alkali-activated fly ash concrete. Iran J Mater Sci Eng 10(1):57–64
20. Fernandez-jimenez A, Palomo JG, Puertas F (1999) Alkali-activated slag mortars Mechanical strength behaviour. Cem Concr Res 29:1313–1321
21. Wang S, Scrivener KL (1995) Hydration products of alkali activated slag cement. Cem Concr Res 25(3):561–571
22. Wang SD, Scrivener KL, Pratt PL (1994) Factors affecting the strength of alkali-activated slag. Cem Concr Res 24(6):1033–1043
23. Bakharev T, Sanjayan JG, Cheng YB (1999) Alkali activation of Australian slag cements. Cem Concr Res 29(1):113–120
24. Krizan D, Zivanovic B (2002) Effects of dosage and modulus of water glass on early hydration of alkali-slag cements. Cem Concr Res 32(8):1181–1188
25. Fernande A, Palomo A (2005) Composition and microstructure of alkali activated fly ash binder: effect of the activator. Cem Concr Res 35:1984–1992

26. Wardhono A, Law DW, Strano A (2015) The strength of alkali-activated slag/fly ash mortar blends at ambient temperature. Procedia Eng 125:650–656
27. Duran Atiş C, Bilim C, Celik Ö, Karahan O (2009) Influence of activator on the strength and drying shrinkage of alkali-activated slag mortar. Constr Build Mater 23(1):548–555
28. Palankar N, Shankar AUR, Mithun BM (2015) Air-cured Alkali activated binders for concrete pavements Air-Cured Alkali Activated Binders for Concrete Pavements. Int J Pavement Res Technol 8(July):289–294
29. Karim MR, Zain MFM, Jamil M, Lai FC (2015) Development of a zero-cement binder using slag, fly ash, and rice husk ash with chemical activator. Adv Mater Sci Eng 2015(247065):1–14
30. Aliabdo AA, Abd Elmoaty AEM, Salem HA (2016) Effect of cement addition, solution resting time and curing characteristics on fly ash based geopolymer concrete performance. Constr Build Mater 123:581–593
31. Assi L, Carter K, (Eddie) Deaver E, Anay R, Ziehl P (2018) Sustainable concrete: building a greener future. J Clean Prod 198:1641–1651
32. Gambhir ML (2013) Concrete technology : theory and practice, 5th edn. McGraw Hill Education (India) Private Limited, New Delhi
33. Palankar N, Ravi Shankar AU, Mithun BM (2015) Studies on eco-friendly concrete incorporating industrial waste as aggregates. Int J Sustain Built Environ 4:378–390
34. Rashad AM (2013) A comprehensive overview about the influence of different additives on the properties of alkali-activated slag—a guide for Civil Engineer. Constr Build Mater 47(2013):29–55
35. Ismail I et al (2012) Influence of fly ash on the water and chloride permeability of alkali-activated slag mortars and concretes. Constr Build Mater 48:1187–1201
36. Mithun BM, Narasimhan MC (2017) Performance of alkali activated slag concrete mixes incorporating copper slag as fine aggregate. National Institute of Technology Karnataka, Surtathkal
37. El-Didamony H, Amer AA, Abd Ela-Ziz H (2012) Properties and durability of alkali-activated slag pastes immersed in sea water. Ceram Int 38(5):3773–3780
38. Palankar N (2016) Performance of alkali activated concrete mixes with steel slag as coarse. National Institute of Technology Karnataka Surtathkal
39. Bakharev T, Sanjayan JG, Cheng YB (2003) Resistance of alkali-activated slag concrete to acid attack. Cem Concr Res 33(10):1607–1611
40. Palankar N, Ravi Shankar AU, Mithun BM (2016) Durability studies on eco-friendly concrete mixes incorporating steel slag as coarse aggregates. J Clean Prod 129:437–448
41. Lee MK, Barr BIG (2002) An overview of the fatigue behaviour of plain and fibre reinforced concrete. Cem Concr Compos 26:299–305
42. Goel S, Singh SP (2014) Fatigue performance of plain and steel fibre reinforced self compacting concrete using S-N relationship. Eng Struct 74:65–73
43. Mithun BM, Narasimhan MC, Palankar N, Ravishankar AU (2015) Flexural fatigue performance of alkali activated slag concrete mixes incorporating copper slag as fine aggregate. SSP-J Civ Eng 10(1):7–18
44. Zidol A, Tognonvi MT, Tagnit-hamou A (2017) Effect of glass powder on concrete sustainability. New J Glas Ceram 7:34–47
45. Islam GMS, Rahman MH, Kazi N (2016) Waste glass powder as partial replacement of cement for sustainable concrete practice. Int J Sustain Built Environ 6(1):37–44
46. Schwarz N, Cam H, Neithalath N (2008) Influence of a fine glass powder on the durability characteristics of concrete and its comparison to fly ash. Cement Concr Compos 30:486–496
47. Zhang L, Yue Y (2018) Influence of waste glass powder usage on the properties of alkali-activated slag mortars based on response surface methodology. Constr Build Mater 181:527–534
48. Redden R, Neithalath N (2013) Microstructure, strength, and moisture stability of alkali activated glass powder-based binders. Cem Concr Compos 45:46–56
49. Tashima MM, Soriano L, Borrachero MV, Monzó J, Cheeseman CR, Payá J (2012) Alkali activation of vitreous calcium aluminosilicate derived from glass fiber waste. J Sustain Cem Mater 1(3):83–93

50. Pascual AB, Tognonvi MT, Tagnit-Hamou A (2014) Waste glass powder based alkali activated mortar. Int J Res Eng Technol 03(13):15–19
51. Banjare D, Bumanis G, Korjakins A (2014) New porous material made from industrial and municipal waste for building application. Mater Sci 20(3):333–338

Progressive Collapse of Steel-Framed Structures

T. Anusha and H. G. Nahushananda Chakravarthy

Abstract Progressive collapse is the phenomena in which the local failure of a primary structural member results in partial or total structural system damage, without any proportionality between the initial and final failure. In the present study, multi-story structure is considered. The modeling and analysis are carried out using SAP2000 software. The different bracing systems such as X, V, K and diagonal bracing are used to analysis the structural behavior. The structure is verified for column removal at 3 different locations such as corner, center and interior. It is observed that maximum vertical deflection is found in interior column removal case and least in corner column removal. Interior column case was observed to be most critical case in progressive collapse. There is a raise in axial load and the load distribution to the adjacent columns for column removal at location 1, 2, 3 is around 30%, 25%, and 20%, respectively. The X bracing system performs well in case of progressive collapse.

Keywords Progressive collapses · Bracing system · Column removal · SAP2000

1 Introduction

Civil engineering methods have evolved through the years, when little attention is paid to progressive failures. Progressive resistance to collapse was not a design idea, before 1968 the collapse of the Ronan Point building. Traditional building structural designs and plans show a technological advancement which is not suitable to resist disproportionate failure. Unfortunately, the failure of the Alfred Murrah Federal Building has demonstrated that the possibility of progressive, and catastrophic collapse is surprisingly possible and needs to be resolved. Due to accidents and building errors, a growing number of buildings are demolished locally. Progressive collapse is expected to spread through the building and lead to a catastrophic

T. Anusha (✉) · H. G. Nahushananda Chakravarthy
Department of Civil Engineering, Siddaganga Institute of Technology, Tumkuru, India

H. G. Nahushananda Chakravarthy
e-mail: chakravarthyn@sit.ac.in

L. Nandagiri et al. (eds.), *Sustainability Trends and Challenges in Civil Engineering*, Lecture Notes in Civil Engineering 162, https://doi.org/10.1007/978-981-16-2826-9_20

loss causing multiple casualties. Progressive collapse is defined as chain reaction of failures caused by the instantaneous failure of one or few gravity load-carrying components. Once the load-bearing portion fails, it could be helpful for the device to have an alternative load-bearing path and to transfer the loads borne by that component to adjoining components. Dynamic internal forces in neighboring components rise as an outcome of the release of internal energy due to the loss of a component.

In his research, Brian I. Song et al. [1] studied the progressive failure capacity of a steel-frame building is analyzed through computer simulations and field experiments. Using SAP2000, a computational simulation of the structure was performed. In the first floor, four columns have been eliminated from the system to examine the reallocation of load within the system. Field experiment data were used to compare and validate simulations and mathematical modeling. Simplified research methods, including sudden replacement of critical columns at specified structural locations, are suggested in the progressive failure design guideline. The findings revealed that for both 3-dimensional and 2-dimensional models, the nonlinear dynamic assessments provide lesser DCR values and vertical displacements than for linear static analysis. Kapil Khandelwal et al. [2] this study is carried out using an alternate path method to ten-floored structure. In this method, critical columns and neighboring braces are immediately excluded from the analysis process and the capability of the structure to tolerate component losses is effectively observed. The loss of the member should then imply a condition in which the component is affected by a drastic event or an unusual load case. Those results suggest that the unique concentrically braced structure is largely susceptible to the progressive collapse than the eccentrically braced structure, while several systems gain from the placement of the seismically built structure on the periphery of the building. In this review, Liu-Lian Li et al. [3] introduce the actions of progressive failure of steel structures under a column elimination case by examining the FE model. First finite eliminate models were validated against the studies of three steel-frame structures regarding dynamic response and collapse modes. Using FE models, the critical load and robustness of steel frames with a number of different mechanical and geometrical factors were calculated. The findings found that critical load and robustness are lower in strong column-weak beam structure than in strong beam-weak column structure. Critical load and robustness are lower in the side-by-side inner column elimination case. On the opposite, the scenario of middle column elimination, robustness and critical load is stronger. In order to explain the gradual collapse of high-rise structures with composite steel structures, Rohola Rahnavard et al. [4] in his study takes into account 3-D modeling using the FE approach. Responses of the system under two conditions of elimination of columns are investigated by nonlinear functional analysis. Two types of lateral resistance structures were chosen for analysis and evaluation. The arrangement comprises of regular and irregular layouts of plans. The behavior of the system is evaluated in depth, and measures to mitigate a progressive failure have been suggested. Relative to corner column elimination case, side column elimination scenario was more susceptible to progressive collapse in the centrally braced moment frame systems.

2 Objective of Work

To analyze the performance of braced structure, when column is eliminated at various location in steel structure.

3 Methodology

3.1 Modeling of the Building

In this analysis, 10 floor steel-framed buildings have been chosen. The building consists of 5 bays of six meters each in both transverse Y-direction and longitudinal X-direction. The standard floor-to-floor height is four meters. On perimeter beams, glazing with a thickness of 25 mm is provided. Column and beam section are made of I Section, and channel section is used for bracings. Slab thickness considered is 200 mm. For modeling and analysis of the structures, SAP2000 Software is used. Material properties and design factors are considered shown in Table 1. Five steel-framed structures such as without bracing, with V bracing, X bracing, diagonal bracing, and K bracing are considered with 3 column removal scenario such as corner, middle, and interior. Elevation of models considered is as shown in Fig. 1.

Table 1 Material properties and design factors

S. No.	Description	Data
1	Seismic zone	III
2	Seismic zone factor (Z)	0.16
3	Damping ratio	0.05
4	Soil type	Hard soil
5	Height of the building	40 m
6	Story to story height	4.0 m
1	Span length	6 m
8	Column used	2ISHB 450
9	Beam used	ISMB 350
10	Bracing used	ISMC 250
11	Thickness of slab	200 mm
12	Floor finish	1.0 kN/m^2
13	Live load	3.0 kN/m^2
14	Glass facade	2.5 kN/m
15	Grade of concrete	M30
16	Grade of structural steel	Fe 350

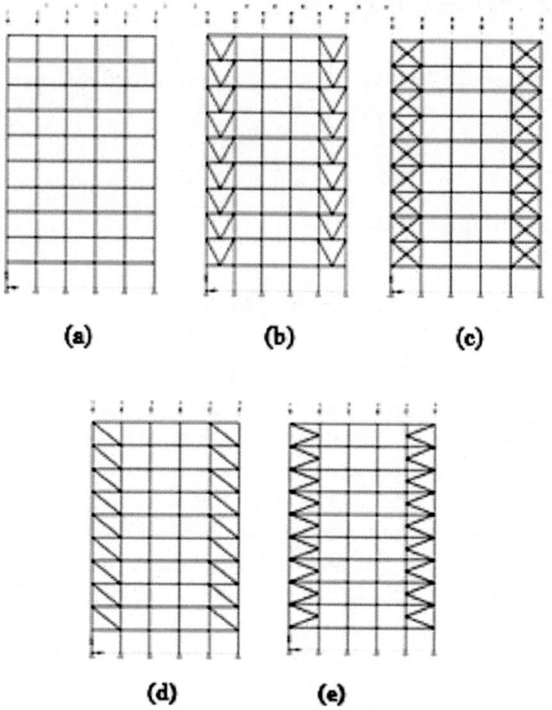

Fig. 1 Elevation of the models

In Fig. 1a–e refer to building without bracing, V bracing, X bracing, diagonal bracing, and K bracing, respectively.

3.2 GSA Guidelines

The General Service Administration (GSA) [5] includes the comprehensive methods and performance standards required to determine the sensitivity of new and current structures to progressive failure. As per GSA specifications, the structure is analyzed for the instantaneous failure of a column. In this study, the following analytical cases are considered for conventional or symmetrical framed systems.

Twenty structural models having 10 floors are considered in this study, mentioned in Table 2. Model A1, A2, A3 are structured with column elimination at position 1, 2, 3 in building A. Models B1, B2, B3 are structured with column elimination at position 1, 2, 3 in building B. Models C1, C2, C3 are structured with column elimination at position 1, 2, 3 in building C. Models D1, D2, D3 are structured with column elimination at position 1, 2, 3 in building D. Models E1, E2, E3 are structured with column elimination at position 1, 2, 3 in building E.

In Fig. 2, position 1, 2, 3 represent the corner, edge, and interior location of column removal at ground story.

Table 2 Building specification

Models	Description
Building A (Model A)	Building without bracing
Building A1 (Model A1)	Building without bracing and column removal at location 1
Building A2 (Model A2)	Building without bracing and column removal at location 2
Building A3 (Model A3)	Building without bracing and column removal at location 3
Building B (Model B)	Building with V bracing
Building B1 (Model El)	Building with V bracing and column removal at location 1
Building B2 (Model B2)	Building with V bracing and column removal at location 2
Building B3 (Model B3)	Building with V bracing and column removal at location 3
Building C (Model C)	Building with X bracing
Building CI (Model CI)	Building with X bracing and column removal at location 1
Building C2 (Model C2)	Building with X bracing and column removal at location 2
Building C3 (Model C3)	Building with X bracing and column removal at location 3
Building D (Model D)	Building with diagonal bracing
Building D1 (Model D1)	Building with diagonal bracing and column removal at location I
Building D2 (Model D2)	Building with diagonal bracing and column removal at location 2
Building D3 (Model D3)	Building with diagonal bracing and column removal at location 3
Building E (Model E)	Building with K bracing
Building E1 (Model El)	Budding with K bracing and column removal at location 1
Building E2 (Model E1)	Building with K bracing and column removal at location 2
Building E3 (Model E3)	Building with K bracing and column removal at location 3

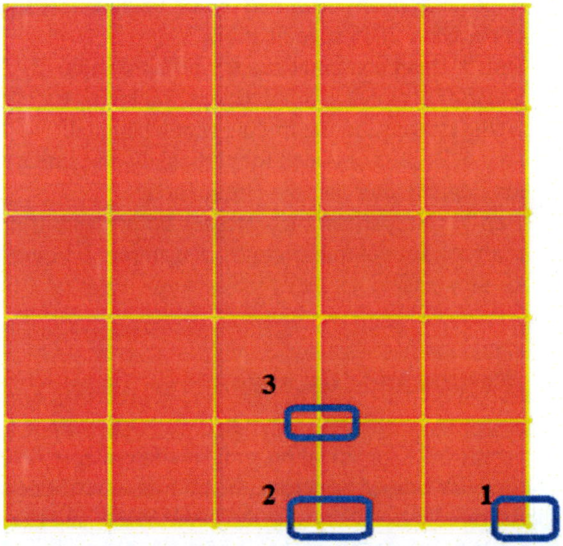

Fig. 2 Column removed position at ground floor

3.3 *Analysis Procedure*

Demand capacity ratios is defined as the ratio of structural member force acting on the member to expected strength of the member. The DCR of each member is determined from the below equation.

$$\mathrm{DCR} = Q_{\mathrm{ud}}/Q_{\mathrm{ce}}$$

where

Q_{ud} Structural member force acting on the member
Q_{ce} Expected ultimate capacity in member.

Acceptance criteria

In order to avoid failure of the structure, the DCR values in each structural element should not be greater than, 2 for symmetrical structural configuration, and 1.5 for unsymmetrical structural configuration.

Structural elements with DCR values beyond the above limits will not have sufficient ability to effectively redistribute loads and are considered to have collapsed, and as a result of that whole system may collapse..

Analysis steps

Step 1. The building is designed according to IS 1893(Part 1) 2016 load combinations in SAP2000, and linear static analysis is performed.
Step 2. A column is eliminated from the location under evaluation, and linear static analysis is performed out on braced steel structure with load combinations as per GSA 2016 specifications.
Step 3. Load combinations are inserted in the SAP2000. Under each scenario, a SAP2000 computer simulation is conducted for various column elimination cases on the system and the findings are evaluated.
Step 4. Results obtained for vertical displacement, axial load, and DCR of beams for element after column elimination.
Step 5. Comparison of outcomes before and after elimination of the column is analyzed for column removal in different bracing structures at various locations for progressive failure of the system.

4 Results and Discussion

As per the GSA specification, progressive collapse (PC) analysis was done. In the case of a regular framed structure, the elimination of one column at the corner (location 1), the middle (location 2) at the outer frame, and the middle of the immediate inner frame at the center (location 3) of the building is performed in accordance with the GSA specifications.

4.1 Axial Loads on Columns

If the columns are eliminated, the axial load won't disappear. It will be redistributed in neighboring columns. The elimination of column leads to rise of load in the adjacent columns.

Location 1 (Corner Column)

Axial load on adjacent column for column removal at location 1 is seen in Fig. 3. The percentage rise in axial load is around 30%.

Location 2 (Middle Column-Exterior)

The elimination of the column at position 2 is more vulnerable to progressive failure and is seen in Fig. 4. The percentage rise in axial load is around 25%.

Location 3 (Middle Column-Interior)

All columns experience similar behaviors in the event of column elimination at position 3, as seen in Fig. 5. It is noticed that in frame 2, and the bracing effect is not noticed. The percentage rise in axial load is around 18 to 20% for all columns.

Fig. 3 Axial Load on adjacent column for removal of column at Location 1

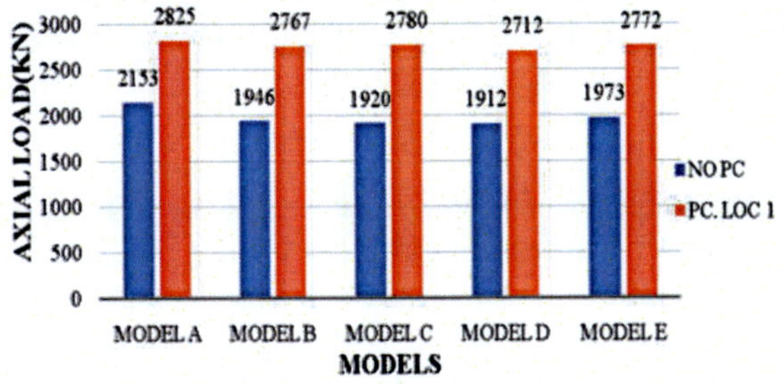

Fig. 4 Axial load on adjacent column for removal of column at Location 2

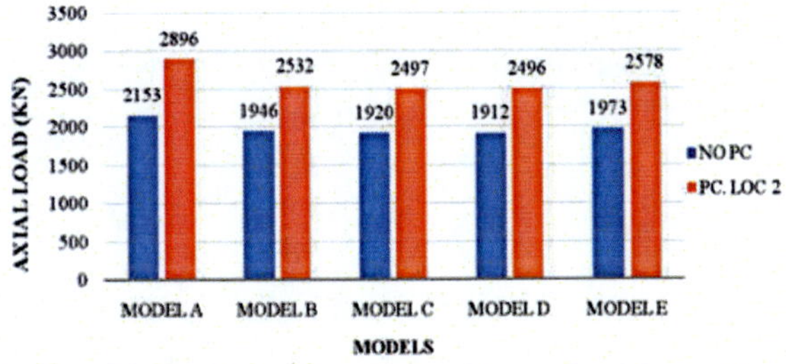

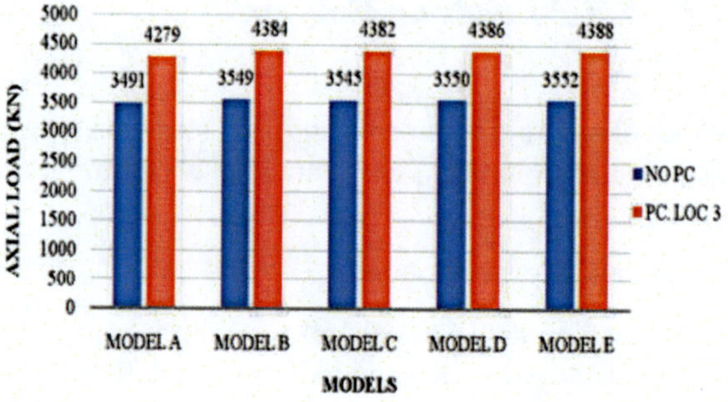

Fig. 5 Axial load on adjacent column for removal of column at Location 3

4.2 *Deflection in Beams*

The removal of the column triggers the beam deflection to rise tremendously. Deflection of beams for different column removal cases is explained below.

Location 1 (Corner Column)

Beam deflection for column removal at position 1 is as seen in Fig. 6. Model A has the maximum deflection of 37 mm. However, by having different bracings, the deflection can be managed. For models B, C, D and E, the deflection of beams increased by 7.5 times, 8.8 times, 7.55 times, and 8.2 times, respectively. Models B and D are better at reducing deflection efficiency.

Location 2 (Middle Column-Exterior)

The deflection of beams at position 2 for column elimination is seen in Fig. 7. It is found from the graph that the bracings do not help to minimize the deflection. For models of bracing systems, the column elimination at position 2 reveals that the deflection in beam is increasing. For structure with and without bracings, the percentage rise is more than 40%.

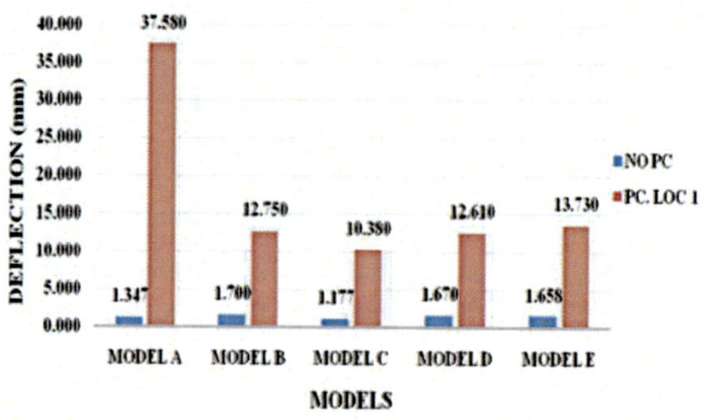

Fig. 6 Deflection of beam at joint 29 for removal of column at Location 1

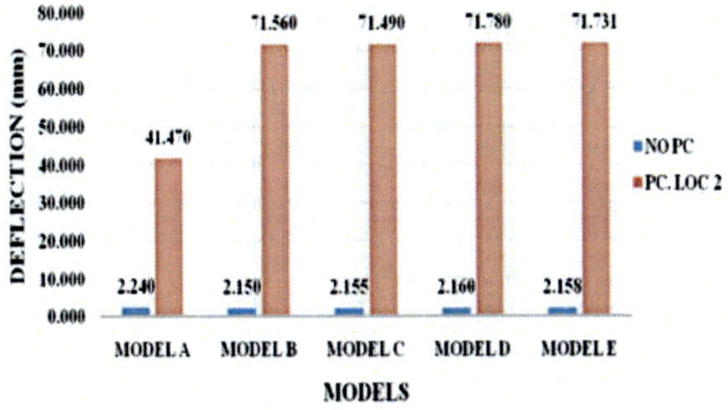

Fig. 7 Deflection of beam at joint 15 for removal of column at Location 2

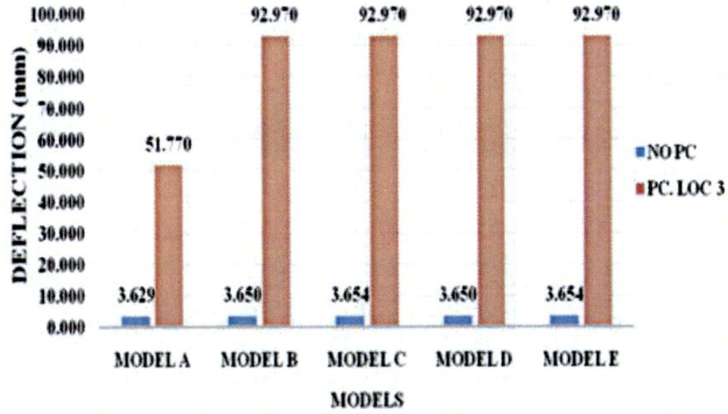

Fig. 8 Deflection of beam at joint 18 for removal of column at Location 3

Location 3 (Middle Column-Interior)

The deflection of the beam for removal of the column at position 2 is seen in Fig. 8. For models of column elimination at position 3, the deflection is increasing as the area affected is more. Maximum deflection of 92.97 mm is observed. The effect of type bracing is negligible on deflection of beam at location 3 and as a result deflection is same.

4.3 DCR Ratios for Beams

Location 1

The DCR value of beams bottom 3 floors is shown in given in Table 3. DCR ratio of beams at floor 1 is as shown in Fig. 9. In model B1 at floor 1, the maximum raise of the DCR ratio in beam is observed. Model C with X bracing is more efficient than other structures.

Table 3 DCR ratio in beams for removal of column at location 1

Location 1		Model A	Model A1	% variation
Floor 3	EI 369	0.397	0.B64	54.0 5%
Floor 2	EI 309	0.414	0.963	57.01%
Floor 1	B29	0.502	1.003	49.95%
		MODEL E	Model E1	% variation
Floor 3	EI 369	0.7B4	0.B69	9.78%
Floor 2	EI 309	0.B22	1014	18.93%
Floor 1	B29	0.90	1.416	36.44%
		Model C	Model C1	% variation
Floor 3	EI 369	0.112	0.123	8.94%
Floor 2	EI 309	0.113	0.1:7	29.94%
Floor 1	E29	0.300	0.321	6.54%
		Model D	Model D1	% variation
Floor 3	EI 369	0.174	0.240	27.50%
Floor 2	EI 309	0.220	0.306	28.10%
Floor 1	B29	0.414	0.522	20.69%
		Model E	Model E1	% variation
Floor 3	EI 369	0.200	0.221	9.50%
Floor 2	EI 309	0.233	0.292	20.21%
Floor 1	B29	0.419	0.520	19.42%
			Max	57.01%

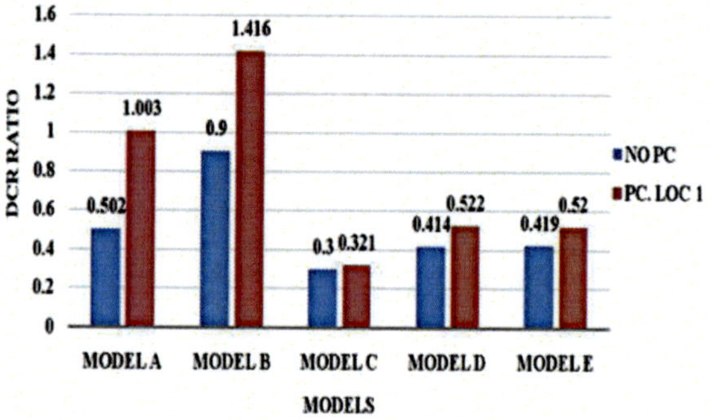

Fig. 9 DCR ratio of beam 29 for removal of column at Location 1

Location 2

The DCR value of beams bottom 3 floors is shown in given in table 4. DCR ratio of beams at floor 1 is as shown in Fig. 10. The removal of the column at position 2 would not benefit from the incorporation of the bracing system. After column

Table 4 DCR ratio in beams for removal of column at Location 2

Location 2		Model A	Model A2	% variation
Floor 3	B137B	0.345	0.994	65.29'%
Floor 2	B131B	0.374	1.091	65.72%
Floor 1	B27	0.460	1.123	59.04%
		Model B	Model B2	% variation
Floor 3	B137B	0.237	1 612	3.5.30%
Floor 2	B131B	0.228	1.664	86.30%
Floor 1	B27	0.388	1.776	78.15%
		Model C	Model C2	% variation
Floor 3	B137B	0.202	1.577	87.19%
Floor 2	B131E	0.185	1.621	88. 59%
Floor 1	B27	0.352	1.737	79.74%
		Model D	Model D2	% variation
Floor 3	B137B	0.265	1.641	83.85%
Floor 2	B131B	0.270	1.707	84.18%
Floor 1	B27	0.436	1.823	76.08%
		Model E	Model E2	% variation
Floor 3	B137E	0.273	1.648	83.43%
Floor 2	B131B	0.285	1.721	83.44%
Floor 1	B27	0.452	1.838	75.41%
			Max	88.59%

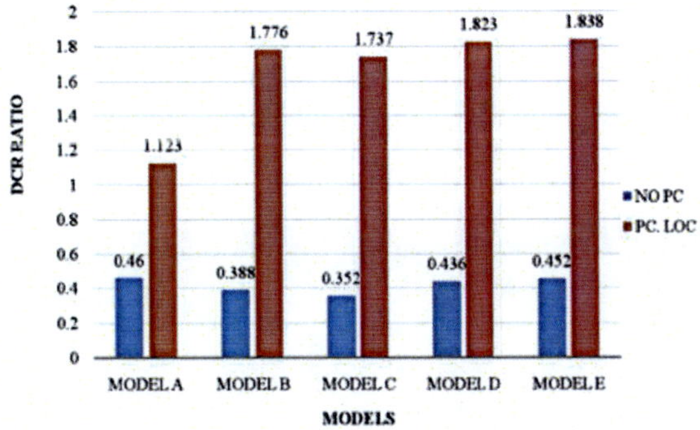

Fig. 10 DCR ratio of beam 27 for removal of column at Location 2

removal, percentage increase of DCR ratio in the beams for bracing structure is more.

Location 3

The DCR value of beams bottom 3 floors is shown in given in Table 5. DCR ratio of beams at floor 1 is as shown in Fig. 11. All beams fail in floor 1 after removal of column at location 3.

Table 5 DCR Ratio in Beams for Removal of Column at Location 3

Locations		Model A	Model A3	% variation
Floor 3	B1373	0.325	1.341	7.5.76%
Floor 2	B1 313	0.353	1.479	76.13%
Floor 1	B32	0.439	1.524	71.19%
		Model B	Model B3	% variation
Floor 1	B1373	0.198	1.976	89.98%
Floor 2	B1313	0.131	2.064	90.75%
Floor1	B32	0.350	2.143	83.67%
		Model C	Model C3	% variation
Floor 3	B1373	0.163	1.975	91.75%
Floor 2	B1313	0.148	2.064	92.83%
Floor 1	B32	0.315	2.138	8.5.27%
		Model D	Model D3	% variation
Floor 3	B1373	0.228	1.995	88.57%
Floor2	B1313	0.235	2.087	88.74%
Floor 1	B32	0.401	2.193	81.71%
		Model E	Model E3	% variation
Floor 3	B1373	0.235	1.975	88. 10%
Floor 2	El 313	0.248	2.083	88. 09%
Floor 1	B32	0.414	2.205	81.22%
				92.83%

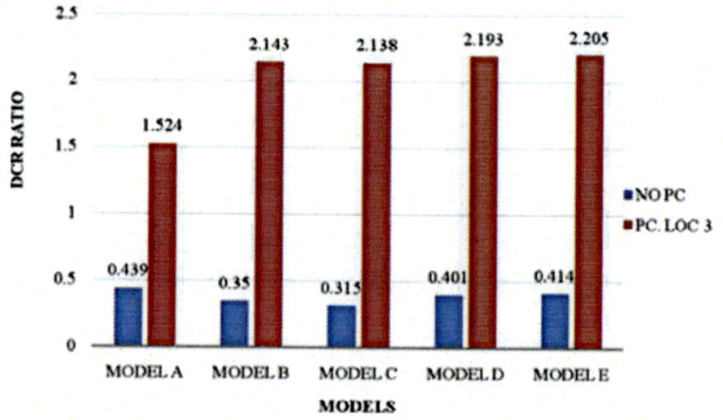

Fig. 11 DCR ratio of beam 32 for removal of column at Location 3

5 Conclusion

(1) There is a raise in axial load, and the load distribution to the adjacent columns for column removal at location 1, 2, 3 is around 30%, 25%, and 20%, respectively.

(2) Maximum beam deflection is observed for removal of columns at location 3, i.e., 92.97 mm, followed by removal of columns at location 2. The lower deflection is observed at location 1 having X bracing, i.e., 10.38 mm.

(3) DCR ratio value of beams exceeds the acceptance criteria in interior column removal. The interior column removal case was found to be more critical in the event of progressive collapse. X bracing proves efficient among other bracing system.

References

1. Song BI, Sezen H (2003) Experimental and analytical progressive collapse assessment of a steel frame building. Eng Struct 56:664–672
2. Khandelwala K, El-Tawila S, Sadek F (2009) Progressive collapse analysis of seismically designed steel braced frames. J Constr Steel Res 65:699–708
3. Li L-L, Li G-Q, Jiang B, Yong Lu (2018) Analysis of robustness of steel frames against progressive collapse. J Constr Steel Res 143:264–278
4. Rahnavard R, Fard FFZ, Hosseini A, Suleimand M (2018) Nonlinear analysis on progressive collapse of tall steel composite buildings. Case Stud Constr Mater 8:359–379
5. General Services Administration (2016) Progressive collapse analysis and design guidelines for new federal office buildings and major modernization projects, GSA

Durability Studies Concerning Permeability on High Performance Concrete Using Artificial Sand as Fine Aggregate

Prashant U. Narale and H. S. Jadhav

Abstract The world scenario, in last 10–15 years, high performance concrete (HPC) entered in the construction field, in particular construction of long span bridges and flyover, high rise building, atomic power plant, prestressed concrete, metro rail project. In this present work, developed optimum mix of M80 grade high strength concrete using fully replacement of artificial sand as fine aggregate and fly ash, silica fume as supplementary cementaneous material with incorporation of super plasticizer by Indian standard method. Also conducted durability test of developed M80 grade concrete includes resistance to chloride attack and impermeability of water. Trial mixes (TM) are designed with water cement ratio 0.26, poly carboxylate ether-based super plasticizer and various percentage addition of silica fume (2.5,5.0,7.5,10.5,12.5), fly ash (5.0,10.0,15.0,20.0) for to obtain optimum mix proportion of M80 grade concrete. Target compressive strength is achieved at 1% dosage of super plasticizer, 5% addition silica fume and artificial sand below 300 micron, which is replaced by fly ash, with well-graded coarse aggregate. Average chloride ion passed through test specimen is 181.33 Coulombs, which is in very low permeability class (100–1000). The maximum water penetration depth of developed M80 grade concrete is 11.5 mm, which is within the permissible limit (< 25 mm) as per Ministry of Road and Surface Transport of India. It is classified as the concrete "impermeable" under aggressive conditions. The present study concluded that artificial sand is substitute material to natural sand as fine aggregate in the presence of silica fume with poly carboxylate ether-based superplasticizer can be suitably used in making eco-friendly HPC.

Keywords High performance concrete · Artificial sand · Silica fume · Fly ash · Super plasticizer · Rapid chloride penetration test · Water penetration test

P. U. Narale (✉) · H. S. Jadhav
Department of Civil Engineering, Rajarambapu Institute of Technology, Shivaji University, Kolhapur, Maharashtra 415414, India

H. S. Jadhav
e-mail: hanmant.jadhav@ritindia.edu

L. Nandagiri et al. (eds.), *Sustainability Trends and Challenges in Civil Engineering*, Lecture Notes in Civil Engineering 162, https://doi.org/10.1007/978-981-16-2826-9_21

1 Introduction

Current scenario of construction industry for development of concrete natural sand is used commonly as fine aggregate. About 30–35% of concrete volume is occupied by sand. Due to infrastructure development, the natural sand resources depleting day by day and disturbing nature of ecosystem of our country. Also, government had restricted quarrying of sand from natural resources. It is necessary to find another alternative to river sand, considering all aspect of high performance concrete (HPC) [1].

The use of artificial sand in concrete has desirable benefits such as increase in the strength [2, 9]. Unacceptable rates of deterioration of recently built concrete structures indicate that the strength criterion is no longer adequate to judge the performance of concrete. It must fulfil all durability aspect. Thus, the concept of HPC has been evolved [3].

The requirement of HPC includes high compressive strength, durability such as impermeability, resistance to chemical attacks (sulfate, chloride, etc.) and resistance to temperature [4]. From material point of view, HPC becomes possible mainly due to use of mineral and chemical admixtures [10–14].

HPC will extend the service life of structures in aggressive environments. Addition of chemical admixtures such as super plasticizer improves the properties of plastic concrete with regard to workability and segregation.

2 Methodology

Various trial mixes (TM) were designed to develop optimum mix proportion of M80 grade concrete incorporating addition of supplementary cementaneous material and super plasticizer based on IS 10262:2019 guidelines [5]. Durability tests were conducted includes rapid chloride penetration test and water impermeability test.

3 Materials and Mix Proportioning

3.1 Materials

1. *Cement:* 53 grade OPC cement confirming to IS 12269:1987 was used important physical properties of cement are shown in Table 1.

2. *Fine aggregate (FA):* Artificial sand of grading zone-II as per IS 383:1987 was used [6]. Physical properties of river and artificial are shown in Table 2.

Table 1 Physical properties of cement

Comp. strength (MPa)	68.5
Initial setting time (min)	40
Soundness (mm)	1.0
Final setting time (min)	365

Table 2 Physical properties of river and artificial sand

Properties	Krishana river sand	Artificial sand Walawa, Sangli
Specific gravity	2.62	2.72
Water absorption (%)	2.80	1.64
Fineness modulus	2.5	4.42

Table 3 Properties of CA

Fineness modulus	5.60
Water absorption (%)	1.40
Specific gravity	2.78

3. *Coarse aggregate (CA)*: Maximum 12.5 mm size of coarse aggregate used for TM1-TM9 as per IS 10262:2019 [6] and fractions of CA and FA of obtained mix proportion of TM10 was used. Physical properties of CA are shown in Table 3.
4. *Water*: Drinking water was considered for making concrete.
5. *Silica fume (SF)*: Confirming to ASTM C 1240–2000, Elkem micro-silica of 920D and specific gravity of 2.2 was used ("ELKEM industry, Mumbai).
6. *Fly ash*: The fly ash of type C was used, in that the CaO present is more than 12%. Properties of fly ash are shown in Table 4.
7. *Chemical admixture*: Poly carboxylate ether (PCE) based "Master Glenium Sky 8521" was used (BASF private limited, Mumbai). Properties of super plasticizer shown in Table 5.

Table 4 Properties of fly ash

Shape	Spherical
Bulk density (kg/m^3)	540–860
Specific gravity	2.20
Size	5–300 μ

Table 5 Properties of used superplasticizer

Relative density	1.11 at 25 °C
pH	> 6.0 at 25 °C
Aspect	Light brown liquid
Chloride ion content	< 0.20%

3.2 Mix Proportioning

As per IS 10262:2019 guideline, various trial mixes were made as shown in Table 6. Details of aggregate gradation of trial mix 10 as shown in Table 7. Figure 1 shows the image of ingredients of proposed concrete grade.

4 Test on Concrete

4.1 Slump Test

A good quality and required strength of concrete is formed by properly mixing ingredients at standard temperature (27 ± 2 °C) and humidity (50–60%) [5]. Concrete mixing with slump test to control on workability shown in Fig. 2.

4.2 Compression Strength

After 7 and 28 days curing period, the concrete cube tested under a compression testing machine of 300-ton capacity. The crushing loads computed based on the average compressive strength of concrete cube. Figure 3 shows the compressive strength test.

4.3 Durability Test

Water Penetration Test (WPT) and Rapid Chloride Penetration Test (RCPT) on developed high strength of M80 grade concrete are conducted to assess the suitability regarding durability point of view. Figure 4 shows the test specimen required for RCPT and water penetration test.

(1) *Rapid Chloride Penetration Test*: The rapid chloride penetration test was carried out as per ASTM C 1202–12. This test method determines electrical current flow through 50 mm thick slices of 100 mm diameter specimen during a 6-h period. After 28 days curing specimens were used. A potential difference

Table 6 Various trial mixes to develop M80 grade concrete

	Trial mix no.										
	Control Mil	TH1	TM2	TM3	TM4	TM5	TM6	Material	TM8	TH9	TM10
Cement (kg/m^3)	573	573	573	573	573	573	573	573	573	573	573
Water (kg/m^3)	149	149	149	149	149	149	149	149	149	149	142
FA (kg/m^3)	802.1	790	770	769	760	755	760	732.5	646.9	624.5	770
CA (kg/m^3)	1020.9	1011	1003	978.8	968	960	968	955	843.6	814.5	1002
SP (kg/m^3)	0.75	0.9	1.0	1.1	1.15	1.2	1	1	1	1	1
SF (kg/m^3)	0	14.32 (2.5%)	28.6 (5%)	42.97 (7.5%)	57.3 (10%)	71.62 (12.5%)	28.65 (5%)	28.65 (5%)	28.65 (5%)	28.65 (5%)	28.65 (5%)
Fly ash (kg/m^3)	0	0	5	0	0	0	28.65 (5%)	57.3 (10%)	85.95 (15%)	114.6 (20%)	

Table 7 Detailed gradation of CA and FA for TM10

CA	Fraction I—20% [20–12.5 mm size] Fraction II—60% [12.5–10 mm size] Fraction III—20% [below 10 mm size]
FA	Fraction I—90% [4.75 mm–300 μ size] Fraction II—10% [below 300 μ replaced by fly ash]

Fig. 1 Ingredient of concrete for TM10

Fig. 2 Concrete mixing with slump test to control on workability

Fig. 3 Compressive strength test setup

Fig. 4 Test specimen for RCPT and WPT

of 60 V kept constant across the ends of the specimen. Filled the one side of the specimen with 3.0% Sodium Chloride (NaCl) solution and the other side of the cell filled with 0.3 N Sodium Hydroxide (NaOH) Solution. Turned power supply on and recorded current with at every 30 minutes [8]. Rapid chloride penetration test is shown in Fig. 5.

(2) *Water Penetration Test*: This test was performed as per German standard DIN 1048:1991.

The specimen cast and cured in a standard procedure, is subjected to prescribed water pressure. After prescribed time of application of pressure, the specimen is broken and the depth of water penetration is measured by scale [4].

The depth of water penetration gives an indication of the qualitative assessment of permeability of concrete. Water penetration test set up is shown in Fig. 6.

Fig. 5 Rapid chloride penetration test setup

Fig. 6 Water penetration test setup

Table 8 Super plasticizer dosage for various addition of silica fume for 100 mm slump

Silica Fume (%)	Superplasticizer (%)
0.0	0.75
2.5	0.90
5.0	1.0
7.5	1.10
10.0	1.15
12.5	1.20

5 Results and Discussion

5.1 Slump Test

For very low w/c ratio (i.e., 0.26), PCE base super plasticizer was used and optimum dosage of super plasticizer was found to fix slump of 100 mm as shown in Table 8 and Fig. 7.

5.2 Compression Strength

The 7 and 28 days results of compressive strength of various trial mix for M80 are summarized in tabular and graphical form as shown in Table 9 and Fig. 8, respectively [1].

Considering only silica fume (TM2) 5% gives maximum compressive strength (87.95 N/mm^2) which is nearly equal to target strength (89.9 N/mm^2). But there is

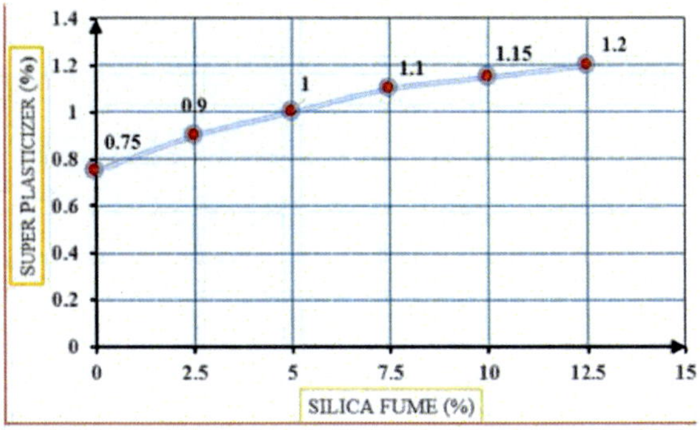

Fig. 7 Superplasticizer dosage percentage versus silica Fume percentage addition

Table 9 7 days and 28 days compressive strength results

Trial mixes	Silica fume (%)	Fly ash (%)	7 days comp. strength (N/mm^2)	28 days comp. strength (N/mm^2)
CM	0.0	0.0	67.33	76.44
TM1	2.5	0.0	74.00	83.33
TM2	5.0	0.0	77.70	87.95
TM3	7.5	0.0	73.33	84.66
TM4	10.0	0.0	70.66	82.00
TM5	12.5	0.0	66.66	78.66
TM6	5.0	5.0	75.75	83.33
TM7	5.0	10.0	74.00	82.00
TM8	5.0	15.0	70.66	79.33
TM9	5.0	20.0	66.66	75.77
TM10	5.0	–	78.66	90.22

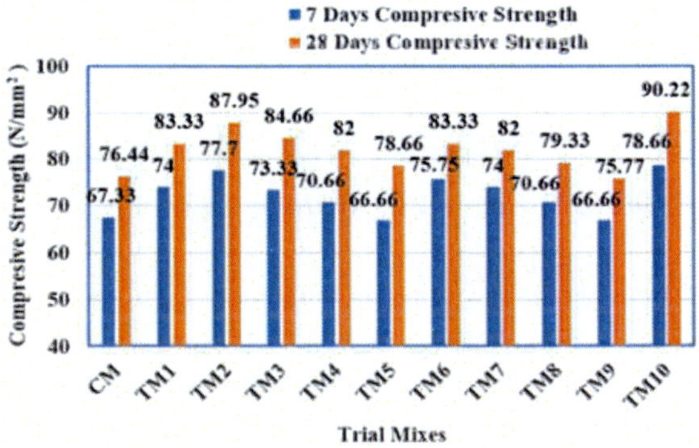

Fig. 8 Chart—Trial mixes versus compressive strength at 7 days and 28 days curing

Fig. 9 Air void in hardened concrete at surface and interior

air void problem in hardened concrete at surface and interior also, shown in Fig. 9. It affects durability aspect.

To overcome this problem and additional more strength with optimized 5% SF, added fly ash, in various percentage (5%, 10%, 15%, and 20%) with respective trail mixes TM6–TM9 are carried out.

From Fig. 8 observed that, with minimum fly ash content and after removing dust in artificial sand, well finished concrete surface with negligible air void and compressive strength up to 90 N/mm^2 which is greater than target strength (89.9 N/mm^2).

Hence, TM10 is optimum mix proportion of M80 grade concrete. This developed M80 high grade concrete was used to check important durability requirement.

5.3 Durability Test

Rapid Chloride Penetration Test: The aggregate charge flows in coulombs, related to the resistance of the concrete specimen to chloride ion penetration. The current is recorded for every 30-min intervals for 6 h and following Eq. 1 is used to calculate the total charge passed over the test specimen.

$$Q = 900\ X(I_0 + 2I_{30} + 2I_{60} + 2I_{90} + 2I_{120} + \cdots + 2I_{300} + 2I_{330} + I_{360}) \quad (1)$$

where,

Q = Charge passes through each cell (coulombs)
I_0 = Initial Current value in amperes immediately
I_t = Current value in amperes at t minutes.

Qualitative indication of the chloride ion carried out as per ASTM C1202 shown in Table 10.

The results of RCPT test of concrete specimens at the age of 28 days of developed M80 grade concrete after computing from Eq. 1 shown in Table 11.

Table 10 Chloride ion permeability rating (ASTM C 1202)

Chloride ion permeability class	Charges passed (Coulombs)
High	< 4000
Moderate	2000 to 4000
Low	1000 to 2000
Very low	100 to 1000
Negligible	< 100

Table 11 Rapid chloride permeability test result

Trial Mix ID	Charges passed (coulombs) through test specimens			Average charge passed (Coulombs)	Penetrability class as PER (ASTM C-1202)
	1	2	3		
TM 10	184	170	190	181.33	Very low (100–1000)

As per rapid chloride penetration test result of developed M80 grade concrete average charge passed through test specimen is 181.33 (Coulombs) which is in very low permeability class. The reduction in the magnitude of charge passed could be because of the consumption of calcium particles in the gel pore liquids and the improvement of tightened irregular and pore structures. The high conductivity way for the ions will be decreased.

Water Penetration Test: Depth of water penetration is measured by using scale after the test by broken tested cube. Specimens of 150 mm cube sizes were used. This method was conducted under a pressure of 0.50 N/mm^2 continuously for 48 h. After the pressure is released the cube is split with the help of compression testing machine and depth of penetration of water is marked and measured as shown in Table 12.

Ministry of Road and Surface Transport of India (MORST) prescribed that the depth of water penetration of good concrete should not more than 25 mm [4].

As a qualitative assessment of concrete, penetration depth up to 50 mm can be classified as the concrete which is "impermeable", a depth not more than 30 mm can be classified as the concrete "Impermeable" under aggressive conditions.

Maximum water penetration depth of developed M80 grade concrete is 11.5 mm which is within the permissible limit (25 mm) as per MORST.

Table 12 Water penetration test result considering penetration depth

Trail mix ID	Water penetration depth through test specimen (mm)			Max value (mm)	Remark (As per MORST)
	1	2	3		
TM10	11	10.5	11.5	11.5	Penetration depth is Permissible (<25 mm)

As per WPT theory penetration depth 11.5 mm is less than 30 mm, therefore it is classified as the concrete "Impermeable" under aggressive conditions.

6 Conclusions

Following are the major conclusions that were drawn,

- Optimum mix proportion of M80 grade concrete is TM10 with SF (5%) and artificial sand below 300 micron which is nearly equal to 10% is replaced by fly ash, with good gradation of coarse aggregate of 20 mm maximum sizes.
- Maximum compressive strength 90.22 N/mm^2. It is possible to achieve higher grade concrete using silica fume as mineral admixture and artificial sand without dust.
- Use of SF in concrete reduces the workability. By using PCE base superplasticizer (HRWR), it is possible to increase workability and high earlier strength at lower w/c ratio.
- As per rapid chloride penetration test result of M80 grade average charge passed through test specimen is 181.33 (Coulombs), which is in very low permeability class (100–1000 Coulombs).
- Maximum water penetration depth of M80 grade concrete is 11.5 mm, which is within the permissible limit (< 25 mm) as per MORST. It is classified as the concrete "Impermeable" under aggressive conditions.
- The reduction in the magnitude of water penetration depth and charge passed could be because of the consumption of calcium particles in the gel pore liquids and the pozzolanic responses of silica fume.
- The eco-friendly high performance concrete with artificial sand is substitute material to natural sand as fine aggregate in the presence of silica fume with poly carboxylate ether based superplasticizer can be effectively used in the construction field to minimize the environmental pollution.

Acknowledgements We would like to thank Department of Civil Engineering of RIT, Sangli and D. Y. Patil College of Engineering and Technology, Kolhapur, Maharashtra, India, provided laboratory facility for research work.

References

1. Narale PU, Jadhav HS (2020) Development of high strength concrete using artificial sand as fine aggregate. IRJET 07(08):1603–1608. e-ISSN: 2395-0056, p-ISSN: 2395-0072
2. Mane KM, Kulkarni DK, Joshi AA (2017) Strength and workability of concrete with manufactured sand. Int J Eng Res Technol 10(1). ISSN 0974-3154
3. Reddy MVS et al (2012) Durability aspects of high-performance concrete containing supplementary cementing materials. IJSCER 1(1):51–59, ISSN 2319 – 6009

4. Shetty MS, Jain AK (2019) Concrete technology- theory and practice. S Chand Publication, Eighth Edition
5. IS: 10262:2019 Recommended Guidelines for Concrete Mix Design. Indian Standard Institution, New Delhi, India, second revision
6. IS: 383–1970, Specification for coarse and fine aggregate from natural sources for concrete. Bureau of Indian Standards, New Delhi, India
7. IS: 9103:1999, Concrete Admixtures—specification. Bureau of Indian Standards, New Delhi, India
8. ASTM C 1202–12, Standard test method for electrical indication of concrete's ability to resist chloride ion penetration.*ASTM International,* West Conshohocken, USA
9. Khartode RR, Kulkarni DB (2016) Mix proportioning of HSC using manufactured sand. Int J Sci Res 5(7). ISSN: 2319-7064
10. Parvathy Karthika A, Gayathri V (2018) Experimental studies on durability aspects of high strength concrete using fly ash and alccofine. Int J Recent Technol Eng (IJRTE) 7(4S). ISSN: 2277–3878
11. Perumal K, Senthilkumar S, Sekar K (2018) Experimental investigations on strength characteristics of high-performance concrete using silica fume and superplasticizer. Int J Trend Sci Res Dev (IJTSRD) 2(6). ISSN No: 2456 – 6470
12. Sheikh KM, Joshi MM (2018) Experimental study of high strength concrete (M70) using manufactured sand. Int J Eng Sci Comput 8(4)
13. Shanmugapriya T, Uma R (2015) Strength and durability studies on high performance concrete with manufactured sand as fine aggregate. Int J Appl Eng Res 10(2):1919–1924. ISSN 0973-4562
14. Sri Hari M, Sai Ram Goud MH et al (2019) Experimental examination on blended concrete by incorporating fly ash and silica fume. Int J Innov Technol Explor Eng (IJITEE) 9(2). ISSN: 2278-3075

Shoreline Change Monitoring of Karwar Coast of Karnataka, India, Using Sentinel-2 Satellite

Sheetal Mutagi, Arunkumar Yadav, and Chandrashekarayya G. Hiremath

Abstract Karnataka coastal area stretches across 320 km approximately. Karwar coast is found to be a famous tourist place and undisturbed beaches in Karnataka. Change in coastal conditions due to port activities, seasonally varying discharge and sediment transport from estuarine, and threat caused by different factors have revolved into the dynamic region; the primary aim of the study is to monitor the change in shoreline using the Sentinel-2 data for a period of 2016–2020. Sentinel-2 data is free to access and has a spatial resolution of 10 × 10 m. Multi-temporal shorelines are extracted using ArcGIS 10.4 tool and rate of change of shorelines are calculated using Digital Shoreline Analysis System tool which includes statistical approaches like linear regression rate, end point rate to identify erosion and accretion rates and comparisons are made. The results indicate that Rabindranath Tagore Beach has erosion of −17.87 m/ yr (highest value) and −0.03 m/yr (lowest value) and accretion of 3.11 m/yr (highest value) and 0.03 m/yr (lowest value). The average shoreline change rate of 6.00 m/yr (LRR) and 5.83 m/ yr (EPR). The results were found to be more precise because of the high-resolution Sentinel-2 data used.

Keywords Sentinel-2A · ArcGIS · Shoreline change · End point rate · Linear regression rate

1 Introduction

The shoreline is complex and unique features are clearly defined as a line/interface where land and water interconnect. The shoreline is variable features easy to define but needs vast knowledge and difficult to delineate the exact position for monitoring change because of the tide and everlasting effects such as changes in the level of the sea, residual transport, concerning space and time continuous changing of

S. Mutagi (✉) · C. G. Hiremath
Department of Water and Land Management, Visvesvaraya Technological University, Belagavi, Karnataka, India

A. Yadav
Department of Civil Engineering, CMR Institute of Technology, Bangalore, Karnataka, India

L. Nandagiri et al. (eds.), *Sustainability Trends and Challenges in Civil Engineering*, Lecture Notes in Civil Engineering 162, https://doi.org/10.1007/978-981-16-2826-9_22

shorelines positions [1]. Shorelines are dynamic and complex due to factors such as geological, meteorological, hydrological, human interruptions, vegetation, policies of agriculture, and land use. Transition of coastal zones are due to the natural factors like sea-level rise, wind and storms, near-shore currents, climate change, global warming and slope processes and additionally affected by human factors like port activities, industrialization, discharge of effluents by large industries and power plants, sewage discharge, coastal engineering, dams or reservoirs construction, activities of dredging, claiming of land, mining and water withdrawals that indirectly affect natural conditions [2]. Coastal areas with river mouth are complex and important habitats, and their estuarine morphology such as sediment transport also varies seasonally due to a rise in sea level and change in shoreline configuration [3]. The advantages of using any particular method for detecting shoreline change rates are dependent on data sources and resources available. Earlier use of Toposheets is now advanced with high-resolution satellite imageries, remote sensing, and different Geographical Information System approaches that showed improvements in the examination of coastal geomorphologic, determination of shorelines in a semi-automatic way, and identification of comparative changes among coastal components. The study analyzed for 55 years was helpful for the further survey of its historic evolution and future trends prediction [4]. Nowadays, the use of Toposheets is called old methods which are not accurate and which gives errors, but T-sheets are the best and only source of historical shoreline data from any areas, and uncertainty can be removed when used in the GIS framework [5]. Over the past years, Remote Sensing and Geographic Information System technology has developed with different methods for solving coastal problems. It has provided a way to generate information required for monitoring and provide analysis to study about previous years and predict and visualize future changes [6]. Prediction of future shoreline positions, estimation of change rates, and prediction of shoreline changes accurately can be done cost-effectively by combined new and old techniques of satellite image analysis and statistics using satellite image data of higher resolution at smaller intervals [7]. Coastal processes and riverine inputs were the cause of different patterns and the presence of changing trends along the coast that sometimes behaves as stable, erosive, accretion zones. Different satellite imagery and statistical methods can be used for determination of shoreline position change [8]. The shoreline change has its effects, but not visible for easy visual interpretation, but a false-color composite of RGB has the limitation of not delineating of land and water boundary correctly, and hence, different techniques like band rationing technique, enhancement of image, supervised and unsupervised classification, principle component analysis (PCA), normalized difference water index (NDWI) can be considered [9]. Detection of shorelines is possible by manual, semi-automatic/ automatic methods. Automatic methods are namely histogram equalization, adaptive thresholding techniques, Image segmentation, algorithms were developed to know about accurate position [10, 11]. Because of shoreline change, the coastal areas are affected in terms of tourism, activities of a fishery, loss of land, and infrastructure damages, which mark the urgency of analysis for shoreline change taking place. Mitigation efforts for the coastal region need to be implemented through structural mitigation/ non-structural mitigation [12]. Dynamic activities like erosion and accretion in shoreline changes were analyzed using RS and GIS techniques, and mapping of shoreline is done to gain information about the changes occurring [13]. Updating of the shoreline change maps and management of natural resources, the study on shoreline change is very important and plays a vital role in providing useful information on coastal landform dynamics [14]. Due to loss of human property, life, and damage to the ecosystem caused by shoreline change, monitoring and mapping of past and ongoing periods is used as an emergency measure to prevent further loss caused by erosion [15]. Shoreline mapping is an important aspect and needs to be updated every year for coastal management and creating a sustainable coastal environment. In the future, society must be educated about the coastal problems and their importance of participation in creating a sustainable and safe environment. The current study was carried out to monitor and compare change rates, erosion, and accretion rates for the period of 2016–2020 using Sentinel-2A satellite imagery data. Due to free access to data, it has 10 × 10 m spatial resolution Sentinel-2 satellite imagery data is used which is better than Landsat TM satellite imagery.

2 Study Area

Karwar is the coastal city in the state of Karnataka which lies on the west coast of southern India situated on the banks of the Kali River which flows west to the Arabian sea. Karwar coast has Rabindranath Tagore Beach which is towards the south. The current study area lies between 14°48′00″ and 14°52′30″ North latitude and 74°06′00″–74°08′00″ East longitude shown in Fig. 1. The whole Karwar city is built on many ridges and due to the influence of meeting of the Arabian Sea and Kali river in past, beach ridges are slowly vanishing along the coast [16]. Karwar coast has seven islands which are significant ecologically with rocky and sandy shores and is diverse in a habitat that ranges from the open shoreline to sandy beaches near the estuarine environment. It was observed that Kali Estuary has nutrient-rich water content influencing primary nutrient product and salinity, resulting in low salinity in regions close to it [17].

3 Data and Methods

3.1 Shoreline Data

This study used Sentinel-2A multi-temporal satellite data for the period 2016–2020 to observe shoreline change. Table 1 gives the details of the type of satellite imagery data used and the time they were acquired.

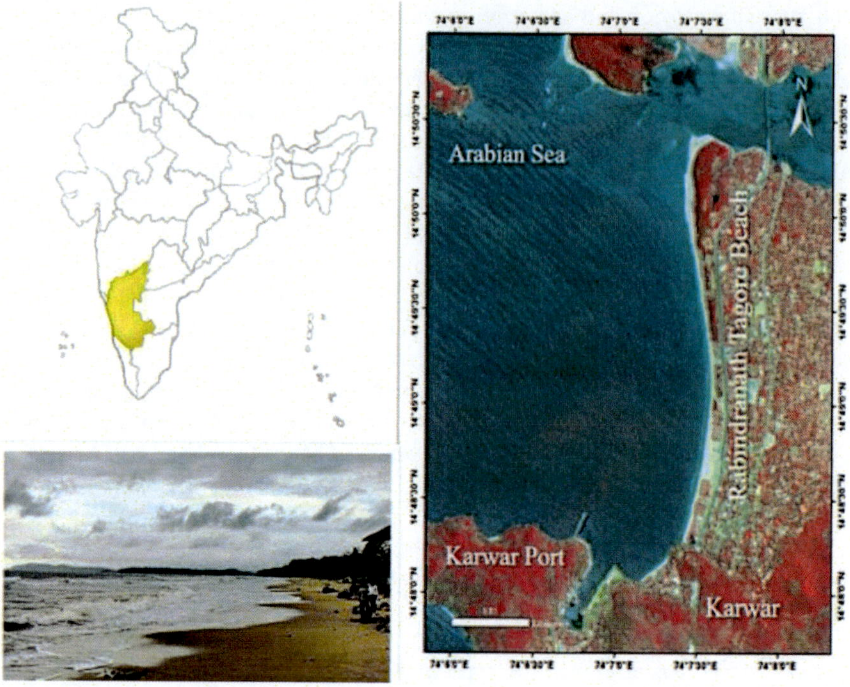

Fig. 1 Study area—Rabindranath Tagore Beach

Table 1 Source of shoreline data

Data	Period	Acquisition time	High tide	Low tide
Sentinel-2A	March 8th, 2016	05:37:57	1.8	0.1
Sentinel-2A	May 20th, 2017	05:42:08	0.7	1.7
Sentinel-2A	May 25th, 2018	05:34:19	0.3	1.7
Sentinel-2A	May 25th, 2019	05:37:18	0.5	1.7
Sentinel-2A	May 25th, 2020	05:38:21	1.9	0.9

Source https://earthexplorer.usgs.gov/

3.2 Image Processing

Image preliminary processing is the pre-operation done to remote sensing data and its main aim is to improve the data by intensifying important features. Preprocessing was carried out for Sentinel-2A satellite image data. The layer-stacking technique is used to stack images/bands having the same spatial resolution of 10 m. For Band 8 (Near-IR), 4 (red), 3 (green), and 2 (blue), the false-color composite was created using ArcGIS 10.4.

3.3 Shoreline Change Detection Analysis

Shorelines were extracted by digitizing manually by ArcGIS 10.4 tool using Sentinel-2A satellite images from 2016 to 2020. Figure 2 shows a flowchart of shoreline change analysis. After manual digitization of shorelines for different periods, the extracted shoreline features are put together in ArcGIS 10.4 with necessary attribute fields such as SHAPE (Polyline), ObjectID, SHAPE Length, Date, and Uncertainty values in shoreline feature class within a personal geodatabase. For the calculation of rate of change, multiple shoreline positions are extracted as individual shapefiles and later merged into single file. Digital Shoreline Analysis System (DSAS) version 4.0 is

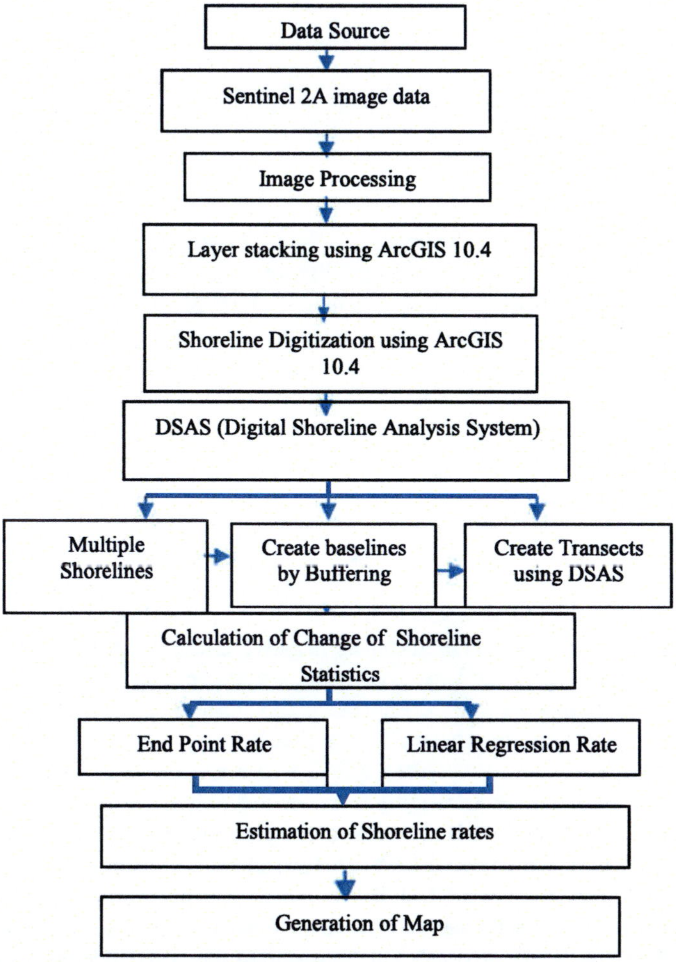

Fig. 2 Flowchart showing shoreline change analysis

an ArcMap extension tool used to calculate the rate of change statistics by casting transects of shorelines from multiple shoreline positions [18]. The baseline act is a starting point for the perpendicular casting of transects created by buffering which follows the direction of the shape of shoreline either onshore, offshore, or mid-shore. The distance between shoreline and baseline was kept 200 m apart. Computation of shoreline change statistics can be done by weighted linear regression, shoreline cover envelope, net shoreline movement, end point rate, and linear regression rate. In this present study, use of linear regression rate (LRR) and end point rate (EPR) statistical approach is used [16]. Calculation of rate of change is done by dividing the shoreline change distance by the time passed between the oldest and the youngest shorelines with the end point rate (EPR) approach. The shoreline movement and estimation of change rate is by Linear Regression Rate (LRR) method using least squares regression lines from total shoreline positions along the transects.

4 Field Investigation

During field investigation, the photographs of study area Rabindranath Tagore Beach subjected to erosion and accretion are taken. Figure 8a and b shows the erosion pattern observed in the central part of the coast. Figure 8c–e shows the accretion pattern observed near the Kali River Estuary and near the port. Most similar results are seen when the comparison is made between obtained results from analysis projected in terms of the graph as shown in Figs. 3 and 4 and field photographs.

5 Results and Discussion

To monitor the change rate of extracted multiple shorelines from 2016 to 2020, the digital shoreline analysis system is used. In this study, the end point rate and linear regression rate methods are utilized to identify the change of rate, erosion, and accretion where erosion is indicated by negative values and accretion by positive values. Rates were calculated by transects cast perpendicular to both shorelines and baseline at 100 m spacing and 200 m transect distance. Figures 3 and 4 show the LRR and EPR map of Rabindranath Tagore Beach.

From previous studies, it was observed that Karwar coast was subjected to both accretion and erosion patterns ranging from low erosion to moderate accretion and high accretion was found on both sides of the Kali River in the year 1973–2006 [6]. The northern part of the Kali River was subjected to erosion and the accretion that happened in the south of the Kali River [19]. According to recent years of study, Rabindranath Tagore Beach resulted in a change from an accretion zone to an erosion zone and experienced low erosion to high erosion in the year 2015–2017 [3].

In the current study for period 2016–2020, from LRR, the erosion is −17.87 m/yr (highest value) and −0.03 m/yr (lowest value) and accretion is 3.11 m/yr (highest

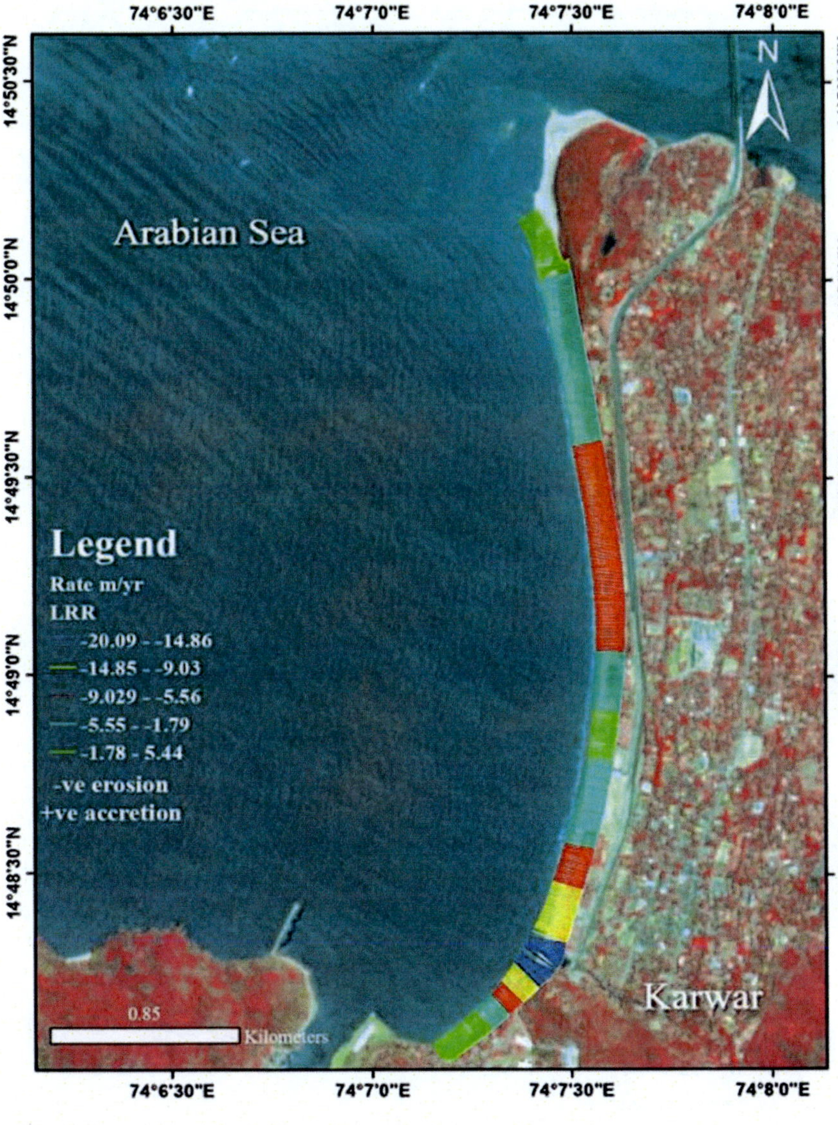

Fig. 4 EPR map of the study area

value) and 0.03 m/yr (lowest value). Similarly, from EPR, the erosion rate is observed as −20.09 m/yr (highest value) and −0.1 m/yr (lowest value) and accretion as 5.44 m/yr (highest value) and 0.25 m/yr (lowest value). It is observed that the study area has high erosion when compared to accretion. High erosion was observed during pre-monsoon and high accretion patterns through the post-monsoon season. Figures 5 and 6 show the EPR and LRR graphs of the rate of change of the shoreline. A compar-

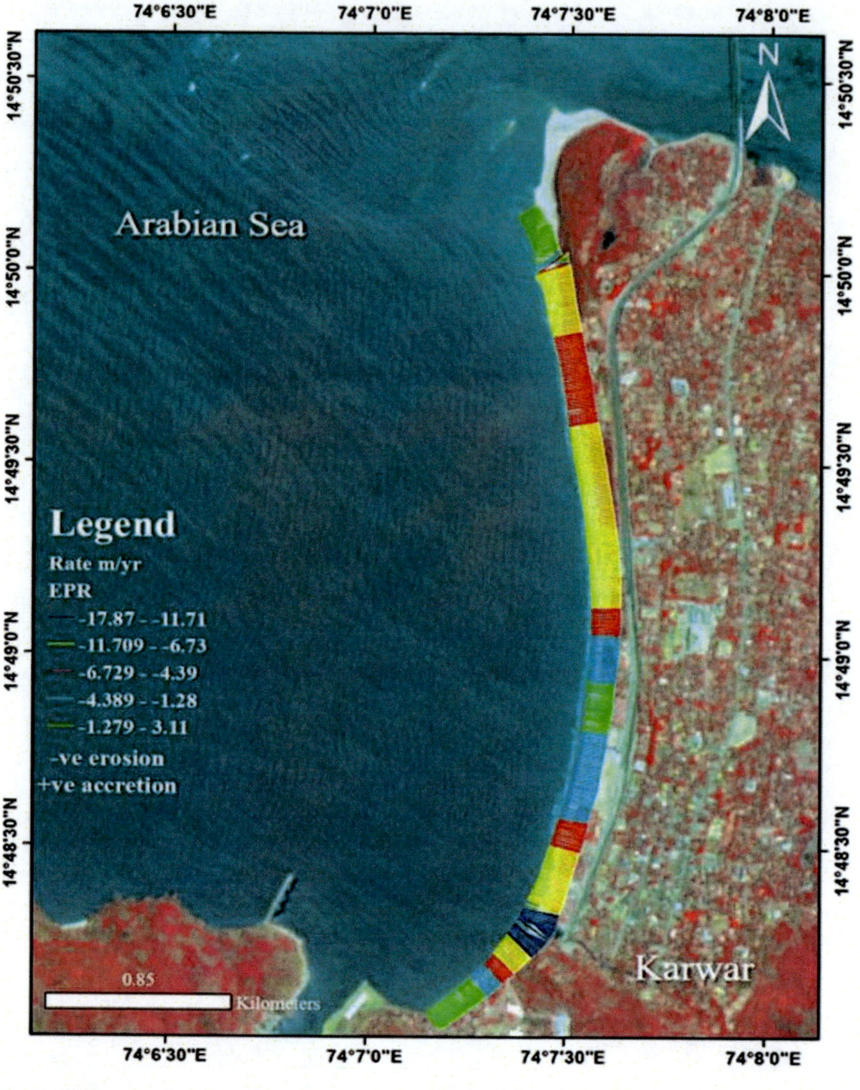

Fig. 3 LRR map of study area

ison between LRR and EPR graphs is made as shown in Fig. 7, negative values in the graph indicating erosion and positive values indicating accretion. When compared, both the methods showed results very close to each other. There is a cumulative effect of different factors and not a single factor on shoreline change, likewise the study area is between the Kali estuary and port, due to port activities and discharge from estuary causes, change in sediment discharges that vary seasonally converted region

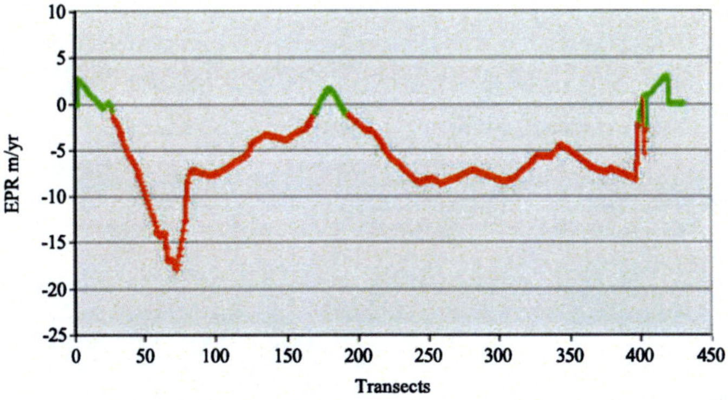

Fig. 5 EPR graph of rate of change of shoreline

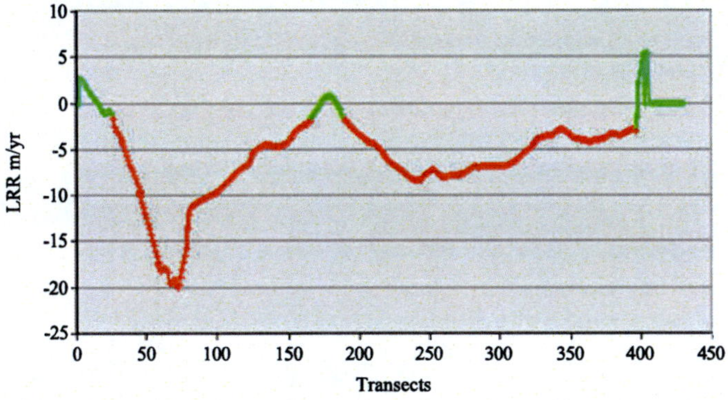

Fig. 6 LRR graph of rate of change of shoreline

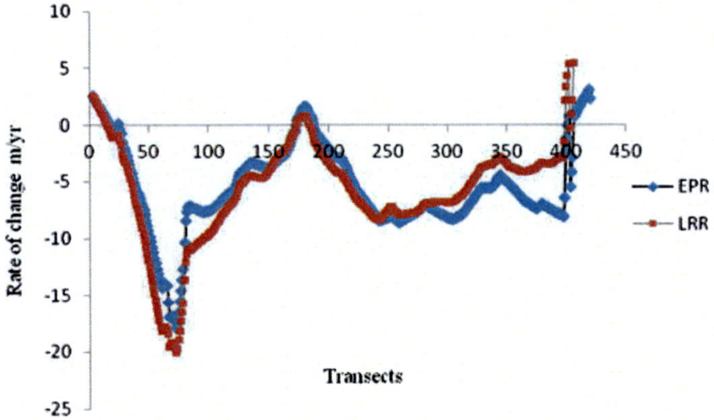

Fig. 7 Comparison graph between LRR and EPR

Table 2 Comparison of the results of two satellite image data

Satellite data	EPR m/ yr		LRR m/ yr		Avg. shoreline rate m/yr	
	Accretion	Erosion	Accretion	Erosion	EPR	LRR
Landsat-8	15.13	−11.71	20.16	−9.29	0.004	1.67
Sentinel-2A	5.44	−20.09	3.11	−17.87	5.83	6.00

Fig. 8 **a, b** Erosion at middle part of coast, **c, d** Accretion observed near port. **e** Accretion near Kali Estuary

into a dynamic zone observed over years. High erosion was observed during pre-monsoon and high accretion during the post-monsoon season. The average shoreline change rates are 6.00 m/yr (LRR) and 5.83 m/yr (EPR). It is observed that the study area has high erosion when compared to accretion. Also, the results from Landsat-8 for the period of 2013–2017 [3] and Sentinel-2A for the period of 2016–2020 for Rabindranath Tagore are shown in Table 2.

6 Conclusions

The present study focused on Sentinel-2 data collected from 2016 to 2020 for shoreline change detection which is the novelty of the present study. Sentinel-2 data was used because of its free access and high resolution. The availability of this data is from the year 2016, and because of high-resolution imagery data, the results obtained in this study are found to be more precise compared to Landsat satellite data. The results obtained from EPR showed that the erosion rate is observed as −20.09 m/yr (highest value) and −0.1 m/yr (lowest value) and accretion as 5.44 m/yr (highest value) and 0.25 m/yr (lowest value) with average shoreline rate of change as 6.00 m/yr (LRR) and 5.83 m/yr (EPR). This study reveals that Rabindranath Tagore Beach is subjected to high erosion when compared to accretion and has progressed from low erosion to high erosion areas from the past to present year. Results obtained must be used as a guide to form preventive measures to reduce shoreline changes. From the current study, the analysis of shoreline obtained can be helpful for the scientists, coastal managers, stakeholders, and coastal people, Policymakers have a guide to form preventive measures against erosion zones and create a sustainable environment along the Karwar coast.

References

1. Moore LJ (2000) Shoreline mapping technique. J Coast Res 16(1):111–124
2. El Mrini A, Maanan M, Anthony EJ, Taaouati M (2012) An integrated approach to characterize the interaction between coastal morph dynamics, geomorphologic setting and human interventions on the Mediterranean beaches of northwestern Morocco. Appl Geogr 35(1–2):334–344
3. Yadav A, Dodamani BM, Dwarakish GS (2018) Shoreline analysis using Landsat-8 satellite image. ISH J Hydraulic Eng
4. Qiao G, Mi H, Wang W, Tong X, Li Z, Li T, Liu S, Hong Y (2018) 55-year (1960–2015) spatiotemporal shoreline change analysis using historical DISP and Landsat time-series data in Shanghai. IJAEOG 68
5. Langley SK, Alexander CR, Bush DM, Jacksontt CW (2003) Modernizing shoreline change analysis in Georgia using topographic survey sheets in a GIS environment. J Coastal Res
6. Umar Z, Akib WA, Ahmad A (2013) Monitoring shoreline change using Remote sensing and GIS: a case study of Padang coastal area, Indonesia. In: 2013 IEEE 9th international colloquium on signal processing and its applications, Kuala Lumpur, pp 280–284
7. Maiti S, Bhattacharya AK (2009) Shoreline change analysis and its application to prediction: a remote sensing and statistics-based approach. Mar Geol 257:11–23
8. ChenthamilSelvan S, Kankara RS, Rajan B (2014) Assessment of shoreline changes along Karnataka coast, India using GIS & Remote sensing techniques. Indian J Geo-Marine Sci 43(7):1286–1291
9. Salghuna NN, Aravind Bharathvaj S (2015) Shoreline change analysis for northern part of the Coromandel coast. Aquat Procedia 4:317–324
10. Aedla R, Dwarakish GS, Reddy DV (2015) Automatic shoreline detection and change detection analysis of netravati-gurpur River mouth using histogram equalization and adaptive thresholding techniques. Aquat Procedia 4:563–570
11. Kuleli T, Guneroglu A, Karsli F, Dihkan M (2011) Automatic detection of shoreline change on coastal Ramsar wetlands of Turkey. Ocean Eng 38(10):1141–1149
12. Mutaqin BW (2017) Shoreline changes analysis in Kuwaru coastal area, Yogyakarta, Indonesia: an application of the digital shoreline analysis system (dsas). Int J Sustain Dev Plan 12(7):1203–1214

13. Raj N, Gurugnanam B, Sudhakar V, Glitson Francis (2019) Estuarine shoreline change analysis along with The Ennore river mouth, southeast coast of India, using digital shoreline analysis system. Geodesy Geodyn 10(3):205–212
14. Kannan R, Ramanamurthy MV, Kanungo A (2016) Shoreline change monitoring in Nellore Coast at East Coast Andhra Pradesh District using remote sensing and GIS. J Fisheries Livest Prod 4:161
15. Jayakumar K, Malarvannan S (2016) Assessment of shoreline changes over the Northern Tamil Nadu Coast, South India using WebGIS techniques. J Coast Conserv 20(6):477–487
16. Hanamgond PT, Mitra D (2007) Dynamics of the Karwar coast, India, with special reference to study of tectonics and coastal evolution using remote sensing data. J Coastal Res Spec 1(50):842–847
17. Sujatha TR (1987) Ecology of the intertidal macrobenthos of Karwar beach
18. Thieler ER, Himmelstoss EA, Zichichi JL, Ergul A (2009) The Digital Shoreline Analysis System (DSAS) version 4.0-An ArcGIS extension for calculating shoreline change. U.S. Geological Survey
19. Naik D, Kunte PD (2016) Impact of port structures on the shoreline of Karnataka, West Coast, India. Int J Adv RS GIS 5(5):1726–1746

Quantification of Traffic Congestion Based on Vehicular Speed Under Heterogeneous Flow Conditions Using Fuzzy Inference Model

A. Vijayakumar, S. Varadarajan, and V. Chitti Babu

Abstract On transport networks, traffic congestion is a phenomenon that arises as usage increases and is marked by slower speeds, longer travel times and increased queuing of automobiles. In plain terms, "the ability of a vehicle to move forward in a traffic state" defines congestion. Traffic congestion in smart cities has been a major concern. In heterogeneous flow like India, congestion impacts the movement of people both in perception and in reality that leads to consumption of time, energy and also leads to the pollution. In order to save precious human life, eliminate road accidents and the essence factor called time, it is essential to ensure a proper measure for traffic congestion. Earlier there were several attempts made to develop different approaches for congestion analysis. At present the congestion levels of ten different road stretches of Visakhapatnam city within the central business district (CBD) area. The key purpose of this analysis is to introduce a scalable fuzzy logic traffic flow model capable of optimally forecasting traffic to define the city's congestion levels by taking into account parameters such as vehicle volume, average speed and road speed limits by using MATLAB and to generate the desired congestion index of the specified study stretches. This research sets the framework for the forecast, early warning and constructive alleviation of traffic congestion

Keywords Level of Service · MATLAB · Membership functions · Fuzzy logic · Linguistic variables

A. Vijayakumar
Department of Civil Engineering, GMR Institute of Technology, Rajam, Andhra Pradesh, India

S. Varadarajan (✉)
Department of Civil Engineering, Aditya Institute of Technology and Management, Tekkali, Andhra Pradesh, India

V. Chitti Babu
Department of Mechanical Engineering, Aditya Institute of Technology and Management, Tekkali, Andhra Pradesh, India

L. Nandagiri et al. (eds.), *Sustainability Trends and Challenges in Civil Engineering*, Lecture Notes in Civil Engineering 162, https://doi.org/10.1007/978-981-16-2826-9_23

1 Introduction

Urban traffic congestion has become a crucial concern that impacts not just the everyday life of people but also the healthy growth of society and economy in countries like India, which have mixed traffic conditions, congestion cannot be totally eliminated, but it could be stabilized to some extent by taking proper enforcement and mitigates measures. Irregular planning, intolerable road capacity, private transit encouragement, tremendous growths of vehicles are the main source of traffic congestion. Countries like China, Singapore, London and US has introduced a concept called "congestion charging" to control congestion. Our country is making some flexible policies to implement the same initially in some of the metro cities. Figure 1 shows how the heterogeneous flow impact congestion.

Today, urban centers generate dynamic issues with transportation in everyday life. The problem of congestion cannot be avoided just by building bridges, motorways or by increasing road capacity alone, but also by ensuring sufficient channeling. It is

Fig. 1 Heterogeneous flow

important to develop a transportation management technology framework to monitor the traffic phenomena.

Traffic management systems have a significant effect on traffic issues, leading to increase the flow of traffic and decrease congestion. There is currently no unified and fixed assessment measure for determining the circumstances of traffic operations. In fact, in different areas, there are different assessment steps [1]. In 1985, the highway capacity manual [HCM] first proposed using the level of service as a road performance assessment index.

The level of service was specified in the United States by six grades and in Japan by three grades. In China, the Ministry of Public Security has chosen the average city road speed as an assessment metric to characterize road traffic conditions [2]. A Congestion Index was developed by Levinson and Lomax that revealed variations between real and desired travel times for various types of roads. The definition of a delay rate index (DRI) was derived on the basis of the 1985 and 1994 HCM, LOS, speed, delay rate and other relationships [3]. In 2006, the Washington State Transportation Department issued a congestion report that described the congestion assessment index as the average peak traffic time [4]. Yan and Liu performed a report based on the Beijing Expressway Network Speed Efficiency Index in which the author took various road segments for report and measured the congestion index of the road segment and the congestion index of the road network with the speed index [5]. Vaziri et al. developed Congestion indices with restricted sampling speed and flow rate information have been established. In simple and rational models, the congestion indices were related to travel speed, travel rate, delay rate, travel rate ratio and delay ratio variables, respectively. Relevant congestion levels were measured by the index values under free and capacity flow conditions to calibrate the congestion index models. For reference, index models were built based on the US Highway Capacity Manual, direction and speed flow rate information [6]. Hamad et al. established a fuzzy inference model with travel speed inputs, free-flow speed, and very low-speed proportions in the overall travel time. Using real-world evidence, the method was illustrated. The findings were similar to those of the manual of highway capability [7]. Das by using MATLAB, assessed congestion modeling under mixed traffic conditions. For the feasibility of study, the author considered six kilometre of stretch dividing it into three segments of 2 km each [8]. Surendra and Parbat developed congestion modeling on the urban road network to obtain an output Parameter Congestion Index using the fuzzy inference system [9].

2 Study Area

"The City of Destiny", Visakhapatnam is situated in the state of Andhra Pradesh, with an area of 550 km^2, and it is the second largest city in the province. Popularly known as the second capital of Andhra Pradesh. Vishakhapatnam is also one of India's fastest growing cities. Due to an abundant increase in urban population, Visakhapatnam city was taken as the study area. Data collection was mainly focused

Fig. 2 Study area

on the central business district (CBD) due to the tremendous increase in vehicles. Figure 2 shows the map of the study area. Ten mid-block areas like Maddilapalem, Gajuwaka, Dabagardens, Gopalapatnam, Duvvada, Kancharapalem, Marripalem, Dwarakanagar, Akkayapalem and Old Gajuwaka are taken in the city for study.

A videographic survey of one hour is carried out at each stretch during the peak hour of that particular locality. A 50 m stretch is taken at each study area by ensuring the study stretch is of gradient free and far from intersection to generate accurate vehicle average speeds by collecting data at that particular locality regarding vehicle volume and the class of vehicle using that particular roadway facility.

3 Development of Fuzzy Inference System (FIS) for Congestion

3.1 Fuzzy Logic

In 1965, Lotfi Zadeh invented the word fuzzy logic. It is a form which purely depends on degree of freedom in which truth of variables may be any real number between 0 and 1, i.e., it cannot be expressed as true or false exactly, but says it as partially true. The membership functions (MF) are the integrated functions which constitute the body of the fuzzy system model. It should be noted that the membership function defines the Fuzzy set having a range of values in the horizontal axis. However, the only constraint for a membership function is in the vertical axis that must be scaled between 0 and 1.

Fuzzy inference system (FIS) is a process which purely depends upon human based knowledge by framing some of the rules in the fuzzy inference engine (FIE). One of the main advantages of this system is that it doesn't promote any error as this is entirely based on human sense. Initially the process of model consists of aggregating the three input data, designing the output variable, framing the rules based on IF–THEN condition and then finally determining the Congestion Index (CI).

3.2 *Input Parameters*

Congestion in road traffic is affected by multiple factors such as vehicular volume, ehicular average speed, road speed limit. All above three parameters combine together to represent the congestion levels within the constraints of rule base.

The three inputs and one output with respect to their linguistic variables and membership functions are tabulated as shown in Table 1. There are no complicated and fast rules on the membership function selection because it really depends on the applications where the fuzzy structures are to be applied. Using a mnemonic acronym that called, "Fuzzy D.I.S.C." these applications can usually categorized.

Table 1 INPUTS and membership functions

S. No.	Input variable	Fuzzy sets	Membership function type
1	Vehicular volume	Too low	trimf
		Low	trimf
		Moderate	trapmf
		High	trimf
		Very high	trapmf
2	Vehicular average speed	Too low	trimf
		Low	trimf
		Moderate	trimf
		High	trapmf
3	Road speed limit	Low	trimf
		Moderate	trimf
		High	trapmf

Table 2 Outputs and membership functions

S. No.	Output variable	Fuzzy sets	Membership function type
1	Congestion index	Low	trimf
		Moderate	trapmf
		High	trapmf
		Very high	trimf

- Decision Maker (e.g., Multiple-criteria decision analysis)
- Identifier (e.g., System identification)
- Scheduler (e.g., Machine scheduling)
- Controller (e.g., Adaptive and supervisory control of nonlinear systems).

3.3 Output Parameters

The linguistic variables of the output parameter with respect to the inputs are given in Table 2.

The graphs of the Input and Output parameters that are designed and executed are shown in results.

3.4 Rule Base of the Model

The inference system parallels the method of human reasoning in the fuzzy logic controller. This is where artificial intelligence is connected with fuzzy logic science. The rules are entirely based on human knowledge, and hence it leads to perfect reasoning.

Based on our inputs given by using the IF-THEN condition, the rules can be specified. The antecedent is considered the part of the law following IF, and the consequent is considered the part following THEN. The process of integrating three inputs and deriving a congestion calculation based on natural language is achieved using a fuzzy law that is manually crafted as follows.

1. IF (Vehicular Volume is too high) AND (Vehicular Average Speed is too low) AND (Road Speed Limit is low), THEN (Congestion Index is very high)
2. IF (Vehicular Volume is too high) AND (Vehicular Average Speed is low) AND (Road Speed Limit is low), THEN (Congestion Index is high)
3. IF (Vehicular Volume is high) AND (Vehicular Average Speed is moderate) AND (Road Speed Limit is moderate), THEN (Congestion Index is moderate).

If the degree of congestion as defined in Figs. 1, 2, 3 and 4 is too high, high, moderate, and low and too low, each represents the degree of congestion. In this method, 57 out of 66 rules are executed to get the desired output, as seen in Fig. 3.

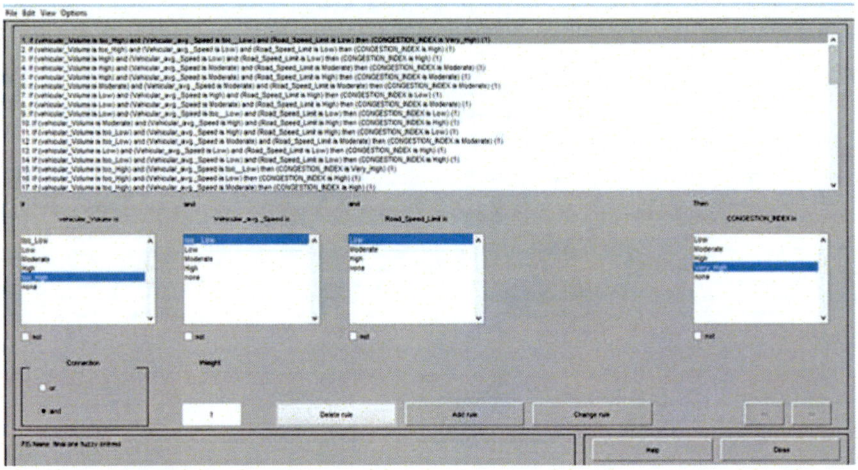

Fig. 3 Rule base in FIS

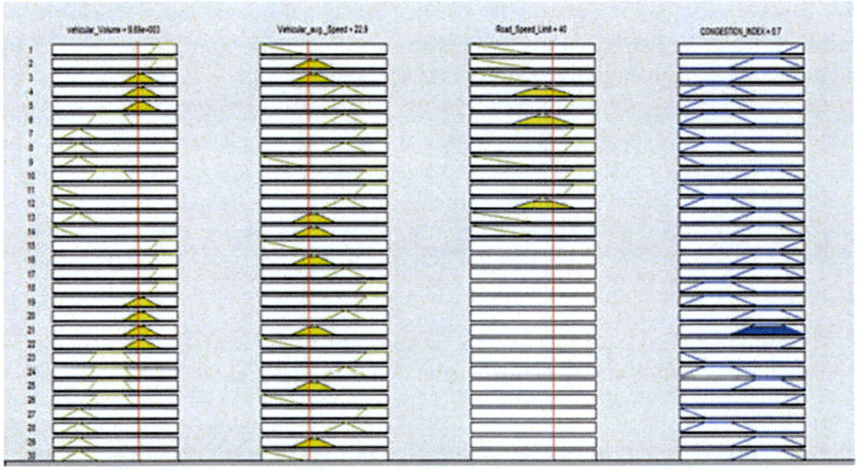

Fig. 4 Generated congestion index

3.5 Fuzzy Inference Framework Implementation to Produce Congestion Index

To compute and output the traffic congestion index, the manually tuned Mamdani-type fuzzy inference method (FIS) is used. In designing fuzzy models, Mamdani FIS is the most used method. The Mamdani architecture used in this paper is constructed with three inputs, one output parameter and fifty-seven fuzzy rules to predict road traffic congestion. Figure 3 depicts how the fuzzy rules are executed. Once all the

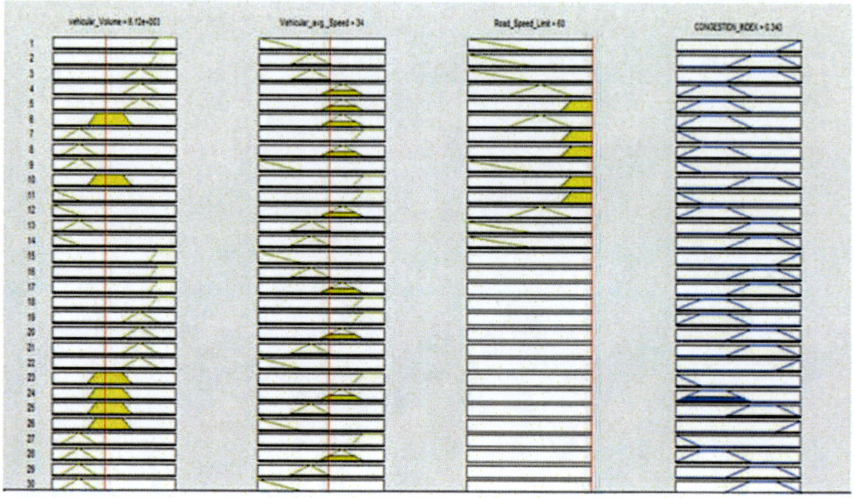

Fig. 5 Generated congestion index

rules are framed, then the fuzzy inference engine is made to run and execute the program that was given to the system in MATLAB.

On entering observational data to the fuzzy inference system of ten considered mid-block sections, the system generates the Congestion Indices of study area. The congestion indices are scaled between 0 and 1, where the indices closer to 0 and 1 indicate free-flow traffic state and a critical congested state, respectively.

Figures 4 and 5 were generated from the fuzzy inference system. In the below figures, the vertical numbering of each segment from each input shows the rule that is framed to run the output.

- Critical congested state generated at Gopalapatnam roadway link as 0.70 ~ 1
- Free traffic flow identified at Kancharapalem roadway link as 0.34 ~ 0.

As mentioned previously, the inference obtained is a number between 0 and 1. In terms of congestion index, this value provides details on the current condition for each and every point of vehicle volume, vehicular average speed and road speed limit. Output shows the corresponding congestion measure for each and every point according to the rule base framed. It is to be noticed that limited number of inputs should be taken in FIS system as the rule base becomes complex in deriving the accurate results for more input parameters.

4 Results

The input parameters developed are vehicle volume, vehicle average speed and road speed limit. The graphs are as shown in Figs. 6, 7, 8, and 9.

Table 3 shows how all the congestion indices of the study network are varying and below are the figures of the developed input and output parameter graphs in

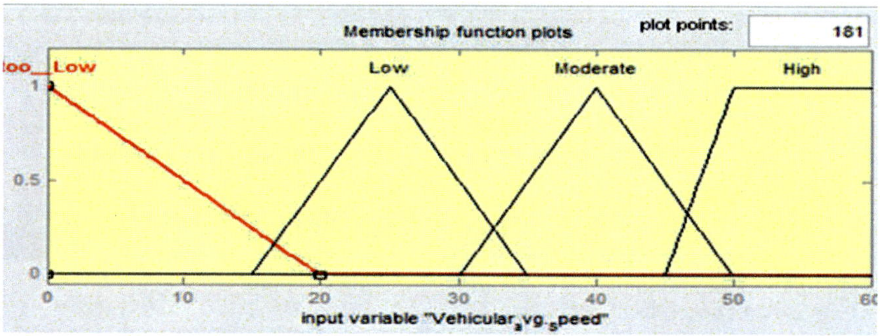

Fig. 6 Vehicular average speed

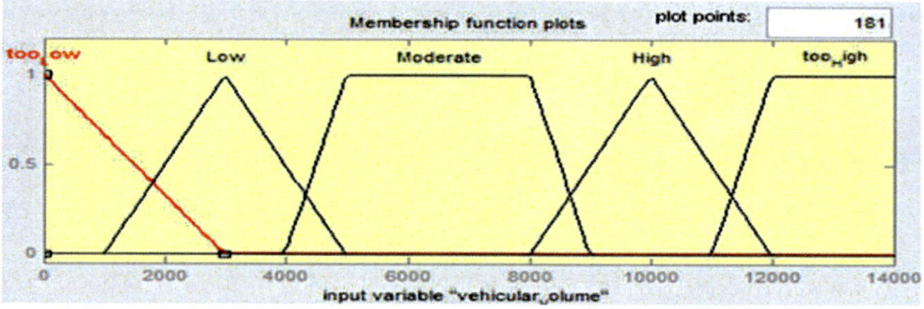

Fig. 7 Vehicular volume

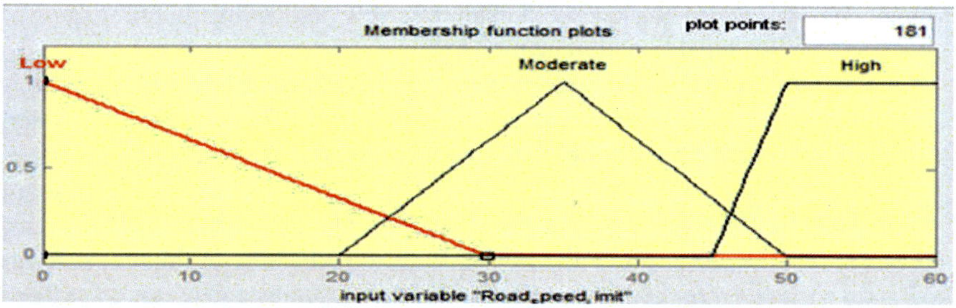

Fig. 8 Road speed limit

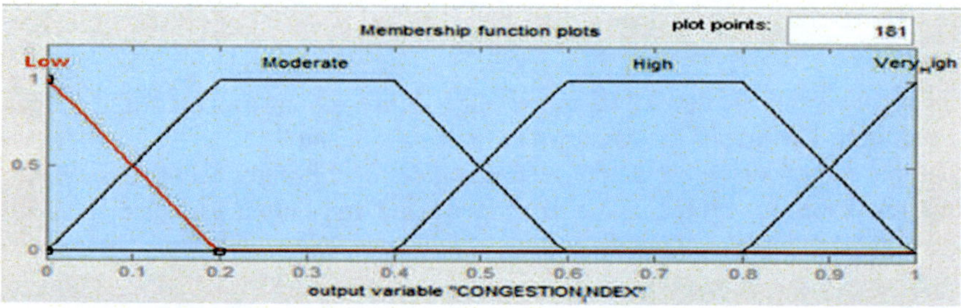

Fig. 9 Congestion index

Table 3 Generated congestion indices

S. No.	Roadway link	Vehicle volume (per h)	Vehicle average speed (kmph)	Road speed limit (km)	Obtained congestion index
1	Maddilapalem	12,887	32.01	60	0.62
2	Kancharapalem	*6115*	*34.05*	*60*	*0.34*
3	Akkayyapalem	4499	20.71	40	0.49
4	Daba Gardens	6582	23.82	40	0.52
5	Gajuwaka	5611	26.70	60	0.48
6	Gopalapatnam	*9688*	*22.94*	*40*	*0.70*
7	Old Gajuwaka	5557	26.22	40	0.51
8	Duvvada	5545	21.71	60	0.47
9	Marripalem	5420	33.12	40	0.39
10	Dwarakanagar	3868	29.32	40	0.55

The idea behind providing italics for two stretches (Kancharapalem & Goplapatnam) in Table 3 is to portray the generated low and higher congestions levels.

fuzzy inference system (FIS) in order to generate the congestion index (CI) by fuzzy inference engine (FIE) as in Fig. 9. Observation shows that study stretch running of vehicular volume of nearly 10000 vehicles per hour is said to be in a maximal congested state. It is also noticed that the same stretch where the maximum congestion is generated in bus rapid transit system (BRTS).

Here, it is observed that the travelers were not using the BRTS roadway facility in a proper manner, i.e., the traffic regulations were not at all, followed by the travelers and all type of vehicle travelers themselves tend to use this BRTS facility in a mixed way. No proper guidelines were even, followed by the traffic department to regulate the roadway facility meant only for the buses. Certain enforcement measures and initiatives are to be taken up by the traffic department leading to the proper usage of this BRTS facility so that the congestion rate occurred as above can be reduced and excellent results-like reduction in travel time, vehicle maintenance cost, lower consumption of fuels can be occurred.

Also, by proper usage of this BRTS facility, there is a great chance of the travelers to change their route choice behavior. This further enhances the public to use this path as the travel time and also the fare is low rather than using their own vehicles at risk which costs more in all aspects. Hence this eventually leads to the encouragement of public transit.

4.1 Vehicular Average Speed

See Fig. 6.

4.2 Vehicular Volume

See Fig. 7.

4.3 Road Speed Limit

See Fig. 8.

4.4 Congestion Index

See Fig. 9.

5 Conclusion

The suggested fuzzy congestion model is extended to the study areas real-world road network. The needed data is gathered mostly around the central business district (CBD) area in the city, which creates more travel times depending on the requirement. The obtained Congestion Index on the major roads of the city should be studied carefully and some mitigate measures are to be taken up in the congestion prone area. By looking at this scenario, we can say that the Kancharapalem roadway link possess free flow state, whereas Gopalapatnam roadway link generates higher congestion effect and some remedial measures have to be taken by anticipating future trends.

References

1. Vaziri M et al (2007) modeling highway congestion index for a developing country: the Iran experience. ScientiaIranica 14(1):1–10
2. Ministry of Public Security (2000) Urban traffic management evaluation indicators system (printed in 2012)
3. Goldstein et al (1991) U.S. Patent No. 5,029,164. Washington, DC: U.S. Patent and Trademark Office
4. Krause B et al (1996) Intelligent highway by fuzzy logic: congestion detection and traffic control on multi-lane roads with variable road signs. In: Proceedings of the Fifth IEEE international conference on fuzzy systems, vol. 3. IEEE, pp 1832–1837
5. He F et al (2016) A traffic congestion assessment method for urban road networks based on speed performance index. Procedia Eng 137:425–433
6. Narayanan R et al (2003) Quantification of congestion using fuzzy logic and network analysis using GIS. In; Map India Conference
7. Manual HC (2000) Transportation research board. National Research Council, Washington, DC, p 113
8. Das AK et al (2016) Traffic congestion modeling under mixed traffic conditions through fuzzy logic approach. Global Res Dev J
9. Kukadapwar et al (2015) Modeling of traffic congestion on urban road network using fuzzy inference system
10. Padhy NP et al (2015) Soft computing: with MATLAB programming. Oxford University Press
11. Hamad K, Kikuchi S (2002) Developing a measure of traffic congestion: fuzzy inference approach. Transp Res Record J Transp Res Board 1802:77–85
12. Levinson H, Lomax T (1996) Developing a travel time congestion index. Transp Res Record J Transp Res Board 1564:1–10
13. Lin FB (1989) Applications of 1985 highway capacity manual for estimating delays at signalized intersections. Transportation Research Record, (1225)
14. Rothenberg MJ (1985) Urban congestion in the United States-what does the future hold. ITE J 55(7):22–39; Clerk Maxwell J (1892) A treatise on electricity and magnetism, 3rd edn, vol 2. Clarendon, Oxford, pp 68–73

An Analytical Approach to Analysis of Concrete Overlay (White Topping) Over Flexible Pavement (Hot Mix Asphalt) Using ANSYS Software

M. S. Nagakumar, N. Ajay, and Sharu Elishuba John

Abstract Concrete overlays offer enormous potential as a rehabilitation strategy for Indian roads, whereas white topping offers a new advanced technique which will extend the life of a deteriorated existing bituminous pavement. In the present work, white topping is considered as the concrete overlay over 150 mm thick hot mix asphalt (HMA) pavement and has been modeled and analyzed by using ANSYS software. The response of the pavement was computed in terms of surface deflections and stresses at interior, corner, and edge, along with the effect of temperature, dowel bars, and slab thickness which were considered as design parameters. Further, parametric studies were carried out on varying thickness of slabs (75, 100, 260, 280, and 300 mm) to identify the significant design parameters of the pavement model. Validation of the finite element (FE) analysis results was carried out using Bradbury's equation for temperature stress and Ioannides modified solution of Westergaard's closed-form formulas for response to wheel loading. The results show that bonding condition, temperature, and thickness of the overlay are critical for the design of rigid overlays.

Keywords ANSYS · Finite element model · Overlay · White topping

1 Introduction

In India, most of the roads are flexible pavements. Structural failures are common in the flexible pavements due to accumulated traffic loads, adverse climatic situations, inadequate ride quality, inaccurate service life prediction, and deterioration due to over loading. These problems may be overcome by adequate maintenance. But the proper pavement maintenance is expensive, and pavement may have inadequate structural ability for the resisting the traffic loads [1]. Several methods are available to maintain the flexible pavement [1], and among them, asphalt overlay on

M. S. Nagakumar · S. E. John
Department of Civil Engineering, RVCE, Bangalore, India
e-mail: nagakumar@rvce.edu.in

N. Ajay (✉)
RASTA-Center for Road Technology, Bangalore, India

L. Nandagiri et al. (eds.), *Sustainability Trends and Challenges in Civil Engineering*, Lecture Notes in Civil Engineering 162, https://doi.org/10.1007/978-981-16-2826-9_24

existing surface is most common method adopted in India [1]. Asphalt overlay represents common practice used for flexible pavement rehabilitation and maintenance of pavement. It will increase the structural support and the serviceability of the existing pavement [2]. The overlay design is mainly based on the thicknesses and stiffness of the existing asphalt, base, and sub-base layers [2]. The PCC overlay on an existing bituminous pavement is termed as white topping [2].

The white topping is comprised of cement concrete slabs laid above the bituminous concrete or hot mix asphalt (HMA) [3]. To make the interface between the two layers smooth, a separation film is set between the current bituminous concrete or HMA and cement concrete. However, this film is missing in case of ultra-thin and thin white topping [3]. To facilitate thermal expansion and contraction of the slab and to continue construction after a break, different types of joints are provided in white topping [4]. Dowel bars are given over the transverse joint to transfer load from one panel to another in conventional white topping.

A few performance studies were carried on the thickness of white topping, bituminous layer, modulus of bituminous layer, temperature differential, etc. [5, 6]. It is a challenge for any highway engineer to design the PCC overlay system because of factors such as restricted boundary conditions, non-availability of accurate field data, less accurate conversion formula for temperature variation, and time-consuming calculations [5, 6]. In addition to this, collection of necessary data, investment of money, time, requirements of reliable and comprehensive data, and the conventional methods like manual calculation which may lead to error [1]. The "finite element method (FEM)" approach can be used to overcome the above drawbacks and help in better prediction of PCC overlay behavior [1].

2 Literature Review

The FE method is used to solve the problems with complex geometries, loading, and material properties numerically. Huang and Wang [7] analyzed two-dimensional concrete slab using FEM based on Kirchhoff's thin plate theory [8–12]. Since 1970s, various software's were developed to simulate the concrete pavements by many researchers [13, 14]. To overcome the limitations in the two-dimensional FE model, the researchers developed three-dimensional FE models to capture the behavior of concrete pavement with different pavement configuration [13, 14]. Cheung and Zienkiewicz [15] were analyzed the three-dimensional rigid pavement using the FEM. They concluded that, three-dimensional model can provide better realistically analysis of pavement when compared to two-dimensional model.

In the present work, white topping is considered as the concrete overlay over existing 150 mm thick HMA pavement and has been modeled using FEM (Three-dimensional model). This was analyzed by using ANSYS software [16].

The response of the pavement was computed in terms of surface deflections and stresses at interior, corner, and edge, along with the effect of temperature, dowel bars, and slab thickness which were considered as design parameters. Further, parametric studies were also carried out on varying thickness of slabs (75, 100, 260, 280, and 300 mm) to identify the significant design parameters of the pavement model. Validation of the FE analysis results was carried out using Bradbury's equation for temperature stress and Ioannides modified solution of Westergaard's closed-form formulas for response to wheel loading.

3 Finite Element Analysis

The modeling and analysis of white topping overlay on HMA have been performed with finite element analysis using ANSYS [12]. The element description used in ANSYS is given in Table 1.

In conventional white topping, dowel bars are used to connect the slab panel to the adjacent slab panel transferring the loads to that slab. The load is transferred through the dowel bars to the adjacent slab. In modeling, dowel bars were modeled as 3D beam elements (BEAM-4). At each transverse joint, dowel bars of 32 mm diameter and 500 mm long spaced at 200 mm center to center at mid depth of slab are placed. The models were analyzed both in bonded and un-bonded conditions (The bottom of sub-grade is assumed as fixed condition in all directions). The pavement model is a five-layer system (Fig. 1) consisting of concrete layer, existing asphalt layer, sub-base layer, and sub-grade layer.

Figure 2 shows schematic arrangement of three adjacent white topping overlay slabs. The stresses and deflections in the white topping overlays were analyzed for various temperature differential starting from 5 to 20°C to find out the effect of temperature on the performance of the pavement. Comparison of bonded and un-bonded conditions of the pavement layers was analyzed to determine its importance.

Figure 3 shows FE mesh of pavement system and dowel bars at transverse joint. The FE model is validated by computing stresses and deflections at edge, corner and interior loading positions and comparing the same using closed-form formulas.

Table 1 Element description used in ANSYS

Members	Elements
Concrete slab	SOLID65
Existing asphalt, granular, and sub-grade layer	SOLID45
Contact element between concrete slab and existing asphalt layer	TARGE 170 and CONTAC174
For determining the temperature stress	SOLID5

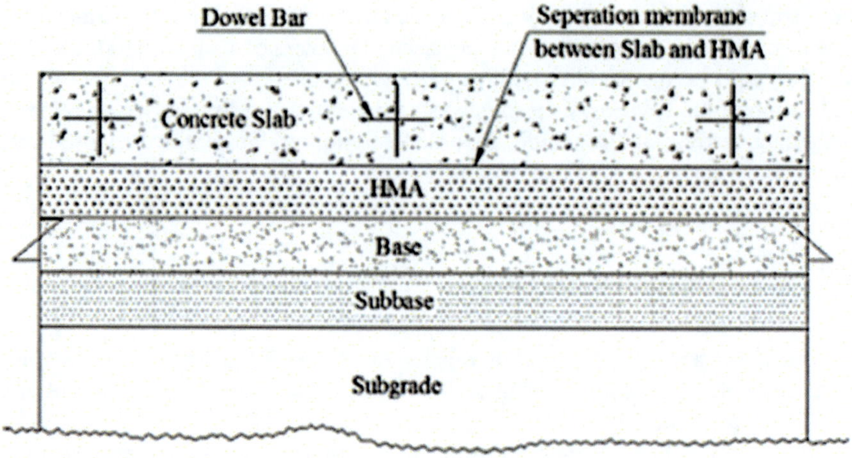

Fig. 1 Schematic arrangement of conventional white topping

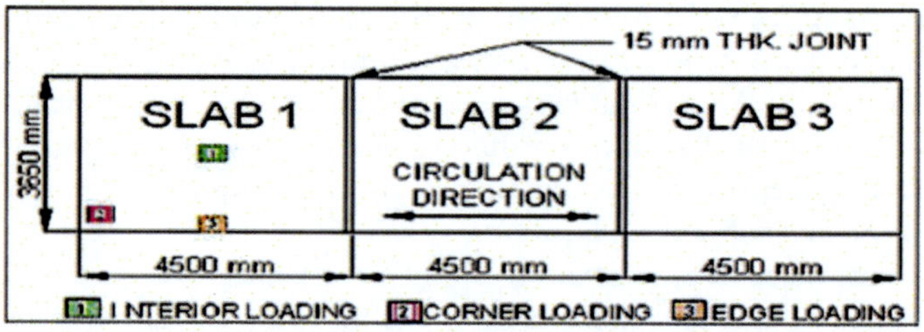

Fig. 2 Schematic arrangement of three adjacent white topping overlay slab

Fig. 3 White topping overlay slab and dowel bar FE model

4 Parametric Study

In the present study, 4500 mm × 3650 mm concrete slab with different thicknesses of slab from 75 mm, 100 mm, 150 mm, 200 mm, and 250 mm over existing 150 mm thick asphalt-treated base (HMA) were considered. The concrete properties and HMA properties are given in Table 2. The parameter's considered were single-axle load (dual tire) $P = 100$ kN, 320 mm thickness of overlay slab, the tire contact area is 139 × 196 mm, different E-value (275,1000,2100), tire pressure (0.70, 0.80, 1.0 MPa), and temperature gradient (lowest considered 20 °C and highest 42 °C).

Westergaard's modified equations (Ioannides et al. and Bradbury equation) were used to verify the finite element model [11].

For Stresses:

Interior:

$$\sigma = \left[3P\frac{1+\mu}{2\pi h^2}\right]\left[\ln\left(\frac{2l}{a}\right) + 0.5 - \gamma\right] + \left[3P\frac{1+\mu}{64h^2}\right]\left(\frac{a}{l}\right)$$

Corner:

Table 2 Properties of concrete and HMA considered for analysis

Concrete	
Grade of concrete (fck)	40 MPa
Modulus of elasticity (Ec)	21,000 MPa
Poisson's ratio (μ)	0.15
Density of concrete	24 kN/m^3
Coefficient of thermal expansion (α)	1.0 × 10^{-5} per °C
HMA	
Modulus of elasticity (base material)	800 MPa
Poisson's ratio (μ) (base material)	0.30
Modulus of elasticity (sub-grade material)	200 MPa
Poisson's ratio (μ) (sub-grade material)	0.35
Modulus of elasticity (steel dowels)	200,000 MPa
Poisson's ratio (μ) (steel dowel)	0.30
Modulus of dowel support (K)	450 MPa/mm
Modulus of sub-grade reaction of HMA, k	0.12 MPa/mm
Modulus of elasticity (AC layer)	275, 1000, 2100 MPa

$$\sigma = \frac{3P}{h^2}\left[1 - \left(\frac{c}{l}\right)^{0.72}\right]$$

Edge:

$$\sigma = \frac{0.803P}{h^2}\left[4\log\left(\frac{a}{l}\right) - 0.034\right]$$

For Deflections:

Interior:

$$\Delta = \frac{P}{8kl^2}\left\{1 + \frac{1}{2\pi}\left[\ln\left(\frac{a}{2l}\right) - 0.673\right]\left(\frac{a}{l}\right)^2\right\}$$

Corner:

$$\Delta = \frac{P}{kl^2}\left[1.205 - 0.69\left(\frac{c}{l}\right)\right]$$

Edge:

$$\Delta = \frac{0.431P}{kl^2}\left[1 - 0.82\left(\frac{a}{l}\right)\right]$$

5 Results and Discussions

5.1 Analysis of Stress and Deflection for Corner, Edge, and Interior Loading

For the analysis of stresses and deflections for corner, edge, and interior loading, the model is developed by considering a thickness of concrete slab as 250 mm and the applied tire pressure is 0.70 MPa. The E-value of slab is considered as 21000 MPa, the load applied is 50 kN, and the temperature considered at top of slab = 20 °C and at bottom of slab = 28 °C, respectively. Full bonding is provided between the overlay and the existing asphalt layer. Figures 4, 5, and 6 show the FE model (250 mm thick) slab along with corner, edge, and interior loading and respective deflection and stress values. A reasonably good agreement has been indicated between the results obtained from proposed model and values obtained from formulas. Similarly, for other thickness of slabs, analyses were carried out for both bonded and un-bonded

Fig. 4 Representation of FE model (250 mm thick slab) along with deflection and stress values under corner loading conditions

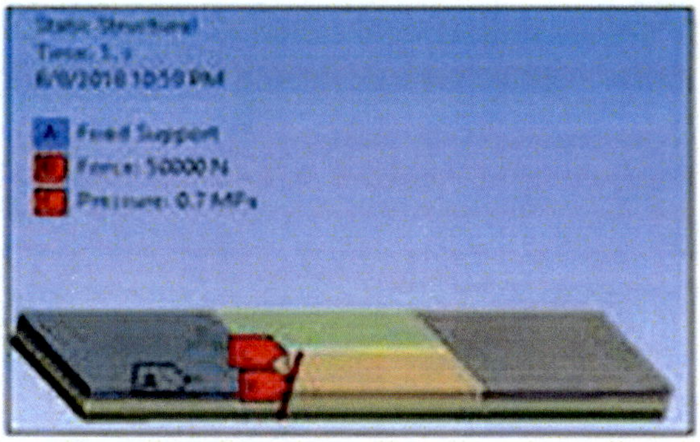

FE Models

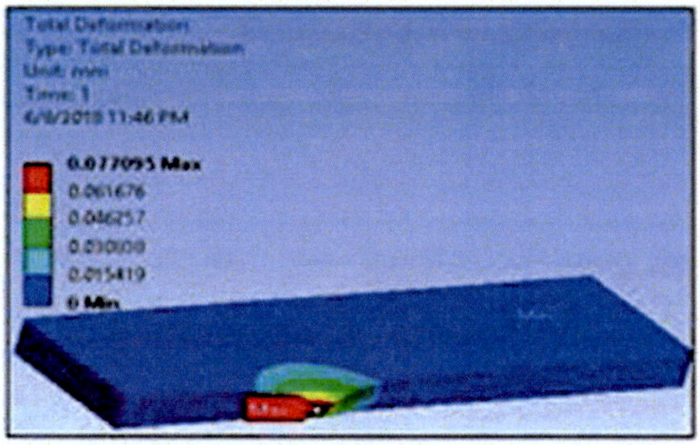

Deflections

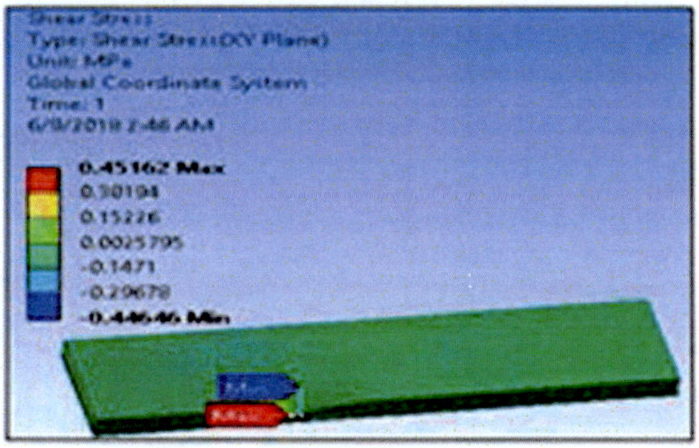

Stresses

conditions between slab and the HMA layer by varying temperature, E-value and tire pressure.

Fig. 5 Representation of FE model (250 mm thick slab) along with deflection and stress values under edge loading conditions

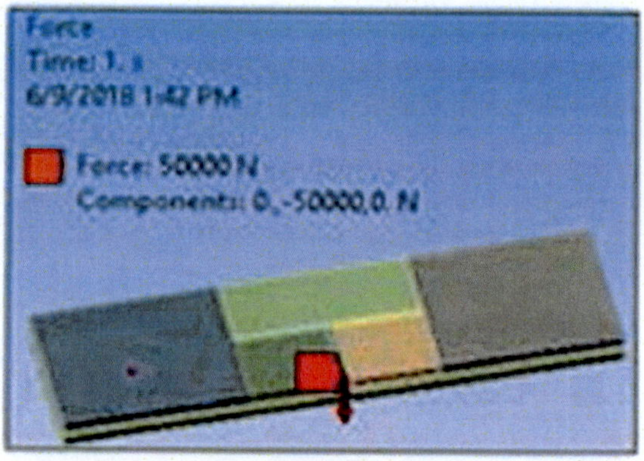

FE Models

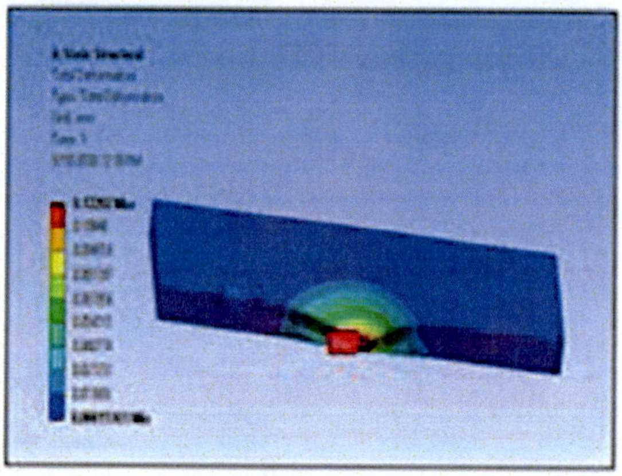

Deflections

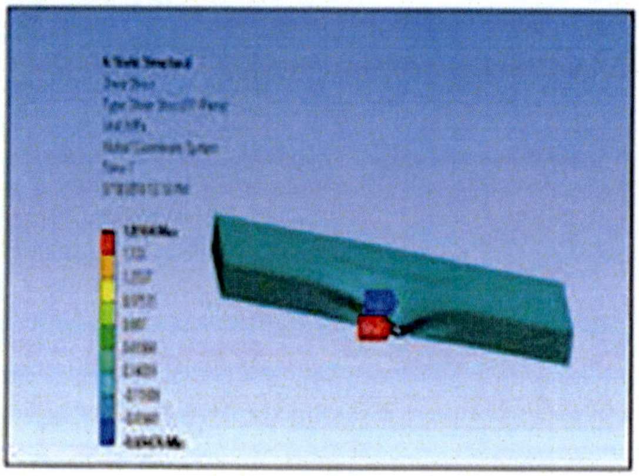

Stresses

Fig. 6 Representation of FE model (250 mm thick slab) along with deflection and stress values under interior loading conditions

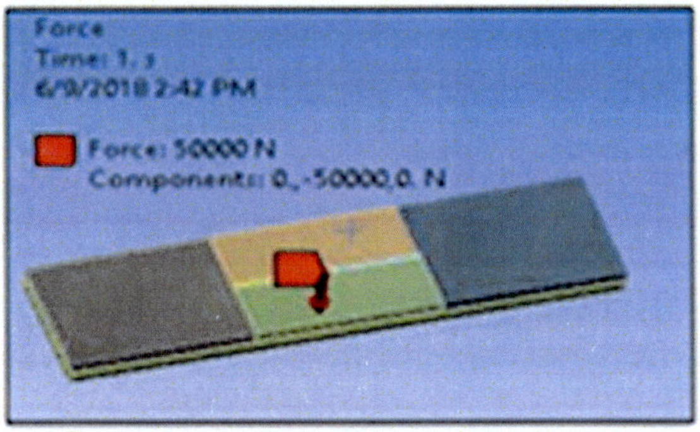

FE Models

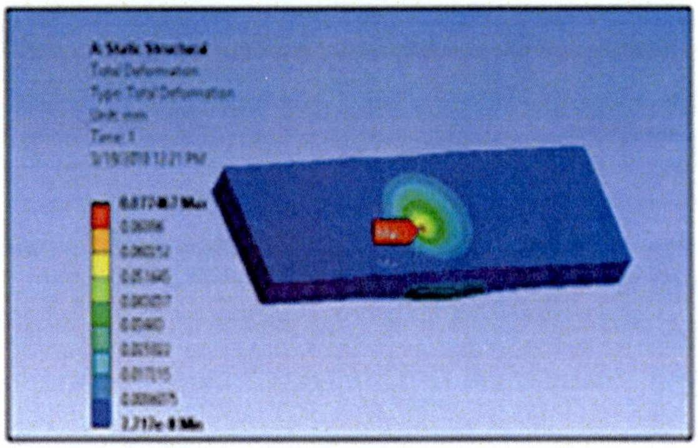

Deflections

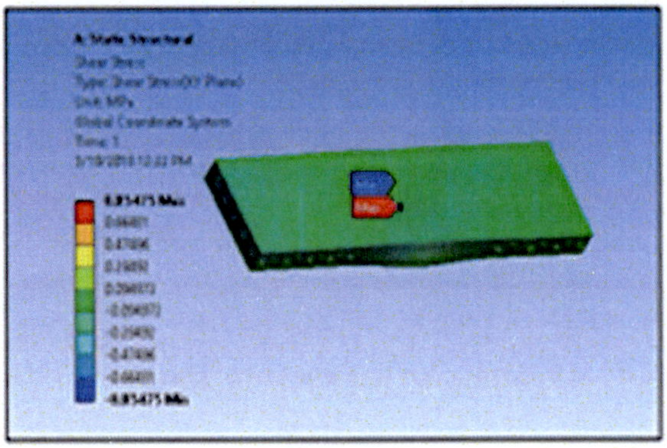

Stresses

5.2 *Vaildation for FE Model*

As seen from Table 3, the edge loading is critical location on the pavement. Whereas the stress and deflection values are higher for edge loading condition when compare to corner and interior loading conditions.

Table 3 Comparisons of results for stresses and deflections using FE model and Ioannides Equation

Condition	Stresses (MPa)		Deflections (mm)	
	FE model	Equations	FE model	Equations
Corner	0.508	0.63	0.070	0.036
Interior	0.750	0.53	0.075	0.055
Edge	1.800	0.75	0.130	0.150

Table 4 Temperature versus deflection for various thicknesses for both bonded and un-bonded

Temperature differential (°C)	Slab thickness (mm)									
	Deflection (mm)—Bonded					Deflection (mm)—Un-bonded				
	75	100	150	200	250	75	100	150	200	250
5	0.54	0.40	0.34	0.2	0.07	1.07	0.77	0.55	0.38	0.29
8	0.73	0.52	0.19	0.19	0.12	1.66	0.84	0.53	0.51	0.31
12	0.91	0.62	0.54	0.57	0.52	1.97	1.03	0.86	0.85	0.79
15	1.17	0.82	0.68	0.60	0.59	2.21	1.22	1.03	0.92	0.81
18	1.47	0.93	0.88	0.79	0.70	2.30	1.38	1.27	1.18	1.01
20	2.35	1.58	1.37	1.01	0.91	3.65	2.51	2.02	1.23	0.94

5.3 Variation in Deflection with Respect to Temperature

In this analysis, the deflection was computed for various values of the temperature and various overlay thickness of 75 mm, 100 mm, 150 mm, 200 mm, and 250 mm. Table 4 shows the deflection in bonded and un-bonded condition when load is applied at the edge of the slab and the result of the same is plotted in Fig. 7.

Figure 7 shows the variation in deflection with respect to the temperature differential. The white topping thicknesses increases as the deflection value decrease, while comparing the bonded white topping that is 18% increase in deflection values and for un-bonded white topping 30% increase in deflection values.

5.4 Variation in Stress with Respect to Temperature

In this analysis, the stresses were computed for various values of the temperature and various overlay thickness of 75 mm, 100 mm, 150 mm, 200 mm, and 250 mm. Table 5 shows the stress in bonded and un-bonded condition when load is applied at the edge of the slab and the result of the same is plotted in Fig. 8.

As seen from Fig. 8, the shear stress increases with decrease in thickness of the overlay. As temperature on top slab increases, the shear stress also increases. White topping for different thicknesses stresses varied from 30 to 40%. The highest value of shear stress found is 3.95 MPa for a temperature differential of 20 °C which is

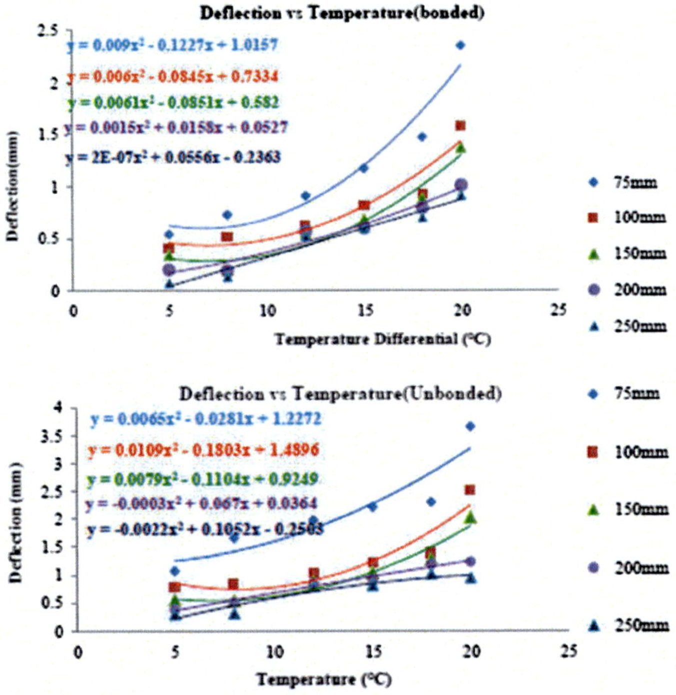

Fig. 7 Relation between temperature versus deflection for various thicknesses for both bonded and un-bonded conditions

Table 5 Temperature versus stresses for various thicknesses for both bonded and un-bonded

Temperature differential (°C)	Slab thickness (mm)									
	Deflection (mm)—Bonded					Deflection (mm)—Un-bonded				
	75	100	150	200	250	75	100	150	200	250
5	1.23	1.20	0.88	0.70	0.62	1.89	1.68	1.47	1.26	1.05
8	1.33	1.14	0.97	0.76	0.51	2.08	1.84	1.6	1.36	1.12
12	1.55	1.38	1.14	0.89	0.66	2.44	2.17	1.9	1.63	1.36
15	1.93	1.79	1.62	1.43	1.33	2.98	2.63	2.28	1.93	1.64
18	2.27	2.17	1.93	1.77	1.69	3.37	3.07	2.77	2.47	2.17
20	2.89	2.55	2.24	1.97	1.83	3.95	3.59	3.23	2.87	2.51

due to the lack of bonding between the overlay and the AC layer. With bonding, the shear stress reduced by 36%.

Fig. 8 Relation between temperatures versus stresses for various thicknesses for both bonded and un-bonded conditions

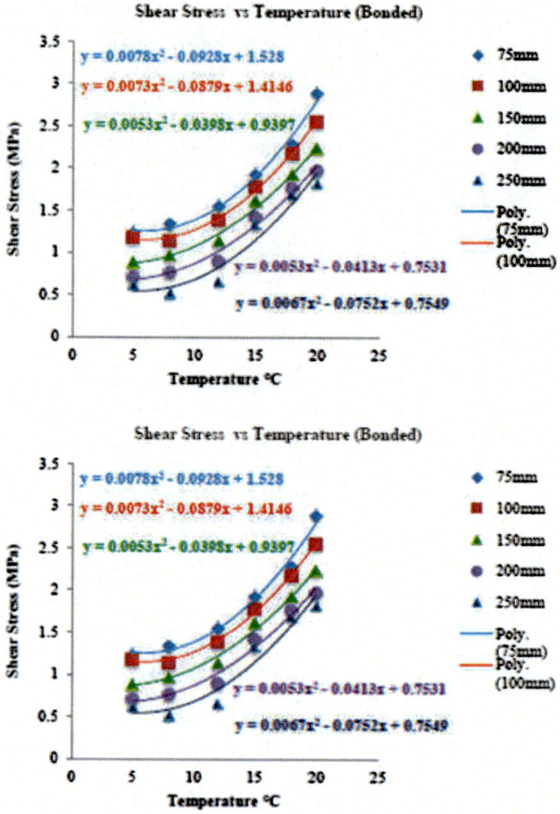

5.5 *Variation in Deflection with Respect to E-Value of AC Layer*

The deflection was computed for variations in the E-value (275, 1000, 2100) of asphalt layer and various overlay thickness of 75 mm, 100 mm, 150 mm, 200 mm, and 250 mm for both bonded and un-bonded conditions.

As seen from Fig. 9, as the AC modulus increases the deflection decreases. The maximum variation in deflection was found to be in the range of 10–12% which implies that the effect of modulus of elasticity of HMA is not significant on the white topping.

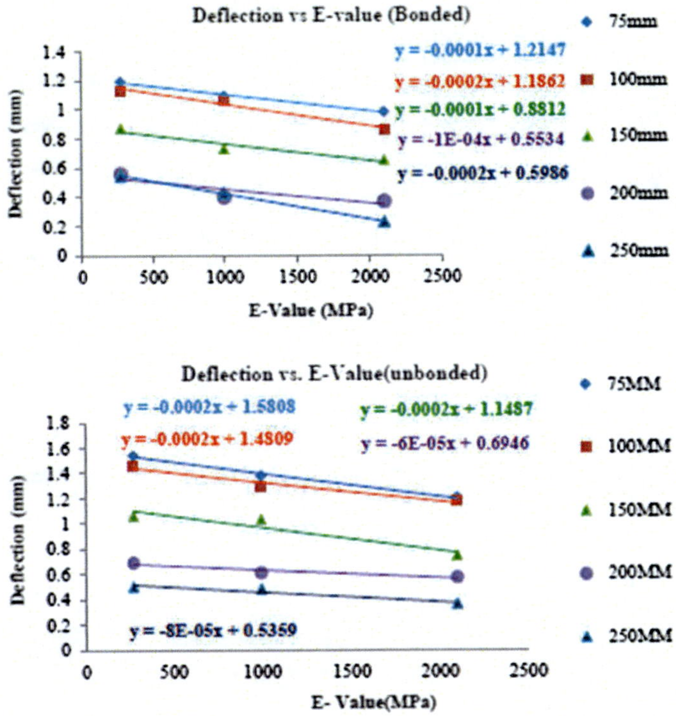

Fig. 9 Variations in deflection with respect to E-value for various overlay thickness for both bonded and un-bonded conditions

5.6 Variation in Stress with Respect to E-Value of AC Layer

The stresses on slab were computed for varying E-value of asphalt layer and various overlay thickness of 75 mm, 100 mm, 150 mm, 200 mm, and 250 mm for both bonded and un-bonded conditions.

As seen from Fig. 10, the stress reduces for increasing in the modulus of elasticity of the AC layer. This is due to increasing the stiffness. There is only 6–8% difference in stresses with the variation of E-value of the asphalt layer. The average value of stress with increase in stiffness on white topping is 0.55 for bonded condition and 0.60 for un-bonded condition, respectively.

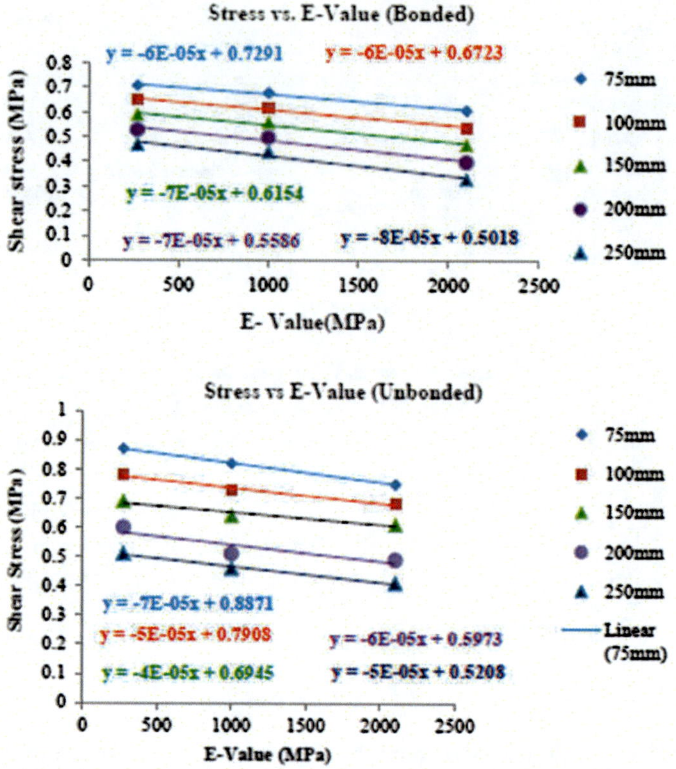

Fig.10 Variations of stresses with respect to E-value on various overlay thickness for both bonded and un-bonded conditions

5.7 Variation in Deflection with Respect to Tire Pressure on Slab

Deflections on slab were computed for varying tire pressure and various overlay thickness of 75 mm, 100 mm, 150 mm, 200 mm, and 250 mm for both bonded and un-bonded conditions. As seen from Fig. 11 as the tire pressure increases from 0.7 to 1.0, the deflection values increase. It is more affected on thin white topping owing to its less thickness. The results also show that there is some significant increase in the deflection for un-bonded pavement rather than bonded pavement, which implies the necessity of bonding between overlay and existing AC layer.

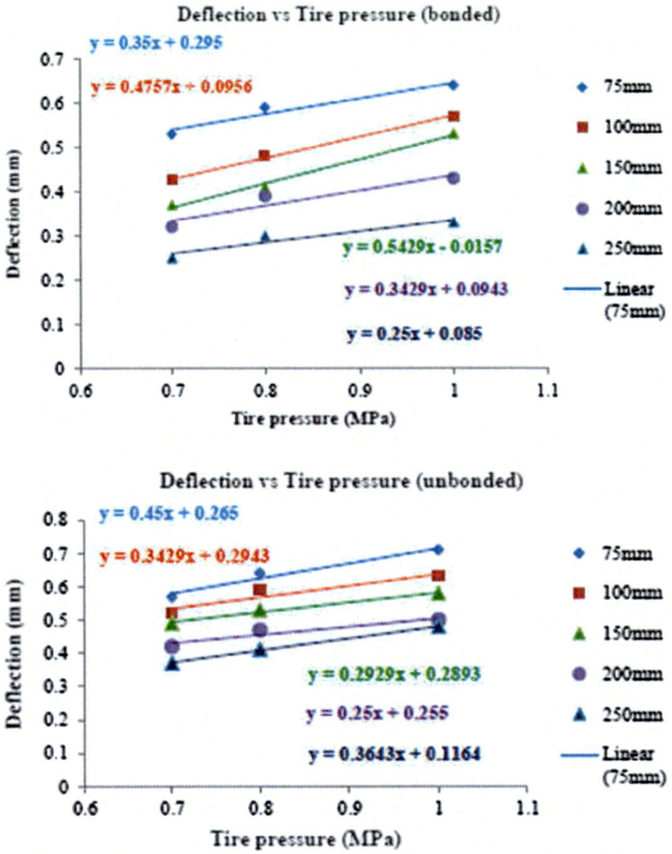

Fig. 11 Variations of deflection with respect to tire pressure on various overlay thickness for both bonded and un-bonded conditions

5.8 Variation in Stress with Respect to Tire Pressure on Slab

Stresses on slab were computed for varying the tire pressure and various overlay thickness of 75 mm, 100 mm, 150 mm, 200 mm, and 250 mm for both bonded and un-bonded conditions.

As seen from Fig. 12, the stresses were found to increase as tire pressure increases for un-bonded pavement under the same overlay thickness. As overlay thicknesses increases, the stresses were found to decrease for the same tire pressure.

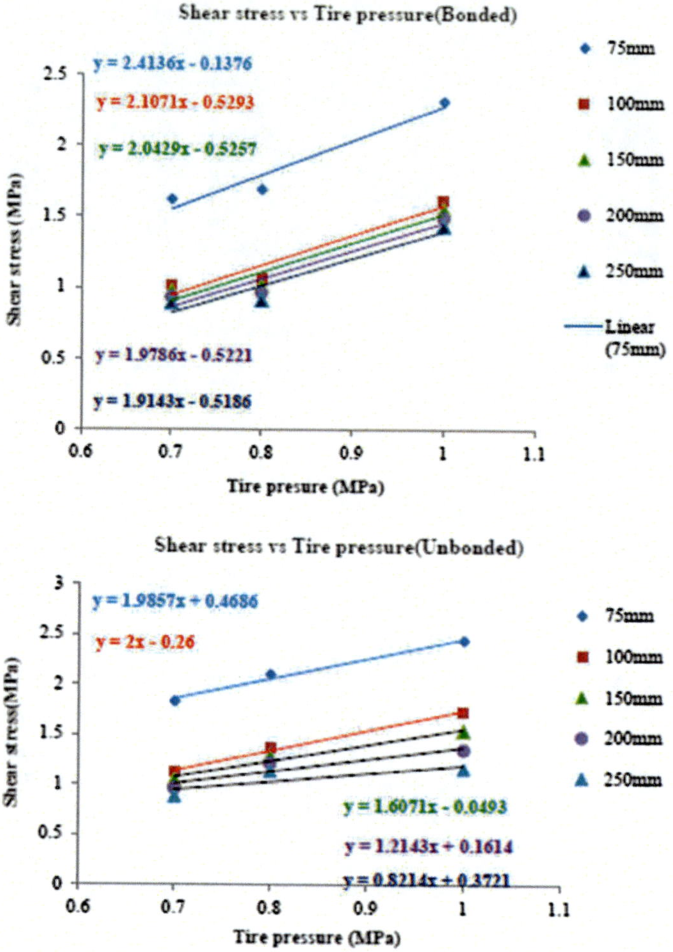

Fig. 12 Variations of stress with respect to tire pressure on various overlay thickness for both bonded and un-bonded conditions

6 Conclusion

A three-dimensional FE model of conventional white topping for bonded and un-bonded conditions have been developed using ANSYS by considering static wheel loading, temperature, tire pressure, and modulus of elasticity for various thicknesses. These models can be used for computing stresses and deflections in conventional white topping. The model has been validated by computing stresses and deflections at edge, corner, and interior loading positions. Also, these models have been validated using closed-form formulas like Ioannides et al. and Bradbury. The results show that bonding condition, temperature, and thickness of the overlay are critical for the design of rigid overlays. Also, sufficient stiffness of the existing asphalt layer should be ensured before laying the overlay. There is good agreement between values obtained from three-dimensional FE model and as well as closed-form formulas results.

Acknowledgements The authors sincerely acknowledge the Department of RV College of Engineering and under-graduate students for their support.

References

1. Singh SG (2014) Overlay of flexible pavements: an ANN approach. Project Report, National Institute of Technology Rourkela
2. Jinwoo A, BooHyun N, Kim J (2014) Numerical analysis of reflective cracking in an asphalt concrete overlay over a flexible pavement. In: Design, analysis, and asphalt material characterization for road and airfield pavements GSP, 246. ASCE, pp 68–73
3. IRC:SP-76-2008, Tentative guidelines for conventional, thin and ultra-thin white-topping. Indian Roads Congress, New Delhi, India
4. Harrington D, Fick G (2014) Guide to concrete overlays, 3rd edn. National Concrete Pavement Technology Center Iowa State University, American Concrete Pavement Association May
5. Rasmussen RO, Rozycki DK (2004) Thin and ultra-thin white topping a synthesis of highway practice. NCHRP Synthesis 338, Transportation Research Board, pp 3–12
6. Mei X, Lin S, Li X, Tian X (2015) Analysis on the rigid-flexible composite base of asphalt pavement of an expressway in Guangdong. In: New Frontiers in Road and Airport Engineering, ASCE, pp 187–192
7. Huang YH, Wang ST (1973) Finite element analysis of concrete slabs and its implications on rigid pavement design. Highway Res Rec 466:55–79
8. Venkatasubramanian V (1964) Investigation on temperature and friction stresses in bonded cement concrete pavements. Ph.D. Thesis, Indian Institute of Technology, Kharagpur
9. William GW, Shoukry SN (2001) Three-dimensional finite element analysis of temperature-induced stresses in dowel jointed concrete pavements. Int J Geo-mech 1(3):291–307
10. Davids WG, Wang Z, Turkiyyah G, Mahoney JP, Bush D (2003) Three-dimensional finite element analysis of jointed plain concrete pavement with EverFE. Transportation Research Record, National Research Council, Washington D.C., 1853, pp 92–117
11. Samir Shoukry N, Gergis William W, Mourad Riad Y (2001) Finite element modeling of rigid pavements. In: Proceedings of the 2nd international symposium on maintenance and rehabilitation of pavements and technological control, Auburn, Alabama, USA, July 29–August 1
12. Zokaei M, Fakhri M, Rahiminezhad S (2017) A parametric study of jointed plain concrete pavement using finite element modeling. Mod Appl Sci 11(11):75–84
13. Ali B, Sadek M, Shahrour I (1998) Finite-element analyses of flexible pavements. J Transp Eng, ASCE 124(5):491–499
14. Nishizawa T, Fukute T, Kokubun S (1999) Study of a method for analyzing the mechanical behavior of composite pavement. Transp Res Rec 1684:101–108
15. Cheung Y, Zinkiewicz O (1965) Plates and tanks on elastic foundations—an application of finite element method. Int J Solids Struct 1(4):451–461
16. ANSYS (2019) User's Manual. ANSYS, Inc., Canonsburg, PA, USA

Investigations on Slag-Fly Ash-Glass Powder Based Ecofriendly Interlocking Paver Blocks

Shriram Marathe, I. R. Mithanthaya, and S. K. Susmitha

Abstract This interrogation presents an laboratory exploration on sustainable interlocking concrete paving blocks (ICPB) produced using alkali activated cementitious materials. The study involves comparison of mechanical properties such as compressive, split-tensile, flexural strengths, and scratch resistance of ICPBs produced using one type of "alkali activated concrete" (AAC) with that of the conventional OPC based paver blocks obtained from field. The AAC is systematically developed using GGBS, fly-ash and powdered waste glass as binding ingredients. The investigations were extended to learn the compression and tensile behavior of geo-grid reinforced alkali activated ICPBs introduced in two layers. The cost analysis was also carried-out to compare the AAC with OPC based ICPBs. The experimental results reveal that the use of alkali activated ICPBs are proved be the sustainable replacement to OPC based conventional ICPBs. Thus, alkali activated concrete ICPBs must be commercialized and produced in a large scale for the sustainable construction of interlocking pavement system.

Keywords Alkali Activation · Fly-ash · Glass powder · Slag · Interlocking Concrete Paver Block (ICPB)

1 Introduction

In India, the industrial production of OPC is more than 300 Mt [1], whereas the production of the industrial wastes that can be used as a substitute to OPC is more than the double this amount. The explore for new environmental friendly building material which will match the durability of primordial OPC based concretes that can be effectively produced using industrial waste materials has motivated the study on non-conventional cementitious systems over the preceding three decades. Hence, there is a demand for an innovative and sustainable concrete technology which

S. Marathe · I. R. Mithanthaya · S. K. Susmitha (✉)
Dept. of Civil Engineering, NMAM Institute of Technology, Nitte, Karkala, India

I. R. Mithanthaya
e-mail: irmithanthaya@nitte.edu.in

L. Nandagiri et al. (eds.), *Sustainability Trends and Challenges in Civil Engineering*, Lecture Notes in Civil Engineering 162, https://doi.org/10.1007/978-981-16-2826-9_25

utilizes less energy and generates fewer CO_2 without compromising the strength and durability performances. Keeping these facts in mind, researchers are trying to develop alternative materials which utilizes industrial by-product materials (wastes), emit little CO_2 and consume less energy, while compared to OPC. Thus, the call for the day is, hence 'Sustainable Development' which demands the fresh concrete technology that utilize fewer natural resource, minimum energy as well as generates lesser amount of CO_2, with no compromise on the strength and durability properties [2–4].

Alkali activated cements are considered as new generation binders, which have a potential to give a sustainable replacement to OPC. There is an urge to limit the usage of the OPC in infrastructural development due to its adverse environmental issues and it's structural limitations [5]. Because of low-internal energy and low-carbon dioxide emission associated with production of Alkali-activated binder (AAB), this new material can be taken as more eco-friendly than the conventional portland cement concretes [4, 6].

These AABs are broadly talked about and encouraged as constituent of the recent and upcoming tool box of 'sustainable cementing binder systems'. Alkali-activation is a common term that is concerned to the reaction of a solid alumino-silicate (called the 'precursor') under alkaline conditions (induced by the 'alkali activator'), to produce a hardened binder which is based on a combination of hydrous alkali-alumino silicate and/or alkali-alkali earth-alumino silicate phases [5, 7, 8].

Interlocking concrete paving blocks (ICPBs) have been comprehensively utilized as a expert pinpointing method for granting pathway in regions where conservative method of construction are fewer long-lasting owing to a lot of functioning and ecological restraints. The strengths, stability, and visually pleasing exterior of ICPB have made them suitable for various range of applications. 10 years ago ICPB technology was introduced in India. They were used in footpath, parking spot, and garden, etc., whereas nowadays they are used for various purpose. The major components of ICPB are shown in Fig. 1. They comprise of base materials, pavers, bedding sand, jointing sand, sealer, and geotechnical fabric and edge restraints. For better architectural view, ICPBs are accessible different sizes, shapes, and colors.

In this current inspection, the attempt was to develop eco-friendly ICPBs of desirable quality by using developed alkali activated binders instead of using OPC binder.

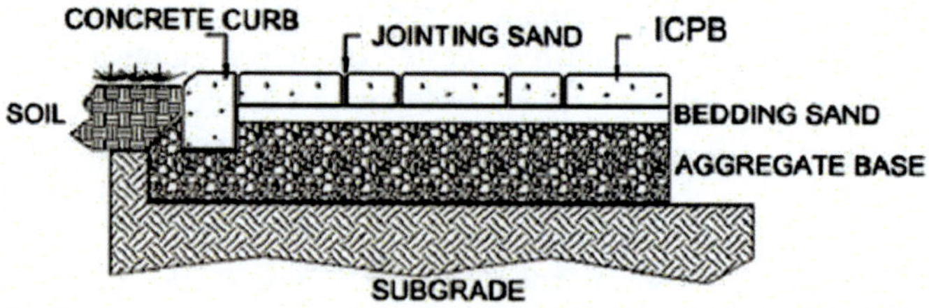

Fig. 1 Typical cross section of ICPB pavement section [9]

IRC guidelines [10] suggest that for paving applications, the minimum compressive strength of individual block must not be less than 30 MPa. IS:15658 [11] states that the grade of the concrete and the minimum thickness of the ICPB is decided for the manufacture of paver blocks based on the expected traffic conditions. For non-traffic category road applications the minimum 28 day compressive strength must be 30 MPa with a minimum thickness of 50 mm; for the light traffic category applications, i.e., average daily traffic (ADT) upto 150 HCVs (Commercial vehicles with laden weight > 30 kN) minimum strength must be 35 MPa with a minimum thickness of 60 mm; for the medium traffic category applications, i.e., ADT from 150 to 450 HCVs minimum strength must be 40 MPa with a minimum thickness of 80 mm; for the heavy traffic category applications i.e., ADT from 450 to 1500 HCVs minimum strength must be 50 MPa with a minimum thickness of 100 mm; and, for the very heavy traffic category applications, i.e., ADT > 1500 HCVs minimum strength must be 55 MPa with a minimum thickness of 120 mm.

2 Materials and Methodology

In this study, the mechanical performance of field ICPBs was compared with that of the developed alkali activated concrete ICPBs. For this purpose, the conventional OPC binder based ICPBs were obtained directly from the local paver block manufacturing company located in Karkala Taluk. The ICPB mix proportion (i.e., field mix) by the manufacturing company was developed for design strength of 40 MPa. Further, two ICPB mix-proportions were developed for AAC mixes. For AAC mixes, three binding ingredients were used namely, ground granulated blast furnace slag (GGBS) of relative density 2.89; Fly Ash (FA) with specific gravity 2.32, and powdered waste glass (GP) with specific gravity 2.45. The graded stone crusher dust of zone-II with relative density of 2.61 is utilized as fine aggregates and 10 mm down gravel crushed stones (with $G = 2.72$) are used as coarse aggregates.

Netlan geo-grid was used in the present investigation as a reinforcement material for the preparation of AAC reinforced paver blocks. The tests on the geo-grid material are carried out as indicated by relevant ASTM standards. Unit weight test was conducted using Archimedes principle and it was found to be 0.85 g/cm^3. Five geo-grid specimens of dimension 10 cm × 10 cm were taken and weighed. The average weight per m^2 of the geo-grid material was found to be 7.545 N (i.e., 769.4 gsm). The tensile strength of the geo-grid specimen was tested in a universal testing machine (UTM) and the ultimate strength of the specimen was found to be 1.061 kN/m. The allowable strength is 0.43 kN/m.

The mix proportion design for AAC-ICPBs was build-up on the method recommended for the mix design of the conventional Portland Cement based concrete mix-proportion design BIS code IS: 10262 [12]. Here in research work, the mix-design was prepared with a total binder content of 440 kg/m^3, with "water-to-binder ratio" of 0.40 and "fine aggregate-to-coarse aggregate ratio" of 0.38:0.62. All the mixes were developed to attain a design slump range of 0–25 mm. Then, the systematic modifications were done based on the literature [6, 13] to obtain the proportioning of AAC mixes to have same binding content (440 kg/m^3) and "water-to-binder ratio" (0.4) as same as the conventional PQC mix. Alkali-activated concrete mixes were made using binders, i.e., ground granulated blast furnace slag: FA in the ratios of 75:25 (M-0 mix). "Total water content" in activator solution for the AAC mixes, is the total water which is present in sodium silicate solution and the additional water which is add on to prepare the mix, to obtain design water content. The alkali activator solutions were balanced separately for each type of mix to provide 4% dosage of Na_2O (by weight of binder), of Ms value 1.25. No super-plasticizers were used in the preparation of AAC mixes.

First reference alkali activated ICPB elements were prepared without using glass powder (M-0 mix); afterward, the AAC ICPBs with 15% dosage of glass powder was considered (i.e., M-15 mix with 75% GGBS + 10% FA + 15% GP). Afterward, studying their compressive, split tensile strengths (Tensile Splitting Strength) and flexural strengths as per IS 15658-2006 [11]. The investigation was extended to study about the mechanical behaviors of reinforced alkali activated ICPB by incorporating geo-grid reinforcement. Comparison was done between results obtained from the test and Portland cement concrete based ICPBs obtained from the field.

Initially, the alkali solution was prepared by mixing calculated quantity of sodium hydroxide and liquid sodium silicate solution mixed to adequate quantity of water essential to maintain a w/c ratio of 0.20. This alkali solution was prepared and kept in a closed plastic container minimum before 24 h of actual casting works. The additional water is added to bring the concrete mix water/binder ratio of 0.40 during the mixing of ingredients. Details of the AAC paver block mix proportion are given in Table 1.

The ICPB used in the present investigation is under the classification of "category A-Type C" as per IRC [10] having a plan area of 32,500 mm^2 with a depth of 100 mm (mean length of 250 mm, mean width of 120 mm, measured in accordance with annexure B of IS:15658-2006 [11]). The preparation sequence of the ICPB involves placing of batched requisite amount of coarse aggregate, fine aggregates (quarry dust) and cementing materials (Alkali activated binders) in drum-type machine mixer and mixing it for about 3 min (i.e., till dry mix gets blended well). Later, the required

Table 1 Mix proportion details for AAC ICPB in kg per m^3

Mix ingredients	Mix ID	
	AAC M-0	AAC M-15
GGBS	330	330
Fly ash	110	44
Glass powder	0	66
Additional water	88	88
Alkaline solution	129.85	129.85
Coarse aggregates	1103	1105
Fine aggregates	669	670

amount of alkaline activator solution and water is poured to this dry mix and green concrete mix is prepared.

The ICPBs were prepared by filling the green concrete mix in the standard ICPB rubber molds. Then, the filled molds were vibrated and compacted in three layers by placing it in vibratory table, and then the surface is finished. The production sequence of ICPBs is shown in Fig. 2. After a day the blocks were de-moulded and air cured in room temperature as shown in Fig. 3. After curing for 7 and 28 days the ICPBs were tested for their strength and abrasion resistance.

Further, two type of reinforced ICPBs were prepared, one with single layer of reinforcement by providing geo-grid reinforcement layer at a depth of 60 mm; and the other with double layer of reinforcement by providing two geo-grid reinforcement layers, respectively, at a depth of 60 mm and 30 mm.

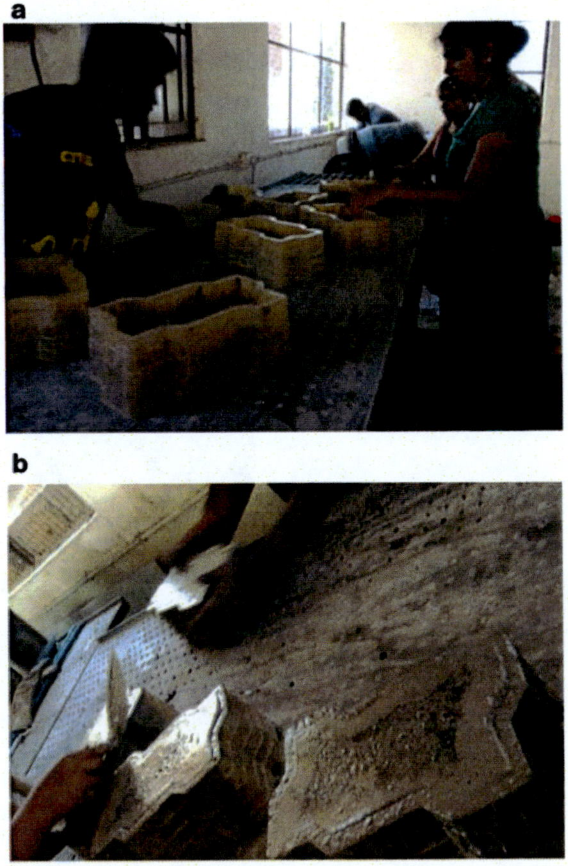

Fig. 2 **a** Production sequence of ICPBs. **b** Finishing of produced ICPBs

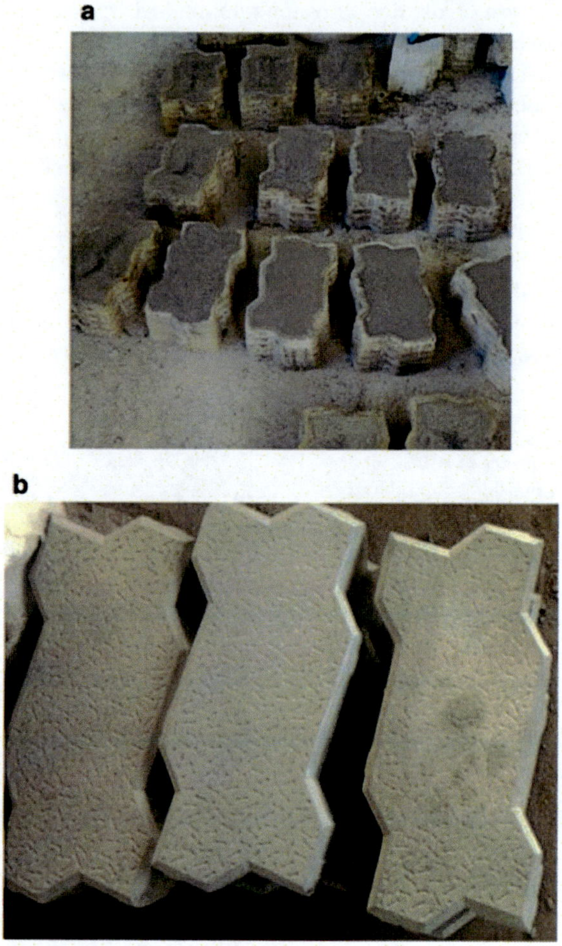

Fig. 3 **a** Air-curing sequence of ICPBs. **b** Air-cured ICPB specimens

3 Results and Discussions of Tests on ICPBs

The compressive, split-tensile, and flexural strengths testing on ICPBs were conducted in the Civil Engineering laboratory is shown in Fig. 3. The failure pattern of ICPBs during its test for compression is shown in Fig. 4, split-tension and flexure strength can also seen in Fig. 5.

Results of compression, split-tensile, and flexural strengths of the ICPBs are, respectively, indicated in Figs. 6, 7, and 8. The guidelines suggested by Indian Standard Code of practice: IS:15658-2006 [11] is carefully followed and the formulae used for the determination of the various strength parameters were also taken from the same guidelines. The averages of three readings were represented as the results in

Fig. 4 **a** Compressive strength testing of ICPBs. **b** Compressive strength testing and failure patterns

every case. Apparent compressive strengths of ICPBs were obtained by the conventional method, i.e., the highest compressive load is divided by the area of plan. The corrected compressive strengths of ICPBs are determined by multiplying the obtained apparent compressive strengths by the appropriate correction factor based on thickness and type (i.e., plain or chamfered edge) of ICPBs (as suggested by the Annex D of the code IS:15658-2006).

In the present investigation, thickness of the paver block used is 100 mm and the edge was chamfered. The multiplication factor used to determine the corrected compressive strength was 1.24 as per the code.

Figure 6 indicates that corrected compressive strengths of the ICPBs which shows that the AAC ICPBs performs better with higher compressive strengths results when matched with that of the field OPC based ICPBs. The mix AAC-M-0 shown a 28 day compressive strength value 49% higher than that of the reference field mix; whereas

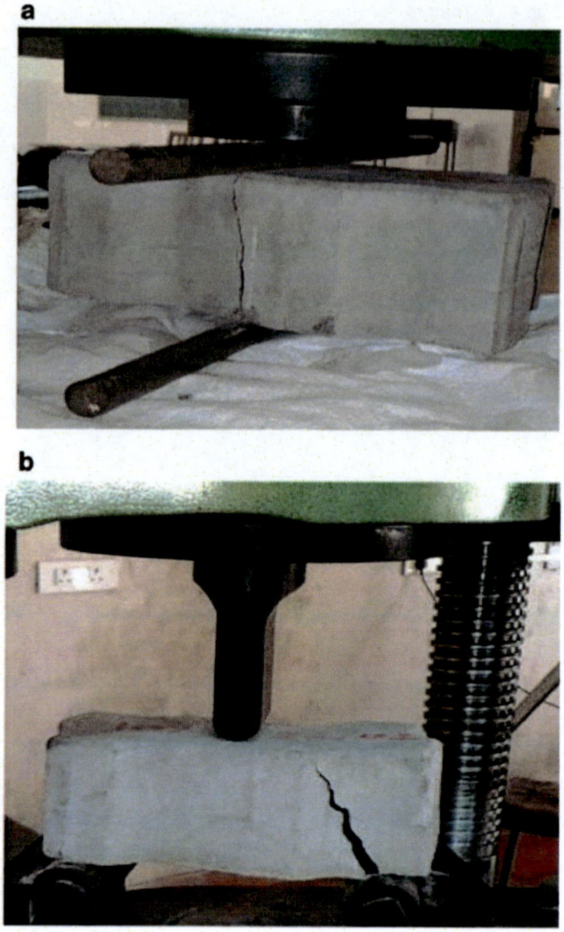

Fig. 5 **a** Split tensile strength testing of ICPB. **b** Flexural strength testing of ICPB

optimum AAC-M-15 shown a value 60% higher than that of the field mix. While compared with the AAC M-0 mix, AAC M-15 mix showed 7% higher compressive strength value. However, the effect of reinforcement is very less on the compressive strength of ICPB mixes, where it indicated a marginal increase in strength, which can be considered to be insignificant. The split tensile strength results of ICPBs were shown in Fig. 7 indicates that the AAC-M-0 ICPB mix is having split tensile strength 91% more than that of the field mix. The AAC-M-15 shows a value which is 101% greater than that of the field mix. Introduction of single layer of geo-grid reinforcement lead to an raise in tensile strengths of about 6%; whereas the introduction of double layer of geo-grid reinforcement lead to an increased tensile strength of about 12% when compared with the un-reinforced AAC-M-15 ICPB mix. The average value of unit weight and the average % water absorption of the ICPBs were indicated in table 2. The results shows that the unit weight of the mixes were almost same for

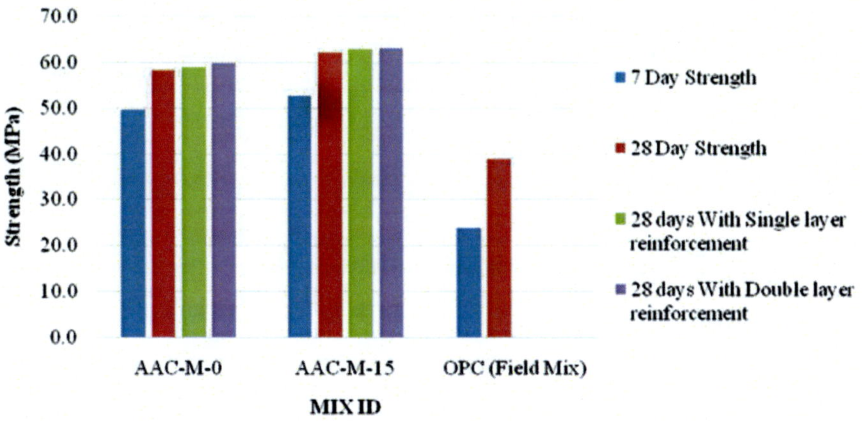

Fig. 6 Results of compression Test on ICPBs

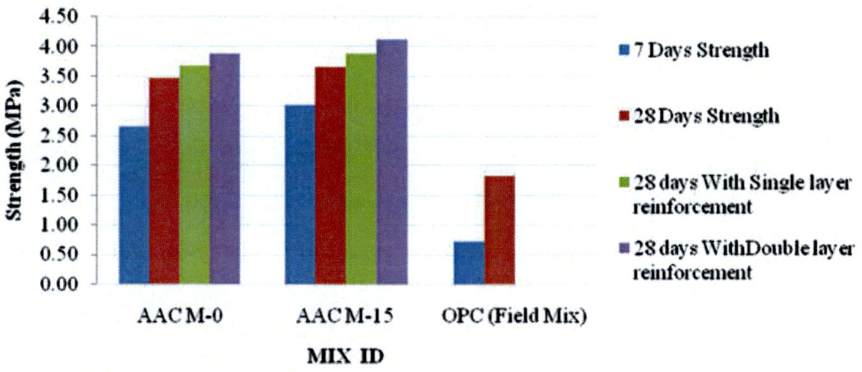

Fig. 7 Results of split tensile strength test of ICPBs

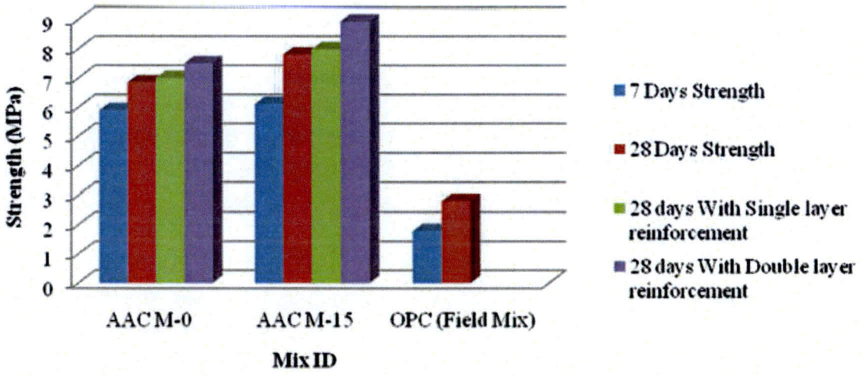

Fig. 8 Results of flexural strength test on ICPBs

Table 2 Density and water absorption results of ICPBs

Mix ID	Test parameter	
	Unit weight (kg/m^3)	Water absorption (%)
AAC-M-0	2342	3.13
AAC-M-15	2382	3.32
OPC (field mix)	2320	6.58

both the field (OPC) mixes and AAC mixes. The water absorption of all the mixes was less than 7% which is desirable; however, the values are less for AAC mixes when compared with that of the field mixes.

The flexural strength of the ICPB depicted in Fig. 8 shows the similar trend as that of the split tensile strength results. The AAC M-0 ICPB mix shows the flexural strength value of about 144% greater than that of the OPC field mix. Whereas AAC-M-15 mix shows a value which is 178% greater than that of the field OPC ICPB mix. Introduction of single layer of geo-grid reinforcement lead to an increased flexural strength of about 14%; whereas the introduction of double layer of geo-grid reinforcement lead to an increased flexural strength of about 14% when compared with the un-reinforced optimum AAC-M-15 mix.

The abrasion test on prepared ICPB specimen was carried-out in-according to IS:15658 [11] to measure the abrasion resistance of the paver block. The specimen for scratch test was obtained from cutting the concrete paving block. The size and shape of test specimen is 7.06 × 7.06 cm (i.e., 50 cm^2 in area) and square, respectively. The preparation of ICPB sample by cutting using marble cutting machine for this test is shown in Fig. 9.

The testing of ICPB specimen for its scratch resisting ability was done by tile abrasion testing machine. The testing set-up is indicated in Fig. 10. The test will be conducted once it's cured for 7 and 28 days. The results of this test are depicted in Fig. 11. The results show that the abrasion value for both the mixes is more than 2.0 mm at the age of 7 days of curing. After 28 days of curing the results were within 2.0 mm which are within the desirable limits as mentioned in Indian standard code specification. However, the AAC mixes were performing better than the OPC based field mixes.

At 28 days of curing, the the AAC-M-0 mix had shown 38.5% better abrasion resistance value when compared it with the OPC filed ICPB. Whereas the AAC-M-15 mix had shown 41.6% better abrasion resistance value when compared it with the OPC filed mix. The optimum AAC-M-15 mix had shown a abrasion resistance value 5% better than that of the AAC-M-0 mix.

The cost analysis of the two types of ICPBs (i.e., OPC field mix and AAC mix) was carried out which shows that for the given strength and thickness the OPC based ICPBs will cost about Rs. 50 per square feet in the local market. Whereas the AAC paver blocks will cost about Rs. 45 per square feet if it is produced in a large scale. Thus, the analysis of overall mechanical performance and the cost analysis show that the AAC binder based ICPBs are far better engineering material than that of the OPC based field ICPBs.

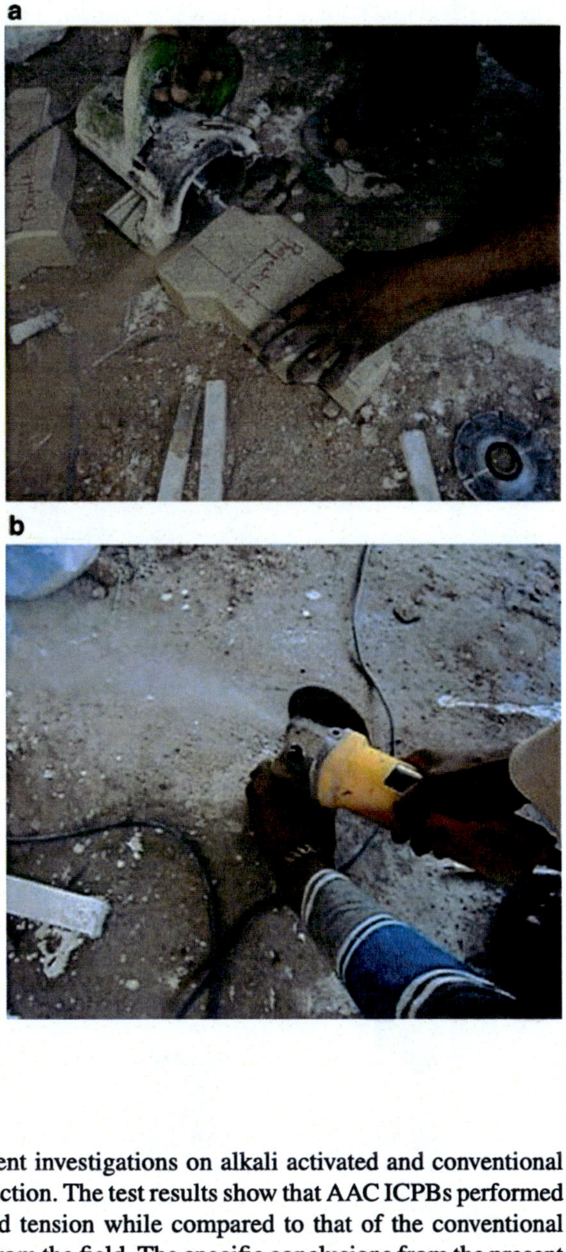

Fig. 9 **a** Preparation of ICPB specimen for Abrasion Test. **b** Preparation of ICPB specimen for Abrasion Test

4 Conclusions

The outcomes from the present investigations on alkali activated and conventional ICPBs are presented in this section. The test results show that AAC ICPBs performed far better in compression and tension while compared to that of the conventional OPC based blocks obtained from the field. The specific conclusions from the present investigations were listed in here:

- The results of compression test revealed that, there is a 49% increase in AAC compressive strengths values than that of the reference OPC based field mix;

Fig. 10 **a** Testing of ICPB specimen in Abrasion Machine. **b** Testing of ICPB specimens in Abrasion Machine

a

b

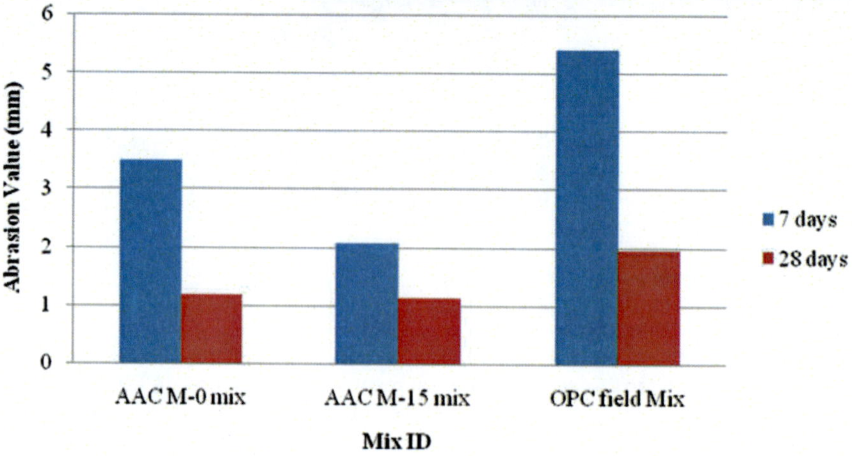

Fig. 11 Abrasion test results on ICPB specimen

whereas optimum AAC-M-15 shown a value 60% more than the field mix. All compressive strengths results satisfy the desirable requirements as per IRC guidelines.
- There was a marginal raise in the tensile strengths values when reinforcement incorporation was done in ICPBs.
- The analysis of cost indicates that the alkali activated ICPBs were about 10% cheaper (when compared with market rate per square feet) when compared to that of the field OPC based ICPBs.
- The abrasion test of ICPBs on the optimum AAC mix had shown 41.6% better abrasion resistance value when compared it with the OPC filed ICPB mix, which indicate that the AAC specimens are better than that of the OPC specimens.

Thus, from overall results it can be affirmed that the AAC paver blocks are far better considering economy and strength aspects. However, the studies can be further extended to investigate the durability aspects of these paver blocks subjected to various extreme environments such as acid, sulfate, corrosion resistance etc.

The experimental results reveal that the use of alkali activated ICPBs are considered to be sustainable replacement to OPC based conventional ICPBs. Thus alkali activated concrete ICPBs must be commercialized and produced in a large scale for the sustainable construction of interlocking pavement system.

Acknowledgements All the experiments were performed using the facilities in the Civil Engineering department of NMAMIT, Nitte. The authors wish to thank students and the authorities of NMAMIT for their kind support and encouragement. Funding provided by "Nitte Education Trust" i.e., "NMAMIT Research Fund" on the funded research project titled as "Study and to Develop Cost-Effective and Green Blocks Using Industrial Waste Materials" is gratefully accepted.

References

1. Andrew RM (2018) Global CO_2 emissions from cement production, 1928–2017. Earth Syst Sci Data 10(4):2213–2239. https://doi.org/10.5194/essd-10-2213-2018
2. Singh B, Ishwarya G, Gupta M, Bhattacharyya SK (2015) Geopolymer concrete: a review of some recent developments. Constr Build Mater 85:78–90. https://doi.org/10.1016/j.conbuildmat.2015.03.036
3. Pacheco-Torgal F, Castro-Gomes J, Jalali S (2008) Alkali-activated binders: a review. Constr Build Mater 22(7):1305–1314. https://doi.org/10.1016/j.conbuildmat.2007.10.015
4. Yang KH, Song JK, Song K (2013) Assessment of CO_2 reduction of alkali-activated concrete. J Clean Prod 39:265–272. https://doi.org/10.1016/j.jclepro.2012.08.001
5. Pacheco-Torgal F, Castro-Gomes J, Jalali S (2008) Alkali-activated binders: areview. Part 1. Historical background, terminology, reaction mechanisms and hydration products. Constr Build Mater 22(7):1305–1314. https://doi.org/10.1016/j.conbuildmat.2007.10.015
6. Mithun BM, Narasimhan MC (2017) Performance of alkali activated slag concrete mixes incorporating copper slag as fine aggregate. National Institute of Technology Karnataka Surtathkal
7. Provis JL (2017) Alkali-activated materials. Cem Concr Res 114:40–48. https://doi.org/10.1016/j.cemconres.2017.02.009

8. Pacheco-Torgal F, Castro-Gomes J, Jalali S (2008) Alkali-activated binders: a review. Part 2. About materials and binders manufacture. Constr Build Mater 22(7):1315–1322. https://doi.org/10.1016/j.conbuildmat.2007.03.019
9. Design Considerations for Interlocking Concrete Pavements. Boston: Unilock Specifications, Boston, USA, 2019
10. IRC:SP-63:2004 (2004) Guidelines for the use of interlocking concrete block pavement, 1st edn. Bureau of Indian Standards, New Delhi
11. IS:15658-2006 (2006) Indian standard precast concrete block for paving-specification, 1st edn. Bureau of Indian Standards, New Delhi, 2006.
12. IS:10262 (2009) Concrete mix proportioning guidelines. Bureau of Indian Standards, New Delhi, pp 1–11 [Online]. Available: http://mis.wbprd.gov.in/Engineering/Codes/IS10262.pdf
13. Palankar N (2016) Performance of alkali activated concrete mixes with steel slag as coarse. National Institute of Technology Karnataka Surtathkal

Pushover Analysis of Irregular Steel Structure with Varying Irregularity Ratios

B. Jayaram Nayak and Kiran Kamath

Abstract In this paper, the seismic performance of irregular steel structure with varying irregularity ratio has been investigated. For the study purpose, two different models with vertical geometric irregularity and three different models with plan irregularity according to IS 1893 (part 1)-2002 have been considered. The irregularity ratio, A/L (where A is offset, and L is base width), has been varied from 0.2 to 0.8. Irregular structures have been modelled using ETABS, a finite element software, and plastic hinges are assigned to incorporate the inelastic seismic behaviour of structures. Performance of thirteen irregular structures has been compared with regular bare frame structure in terms of base shear carrying capacity, roof displacement and performance point, using pushover analysis. Two slab models with shell element and membrane element considered for all structural models. The results indicate that as irregularity ratio increases, base shear carrying capacity and performance point of irregular structure decrease. Structural models considered slab as shell obtained higher seismic performance compared to slab as membrane.

Keywords Irregularity ratio · Plan irregularity · Pushover analysis · Seismic behaviour · Vertical geometric irregularity

1 Introduction

An earthquake is a natural phenomenon which induces seismic wave causing ground motion. Due to this, lateral forces will act on structures which in turn cause severe damage or collapse of structures. When the structure is under seismic excitations, elements in structure reach its inelastic zone. For seismic evaluation, nonlinear behaviour of structure in inelastic zone plays significant role. Hence, elastic behaviour

B. Jayaram Nayak (✉)
Department of Civil Engineering, Moodlakatte Institute of Technology, Moodlakatte, Kundapura, Udupi, Karnataka, India

K. Kamath
Department of Civil Engineering, Manipal Institute of Technology, Manipal Academy of Higher Education, Manipal, Udupi, Karnataka, India
e-mail: kiran.kamath@manipal.edu

L. Nandagiri et al. (eds.), *Sustainability Trends and Challenges in Civil Engineering*, Lecture Notes in Civil Engineering 162, https://doi.org/10.1007/978-981-16-2826-9_26

of structure is not sufficient to analyse and design the structures [1]. During seismic excitations, considering inelastic behaviour with elastic behaviour, the real behaviour of structures can be studied.

In the last decade, pushover analysis is used to study performance of structure under seismic exaction. Pushover analysis is a sequential analysis method to study the inelastic behaviour of structures, when the structure pushed by providing monotonically increasing lateral force until a predefined target roof displacement is reached or till collapse of structure [2]. Also, pushover analysis provides details regarding capacity curves and demand curves which represent ability of structure to resist the lateral loads and earthquake ground motion, respectively [3]. Performance point can be obtained by superimposing capacity curves and demand curves on each other and is analysed with respect to base shear and roof displacement. Henceforth, pushover analysis gives accurate predication of structural performance for demand seismic loading.

Irregular structures have been commonly used due to site restriction, various functional requirements and architectural demands. Researchers have shown that irregular structure attracts more seismic forces compared to regular structures [4]. In an earlier research, authors have study on the amount of eccentricity of structures effecting the seismic behaviour. Results indicate that as eccentricity increases, torsion increases. Due to which accuracy of seismic response is reduced [5]. Authors of the researcher studied seismic response considering various irregularities type of structural configurations. Results obtained show that seismic response of structure varies with the irregularities type [6]. When plan irregularity and vertical irregularity structures on a sloping ground are considered, the vertical irregularity structures are more critical in seismic performance [7]. Also, various researches have been carried out in order to study the seismic performance of steel frame structures. Results show that steel structures have high seismic performance [8–10]. However, limited studies have been carried out in the effect irregularity ratio on irregular steel structures for seismic evaluation. So, in this study, using pushover analysis, the effect of irregularity ratio on vertical geometric irregularity and plan irregularity steel structures is considered.

2 Description of Models

A 10-story structure with height of each story 3 m is considered as regular bare frame structure. The plan of dimension 15 m × 15 m of bare frame structure considered in the study is given in Fig. 1. The reinforced concrete slab considered is of 150 mm thick with M 20 grade of concrete.

The dead and live loads considered on structure are based on IS 875 (part I) [12] and IS 875 (part II) [13], respectively. Live load considered on all floors is 3 kN/m^2 and on roof is 1.5 kN/m^2; dead load on floor is 1 kN/m^2 on all structural models. For seismic evaluation, structural models considered are situated in seismic zone III with response reduction factor as four. All the structural configurations are

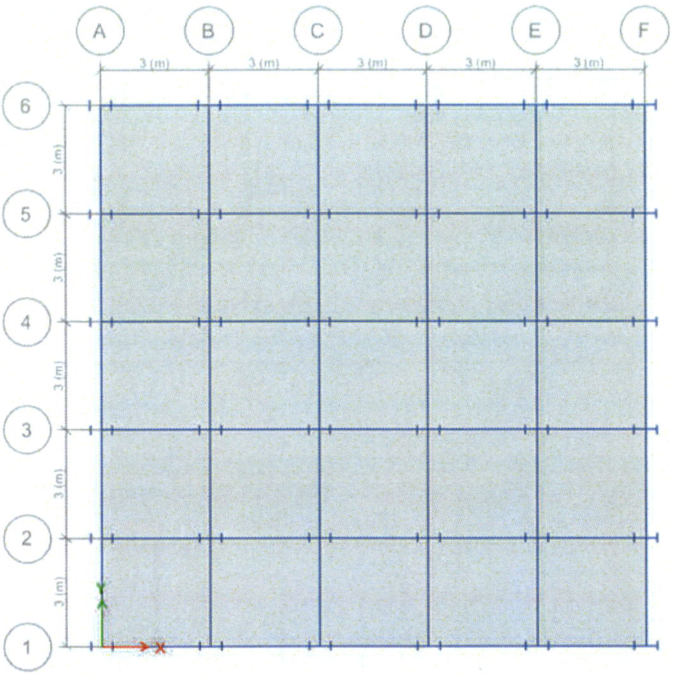

Fig. 1 Plan of the building

having importance factor of one with soil type medium. All the structural models are considered fixed at the base. The seismic excitations on all models are considered by application of earthquake loading in X-direction and Y-direction.

Thirteen irregular models are considered by varying irregularity ratio. The irregularity ratio is varied by varying the offset (A) and keeping the base width (L) constant.

As per IS 1893 (part 1)-2002 [11], Fig. 2 shows type (i) and type (ii) structures which are two types of vertical geometric irregularities. Figure 3 shows type (iii), type (iv) and type (v) structures which are three types of plan irregularities.

Figures 4, 5 and 6 show the variation in irregularity ratio of type (i) structure. Figures 7 and 8 show the variation in irregularity ratio of type (ii) structure. Figures 9, 10 and 11 show the variation in irregularity ratio of type (iii) structure. Figures 12, 13 and 14 show the variation in irregularity ratio of type (iv) structure. Figures 15 and 16 show the variation in irregularity ratio of type (v) structure. Model numbering for irregular structures is done from Figs. 4, 5, 6, 7, 8, 9, 10, 11, 12, 13, 14, 15 and 16.

Using ETABS, a finite element software, beams and columns are modelled as frame elements with slab considered as membrane element and shell element. The structural models both regular and irregular structures are designed according to IS 800 (2007) [14]. The designed steel section used for beams and columns is ISMB 200

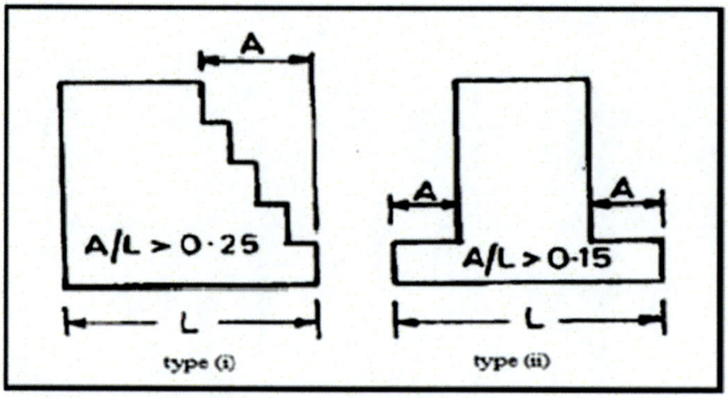

Fig. 2 Two types of vertical geometric irregularity as per IS 1893 (part 1)—2002

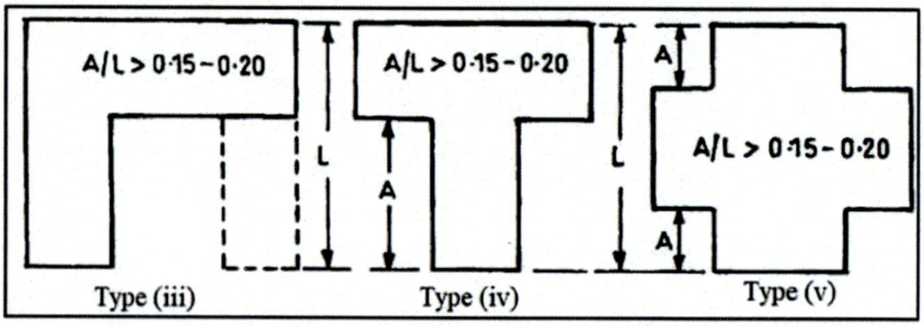

Fig. 3 Three types of plan irregularity as per IS 1893 (part 1)—2002

and ISWB 600-2, respectively, with grade of steel used is Fe 250. The auto hinges for incorporation of inelastic behaviour of structures, M3 and P-M2-M3 hinges, are assigned to beams and columns, respectively, according to ASCE 41-13 [15]. For seismic performance evaluation, the base shear carrying capacity, roof displacement and performance point are considered in both the directions of applied earthquake loading, i.e. X- and Y-direction.

3 Results and Discussions

The results obtained from nonlinear pushover analysis are discussed here. Figures 17, 18, 19, 20, 21, 22, 23, 24, 25 and 26 show the pushover curves comparing the irregular structural models with bare frame structural model.

From Figs. 17, 18, 19, 20, 21, 22, 23, 24, 25 and 26, it is clear that structures with slab considered as shell element have shown higher seismic base shear capacity with least roof displacement compared to structures with slab considered as membrane

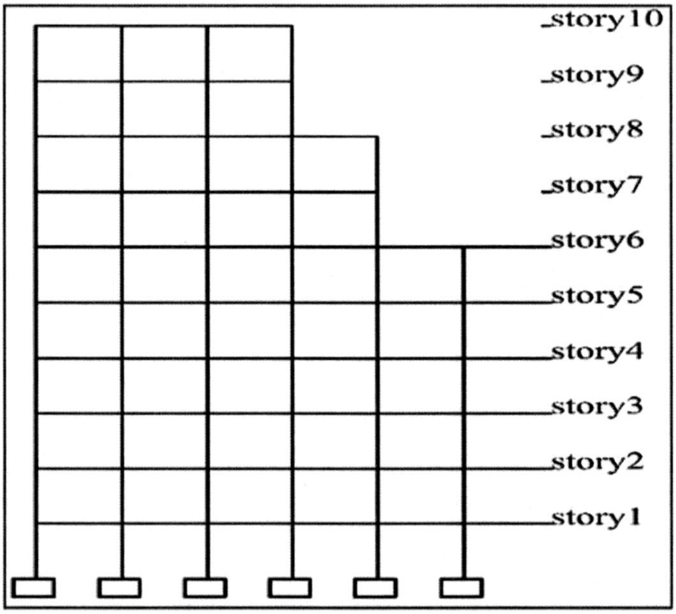

Fig. 4 Model 1, frame showing type (i) irregular structure with irregularity ratio of 0.4

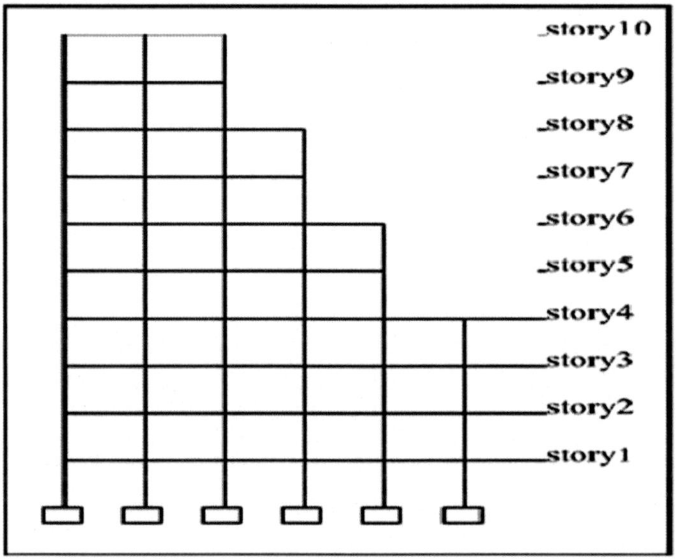

Fig. 5 Model 2, frame showing type (i) irregular structure with irregularity ratio of 0.6

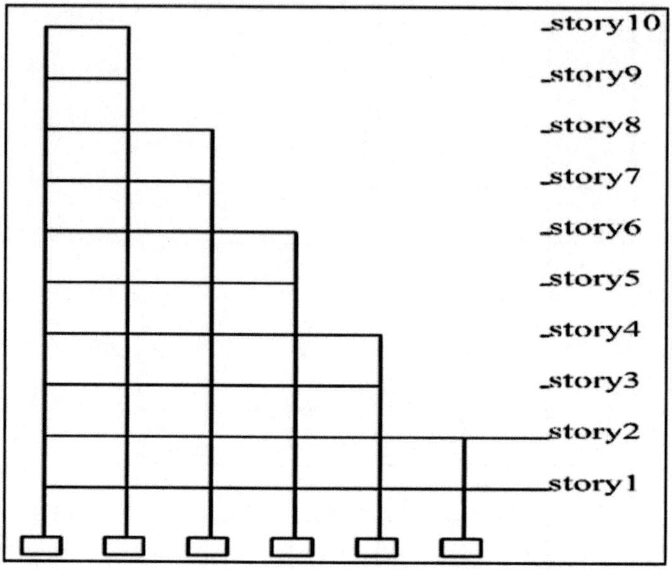

Fig. 6 Model 3, frame showing type (i) irregular structure with irregularity ratio of 0.8

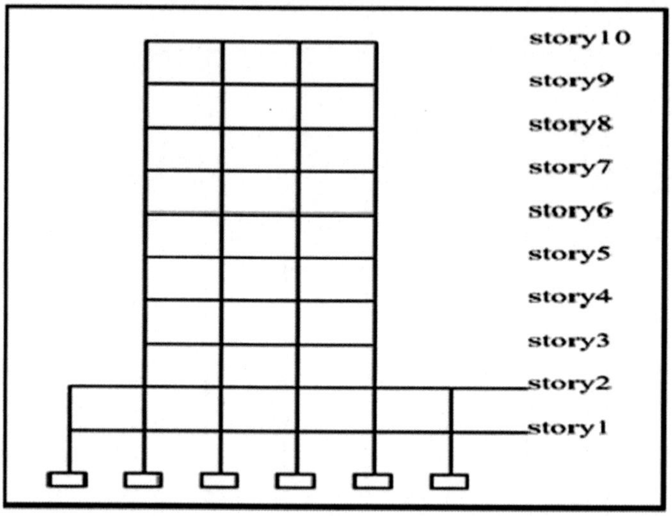

Fig. 7 Model 4, frame showing type (ii) irregular structure with irregularity ratio of 0.2

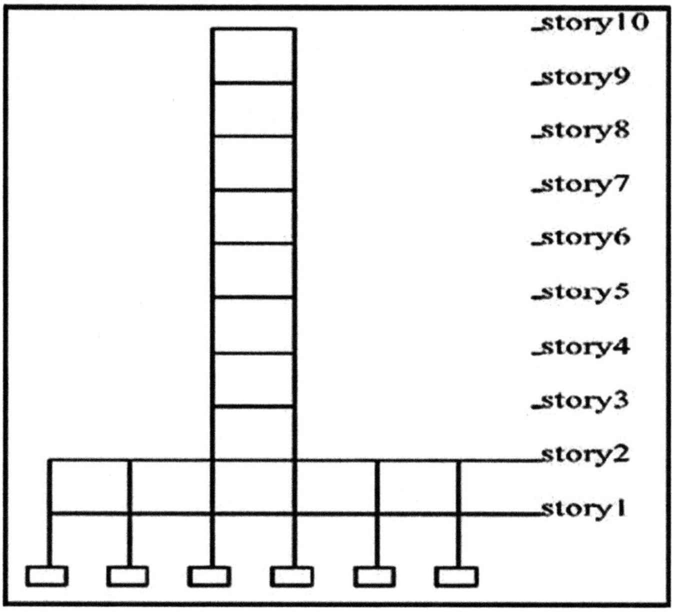

Fig. 8 Model 5, frame showing type (ii) irregular structure with irregularity ratio of 0.4

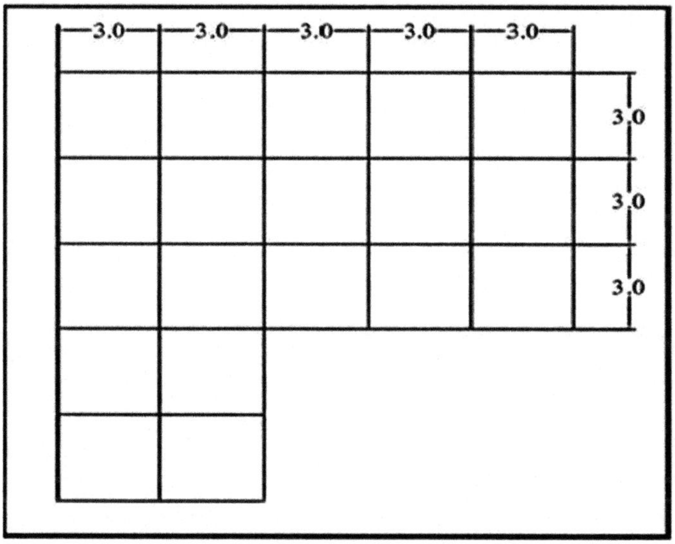

Fig. 9 Model 6, frame showing type (iii) irregular structure with irregularity ratio of 0.4

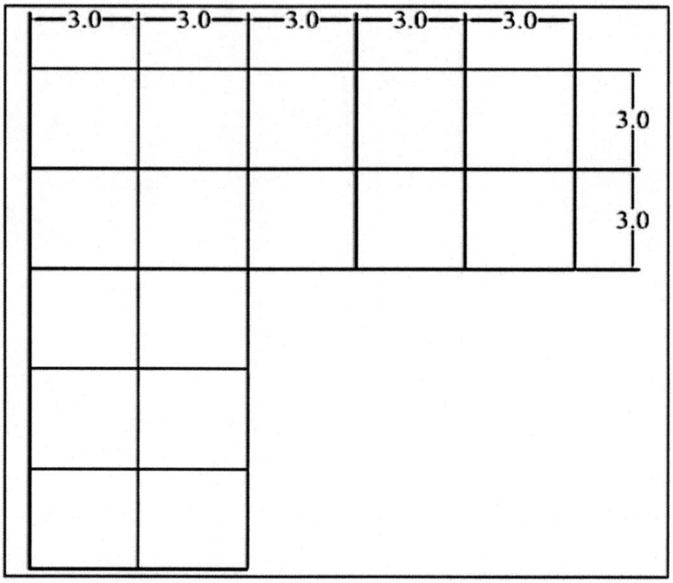

Fig. 10 Model 7, frame showing type (iii) irregular structure with irregularity ratio of 0.6

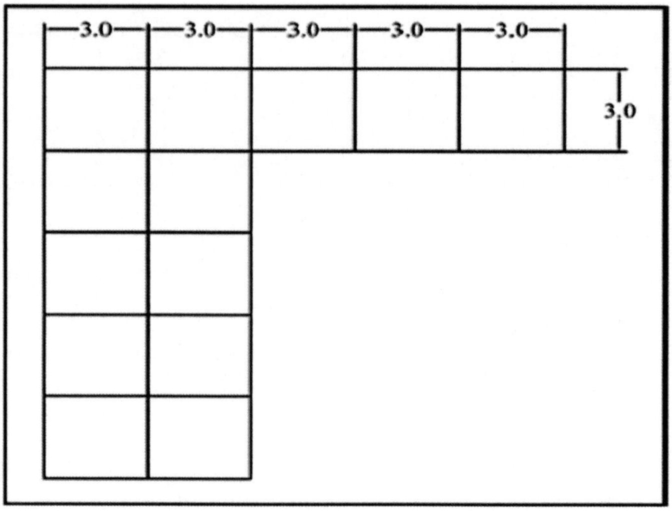

Fig. 11 Model 8, frame showing type (iii) irregular structure with irregularity ratio of 0.8

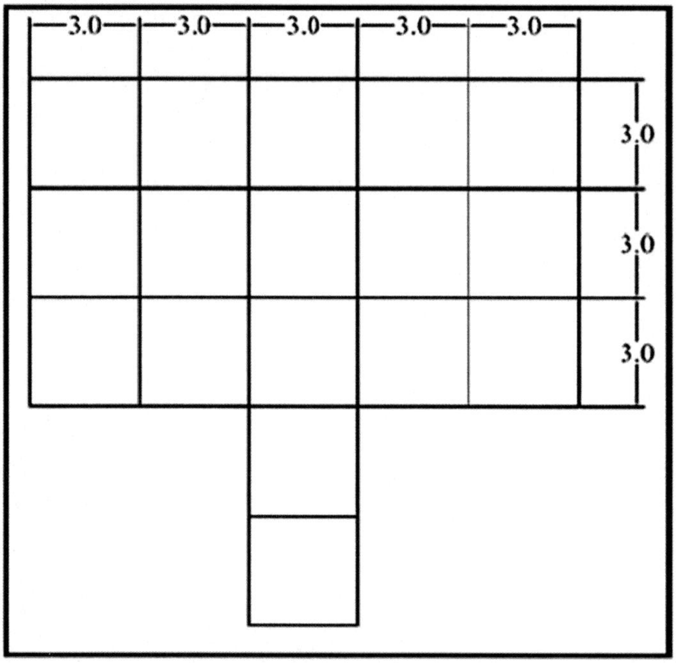

Fig. 12 Model 9, frame showing type (iv) irregular structure with irregularity ratio of 0.4

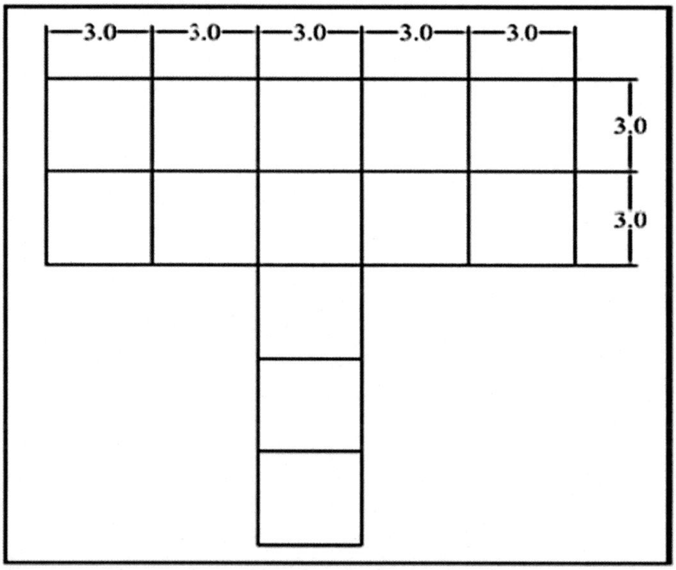

Fig. 13 Model 10, frame showing type (iv) irregular structure with irregularity ratio of 0.6

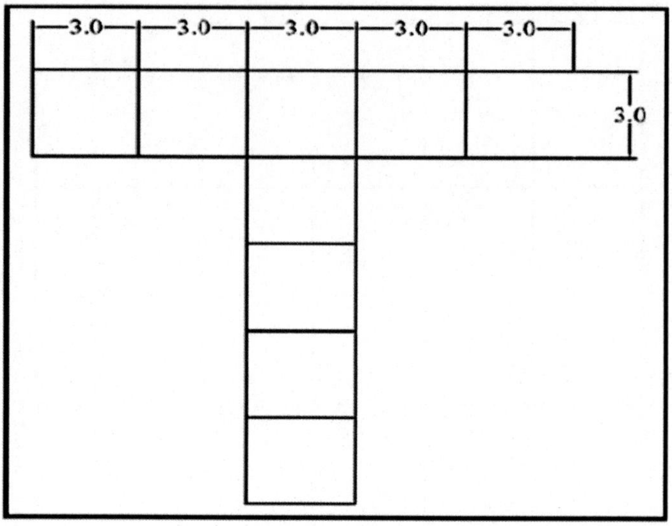

Fig. 14 Model 11, frame showing type (iv) irregular structure with irregularity ratio of 0.8

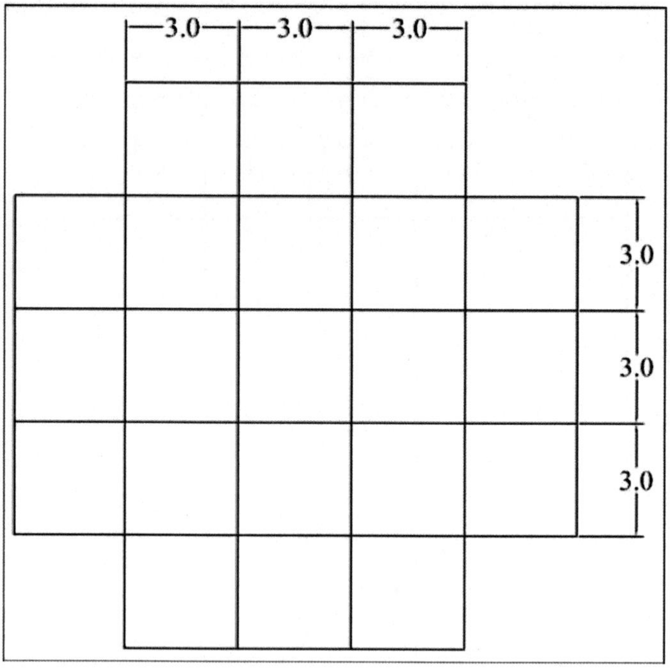

Fig. 15 Model 12, frame showing type (v) irregular structure with irregularity ratio of 0.2

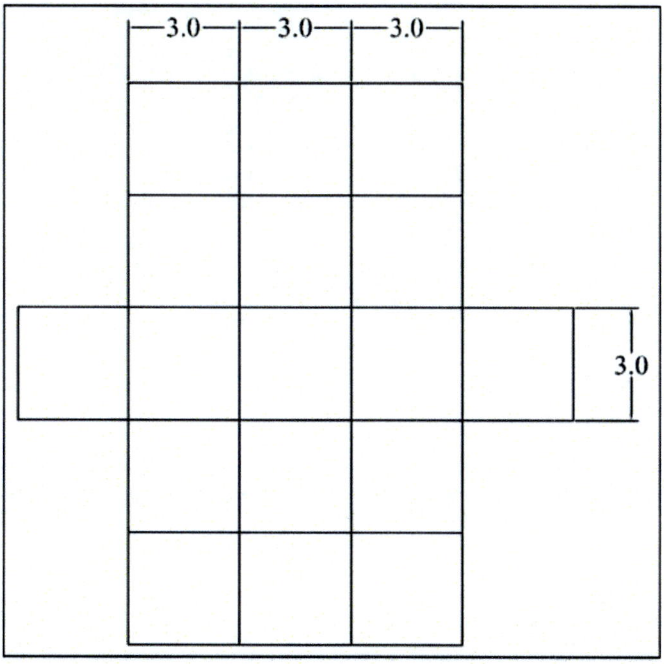

Fig. 16 Model 13, frame showing type (v) irregular structure with irregularity ratio of 0.4

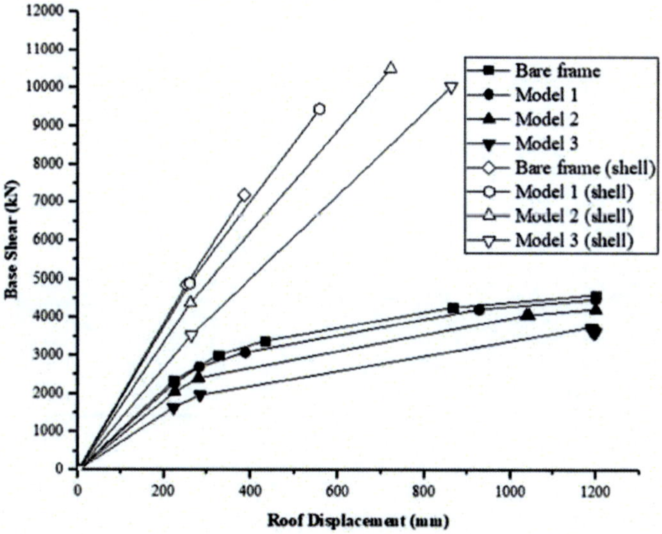

Fig. 17 Pushover curves, comparing vertical geometric irregularity of type (i) (Fig. 2) structures with regular structure in X-direction of earthquake

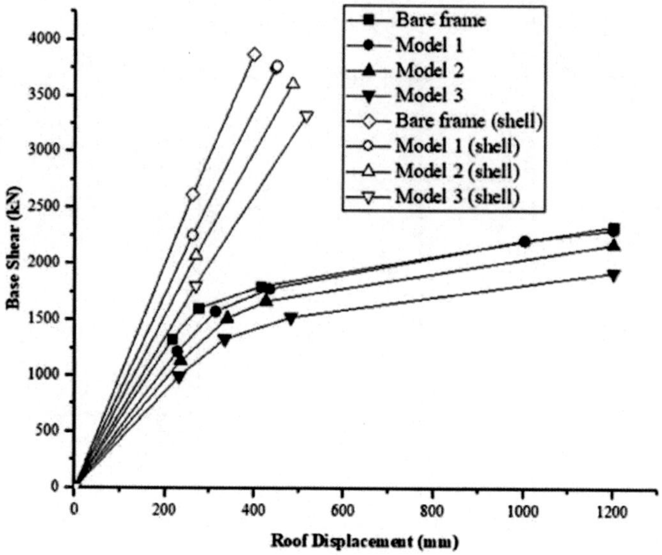

Fig. 18 Pushover curves, comparing vertical geometric irregularity of type (i) structures with regular structure in Y-direction of earthquake

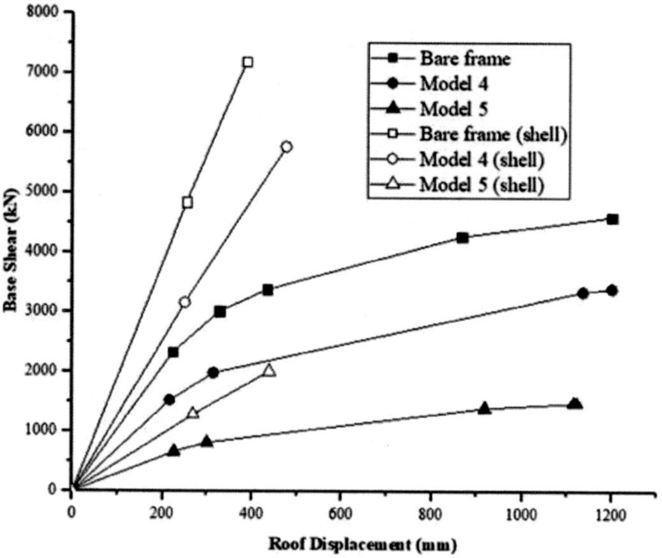

Fig. 19 Pushover curves, comparing vertical geometric irregularity of type (ii) (Fig. 2) structures with regular structure in X-direction of earthquake

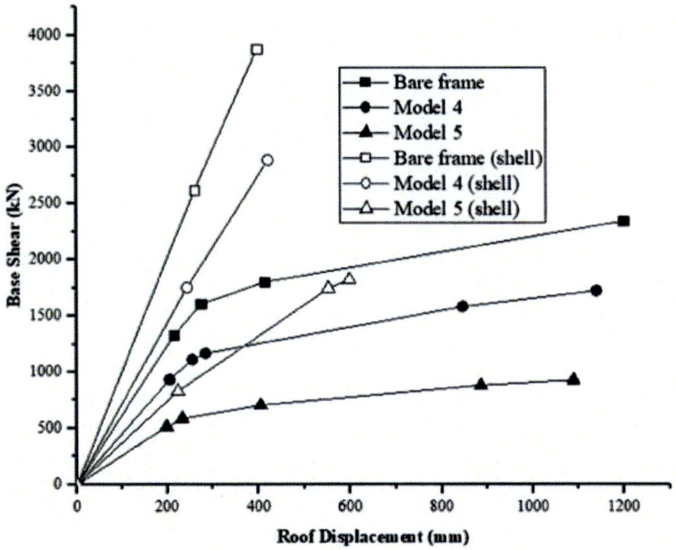

Fig. 20 Pushover curves, comparing vertical geometric irregularity of type (ii) structures with regular structure in Y-direction of earthquake

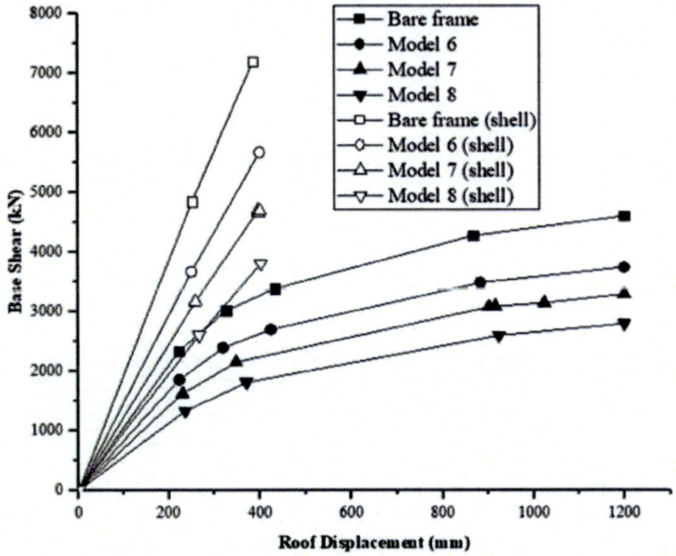

Fig. 21 Pushover curves, comparing plan irregularity of type (iii) (Fig. 3) structures with regular structure in X-direction of earthquake

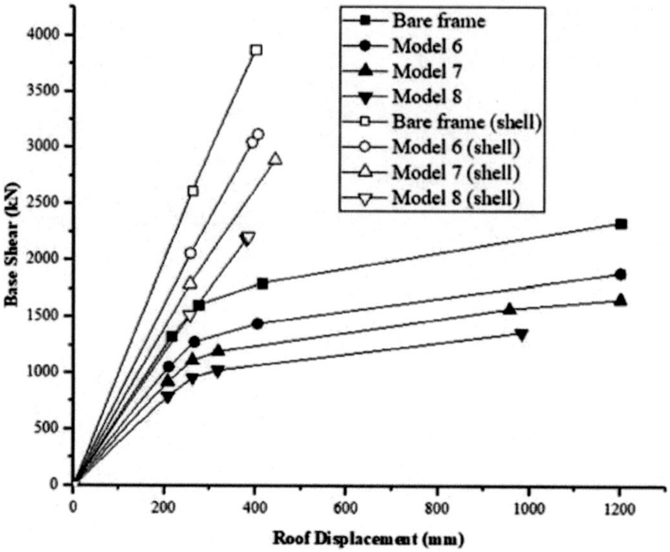

Fig. 22 Pushover curves, comparing plan irregularity of type (iii) structures with regular structure in Y-direction of earthquake

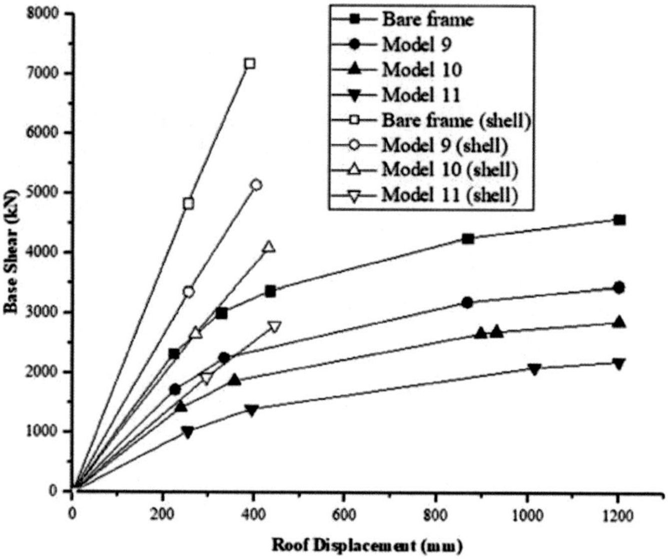

Fig. 23 Pushover curves, comparing plan irregularity of type (iv) (Fig. 3) structures with regular structure in X-direction of earthquake

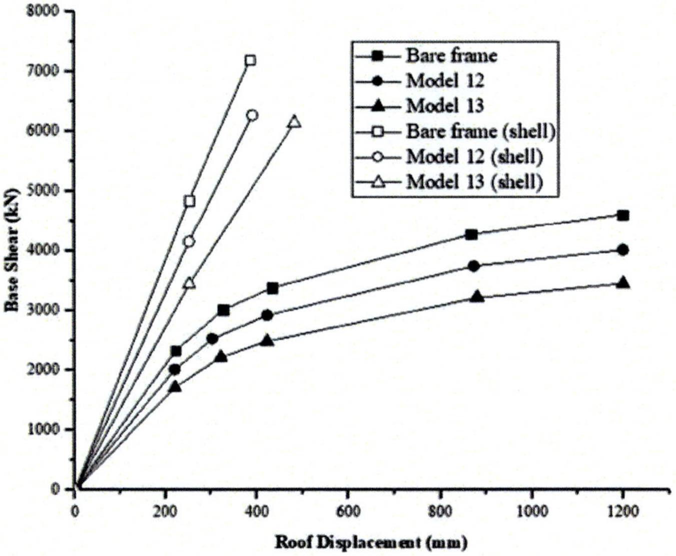

Fig. 24 Pushover curves, comparing plan irregularity of type (iv) structures with regular structure in Y-direction of earthquake

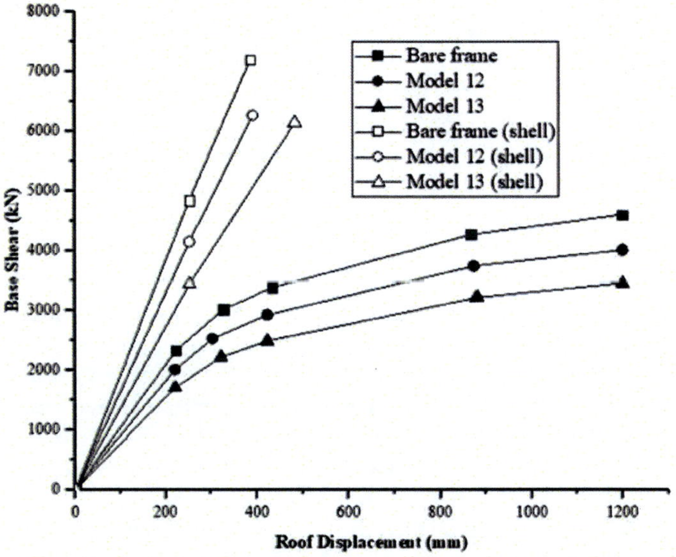

Fig. 25 Pushover curves, comparing plan irregularity of type (v) (Fig. 3) structures with regular structure in X-direction of earthquake

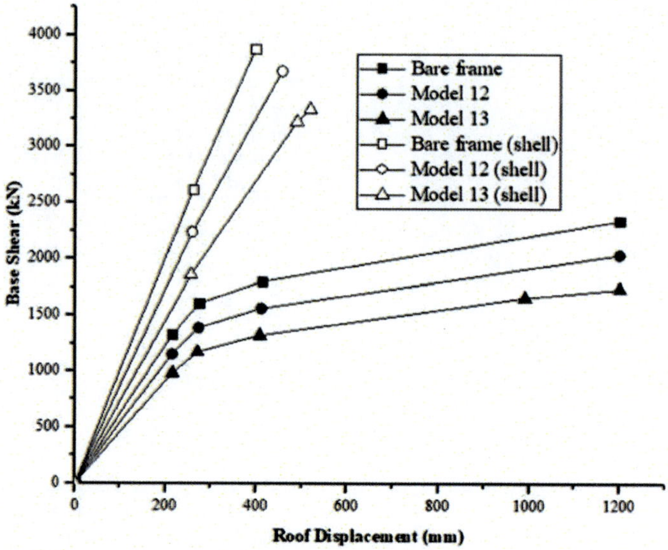

Fig. 26 Pushover curves, comparing plan irregularity of type (v) structures with regular structure in Y-direction of earthquake

element. Stiffness of structures with slab as shell element is higher compared to structures with slab as membrane element.

As shell element will resist seismic forces compared to the membrane element, for seismic analysis, considering shell elements for slab gives accurate values compared to the overestimated values of the membrane elements for slab in structures.

From Figs. 17, 18, 19, 20, 21, 22, 23, 24, 25 and 26, it evident that regular bare frame structure has higher seismic base shear capacity compared to irregular structures in both the directions of earthquake.

Also as irregularity ratio increases, base shear capacity decreases in both the directions of earthquake and in all the cases of vertical geometric irregularity and plan irregularity with slab considered as shell and membrane elements.

In both the cases, vertical geometrical irregular structures with type 1 (Models 1 to 3) have shown higher seismic base shear capacity compared with least roof displacement compared to all other structures considered in the study. In plan irregular structures, the type 5 (Model 12 and Model 13) has shown higher base shear capacity compared all other plan irregular structures in the direction.

Table 1 indicates the structural models with slabs considered as shell element have higher base shear capacity than slabs considered as membrane elements. However, it can be seen in Table 2 the roof displacement decreases for structural models with slab considered as shell elements compared to slabs considered as membrane element. Also, it is evident that in irregularity ratios increase, percentage base shear increase has been decreased in both vertical geometrical irregular structures (from Models 1 to 5) and plan irregularity (from Models 6 to 13).

Table 1 Comparative percentage increase in base shear in structural models of slab as shell elements to membrane elements

Structural models	Percentage increase in Base Shear in X-direction (%)	Percentage increase in Base Shear in Y-direction (%)
Bare frame	36.07	39.56
Model 1	52.59	38.77
Model 2	59.83	39.70
Model 3	63.03	42.01
Model 4	41.09	40.35
Model 5	26.52	49.14
Model 6	34.08	10.90
Model 7	30.11	42.62
Model 8	26.61	38.16
Model 9	32.82	43.68
Model 10	29.77	46.17
Model 11	21.06	41.28
Model 12	35.96	44.57
Model 13	43.81	47.95

Table 2 Comparative percentage decrease in roof displacement in structural models of slab as shell elements to membrane elements

Structural models	Percentage decrease in roof displacement in X-direction (%)	Percentage decrease in roof displacement in Y-direction (%)
Bare frame	210.53	201.72
Model 1	114.78	167.27
Model 2	66.06	147.64
Model 3	37.86	132.99
Model 4	152.25	170.18
Model 5	155.37	81.84
Model 6	199.42	197.13
Model 7	201.49	171.25
Model 8	197.32	155.52
Model 9	197.57	162.65
Model 10	178.17	136.09
Model 11	169.99	156.60
Model 12	205.94	162.31
Model 13	147.98	131.05

Table 3 Base shear and displacement at performance point in X-direction

Types of structural model	Performance point (slab as membrane)		Performance point (slab as shell)	
	Base Shear (kN)	Displacement (mm)	Base Shear (kN)	Displacement (mm)
Bare frame	3865.16	368.79	5244.69	273.39
Model 1	3639.88	360.22	4976.94	265.75
Model 2	3196.81	354.53	4383.10	262.62
Model 3	2552.29	348.74	3483.06	260.88
Model 4	2481.46	351.81	3341.52	263.18
Model 5	961.55	328.74	1230.61	257.07
Model 6	3030.24	365.72	4044.73	276.72
Model 7	2619.79	373.43	3453.11	283.34
Model 8	2206.91	393.84	2859.80	293.96
Model 9	2760.60	362.50	3669.03	278.46
Model 10	2206.94	372.26	2877.02	292.25
Model 11	1643.24	409.63	2079.18	318.54
Model 12	3313.04	363.78	4465.89	271.49
Model 13	2761.46	356.95	3673.32	269.85

From Tables 3 and 4, it can be seen that, at performance point, base shear capacity of regular bare frame structure is greater than all irregular structures. Also, as irregularity ratio increases, at performance point, base shear capacity and displacement decrease in all the cases and in both directions. The base shear capacity of structures with slab as shell element is higher compared to structures with slab as membrane element at performance point.

The vertical geometrical irregular structure of type 1 (Models 1 to 3) has shown higher base shear capacity at performance point compared to other irregular structural models considered. Higher base shear indicates higher seismic performance of structures.

From Tables 3 and 4, it can be observed that for bare frame structures the value of base shear and displacement at performance point has higher value in X-direction when compared to Y-direction due to sectional properties of structural members. The structural models considered in the present study have higher moment of inertia for beams and columns along X-direction than in Y-direction, hence the performance point for seismic analysis along X-direction is higher than that along Y-direction.

Table 4 Base shear and displacement at performance point in Y-direction

Types of structural model	Performance point (slab as membrane)		Performance point (slab as shell)	
	Base Shear (kN)	Displacement (mm)	Base Shear (kN)	Displacement (mm)
Bare frame	3010.99	487.34	3864.55	384.17
Model 1	2829.74	526.64	3626.35	418.05
Model 2	2495.44	516.19	3183.31	412.20
Model 3	2057.33	477.22	2600.91	385.33
Model 4	2030.35	449.27	2573.38	258.07
Model 5	954.54	369.85	1137.46	307.10
Model 6	2363.28	468.39	3009.01	374.44
Model 7	2041.52	458.85	2584.44	369.10
Model 8	1720.31	450.24	2158.53	363.83
Model 9	2152.02	471.53	2730.70	375.75
Model 10	1720.90	458.09	2163.45	368.33
Model 11	1287.26	436.65	1593.57	356.46
Model 12	2582.15	480.54	3294.47	380.89
Model 13	2152.13	471.34	2728.12	375.96

4 Conclusion

Based on the result and discussion, it can be concluded that, as irregularity ratio increases, base shear carrying capacity and performance point of irregular structure decrease. Regular bare frame structural model showed higher seismic performance in both X- and Y-direction compared to all irregular structural models considered. The vertical geometrical irregularity models of type (i) have higher seismic performance compared to all the irregular configurations considered in the study. Also, slab considered as shell element has higher seismic performance compared to structures with slab considered as membrane element.

References

1. Kalibhat MG, Kamath K, Prasad SK, Ramya RP (2012) Seismic performance of concentric braced steel frames from pushover analysis. J Mech Civil Eng, 67–73
2. Kamath K, Anil S (2016) Fragility curves for low-rise, mid-rise and high-rise concrete moment resisting frame building for seismic vulnerability assessment. Int J Civil Eng Technol 08(03):510–519
3. ATC 40 (1996) Seismic evaluation and retrofit of concrete buildings. America Society of Civil Engineers, Reston, VA

4. Yashar Y, Maryam P (2012) Investigation the 3D-pushover analysis of unsymmetrical concrete structure. In: 15th world conference on earthquake engineering
5. Milind VM (2015) Pushover analysis of a structure with plan irregularity. J Mech Civil Eng 12(04):46–55
6. Rai CB, Ricardo (2015) Pushover analysis of asymmetrical three dimensional building frames. J Civil Eng Manage, 03–12
7. Babu NJ, Balaji KVJD, Gopalaraju SSSV (2012) Pushover analysis of unsymmetrical structure on sloping ground. Int J Civil Struct Environ Infrastruct Eng Res Dev 02(04):45–54
8. Magudeaswaran P, Dinesh A, Eswaramoorthi P (2016) Pushover analysis of steel frame. Int J Adv Eng Technol 07(02)
9. Kamath K, Shruthi, Shashikumar R (2015) Comparative study on concentric steel braced frame structure due to effect of aspect ratio using pushover analysis. Int J Sci Res Eng Technol 04(03):247–252
10. Padmakar M (2013) Pushover analysis of steel frame. NITK Rourkela Project Thesis
11. IS 1893 (Part-I) (2002) Criteria for earthquake resistant design of structure. Bureau of Indian Standards, New Delhi, India
12. IS 875 (Part-I) (1987) Code of practice for design loads for buildings and structures-Dead loads. Bureau of Indian Standards, New Delhi, India
13. IS 875 (Part-II) (1987) Code of practice for design loads for buildings and structures-Imposed loads. Bureau of Indian Standards, New Delhi, India
14. IS 800 (2007) Code of practice in general construction in steel. Bureau of Indian Standards, New Delhi, India
15. ASCE/SEI 41-13 (2014) Seismic evaluation and retrofit of existing buildings. America Society of Civil Engineers, Reston, VA

Studies on the Behavior of Gabion Wall Subjected to Lateral Monotonic Loading

N. R. Chirdeep, N. C. Balaji, Rohith Jain, and G. S. Suresh

Abstract This paper presents the experimental investigation results on the behavior of gabion wall subjected to lateral monotonic loading. Behavior of reinforced concrete retaining wall due to lateral loading could be significantly improved by replacing concrete wall by gabion wall. Diameter of mesh wire, spacing between the wires and void ratio of infill material and tensile strength of the wire significantly influence the behavior of gabion walls. Single twisted wire mesh was used as outer cage, and stones and soil were used as infill materials for the construction of gabion wall. Concrete reaction retaining wall was constructed in the laboratory to support the gabion wall and to resist the reaction of load applied on the gabion wall. The main objective of experimental work is to study the behavior of gabion wall having soil and stone as infill material subjected to lateral loading. Lateral monotonic loading was applied, and the strain in the mesh wire was measured using vernier caliper for each load increment. The test results included experimental load-strain curve, load-deformation curve, H/Δ ratio, and moment–curvature characteristics. Gabion wall having stone as infill material showed an increase in resistance to deformation ranging from 30 to 40% compared to gabion wall having soil as infill material.

Keywords Gabion wall · Lateral monotonic loading · Deformation characteristics · Single twisted wire mesh

1 Introduction

The gabion is defined as a wire mesh container of different sizes, unvaryingly divided into internal compartments, unified with other comparable components and filled with stone at the job site to form elastic, permeable uniform structures for earth retaining, and erosion control purposes, for example, retaining walls, sea walls, channel facings, revetments, offshore bunds, dykes, and weirs [1] The different

N. R. Chirdeep · N. C. Balaji (✉) · G. S. Suresh
Department of Civil Engineering, The National Institute of Engineering, Mysuru, India

R. Jain
Department of Civil Engineering, MIT-Thandavapura, Mysuru, India

L. Nandagiri et al. (eds.), *Sustainability Trends and Challenges in Civil Engineering*, Lecture Notes in Civil Engineering 162, https://doi.org/10.1007/978-981-16-2826-9_27

components of the gabion box are double-twisted wire mesh, selvedge wire, edge wire, and lacing wire. The tensile strength of wire used for lacing wire, double-twisted mesh, and stiffener will be 350–550 MPa at a minimum elongation of 10% [2]. The experiment was performed on a gauge length of test specimen as 200 mm. In recent years, the interest in gabion structures is increasing drastically. This curiosity has been produced by the gabion's ecologically aesthetic structures, but additional interest is being established by the versatility and advantages of gabion structure. Gabion structures are essentially cages made up of wire mesh which is filled with stones, rocks, or sand. A gabion wall is a retaining wall made up of stones or rocks stacked layer by layer and confined inside a tied wire mesh. They can be stacked in any shapes from stepped walls to battering wall. As they dissipate energy from the flowing water, they allow water to pass through them, but do not get washed away. As time passes, the voids between the stones may get filled with silt, vegetation, etc. This may increase the strength of the gabion wall in some cases by reinforcing it with the filled debris. One of the main aspect of a gabion retaining wall is its durability. The durability of the gabion wall mainly depends on the lifespan of the wire rather than the durability of the filler material. The wire mesh can be made up of galvanized steel wire or stainless steel wire whose life expectancy can reach up to 60 years. The filler material should chiefly comply with the dimensional and structural characteristics. Gabion walls are protective assemblies used to design, construct, and conserve for providing technical solution. Gabions are being used as building material for various construction works, because of their excellent engineering adaptability, outstanding strength and dependability. They are used in many construction works, such as, soil erosion protection, road construction, retaining walls, bridge protection, river training, etc.

A broad literature assessment was shown to study the experimental work carried out on gabion walls. In the experimental study conducted by Lin and Yang [3], the extreme strains of reinforcements in the fifth layer and the third layer are about 0.90% and 0.78%, respectively, under the maximum load $q = 250$ kPa. As per Saravanapriya S [4], the outcome indicates that the lateral deformation for the geogrid gabion wall and geogrid gabion wall (partial replacement with sand) was reduced by 20% and 55%, respectively, when related to conventional gabion wall. Ramli and Karasu [5] conducted tests for two types of gabion wall, which varied in its shape. Based on the observations, it was concluded that the hexagonal shape gabion wall has better structural integrity than the conventional rectangular gabion wall due to their special interlocking property. It was ascertained from the results obtained by Jiang and Wang [6], and the deformation occurs at center of gabion was found to be maximum, and it was increasing laterally. As per the study conducted by Yang et al. [7], the tensile strength of 2.7 mm diameter wire mesh was 1.4 times greater than 2.2 mm diameter wire mesh. Employing woven waste tire stripes as replacement to steel wire in their study, Apriyono et al. [8] concluded that the strength of the gabion models used in their study is 83.197, 59.426, and 62.397 kPa for the wire, nail, and glue connector.

Table 1 Different variables considered for the test

Specimen/variable	Dia of wire mesh in mm	Wire mesh opening size in mm	Gabion box size ($l \times b \times h$) in m	Shape of opening	Filler material
Specimen 1	2.5	80 × 80	1 × 0.3 × 0.4	Single twisted rhombus	Stones
Specimen 2	2.5	80 × 80	1 × 0.3 × 0.4	Single twisted rhombus	Soil

2 Experimental Program

The gabion walls were tested under lateral monotonic loading with fixed support at the bottom of the wall. In order to interpret the behavior of gabion wall subjected to lateral monotonic loading, performance on deformation resistance of gabion wall is to be evaluated. To analyze the deformation characteristics of gabion wall for both stones and soil as infill material, the strain, deformation and moment–curvature characteristics were obtained. Further to evaluate the brittleness of the material, H/Δ values were used. The gabions used for this study was classified into two types of specimens with respect to the infill material used. A single twisted steel wire mesh of 2.5 mm diameter was used for the study. The dimensions of the gabion wall used were same for both the cases. But, the infill materials used were both soil and stones. A total of three trials was conducted for gabion wall having stone and soil as infill materials. As shown in Table 1.

3 Details of Gabion Wall

The size of the gabion wall selected for the study was 1000 mm × 300 mm as length and width, respectively. Gabion wall was constructed upto 400 mm height, and the infill material was varied. Single twisted gabion wire mesh of 2.5 mm diameter and opening size of 80 mm × 80 mm was used. The gabion wall was set up by adopting guidelines given by IS 16014:2018.

3.1 Materials for Constrcution of Gabion Wall

Single twisted steel mesh is a wire frequently used in many fields in practice. It is made of thin, elastic galvanized steel wire, with rhombus shaped openings. Mesh size of 80 mm × 80 mm was used for the study. The diameter of the wire mesh was 2.5 mm. Before conducting experiments, a wire sample from the mesh was tested in UTM to determine tensile strength. The yield strength and ultimate strength of the wire sample were found to be 352 Mpa and 587 Mpa, respectively. Stones of higher unit

Table 2 Test results on soil

Properties	Results	IS Conforming
Specific gravity	2.47	IS 2720 Part (3)
Fineness modulus	2.72	IS 2720 Part (4)
Bulk density (kN/m^3)	18.86	IS 2720 Part (29)
Dry density (kN/m^3)	14.92	IS 2720 Part (29)
Grade of soil	Medium to well graded	
Type of soil	Sandy soil	

weight are desirable. To make sure the toughness of the gabion assembly, the stone should have high opposition to weathering and erosion and maximum compression strength. The suitable size for infilling stone differs from 1.5 to 2 times the length of wire mesh, and the stone should be sufficiently large to evade its escape over the opening of wire mesh. Specific gravity of stones is 2.75, compressive strength was 62 Mpa, and density of stones was 27 kN/m^3. The size ranges from 150 to 200 mm. Soils are also used to fill gabion wall. The properties of the soil are discussed in Table 2. LDPE sheets were used as outer covering to fill the soil inside the gabion wire mesh. The soil was compacted using rammer after filling it in layers to achieve maximum density and minimum void ratio.

3.2 *Specifications of Equipment Used for Testing*

There were many equipment's used to test gabion wall. The details of those equipment's have been discussed in this section (Figs. 1 and 2).

- Mechanical jack of stroke 300 mm and maximum capacity of 30 kN was used to apply the load on gabion wall. The increase in loading was gradual. Mechanical jack uses a screw thread for lifting or pushing heavy objects is as shown in Fig. 3. A steel rod was used to rotate the screw and apply the load.

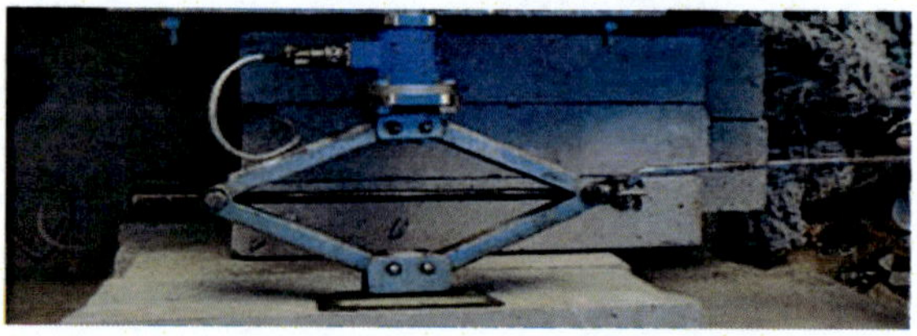

Fig. 1 Mechanical Jack

Fig. 2 Load cell

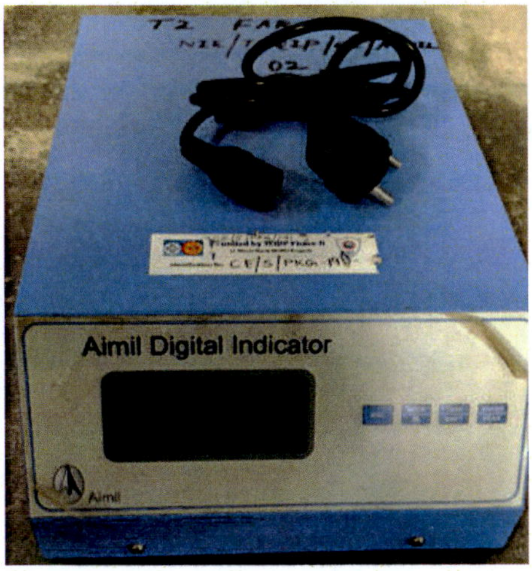

Fig. 3 Digital indicator

- The samples were placed on the concrete bed and were subjected to require loading using load cell. A load cell of capacity 100 kN was used to measure the load applied on the gabion wall using mechanical jack. The face of the load cell has a spherical surface which helps in making contact with the face of the gabion wall. It is very sensitive, having least count of 0.01 kN. The load cell used to conduct experiment is as shown in Fig. 4.

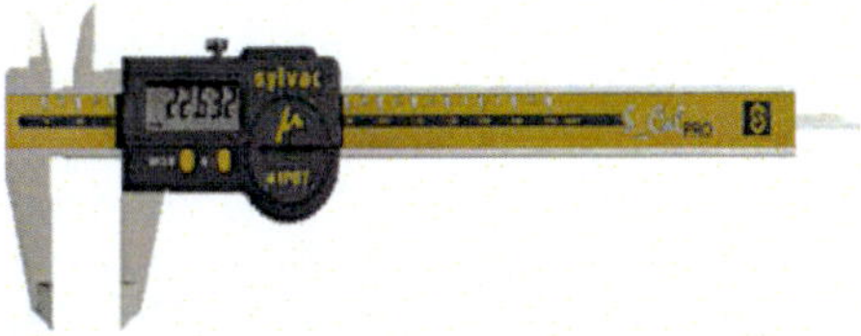

Fig. 4 Vernier Caliper

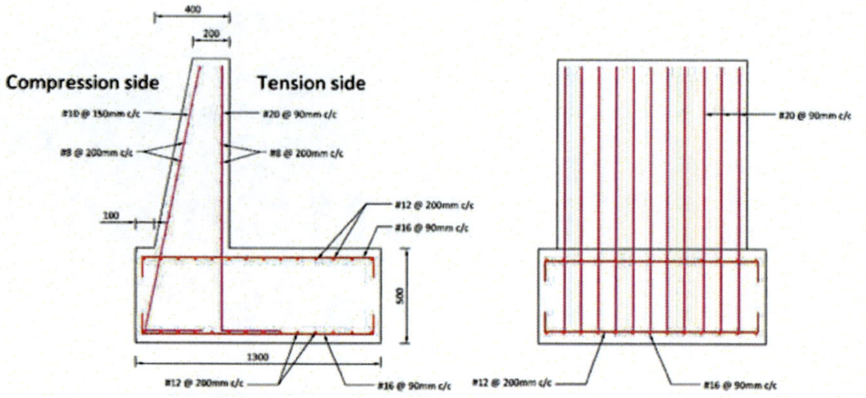

Fig. 5 Detailing of reaction retaining wall

- A digital indicator was used to display the load applied on the gabion wall as shown in Fig. 5.
- A Vernier Caliper was used to measure the deformation on the gabion wire mesh as shown in Fig. 6. After loading the gabion wall for certain load increment, the change in length and diameter of the wire at 100 mm from bottom of gabion wall was measured using Vernier Caliper.
- Later, this data was used to measure the strain at that point. Dial gauges were also used to note down the deformation occurring at the top of the gabion wall.

Fig. 6 Concrete reaction retaining wall

4 Test Setup

4.1 Construction of Reaction Retaining Wall

Before conducting experiments, a thorough study has been conducted to decide the platform for construction of gabion wall and also the loading surface which has to resist the reaction of the mechanical jack. After detailed study, a concrete retaining wall was designed and constructed in such a way that it has to resist the reaction force and also the rotation at the base of the gabion wall. A concrete retaining wall was constructed first which rests firmly on the ground. The design of concrete retaining wall was carried out to resist lateral load of 400 kN at the top. The dimensions of retaining wall were decided based on the results obtained from the design. Detailing of retaining wall is as shown in Figs. 7 and 8. 10 bolts of 20 mm diameter were inserted 500 mm into the concrete bed of retaining wall perpendicular to heel, along the length of the wall. These bolts were used to fix the base of the gabion wall on the ground, so as to achieve fixity at base of the gabion wall.

Fig. 7 Testing of gabion wall specimen 1

Fig. 8 Testing of gabion wall specimen 2

4.2 Construction of Gabion Wall

The gabion wall specimens were tested on the heel portion of the reaction retaining wall constructed which has maximum load capacity of 400 kN. The gabion wall specimens were tested under lateral monotonic loading, and they were fixed at the bottom with a span of 1000 mm. Figures 9 and 10 show experimental setup for gabion wire mesh. Deformation at the top, type of failure, ultimate load, and strain measurement at 100 mm from bottom was measured for respective loading increments which was carefully observed and recorded. Failure of gabion wall is as shown in Figs. 11 and 12. The steps followed for the construction of gabion wall is discussed below.

- Preparing the gabion wire mesh and placing it on bolts
- Fixing the edges of gabion wire mesh and filling it with infill materials
- Fixing plates, studs, and dial gauges
- Testing of gabion wall
- Observing failure of gabion wall.

Fig. 9 Failure of specimen 1

Fig. 10 Failure of specimen 2

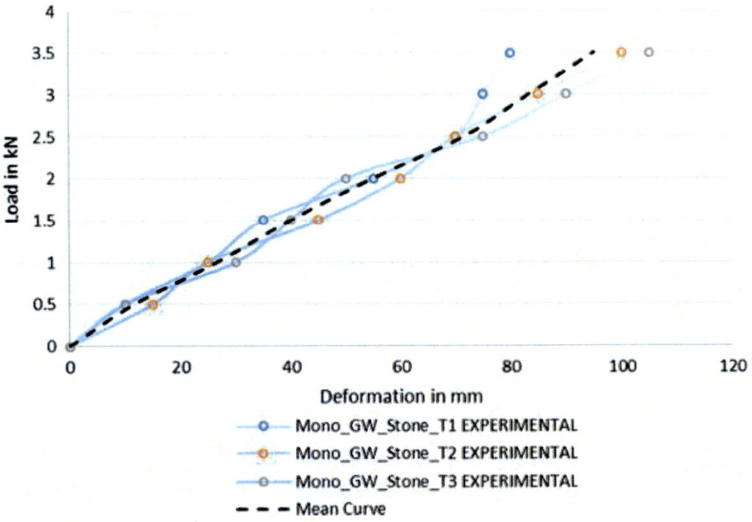

Fig. 11 Load versus deformation curve for specimen 1

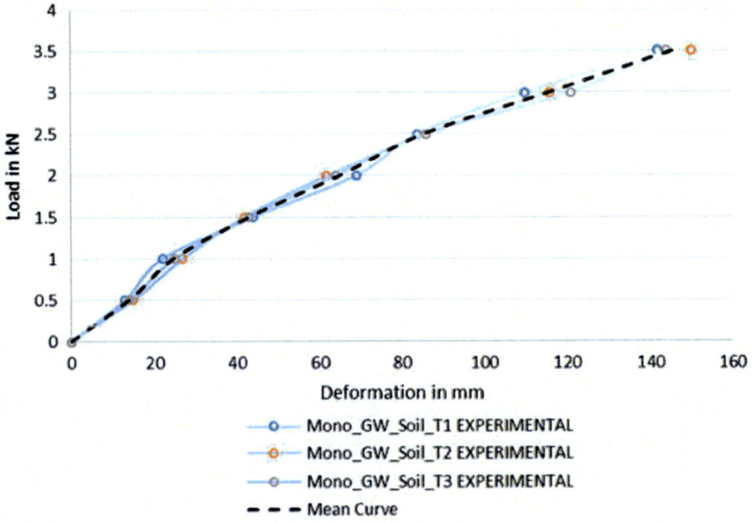

Fig. 12 Load versus deformation curve for specimen 2

5 Results and Discussions

5.1 *Load Deformation Behavior*

The behaviors of gabion wall were studied with load-controlled testing, under lateral monotonic loading. Vernier Caliper was utilized for strain measurement. In monotonic loading, the loads are applied gradually up to the ultimate load, and strain is measured until the failure load. Investigations carried to arrive at behavior of gabion wall for both stone and soil as infill materials are presented below.

5.1.1 Gabion Wall Having Stone as Infill Material

The gabion wall having stone as infill material shows linear variation of load versus deformation curve. The failure of gabion wall was decided by deformation value. There was excess deformation for smaller value of loading. Maximum load being 3.5 kN and maximum deformation obtained for all the three trials were 80 mm, 100 mm, and 105 mm, respectively. So, average deformation was found to be 95 mm. Stones deform lesser compared to soil for similar value of loading.

5.1.2 Gabion Wall Having Soil as Infill Material

Initially, there is slight increase in deformation for the given loading upto 1 kN. After that, the curve increases sharply with the given loading upto 3.5 kN. Maximum deformation obtained for all three trials were 142 mm, 150 mm, and 144 mm, respectively. Maximum deformation obtained for gabion wall having soil as infill material was 1.5 times more than that for gabion walls having stone as infill material. Since soil has less density compared to stones, the resistance to deformation obtained for soil will be less compared to stones.

5.2 *Load Strain Behavior*

5.2.1 Gabion Wall Having Stone as Infill Material

The load strain behavior follows curvilinear pattern. Initially, the slope of the curve increases gradually. Later on, there was a sudden change in the curve which has sharp slope. This indicates that the stones were being adjusted or there was a slight movement in the stones to fill the void ratio inside the gabion wall when the load was applied initially. This process occurred upto 2 kN. Later, the stones inside the gabion wall were packed well. Therefore, it shows marginal increase in the slope indicating there was a slight increase in strain value for larger value of load. The gabion wall

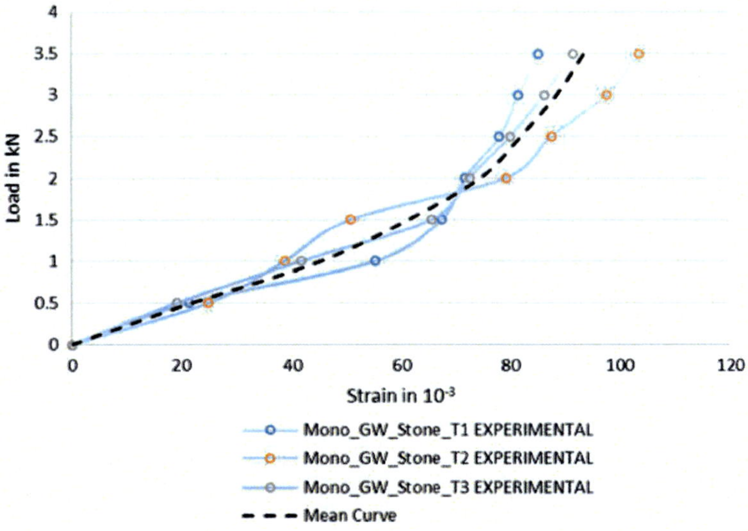

Fig. 13 Load versus strain curve for specimen 1

resists the load after the stones are packed well. As shown in graph, the strain value increases twice as the previous value for each load increment upto 2 kN. Later on, the strain value was increased by 20% for each load increment. Maximum load being 3.5 kN and maximum average tensile strain obtained for this load was 0.09344.

5.2.2 Gabion Wall Having Soil as Infill Material

The load strain behavior follows linear pattern. The curve shows almost linear behavior for all the three trials. The strain values for each load increment were increased by 64% from 0.5 to 1 kN. For next increment, the strain was increased by 37.5% for 1.5 kN and for 2 kN, and the strain value was increased by 21%. This pattern continues upto 3.5 kN, where strain value was increased by 14%. Maximum load being 3.5 kN and maximum average tensile strain obtained for this load was 0.318. Therefore, maximum tensile strain for specimen 2 was 70.6% more than that of specimen 1 (Figs. 13 and 14).

5.3 Moment Curvature Characteristics

5.3.1 Gabion Wall Having Stone as Infill Material

This curve shows same variation as load deformation curve. This relationship provides information about the ductility, stiffness, and strength of the material.

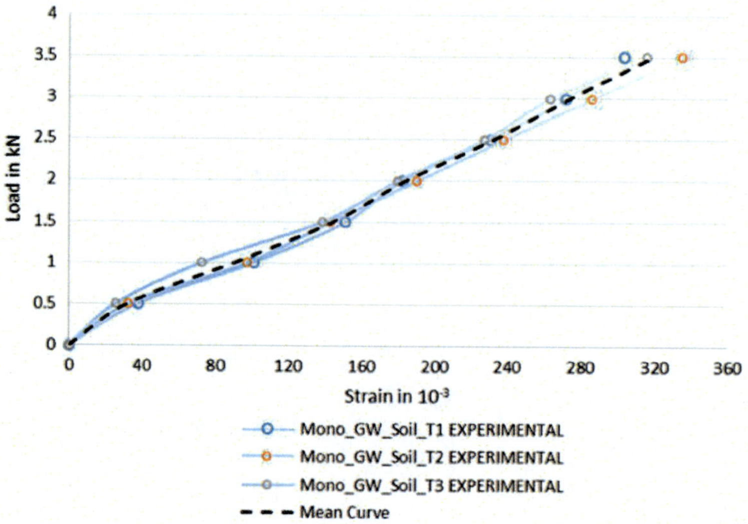

Fig. 14 Load versus strain curve for specimen 2

Maximum moment obtained for the gabion wall was 1.05 kN m, and maximum angle of rotation of cross-section under bending of gabion wall was found to be 17.55.

5.3.2 Gabion Wall Having Soil as Infill Material

This curve shows same variation as load deformation curve. Maximum moment obtained for the gabion wall was 1.05 kN m, and maximum angle of rotation of cross-section under bending of gabion wall was found to be 25.84. The curvature obtained for specimen 2 is 32% more than that of specimen 1 (Figs. 15 and 16).

5.4 H/Δ Ratio

The maximum value of H/Δ ratio was found to be 4.21 and 2.76 for specimen 1 and specimen 2, respectively, for 3.5 kN. This value provides information about the stiffness of the material. The H/Δ ratio for specimen 1 is 34.44% more than that of specimen 2. Specimen 1 is stiffer and offers higher resistance to deformation when compared with specimen 2. Specimen 2 is more ductile when compared to specimen 1 (Fig. 17).

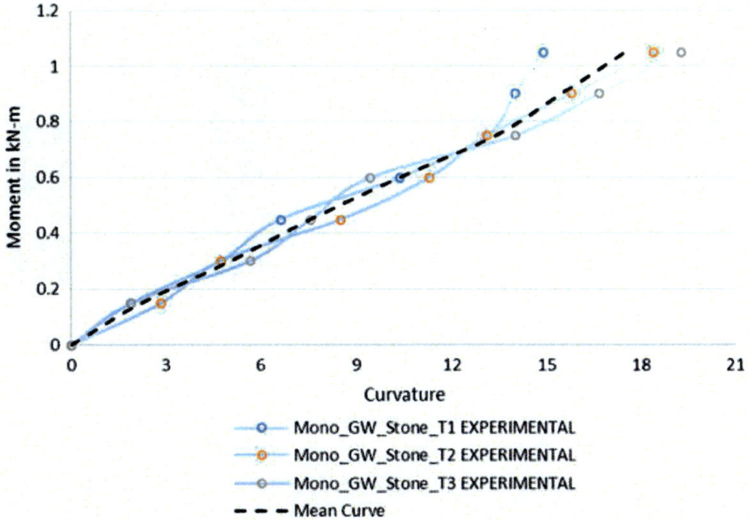

Fig. 15 Moment curvature characteristics for specimen 1

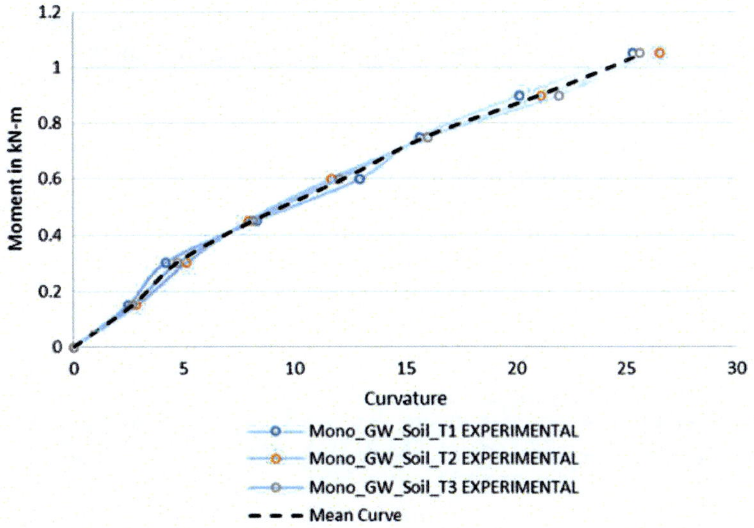

Fig. 16 Moment curvature characteristics for specimen 2

6 Conclusion

Following observations were made for monotonic loading through experimental approach:

Fig. 17 H/Δ ratio for specimen 1 and specimen 2

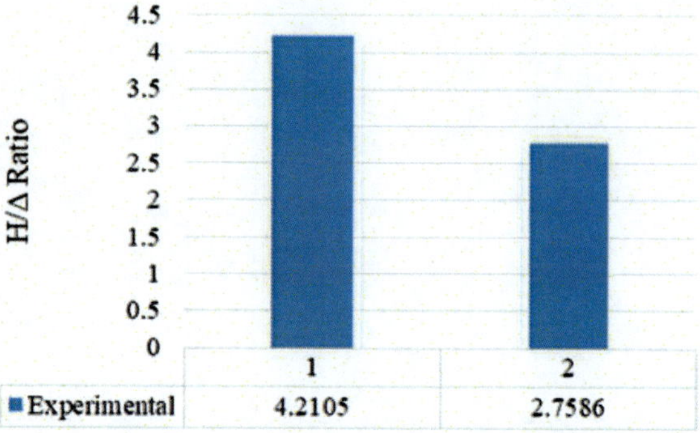

- Load versus deformation curve for specimens with soil and stone as infill materials followed typical curvilinear pattern.
- The initial variation in load versus strain curve for stone as infill material shows that filling of void ratio occurred upto 1.5 kN loading, and further variation shows that the material got stiffer.
- Load versus strain curve for specimens with soil and stone as infill materials followed typical curvilinear pattern.
- Moment versus curvature diagram for specimens with soil and stone as infill materials showed same behavior as load versus deformation curve.
- The variation in deformation, strain values in successive trials is attributed to sensitivity of instrumentation and experimental errors.
- Maximum deformation of 150 mm and maximum strain of 0.335 were obtained in 2nd trial for specimen having soil as infill material.
- Soil as infill material does not offer resistance to deformation when compared to stones.
- H/Δ ratio was maximum for specimen having stone as infill material, which indicates that this specimen was stiffer when compared to specimen having soil as infill material.
- The deformation value of gabion wall having stone as infill materials was increased by 34.45% after replacing it with soil.
- There was an increase in strain value of 70.6% in gabion wall having soil as infill material when compared with stone as infill material.

References

1. IS 16014:2018. Mechanically woven, double twisted, hexagonal wire mesh Gabions, revet mattresses and rock fall netting
2. ASTM A975-97. Standard specification for double–twisted hexagonal mesh Gabions and Revet Mattresses

3. Lin YL, Yang GL, Li Y, Zhao LH (2010) Engineering behaviors of reinforced gabion retaining wall based on laboratory test. J Cent South Univ Technol 17(6):1351–1356
4. Saravanapriya S (2018) Experimental investigation on improvement in strength characteristics of gabion wall. Int J Civil Eng Technol 9(9):628–641
5. Ramli M, Karasu T, Dawood ET (2013) The stability of gabion walls for earth retaining structures. Alex Eng J 52(4):705–710
6. Jiang Y, Wang X (2011) Stress-strain behavior of gabion in compression test and direct shear test. In: ICTE 2011, pp 1457–1462
7. Yang GL, Huang XJ, Lin YL (2010) Test study on engineering properties of gabion structures. In: Advances in environmental geotechnics Springer, Berlin, Heidelberg, pp 805–811
8. Apriyono A, Sumiyanto S, Yanto Y, Iswahyudi S, Pamungkas RW (2019) Analysis of different stripe connectors for a Gabion Wall based on Woven Waste tire stripes. Civil Eng Dimension 21(2):66–69

Exploring the Application of Data Envelopment Analysis in the Evaluation of Public Transport Organizations

P. Praveen Kumar, Varghese George, and Raviraj H. Mulangi

Abstract The work provides details on a study on performing a comparative analysis of productivity and efficiency of selected public transport organizations in India. The *data envelopment analysis* (*DEA*) method was used in this study using subroutines specially developed in the open-source *R* programming environment used in statistical analyses. The *DEA* approach is a commonly adopted non-parametric method of analysis for the measurement of efficiency-related characteristics. Based on the availability and consistency of information, selected input and output data on 27 State Road Transport Undertakings (SRTUs) were used in this study. The *scores of efficiencies* were then computed using an input-oriented *DEA* model with *constant return to scale*. Finally, the analysis was used to distinguish between *efficient* and *inefficient* SRTUs based on *weights* assigned to *peer units*. The results of this study indicate that only 10 out of the 27 SRTUs performed well. It was also found that the performance parameters could be improved by reducing the quantity of input-related parameters in order to maintain the same level of output.

Keywords SRTU · DEA · Efficiency · Non-parametric approach · Benchmarking · Bus Transport

P. P. Kumar (✉) · V. George · R. H. Mulangi
Department of Civil Engineering, National Institute of Technology Karnataka, Surathkal Mangalore 575025, India

R. H. Mulangi
e-mail: ravirajmh@nitk.edu.in

L. Nandagiri et al. (eds.), *Sustainability Trends and Challenges in Civil Engineering*, Lecture Notes in Civil Engineering 162, https://doi.org/10.1007/978-981-16-2826-9_28

1 Introduction

The *data envelopment analysis* (*DEA*) approach is a commonly adopted non-parametric method of analysis for the measurement of efficiency-related characteristics of business organizations. This method was originally seen to be applied to assess the performance of the banking sector, and other economic sectors including the health-care sector. The present study provides details on the application of the *DEA* approach in performing a comparative analysis of productivity and efficiency of selected public transport organizations in India. The focus of the present study is on performing a comparative study of the performance of three important input variables such as, *total operating cost* (TOC), *average fleet operated* (AFO), and *total staff employed* (TSE) categorized as part of service inputs, and three output variables such as, *total revenue* (TR), *total passenger-km* (TPKM), and *total passengers carried* (TPC) classified under *service consumption*. The *data envelopment analysis* (*DEA*) method was used in this study using sub-routines specially developed in the *R* programming language commonly used for statistical analyses.

The theoretical foundations for the application of the *DEA* approach were laid by Charnes et al. [1] while Coll-Serrano et al. [2] developed a package named '***deaR***' using the *R programming language* [3] for *DEA* analysis and comparison of the performances of various organizations in general. The *R programming language* is generally used in developing tools for performing statistical analyses.

Based on the availability and consistency of information, selected input and output data on 27 State Road Transport Undertakings (SRTUs) were used in this study. The *scores of efficiencies* were then computed using an input-oriented *DEA* model with *constant return to scale*. Finally, the analysis was used to differentiate between *efficient* and *inefficient* SRTUs based on the *weight to peer units*. The results of this study indicate that only 10 out of the 27 SRTUs performed well. It was also found that the performance parameters could be improved by reducing the quantity of input-related variables in order to maintain the same level of output.

2 Basic Studies and Literature Review

(a) *Theoretical Foundations of the DEA Model*

Efficiency of various organizations can be evaluated based on parametric approaches such as, the least-square regression method [4, 5], and the stochastic frontier analysis approaches that include the Cobb–Douglas or translog frontier analysis method [6], and the semi-parametric approaches such as, the tobit regression method, and the truncated regression method [7], and the non-parametric approaches such as the data envelopment analysis (DEA) method [1]. The most suited method for analysis is adopted based on the availability, type of data, and the consistency of data. Parametric techniques are adopted when data collection involves randomness as in a sample survey, where there are chances of occurrence of measurement errors. On the other hand, a non-parametric approach such as the *DEA* method could be adopted where random-errors are expected to be the least [8].

The objective of the *DEA* approach proposed by Charnes et al. [1] is to maximize the efficiency of the organization or *decision-making unit* (DMU) under study. This approach focuses on maximizing the *normalized* ratio of the output to the input of selected characteristics. This approach was based on the measurement of productive efficiency [9]. The *DEA* model can be formulated as an *input oriented*, *output oriented*, or a combined approach. The original approach adopted by Charnes et al. [1] in the CCR model as it later came to be known was based on inputs to the *decision-making unit* with a specific rate of return (or constant returns to scale). The measure of efficiency obtained using the CCR model is called as the *overall technical efficiency* (OTE).

Banker et al. [10] developed an alternative model that came to be known as the BCC model, where varying rates of returns (or variable returns to scale) could be studied. The measure of efficiency obtained using the BCC model is called as the *pure technical efficiency* (PTE).

In the later stages, various modified forms of the *DEA* approach evolved. The *bootstrap DEA* developed by Sadjadi and Omrani [11], the *fuzzy DEA* developed by Bray et al. [12] that could handle uncertain and partially erroneous data, and the *robust optimization model* or *robust DEA* developed by Landete et al. [13], are some of the examples.

(b) *Measurement of Efficiency in Public Transport Organizations*

A number of studies were performed on the application of the DEA approach in evaluating the performance of public transport organizations in India.

Agarwal et al. [14] performed investigations on determining the technical efficiency of 35 State Road Transport Undertakings (SRTUs) in India using the *DEA* approach for data on performance of bus transport organizations in the 2004–05.

Saxena and Saxena [15] applied the technique based on the*DEA* approach in comparing the technical and managerial efficiencies of 25 SRTUs in India based on data compiled over a three-year period between 2002 and 2005. The scale efficiencies were then computed as the ratio of the efficiency scores obtained using the CCR model and the BCC model.

Kumar [16] adopted the use of DEA models such as the CCR model, the BCC model, and the Andersen and Petersen's super-efficiency model [17] to compute the efficiency scores of public transport organizations. The studies also focused on the factors that resulted in the variation of performance among various organizations using the Tobit analysis. Mulangi et al. [18] evaluated the performance of 13 divisions of Karnataka State Road Transport Corporation using the principal component analysis (PCA) approach and the DEA method.

It may be noted that if a number of high performing DMUs are included in the reference set of organizations studied, then the DEA analysis can ensure a smoother frontier line of points. However, in case the number of high performing DMUs is lesser, then the frontier points appear to be jagged. Bose and Patel [19] proposed the use of artificial neural networks (ANNs) in generating additional frontier points that assisted in smoothening the frontier points. A two-step approach was used in this study, where the data on the functioning of 12 hospitals was adopted based on CCR model.

3 Methodology

In general, *efficiency* can be defined as the ratio of output to input. A *decision-making unit* (DMU) or an organization is considered to be efficient if a higher output can be attained at a lower input. When two or more input and output parameters are involved, then the efficiency can be compared using a composite efficiency factor as given in (1a) and (1b).

$$\text{Efficiency} = \text{Weighted Outputs/Weighted Inputs} \quad (1a)$$

Or,

$$\text{Combined Efficiency Factor} = (u_1Y_1 + u_2Y_2 + \ldots + u_sY_s)/(v_1X_1 + v_2X_2 + \ldots + v_mX_m) \quad (1b)$$

where $Y_1, Y_2, \ldots Y_s$ indicate the 's' output variables, $u_1, u_2, \ldots u_s$ indicate the weights assigned to the 's' output variables; $X_1, X_2, \ldots X_m$ indicate the 'm' input variables; and $v_1, v_2, \ldots v_m$ indicate the weights assigned to the 'm' input variables.

However, when comparing parameters with different measurement units, for various organizations, *DEA* model can be used as part of the *multi-criteria decision-making model* (MCDM). The *DEA* model proposed by Charnes et al. [1] is capable of handling multiple input and output variables for the comparison of performance of DMUs. The model incorporates the use of different weights for different input and output variables since managers do not give equal weightage to different types of input and output variables. In this approach, a *linear programming* (LP) model is formulated for each of the DMUs, where the *combined efficiency factor* is maximized as shown in (2a), subject to constraints as in (2b) to (2d) given below:

$$\text{Maximize} = (u_1Y_1 + u_2Y_2 + \ldots + u_sY_s)/(v_1X_1 + v_2X_2 + \ldots + v_mX_m) \quad (2a)$$

Such that,

$$(u_1Y_1 + u_2Y_2 + \ldots + u_sY_s)/(v_1X_1 + v_2X_2 + \ldots + v_mX_m) \leq 1 \tag{2b}$$

$$u_1, u_2, \ldots u_s \geq 0 \tag{2c}$$

$$v_1, v_2, \ldots v_m \geq 0 \tag{2d}$$

where $Y_1, Y_2, \ldots Y_s$ indicate the 's' output variables, $u_1, u_2, \ldots u_s$ indicate the weights assigned to the 's' output variables; $X_1, X_2, \ldots X_m$ indicate the 'm' input variables; and $v_1, v_2, \ldots v_m$ indicate the weights assigned to the 'm' input variables.

Since the objective function as expressed in (2a) is fractional in nature, it is required to reformulate the objective function of the LP problem as in (3a). Here, the denominator of the objective function is assumed to be equal to 1, and an additional constraint is provided as in (3c). Thus, the constraint given by (2b) can be reformulated as shown in (3b). The modified primal form of the LP problem for the qth *decision-making unit* (DMU) can be expressed according to CCR model as follows,

$$\text{Maximize}\, h_q = \sum_{r=1}^{s} u_r Y_{rq} \tag{3a}$$

Such that,

$$\sum_{r=1}^{s} u_r Y_{rq} - \sum_{i=1}^{m} v_i X_{iq} \leq 0 \quad q = 1, 2, \ldots, n \tag{3b}$$

$$\sum_{i=1}^{m} v_i X_{iq} = 1 \tag{3c}$$

$$u_1, u_2, \ldots u_s \geq 0 \tag{3d}$$

$$v_1, v_2, \ldots v_m \geq 0 \tag{3e}$$

where h_q represents the *combined efficiency factor* for the qth decision-making unit.

The methodology can be explained using a simple demonstration. Let us consider four DMUs, namely 1, 2, 3, and 4 with inputs and outputs as listed in Table 1:

Here, 's' number of outputs measured = 3, and 'm' number of inputs measured = 2. The basic input-oriented CCR model for the first DMU can be formulated using (4a) to (4e) corresponding to (3a) to (3e) as given below:

$$\text{Maximize}\, h_1 = \sum_{r=1}^{3} u_r Y_{r1} \tag{4a}$$

$$=> (u_1 * 5) + (u_2 * 50) + (u_3 * 10)$$

Table 1 A sample problem for the DEA approach

DMU (q)	Inputs (m)		Outputs (s)		
	(X_1)	(X_2)	(Y_1)	(Y_2)	(Y_3)
1	1	25	5	50	10
2	0.8	22	4	45	8
3	0.6	15	3	55	6
4	1	27	4	40	8

Such that,

$$\sum_{r=1}^{3} u_r Y_{rq} - \sum_{i=1}^{2} v_i X_{iq} \leq 0 \tag{4b}$$

Or,

$$[(u_1 * 5) + (u_2 * 50) + (u_3 * 10)] - [(v_1 * 1) + (v_2 * 25)] \leq 0$$

$$[(u_1 * 4) + (u_2 * 45) + (u_3 * 8)] - [(v_1 * 0.8) + (v_2 * 22)] \leq 0$$

$$[(u_1 * 3) + (u_2 * 55) + (u_3 * 6)] - [(v_1 * 0.6) + (v_2 * 15)] \leq 0$$

$$[(u_1 * 4) + (u_2 * 40) + (u_3 * 8)] - [(v_1 * 1) + (v_2 * 27)] \leq 0$$

Also,

$$\sum_{i=1}^{2} v_i X_{i1} = 1 \tag{4c}$$

Or,

$$[(v_1 * 1) + (v_2 * 25)] = 1$$

$$u_1, u_2, u_3 \geq 0 \tag{4d}$$

$$v_1, v_2 \geq 0 \tag{4e}$$

Similarly, we need to formulate an LP problem for other 3 DMUs. Thus, there will be 4 objective functions for 4 DMUs along with the associated constraint equations.

The solution to the LP problems formulated above can be obtained as shown in Table 2 using *solver* sub-routine in *Excel* spreadsheet application. The solutions can also be computed using other programming tools including MATLAB.

Here, it may be observed that the efficiency is measured in a scale varying between 0 and 1.0. Thus, DMUs such as 1, 2, and 3 having values approximately equal to 1.0 may be considered to be highly efficient.

A dual form of the above mentioned primal LP formulation expressed vide (3a) to (3e) can also be derived as demonstrated by Charnes et al. [1]. In the primal LP problem, since the objective function is required to be maximized, the dual problem can be formulated to perform a minimization operation. The solutions to the primal and dual problems will be identical as both represent the same optimal conditions. This approach is adopted if there are more number of constraints in the primal LP problem, so as to simplify the computations.

For the primal LP problem formulated above vide (3a) to (3e), the dual LP problem with slack variables can be formulated as expressed in (5a) to (5d) as demonstrated in CCR model.

$$\text{Minimize } \theta - \varepsilon\left(\sum_{r=1}^{s} S_r^+ + \sum_{i=1}^{m} S_i^-\right) \tag{5a}$$

Such that,

$$\sum_{q=1}^{n} \lambda_q X_{iq} + S_+^- = \theta X_{iq} \quad i = 1, 2, .., m \tag{5b}$$

$$\sum_{q=1}^{n} \lambda_q Y_{rq} - S_r^+ = Y_{r\,q} \quad r = 1, 2, .., \text{s} \tag{5c}$$

$$\lambda_q, S_i^-, S_r^+ \geq 0 \tag{5d}$$

where x_{iq} = amount of input '*i*' used by the *q*th DMU, y_{rq} = amount of output '*r*' produced by the *q*th DMU; m = the number of inputs; s = the number of outputs; and n = the number of DMUs.

In the above formulation, θ and λ_q (where $q = 1, 2, \ldots n$) are the dual variables of the primal LP problem. Also, S_i^- and S_r^+ are the 'Slack variables' (additional variables) added to the model in order to convert inequality constraints to equality

Table 2 Summary of solutions to the sample problem for the DEA approach

DMU (q)	Efficiency (h_q)
1	1
2	1
3	1
4	0.8

constraints. The *slack variables* indicate the additional increase in outputs (S_r^+), and/or the decrease in inputs (S_i^-) required to be implemented for a decision-making unit to become more efficient. 'ε' is a '*Non-Archimedean infinitesimal*' constant which is smaller than any positive real number, and is normally assumed as 10^{-6}.

A. *Data Collection and Preprocessing*

Bus transport organizations in India are mostly owned and operated by State Road Transport Undertakings (SRTUs). These provide passenger transport services for rural, inter-city, and urban areas. Central Institute of Road Transport (CIRT) in Pune collects data on financial and physical performances submitted by the SRTUs, compiles the data, and publishes the summarized findings. The transport research wing of the Ministry of Road Transport and Highways (MoRTH) also compiles and publishes details on the physical and financial performance of SRTUs annually. The information compiled as part of *financial performance* includes the total revenue, total cost, and details on other cost components, while the data compiled as part of *Physical Performance* includes details on fleet-utilization, capacity-utilization, manpower productivity, and other related operational information.

In the year 2014–15, forty-six SRTUs reported details on financial and physical performance to the transport research wing of Ministry of Road Transport & Highways, Government of India [20]. Out of the forty-six SRTUs, some of the data pertaining to a number of SRTUs were found to be missing. Such data were eliminated, and the data reported by twenty-seven SRTUs were finally selected for analysis in the present study. The final set of data used in this study is provided in Table 3.

In the present study, it was proposed to compare the performance of bus transport organizations using the DEA approach considering three important input and three important output variables. These variables form part of the set of essential input and output indicators identified by a number of researchers [21–24].

The variables such as *total operating cost* (TOC), *average fleet operated* (AFO), and *total staff employed* (TSE) can be categorized under *service inputs*, while *total revenue* (TR), *total passenger-km* (TPKM), and*total passengers carried* (TPC) can be classified under *service consumption.*

In a DEA approach, the number of DMUs selected should be greater than or equal to three times the total number of input and output variables considered, or the number of DMUs must be greater than product of the number of input and output variables in order to ensure reliable results [25]. This criterion was satisfied while formulating the problem for the present study using data on 27 SRTUs.

4 Results and Discussion

The computations related to the DEA analysis can be performed using the '*deaR*' open-source package, a GPL software sub-routine developed by Vicente Coll-Serrano et al. [2] as part of the open-source *R* programming environment [3]. *R*

Table 3 An overview of existing performance of 27 SRTUs in India

Sl. No.	Names of SRTUs	TOC *(10^5 Rs)*	AFO (Number)	TSE (Number)	TR *(10^5 Rs)*	TPKM *(10^5)*	TPC *(10^5)*
1	Ahmedabad Municipal Transport Service (AMTS)	35,413.26	719	4971	13,011.4	15,851.6	2024.37
2	Andhra Pradesh State Road Transport Corporation (APSRTC)	540,280.1	12,023	61,806	480,768.1	548,032.3	23,192.1
3	Brihanmumbai Electricity Supply and Transport (BEST) undertaking	235,503	3636	35,705	150,855.9	90,716	12,216
4	Bangalore Metropolitan Transport Corporation (BMTC)	232,174.8	6014	36,474	225,684.4	217,253.9	19,366.9
5	Bihar State Road Transport Corporation (BSRTC)	2348.66	85	749	1562.75	2674.7	25.36
6	Calcutta State Transport Corporation (CSTC)	23,191.2	401	4799	7240.52	6306	616
7	Chandigarh Transport Undertaking (CTU)	18,139.97	378	2102	11,107.08	15,441	478.15
8	Delhi Transport Corporation (DTC)	510,467.7	4180	32,864	111,321.1	136,064.6	14,187.28
9	Gujarat State Road Transport Corporation (GSRTC)	299,568.7	6683	39,257	286,218.5	333,368.2	7734.65
10	Jammu and Kashmir State Road Transport Corporation (JKSRTC)	8372.61	287	1720	5341	6065.75	45.69

(continued)

Table 3 (continued)

Sl. No.	Names of SRTUs	TOC ($10^5 Rs$)	AFO (Number)	TSE (Number)	TR ($10^5 Rs$)	TPKM (10^5)	TPC (10^5)
11	Kadamba Transport Corporation (KTS)	16,581.87	361	2054	14,388.58	9523.52	345.08
12	Karnataka State Road Transport Corporation (KnSRTC)	324,818	7572	37,326	320,468.9	354,273.7	10,129.75
13	Maharashtra State Road Transport Corporation (MSRTC)	764,967	16,702	107,500	725,866	523,753.8	24,556.91
14	Meghalaya Transport Corporation (MTC)	1360.98	42	286	1179.95	399.09	3.82
15	Metropolitan Transport Corporation Chennai (MTC-CNI)	159,599.3	3187	25,219	137,651.8	189,963	18,120
16	Mizoram State Transport (MZST)	2383.36	18	418	231.45	229.15	0.48
17	North Bengal State Transport Corporation (NBSTC)	18,110.4	501	3001	8770.19	12,000.51	763.68
18	North Eastern Karnataka Road Transport Corporation (NEKnRTC)	146,413.9	3795	18,412	144,901.2	154,800.5	4891
19	North Western Karnataka Road Transport Corporation (NWKnRTC)	174,668.1	4507	23,573	171,583.9	179,082	8303.75
20	Rajasthan State Road Transport Corporation (RSRTC)	214,750.6	4324	20,551	183,692.9	223,386.3	3582.29

(continued)

Table 3 (continued)

Sl. No.	Names of SRTUs	TOC *(10^5Rs)*	AFO (Number)	TSE (Number)	TR *(10^5Rs)*	TPKM *(10^5)*	TPC *(10^5)*
21	South Bengal State Transport Corporation (SBSTC)	18,967.54	403	2148	15,242.53	18,923.67	1020.04
22	Tamil Nadu State Transport Corporation. Ltd., Coimbatore (TNSTC-CBE)	154,469.6	3061	18,964	126,100.7	228,832.2	9014.78
23	Tamil Nadu State Transport Corporation. Ltd., Kumbakonam (TNSTC-KUM)	182,563.2	3622	23,198	164,740.1	275,548	10,748.46
24	Tamil Nadu State Transport Corporation. Ltd., Madurai (TNSTC-MDU)	120,836.4	2381	14,850	106,235.8	187,390.6	6563.54
25	Tamil Nadu State Transport Corporation. Ltd., Salem (TNSTC-SLM)	117,933.4	2212	14,326	90,742.01	173,098.1	5580.85
26	Tamil Nadu State Transport Corporation. Ltd., Villupuram (TNSTC VPM)	177,126.8	3352	22,573	161,852.4	293,359.9	8450.15
27	Uttar Pradesh State Road Transport Corporation (UPSRTC)	327,306.7	9128	24,904	327,554.6	283,054	5440

is a programming language available under the free software environment for statistical computing and graphical representation. This programing environment was originally developed by the R Foundation for Statistical Computing, and is presently supported by them [3].

The '*deaR*' open-source package [2] supported by the open-source *R* programming environment was developed to handle classical formulations of DEA models in addition to fuzzy-based DEA models. The algorithm is programmed to perform the analysis in a two-stage process. In the first phase, the value of the efficiency scores

(θ) was optimized, followed by the second phase where the algorithm minimizes the quantity used as input based on the optimized value of θ in order to see how the performance of inefficient DMUs can be enhanced.

In the present study, the *input-oriented CCR model* was adopted, where the algorithm minimizes the quantity used as input. The details on the optimized values of λ_q assigned as coefficients to the input and output parameters generated by the '*deaR*' open-source package [2] are shown in Appendix 1. The cells in the last column of the table provide the sum of the λ_q values for each of the SRTUs. These values indicate the level of improvement required to be undertaken when compared to efficient DMUs.

Of the 27 SRTUs, 10 SRTUs were found to be efficient since these possessed *efficiency scores* equal to 1. These SRTUs are considered to define the benchmark for best practices, and together form the *peer group* in DEA terminology as examples to be followed. These SRTUs are classified as *efficient frontier* organizations that ensure effective utilization of resources. See Fig. 1.

The 10 exemplary SRTUs that constitute the peer group include: Bangalore Metropolitan Transport Corporation (*BMTC*), Karnataka State Road Transport Corporation (*KnSRTC*), Metropolitan Transport Corporation Chennai (*MTC-CNI*), North Eastern Karnataka Road Transport Corporation (*NEKnRTC*), Rajasthan State Road Transport Corporation (*RSRTC*), Tamil Nadu State Transport Corporation Ltd., Coimbatore (*TNSTC-CUM*), Tamil Nadu State Transport Corporation Ltd., Kumbakonam (*TNSTC-KUM*), Tamil Nadu State Transport Corporation Ltd., Madurai (*TNSTC-MDU*), Tamil Nadu State Transport Corporation Ltd., Villupuram (*TNSTC-VPM*), and Uttar Pradesh State Road Transport Corporation (*UPSRTC*).

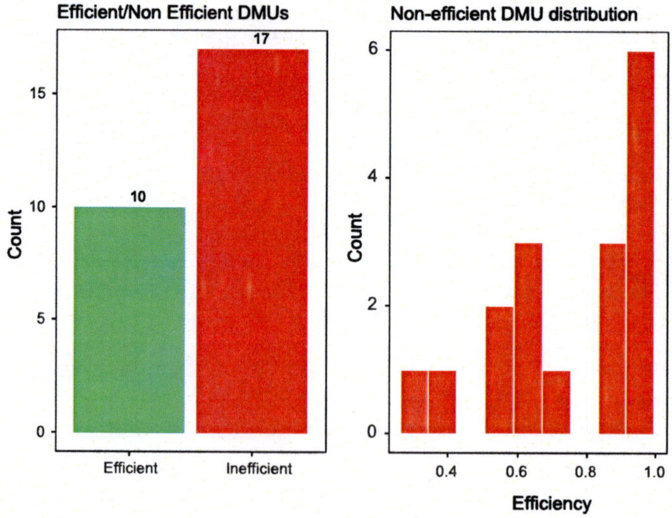

Fig. 1 Details on efficient and inefficient DMUs, and the efficiency score distribution

Table 4 Efficiency scores and returns to scale values based on the DEA analysis for SRTUs

S. No.	SRTU[a]	Eff. score	'λ_q' sum	RTS	S. No.	SRTU[a]	Eff. score	'λ_q' sum	RTS
1	AMTS	0.56678	0.1117	IRS	15	MTC-CNI	1	1	CRS
2	APSRTC	0.94613	2.3985	DRS	16	MZST	0.2663	0.0014	IRS
3	BEST	0.89596	0.9914	IRS	17	NBSTC	0.52659	0.051	IRS
4	BMTC	1	1	CRS	18	NEKnRTC	1	1	CRS
5	BSRTC	0.72049	0.009	IRS	19	NWKnRTC	0.99817	1.0469	DRS
6	CSTC	0.3922	0.048	IRS	20	RSRTC	1	1	CRS
7	CTU	0.65793	0.0606	IRS	21	SBSTC	0.92248	0.0815	IRS
8	DTC	0.61316	0.8023	IRS	22	TNSTC-CBE	1	1	CRS
9	GSRTC	0.98502	1.0816	DRS	23	TNSTC-KUM	1	1	CRS
10	JKSRTC	0.64877	0.0174	IRS	24	TNSTC-MDU	1	1	CRS
11	KTS	0.90292	0.0596	IRS	25	TNSTC-SLM	0.94668	0.7606	IRS
12	KnSRTC	1	1	CRS	26	TNSTC-VPM	1	1	CRS
13	MSRTC	0.98629	2.9751	DRS	27	UPSRTC	1	1	CRS
14	MTC	0.86633	0.0036	IRS		–	–	–	–

[a]The full names of the SRTUs are provided in Table 3

RTS Returns to Scale; *CRS* Constant Returns to Scale; *IRS* Increasing Returns to Scale; and *DRS* Decreasing Returns to Scale

The remaining 17 SRTUs have an *efficiency score* of lesser than 1, indicating that these SRTUs are not operated efficiently.

In the case of *APSRTC*, the efficiency score is 0.94613 and the $\lambda_q sum$ value is 2.3985 as observed in Appendix 1, and in Table 4. Since the $\lambda_q sum$ value is greater than 1.0, the SRTU is said to operate at *decreasing return to scale* (DRS) according to Banker et al. [10] and Seiford and Zhu [26].

The $\lambda_q sum$ value of is 2.3985 for *APSRTC* was computed based on the benchmark performances of Karnataka State Road Transport Corporation (*KnSRTC*), Metro Transport Corporation Ltd. Chennai (*MTC-CNI*), Tamil Nadu State Transport Corporation. Ltd., Villupuram (*TNSTC-VPM*), and Uttar Pradesh State Road Transport Corporation (*UPSRTC*) with λ_q values of 0.1397, 0.43, 1.3411, and 0.487, respectively, resulting in a summed up value of 2.3985 as shown in Appendix 1.

This indicates that if the performance of *APSRTC* must be improved, then the targeted total operating cost (TOC) of this SRTU must be equal to the sum of 0.1397 × TOC for *KnSRTC*, 0.43 × TOC for *MTC-CNI*, 1.3411 × TOC for *TNSTC-VPM*, and 0.487 × TOC for *UPSRTC* which is equal to 511,176.1 units in place of the existing value of 540,280.1 units as shown in Appendix 2, and Table 3, respectively, which are attained as part of the output generated by the '*deaR*' Program. In a similar way, the inputs such as the average fleet operated (AFO), and the total staff employed (TSE) must be reduced. Thus, the improvement in performance for *APSRTC* can be attained for each input variable as the sum of the product of the λ_q values, and the corresponding input values of the benchmarked organizations. This is expressed as,

$$X_{iq-\text{target}} = \sum_{q=1}^{n} \lambda_q X_{iq} \tag{6a}$$

where $X_{iq\text{-target}}$ = the target input i for the qth DMU, X_{iq} = actual input i for the qth DMU, and λ_q = dual variables of the primal LP problem.

The required improvements to the SRTUs can also be found in another manner based on the efficiency scores and the values of slack variables. In the case of *BEST*, the *efficiency score* is 0.89596 and the λ_q*sum* value is 0.9914 as observed in Appendix 1, and in Table 4. Since the λ_q*sum* value is lesser than 1.0, the SRTU is said to operate at *increasing return to scale* (IRS) according to Banker et al. [10] and Seiford and Zhu [26].

Here, the performance of an inefficient bus transport organization such as *BEST* can be improved by proportionally reducing inputs as part of the *input-oriented CCR model* such that the output increases proportionately. Since the desired efficiency score should have been 1.00, the improvement in performance of *BEST* can be achieved by reducing the inputs by 10.404% (obtained as 100 * [1 − 0.89596]). Additionally, the values of the input-related slack variables need to be reduced.

This indicates that if the performance of *BEST* must be improved, then the targeted total operating cost (TOC) of this SRTU must be equal to 0.10404 times the existing value of TOC of 235,503 units for *BEST* minus the value of 42,353.5, and the slack variable value for TOC for *BEST* as shown in Table 3 and in Appendix 3. This computes to the target value of 168,648.05 units of TOC for BEST as shown in Appendix 2.

In a similar way, the inputs such as the average fleet operated (AFO), and the total staff employed (TSE) must be reduced. Thus, the improvement in performance for *BEST* can be attained for each input variable as the product of efficiency score and the input variable value minus the slack value for the corresponding input. This is expressed as,

$$X_{iq-\text{target}} = \left(\theta * X_{iq}\right) - S_i^- \tag{6b}$$

where $X_{iq\text{-target}}$ = the target input i for the qth DMU, X_{iq} = actual input i for the qth DMU, $\theta*$ = efficiency score of the qth DMU, and S_i^- = value of the corresponding input slack variable.

In the case of output variables, it is required to add the slack value for the corresponding output. This is expressed as,

$$Y_{rq-\text{target}} = Y_{rq} + S_r^+ \tag{7}$$

where $Y_{rq\text{-target}}$ = target output '*r*' for the qth DMU; Y_{rq} = actual output '*r*' for the qth DMU; and S_i^+ = value of the corresponding output slack variable.

In the case of *BMTC*, the efficiency score is 1.0 and the $\lambda_q sum$ value is 1.0 as observed in Appendix 1, and in Table 4. Since the $\lambda_q sum$ value is equal to 1.0, the SRTU is said to operate at *constant return to scale* (CRS) according to Banker et al. [10] and Seiford and Zhu [26].

Usually, during short-term periods such as during certain seasons of the year, organizations may seem to operate at *increasing returns to scale* (IRS) or at *decreasing returns to scale* (DRS). Organizations that undergo *decreasing returns to scale* experience rising average costs, while organizations with *increasing returns to scale* are characterized by reduction in average costs.

However, according to micro-economic theory, it is required for organizations to operate at a *constant return to scale* (CRS) ensuring productivity over long periods by minimizing costs and by maximizing revenues [27]. This implies that in the long run, the organizations are expected to operate at *constant return to scale* (CRS) by either expanding the services, or by curtailing unprofitable services so as to compete with other similar firms.

Out of the seventeen SRTUs that operate at lower efficiencies, it is found that thirteen SRTUs are operating at *increasing returns to scale* (IRS), while four SRTUs are found to operate at *decreasing returns to scale* (DRS). Since the thirteen SRTUs are operating at increasing returns to scale (IRS), it can be said that the scale of operations of these SRTUs can be further expanded. On the other hand, the four SRTUs that are seen to operate at *decreasing returns to scale* (DRS) can be made more efficient by curtailing services offered to unproductive routes.

On further examination of Appendix 1, it can be seen that *TNST-VPM* or Tamil Nadu State Transport Corporation-Villupuram has been considered as a benchmark for comparison of efficiencies of fourteen comparatively inefficient SRTUs, while *MTC-CNI* (Metro Transport Corporation Chennai) has been considered as a benchmark of performance by eight inefficient SRTUs. Figure 2 provides a graphical representation of the ranking of efficient SRTUs based on the referenced benchmarks. Figure 3 provides a graphical output generated by the *deaR* program representing inefficient SRTUs in red benchmarked against efficient SRTUs in green.

5 Conclusion

The overall technical efficiency of 27 State Road Transport Undertakings (SRTUs) operating in India was determined using the DEA approach. For this purpose, the primary input oriented DEA model with constant returns to scale was adopted. It is evident from the results that:

- More than half of SRTUs are operating Inefficiently.
- Among the Efficient SRTUs, few are 'Marginal Efficient SRTU' which it may lose if necessary measures are not taken to improve its output.
- The inefficient SRTUs need to improve its performance. This can be achieved either by improving its outputs are by downsizing its size.

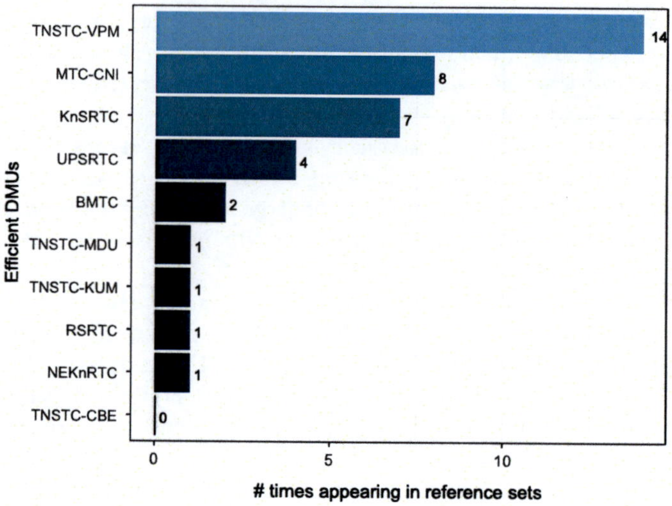

Fig. 2 Ranking of efficient SRTUs based on referenced benchmarks

Based on the analysis performed and results generated and discussed in the above section vide Table 3 and Appendix 2, it can be seen that, for example, the performance of APSRTC may be improved by reducing the *AFO*, from 12,023 units to 11,375 units and reducing the *TSE* from 61,806 units to 58,477 units. This will result in reduction in *TOC* from 540,280.1 to 511,176 lakhs rupees. These values were arrived at by DEA analysis in *R programming language* [3] using *deaR* package [2] based on comparison with KnSRTC, MTC-CNI, TNSTC-VPM and UPSRTC, as inferred based on Appendix 1. Similar observation may be made on other 17 non-performing SRTUs.

With the Efficiency measurement, the management can take right decision to be a self-reliant and professional organization to improve its financial performance judiciously. In the process of improving its efficiencies, SRTU should not compromise its primary objective of serving the needs of commuters. A balanced approach to maintain the equilibrium of efficiency and effectiveness should be considered in the process.

The DEA method is one of the non-parametric approaches to determine efficiency of an organization with multiple input and output variables. This approach is widely used in banking sector, and, also in Hospital sector, and other organization, where performance has to evaluated. DEA analysis is capable of indicating the parameters that the management needs to focus on so as to improve the operational efficiency.

The above section provided details on the analysis approach adopted using DEA method. However, it may be also observed that similar studies on performance analysis of SRTUs were made by various researchers mentioned in the literature, and also, using capabilities of RDBMS and expert system as demonstrated by George et al. [28, 29].

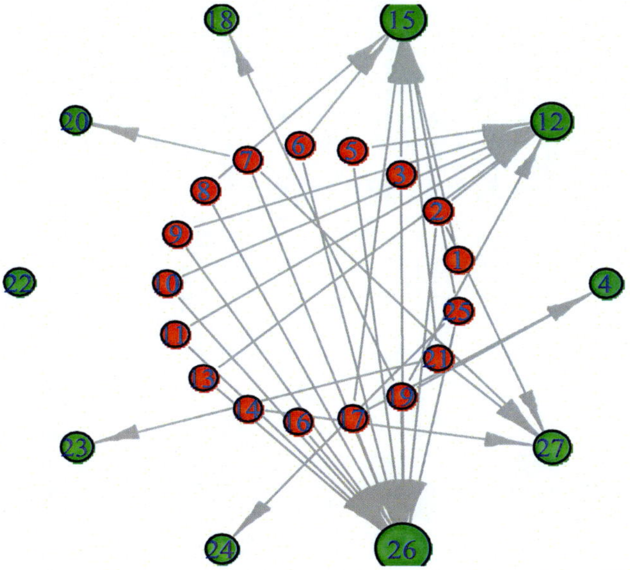

1 = AMTS	8 = DTC	15 = MTC-CNI	22 = TNSTC-CBE
2 = APSRTC	9 = GSRTC	16 = MZST	23 = TNSTC-KUM
3 = BEST	10 = JKSRTC	17 = NBSTC	24 = TNSTC-MDU
4 = BMTC	11 = KTS	18 = NEKnRTC	25 = TNSTC-SLM
5 = BSRTC	12 = KnSRTC	19 = NWKnRTC	26 = TNSTC-VPM
6 = CSTC	13 = MSRTC	20 = RSRTC	27 = UPSRTC
7 = CTU	14 = MTC	21 = SBSTC	

Fig. 3 An output by the deaR program representing inefficient SRTUs in red benchmarked against efficient SRTUs in green

Appendix 1

Summary of the Optimized Values of λ_q Assigned as Coefficients to the Input and Output Parameters Generated by the 'deaR' Program.

S. No.	SRTUs	1	2	3	4	5	6	7	8	9	10	11	12	13	14	15	16	17	18	19	20	21	22	23	24	25	26	27	$\lambda_{q\ sum}$
1	AMTS	0	0	0	0	0	0	0	0	0	0	0	0	0	0	0.1117	0	0	0	0	0	0	0	0	0	0	0	0	0.1117
2	APSRTC	0	0	0	0	0	0	0	0	0	0	0	0.1397	0	0	0.43	0	0	0	0	0	0	0	0	0	0	1.3411	0.4877	2.3985
3	BEST	0	0	0	0	0	0	0	0	0	0	0	0	0	0	0.3969	0	0	0	0	0	0	0	0	0	0	0.5945	0	0.9914
4	BMTC	0	0	0	1	0	0	0	0	0	0	0	0	0	0	0	0	0	0	0	0	0	0	0	0	0	0	0	1
5	BSRTC	0	0	0	0	0	0	0	0	0	0	0	0.0007	0	0	0	0	0	0	0	0	0	0	0	0	0	0.0083	0	0.009
6	CSTC	0	0	0	0	0	0	0	0	0	0	0	0	0	0	0.0218	0	0	0	0	0	0	0	0	0	0	0.0262	0	0.048
7	CTU	0	0	0	0	0	0	0	0	0	0	0	0	0	0	0	0	0	0	0	0.0013	0	0	0	0	0	0.0516	0.0077	0.0606
8	DTC	0	0	0	0	0	0	0	0	0	0	0	0	0	0	0.766	0	0	0	0	0	0	0	0	0	0	0.0363	0	0.8023
9	GSRTC	0	0	0	0	0	0	0	0	0	0	0	0.7008	0	0	0	0	0	0	0	0	0	0	0	0	0	0.3808	0	1.0816
10	JKSRTC	0	0	0	0	0	0	0	0	0	0	0	0.016	0	0	0	0	0	0	0	0	0	0	0	0	0	0.0014	0	0.0174
11	KTS	0	0	0	0	0	0	0	0	0	0	0	0.0299	0	0	0	0	0	0	0	0	0	0	0	0	0	0.0297	0	0.0596
12	KnSRTC	0	0	0	0	0	0	0	0	0	0	0	1	0	0	0	0	0	0	0	0	0	0	0	0	0	0	0	1
13	MSRTC	0	0	0	0	0	0	0	0	0	0	0	1.5404	0	0	0	0	0	0	0	0	0	0	0	0	0	1.4347	0	2.9751
14	MTC	0	0	0	0	0	0	0	0	0	0	0	0	0	0	0	0	0	0	0	0	0	0	0	0	0	0	0.0036	0.0036
15	MTC-CNI	0	0	0	0	0	0	0	0	0	0	0	0	0	0	1	0	0	0	0	0	0	0	0	0	0	0	0	1
16	MZST	0	0	0	0	0	0	0	0	0	0	0	0	0	0	0	0	0	0	0	0	0	0	0	0	0	0.0014	0	0.0014
17	NBSTC	0	0	0	0.0148	0	0	0	0	0	0	0	0	0	0	0.0177	0	0	0	0	0	0	0	0	0	0	0.0185	0	0.051
18	NEKnRTC	0	0	0	0	0	0	0	0	0	0	0	0	0	0	0	0	0	1	0	0	0	0	0	0	0	0	0	1

(continued)

(continued)

S. No.	SRTUs	1	2	3	4	5	6	7	8	9	10	11	12	13	14	15	16	17	18	19	20	21	22	23	24	25	26	27	$\lambda_{q\ sum}$
19	NWKnRTC	0	0	0	0.2132	0	0	0	0	0	0	0	0.0147	0	0	0	0	0	0.8133	0	0	0	0	0	0	0	0.0057	0	1.0469
20	RSRTC	0	0	0	0	0	0	0	0	0	0	0	0	0	0	0	0	0	0	0	1	0	0	0	0	0	0	0	1
21	SBSTC	0	0	0	0	0	0	0	0	0	0	0	0	0	0	0.0314	0	0	0	0	0	0	0	0.0337	0	0	0	0.0164	0.0815
22	TNSTC-CBE	0	0	0	0	0	0	0	0	0	0	0	0	0	0	0	0	0	0	0	0	0	1	0	0	0	0	0	1
23	TNSTC-KUM	0	0	0	0	0	0	0	0	0	0	0	0	0	0	0	0	0	0	0	0	0	0	1	0	0	0	0	1
24	TNSTC-MDU	0	0	0	0	0	0	0	0	0	0	0	0	0	0	0	0	0	0	0	0	0	0	0	1	0	0	0	1
25	TNSTC-SLM	0	0	0	0	0	0	0	0	0	0	0	0	0	0	0.0039	0	0	0	0	0	0	0	0	0.4681	0	0.2886	0	0.7606
26	TNSTC-VPM	0	0	0	0	0	0	0	0	0	0	0	0	0	0	0	0	0	0	0	0	0	0	0	0	0	1	0	1
27	UPSRTC	0	0	0	0	0	0	0	0	0	0	0	0	0	0	0	0	0	0	0	0	0	0	0	0	0	0	1	1

Appendix 2

Target Values of Input and Output Variables to be Attained to Improve the Efficiency of SRTUs as Determined Using the 'deaR' Program.

S. No.	SRTUs	TOC (10^5 Rs)	AFO	TSE	TR (10^5 Rs)	TPKM (10^5)	TPC (10^5)
1	AMTS	17,830.46774	356.0522732	2817.47169	15,378.48868	21,222.7041	2024.37
2	APSRTC	511,176.1004	11,375.33994	58,476.60819	480,768.07	662,640.0204	23,192.1
3	BEST	168,648.0517	3257.71461	23,429.45305	150,855.93	249,797.1581	12,216
4	BMTC	232,174.81	6014	36,474	225,684.43	217,253.9	19,366.9
5	BSRTC	1692.1962	33.01612711	212.8193117	1562.75	2674.7	76.99175789
6	CSTC	8118.788121	157.2706949	1140.870063	7240.52	11,827.73075	616
7	CTU	11,934.88086	248.6985903	1382.9747	11,107.08	17,603.25173	482.4236559
8	DTC	128,688.1015	2563.026053	20,138.0016	111,321.11	156,166.8399	14,187.28
9	GSRTC	295,082.5098	6582.918972	34,753.83787	286,218.46	359,984.6597	10,316.69449
10	JKSRTC	5431.921584	125.530634	627.3300791	5341	6065.75	173.5256193
11	KTS	14,972.06779	325.9533738	1786.357421	14,388.58	19,303.3777	553.7853716
12	KnSRTC	324,817.95	7572	37,326	320,468.94	354,273.66	10,129.75
13	MSRTC	754,479.0005	16,473.18652	89,882.9975	725,866	966,608.2861	27,727.34444
14	MTC	1179.057206	32.88180121	89.71169779	1179.95	1019.645636	19.59651606
15	MTC-CNI	159,599.32	3187	25,219	137,651.82	189,963	18,120
16	MZST	253.2925826	4.793383007	32.27954493	231.45	419.5066133	12.0837725
17	NBSTC	9536.841055	207.4388191	1403.735896	8770.19	12,000.51	763.68
18	NEKnRTC	146,413.91	3794.99999	18,411.99996	144,901.15	154,800.5202	4891.000001

(continued)

(continued)

S. No.	SRTUs	TOC (10^5 Rs)	AFO	TSE	TR (10^5 Rs)	TPKM (10^5)	TPC (10^5)
19	NWKnRTC	174,348.8662	4498.761965	23,426.68879	171,583.93	179,082	8303.75
20	RSRTC	214,750.58	4324	20,551	183,692.89	223,386.27	3582.290001
21	SBSTC	16,527.7362	371.7594271	1981.486971	15,242.53	19,887.32208	1020.04
22	TNSTC-CBE	154,469.63	3061	18,964	126,100.73	228,832.2	9014.78
23	TNSTC-KUM	182,563.19	3622	23,198	164,740.11	275,547.98	10,748.46
24	TNSTC-MDU	120,836.35	2381	14,850	106,235.8	187,390.55	6563.54
25	TNSTC-SLM	108,289.3194	2094.064537	13,562.19193	96,962.61886	173,098.12	5580.85
26	TNSTC-VPM	177,126.83	3352	22,573	161,852.37	293,359.86	8450.15
27	UPSRTC	327,306.71	9128	24,904	327,554.55	283,054	5440

Appendix 3

Slacks Generated by the 'deaR Program Indicating the Quantum of Inputs to be Reduced for Better Performance for Selected SRTUs.

S. No.	SRTUs	TOC (10^5 *Rs*)	AFO	TSE	TR (10^5 *Rs*)	TPKM (10^5)	TPC (10^5)
1	AMTS	2241.118967	51.46374883	0	2367.088678	5371.104	0
2	APSRTC	0	0	0	0	114,607.8	0
3	BEST	42,353.50534	0	8560.838522	0	159,081.2	0
4	BMTC	0	3.26175E–09	5.32136E–09	0	7.97E–07	0
5	BSRTC	0	28.22589046	326.8309372	0	0	51.63175789
6	CSTC	976.7134764	0	741.2797249	0	5521.731	0
7	CTU	0	0	0	0	2162.252	4.273655932
8	DTC	184,312.3926	0	13.02428546	0	20,102.27	0
9	GSRTC	0	0	3915.270326	0	26,616.43	2582.044492
10	JKSRTC	0	60.66715794	488.559127	0	0	127.8356193
11	KTS	0	0	68.23601359	0	9779.858	208.7053715
12	KnSRTC	0	0	0	0	2.4E–06	4.13628E-08
13	MSRTC	0	0	16,143.13506	0	442,854.5	3170.434435
14	MTC	0	3.504040343	158.0585566	0	620.5556	15.77651606
15	MTC-CNI	7.25845E–06	0	7.31036E-07	2.60154E-06	6.45E–06	0
16	MZST	381.3939353	0	79.03346045	0	190.3566	11.6037725
17	NBSTC	0	56.38513669	176.5748654	0	0	0
18	NEKnRTC	0	1.10015E-05	4.22219E-05	0	0.000131	0

(continued)

(continued)

S. No.	SRTUs	TOC *(10^5 Rs)*	AFO	TSE	TR *(10^5 Rs)*	TPKM *(10^5)*	TPC *(10^5)*
19	NWKnRTC	0	0	103.2237471	0	0	0
20	RSRTC	5.12233E–06	0	0	0	8.84E-06	7.7562E-07
21	SBSTC	969.4394905	0	0	0	963.6521	0
22	TNSTC-CBE	2.60857E–06	5.53859E–08	0	4.07414E–06	0	0
23	TNSTC-KUM	1.93385E–08	0	0	0	6.59E–07	0
24	TNSTC-MDU	0	0	0	1.04319E–07	2.29E–07	0
25	TNSTC-SLM	3356.337271	0	0	6220.608855	0	0
26	TNSTC-VPM	0	−1.22929Ev09	5.6224E–10	6.22801E–08	1.24E–07	0
27	UPSRTC	2.51068E–07	9.48674E–09	0	0	1.26E–06	4.71907E–08

References

1. Charnes A, Cooper WW, Rhodes E (1978) Measuring the efficiency of decision making units. Eur J Oper Res 2(6):429–444
2. Coll-Serrano V, Bolos V, Benitez Suarez R (2020) deaR: conventional and Fuzzy Data envelopment analysis
3. R Core Team (2019) "R: a language and environment for statistical computing." R Foundation for Statistical Computing, Vienna, Austria
4. Merewitz L (1977) On measuring the efficiency of public enterprises: Bus operating companies in the San Francisco Bay area. Transp (Amst) 6(1):45–55
5. Alexandersson G, Hultén S, Fölster S (1998) The effects of competition in Swedish local bus services. J Transp Econ Policy 32(2):203–219
6. Battese GE, Coelli TJ (1995) A model for technical inefficiency effects in a stochastic frontier production function for panel data. Empir Econ 20(2):325–332
7. Simar L, Wilson PW (2007) Estimation and inference in two-stage, semi-parametric models of production processes. J. Econom. 136(1):31–64
8. Asmare E, Begashaw A (2018) Review on parametric and nonparametric methods of efficiency analysis. Biostat Bioinforma 2(2):1–7
9. Farrell MJ (1957) The measurement of productive efficiency. J R Stat Soc 120(3):253–290
10. Banker RD, Charnes A, Cooper WW (1984) Some models for estimating technical and scale inefficiency in data envelopment analysis. Manage Sci 30(9):1078–1092
11. Sadjadi SJ, Omrani H (2010) A bootstrapped robust data envelopment analysis model for efficiency estimating of telecommunication companies in Iran. Telecomm Policy 34(4):221–232
12. Bray S, Caggiani L, Ottomanelli M (2015) Measuring transport systems efficiency under uncertainty by fuzzy sets theory based data envelopment analysis: theoretical and practical comparison with traditional DEA model. Transp Res Procedia 5:186–200
13. Landete M, Monge JF, Ruiz JL (2017) Robust DEA efficiency scores: a probabilistic/combinatorial approach. Expert Syst Appl 86:1339–1351
14. Agarwal S, Yadav SP, Singh SP (2010) DEA based estimation of the technical efficiency of state transport undertakings in India. Oper Res Soc India 47(3):216–230
15. Saxena P, Saxena RR (2010) Measuring efficiencies in Indian public road transit: a data envelopment analysis approach. Opsearch 47(3):195–204
16. Kumar S (2011) State road transport undertakings in India: technical efficiency and its determinants. Benchmarking An Int J 18(5):616–643
17. Andersen P, Petersen NC (1993) A procedure for ranking efficient units in data envelopment analysis. Manage Sci 39(10):1261–1264
18. Mulangi RH, Sitharam TG, Hanumanthappa D, Siddeshwar H (2014) Performance evaluation of state transport undertakings using non-parametric techniques. Indian J Transp Manag 40(2):196–211
19. Bose A, Patel GN (2015) 'NeuralDEA'—a framework using neural network to re-evaluate DEA benchmarks. Opsearch 52(1):18–41
20. Ministry of Road Transport and Highway (2016) Review of the performance of state road transport undertakings for 2014–2015
21. Fielding GJ, Brenner ME, Faust K (1985) Typology for bus transit. Transp Res Part A Gen 19(3):269–278
22. Fielding GJ, Babitsky TT, Brenner ME (1985) Performance evaluation for bus transit. Transp Res Part A Gen 19(1):73–82
23. Wright AA, Thiriez S (1987) Bus services: reducing costs, raising standards
24. Karlaftis MG (2004) A DEA approach for evaluating the efficiency and effectiveness of urban transit systems. Eur J Oper Res 152(2):354–364
25. Cooper WW, Seiford LM, Tone K (2007) Data envelopment analysis a comprehensive text with models, applications, references and DEA-solver software. Springer, New York

26. Seiford LM, Zhu J (1999) An investigation of returns to scale in data envelopment analysis. Omega 27(1):1–11
27. Gelles GM, Mitchell DW (1996) Returns to scale and economies of scale: further observations. J Econ Educ 27(3):259–261
28. George V, Dhingra S, Sikdar P (1998) The role of expert systems, and rdbms strategies in a DSS for urban bus transport management. In: 4th design and decision support systems conference in architecture and urban planning (DDSS-4)
29. George V, Dhingra S, Sikdar P (1996) An advanced RDBMS approach for the design of an information system for performance evaluation in urban public transport. In: International Workshop on Transport Planning and Management (TPMDC-96)

Investigation, Design and Construction Methodology of 120 m High IKOCP Dumping Yard, Singareni Collieries

Srikanth Emmadi, G. V. Ramana, and P. S. Prasasd

Abstract The Singareni Collieries Company Limited (SCCL) is a government-based coal mining company jointly owned by the Telangana Government and Government of India. The Singareni coal reserves stretch over an area of 350 km on Pranahita—Godavari Valley of Telangana with a proven geological reserve aggregating to whopping 8791 million tonnes of coal. SCCL is currently operating 29 underground mines and 17 opencast mines in four districts of Telangana. Out of 17 open cast mines, Indaram Khani Open Cast Project (IKOCP) is one of the open cast mine operating by SCCL. Huge quantity of waste/debris (Sandstone boulders, clay and some other rock lumps) is also generated during mining of coal. The SCCL is proposed to construct a 120 m high IKOCP dumping yard in 1.2 km × 1.0 km of Srirampur Region, near IKOCP, Telangana. Nine bore holes were drilled up to a depth of 7 m at different locations in the proposed dumping yard area. The standard penetration test (SPT) was also carried out during subsoil investigation. The site was investigated and the slope stability analysis of the proposed dumping yard under different conditions was carried out by using GEO 5 software. The present study is intended to highlight the outcome of the study along with suitable design and construction methodology of 120 m high dumping yard based on the field and laboratory investigations.

Keywords Mine overburden · Shear strength parameters · Opencast mine · Dumping yard · Slope stability

---

S. Emmadi (✉) · G. V. Ramana
Department of Civil Engineering, National Institute of Technology, Warangal, India

G. V. Ramana
e-mail: gvramana@nitw.ac.in

P. S. Prasasd
Geotechnical Engineering Division, CSIR-Central Road Research Institute, New Delhi, India

L. Nandagiri et al. (eds.), *Sustainability Trends and Challenges in Civil Engineering*, Lecture Notes in Civil Engineering 162, https://doi.org/10.1007/978-981-16-2826-9_29

1 Introduction

Due to accelerated industrialisation and extensive growth in population, the energy demand is increased rapidly which forces extraction of coal from open cast mines to serve energy needs. Generally, extraction of coal from mines by open cast mining is relatively better option than underground methods. For successful operation of open cast mine, proper management of overburden mine waste is very essential otherwise it may lead to instability problems. This might affect production of mining operation and safety of human life. For example, pit dumps minimise reusing of overburden material and are effective in utilisation of existing area [1, 2]; however, failure of dump may interrupt quarrying operations, hazard personnel and destroy machinery [2]. In case of material handling and land use, overburden dumps are less effective and overcome the effects of mining operations in case of failure of dump [3, 4]. Many case studies in India reveal that dump failure results in significant damage to quarrying properties, hazard to human life and interrupts coal production [5, 6]. There are dump slope failures occurred at Singareni colliery during December 2009 [7]. In order to provide safety of personnel and machinery, the waste dump should have a proper design [6, 8]. The study area is an open cast mine in Indaram Khani Open Cast Project (IKOCP) and it is operating by SCCL. The SCCL main function is to search and produce the coal deposits along the Godavari valley coal field. Huge quantity of waste/debris (Sandstone boulders, clay and some other rock lumps) is also generated during mining of coal has to be accommodated in minimal area in order to restrict the wastage of useful agricultural land. Hence SCCL is proposed to construct a 120 m high IKOCP dumping yard in the area of 1.2 km $\times$ 1.0 km of Srirampur Region, near IKOCP, Singareni Collieries, Telangana, as the waste materials placed in the form of dumps to greater heights. Hence, there is a need to study the geotechnical properties and parameters of dump materials along with foundation soil which governs the design of stability of Dump Yard.

In the present study, field and laboratory investigation has been carried out to evaluate the representative geotechnical properties (shear strength parameters; unit weight) of both proposed dumping yard foundation soil and excavated mine overburden (OB). Slope stability software (GEO 5) has been used to assess the local and global stability of the overburden dump fill. The final slope of the dump yard (overburden dump fill) of 120 m high has been proposed based on the factor of safety value.

2 Geotechnical Investigation on Proposed Site

A total of nine bore holes (BH-1, BH-2, BH-3, BH-4, BH-5, BH-6, BH-7, BH-8 and BH-9) were drilled at different locations in the proposed dumping yard area to find the subsoil strata and its properties of foundation soil. The depth of each bore hole varies from 3.5 to 7.0 m. The location of the bore holes was selected by the

Fig. 1 SPT test on the proposed original foundation site

project authorities at the proposed original foundation site of IKOCP OB dumping yard Srirampur Region Singareni Collieries Company Limited.

2.1 *Field Tests*

Standard Penetration Test at different levels has been carried out in all bore holes (Fig. 1) and the results are presented in Table 1. Undisturbed soil samples were also collected from each bore hole by using 38 mm diameter tubes. The typical disturbed soil samples from SPT split spoon sampler tube of BH-2 (Label Name: SPT-02 IKOCP) is presented in Fig. 2.

2.2 *Sample Preparation and Laboratory Tests*

A total of 18 representative soil samples (9 Nos. disturbed soil samples and 9 Nos. undisturbed soil samples) from 9 bore holes (SPT-1 to SPT-9) were collected and conducted grain size analysis, Atterberg limits and shear strength characteristics as per procedures outlined in relevant Indian Standards [9–15] and other standard procedures.

(a) *Physical characteristics of field samples*

The representative samples collected from 9 boreholes were subjected to grain size distribution analysis and consistency limits tests and the results are shown in Table 1. The grain size distribution analysis has exposed that the majority of the soil samples

Table 1 SPT value and physical characteristics of foundation soil

BH. No	SPT value (Ncor)	Depth (m)	Grain size analysis				Atterberg limits			IS soil classification
			< 0.002 mm	0.002–0.075 mm	0.075–4.75 mm	> 4.75 mm	LL	PL	PI	
			Clay	Silt	Sand	Gravel				
1	5–40	1.2–3.45	56	37	7	0.0	68.0	23.0	45.0	CH
2	8–41	1.5–6.0	22	43	35	0.0	38.0	20.0	18.0	CI
3	9–31	1.2–3.9	62	30	8	0.0	70.0	24.0	46.0	CH
4	10–27	1.1–3.35	59	38	3	0.0	74.0	23.0	51.0	CH
5	10–38	1.5–6.0	8	46	44	2.0	0.0	0.0	0.0	ML
6	10–25	1.5–3.4	15	46	39	0.0	30.0	16.0	14.0	CL
7	10–31	1.5–3.45	16	51	32	1.0	30.0	16.0	14.0	CL
8	16–42	1.5–5.55	64	30	6	0.0	78.0	23.0	55.0	CH
9	16–46	1.5–6.9	71	24	5	0.0	76.0	22.0	54.0	CH

Fig. 2 Bore hole number 2 (BH-2)

contain primarily clay fraction then followed by silt and sand (including fine, medium and coarse) fractions barring few exceptions. From the Atterberg Limits tests, it was observed that liquid limit values of the 9 No's of representative soil samples vary from 30 to 78%. Barring one sample SPT-5 (which exhibited non-plastic nature), the remaining 8 representative soil samples exhibited plasticity characteristics and the plasticity index values varied from 14 to 55%. The subsoil samples have been classified as per IS soil classification system and the results are shown in Table 1.

(b) *Triaxial test*

Eight chosen soil samples (Undisturbed) were subjected to Triaxial Shear Test (UU test). The soil samples were sheared at three various confining pressures of 1.0, 1.5 and 3.0 kg/cm^2 respectively. The values of shear strength parameters, i.e. cohesion (c) and angle of internal friction (φ) of the tested soil samples of the areas are shown in Table 2. The values of shear strength parameters, i.e. cohesion (c) and angle of

Table 2 Foundation investigations at the original ground site of IKOCP

SI. No	Field No.	Depth (m)	In-situ bulk density kN/m^3	In-situ dry density kN/m^3	Natural moisture content %	IS soil classification	Shear strength parameters (UU test)	
							c (kPa)	ϕ
1	SPT1	1.2–3.45	19.1	15.8	21.07	CH	32	11
2	SPT2	1.5–6.0	15.8	14.4	9.89	CI	36	12
3	SPT3	1.2–3.9	18.0	15.2	18.25	CH	42	9
4	SPT4	1.1–3.35	18.7	16.0	17.13	CH	30	11
5	SPT5	1.5–6.0	17.0	15.8	7.75	ML	44	17
6	SPT6	1.5–3.4	19.1	17.1	11.65	CL	23	13
7	SPT7	1.5–3.45	17.6	16.3	7.90	CL	26	11
8	SPT8	1.5–5.55	21.0	17.6	19.08	CH	35	9
9	SPT9	1.5–6.9	21.5	18.2	18.38	CH	46	10

internal friction (φ) of the tested samples (i.e. 8 No's) from SPT-1 to SPT-4 and SPT-6 to SPT-9 were found to vary from 23 to 46 kPa and 9° to 13° respectively. In similar lines, the values of total shear strength parameters of SPT-5, cohesion (c) and effective angle of internal friction (φ) were found to 44 kPa and 17° respectively.

3 Stability Analysis of Dumping Yard

The shape of the dumping yard mainly depends on the nature and geography of the area available for dump. Since the available land is almost plain, heaped fill dumping construction method is proposed. The area available for the proposed dumping yard is 1.2 km × 1.0 km. The height of the proposed dump fill is 120 m. The slope of the proposed dumping fill is 23.20. Berm has been provided at every 15 m high intervals. The average slope of the dumping fill with berms is 17.50. The top view and cross section of the proposed dumping yard is shown in Fig. 3. Since dumping yard is to be constructed on clayey soil, the designer has to ensure the internal and global stability of the dump fill. Failure of dumping yard may be due to either inadequate bearing capacity or due to deep seated shear failure. The failure of dumping yard may also take place due to loose dump fill / loose dumped slope upon saturation due to rain. The main objective of the stability analysis is to ensure that the proposed dumping yard does not face any risk due to shear strength parameters of the dumped fill material or drainage issue.

Geo5 software is a tool to provide analytical solutions for various geotechnical applications like design of abutments, cantilever wall, gabion wall, gravity wall and masonry wall. The Geo 5 software is very much useful for design of embankments, stability analysis of normal slopes and steep slopes with soil nailing technique. This software can also be used for settlement calculations and rock slope stability. In the current study stability analysis of the dumping yard has been carried out with GEO 5 (slope stability) software. The stability analysis (verified methodology) has been carried out using classic way, i.e. the stability of dumping yard can be corroborated according to theory of limit or FoS. Bishops method (circular slip surface) is used for analysis of slopes of dumping yard. The foundation soil and overburden dump properties considered for stability analysis are presented in Table 3. Different slope angle of the dump fill has been considered for stability analysis. In the present study, only optimised design cross section (The slip circle with lowest factor of safety) is presented. The results of the stability analysis for different conditions (effect of groundwater table, saturation and earthquake forces) are presented in Table 4. The results of the stability analysis (Table 4) clearly shows that the overburden dump fill slope is stable in all conditions except under saturated condition. Results of the stability analysis indicated that proper drainage arrangements are required for entire slope and top of the dump yard so that the rainwater can be disposed of immediately without saturating the slope.

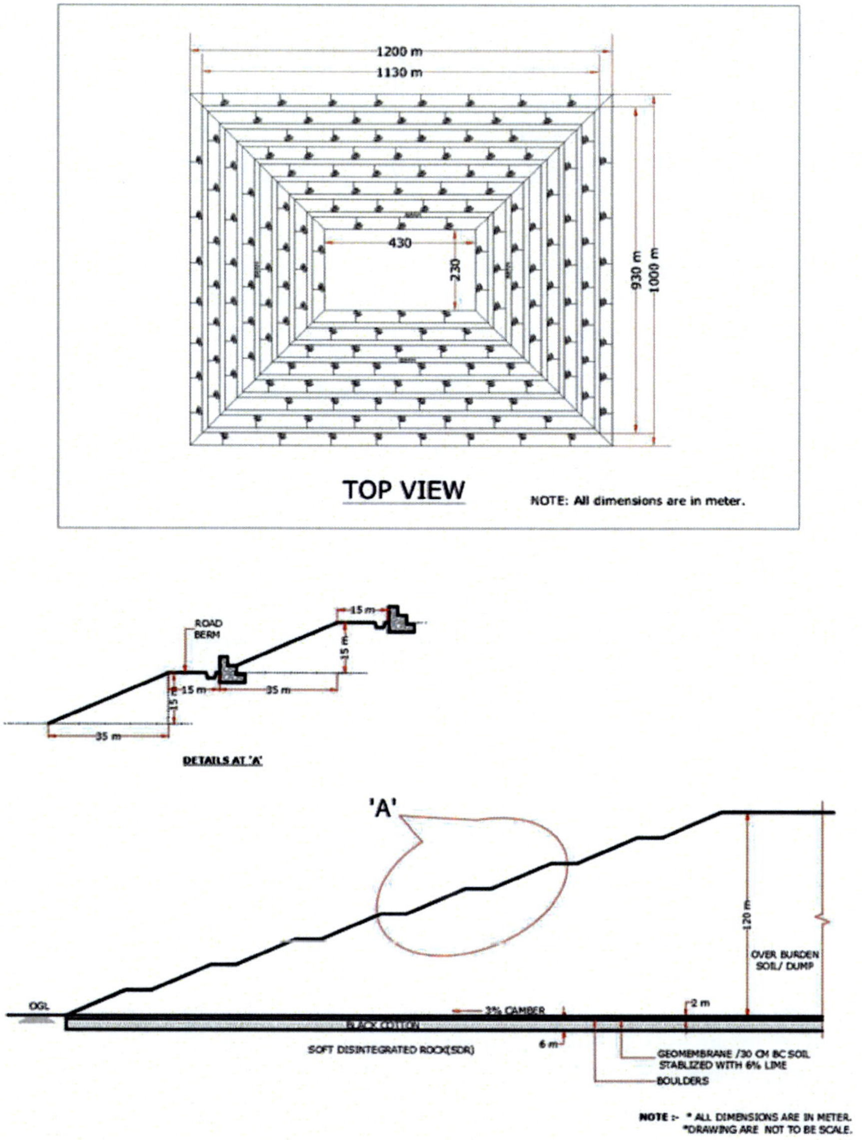

Fig. 3 Top view and cross section of the proposed dumping yard

4 Methodology for Construction of 120 m High Dumping Yard

Before dumping yard construction, it is proposed to first remove/excavate approximately 2 m of top soil. Top 20 cm soil should be stacked separately for dump yard

Table 3 Material properties considered for stability analysis

	γ_b (kN/m^3)	c (kPa)	φ
Soft disintegrated rock	19	40	20°
Clayey soil	18	35	9°
Boulders	19	0	32°
Overburden (OB) fill	18	15	25°

reclamation. Remaining 180 cm soil also stacked separately. This soil will also be used as a soil cover for the slope of dumping yard for each stage. After removal of top soil, the soil should be properly cambered for drainage purpose (Fig. 3). The base soil should be properly compacted up to 95% of maximum dry density with pneumatic roller/ sheep foot roller. 25 mm thick sand bedding layer should be spread evenly over the foundation soil to serve as cushion for geomembrane laying and this sand layer should be sprinkled with water and rolled. Geomembrane should be laid over the compacted ground. The geomembrane layer would act as separator between filled soil and Foundation soil. The geomembrane would act as an impermeable sheet and helps in preventing water seeps into foundation soil. The specification of geomembrane should satisfy according to IRC [16].

Over the geomembrane, a 10 cm sand and 190 cm stone/boulder soling should be provided. This will act as a foundation for proposed overburdening dumping fill. Construction of the new dumping yard should be taken up layer wise above the prepared base. In general the waste mine materials usually back dumped in the dumping yard with end tipping method. After dumping the waste mine material, it is generally levelled with a grader. In general the layer is in a very loose condition and these layered materials usually compacted through dumpers while transporting the dumping material (Natural compaction). Some compaction will also take place in due course of time due to further movement of vehicles. At every 15 m, interval height berm should be provided. This berm will be useful for stability of an embankment, drainage can be provided on berm and also can be used as a road for transporting the overburden dump fill. In general, overburden dump fill slopes remain denuded. Hence, the side slopes are very much prone for erosion due to rainfall and wind. The embankment side slopes need protection against erosion. The Gabion toe wall plays an important role in reducing the embankment erosion. The erosion control measures should be applied up to top of the embankment. To ensure that gabion toe wall does not bulge / gets displaced during embankment construction, it is suggested that gabion toe wall construction can be taken up subsequently after embankment construction by suitably trimming the compacted embankment and using a plate compactor. Non-woven polymeric geotextile should be placed all along behind the gabion toe wall.

After constructing the 120 m high embankment (overburden dump fill), the embankment side slopes should be properly protected by providing 300–400 mm (200 mm thick soil cover + 150–200 mm high geocell filled up with soil) thick layer of soil cover (use already excavated and stacked soil for this purpose). This technique helps not only in case of erosion due to run-off water but also in case of wind

Table 4 Stability analysis of proposed dumping fill

S. No		FoS	Remarks
1	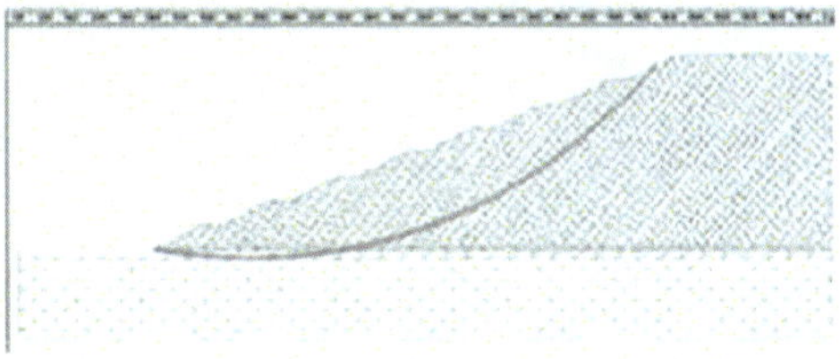	1.54	The Dump fill is in dry condition and Water table far below the ground level
2	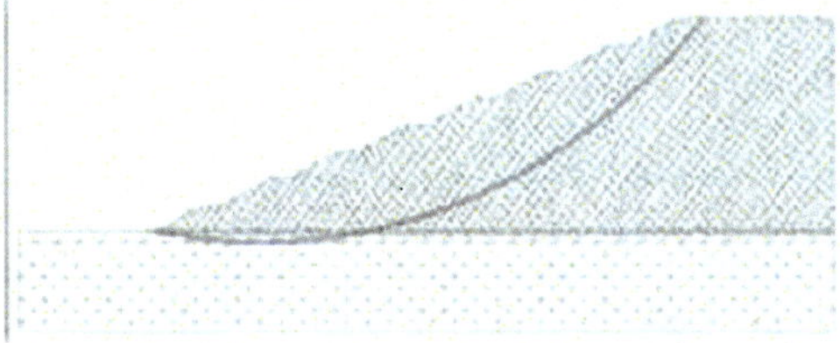	1.53	The Dump fill is in dry condition and Water table is at ground level
3		0.96	The Dump fill is in saturation condition
4	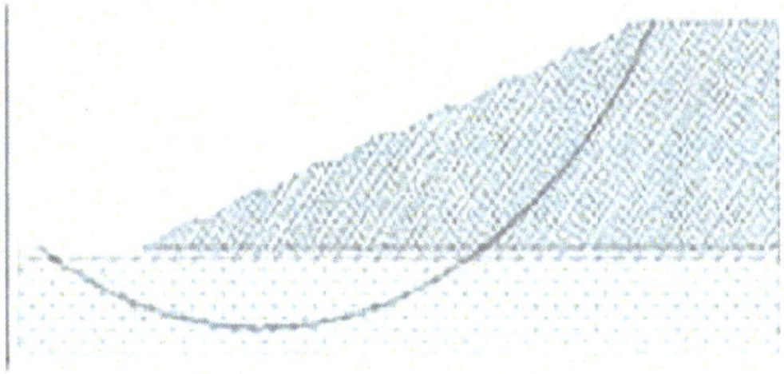	1.53	The Dump fill is in dry condition and Water table is at ground level
5	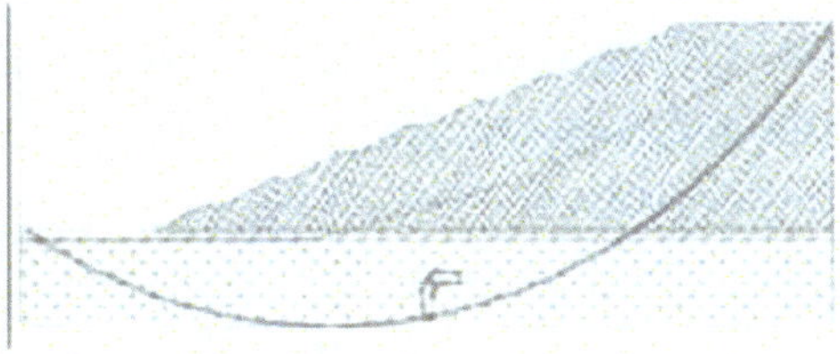	1.11 (with earthquake factors)	The dump fill is in partially saturated condition

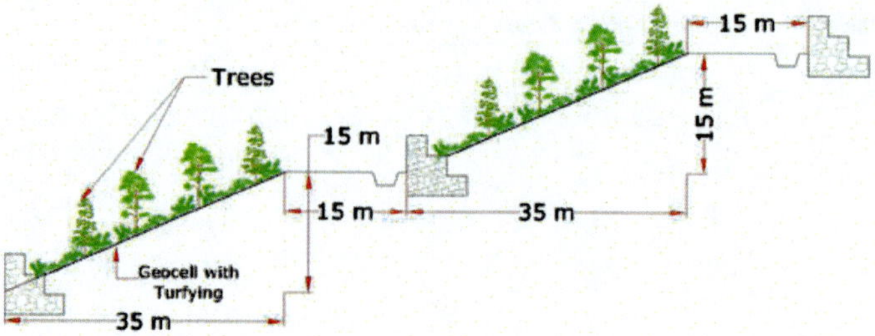

Fig. 4 Embankment slope protected with gabion, drainage, soil cover with geocell with turfing and plantation

erosion. Side slopes of embankment covered with geocell filled up with soil suitable for turfing and tree plantation. Local grass/creepers and plants can be selected based on the expert opinion of agricultural scientist. Figure 4 presents the embankment slide slopes protected with gabion, drain, geocell with turfing and trees.

5 Conclusions

The SCCL is proposed to construct a 120 m high IKOCP dumping yard. The proposed dumping yard has been designed based on the field investigation and stability analysis. The conclusions are drawn from the present study as follows.

- The slope angle of the dumping yard should be less than or equal to angle of repose of overburden fill material.
- The stability of dumping yard is safe in all conditions except under full saturation condition. To avoid full saturation of dumping yard due to rains, longitudinal and transverse (Garland) drains should be provided all along the dumping yard.
- Since the foundation soil is weak (clayey soil), before laying the next layer of fill, the overburden dump fill should be dumped and spread uniformly over the entire proposed area.
- To prevent the rill and gully erosion against wind and rain, the side slopes of the dumping yard should be protected with gabion, drainage, soil cover with geocell with turfing and plantation.
- Since the height of the dumping yard is too high (120 m) and back dumping the overburden fill with end tipping method, the dumping yard should be properly monitored with settlement gauges, piezometers and inclinometers.

References

1. Kainthola A, Verma D, Gupte SS, Singh TN (2011) A coal mines dump stability analysis- A case study. Int J Geomaterial 1:1–13
2. Richards BG, Coulthard MA, Toh CT (1981) Analysis of slope stability at Goonyella Mine. Can Geotech J 8:179–194
3. Dawson RF, Morgenstern NR, Stokes AW (1998) Liquefaction flowslides in Rocky Mountain coal mine waste dumps. Can Geotech J 35:2
4. Roberson (1985) A mine waste disposal: an update on geotechnical and geohydrological aspects. In: Steffen, Robertson, Kirsten (eds) Vancouver, Canada
5. Speck RC, Huang SL, Kroeger EB (1993) Large-scale slope movements and their effect on spoil-pile stability in Interior Alaska. Int J Surf Min Reclam 7(4):161–166
6. Sharma A, Rai R, Srivastava BK (2011) Stability analysis of internal mine waste dump by numerical simulation. Indian J Min Engg 50(11):232–38
7. Poulsen B, Khanal M, Rao AM, Adhikary D, Balusu R (2014) Mine overburden dump failure: a case study. Geotech Geol Eng 32(2):297–309. https://doi.org/10.1007/s10706-013-9714-7
8. Hancock GR, Turley E (2006) Evaluation of proposed waste rock dump designs using the SIBERIA erosion model. Environ Geol 49(5):765–779
9. IS: 2720-Part 2: Methods of test for soils—determination of water content. Bureau of Indian Standards, New Delhi (1983)
10. IS 2720-Part 8: Methods of test for soils—determination of water content—dry density relation. Bureau of Indian Standards, New Delhi (1983)
11. IS: 1498: Classification and identification of soils for general engineering purposes. Bureau of Indian Standards, New Delhi (1970)
12. IS: 2720-Part 4: Methods of test for soils grain size analysis. Bureau of Indian Standards, New Delhi (1985)
13. IS: 2720-Part 5: Methods of test for soils determination of liquid and plastic limit. Bureau of Indian Standards, New Delhi (1985)
14. IS: 2720-Part 12: Method of test for soil; Triaxial shear test, Bureau of Indian Standards, New Delhi (1985).
15. IS: 2720-Part 13: Method of test for soil; direct shear test; Bureau of Indian Standards, New Delhi (1981)
16. IRC34-2011. Recommendations for Road Construction in Areas Affected By Water Logging Flooding; Indian Roads Congress (IRC), Ministry of Road Transport & Highways (MORTH, formerly MOST)

Stresses in Masonry Cylindrical Shells Using Realistic Boundary Conditions

P. Subrahmanya V. Bhat, M. V. Renukadevi, and K. S. Jagadish

Abstract Masonry shells have been used as roofing since ancient times. Cylindrical and spherical shells are most common among them. They have the advantage of simplicity of construction without using elaborate support systems. Attempts have been made to use theory of shells in designing modern masonry shells. While analyzing circular cylindrical shells, the shear stress at the longitudinal boundary is frequently assumed to be zero. An earlier study by Yogananda using this assumption showed large tensile membrane stresses in the longitudinal direction. Such a boundary condition is unrealistic as the wall support is never very smooth. It appears that no axial deformation along the longitudinal boundaries is more realistic instead. Hence, the alternative condition has been considered in this paper using Flugge's approach. The analysis results showed that the longitudinal membrane stresses are no longer tensile along the boundaries. The bending moments also show a reducing trend. Thus, the new boundary condition is well suited for designing masonry cylindrical shells. However, the diagonal tension due to in plane shear stress is increased marginally as the boundary condition is altered. The masonry shell is generally provided with a thin layer (4 cm) of cement concrete with nominal steel reinforcement both for waterproofing and for resisting settlement creeps. This will also help in resisting some of the small tension value which may occur.

Keywords Masonry shells · Cylindrical shells · Boundary conditions

1 Introduction

The use of masonry for domes and vaults has been known all over the world for thousands of years. Nubian vaults of Africa, resurrected by Hassan Fathy [1, 2],

P. S. V. Bhat (✉) · M. V. Renukadevi
Department of Civil Engineering, RVCE, Bengaluru, India

M. V. Renukadevi
e-mail: renukadevimv@rvce.edu.in

K. S. Jagadish
Department of Civil Engineering, IISc, Bengaluru, India

L. Nandagiri et al. (eds.), *Sustainability Trends and Challenges in Civil Engineering*,
Lecture Notes in Civil Engineering 162, https://doi.org/10.1007/978-981-16-2826-9_30

represent perhaps the oldest examples. Subsequently, the Pantheon of Rome, Hagia Sophia of Istanbul are well known examples of large masonry domes. In the Indian context, traditional Jain temples used corbelled domes. Later, during the Islamic rule in India, arcuate domes became common. In the Vijayanagar period, the domes used by the Muslim rulers of Bahmani Kingdom influenced masonry dome and vault construction for secular and temple structures even by Hindu rulers. It must, however, be noted that the earlier domes, and vaults were essentially designed as architectural novelties using empirical means for their structural design. It is only in recent times that Heyman [3] and Grimm [4] recognized such structures as masonry shells. This means that structural theories of shells may now be used to understand and design masonry shells. Yogananda [5] attempted a stress analysis of masonry cylindrical shell roofs using Flugee's and Donnel theory [6, 7] to calculate stresses in various shells.

Yogananda considered cylindrical shells according to their lengths and span-rise ratios. Accordingly, there types of lengths, namely short, moderate, and long shells are considered. In the span-rise ratio category, he considered shallow, moderately shallow, and non-shallow shells. The basis for the selection of shell classification is as below.

2 Shell Parameters

Three dimensions of a cylindrical shell are important in analyzing the stresses. They are the span, rise, and length of the shell. The relevant aspect ratios are the rise/span and the length/span. Vlasov [8] defines a shallow shell is the one whose rise/span ratio is less than 0.2. The maximum value possible for this ratio in a circular cylindrical shell is 0.5 (semicircular cylinder). Shells with a rise/span ratio close to 0.5 (>0.4) is termed as non-shallow. The other shells with rise/span ratios between 0.2 and 0.4 are considered as moderately shallow. Regarding the length/span ratios, a shell may be considered as short if the ratio is unity. It may be considered long if the ratio is three or more. The shell may be considered moderately long if the ratio is in between.

3 Numerical Studies

For comparison of results, the shells considered by Yogananda are considered here also. The specific dimensions of shell considered are shown in Table 1. Here, the length of the shell is 'l', its rise is 'R', span 'B', radius is 'a', and 'ϕ_0' is the semi vertical angle from crown to longitudinal boundary as shown in the Fig. 1. In shell designation, MS refers to moderately shallow and NS refers to non-shallow. The first letter of suffix indicates the length of the shell and the remaining the span.

The membrane and bending stress resultants of a typical shell element are shown in Fig. 2. Yogananda considered the following boundary conditions for the analysis:

Table 1 Various circular cylindrical shells

Shell designation	Length ('l' m)	Span ('B' m)	Rise ('R' m)	Radius ('a' m)	Half angle (ϕ_0°)
MS33	3.0	3.0	1.0	1.625	67.4°
MS63	6.0	3.0	1.0	1.625	67.4°
MS93	9.0	3.0	1.0	1.625	67.4°
MS645	6.0	4.5	1.5	2.437	67.4°
MS945	9.0	4.5	1.5	2.437	67.4°
NS33	3.0	3.0	1.5	1.5	90°
NS63	6.0	3.0	1.5	1.5	90°
NS93	9.0	3.0	1.5	1.5	90°
NS645	6.0	4.5	2.25	2.25	90°
NS945	9.0	4.5	2.25	2.25	90°

Fig. 1 Shell geometry

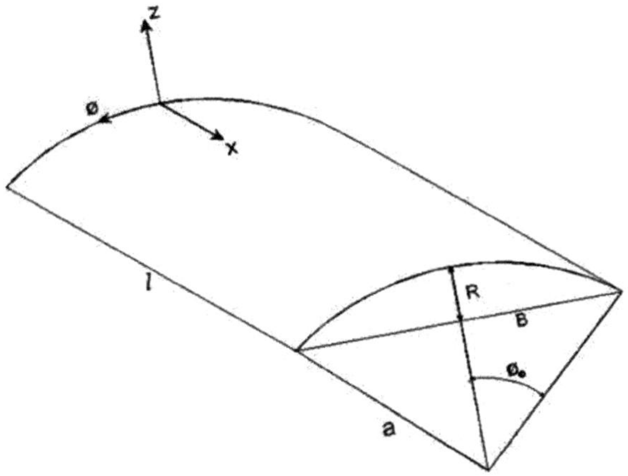

Along longitudinal boundaries, i.e.,
at $\phi = \pm\phi_0$ $N_{x\phi} = v = w = M_\phi = 0$ and
Along the curved boundaries, i.e.,
at $x = 0$ *and* l, $u = v = w = M_x = 0$.

These boundary conditions are referred to as old boundary conditions (OBC) further.

Using these conditions, the stresses in the shells for dead and live load of 2.05 kN/m^2 were obtained by neglecting the Poisson's ratio. Yogananda found that the N_x values at midspan of non-shallow shells could go up to 8 kN/m to 16 kN/m leading to tension values ranging from 0.11 to 0.23 N/mm^2. These values are not admissible for unreinforced masonry. Apparently, these large tension values are due to the boundary condition $N_{x\phi} = 0$ at $\phi = \pm\phi_0$. This is an unrealistic boundary condition as the top

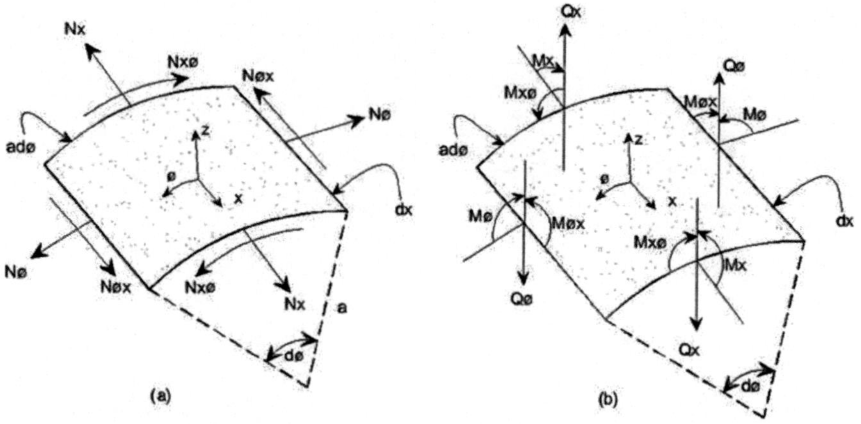

Fig. 2 **a** Membrane, **b** Bending and shear stress resultants

of the wall on which the shell rests is never smooth. It appears that the boundary condition $u = 0$ is more realistic in the longitudinal boundaries.

Now the modified boundary conditions considered in the present study are:

Along longitudinal boundaries, i.e.,

at $\phi = \pm\phi_O u = v = w = M_\phi = 0$ and.

Along the curved boundaries, i.e.,

at $x = 0$ and l, $u = v = w = M_x = 0$.

Which is referred to as new boundary conditions (NBC) now onwards.

The various shells are now analyzed for the new boundary conditions using Flugge's equations, and the results are compared using the stress resultants obtained by Yogananda. The thickness of shell considered is 7 cm.

4 Cylindrical Shell Analysis

Complete solution to cylindrical shell problem consists of two parts. The first one is membrane or particular solution, which will account for the applied loads. The latter is the bending solution, which will account for the edge or boundary conditions. The membrane state of stress alone will results in free deformation of the shell along the boundaries. The bending solution will counteract this depending upon the edge conditions superimposed. So boundary conditions will play vital role in the bending analysis, miss interpretation of the same will lead to abnormal results.

For obtaining the membrane solution, the load '*p*' acting on the surface of the shell is expressed as single trigonometric series function as below.

$$p(x) = \sum_{n=1,3,5..}^{\infty} \frac{4p}{n\pi} \sin\left(\frac{\lambda x}{a}\right) \quad (1)$$

where $\lambda = n\pi a/l$.

Using (1) in membrane equilibrium equations and with the help of basic stress strain relations, expressions for membrane stress resultants and displacements are obtained.

For the bending solution, Flugge's approach carried out by Yogananda [5] is followed. The stresses are expressed in terms of strains and strains in terms of displacements. The expressions for stress resultants in terms of displacements are obtained by integrating the stresses thought the thickness of the shell. These stress resultants are used to transform the bending equilibrium equations having ten unknowns to three equations with only three unknowns, i.e., with displacement components u, v, and w. Considering $u = Ae^{m\phi}$, $v = Be^{m\phi}$, and $w = Ce^{m\phi}$, the equilibrium equations can be written in matrix form as below.

$$\begin{bmatrix} \lambda^2 - 0.5(1+k)m^2 & -0.5m\lambda & -k\left\{\lambda^3 + 0.5m^2\lambda\right\} \\ -0.5m\lambda & m^2 - 0.5\lambda^2 - 1.5\lambda^2 k & m + 1.5m\lambda^2 k \\ -k\left\{\lambda^3 + 0.5m^2\lambda\right\} & m + 1.5km\lambda^2 & 1 + k\left(\lambda^4 + m^4 - 2\lambda^2 m^2 + 2m^2 + 1\right) \end{bmatrix} \begin{Bmatrix} A \\ B \\ C \end{Bmatrix} = 0 \quad (2)$$

where $k = h^2/12a^2$, $h =$ thickness of shell.

For the nontrivial solution, the determinent of the coeffecient matrix should vanish, i.e.,

$$\begin{vmatrix} \lambda^2 - 0.5(1+k)m^2 & -0.5m\lambda & -k\left\{\lambda^3 + 0.5m^2\lambda\right\} \\ -0.5m\lambda & m^2 - 0.5\lambda^2 - 1.5\lambda^2 k & m + 1.5m\lambda^2 k \\ -k\left\{\lambda^3 + 0.5m^2\lambda\right\} & m + 1.5km\lambda^2 & 1 + k\left(\lambda^4 + m^4 - 2\lambda^2 m^2 + 2m^2 + 1\right) \end{vmatrix} = 0 \quad (3)$$

The resulting characteristics equation in terms of m is as follows,

$$m^8 - 2\left(2\lambda^2 - 1\right)m^6 + \left(6\lambda^4 - 8\lambda^2 + 1\right)m^4 - 2\lambda^2\left(2\lambda^4 - 3\lambda^2 + 2\right)m^2 + \left(\frac{\lambda^4}{k} + \lambda^8\right) = 0 \quad (4)$$

The eight roots of the equation lead to eight solutions of the form constant x $e^{mj\phi}$, $j = 1,2, 0.8$ with eight constants. The resulting solution need to be solved along with membrane solution using the appropriate boundary conditions at $\phi = \pm\phi_0$. Since both geometry and loading is symmetry, all the anti-symmetric terms of the series can be omitted. The series solution can be finally truncated depending on the accuracy of the solution required. In the present analysis, six terms are considered namely $n = 1, 3, 5, 7, 9$, and 11.

5 Stresses in Circular Cylindrical Shells

The results for the NBC at $\phi = \pm \phi_0$ have been compared with earlier results of Yogananda. Comparison have been made for the stress resultants M_ϕ, N_ϕ, N_x, and $N_{x\phi}$. Figures 3, 4, 5, 6, 7, 8, 9, 10, 11, 12, 13, 14, 15, 16, 17, and 18 shows the comparisons. The comparison have been made for moderately shallow and non-shallow shells.

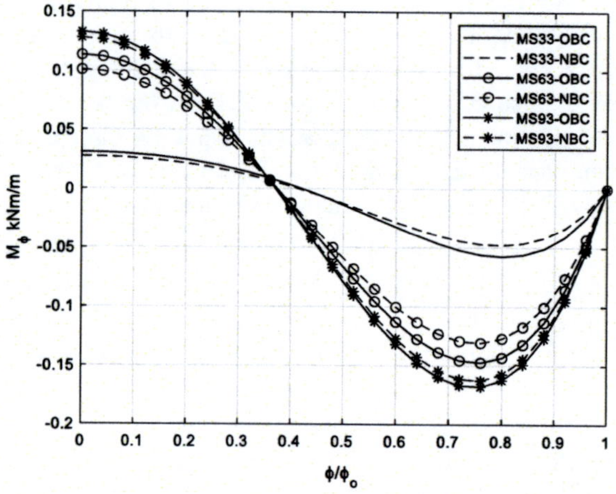

Fig. 3 Comparison of M_ϕ at $x = l/2$ for OBC/NBC 3m span MS shells

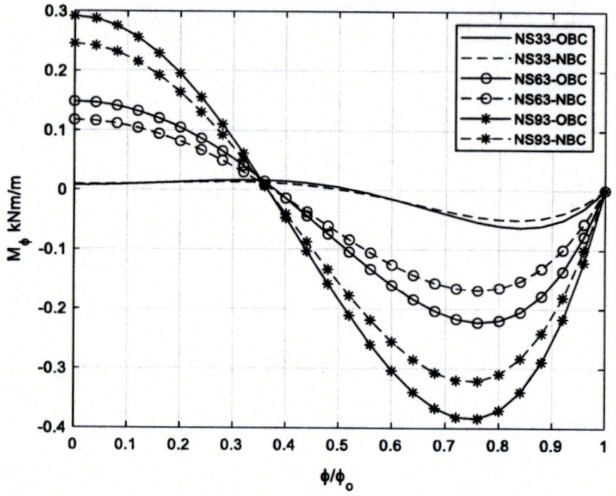

Fig. 4 Comparison of M_ϕ at $x = l/2$ for OBC/NBC 3m span NS shells

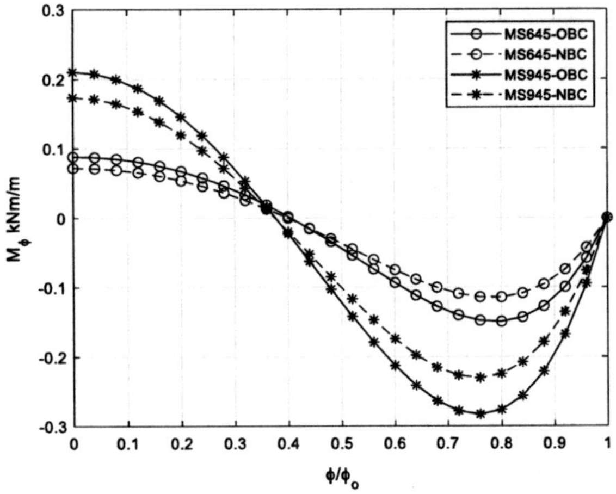

Fig. 5 Comparison of M_ϕ at $x = l/2$ for OBC/NBC 4.5m span MS shells

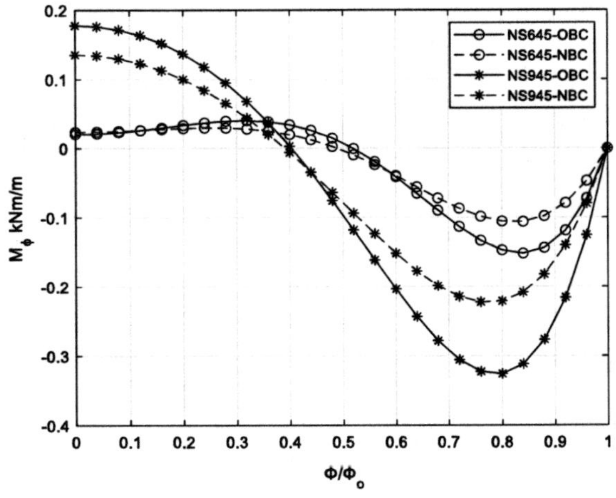

Fig. 6 Comparison of M_ϕ at $x = l/2$ for OBC/NBC 4.5m span NS shells

5.1 Comparison for M_ϕ

The maximum bending moment M_ϕ, generally occurs when $x = l/2$ (mid length) and near $\phi/\phi_o = 0.75$. The maximum value is a negative moment. The positive maximum moment occurs at the crown, when $\phi/\phi_o = 0$, but this value is slightly smaller than the maximum negative moment. Hence, the negative moment will lead to maximum flexural tension. It may also be observed that pattern of moment is quite same for both the boundary conditions. However, the moments for the NBC are invariably smaller than for OBC. The difference is less pronounced when the shell lengths are smaller and are largest when the length of shell is twice or thrice the span.

Fig. 7 Comparison of N_ϕ at $x = l/2$ for OBC/NBC 3m span MS shells

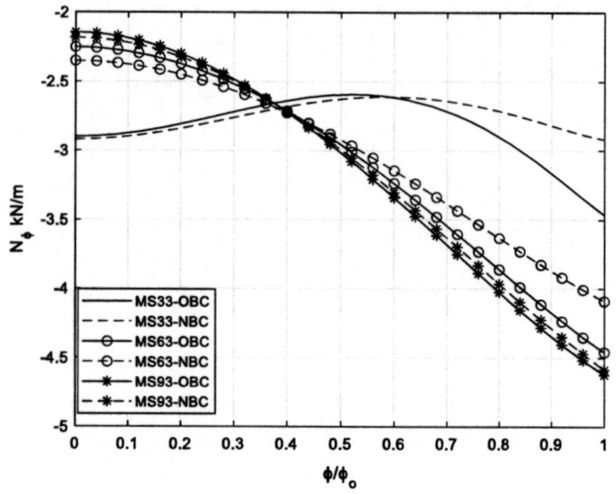

Fig. 8 Comparison of N_ϕ at $x = l/2$ for OBC/NBC 3m span NS shells

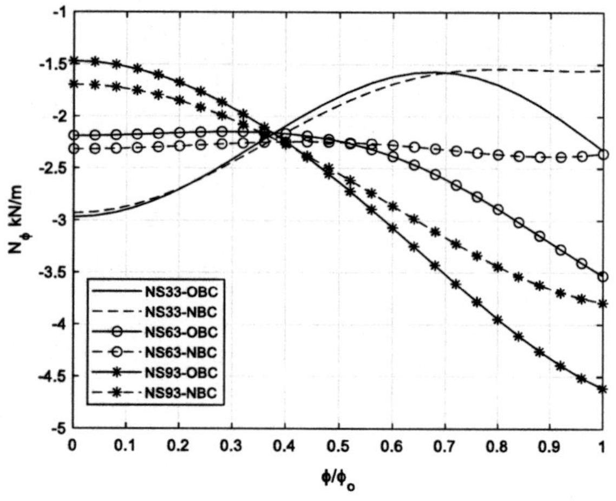

5.2 *Comparison for N_ϕ*

This stress resultant is always compressive for both the boundary conditions. The maximum value occurs often at $\phi = \pm\phi_0$. When the boundary condition is u = 0, the N_ϕ values are generally smaller near the longitudinal boundaries and greater near the crown. The trend is somewhat different for shells with span 3 m and length 3 m and for shells when the span is 4.5 m and length 6/9 m. When the span is 3 m, the maximum value is at $\phi = \pm\phi_0$, except for the cases NS33 and MS33. In these cases, the maximum compression is at $\phi = 0$. Since the shell is square in plan, the design load is distributed more or less equally on all the sides, and the arch action due to curvature is not felt at midspan. The moment variation due to arch action starts showing as the length increases beyond twice the span. The maximum compression

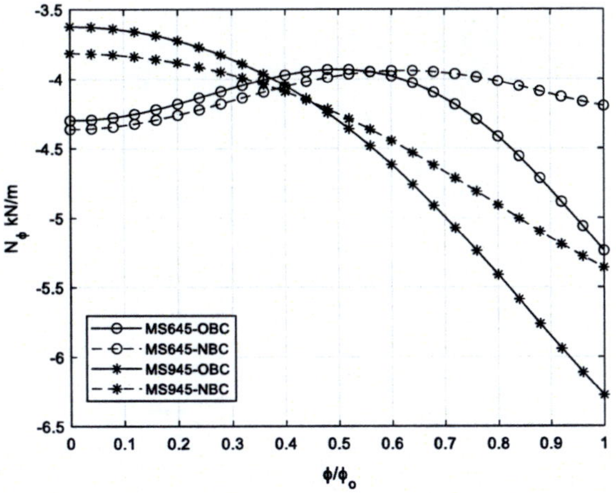

Fig. 9 Comparison of N_ϕ at $x = l/2$ for OBC/NBC 4.5m span MS shells

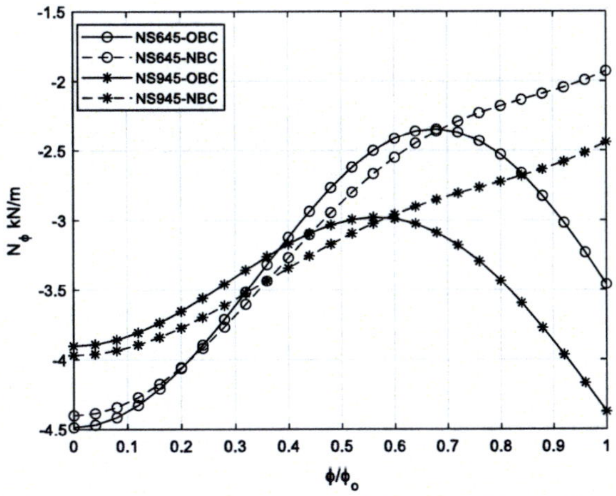

Fig. 10 Comparison of N_ϕ at $x = l/2$ for OBC/NBC 4.5m span NS shells

N_ϕ also occurs at $\phi = 0$ for shells NS645 and NS945 for NBC. The resulting compressive stresses are fairly low, and the maximum among all cases considered is in the range of 0.6–0.9 N/mm^2. A brick masonry with a brick strength of 3.5 N/mm^2 can easily withstand this stress when the mortar is CM 1:6. Since most of the case mortar of CM 1:4 is recommended for masonry shells, the issue of compressive strength is not problemative.

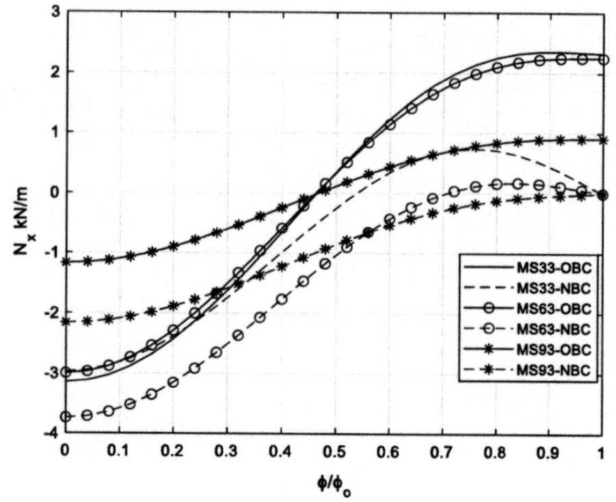

Fig. 11 Comparison of N_x at $x = l/2$ for OBC/NBC 3m span MS shells

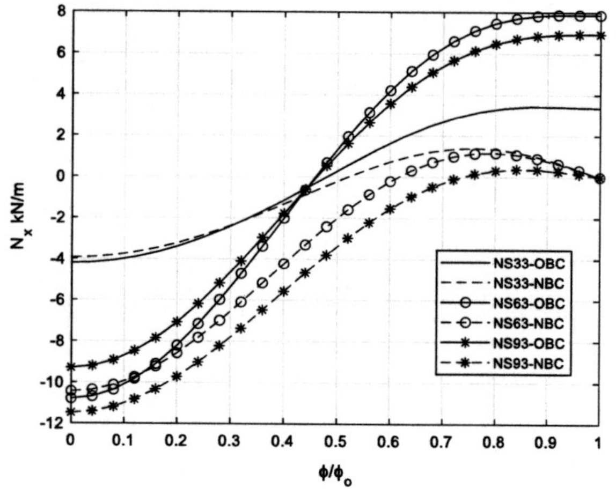

Fig. 12 Comparison of N_x at $x = l/2$ for OBC/NBC 3m span NS shells

5.3 Comparison for N_x

The comparison between the results for two boundary conditions throws up an interesting result. In the analysis of Yogananda, the N_x value is generally tensile as ϕ/ϕ_0 is more than 0.5. However, when boundary condition is $u = 0$, the tension values are practically suppressed, N_x being zero at $\phi = \pm \phi_0$. It can be observed that the shell MS93 is purely compressive without axial tension along the longitudinal direction. N_x attains a small tension value for all other shells near $\phi/\phi_o = 0.75$ which vanishes toward crown where it is compressive for all the shells. This shows that when the boundary condition is realistic $u = 0$, the large tensile stresses are not there. It may also be observed that even when $N_{x\phi} = 0$, the tensile stress is mostly in the range

Fig. 13 Comparison of N_x at $x = l/2$ for OBC/NBC 4.5m span MS shells

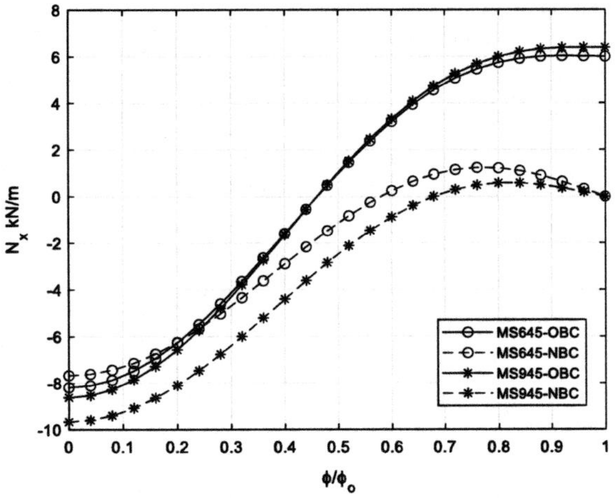

Fig. 14 Comparison of N_x at $x = l/2$ for OBC/NBC 4.5m span NS shells

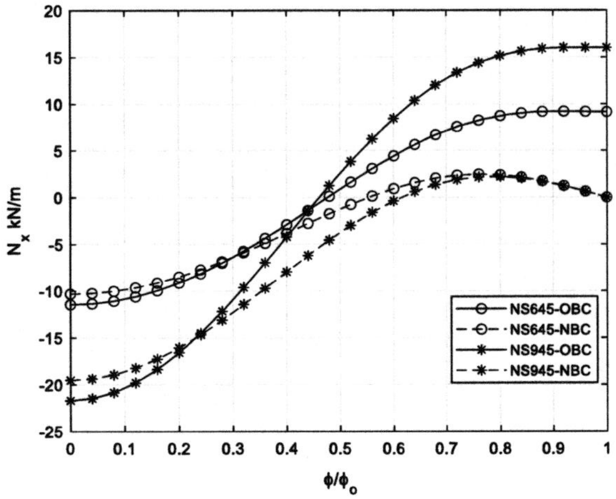

of 0.1–0.2 N/mm^2. The largest value is for the shell NS945, and all the other shells have lower tensile values.

5.4 Comparison for $N_{x\phi}$

The shear stress $N_{x\phi}$ shows significantly different variation in the two boundary conditions. When $u = 0$ at $\phi = \pm\, \phi_O$, $N_{x\phi}$ is no longer zero and attain a large value ($x = 0$ or l). The peak value is usually around $\phi/\phi_o = 0.6$. This peak value is marginally more than the value at $\phi = \pm\, \phi_O$. Again, the peak value for $u = 0$ is invariably more than the peak value for the boundary condition $N_{x\phi} = 0$ used

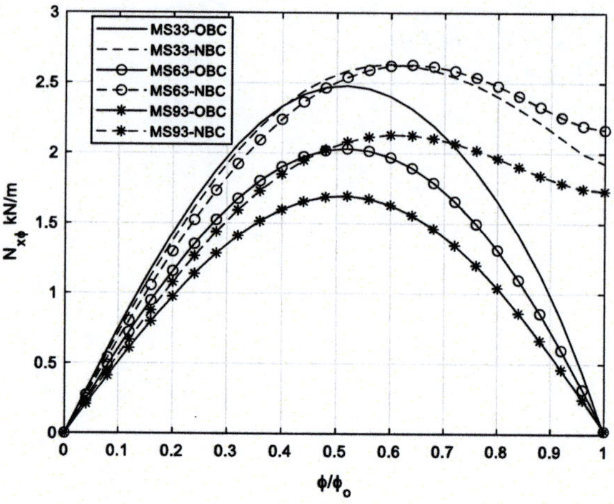

Fig. 15 Comparison of $N_{x\phi}$ at $x = 0$ for OBC/NBC 3m span MS shells

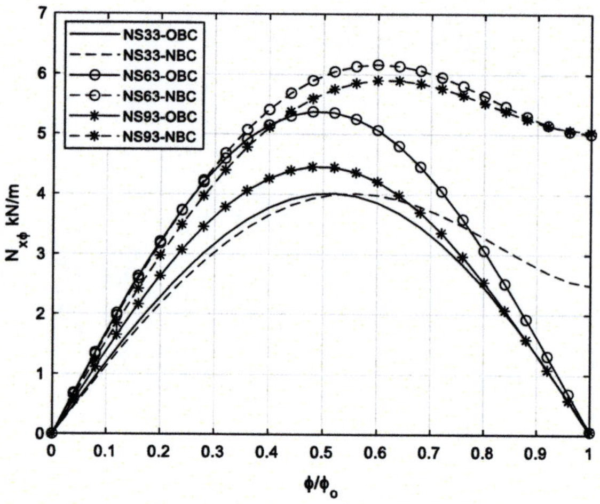

Fig. 16 Comparison of $N_{x\phi}$ at $x = 0$ for OBC/NBC 3m span NS shells

by Yogananda. The largest shear value of around 11.11 kN/m occurs for the shell NS945. It must be recognized that this peak value of membrane shear leads to a diagonal tension near the boundary.

6 Values of Stress Resultants and Corresponding Stresses for Various Shell Dimensions

The critical values of stress resultants for the various shells considered are presented in Table 2.

Fig. 17 Comparison of $N_{x\phi}$ at $x = 0$ for OBC/NBC 4.5m span MS shells

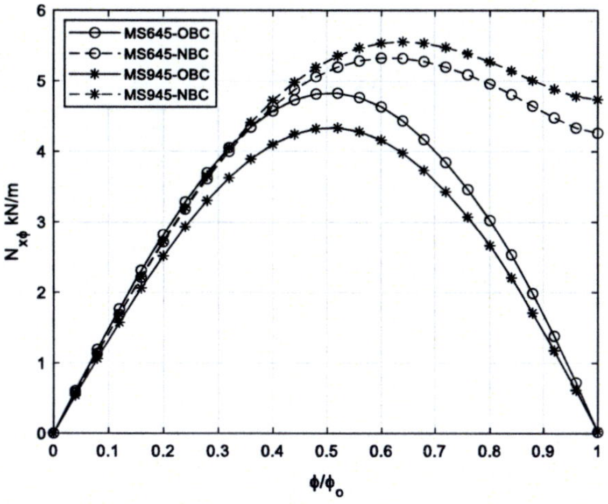

Fig. 18 Comparison of $N_{x\phi}$ at $x = 0$ for OBC/NBC 4.5m span NS shells

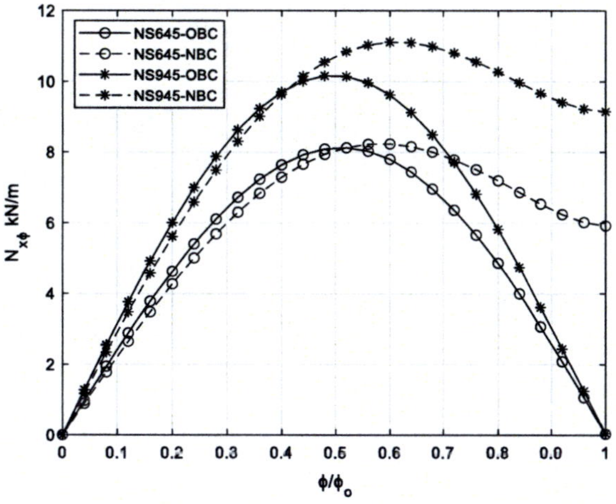

The stresses due to these stress resultants are presented in Table 3. With reference to M_ϕ, a flexural stress from $6M_\phi/bh^2$ is presented in form of N/mm^2. The other stress resultants lead to direct stresses in terms of N/mm^2. It may be noted that the largest flexural tensile stresses occur in the non-shallow shell NS93, and the value is 0.393 N/mm^2. For masonry, this value is rather high. It is possible that the presence of axial compression N_ϕ may partially offset this flexural tension. Also, a thin layer of concrete with minimum reinforcement may be provided over the masonry shell to bear the small tension values. But still such shell proportions can be avoided.

The stresses N_x are by and large quite low. The maximum value is seen in shell NS645, and the value is 0.035 N/mm^2. According to the measurements by Mathana [9] and Raghunath [10], the masonry tensile strength is about 0.166 N/mm^2 for

Table 2 Critical stress resultants for NBC

Stress resultants	$x = l/2,\ \phi/\phi_0 = 0.75$		$x = l/2,\ \phi/\phi_0 = 0.75$	$x = 0$ or $l,\ \phi/\phi_0 = 0$
Shell type	M_ϕ kNm/m	N_ϕ kN/m	N_x kN/m	$N_{x\phi}$ kN/m
MS33	0.046	−2.673	0.738	2.634
MS63	0.131	−3.509	0.105	2.624
MS93	0.163	−3.799	−0.192	2.130
NS33	0.044	−1.551	1.395	3.985
NS63	0.168	−2.340	1.111	6.163
NS93	0.322	−3.307	0.093	5.900
MS645	0.113	−3.971	1.221	5.327
MS945	0.230	−4.786	0.447	5.531
NS645	0.105	−2.242	2.463	8.232
NS945	0.221	−2.774	2.124	11.115

Table 3 Critical stresses for NBC

Stress resultants	$x = l/2,\ \phi/\phi_0 = 0.75$		$x = l/2$ $\phi/\phi_0 = 0.75$	$x = 0$ or $l,\ \phi/\phi_0 = 0$
Shell type	M_ϕ N/mm^2	N_ϕ N/mm^2	N_x N/mm^2	$N_{x\phi}$ N/mm^2
MS33	0.057	−0.038	0.011	0.038
MS63	0.160	−0.051	0.002	0.038
MS93	0.199	−0.055	−0.003	0.030
NS33	0.055	−0.022	0.020	0.057
NS63	0.206	−0.034	0.016	0.088
NS93	0.393	−0.048	0.002	0.084
MS645	0.140	−0.057	0.018	0.076
MS945	0.282	−0.069	0.007	0.079
NS645	0.120	−0.032	0.035	0.118
NS945	0.272	−0.040	0.031	0.159

CM1:6, stresses parallel to the bed joints. Hence, the shell stresses are well within the limit. However, according to Yogananda's boundary conditions, the tensile stress is 0.24 N/mm^2. Thus, the use of realistic boundary condition $u = 0$ leads to satisfactory stresses in masonry shells.

It is interesting to note that the shell MS93 is under pure compression without axial tension along the longitudinal direction. Also, shells like MS63, NS93, MS945 show marginal tension values. So it is clear from the results that it is possible to select the shell proportions such that the direct tension value N_x, which is critical for masonry design can be avoided or minimized.

The shear stress resultant $N_{x\phi}$, however, leads to larger diagonal tension going up to 0.159 N/mm^2(NS945).

7 Discussion of Results

The change of boundary condition $N_{x\phi} = 0$ at $\phi = \pm\phi_0$ to $u = 0$ at $\phi = \pm\phi_0$ generally leads to less value of $M_\phi N_\phi$ and N_x. In fact, the tension values of N_x in the Yogananda boundary condition are practically eliminated. The values of $N_{x\phi}$ near the boundaries $x = 0$ and $x = l$, $x = 0$, however, get enhances by about ten percent.

Increase in the length of the shell generally leads to larger bending moments near the midspan. It would be ideal to keep the length of the shell as less than twice the span. In general, the value of the new boundary condition leads to more favorable stresses in masonry shell proportions considered here.

The large flexural tension values occur in shells NS93 and NS945 for the boundary condition $N_{x\phi} = 0$ at $\phi = \pm\phi_0$. The values figure out to be 0.47 N/mm^2 and 0.4 N/mm^2, respectively. When the boundary condition is changed to $u = 0$, the flexural tension values reduced to 0.39 N/mm^2 and 0.27 N/mm^2. Even these values are on the higher side, it is desirable to restrict the length of 3 m span shell to twice the span and 4.5 m span shell to 1.3 times the span. Even then, these shells will have small value of flexural tension. The presence of axial compression due to N_ϕ means such flexural tensions are not serious, especially when a mortar of CM 1:4 is used for the shell construction.

The critical stress in the N_x value which becomes tensile as ϕ/ϕ_0 is increased beyond 0.5 when the boundary condition is $N_{x\phi} = 0$ at $\phi = \pm\phi_0$. The tension value reachs 8 kN/m for shell NS63 and 16 kN/m for shell NS945. The corresponding stresses are 0.11 N/mm^2 and 0.23 N/mm^2, respectively. These tension values are certainly not desirable for unreinforced masonry. However, when the boundary condition is the more realistic $u = 0$ at $\phi = \pm\phi_0$, the tension value reduces to 0.014 N/mm^2 and 0.036 N/mm^2. These tension values are well below the permissible stress values. Again, a masonry shell is always covered by a thin layer of reinforced concrete (4 cm thick with nominal steel) to take care of settlement stresses. This additional feature will also be able to provide any small tension values.

The stress $N_{x\phi}$ in the shell, however, is increased with the boundary condition $u = 0$. This leads to a diagonal tension when $x = 0$ and $x = l$ for $\phi/\phi_0 = 0.6$. This tension is generally small except for the shell NS645 and NS945. Again, the layer of reinforced concrete can take care of longitudinal tension values. The stress values are well within the permissible limits for other shell proportions considered here.

8 Conclusion

This paper examines the stresses in masonry shells as the edge boundary condition $N_{x\phi} = 0$ at $\phi = \pm\phi_0$ is changed to $u = 0$ at $\phi = \pm\phi_0$. The results of earlier study by Yogananda [5] have been compared with the new results, utilizing Flugge's theory for circular cylindrical shells. Shells of span 3 m and 4.5 m have been considered. For 3 m span shells, lengths of 3, 6, and 9 m have been considered. For 4.5 m span shells, lengths of 6 m and 9 m have been considered. Regarding the midspan rise, moderately shallow and non-shallow shells have been considered.

- There is a drastic reduction in the flexural tension values for the new boundary condition. The values are quite high for long shells. Even though the presence of axial compression N_ϕ at the same location is beneficial in reducing the effect of this tension, the stresses values are quite high for masonry. So, such shell proportions can be avoided.
- There is a significant reduction in the direct tension values for new boundary condition. The largest value is observed for non-shallow shells. It is interesting to note that as the length of the shell increases, the tension value decreases. However, the tension value increases with span.
- The diagonal tension due to $N_{x\phi}$ is, however, increased marginally as the boundary condition is changed from $N_{x\phi} = 0$ to $u = 0$. A thin layer of concrete with nominal reinforcement will help in resisting any small tension which may occur.
- The study reveals that it is possible to select the shell proportions such that the stress values are within permissible limit.
- New boundary condition leads to more favorable stresses in masonry cylindrical shells.

References

1. Fafthy H (1973) Architecture for the poor. The University of Chicago Press, Chicago & London
2. Fathy H (1986) Natural, energy and Vernicular architecture. The University of Chicago Press, Chicago & London
3. Heyman J (1967) On shell solutions for masonry domes. Int J Solids Struct 3:227–241
4. Grimm CT (1975) Brick masonry shells. J Struct Div, ASCE 101:79–95 (Jan)
5. Yogananda MR (1991) The analysis of unreinforced brick masonry vaults. Department of Civil Engineering, IISc, Bengaluru
6. Flügge W (1973) Stresses in shells, 2nd edn. Springer-Verlag, Berlin Heidelberg, New York
7. Donnel LH (1933) Stability of thin walled tubes under torsion. National Advisory Committee on Aeronautics, No. 479
8. Vlasov VZ (1958) Allgemeine Schalentheorie und ihre anwendung in der Technik Academie-Verlag, Berlin
9. Matthana MH (1996) Strength of brick masonry and masonry walls with openings. Ph.D. Thesis, Department of Civil Engineering, IISc, Bengaluru (Oct)
10. Raghunath S (2012) Static and dynamic behaviour of brick masonry with containment reinforcement. Ph.D. Thesis, Department of Civil Engineering, IISc, Bengaluru (Jan)

Experimental Investigations on Utilization of Bagasse Ash in Adobe Bricks

Manish S. Dharek, K. S. Sreekeshava, Jagadish Vengala, Kilabanur Pramod, Prashant Sunagar, and M. V. Shivaprakash

Abstract Utilizing the industrial and agricultural wastes as a active raw material has been the prime focus for Researchers around the globe. Utilizing these by-products/waste materials can be beneficial in terms of cost savings in addition to addressing the issue of sustainability and reducing the greenhouse gas emissions. The present work was aimed at developing eco-friendly unfired bricks termed as adobe bricks by utilizing agro waste from sugar mills which is termed as sugarcane bagasse ash. The study involved investigating the performance of adobe bricks mixed with sugarcane bagasse ash stabilized with cement. Locally available soil from North part of Bengaluru was used as part of this study. The adobe bricks were casted by varying the sugarcane bagasse ash in the range 0–60% at increments of 10%. Engineering properties of these adobe bricks such as compressive strength, water absorption, dry density, flexural strength were studied as per codal provisions. Mix with 30% blending of sugarcane bagasse ash in the adobe bricks showed good results and satisfied the codal provisions as per Indian standards.

Keywords SCBA · OPC · Compressive strength · Sun-dried bricks

M. S. Dharek (✉)
Department of Civil Engineering, BMS College of Engineering Bengaluru, Bengaluru, India

K. S. Sreekeshava
Department of Civil Engineering, Jyothy Institute of Technology, Bengaluru, India

J. Vengala
Department of Civil Engineering, PVP Siddhartha Institute of Technology, Vijayawada, India

K. Pramod
Ionidea Educational Software private limited, Bengaluru, India

P. Sunagar
Department of Civil Engineering, Ramaiah Institute of Technology, Bengaluru, India

M. V. Shivaprakash
Department of Civil Engineering, MITS, Madanpalle, India

L. Nandagiri et al. (eds.), *Sustainability Trends and Challenges in Civil Engineering*, Lecture Notes in Civil Engineering 162, https://doi.org/10.1007/978-981-16-2826-9_31

1 Introduction

Use of adobe blocks for dwelling has been in practise from ancient days especially in the American and middle east regions. "Adobe" a Spanish word has been derived from old Arabic word "al-tob" dating from the Moorish occupation in Spain. Adobe block/bricks for construction has been adapted universally due to changes in economic and social reasons. Quite a good amount of investigations in order to provide engineering solutions have been done on utilization of adobe as a building blocks. Earth has been the most broadly known and richly accessible material for human culture to utilize it in development. From the times of Egyptian and Mesopotamian, earth is primary part of any development in its various structures. Nearby accessibility of mud makes its utilization invaluable, and simplicity of handling makes it one of the most energy proficient building material, savvy and it is a truly dependable material for any working by and large and minimal effort structures specifically. Generally, mud has been widely utilized for building development in India and somewhere else. Customary earth development innovation has gone through an extensive change that upgrades earth's strength and quality as a development material for minimal effort structures. Such techniques incorporate slammed earth and machine packed settled earth blocks. The development practises of today vigorously rely upon materials like concrete blocks, bricks, and different metals like steel, aluminium and so on [1–3]. These are energy escalated materials which devour a great deal of energy and hence the creation of these building blocks negatively affects climate [4, 5]. Since these materials can be created distinctly specifically territories there is a need to move these materials to the site where they are to be utilized and again bringing about utilization of energy, so it is very obvious that these materials contain parcel of typified energy in them. Thus, it is essential to deliver building materials which burn-through less energy and can be utilized for development [6].

2 Literature Review

Studies on utilization of quarry dust (QD), lime (L) and sugarcane bagasse ash (SCBA) were carried out by [7]. The physical properties of quarry dust and lime were determined using the laboratory test methods. SCBA–quarry dust (QD) lime (L) combination blocks were designed and developed in different mix proportions. A compressive strength of 6 MPa was observed. Particle size distribution analysis of SCBA was carried out using the hydrometer test. Chemical analysis of SCBA was done using energy-dispersive x-ray fluorescence spectrometer. The blocks with 20% addition of lime to SCBA and quarry dust exhibited a compressive strength of up to 6.59 MPa, which is almost double that of the conventional clay blocks (3.5 MPa). It was also observed that masonry bonding of SCBA– QD–L blocks is stronger compared to commercially available fly ash and burnt clay blocks. Research carried out by Pusit et al. [8] focussed on the feasibility of using agricultural by-products such as rice husk and bagasse ash in adobe block mixtures with a percentage replacement of 0, 1, 2, 3 and 6% by weight of materials. The study explored the level at which replacing rice husk and bagasse in adobe block affect the optimal compressive strength, shrinkage, thermal conductivity and moisture absorption. Bagasse and rice husk fibre can improve the compressive strength of adobe block. Bagasse was more effective than rice husk in all mix proportions. The proportion of clay 84%, sand 10%, bagasse 6% gives compressive strength as 3 MPa. Prabhu et al. [9] studied the effect of addition of bagasse ash on the compressive strength and water absorption of bricks.

Studies on effect of addition of aluminium sulphate on the compaction and mixing characteristics of bricks were also studied as part of this work. Mix with 20% addition of bagasse ash gave compressive strength of 6.3 MPa when compared with conventional clay brick (3.5 MPa). The water absorption values were also found to reduce with increase in bagasse ash content in the brick. It was observed that addition of aluminium sulphate improved compaction. Studies on bricks with binary and ternary mixtures containing clay, bagasse ash, fly ash and silica fume was taken up by Ignacio et al. [10]. Characterization of ternary and binary mixtures were carried out. Studies concluded that mixture with clay, bagasse ash and silica fume showed good results in terms of compressive and bending strength. Based on the above literature, it was seen that sugarcane bagasse ash be effectively used as an active ingredient. The objectives of the present research have been summarized below.

- Testing of soil, bagasse ash for its engineering properties.
- Preparation of optimal mix to produce an adobe brick of required strength incorporating the bagasse ash.
- Evaluation of engineering properties of adobe brick viz. compressive strength, flexural strength, dry density, water absorption in comparison with normal burnt clay brick.

3 Characterization of Materials

The constituents used for casting the adobe bricks included soil, sugarcane bagasse ash, ordinary portland cement, water. The details of the tests carried on these constituents are summarized below.

3.1 Soil

Soil used in the present study was sourced from North part of Bengaluru. Tests were conducted on the soil as per IS 2720-1983 [11]. The details of the physical properties of soil is shown in Table 1.

Table 1 Physical properties of soil

Serial No	Property	Result obtained
1	Soil classification	Silty clay
2	Specific gravity	2.67
3	Liquid limit and plastic limit	23.29%, 17.79%
4	Plasticity index	5.51
5	Water content	12.5%
6	Unconfined compressive strength	0.132 MPa
7	Maximum dry density Optimum moisture content	1.86 g/cc 16.5%

3.2 *Sugarcane Bagasse Ash (SBA)*

Sugarcane bagasse ash, a by-product obtained from sugar manufacturing factories is produced by burning the sugarcane bagasse. The bagasse ash for the present work was collected from sugarcane manufacturing plant at Doddagarudanahalli, Mandya district, Karnataka. The ash was brought to the lab and sieved through 600 μ sieve. The specific gravity of the SBA was found to be 2.35and blaine's fineness was 938 m^2/kg.

3.3 *Ordinary Portland Cement (OPC)*

Ordinary portland cement-53 Grade conforming to IS 12269-2013 [12] was used in this study which was procured locally from Bengaluru. The results of tests conducted on cement are shown in Table 2

Table 2 Physical properties of ordinary Portland cement-53 Grade

S. No.	Property	Result obtained	Remarks
1	Specific gravity	3.12	Range 3.05–3.15
2	Consistency	28%	–
3	Setting time-initial	43 min	≥ 30 min
4	Setting time-final setting	568 min	≯ 600 minutes
5	Soundness by Le chatliers method	7.0 mm	≯ 10 mm
6	Compressive strength	56.3 MPa @ 28 days	≥ 53 MPa

4 Methodology Adopted

Wooden moulds in a gang mould pattern was planned for casting the adobe bricks of size 190 mm × 90 mm × 90 mm which is the size recommended in IS 1077-1992 [13]. The gang moulds to accommodate five adobe bricks at a time were fabricated, details of which are shown in Fig. 1.

The blend had to kneaded well in order to obtain a reliable blend. The optimum content of water required for making adobe was decided by conducting a field trials through the method of rolling the soil into smooth balls between the hands which should not disintegrate when dropped from 0.5 m elevation. The inner surface of mould was oiled in order to prevent the sticking of soil inside the mould. The blend was filled into the precast wooden mould of dimensions 190 mmx 90 mm x 90 mm and later compacted immediately by dropping the ball from an elevation of around 0.5 m. This was done to ensure soil got compacted inside the mould without formation of voids. These were demoulded after 24 h from the moulds and subjected to wet curing through sprinkling of water on a regular basis for seven days and stacked for a minimum duration of about seven days later after curing as shown in Fig. 2. The main purpose of curing of these adobe bricks was to ensure they gain sufficient strength and undergo minimal shrinkage if any so that these adobe bricks do not shrink at a later stage.

The numerical value in parenthesis in Table 4 indicates the coefficient of variation.

The details of the mix proportion by weight which were studied as part of the present investigations are shown in Table 3. The cement content of 10% was kept uniform for all mixes.

Fig. 1 Details of mould (left) and mould fabrication process (right)

Fig. 2 Demoulded adobe bricks (right) and curing of adobe bricks (right)

Table 3 Details of the mix proportion for adobe bricks

Mix type	% of soil	% of SBA	% of cement
A	100	0	10
B	90	10	10
C	80	20	10
D	70	30	10
E	60	40	10
F	50	50	10

5 Tests on Adobe Bricks

The assessment of engineering properties of these adobe bricks such as compressive strength, water absorption, dry density, flexural strength were carried out as per IS 3495-1992-Part I and Part II [14]. The details are shown in Table 4.

Table 4 Details of engineering properties of adobe bricks

Mix type	Mean compressive strength	Mean water absorption	Mean dry density	Mean flexural strength
A	1.80 MPa (15.2%)	21.80% (8.9%)	1.55 g/cc (15.4%)	0.19 MPa (13.6%)
B	2.45 MPa (18.7%)	19.40% (12.2%),	1.53 g/cc (14.2%)	0.27 MPa (15.9%)
C	3.50 MPa (21.3%)	17.40% (13.9%)	1.49 g/cc (9.8%)	0.34 MPa (22.6%)
D	4.12 MPa (12.3%)	18.20%, (16.8%)	1.41 g/cc (16.1%)	0.45 MPa (15.4%)
E	3.12 MPa (13.9%)	23.60% (15.7%)	1.38 g/cc (17.0.2%)	0.30 MPa (13.9%)
F	2.30 MPa (18.6%)	20.20%, (14.5%)	1.36 g/cc (13.2%)	0.26 MPa (18.6%)

6 Results and Discussions

The variation of mean compressive strength for varying percentage of SBA is shown in Fig. 3. The maximum compressive strength of 4.2 MPa is obtained for mix D. Hence, when the soil is replaced with SBA, the pozzolana reacts completely with Ca(OH)2 resulting from hydration of OPC and hence it is the optimum. The range of compressive strength obtained was between 1.80 and 4.12 MPa.

The variation of mean water absorption for varying percentage of SBA is shown in Fig. 4. The range of water absorption was between 17.40 and 23.60%. Mix C showed the least value of water absorption, i.e. 17.40%. The water absorption of Mix D was also very close to Mix C, i.e. 18.20%.

The variation of mean dry density for varying percentage of SBA is shown in Fig. 5. The range of dry density was between 1.36 and 1.55 g/cc. Mix A showed maximum dry density, whereas mix F showed the least dry density. The dry density valued dipped sharply for Mix E and Mix F.

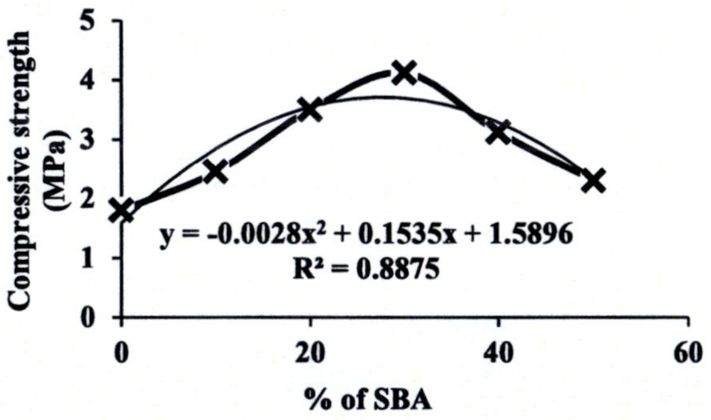

Fig. 3 Variation of compressive strength of adobe bricks versus % of SBA

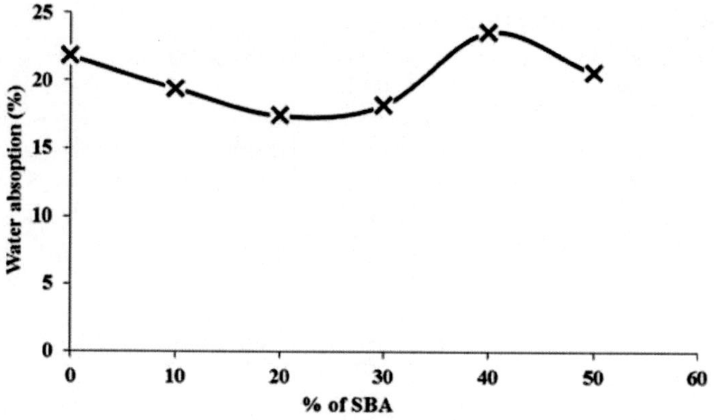

Fig. 4 Variation of water absorption of adobe bricks versus % of SBA

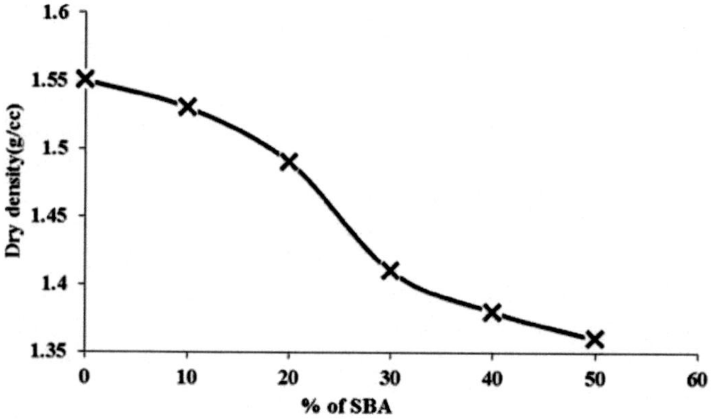

Fig. 5 Variation of dry density versus % of SBA

The variation of mean flexural strength for varying percentage of SBA is shown in Fig. 6. The maximum compressive strength of 0.45 MPa was obtained for mix D.

The range of flexural strength obtained was between 0.19 and 0.45 MPa. The optimal level of SBA from point of view of strength was found to be at 30%.

7 Conclusions of the Present Study

The following broad conclusions are drawn from the present study on utilization of bagasse ash in adobe bricks.

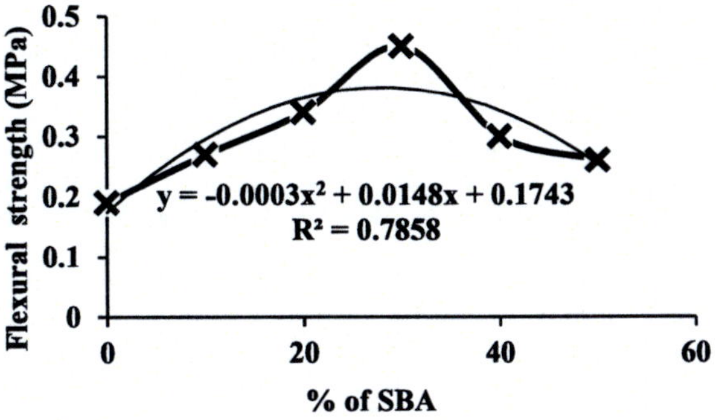

Fig. 6 Variation of Flexural strength of adobe bricks versus % of SBA

- The minimum compressive strength prescribed as per IS 1077-1992 [13] for clay bricks is 3.50 MPa. In the present study, compressive strength of mix D (30% SBA) was 4.20 MPa satisfying the codal provisions
- The maximum water absorption prescribed as per IS 1077-1992 [13] for clay bricks is 20%. In the present study mix B, C, D satisfied this criteria. However from strength point of view, mix C was found to be optimum
- The dry density values as per literature [15] is in the range 1.4–1.8 g/cc for bricks. In the present study, Mix A, B, C, D satisfied this criteria.
- The flexural strength as per literature [15] is in the range 0.3–0.4 MPa for table moulded bricks. In the present study, Mix C and Mix D satisfied this criteria
- Based on the results obtained, Mix C with 30% SBA was found to be optimal thus clearly showing the effectiveness of SBA as an active ingredient in production of adobe bricks.

References

1. Dharek MS, Raghunath S, Sunagar P, Bhashyam AH, Sreekeshava KS (2020) Stress strain characteristics of reinforced hollow concrete block masonry melded with mesh reinforcement. IOP Conf Ser Mater Sci Eng 814:012040
2. Dharek MS, Raghunath S, Ashwin CP (2021) Experimental behaviour of unreinforced and reinforced concrete block masonry walls under uniaxial compression. Mat Today Proc 46:2462–2467. https://doi.org/10.1016/j.matpr.2021.01.398
3. Sumalatha J, Dharek MS, Niranjan GH, Sunagar P, Chaurasiya AK (2020) Development of sustainable building blocks with tyre waste, flyash and lime. Int J Civil Eng Technol 11(5):93–104. http://www.iaeme.com/IJCIET/issues.asp?JType=IJCIET&VType=11&IType=5
4. Vengala J, Raju S, Shiva B, Manjunatha LR, Yogananda MV (2020) Use of GGBS in manufacturing of solid concrete blocks. In: Babu K, Rao H, Amarnath Y (eds) Emerging trends in civil engineering. Lecture notes in civil engineering, vol 61. Springer, Singapore, pp 177–195. https://doi.org/10.1007/978-981-15-1404-3_16
5. Vengala J, Mangloor S, Goud TKC (2019) Performance of autoclaved aerated concrete blocks under varying temperatures. Int J Recent Technol Eng (IJRTE) 7(6C2). ISSN: 2277-3878. https://www.ijrte.org/wp-content/uploads/papers/v7i6c2/F11120476C219.pdf

6. Dharek MS, Raghunath S, Ashwin CP (2019) Strength and elastic properties of grouted reinforced hollow concrete block masonry prisms. In: Chapter 8, Proceedings of AICERA—2019. CRC Press, Taylor and Francis publications, pp 42–47
7. Madurwar M, Mandavgane S, Ralegaonkar R (2014) Development and feasibility analysis of bagasse ash bricks. J Energy Eng ASCE 14(2014):1–16
8. Lertwattanaruk P, Choksiriwanna J (2011) The physical and thermal properties of adobe brick containing bagasse for earth construction. Int J Build Urban, Inter Landscape Technol (BUILT) 14:53–62
9. Prabhu P, Ramesh S, Archana M (2019) An experimental study on bricks by partial replacement of bagasse ash. JRMT I, 201(1963):1–13
10. Maza-Ignacio OT, Jiménez-Quero VG, Guerrero-Paz J, Montes-García P (2020) Recycling untreated sugarcane bagasse ash and industrial wastes for the preparation of resistant, lightweight and ecological fired bricks. Const Build Mat 234:117314. https://doi.org/10.1016/j.conbuildmat.2019.117314
11. IS 2720-Part 1 to Part 17 (1983) Method of test for soils. Bureau of Indian Standards, New Delhi, India
12. IS 12269 (2013) Ordinary Portland cement-53 grade specification. Bureau of Indian Standards, New Delhi, India
13. IS 1077 (1992) Common burnt clay building bricks—specification. Bureau of Indian Standards, New Delhi, India
14. IS 3495- Part 1 and Part 2 (1992) Method of tests of burnt clay building bricks. Bureau of Indian Standards, New Delhi, India
15. Keshava M, Vijayendra KV, Raghunath S (2010) Strength efficiency of commonly used block work Masonry. IJEE, vol 3, pp 586–598
16. Dharek MS, Sunagar P, Sreekeshava KS, Nagashree B, Thejaswi P, Kilabanur P, Nruthya K, Satish Chandra C (2021) Experimental investigations on strength performance of the brick produced by blending demolished waste with pozzolanic materials. In: Advances in sustainable construction materials. Springer, Singapore, pp 573–583. https://doi.org/10.1007/978-981-33-4590-4_54

Analysis of Storm Water Management Model for Yediyur Lake, Bangalore

K. Tejaswini, H. S. Yashaswini, B. N. Skanda Kumar, Vibha Ramesh, C. Chandre Gowda, B. E. Bhojaraja, S. R. Srilakshmi, and S. Harish Kumar

Abstract Lack of organized planning may raise critical issues in management of water resources. The encroachment of wetlands and construction of any infrastructure in the natural flood zones severely affect them during monsoon. So, a systematic and technical supervision is required to manage the storm water drains in the urban zones. In this study, a micro-catchment in Bangalore (Yediyur Lake), Karnataka, India, was analyzed in terms of quality and quantity using storm water management model. The size of drains and conduits was designed and simulated for 10 year return period rainfall event. The impact of excess volume at each node was checked. The best management practices for urban surplus water management have been portrayed in order to mitigate the flash floods and also to reduce the contamination of the runoffs.

Keywords Storm water management model · Flood · Rainfall event · Contamination

1 Introduction

In urban areas, most of the surface area is covered by impermeable layer such as parking spaces, built-up areas, roads, and pavement blocks. The impermeable surface increases the direct runoff and reduces the seepage of water in to the ground. These layers are the main cause for drastic hydrological response such as urban floods. So, it is essential to model the hydrological conditions accurately to assess the future risks and to devise the strategies for mitigation. Urban floods are caused

K. Tejaswini · H. S. Yashaswini · Vibha Ramesh · S. R. Srilakshmi · S. Harish Kumar (✉)
Department of Civil Engineering, Jyothy Institute of Technology, VTU, Bengaluru, India

B. N. Skanda Kumar · C. Chandre Gowda
Centre for Incubation, Innovation, Research and Consultancy, Jyothy Institute of Technology, VTU, Bengaluru, India
e-mail: skanda.kumar@jyothyit.ac.in

B. E. Bhojaraja
Department of Civil Engineering, NMAM Institute of Technology, VTU, Nitte, Karkala, Udupi, India

L. Nandagiri et al. (eds.), *Sustainability Trends and Challenges in Civil Engineering*, Lecture Notes in Civil Engineering 162, https://doi.org/10.1007/978-981-16-2826-9_32

by augmented development activity and unplanned expansion of infrastructure due to the increase in population. Expansion of city areas depends on energy-intensive water sources (distant rivers or aquatic body) to meet their water demand. Storm water originating from precipitation events is often stored in ponds and puddles which serves as drinking water source for urban catchments.

The pace of urbanization is increasing globally, putting more pressure on local water quality [1]. Release of urban and industrial wastewater contributes to the poor water quality in a number of ways. During monsoons, urban storm water runoff results in increase of microorganisms in the drain at outfall discharges [2]. The correlation between water quality and urban planning is significant in safeguarding urban ecosystems such as lakes [3]. The quality of Bellandur Lake was severely affected due to high-level physio-chemical concentrations by industrial discharges [4]. Lakes should be monitored regularly to reduce the deterioration of quality from source to point [4]. The restoration approaches substantially improve lakes condition and health of aquatic dwellers [5].

Storm water detention basins control the flood and are considered major part of best management practice during peak storm water flows [6]. The multi-criteria analysis (MCA) approach was carried out for urban storm water drainage management [7]. Many legal and administrative measures were imposed in developing the natural water flow routes through restoring the tank bunds [8]. Nevertheless, the contaminants such as copper and other pollutants deposited on roads from the motor vehicles are carried from overland flow, further transported to aquatic habitats [9]. Consequently, the widespread degradation of water resources caused by pollution from urban runoff combined with the impacts of flooding and poor drainage of runoff has led to a critique of traditional approaches for the design and operation of urban drainage systems [10].

Many model studies along with field and laboratory experiments were carried out to find the biological retention for urban storm water at laboratory scales [11]. Biological retention helped in removal of nitrate and other constituents [11, 12]. In the recent years, mechanisms and design enhancements for nitrogen uptake, denitrification through various storm water (bio-retention and using filters) control measures were reviewed [13]. Artificial neural networks were used to predict storm water quality at urbanized catchments located throughout the United States [14]. Many studies used in evaluation and identification of suitable sites for storm water harvesting assist the planner to prioritize the respective schemes in the corresponding areas with significant impact on the water demand [15]. Linear and integer programming formulation was developed to find the optimal sizes of storm water. The results demonstrated the feasibility of meeting the objective of storm water retention and water quality management by optimal application of storm water control measures [16, 17].

In similar purview of the contemporary moment, flood risk has been heightened because of an intensified alignment between the flow/fixity of capital and storm drains [18]. Due to urban encroachments of open and wetland areas, the impact of the urban flood has directly affected the habitants [19]. Annual average economical loss in India because of flood is 75–80% of total economic loss [20]. It is necessary to develop better disaster management tactics and disaster risk reduction techniques for the better way of life.

Lack of organized planning may raise critical issues in management of water resources. So, a systematic and technical supervision is required to manage the storm water drains in the urban zones. In this study, a micro-catchment in Bangalore (Yediyur Lake), Karnataka, India, was analyzed in terms of quality and quantity using SWMM.

2 Study Area

Yediyur Lake, Jayanagar, Bangalore, India, is selected as study area. The geographic location of the inflow to lake is at 12.93° N latitude and 77.57° E longitude. Jayanagar is located in the center of the Bangalore city. Being one of the oldest lakes in Bangalore, Yediyur Lake is the one of the most visited lakes in the city.

In Bangalore, rainfall peaks during the monsoon starting from last week of May to August and continues with post-monsoon showers later in September–October. Since it is closer to Chennai, so at times it is affected due to depressions in Bay of Bengal. During pre-monsoon, rainfall is intense with shorter spells, and during monsoon, rainfall recurs with longer spells. Previously, the recharge potential of rainfall infiltrating into the ground was good. But in the present time, urbanization has decreased the infiltration capacity with increased built-up cover in the city. Hence, the majority rainwater is carried away by storm water drains giving rise to sudden accumulation of water at the low lying areas.

For over 100 years, rainfall variability ranged from 500 to 1350 mm during 1901–2000 over Bangalore. However, Bangalore witnessed lowest rainfall of 24.3 mm in 1984 and highest rainfall of 247.3 mm in 1998. The average annual rainfall was recorded about 787 mm with around 60 rainy days per year [21].

The urban space of Bangalore city has grown more than 632% with the encroachment of wetlands and floodplains. The rapid urbanization affected the floodplain areas with the loss of natural water bodies for flood storage resulting in the creation of 134 flood-prone areas. Hence, the quantity of storm water inflowing to the sewers has increased with the sewer system getting overburdened. In addition, overflows occur in combined sewers resulting in untreated sewage entering into the streams. The difference in the capacity of wastewater treatment system installed (450 MLD) when compared to that of the domestic water generated (700 MLD) is proof for the magnitude of water available [21].

From the review, it was observed that, it is very important to analyze the storm water discharges in terms of quality and quantity in urban areas. Based on this, the objectives were defined for this study (Fig. 1).

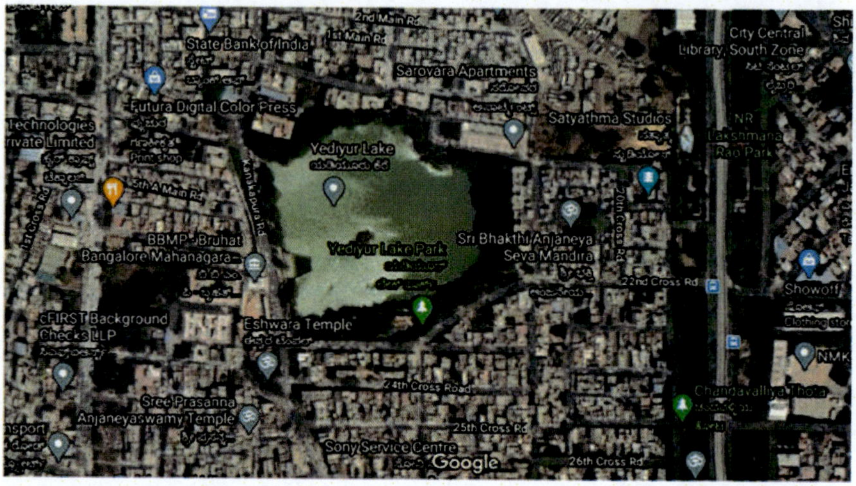

Fig. 1 Yediur Lake, Bangalore, Karnataka (*Source* Google Earth Imagery: May 21st, 2019)

Objectives of the study:

- Qualitative and quantitative analysis to evaluate inundation risks in urban drainage system.
- Providing significant best management practices for the improvement of water quality in storm water drainage systems.

3 Methodology

In the study, Yediyur Lake was considered, and the details of micro-watershed data were collected (stream flow, precipitation and metadata) to analyze through storm water management model (SWMM). ArcGIS was used to do watershed delineation and crop the DEM. Later, it was used in AutoCAD to create a backdrop. After drawing the drainages through the area, then SWMM was used to find the flow volume and determine the contaminate constituents. The quality analysis of the water in drains and the quantity analysis in terms of water volume were determined (Fig. 2). Once the analysis was competed, then the best management practices were suggested.

SWMM was used for modeling the storm water, sewer and other drainage network and analyzing the input–output discharge from them. It was utilized to design, plan and evaluate the controlling facilities, such as storm water drainages and pipe appurtenances pipes.

SWMM was found to be a useful tool for providing cost-efficient solutions to create sustainable infrastructure configuration. It was developed to support national, state and local objectives of storm water management by reducing the runoff through artificial recharge techniques and regulate the damage to natural water bodies. The

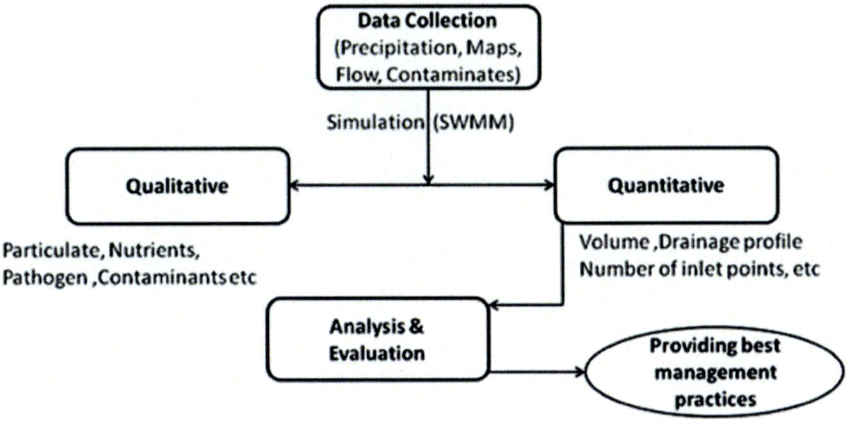

Fig. 2 Methodology flowchart

major applications of SWMM consist of storm water drainage system design, flood plain mapping and protection, impact assessment on the existing sewers and redesign for overflow conditions as well as non-point source modeling of waste load.

4 Results

The analysis of the micro-watershed in terms of quality and quantity was carried. The simulation was carried for a storm event of 6 h. The analysis was done for the maximum precipitation event occurred in during previous 10 years. Totally, 47 sub-catchments and 29 junctions were observed, which were contributing to the lake.

The maximum discharge observed along the sub-watershed was 665.44 m^3/s for 6-h storm. The nodal contaminant inflow has been displayed in Fig. 3. The flooding zones are been highlighted in Fig. 4.

From Fig. 4, it was observed that the carrying capacity of the drains is limited for the extreme event which stimulates the cascading effect and leads to severe damages for the neighboring areas within the vicinity of the lake.

Table 1 show the runoff volume about 0.531 MCM and depth 102 mm at the outlet node for the storm event occurred, excluding the losses. The total volume available as the final storage was about 91,000 m^3.

Table 2 describes the lead content at varying runoff qualities. The standard TDS content recommended for portable water is 4.15 mg/l and lead content is about 0.01 mg/l; hence, treatment is essential at current situation, before it reaches into any animal of plant body. The treatment must be carried for TDS and lead in surface runoff.

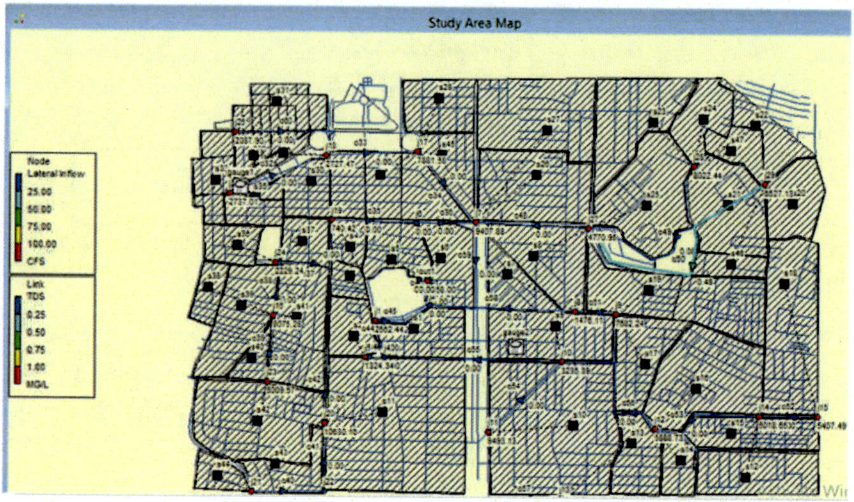

Fig. 3 Contaminant inflow toward the lake

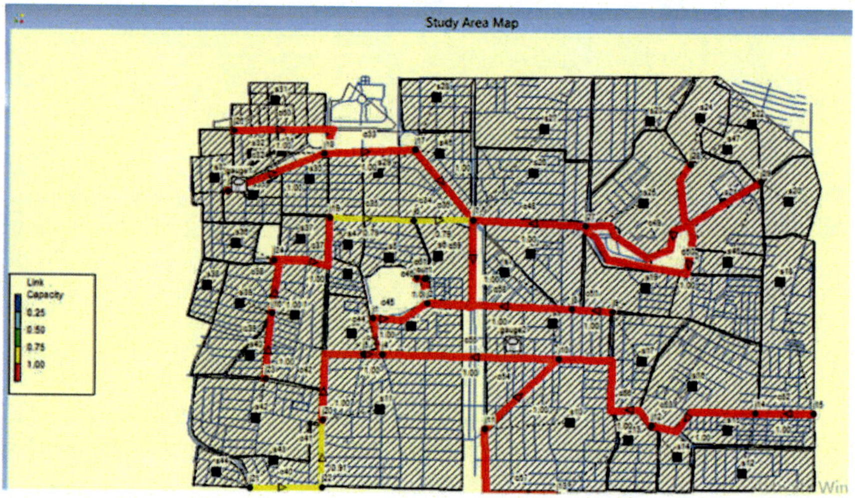

Fig. 4 Flooding zones within the study are for storm event

5 Conclusion

SWMM was analyzed for Yediyur Lake, Bangalore. From the study, it was observed that total volume at the outfall was 68.5 Mm^3 of water for total precipitation of 102 mm (10 year return period) in all the sub-catchments and the maximum flooded hours were 5.97 h. During qualitative analysis, TDS and lead content for the study area

Table 1 Link capacity

	Volume	Depth
Runoff volume continuity	MCM	Mm
Rainfall	0.531	102
Evaporation	0.000	0.000
Infiltration	0.023	6.29
Surface runoff	0.42	119.54
Total storage	0.091	25.82
Continuity error (%)	−1.037	

MCM million cubic meter

Table 2 Water quality and concentrations

Outfall node	out1
Flow frequency (%)	99.86
Average flow (m^3/s)	3.46
Maximum flow (m^3/s)	4.52
Total volume 10^6 (l)	68.5
Total TDS (mg/l)	1260.8
Total lead (mg/l)	0.315

were about 1260.79 mg/l and 0.315 mg/l (Table 2), respectively, and it exceeds the drinking standards (IS 10500:2012) for which treatment is required. Total infiltration in the sub-catchment is 6.29 mm, and peak runoff in the watershed area was 4.52 m^3/s.

From the results, it can be concluded that

- The contaminant concentrations are higher in the basin, so pretreatment at the inlet or prior is essential.
- Flow paths in the drainage network found to be unique at the branches, where the excess rainfall congestion at inlets increased and the uncertainty in peak runoff estimation increased.
- By improving the drainage network (looped), the decrease in the discharge was predicted to occur at a shorter time period with more flow path availability.
- The source of higher TDS and lead is due to the industrial discharges.

Best Management Practices:

From the study, some of the management practices are listed, and recommendations for the future in the areas of structural best management practices include managing:

- Artificial recharge structures.
- Detention and retention basins.
- Wetlands and treatment units.
- Vegetation strips and swales.

- Non-structural best management practices include
- Industrial monitoring.
- Reduction in the fertilizers and pesticides.
- Silt and waste management.
- Discharge maintenance.
- Educational and outreach activities.
- Individual storm and sewer systems.

Acknowledgements The authors wish to thank JIT and CIIRC for the support to carry out the research.

References

1. Hall KJ, Anderson BC (1988) The toxicity and chemical composition of urban storm water runoff. Can J Civ Eng 15(1):98–106
2. Jeng HAC, Englande AJ, Bakeer RM, Bradford HB (2005) Impact of urban storm water runoff on estuarine environmental quality. Estuar Coast Shelf Sci 63(4):513–526
3. Goonetilleke A, Thomas E, Ginn S, Gilbert D (2005) Understanding the role of land use in urban storm water quality management. J Environ Manage 74(1):31–42
4. Pattusamy V, Nandini N, Vijay Kumar M, Bheemappa K (2013) Water quality studies of Bellandur lake, urban Bangalore, Karnataka, India. Int J Adv Res 1(4):77–82
5. Walsh CJ, Fletcher TD, Ladson AR (2005) Stream restoration in urban catchments through redesigning storm water systems: looking to the catchment to save the stream. J N Am Benthol Soc 24(3):690–705
6. Hogan DM, Walbridge MR (2007) Best management practices for nutrient and sediment retention in urban storm water runoff. J Environ Qual 36(2):386–395
7. Martin C, Ruperd Y, Legret M (2007) Urban storm water drainage management: the development of a multicriteria decision aid approach for best management practices. Eur J Oper Res 181(1):338–349
8. Gowda K, Sridhara MV (2007) Conservation of tanks/lakes in the Bangalore metropolitan area. Manage Environ Qual Int J
9. Sandahl JF, Baldwin DH, Jenkins JJ, Scholz NL (2007) A sensory system at the interface between urban storm water runoff and salmon survival. Environ Sci Technol 41(8):2998–3004
10. Gogate GN, Pratap M (2012) Sustainable stormwater management in developing and developed countries: A review. In: International conference on advances in design and construction of structures (ADCS 2012), Bangalore, India
11. Davis AP, Shokouhian M, Sharma H, Minami C (2001) Laboratory study of biological retention for urban storm water management. Water Environ Res 73(1):5–14
12. Davis BS, Birch GF (2009) Catchment-wide assessment of the cost-effectiveness of storm water remediation measures in urban areas. Environ Sci Policy 12(1):84–91
13. Collins KA, Lawrence TJ, Stander EK, Jontos RJ, Kaushal SS, Newcomer TA, Grimm NB, Ekberg MLC (2010) Opportunities and challenges for managing nitrogen in urban storm water: a review and synthesis. Ecol Eng 36(11):1507–1519
14. May DB, Sivakumar M (2009) Prediction of urban stormwater quality using artificial neural networks. Environ Model Softw 24(2):296–302
15. Inamdar PM, Cook S, Sharma AK, Corby N, O'Connor J, Perera BJC (2013) A GIS based screening tool for locating and ranking of suitable storm water harvesting sites in urban areas. J Environ Manage 128:363–370

16. Loáiciga HA, Sadeghi KM, Shivers S, Kharaghani S (2015) Stormwater control measures: optimization methods for sizing and selection. J Water Resour Plann Manage 141 9
17. Ranganathan M (2015) Storm drains as assemblages: the political ecology of flood risk in post-colonial Bangalore. Antipode 47(5):1300–1320
18. Avinash S (2016) Flood related disasters: concerned to urban flooding in Bangalore, India. Int J Res Eng Technol 3:76–83
19. Avinash S, Prasad KL, Reddy GS, Mukund D (2018) Urban flood forecast system-a case study of Bangalore, India. Univ Rev
20. Paul R, Kenway S, McIntosh B, Mukheibir P (2018) Urban metabolism of Bangalore city: a water mass balance analysis. J Ind Ecol 22(6):1413–1424
21. Ramachandra TV, Mujumdar PP (2009) Urban floods: case study of Bangalore. Disaster Dev 3(2):1–98

An Iterative Procedure to Determine Natural Frequencies and Mode Shapes from Discrete and Continuous Approaches

E. Meghana Reddy, N. Srujana, and T. Bhavani

Abstract Free vibration analysis is the preliminary task that a structure must undergo before conducting either linear or nonlinear analysis under external time-dependent loadings. Discrete and continuous approaches are commonly used to perform modal analysis. The results evolved from the eigenvalue problem of the discrete system approach and Euler Bernoulli's equation of continuous system approach are not consistent. An alternative methodology is proposed to determine the preliminary dynamic parameters based on the modal superposition method and Rayleigh quotient method using iterations. In this study, G + 9 structure is analysed for fundamental frequencies and mode shapes by applying conventional approaches. The other conventional approaches are highlighted in this work for their accuracy and straightforwardness in contrary to the computational efforts generally experienced in terms of modelling and analysis of the structure in software tools. The fundamental frequency obtained from the iterative procedure is supported by approximate methods. The proposed approach is suitable for symmetric structures having uniformly distributed mass and stiffness throughout the structure height.

Keywords Discrete system · Continuous system · Rayleigh quotient · Modal superposition method · Natural frequency · Mode shapes

1 Introduction

The natural frequency is essential to understand the system behaviour and able to draw the deformation profile of the structure. For a single degree of freedom (SDOF) system (where the structure is able to deform in any one direction either x or y-axis) with one mass and equivalent stiffness of the column members, identifying the natural frequency is straightforward. But in the case of multi-degree of freedom (MDOF) system, calculation of equivalent stiffness is challenging. Hence, a popular idealization-shear building concept is introduced which is used to determine the natural frequencies of a multi-degree of freedom system. A shear building may be

E. Meghana Reddy (✉) · N. Srujana · T. Bhavani
Department of Civil Engineering, Vardhaman College of Engineering, Hyderabad, India

L. Nandagiri et al. (eds.), *Sustainability Trends and Challenges in Civil Engineering*, Lecture Notes in Civil Engineering 162, https://doi.org/10.1007/978-981-16-2826-9_33

defined as a structure in which there is no rotation at the floor level. The structure behaves like a cantilever beam, and it undergoes deflection only due to shear force. The major assumptions of shear building concepts are: (1) the total mass is located at the floor level (i.e. it transfers the system of infinite number of degrees of freedom into as many degrees as the system masses lumped at floor levels), (2) the beams are rigid when compared to columns (it means, no rotation at the joint of column and beam), (3) the deformation of the structure is independent of the axial forces of the column (which indicates that the beams and floors remain horizontal during vibration). Classical methods and approximate methods are considered under the discrete system approach (lumped masses at floor levels) and are used to calculate the natural frequency of MDOF systems. Another methodology, i.e. continuous system approach at where the mass and stiffness are uniformly distributed throughout the length of the structure, and those systems are also known as "distributed parameter systems". The discrete system consists of a finite number of degrees of freedom (DOF), whereas in a continuous system infinite number has infinite DOF because the displacement of every point in the elastic body is specified by an infinite number of coordinates. The displacement is dependent on two variables $\Phi(x)$ and $y(t)$. As a result, the motion of the continuous system is governed by partial differential equations to be satisfied over the entire domain of the system subjected to boundary conditions and initial conditions. In the case of a building, the structure is considered as cantilever system (i.e. one end is fixed at ground level and another end is free at the top floor).

Late 80's the computer algorithms were developed to analyse indeterminate structures using numerical methods. Computational efforts were greatly reduced, but all these algorithms are based on flexibility and stiffness matrix methods. Kamgar and Reza [1] developed an approximate method for estimating the natural frequency of tube framed structure using Timoshenko's beam model in which flexural and shear deformations are considered. The natural frequencies and mode shapes are calculated based on the flexural and shear rigidities along with the effects of rotational inertia. Hamdan and Jubran [2] used base beam equation to calculate the mode shape functions which satisfy all the geometric and natural boundary conditions. The mode shape functions are used in conjunction with Galerkin's method to obtain the response of the cantilever. Malekinejad and Reza [3] presented an approximate formula for dynamic response of tubular tall building structures on the basis of D'Alembert's principle. Using the principle and applying the compatibility conditions on the deformation of the tubes, the governing dynamic equation of the tubular structure's motion is derived. Then, natural boundary conditions of the parallel cantilevered flexural–shear beams are derived, and by using Rayleigh–Ritz method, the value problem is solved, and trivial and nontrivial solutions are derived, which can be used for calculating natural frequencies and mode shapes of tubular structures. Roh et al. [4] developed a model, based on eigenvalues and eigenvectors which named as "frequency adaptive lumped mass stick model" has only a small number of stick elements and nodes to provide the same natural frequencies of the structure and is applied to a nuclear containment building. Cunha and Sampaio [5] studied the influence of a discrete element in the nonlinear dynamics of a continuous mechanical system subject to randomness in the model parameters. The mechanical system consists of an elastic bar that is fixed at one end (right side), and other the end is attached to a couple of springs (one linear and another nonlinear) and lumped mass. In this system, I was subjected to Gaussian white-noise distributed external force. Malekinejad et al. [6] proposed a discrete–continuous approach for free vibration analysis of the combined system of framed tube, shear core, and outer rigger belt truss. The structure is discretized at each floor of the building as a series of lumped masses placed at the centre of shear core. Sohani and Eipakchi [7] worked on beam theories proposed by Euler–Bernoulli and Timoshenko for the analysis of free vibrations. Using perturbation technique, the behaviour of beams with arbitrary varying cross sections is studied. The beam equations are solved using Wentzel, Kramers, Brillouin approximation. Liu [8] verified variational iteration technique with the domain decomposition method to solve free vibration equations of Euler–Bernoulli beam. The tested method is found to be efficient in solving uniform Euler–Bernoulli beam problems and solution converges rapidly.

A minimum study is addressed in the literature regarding the iteration procedures involved in classical methods to estimate the fundamental frequency of the structures. The present study introduces an alternative iterative procedure by using natural frequency obtained from any approach and with the mode shapes corresponding to the natural frequency obtained from the mode superposition method. The proposed methodology is limited to flexural modes especially the fundamental mode and corresponding mode shape.

2 Methodology

The proposed methodology is carried out in three levels. Initially, solving the eigenvalue problem from the modal superposition method using mass and stiffness matrix for identifying the angular frequencies and mode shapes. Next, estimate the mode shapes of the structure with the idealization of a vertical cantilever beam using the continuous approach. Consider Rayleigh quotient method to calculate the fundamental frequency. At this level, the fundamental frequencies from the two approaches must be matched well. In case of a discrepancy with frequencies, the iterative procedure must be considered. In the iteration procedure, the frequency obtained from the continuous approach is substituted in the mode superposition method to identify the mode shape corresponding fundamental frequency. Evaluate natural frequency using mode shapes by applying the Rayleigh quotient method. Continue the procedure till the natural frequency converges. The mode shape related to the converged natural frequency is considered as the exact shape profile of the frequency. A case study of the G + 9 structure is presented in the study for the identification of fundamental frequency from the conventional approaches.

2.1 Description of the Structure

A 30 m height of structure having length and width as 12 m (3bays) and 8 m (2bays) with a floor height of 3 m having column sizes 600 mm × 600 mm for bottom five storey and 400 × 400 mm for remaining storeys are considered. Beam dimensions are 300 mm × 400 mm, and the thickness of the slab is 150 mm. The shear building assumed the G + 9 structure with each level idealization is shown in Fig. 1.

Fig. 1 Idealized G + 9 building

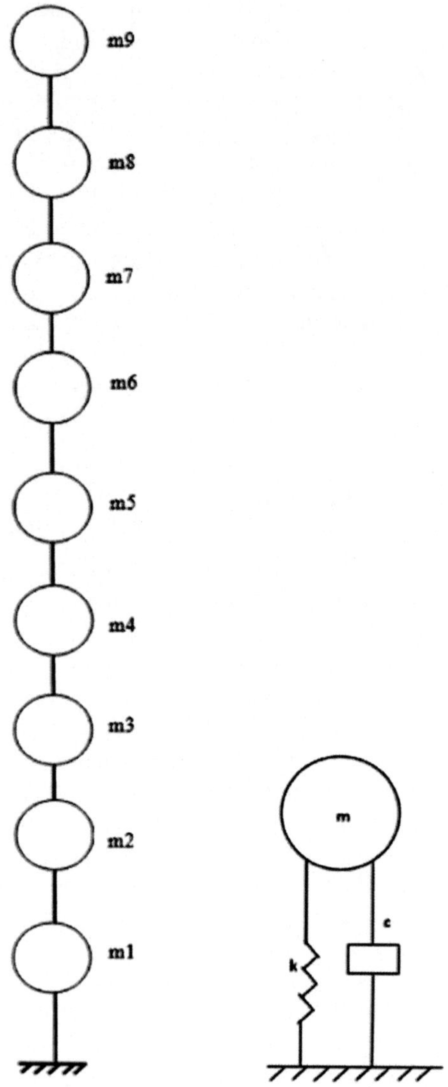

Table 1 Global mass matrix of G + 9 structure

88800	0	0	0	0	0	0	0	0	0
0	88800	0	0	0	0	0	0	0	0
0	0	88800	0	0	0	0	0	0	0
0	0	0	88800	0	0	0	0	0	0
0	0	0	0	79800	0	0	0	0	0
0	0	0	0	0	70800	0	0	0	0
0	0	0	0	0	0	70800	0	0	0
0	0	0	0	0	0	0	70800	0	0
0	0	0	0	0	0	0	0	70800	0
0	0	0	0	0	0	0	0	0	63000

2.2 *Preliminary Data Collection*

(a) *Calculation of mass matrix*: Lumped masses are considered at the individual floor levels by calculating the dead loads of half of the column above and below the floor level including self-weight of beams and slab. The global mass matrix is framed from the lumped mass located at storey levels is shown in Table 1.

(b) *Calculation of stiffness matrix*: Stiffness at each floor is calculated with the assumptions of the floor is rigidly connected (in a way of fixed) for one case and floor rotation is allowed for the other case. In one case, the shear building concept is adopted that is rotation is not allowed at the floor level. The stiffness of each storey is calculated by assuming each storey columns as a cantilever bar (one end fixed and another end is free). The global stiffness matrix is shown in Table 2.

In another case, the rotation is allowed at the floor level, i.e. stiffness is calculated by assuming ground floor columns as the cantilever ($K_1 = 12EI/L^3$) and other floor columns as simply supported ($K_2, K_3 \ldots K_{10} = 3\ EI/L^3$). The stiffness matrix is framed as shown in Table 3.

The discrete approach is utilized in the classical method and approximate method of estimating natural frequencies and mode shapes. Approximate methods have proven to be efficient in estimating the fundamental natural frequency of the structure.

2.3 *Conventional Approach (Eigenvalue Problem)*

The structure is idealized as spring mass system connected in series (Fig. 1). D Alembert's principle is adopted to write the equilibrium equation of motion in free vibration condition. The undamped equation of motion represented globally is shown in "(1)". Damping characteristics are neglected in the study due to the basic definition

Table 2 Global stiffness matrix of G + 9 structure (Rotation is not allowed)

3.15E + 09	−1.60E + 09	0	0	0	0	0	0	0
–	3.15E + 09	−1.58E + 09	0	0	0	0	0	0
0	−1.610E + 09	3.15E + 09	−1.60E + 09	0	0	0	0	0
0	0	−1.58E + 09	3.15E + 09	−1.60E + 09	0	0	0	0
0	0	0	−1.60E + 09	1.89E + 09	−3.10E + 08	0	0	0
0	0	0	0	−3.10E4 + 08	6.23E + 08	−3.10E + 08	0	0
0	0	0	0	0	−3.10E + 08	6.23E + 08	−3.10E + 08	0
0	0	0	0	0	0	−3.10E + 08	6.23E + 08	−3.10E + 05
0	0	0	0	0	0	0	−3.10E + 08	3.12E + 08

of natural frequency as a function of equivalent stiffness and equivalent mass.

$$[\mathrm{M}]\ddot{u}(t) + [\mathrm{K}]u(t) = 0 \tag{1}$$

where [M] is the global mass matrix, [K] is the global stiffness matrix, and $u(t)$ is the shape vector or displacement vector of response.

Equation (1) is written in form of the matrix

$$[\mathrm{M}] \times \{\ddot{u}\} + [\mathrm{K}] \times \{u\} = \{0\} \tag{2}$$

The characteristic equation or frequency equation is

$$\left|[\mathrm{K}] - \omega_j^2[\mathrm{M}]\right| = 0 \tag{3}$$

The eigenvalues are estimated by using the algorithms of Matlab. The fundamental frequencies (ω^2) of G + 9 structure in both the stiffness cases of rotation allowed and rotation not allowed are shown in Table 4. The natural frequency of 8.2 rad/s is considered in the study. The normalized mode shape values corresponding graph (height of storey is on the x-axis and corresponding mode shape values on y-axis) are shown in Fig. 2.

Table 3 Global stiffness matrix of G + 9 structure (Rotation is allowed)

1.97E + 09	−3.94E + 08	0	0	0	0	0	0	0	0
−3.94E + 08	78872083	−3.9E + 08	0	0	0	0	0	0	0
0	−3.94E + 08	7.89E + 08	−3.94E + 08	0	0	0	0	0	0
0	0	−3.9E + 08	7.89E + 08	−3.94E + 08	0	0	0	0	0
0	0	0	−3.94E + 08	472258561	−7.8E + 07	0	0	0	0
0	0	0	0	−77898319	1.56E + 08	−7.8E + 07	0	0	0
0	0	0	0	0	−7.8E + 07	1.56E + 08	−7.8E + 07	0	0
0	0	0	0	0	0	−7.8E + 07	1.56E + 08	−7.8E + 07	0
0	0	0	0	0	0	0	−7.8E + 07	1.56E + 08	−7.8E + 07
0	0	0	0	0	0	0	0	−7.8E + 07	7.8E + 07

Table 4 Fundamental frequencies

Rotation conditions at floor level	Fundamental natural frequency
Lumped mass-rotations not allowed	23 rad/s
Lumped mass-rotations allowed at	8.2 rad/s

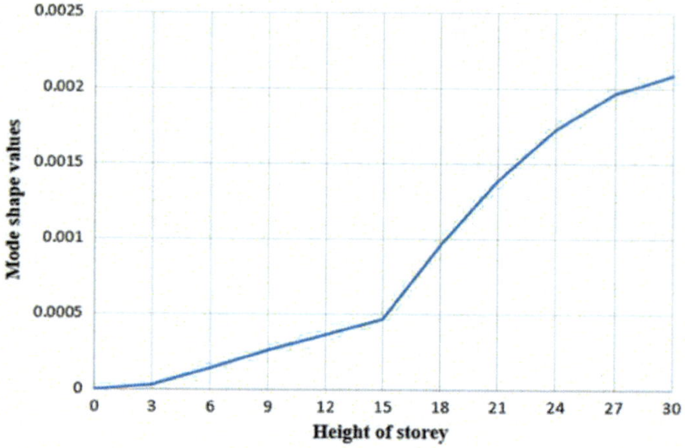

Fig. 2 The first mode shape in eigenvalue problem

2.4 Continuous Approach (Euler Bernoulli's)

In the continuous approach (Paz [9]), the mass and stiffness are uniformly distributed throughout the length of the structure. By using the Euler–Bernoulli theory (i.e. plane of the section remains plane during flexure) an approximate equation is derived. From the solution of this approximate equation, the mode shapes are calculated.

$$EI\left(\partial^4 y/\partial x^4\right) + m\left(\partial^2 y/\partial t^2\right) = 0 \tag{4}$$

The deflections due to shear force and inertia force caused by the rotation of column cross sections are neglected. The solution to the above equation is found by the variable separable method. The approximate solution is given in function of position $\Phi(x)$ and function of time $f(t)$.

$$\text{i.e. } y(x, t) = \Phi(x) \times f(t).$$

$$\Phi^{\text{IV}}(x) - a^4\Phi(x) = 0 \tag{5}$$

Let, $\Phi(x) = A\sin(ax) + B\cos(ax) + C\sinh(ax) + D\cosh(ax)$

x	Mode shape Φ(x)
0	0
3	0.033547067
6	0.127743015
9	0.27297045
12	0.459780413
15	0.679069962
18	0.922311243
21	1.181820815
24	1.451058773
27	1.724948213
30	2.000207019

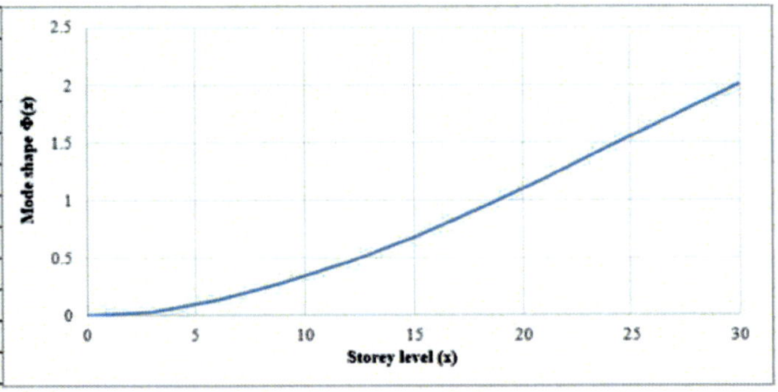

Fig. 3 Mode shape of first natural frequency

Where A, B, C, and D stands for the constants of integration, these four define the shape and amplitude of the beam in free vibration. The constants are evaluated by considering the boundary conditions at the end of the beam. The frequency equation is shown in "(6)"

$$\cos(a_n L) \times \cosh(a_n L) + 1 = 0 \tag{6}$$

Roots of the equation corresponds to the successive natural frequencies. The first natural frequency is given by,

$$\omega_n = \frac{(a_n L)^2}{\sqrt{\frac{EI}{mL^4}}} \text{ and normal mode is } \Phi(x).$$

$$\Phi(x) = (\cosh a_n x - \cos a_n x) - \frac{(\cos a_n L + \cosh a_n L)}{(\sin a_n L + \sinh a_n L)}(\sinh a_n x - \sin a_n x) \tag{7}$$

where x is the level height from the base, $a_n = \frac{\sqrt{3516}}{L}$, L is length of the structure.

The first mode shape corresponding the storey level using "(7)", i.e. $X = 0, 3, 6, \ldots 0.30$ is shown in Fig. 3.

2.5 Rayleigh Quotient Method

Using the mode shape values of first natural frequency, equivalent mass and the equivalent stiffness matrices are calculated. Rayleigh quotient method is adopted to identify the fundamental frequency of the structure. The equivalent mass and stiffness matrices are shown in "(8)" and "(9)".

$$[\mathrm{M}]_{1\times1} = \{\Phi\}^{\mathrm{T}}_{1\times10} \times [\mathrm{m}]_{10\times10} \times \{\Phi\}_{10\times1} \tag{8}$$

$$[K]_{1\times1} = \{\Phi\}^{T}_{1\times10} \times [k]_{10\times10} \times \{\Phi\}_{10\times1} \quad (9)$$

The natural frequency is calculated from the below formula comes out to be 9.32 rad/s.

$$\omega = \sqrt{([K]/[M])} \quad (10)$$

2.6 Numerical Analysis

G + 9 structure is modelled in the ETABS, and modal analysis is performed. The plan and isometric view are shown in Fig. 4. The characteristic strength given as material properties of concrete is 30 N/mm^2. Sectional properties are assigned, for bottom five storey columns as 600 × 600 mm, top five storey columns as 400 × 400 mm, beam size is 300 × 400 mm, and the thickness of slab as 150 mm. Slab is modelled as diaphragm using shell 8-noded elements.

First natural frequency is identified at 6.32 and 7.53 rad/s at first two natural frequencies. First and second modes corresponding frequencies are shown in Fig. 5.

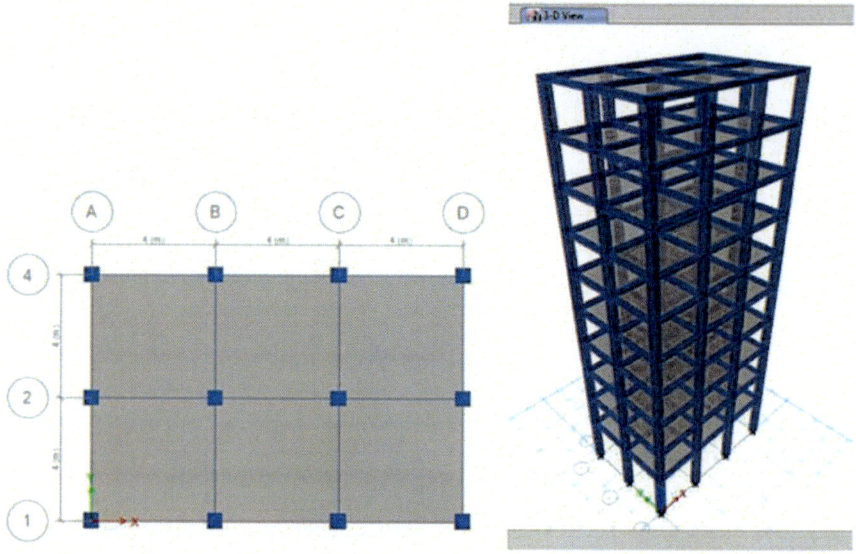

Fig. 4 Plan and isometric view of G + 9 building

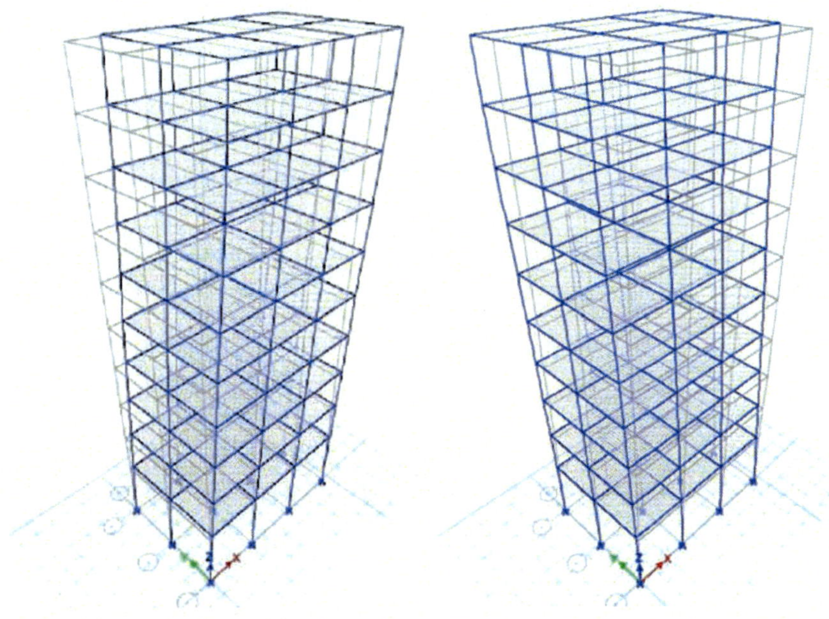

Fig. 5 First and second natural frequencies of the building

2.7 Approximate Methods

Modified Stodola and Holzar's methods are considered for the study of natural modes. In modified Stodola, mode shape is assumed at the initial stage and later it is adjusted iteratively until an approximate of the mode shape is achieved.

$$\{u\} = \omega_j^2[\mathrm{A}]\{u\} \tag{11}$$

where $[A] = [f]\,[M]$ is called dynamic equation, $[f]$ is the flexibility matrix, i.e. inverse of stiffness matrix. Table 5 describes the iterations for the mode shape convergence. The natural frequency is calculated at the 6th iteration.

The value of natural frequency is calculated as,

$$\omega^2 = 1/0.0147 = 68.0272$$
$$\omega = \surd 68.0272 = 8.24\,\mathrm{rad/s}$$

The natural frequency is identified as 8.24 rad/s. The converged mode shape is shown in Fig. 6.

Holzar's method is an iterative procedure with the assumed natural frequency at a minimum value. The first natural frequency of 8.2 rad/s is estimated at converged

Table 5 The converged mode shape evaluation using Stodola method

1st iteration	(u)1	(uc)1	2nd iteration	(uc)l = (u)2	(uc)2	3rd iteration	(uc)2 = (u)3	(uc)3
	0.1	0.000256555		0.016807456	0.000211336		0.014173068	0.006204466
	0.2	0.001260255		0.082562109	0.001052895		0.070611525	0.001019137
	0.3	0.002218923		0.145366425	0.001875864		0.125803201	0.001817909
	0.4	0.003 110,637		0.203745234	0.002656099		0.173799675	0.002588353
	0.5	0.003911081		0.256223368	0.003410457		0.228719361	0.003318536
	06	0.007454162		0.488338192	0.006916287		0.463834873	0.006780784
	0.7	0.010451915		0.684727517	0.009978278		0.659184685	0.009821463
	0.8	0.012813455		0.839437094	0.012417936		0.83279825	0.012253935
	09	0.014447894		0.945512671	0.014094649		0.94524555	0.013929496
	1	0.015264343		1	0.014911097		1	0.014745945

4th iteration	(uc)3 = (u)4	(uc)4	5th iteration	(uc)4 = (u)5	(uc)5		6th iteration	(uc)5 = (u)6
	0.013865399	0.000203424		0.013824328	0.006203269			0.013813492
	0.069113067	0.001013998		0.C68909455	0.001013234			0.06888084
	0.123281976	0.00180901		0.122936982	0.001807682			0.122888347
	0.175529S35	0.002576262		0.175077991	0.002574447			0.175013977
	0.225047371	0.003303988		0.224532961	0.06330179			0.224459574
	0.459840583	0.006757564		0.459231583	0.006753947			0.459141265
	0.666044993	0.009793201		0.65552787	0.009788713			0.665448626
	0.831003716	0.012223485		0.83068548	0.012218607			0.836635322
	0.944632309	0.013898488		0.944515636	0.013893504			0.944496837
	1	0.014714937		1	0.014709953			1

Storey level (m)	Mode shape {u}
0	0
3	0.013818492
6	0.06888084
9	0.122888347
12	0.175013977
15	0.224459574
18	0.459141265
21	0.665448626
24	0.830635322
27	0.944496837
30	1

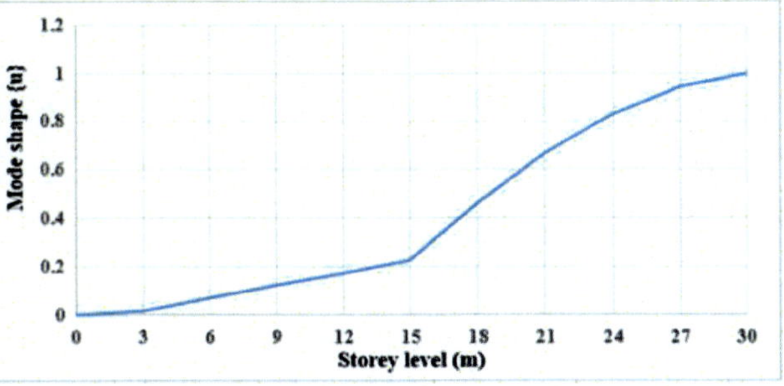

Fig. 6 Mode shape of the building using Stodola method

mode shape values (shown in Table 6). The mode shapes derived from Holzar's method are shown in Fig. 7.

The frequencies extracted from conventional method, approximation method, discrete continuous approach, numerical method and proposed iterative approach are shown in Table 7. For the superposition method of discrete approach, fundamental frequency in the case of rotation not allowed is presented.

2.8 *Iterative Procedure*

An iterative procedure is considered to calculate the approximate frequency from the discrete approach and continuous approach. The iteration is followed in two steps. Step 1: the natural frequency obtained from the continuous approach, and numerical analysis is substituted separately in the eigenvalue problem to retrieve mode shapes. Step 2: using the mode shapes, Rayleigh quotient frequency is determined. The Rayleigh quotient frequency is substituted in the eigenvalue problem to get the modified mode shapes. Repeat the steps till the natural frequency converges. At the 3rd iteration, the value converged to 8.24 rad/s. The normalized mode shapes are shown in Fig. 8.

3 Summary and Discussions

- In the discrete approach, the conventional method with the assumption of allowed rotation, the natural frequency is obtained as 8.24 rad/s. But in the case of restricted rotation, the natural frequency of the structure is observed to be higher than 8.24 rad/s. This is due to the provision of considering the stiffness in the system by not allowing the rotation. Hence, the natural frequency of 23 rad/s is evident. In the first case, stiffness for all the columns is taken as $12EI/L^3$ for all the storeys

Table 6 Mode shape values for assumed frequency

Assumed frequency	Position	*mi*	Frequency^2 * mi	*xi*	Frequency^2 * mi * xi	Sum (frequency^2 * mi * xi)	*kij*	Sum (frequency^2 * mi * xi)/*kij*
1st iteration								
0.5								
0.25	1	63600	15900	1	15900	15900	77898319.29	0.000204112
	2	70800	17700	0.999795888	17696.38721	33596.38721	77898319.29	0.000431285
	3	70800	17700	0.999364603	17688.75347	51285.14068	77898319.29	0.00065836
	4	70800	17700	0.9987063243	17677.10049	68962.24117	77898319.29	0.000885285
	5	70800	17700	0.997820957	17661.43094	86623.67212	77898319.29	0.00111201
	6	79800	19950	0.996708948	19884.34351	106508.0156	394360241.4	0.000270078
	7	88800	22200	0.99643887	22120.94291	128628.9585	394360241.4	0.000326171
	8	88800	22200	0.996112699	22113.70191	150742.6604	394360241.4	0.000382246
	9	88800	22200	0.995730452	22105.21604	17287.8765	394360241.4	0.000438299
	10	88800	22200	0.995292153 0.995168571	22095.4858	194943.3623	1577440966	0.000123582
20th iteration	*Position*	*mi*	*Frequency*^{2} *** mi**	*xi*	*Frequency*^{2} *** mi * xi**	*Sum (frequency*^{2} ** mi *)xi*	*kij*	*Sum (frequency*^{2} ** mi * xi)/kij*
10								
100	1	63600	6360000	1	6360000	6360000	77898319.29	0.081644894
	2	70800	7080000	0.918355106	6501954.152	12861954.15	77898319.29	0165112088
	3	70800	7080000	0.753243018	5332960.563	18194914.72	77898319.29	0.233572622
	4	70800	7080000	0.519670396	3679266.402	21874181.12	77898319.29	0.280804276
	5	70800	7080000	0.23886612	1691172.13	23565353.25	77898319.29	0.302514271

(continued)

Table 6 (continued)

Assumed frequency	Position	*mi*	Frequency^2 * mi	*xi*	Frequency^2 * mi * xi	Sum (frequency^2 * mi * xi)	*kij*	Sum (frequency^2 * mi * xi)/*kij*
	6	79800	7980000	−0.063648151	−507912.2422	23057441.01	394360241.4	0.058467966
	7	88800	8880000	−0.122116116	−1084391.112	21973049.9	394360241.4	0.055718218
	8	88800	8880000	−0.177834334	−1579168.888	20393881.01	394360241.4	0.051713836
	9	88800	8880000	−0.229548171	−2038387.755	18355493.26	394360241.4	0.046544989
	10	88800	8880000	−0.27609316	−2451707.261	15903786	1577440966	0.010082017
			−0.286175177					

Natural frequency (rad/sec)	xij
0.5	0.994797825
1	0.979259651
1.5	0.953589637
2	0.918125069
2.5	0.873331928
3	0.81979877
3.5	0.758228975
4	0.689431476
4.5	0.614310077
5	0.533851489
5.5	0.449112234
6	0.361204582
6.5	0.271281691
7	0.180522147
7.5	0.09011408
8	0.042989846
8.5	-0.084943897
9	-0.167314415
9.5	-0.244806045
10	-0.316421226

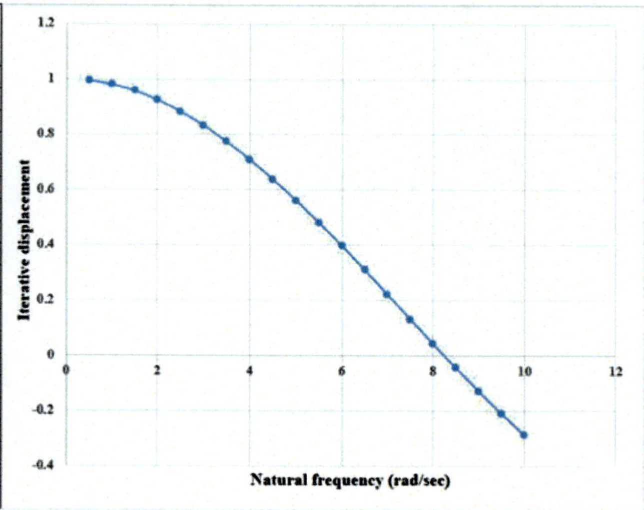

Fig. 7 First natural frequency using Holzar's method

Table 7 Fundamental frequencies from various methods

Method of analysis	Fundamental frequency (rad/s)
Mode superposition method	8.24 23 (rotation not allowed)
Continuous approach	9.32
Stodola method	8.24
Holzar's method	8.20
Numerical method	6.32
Proposed iterative approach	8.20

Storey level x (m)	Mode shape Φ(x)
0	0
3	2.87865E-05
6	0.000143492
9	0.000256001
12	0.00036459
15	0.000467599
18	0.000956512
21	0.001386324
24	0.001730472
27	0.001967696
30	0.002083334

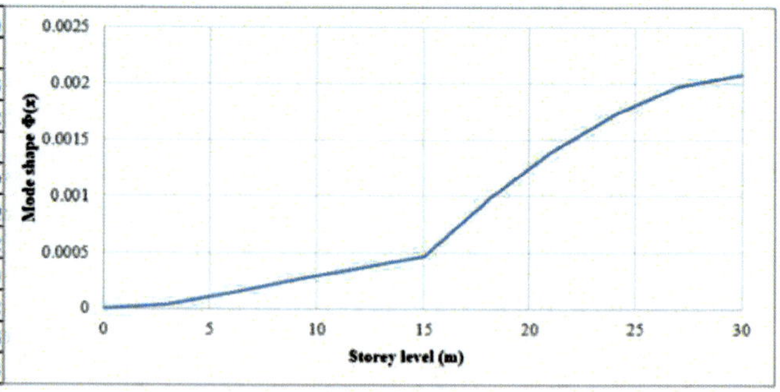

Fig. 8 Mode shape of the structure at the converged first frequency

but in other case, the stiffness is taken as $12EI/L^3$ for the fixed base and $3EI/L^3$ for other storeys.

- From the approximate methods, the structural natural frequency of 8.2 rad/s is observed in the Stadola method and Holzer's method.
- In the continuous system, Euler–Bernoulli's bending theory is adopted in the discrete–continuous approach. The structure is idealized as a cantilever structure with uniform distributed mass and stiffness. The mode shape of the first mode is identified using shear building idealization. First natural frequency is determined by adopting Rayleigh quotient method. The natural frequency value obtained is 9.32 rad/s which is almost 10% greater than the value calculated from the conventional approach (i.e. eigenvalue problem second case). Because the structure is very stiff with the basic Euler's assumption of the vertical cantilever. Hence, high stiffness is drawn to the high natural frequency. One other limitation of the approach is structured with varying sections cannot be modelled.
- In ETABS, the natural frequency obtained from this analysis is 6.32 rad/s which is 23% lesser than the value obtained from the conventional approach.
- The iterative procedure adopted in the study from the values derived from the discrete, continuous, and numerical approaches, the natural frequency of 8.2 rad/s is converged at 3rd iteration.

4 Conclusions

- With the shear building assumption of the structure having lumped mass at intermediate levels and restrained floor rotations, the natural frequency of the structure is identified as 23 rad/s, which is very high for the structure. This is due to the assumption by allowing the structure to behave rigidly with the base. Since the structure has ten degrees of freedom, each degree of freedom is designed to move rigidly and relatively along with the below floor level, starting from the first mass connected at the first-floor level near the base of the structure to the top floor mass. The above idealization may work well with the SDOF system.
- With the assumption of lumped mass and allowed floor rotations, by means of making the structure flexible at the floor levels except the floor connected to the base of the structure. Therefore, the natural frequency is reduced to 8.2 rad/s.
- Approximate methods are in good pace with the structure assumed of lumped mass and with allowed floor rotations. It is maintained 99% accuracy.
- The fundamental frequency obtained from the continuous approach is 94% accurate with the complete discrete method with the assumption of allowed floor rotations. This is due to the consideration of change in column sections from 6th storey of the building which is contradicting the Euler-Bernoulli's approach considered for the study. The profile of the mode shape has not accommodated the changes in mass and stiffness. Hence, due to the limitation, this method is not suitable to be considered in the iteration procedure of identifying the first natural frequency. However, the natural frequency from the discrete–continuous

approach is used in the proposed iterative procedure, but the value converged at 8.2 rad/s. Timoshenko beam theory can be considered to validate the approximate methods for the structures having varying cross sectional elements with allowed rotations, and also the Timoshenko beam theory possibly addresses the torsional modes and mode shapes in terms are shear effects.

- The iteration procedure is efficient in estimating the natural frequency obtained from numerical method, Rayleigh quotient method, and eigenvalue problem.
- The resonant frequency obtained from iteration procedure is highly accurate to the frequency calculated from approximate methods.
- The proposed approach results accurate fundamental frequency for symmetric structures. For non-symmetric structures, the approach can be successful provided appropriate stiffness at each floor level is assumed.

References

1. Kamgar R, Reza R (2010) A simple approximate method for free vibration analysis of framed tube structure. Struct Design Tall Spec Build 2(22):217–234
2. Hamdan MN, Jubran BA (1991) Free and forced vibration of a cantilever beam carrying a concentrated mass. J King Abdulaziz Univ Eng Sci 3:71–83
3. Malekinejad M, Rahgozar R (2012) An analytical model for dynamic response of analysis of tubular tall buildings. Struct Des Tall Special Build 23(1)
4. H Roh, H Lee, JS Lee (2013) New lumped-mass-stick model based on model characteristics of structure: development and application to a nuclear containment building. Earthquake Eng Eng Vibr 12:307–317
5. Cunha A, Sampio R (2014) On the nonlinear stochastic dynamics of a continuous system with discrete attached elements. Appl Math Model 809–819
6. Malekinejad M, Rahgozar R, Malekinejad A, Rahgozar P (2016) A continuous-discrete approach for evaluation of natural frequencies and mode shapes of high-rise buildings. Int J Adv Struct Eng 8:269–280
7. Sohani F, Eipakchi HR (2018) Analytical solution for model analysis of Euler–Bernoulli and Timoshenko beam with an arbitrary varying cross-section. Math Models Eng 4(3):164–174
8. Liu Y (2018) A numerical method for solving free vibration of Euler-Bernoulli beam. World Acad Sci Eng Technol 65
9. Paz M (2019) Structural dynamics theory and computation. Springer

A Novel Approach to Utilize Fly Ash to Enhance Compressive Strength of Diesel-Soaked Concrete

Sneha S. Bandekar, Chidanand Patil, Girish S. Kulkarni, and K. B. Prakash

Abstract Petroleum, popularly known as black diamond, has been the major cause of the renaissance in civilization history. Petroleum products adversely affect the environment. Concrete is a prominent construction material. Concrete finds applications in the construction of petroleum storage areas, garages, and petrol bunks. This study attempts to study the impact of diesel on concrete structures. The concrete properties were enhanced by partially replacing the costly cement with solid waste from thermal power stations "fly ash". Concrete cubes were cast, cured in water and diesel. The concrete samples were tested for strength, age, and durability. Trials were conducted with different blends of fly ash to partially replace cement, enhance the quality of concrete, and reduce concrete cost. Mineralogical and morphological analysis of concrete can be done using scanning electron microscope (SEM) and X-ray diffraction (XRD), which indicates the presence of phases that positively affect compressive strength.

Keywords SEM · XRD · Compressive strength · Diesel-soaked concrete

1 Introduction

Usage of concrete for construction purposes has become inevitable. Resisting of loads is a unique property anticipated in concrete. The sorptivity, permeability, and diffusivity are some of the transport material properties determining the durability

S. S. Bandekar (✉)
Department of Chemical Engineering, KLE DR. M.S.Sheshgiri College of Engineering & Technology, Udyambag, Belgaum, Karnataka 590008, India

C. Patil
Department of Civil Engineering, KLE DR. M.S.Sheshgiri College of Engineering & Technology, Udyambag, Belgaum, Karnataka 590008, India

G. S. Kulkarni
Department of Technology, Shivaji University Kolhapur, Kolhapur, Maharashtra, India

K. B. Prakash
Govt. Engineering College Haveri, Haveri, Karnataka, India

L. Nandagiri et al. (eds.), *Sustainability Trends and Challenges in Civil Engineering*, Lecture Notes in Civil Engineering 162, https://doi.org/10.1007/978-981-16-2826-9_34

and age of concrete. Concrete undergoes degradation when it comes in contact with petroleum products. The structural mortification of concrete due to petroleum products spillage has gained paramount importance. Spillage of petroleum products on concrete is a widespread scenario in storage tanks, automobile service stations, and garages.

Concrete linings are used in the boreholes of oil extraction plants. Petroleum products are stored in steel or concrete tanks [1, 2]. Due to corrosion problems, concrete tanks are favored for storing petroleum products. A significant breakdown in compressive strength by 18–90% on concrete was found due to crude oil coming into contact with it [3]. Dynamic modulus of concrete was analyzed for concrete specimens soaked in crude oil over three months. It was found that the increase in the dynamic modulus in the case of crude oil-soaked specimens was less when compared to water-saturated ones [4]. The compressive strength was found to be decreased for concrete specimens soaked in crude oil [3]. Petroleum products cause adverse effects on concrete and thus start degrading. The percolation of petroleum products into concrete poses a threat to underground water. This culminates in the cracking of concrete structures.

1.1 Fly Ash with Concrete

To extract the maximum reimbursement to tackle economic and environmental issues, the addition of fly ash as a replacement material is crucial to concrete. Since fly ash is an industrial waste, it has the overhead of disposal. So, it can be efficiently be exploited as a replacement agent in concrete and thus help minimize the issues related to environmental pollution [5–7]. Studies reveal the significance of fly ash morphological size on the strength of mortar [8]. Positive results were found with replacements of concrete compressive strength with fly ash [9–11], and relations between the two parameters were determined.

Fly ash has been used exclusively along with other pozzolanic materials like palm oil fuel, silica fumes, and burnt clay to enhance the strength properties of concrete. Though much research is laid on enhancing the compressive strength of concrete using fly ash and other additives, the proper proportion of only fly ash selection needs much consideration. Also, this has to be done on all the classes of petroleum products. This helps in optimum disposal of fly ash on a large-scale basis. The proposed work is based on the proper selection of fly ash proportion to enhance the compressive strength characteristics of concrete coming in contact with diesel. The proportion of fly ash selected for the study was 0% to 40% to test the mechanical behavior of Diesel soaked concrete.

Table 1 Physical properties of ordinary Portland cement

Particulars	Parameters	Experimental result	As per standard
Fineness		308 m^2/kg	Not > 225 m^2/kg
Compressive strength	3 days	36	Not > 23
	7 day	46	Not > 33
	28 day	60	Not > 43
Soundness	By Le-Chatelier Exp. (mm)	1.5 mm	Not < 10 m
	By autoclave Exp. (%)	0.070	Not < 0.8
Setting time (minutes)	Initial	180	Not < 30
	Final	250	Not > 600

1.2 Research Significance

The proposed work aims at inferring the performance of concrete soaked in diesel for over 3 months. The study involves assessing the degrading effect of diesel on concrete versus diesel-free concrete kept under experimental surveillance for the same period. In this consideration, the chemical characteristics of the above specimens were premeditated via XRD and SEM analysis. It also deals with finding the right proportion of fly ash to be used as a cement replacement material to sustain diesel's harmful impacts on concrete.

1.3 Materials and Method

1.3.1 Cement

Ordinary Portland cement, 43 grade conforming IS: 8112-19,898, contains CaO 60–67%, SiO_2 17–25%, Al_2O_3 3–8%, Fe_2O_3 0.5–6%, MgO 0.5–4%, Na_2O_2%–3.5%, and SO_3 2–3.5%. Table 1 gives the physical properties of Portland cement.

1.3.2 Fly Ash

Fly ash used was obtained from Raichur Thermal Power Station, Shaktinagar, containing silica (SiO_2): 38–63%, Iron oxide (Fe_2O_3): 3.3–6.4%, alumina (Al_2O_3): 27–44%, titanium oxide (TiO_2): 0.4–1.8%, potassium oxide (K_2O): 0.04–0.9%, Calcium oxide (CaO): 0.2–8.0%, magnesium oxide MgO: 0.01–0.5%, Phosphorus pentaoxide (P_2O_2), sulfate (SO_4) and disodium oxide (Na_2O) together having 0.07–0.43%, as shown in Table 2 [7, 11].

Table 2 Chemical composition of fly ash

Chemical components	%
Iron oxide (Fe_2O_3)	6.28
Alkalies ($Na_2O + K_2O$)	0.46
Calcium oxide (CaO)	3.62
Magnesium oxide (MgO)	0.34
Titanium oxide (TiO2)	0.31
Silica (SiO_2)	56.88
Alumina (Al_2O_3)	27.65

1.3.3 Sand

Sand was collected from local vendors which has specific gravity 2.64, and coarse aggregates of (20 mm and down) 2.90 were used. Table 3 gives the sieve analysis of sand, and Table 4 presents the physical properties of sand particles.

Table 3 Presents sieve analysis of sand particles as per IS standards (IS: 383-1970)

IS sieve size in μm	Weight retained (g)	Cumulative weight retained (g)	Cumulative % weight retained	Cumulative % passing	Grading zone II
10	0	0	0	100	100
4.75	5	1	5	95	90–100
2.36	44	45	9	91	75–100
1.18	30	75	15	85	55–90
600	50	125	45	55	35–59
300	185	310	72	28	8–30
150	120	430	92	8	0–10
Pan	70	500		–	–
Total	504	–	238	–	–

Table 4 Presents physical properties sand particles as per IS standards (IS: 2386-1963)

Parameters	Results	Permissible limit as per IS: 2386-1963
Specific gravity	2.63	Should be between the limits 2.6–2.7
Moisture content	0.65%	–
Organic impurities	Colorless	Colorless/straw color/dark color
Silt content	0.7%	Should not be more than 6–10%
Bulking of sand	16%	Should not be more than 40%

Table 5 Presents properties of deisel as per IS standards (ASME)

Product	Parameters	Value
Diesel	Density at 293 K	840 kg/m^3
	Kinematic viscosity at 313 K	3.161 mm^2/s
	Flashpoint	341.15 K

1.3.4 Water

Water used for concrete mixing, specimen preparation and curing is taken from local available sources, i.e. Udyambag, Belagavi Municipal water supply, as water aided an essential key role in the hydration of cement which resulted in the setting and hardening of the concrete as per Indian Standards (IS 516-1959).

1.3.5 Diesel

Petroleum products like diesel from automobile service stations are taken as a soaking medium. Prepared concrete specimens are soaked in diesel. After a specific curing time, strength was noted down (Table 5).

1.3.6 Preparation of Specimens

The concrete was designed for M 30 grade as per IS: 10262-2009, which yields a proportion of 1:1.76: 3.15 with a water/cement ratio of 0.49, and various percentages of fly ash were added in the range of 0–40%. The mix was designed to achieve a nominal 28 days concrete compressive strength of 30 N/mm^2. Then, the concrete mix was placed in the molds to cast the specimens. Standard cube specimens of 150 mm × 150 mm × 150 mm were cast. Hand compaction, along with the vibrations from the vibrating table, was employed in preparing the specimens. This bequeathed fine finishing to the specimens, it was left for 24 h without disturbing it, and it was demolded and water cured for 28 days. The specimens were then categorized into two groups (Group A and B). Group B specimens were kept for water curing and will be continued without any disturbance for another 90 days. At the same time, Group A specimens were soaked in diesel. Then, as per IS standards, Group A and Group B, specimens were tested for their respective compressive strengths. In Group A specimens, the compressive test was performed on the 14th, 28th, 40th, 52nd, and 90th day of the diesel curing to monitor the consequences of diesel on concrete. Group B specimens were tested on the 28th and 90th day of the water curing process as the change in compressive strength in this context is significantly less.

To scrutinize the impact of diesel on the strength properties of concrete, the subjection of the specimens to the chemical and morphological tests was of paramount importance. So, X-ray diffraction test (XRD) and scanning electron microscope test (SEM) analyses were conducted in this regard.

1.3.7 Experimental Procedure

The present study deals with the utilization of fly ash along with two variants in the compressive strength analysis of the host specimen (one water cured and the other soaked in diesel), which can be better explained by the below-mentioned steps.

Totally, 189 concrete cubes with different fly ash replacements were prepared and cured in water for 28 days. On the 28th day, three cubes of each of the nine samples (varying in percentages in the interval of 5 from 0 to 40% of fly ash) were prepared and tested for their compressive strength. Twenty-seven out of 189 specimens were prepared and tested. 135 specimens were left for diesel curing (Group A), while the remaining twenty-seven specimens were cured in water (Group B). At mentioned time intervals (14, 28, 40, 52, 90 days), Group A specimen's compressive strengths were analyzed. In the case of Group B specimens, the compressive test was performed on specimens cured only for 28th and 90th day. The specimens of both Groups A and B compressive test results were compared against their corresponding time intervals. In the case of XRD and SEM, powdered samples were used for the analysis of compressive strength. The entire fly ash concrete cube was used with its shape intact.

2 Results and Discussions

The compressive strength of Group A specimens is as shown in Fig. 1, which indicates the strength results of diesel-soaked concrete at fixed time intervals. The various percentages of fly ash used in concrete and the period are shown along the X-axis with their corresponding compressive strengths plotted against Y-axis. The compressive strength increased during the early stages of soaking. However, the rate of increase

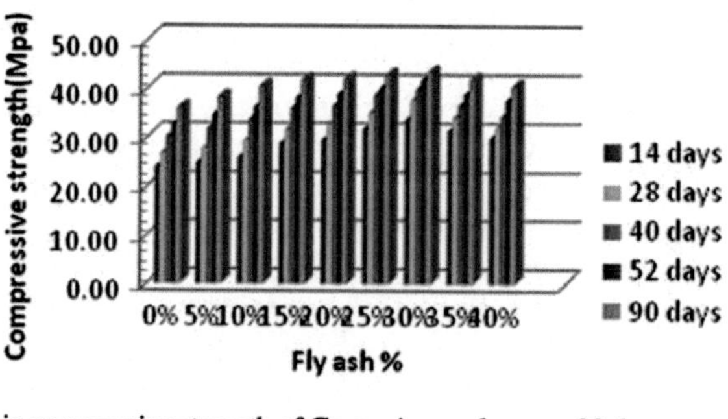

Fig. 1 Variation in compressive strength of Group A samples over 90 days

decreases gradually along with fly ash content up to 25% replacement of fly ash. It increases drastically at 30% fly ash content. The decrease in compressive strength followed the same trend before 30% of fly ash up to 40% fly ash contents. Figure 2 shows the compressive strengths of the three types of specimens. Where fd-28-f gives concrete specimens containing fly ash and water cured for 28 days, fd-90-f gives concrete specimens containing fly ash and water cured for 90 days, and fd-90-d gives concrete specimens containing fly ash and soaked in diesel for 90 days. The varying percentages of fly ash used are shown along X-axis, and the compressive strengths are shown along Y-axis. The compressive strengths of reference concrete were found to decrease from the 28th day to the 90th day. In contrast to the 30% replacement of fly ash, other specimen compressive strengths were relatively low. The deterioration of concrete in diesel over some time can be seen. Figure 3 represents the percentage variation of the compressive strengths of both Group A and B specimens. The fly ash compositions used are represented on the X-axis with the corresponding variations of compressive strengths.

The percentage increase or decrease in compressive strength concerning the initial reference mix shows the amount of degradation undergone by the mix under diesel contact. It can be observed that positive variation concerning 30% fly ash mix in concrete indicates the lower deteriorating effect on the specimen due to diesel. The

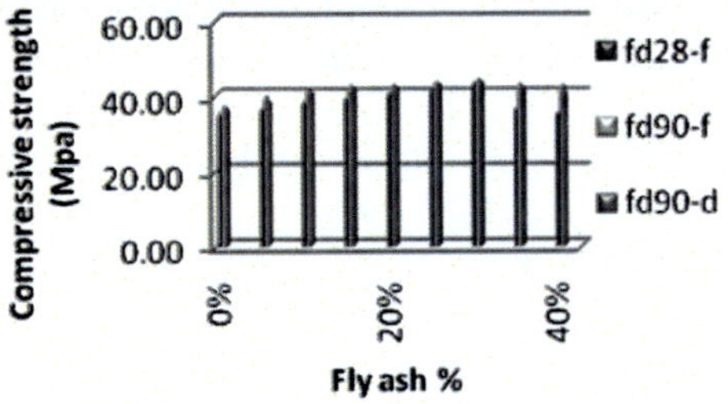

Fig. 2 Variation of compressive strength of fly ash mixed with and without diesel-soaked concrete

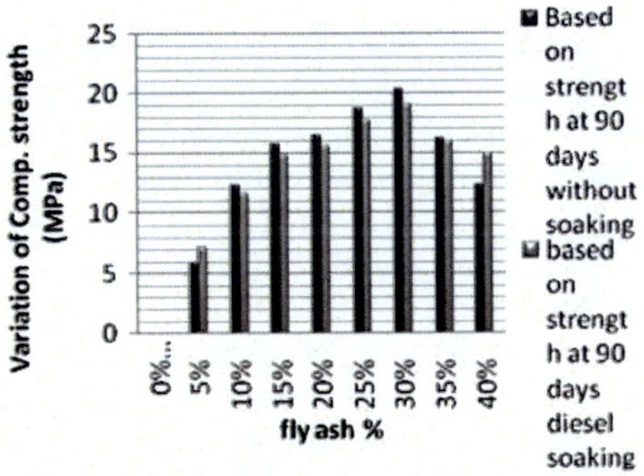

Fig. 3 Percent variation of diesel concrete strength based on that of diesel-free concrete

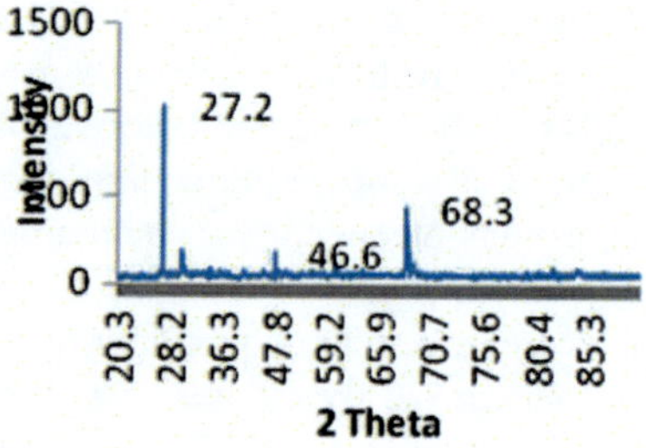

Fig. 4 XRD results of concrete before diesel soaking

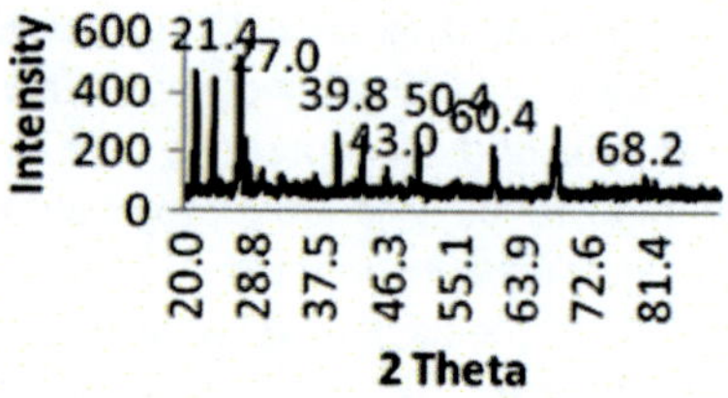

Fig. 5 XRD results of concrete after diesel soaking

figures mentioned above aid in examining the concrete strength configurations due to diesel. To determine the chemical behavior of diesel on concrete (with and without diesel soaking), prepared concrete as per different mediums was characterized by the XRD and SEM. Figures 4 and 5 depict the diffractograms, which analyze the presence of aluminum silicate, quartz (SiO_2), $CaSO_4$ (andrite), and free CaO compounds. The peak position at 2ɵ values of 27.0, 50.4, and 60.4 was of aluminum–silicon oxide which matched with standard JCPDS card No 79-1455. Aluminum silicate reacts with water to give hydration product of aluminum silicate hydrate A-S-H, this is also responsible for early strength in concrete [12]. The peak position at 2θ values of 39.8 and 43.0 represents the crystal system of zinc chromium sulfide ($ZnCr_2S_4$), which matches with JCPDS card No. 88-0886. Cement is known to have a trace composition of chromium and sulfate and zinc usually from the fly ash components. The mineral zinc chromium sulfide would have shown its presence due to the reaction of cement and fly ash. Two interpretations could be drawn from this presence of peak. Either calcium sulfate has not fully reacted which shows the presence of an unreacted gypsum peak. Our earlier findings have shown the presence of calcium hydroxide at peak position 39.8. This free calcium oxide could react with the sulfate which originates from petroleum products which in turn reacts with calcium hydroxide to give C-S–H peak. It is known that more amount of calcium sulfate in cement leads to the deterioration of concrete. This could be one of the reasons for the reduction in strength which was soaked in deisel.

The phase shifts in these figures are attributed to the decrease in compressive strengths. The readings shown in Fig. 4 correspond to the fly ash concrete subjected to water curing. Note the intensities of all the compounds mentioned above. The readings shown in Fig. 5 correspond to the fly ash concrete subjected to diesel

Fig. 6 SEM results of concrete before diesel soaking

soaking for 90 days. The intensities of all the components are reduced considerably when compared to the former figure. The presence of additional peaks (intensities) is attributed to additional compounds. The peak values of the above-mentioned compounds are reduced here. Maximum reduction of SiO_2 takes place in diesel-soaked concrete. The micrographs show that the materials are heterogeneous and contain a large proportion of unreacted fly ash.

On similar grounds, the SEM analysis results are in good agreement with the above XRD results (Fig. 7). The instrument specification for SEM testing is JEOL JSM-6360, vacuum system-rotary up to 10^{-2} Torr. Micrographs were taken on the 90th day for both the Group A and B samples. Figures 6 and 7 are SEM images of before and after Diesel soaked concrete. In the case of Group B samples, Fig. 7 indicates the aggressive effect of diesel on concrete. It has been observed from the SEM images of different resolutions that the microstructure of concrete soaked in diesel shows the less dense hydration. Few rod-like structures were noticed in Fig. 7 SEM photographs, and this may lead to the ettringite formation which leads to the cracking of concrete. Small ettringite formation may lead to expanding the concrete structure, and due to this, there may be a chance of deterioration of concrete. These results act as an archetype to emphasize that diesel has an aggressive impact on concrete, which degrades its compressive strength.

3 Conclusion

As a summary, the following conclusions can be drawn from the proposed implementation:

The mechanical behavior of concrete (compressive strength) when soaked in diesel found that there is a reduction in the strength properties by 2.40%, and the magnitude of the reduction varies with diesel.

Fig. 7 SEM results of concrete after diesel soaking

The studies on material characterization by X-ray diffraction (XRD) and material morphology by scanning electron microscope (SEM) have shown that the addition of fly ash in the concrete leads to the formation of dense microstructure which can resist the deterioration effects of deisel in concrete. The study dealt with assessing the degrading impact of diesel on fly ash concrete with the aid of XRD, SEM, and compressive strength analysis. To reduce the deteriorating effect of diesel on concrete, the results obtained showed that 30% replacement of fly ash bequeathed optimum compressive strength. Thus, it serves as an effective paradigm, and industrial waste materials can be used to reduce pollution.

4 Future Scope

The obtained results can be effectively be used in building structures that are prone to come in contact with diesel. Also, the proposed methods can be used to study the effect of other petroleum products on fly ash mixed concrete. This can bestow better compressive strengths to the concrete structures and help them to overcome the degrading effect of petroleum products. Furthermore, this creates an urge to effectively utilize solid wastes, thereby overcoming environmental pollution.

Acknowledgements The authors are deeply grateful to the Principal, KLE Dr. M. S. Sheshgiri, College of Engineering and Technology, Belgaum, and Department of Technology, Shivaji University, Kolhapur, for all the crucial amenities provided for the experimentation.

References

1. Nuruddin F, Shafiq N, Kamal NM (2008) Microwave incinerated rice husk ash (MIRHA) concrete: a new material in the construction industry

2. Svintsov AP, Gamal TS, Shumilin EE (2017) Effect of mineral and vegetable oil on deformation properties of concrete. RUDN J Eng Res 18(2):245–253
3. Ajagbe WO, Omokehinde OS, Alade GA, Agbede OA (2012) Effect of crude oil impacted sand on compressive strength of concrete. Constr Build Mater 26(1):9–12
4. Matti MA (1983) Effect of oil soaking on the dynamic modulus of concrete. Int J Cem Compos Lightweight Concrete 5(4):277–282
5. Gopalan MK, Hague MN (1985) Design of fly ash concrete. Cem Concr Res 15(4):694–702
6. Chang YM, Lin LK, Chang TC (1997) A preliminary assessment on the reuse of fly ash emitted from fluidized bed incineration of sludge cake. Resour Conserv Recycl 20(4):245–266
7. Ondova M, Stevulova N, Estokova A (2012) The study of the properties of fly ash based concrete composites with various chemical admixtures. Procedia Eng 42:1863–1872
8. Erdoğdu K, Türker P (1998) Effects of fly ash particle size on strength of Portland cement fly ash mortars. Cem Concr Res 28(9):1217–1222
9. Pofale AD, Deo SV (2010) A long-term comparative study of a concrete mix design procedure for fine aggregate replacement with fly ash by minimum voids method and maximum density method. KSCE J Civ Eng 14(5):759–764
10. Siddique R (2003) Effect of fine aggregate replacement with Class F fly ash on the mechanical properties of concrete. Cem Concr Res 33(4):539–547
11. Chindaprasirt P, Jaturapitakkul C, Sinsiri T (2005) Effect of fly ash fineness on compressive strength and pore size of blended cement paste. Cement Concr Compos 27(4):425–428
12. Ozyildirim C, Carino NJ (2006) Concrete strength testing. Significance of tests and properties of concrete and concrete-making materials. In: Lamond JE, Pielert JH (eds) ASTM STP D, vol 169, pp 125–140

Studies on the Effect of Frequency Content of Earthquakes on Structures with Infill

A. R. Avinash, A. Krishnamoorthy, and Kiran Kamath

Abstract Most of the modern buildings are constructed using reinforced cement concrete, and these structures commonly have infill walls. Though the mass of infill walls is accounted for in the conventional analysis, the stiffness is usually ignored. One of the reasons for ignoring the stiffness of the infill is the difficulty involved in the modelling. However, such an assumption might not yield realistic results. In this paper, the effect of modelling the infill on the seismic response of a fixed base plane frame structure is considered. Further, an attempt is also made to understand the behaviour of frames under earthquakes of different frequency content. Results show that the base shear response of the fixed base structure is sensitive to the frequency content of the earthquake. However, other responses such as roof displacement and roof acceleration are not much sensitive to the frequency content of the earthquakes.

Keywords Frequency content · Infill · Time history · Plane frame · Storey shear · Base shear

1 Introduction

Reinforced concrete (RC) multi-storey buildings are quite common nowadays. Masonry infilled walls are used widely in these structures as partition walls due to low cost, excellent heat resistant properties and ease of availability [1]. Under the action of gravity loads, these masonry infills do not offer any structural advantage. Also, due to their brittle behaviour, usually, infills are treated as non-structural

A. R. Avinash (✉) · A. Krishnamoorthy · K. Kamath
Department of Civil Engineering, Manipal Institute of Technology, Manipal Academy of Higher Education, Manipal 576104, India
e-mail: avinash.ar@manipal.edu

A. Krishnamoorthy
e-mail: moorthy.mit@manipal.edu

K. Kamath
e-mail: kiran.kamath@manipal.edu

L. Nandagiri et al. (eds.), *Sustainability Trends and Challenges in Civil Engineering*, Lecture Notes in Civil Engineering 162, https://doi.org/10.1007/978-981-16-2826-9_35

elements. However, the addition of such infill adds a considerable mass and stiffness to the structure. During a seismic event, the infill walls also act as structural elements and significantly alter the response of the structure [2]. Despite this, most of the times, the effect of infill is neglected in the conventional analysis. One of the main reasons behind ignoring the infill effect is the complex behaviour of infill and the difficulty involved in the modelling. However, to reduce the computational effort, many researchers have considered several simplified models such as diagonal compression elements to incorporate the stiffness effects of infill. Some researchers used single diagonal compression element [3, 4]. Others have tried two diagonal struts and three diagonal struts in their models [5, 6]. However, it has been reported that behaviour of masonry is very complicated due to the participation of higher modes [7], and hence, such simplified methods may not produce satisfactory results. In this context, in the present work, in order to obtain a realistic response of structure with infill, the infill is modelled as continuum using finite element method (FEM).

The characteristics of earthquake significantly influence the response of the structure during a seismic event. One of the important characteristics of an earthquake is the frequency content. The response of a structure which is subjected to a high-frequency, short-duration earthquake is different than the structure subjected to a low-frequency, long-duration earthquake. Several researchers have tried to fix a range to classify earthquakes based on such frequency contents. The frequency content of an earthquake can be expressed with a value which is a ratio of peak ground acceleration to peak ground velocity. A ratio of 0.8 or lesser indicates a low-frequency content and any value more than 1.2 indicates a high-frequency content, whereas values in between indicate a medium frequency content [8]. In this study, the effect of frequency content on the response of the infilled structure is also studied.

2 Details of the Structure and Ground Motion parameters

2.1 Description of the Structure

The structure considered for the study is a RC space frame building as shown in Fig. 1a. As shown in the figure, the building is of four storeys with each storey being 3 m in height. The building also has a plinth level of height 1 m. The details of geometrical and material properties are shown in Table 1. Since the frame is symmetric, it is idealized as a two-dimensional (2D) plane frame structure as shown in Fig. 1b. Gravity loads on the frame include dead load and the live load. Dead loads include self-weight of the slabs, beams, columns and floor finish. A floor finish load of 1 kN/m^2 is considered for all floors. A live load of 2 kN/m^2 is considered for all floors except roof slab.

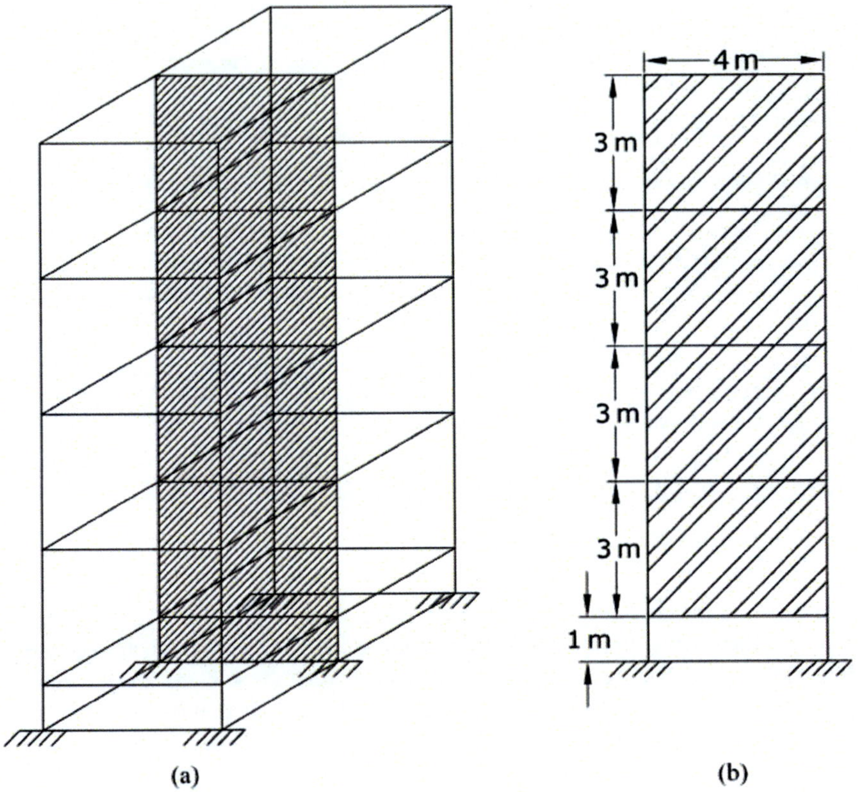

Fig. 1 Idealization of a space frame into a plane frame **a** space frame **b** plane frame

Table 1 Geometrical and material properties of the structure

Properties of the frame		Properties of the masonry infill	
Size of columns	0.23 m × 0.60 m	Type of infill	Brick masonry
Size of beams	0.23 m × 0.45 m	Thickness of infill	0.3 m
Modulus of elasticity	27,386 MPa	Modulus of elasticity	2068.4 MPa
Unit weight	25 kN/m^3	Unit weight	18 kN/m^3
Poisson's ratio	0.2	Poisson's ratio	0.3

2.2 *Description of the Earthquake Records Used for the Study*

Four strong earthquake ground motions are considered for the study. These earthquakes are chosen based on their frequency content. Details of these earthquakes are given in Table 2. As seen from the Table 2, Chi-Chi and Imperial Valley have low frequency content. El-Centro earthquake has intermediate frequency content,

Table 2 Details of the earthquake ground motions considered for the study

Earthquake	Peak ground acceleration (g)	Station	Magnitude	Frequency content	Category
Chi-Chi	0.31	CWB TCU075	7.3	0.29	Low
Imperial valley	0.44	USGS 5158	6.6	0.40	Low
El-Centro	0.30	USGS 0117	6.7	0.81	Intermediate
Kobe	0.35	Kakogawa CUE	6.9	1.25	High

whereas Kobe earthquake has a high frequency content. The accelerograms of these earthquakes are shown in Fig. 3. The peak ground acceleration (PGA) in terms of 'g' (acceleration due to gravity) is shown in these figures.

3 Modelling Details of the Structure

The structure is the arrangement of beams and columns and infill walls. The beams and columns of the structure are discretized as two-noded plane frame elements (Fig. 2a) and infill is modelled as four-noded quadrilateral plane stress elements (Fig. 2b). Frame elements have two end nodes with three degrees of freedom at each node (Fig. 2c). These degrees of freedom represent two translations along two axes (x and y) and one rotation about the 'z' axis. Plane stress elements have four nodes with two degrees of freedom (one horizontal translation and another vertical translation) at each node (Fig. 2c). In general, the stiffness matrix for each element is obtained

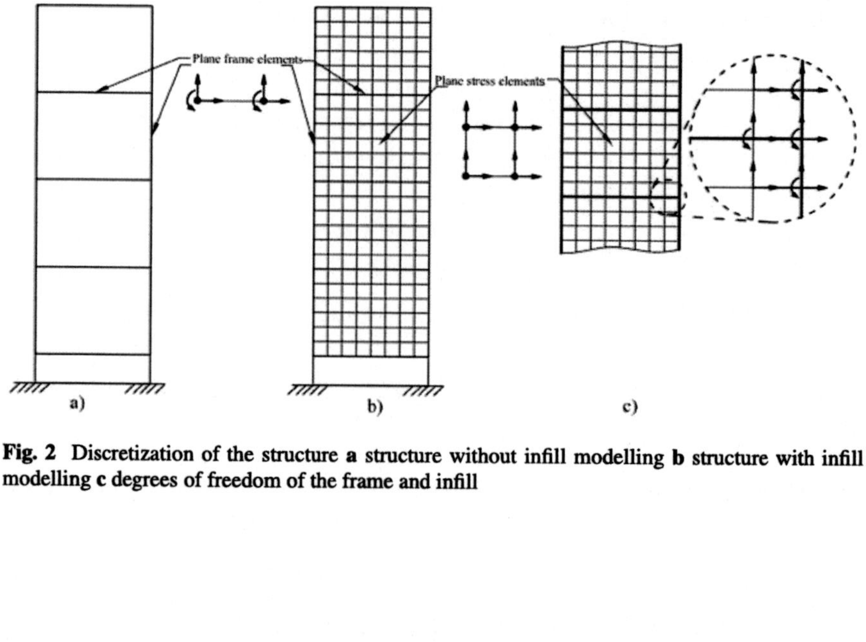

Fig. 2 Discretization of the structure **a** structure without infill modelling **b** structure with infill modelling **c** degrees of freedom of the frame and infill

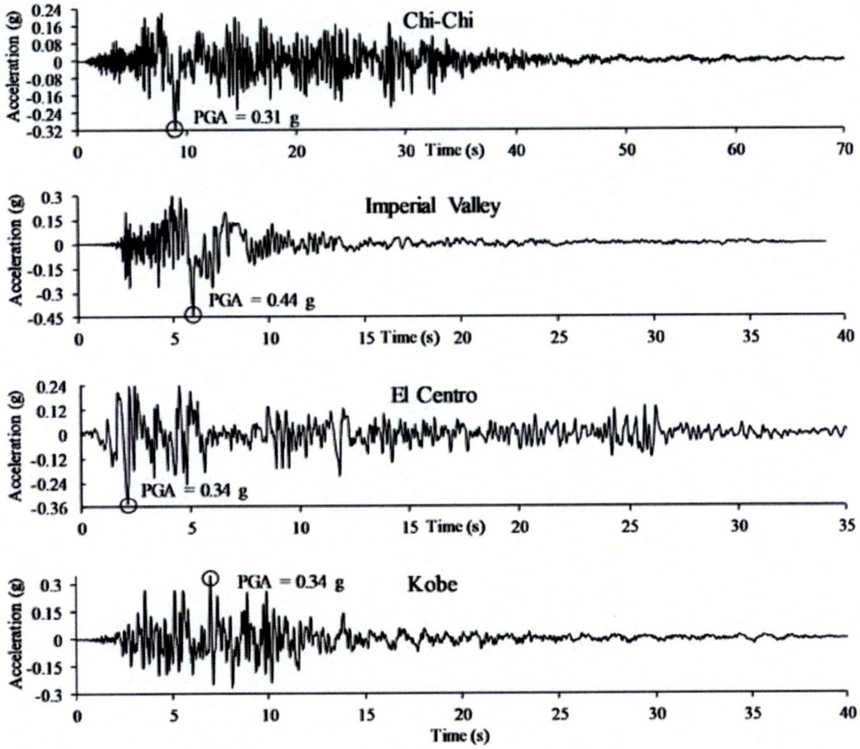

Fig. 3 Accelerograms of various earthquakes considered for the study

using the equation,

$$[K] = \int_v [B]^T[C][B]\mathrm{d}v \tag{1}$$

where
$[B]$ = matrix relating strain to displacement
$[C]$ = constitutive matrix relating stress to strain given by,

$$\frac{E}{1-\mu^2}\begin{bmatrix} 1 & \mu & 0 \\ \mu & 1 & 0 \\ 0 & 0 & \frac{1-\mu}{2} \end{bmatrix}$$

where
μ = Poisson's ratio of the material
E = modulus of elasticity of the material
Similarly, the mass matrix can be obtained using the equation,

$$M = \rho \int_{v} [N]^T [N] \mathrm{d}v \tag{2}$$

where

$\rho =$ mass density of the material

$N =$ shape function of the element

The stiffness matrix of the frame element in local direction is given by,

$$[K_{\mathrm{fl}}] = \begin{bmatrix} \frac{EA}{L} & 0 & 0 & -\frac{EA}{L} & 0 & 0 \\ 0 & \frac{12EI}{L^3} & \frac{6EI}{L^2} & 0 & -\frac{12EI}{L^3} & \frac{6EI}{L^2} \\ 0 & \frac{6EI}{L^2} & \frac{4EI}{L} & 0 & -\frac{6EI}{L^2} & \frac{2EI}{L} \\ -\frac{EA}{L} & 0 & 0 & \frac{EA}{L} & 0 & 0 \\ 0 & -\frac{12EI}{L^3} & -\frac{6EI}{L^2} & 0 & \frac{12EI}{L^3} & -\frac{6EI}{L^2} \\ 0 & \frac{6EI}{L^2} & \frac{2EI}{L} & 0 & -\frac{6EI}{L^2} & \frac{4EI}{L} \end{bmatrix}$$

where

$A =$ cross-sectional area of the frame element

$L =$ length of the frame element

$I =$ second moment of area of the element

The consistent mass matrix of the frame element is given by,

$$[M_{\mathrm{fl}}] = \frac{\rho AL}{420} \begin{bmatrix} 140 & 0 & 0 & 70 & 0 & 0 \\ 0 & 156 & 22L & 0 & 54 & -13L \\ 0 & 22L & 4L^2 & 0 & 13L & -3L^2 \\ 70 & 0 & 0 & 140 & 0 & 0 \\ 0 & 54 & 13L & 0 & 156 & -22L \\ 0 & -13L & -3L^2 & 0 & -22L & 4L^2 \end{bmatrix}$$

The stiffness matrix $[K_{\mathrm{fl}}]$ and mass matrix $[M_{\mathrm{fl}}]$ from local direction are transformed into the global direction using the equations,

$$\left[K_{\mathrm{gf}}\right] = [T]^T [K_{\mathrm{fl}}][T] \tag{3}$$

$$\left[M_{\mathrm{gf}}\right] = [T]^T [M_{\mathrm{fl}}][T] \tag{4}$$

where

$[T] =$ transformation matrix

The overall mass matrix $[M_{\mathrm{s}}]$ and stiffness matrix $[K_{\mathrm{s}}]$ of the structure are obtained by assembling mass and stiffness matrix of each element by direct stiffness method. The masonry is modelled using plane stress element with two translational degrees of freedom u and v. Thus, the stiffness of a frame element corresponding to the two translational displacements are added to the stiffness of plane stress element at the nodes where the frame elements are connected to the plane stress elements. The

Table 3 Validation of the developed model considering El-Centro earthquake

Time period (s)				Peak roof displacement (m)			
Without infill modelling		With infill modelling		Without infill modelling		With infill modelling	
Program	SAP2000	Program	SAP2000	Program	SAP2000	Program	SAP2000
0.49	0.50	0.19	0.19	0.072	0.076	0.016	0.011

dynamic equation of equilibrium of the structure in matrix form is given as,

$$[M_s]\{\ddot{u}\} + [C_s]\{\dot{u}\} + [K_s]\{u\} = \{F(t)\} \quad (5)$$

where

$$[C_s] = \alpha[M_s] + \beta[K_s] \quad (6)$$

where

$[C_s]$ = the damping matrix which is obtained by considering Rayleigh type damping

α, β = Rayleigh constants which are obtained by knowing the damping ratio for each mode.

u, $\dot{u}$ and $\ddot{u}$ are the relative displacement, velocity and acceleration vectors, respectively. $F(t)$ is the nodal load vector given by,

$$\{F(t)\} = -[M_s]\{I\}\ddot{u}_g(t) \quad (7)$$

where

$\{I\}$ = is influence vector

$\ddot{u}_g(t)$ = ground acceleration

A program is written in *C* language to analyse the plane frame building. For the validation purpose of the method, the model is developed and analysed in SAP2000 (evaluation version), a commercial structural analysis and design software package. These models are subjected to the El-Centro earthquake (Fig. 3). The comparison of the time period of the structures and peak roof displacement results are shown in Table 3. As seen from the table, the time period of the model obtained from the proposed analysis matches closely with the SAP2000 results. Further, peak roof displacement of the models also shows agreeable results.

4 Results and Discussions

The structure without considering the infill effect is subjected to above-mentioned earthquakes and the responses of the structure are obtained in terms of base shear, storey shears, roof displacement and roof accelerations. Similarly, the structure with infill modelling is also subjected to the same earthquakes and results are obtained. The results of the structure without infill modelling are compared with the structure modelled by considering the effect of infill.

Figure 4 shows the time history response of base shear of structures subjected to different earthquakes. It can be seen that for all earthquake inputs considered in the study, the base shear of the structure is increased when the infill is modelled. A comparison is made to study the effect of modelling infill on the peak base shear in Fig. 5. The figure shows the base shear response in the increasing order of the frequency content. The increase in peak base shear is prominent for the El-Centro earthquake which has an intermediate frequency content.

The storey shear response of the structures can be seen in Fig. 6. Even though the storey shears of structure with infill modelling is high at the plinth level, a considerable reduction can be seen at all the upper storeys. This reduction is considerable for earthquakes of low frequency content. However, it is not so significant for the intermediate and high frequency content earthquakes.

The effect of infill modelling on the roof displacement is seen in the time history plot shown in Fig. 7 for various earthquakes with different frequency content. In

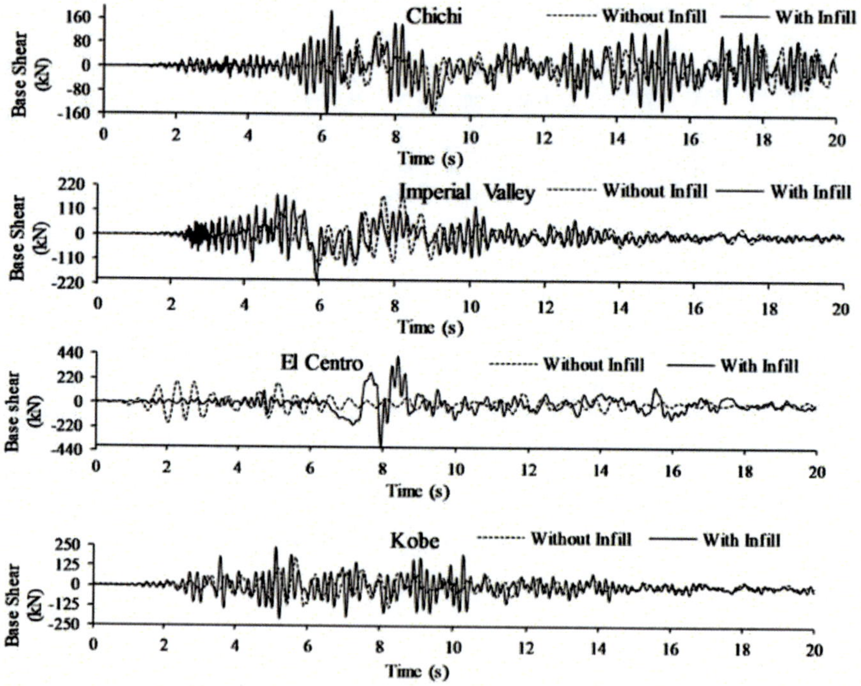

Fig. 4 Time history responses of base shear for the structure with and without infill effect corresponding to earthquakes of various frequency contents

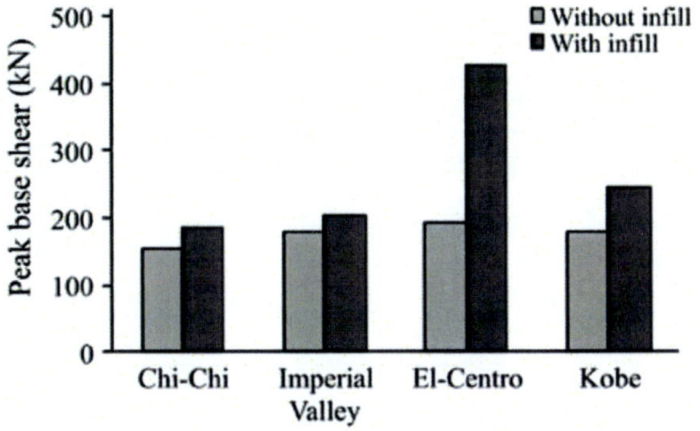

Fig. 5 Peak base shear for the structure with and without infill modelling corresponding to earthquakes of various frequency contents

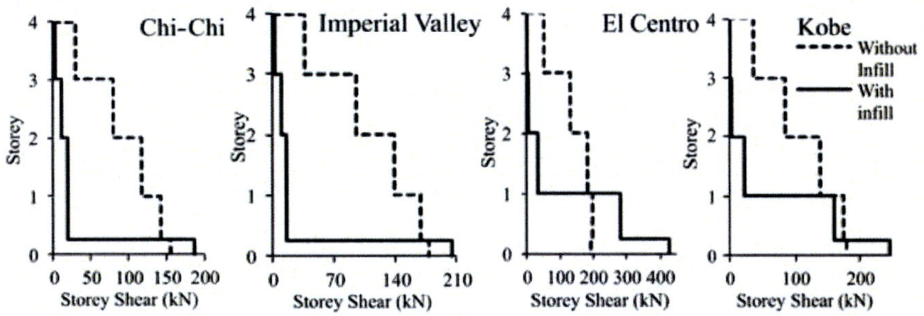

Fig. 6 Peak storey shears for the structures without infill modelling and with infill modelling corresponding to earthquakes of various frequency contents

all the cases, the roof displacement reduces significantly when infill modelling is considered. It can also be seen from Fig. 8 that the peak roof displacement reduction is similar to all earthquake inputs. Thus, the frequency content of the earthquake does not have a notable effect on roof displacement.

Figure 9 illustrates the effect of modelling infill on roof accelerations. It is evident from this time history response that with the infill modelling, there is a significant reduction in roof acceleration. It can be observed from Fig. 10 that the peak roof acceleration reduction is not significant in case of Imperial Valley earthquake which has a low frequency content. This indicates that the effect of infill on roof acceleration is influenced significantly by the frequency content of earthquake.

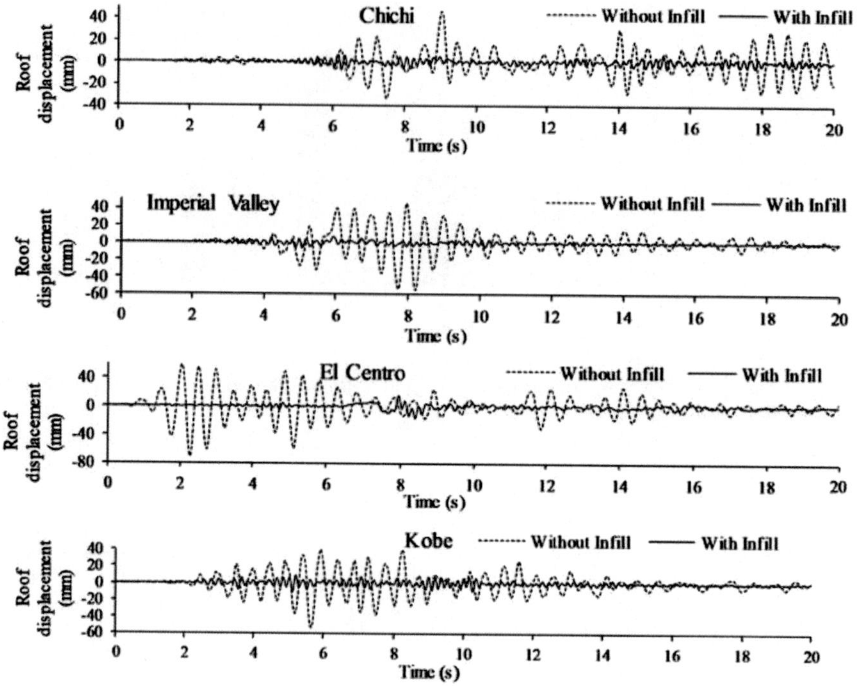

Fig. 7 Time history response of roof displacement for the structure with and without infill modelling for earthquakes with different frequency contents

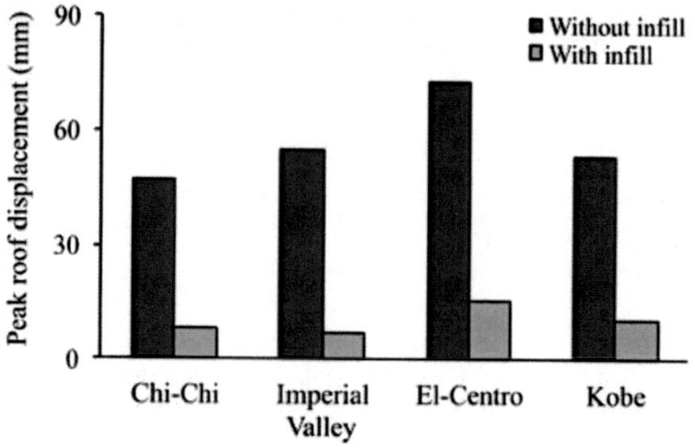

Fig. 8 Peak roof displacement response for the structure with and without infill modelling for earthquakes with different frequency contents

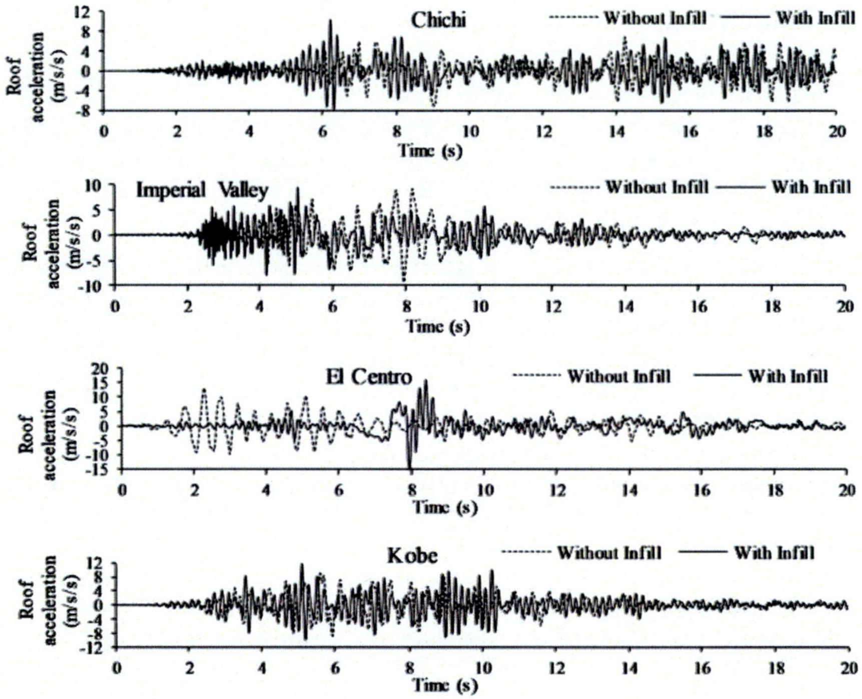

Fig. 9 Time history response of roof acceleration for the structure with and without infill modelling for earthquakes with different frequency contents

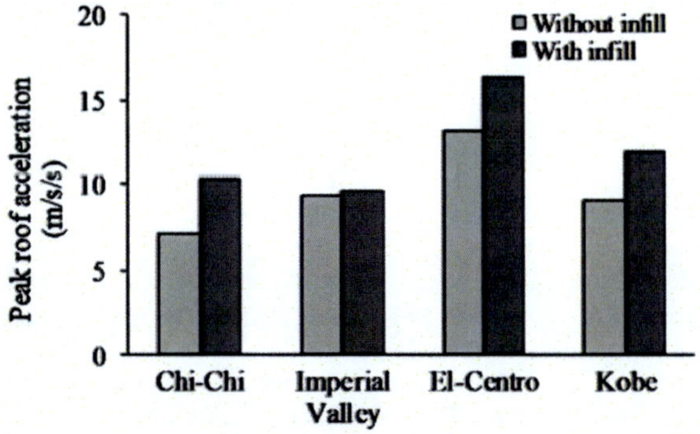

Fig. 10 Peak roof acceleration response for the structure with and without infill modelling for earthquakes with different frequency contents

5 Conclusions

Masonry infills when used in structures have a tendency to significantly modify the response of the structure. Thus, realistic modelling and analysis of the infill is important from both the safety and economy point of view. Hence, in the proposed study, the effect of infill on modelling of structure is studied. Following are the conclusions of the proposed study:

- It is evident from the study that ignoring the effect of infill in modelling has resulted in gross underestimation of base shears. Hence, consideration of infill in modelling is required to obtain realistic response.
- Storey shear results indicate that neglecting the effect of infill in modelling of structure results in significant overestimation of forces in upper storeys. In view of this, with a proper approach, it is possible to achieve some economy in construction by reducing the member sizes of upper storeys.
- Roof displacement reduces significantly when the effect of infill is considered in modelling. This again indicates an overestimation of the roof displacements in conventional analysis.
- Further, roof acceleration plots indicate a similar overestimation of forces when infill effect is ignored in modelling.
- The effect of infill on base shear response is sensitive to frequency content of the earthquake.
- The story shear reduces when the effect of infill is considered in modelling of structure for an earthquake with low frequency content, whereas for other earthquakes with intermediate or high frequency contents, the effect of considering infill in modelling is not much significant for storey shear responses.
- The effect of frequency content does not appear to be significant for roof displacement and roof accelerations.

References

1. Kaushik HB, Rai DC, Jain SK (2007) Stress-strain characteristics of clay brick masonry under uniaxial compression. J Mater Civ Eng 19(9):728–739
2. Flanagan RD, Bennett RM (1999) In-plane behavior of structural clay tile infilled frames. J Struct Eng 125(6):590–599. https://doi.org/10.1061/(ASCE)0733-9445(1999)125:6(590)
3. Cavaleri L, Di Trapani F (2014) Cyclic response of masonry infilled RC frames: experimental results and simplified modeling. Soil Dyn Earthq Eng 65:224–242
4. Rodrigues H, Varum H, Costa A (2010) Simplified macro-model for infill masonry panels. J Earthq Eng 14(3):390–416
5. El-Dakhakhni WW, Elgaaly M, Hamid AA (2003) Three-strut model for concrete masonry-infilled steel frames. J Struct Eng (NY) 129(2):177–185

6. Crisafulli FJ, Carr AJ (2007) Proposed macro-model for the analysis of infilled frame structures. Bull New Zealand Soc Earthq Eng 40(2):69–77
7. Priestley MJN, Crosbie RL, Carr AJ (1977) Seismic forces in base-isolated masonry structures. BNZSEE 10(2):55–68
8. Zhu TJ, Heidebrecht AC (1988) Effect of peak ground a/v ratio on structural damage. J Struct Eng 114(5):1019–1037

Soil Structure Interaction Analysis of Tanks Filled with Fluid Subjected to Near-Fault Earthquakes

M. Chaithra and A. Krishnamoorthy

Abstract When the tanks with fluid are analysed conventionally, soil flexibility is not considered, and the tank base is assumed to be fixed. But, for practical analysis of tank resting on soil, the interaction between the tank, soil and foundation systems has to be considered. In the present study, the tank supported on the soil is studied considering the influence of soil. The tank, foundation and soil systems are discretized using finite element method. The tank is modelled using frame elements with two nodes, and foundation and soil are modelled using plane strain elements of four nodes. Pressure formulation approach is used to model the mutual influence between fluid and tank. The tank is subjected to two near-fault earthquakes. The analysis is used to investigate the effect of soil on the response of the tank. The various response parameters considered for the study are base shear, variation of fluid pressure on tank wall and displacement at top. Three soil types with different values of flexibility are considered in order to evaluate the effect of SSI. It is concluded from the analysis that the soil flexibility considerably influences the response of tank.

Keywords Tanks · Soil structure interaction · Seismic · Analysis

1 Introduction

Liquid storage tanks have played a vital role in daily human activities. These tanks must perform well and should remain functional even after a major seismic event. If forces on the tank are not estimated properly, it will lead to inappropriate design and may fail during a seismic event. Such inadequate designs have attributed to failure of storage tanks filled with fluid during past earthquakes. Such failures created a lot of interest in safeguarding the tanks against dynamic forces. The dynamic

M. Chaithra (✉) · A. Krishnamoorthy
Department of Civil Engineering, Manipal Institute of Technology, Manipal Academy of Higher Education, Manipal 576104, India
e-mail: chaithra.mitthur@manipal.edu

A. Krishnamoorthy
e-mail: moorthy.mit@manipal.edu

L. Nandagiri et al. (eds.), *Sustainability Trends and Challenges in Civil Engineering*, Lecture Notes in Civil Engineering 162, https://doi.org/10.1007/978-981-16-2826-9_36

behaviour of fluid-filled tanks subjected to earthquakes is different compared to the response of common structures such as bridges or buildings. This difference is obtained due to the influence of pressure on tank walls. The presence of liquid in storage tanks has a considerable effect on the dynamics of structure. When the tank is subjected to seismic force, the motion of the structure restricts the motion of fluid which in turn loads the structure. This mutual influence between the structure and fluid is known as fluid structure interaction (FSI). Also, when the tanks with fluid are analysed conventionally, soil flexibility is not considered, and the tank base is assumed to be fixed. However, various studies indicated that, apart from the tank stiffness, the actual response of the tank also depends on the foundation stiffness and supporting soil stiffness. Soil conditions have a great influence on the damage of structures during ground motion. SSI can affect the behaviour of the structure based on the relative stiffness of the structure and soil. Hence, it is required to consider the interaction between tank and soil system for the realistic analysis. The process in which the soil and structure has mutual influence on their response is known as soil–structure interaction (SSI). Such analysis, considering fluid–structure–soil interaction is required to achieve a safe and economical structural design of the tank.

Housner [1] proposed a simplified analytical solution for circular and rectangular tanks under horizontal ground excitations assuming the tank base to be fixed without considering the soil flexibility. Kim et al. [2] presented a method for a cylindrical tank by considering the effects of the liquid and soil system in the frequency domain. Kianoush and Ghaemmaghami [3] also studied the effect of SSI on a tank filled with fluid for different soil types. Everstine [4] has reviewed different finite element formulations to solve FSI problems.

It may be noted from the literature that the study on the effect of both SSI and FSI on dynamic behaviour of liquid storage tanks is limited. Many literatures are available on investigation of seismic responses of circular storage tanks. However, rectangular tanks are also used for storage of fluids, and hence, vibratory response of these tanks is also of practical interest. Chaithra et al. [5] studied the influence of SSI on a shallow tank filled with fluid. In this, the effect of SSI is considered for Chi-Chi earthquake. Since, the type of earthquake has strong influence on response of tank and on SSI effect, in the present study, the effect of SSI is evaluated for two near-fault earthquakes. Also, type of tank has great influence on the response of the tank. Hence, a tall tank model resting on three types of soil is considered.

A rectangular water tank is considered in the present study. The tank is subjected to two near-fault earthquakes, viz. Imperial Valley and Loma Prieta. The dynamic behaviour of the tank is obtained in terms of displacement at top, variation of fluid pressure and base shear by considering FSI as well as SSI. The results obtained are then compared with the response of tank without considering SSI, i.e. rigid base structure.

2 Methodology

The tank, fluid and soil are discretized using finite elements.

2.1 Modelling the Tank and Soil

The tank is modelled using two-noded frame element having one rotational and two translational degrees of freedom at each node, and four-noded plane strain element is used to model the foundation and soil system. The equation of motion for the tank and soil is given by,

$$Kq + M\ddot{q} + C\dot{q} - \int_s \left(N^T \tilde{N} ds\right) p = f(t) \tag{1}$$

where

K	$\int_v B^T DB\, \mathrm{d}V$, stiffness matrix
M	$\rho \int_v N^T N\, \mathrm{d}V$, mass matrix
q, $\dot{q}$, $\ddot{q}$	vectors of displacement, velocity, acceleration at nodes
D	constitutive matrix relating stress to strain
ρ	mass density of material
N	shape function defining displacement of structure
$\tilde{N}$	shape function defining pressure distribution in fluid
C	damping matrix and is obtained using Rayleigh's method as,

$$C = \alpha M + \beta K \tag{2}$$

where α and β are Rayleigh constants.
The force due to earthquake, $f(t)$ is given by,

$$f(t) = -MI\ddot{u}_g(t) \tag{3}$$

where

I	influence vector
$\ddot{u}_g(t)$	ground acceleration

As observed, the loads due to earthquake ground acceleration are applied at the corresponding degrees of freedom. It may be noted that $\ddot{u}_g(t)$ in (3) is the ground acceleration which can be obtained from the accelerogram of earthquake at various time interval.

2.2 Modelling the Fluid

Fluid is modelled as four-noded rectangular element, and Eulerian pressure formulation approach [6] is used to model the equation of motion for fluid. Sloshing is neglected in the analysis, and hence, the pressure along the top nodes of fluid is considered as zero. The motion at the interface is prescribed by the movement of the structure. Hence, the equation of motion for fluid is given by,

$$Hp + G\ddot{p} + \int_s \left(\tilde{N}^T \rho N \mathrm{d}s\right)\ddot{q} = 0 \tag{4}$$

where

H $\int_v \nabla^T \tilde{N}^T \nabla \tilde{N} \,\mathrm{d}V$
G $1/g \int_s N^T N \,\mathrm{d}s$
ρ density of fluid
p fluid pressure
$\ddot{q}$ acceleration of the structure
g acceleration due to gravity

Equations (1) and (4) can be combined and written in matrix form as,

$$\begin{bmatrix} K & -S \\ 0 & H \end{bmatrix}\begin{bmatrix} q \\ p \end{bmatrix} + \begin{bmatrix} C & 0 \\ 0 & 0 \end{bmatrix}\begin{bmatrix} \dot{q} \\ \dot{p} \end{bmatrix} + \begin{bmatrix} M & 0 \\ \rho S^T & G \end{bmatrix}\begin{bmatrix} \ddot{q} \\ \ddot{p} \end{bmatrix} = [f(t)] \tag{5}$$

where

$S = \int_s \left(N^T N \,\mathrm{d}s\right)$, is the interaction matrix for both structure and fluid and is obtained from the boundary condition imposed between the tank wall and liquid.

It may be noted that the (5) combines the structure and fluid and is known as coupled fluid–structure interaction equation.

2.3 Determination of Seismic Response

Newmark's method is considered to obtain the response of the tank system in the incremental form. In this method, response at time 't' is obtained first and is used to obtain the response at time '$t + \Delta t$'. Constant average acceleration method is used by considering $\beta = ¼$ and $\gamma = ½$.

3 Numerical Analysis

The response of the tank filled with fluid is studied for real earthquakes. The response quantities considered for the study include displacement at the top of tank, base shear and variation of fluid pressure on tank wall. The effect of considering SSI on the response of tank is studied.

3.1 Details of Structure, Fluid, and Soil System Considered for the Analysis

The finite element model of tank, fluid and soil is shown in Fig. 1. Material properties and geometric properties considered for the problem are as follows:

Tank

Tank height—11.2 m
Tank width—19.6 m
Tank wall thickness—1.2 m
Mass density of tank wall material—2.3 kNs^2/m^4

Fluid

Bulk modulus—2.0684×10^6 kN/m^2
Mass density—1.0 kNs^2/m^4

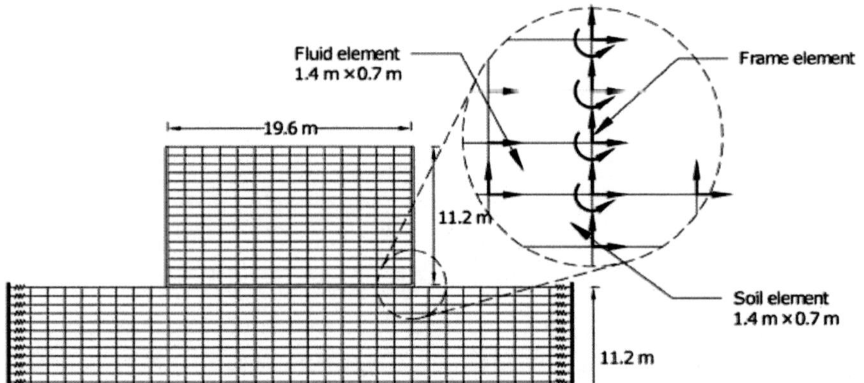

Fig. 1 Finite element discretization of soil, fluid and tank system

Modulus of elasticity of soil

Hard soil—5,000,000 kN/m^2
Medium soil—50,000 kN/m^2
Soft soil—5000 kN/m^2
Poisson's ratio—0.3

The properties of the tank model used for the present study are similar to the tank model considered by Kianoush and Ghaemmaghami [6]. The tank resting on soil is subjected to Imperial Valley and Loma Prieta earthquake ground motions, and responses are obtained in terms of base shear and displacement at top at various time intervals. To study the influence of soil on the tank, the results obtained from the analysis of the tank resting on soil are compared with the results obtained from the analysis of isolated tank without considering SSI (tank fixed at base).

3.2 *Time History Response*

The SSI effect on response of tank subjected to Imperial Valley and Loma Prieta earthquakes is studied at different time intervals during the ground motion. The accelerograms for the earthquakes with peak ground acceleration (PGA) is shown in Fig. 2. The time history response for the tank with fixed base and tank supported on various types of soil in terms of base shear is shown in Fig. 3 for Imperial Valley earthquake and in Fig. 4 for Loma Prieta earthquake. Figures 5 and 6 show the time history response for the tank with fixed base and tank supported on three types of soil in terms of displacement at the top for Imperial Valley earthquake and Loma Prieta earthquake, respectively.

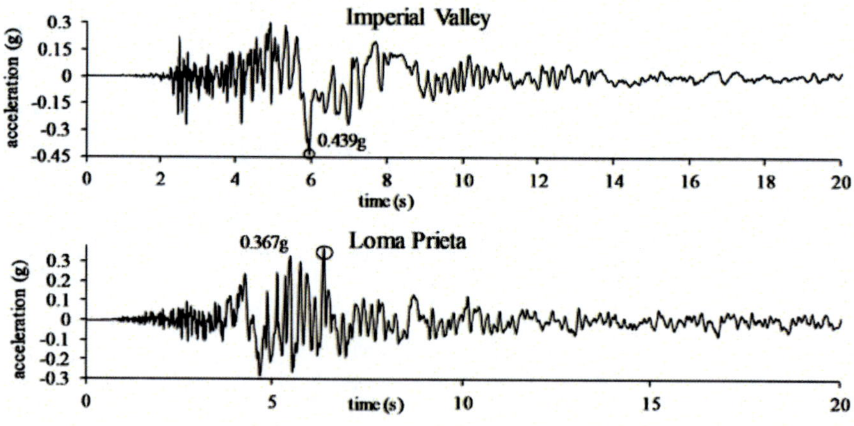

Fig. 2 Accelerograms of earthquake ground motions considered for the study

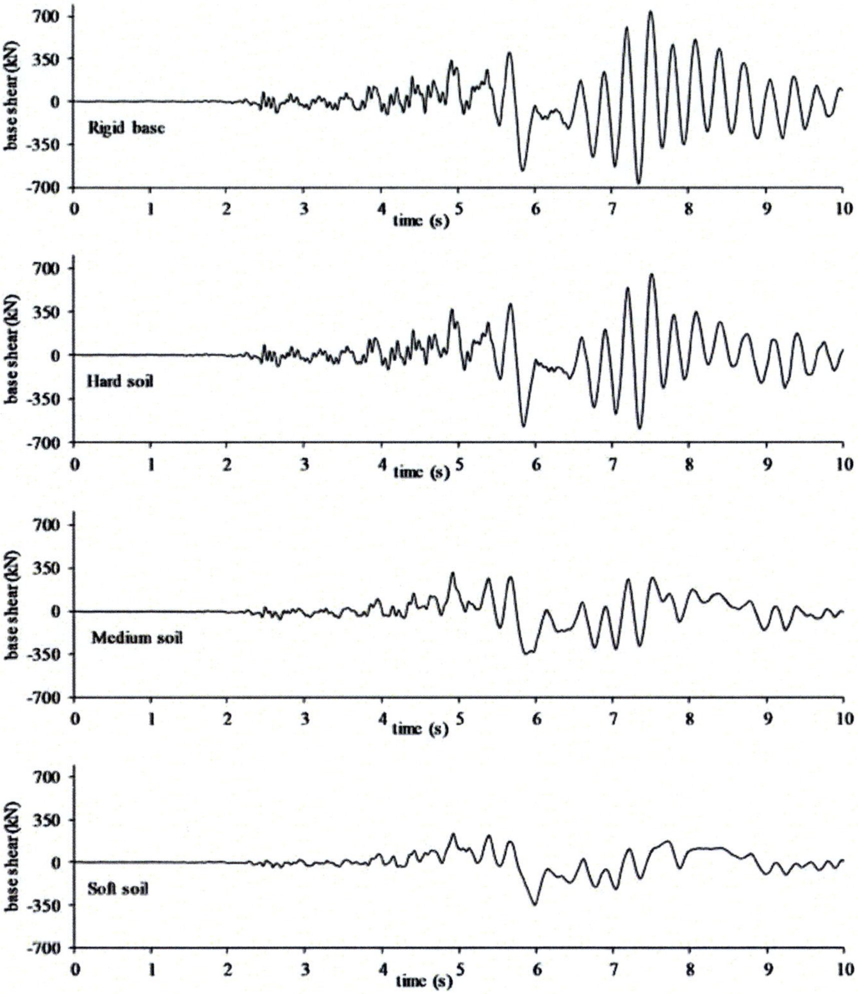

Fig. 3 Time history response of base shear of tank on various soil types for Imperial Valley earthquake

The peak response of the tank resting on various soil types is given in Table 1. It can be observed from Figs. 3, 4 and Table 1 that the SSI influences the base shear of the tank. The percentage reduction of base shear for tank resting on soft, medium and hard soil is 12.2%, 52.5% and 52.8%, respectively. The effect of SSI is significant for the tank resting on soft and medium soil and reduces as the soil stiffness increases. The base shear is maximum for tank resting on hard soil compared to soft and medium soil for both earthquakes. It can be observed from Figs. 5, 6 and Table 1, that, the top displacement of the tank is affected considerably due to SSI. The percentage reduction of top displacement for tank resting on soft, medium and hard soil is

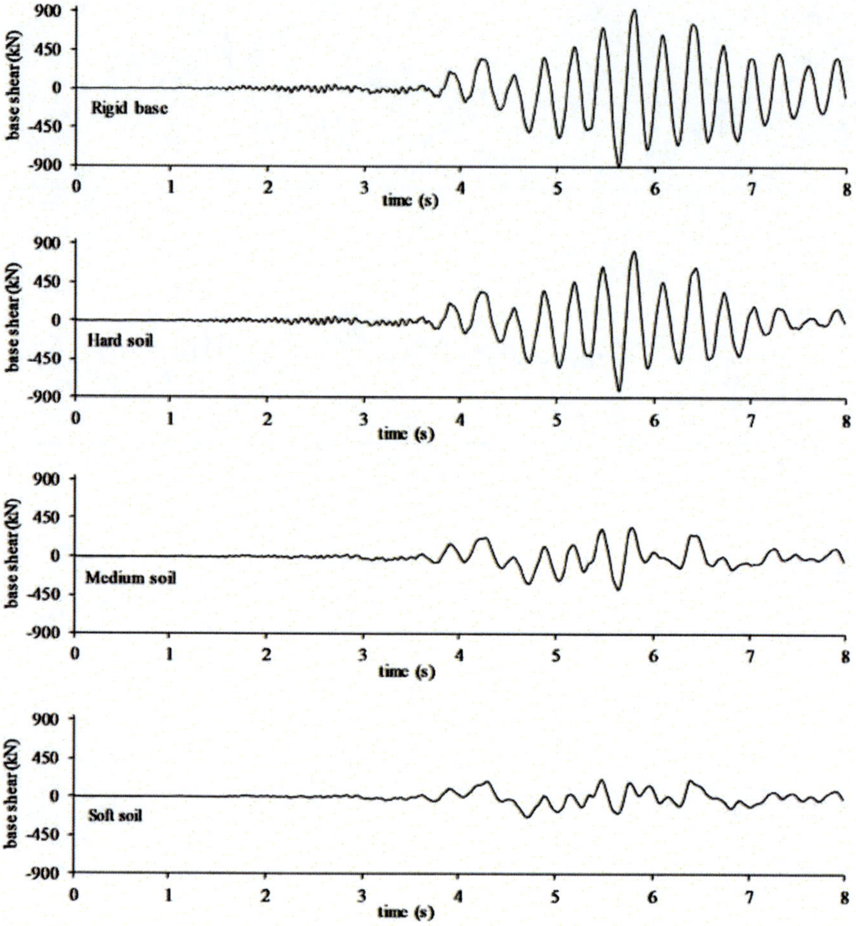

Fig. 4 Time history response of base shear of tank on various soil types for Loma Prieta earthquake

13.8%, 62.3% and 66.1%, respectively. The effect is significant for the tank resting on soft soil and reduces as the stiffness of soil increases. However, the response may decrease or increase due to the shift in natural frequency of the tank and soil system when SSI is considered. For the present case, the response of the tank reduces as the soil stiffness decreases.

3.3 *Effect of SSI on Fluid Pressure*

Figure 7 shows the variation of pressure at fluid-wall interface along height of the wall for the tank supported on three types of soil subjected to Imperial Valley and

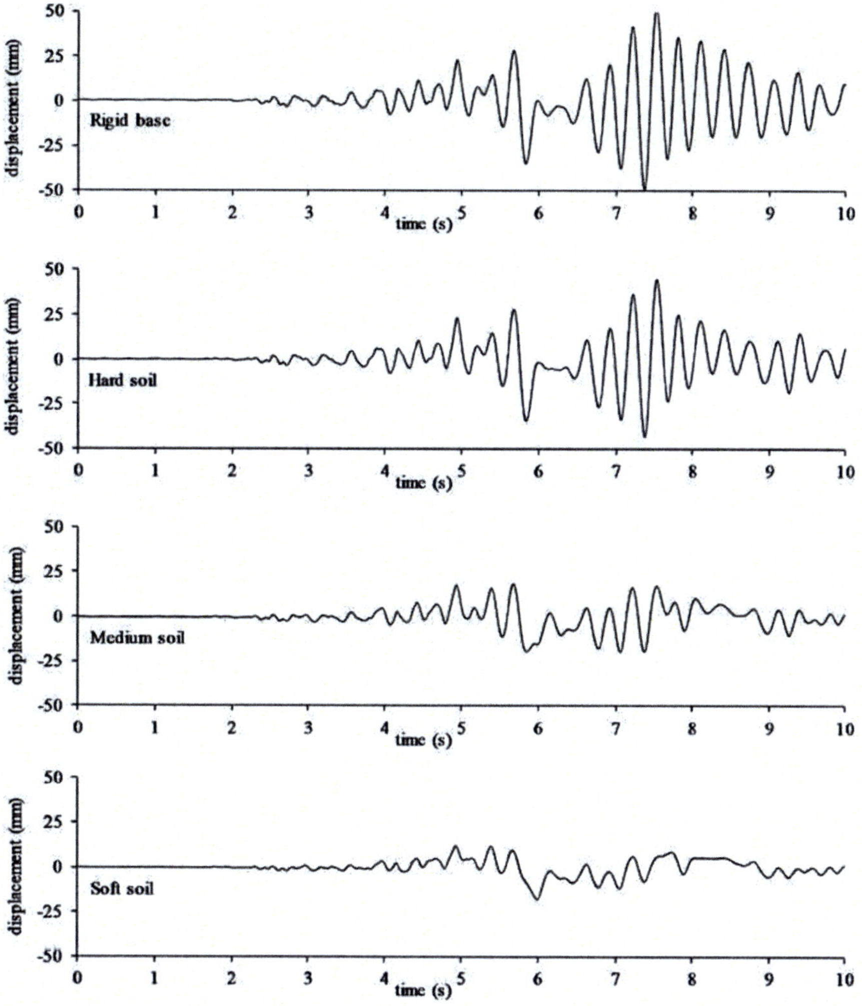

Fig. 5 Time history response of top displacement of tank on various soil types for Imperial Valley earthquake

Loma Prieta earthquake. It can be observed that the pressure is not affected much due to SSI for the tank on hard soil whereas, the pressure decreases due to SSI on medium and soft soil.

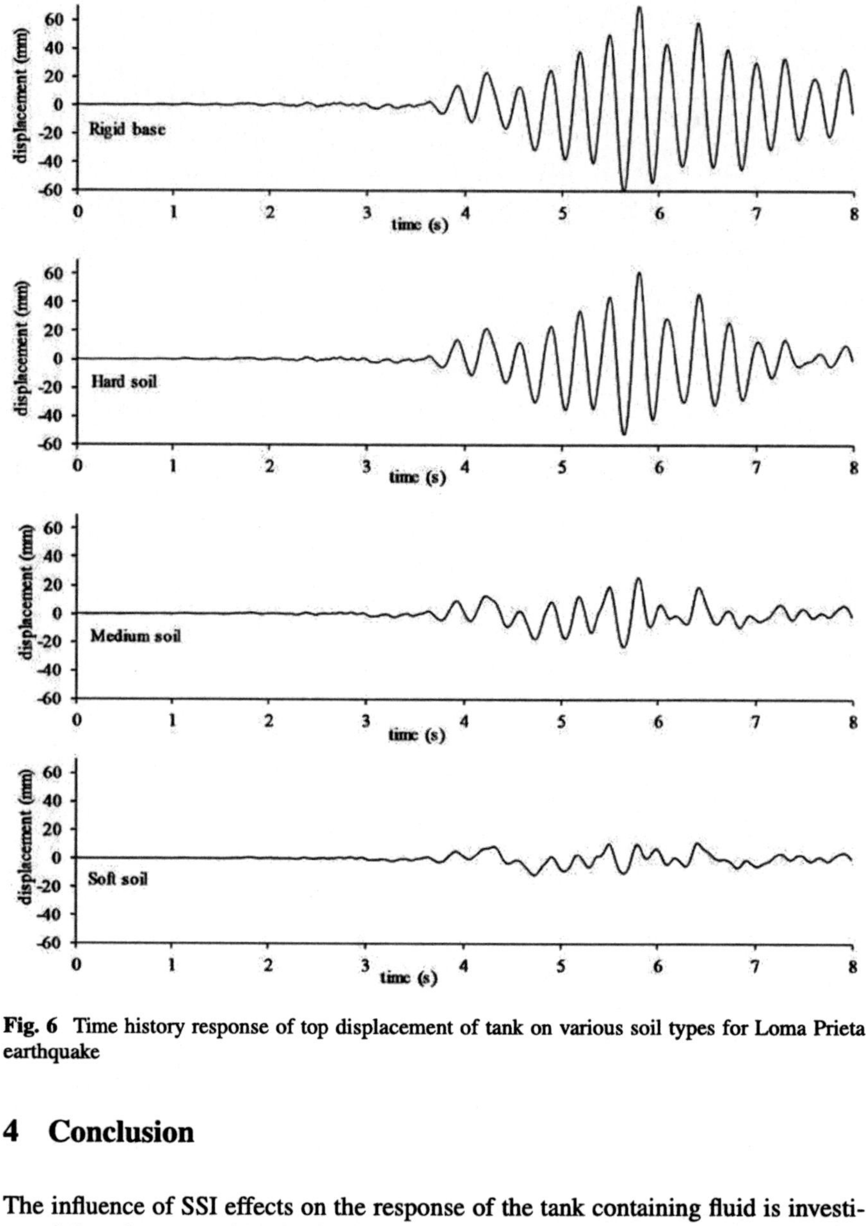

Fig. 6 Time history response of top displacement of tank on various soil types for Loma Prieta earthquake

4 Conclusion

The influence of SSI effects on the response of the tank containing fluid is investigated. In order to consider the flexibility of soil on SSI, three types of soils, viz. soft, medium and hard, are considered for the analysis. From the study, it is concluded that SSI does not have much effect on response of tank resting on hard soil. However, SSI affects the response of the tank and fluid significantly when resting on soft soil. For the two near-fault earthquakes considered for the study, the effect of SSI is observed

Table. 1 Peak response of the tank supported on various soil types

Earthquake	Base shear (kN)					Displacement (mm)				
	Rigid base	Hard soil	Medium soil	Soft soil	Percentage reduction	Rigid base	Hard soil	Medium soil	Soft soil	Percentage reduction
Imperial Valley	741.605	650.973	351.926	350.13	53	51.821	44.653	19.544	17.588	66
Loma Prieta	941.313	823.398	372.652	225	76	70.887	62.221	26.434	11.68	84

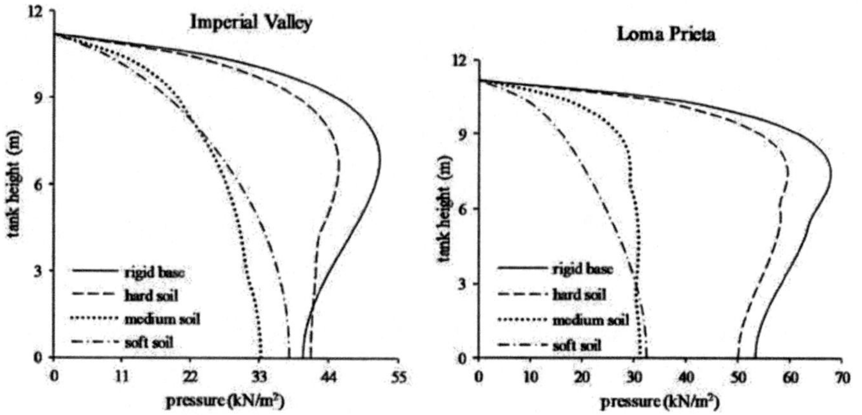

Fig. 7 Pressure variation of fluid on tank wall

to be beneficial for all the responses such as top displacement, base shear and pressure for the tank with fluid. Hence, considering the SSI effect is more important when the tank is resting on soft soil rather than on hard soil.

References

1. Housner GW (1963) The dynamic behavior of water tanks. Bull Seismol Soc Am 53(2):381–387
2. Kim JM, Chang SH, Yun CB (2002) Fluid-structure-soil analysis of cylindrical liquid storage tanks to horizontal earthquake loading. Struct Eng Mech 13(6):615–638
3. Kianoush MR, Ghaemmaghami AR (2011) The effect of earthquake frequency content on the seismic behavior of concrete rectangular liquid tanks using the finite element method incorporating soil–structure interaction. Eng Struct 33:2186–2200
4. Everstine GC (1997) finite element formulations of structural acoustics problems. Comput Struct 65(3):307–321
5. Chaithra M, Krishnamoorthy A, Naurin Nafisa PM (2015) Soil structure interaction analysis of tanks filled with fluid. In Proceeding international conference on computational methods in engineering and health sciences, Malaysia, December 2015
6. Zienkiewicz OC, Bettess P (1978) Fluid-structure dynamic interaction and wave forces. An introduction to numerical treatment. Int J Numer Meth Eng 13(1):1–16

Analysis of Interaction-Dynamics Between Vehicle and Bridge

S. Akhila and Durgaprasad Janjanam

Abstract Past examinations on the deducing the (bridge frequency (B-f) in an indirect way from the dynamic reaction of a moving vehicle were centered fundamentally around the possibility of the procedure. Not many examinations are done to contemplate the impacts of important unique factors of the framework on the vertical reaction of the moving vehicle. To comprehend such a space, the important factors will be analyzed not just for assessing their comparative impact, however for enhancing the likelihood of effectively recognizing the B-f of requirement. Different amplitude ratio (AR's) will be characterized and communicated as far as the important factors, which serve as valuable markers for assessing the likelihood of effective recognition of the B-f from the moving automobile. For the case examined, where it is possible that a couple of B-f to be indicated, it was indicated that the preliminary AR assumes a job that's a higher priority when compared to others in fruitful recognizable proof of the B-f. Recommendations are done for utilization of strategy in the discipline.

Keywords Bridge frequency · Natural frequency · Acceleration amplitude ratio (AAR) · Driving frequency · Amplitude ratio

1 Introduction

For whatever length of time that the estimation of bridge frequency (B-f) is considered, one conventional methodology is to attach various vibration sensors, for example, accelerometers, on suitable areas on the connect brace and to document the dynamic reactions of the bridge under different vibration origin by means of an information obtaining framework. The frequencies of the connection could be differentiated from the documented reaction by a Fourier range (F-R) examination. Such a methodology is eluded as the direct method, because it depends for most

S. Akhila (✉) · D. Janjanam
Department of Civil Engineering, Nitte Meenakshi Institute of Technology, Bengaluru, India
e-mail: akhila3neel@gmail.com

D. Janjanam
e-mail: durgaprasad.j@nmit.ac.in

L. Nandagiri et al. (eds.), *Sustainability Trends and Challenges in Civil Engineering*, Lecture Notes in Civil Engineering 162, https://doi.org/10.1007/978-981-16-2826-9_37

part on the bridge reaction. Even though an extensive scope of dynamic properties shall be observed for the bridge, the organization and support of the vibration sensor and statistics lumberjacks are for the most part expensive, labor-intensive, and time-consuming, despite the fact that remote information transmission is turning out to be well known these days.

As of late, another idea for estimating the B-f alluded to as the indirect method. The principle thought of unintended approach is to deduce B-f from the dynamic reaction of an automobile moving on the bridge, instead of that from the bridge. Passing automobile assumes double job of vibration agitator to the bridge and communication recipient of the bridge reaction.

In order to get a handle on important factors included, the bridge and automobile were considered as a simple beam referring a sprung mass, individually, in the hypothetical investigations. There was no exertion done to consider parameters, for instance, the automobile damping, the underlying dynamic reaction of whichever the automobile or bridge, and so forth. Hypothetically talking, every one of these components can influence the automobile reaction to a few degrees. To upgrade the common applicability of the method, one should understand all the more altogether the general impact of each factor of the V-B (vehicle-bridge) framework as to the likelihood of effective recognition of the B-f unconfined to the principle mode.

The target of the current project is to assess in what way every one of the important factors of the V-B framework might influence its dynamic reaction of a moving automobile. Such data will be valuable for tuning the trial automobile, to lessen the amount of futile preliminary runs during the ground test.

To the finish, the investigative answer for the dynamic reaction of a moving automobile on the bridge shall initially be inferred. At that point, amplitude ratio (AR) shall be characterized and communicated in series of important factors that fill in as accommodating markers for assessing the likelihood of effective recognition of the B-f of concern. A case, where it is possible that a couple of B-f are wanted, shall be concentrated to outline the impact of the factors considered.

2 Literature Survey

The issue of an automobile going on a bridge is generally experienced in the conveyance facilities, for example, high-way spans, railway spans, airplane/runway bridges in air terminals, etc. At the point when an automobile moves above a bridge, convinced effect or dynamic enhancement affect will be originated above the bridge, which should be considered in the design of bridges. Consideration paid to this topic goes back to crafted by Willis [1] and Stokes [2] in the mid-nineteenth hundred. In this viewpoint, most of the inscription has been given to inspection of the bridge vibrations using the supposed (1) moving load, (2) moving mass, and (3) moving sprung mass model for the automobiles.

2.1 Moving Load

As stated by Timoshenko [3], the imperfect balance of the locomotive driving-wheels is the primary origin of impact effect in bridges of lengthy span inferred that forced vibration delivered by need of equalization in trains might be of functional significance.

2.2 Moving Mass

A numerical-analytical tactic is given by Akin and Mofid [4], to describe the action of beams through various limit settings and to transport a momentous mass. Analyzing the response of beams subjected to moving mass requires resolving partial distinguish equations that are generally agreed by abridging assumptions, adequately intricate. As a general rule, these expectations include fixed limit states or neglect the momentous inertia of mass or beam mass. The apathy of a moving mass can be diminished now and then and may be safely ignored. Be that as it may, disregarding the inertia in cases with a huge moving mass may cause an error of 20–80%, depending on parameters such as frame adaptability, mass speed, and uniform components. An investigative numerical technique is implemented that can be used to determine the complex behavior of beams, conveying a moving mass, with different boundary settings. The modification of the normal prevailing equation into a new, solvable sequence of ordinary differential equations is shown in this paper. In addition, the paper shows that the return of structures due to moving mass, which has been routinely overlooked before, must be properly taken into account because it sometimes differs entirely from the moving force model.

2.3 Moving Sprung Mass Models

The dynamic relation between a moving car and the assistant bridge is contemplated [5]. It is asserted that reasonably reliable findings could be achieved by evaluating only the primary mode for bridge and vehicle reactions together. Two frequency classes, i.e., the moving frequency of the vehicle and the N-f of the bridge, administer the displacement, velocity rate, and velocity of the bridge at different numbers. Then again, three distinct frequencies that occur as driving frequencies, automotive frequency, and B-f reflect the automotive reactions. All three frequency sets can be identified from the vehicle reaction, directed at the situation in which 0 damping and even roadway exterior are presumed for the bridge. In the off case that the results of damping and intermittent paving are taken into consideration, the dominant B-f can still be identified from the automotive response. Additional hypothetical and forensic analysis both are necessary in this respect.

3 Analytical Investigation

3.1 Construction of the Theory

For current purposes, consider the V-B model, as seen in Fig. 1. The automobile is modeled as a m_v of lumped mass in kg, backed by a spring of stiffness k_v in N/m and scramble pot coefficient of damping c_v in N s/m, which passes through a clear beam of length L in m at speed "v" in m/s. In Fig. 1, as mentioned elsewhere, the easy beam showed up would be considered as the system for the necessary equations to be inferred. With the ultimate objective of speculation, the vast majority of well-thought-out dynamic features would be expressed in the wording of dimensionless variables. The bar is thought to be a B-E (Bernoulli Euler) type with consistent *c/s* and even roadway.

The trial automobile is displayed as a SDOF sprung mass. The equation of motion as for the trial automobile can be composed as [4, 6]

$$m_v\ddot{q}_v + c_v\dot{q}_v + k_v(q_v - u_b|_{x=vt}) = 0 \tag{1}$$

where q_v and u_b (x, t) mean, respectively, the vertical displacements of the automobile and the beam.

Prior to entering the bridge, the vehicle is expected to vibrate with steady amplitude A_{v0}. Thusly, the preliminary state of the automobile can be composed as [6]

$$q_v(0) = A_{v_0} \tag{2}$$

$$q_v(0) = 0 \tag{3}$$

We expect the beam to be programmed to vibrate by various means prior to the trial car's arrival above the bridge. The movement of the bridge can scarcely be

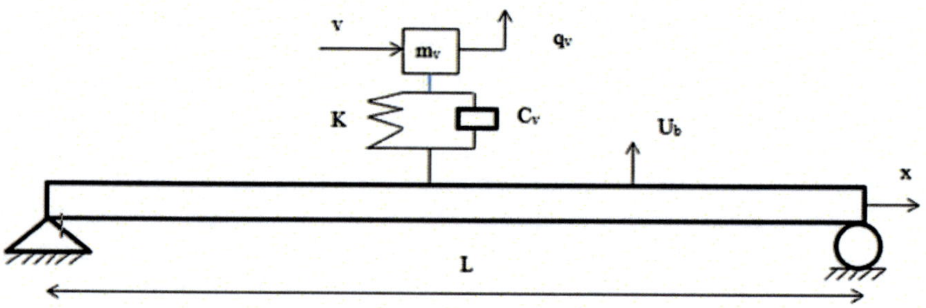

Fig. 1 Automobile modeled as a sprung mass moving through a simple beam

affected by the moving vehicle in a situation where the mass of the automobile is considerably smaller than that of the bridge. In addition, because of the moderately limited operating time of the trial vehicle, the V-B interface is negligible. Along these lines, the vertical motion of the bridge can be sensibly thought to be harmonic motion with steady frequencies and amplitudes across the automobile's time frame, given as follows [6],

$$u_b(x,t) = \sum_n A_n \sin\frac{n\pi x}{L} \sin\omega_{b,n} t \tag{4}$$

where the nth modular B-f and its equivalent amplitude are meant by $\omega_{b,n}$ and A_n, and the word $\sin(n\pi x/L)$ means the nth modular bridge form.

Substituting Eq. (4) $u_b(x, t)$ into Eq. (1)

$$m_v\ddot{q}_v + c_v\dot{q}_v + k_v(q_v - \sum_n A_n \sin\left(\frac{n\pi L}{L}\right) \sin\omega_{b,n} t = 0 \tag{5}$$

$$\ddot{q}_v + 2m_v\omega_v\xi_v\dot{q}_v + \omega_v^2 q_v = \frac{\omega_{v,n}^2}{2}\sum_n A_n \cos\omega_{b,n} t\left(1 - \frac{n\pi L}{\omega_{b,n} t}\right)$$
$$- \cos\omega_{b,n} t\left(1 + \frac{n\pi L}{\omega_{b,n} t}\right) \tag{6}$$

The parameters $\omega_{bl,n}$ and $\omega_{br,n}$, individually, are the L-R-shifted frequencies of the nth mode of the bridge[5], characterized as

$$\omega_{bl,n} = \omega_{b,n}(1 - S_n) \tag{7}$$

$$\omega_{br,n} = \omega_{b,n}(1 + S_n) \tag{8}$$

where S_n is a dimensionless speed parameter [5], characterized as the ratio of n occurrences the driving frequency $\pi v/L$ of automobile to the nth frequency $\omega_{b,n}$ of the bridge [6],

$$S_n = \frac{n\pi v/L}{\omega_{b,n}}, \quad \text{where } v = \frac{x}{t} \tag{9}$$

$$\ddot{q}_v + \ddot{2}\xi_v\dot{q}_v + w_v^2 q_v = \frac{\omega_v^2}{2}\sum_n A_n(\cos\omega_{bl,n} t - \cos\omega_{br,n} t) \tag{10}$$

Remarkable from Eq. (10) is that the automobile's dynamic response can be seen as remaining entirely energized by the progression of bridge vibrations with L-R-shifted nth mode frequencies. The reply to Eq. (10), the homogeneous structure, q_{vh}, and separate arrangements, q_{vp1} and q_{vp2}, can be communicated as a whole, i.e.,

$$q_v = q_{vh} + q_{vp1} + q_{vp2} \tag{11}$$

Here, q_{vp1} and q_{vp2} are the particular answers, each relating to L-R-shifted bridge vibration notions. Every one of them needs to fulfill a nonhomogeneous differential condition considered as shown below [7]

$$\ddot{q}_{vp1} + 2\xi_v\omega_v\dot{q}_{vp1} + \omega_v^2 q_{vp1} = \frac{w_v^2}{2}\sum_n A_n \cos\omega_{bl,n}\, t \tag{12}$$

$$\ddot{q}_{vp2} + 2\xi_v\omega_v\dot{q}_{vp2} + \omega_v^2 q_{vp2} = \frac{-w_v^2}{2}\sum_n A_n \cos\omega_{br,n}\, t \tag{13}$$

Following the traditional strategy, the homogeneous solution q_{vh} to Eq. (10) can be gotten as because the system is under damped the complimentary solution is [7]

$$q_{vh} = (c_1 \sin\omega_{vd}t + c_2\cos\omega_{vd}t)e^{-\xi v\omega_v t} \tag{14}$$

where ω_{vd} is the damped frequency of the automobile, interpreted as

$$\omega_{vd} = \omega_v\sqrt{1-\xi_v^2} \tag{15}$$

and c_1 and c_2 are the coefficient to be ascertained from the preliminary states. At the same time, the particular answers can be acquired as

Consider Eq. (12)

$$\ddot{q}_{vp1} + 2\xi_v\omega_v\dot{q}_{vp1} + \omega_v^2 q_{vp1} = \frac{w_v^2}{2}\sum_n A_n \cos\omega_{bl,n}\, t \tag{16}$$

The particular solution is obtained as [7]

$$\begin{aligned} q_{vp1} = \sum_n & \frac{A_n/2}{\left[\left(1-\mu_{l,n}^2\right)^2 + \left(2\xi_v\mu_{l,n}\right)^2\right]} \\ & * \left[\left(2\xi_v\mu_{l,n}\right)\sin\omega_{bl,n}t + \left(1-\mu_{l,n}^2\right)\cos\omega_{bl,n}t\right] \end{aligned} \tag{17}$$

where $\mu_{l,n}$ is a dimensionless parameter characterized as the nth mode to vehicle frequency (V-f) ratios of L-S B-f; i.e., similarly, for Eq. (13)

$$\begin{aligned} q_{vp2} = \sum_n & \frac{-A_n/2}{\left[\left(1-\mu_{r,n}^2\right)^2 + \left(2\xi_v\mu_{r,n}\right)^2\right]} \\ & * \left[\left(2\xi_v\mu_{r,n}\right)\sin\omega_{br,n}t + \left(1-\mu_{r,n}^2\right)\cos\omega_{br,n}t\right] \end{aligned} \tag{18}$$

where $\mu_{r,n}$ are dimensionless parameters characterized as the nth mode to V-f ratios of R-S B-f; i.e.,

$$\mu_{r,n} = \frac{\omega_{br,n}}{\omega_v} \tag{19}$$

the above-mentioned two parameters $\mu_{l,n}$ and $\mu_{r,n}$ can be associated to μ_n as [6]

$$\mu_n = \frac{\omega_{b,n}}{\omega_v} \tag{20}$$

$$\mu_{l,n} = \mu_n(1 - S_n) \tag{21}$$

$$\mu_{r,n} = \mu_n(1 + S_n) \tag{22}$$

$$c_1 = \frac{A_{v_0}\xi_v}{\sqrt{1-\xi_v^2}} \times \left\{1 - \sum_i \frac{1}{2R_i}\left[\frac{1+\mu_{l,i}^2}{(1-\mu_{l,i}^2)^2 + (2\xi_v\mu_{l,i})^2} - \frac{1+\mu_{r,i}^2}{(1-\mu_{r,i}^2)^2 + (2\xi_v\mu_{r,i})^2}\right]\right\} \tag{23}$$

$$c_2 = A_{v0} \times \left\{1 - \sum_j \frac{1}{2R_j}\left[\frac{1+\mu_{l,j}^2}{\left(1-\mu_{l,j}^2\right)^2 + (2\xi_v\mu_{l,j})^2} - \frac{1-\mu_{r,j}^2}{\left(1-\mu_{r,j}^2\right)^2 + (2\xi_v\mu_{r,j})^2}\right]\right\} \tag{24}$$

$$q_v = (c_1 \sin\omega_{vd}t + c_2\cos\omega_{vd}t)e^{-\xi_v\omega_v t} + \sum_n (d_n \sin\omega_{bl,n}t + e_n\cos\omega_{bl,n}t + f_n\sin\omega_{br,n}t + g_n\cos\omega_{br,n}t) \tag{25}$$

where c_1 and c_2, the coefficients of the terms $\sin\omega_{vd}t$ and $\cos\omega_{vd}t$, d_n, e_n, f_n, and g_n, the coefficient of the terms $\sin\omega_{bl,n}t$, $\cos\omega_{bl,n}t$, $\sin\omega_{br,n}t$, and $\cos\omega_{br,n}t$, can be acquired by evaluation with Eqs. (17) and (18); i.e.,

$$d_n = \frac{A_n}{2}\frac{2\xi_v\mu_{l,n}}{\left[(1-\mu_{l,n}^2)^2 + (2\xi_v\mu_{l,n})^2\right]} \tag{26}$$

$$e_n = \frac{A_n}{2}\frac{1-\mu_{l,n}^2}{\left[(1-\mu_{l,n}^2)^2 + (2\xi_v\mu_{l,n})^2\right]} \tag{27}$$

$$f_n = \frac{A_n}{2} \frac{-2\xi_v \mu_{r,n}}{[(1-\mu_{r,n}^2)^2 + (2\xi_v \mu_{r,n})^2]} \tag{28}$$

$$g_n = \frac{A_n}{2} \frac{-(1-\mu_{r,n}^2)}{[(1-\mu_{r,n}^2)^2 + (2\xi_v \mu_{r,n})^2]} \tag{29}$$

Take in that the vertical displacement of automobile in Eq. (25) is overwhelmed by three noteworthy gatherings of frequencies. Two gatherings remain identified with the normal B-f, i.e., the L-R-shifted B-f of the *n*th mode, $\omega_{bl,n}$ and $\omega_{br,n}$, and the third gathering is identified with characteristic frequency of the automobile, i.e., the damped frequency ω_{vd} of the automobile. Precisely, the L-R-shifted B-f are the B-f (ω_n) prevailing in the dynamic reactions of the transient automobile yet decreased or amplified by proportions $(1 - S_n)$ and $(1 + S_n)$, separately, as showed in Eqs. (7) and (8). Genuinely, the vibration of the bridge will be conducted to the voyaging automobile with its frequencies moved by specific sums, as demonstrated by the relationships with L-R-moved B-f, because of the coupling among the automobile and bridge. The additional perception is that the parameter associated with ω_{vd} of the automobile will decay because of essence of the automobile damping.

In handy relevance, it is simpler to quantify an acceleration reaction, as opposed to the displacement reaction, of the voyaging automobile [5]. To the finish, the perpendicular quickening reaction of the automobile shall be gotten by differentiate the displacement reaction in Eq. (25) w.r.t "t" twice, [6]

$$\begin{aligned}\ddot{q}_v = {} & \left(-c_1\omega_{vd}^2 \sin \omega_{vd}t - c_2\omega_{vd}^2 \cos \omega_{vd}t\right)e^{-\xi_v\omega_v t} + (c_1\omega_{vd} \cos \omega_{vd}t - c_2\omega_{vd} \sin \omega_{vd}t) \\ & e^{-\xi_v\omega_v t} * \left(\xi_v\omega_v\right) + (c_1\omega_{vd} \cos \omega_{vd}t - c_2\omega_{vd} \sin \omega_{vd}t)e^{-\xi_v\omega_v t}(-\xi_v\omega_v) \\ & + (c_1 \sin \omega_{vd}t + c_2 \cos \omega_{vd}t)e^{-\xi_v\omega_v t}\left(-\xi_v^2\omega_v^2\right) \\ & + \sum_n (d_n(-\omega_{bl,n}^2 \cos \omega_{bl,n}t - e_n\omega_{bl,n}^2 \sin \omega_{bl,n}t - f_n\omega_{br,n}^2 \cos \omega_{br,n}t \\ & - g_n\omega_{br,n}^2 \sin \omega_{br,n}t)\end{aligned} \tag{30}$$

where $\tilde{\tilde{c}}_1$ and $\tilde{\tilde{c}}_2$, the coefficients of the notion sin ω_vd*t* and cos ω_vd*t*, can be related to c_1 and c_2, respectively, as follows:

$$\tilde{\tilde{c}}_1 = \xi_v^2\omega_v^2 c_1 - \omega_{vd}^2 c_1 + 2\xi_v\omega_v\omega_{vd}c_2 \tag{31}$$

$$\tilde{\tilde{c}}_2 = \xi_v^2\omega_v^2 c_2 - \omega_{vd}^2 c_2 - 2\xi_v\omega_v\omega_{vd}c_1 \tag{32}$$

On the other hand, the coefficients of the terms sin $\omega_{bl,n}t$, cos $\omega_{bl,n}t$, sin $\omega_{br,n}t$, and cos $\omega_{br,n}t$ terms, i.e., $\tilde{\tilde{d}}_n$, $\tilde{\tilde{e}}_n$, $\tilde{\tilde{f}}_n$, $\tilde{\tilde{g}}_n$ can be associated to d_n, e_n, f_n, and g_n, respectively, as follows [6]:

$$\tilde{\tilde{d}}_n = -\omega_{bl,n}^2 d_n \tag{33}$$

$$\tilde{\tilde{e}}_n = -\omega_{bl,n}^2 e_n \tag{34}$$

$$\tilde{\tilde{f}}_n = -\omega_{br,n}^2 f_n \tag{35}$$

$$\tilde{\tilde{g}}_n = -\omega_{br,n}^2 g_n \tag{36}$$

$$\ddot{q}_v = (\tilde{\tilde{c}}_1 \sin \omega_{vd} t + \tilde{\tilde{c}}_2 \cos \omega_{vd} t) e^{-\xi_v \omega_v t} + \sum_n (\tilde{\tilde{d}}_n \sin \omega_{bl,n} t + \tilde{\tilde{e}}_n \cos \omega_{bl,n} t$$
$$+ \tilde{\tilde{f}}_n \sin \omega_{br,n} t + \tilde{\tilde{g}}_n \cos \omega_{br,n} t) \tag{37}$$

Once more, from Eq. (37), it very well may be seen that the acceleration reaction of the voyaging automobile is ruled by a similar three gatherings of frequencies, i.e., $\omega_{bl,n}$ and $\omega_{br,n}$, and ω_{vd}. As well, the acceleration reaction related with ω_{vd} of the automobile also degenerates in an exponential way because of the automobile damping.

Meanwhile a large portion of the reactions concerned of the acceleration kind, the term "acceleration amplitude" will be utilized when an inconstant is recently presented, yet might be alluded to simply as "amplitude" for brusqueness a short time later.

3.2 *Amplitude Ratios in Fourier Transform*

Planned for looking for an ideal condition for extricating the B-f, the AR's among different frequencies would be examined in this segment. Amplitudes comparing $\omega_{bl,n}$ and $\omega_{br,n}$, individually, of the bridge can also be determined by the acceleration reaction of the automobile in Eq. (37) as

$$\widetilde{\widetilde{A}}_{bl,n} = \sqrt{\tilde{\tilde{d}}_n^2 + \tilde{\tilde{e}}_n^2} = \frac{A_n \omega_{bl,n}^2}{2\sqrt{\left(1-\mu_{l,n}^2\right)^2 + \left(2\xi_v \mu_{l,n}\right)^2}} \tag{38}$$

$$\widetilde{\widetilde{A}}_{br,n} = \sqrt{\tilde{\tilde{f}}_n^2 + \tilde{\tilde{g}}_n^2} = A_n \frac{\omega_{br,n}^2}{2\sqrt{\left(1-\mu_{r,n}^2\right)^2 + \left(2\xi_v \mu_{r,n}\right)^2}} \tag{39}$$

As this can be seen, the amplitudes $\widetilde{\widetilde{A}}_{bl,n}$ and $\widetilde{\widetilde{A}}_{br,n}$ of the altered B-f are equivalent to the preliminary amplitude A_n of B-f of the nth approach incremented by the proportion notions on right side of Eqs. (38) and (39).

Accordingly, the amplitude related with ω_{vd} can be determined from the automobile acceleration in Eq. (37) alongside the meaning of ω_{vd} in Eq. (15) as

$$\widetilde{\widetilde{A_{vd}}} = \sqrt{\widetilde{\widetilde{c_1^2}} + \widetilde{\widetilde{c_2^2}}} \tag{40}$$

So as to effectively deduce the B-f $\omega_{b,n}$, it is ideal that the amplitudes $\widetilde{\widetilde{A_{vd}}}$ related with the automobile be kept as little as could be expected, while $\widetilde{\widetilde{A_{bl,n}}}, \widetilde{\widetilde{A_{br,n}}}$ related with B-f ought to be made as big as could reasonably be expected. The likelihood of effectively extracting the bridge associated frequencies relies principally upon the deceivability of the pinnacles related thru these frequencies in an F-R, the overall amplitude of the focused frequency as for the others in the span [6].

$$\widetilde{\widetilde{A_{vd}}} = \omega_v^2 \sqrt{c_1^2 + c_2^2} \tag{41}$$

In the initial place, considering the AR of the L-S (left shifted) B-f of the nth mode to R-S (right shifted) one, $\widetilde{\widetilde{A_{bl,n}}}/\widetilde{\widetilde{A_{br,n}}}$ by the articulations for $\widetilde{\widetilde{A_{bl,n}}}$ and $\widetilde{\widetilde{A_{br,n}}}$ in Eqs. (38) and (39) and the denotations of $\omega_{bl,n}$ and $\omega_{br,n}$ in Eqs. (7) and (8), it very well may be inferred that

$$\frac{\widetilde{\widetilde{A_{bl,n}}}}{\widetilde{\widetilde{A_{br,n}}}} = \left(\frac{1 - S_n}{1 + S_n}\right)^2 \sqrt{\frac{\left[1 - \mu_n^2(1 + S_n)\right]^2 + \left[2\xi_v \mu_v (1 + S_n)\right]^2}{\left[1 - \mu_n^2(1 - S_n)\right]^2 + \left[2\xi_v \mu_v (1 - S_n)\right]^2}} \tag{42}$$

Figure 2 shows the contour of the AR $\widetilde{\widetilde{A_{bl,n}}}/\widetilde{\widetilde{A_{br,n}}}$ with respect to the S_n and nth μ_n with $\xi_v = 10\%$ [8]. The contour line set apart as 1 is important since it speaks to the limit between the two regions with $\widetilde{\widetilde{A_{bl,n}}}$ more prominent or smaller than $\widetilde{\widetilde{A_{br,n}}}$.

Secondly, considering the AR α_n of the top acceleration of the damped automobile frequency to the big one of the top related with the nth bridge shifted frequencies [6],

$$\alpha_n = \frac{\widetilde{\widetilde{A_{vd}}}}{\max\left(\widetilde{\widetilde{A_{bl,n}}}, \widetilde{\widetilde{A_{br,n}}}\right)} \tag{43}$$

$$\alpha_n = \frac{\omega_v^2 2\sqrt{\left(1 - \mu_{l,n}^2\right)^2 + \left(2\xi_v \mu_{l,n}\right)^2}}{A_n \omega_{bl,n}^2}$$

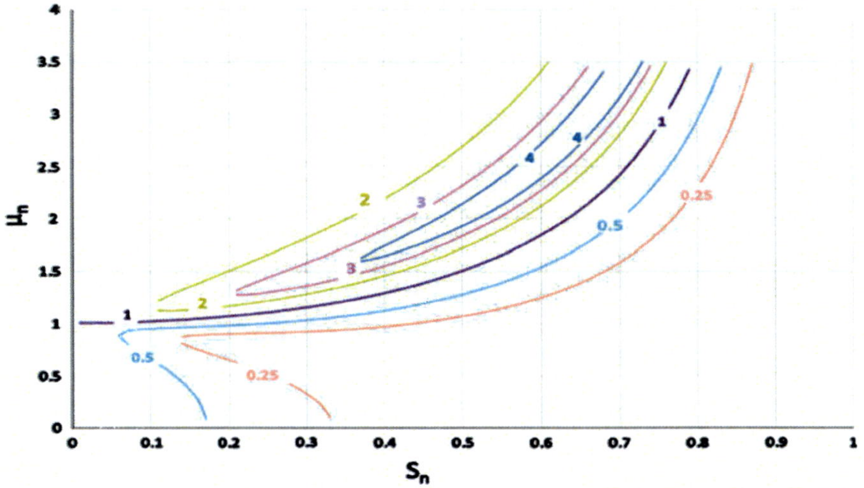

Fig. 2 Graph for the AR $\widetilde{\widetilde{A_{bl,n}}}=\widetilde{\widetilde{A_{br,n}}}$ versus S_n and nth bridge/vehicle frequency ratio μ_n

$$* \sqrt{\begin{aligned}&\left(\frac{A_{v0}\xi_v}{\sqrt{1-\xi_v^2}}\{1-\sum_i \frac{1}{2\tilde{\tilde{R}}_i}[\frac{1+\mu_{l,i}^2}{(1-\mu_{l,i}^2)^2+(2\xi_v\mu_{l,i})^2}-\frac{1+\mu_{r,i}^2}{(1-\mu_{r,i}^2)^2+(2\xi_v\mu_{r,i})^2}]\}\right)^2\\&+\left(A_{v0}\{1-\sum_j \frac{1}{2\tilde{\tilde{R}}_j}[\frac{1+\mu_{l,j}^2}{(1-\mu_{l,j}^2)^2+(2\xi_v\mu_{l,j})^2}-\frac{1-\mu_{r,j}^2}{(1-\mu_{r,j}^2)^2+(2\xi_v\mu_{r,j})^2}]\}\right)^2\end{aligned}} \tag{44}$$

$$\alpha_n = \frac{\tilde{\tilde{R}}_n 2\sqrt{(1-\mu_{l,n}^2)^2+(2\xi_v\mu_{l,n})^2}}{(1-S_n)}$$

$$* \sqrt{\begin{aligned}&\left(\frac{\xi_v}{\sqrt{1-\xi_v^2}}\{1-\sum_i \frac{1}{2\tilde{\tilde{R}}_i}[\frac{1+\mu_{l,i}^2}{(1-\mu_{l,i}^2)^2+(2\xi_v\mu_{l,i})^2}-\frac{1+\mu_{r,i}^2}{(1-\mu_{r,i}^2)^2+(2\xi_v\mu_{r,i})^2}]\}\right)^2\\&+\left(\{1-\sum_j \frac{1}{2\tilde{\tilde{R}}_j}[\frac{1+\mu_{l,j}^2}{(1-\mu_{l,j}^2)^2+(2\xi_v\mu_{l,j})^2}-\frac{1-\mu_{r,j}^2}{(1-\mu_{r,j}^2)^2+(2\xi_v\mu_{r,j})^2}]\}\right)^2\end{aligned}} \tag{45}$$

The AR α_n offers a significant record for assessing the likelihood of effectively recognizing the bridge interrelated frequencies in F-R of acceleration reaction of a moving automobile. The greater the α_n is, the lower the likelihood of effective recognition identification of the B-f.

This could be effortlessly comprehended, since a pinnacle that is excessively little analogized with the neighboring pinnacle might be covered up by the others when they're excessively near, or even adulterated by the commotions and rendered undetectable.

$$\widetilde{\widetilde{R}}_n = \frac{\omega_v^2 A_{v_0}}{\omega_{b,n}^2 \, A_{b,n}} \tag{46}$$

the AR α_n for the V/B frequency could be inferred as

$$\alpha_n = 2\sqrt{\alpha_{c1n}^2 + +\alpha_{c2,n}^2} \tag{47}$$

where the terms $\alpha_{c1,n}$ and $\alpha_{c2,n}$ are given first box [6]

$$\alpha_{c1,n} = \left\{ \begin{array}{l} \widetilde{\widetilde{R}}_n \frac{2\sqrt{\left(1-\mu_{l,n}^2\right)^2+\left(2\xi_v\mu_{l,n}\right)^2}}{(1-S_n)^2} \\ * \frac{\xi_v}{\sqrt{1-\xi_v^2}} \left\{ 1 - \sum_i \frac{1}{2\tilde{R}_i} \left[\frac{1+\mu_{l,i}^2}{\left(1-\mu_{l,i}^2\right)^2+\left(2\xi_v\mu_{l,i}\right)^2} - \frac{1+\mu_{r,i}^2}{\left(1-\mu_{r,i}^2\right)^2+\left(2\xi_v\mu_{r,i}\right)^2} \right] \right\} \\ \text{if } \widetilde{\widetilde{A_{bl,n}}} \geq \widetilde{\widetilde{A_{br,n}}} \\ \widetilde{\widetilde{R}}_n \frac{2\sqrt{\left(1-\mu_{r,n}^2\right)^2+\left(2\xi_v\mu_{r,n}\right)^2}}{(1-S_n)^2} \\ * \frac{\xi_v}{\sqrt{1-\xi_v^2}} \left\{ 1 - \sum_i \frac{1}{2\tilde{R}_i} \left[\frac{1+\mu_{l,i}^2}{\left(1-\mu_{l,i}^2\right)^2+\left(2\xi_v\mu_{l,i}\right)^2} - \frac{1+\mu_{r,i}^2}{\left(1-\mu_{r,i}^2\right)^2+\left(2\xi_v\mu_{r,i}\right)^2} \right] \right\} \\ \text{if } \widetilde{\widetilde{A_{bl,n}}} \leq \widetilde{\widetilde{A_{br,n}}} \end{array} \right\} \tag{48}$$

$$\alpha_{c2,n} = \left\{ \begin{array}{l} R_n \frac{2\sqrt{\left(1-\mu_{l,n}^2\right)^2+\left(2\xi_v\mu_{l,n}\right)^2}}{(1-S_n)^2} \\ * \left\{ 1 - \sum_i \frac{1}{2\tilde{R}_i} \left[\frac{1-\mu_{l,i}^2}{\left(1-\mu_{l,i}^2\right)^2+\left(2\xi_v\mu_{l,i}\right)^2} - \frac{1-\mu_{r,i}^2}{\left(1-\mu_{r,i}^2\right)^2+\left(2\xi_v\mu_{r,i}\right)^2} \right] \right\} \\ \text{if } \widehat{A_{bl,n}} \geq A_{br,n} \\ R_n \frac{2\sqrt{\left(1-\mu_{r,n}^2\right)^2+\left(2\xi_v\mu_{r,n}\right)^2}}{(1-S_n)^2} \\ * \left\{ 1 - \sum_i \frac{1}{2\tilde{R}_i} \left[\frac{1+\mu_{l,i}^2}{\left(1-\mu_{l,i}^2\right)^2+\left(2\xi_v\mu_{l,i}\right)^2} - \frac{1-*\mu_{r,i}^2}{\left(1-\mu_{r,i}^2\right)^2+\left(2\xi_v\mu_{r,i}\right)^2} \right] \right\} \\ \text{if } \overbrace{\widehat{A_{bl,n}}} \leq \overline{A_{br,n}} \end{array} \right\} \tag{49}$$

As it can be seen, the two term $\alpha_{c1,n}$ and $\alpha_{c2,n}$ rely upon the ξ_v of the automobile, $\widetilde{\widetilde{R_{i(j)}}}$, S_n, $\mu_{l,i(j)}$ and $\mu_{r,i(j)}$. Since the in-between parameters $\mu_{l,i(j)}$ and $\mu_{r,i(j)}$. as mentioned in Eqs. (21) and (22) are associates of $S_{i(j)}$ and $\mu_{i(j)}$ the two terms $\alpha_{c1,n}$ and α_{c2}, rely completely upon the following four parameters: ξ_v, $\widetilde{\widetilde{R_{i(j)}}}$, $S_{i(j)}$, $\mu_{i(j)}$. Accordingly, the α_n, that is a function of $\alpha_{c1,n}$ and $\alpha_{c2,n}$, can be completely controlled by these four parameters: ξ_v, $\widetilde{\widetilde{R_{i(j)}}}$, $S_{i(j)}$, $\mu_{i(j)}$.

The directly above figuring demonstrates that a colossal quantity of components may influence α_n for the V/B frequency. Unmistakably, the deducing of B-f from the automobile reaction is a muddled physical issue. To examine the crucial parameters included, we will study underneath a special case, apiece consisting of a couple of B-f of interest.

3.3 Case I: One Frequency of the Bridge Involved

Considering the case where a particular basis pushes the bridge and sets it in motion for a solitary vibration mode. Let us agree that the various modes are not invigorated, with the exception of this vibration mode, or animated to a neck and neck so low that they can be forgotten for all intents and purposes. Meanwhile, in repetition, the bridge's preliminary mode is almost definitely to be stimulated; here, we intend to take the preliminary mode as the one to demonstrate the interrelated results. The scene is like that. When the test science reaches the bridge with a movement of ω_v and A_{v0}, the bridge is formed in the motion of the preliminary mode with $\omega_{b,1}$ and A_1.

Acceleration of the automobile's input by its acceleration on the bridge, as seen from Eq. 37 with $n = 1$, will be dominated by three important frequencies will be regulated by by $n = 1$: L-R-shifted B-f of the preliminary mode, $\omega_{bl\,n}$ and $\omega_{br,n}$ and ω_{vd}.

Aimed at the current case, α_1, can be acquired from Eq. (47) where the terms $\alpha_{c1,1}$ and $\alpha_{c2,1}$, shall be gotten from the conditions $n = i = j = 1$.

As indicated by the earlier discussion, α_1 is reliant on the following three non-dependent inconstant: $\widetilde{\widetilde{R}}_1$, S_1, and μ_1, that is, α_1 ($\widetilde{\widetilde{R}}_1$, S_1, μ_1) the ξ_v is set as 10% and is accordingly it is not a non-dependent inconstant.

Figure 3 previews the AR α_1 regarding the speed boundary S_1 and μ_1 for the three beginning $\widetilde{\widetilde{R}}_1$= 2, 5, and 10, alongside the contours contrived in Fig. 4a–c. It is understood that the patterns of the AR α_1 are comparative regardless if the beginning $\widetilde{\widetilde{R}}_1$ is little ($\widetilde{\widetilde{R}}_1$=2) or huge ($\widetilde{\widetilde{R}}_1$=10). All in all, the acceleration amplitude ratio (AAR) α_1 declines steadily as the f-R μ_1 increments from 0 to 1 and arrives at its lowest around the curve $\mu_1 = 1$. In any case, α_1 increments quickly as μ_1 increments from 1 and thereafter, with the exception of within the V-shape span focused by a conspicuous bend.

Subsequently, a lower value of the α_1 infers a superior resolute for the B-f; it is envisioned that the B-f could be all the additional effortlessly deduced if the μ_1 happens to be situated in the < 1 or if the related boundaries occur to be found within the V-shape range focused by the conspicuous bend for $\mu_1 > 1$.

Such a perception additionally clarifies the motivation behind where for the B-f of complex modes cannot be deduced in a path as simple as that for the principal mode due to the comparatively better numbers of μ_n for frequencies of complex modes.

From a pragmatic perspective, there is a complex likelihood of accomplishment in recognizing the B-f when α_1 is decreased, and ideally once it is lower than the preliminary $\widetilde{\widetilde{R}}_1$. To the extent the reduction of α_1 is concerned, the inconstant $\mu_{1,P}$ (i.e., B/V f-R) is an unmanageable obscure by and by because of the factual that

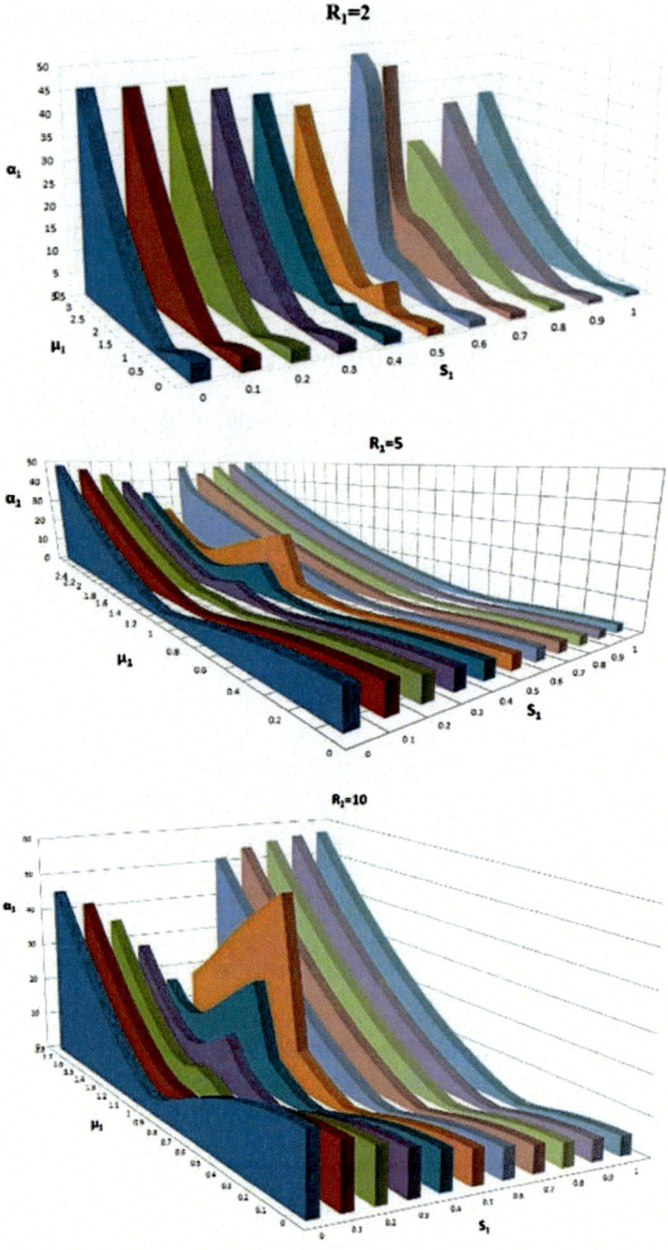

Fig. 3 α_1 versus S_1 and μ_1 (note $\widetilde{\widetilde{\widetilde{R}}}_1 = R_1$)

(1) both the BVF are constant and can scarcely be altered; (2) the B-f are obscure previously the evaluation is directed.

On the way to manage the unmanageable character of μ_1, we will split its span into N_p parts, supposing equivalent likelihood of event for each parts and signify the value of the μ_1 on the pth part by $\mu_{1,p}$. Likewise, the likelihood of diminishing the α_1 is [6]

$$P_{(\alpha_{1\leq}R_1)}\left(\widetilde{\widetilde{R}}_1, S_1\right) = \frac{\sum_{p=1}^{N_P} \phi_P(\widetilde{\widetilde{R}}_1, \mu_{1,p}, S_1)}{N_P} \tag{50}$$

where

$$\phi_p(\widetilde{\widetilde{R}}_1, \mu_{1,p}, S_1) = \begin{cases} 1, & \text{if}\, \alpha_1(\widetilde{\widetilde{R}}_1, \mu_{1,p}, S_1) \leq \widetilde{\widetilde{R}}_1 \\ 0, & \text{if}\, \alpha_1(\widetilde{\widetilde{R}}_1, \mu_{1,p}, S_1) > \widetilde{\widetilde{R}}_1 \end{cases} \tag{51}$$

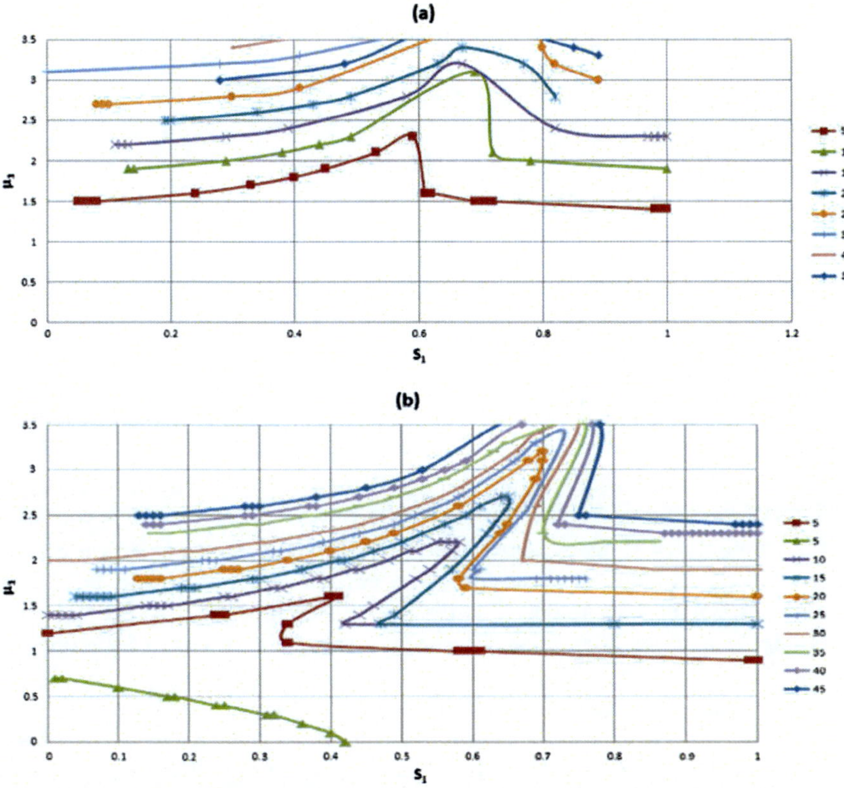

Fig. 4 **a** Graph for α_1 with respect to S_1 and μ_1: $\widetilde{\widetilde{R}}_1 = 2$. **b** Graph for α_1 with respect to S_1 and μ_1: $\widetilde{\widetilde{R}}_1 = 5$. **c** Graph for α_1 with respect to S_1 and μ_1: $\widetilde{\widetilde{R}}_1 = 10$

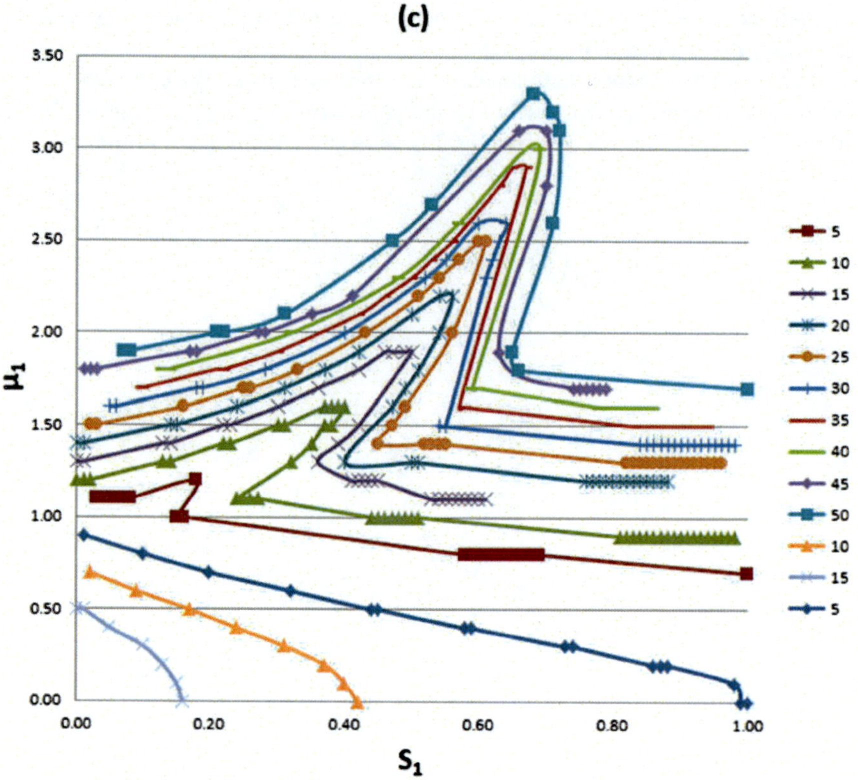

Fig. 4 (continued)

The likelihood of decreasing α_1 has been contrived in Fig. 5 concerning S_1 for preliminary $\widetilde{\widetilde{R_1}} = 2$, 5, and 10. As can be seen, the likelihood of decreasing α_1 gets high as S_1 increments from nil to a specific value (e.g., 0.42 for the bend $\widetilde{\widetilde{R_1}} = 2$, 0.3 for the bend $\widetilde{\widetilde{R_1}}= 5$, and 0.25 for the bend $\widetilde{\widetilde{R_1}}=10$) and afterward remains constant for additional increment in S_1. Such a perception is gotten beneath the idyllic condition that the preliminary $\widetilde{\widetilde{R_1}}$ stays constant as S_1 builds that needs to be affirmed from the ground trial. Another apprehension is that, as the automobile velocity builds, the moving duration of the automobile on the bridge is diminished, as is the hiatus for recording the automobile reaction. These will create a cut-off on the highest speed permitted.

The decrease of the α_1 is not at all assurance of effectively recognizing the B-f, except if this tends to decrease to an admissible level. Hypothetically, the likelihood of effectively recognizing the B-f can be characterized as the ace for appearance of α_1 to be lesser than a threshold t_α. Since α_1 is dependent on $\widetilde{\widetilde{R_1}}$, S_1 (speed boundary), and μ_1, the likelihood of effectively recognizing the B-f is likewise an element of these three non-dependent factors. For the current purposes, we will segregate the

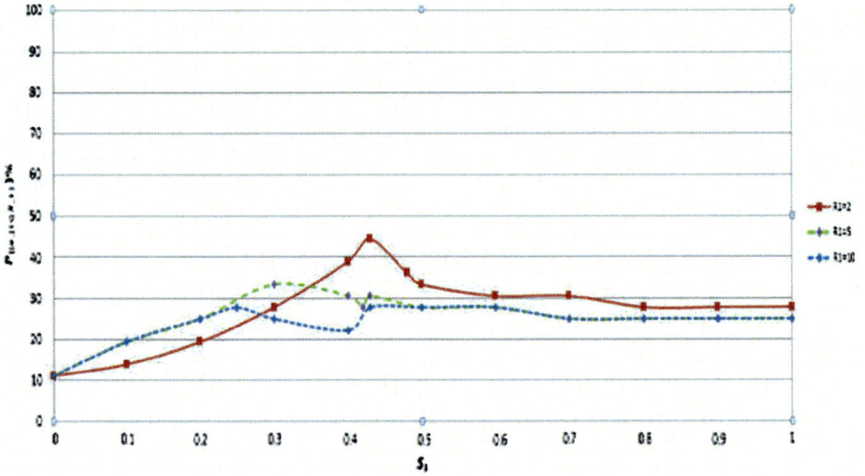

Fig. 5 Probability of reducing α_1 versus S_1 (note: $\widetilde{\widetilde{R}}_1 = R_1$)

span of S_1 and μ_1 into N_q and N_p equivalent parts, individually, supposing equivalent likelihood of appearance for apiece.

Let us indicate the value of the inconstant μ_1 on the pth portion and, that of S_1 on the qth portion by $\mu_{1;p}$ and $S_{1,q}$, individually. In understanding, the likelihood of effectively recognizing the B-f can be characterized as [6]

$$P_{(\alpha_1 \le t_\alpha)}\left(\widetilde{\widetilde{R}}_1\right) = \frac{\sum_{p=1}^{N_P} \sum_{q=1}^{N_q} \phi_{pq}\left(\widetilde{\widetilde{R}}_1, \mu_{1,p}, S_{1,q}\right)}{N_P \times N_q} \tag{52}$$

where

$$\phi_{pq}(\widetilde{\widetilde{R}}_1, \mu_{1,p}, S_{1,q}) = \begin{cases} 1, & \text{if}\, \alpha_1\,(\widetilde{\widetilde{R}}_1, \mu_{1,p}, S_{1,q}) \le t_\alpha \\ 0, & \text{if}\, \alpha_1\,(\widetilde{\widetilde{R}}_1, \mu_{1,p}, S_{1,q}) > t_\alpha \end{cases} \tag{53}$$

In this, the threshold limit t_α is prudently set 5, suggesting that the pinnacle amplitude of the bridge related frequency ought to be at least be 0.2 times that of the damped V-F in the F-R to ensure its discernibility.

Now, we might want to include that in spite of the fact that the automobile speed is manageable practically speaking, the presumption of S_1 as an unmanageable parameter is as yet prudent, since, as indicated by the meaning of speed boundary the B-f associated with the divisor are obscure preceding the volume. Hence, one can regulate the pattern of the speed parameter (as it is relative to automobile velocity), yet could scarcely get a handle on its value prior the trial is made. Accordingly, the velocity parameter is viewed as unmanageable.

It is understood that reverberation brought about by the fortuitous event of the bridge and automobile frequencies is hard to deduce the B-f utilizing the strategies introduced thus. Therefore, such resonance conditions will be prohibited from the current contemplations. All the more accurately, we will prohibit the state of definite fortuitous event as of the bridge and automobile frequencies, as suggested by $\mu_{bl,1} = 1(\text{or } \omega_{bl,1} = \omega_v)$ and $\mu_{br,1} = 1(\text{or } \omega_{br,1} = \omega_v)$, yet consider the cases when $\mu_{bl,1}$ and $\mu_{br,1}$ are moderately prominent or < 1, as far as the automobile and B-f can be systematically isolated.

Then again, the reverberation of the bridge with the automobile will show itself in the fairly huge reaction of the test automobile when it goes inside the bridge, contrasted with the non-reverberating cases because of the fact that the reverberation occurrence is identified, the B-f is recognized to be equivalent to the V-f, and the objective of deducing the B-f is correspondingly made.

Figure 6 displays the likelihood of effectively recognizing the B-F regarding preliminary $\overset{\approx}{R}_1$. As can be seen, the likelihood of effective recognition diminishes radically from its 100% to about half as $\overset{\approx}{R}_1$ increments from 0 to 1 and afterward diminishes steadily from about half to 10% as $\overset{\approx}{R}_1$ increments from 1 to 10.

The overall conclusion is that building $\overset{\approx}{R}_1$ will bring about a diminishing of the likelihood of effective recognition. In this manner, in the event that one can possess control over the preliminary $\overset{\approx}{R}_1$ to be as low as possible, either by intensifying the preliminary amplitude of the bridge (say, by exploiting existing traffic) or by lessening the preliminary amplitude of the automobile, all things considered, the B-f can be effectively recognized.

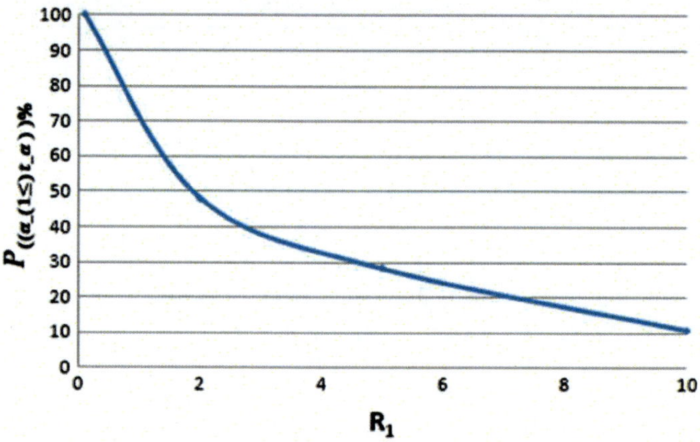

Fig. 6 Probability of effectively identifying the bridge frequency versus the preliminary acceleration amplitude ratio $\overset{\approx}{R}_1$ (note: $\overset{\approx}{R}_1 = R_1$)

3.4 Case II: Two Frequencies of the Bridge Involved

We will reflect the condition in the current section when the extension is invigorated by unique bases and set to travel with 2 vibration modes. It is assumed that all the additional modes of the bridge are not invigorated for the present tenacities or are invigorated to so little a degree that they can be overlooked. Usually, the original 2 modes are the ones that are well on the way to being revitalized. In these lines, we will take the original 2 bridge methods as the ones for analysis in the current chapter. The bridge is thus assumed to be placed in vibration mainly in the modes I and II, each with and A_1 and A_2, when the car enters the bridge with a vibration of and amplitude.

By setting $n = 1$–2 in Eq. (25), the acceleration feedback of the automobile during its entrance on the bridge will be influenced by five significant frequencies, i.e., $\omega_{bl,1}$ and $\omega_{br,1}$ $\omega_{bl,2}$ and $\omega_{br,2}$, and ω_{vd}. Their linking amplitudes in the F-R are signified as $\tilde{\tilde{A}}_{bl,1}$, $\tilde{\tilde{A}}_{br,1}$, $\tilde{\tilde{A}}_{bl,2}$, $\tilde{\tilde{A}}_{br,2}$ and $\tilde{\tilde{A}}_{vd}$, individually. Like Case I, one is keen on of the damped V-B to the I and II bridge-related frequencies, separately, got from Eq. (47) as [6]

$$\alpha_n = 2\sqrt{\alpha_{c1,n}^2 + \alpha_{c2,n}^2} \quad \text{For } n = 1, 2 \tag{54}$$

where the terms $\alpha_{c1,n}$ and $\alpha_{c2,n}$ can be got from the equations by considering $i = j = (1, 2)$.

For the current situation with two vibrational modes included, $N_i = 2$, α_1 and α_2 rely upon six factors, $\tilde{\tilde{R}}_1$, $\tilde{\tilde{R}}_2$, S_1, S_2, μ_1 and μ_2. Among these factors, just five (=$2N_i + 1$) are autonomous. To improve the accompanying conversation, we will present additional dimensionless parameter ν_{21} to demonstrate the ratio of the II/I frequency of the bridge,

$$\nu_{21} = \frac{\omega_{b.2}}{\omega_{b,1}} \tag{55}$$

Similarly, the co-relation between the speed parameters S_1 and S_2 w.r.t the I and II B-f, respectively, can be provided as

$$S_2 = \frac{2}{\nu_{21}} S_1 \tag{56}$$

and that among the, μ_1 and μ_2 is

$$\mu_2 = \nu_{21}\mu_1 \tag{57}$$

Along these lines, the α_1 and α_2 are elements of the five free factors $\widetilde{\widetilde{R}}_1$, $\widetilde{\widetilde{R}}_2$, S_1, μ_1, and ν_{21}, i.e., $\alpha_{1(2)} = \alpha_{1(2)}(\widetilde{\widetilde{R}}_1, \widetilde{\widetilde{R}}_2, S_1, \mu_1, \nu_{21})$.

To lessen the quantity of free factors, the accompanying information is viewed as reasonable for functional issues.

- The preliminary R for the I mode is taken as $\widetilde{\widetilde{R}}_1 = 2$;
- For the subsequent mode, it is occupied as $\widetilde{\widetilde{R}}_2 = 5$, in view of the thought that the AAR of the initial type is additional prominent than that of the subsequent type;
- The II/I bridge f-R is taken as $\nu_{21} = 2$ (the II B-f is about twice that of the preliminary B-f in the field trial of Ref. [8]).

As for the speed limit S_1 and the preliminary mode, the subsequent ARs of the I and II modes were built in Fig. 7, next to their description in Fig. 8a and b. The trend is similar to that of the current case, but more complex than that of Case I. Due to the presence of two reverberation conditions for the corresponding mode, the dispersion of the current case is more confused, apiece with $\mu_{bl,2} = 1(\omega_{bl,2} = \omega_v)$ or $\mu_{br,2} = 1(\omega_{br,2} = \omega_v)$ in totaling to the two reverberation conditions for the preliminary mode, with $\mu_{bl,1} = 1$ and $\mu_{br,1} = 1$, as referenced in Case I.

The perceptions for the current case can be watched as an augmentation of the outcomes drawn for Case I, which are specified as follows. (1) Under certain conditions (e.g., for $S_1 < 0{:}4$ for α_1 or $S_2 < 0{:}3$ for α_2), the rise in the automobile speed can prompt an increment in the likelihood of diminishing α_1 or α_2 to a level littler than the preliminary $\widetilde{\widetilde{R}}_1$ or $\widetilde{\widetilde{R}}_2$, individually. In any case, the decrease of α_1 or α_2 need not ensure benefit in recognizing the B-f, except if it very well may be diminished to an adequate level from real contemplations. (2) Both the α_1 and α_2 decline steadily as there μ_1 *and* μ_2 . Increment from 0 to the reverberation ($\mu_{b.1} = 1$ *and* $\mu_{b,2} = 1$), and afterward increment quickly as μ_1 *and* μ_2 increment further. The technique that the estimation of α_2 is always higher than of α_1 also acknowledge with the hypothesis for Case I, in that frequencies of advanced modes cannot be separated in a manner that is as simple as that for the principal mode due to their complex B/V f-Rs.

Once more, to handle the intractable character of the initial μ_1 and the primary S_1, we will partition the spans of μ_1 and S_1 into N_p and N_q equivalent portions, correspondingly. Expecting the likelihood of happening for apiece part to be equivalent, we will designate the estimate of the inconstant μ_1 on the pth portion and that of S_1 on the qth portion by $\mu_{1,p}$ and $S_{1,q}$, respectively.

As needs be, the likelihood of lessening α_1 or α_2 can be characterized as [6]

$$P_{\left(\alpha_{1(2)} \leq \widetilde{\widetilde{R}}_{1(2)}\right)}(\widetilde{\widetilde{R}}_1, \widetilde{\widetilde{R}}_2, \nu_{21}) = \frac{\sum_{p=1}^{N_p} \sum_{q=1}^{N_q} \phi_{pq}\left(\widetilde{\widetilde{R}}_1, \widetilde{\widetilde{R}}_2, S_{1,q}, \mu_{1,p}, \nu_{21}\right)}{N_p * N_q}. \quad (58)$$

where

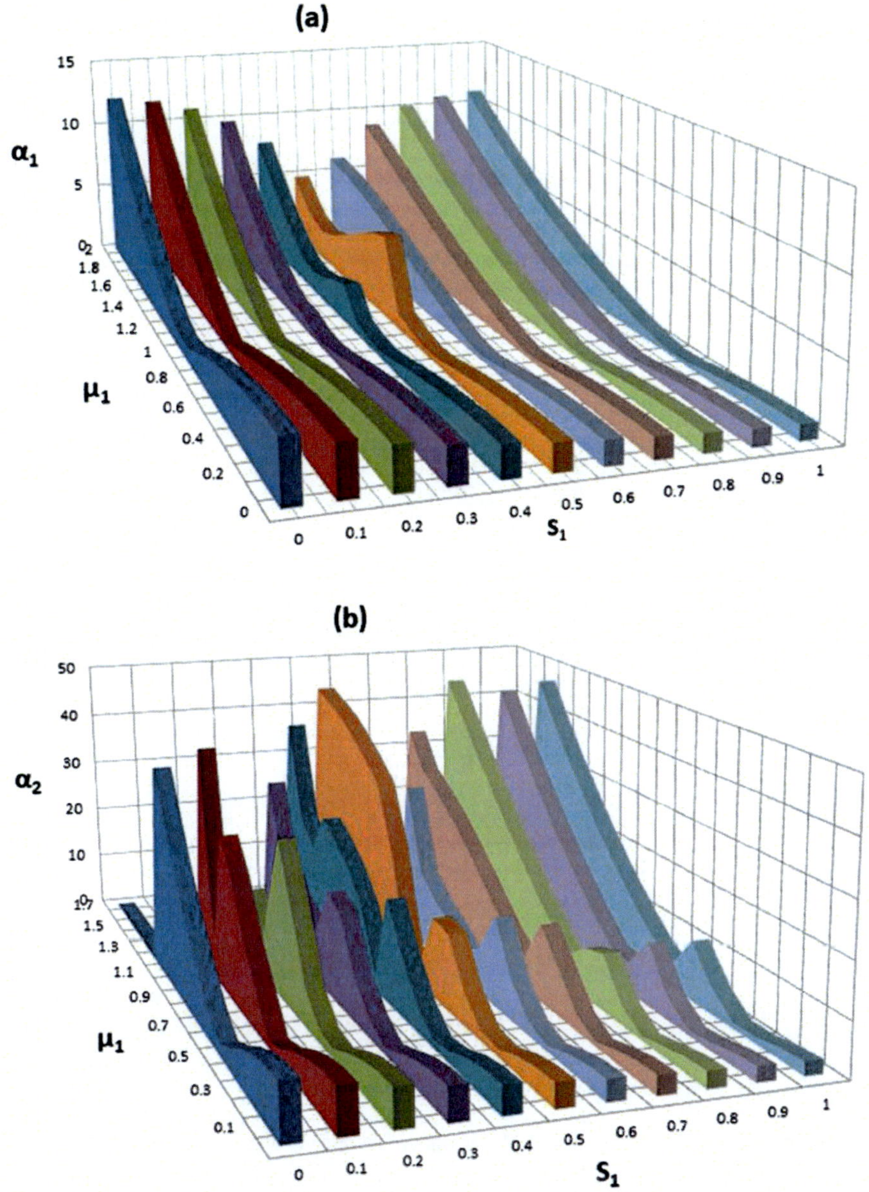

Fig. 7 α_1; α_2 versus S_1 and μ_1

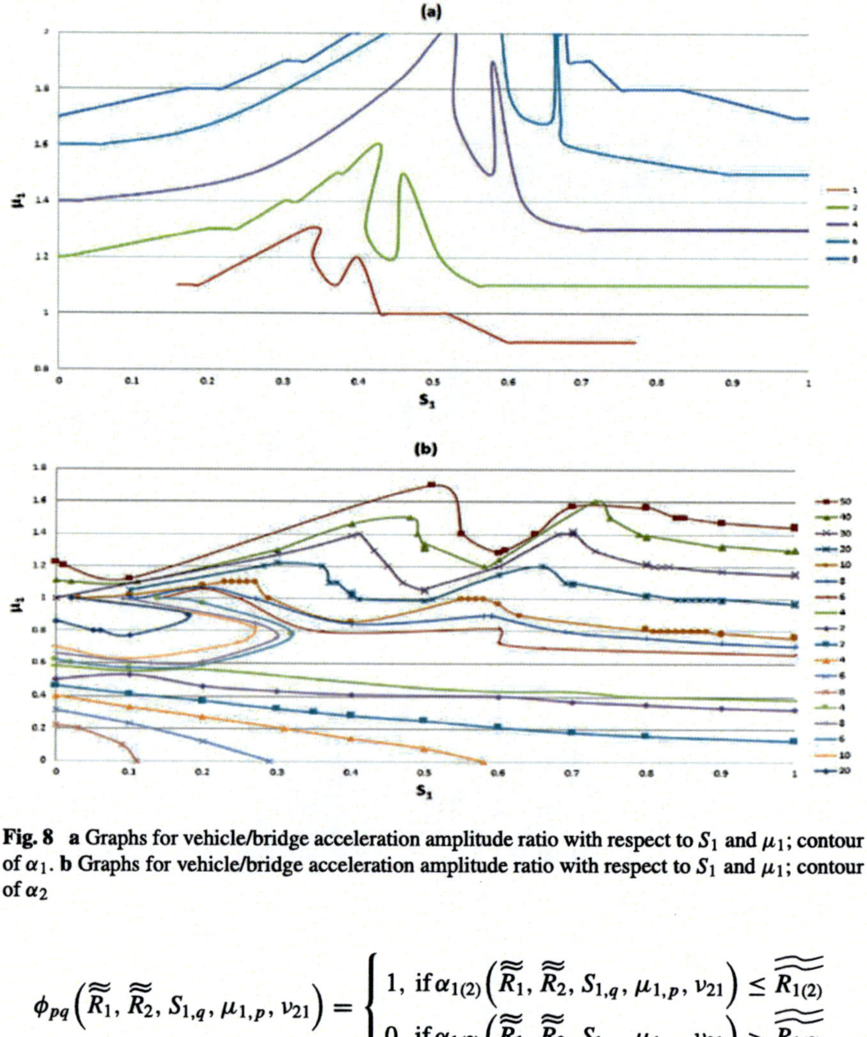

Fig. 8 **a** Graphs for vehicle/bridge acceleration amplitude ratio with respect to S_1 and μ_1; contour of α_1. **b** Graphs for vehicle/bridge acceleration amplitude ratio with respect to S_1 and μ_1; contour of α_2

$$\phi_{pq}\left(\widetilde{\widetilde{R}}_1, \widetilde{\widetilde{R}}_2, S_{1,q}, \mu_{1,p}, \nu_{21}\right) = \begin{cases} 1, \text{ if } \alpha_{1(2)}\left(\widetilde{\widetilde{R}}_1, \widetilde{\widetilde{R}}_2, S_{1,q}, \mu_{1,p}, \nu_{21}\right) \le \widetilde{\widetilde{R_{1(2)}}} \\ 0, \text{ if } \alpha_{1(2)}\left(\widetilde{\widetilde{R}}_1, \widetilde{\widetilde{R}}_2, S_{1,q}, \mu_{1,p}, \nu_{21}\right) > \widetilde{\widetilde{R_{1(2)}}} \end{cases} \tag{59}$$

Figure 9 shows the likelihood of lessening α_1 and α_2 versus ν_{21} for preliminary $\widetilde{\widetilde{R}}_1 = 5$ (I mode) and $\widetilde{\widetilde{R}}_2 = 5$ (II mode). It is seen that the likelihood of lowering the α_1 remains nearly invariable as ν_{21} increments, while that of lowering α_2 reduces as ν_{21} increments. As indicated by Eq. (57), the ν_{21} is directly proportional to μ_2. Consequently, it very well may be gathered that the greater μ_2 is, the lesser the likelihood of lowering the second AAR. This added affirms the hypothesis for Case I about the complication for deducing the B-f of greater modes. From a practical perspective, in the event that the II B-f is a lot higher than the I, at that point the

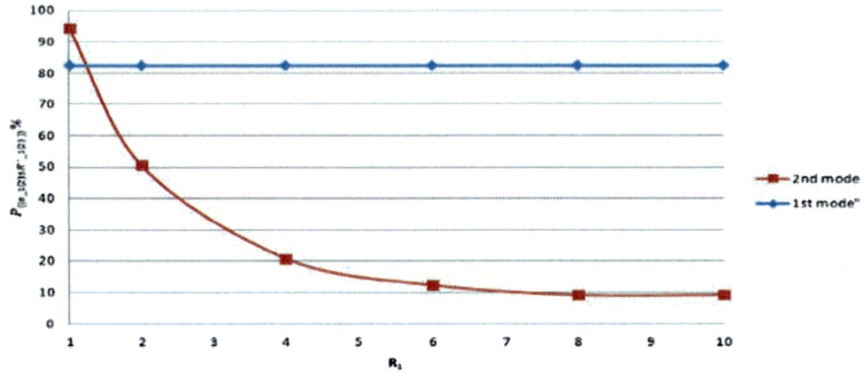

Fig. 9 Probability of reducing α_1 and α_2

likelihood of lowering the II AAR will be a lot lower, inferring that the likelihood of effectively recognizing the II B-f will be a lot of lower also. Such a perception offers a few hints on the intrinsic complication related with the recognizing of the subsequent or higher modes of frequencies for the bridge.

As a general rule, ν_{21} is also an unmanageable obscure other than the two inconstant μ_1 and S_1. By a similar token, we can segregate the span of ν_{21} into N_r parts, with equivalent likelihood of appearance for each part, and indicate the value of the inconstant ν_{21} on the rth part by$\nu_{21,r}$. In this manner, the likelihood of effectively recognizing the B-f, for α_1 and α_2 to be lower than the threshold $t_{\alpha1}$ and $t_{\alpha2}$, individually, can be characterized as [6]

$$P_{\left(\alpha_{1(2)} \leq t_{\alpha_{1(2)}}\right)}\left(\widetilde{\widetilde{R}}_1, \widetilde{\widetilde{R}}_2\right) = \frac{\sum_{p=1}^{N_p} \sum_{q=1}^{N_q} \sum_{r=1}^{N_r} \phi_{pqr}(\widetilde{\widetilde{R}}_1, \widetilde{\widetilde{R}}_2, \mu_{1,p}, S_{1,q}, \nu_{21,r})}{N_p \times N_q \times N_r} \tag{60}$$

where

$$\phi_{pqr}\left(\widetilde{\widetilde{R}}_1, \widetilde{\widetilde{R}}_2, \mu_{1,p}, S_{1,q}, \nu_{21,r}\right) = \begin{cases} 1, \text{ if} \alpha_{1(2)}\left(\widetilde{\widetilde{R}}_1, \widetilde{\widetilde{R}}_2, \mu_{1,p}, S_{1,q}, \nu_{21,r}\right) \leq t_{\alpha_{1(2)}} \\ 0, \text{ if} \alpha_{1(2)}\left(\widetilde{\widetilde{R}}_1, \widetilde{\widetilde{R}}_2, \mu_{1,p}, S_{1,q}, \nu_{21,r}\right) > t_{\alpha_{1(2)}} \end{cases} \tag{61}$$

Thus, both threshold values t_{α_1} and t_{α_2} are established as 5. Like Case I, reverberation circumstances are avoided since the computation of the likelihood for effectively recognizing the B-f Fig. 9 shows the likelihood of effective recognition of the B-f of the I and II modes regarding $\widetilde{\widetilde{R}}_1$ and $\widetilde{\widetilde{R}}_2$. It is seen that likelihood of effectively recognizing the first B-f is firmly associated with the preliminary $\widetilde{\widetilde{R}}_1$ of the primary mode, however for the most part superfluous to the ratio $\widetilde{\widetilde{R}}_2$ of the subsequent mode.

The opposite is likewise valid for the subsequent B-f, that is, the likelihood for effectively recognizing the II B-f is unequivocally associated with the preliminary $\overset{\approx}{R}_2$ of the subsequent mode, yet for the most part nondependent of the ratio $\overset{\approx}{R}_1$ of the primary mode. The likelihood of effectively recognizing the B-f of the initial mode arrives at 100% when the preliminary AR $\overset{\approx}{R}_1 < 3$. It reduces steadily from 100 to about 40/100 as $\overset{\approx}{R}_1$ increments from 3 to 10. For the subsequent mode, the likelihood of effective recognition can scarcely arrive at 100% except if $\overset{\approx}{R}_2$ is fairly less, say 0.1. It reduces steadily from 100 to about 15/100 as $\overset{\approx}{R}_2$ increments from 0.1 to 10. Generally, the likelihood of effectively recognizing the B-f of either the initial or subsequent mode is inversely proportional to the corresponding preliminary AR (e.g., $\overset{\approx}{R}_1$ for the primary mode and $\overset{\approx}{R}_2$ for the subsequent mode). The outcomes procured for Case I bolster the perception made in this regard, if the movement of two modes is viewed as the blend of two individual movements, with the association consequence between $\overset{\approx}{R}_1$ and $\overset{\approx}{R}_2$ disregarded. In this way, to build the likelihood of effectively recognizing the B-f of a specific mode practically, it is proposed that the preliminary automobile/bridge AAR related with that specific method be decreased, in the method of either enhancing the AAR of the bridge movement (state, by profiteering running traffic) or lowering that of the automobile movement (state, by permitting the automobile to enter the bridge in a smooth manner) (Fig. 10).

4 Conclusion

That enhance the likelihood of effectively recognizing the B-f utilizing a trial automobile, the impacts of a few important boundaries on the dynamic reaction of the automobile moving over the bridge are examined. In the initial place, the dynamic reaction of the trial automobile with preliminary vibrations going on a bridge under vibration of specific modes is deduced. The acceleration reaction of the passing automobile is governed by three set of frequencies: L-S B-f, R-S B-f, and V-f, as can be visualized from the Fourier spectrum. The key attention in effective recognition is the discernibility of the pinnacles associated with the B-f of interest and that which depends on the comparative scales of these pinnacles regarding the V-f.

In the present report, the AAR of the V/B frequencies is presented, which proposals a list for assessing the likelihood of effective recognition. The littler V/B AAR is, the greater the likelihood of effectively recognizing the B-f. As for the situations where just a couple of B-f is included, the V/B AAR is appeared to rely on the accompanying factors: (1) starting V/B AAR; (2) speed boundaries; (3) B/V f-R; and (4) II/I bridge f-R (for Case II alone).

Out of these inconstant, the preliminary V/B AAR has all the earmarks of being generally significant. The lower the ratio of every particular mode is, the more prominent the likelihood of effectively recognizing the B-f of that specific mode. The likelihood of effective recognition might arrive at 100% relying on the prerequisite that

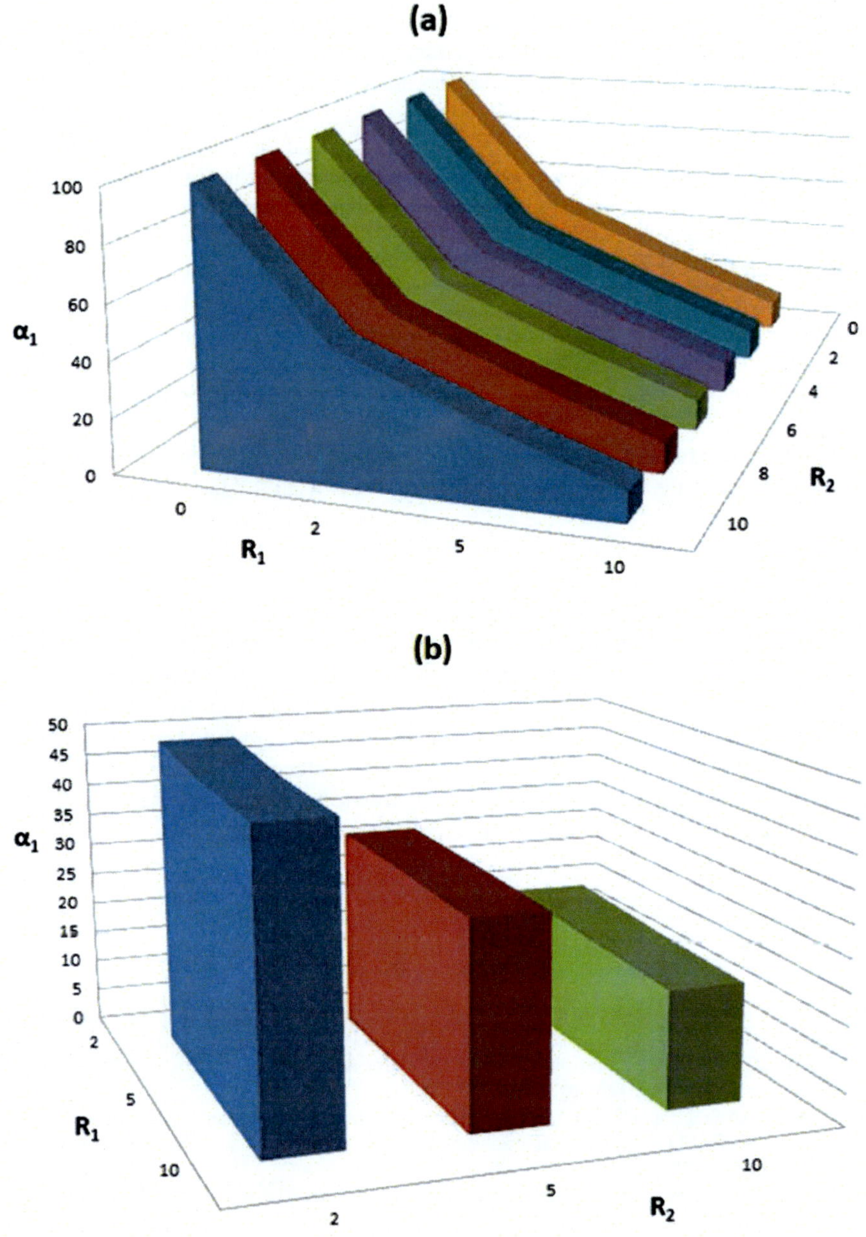

Fig. 10 Probability of effectively identifying bridge frequencies versus initial amplitude ratio: **a** first mode, **b** second mod

the preliminary AAR < 3. To enhance the likelihood of effectively recognizing the B-f, it has been proposed that the preliminary AAR of each type be lessened however much as possible, whichever by enhancing the acceleration amplitude of the bridge or by lowering that of the automobile.

Regarding the speed boundary, it is seen that within the scope of 0 to 4/10, the greater the speed boundary is, more the likelihood of diminishing the V/B AAR. When speed boundary surpasses this span, no extra improvement can be done by speeding up the vehicle. Yet, decrease of this ratio ought not ensure accomplishment of-Recognition, except if the ratio is maintained at a value less than the threshold value. In addition, a ground test needs the travel duration of the automobile on bridge to be lengthy enough for information gathering. This shall surely put a limiting point on automobile speed.

The B/V f-R's are unmanageable factors practically speaking, so it does not mean a lot to discuss their impact on frequency recognition. Despite the fact that the II/I bridge f-R is additional unmanageable factor, it is an inconstant identified with the characteristics of the bridge and might generally assessed for specific bridges. More prominent the II/I bridge f-R is, lesser the likelihood for lowering the V/B AAR. Such a perception proposals a few hints on complication associated with recognizing the second or greater frequencies of specific bridges.

Current examination for deducing the B-f dynamic reaction of a moving automobile depends for the most part on the systematic detailing and analytical framing. Outcomes acquired proposal a hypothetical guide for deducing the B-f practically. Albeit a revaluation of the invariable values suggested by the connected past works upholds the legitimacy of current hypothesis, further in-ground tests are necessary to confirm the conclusion in this, while giving a criticism to fine tune the hypothesis utilized. At long last, the conclusions drawn herein are dependent upon the information and hypothesis utilized in the modeling.

References

1. Willis R (1849) Appendix to the report of the commissioners appointed to inquire into the application of iron to railway structures. H.M. Stationary Office, London, UK
2. Stokes GG (1849) Discussion of a differential equation relating to the braking of railway bridges. Trans Camb Philos Soc 8(5):707–735
3. Timoshenko SP (1922) On the forced vibrations of bridges. Philos Mag Ser 6(43):1018–1019
4. Akin JE, Mofid M (1989) Numerical solution for response of beams with moving mass. J Struct Eng ASCE 115(1):120–131
5. Yang YB, Lin CW (2005) Vehicle-bridge interaction dynamics and potential applications. J Sound Vibration 284:205–226
6. Yang YB, Chang KC (2009) Extracting the bridge frequencies indirectly from a passing vehicle: parametric study. Eng Struct 31:2448–2459
7. Damodarasamy SR, Kavitha S (2015). Basics of structural dynamics and aseismic design
8. Lin CW, Yang YB (2005) Use of a passing vehicle to scan the fundamental bridge frequencies: an experimental verification. Eng Struct 27:1865–1878

The Influence of Ceramic Ball as a Fine Aggregate in Concrete

Pradeep Karanth, Sabyath P. Shetty, Thushar S. Shetty, and Sushanth S. Bhandary

Abstract River sand, a non-renewable resource, is used in concrete as a fine aggregate, which is called filler material that fills the voids in concrete. This paper provides outcomes of compressive strength of concrete by using the ceramic ball as substitutes of fine aggregate that is compared with nominal mix concrete. Here, the ceramic ball is a commercial waste product produced in industries. The report includes the chemical test of the substitute, and it is having 73% of silica and 0.7% of alumina. Zone I and Zone II are selected as fineness modulus of sand for calculation of nominal mix design and partial mix design. As per the 28th-day compressive strength result, 30% replacement of fine aggregate gives a good result in comparison with a nominal strength of concrete, and hence, this waste material can be used as fine aggregate in concrete production. The main objective of this project is to find ways of disposing of waste products and to reduce the use of sand in the production of concrete as a fine aggregate.

Keywords Compressive strength · Ceramic ball · Concrete · Molecular sieve

1 Introduction

The concrete is usually obtained by mixing cement, fine aggregate, coarse aggregate (gravel), and water using the correct proportion of mix as per the codebook. The material that is used as fine aggregate and coarse aggregate is natural elements that affect the nature of the earth especially the environment of the river by extraction

P. Karanth (✉) · S. P. Shetty · T. S. Shetty · S. S. Bhandary
Department of Civil Engineering, NMAMIT-Nitte, Karkala, India
e-mail: pradeepkaranth@nitte.edu.in

S. P. Shetty
e-mail: shettysabyath@nitte.edu.in

T. S. Shetty
e-mail: shettythushar@nitte.edu.in

S. S. Bhandary
e-mail: Sushanth.bhandary@nitte.edu.in

L. Nandagiri et al. (eds.), *Sustainability Trends and Challenges in Civil Engineering*, Lecture Notes in Civil Engineering 162, https://doi.org/10.1007/978-981-16-2826-9_38

Fig. 1 Ceramic balls

of sand. Over extraction of sand will increase the river's ability to erode, this can provide a threat to nearby infrastructure, and breeding of aquatic life is also affected. For the construction industry, it has become difficult to obtain sand as fine aggregate as there is a ban on sand mining, and this increases the cost rate of sand day by day. This project used ceramic balls as a replacement for fine aggregate to solve this problem, to find out if this waste can be used as a fine aggregate instead of sand. Ceramic balls as shown in Fig. 1 are the commercial waste produced in the industry which makes the industrial very difficult to dispose of.

Figure 2 shows the flowchart diagram of methodology. In this project, waste material was collected from the industry and carried out for a particular chemical test that is silica content and alumina content test. The waste material is first crushed to a small particle with the help of crushing equipment. For the calculation of mix design, the following test is been conducted as per IS 10262:2009, and they are sieve analysis of partially mixed sand with waste material in percentage varying from 0 to 50% to select a suitable zone. Water absorption test was conducted on the sand, aggregate, and waste material, and a specific gravity test was conducted on cement, sand, aggregate, and waste material. As per the M30 grade mix design, cubes are cast by partially replacing sand with the waste material in varying percentages from 0 to 100%, and then, compressive strength test for the 7th day and 28th day is tested. The following are the lab test conducted in detail.

1.1 Chemical Test

This test is conducted to check whether the waste sample has binding component or it can be the substitute of fine aggregate or not. The chemical tests that were conducted are determination of silicon dioxide to find out silica content and determination of the combined oxides (Al_2O_3 and Fe_2O_3) to find out alumina content. Following are the procedure of test conducted.

- Take 2 g of sample in 100 ml beaker.

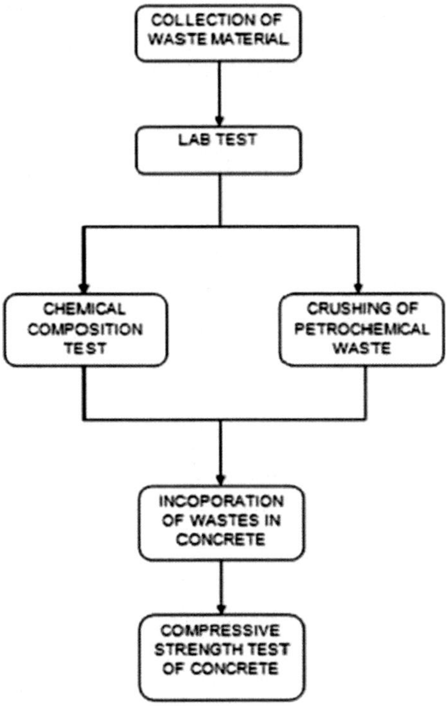

Fig. 2 Flowchart of methodology

- 20 ml of concentrated HCL was applied and then heated to dryness on the electric hotplate.
- An additional 12 ml of HCL, 60 ml of distilled water have been added and heated up to a boiling point.
- The hot solution was cleaned with less filter paper than ash.
- The precipitate was washed with 60 ml of hot distilled water.
- The precipitate and filter paper were transmitted to a clean vessel.
- For 50 min, the vessel and its innards were burned to 800 °C, and the vessel was left in the dryer to cool and then measured.

Calculation:

$$\text{Percentage of Silicon dioxide} = \frac{\text{Weight of SiO}_2}{\text{Weight of sample}} \times 100\%$$

Identification of combined oxides

- 2 g of ammonium chloride was added to the remaining filtrate, and drops of the indicator of methyl red were heated to boil.
- The ammonia solution has been applied continuously until the color changes to yellow.

- The beaker was left behind for 10 min.
- The solution was filtered through ashless filter paper.
- Precipitate and filter paper washed with a 2% ammonium nitrate solution.
- The precipitate and filter paper were transmitted to a clean crucible.
- The crucible and its contents were burned at an 800 °C for 50 min. It was left to cool in the dryer for minutes and then weighted to assess the combined oxides.

Determination procedure of ferric oxide percentage:

- 2 g of the sample was taken in a 100 ml beaker, and 10 ml of concentrated HCL was applied to the beaker and was well controlled until the green color had disappeared.
- Then, 20 ml of distilled water was applied, and the solution moved quantitatively to a 100 ml volumetric flask.
- Then, a beaker was poured in and left to cool down.
- Add 10 ml of buffer solution and 2 ml of salicylic acid.
- The mixture was titrated until the end point against the standard EDTA solution.
- The titration has been repeated for up to two consecutive readings.
- The number of moles and the weight in grams was measured by titration, and the percentage was then estimated.

Percentage of Aluminum Oxide:

$$\text{Percentage of aluminum oxide} = \text{percentage of combined oxide} - \text{percentage of ferric oxide.}$$

From the test, it is observed that silica content and alumina content in ceramic balls are 93 and 0.7%. Since ceramic balls are having 93% of silica content which is more than 50%. Hence, ceramic ball can be used as substitutes of fine aggregate.

1.2 *Sieve Analysis*

This test is conducted to select the zone of fineness modulus of partial replacement of sand with waste sample so that it can be used to calculate mix design which is required for casting concrete cubes. This test is done on partial replacement of sand with waste sample having sand replacement varying from 0 to 50%. The following are the procedure carried out for the test.

- 500 g of sand replaced by waste material by percentage was taken.
- The material was sieved by standard IS sieve.
- Retained weight on respective sieve, including pan, was noted and tabulated as shown in Table 1, and suitable zone is selected.

Table 1 shows the report of sieve analysis conducted on partially replaced sand with waste sample. In the table, the * mark tells that the value of cumulative

Table 1 Sieve analysis report of 0–50% replacement of sand with waste sample

Sieve	Weight retained (g)	Cumulative weight retained (g)	Cumulative percentage weight (%)	Cumulative percentage passing (%)	Zone I (%)	Zone II (%)	Zone III (%)	Zone IV (%)
0% replacement of sand with sample								
10	0	0	0	100	100*	100*	100*	100*
4.75	6	6	1.2	98.8	90–100*	90–100*	90–100*	95–100*
2.36	24	30	6	94	60–95*	75–100*	85–100*	95–100*
1.18	136	166	33.2	66.8	30–70*	55–90*	75–100*	90–100
0.6	83	249	49.8	33	16–34*	35–59	60–79	80–100
0.3	72	321	64.2	18	05–20*	8–30*	12–40*	15–50*
0.15	106	427	85.4	9.6	0–10*	0–10*	0–10*	0–15*
Pan	73	500	100	0	Fineness modulus value is 2.39			
10% replacement of sand with sample								
10	0	0	0	100	100*	100*	100*	100*
4.75	10	10	2	98	90–100*	90–100*	90–100*	95–100*
2.36	33	43	8.6	91.4	60–95*	75–100*	85–100*	95–100
1.18	78	121	24.2	75.8	30–70	55–90*	75–100*	90–100
0.6	117	238	47.6	52.4	16–34	35–59*	60–79*	80–100
0.3	219	457	91.4	8.6	05–20*	8–30*	12–40	15–50
0.15	35	492	98.4	1.6	0–10*	0–10*	0–10*	0–15*
Pan	8	500	100	0	Fineness modulus value is 2.72			
20% replacement of sand with sample								
10	0	0	0	100	100*	100*	100*	100*
4.75	10	10	2	98	90–100*	90–100*	90–100*	95–100*
2.36	31	43	8.6	91.4	60–95*	75–100*	85–100*	95–100
1.18	93	136	27.2	72.8	30–70	55–90*	75–100*	90–100
0.6	114	254	50.8	49.2	16–34	35–59*	60–79	80–100
0.3	198	452	90.4	9.6	05–20*	8–30*	12–40	15–50
0.15	33	485	97	3	0–10*	0–10*	0–10*	0–15*
Pan	15	500	100	0	Fineness modulus value is 2.76			
30% replacement of sand with sample								
10	0	0	0	100	100*	100*	100*	100*
4.75	5	5	1	99	90–100*	90–100*	90–100*	95–100*
2.36	42	47	9.4	90.6	60–95*	75–100*	85–100*	95–100
1.18	83	130	26	74	30–70	55–90*	75–100	90–100
0.6	120	250	50	50	16–34	35–59*	60–79	80–100
0.3	179	429	85.8	14.2	05–20*	8–30*	12–40*	15–50
0.15	36	465	93	7	0–10*	0–10*	0–10*	0–15*

(continued)

Table 1 (continued)

Sieve	Weight retained (g)	Cumulative weight retained (g)	Cumulative percentage weight (%)	Cumulative percentage passing (%)	Zone I (%)	Zone II (%)	Zone III (%)	Zone IV (%)
Pan	35	500	100	0	Fineness modulus value is 2.652			
40% replacement of sand with sample								
10	0	0	0	100	100*	100*	100*	100*
4.75	8	8	1.6	98.4	90–100*	90–100*	90–100*	95–100*
2.36	48	56	11.2	88.8	60–95*	75–100*	85–100*	95–100
1.18	85	141	22.8	77.2	30–70	55–90*	75–100*	90–100
0.6	105	246	49.2	50.8	16–34	35–59*	60–79	80–100
0.3	170	416	83.2	16.8	05–20*	8–30*	12–40*	15–50*
0.15	63	479	95.8	4.2	0–10*	0–10*	0–10*	0–15*
Pan	21	500	100	0	Fineness modulus value is 2.63			
50% replacement of sand with sample								
10	0	0	0	100	100*	100*	100*	100*
4.75	9	9	1.8	98.2	90–100*	90–100*	90–100*	95–100*
2.36	19	28	5.6	94.4	60–95*	75–100*	85–100*	95–100
1.18	105	133	26.6	73.4	30–70	55–90*	75–100	90–100
0.6	158	291	48.2	41.8	16–34	35–59*	60–79	80–100
0.3	130	421	84.2	15.8	05–20*	8–30*	12–40*	15–50*
0.15	72	493	98.6	1.4	0–10*	0–10*	0–10*	0–15*
Pan	7	500	100	0	Fineness modules value is 2.75			

percentage passing is satisfying the value of the zone with respect to sieve sizes. It is observed that 0% replacement satisfies Zone I and the remaining, i.e., 10–50% replacement satisfies Zone II.

1.3 Specific Gravity

Specific gravity is the proportion of an object's mass to a reference liquid. In this project, specific gravity test was 'ball'. The main reason to conduct this test is to check whether ceramic ball is having same density as fine aggregate is having for better bonding in concrete and it is required for calculation of mix design. Here, water is used as reference liquid to test specific gravity of fine aggregate, coarse aggregate, and ceramic ball. Whereas kerosene is used as reference liquid to test specific gravity of cement. Table 2 gives the result of specific gravity test.

Table 2 Specific gravity test result

Specific gravity test result	
Sample	Specific gravity
Sand	2.55
Coarse aggregate	2.83
Cement	3.08
Ceramic ball	2.75

Fig. 3 Shows the workability of fresh concrete by using the Slump test

1.4 Slump Test

As per IS 456:2000, the slump to be obtained during the test is 100 mm for heavily reinforced section in slabs, beams, walls, and columns. This is required to obtain ease in workability. Soon after the slump is obtained, the concrete is casted in the cube (Fig. 3).

2 Process of Concrete Casting

The concrete is cast for M30 grade, and the mix proportion as per calculation is 1.00:1.71:2.98. The quantity of water used to mix concrete materials is 3.9 L. The mold used for casting of concrete cube is 15 cm * 15 cm * 15 cm. Three cubes each are cast for 7 days and 28 days for 0% replacement of sand with the ceramic ball up to 100% replacement of sand with ceramic ball. The cubes are vibrated manually with the help of a tamping rod. The cubes were demolded after 24 h and held for 7 days and 28 days for curing, and the compressive strength of the cubes was measured on that day by the UTM.

3 Compressive Strength Test

For nominal cubes and partly substituted cubes, the compressive strength test is carried out for the 7th and 28th days. Since the cube is casted for M30 grade, as per IS 10262-2009, the 7th day compressive strength for nominal and partially replaced cubes should obtain 65% of the compressive strength of concrete mix (M30) that is 19.5 N/mm^2, and 28th day compressive strength should obtain 100% compressive strength of concrete mix that is 30 N/mm^2 (Figs. 4 and 5).

Table 2 shows the compressive strength test result of concrete cubes for 7th day and 28th day, and Figs. 6 and 7 show line chart diagram of 7th day and 28th day compressive strength. From 28th day line chart diagram, it is observed that 50% replacement and 100% replacement obtain minimum strength, and hence, this waste sample can be used as fine aggregate in construction industry for 50 and 100% replacement.

Fig. 4 Compressive test machine

Fig. 5 Shows the Compression testing of the Concrete cube

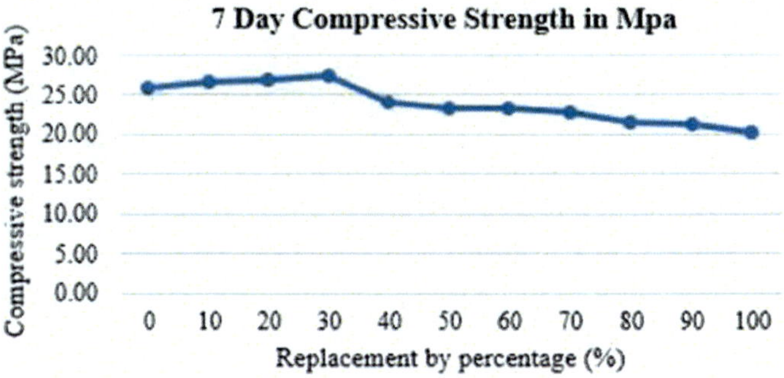

Fig. 6 Line chart of 7th day compressive strength

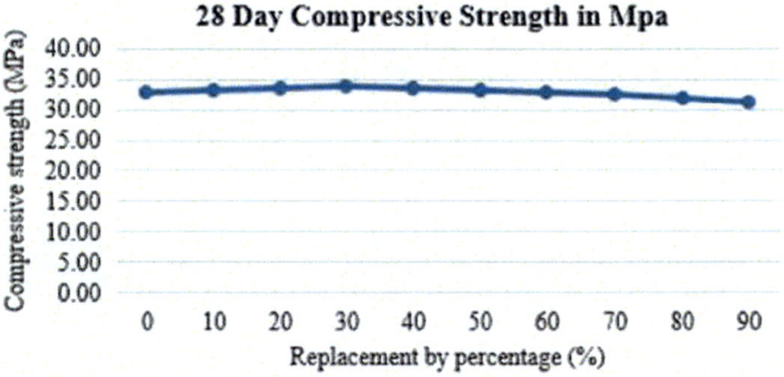

Fig. 7 Line chart of 28th day compressive strength

4 Calculation Procedure of Mix Proportion as per IS 10262:2009

Mix proportion is calculated as per above test result of sieve analysis, specific gravity test and slump test.

Data for M30 grade concrete mix proportion

- Note down complete data for proportion like grade, type of cement, etc.
- (2) Calculate the target strength for mixing by the following equation.

$$f'ck = fck + 1.65s$$

where

s = standard deviation, i.e., 5 N/mm^2.

- Selection of the water–cement ratio: As per IS 456:2000 from Table 3, the respective water–cement ratio should be selected.

Table 3 Compressive strength test result of cube for 7th day and 28th

Compressive strength test of the cubes after 7 days and 28 days of curing treatment											
Replacement by percentage (%)	0	10	20	30	40	50	60	70	80	90	100
7th day strength in MPa	25.92	26.50	26.70	26.74	24.82	24.82	24.82	24.82	25.80	22.12	20.00
28th day strength in MPa	33.04	33.28	33.50	33.85	33.62	33.25	32.90	32.40	31.75	31.28	30.25

- Water content selection: As per IS 10262:2009 from Table 2, the respective water content should be selected.
- With respect to water–cement ratio and water content, cement content should be selected from Table 5 of IS 456:2000.
- The proportion of gross aggregate volume and fine aggregate content can be obtained from Table 3 (with respect to zone for water–cement ratio of 0.5).
- Mix calculations: In this calculation, the volume of concrete should be taken as 1 m^3, and the volume of cement, the volume of water, the total volume, and the volume of the fine aggregate and the gross aggregate should be calculated.
- Note down the calculated mix proportion for castings such as cement, water quantity, total volume in aggregate, and gross aggregate mass.

5 Conclusion

- The ceramic ball is having high silica content (73%), and hence, it is used as a substitute for sand.
- From sieve analysis, Zone I is selected for nominal mix design, and Zone II is selected for partial mix design calculation.
- Specific gravity for the ceramic ball is 2.75 and sand is 2.55 which is nearly similar, and hence, the ceramic ball can be used as a substitute for sand in concrete.
- The characteristic strength of the M30 concrete mix is 38.25 MPa. The strength in 30% replacement is 33.85 MPa. Hence, waste samples can be used to replace 30% sand in fine aggregate.
- On the 28th day, the compressive strength of the concrete with 100 percent substitution of sand with the ceramic ball as an aggregate is 30.25 MPa, which is slightly lower than that of Nominal mix concrete (With 0 percent replacement). As a result, this waste sample can also be used to replace sand completely.

References

1. Akhtar JN, Ahamad T, Akhtar MN, Abbas H (2014) Influence of fibres and fly ash on mechanical properties of concrete. Am J Civil Eng Archit 2(2):64–69
2. Nataraja MC, Dileep Kumar PG, Manu AS, Sanjay MC (2013) Use of granulated blast furnace slag as fine aggregate in cement mortar. Int J Struct Civ Eng Res 2(2):61–68
3. Dimov D, Amit I, Gorrie O, Barnes MD, Townsend NJ, Neves AIS, Withers F, Russo S, Craciun MF (2018) Ultrahigh performance nanoengineered graphene–concrete composites for multifunctional applications. Adv Funct Mater 28

Comprehensive Analysis of Outrigger System for High Rise Structures Subjected to Wind and Earthquake Loadings

N. M. Priyanka, D. T. Abhilash, H. A. Ajay, H. S. Mohan, and S. Apoorva

Abstract Tall building development has been rapidly increasing worldwide introducing new challenges that need to be met through engineering judgment. High rise buildings are very vulnerable to earthquake and wind loads on regular basis. Hence, to make high rise buildings safe against lateral loads, different types of structural systems are used. One of these structural systems is outrigger system. This project aims to show the analysis of a 60 storied three-dimensional building located at Delhi using ETABS 2016 software. Analysis of eight different models with outriggers and belt truss configuration is carried out by response spectrum method for seismic zone IV, to understand the performance for various parameters like maximum storey displacement, maximum storey drift and storey shear. The present work contains a comparative study on regular building with wall outriggers and braced outriggers. The results showed the reduction in responses by using outrigger and belt truss system.

Keywords Outriggers · Belt truss · Belt wall · Bracing

1 Introduction

People have been fascinated by high rise buildings from the olden days [1]. People from countryside regions are voyaging in large numbers to bigger towns because of increase in employment opportunities and infrastructure amenities [2]. Due to this, cities come to be compactly inhabited and rate of property has been increasing

N. M. Priyanka (✉) · H. S. Mohan
Department of Civil Engineering, Navkis College of Engineering, Hassan, India

D. T. Abhilash
Department of Civil Engineering, Adichunchanagiri Institute of Technology, Chikkamagaluru, India

H. A. Ajay
Department of Civil Engineering, Presidency University, Bengaluru, India

S. Apoorva
Department of Civil Engineering, JSS Academy of Technical Education, Bengaluru, India

L. Nandagiri et al. (eds.), *Sustainability Trends and Challenges in Civil Engineering*, Lecture Notes in Civil Engineering 162, https://doi.org/10.1007/978-981-16-2826-9_39

which leads to the growth of skyscraper [2]. From the olden pyramidical period to today's technical period, supremacy and prosperity of the nation have been conveyed through skyscraper. Thus, we can say that there is a competition in mankind to have the world's tallest building. From a structural engineer's point of vision, a skyscraper may be defined as one that because of its tallness is exaggerated by the movements of horizontal forces due to earthquake or wind and plays important criteria in its structural design. There are several reasons for the demand of tall buildings in bigger towns. Some of the most important reasons are scarcity of land in urban areas; due to rapid industrial development, there is increasing demand for domestic and commercial space, economic growth of the city and technological advancements [3]. Ground motion due to earthquake can be happened any place in the world, and the risk involved with skyscraper due to high intensity earthquake is more, and special attention should be given in its design and analysis, as skyscraper often have thousands of residents [4]. The structure should be designed to perform well during an earthquake and ensure that the structure remains safe without producing major damage to the structure [5].

1.1 Outrigger Structure

It is one of the most effective lateral load resisting systems and has been used for four decades [6]. Outrigger system connects the central core to the external peripheral columns in the case of conventional outrigger systems, and in the case of virtual outrigger systems, there is no direct connection between central core and outermost columns, but it connects outermost columns together so that it will help to reduce the responses [7, 8]. It may be positioned centrally or at one side of the structure [3]. Outrigger system concept has been developed from sailing ship industry. In sailing ship industry, outriggers have been used for many years in order to resist the wind [7]. When the building is exposed to earthquake, wind loads rotational moments at the core will be transferred to outermost columns by joining columns and core and then transferred to foundation; hence, overall deflection of the building will be reduced [2, 3, 6].

2 Objectives of the Present Work

Following are the objectives of the current project adopted from the various literature review done.

- To study and understand the concept of outrigger structural system and to ascertain the optimum conditions for buildings at large height.

- To analyse the important structural response factors such as storey shear, maximum storey displacement and maximum storey drift for dissimilar structural configurations.
- To study the performance of the outrigger system with and without belt truss system/belt wall and to compare wall outriggers with braced outriggers.

3 Methodology

Following steps have been adopted as the methodology process for achieving the anticipated objectives.

- Structure parameters such as floor plans, outrigger type, tower dimensions and number of outrigger levels have been worked out.
- Ground conditions and foundation type are assumed prior to modelling of the structure.
- Drafting and modelling of the tower were done on ETABS using corresponding materials selection. IS codes for concrete and steel are adopted for the analysis.
- Loads are applied for gravity, wind and earthquake condition.
- The analysis was run on the model in the ETABS software. For seismic loads, dynamic analysis was carried out by using response spectrum method. The performance of the structure is checked on parameters such as maximum storey drift, maximum storey displacement and storey shear.
- Results are obtained, and the behaviour of the model is studied, and necessary adjustments are made to give optimum conditions for the building to stay safe. The analysis results are further represented in graphical and tabular forms.

3.1 Details of the Model

- Structure—SMRF
- No. of stories—60
- Storey height—3.5 m
- Total height of the building—210 m
- Grade of concrete—M60 for columns, M40 for beams and M30 for slabs
- Grade of steel—HYSD Fe 500
- Density of concrete—25 kN/m^3
- Density of brick masonry—19 kN/m^3
- Thickness of RC slab—150 mm

- Beam size (bracing beams and all other beams)—230 mm × 600 mm
- Thickness of RC shear wall—300 mm
- Thickness of outrigger wall—500 mm
- Thickness of brick masonry wall—230 mm
- Columns

1–15	Stories	900 mm × 900 mm
16–30	Stories	750 mm × 750 mm
31–45	Stories	600 mm × 600 mm
46–60	Stories	450 mm × 450 mm

- Floor finishes—1.5 kN/m^2
- Live load on floors—3 kN/m^2
- Live load on roof—1 kN/m^2
- Parapet load on top floor = 0.23 × 1 × 19 = 4.37 kN/m
- Wall load on beams = (3.5–0.6) × 0.23 × 19 = 12.673 kN/m Earthquake live load on slab according to clause 7.3.1 and 7.3.2 of IS 1893 (Part-1)—2002 is calculated as:
- Roof (clause 7.3.2) = 0
- Floor (clause 7.3.1) = 0.5 × 3 = 1.5 kN/m^2
- Location—Delhi
- Seismic zone—IV
- Seismic intensity—Severe
- Seismic zone factor (Z)—0.24
- Response reduction factor (R)—5
- Importance factor (I)—1
- Type of soil—medium
- Structure—SMRF
- Damping—5% (for RC structure)
- Supports at base—fixed
- Diaphragm—rigid
- Basic wind speed—47 m/s (for Delhi)

Following models have been studied and analysed in the present study.

Model 1—A bare frame model.
Model 2—Building with core wall.
Model 3—Building with core wall and single outrigger with peripheral belt wall at 30th storey.
Model 4—Building with core wall and double outrigger with peripheral belt wall at 20th and 40th storey.
Model 5—building with core wall and triple outrigger with peripheral belt wall at 15th, 30th and 45th storey (Fig. 2).
Model 6—building with core wall and triple outrigger without peripheral belt wall at 15th, 30th and 45th storey.

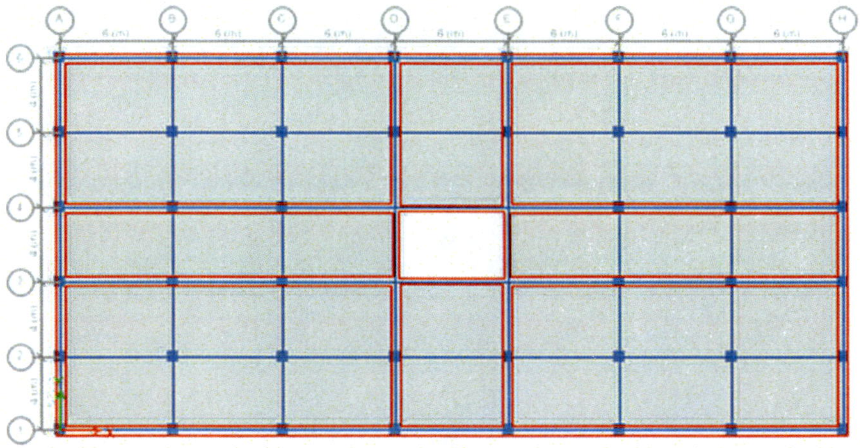

Fig. 1 Plan of the model at outrigger level

Model 7—building with core wall and braced outriggers with peripheral belt truss (X bracings) at 15th, 30th and 45th storey (Fig. 4).
Model 8—building with core wall and braced outriggers without peripheral belt truss (X bracings) at 15th, 30th and 45th storey.

4 Results and Discussions

Optimum location of the outrigger system is of major concern in the present study. It was found out by placing the outrigger at different locations of the structure. Based on some literature survey, it was found that for one outrigger structure placing the outrigger at mid height of the structure will be efficient, and for two outrigger structure, it would be better to place the outrigger systems at 2 h/3 and h/3 from bottom of the building. Optimum position of the multi outrigger system was found out by placing the outrigger systems at different locations of the building.

4.1 Results of Bare Frame Model (M1)

- Maximum storey displacement (Fig. 6)

Maximum storey displacement was recorded under the action of wind loads along Y direction. The maximum storey displacement at the topmost storey of the structure was found to be 1327.56 mm due to WLY direction, 549.446 mm due to WLX

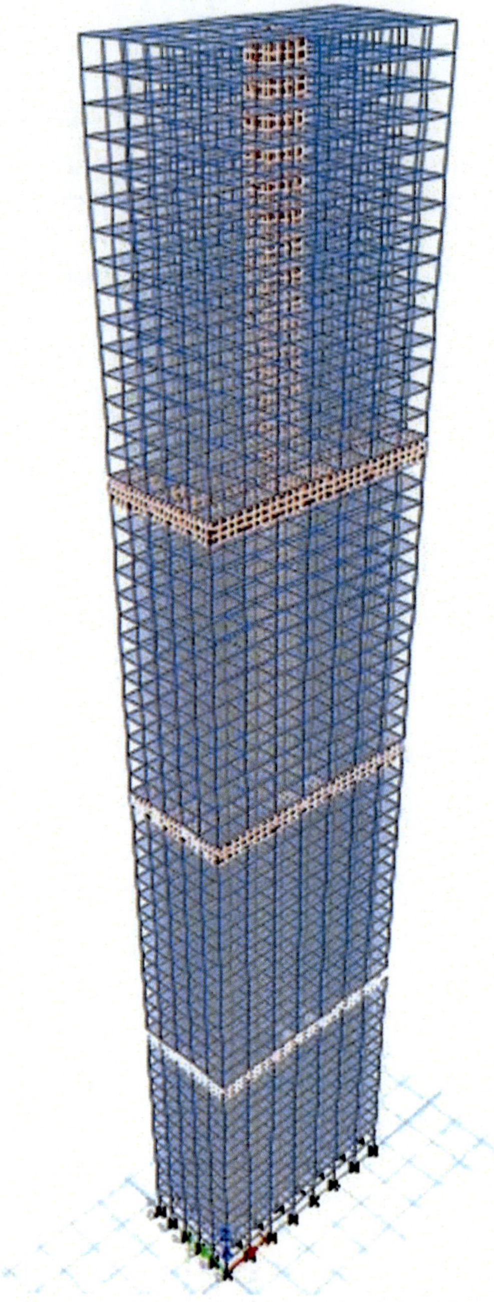

Fig. 2 3D view of the building with core wall and triple outrigger with peripheral belt wall at 15th, 30th and 45th storey

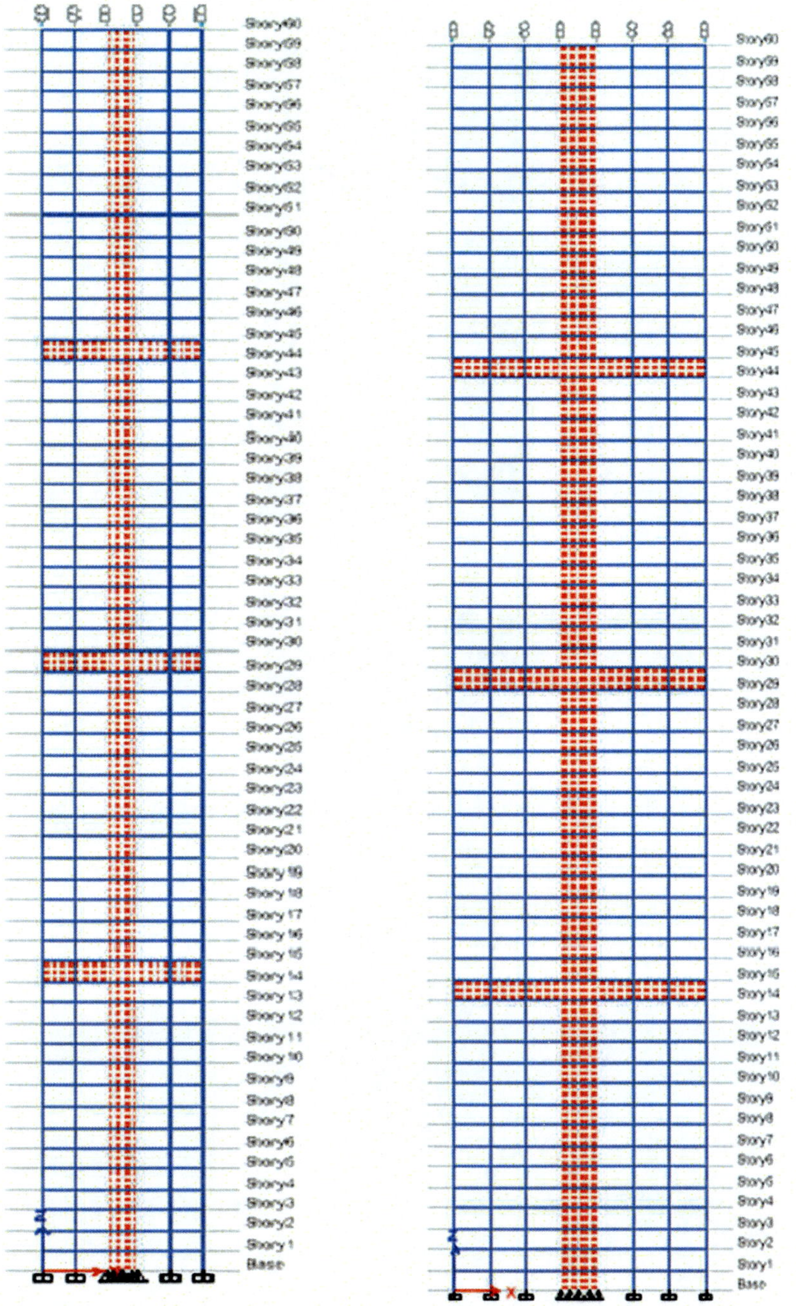

Fig. 3 Elevation views of the building with core wall and triple outrigger with peripheral belt wall at 15th, 30th and 45th storey

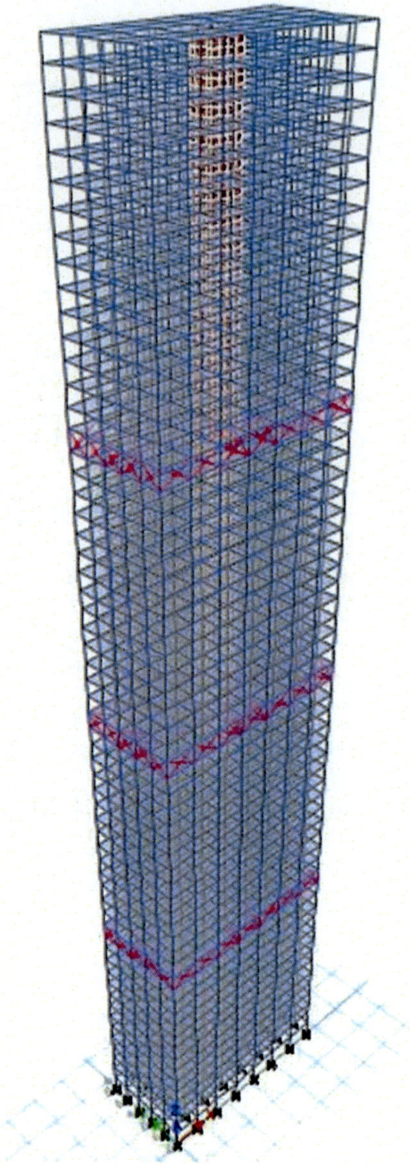

Fig. 4 3D view of the building with core wall and braced outriggers with peripheral belt truss (X bracings) at 15th, 30th and 45th storey

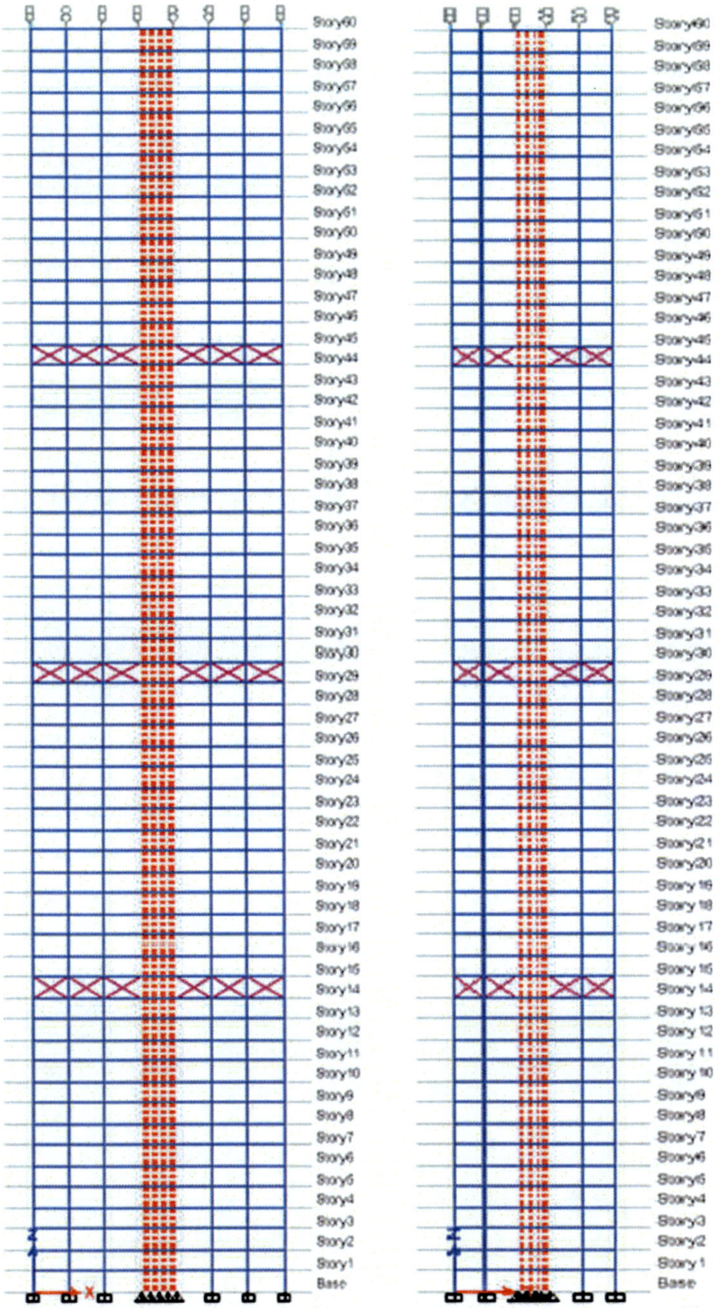

Fig. 5 Elevation views of the building with core wall and braced outriggers with peripheral belt truss (X bracings) at 15th, 30th and 45th storey

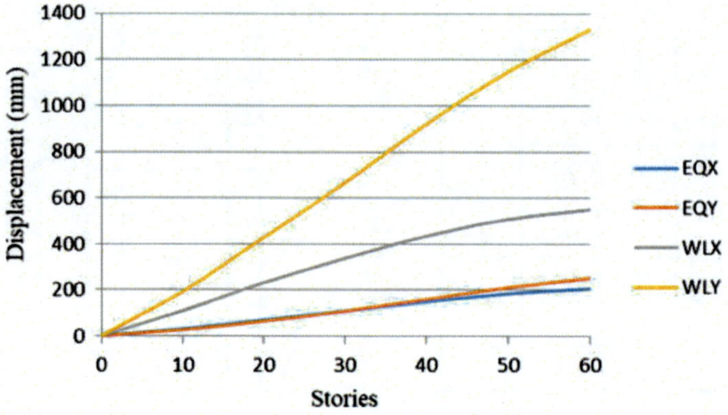

Fig. 6 Variation in maximum storey displacement for M1 under earthquake and wind loads

direction, 209.098 due to EQX direction and 252.885 mm due to EQY direction, respectively.

- Maximum storey drift (Fig. 7)

Maximum storey shear value was recorded under the action of wind loads along *Y* direction. The value of maximum storey drift at 40th storey of the building was found to be 0.00693 mm due to WLY direction, 0.00363 mm at 10th storey due to WLX direction, 0.00148 mm at 40th storey due to EQY direction and 0.00115 mm at 40th storey due to EQX direction, respectively.

- Storey shear (Fig. 8)

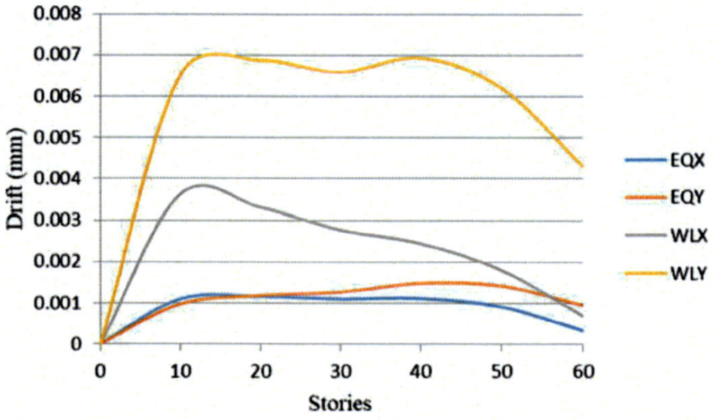

Fig. 7 Variation in maximum storey drift for M1 under earthquake and wind loads

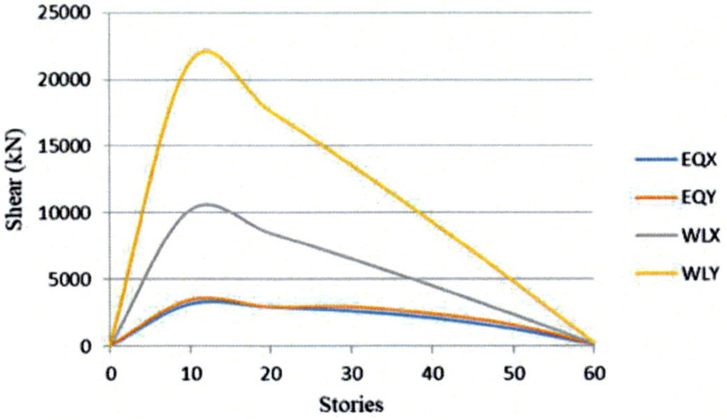

Fig. 8 Variation in storey shear for M1 under earthquake and wind loads

Maximum storey shear value was recorded under the action of wind loads along Y direction. The recorded value is 21272.25 kN at the bottommost storey and 235.15 kN at the topmost storey because of wind loads along *Y* direction. Storey shear value was found to be 3128.8 kN at the bottommost storey and 128.438 kN at the topmost storey due to earthquake loads.

4.2 *Results of Building with Core Wall and Triple Outrigger with Peripheral Belt Wall at 15th, 30th and 45th Storey (M5)*

- Maximum storey displacement (Fig. 9)

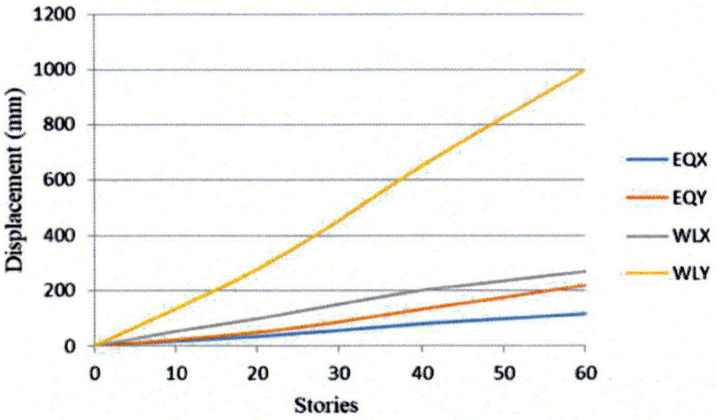

Fig. 9 Variation in maximum storey displacement for M5 under earthquake and wind loads

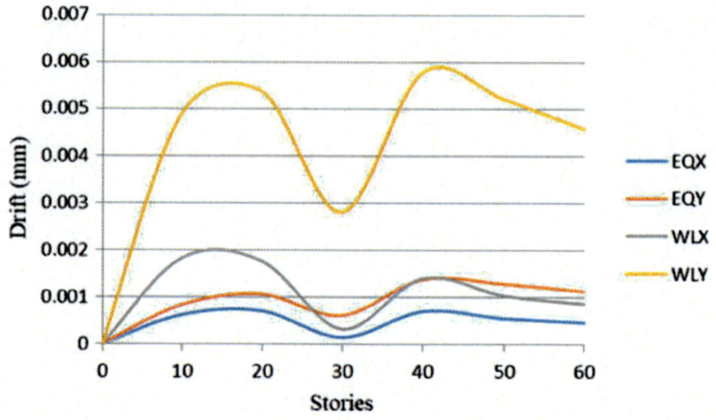

Fig. 10 Variation in maximum storey drift for M5 under earthquake and wind loads

The maximum storey displacement at the top was found to be 994.504 mm due to WLY direction, 266.795 mm due to WLX direction, 215.882 mm due to EQY direction and 116.524 mm due to EQX direction. There is a reduction of 33.48%, 22.76%, 14.43% and 7.24% of storey displacement due to wind loads when compared to storey displacement of M1, M2, M3 and M4 respectively.

- Maximum storey drift (Fig. 10)

The maximum storey drift value was seen to be 0.00578 mm at 40th storey due to WLY direction, 0.00184 mm at 10th storey due to WLX direction, 0.00138 mm at 40th storey due to EQY direction and 0.00071 mm at 20th storey due to EQX direction. There is a reduction of 73.62% of drift due to wind loads when compared to drift of M1.

- Storey shear (Fig. 11)

The recorded value is 3457.45 kN at the lowermost storey and 127.12 kN at the top most storey due to earthquake loads. Storey shear value was found to be 21272.25 kN at the bottommost storey and 235.15 kN at the topmost storey because of wind loads along *Y* direction. There is an increment of 10.5% of storey shear when compared to storey shear of M1 due to earthquake loads. But there is no considerable change in the storey shear because of wind loads by the introduction of triple outriggers in the building.

4.3 Comparative Results of All the Models in Graphical and Tabular Form

- Results for maximum storey displacement (Fig. 12)

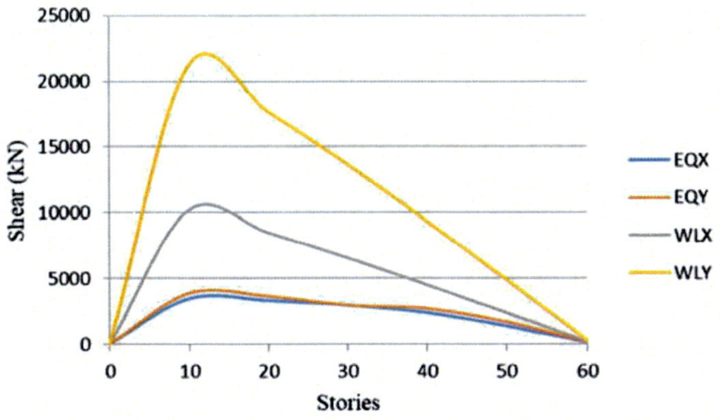

Fig. 11 Variation in storey shear for M5 under earthquake and wind loads

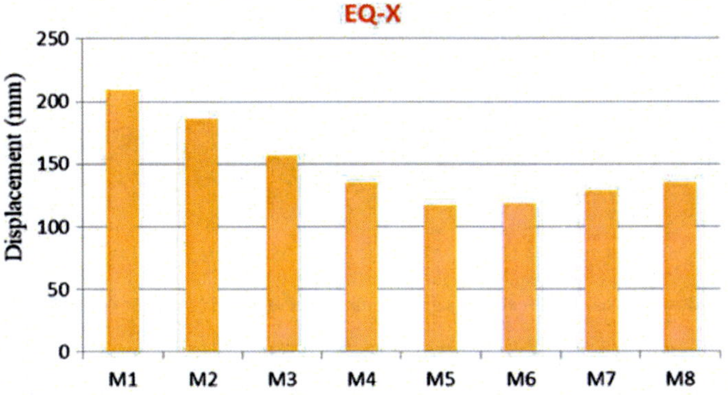

Fig. 12 Variation in maximum storey displacement under earthquake loads along *X* direction for each model

When triple outriggers with belt wall were placed at 15th, 30th and 45th storey, maximum storey displacement has been reduced to 116.524 mm. This was a reduction of 44.27%. When triple outriggers were placed without belt wall, there is a reduction of 43.45% of maximum storey displacement (Table 1).

From the above results, it can be seen that model 1 has maximum storey displacement of 209.098 mm at the top most storey; by the introduction of concrete core wall, there is a reduction of 11.16% of displacement (Fig. 13).

When a single outrigger with belt wall was placed at 30th storey, displacement has been reduced to 377.403 mm. When double outriggers with belt wall were placed at 20th and 40th storey displacement was reduced to 317.308 mm (Table 2).

When triple outriggers with belt wall were placed at 15th, 30th and 45th storey maximum storey displacement has been reduced to 266.795 mm. This was a reduction

Table 1 Maximum storey displacement because of earthquake loads along *X* direction

Model number	Maximum storey displacement in mm	% reduction
1	209.098	–
2	185.76	11.16
3	157.088	24.87
4	135.038	35.41
5	116.524	44.27
6	118.234	43.45
7	128.293	38.65
8	134.954	35.45

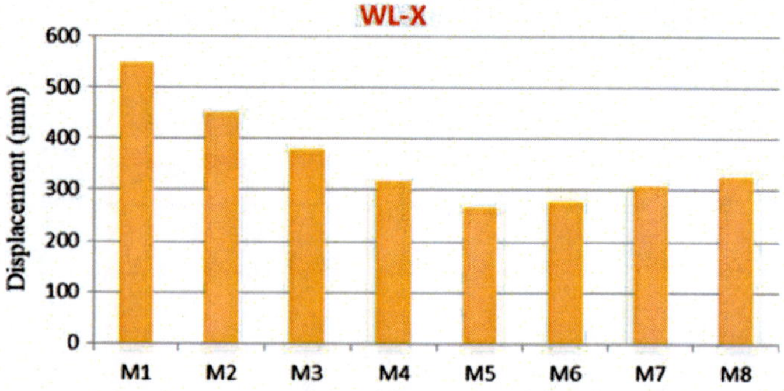

Fig. 13 Variation in maximum storey displacement under wind loads along *X* direction for each model

Table 2 Maximum storey displacement because of wind loads along *X* direction

Model number	Maximum storey displacement in mm	% reduction
1	549.446	–
2	451.511	17.82
3	377.403	31.31
4	317.308	42.23
5	266.795	51.44
6	378.747	49.27
7	307.859	43.97
8	325.974	40.67

of 51.44%. When triple outriggers were placed without belt wall, there is a reduction of 49.27% of maximum storey displacement.

- Results for maximum storey drift (Fig. 14)

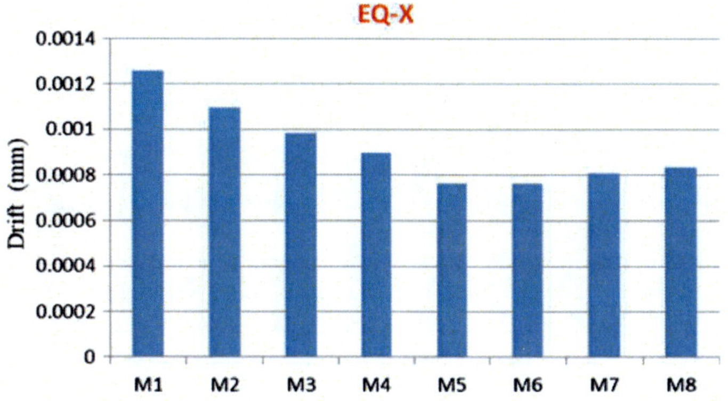

Fig. 14 Variation in maximum storey drift under earthquake loads along *X* direction for each model

By the introduction of concrete core wall, there is a reduction of 12.73% of storey drift. When a single outrigger with belt wall was placed at 30th storey, drift has been reduced to 0.000981 mm. When double outriggers with belt wall were placed at 20th and 40th storey, drift was reduced to 0.000897 mm (Table 3).

When triple outriggers with belt wall were placed at 15th, 30th and 45th storey maximum storey drift has been reduced to 0.000763 mm. This was a reduction of 39.29%. When triple outriggers were placed without belt wall, there is a reduction of 39.46% of maximum storey drift (Fig. 15).

When a single outrigger with belt wall was placed at the 30th storey, drift has been reduced to 0.002572 mm. When double outriggers with belt wall were placed at 20th and 40th storey, drift was reduced to 0.002361 mm (Table 4).

When braced outriggers were placed in place of wall outriggers storey, drift was seen to be 0.002078 mm, and it can be clearly seen that braced outriggers are less effective when compared to maximum storey drift of wall outriggers.

- Results for storey shear (Fig. 16)

Table 3 Maximum storey drift because of earthquake loads along *X* direction

Model number	Maximum storey drift in mm	% reduction
1	0.001257	–
2	0.001097	12.73
3	0.000981	21.96
4	0.000897	28.64
5	0.000763	39.29
6	0.000761	39.46
7	0.000807	35.79
8	0.000835	33.57

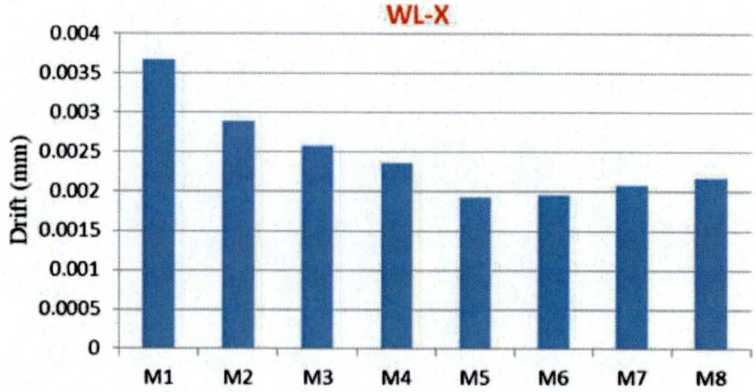

Fig. 15 Variation in maximum storey drift under wind loads along *X* direction for each model

Table 4 Maximum storey drift because of wind loads along *X* direction

Model number	Maximum storey drift in mm	% reduction
1	0.00367	–
2	0.00289	21.25%
3	0.002572	29.92%
4	0.002361	35.67%
5	0.001918	47.74%
6	0.001945	47%
7	0.002078	43.38%
8	0.002168	40.93%

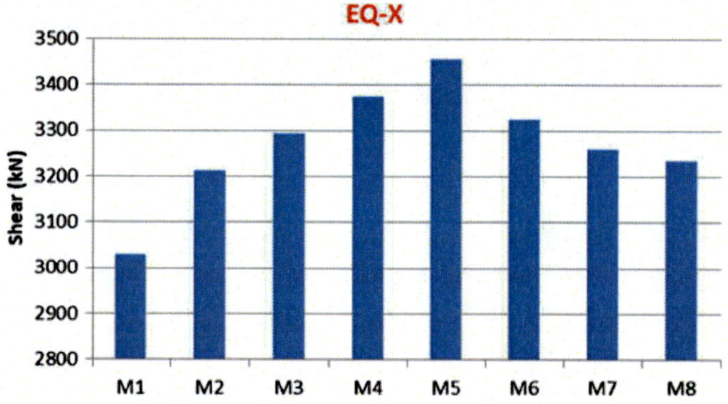

Fig. 16 Variation in storey shear under earthquake loads along *X* direction for each model

Table 5 Storey shear because of earthquake loads along x direction

Model number	Storey shear in kN	% reduction
1	3128.8	–
2	3213.14	2.69
3	3294.51	5.29
4	3375.86	7.89
5	3457.46	10.50
6	3324.58	6.25
7	3259.78	4.18
8	3234.39	3.37

By the introduction of concrete core wall, there is an increment of 2.69% of storey shear. When a single outrigger with belt wall was placed at the 30th storey, storey shear has been increased to 3294.51 kN. When double outriggers with belt wall were placed at 20th and 40th storey, storey shear was increased to 3375.86 kN (Table 5).

- Performance of the building increases with increase in the number of outrigger system, and use of belt truss/belt wall with outriggers is more effective than only outriggers.
- Wall outriggers reduce the responses more when compared to braced outriggers.
- Single outrigger at 30th storey reduces the maximum storey displacement by 31.31% and maximum storey drift by 29.91%.
- Double outrigger at 20th and 40th storey reduces the maximum storey displacement by 42.23% and maximum storey drift by 35.67%.
- Triple outrigger at 15th, 30th and 45th storey reduces the maximum storey displacement by 51.44% and maximum storey drift by 47.74%.
- Triple outrigger without belt wall at 15th, 30th and 45th storey reduces the maximum storey displacement by 49.27% and maximum storey drift by 47%.
- Braced outrigger system with external belt truss at 15th, 30th and 45th storey reduces the maximum storey displacement by 43.97% and maximum storey drift by 43.38%.
- Braced outrigger without peripheral belt truss at 15th, 30th and 45th storey reduces the maximum storey displacement by 40.67% and maximum storey drift by 40.93%.
- The value of storey shear has been increased to very small extent by the introduction of lateral load resisting systems since it increases with increase in the weight of the structure, but this increment is within the acceptable limits; hence, it is of less concern.

When triple outriggers with belt wall were placed at 15th, 30th and 45th storey, storey shear has been increased to 3457.46 kN. This was an increment of 10.5%. When triple outriggers were placed without belt wall there was an increment of 6.25%

of storey shear. When braced outriggers were placed in place of wall outriggers storey shear was seen to be 3259.7752 kN.

5 Conclusion

- Performance of the building under the action of lateral loads increases with the usage of outrigger system when compared to structure without outrigger.

References

1. Raj Kiran Nanduri PMB, Suresh B, Ihtesham Hussain MD (2013) Optimum position of outrigger system for high rise reinforced concrete buildings under wind and earthquake loadings. Am J Eng Res 02(08):76–89
2. Sharma P, Singh G (2018) Dynamic analysis of outrigger systems in high rise buildings against lateral loading. Int J Civ Eng Technol 9(8):61–70
3. Shah NK, Gore NG (2016) Review on behavior of outrigger system in high rise building. IJEDR 03(06)
4. Herath N, Haritos N, Mendis P (2009) Behaviour of outrigger beams in high-rise buildings under earthquake loads. In: Proceedings of Australian earthquake engineering society conference, 2009
5. Rutenberg A, Tal D (1987) Lateral loads response of belted tall building structures. Eng Struct 9(1):53–67
6. Sundar RS, Gore NG (2017) A study on tall RC structure with outrigger system subjected to seismic and wind loading. IJERT 6(02)
7. Gadkari AP, Gore NG (2016) Review on behavior of outrigger structural system in high-rise building. IJEDR 4(2)
8. Gatti, Ishak (1997) Structural behaviour of tall buildings stiffened by outrigger thin wall cores. Multi-purpose high-rise towers and buildings, pp 2–6

Possible Risk on Human Health and Agricultural Land Associated with Application of Humanure

S. Srirashmi, R. Varshini, D. Istalingamurthy, and K. S. Lokesh

Abstract Ecological sanitation can be used in water scare places, and this type of sanitation the end product obtained is humanure which can be applied to agricultural fields to improve the fertility of soil. Ground water contamination, soil pollution can be avoided by this type of sanitation. The study area is Mosarahalla (Singapatna) in H.D. Kote taluk, Mysuru. This village has over 57 eco-san toilets installed in 2006. The decomposed manure is removed once in a year and applied to fields during pre-monsoon season in bear hands. The soil and water samples were collected from different sources of village and was analysed. The chemical parameters like pH, fluoride, total iron, Calcium hardness, Total hardness, chloride and electric conductivity of water was analysed and a comparative study was made as all the sample were in Limits as specified. Soil samples were analysed for pH, electrical conductivity, organic carbon, potash, available nitrogen, available potassium. A questionnaire survey was made to understand the health of the villages. Most of them suffered from skin infection, gastro intestinal problem and also respiratory related issues. Microbial analysis was done for soil and showed that the presence of certain fungi may be responsible for the disease caused among the villagers. The results of analysis of soil and water sample showed that increases chloride content, moderately hard water may be the reason for skin infection and cardiovascular diseases. Microbial analysis showed that certain fungi present in soil samples are carcinogen and potent to cause respiratory disorders. Proper treatment like reverse osmosis can be done to decrease the increased levels of chloride and hardness. Taking proper measure during to application of humanure and proper storage with application of fungicides, the risk of getting effected by humanure can be avoided.

Keywords Humanure · Chemical characteristics of water · Microorganism · Eco-san toilets

S. Srirashmi · R. Varshini (✉)
Department of Civil Engineering, The National Institute of Engineering (Autonomous Under VTU), Mysuru, India

D. Istalingamurthy · K. S. Lokesh
Department of Environmental Engineering, JSS Science and Technology University, Mysuru, India

L. Nandagiri et al. (eds.), *Sustainability Trends and Challenges in Civil Engineering*, Lecture Notes in Civil Engineering 162, https://doi.org/10.1007/978-981-16-2826-9_40

1 Introduction

1.1 General Introduction

Eco-San toilets are composting toilets in which microorganisms treats human excreta in biological process forming compost as end product that can be applied for agricultural land thus "closing the loop" by putting back the nutrients to soil [1]. These toilets are low-cost waterless toilets constructed using locally available materials. When the pit of eco-san toilets get completely filled after about ten to twelve months, the faeces get composted completely and can be used as manure for agricultural fields. The end product of the eco-san toilets is called as humanure. Humanure regenerates the fertility of cultivated land [2]. Using humanure obtained from Eco-San toilets have many advantages in context with artificial fertilizers Ground water contamination and soil pollution that is caused by practice of open defecation [3] can be avoided by use of Eco-San toilets. According to a report in 2018, over 2 billion people live in countries experiencing high water stress. When compared to flushing system of sanitation these toilets use minimal amount or no water hence, it is generally recommended to be used in water scarce places. There are different types of ecological sanitation used worldwide some of them are safichoo, bio-digestor toilets, Eco-san toilets, tent toilets, bamboo toilets, two pit compost, nano membrane toilets, etc.

The Possible Risk Associated with Use of Humanure

When the humanure is exposed to water and air there are possibility of getting infected from pathogens like *E. coli*, listeria and cryptosporidium. The toxic in the humanure can cause respiratory problems [4] in cattle and humans and also effect the reproduction system in cattle. Improperly treated manure can be threat to aquatic life. There are estimated risk of acute gastrointestinal infection due to the presence of aerosolized pathogens. The humanure may also be contaminate with pathogenic bacteria and viruses and also medicines taken for few diseases this can cause nausea, eye irritation, headaches, chronic cough, increased cardiopulmonary risk, asthma, etc.

1.2 Indian Scenario

India is a welfare state according to the constitution. Providing basic facilities is one of the important responsibilities of a democratic country like India. In the post-independence period, the Government of India has been concentrating on various development activities in the country. Most Indians depend on on-site sanitation facilities which means mainly pit latrines in rural areas. In rural areas, the government has been promoting community-led sanitation approaches such as the total sanitation campaign, with some success. The main objectives of TCS included, bring about an improvement in the general quality of life in rural areas. Accelerate sanitation coverage in rural areas generally felt the demand for sanitation facilities through awareness creation and health education. Cover schools/Anganwadi in rural areas with sanitation facilities and promote hygiene education and sanitary habits among students. If we take the example of Mumbai city, an average of 81 persons shares a single toilet. In some places, this figure rises to an eye-watering 273. Even the lowest average is still 58, according to local municipal authority figures. In many urban and semi-urban pockets, it is still a common sight to see people squatting by roads and railway tracks or along the coast, openly defecating in the city where some of the world's richest people live. The level of investment in water and sanitation by international standards has increased in size during the 2000s. For example, in 1980 rural sanitation coverage was estimated at 1% and reached 95% in 2018. Also, the share of Indians with access to improved sources of water has increased significantly from 72% in 1990 to 88% in 2008. In India in 2019, 98.9% have access to "at least basic sanitation". Between 2014 and 2018, the NDA Government in India built around 92.2 million toilets all across India, due to which the basic sanitation coverage went up from 38.7% in October 2014 to 98.9% in February 2019.

The circular sanitation economy (close the loop form) recovers toilet resources for water, energy and agriculture are pilot phase around the country but they are worth of about 14 billion dollars today and could be worth of over 25 billion dollars by 2021 [5]. The government of India is promoting the production and co-marketing of compost through existing fertilizer companies. This study finds that more and more farmers are shifting usage of chemical fertilizers to compost or humanure. In some cases, it is found that the cultivation and input cost can be cut by 70% with organic farming. India is home *to 30% of the total organic producers* in the world, but accounts for just 2.59% (1.5 million hectares) of the total organic cultivation area of 57.8 million hectares, according to the *World of Organic Agriculture 2018* report.

2 Objectives

2.1 Major Objective

To study the effect of application of humanure on human health and agriculture.

2.2 Specific Objective

i. To analyse the chemical characteristics of water collected from various sources
ii. To analyse nutrients, present in different soil samples
iii. To identify the microorganisms mainly bacteria and fungi present in humanure and various soil sample

iv. To identify the possible health risk associated with microorganisms present in humanure through microbial examination and questionnaire.

3 Materials and Methodology

3.1 Study Area

Mysore district falls in the survey of India degree sheet Nos. 48P, 57D, 57H and 58A. The district is bounded by north latitudes 110°45′–120°40′ and east longitudes 750°59′–770°05′ covering an area of 6269 km^2. Mysore district is divided into 7 taluks namely HD Kote, Hunsur, K.R.Nagar, Mysore, Nanjangud, Periyapatna, T. Narasipur for administrative purposes. The study area is a village named Mosarahalla in H.D. Kote taluk.

The study area is a water scare village in which 57 ecological sanitation toilets are installed in 2006 under MYRADA scheme and zillah panchayat with the cost of each toilet being around Rs. 11,000/-. These toilets consist of a trench over with ceramic toilet is placed, the wall is of burnt brick masonry with 12-mm thick plastering on both side and asbestos sheet over top. The excreta is collected in the pit and charcoal ash is mixed which helps for further treatment and transformation of excreta into humanure. Humanure is removed once in every two or one year. The manure that is obtained from eco-san toilets are applied for fields every one year or two year during the pre-monsoon period because humanure is very concentrated and require more for dilution. The major crops cultivated in Mosarahalla Village are ragi, jowar, cotton, banana and groundnut.

Features of Study Area

The taluk is connected by weather roads thus making it accessible throughout the year. The River Kabini which is one of the major tributaries of the River Cauvery flows in this region. It receives an annual rainfall during South-West monsoon. There are 6 rain gage stations in H.D Kote which receives a normal rainfall of 904 mm (1901–70) and actual rainfall of 1171.1 mm (2005). H.D Kote taluk in the southern parts of the district has higher elevation ranging from 2200 to 3150 m AMSL. Mosarahalla has red loamy soil. In the yield wise distribution of aquifers, it is seen that Mosarahalla of H.D. Kote taluk falls under small patches of land with a yield less than one liter per second (lps).

3.2 Sample Collection

Ecological sanitation can be the method used to "close the loop" and put the nutrients back to soil. In Mosarahalla in H.D Kote taluk has eco-san toilets installed in 2006.

The end product of these toilet is humanure which is composite of human excreta, charcoal ash. This humanure is applied to field in bare hand without any gloves.

This study concentrates on the effect of usage of humanure on human health and agricultural crop through conducting microbial test on soil to understand the number of microbes (bacteria and fungi) present in them and their effect on human and agriculture

i. Collection of Water Sample

2-L cans were used to collect the water samples. Six water samples from different source was collected. The cans were rinsed with the source water first. Sampling time and data was recorded on each sample. Water samples were analyzed for various chemical properties.

The chemical properties the water samples were analyzed for *pH, fluoride content, total iron, chloride content, calcium hardness and electrical conductivity* (Table 1).

ii. Collection of Soil and Humanure Sample

Eight representative soil samples were collected from different plots among which three samples were collective sample collected by quadrant method and one humanure sample was collected on 25th Jan, 2020. Five-kilogram polythene bags were used to collect the sample. The samples were analysis for chemical characteristics and soil nutrients. Chemical properties soil was analyzed for was *pH and electrical conductivity. Organic carbon, available nitrogen, available phosphorous and available potash were the nutrients analysed in samples* (Table 2).

iii. Microbial test

Microbiological test was conducted on humanure and soil samples to identify the bacteria and fungi present in the sample. The results from the experiments were helpful to justify the objective of the study.

iv. Questionnaire survey

Questionnaire survey was done to understand the condition of health of villagers. Local temple used as a central meeting point for all the respondents, where the questionnaires were individually administered. There are 2 street in the village and

Table 1 Source of water samples collected

Sample	Source
S1	Drinking water
S2	Water used for toilets
S3	Bore well water
S4	Surface water
S5	Overhead tank water
S6	Surface water near humanure vegetation

Table 2 Source of soil samples collected

Sample	Collection plot
S1	Humanure sample
S2	Sample near surface water
S3	Vegetation soil with no humanure
S4	Sample near humanure toilets
S5	Humanure and other manure applied soil
S6 (a, b, c)	Humanure applied soil sample
S7	Sample in village street

about 57 eco-san toilets are being installed in 2006 by MYRADA plan and zillah panchayat. This questionnaire survey is a quantitative method used for descriptive purpose in this study.

The following questions were asked during the questionnaire survey

1. Name of the person
2. If humanure was used in by them?
3. Number of family members with medication
4. Any diseases like B.P, Diabetes, TB, Skin infections, amoebiasis etc.
5. Age group of the family members
6. Frequency of usage of toilets
7. Medication taken for Specific Diseases
8. Since how many years the medicines are taken?

All the respondents where the residents of the village. The above questions were asked for 13 residents and the respond was noted for further use.

4 Results and Discussion

4.1 Results on Water Sample Analysis

All the results are verified with respect to IS standard code IS:10500-2012 (Tables 3 and 4).

4.2 Result of Soil Sample Analysis

Chemical parameter and soil nutrient (Figs. 1, 2, 3 and 4).

Microbial test

i. Sample name: Humanure sample

Table 3 IS:10500-2012 permissible and desirable limits for drinking water

Parameter	Desirable limits	Permissible limit
pH	6.5–8.5	No relaxation
Fluoride	1.00 mg/l	1.5 mg/l
Total iron	0.30 mg/l	No relaxation
Chloride	250 mg/l	1000 mg/l
Total hardness	300 mg/l	600 mg/l
Calcium hardness	75 mg/l	200 mg/l
conductivity	< 1.1 micro mhos/cm	–

- Bacteria (Fig. 5)

  The number of bacteria present in a gram of humanure sample analysed,
  For 10^{-5} dilution $= \frac{77}{0.1\times10^{-5}}$
  $= 77 \times 10^6$ bacteria per gram of soil sample
  For 10^{-6} dilution $= \frac{100}{0.1\times10^{-6}}$
  $= 10 \times 10^8$ bacteria per gram of soil sample.

- Fungi (Fig. 6)

  For 10^{-3} dilution $= \frac{46}{1\times10^{-3}}$
  $= 46 \times 10^3$ colonies of fungi are present in per gram of soil sample.
  The fungi present in humanure sample are

- *Aspergillus niger*
- *Aspergillus parasiticuss*
- Penicillium

4.3 Discussion

According to drinking water standards of IS:10500-2012 the permissible limit for chloride content is 250 mg/l. The results of the water samples analysed showed that the chloride content is range of 430–476 mg/l. The chloride content is responsible for maintaining fluids in body. When chloride levels are moderately high there are no symptoms of hyperchloremia, however with long term exposure Cl^- can cause a range of symptoms.

- Fertilizers and humanure from eco-san toilets are applied to field from past 10–15 years. Potassium chloride is the salt in most of the commonly used fertilizer as a source of potassium. Potassium increases the soil fertility and improves the yield however, the use of these kinds of fertilizers will leach into rivers and underground water and can results in increase chloride content.

Table 4 Data obtained from questionnaire survey conducted among the residence of village

S. No.	Name of the person	Usage of humanure	No. of family member	Recent history of illness	Number of people with medication	Any long-term disease in family members	Age group	Frequency of toilets used (max)	Medication taken for specific disease	Since how many year (years)
1	Suresh	Yes	4	Wife was operated for thyroid 5 years ago	0	–	31–40 and 11–20 years	2	–	–
2	Putta Naika	Yes	10	A man had severe stomach ace 1 year ago	0	–	> 61, 31–40 and 21–30	2	–	–
3	Chandra Naika	No	3	Burp frequently	1	BP for wife	> 61, 31–40 ad 21–30	2	–	–
4	Sankus Naika	No	2	Father—heart surgery Mother—frequent sneeze	1	Heart problem	> 61 and 51–60	2	Tablets for heart problem	3
5	Siddharaju Chinpatna	Yes	20	–	1	Skin infection	41–50 and 51–60	2		1
6	Madevamma	No	4	–	1	Blood pressure	61 and above	2	Amlodipine tablets	> 5
7	Puttaswamy	Yes	3	–	1	Nose infection	> 60 and 51–60	1	Vicks action 500	4
8	Swamesha	Yes	5	Father—cancer from 10 year Son—dialysis from 3 years	2	Cancer and kidney failure	> 61 and 31–40	1	12 different tablets	10 (causer) and 3 (dialysis)

(continued)

Table 4 (continued)

S. No.	Name of the person	Usage of humanure	No. of family member	Recent history of illness	Number of people with medication	Any long-term disease in family members	Age group	Frequency of toilets used (max)	Medication taken for specific disease	Since how many year (years)
9	Mara Naika	Yes	5	–	1	Asthma	>61	2	Asthalin-salbutanal inhalation 100 mg/dose	5
10	Sanna Swamy	Yes	4	–	–	–	31–40, 21–30 and 0–10	1	–	–
11	Govinda Naika	No	6	Frequency headache	1	Headache	21–30, 31–40 and 41–50	2	Amla juice	12
12	Channa Naika	No	3	–	–	–	> 61, 21–30 and 41–50	2	–	–
13	Devamma	Yes	5	–	1	Gastric	> 61, 51–60 and 21–30	2	Locally available tablets	20

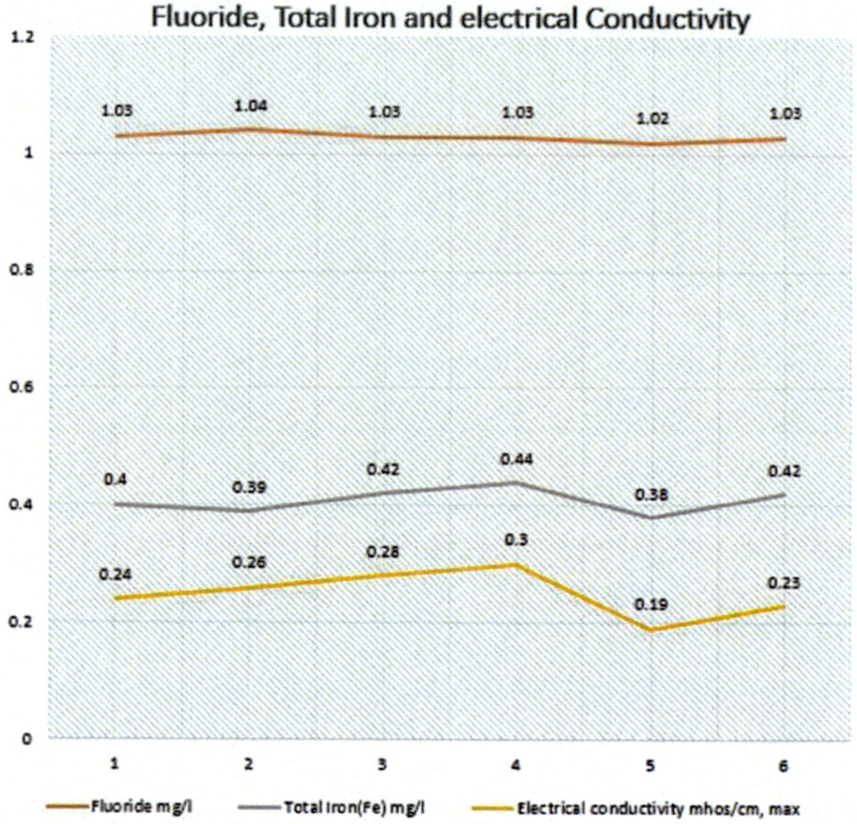

Fig. 1 Result of fluoride, total iron and electrical conductivity in water

- The samples were collected during end of January. If the evaporation losses are greater than the precipitation this may also lead to increase the chloride content in the surface water bodies.
- Leaching of human waste from septic tank and also from open defecation practiced in few parts of village might have led to increase the chloride content over period of time [6].

4.4 Discussion on Calcium and Total Hardness Exceeding Desirable Limit

Water which does not produce lather with soap is called hard water. The hardness does not have a very significant importance with respect to health, but the hardness adheres to the surface of tubs, sinks, utensils, taps etc. it also affects the texture of

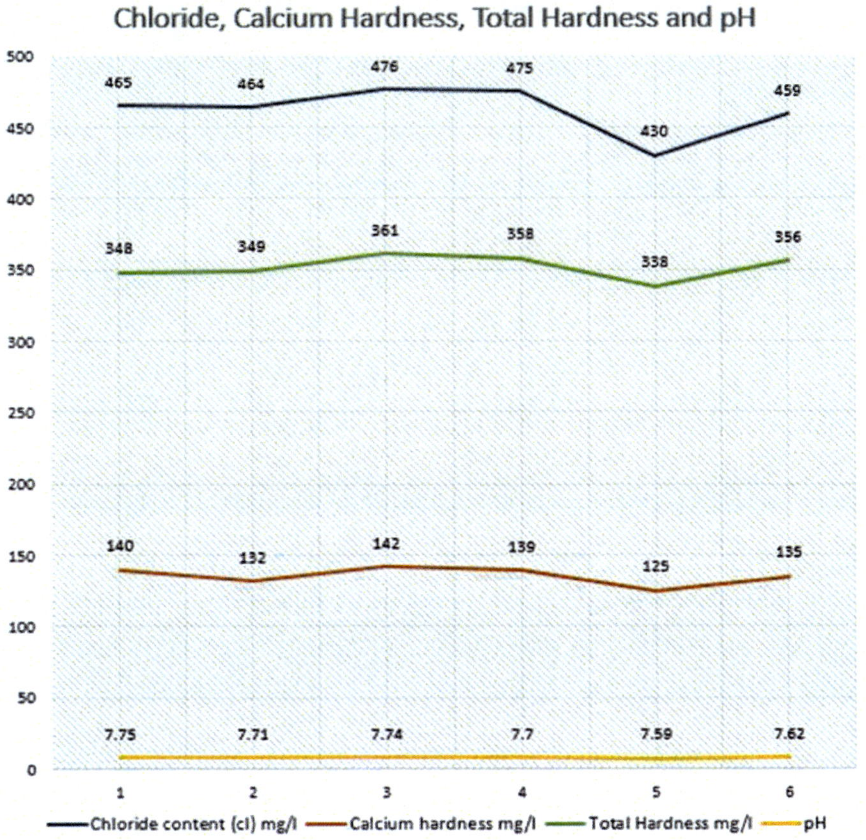

Fig. 2 Result of chloride, calcium hardness and total hardness in water sample

skin and kills natural antibodies on surface of skin and make vulnerable to infections. Water with hardness of 61–120 mg/l as 'Ca' is considered as soft. 121–180 mg/l as moderately hard, more than 180 mg/l as hard.

- Prolong intake of very hard water may be a cause for formation of kidney stones as 3/4th of the kidney stones [7] are composed of calcium salts in form of oxalate and less in form of calcium phosphates.
- The calcium hardness and total hardness of all the samples analysed are in the range of 125–142 mg/l and 338–361 mg/l, respectively, which tell that is it moderately hard.

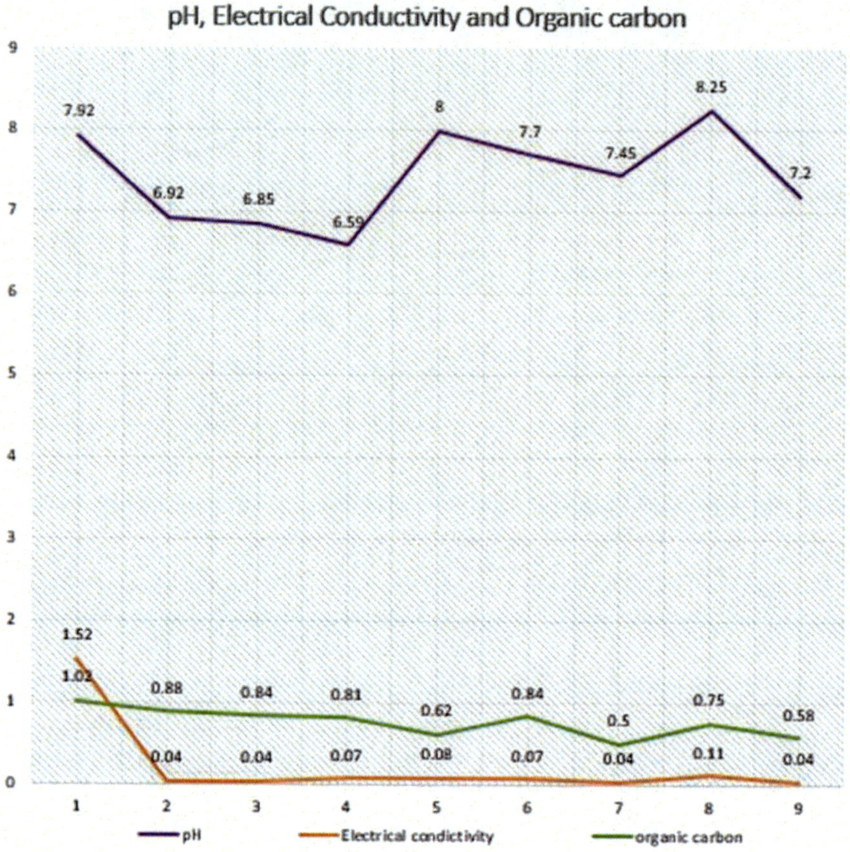

Fig. 3 Result of pH, electrical conductivity and organic carbon in soil sample

4.5 *Discussion on Questionnaire Survey*

The questionnaire survey was done with the residents of villages to understand the condition of the heath. During the questionnaire survey it was found that most of the residence suffered gastrointestinal problems, respiratory problems, headache, high blood pressure and skin infection. The probable reason for these health issue may be

- The increased chloride content can lead to hyperchloremia which causes a wide range of symptoms like high blood pressure, fluid retention, numbness, confusion also muscle spam
- The increase in hardness make skin vulnerable for skin infection
- The application of humanure in bare hands can causes respiratory problems, skin problems, nausea and also it may be due to aerosol microorganism present in environment.

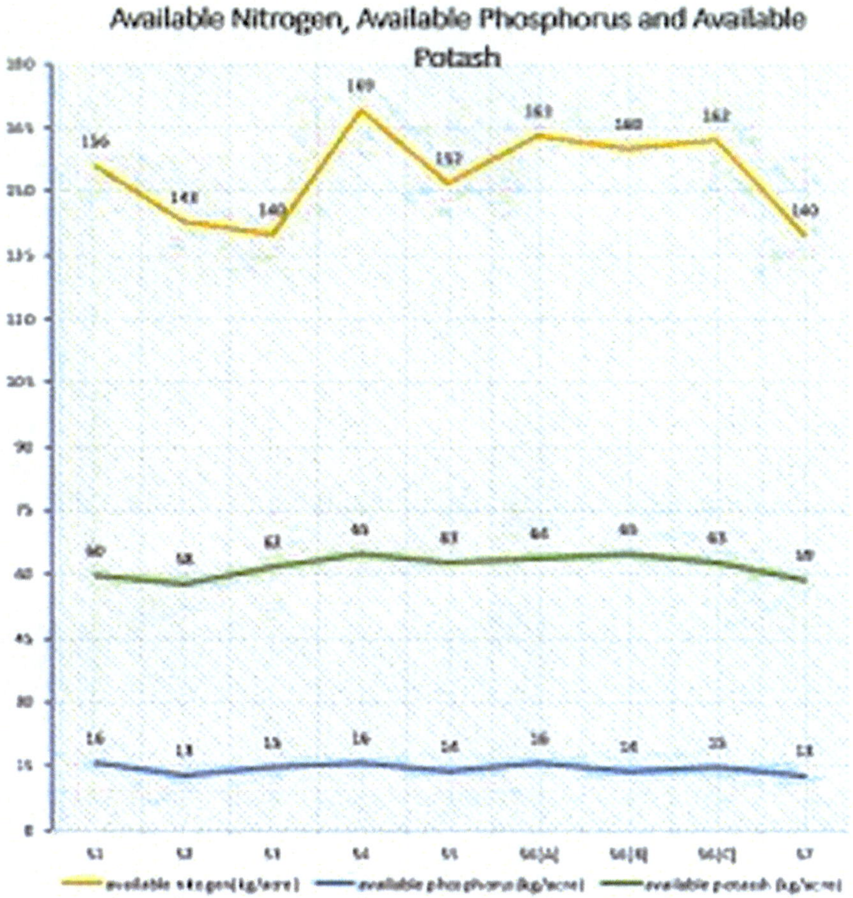

Fig. 4 Result of available nitrogen, available phosphorus and available potash in soil sample

- Few crops like ground nuts and corn has ability to get contaminated by toxins produced by fungi (mycotoxin). Few fungi like Penicillium and *Aspergillus parasiticuss* produces a mycotoxin called as aflatoxin. This aflatoxin is a potential carcinogenic. This may also be a reason for kidney failure and cancer caused among a resident.

4.6 Discussion on Humanure Sample

From the analysis of humanure sample it is found that the levels of organic carbon, conductivity, and pH are high and available nitrogen, Phosphorus and Potash are in

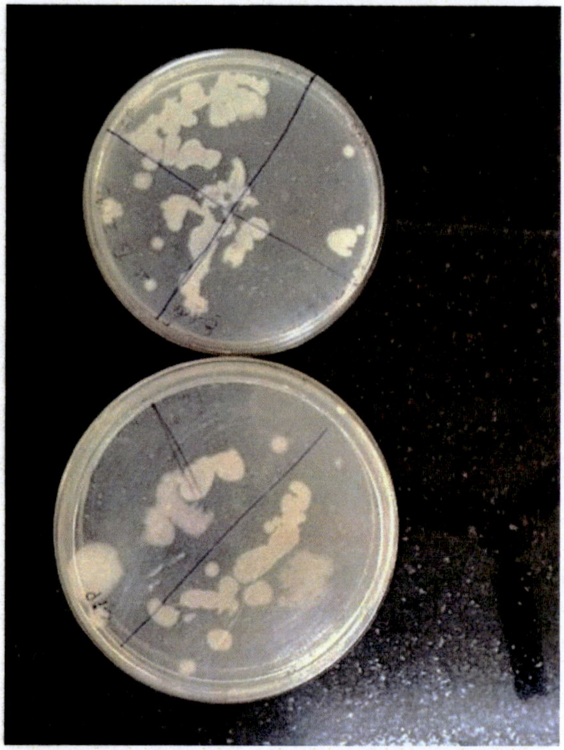

Fig. 5 Developed bacterial colonies

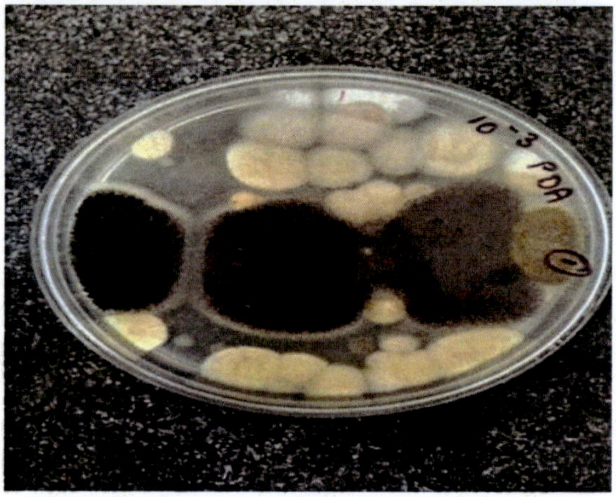

Fig. 6 Developed fungi in humanure soil

medium range. The pH of humanure sample was found to be 7.92 which is slightly high (alkaline). The use of alkaline pH can disrupt the ability of plants to absorb certain nutrients. The probable reason for alkaline pH in humanure sample are

- Human stool is alkaline in pH thus making the humanure slightly alkaline
- The pH of any compost will be slightly acidic initially and during the course of time due to microbial activity the pH of humanure becomes Alkaline, this may also be the reason for alkaline pH in our sample.

The electrical conductivity of the humanure is also high as 1.52. The reason of high conductivity may be due to

- Human excreta have bicarbonates, chlorides and salts like sodium and potassium in it hence, the presence of these ions can rise the conductivity of humanure sample.

4.7 Discussion on Increased pH in Soil Samples

The pH of the soil samples varies from 6.59 to 8.25. The pH of the soil influences the intake of micro and macro nutrients by plants. It is found that plants can absorb nutrients easily from slightly acidic soil. The possible cause for rise in pH may be due to

- The water used is moderately hard which has ions like carbonated and bicarbonates. The presence of carbonated and bicarbonates in soil can rise the pH of soil making it Alkaline in nature.
- The study area has pockets of limestone. The weathering of such rocks can also rise the pH of the soil.
- Humanure is applied to field this may also be a reason for increase in pH [8].

4.8 Discussion on Increased Organic Carbon in Soil

Organic carbon has important role to improve the soil fertility, better water holding capacity and also effect the microbial metabolic activity. The reason for increased organic carbon in soil sample are

- Coal ash is added for treatment of humanure this will increase the organic carbon in soil.
- The erosion of top soil from elevated field can also lead to increase in organic carbon due to deposition.
- Sample 4, Sample near Eco-san toilets has carbon content of 0.81 which is slightly high. The leaching of humanure might be the reason for this.

- Vegetation where humanure was applied has high organic carbon thus this is because of application of humanure for better fertility of soil and to obtain good yield [9].

5 Conclusions

Humanure is end product of the eco-san toilets. The humanure when applied to agricultural field increases the fertility [10], and humus content of the soil which improves the yield of crops [11]. The application of humanure in bare hands and improper storage can be peril for human health. Many researches have shown the effect of application of humanure on soil, water and human health.

In Mosarahalla, H. D. Kote taluk there are about 90 houses in 2 street of village and 57 eco toilets are installed in the year 2006. The water and soil samples were collected from different sources and places in the village. A questionnaire survey was done to understand the status of health among the villagers. The analysis of water samples revealed that the chloride content was high and water was moderately hard. The possible reasons for increase in chloride content are maybe due to geological setting, use of fertilizer and humanure to fields [12] and also may be due to leaching of wastes from septic tanks. The increase in hardness is may be due to geological setting. H. D. Kote has pockets of dolomitic limestone, calcite and high calcium lime stone which might be the cause for increase in hardness.

The microbial test of the samples showed the presence of bacterial colonies and few fungi like *Aspergillus parasiticuss*, Penicillium and *Aspergillus niger*. The questionnaire survey results show many residents suffered from dermatitis, kidney failure [7, 13], headache, high blood pressure, respiratory problems and cancer. The fungi present in the soil sample produces a mycotoxin named aflatoxin which is a potential carcinogenic. Aspergillus are also responsible for causing respiratory problems and polyps in nose. The water with excess of chloride can result in cardiovascular disease, hyperchloremia and nausea [14]. Hardwater can kill the natural antibodies present on surface of skin making it vulnerable for skin infections [15]. Humanure sample and vegetation where humanure was applied has oil with alkaline nature this can be treated by adding proper amount of gypsum to soil and also by growing leguminous plants [16] in crop rotation. The use of humanure in bare hands can also result in respiratory problems and skin infection however, prolonged contact of humanure containing those fungi can be carcinogenic.

Excess chloride content and hardness can be removed by treating the water through reverse osmosis. Reverse osmosis can potentially reduce the concentration of dissolved solids and salts. The risk associated with humanure can be controlled by taking personal protective measures [PPM] [17] like wearing rubber or latex gloves during the application, wearing mask to avoid inhalation of air bound microorganisms and wear gumboots. Consuming food crops grown in such field (corn, house gram and groundnut) [15] can also be a risk, however, cleaning in fresh running water and cooking in high pressure vessel can kill the microbes present. The fungi present in the humanure if treated with fungicide and baking soda diluted with water can be applied to fields to kill the fungus in soil.

References

1. Jenkins JC The humanure handbook
2. Wang B et al Soil productivity and structure of bacterial and fungal communities in unfertilised arable soil
3. Harilal V An assessment of the viability and sustainability of the use of humanure for household agriculture purposes. J Case Study Cotton Lands eThekwini Municipality
4. Komnitsas K et al Assessment of human health and ecosystem risk due to agricultural waste compost applications on soils
5. Toilet board coalition. The sanitation economy in India
6. Hunt M et al Chlorides in fresh water
7. Shuster J et al (1982) Water hardness and urinary stone disease
8. Phuc PD et al Human health implications by humanure. J Case Study Nghean Province Vietnam
9. Li X et al Short term response of soil microbial community to field conversion from dry land to paddy land under the land consolidation process in North China
10. Jenkins J (2015) Thermophilic composting as a sanitation alternative, Haiti. J Case Study Curr Bus Model Sanergy Kenya
11. Urra J et al Potential benefits and risks for soil health derived from the use of organic amendments in agriculture
12. Effebi KR et al Physicochemical and microbiological characterisation of human faeces and wine from composting toilets in Abidjan
13. Bellizzi V (1999) Effects of water hardness on urinary risk factor for kidney stone in patients with idiopathic nephrolithiasis
14. Shuster J et al (1982) Water hardness and urinary stone disease. J Urol
15. Deepa N, Sreenivasa MY A review on its global occurrence, epidemiology, toxicity and detection
16. Bassil RJ et al Antibiotic uptake by plants from manure-amended soils
17. Gajurel DR et al Sanitation and associated hygienic risk

An Investigation of Coefficient of Torsional Irregularity for Irregular Buildings in Plan

K. K. Sneha and Janjanam Durgaprasad

Abstract In many of the structures in the present scenario, most of the buildings have irregular configurations in both plan and elevation of the structure. This irregular configuration in future may subject to damage due to earthquakes. Structures experience lateral deflections under earthquake loads. Asymmetric distribution of mass, rigidity and strength due to irregularity in the design of the building when the building is exposed to translations, and unnecessary floor rotations are the key causes of serious harm to the structures. In most of the research works, it is repeatedly confirmed that irregular structure suffers more damage than regular structure. The code proposed torsional irregularity coefficient does not include eccentricity in the direction of excitation of earthquake. The objective of this work is, first of all, to analyze the requirements for excessive structural irregularity and, secondly, to address the possible irregularity provisions of the code. A parametric investigation is conducted on six types of asymmetrical normal buildings with different floor numbers and structural wall location and six types of unusual L-shaped buildings with different story numbers to accomplish this task. Based on this analysis, the maximum irregularity coefficient found in 1 story structure and torsional irregularity coefficients are found to increase as the number of story in most structures decreases. Also, floor rotation increases with respect to number of floors, as the story numbers increases, floor rotation reach the maximum value.

Keywords Torsional irregularity coefficient · Rotation of floor · Irregular buildings

K. K. Sneha (✉) · J. Durgaprasad
Department of Civil Engineering, Nitte Meenakshi Institute of Technology, Bangalore, Karnataka, India

J. Durgaprasad
e-mail: durgaprasad.j@nmit.ac.in

L. Nandagiri et al. (eds.), *Sustainability Trends and Challenges in Civil Engineering*, Lecture Notes in Civil Engineering 162, https://doi.org/10.1007/978-981-16-2826-9_41

1 Introduction

Earthquake is one of the most dangerous and devastating natural disasters from which it was very difficult to save our engineering properties. The reliability of the build environment's seismic strength has to be checked and many analytical procedures have to be developed in order to ensure that the structure is resistant to the earthquake.

Several guidelines are being updated on this topic all over the world. The behavior of the building depends on many factors during an earthquake, such as stiffness, sufficient lateral strength, ductility, simple and periodic building configurations. More harm is suffered by buildings with irregular geometry and nonuniformly distributed mass and stiffness. The irregularity in plan or elevation of the building causes the structures to get highly effected by seismic forces. The irregularities of the building are of many types; mainly they are branched into two groups, they are plan irregularities and vertical irregularities.

In this work, an investigation is done on torsional irregularity which is a type of plan irregularity. There are numerous studies exploring different aspects of the torsional irregularity of the building. Ozmen et al. [2] considered certain group of buildings and performed seismic analysis using ETABS. If the eccentricity decreases, they find that the structure is subject to torsion as the structure's torsional rigidity decreases. On the contrary, as the eccentricity increases, torsional rigidity of the structure is found to be increased. Kewalramani and Syed [3] performed analysis on symmetrical and asymmetrical reinforced concrete buildings of uniform panel dimension. They have found that interstory drift is maximum for the square shape building due to inadequacy. Also, base shear results were found maximum for square shape building indicating that structure is stiffer for seismic response. They concluded that significant variation in values of torsion-related parameters was observed due to asymmetry in plan and the presence of stiff elements within the structure. Ilerisoy [4] an investigation of structural irregularities has been done in different earthquake codes of certain countries. The seismic codes of different countries having various seismic regulations were compared and the results of plan irregularities are analyzed. Hussain and Tengli [5] in their work, regular and irregular building has been used for analysis purpose which is provided with a certain percentage of irregularities with constant height. Results of the study indicate that there is an increase in shear forces in columns due to torsion in irregular structures. They concluded that the planned structural irregularities had a major impact on the structure's seismic response, especially in terms of base shear and displacement. Ozhendekci and Polat [6] concluded that code proposed torsional irregularity parameter does not include the eccentricity in the direction parallel to excitation due to seismic forces. They proposed new parameter, effective modal masses to define torsional irregularity.

In this work, asymmetric regular and L shaped irregular buildings have been considered. Six types of asymmetric regular building have been chosen with varying shear wall positions in a rectangular plan. On the other hand, six types of asymmetric irregular building of L shape are chosen. Both the types of models are varied with number of floors, i.e., 10 floors, 5 floors, and 1 floor buildings are considered in this work. The outcome of this work is to determine the reasons for excessive torsional behavior of buildings and to address the provisions provided for torsional irregularity in codes. Based on the studies, it is found that torsional irregularity coefficient, eccentricity, and rotation of the floor are the parameters that defines the severity of torsional effect on the building.

Equivalent stiffness method is used to determine the eccentricity and angle of rotation of the floor. The torsional irregularity coefficient of the building is determined using computer analysis. The provisions of IS 1893: 2016 (part 1) Indian Standard, 'Criteria for Earthquake Resistant Design of Structures', are followed throughout the investigation of this work.

2 Torsional Irregularity of Structure

Torsional irregularity is one of the types of plan irregularity of the building. The torsional behavior in plan asymmetric building is one of the frequent causes for failure during strong ground motions and structural damage. The unsymmetrical distribution of mass and stiffness along the height of the building is one of the most common reasons for the torsion in buildings. Seismic forces are required to act at the floor center of mass (CM) and lateral force resistance acts at the floor diaphragm center of rigidity (CR). Hence, if CM coincides with CR of floor diaphragms, only translational motion gets produced without any rotation when lateral force acts to the structure. Hence, the distance between the CM and CR should be less to prevent the building from rotation. In Fig. 1, it can be observed that the CM and CR are at the same location. Due to the effect of lateral forces, eccentricity gets produced which leads to torsional moment in each floor. The larger the eccentricity between the

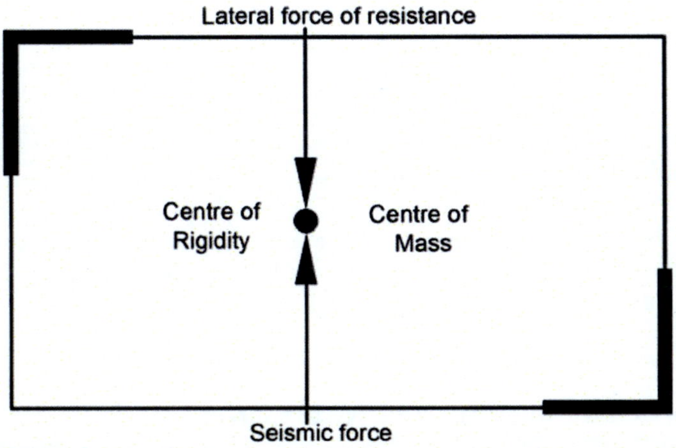

Fig. 1 Regular building

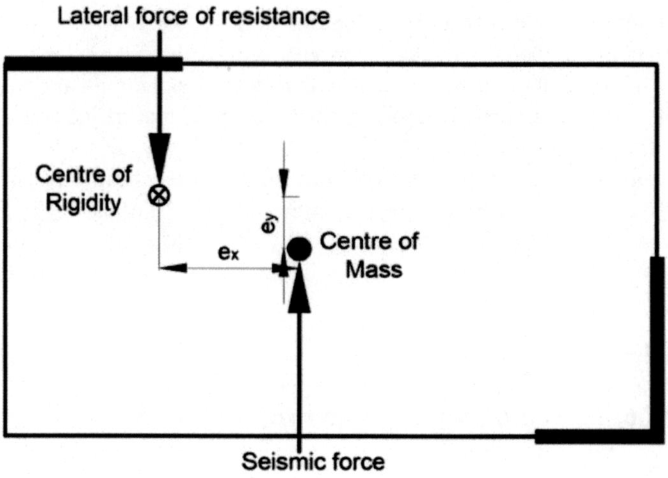

Fig. 2 Building with eccentricity

center of rigidity and center of mass, the larger will be the torsional effects. Figure 2, shows a floor diaphragm where the CM and CR did not coincide. This leads to larger displacements and drifts in the building.

We can say that the eccentricity which is produced due to the center of mass and rigidity is the main reason for torsional behavior of building when it is excited by the seismic forces. In any particular floor diaphragm of the structure, the eccentricity between the CM and the CR may be caused by the asymmetric distribution of the stiff lateral force resisting elements with respect to the center of gravity of the floor, or by the asymmetric existence of large masses with respect to the structure's CR. The severity of torsional irregularity is presented by torsional irregularity coefficient.

3 Torsional Irregularity Coefficient

3.1 Definition

The ratio of the maximum story drift at one end of the structure to the average displacement drift at two ends of the structure is the torsion irregularity coefficient defined by the code.

3.2 Torsional Provisions in Seismic Codes

In general, the torsional irregularity coefficient represented in most of the codes as

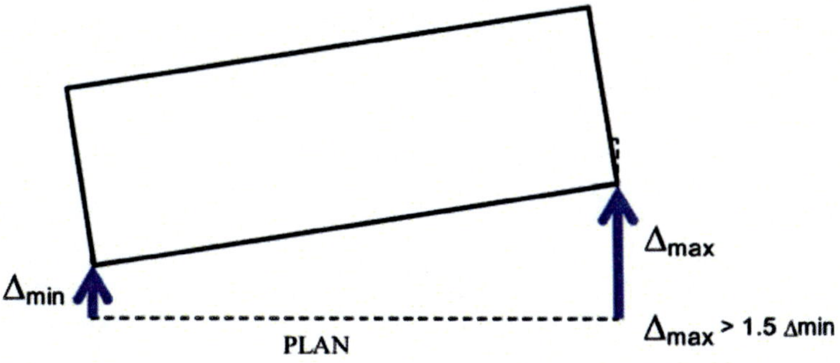

Fig. 3 Maximum and average displacements [14]

$$\eta_t = \frac{\delta_{\max}}{\delta_{\text{avg}}} \tag{1}$$

where

$\delta_{\max}$ = Maximum drift of story at one of the structures.
δ_{avg} = Average drift of two end of the structures.

As per IS 1893:2016 (part 1) Indian standard code, if

- $\eta_t < 1.5$, torsional irregularity does not exist in the structure.
- $\eta_t > 1.5$, torsional irregularity exists and the building configuration need to be revised (Fig. 3).

4 Equivalent Stiffness Method

The eccentricity and angle of rotation of the floor which has been excited by the seismic forces are determined by using equivalent stiffness method. For complex structures, the concept of equivalent bending (*K*) and torsional stiffness (*J*) are used to estimate the angle of twist. Torsional rigidity (stiffness) is a characteristic feature of the material that indicates how rigid it is and its resistance when twisted at an angle per grade change.

Following are the major steps need to be followed to determine the angle of rotation of each story of building as per equivalent stiffness method.

4.1 Centre of Rotation

If an earthquake occurs, the rotation of buildings is expected about their center of rigidity, which can also be known as center of rotation. Select the proper coordinate system, say X, Y, Z for the required model. Mark all vertical members and their stiffness K_1, K_2, K_3, ..., K_n on the plan XY. The location of center of rotation can be obtained using,

$$\bar{X} = \frac{\sum K_i x_i}{\sum K_i} \tag{2}$$

$$\bar{Y} = \frac{\sum K_i y_i}{\sum K_i} \tag{3}$$

K_i = Rigidity of member against shear force.
E = Modulus of elasticity of member.
I = Moment of inertia of the section.
x_i = distance of structural member from origin of building to center of gravity of structural member in x-direction.
y_i = distance of structural member from origin of building to center of gravity of structural member in y-direction.

4.2 Lateral Load

The total lateral force acting on each story is calculated to determine the torsional moment. The total lateral force is calculated as,

$$Q_{ix} = \left(\frac{W_i h_i^2}{\sum_{j=1}^{n} W_j h_j^2} \right) V_B \tag{4}$$

Q_{ix} = Total lateral force at floor i in x-direction.
W_i = Seismic weight of floor i.
h_i = Height of floor i from the base.
V_B = Base shear along principal direction of building.

4.3 Torsional Moment

The more the distance between the Centre of Rigidity(CR) and Centre of Mass(CM), eccentricity will get generated and the structure is forced to twist around the center of rigidity and subjected to great torsional moment. Eccentricity, which is the distance between the CR and CM, induces a moment that creates additional forces on the building. The torsional moment can be calculated as,

$$T = W_x \times \text{Eccentricity} \tag{5}$$

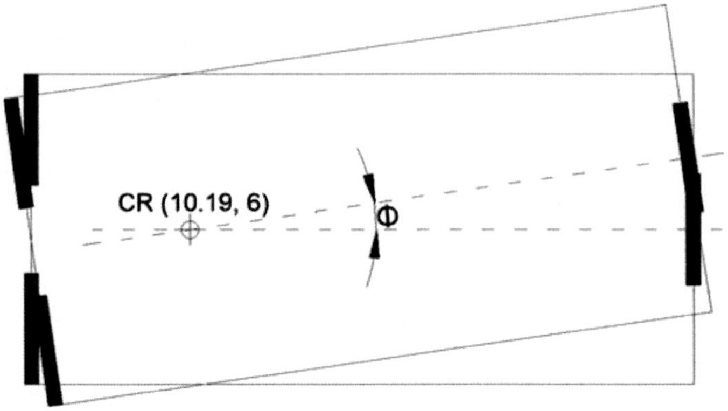

Fig. 4 Angle of rotation

T = Torsional moment.
W_x = Lateral force acting in x-direction.
Eccentricity = Spacing between the CM and CR.

4.4 *Angle of Rotation*

The building is expected to twist or rotate at center of rotation at some angle due to torsional moment. The angle of twist or rotation (θ) can be calculated as,

$$\theta = \frac{T}{J} = \frac{T}{\sum K_i r_i^2} \tag{6}$$

θ = Angle of rotation.
K = Stiffness of each vertical member.
r = Perpendicular distance of each vertical member from the center of rotation.

5 Model Parameters of Building

5.1 *Asymmetrical Regular Building*

Six standard regular asymmetric models, i.e., A, B, C, D, E, and F, have been taken into account and consist of frames and walls. All the models are composed of 30 × 12 m^2, five bays along x-direction and three bays along y-direction. The models are provided with varying shear wall position. The position of shear walls is arranged in such a way that they are symmetrical along x-axis. The schematic floor plans of asymmetrical regular buildings are shown in Fig. 5.

Column size 600 × 300 mm, beam size 300 × 500 mm, thickness of slab 125 mm, thickness of wall 200 mm, live load 2 kN/m^2, floor finish 2 kN/m^2, zone factor, $Z = 0.16$ (zone III), site condition II (medium soil), response reduction factor, $R = 5$ (SMRF), importance factor, $I = 1.5$.

5.2 L Shaped Irregular Building

The typical structures have ten bays in x-direction and ten bays in y-direction, and each bay is placed at the distance of 5 m. Six types of models have been considered L1, L2, L3, L4, L5, and L6. The schematic floor plans of L shaped regular buildings are shown in Fig. 6.

The model parameters of this L shaped irregular buildings are same as that of asymmetrical regular buildings.

Using equivalent stiffness method, the parameters center of rotation, lateral force, torsional moment, and angle of twist have been found out. From the obtained results of this parameter, discussion has been done regarding the code proposed torsional irregularity coefficient for the buildings.

Tables 1, 2 and 3 show the results of equivalent stiffness method that is obtained for type A, 10 story, 5 story, and 1 story structures. Similarly, the method is followed for other types of buildings.

6 Results and Discussions

6.1 Asymmetrical Regular Building

The main parameter which defines the torsional irregularity is torsional irregularity coefficient, nothing but the ratio of maximum displacement by average displacement.

Table 4 shows the maximum horizontal displacement limit for all 10 story, 5 story, and 1 story asymmetric regular buildings.

Table 5 shows the torsional irregularity coefficient of 10 story asymmetric regular buildings. This coefficient is determined for 5 story and 1 story structures too. The results are obtained by seismic analysis of the building using computer analysis.

The variation of maximum torsional irregularity coefficient of all 10, 5, and 1 story structures are represented in terms of graph as shown in Fig. 7.

From Fig. 7, it is noted that the maximum torsional irregularity coefficient value is found to be in 1 story structure in all the asymmetrical regular structure.

Fig. 5 Schematic floor plans of asymmetrical regular building [2]

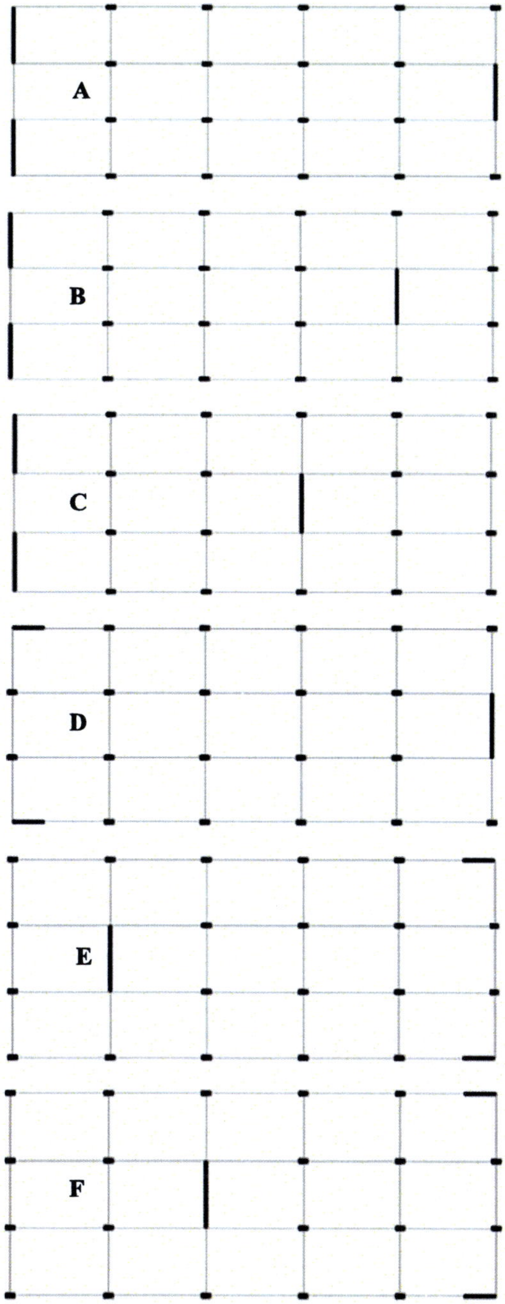

Fig. 6 Schematic floor plans of L shaped irregular building [1]

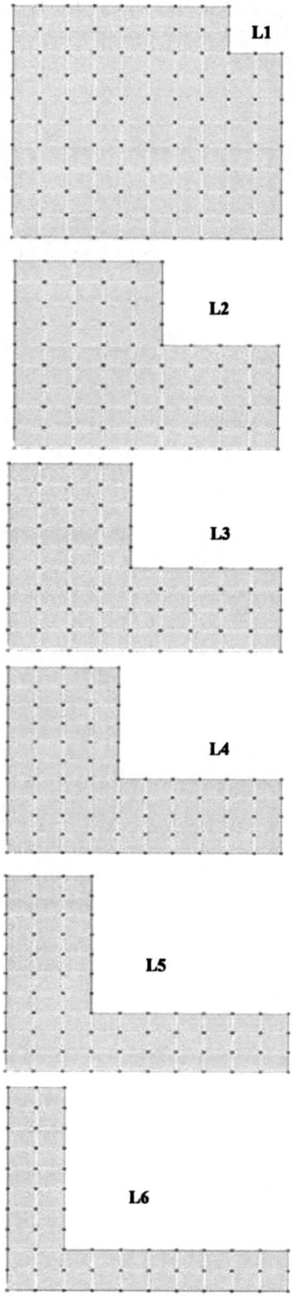

Table 1 Angle of rotation obtained by equivalent stiffness method for type A, 10 floor structure

Story No.	Type A, 10 floor structure					
	Center of rotation (m)	Lateral force (kN)	Eccentricity (m)	Torsional moment (kN m)	$K * r_i^2$	Angle, θ (radian)
10	10.19	468.40	4.89	2917.197	1,782,828	0.00163
9	10.19	605.07	4.89	4351.141	2,438,175	0.00178
8	10.2	483.58	4.88	3472.697	3,456,906	0.00100
7	10.2	375.70	4.87	2697.960	5,128,633	0.00052
6	10.2	281.41	4.87	2020.880	8,068,145	0.00025
5	10.21	200.72	4.86	1439.447	13,729,534	0.00010
4	10.21	133.64	4.86	958.349	26,085,053	3.6E−05
3	10.22	80.151	4.85	573.970	58,399,366	9.8E−06
2	10.26	40.26	4.81	286.699	1.7E+08	1.6E−06
1	10.44	14.039	4.63	97.447	7.9E+08	1.2E−07

Table 2 Angle of rotation obtained by equivalent stiffness method for type A, 1 floor structure

Story No.	Type A, 5 floor structure					
	Center of rotation (m)	Lateral force (kN)	Eccentricity (m)	Torsional moment (kN m)	$K * r_i^2$	Angle, θ (radian)
5	10.19	115.40	4.89	565.276	1,979,708.61	0.000285
4	10.19	137.98	4.89	675.007	2,839,058.14	0.000237
3	10.2	77.617	4.88	378.774	4,442,440.82	8.52E−05
2	10.2	34.496	4.87	168.310	8,558,086.05	1.96E−05
1	10.2	8.624	4.87	42.072	36,028,895.2	1.16E−06

Table 3 Angle of rotation obtained by equivalent stiffness method for type A, 1 floor structure

Story No.	Type A, 1 floor structure					
	Center of rotation (m)	Lateral force (kN)	Eccentricity (m)	Torsional moment (kN m)	$K * r_i^2$	Angle, θ (radian)
1	10.2	8.624	4.87	42.072	36,028,895.2	1.16E−06

Table 4 Maximum horizontal displacement limit (mm) for asymmetrical regular building

Story No.	Type of structure					
	A	B	C	D	E	F
10 story	83.382	128.135	132.668	217.891	189.138	169.004
5 story	33.654	40.648	9.803	83.692	68.48	48.476
1 story	0.476	0.672	0.432	3.426	2.317	0.07

Table 5 Torsional irregularity coefficient for 10 story asymmetrical regular building

Story No.	Type of structure					
	A	B	C	D	E	F
1	1.439	1.618	1.997	1.769	1.866	1.545
2	1.433	1.602	1.986	1.746	1.812	1.484
3	1.428	1.588	1.973	1.725	1.765	1.439
4	1.424	1.574	1.958	1.707	1.728	1.409
5	1.42	1.562	1.943	1.691	1.699	1.387
6	1.416	1.549	1.929	1.679	1.675	1.371
7	1.412	1.537	1.916	1.667	1.655	1.359
8	1.408	1.525	1.904	1.656	1.636	1.349
9	1.404	1.513	1.894	1.644	1.617	1.338
10	1.401	1.512	1.883	1.63	1.595	1.327

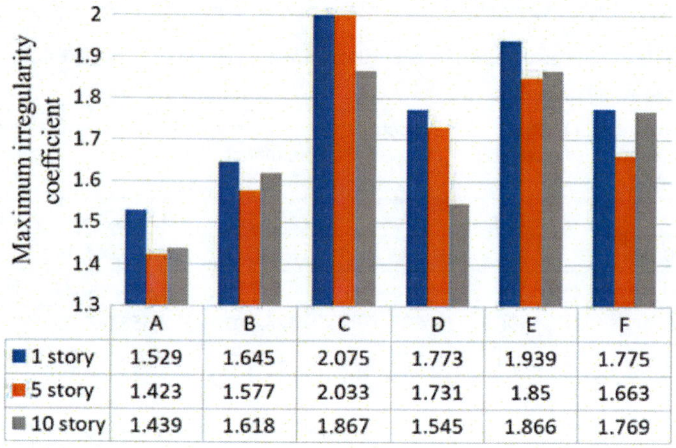

Fig. 7 Maximum torsional irregularity coefficient of asymmetric regular buildings

Similarly, the variation of angle of rotation of all 10 story structures are represented in terms of Table 6 and the maximum angle of rotation of all 10, 5, and 1 story structure are presented in graph as shown in Fig. 8.

The rotation is maximum in the top floors in all the types of the model. The type C structure shows higher rotation compare to all the models. The 10 story structures show maximum rotation than the 1 story structure.

Table 6 Angle of rotation for 10 story asymmetrical regular building

Story No.	Type of structure					
	A	B	C	D	E	F
1	5.0E−08	1.1E−07	2.1E−07	7.7E−08	4.8E−08	4.26E−09
2	9.7E−07	2.2E−06	4.4E−06	1.4E−06	1.1E−06	1.41E−07
3	6.4E−06	1.3E−05	2.9E−05	9.6E−06	7.9E−06	1.05E−06
4	2.5E−05	5.3E−05	0.00011	3.8E−05	3.2E−05	4.41E−06
5	7.6E−05	0.00015	0.00035	0.00011	9.7E−05	1.34E−05
6	0.0001	0.00039	0.00086	0.00027	0.00024	3.34E−05
7	0.0004	0.00083	0.00184	0.0005	0.00051	7.45E−05
8	0.0007	0.00162	0.00358	0.00116	0.0010	0.000149
9	0.0014	0.00291	0.00645	0.00208	0.0018	0.000282
10	0.0014	0.00299	0.00684	0.00331	0.0029	0.000522

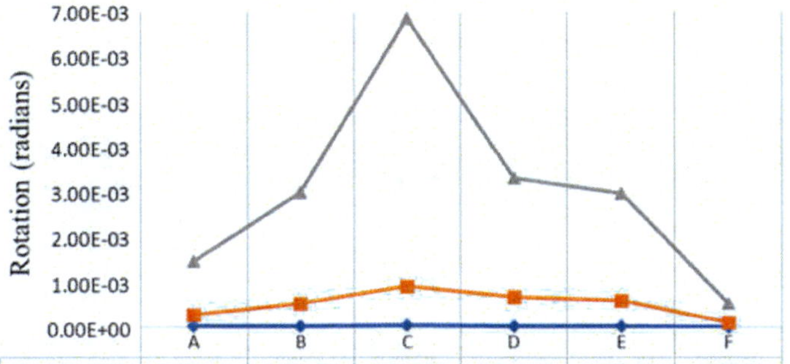

	A	B	C	D	E	F
1 STOREY	1.00E-05	2.00E-05	3.60E-05	2.70E-05	2.00E-05	4.00E-06
5 STOREY	0.00028	0.00052	0.00091	0.00067	0.0006	0.0001
10 STOREY	0.00146	0.00299	0.00684	0.00331	0.00296	0.00052

Fig. 8 Maximum angle of rotation of asymmetric regular buildings

6.2 L Shaped Irregular Building

The maximum horizontal displacement limit for 10 story, 5 story, and 1 story L shaped irregular buildings are shown in Table 7.

Table 7 Maximum horizontal displacement limit (mm) for L shaped irregular building

Story No.	Type of structure					
	L1	L2	L3	L4	L5	L6
10 story	66.835	74.967	85.681	78.557	81.923	90.475
5 story	0.036	0.944	29.706	25.731	26.915	29.425
1 story	0.507	0.034	0.577	0.025	0.595	0.02

The maximum torsional irregularity ratio and angle of rotation of all L shaped irregular buildings for 10 story structures has been shown in Table 8.

Figure 9 shows the graphical representation of the maximum torsional irregularity coefficient for 1 story, 5 story, and 10 story of all L shaped irregular structures. The 1 story structure shows higher irregular coefficient in type L1 and 10 story structure shows the lesser value. Except L1, rest of the structures shows higher irregular coefficient value in 10 story structure.

The floor rotation of L shaped irregular structures is shown in Table 9. Maximum rotation is found in top floors and structure type L6 shows the maximum rotation and type L1 shows the lesser one.

Table 8 Torsional irregularity coefficient for 10 story L shaped irregular building

Story No.	Type of structure					
	L1	L2	L3	L4	L5	L6
1	1.1	1.121	1.136	1.2	1.19	1.245
2	1.096	1.12	1.138	1.188	1.199	1.262
3	1.094	1.119	1.139	1.184	1.204	1.272
4	1.092	1.119	1.14	1.182	1.208	1.279
5	1.091	1.119	1.141	1.182	1.21	1.283
6	1.091	1.119	1.142	1.181	1.212	1.286
7	1.09	1.119	1.142	1.181	1.213	1.289
8	1.09	1.119	1.142	1.182	1.214	1.292
9	1.089	1.118	1.143	1.182	1.216	1.294
10	1.089	1.118	1.143	1.183	1.217	1.297

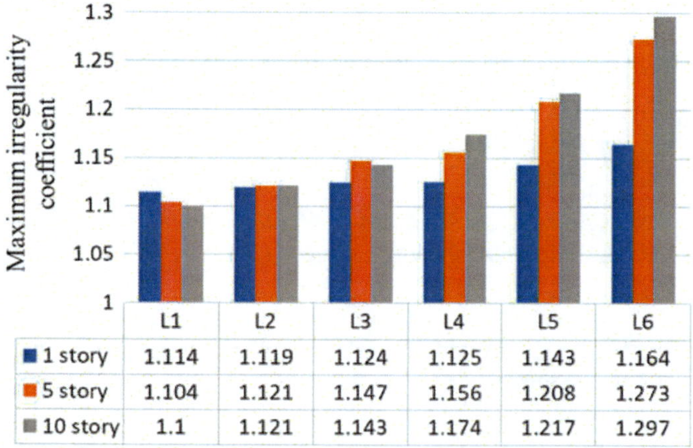

Fig. 9 Maximum torsional irregularity coefficient of L shaped irregular buildings

Table 9 Angle of rotation for 10 story L shaped irregular building

Story No.	Type of structure					
	L1	L2	L3	L4	L5	L6
1	5.4E−10	1.1E−09	1.1E−09	4.4E−09	3.3E−09	1.4E−08
2	1.7E−08	3.5E−08	3.4E−08	1.4E−07	1.1E−07	4.48E−07
3	1.3E−07	2.7E−07	2.5E−07	1.1E−06	8.2E−07	3.40E−06
4	5.5E−07	1.1E−06	1.1E−06	4.5E−06	3.4E−06	1.43E−05
5	1.6E−06	3.4E−06	3.3E−06	1.4E−05	1.1E−05	4.38E−05
6	4.2E−06	8.6E−06	8.2E−06	3.4E−05	2.6E−05	0.000108
7	9.1E−06	1.8E−05	1.7E−05	7.4E−05	5.6E−05	0.000235
8	1.7E−05	3.6E−05	3.5E−05	0.00014	0.00012	0.000459
9	3.1E−05	6.5E−05	6.1E−05	0.00026	0.00021	0.000828
10	3.7E−05	7.8E−05	7.4E−05	0.00032	0.00029	0.00191

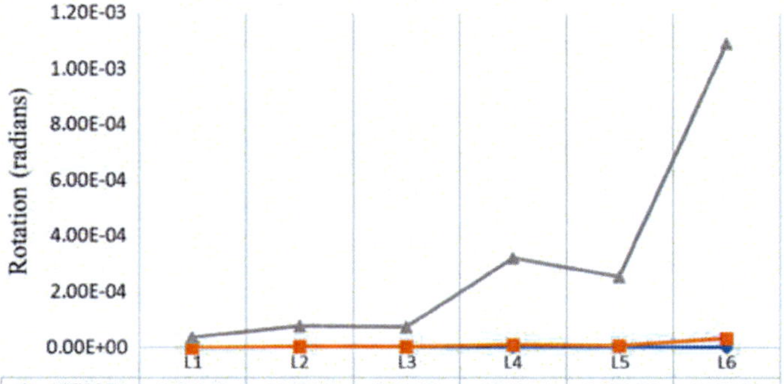

	L1	L2	L3	L4	L5	L6
1 STOREY	4.82E-08	9.98E-08	9.59E-08	4.04E-07	3.20E-07	1.36E-06
5 STOREY	1.16E-06	2.44E-06	2.32E-06	1.01E-05	7.99E-06	3.41E-05
10 STOREY	3.72E-05	7.81E-05	7.42E-05	0.00032462	0.00025568	0.00108986

Fig. 10 Maximum angle of rotation of L shaped irregular buildings

Figure 10 shows the maximum floor rotation of L shaped irregular buildings. It is seen that all the 10 story structure shows maximum rotation and 1 story structure shows the lesser rotation.

6.3 *Evaluation of Torsional Irregularity Coefficient and Rotation of Floor*

Figures 11, 12, 13, and 14 in this section represents the torsional irregularity coefficient and rotation of floor along the height of the structure.

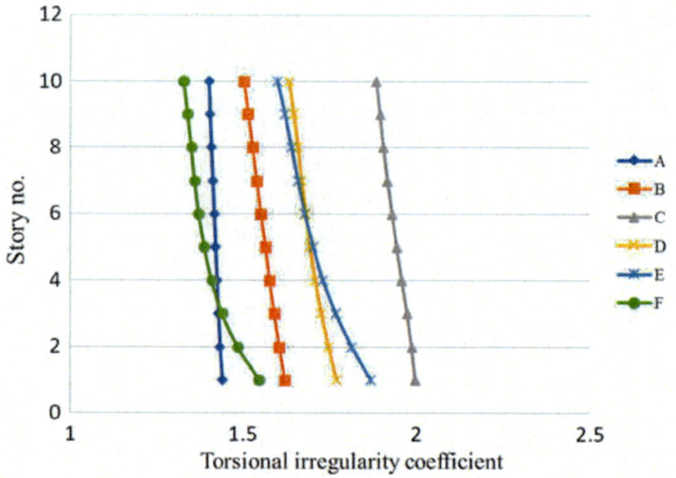

Fig. 11 Torsional irregularity coefficient of 10 story asymmetric regular buildings

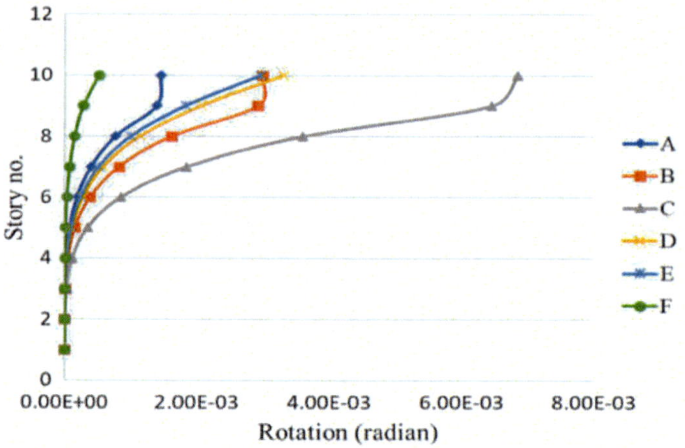

Fig. 12 Angle of rotation of 10 story asymmetric regular buildings

- Asymmetrical regular building
- L Shaped irregular building

The torsional irregular coefficient together with floor rotation has been shown in this section for both the types of models. In both types of structures, the angle of rotation is maximum in top story. The type C shows maximum rotation in asymmetrical regular buildings, and type L6 shows maximum angle of rotation in L shaped irregular buildings. But if we observe the torsional irregularity coefficient in both the groups of structure types (Figs. 11 and 13), in asymmetrical regular

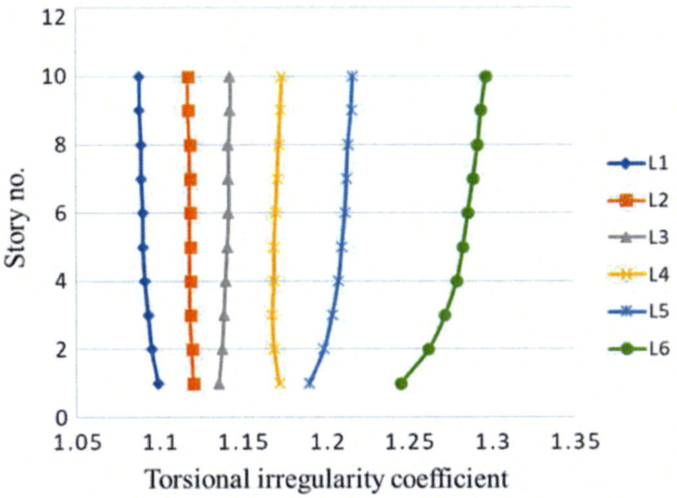

Fig. 13 Torsional irregularity coefficient of 10 story L shaped irregular buildings

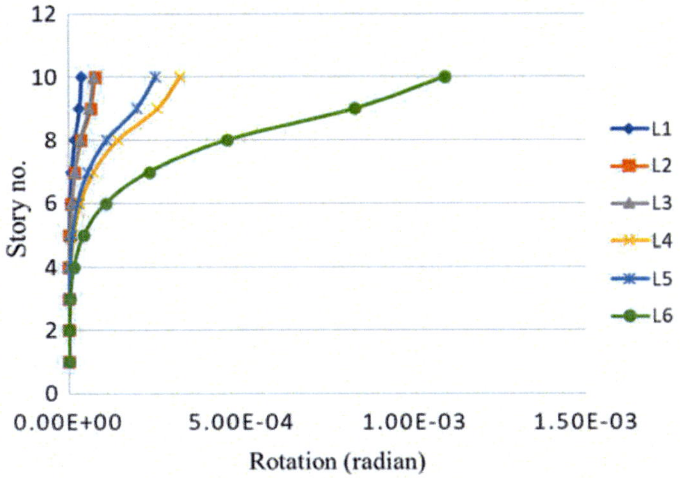

Fig. 14 Angle of rotation of 10 story L shaped irregular buildings

buildings, coefficient value is found to be maximum in bottom story and gradually decreased along the height of the building. But in L shaped irregular structures, the results are quite reversible that coefficient values are maximum in top story. The angle of rotation of floor obtained from equivalent stiffness method of all structures has been represented graphically, it can be observed that,

Maximum rotation can be seen in top floor of the structure and the rotation increases gradually as the number of floor increases.

The angle of rotation is decreased for 1 story structures of all types of building, whereas torsional irregularity coefficient increased for one story structures.

The angles which are obtained for both the types of structures is almost zero only in lower floors. If we observe the angles in top floors, the results are not almost zero and there is rotation. In the similar way, angles in radians of one story structures of both asymmetric and L shaped irregular structures are almost zero which indicates that building is safer from rotation. Lower the angle of rotation, lesser will be the torsion effect and hence safer the building. As the number of floors increases, angle of floor rotation increases simultaneously.

The above observations for torsional irregularity coefficient and angle of rotation of floor are quite contradictory. We can clearly observe that angle of rotation of floor can be considered the indicator of the torsional behavior compared to the code proposed torsional irregularity coefficients of the structures [2]. It can also be seen that both the coefficients of floor rotation and torsion irregularity are inversely proportional to each other. It can be inferred as follows, by analyzing the observations made in the above:

- Regulations for torsional irregularity coefficients which are defined in code books are not practical about the torsional features of any of their structures.
- The torsional irregularity coefficient regulations defined in the codes should be updated.

6.4 Proposal of New Definition for Torsional Irregularity

From the results of the torsional irregularity coefficient and floor rotation studied above, we can note that the rotation of the floor more realistically reflects the structure's torsional behavior. Therefore, a new definition of the coefficient of torsional irregularity based on floor rotation should be proposed. Along with the new definition, similar to the limitations of drift which is provided in the code regulations, an upper bound for angle of rotation of floor need to be proposed in the standard codes [2]. By considering the floor rotation θ of above models, an appropriate limitation for θ can be,

$$\theta_i \leq 0.0012$$

7 Conclusion

In this work, an investigation is carried out on six types of asymmetric regular building by considering shear walls in different positions and L shaped irregular building. All the buildings are varied with story numbers. The irregularity in plan and asymmetric placement of the structural walls of these structures will have major effects on the demands for seismic response due to the lateral-torsional coupled behavior.

The torsional behavior of the building is one of the most severe factors, which is the reason for extreme damage to the structural elements of the building. A significant parameter which measures the extent of the torsional effect on the structure is the coefficient of torsional irregularity. A significant parameter which measures the extent of the torsional effect on the structure is the coefficient of torsional irregularity. Torsional irregularity is described in most of the standard codes by considering the displacement of buildings. It can be clearly observed in this analysis that the torsional irregularity coefficient does not realistically characterize the torsion behavior of the structures.

From the obtained results, it is seen that single story structures were showing higher torsional irregularity coefficient value in asymmetrical regular buildings, and in most of the L shaped irregular buildings, torsional irregularity coefficient shows the minimum value. It is clearly seen that the angle of rotation of floor of all the buildings is increased as the number of floors is increased. The rotation of floor decrease in proportion to number of floors, i.e., for single story structure, floor rotation is minimum. Floor rotations of the structure attain maximum value when the shear walls are placed near the center of mass in asymmetric regular models [2]. By considering all these behaviors of the structures, the code proposed torsional irregularity coefficient does not represent torsion behavior reasonably. The torsional or twisting behavior of the structures is defined realistically by the rotation of floor. Hence, it is required to propose the new amendments for torsional irregularity based on floor rotations. By taking into account the floor rotation of the above examples, a suitable constraint for angle of rotation θ could be,

$$\theta_{\mathrm{i}} \leq 0.0012$$

Acknowledgements The first author gratefully acknowledges and would like to forward deepest appreciation and gratitude to advisor Dr. Durga Prasad J, Professor, Nitte Meenakshi Institute of Technology Yelahanka, for his patience and constructive advice throughout the thesis work.

References

1. Abdel Raheem SE, Ahmed MMM, Ahmed MM, Abdel-shafy AGA (2018) Evaluation of plan configuration irregularity effects on seismic response demands of L-shaped MRF buildings. Bull Earthq Eng 3845–3869 (in press)
2. Ozmen G, Girgin K, Durgun Y (2014) Torsional irregularity in multi-story structures. Int J Adv Struct Eng 121–131 (in press)
3. Kewalramani MA, Syed ZI (2020) Seismic analysis of torsional irregularity in multi-storey symmetric and asymmetric buildings. Eurasian J Anal Chem 13:3553–3558 (in press)
4. Ilerisoy ZY (2019) Discussion of the structural irregularities in the plan for architectural design within the scope of earthquake codes. Period Polytech Archit 50–62 (in press)

5. Hussain SM, Tengli SK (2018) Study on torsional effects of irregular buildings under seismic loads. Int J Appl Eng Res 13(7):55–60. ISSN 0973-4562 (in press)
6. Ozhendekci N, Polat Z (2008) Torsional irregularity of buildings. In: 14th world conference on earthquake engineering, Beijing, China, Oct 2008, pp 12–17 (unpublished)
7. Horizontal structural irregularities, seismic loads: guide to the seismic load provisions of ASCE 7–10 (2015) Am J Struct Eng (ASCE) (in press)
8. Goel RK, Chopra AK (1992) Seismic code analysis of buildings without locating centers of rigidity. Am J Struct Eng (ASCE) 3039–3055 (in press)
9. Patil SS, Mujawar AG, Mali PA, Katti MR (2017) A study of torsional effect on multi-storied building with plan-irregularity. Int J Adv Res 1625–1632 (unpublished)
10. Tso WK, Dempsey KM (1980) Seismic torsional provisions for dynamic eccentricity. Earthq Eng Struct Dyn 275–289 (in press)
11. Humar JL, Kumar P (1998) Torsional motion of buildings during earthquakes. I. Elastic response. Can J Civil Eng 898–916 (in press)
12. Wakchaure MR, Nagare YU (2013) Effect of torsion consideration in analysis of multi storey frame. Int J Eng Res Appl (IJERA) 3(4):1828–1832. ISSN: 2248-9622 (in press)
13. Mehana MS, Mohamed O, Isam F (2019) Torsional behaviour of irregular buildings with single eccentricity. IOP Conf Ser: Mater Sci Eng (in press)
14. IS 1893(Part 1):2016. Criteria for earthquake resistant design of structures (sixth revision)
15. Adarsh A, Rajeeva SV (2017) Torsional irregularities in multi-storeyed structures. Int J Res Sci Innov (IJRSI) IV(IX). ISSN 2321-2705 (in press)
16. Chambers J, Kelly T (2004) Nonlinear dynamic analysis—the only option for irregular structures. In: 13th world conference on earthquake engineering, Vancouver, BC, Canada Paper No. 1389, 1–6 Aug 2004 (unpublished)
17. Chopra AK, Goel RK (1991) Evaluation of torsional provisions in seismic codes. Earthq Eng Struct Dyn 27(2):173–185 (unpublished)
18. Soni AG, Agrawal DG, Pande AM (2015) Effect of irregularities in buildings and their consequences. Int J Mod Trends Eng Res (IJMTER) 02(04). ISSN (Online): 2349–9745; ISSN (Print): 2393-8161 (in press)
19. Anagnostopoulos SA, Kyrkos MT, Stathopoulos KG (2013) Earthquake induced torsion in buildings: critical review and state of the art. In: The 2013 world congress on advances in structural engineering and mechanics, Jeju, Korea, 8–12 Sept 2013 (unpublished)
20. Chopra AK, De La Llera JC (1996) Accidental and natural torsion in earthquake response and design of buildings. In: Eleventh world conference on earthquake engineering, Paper No. 2006 (unpublished)
21. Duan XN, Chandler AM (1997) An optimized procedure for seismic design of torsionally unbalanced structures. Earthq Eng Struct Dyn 26:737–757 (unpublished)
22. Zheng N, Yang Z, Shi C, Chang Z (2004) Analysis of criterion for torsional irregularity of seismic structures. In: 13th world conference on earthquake engineering, Vancouver, BC, Canada Paper No. 1389, 1–6 Aug 2004 (unpublished)
23. Babu DV, Babitha Rani H, Pavan Siva Kumar Ch (2017) Assessment of the torsion effect on buildings under lateral seismic ground motion. Int J Eng Technol Sci Res (IJETSR) 4(9). ISSN: 2394-3386 (unpublished)
24. Gokdemir H, Ozbasaran H, Dogan M, Unluoglu E, Albayrak U (2013) Effects of torsional irregularity to structures during earthquakes. Eng Fail Anal 713–717 (unpublished)
25. Kuang Y-p, Jiang X-l, Jiang N (2018) Inelastic parametric analysis of seismic responses of multistorey bidirectional eccentric structure. Hindawi Shock Vib 2018:Article ID 7023205, 20 pp (in press)
26. Mahdi T, Soltan Gharaie V (2011) Plan irregular RC frames: comparison of pushover with nonlinear dynamic analysis. Asian J Civ Eng (Build Hous) 12(6):679–690 (in press)

Regionalization of Flow Duration Curves for West Flowing Rivers of India

Chandrashekarayya G. Hiremath and Lakshman Nandagiri

Abstract A flow duration curve (FDC) by providing a concise description of the temporal variability of streamflow from a river basin is extremely useful as a hydrological signature and also in the design of water resources projects. However, long-term historical measured streamflow records are essential to derive the FDC, and hence, a major challenge has been to develop methods to estimate the FDC for ungauged river basins. Also, for the derived FDC to provide an accurate representation of the variability of natural flows, no regulations/abstractions must exist upstream of the gauging station. Therefore, the present study was being taken up to develop a methodology for prediction of FDCs in ungauged river basins located in the hydrologically homogeneous West flowing rivers of India using streamflow records of 14 unregulated river basins. FDCs were derived for each basin using frequency analysis, and flow quantiles at specific exceedance probabilities were extracted. Subsequently, the regionalization approach involving transfer of hydrological information from gauged basins to ungauged basins was adopted by developing regression models relating flow quantiles to easily derived basin physical characteristics. The performance of the developed regression models was evaluated using validation datasets and found to yield satisfactory results. Results of this study will permit estimation of flow quantiles and FDCs in ungauged basins located within West flowing rivers of India.

Keywords Flow duration curves · Regionalization · Regression analysis

1 Introduction

Water is a vital resource, and its management is essential for the economic development of all nations. River discharge is an important hydrological data essentially required for analysis and design of water resources projects viz., canals, reservoirs and hydropower projects, irrigation, water supply, navigation, recreation and for

C. G. Hiremath (✉) · L. Nandagiri
Department of Water Resources & Ocean Engineering, National Institute of Technology, Surathkal, Mangalore, India

L. Nandagiri et al. (eds.), *Sustainability Trends and Challenges in Civil Engineering*, Lecture Notes in Civil Engineering 162, https://doi.org/10.1007/978-981-16-2826-9_42

assessment of drought and flood risks [1, 2]. Although there is significant advancement in hydrological research, the issues related to insufficient hydrometric data, together with problems of land use changes and climatic change impacts, resulting in decreased water availability and degradation of ecosystem is prevalent in many developing countries. These issues are challenging, especially during the estimation of runoff from an ungauged or poorly gauged basin [3]. This undermines the water resources planning and management not only at the ungauged site, but also at the river basin level [1].

The non-availability of data in ungauged basins needs hydrological regionalization that allows assessment of parameter values of hydrological predictive tools without calibration [1]. The mechanism that facilitates extrapolation of parameter values from gauged sites to other site at which data are required but are not available is commonly referred as regionalization [4, 5]. Previous research studies have used the river flow information from a flow duration curve (FDC) as a criterion for hydrological regionalization [1, 6–8]. For example—Quimp et al. 1983 estimated FDC for ungauged sites in Philippines by regionalization of FDC from different gauged basins. Yu and Yang 1996 developed regional FDC for Southern Tailwan by adopting multivariate statistical analysis of flow records from 34 sites [7].

FDC is a tool for representing the relationship between the river discharge value and their associated exceedance probabilities for a particular site [6, 9, 10]. The regional FDC works on the supposition that the regions within which the basins are studied will perform in a similar manner. Hence, the flow duration curve for sites, where the flow measurements are unavailable, is derived from the method of regression analysis [11]. The use of multivariate regression techniques to develop the statistical relationship between river flow and geomorphology has been adopted by many researchers for estimating flow values at ungauged sites [2]. A similar kind of regionalization technique is attempted in this study which aims to develop a methodology for estimating the FDC in ungauged river basins located in the hydrologically homogeneous West flowing rivers of India using stream flow records of 14 unregulated river basin stations. This technique involves development of FDCs, and subsequently, regionalization of parameters from gauged to ungauged basins was adopted by developing regression models relating flow quantiles to easily derived basin physical characteristics.

2 Study Area and Data

South Indian Rivers can be categorized into West flowing and the East flowing rivers. The West flowing rivers of India with unregulated flow is selected as the study region. West flowing river flows are seasonal, and this seasonality of flow has resulted in construction of multi-purpose reservoirs in order to manage water supply problems and to protect flood prone deltaic region [13]. All the West flowing rivers drain into the Arabian Sea.

Out of all West flowing Indian Rivers, 14 unregulated river basin station having a combined geographical area of 13,793.857 km^2 was delineated for the study.

Figure 1 displays the location of the identified unregulated river basin gauge station in the West Flowing rivers of India. The details of the selected stations are as given in Table 1.

West flowing rivers are surrounded by Western Ghats Mountains on the east and the Arabian Sea on the west side with physiographic consisting of highland hill slopes on east and the coastal plains on west side. The mean annual rainfall in the study area is around 2400 mm [13, 14]. Majority of river flow occurs during the wet monsoon month from July to August, and the flow is negligible in many of the rivers during

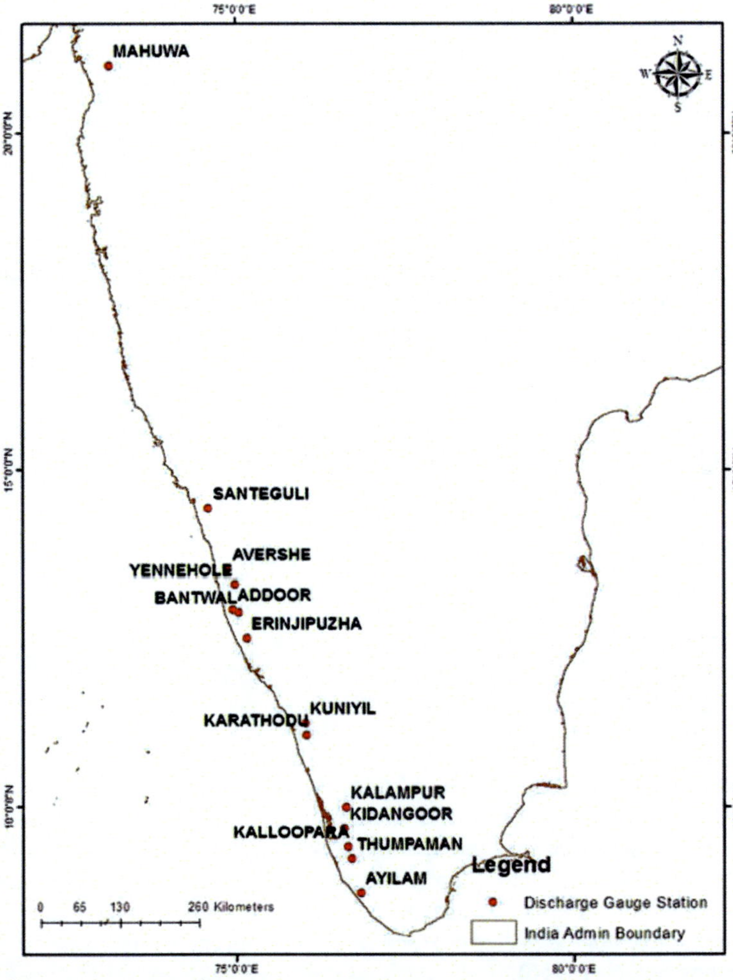

Fig. 1 Map showing the location of selected river basin gauge station

Table 1 List of selected river basin gauge station in West flowing rivers of India

River basin name	S. No.	Gauge station name	Data period	Drainage area (km^2)
West flowing rivers	1	Santeguli	1991–2018	988
	2	Avershe	2003–2018	276
	3	Yennehole	1991–2018	334
	4	Addoor	2004–2018	690
	5	Bantwal	1991–2018	3204
	6	Erinjipuzha	1991–2018	852
	7	Kidangoor	1991–2017	593
	8	Kalloopara	1991–2017	700
	9	Thumpaman	1991–2018	796
	10	Ayilam	1991–2015	533
	11	Kuniyil	1991–2018	1998
	12	Karathodu	1991–2015	770
	13	Kalampur	1991–2017	358
	14	Mahuwa	1992–2017	1701

the dry periods. Daily discharge data from 14 selected gauge stations maintained by CWC, and Government of India were extracted from the water resources information system (WRIS) web-portal. In addition, the delineation of catchment and assessment of its characteristics were carried out by processing the 30 m SRTM—digital elevation model (DEM) information using ArcMap tool. For the analysis, different catchment characteristics were determined which include: catchment area (km^2), basin perimeter (km), length of the basin (km), basin width (km), drainage density (km/km^2), elongation ratio, form factor, shape factor, circulatory ratio, basin slope, maximum and minimum elevation (km).

3 Methodology

3.1 *Flow Duration Curve*

The flow data of the selected gauge station were collected separately and were arranged in descending magnitude. In order to develop the FDC for each basin, the plotting positions (exceedance probability) were assigned to each of the discharge value using the Weibull Plotting Position method as given in Eq. (1). Next the flow values are plotted as ordinate against the associated percentage of the time; the river discharge has exceeded/equalled as abscissa, and smooth curve is obtained by joining the resulting points [15, 16].

$$P = \frac{M}{n+1} 100\% \tag{1}$$

Here, 'P' is the probability of exceedance less than or equal to the magnitude of river flow, 'M' is the number of counts of flow values falling within the specified interval and 'n' is the number of observations. From this study, total nine flow quantiles with their corresponding exceedance probability values viz., 10, 20, 30, 40, 50, 60, 70, 80 and 90% were extracted by interpolation for each of the selected unregulated basin.

3.2 Multiple Regression Models

Multiple regression model is one of the commonly adopted technique for determining the model parameter values at gauged basins by developing a regression relationship between catchment characteristics and the optimal parameters [8]. In order predict the river flow value at ungauged stations through regionalization, the regression equation shown in Eq. (2) has been developed.

$$Q(D) = \beta_0 + \beta_1 Z_1 + \beta_2 Z_2 + \cdots + \beta_n Z_n + \varepsilon \tag{2}$$

where $Q(D)$ is the flow quantile of specific exceedance probability 'D' (10%, 20%, …, 90%) of study catchment; Z_1, Z_2, …, Z_n are the selected catchment characteristics and β_1, β_2, …, β_n are regression co-efficient [12, 15, 17]. The co-efficient contained through regression is utilized to determine the flow quantile values in ungauged basins. This is carried out by substituting the characteristics corresponding to ungauged basin in the developed regression relationship [10, 15]. In the beginning, twelve catchment characteristics were selected for the multiple regression analysis in each of the 14 unregulated stations. Later, the use of stepwise linear regression procedure facilitated the identification of the most significant catchment characteristics for explaining the observed variability in the flow quantiles. The identified significant characteristics for 14 basins include: catchment area (km^2), perimeter of the basin (km), basin length (km), basin width (km), shape factor, elongation ratio, circulatory ratio, form factor and drainage density (km/km^2).

3.3 Model Performance Statistics

The following statistical measures were utilized to assess the performance of the predicted quantiles with that of the observed flow quantiles.

- Coefficient of Determination (R^2):

$$R^2 = \left(\frac{\sum_{i=1}^{N} (O_i - O_a)(S_i - S_a)}{\sqrt{\sum_{i=1}^{N} (O_i - O_a)^2} \sqrt{\sum_{i=1}^{N} (S_i - S_a)^2}} \right) \tag{3}$$

- Root Mean Square Error (RMSE):

$$\text{RMSE} = \sqrt{\frac{\sum_{i=1}^{N} (S_i - O_i)^2}{N}} \; (\ldots\ldots \text{m}^3/\text{s}) \tag{4}$$

- Mean Relative Error (MRE):

$$\text{MRE} = \frac{1}{N} \sum_{i=1}^{N} \left(\frac{S_i - O_i}{O_i} \right) \tag{5}$$

In above equations, 'N' denotes the number of selected basins, 'O' represents the observed flows and 'S' is the estimated flow values, O_a and S_a denotes the average value of observed and simulated flow rates. For an ideal model, the value of RMSE and MRE value should be zero. The value of R^2 ranges between 0 and 1; where 1 signifies better model performance [20].

4 Results and Discussion

To evaluate the variation of the results with respect to changing basin location and characteristics, 14 unregulated river basin gauge station pertaining to the West flowing rivers was used.

4.1 Regional Flow Quantiles

As discussed in Sect. 3.1, the observed flows of the selected river basin stations were plotted against the corresponding time of flow exceedance flow. Figure 2 displays the flow duration curves estimated for the selected river gauge stations. The maximum average flow from all the selected West flowing river basins combined was 229.47 m^3/s, and the minimum average flow value was about 0.108 m^3/s. Total nine flow quantiles (10, 20, 30, 40, 50, 60, 70, 80 and 90%) values were extracted by interpolation for each of the FDC plotted for selected unregulated stations for regression analysis.

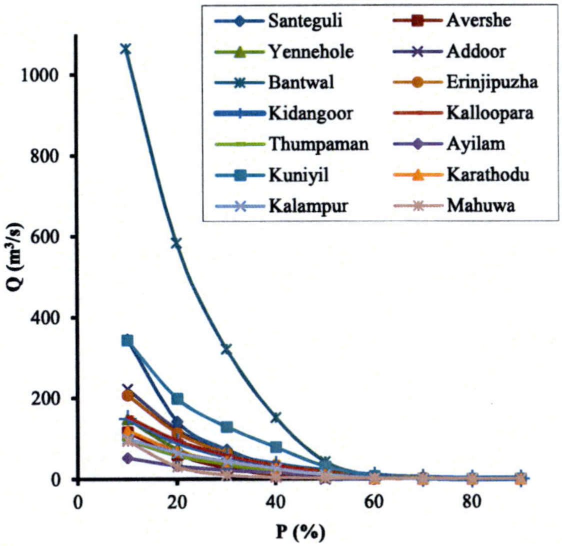

Fig. 2 Flow duration curve for 14 West flowing river dataset

4.2 Multiple Regression Analysis

As discussed previous, FDC is used as a criterion for hydrological regionalization in this study—to predict the river flow information at ungauged basins by utilizing the multiple regression technique. Subsequently, 14 regression relationships were developed (in the form of Eq. 2) by utilizing the flow quantile values and catchment characteristics data available from all the delineated unregulated stations. The significant correlated characteristics were determined using stepwise regression technique. The catchment characteristics, namely the catchment area (km^2) 'A', basin perimeter 'P' (km), basin length 'L' (km), basin width 'W' (km), drainage density 'D_D' (km/km^2), elongation ratio 'R_L', form factor 'FF', shape factor 'SF' and circulatory ratio 'R_c' are found be correlated significantly to flow quantiles associated with 14 West flowing river stations. The regression relationship coefficients for significantly correlated catchment characteristics for each of the flow quantile are shown in Table 2. As per study conducted by Vogel et al. 1999 [2], the regression relationship developed based on more than catchment area alone can estimate the river flow values better than that of runoff maps at the ungauged sites.

The significantly correlated catchment characteristics and co-efficient derived through the regression analysis are utilized to produce flow quantile values in ungauged basins. Flow quantile estimate from multiple regression analysis is compared with those of the observed values, and the performance is evaluated in terms of correlation coefficient, RMSE and MRE. Table 3 shows the performance statistics of each of the flow quantile obtained by multiple regression model developed (Table 2) using the complete dataset of all the 14 West flowing river basin stations. This statistical analysis indicates better prediction performance results for

Table 2 Regression relationship coefficients for each flow quantile of 14 West flowing river dataset[a]

Flow quantile	Regression coefficients									
	β_0	β_1	β_2	β_3	β_4	β_5	β_6	β_7	β_8	β_9
Q_{10}	−4611.94	0.850	−5.765	10.587	−25.070	−8560.98	242.25	−3893.6	13,152.6	−242.89
Q_{20}	−4197.17	0.360	−0.764	−1.971	−13.073	−8013.01	206.77	−439.05	10,534.28	−146.44
Q_{30}	−2275.35	0.116	0.851	−4.864	−4.546	−4472.21	112.27	640.90	5417.08	−85.29
Q_{40}	−853.42	−0.015	1.558	−5.563	−0.930	−1767.07	43.69	1077.89	1772.96	−46.05
Q_{50}	5.82	−0.039	0.950	−2.758	−0.281	−74.90	0.93	596.83	−150.88	−20.78
Q_{60}	266.05	−0.018	0.277	−0.476	−0.582	444.34	−12.89	128.23	−607.79	−10.35
Q_{70}	50.99	−0.002	−0.027	0.293	−0.331	105.56	−2.73	−41.75	−96.57	−4.60
Q_{80}	122.10	0.001	−0.061	0.257	0.016	218.81	−5.71	−50.71	−255.40	−0.72
Q_{90}	51.06	0.001	−0.037	0.134	0.037	94.19	−2.33	−29.43	−105.88	−0.18

[a]Regression relationship, $Q(D) = \beta_0 + \beta_1 A + \beta_2 P + \beta_3 L + \beta_4 W + \beta_5 \mathrm{FF} + \beta_6 \mathrm{SF} + \beta_7 R_c + \beta_8 R_L + \beta_9 D_D$

Table 3 Performance statistics of multiple regression model

Flow quantiles	MRE	RMSE (m^3/s)	R^2
Q_{10}	0.049	29.010	0.993
Q_{20}	0.045	15.095	0.994
Q_{30}	0.027	6.621	0.996
Q_{40}	0.006	3.354	0.996
Q_{50}	75.187	3.523	0.947
Q_{60}	29.806	1.725	0.802
Q_{70}	13.646	0.543	0.914
Q_{80}	−1.255	0.107	0.987
Q_{90}	−1.546	0.082	0.972

most of the flow quantiles. The weakening of model performance with higher statistical values of MRE and RMSE for few of the flow quantiles shows the possibility of data uncertainties while calculating the flow quantile. Correlation coefficient (R^2) quantifies the strength of association between observed and model estimates. The R^2 is found to be better for all the predicted flow quantile in 14 West flowing river dataset except for one flow quantiles—Q_{60}.

Further the performance statistics of MRE and RMSE show a mixed response, with much higher RMSE values in initial three flow quantile—Q_{10}, Q_{20} and Q_{30}. In contrast, the MRE values are low in first four flow quantile and considerably higher in Q_{50}, Q_{60} and Q_{70} indicating the model is over predicting the flows at these three quantiles. MRE results for last two quantiles Q_{80} and Q_{90} are under predicted by the multiple regression model. The performance result from Table 3 reveals that the regression model analysis for study stations provides better estimate as predicted values are closely correlated to the observed flow corresponding to different exceedance probability.

The unsatisfactory model performance in certain flow quantiles can be linked to the varying topography and land-use conditions in the delineated basins of the study area. Overall, the predicted flow quantile using multiple regression model simulation for study area provides much better model performance. Major deviations were detected in only five of the stations including Adoor with minimum value of Q_{10}, Q_{20} and Q_{30}, Kidangoor station with over predicted flow values at Q_{10} and Q_{20}, and under predicted at Q_{50} and Q_{60}, Ayilam station with maximum value of Q_{10} to Q_{30}, Karathodu station with maximum value of Q_{10}, Q_{20} and Q_{30}, and Kalampur station with minimum value of Q_{10} and Q_{20}. Further the observations from catchment characteristics reveal that the drainage density of the above-mentioned five stations is much higher than the rest of stations in West flowing river dataset. The negative values predicted in the study are not practically as the catchment responses are non-negative process [21] and can be set to be zero [22] for plotting the FDC. Figure 3 illustrates the predicted FDC for all the 14 river station dataset which is in agreement with the observed FDC curves.

Fig. 3 Predicted flow duration curve for 14 West flowing river dataset

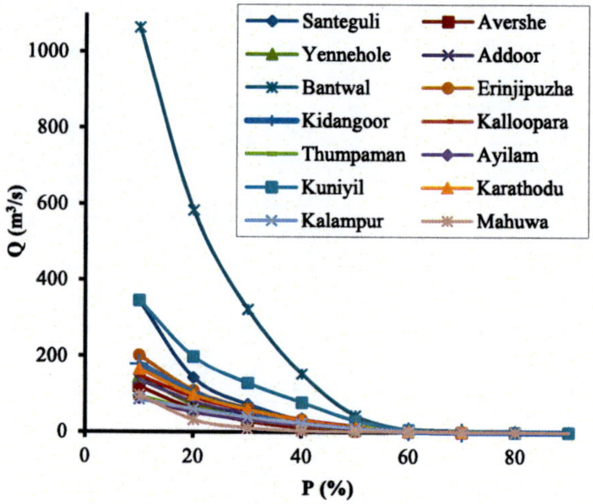

In summary, the results indicate that the performance of the developed regression model for flow prediction has yielded satisfactory results. Such analysis will permit estimation of flow quantiles and FDCs in ungauged basins located within West flowing rivers of India. The non-satisfactory performance statistical value necessitates more comprehensive research work to overcome the uncertainty involved in such prediction.

5 Conclusions

The regionalization concept in predicting the flow quantiles in ungauged basins was evaluated for river basin gauge station located in West flowing rivers of India. The FDC for each of the station was developed by utilizing the historic time-series flow data of 14 unregulated river basin gauge stations, and total of nine flow quantiles (Q_{10} to Q_{90}) were extracted by interpolation. Subsequently, regionalization of parameters from gauged to ungauged basins was adopted by developing regression models relating flow quantiles to easily derived basin physical characteristics. Stepwise regression analysis was used to determine the significant correlated characteristics. Total nine characteristics in 14 West flowing river dataset were found to be significantly correlated. The developed regression co-efficient from all stations combined was utilized to synthesize flow quantile values in presumed ungauged basins. The tested prediction accuracy of the regression model indicates that the flow quantile developed for 14 West flowing river dataset performs very well in predicting the percentile flow values.

However, certain deviations were detected in five locations. Stations including Adoor, Kidangoor and Kalampur under-predicted the flow quantile values, whereas over-predicted flow quantiles were seen in Kidangoor, Ayilam and Karathodu station. This variation was mainly due to difference in catchment characteristics especially the flow quantiles and drainage density of the sites within the study area. Hence, care must be taken while dealing with different river basin having considerable difference in characteristics. This study is an attempt to solve the issues related to prediction of flows in poorly gauged or ungauged basins, and comprehensive research work is needed to reinforce the outcomes of the present study.

Acknowledgements Grateful thanks to the Mr. Nataraja, M, Research Scholar, VTU, Belagavi, Karnataka and Mr. Niranjan S, Research Scholar, National Institute of Technology, Surathkal, Karnataka for their immense help and support provided during this study.

References

1. Masih I, Uhlenbrook S, Maskey S, Ahmad MD (2010) Regionalization of a conceptual rainfall–runoff model based on similarity of the flow duration curve: a case study from the semi-arid Karkheh basin, Iran. J Hydrol 391(1–2):188–201
2. Vogel RM, Wilson I, Daly C (1999) Regional regression models of annual streamflow for the United States. J Irrig Drain Eng 125(3):148–157
3. Sivapalan M, Takeuchi K, Franks SW, Gupta VK, Karambiri H, Lakshmi V, Oki T (2003) IAHS Decade on Predictions in Ungauged Basins (PUB), 2003–2012: shaping an exciting future for the hydrological sciences. Hydrol Sci J 48(6):857–880
4. Merz R, Blöschl G (2004) Regionalisation of catchment model parameters. J Hydrol 287(1–4):95–123
5. Blöschl G, Sivapalan M (1995) Scale issues in hydrological modelling: a review. Hydrol Process 9(3–4):251–290
6. Fennessey N, Vogel RM (1990) Regional flow duration curves for ungauged sites in Massachusetts. J Water Resour Plan Manag 116(4):530–549
7. Isik S, Singh VP (2008) Hydrologic regionalization of watersheds in Turkey. J Hydrol Eng 13(9):824–834
8. Li M, Shao Q, Zhang L, Chiew FH (2010) A new regionalization approach and its application to predict flow duration curve in ungauged basins. J Hydrol 389(1–2):137–145
9. Vogel RM, Fennessey NM (1994) Flow duration curves. I: New interpretation and confidence intervals. J Water Resour Plan Manag 120(4):485–504
10. Yu P, Yang T, Wang Y (2002) Uncertainty analysis of regional flow duration curves. J Water Resour Plan Manag 128(6):424–430
11. Yu PS, Yang TC (1996) Synthetic regional flow duration curve for southern Taiwan. Hydrol Process 10:373–391
12. Chiang S, Tsay T, Nix SJ (2002) Hydrologic regionalization of watersheds. I: Methodology development. J Water Resour Plan Manag 128(1):3–11
13. Panda DK, Kumar A, Mohanty S (2011) Recent trends in sediment load of the tropical (Peninsular) river basins of India. Global Planet Change 75(3–4):108–118
14. Hydrological data (unclassified) book (2018) Hydrological Data Directorate Information System Organization, Water Planning & Projects Wing, Central Water Commission, New Delhi, Government of India
15. Nruthya K, Srinivas V (2015) Evaluating methods to predict streamflow at ungauged sites using regional flow duration curves: a case study. Aquat Procedia 4:641–648
16. Sugiyama H, Vudhivanich V, Whitaker AC, Lorsirirat K (2003) Stochastic flow duration curves for evaluation of flow regimes in rivers. J Am Water Resour Assoc 39(1):47–58

17. Nandagiri L (2010) Regionalization of flow duration quantiles—comparison of multiple least-squares regression and artificial neural networks approaches. In: Proceedings: national conference on 'sustainable water resources management and impact of climate change', BITS-Hyderabad, India, March 2010
18. Efron B (1981) Nonparametric estimates of standard error: the jackknife, the bootstrap and other methods. Biometrika 68(3):589–599
19. Mccuen RH (2005) Accuracy assessment of peak discharge models. J Hydrol Eng 10(1):16–22
20. Domínguez E, Dawson CW, Ramírez A, Abrahart RJ (2010) The search for orthogonal hydrological modelling metrics: a case study of 20 monitoring stations in Colombia. J Hydroinf 13(3):429–442
21. Müftüoğlu R (1984) New models for nonlinear catchment analysis. J Hydrol 73(3–4):335–357
22. Sharma N, Zakaullah M, Tiwari H, Kumar D (2015) Runoff and sediment yield modeling using ANN and support vector machines: a case study from Nepal watershed. Model Earth Syst Environ 1(3):1–8

Comparative Analysis of Precast Prestressed Hollow Core Slabs

B. Rekha and R. Ravindra

Abstract The time-to-market pressure is very high in construction industry. It shortens the project duration and requires faster delivery means. It triggers for having quicker engineering design and analysis solutions for building and its components. The advent of precast prestressed components in construction industry triggers better cost-effective design and analysis schemes for these components to withstand the quality pressure and verification demands. This requires an appropriate modelling mechanism of precast prestressed components. The work undertakes a numerical model for precast prestressed hollow core (PHC) slabs using ANSYS APDL and validated with the results obtained from experimental results. Comparable results justify the suitability of discrete modelling mechanism as substitute to experimental evaluation. It also confirms the usage of contact pairs, meshing and element selection of ANSYS model. The performance evaluation of PHC slabs under single line loading (LL) at mid-span, two line loading (TLL) and uniformly distributed loading (UDL) cases are conducted both experimentally and numerically. The performance in terms of deflection, stress and crack patterns are evaluated in this work. It was also observed that PHC slabs show shear compression failure under TLL and UDL cases. The work is extended with parametric study of core shape of PHC slabs. It was observed that slabs with circular cores (PHC) have higher load carrying capacity with lesser deflection than PHC slabs with square cores (PHSC).

Keywords Hollow core slabs · Hollow square core slabs · Load carrying capacity · Crack pattern · Numerical analysis

B. Rekha (✉) · R. Ravindra
Department of Civil Engineering, RV College of Engineering, Visvesvaraya Technological University, Bengaluru, Karnataka, India
e-mail: rekhab.phdcv@rvce.edu.in

R. Ravindra
e-mail: ravindrar@rvce.edu.in

L. Nandagiri et al. (eds.), *Sustainability Trends and Challenges in Civil Engineering*, Lecture Notes in Civil Engineering 162, https://doi.org/10.1007/978-981-16-2826-9_43

1 Introduction

There is wide acceptance of precast prestressed hollow core (PHC) slabs for floor construction of multi-storey buildings, horizontal/vertical walls or noise barriers [1]. There are experimental studies on PHC slabs mentioned in literature [2–6] and these experiments give vital information on 'shear span to effective depth (*a*/*d*) ratio', shape of core, stress distribution, deflection, vertical shear capacity and ultimate loading capacity. But it is observed by research fraternity [7] that there is a need of numerical analysis of precast systems due to its economical viability than experimental analysis due to the nature of testing.

The numerical studies on PHC slabs were carried out in [8–11]. Brunesi et al. [9] compared PHC slabs for its shear strength capacity experimentally and numerically using nonlinear fracture mechanism-based model. Hoogenboom [10] presented finite element analysis procedure with stress recovery method for PHC slabs and predicted the feasibility of PHC slabs with large opening. Deeb et al. [11] presented an analytical model to study shear behaviour of PHC slabs in ANSYS. They used smeared cracking approach for the concrete and 3D link element for pre-stressing strands. It was observed that the results of experiments match with that of ANSYS results. The proposed work aims to validate numerical model with experimental results. Deflections at mid-span, cracking load, failure load and failure pattern were experimentally measured and compared with analytical results. The analysis also extends by conducting the numerical analysis on effect of changing core shape from circular to square.

2 Methodology

This section provides the approach used and the loading considered during the execution of work. It also indicates the details of equipment and test environment deployed while performing the experiments.

2.1 Approach

The scope of work includes the measurement of flexural strength, deflection and crack patterns of PHC slabs under loading cases experimentally. The results are validated with PHC ANSYS models by comparing these results. The scope of work also includes the parametric study of PHC slabs by changing the circular core shapes to square shape. The following steps are carried out as part of proposed work.

- Preparation of the PHC specimens for conducting experiments.
- Testing of the specimen under defined loading cases using a loading frame.
- Modelling of PHC specimens in ANSYS APDL releases 18.1.

- Validation of ANSYS models by comparing experiment measurements to that ANSYS simulation results.
- Performance evaluation of stress and crack patterns.
- Parametric study on the effect of core shape on PHC slabs.

2.2 Loading Cases

Single line loading (LL) at mid-span, two line loading (TLL) and uniformly distributed loading (UDL) cases are carried out to measure deflection and flexural behaviour of PHC slabs. These slabs are tested on a loading frame for the following load cases (a) LL at mid-span of slab (Fig. 1), (b) TLL (Fig. 2) and (c) UDL (Fig. 3) cases.

Fig. 1 Experimental setup—LL

Fig. 2 Experimental setup—TLL

Fig. 3 Experimental setup—UDL

2.3 *Experimental Setup*

The load is applied through a hydraulic jack and the test data are captured using PC-based data logger using a loading frame unit. The 2000 kN capacity load cell is used to measure load and linear variable differential transducer (LVDT) is used for measuring deflection. The slab is simply supported at ends with a bearing length of 150 mm. The slabs are loaded gradually up to the failure. The loading process is carried out to obtain first crack, cracking load, failure load and deflection. The images of crack pattern are monitored at each stage of loading process and it is photographed for the verification at a later stage.

An I-section is kept at mid-span of slab width in LL arrangement as shown in Fig. 1. Two I-section beams at one-third span of slab width in TLL arrangement as shown in Fig. 2. The load applied is uniformly transferred through these sections in UDL arrangement as shown in Fig. 3. In UDL arrangement, load is imposed on the specimen by placing a six number of 160 mm diameter 40 mm thick mild steel discs, two discs spaced equally along its width and 3 numbers of discs spaced equally along its span length. Above these discs I-sections are placed to apply the load. In all these arrangements—LL, TLL and UDL, another set of I-sections of 150 mm height and 150 mm width are kept for supporting the slabs.

3 Preparation of the Specimen

The experimental specimen consists of 3 PHC slabs. Each of them has dimensions of 1500 mm × 595 mm × 125 mm, with 5 circular voids of 70 mm diameter as shown in Fig. 4. The circular voids are provided symmetrically at mid-depth of slabs. Slabs are reinforced with 6 numbers of pre-stressing strands of 5 mm diameter with Poisson's coefficient of 0.3. They are having a maximum tensile strength of 1595 MPa and Young's modulus of 210 kN/mm^2. For having accelerated curing, steam curing at atmospheric pressure was carried out.

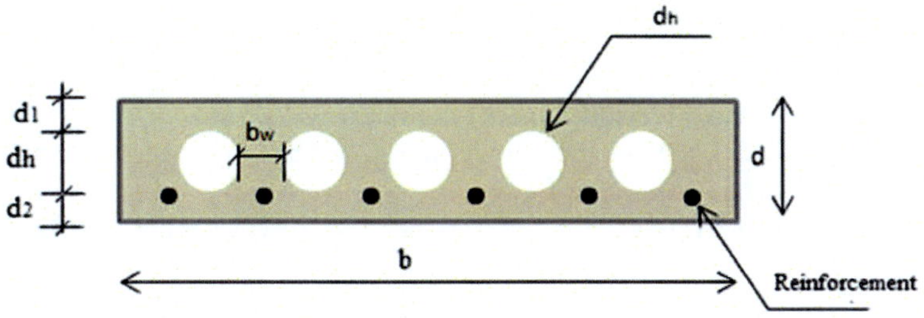

Fig. 4 PHC slab

The geometrical dimensions of the proposed specimens satisfy requirements of IS10297-1982 Clause 4.4.3 [12] and are tabulated in Table 1.

Concrete mix design for M40 grade is made in accordance with IS10262:2009 [13] and the mix proportion details are shown in Table 2.

Table 1 Specimen dimensions

Parameter	Dimensions (mm)	ISO10297-1982 (mm)
Width, b	595	< 2100
Span	1500	–
Depth, $d = \text{span}/30$	125	> 50
Top flange depth, $d1$	31.25	20 (min) or $d/4$
Bottom flange depth, $d2$	20	20 (min) or $d/8$
Hollow core diameter, $d_h = d - (d1 + d2)$	70	–
Web width $b_0 = b - (n * d_h)$, where n is number of voids	245	$\geq b/3$
Net web width, $bw = b_0/n$	40.833	–
a/d ratio	0.1723	–

Table 2 Concrete mix design details

S. No.	Material	Quantity
1	Cement	285 kg/m^3
2	GGBS	122 kg/m^3
3	Fine aggregate	807 kg/m^3
4	Coarse aggregate	1127 kg/m^3
5	Chemical admixture	7.40 kg/m^3
6	Water cement ratio	0.343

4 Numerical Model

ANSYS APDL is used for carrying out nonlinear finite element analysis of PHC slabs. ANSYS model of slab is as shown in Fig. 5. Meshing size of 25 mm [14] is adopted for the slabs considered.

SOLID165, LINK180 and SOLID185 are the element types considered for concrete, pre-stressing strands and steel plates respectively and are given in Table 3.

To avoid uneven loading/stress concentration, steel plates of 10 mm thickness are provided at the loading areas. Slab and steel plate were modelled as contact-pair elements, in which steel plate is assigned as contact and concrete are assigned as target elements. The boundary conditions of slabs are provided such that *Y*-direction (UY=0) movements are restrained on the right-hand side of slab and *X*, *Y* and *Z*

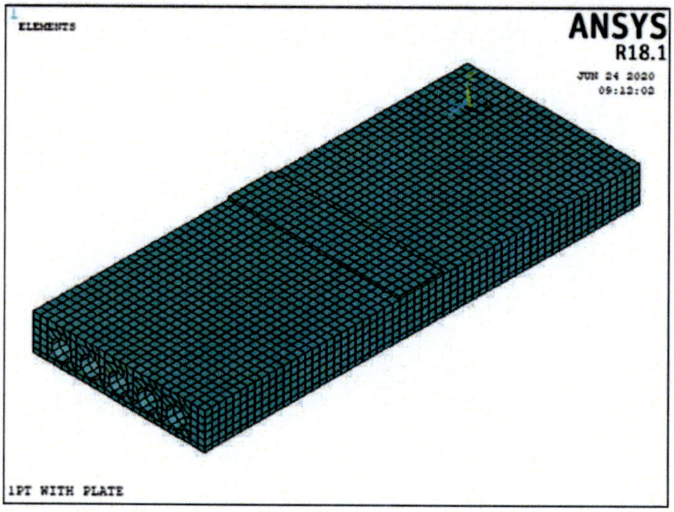

Fig. 5 Meshed PHC slab model

Table 3 Modelling element types

Material	Element type	Remarks
Concrete	SOLID65	To be able to account for the failure modes of concrete cracking in tension and crushing in compression, a special brittle finite element material model has to be used
Pre-stressing strand	LINK180	3D spar uni-axial tension–compression element
Steel plates	SOLID185	Element with high strain and stiffen stress
Contact pairs	CONTA 173 & TARGE 170	Steel plate was taken as the contact element. Concrete was taken as the target

direction (UX = 0, UY = 0 & UZ = 0) movements are restrained on left-hand side of slab. The loading steps are followed the same manner in both experimental and numerical models. The incremental loading steps of 10 kN applied until the first crack is observed and each step was applied with sub-steps of 2, i.e. applied twice. Further on, the load incremented with same incremental loading 10 kN, but with sub-steps of 10 until the failure happens. Load increment was divided into 2 sup-steps. Close to the failure load, increment was divided into 10 sub-steps.

5 Validation of Model

The deflections obtained experimentally and numerically for load cases-LL, TLL and UDL are plotted to validate the performance of PHC slabs. The load deflection plots under LL, TLL and UDL are shown in Figs. 6, 8 and 10 respectively. The plots are depicted for comparison. In experimental evaluation, the ultimate deflection values in slabs at failure load is found to be 11.21 mm, 8.30 mm, 5.20 mm respectively for loads-LL, TLL and UDL respectively. The deflection observed from ANSYS model on slabs under LL, TLL and UDL loading cases are as shown in Figs. 7, 9 and 11 respectively. The ultimate deflection computed by numerical analysis in the slabs at failure load are found to be 10.73 mm, 8.18 mm, 4.84 mm under LL, TLL and UDL respectively. It is observed that numerical analysis shows lesser deflection and higher ultimate load. Initially the deflections at initial stage are same both experimentally and numerically. Small divergence of values in deflection is observed at ultimate loads. The deflection under LL experimentally is similar with ANSYS analysis up to 350 kN as shown in Fig. 6. Further, numerical analysis shows lower deflection trend compared to experimental results may due to uncertainty of concrete after first crack.

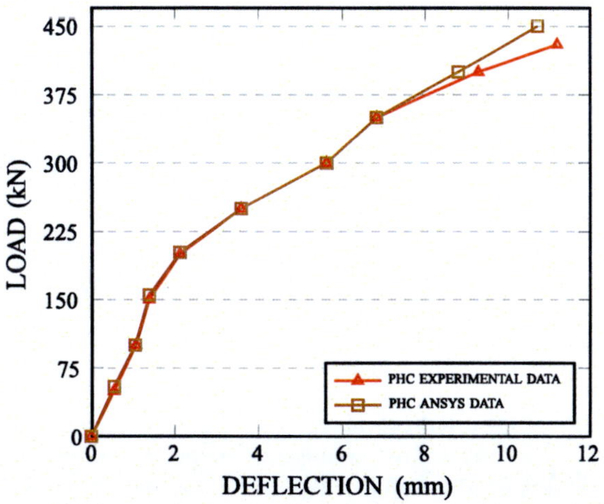

Fig. 6 Load deflection plot under LL

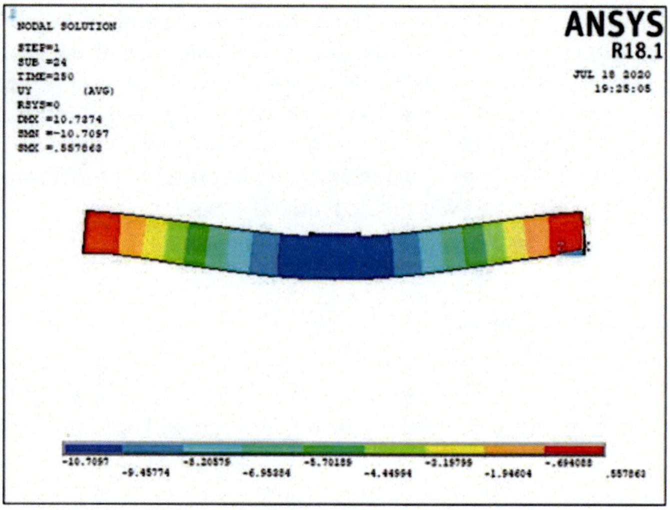

Fig. 7 Deflection plot of PHC slab under LL

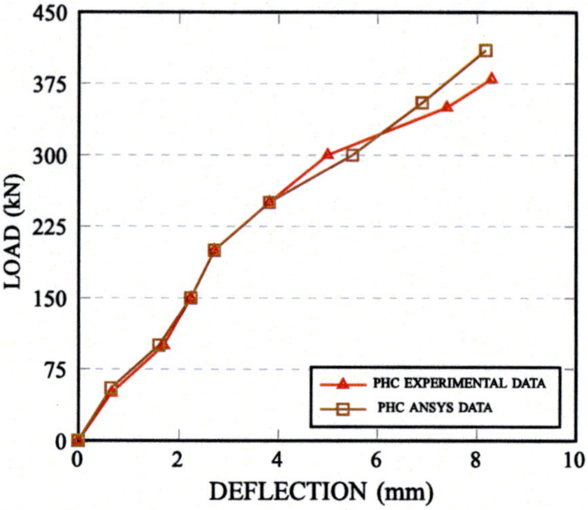

Fig. 8 Load deflection plot under TLL

Experimentally measured deflection of slab under UDL has shown similar trend with ANSYS results up to 300 kN as shown in Fig. 10. Further, numerical analysis shows lower deflection trend compared to experimental results.

Experimentally measured deflection of slab under TLL, have shown similar trend with ANSYS results up to 325 kN as shown in Fig. 8. Further, numerical analysis shows lower deflection trend compared to experimental results.

The comparative analysis of experimental and numerical results of PHC slabs are tabulated in Table 4. The results obtained from ANSYS software are comparable

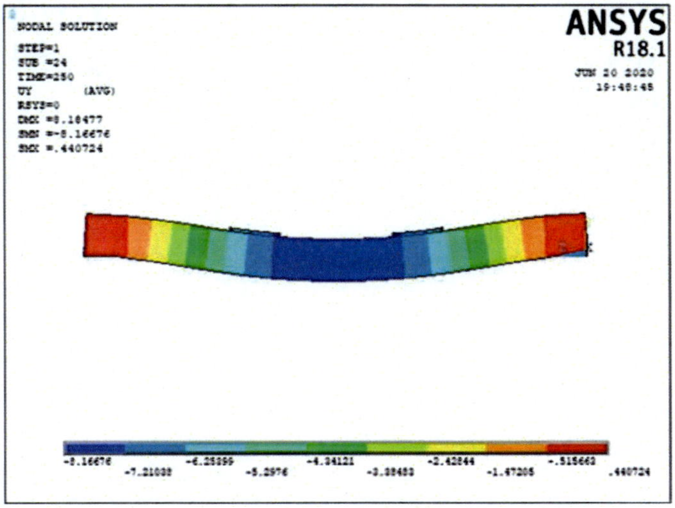

Fig. 9 Deflection plot of PHC slab under TLL

Fig. 10 Load deflection plot under UDL

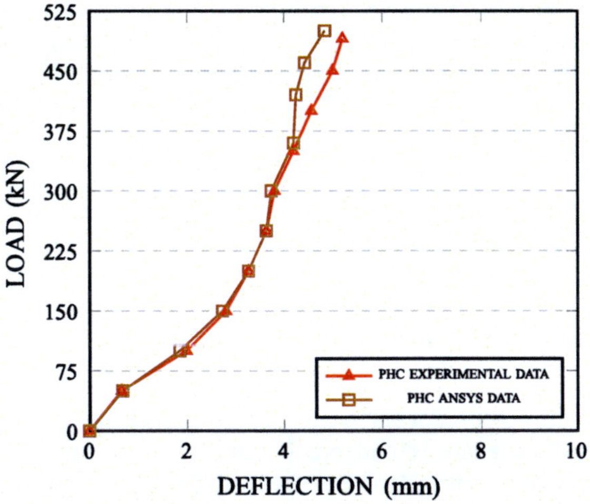

with experimental results. The usage of contact pairs, meshing technique and element selection of ANSYS model provides satisfactory results as of experimental results. This confirms the feasibility of modelling techniques as a cost-effective alternative design and analysis mechanism. It also confirms the suitability of models for the performance evaluation and parametric study of PHC slabs in the present work.

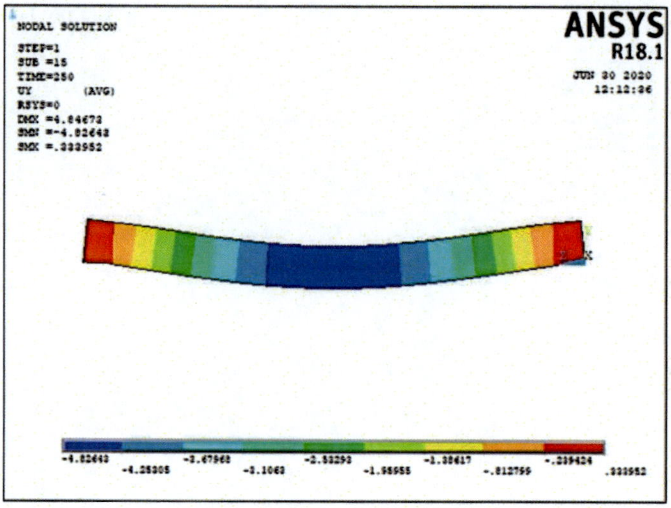

Fig. 11 Deflection plot of PHC slabs under UDL

Table 4 Comparison of results

Loading case	Experimental			Numerical		Mode of failure
	Crack load (kN)	Failure load (kN)	Deflection (mm)	Failure load (kN)	Deflection (mm)	
LL	310	430	11.21	450	10.73	Shear compression
TLL	340	380	8.3	410	8.18	Web shear
UDL	420	490	5.2	500	4.84	Web shear

6 Performance Evaluation

This section details the performance evaluation of PHC slabs. It consists of stress pattern analysis and crack pattern observation. In addition, the crack patterns observed with ANSYS models are compared with patterns obtained during the test experiments.

6.1 Stress Patterns

The purpose of stress distribution analysis is to find the maximum stress points of PHC slabs under the loading of LL, TLL and UDL cases using numerical method. Von Mises stress distribution is based on distortion energy theory, is being considered for the analysis.

The stress patterns observed on the application of LL case are as shown in Fig. 12. It is observed that for LL case, the maximum stress concentration occurred in bottom region of voids and at support region.

The stress pattern observed with the application of TLL case is shown in Fig. 13. When TLL case is applied, it is observed that the maximum stress concentration occurred at web area between cores and at support region. The stress patterns

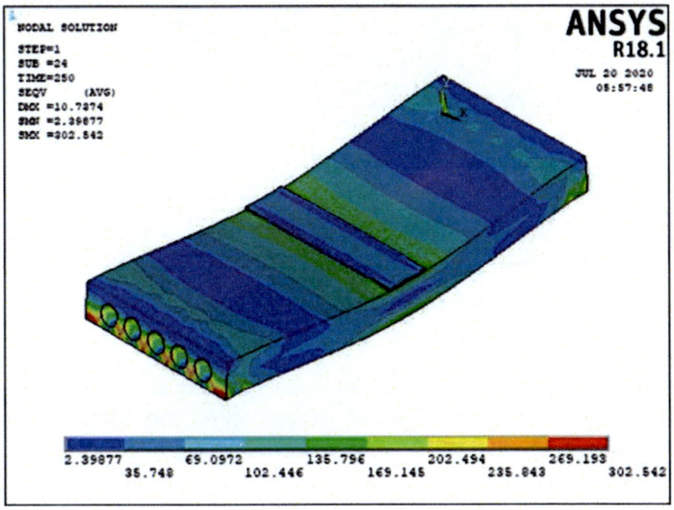

Fig. 12 Stress pattern on PHC slab due to LL

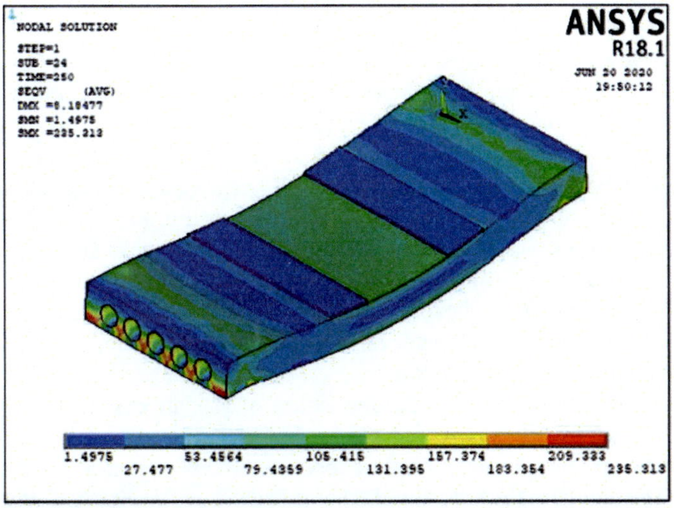

Fig. 13 Stress pattern on PHC slab due to TLL

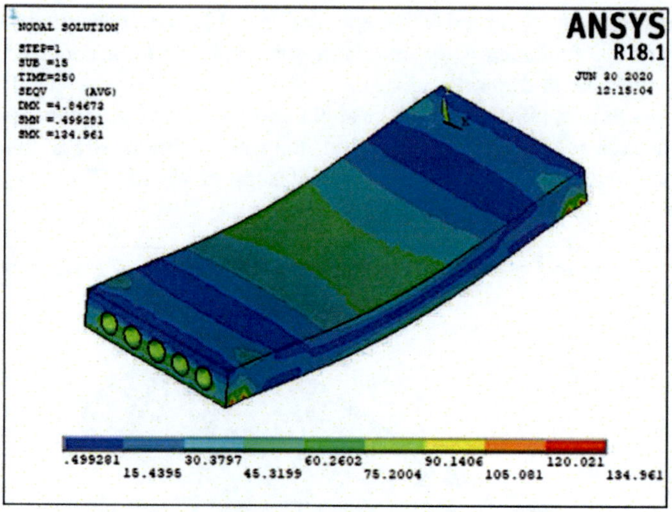

Fig. 14 Stress pattern on PHC slab due to UDL

observed with the application of UDL case on slabs are as shown in Fig. 14. The maximum stress pattern is observed only at support regions when UDL is applied. The stress pattern analysis is useful in understanding the weak zones in a structure subjected to the loading.

6.2 *Crack Patterns*

The behaviour of crack patterns gives an insight into the vital clues of failures and it assists engineers to take remedial measures to prevent major catastrophe. A comparative analysis was conducted between crack patterns observed during experimental study and the results obtained from numerical analysis. Fig. 15 indicates the crack patterns observed on the application of LL case. The visible crack occurred at a load of 310 kN initially and the width of the initial crack was less than 1 mm. At the failure load of 430 kN, the crack width increased to 3 mm. The cracks are observed at the mid-span of the slab, indicates shear compression failure [15]. With the increase in loading, the crack begins at flexural zones, then it spreads to shear zones and later induces into the compression zone. Finally, it ends with crushing of concrete and fails at the application loading point.

The crack patterns observed on the application of TLL are as shown in Fig. 16. The first visible crack in slab is observed at a load of 340 kN and the initial crack width is found to be less than 1 mm. At failure load of 380 kN, the crack width increased to 3 mm. The crack originated from the point of load application, i.e. at 1/3rd span of the slab and extended up to top surface of slab through the cores. According to

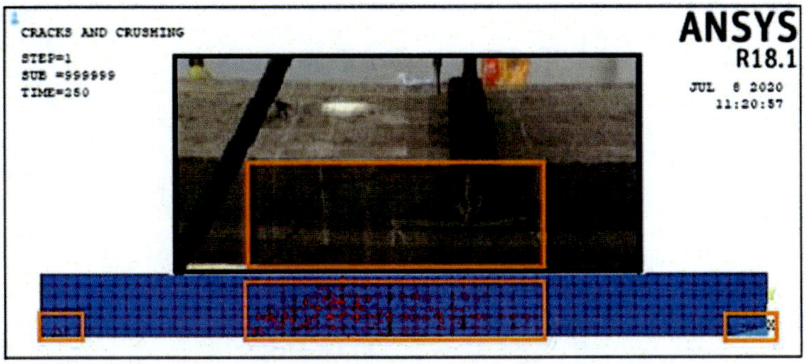

Fig. 15 Crack patterns observed under LL

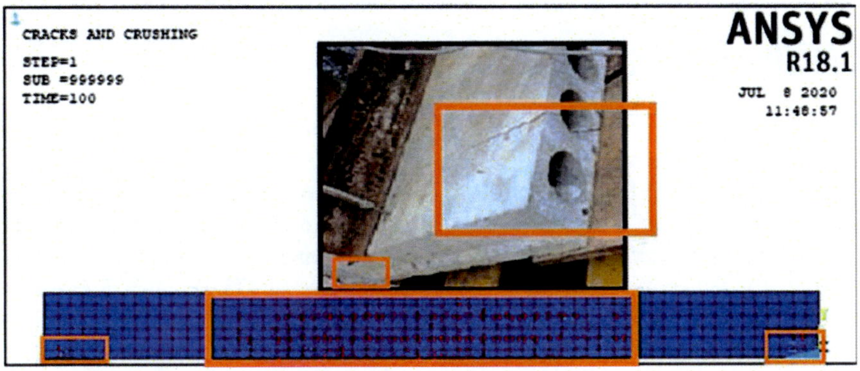

Fig. 16 Crack patterns observed under TLL

FIP recommendations [15], the crack developed is a longitudinal crack at void/core and results in web-shear mode of failure.

The crack patterns observed for UDL case are as shown in Fig. 17. The first visible crack occurred at a load of 420 kN and the initial width is found to be less than 1 mm. At failure load of 490 kN, the crack width increased to 3 mm. It is observed that the crack behaviour and failure mode exhibit the similar behaviour as in case of TLL. It is also observed that the failure modes of PHC are due to web shear [15].

The load at the occurrence of first crack, failure load and failure mode for loading cases of LL, TLL and UDL are shown in Table 4. It is observed that in accordance with ACI 318 [15], the chosen slab with *a*/*d* ratio of 0.1723 failed with web shear mode, under TLL and UDL cases. Whereas it failed with shear compression, when subjected to LL case.

Fig. 17 Crack patterns observed under UDL

7 Effect of Core Shape in PHC Slabs

The core shape of PHC slabs place an important role in load carrying capacity. In order to find the effect of the core shapes on the performance of PHC slab, ANSYS analysis is conducted by changing the core shape from circular to square. The cross-sectional area for both circular and square cores are kept same. The diameter of circular core considered was 70 mm while for a square core each side measured 62 mm. The deflection observed on PHSC slabs under LL, TLL and UDL cases are as shown in Figs. 18, 20 and 22 and the corresponding stress patterns of PHSC slabs under LL, TLL and UDL cases are as shown in Figs. 19, 21 and 23 respectively.

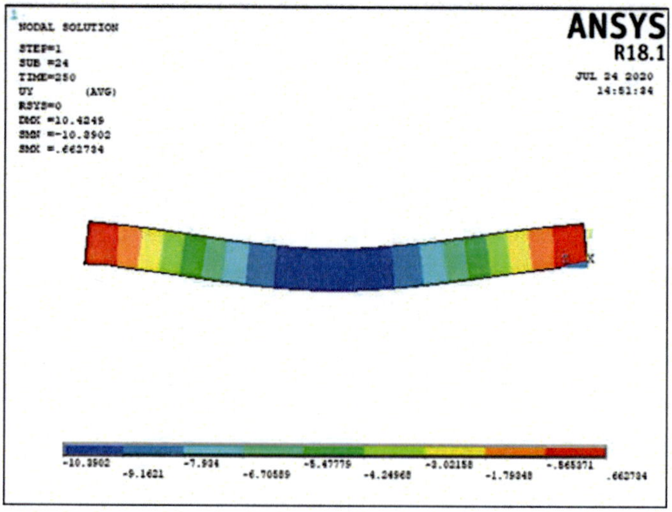

Fig. 18 Deflection of PHSC slab under LL

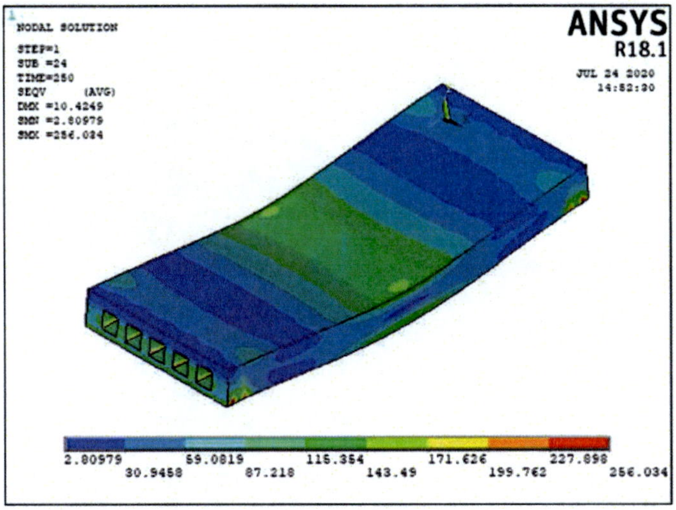

Fig. 19 Stress pattern on PHSC slab due to LL

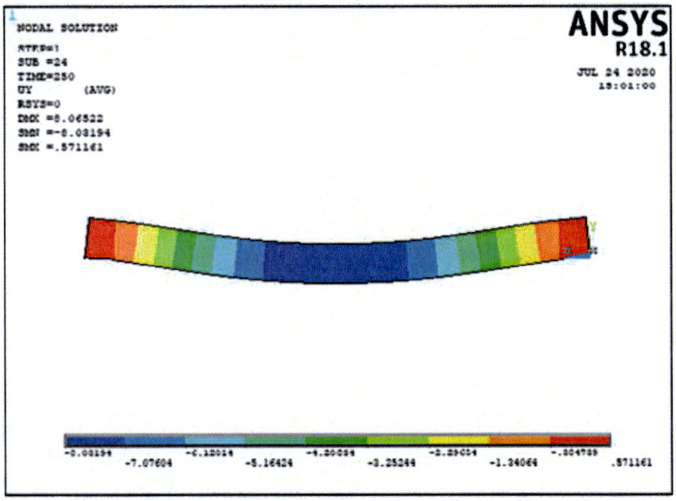

Fig. 20 Deflection of PHSC slab under TLL

The failure load and corresponding deflection are tabulated for circular core and square core slabs as given in Table 5. The load carrying capacity of PHSC slabs are found to be lesser than PHC slabs. An improvement in load carrying capacity of 8.88%, 9.7% and 9.6% was observed in circular core slabs compared to square core slabs when subjected to LL, TLL and UDL cases respectively. It is also observed that there is only marginal increase in deflection in slabs with square core compared

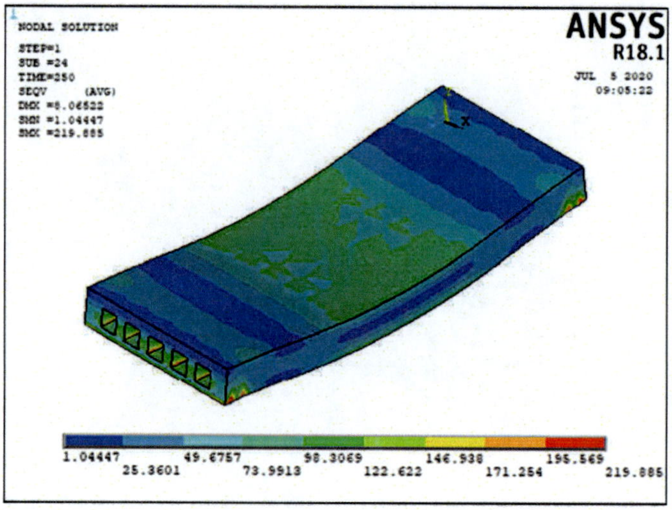

Fig. 21 Stress pattern on PHSC slab due to TLL

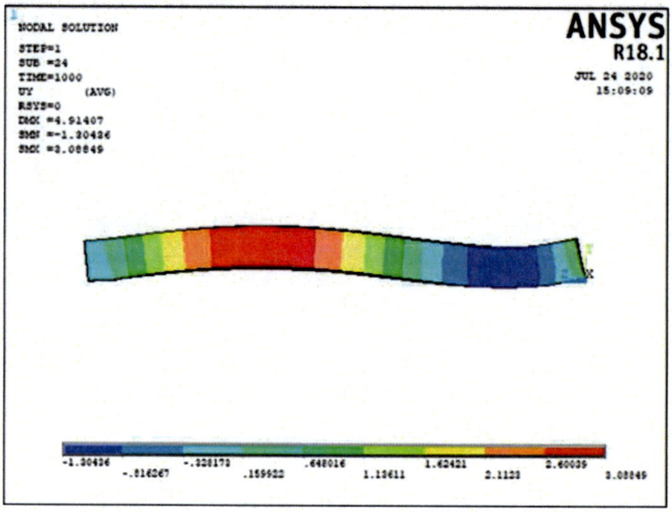

Fig. 22 Deflection of PHSC slab under UDL

to circular core. However, the load carrying capacity for slabs with circular core is higher than the slabs with square cores.

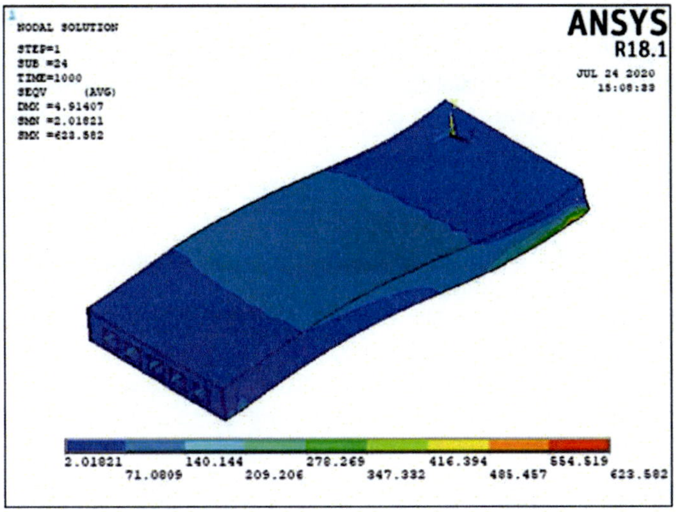

Fig. 23 Stress pattern on PHSC slab due to UDL

Table 5 PHC slab versus PSHC slab

Loading case	Circular core		Square core	
	Failure load (kN)	Deflection (mm)	Failure load (kN)	Deflection (mm)
LL	450	10.16	410	10.43
TLL	410	8.18	370	8.06
UDL	520	4.84	470	4.91

8 Conclusion

Replacement of cast-in-situ construction by precast industry, triggers tremendous research potential in precast structures. The work brings out an appropriate modelling mechanism in ANSYS to perform numerical analysis to assess deflection, stress pattern and crack pattern of PHC slabs. The chosen discrete modelling mechanism provides compatible results to that of experimental evaluation. In this work, performance evaluation of PHC slabs is compared experimentally and numerically. The observations obtained are as follows:

- The range of variation in maximum deflection of PHC slabs between numerical analysis and experimental results were within 9.36%, 1.44% and 6.90% under LL, TLL and UDL respectively.
- It is observed that the maximum stress concentration occurred at support irrespective of loading case applied. In addition, maximum stress concentration occurred at slab core area when LL is applied and at slab web area when TLL is applied.

- PHC slabs failed with shear compression, web-shear and bond-and-anchorage failure modes when subjected to LL, TLL and UDL cases respectively.
- The core shape of PHC slabs does not have much impact on deflection behaviour of slabs. However, there is an improvement of load carrying capacity of 8.88%, 9.7% and 9.6% observed with circular core PHC than square core subjected to LL, TLL and UDL cases respectively.

It was measured that performance of PHC slabs in terms of deflection, crack pattern and failure and observed that experimental measurements have greater agreement with ANSYS simulation results. It is proved the performance evaluation of PHC slabs using numerical analysis is better and cost-effective alternative to test experiments. Considering load carrying capacity, it is preferable to use circular core shape PHC slabs as roofing/flooring/walls/noise barrier components.

References

1. Al-Azzawi AA, Abed SA (2017) Investigation of the behavior of reinforced concrete hollow-core thick slabs. Comput Concr 19(5):567–577
2. Walraven JC, Mercx WM (1983) The bearing capacity for pre-stressed hollow core slabs. Heron (Delft) 28(3):1–46
3. Aswad A, Jacques FJ (1992) Behavior of hollow-core slabs subject to edge loads. PCI J 37(2)
4. Rahman M, Baluch M, Said M, Shazali M (2012) Flexural and shear strength of prestressed precast hollow-core slabs. Arab J Sci Eng 37(2):443–455
5. Haruna SI (2014) Flexural behavior of precast prestressed concrete hollow core slabs with cast-in-place concrete topping
6. Derkowski W, Surma M (2013) Shear capacity of prestressed hollow core slabs on flexible supports. Czasopismo Techniczne. Budownictwo 110(2-B):3–12
7. Rekha B, Ravindra R, Kumar P (2016) Experimental investigations of prestressed hollow core slabs and grouted hollow core slabs. J Struct Eng Manag 3(3):16–21
8. Fallon E, Robinson J (1999) The behaviour of hollowcore floor slabs. In: Innovation in concrete structures: design and construction. Thomas Telford Publishing, pp 507–518
9. Brunesi E, Bolognini D, Nascimbene R (2015) Evaluation of the shear capacity of precast-prestressed hollow core slabs: numerical and experimental comparisons. Mater Struct 48(5):1503–1521
10. Hoogenboom P (2005) Analysis of hollow-core slab floors. Heron 50(3)
11. Deeb AN, Tarkhan M, El-Tehewy E (2012) Finite element modeling of pre-stressed hollow core slabs. In: The 9th international conference on civil and architecture engineering, vol 9. Military Technical College, 2012, pp 1–12
12. Standard I (1982) Is 10297: code of practice for design and construction of floors and roofs using precast reinforced/prestressed concrete ribbed or cored slab unit. Bureau of Indian Standards
13. Standard I (2009) Is 10262: 2009, concrete mix proportioning guidelines. Bureau of Indian Standards
14. Al-Azzawi AA, Abed SA (2006) Numerical analysis of reinforced concrete hollow-core slabs. ARPN J Eng Appl Sci 11(5):9284
15. Committee A (2005) Building code requirements for structural concrete (ACI 318-05) and commentary (ACI 318R-05). American Concrete Institute

Study on Rainfall Trends and Water Requirement for Crops in Bellary District of Karnataka, India

R. Sreedevi and B. R. Ramesh

Abstract The global climate change could have important effects on various weather parameters including rainfall in many countries around the world. The trends of rainfall have a great impact in the existing regions for crop production. This study is carried out to examine the trend of rainfall in Bellary District of Karnataka, India. Rainfall data of 51 rain gauge stations for a period of 40 years (1977–2016) and monthly evapotranspiration (ETo) data of 2007–2016 is considered for the analysis. Rainfall anomaly index (RAI) and IMD method are used to find the trend of rainfall. Meteorological drought severity index (MDSI) is used to assess the severity of drought in the study area. The crop water requirement studies with 10 years whether parameters using Penman Montaith method for crops in Kharif and Rabi season are analysed to assess the necessity of irrigation water for both Kharif and Rabi season. Rainfall anomaly index (RAI) shows the decreasing trend of rainfall, and IMD method shows the occurrence and frequency of various drought. MDSI analysis indicates that the study area is facing moderate drought. Water requirement computations to crops for both Kharif and Rabi season indicate that dry crops can be grown in Kharif season without any irrigation. However, both dry, wet and seasonal crops require irrigation supply in both the seasons.

Keywords Drought · Trend analysis · Precipitation · Crop season · RAI · IMD · MDSI

1 Introduction

Drought is considered as one of the most complex of all the natural hazards, which affects more people than any other hazard. It is widely defined as severe water shortage. Low rainfall and fall in agricultural production are the main causes for drought [1]. The key parameters like rainfall and temperature of the weather system

R. Sreedevi (✉)
Department of CE, East West Institute of Technology, Bengaluru 91, India

B. R. Ramesh
Department of CE, Nitte Meenakshi Institute of Technology, Bengaluru 64, India

L. Nandagiri et al. (eds.), *Sustainability Trends and Challenges in Civil Engineering*, Lecture Notes in Civil Engineering 162, https://doi.org/10.1007/978-981-16-2826-9_44

which are responsible for drought are temporal and spatial variability at different scales. Thus, analysis of their behaviour is of vital importance for understanding of climate variability. Rainfall is vital climate parameter which has direct effect on crop growth and yield under rain fed condition [2]. Any change in the weather parameters like temperature, rainfall, wind pattern over a decade or longer is termed as climate change (IPCC, 2015). In planning of agriculture allied projects, rainfall probability is to be considered as an important factor. Hence, information about rainfall distribution pattern at specified location is quite essential for planning of irrigation system and cropping pattern for the project [3].

In countries like India, which mainly depends on monsoon for agricultural activities, rainfall data is a crucial parameter necessary for planning and design of water resources projects. Study of trends of monsoon in India is very important [4]. To identify the climate change studies using annual series of sub-regions have been attempted by researchers [5]. Precipitation pattern and their relationship with ENSO are conducted by Sarkar and Kafatos [6]. Annual rainfall studies from 1901 to 1960 were conducted by Parthasarathy and Dhar, indicated an increasing trend in and around Central India and a decreasing trend of rainfall in some regions of eastern India [7]. Nandeesha used India Meteorological Department (IMD) method to assess meteorological drought gives overall idea of drought proneness in any area with combination of different severities of drought [8].

2 Study Area

The present study area—Bellary region, in the state of Karnataka has the second largest of arid zone after Rajasthan in terms of total geographical area prone to drought [9]. The district is elongated from south-west to north-east and is located in the eastern part of the Karnataka state and lies between the north latitude 14° 30′:15° 47′ east longitude 75° 40′:77° 11'. Seventy five per cent of the labour force in the district is dependent on agriculture and allied activities. The Tungabhadra perennial river-tributary to Krishna River forms major drainage system in Bellary district (Fig. 1).

3 Methodology

3.1 Rainfall Anomaly Index

Rainfall anomaly index (RAI) of ranking procedure is used to assign magnitudes for precipitation [10]. The range of RAI values is divided into nine categories as shown in Table 1.

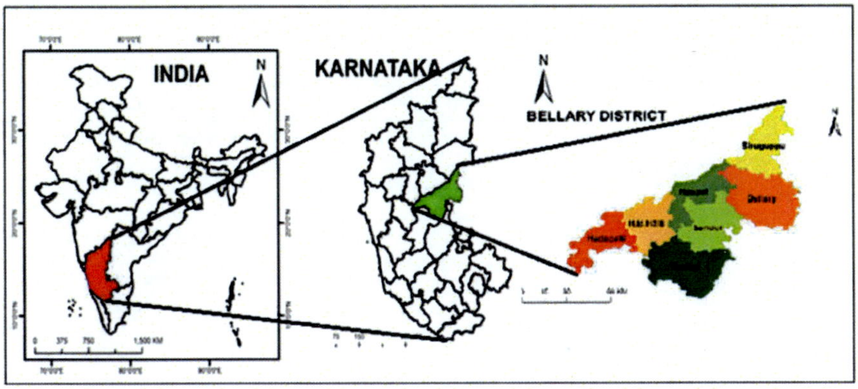

Fig. 1 Bellary district

Table 1 Classification of RAI values

Range of values	Classification
≥ 3.00	Extremely wet
2.00–2.99	Very wet
1.00–1.99	Moderately wet
0.50–0.99	Slightly wet
0.49 to − 0.49	Near normal
− 0.50 to − 0.99	Slightly dry
− 1.00 to − 1.99	Moderately dry
− 2.00 to − 2.99	Very dry
≤− 3.00	Extremely dry

The procedure of ranking is done by arranging the meteorological data in descending order. The top ten values are averaged to form a threshold for positive anomaly, and the last ten values are averaged to form a threshold for negative anomaly. The arbitrary threshold values of + 3 and − 3 have, respectively, been assigned to the mean of the ten most extreme positive and negative anomalies. Nine abnormality classes are then given against a scale of numerical values of the relative rainfall anomaly index [11].

3.2 Indian Meteorological Department (IMD) Method

Significant decrease in precipitation from the normal precipitation in an area is termed as meteorological drought. IMD method is employed for the assessment of drought severity [12]. Percentage deviation of annual rainfall from the long-term annual mean

Table 2 IMD classification of drought

> 0	M_0	No drought
0 to − 25	M_1	Mild drought
− 25 to − 50	M_2	Moderate drought
<− 50	M_3	Severe drought

Table 3 Range of drought severity in MDSI

S. No.	Range	Drought severity
1	1.34–1.64	No drought
2	1.64–1.94	Mild drought
3	1.94–2.24	Moderate drought
4	2.24–2.54	Severe drought

rainfall is employed to assess the drought in this method. The definition of drought according to this method is presented in Table 2.

3.3 *Meteorological Drought Severity Index (MDSI)*

Meteorological drought severity index is a number computed based on the frequency analysis. It is the number of times the precipitation is deviated in a given period with respect to mean rainfall of the region.

The method involves frequency analysis of various classes of drought severity. Weightages 1, 2, 3 and 4 are assigned to drought severity classes 'No drought', 'Mild drought', 'Moderate drought' and 'Severe drought', respectively. The MDSI is found out by multiplying the frequency of each class of drought severity by the corresponding weightage. Four drought severity classes were delineated based on the range of drought severity index [12] as shown in Table 3.

3.4 *Water Requirement and Crop Season*

Crop water requirements (CWR) are the water consumed by crop for their evapotranspiration (ETo). The crop must be disease-free, growing in large fields under non-restricting soil conditions including soil, water and fertility and produce full production potential under the given growing environment (L. S. Pereira, I. Alves, in Reference Module in Earth Systems and Environmental Sciences, 2013). This method needs the climatic and agronomic conditions of the area.

Cropping season in India is classified into two main seasons Viz., Kharif and Rabi based on the monsoon. The Kharif cropping season is from July to October during

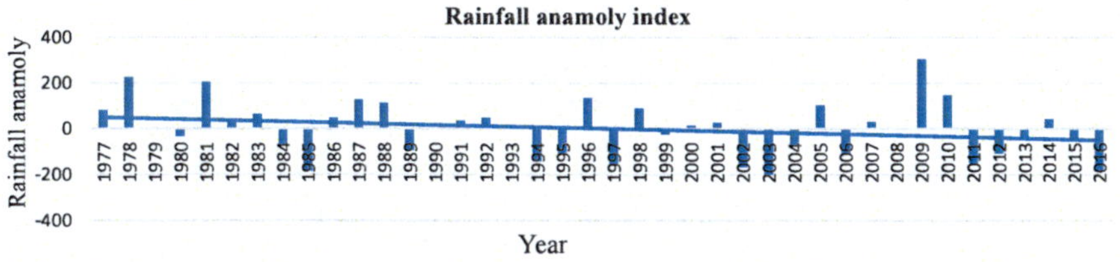

Fig. 2 Rainfall anomaly i

the south-west monsoon, and the Rabi cropping season is from October to March (winter). The crops grown between March and June are called summer crops.

The requirement of crop water for its survival, growth and development is to be applied either naturally by precipitation or artificially by irrigation. Ten year (2007–2016) average value of monthly mean rainfall (cm) for Bellary region is considered, and ETo (cm) computation for different crops and seasons is presented in Table 9.

4 Results

4.1 Rainfall Anomaly Index

Annual rainfall anomaly index can be used to analyse the frequency and intensity of the dry and rainy years. Positive anomaly have their values above average, and negative anomaly have their values below average.

The rainfall anomaly index analysis is conducted for rainfall data between 1977 and 2016. The year wise classification is shown in Fig. 2. Out of 40 year rainfall, 21 years shows positive anomaly, and 19 year shows the negative anomaly. Positive anomalies are more in earlier years, and negative anomalies are more in recent years.

Classification of different categories of rainfall anomalies and the number of years of occurrence is presented in Table 4. Extremely dry and extremely wet years were not observed during the period. Even though it shows one very wet year and zero very dry year, number of moderately dry and slightly dry years are seven each; moderately wet and slightly wet years are 2 each, whereas the near normal years are twenty-one in the forty years of study; therefore, dry years are observed more frequent, and area can be moderately classified as agricultural drought.

4.2 IMD Method

Study area consists of fifty-one rain gauge stations. Drought classification in the region on basis of rain gauge stations is presented in Table 5. The classification

Table 4 Rainfall anomaly index

Range of values	Classification	No. of years (1977–2016)
≥ 3.00	Extremely wet	0
2.00–2.99	Very wet	1
1.00–1.99	Moderately wet	2
0.50–0.99	Slightly wet	2
0.49 to – 0.49	Near normal	21
– 0.50 to – 0.99	Slightly dry	7
– 1.00 to – 1.99	Moderately dry	7
– 2.00 to – 2.99	Very dry	0
≤– 3.00	Extremely dry	0

indicates twenty-three stations with no drought (> 0) conditions, twelve stations are under mild drought (0 to – 25), twelve are under moderate drought (– 25 to – 50) and four stations are facing severe drought (<–50).

Frequency of drought occurrence is also calculated according to year, i.e. from 1977 to 2016. The results are shown in Table 6. Percentage deviation is calculated using the average annual rainfall of all the forty years and the average rainfall of the individual year. In the forty years of study, four year shows 'No drought' conditions, seventeen years are under 'Mild drought', sixteen years are under 'Moderate drought' and three years are facing 'Severe drought'.

The analysis of frequency of drought based on rain gauge stations indicates rate of non-occurrence of drought in the study area is 45% (as the average rainfall of 12 rain gauge stations are more than 500 mm). Frequency of mild and moderate and severe drought is 23% and 24% and 8%, respectively.

The analysis of frequency of drought on basis of years (40 years) indicates severe drought as 8%. There is a drastic change in the non-occurrence of drought, i.e. from 45 to 10%, with increase in chances of occurrence of mild drought from 24 to 42% and the moderate drought from 24 to 40%.

Drought classes according to the rain gauge stations and according to the year are shown through Pie chart, Figs. 3 and 4.

4.3 *Meteorological Drought Severity Index (MDSI)*

Meteorological drought severity index (MDSI) was developed by calculating probability of each drought severity class. Weightages of 1, 2, 3 and 4 were assigned, and the cumulative weighted value is arrived as per MDSI. From the IMD results presented in Tables 5 and 6, the meteorological drought severity status of study area was determined. Four drought severity classes were delineated based on the range of drought severity index as shown in Table 3.

Table 5 Occurrence and frequency of drought according to rain gauge station

Occurrence and frequency of drought								
Name of the district	No drought		Mild drought		Moderate drought		Severe drought	
	Occurrence	Frequency	Occurrence	Frequency	Occurrence	Frequency	Occurrence	Frequency
Bellary	23	0.45	12	0.24	12	0.24	4	0.08

Table 6 Occurrence and frequency of drought according to year

Occurrence and frequency of drought								
Name of the district	No drought		Mild drought		Moderate drought		Severe drought	
	Occurrence	Frequency	Occurrence	Frequency	Occurrence	Frequency	Occurrence	Frequency
Bellary	4	0.1	17	0.425	16	0.4	3	0.075

Fig. 3 Frequency according to rain gauge station

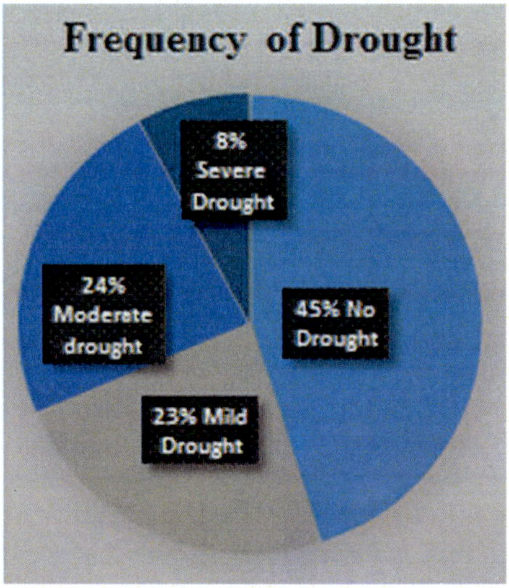

Fig. 4 Frequency according to year

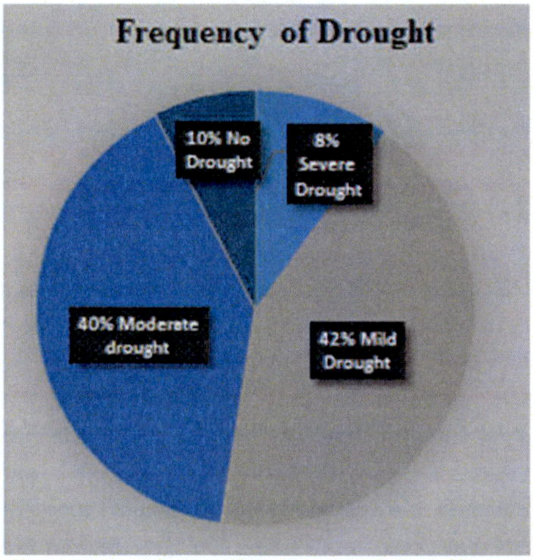

MDSI was calculated by taking the average of fifty-one rain gauge stations with average annual rainfall of 387 mm. Probability of drought severity index is obtained as 45% chances of occurrence of no drought, 24% incidence of mild and moderate drought and 8% existence of severe drought. Through these probability of classes drought severity index is attained as 1.94 with which it can be alleged that drought severity class for the study area is moderate, shown in Table 7.

MDSI was also calculated by taking the average of forty years with average annual rainfall of 387 mm. Probability of drought severity index is obtained as 10% chances of occurrence of no drought, 42.5% incidence of mild, 40% of occurrence

Table 7 MDSI according to rain gauge station

Meteorological Drought Severity Index							
Name of the district	Average annual rainfall (mm)	Probability of drought severity class				Drought Severity Index	Drought severity class
		No drought	Mild	Moderate	Severe		
Bellary	387	0.45	0.24	0.24	0.08	1.94	Moderate

Table 8 MDSI according to year

Meteorological Drought Severity Index							
Name of the district	Average annual rainfall (mm)	Probability of drought severity class				Drought Severity Index	Drought severity class
		No drought	Mild	Moderate	Severe		
Bellary	387	0.1	0.425	0.4	0.075	2.15	Moderate

of moderate drought and 8% existence of severe drought. Through these probability of classes, drought severity index is attained as 2.15 with which it can be alleged that drought severity class for the study area is moderate as shown in Table 8 (Table 9).

4.4 Water Requirement to Crops

Study on water requirement of crops is required to understand the necessity of irrigation to the crops in the region and also to suggest the crop planning suited to the region. The study is conducted for both Kharif and Rabi seasons considering the meteorological data from year 2007 to 2016.

Kharif crops

Paddy (Kharif crop): High rainfall or assured irrigation is essential for areas of rice cultivation. Even though Bellary district faces moderate drought, few villages in the lower rainfall region of the Bellary district grow paddy.

Crop period for medium type of paddy is 120–140 days (4–5 months). Sowing starts in the middle of august, and the harvesting is at the middle or at the end of January. Eto graph in Fig. 5 shows that there is a shortage of water in the months of November and January for paddy in the Kharif season. So there is a requirement of additional supply of water to the crop in the month of scarcity.

Jowar (Kharif crop): This crop is the main food and fodder crop of dry land agriculture. Crop period for Jowar is 110–120 days (4–5 months). Sowing starts in the starting of June, and the harvesting is at middle or at the end of October. ETo graph

Table 9 Average values of MMR and ETo

Month	January	February	March	April	May	June	July	August	September	October	November	December
MMR (cm)	1.4	2.1	4.3	7.0	15.6	13.1	7.7	11.3	14.0	17.4	8.7	2.6
ETo (cm)	5.53	6.905	7.989	3.293	9.551	8.115	6.642	6.189	5.501	5.022	4.34	4.614

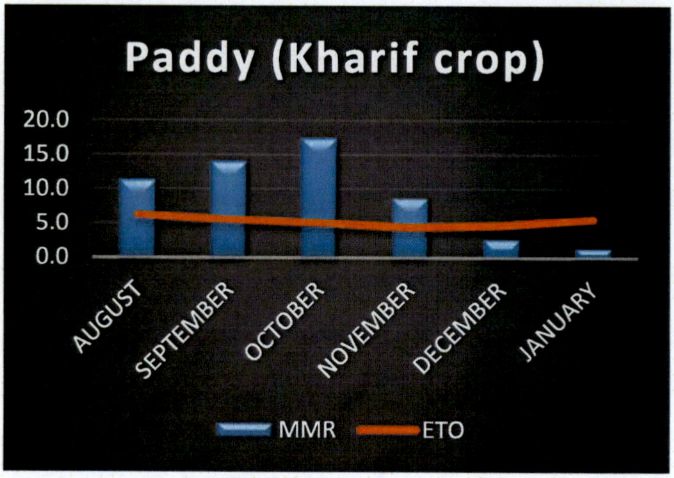

Fig. 5 Monthly water required for paddy

in Fig. 6 shows that no shortage of water for the Jowar crop in the Kharif season. Crop gets sufficient water for the growth and can expect a good production.

Bajra (Kharif crop): This crop is a drought resistant crop which is generally preferred in low rainfall areas and lighter soils. Crop period for bajra is 90–110 days (3–4 months). Sowing of the crop starts in the starting days of June and at the middle of September. ETo graph in Fig. 7 shows that the evapotranspiration in all months

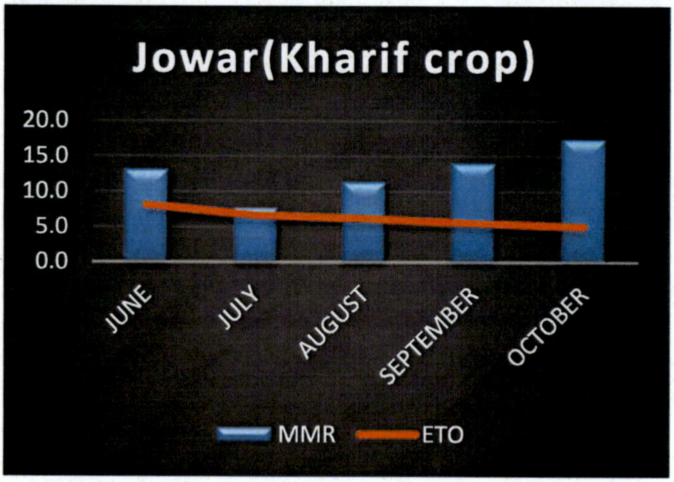

Fig. 6 Monthly water required for jowar

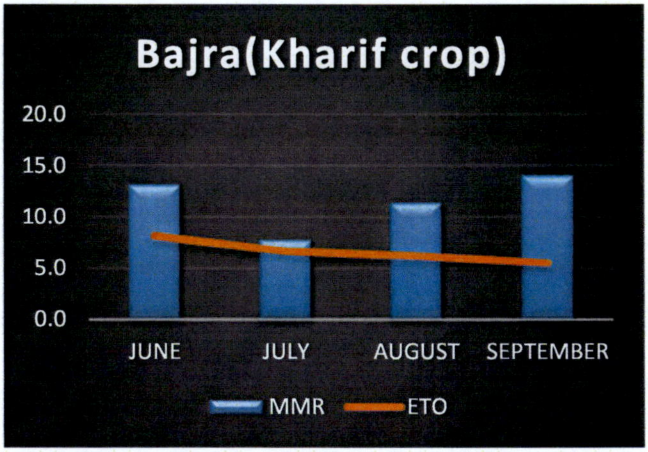

Fig. 7 Monthly water required for bajra

falls below the range of monthly mean rainfall. Crops in the kharif season have sufficient amount water for the expected production rate.

Maize (Kharif crop): This crop is essentially a warm weather crop grown in different regions and is one of the main cereals of the world. Duration of the crop is between 80 and 100 days (4–5 months). Sowing starts from the end of the august, and the harvesting is in the starting or in the middle of the December. ETo graph in Fig. 8

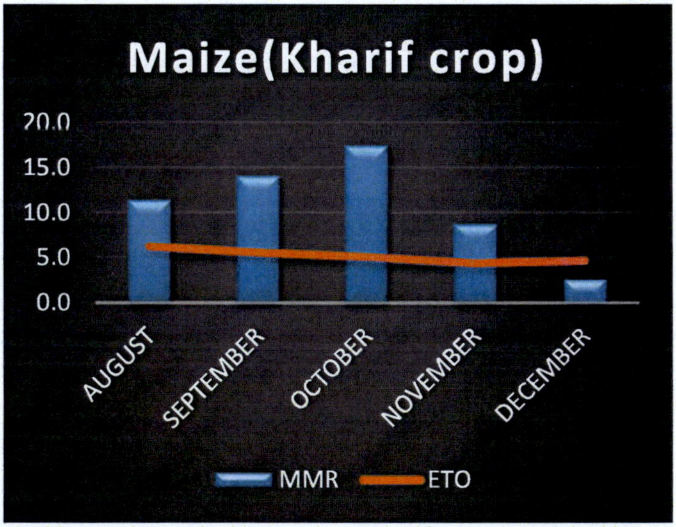

Fig. 8 Monthly water required for maize

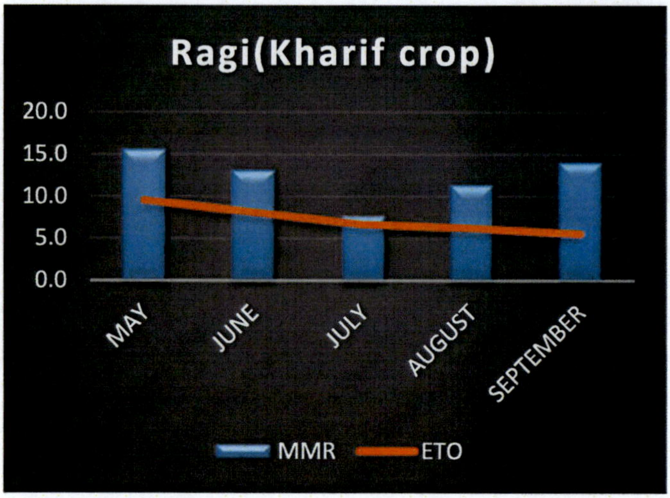

Fig. 9 Monthly water required for ragi

shows that the crop gets required amount of water in all the months of the crop period, expect November. Hence, a little irrigation supply in the month of November is necessary good yield.

Ragi (Kharif crop): This crop is a not season bound crop and hence can cultivated through the year, if moisture is available. It can also be grown as hot weather crop. Crop duration of the Raagi is 100–120 days (4–5 months). Sowing of the crop starts from May, and harvesting can be done at the middle or end of the September. ETo graph in Fig. 9 shows that the range of monthly mean rainfall falls above the evapotranspiration in all months. Hence, there is no additional supply of water is required for the crop.

Tur (Kharif crop): This crop is commonly known as tur in India has the ability to produce high economic yields under soil moisture deficit makes it an important crop in rain fed and dry land agriculture. Crop period of this is 200–210 days (6–7 months). Sowing starts at the starting and middle of the June, and the harvesting is done at the end of the December. ETo graph in Fig. 10 shows that evapotranspiration is below the monthly mean rainfall in this crop period expect November. Hence, extra water is necessary for the crop in this month for expected yield.

Gram (Kharif crop): This crop is semi-spreading verity, suitable for rain fed conditions. Crop period for the grams varies between 140 and 145 days (4–5 months). Sowing of the gram is in the middle of the June, and the harvesting is at the middle or at the end of the October. The rate of evapotranspiration falls within the monthly mean rainfall so that there is no requirement of additional supply of water for the better yield of crop as shown in Fig. 11.

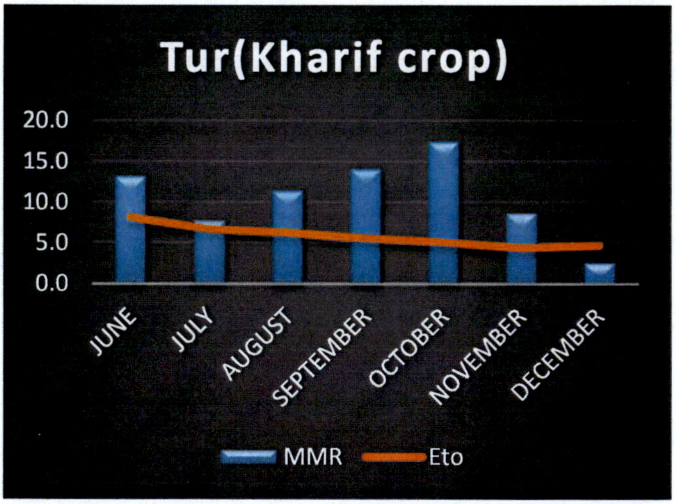

Fig. 10 Monthly water required for tur

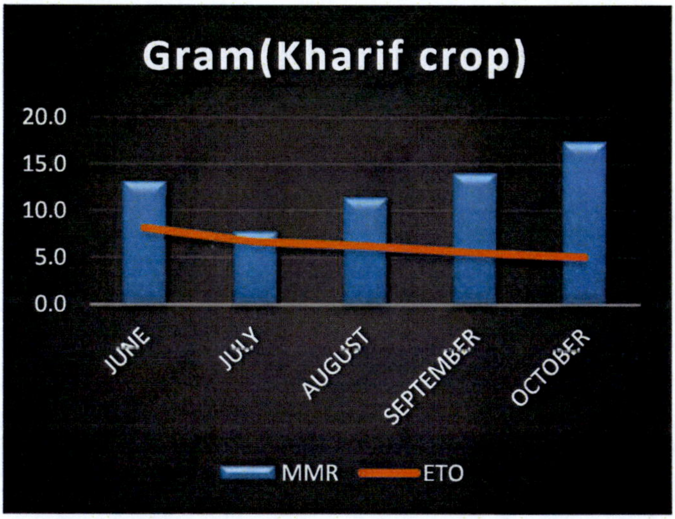

Fig. 11 Monthly water required for gram

Groundnut (Kharif crop): This crop is generally grown as a rain fed Kharif crop. Crop period of the ground nut varies from 100 to 120 days (4–5 months). Sowing and harvesting of the crop are done at the starting of the June and ending of September, respectively. Rainfall in these months is above the evapotranspiration rate as shown in Fig. 12. Good production of the crop can be done without surplus water.

Sunflower (Kharif crop): This used as a silage crop. It can be used as a double crop after early harvested small grains or vegetables, an emergency crop, or in areas with

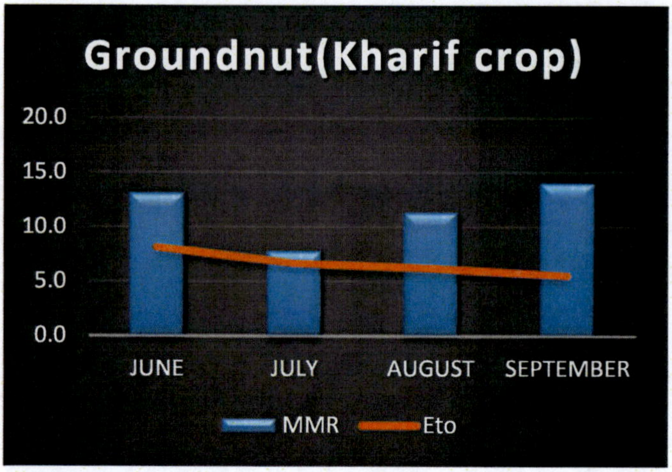

Fig. 12 Monthly water required for groundnut

a season too short to produce mature corn for silage. Crop period of the sunflower varies from 90 to 100 days (3–4 months). Sowing and harvesting of the crop is done at the starting of the July and middle of October, respectively. Rainfall in these months is above the evapotranspiration rate as shown in Fig. 13. Good crop production can be done without surplus water.

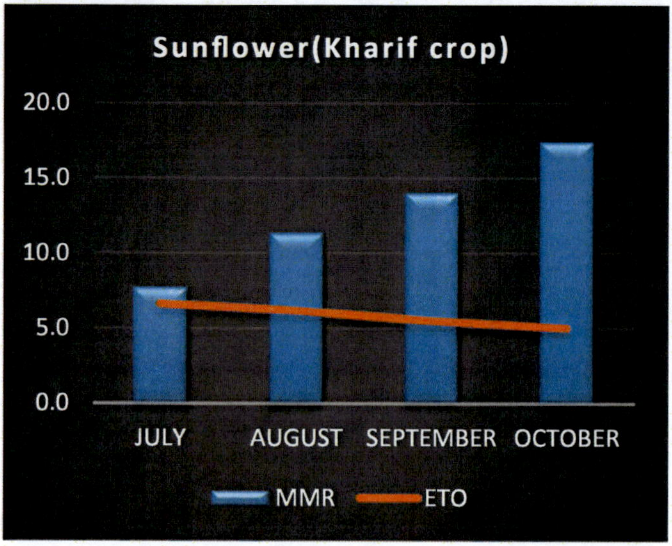

Fig. 13 Monthly water required for sunflower

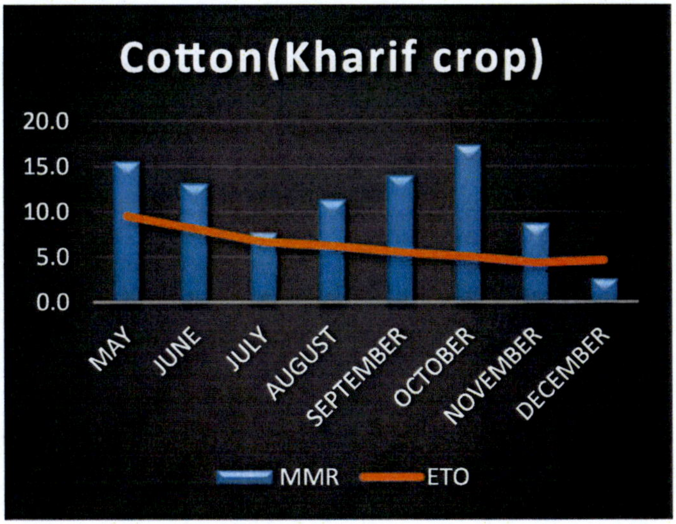

Fig. 14 Monthly water required for cotton

Cotton (Kharif crop): Most of the cotton growing areas are in the high to medium rainfall zones. The sowing seasons vary from region to region. Cotton is a Kharif crop which requires 6–8 months to mature. Figure 14 shows that evapotranspiration is below the monthly mean rainfall in this crop period expect November. Hence, extra irrigation is necessary for the crop in this month.

Rabi crops

Figures 15, 16, 17, 18, 19, 20, 21, 22, 23 and 24 show the monthly mean rainfall and the evapotranspiration rates in the respective months for all the crops in the Rabi season. Rate of evapotranspiration except in the beginning of the season is higher than the rainfall in most of the months for all. Therefore, rainfall alone is not sufficient for the growing of the crops. Even the shortage of water for a single month leads to failure of the crops. So it requires additional irrigation water for the proper production of all the crops.

Seasonal Crop: Sugarcane is one of the seasonal cash crop, but it is also used as livestock fodder. Crop period of sugarcane is 12–18 months. Planting season in tropical regions is from June to august. Crop period of sugarcane experiences the shortage of water in about five months as shown in Fig. 25. If additional water is not supplied during these months, proper yield of the crop and good production cannot be achieved.

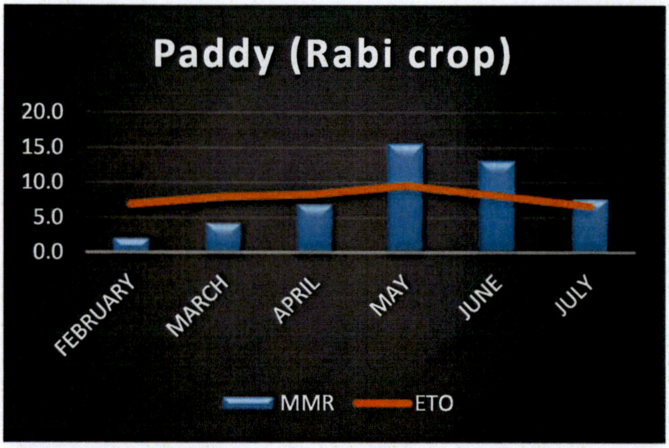

Fig. 15 Monthly water required for paddy

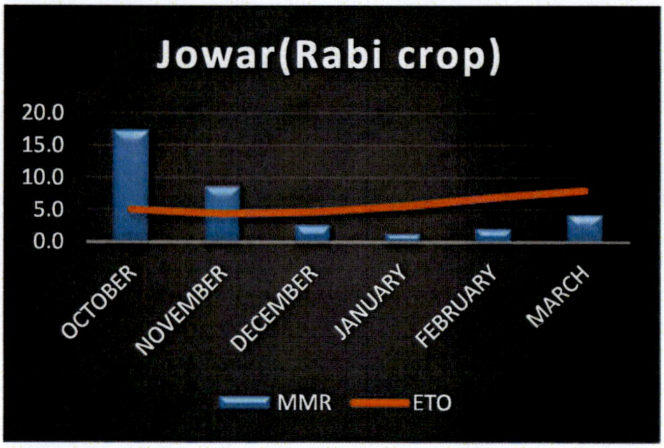

Fig. 16 Monthly water required for jowar

5 Conclusion

This exertion is carried out to study the trends of rainfall in Bellary District from 1977 to 2016, over 51 rain gauge stations. With more negative anomalies in the recent years of study indicates the occurrence of lowest rainfall, and the trend of rainfall is decreasing. According to rainfall anomaly index drought categories, 21 out of 40 years have the normal conditions of rainfall, and in the remaining 19 years, 7 years are under slightly dry and moderately dry condition each, 2 years are under

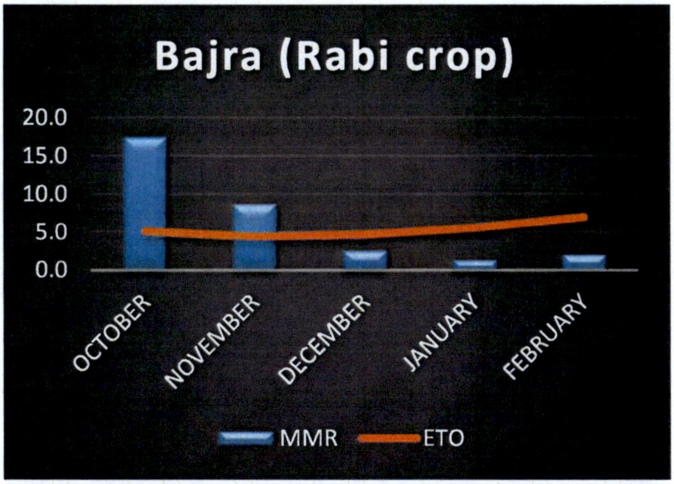

Fig. 17 Monthly water required for bajra

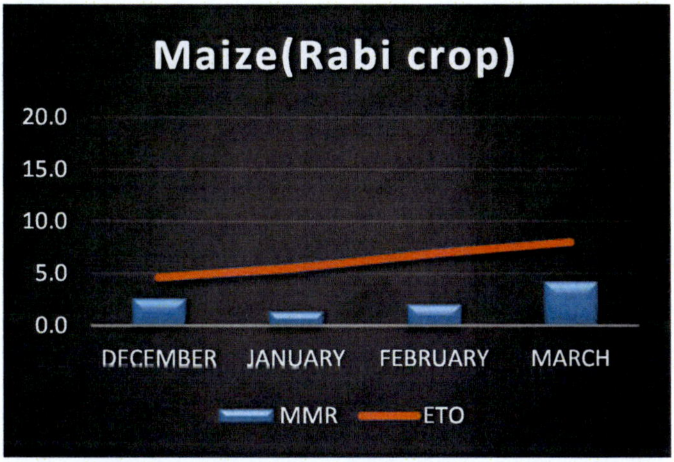

Fig. 18 Monthly water required for maize

slightly wet and moderately wet condition each with 1 year of very wet shows that the study area is facing lowest rainfall and moderate drought.

Occurrence and frequency of drought are analysed according to rain gauge stations and year using IMD method, where occurrence of drought is shown in numbers, and the frequency is given in percentage. The results obtained indicate that Bellary region is experiencing drought. MDSI is another method to locate drought severity class with the drought index. This analysis in the present study indicates that Bellary district is experiencing a moderate drought.

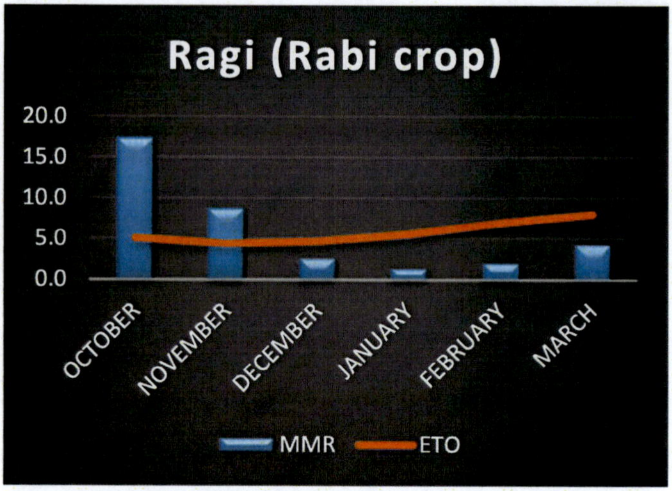

Fig. 19 Monthly water required for ragi

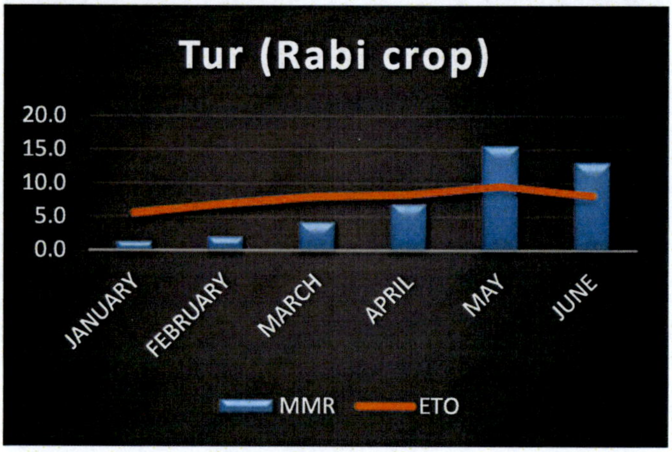

Fig. 20 Monthly water required for tur

Water requirement for crops which are grown in Bellary district is assessed in graphical method for the year 2016. The results show that dry crops will have sufficient water during the Kharif season, but experience insufficiency in Rabi season. Wet crops show irrigation requirement in both the seasons. Seasonal crops also require irrigation for few months. This scarcity of water is due to low rainfall in the study area.

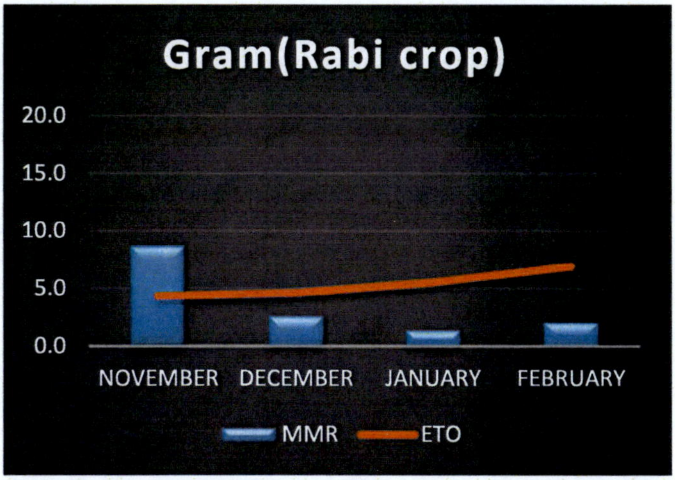

Fig. 21 Monthly water required for gram

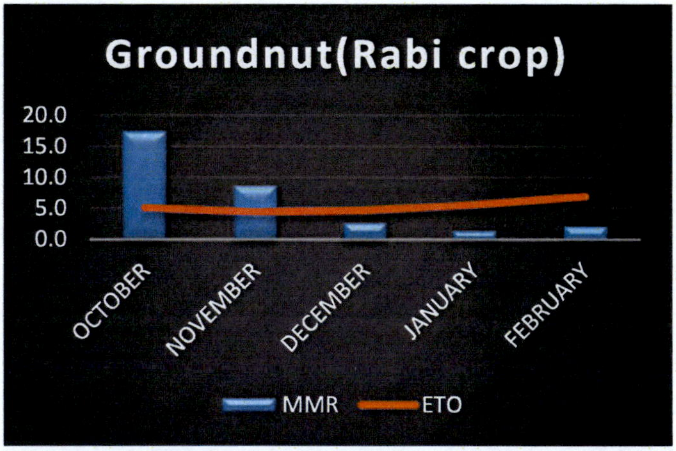

Fig. 22 Monthly water required for groundnut

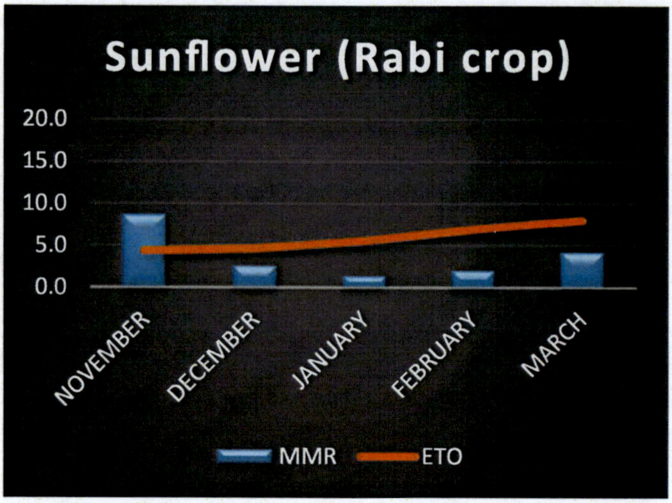

Fig. 23 Monthly water required for sunflower

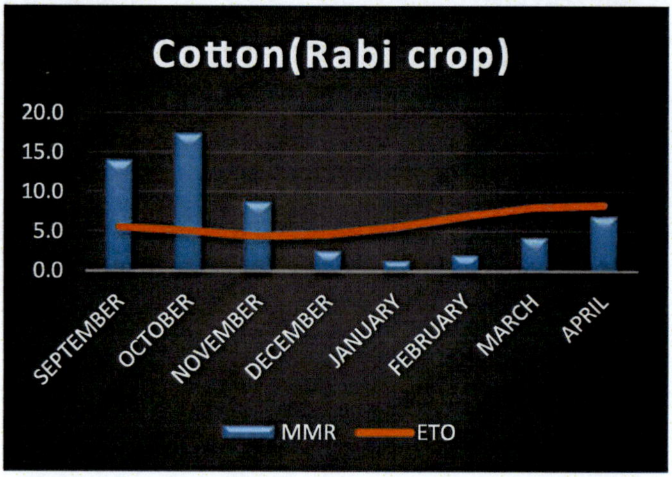

Fig. 24 Monthly water required for cotton

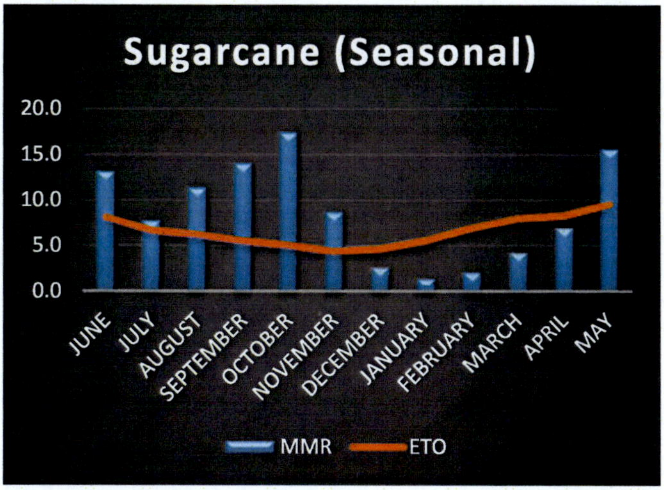

Fig. 25 Monthly water required for sugarcane

References

1. Kant S, Meshram S, Sahu KC (2014) Analysis of rainfall data for drought investigation at Agra U.P. Recent Res Sci Technol 6(1):62–64. ISSN: 2076–5061, http://recent-science.com/
2. Waghaye AM, Rajwade YA, Randhe RD, Kumari N (2018) Trend analysis and change point detection of rainfall of Andhra Pradesh and Telangana, India. J Agrometeorol 20(2):160–163
3. Yadav SK, Rawat S, Gautam S (2018) Variability in rainfall for Aravalli range of Rajasthan, India. Int J Curr Microbiol Appl Sci 7(07). https://doi.org/10.20546/ijcmas.2018.707.505
4. Guhathakurta P, Rajeevan M (2008) Trends in rainfall pattern over India. Int J Climatol 28:1453–1469
5. Joshi MK, Pandey AC (2011) Trend and spectral analysis of rainfall over India during 1901–2000. J Geophys Res: Atmosph 116:D06104
6. Sarkar S, Kafatos M (2004) Inter annual variability of vegetation over the Indian sub-continent and its relation to the different meteorological parameters. Remote Sens Environ 90:26–280
7. Parthasarathy B, Dhar ON (1974) Secular variations of regional rainfall over India. Q J R Meteorol Soc 100:245–257
8. Nandeesha, Ramu (2015) Agricultural drought assessment using GIS and remote sensing applications in different agro climatic zones of Karnataka state. Centre for Geoinformatics Technology, DOS in Geography, Manasagangotri
9. Wani SP, Sarvesh KV, Krishnappa K, Dharmarajan BK, Deepaja SM (2012) Bhoochetana: mission to boost productivity of rainfed agriculture through science-led interventions in Karnataka. ICRISAT, Hyderabad, p 84
10. Van Rooy MP (1965) A rainfall anomaly index (RAI) independent of time and space. Notos 14:43–48
11. Rangarajan S, Thattai D, Cherukuri A, Borah TA, Joseph JK, Subbiah A (2019) A detailed statistical analysis of rainfall of Thoothukudi district in Tamil Nadu (India). Water Resources and Environmental Engineering II, Springer. https://doi.org/10.1007/978-981-13-2038-5_1
12. Irrigation Commission Report Government of India, Ministry of Agriculture and Irrigation, India (1972)
13. India Meteorological Department (2002) Southwest Monsoon End-of-Season Report, India

Mitigation of Seismic Pounding Observed in Adjacent Buildings with Fluid Viscous Damper

Basanagouda I. Patil, Bapugouda B. Biradar, and Rashmi Doddamani

Abstract Seismic pounding between closely spaced building structures is one among the main causes of severe building damages observed in seismically proven regions. Due to the earthquake-induced vibrations, the buildings which are adjacent to each other with dissimilar dynamic characteristics will move out of phase resulting in the collision as there would be no energy dissipation system to accommodate the relative motions. Seismic pounding can be prevented by passive structural control of energy dissipation systems, i.e., dampers. The current study aims to mitigate the seismic pounding observed in the adjacent buildings connected with fluid viscous dampers (FVDs). G+14 and G+9 multistoried adjacent buildings are modeled and analyzed using ETABS 2017. Two stories are connected with one viscous damper. Adjacent buildings with similar height (G+14 storied) and varying height (G+14 & G+9 storied) connected with FVDs subjected to El Centro earthquake are studied. Nonlinear time history analysis is carried out. Considering displacement and seismic gap as the main parameters, adjacent buildings with similar and different height connected with and without FVDs are compared.

Keywords Seismic pounding · Seismic gap · Adjacent buildings · Fluid viscous dampers (FVDs) · Time history analysis (THA)

B. I. Patil · B. B. Biradar · R. Doddamani (✉)
School of Civil Engineering, K. L. E. Technological University, Hubballi, India

B. I. Patil
e-mail: basanagouda.patil@kletech.ac.in

B. B. Biradar
e-mail: bapugouda.biradar@kletech.ac.in

L. Nandagiri et al. (eds.), *Sustainability Trends and Challenges in Civil Engineering*, Lecture Notes in Civil Engineering 162, https://doi.org/10.1007/978-981-16-2826-9_45

1 Introduction

Over the past few decades, the world has experienced numerous devastating earthquakes, resulting in increased loss of human life due to the collapse of the structures. Earthquakes are the major natural hazards that cause severe damages to the structure. Building pounding between the structures has been usually observed in majority of the earthquakes [6]. Due to the earthquake-induced vibrations, the adjacent buildings placed with minimum separation gap will move out of the phase due to their different dynamic characteristics resulting in the collision [5]. Seismic pounding occurs when the two adjacent buildings collide [1]. One of the simplest and most effective ways to prevent pounding is to provide enough gap between the adjacently placed structures. But on occasions, obtaining the required separation is a problem in metropolitan cities because of increasing cost of land, scarcity of land space and detailing the problem [2]. Due to the sudden ground vibrations, a huge amount of energy will be released and imparted to the structures. As a result, the structure undergoes large displacements, drifts which may further lead to the collision of adjacent buildings and hence failure of the structure. Hence, for the safety of the structure, there is a need to dissipate the energy imparted to the structure. So, to counteract these loads, a proper lateral load resisting system should be used to fulfill the design and serviceability criteria [3]. Hence, the substitute to gap provision in the structural design is to reduce the consequence of pounding by reducing sideways motion which can be achieved by use of passive energy dissipation device called "Dampers" which are used to attenuate the unwanted vibrations in the structure. Fluid viscous damper (FVD) is a technique that is used for the stabilization of buildings against lateral forces [8]. Installing FVD within the structures mitigates the structural response and also avoids seismic pounding between the structures [4]. Application of these dampers substantially reduces story displacement and inter-story relative displacements [7]. FVD works on the principle of energy dissipation. Seismic energy is absorbed by silicone-based fluid passing between piston–cylinder arrangements. The passive damper modifies the structural response without using any external power supply. It operates at the temperature ranging from 40 to 70 °C [9] (Fig. 1).

Following two types of damage due to pounding can occur

1. Local damage when there is an impact.
2. Global damage which occurs due to energy and momentum transfer occurred by collision.

Local damage occurs due to collision force, while global damage is based on the dynamic characteristics of both the structures at the time of collision (Figs. 2 and 3).

Seismic gap:

IS 4326 specifies that a minimum gap should be provided between the buildings which are adjacent to each other to avoid collision when earthquake occurs (Table 1).

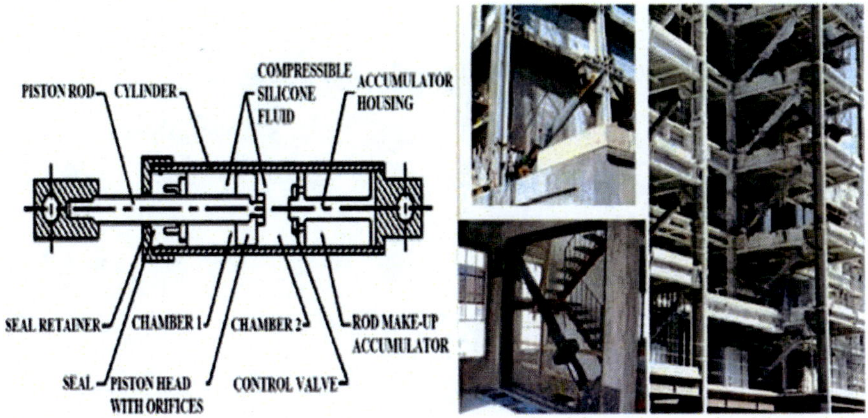

Fig. 1 Fluid viscous dampers installed diagonally in building. *Source* (1) https://www.researchgate.net/figure/Construction-of-fluid-viscous-damper-Constantinou-and-Symans-1992_fig9_257335162. (2) https://www.slideshare.net/NamanKantesaria/naman-act

Fig. 2 Pounding damage during Mexico earthquake 1985. *Source* http://www.smate.wwu.edu/teched/geology/eq-Mexico.html

Structural vibration control aims to produce buildings that can withstand severe dynamic loads such as seismic and wind. This aim is achieved by using dampers. So, the current study includes finding out the effectiveness of fluid viscous dampers installed in multistoried adjacent buildings in reducing seismic pounding.

Fig. 3 Causality of L'Aquila earthquake, Italy 2009. *Source* http://www.airworldwide.com/PublicationsItem.aspx?id=17228

Table 1 Gap width for buildings built adjacent to each other

S. No.	Mode of construction	Gap/story (mm) for design seismic coefficient $\alpha_h = 0.12$
1	Box system/frames with shear walls	15
2	Moment-resistant reinforced concrete frame	20
3	Moment-resistant steel frame	30

Note Minimum gap shall not be less than 25 mm. Seismic gap shall be calculated according to IS 1893 (Part 1)—2016 clause 7.11.3

2 Methodology

For the study, two models are considered for analysis, i.e., model 1 of G+14 storied similar height adjacent buildings; model 2 of 15 and 10 storied different height adjacent structures. The structures are modeled and analyzed using ETABS 2017. Earthquake loads and analyses carrying out in the study are as per IS 1893 part 1:2016. Nonlinear time history analysis is carried out using El Centro earthquake data.

Step 1: Initially, the analysis is carried out considering individual buildings and calculated for the seismic gap.
Step 2: Then, the buildings are placed adjacently with proper seismic gap considered.
Step 3: Analysis is carried out for both model 1 and model 2; seismic pounding between the adjacent buildings is verified.
Step 4: Fluid viscous dampers (FVD) will be installed within the structure if they are subjected to seismic pounding.

Step 5: Considering displacement, story drift, and seismic gap as the main parameters, adjacent buildings with similar height and different height connected with and without fluid viscous dampers are compared.

3 Structural Modeling and Analysis

Model 1: Adjacent buildings with similar height (G+14).
Model 2: Adjacent buildings with different height (G+14 & G+9) (Table 2, Figs. 4, 5 and 6; Graph 1).

Table 2 Section, material properties and loads considered

Building parameters considered		
1	Typical story height	3.2 m
2	Size of columns	500 × 500 mm
3	Size of beams	300 × 500 mm
4	Depth of slab	150 mm
5	Main wall	230 mm
6	Parapet wall	115 mm
Material properties		
1	Grade of concrete	M30
2	Grade of steel	HYSD500
Loads as per IS 875:1983 (part I and II)		
1	Live load	3 kN/m^2
2	Floor finish	1 kN/m^2
3	Dead load of wall	12.42 kN/m
4	Dead load of parapet wall	2.07 kN/m
Seismic loads as per IS 1893:2016		
1	Seismic zone factor	0.24 (zone IV)
2	Importance factor	1
3	Response reduction factor	5
4	Site type	II (medium)
Link type—damper exponential		
1	Mass	44 kg
2	Weight	250 kN

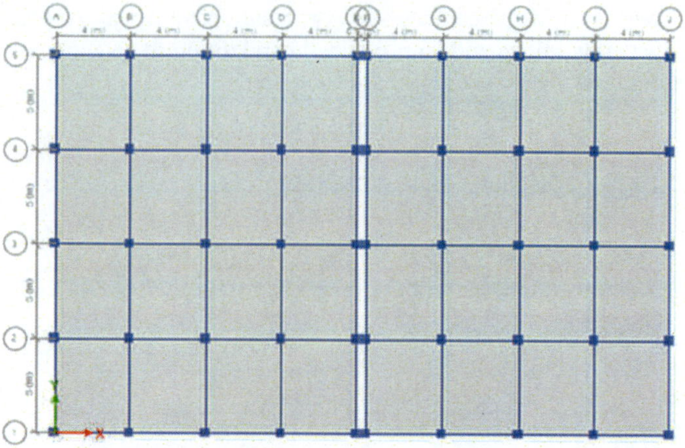

Fig. 4 Plan of G+14 storied similar height adjacent buildings placed with seismic gap of 600 mm

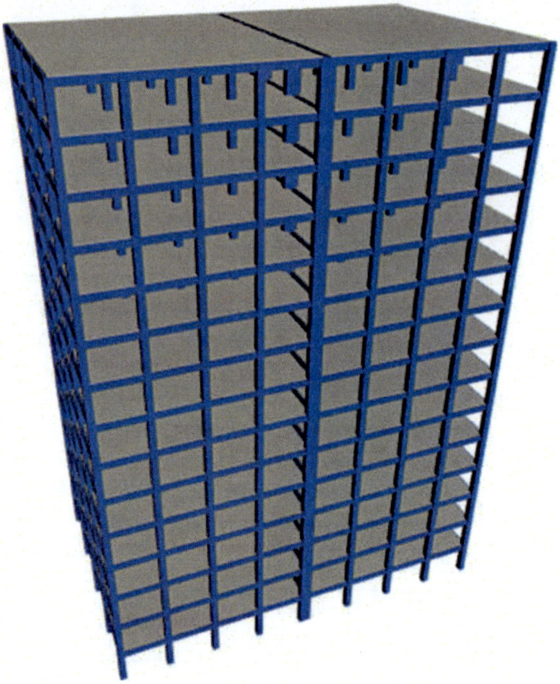

Fig. 5 3D view of G+14 storied adjacent buildings

Fig. 6 3D view of G+14 and G+9 storied adjacent buildings

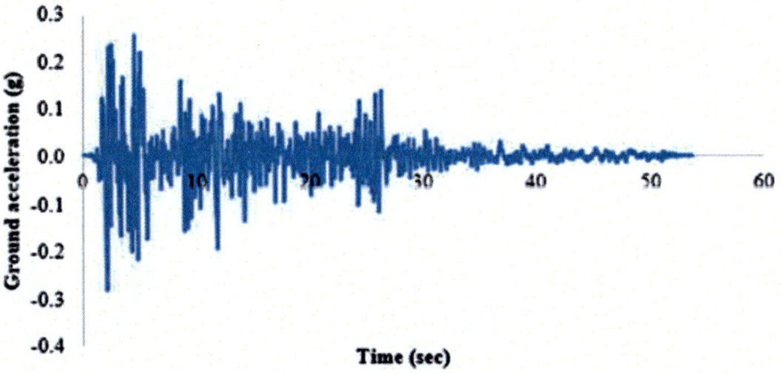

Graph 1 Ground acceleration of El Centro earthquake. *Source* http://peer.berkeley.edu

4 Results and Discussions

See Graphs 2, 3, 4, 5 and 6 and Tables 3 and 4.

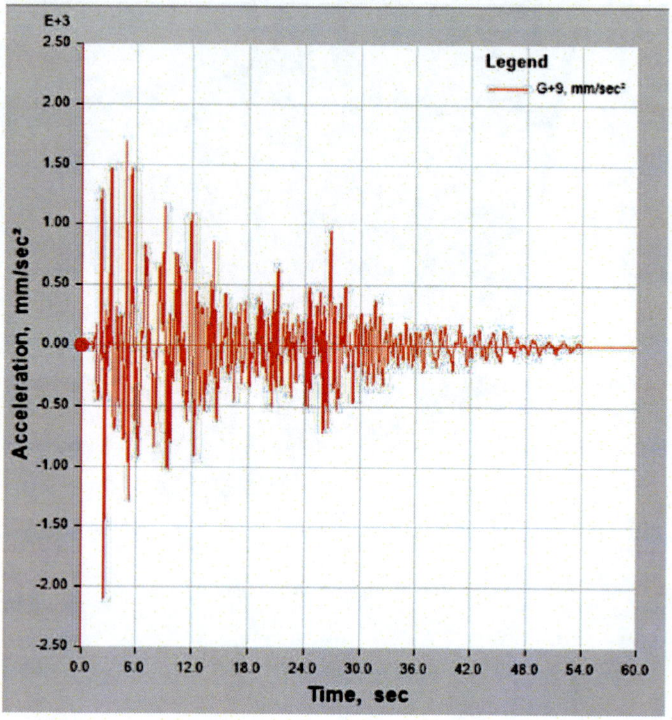

Graph 2 Acceleration of G+9 building

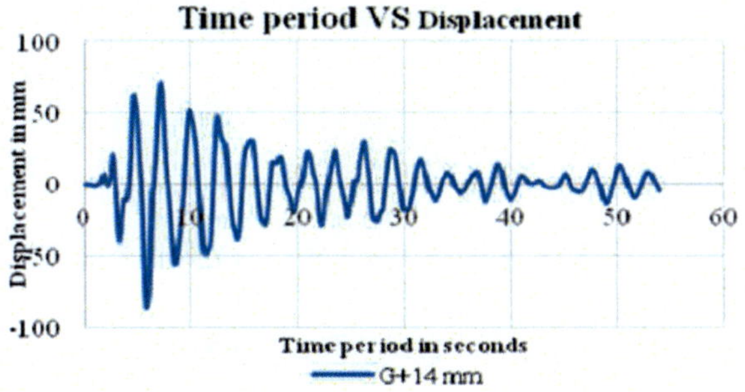

Graph 3 Maximum story displacement along THA-X

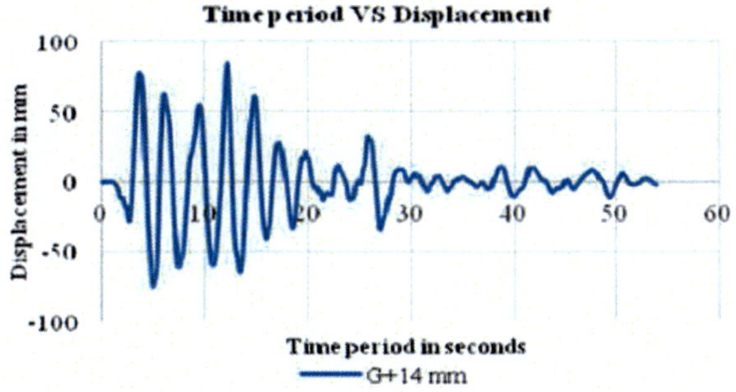

Graph 4 Maximum story displacement along THA-Y

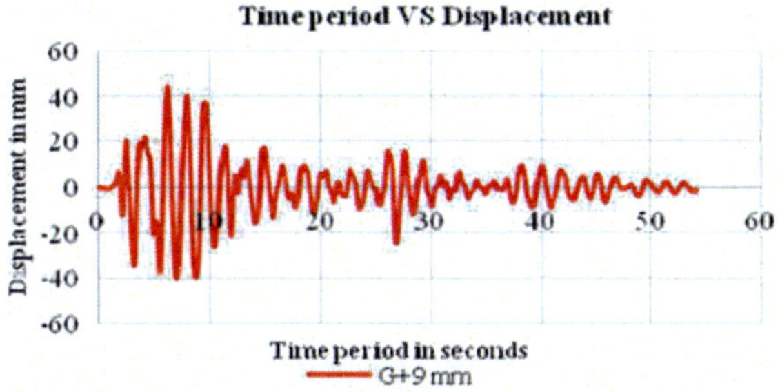

Graph 5 Maximum story displacement along THA-X

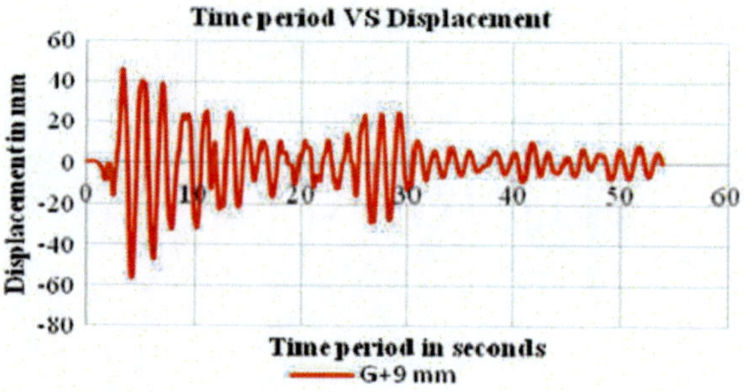

Graph 6 Maximum story displacement along THA-Y

Table 3 Displacements in *x*- and *y*-direction

Model 1: buildings with similar height	Displacement in *x*-direction (THA-X) mm	Displacement in *y*-direction (THA-Y) mm
Building 1: 15 storied	70.970	85.247
Building 2: 15 storied	70.970	85.247
Seismic gap	709.7	852.47
Seismic gap considered	600 mm	

Table 4 Displacements in *x*- and *y*-direction

Model 2: buildings with different height	Displacement in *x*-direction (THA-X), mm	Displacement in *y*-direction (THA-Y), mm
Building 1: 15 storied	70.970	85.247
Building 2: 10 storied	44.446	44.954
Seismic gap	577.08	651.005
Seismic gap considered	500 mm	

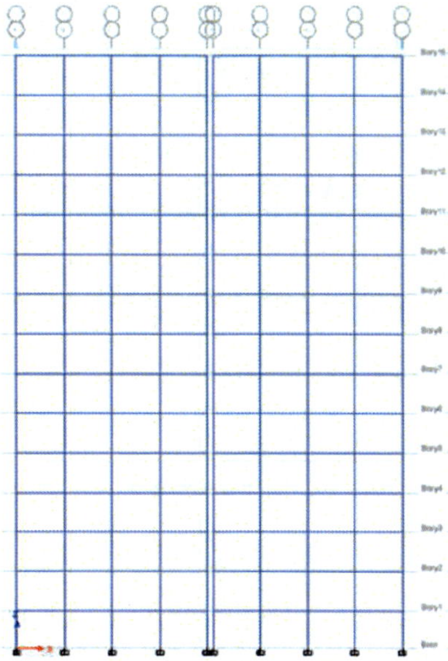

CASE 1

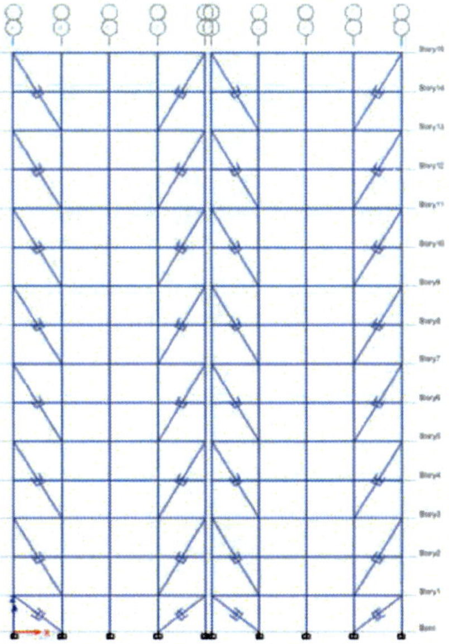

CASE 2

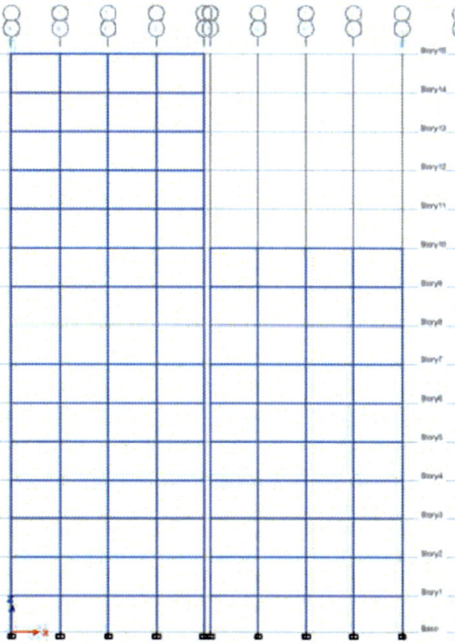

CASE 3

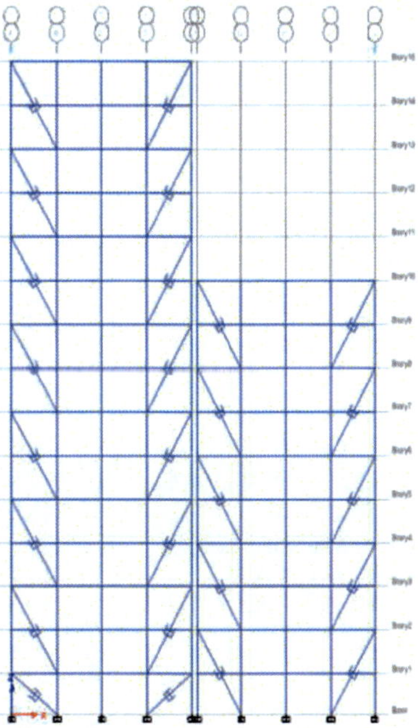

CASE 4

Case 1: Similar height adjacent buildings without FVD.
Case 2: Similar height adjacent buildings with FVD.

Case 3: Different height adjacent buildings without FVD.

Case 4: Different height adjacent buildings with FVD (Graphs 7, 8, 9, 10, 11, 12, 13 and 14; Tables 5, 6, 7, 8, 9 and 10).

- In case of without dampers, the seismic gap obtained is 752.10 mm in *x*-direction and 867.41 mm in *y*-direction which is exceeding the seismic gap considered, i.e., 600 mm. Hence, seismic pounding is observed.
- With the use of FVD, the seismic gap obtained is 445.96 mm in *x*-direction and 335.66 mm in *y*-direction which is within the seismic gap considered. Hence, seismic pounding is prevented.
- The displacement was reduced by 40.70% in *x*-direction and 61.30% in the *y*-direction with the use of FVD.

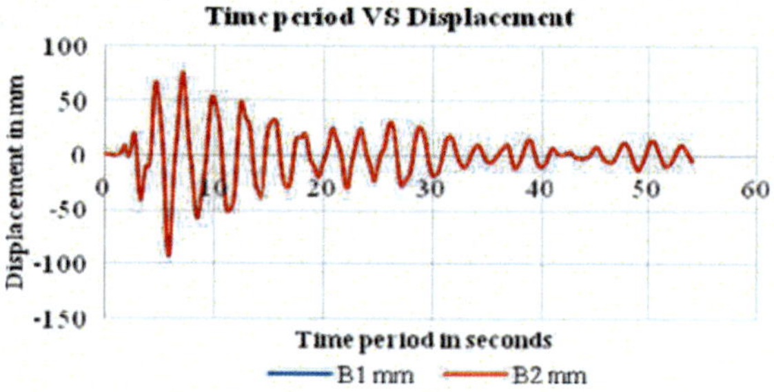

Graph 7 Maximum story displacement of CASE 1 along THA-X

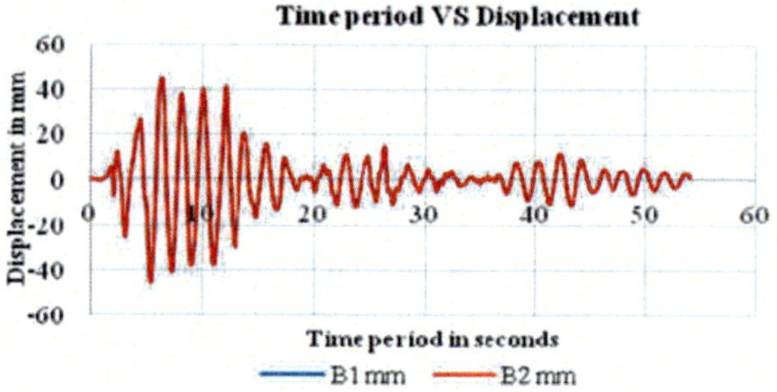

Graph 8 Maximum story displacement of CASE 2 along THA-X

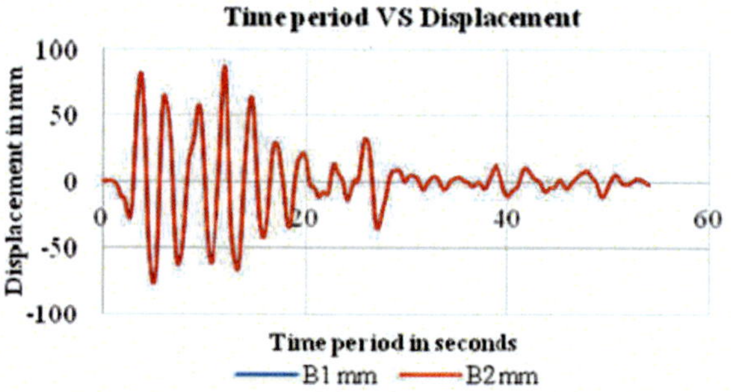

Graph 9 Maximum story displacement of CASE 1 along THA-Y

Graph 10 Maximum story displacement of CASE 2 along THA-Y

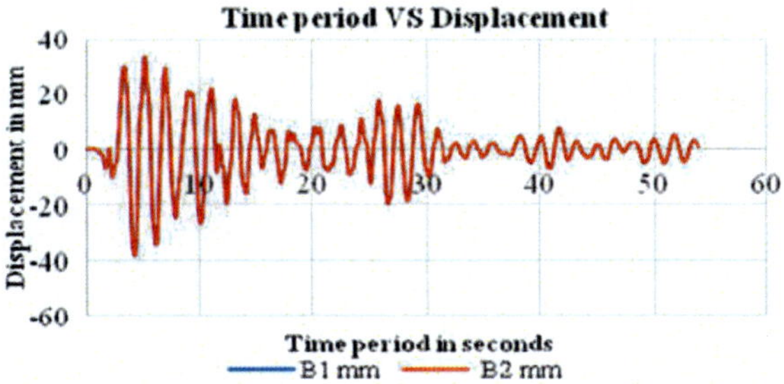

Graph 11 Maximum story displacement of CASE 3 along THA-X

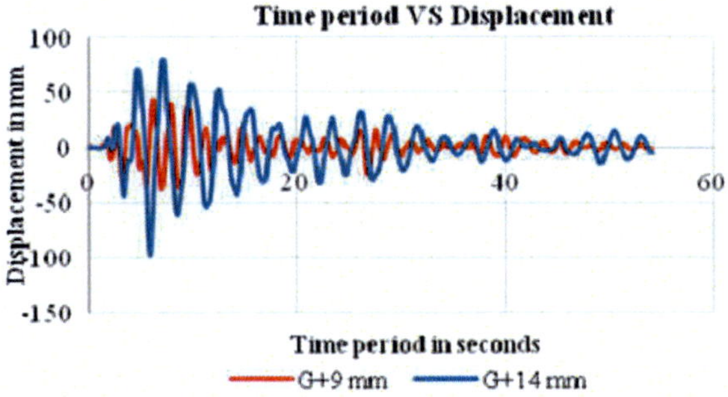

Graph 12 Maximum story displacement of CASE 4 along THA-X

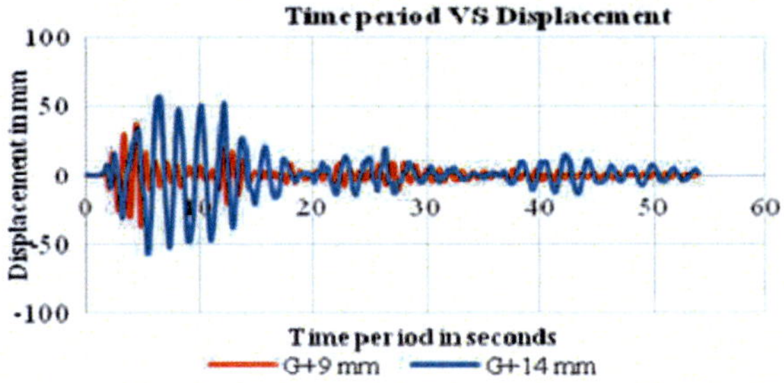

Graph 13 Maximum story displacement of CASE 3 along THA-Y

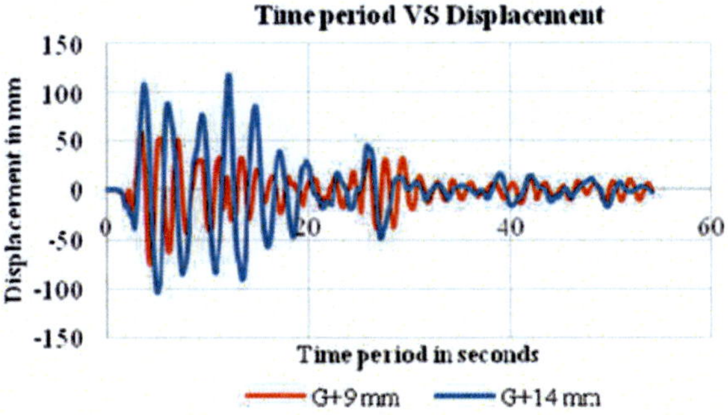

Graph 14 Maximum story displacement of CASE 4 along THA-Y

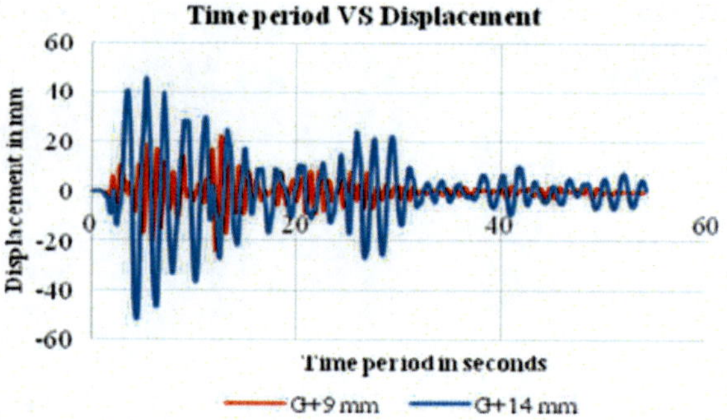

Table 5 % difference of displacements in *x*-direction

	Case 1: without dampers	Case 2: with dampers	Percentage difference %
Displacements (mm) for G+14	75.210	44.596	40.70
Displacements (mm) for G+14	75.210	44.596	40.70
Seismic gap calculation (mm)	752.10	445.96	40.70

Table 6 % difference in displacements in *y*-direction

	Case 1: without dampers	Case 2: WITH dampers	Percentage difference %
Displacement (mm) for G+14	86.741	33.566	61.30
Displacements (mm) for G+14	86.741	33.566	61.30
Seismic gap calculation(mm)	867.41	335.66	61.30

Table 7 Maximum displacements of building corresponding to peak time interval

Maximum displacement obtained at peak time	Case 1	Case 2
THA-X (mm)	75.210	44.596
Time (s)	7.13	6.45
THA-Y (mm)	86.741	33.566
Time (s)	12.09	5.22

Table 8 % difference in displacements in *x*-direction

	Case 3: without dampers	Case 4: with dampers	Percentage difference %
Displacements (mm) for G+14	78.82	56.58	28.21
Displacements (mm) for G+9	42.67	36.80	13.76
Seismic gap calculation (mm)	607.48	466.92	23.14

Table 9 % difference in displacements in y-direction

	Case 3: without dampers	Case 4: with dampers	Percentage difference %
Displacements (mm) for G+14	117.35	59.66	49.16
Displacements (mm) for G+9	45.30	21.77	51.94
Seismic gap calculation (mm)	813.28	407.15	49.94

Table 10 Maximum displacements of building corresponding to peak time interval

Maximum displacement obtained at peak time	Case 3	Case 4
THA-X (mm)	78.820	56.583
Time (s)	7.13	6.45
THA-Y (mm)	117.352	45.304
Time (s)	12.10	5.22

- In case of without dampers, the seismic gap obtained is 607.48 mm in *x*-direction and 813.28 mm in *y*-direction which is exceeding the seismic gap considered, i.e., 500 mm. Hence, seismic pounding is observed.
- With the use of FVD, the seismic gap obtained is 466.92 mm in *x*-direction and 407.15 mm in *y*-direction which is within the seismic gap considered. Hence, seismic pounding is prevented.
- The displacement was reduced by 28.21% in *x*-direction and 51.94% in the *y*-direction with the use of FVD (Graphs 15, 16, 17, 18, 19, 20, 21, 22, 23, 24, 25 and 26).

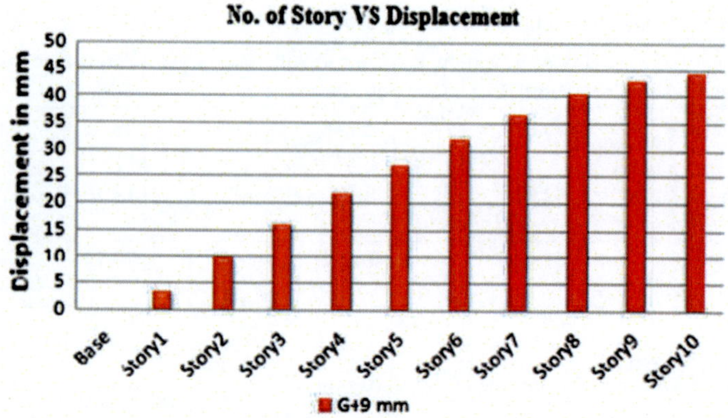

Graph 15 Number of story versus displacement of G+9 along THA-X

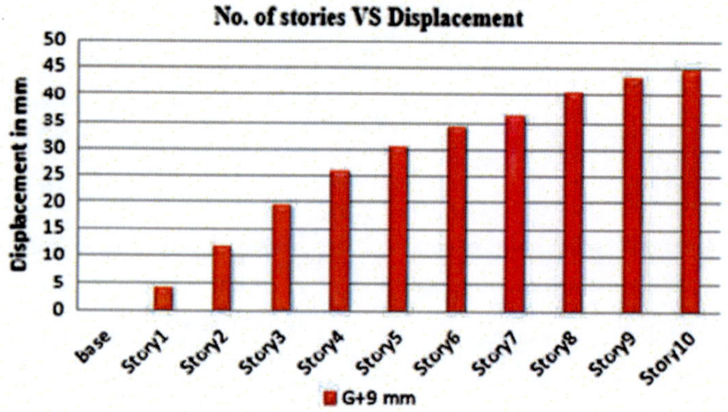

Graph 16 Number of story versus displacement of G+9 along THA-Y

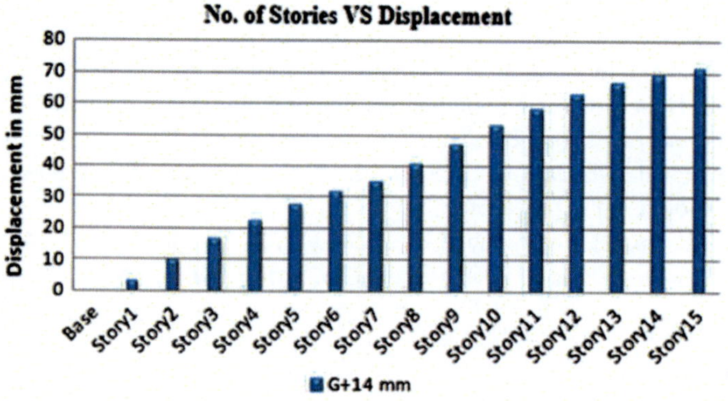

Graph 17 Number of story versus displacement of G+14 along THA-X

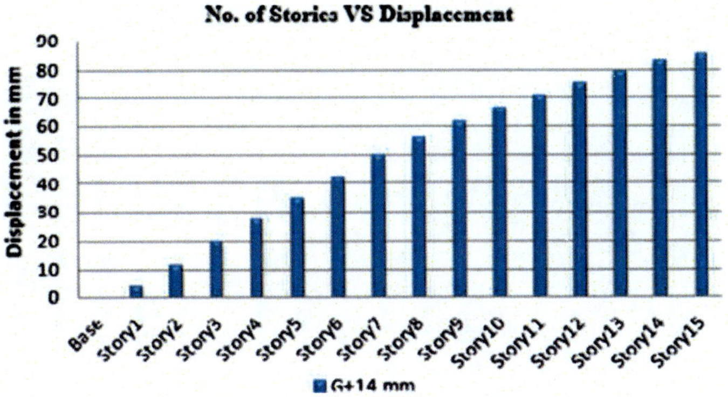

Graph 18 Number of story versus displacement of G+14 along THA-Y

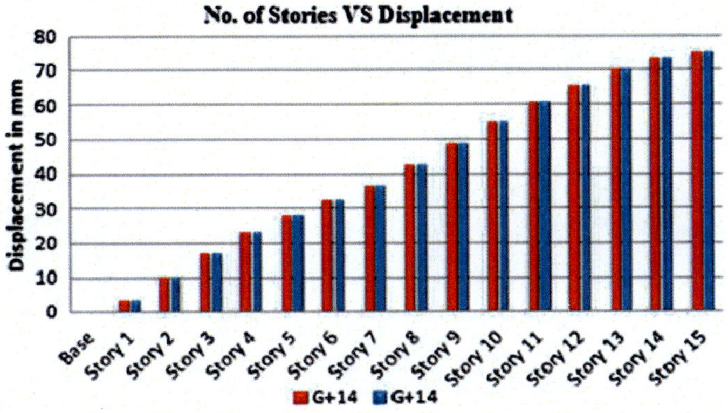

Graph 19 Number of story versus displacement of CASE 1 along THA-X

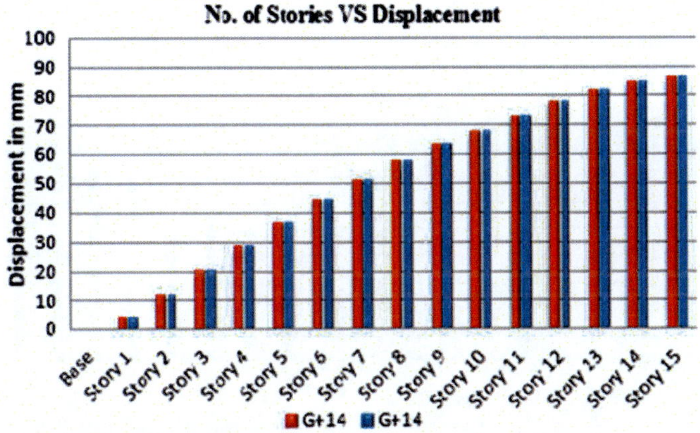

Graph 20 Number of story versus displacement of CASE 1 along THA-Y

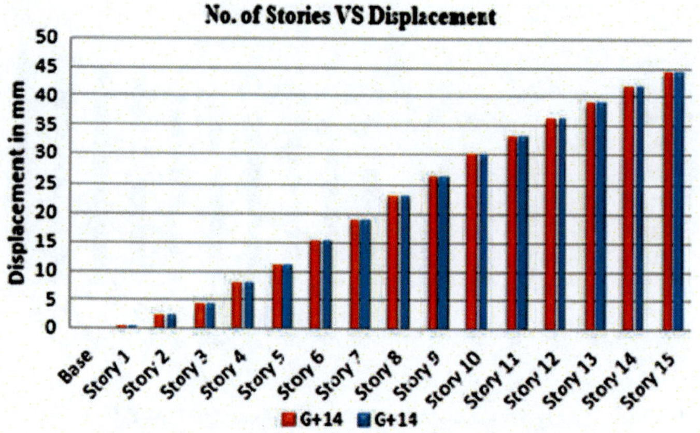

Graph 21 Number of story versus displacement of CASE 2 along THA-X

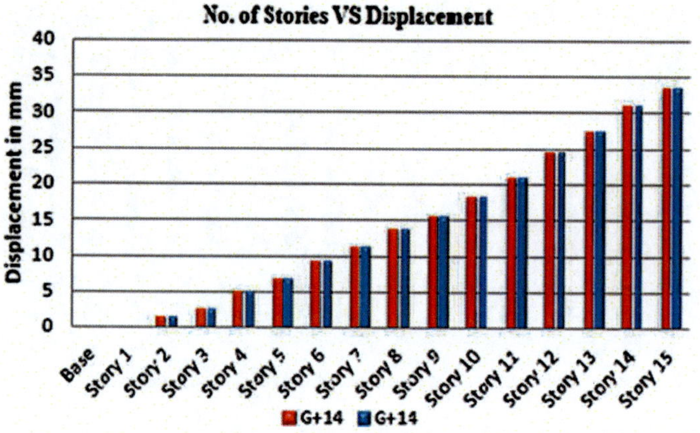

Graph 22 Number of story versus displacement of CASE 2 along THA-Y

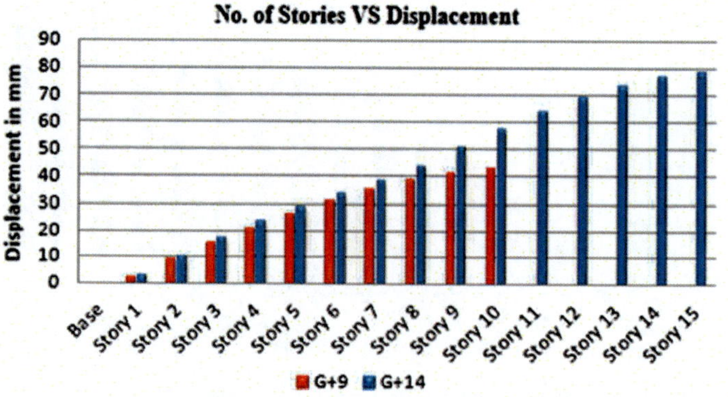

Graph 23 Number of story versus displacement of CASE 3 along THA-X

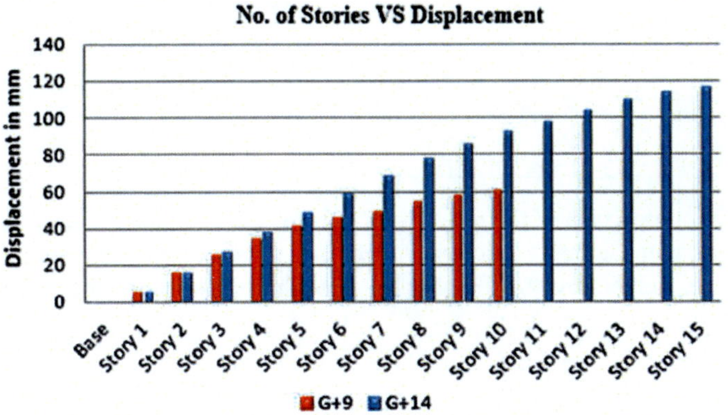

Graph 24 Number of story versus displacement of CASE 3 along THA-Y

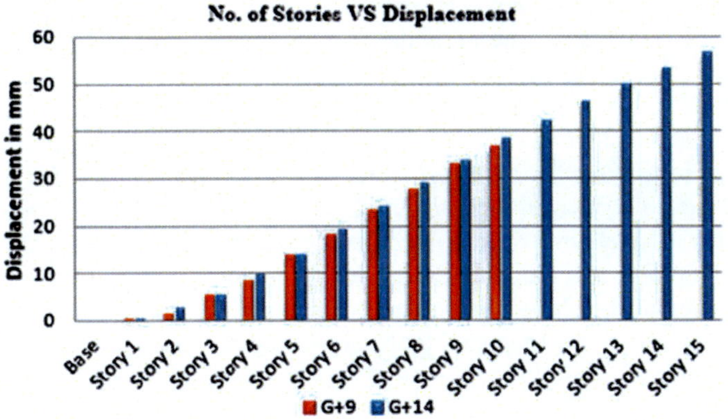

Graph 25 Number of story versus displacement of CASE 4 along THA-X

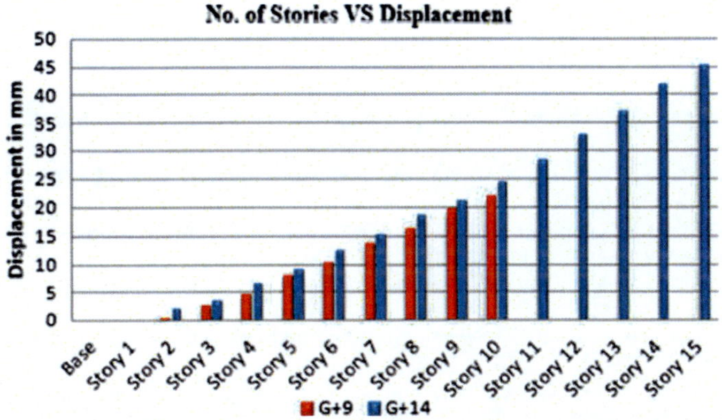

Graph 26 Number of story versus displacement of CASE 2 along THA-Y

5 Conclusions

After the analysis, the following conclusions are drawn as follows:

1. MODEL 1: G+14 storied adjacent buildings (similar height)
   - On comparison of the results of displacement of buildings without dampers and with damper, it is concluded that provision of dampers reduces the displacement by 40.70% along x-direction and 61.30% along y-direction.
2. MODEL 2: G+14 and G+9 storied adjacent buildings (different height)
   - On comparison of the results of displacement of buildings without dampers and with damper, it is concluded that provision of dampers reduces the displacement by 28.21% in x-direction and 51.94% in the y-direction.
3. The displacements in both the models were exceeding the seismic gap considered, i.e., 600 mm in case of model 1 and 500 mm for model 2. Hence, seismic pounding was observed between the adjacent buildings without dampers.
4. With the use of dampers, the seismic pounding effect was reduced by 61.30% in model 1 and about 49.94% in model 2. Hence, after placing the dampers, it was observed that the pounding effect was effectively controlled.
5. Two stories connected with one fluid viscous dampers (FVD) were noticed to be very effective in reducing the seismic responses between the structures and also helpful in avoiding the seismic pounding phenomenon.
6. The similar height adjacent structures having floors at same level will behave similarly as compared to the adjacent structures of different heights.

References

1. Shirole AB (2015) Seismic pounding between adjacent building structures. Int J Innov Res Adv Eng
2. Patel RR et al (2017) Study of seismic pounding effects between adjacent structures. Int J Innov Res Adv Eng
3. Ravindranatha et al (2015) Prevention of seismic pounding between adjacent buildings. Civ Eng Syst Sustain Innov
4. Bhaskararao AV et al (2004) Seismic response of adjacent buildings connected with dampers. In: 13th world conference on earthquake engineering, Canada
5. Mazanoglu K et al (2016) An optimization study for viscous dampers between adjacent buildings. Elsevier
6. Somwanshi MA et al (2017) Seismic pounding and vulnerability to the high-rise building structures. Int Adv Res J Sci Eng Technol
7. Aydin E et al (2017) Application of viscous dampers for prevention of pounding in adjacent reinforced concrete buildings. In: 16th world conference on earthquake, 16WCEE 2017, Santiago, Chile, 9–13 Jan 2017
8. Dona M et al (2018) Effectiveness of fluid-viscous dampers for improved seismic performance of inter-storey isolated buildings. Elsevier

9. Pipinato A (2016) Innovative bridge design handbook: construction, rehabilitation and maintenance. Butterworth-Heinemann
10. IS 1893 (Part 1):2016. Criteria for earthquake resistant design of structure
11. IS 4326: Earthquake resistant design and construction of buildings—code of practice (second revision) (1993)
12. IS 875 (Part 1):1987: Code of practice for design loads of buildings & structures. Part 1. Dead loads. Bureau of Indian standards, New Delhi

Hydrological Review and Dam Break Analysis of Suvaranavathi Dam Using HEC-RAS

B. S. Naveen Kumar, K. Usha, and M. S. Kanchana

Abstract Dam is a hydraulic structure constructed across the river to impound water. Generally dams are used for multipurpose, thus dam stability is essential. Floods resulting from the failure have caused some of the most destructive catastrophes in the past decades (George and Nair Aquat Procedia 4:853–860, 2015). The current study was undertaken to understand the Hydrological Review and Dam Break Analysis (DBA) of Suvarnavathi dam constructed across Suvarnavathi River. An unsteady flow simulation is done using HEC-RAS (Hydrological Engineering Center River Analysis System) model to determine, probable maximum flood, flood travel time and the affected villages. From hydrological review, it is clear that existing spillway capacity is inadequate. Hence dam height must be raised further to a height of 10.1 m. The results obtained from Dam break analysis under overtopping and piping criteria furnished proof to give statement that, 46 villages and a city will be under submergence and rehabilitation cost turns out to be 3440.64 crores under overtopping failure. 23 villages and a city will be under submergence and rehabilitation cost turns out to be 1802.73 crores under piping failure.

Keywords Dam break · Flood · HEC-HMS · HEC-RAS

1 Introduction

Dam is a very important structure. It is defined as a barrier constructed across the river to impound water or to create reservoir which can be used for multipurpose like irrigation, domestic, power generation, and flood mitigation, etc. Dams provide many societal benefits, but some of the most destructive catastrophes in the last two centuries have also caused by floods resulting from the failure of the built dams

B. S. Naveen Kumar (✉) · M. S. Kanchana
Department of Civil Engineering, Hydraulics Branch, National Institute of Engineering, Mysore, Karnataka, India
e-mail: 2018hyd_naveenkumarbs@nie.ac.in

K. Usha
K.E.R.S., Karnataka Mandya, India

L. Nandagiri et al. (eds.), *Sustainability Trends and Challenges in Civil Engineering*, Lecture Notes in Civil Engineering 162, https://doi.org/10.1007/978-981-16-2826-9_46

[1]. Simulation of events of dam breaches and the subsequent floods are important to characterizing and minimizing risks due to possible disruptions of dams. A dam breach can result in a flood wave moving along a valley at fairly high speeds up to tens of meters deep. The effect on developed areas of such a wave can be destructive. Such destructive power, if early notice and evacuation are not feasible, comes as an imminent loss of lives.

A dam can be classified into different types based on function served, material used, structural behavior, hydraulic design and rigidity. Dam Break Analysis (DBA) is an important topic in which peak flood discharge, which can be discharged through spillway can be estimated (adequacy of spillway). In flood condition, if peak discharge is discharged then Inundation maps are very much useful to safeguard the lives and property. The negative effect of natural disaster can be reduced by providing reliable information to the community through the flood inundation maps. By using inundation maps and Emergency Action Plans (EAP) we may reduce risk of life and loss of property in downstream areas. Disaster prevention focused on flood inundation maps and identification of flood risk areas is the key goal of EAP. The category of flood risk is divided into three types: Very high, Medium, Very low risk.

The first collapse of dam was reported in Madhya Pradesh in 1917 when, due to overtopping, the Tigra dam collapsed. The Machu II dam failure in Gujarat is the worst dam failure in India in1979, unleashed floodwater in which about 2000 people died. The breach of Pratapura dam, Vadodara, Gujarat and Nandagaon dam, Nagpur, Maharashtra, both in 2005 triggered severe floods in downstream areas, among the recent dam failures [2]. The above instances of the dam breaches demonstrate the danger posed by big and small dams, is very real. As public understanding of these potential risks increases and tolerance of catastrophic environmental effects and loss of life declines, it is becoming an important necessity rather than a management choice to mitigate and minimize the risk from individual systems.

2 Materials and Methods

2.1 Study Area

The Suvarnavathi Dam is located across Suvarnavathi River near Attigulipura village, Chamarajanagar taluk at a distance of about 16 km away. The longitude and latitude are 77°00′59″ and 11°49′53″, respectively. It has a gross storage capacity of 1.260 Thousand Million Cubic feet (TMC). Filling period starts from the month of June to November, The reservoir submerges 1112 acres of forest, 100 acres of cultivable land. The dam is constructed with masonry spillway and earthen flanks on either side. Length of the dam is about 1170.43 m long and height of the dam is about 25.90 m above the river bed level. Spillway capacity is 23,900 cusecs. It irrigates the downstream area of about 9694 acres.

2.2 Input Data for Modeling

Input data's required for the model are salient feature of the dam these data were fetched from the Karnataka water resource department Web site. Standard project storm (SPS) value of the study area was collected from Indian Metrological Department (IMD). % of 24 h storm rainfall collected from IMD. Reservoir elevation, storage and discharge data were collected from dam authority. Catchment delineation is done to obtain area of the catchment. Catchment area was found to be 392 km^2. ArcGIS software is used to perform catchment delineation in order to extract the catchment parameters. Inflow hydrograph is prepared by using synthetic unit hydrograph principle. Flood routing is done by using Hydrological Engineering Centre Hydrological Modeling System (HEC-HMS) model. Hydrological Engineering Centre River Analysis System (HEC-RAS) model is used to analyze dam breach under different criteria.

2.3 HEC-HMS Model

The hydrologic modeling system is designed to simulate in dendrite watershed systems, precipitation runoff process. It is planned to be applicable to solve the widest possible variety of problems across a wide range of geographic areas. This includes water supply and flood hydrology for large river basins and drainage for small urban or natural watershed runoff. Hydrographs created by the programmer are used directly or in conjunction with other forecasting like future effect of urbanization, design of reservoir spillway, reduction of flood damage, management of floodplain and system operation [3].

The program is a generalized modeling system capable of representing several different watersheds. By splitting the hydrologic cycle into manageable parts and creating boundaries across the watershed of interest, a model of the watershed is built. With a mathematical model, any mass or energy flux in the cycle can then be interpreted. For the representation of each flux, multiple model choices are available in most cases. The objectives of hydrological analysis and engineering judgment are relevant for each mathematical model included in the programmer in various settings and under the watershed [3].

2.4 HEC-RAS Model

HEC-RAS is designed to carry out one-dimensional and two-dimensional hydraulic calculations for the entire natural and constructed canal network. Several river analysis components are used in HEC-RAS method for: (1) steady flow water surface profile computations; (2) one-and two-dimensional unsteady flow simulation; (3)

movable boundary sediment transport computations; and (4) analysis of water quality. A key aspect is that all four components use a shared representation of geometric data and common routines for geometric and hydraulic computation. The system contains many hydraulic design features in addition to these river analysis elements, which can be invoked once the specific water surface profiles are computed [4].

This part of the modeling system for HEC-RAS is capable of simulating one-dimensional and combined one-dimensional unstable flow through a complete network of open channels, floodplains and alluvial fans. In the unstable flow computation module, the unsteady flow portion can be used to perform subcritical, supercritical and mixed flow regime calculations [4].

The unstable flow components special characteristics include: comprehensive dam break analysis of hydraulic structure capabilities; levee breaching and overtopping; pumping stations; navigation dam operations; pressurized pipe systems; automatic calibration features; user-defined rules; and combined modeling of one- and two-dimensional unstable flow [4].

2.5 Boundary Conditions (BC)

A slope of 1 in 1000 is considered. Lateral inflow hydrograph is fed. The maximum inflow was 4179 m^3/s . The maximum gate opening of dam was 6.10 m. The elevation of water at maximum discharge is 748.28 m.

2.6 Empirical Equations

As per Froehlich [5] the breach parameters are B_{avg} andt_f

B_{avg} = Average width of final trapezoidal breach in meter.

B = Final bottom width in meter = $B_{avg} - H_b \times$ Slope.
where

H_b Height of breach in meter.
t_f Breach formation times in hours.
V_w Reservoir volume at the time of failure in 1000 m^3.

Equation (1) is used to calculate width of breach and (Eq. 2) is used to calculate breach formation time for the corresponding breach width (Fig. 1).

$$B_{avg} = 0.23 \times K_m \times V_W^{1/3} \tag{1}$$

where K_m
= 1.5 for Overturning failure.

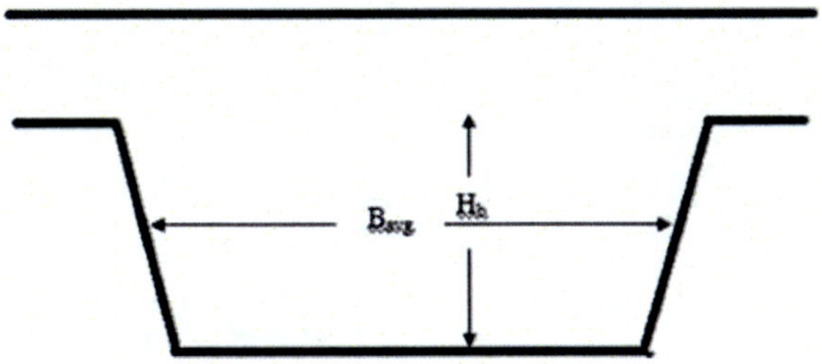

Fig. 1 Typical breach cross section

= 1.0 for other failure modes.

$$t_f = 60 \times \sqrt{\frac{V_w}{g \times H_b^2}} \tag{2}$$

where

g–Gravitational constant 9.81 m/s^2.

2.7 *DBA Simulation by Using HEC-RAS Model*

The clipped digital elevation model (DEM) is inserted into HEC-RAS model. Manning's coefficient is extracted from Arc-GIS and fed into the model. Water storage area, flow area and boundary conditions are fed in to the model. Elevation and storage data is entered. Mesh is created for simulation. Geometric parameters for the dam are entered. The center line terrain profile is shown in Fig. 2. As per Froehlich [5]; practical breach velocity will be less than 4 m per second at the time of breach time. Hence the bottom width of breach and breach formation time is

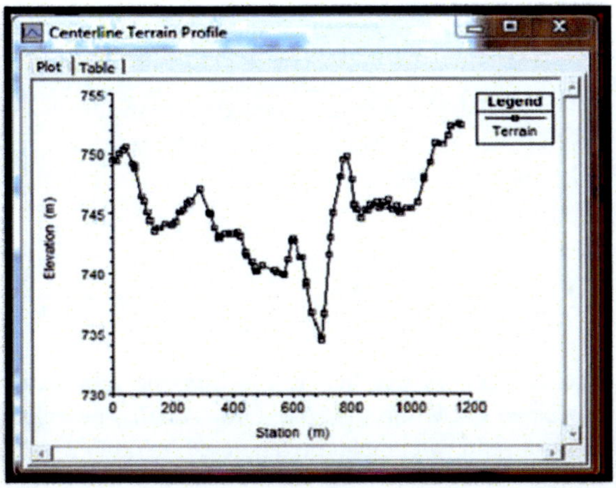

Fig. 2 Centerline terrain profile

Table 1 Breach parameters for overtopping failure and piping failure condition

Breach parameters	Overtopping failure	Piping failure
Bottom width in meter 'B'	138	106
Breach formation time in hours 't_f'	2.401	2.051

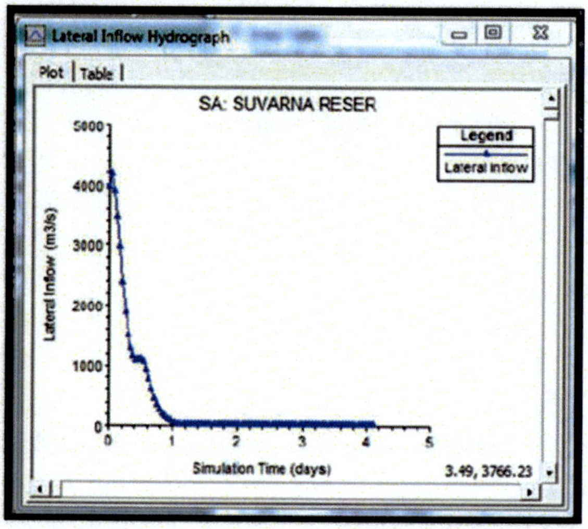

Fig. 3 Inflow hydrograph

changed within the given range so that breach velocity must be within permissible limit. From trial and error procedure for the given ranges of breach bottom width and breach formation time the following values are obtained and shown in Table 1.

The plot between Simulation Time and Lateral Inflow is shown in Fig. 3.

3 Results

After simulation the model gives various results to analyze like breach bottom elevation, breach bottom width, breach outflow, breach side slope, breach velocity, etc. It helps further to prepare Inundation mapping and flood hazard mapping.

3.1 Hydrological Review of Suvarnavathi Dam

The Roughness coefficient for Suvarnavathi catchment area is obtained from processed DEM through Arc-GIS model. The values of manning's coefficient for Suvarnavathi catchment are tabulated in Table 2.

Table 2 Manning's coefficient of Suvarnavathi catchment

S. No.	Roughness coefficients	η Value
1	11	0.034
2	14	0.06
3	20	0.034
4	30	0.034
5	40	0.1
6	50	0.1
7	70	0.11
8	100	0.11
9	110	0.035
10	120	0.035
11	130	0.07
12	140	0.034
13	210	0.04

The left bank elevation is 749.797 m and right bank elevation is 752.408 m, the cross-sectional details are given in Table 3.

Table 3 Elevations of embankment along center line

S. No.	Station	Elevation in m
1	0	749.797
2	75	748.919
3	150	743.789
4	225	745.122
5	300	747.000
6	375	743.289
7	450	741.000
8	525	740.303
9	600	742.854
10	675	734.392
11	750	748.000
12	825	744.633
13	900	745.971
14	975	745.395
15	1050	748.000
16	1125	751.512
17	1170.43	752.408

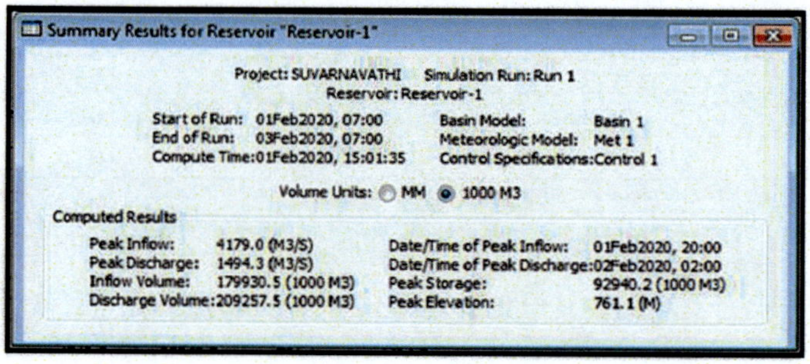

Fig. 4 Flood routed results of Suvarnavathi dam

Catchment of Suvarnavathi River is ungauged catchment, hence synthetic unit hydrograph is extracted as per principles of unit hydrograph [6]. Critical sequencing of rainfall is done by collecting Standard Project Storm (SPS) value from Indian Metrological Department (IMD). The SPS value is distributed into 24 h rainfall distribution. Hourly % rainfall distribution for 12 h is sufficient to find Time Distribution Coefficient (TDC). Critical sequencing of rainfall is done to feed the precipitation values in HEC-HMS model. The data's of elevation of water, storage of water at different levels and discharge of water at different levels are collected from dam authority.

Flood routed results from HEC-HMS model refer Fig. 4; the peak inflow to the reservoir was 4179.00 m^3/s. The Dam spillway was designed for 830.24 m^3/s. The top of dam was at an elevation of 751.30 m. Due to flood, peak discharge was 1494.30 m^3/s. and it is flowing over the dam at an elevation of 761.10 m hence the existing spillway is inadequate, hence raise the dam by a height of 10.1 m including 0.3 m of freeboard.

3.2 *Dam Break Analysis Under Overtopping Failure Criteria*

The breach bottom width has increased from 0 to 138 m in 2 h 25 min. The breach elevation has reduced from 751.19 m to 727.98 m in just 39 min. Hence breach height has restricted to 23.21 m. Breach left side slope and right side slope is constant 1:1.The breach velocity at breach formation time is 3.55 m per second which is less than 4 m per second [5].

The outflow hydrograph is shown in Fig. 5, the maximum flood was found to be 13,475.27 m^3/s. The flood reaches its maximum value after 1 h 29 min from breach start time and reaches its base flow of 19.6 m^3/s after 32 h 46 min. The graph of flood discharge versus time is shown in Table 4.

In the downstream, the river has 45 km length to reach the Cauvery River at its confluence point. The flood water covers the downstream area by a depth ranging from zero to 15 m. The flood water may flow with a velocity of up to 6 m/s.

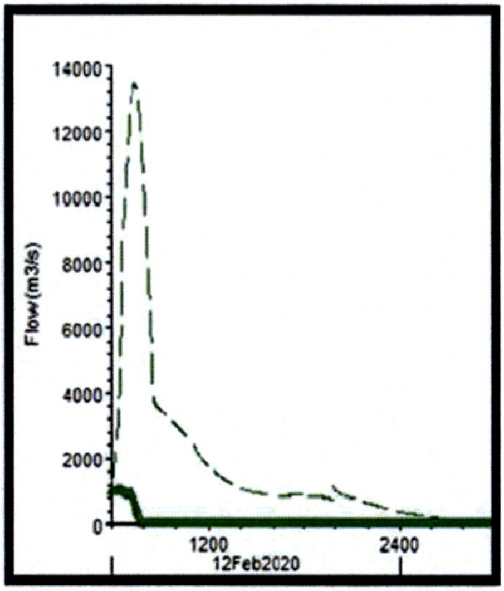

Fig. 5 Outflow flood hydrograph

Table 4 Peak flood result

	Time series	Maximum	Time at max	Volume (1000 m^3)
1	HW stage	752.21	12 Feb 2020 0627	
2	HW stage	738.35	12 Feb 2020 0627	
3	Flow	13,475.27	12 Feb 2020 0627	164,330.33

The flooded water will effect 46 villages and a town which are located in downstream side of the Suvarnavathi Dam. The population data from 2011 census of the flood effected villages are taken from https://www.census2011.co.in/ and the number of families are calculated by considering 5 persons per family [7]. The list of villages effected population and rehabilitation cost is tabulated in Table 5.

Thus, the total cost for rehabilitation of the villages is estimated to be Rs. 3440.64 crores. This has been calculated by considering 15 Lakhs per family. It includes land cost, construction cost and infrastructure cost. From RAS Mapper, it can be analyzed that, time required by flood water to reach the above listed villages. It is listed in Table 6. This information helps the authority to evacuate the people and animal life with their valuables within safe time.

By using RAS mapper (flood inundation map) velocity of flow, depth of flow and time of flood reach at various points can be extracted. Here, only time of flood reach to various places is listed as in above Table 6. RAS mapper showing depth of flood at a point is shown in Fig. 6. The typical RAS mapper showing time of flood reach to Hongalavadi village, Yelandur city and Cauvery River (Confluence point) is shown in Figs. 7, 8 and 9, respectively. The time of flood reach to various villages and city are tabulated by using the typical technique as shown in Figs. 7, 8 and 9.

Table 5 Rehabilitation cost of villages due to flood under OTP failure

S. No	Villages	Population affected	No of families (Approx.)	Rehabilitation cost in lakhs
1	Hongalawadi	398	80	1194.00
2	Karadihalla	70	14	210
3	Kanikere	97	19	291
4	Devarajaura	78	16	234
5	Dollipura	620	124	1860.00
6	Byadamudlu	2355	471	7065.00
7	Basavanapura	857	171	2571.00
8	Hebbasur	5351	1070	16,053.00
9	Boodithittu	920	184	2760.00
10	Kudluru	4054	811	12,162.00
11	Handarakalli	2216	443	6648.00
12	Mallupura	375	75	1125.00
13	Kariyanakatte	4971	994	14,913.00
14	Kempanapura	4353	871	13,059.00
15	Kannegala	1365	273	4095.00
16	Kottamballi	1639	328	4917.00
17	Alur	2733	547	8199.00
18	Amble	6166	1233	18,498.00

(continued)

Table 5 (continued)

S. No	Villages	Population affected	No of families (Approx.)	Rehabilitation cost in lakhs
19	Avalakandalli	1804	361	5412.00
20	D Kandahalli	672	134	2016.00
21	Homma	1195	239	3585.00
22	Dugahatti	1569	314	4707.00
23	Changarahalli	3769	754	11,307.00
24	Dyamkandalli	672	134	2016.00
25	**Yelandur**	8779	1756	26,337.00
26	Y.K. Mole	1608	322	4824.00
27	Honganuru	6539	1308	19,617.00
28	Yeriyur	5129	1026	15,387.00
29	Uppinamole	1217	243	3651.00
30	Chamalapura	2249	450	6747.00
31	Mellagahalli	1860	372	5580.00
32	Komaranapura	2207	441	6621.00
33	Basavapura	3406	681	10,218.00
34	Shivakahalli	1052	210	3156.00
35	Boodithittu	920	184	2760.00
36	Kinkanahalli	1634	327	4902.00
37	Katnavadi	831	166	2493.00
38	T.Hosur	2287	457	6861.00

(continued)

Table 5 (continued)

S. No	Villages	Population affected	No of families (Approx.)	Rehabilitation cost in lakhs
39	Bannisarrige	1428	286	4284.00
40	Mamballi	5911	1182	17,733.00
41	Agara	2866	573	8598.00
42	Teramballi	2256	451	6768.00
43	Uttamballi	1959	392	5877.00
44	Mulluru	5761	1152	17,283.00
45	Hampapura	3116	623	9348.00
46	Dasanapura	830	166	2490.00
47	Harale	2544	509	7632.00

Table 6 Time required by flood water to reach the listed villages

S. No.	Villages	Time required to reach flood water
1	**Hongalawadi**	**47 min**
2	Karadihalla	52 min
3	Kanikere	59 min
4	Devarajaura	1 h 07 min
5	Dollipura	1 h 10 min
6	Byadamudlu	1 h 26 min
7	Basavanapura	1 h 59 min
8	Hebbasur	2 h 07 min
9	Boodithittu	2 h 31 min
10	Kudluru	3 h 11 min
11	Handarakalli	3 h 33 min
12	Mallupura	3 h 43 min
13	Kariyanakatte	4 h 32 min
14	Kempanapura	5 h 01 min
15	Kannegala	5 h 06 min
16	Kottamballi	5 h 17 min
17	Alur	5 h 20 min
18	Amble	5 h 33 min
19	Avalakandalli	5 h 50 min
20	D Kandahalli	5 h 51 min
21	Homma	5 h 56 min
22	Dugahatti	6 h 14 min
23	Changarahalli	6 h 17 min
24	Dyamkandalli	6 h 23 min
25	**Yelandur**	6 h 40 min
26	Y.K. Mole	6 h 51 min
27	Honganuru	7 h 45 min
28	Yeriyur	7 h 54 min
29	Uppinamole	8 h 06 min
30	Chamalapura	8 h 53 min
31	Mellagahalli	8 h 59 min
32	Komaranapura	9 h 23 min
33	Basavapura	9 h 58 min
34	Shivakahalli	9 h 59 min
35	Boodithittu	10 h 04 min
36	Kinkanahalli	10 h 33 min
37	Katnavadi	10 h 35 min

(continued)

Table 6 (continued)

S. No.	Villages	Time required to reach flood water
38	T.Hosur	11 h 43 min
39	Bannisarrige	11 h 56 min
40	Mamballi	12 h 47 min
41	Agara	12 h 49 min
42	Teramballi	13 h 04 min
43	Uttamballi	13 h 25 min
44	Mulluru	13 h 54 min
45	Hampapura	15 h 27 min
46	Dasanapura	17 h 57 min
47	Harale	22 h 51 min

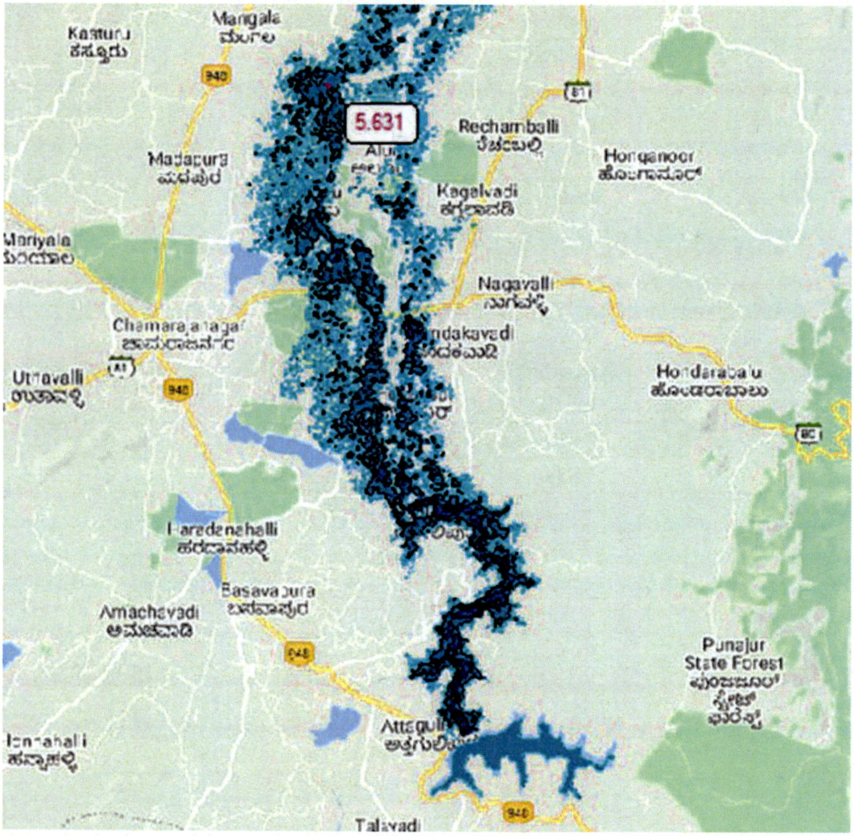

Fig. 6 Typical RAS mapper showing depth of water at a point

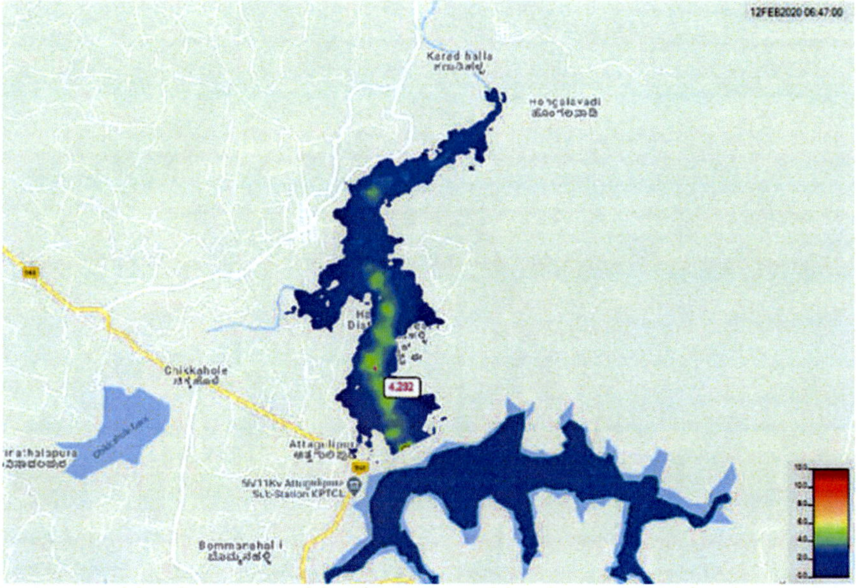

Fig. 7 Typical RAS mapper showing time of flood (47 min) to reach the village Hongalavadi

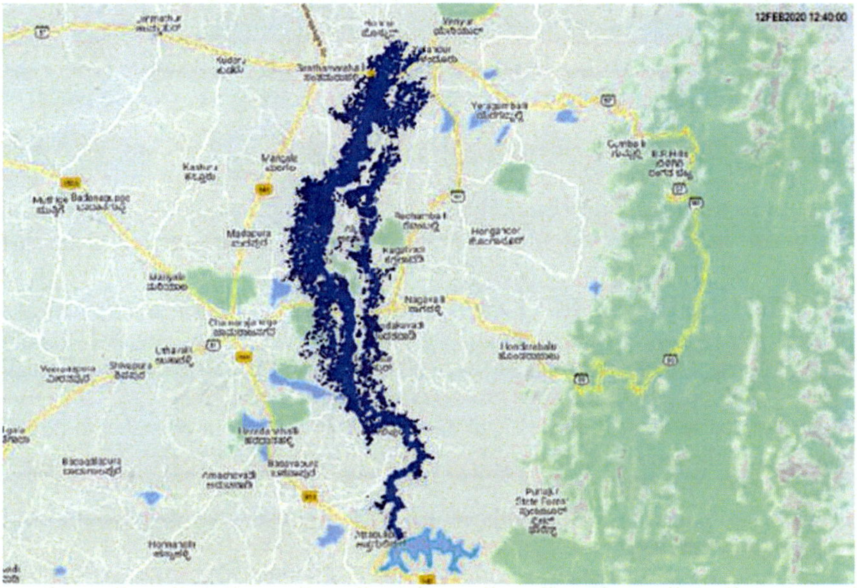

Fig. 8 Typical RAS mapper showing time of flood (6 h 40 min) to reach the city Yelandur

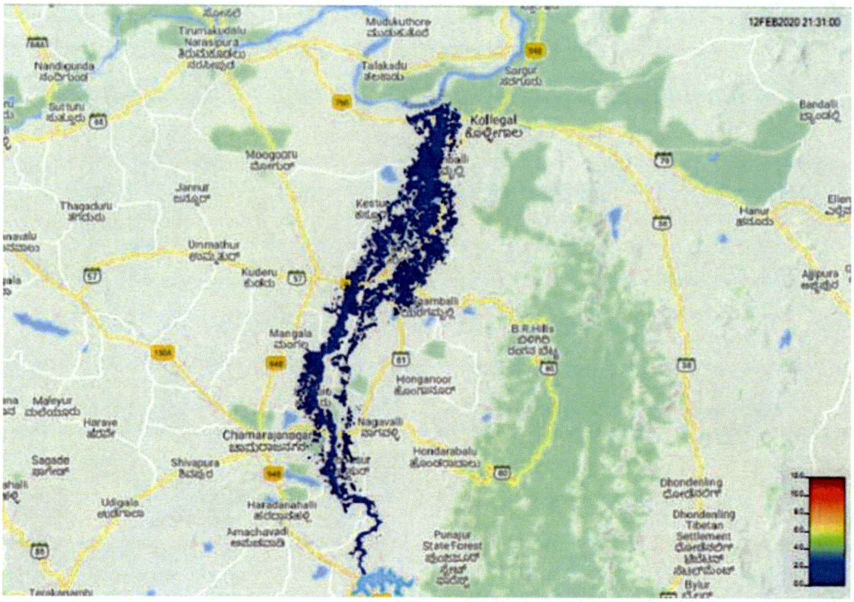

Fig. 9 Typical RAS mapper showing time of flood (15 h 31 min) to reach the confluence point 'Cauvery River'

3.3 Dam Break Analysis Under Piping Failure Criteria

The Outflow hydrograph is shown in Figure-10, the maximum flood was found to be 8059.91 m^3/s. The flood reaches its maximum value after 1 h 38 min from breach start time and reaches its base flow of 19.60 m^3/s after 6 h 24 min. The outflow hydrograph is shown in Fig. 10. The flood values and time of flood reach are summarized in Table 7.

In the down stream, the Suvarnavathi River has to flow 45 km length to reach the Cauvery River at its confluence point. The flood water reaches its confluence point after 4 days. The flood water covers the downstream area by a depth ranging from 0 to 8 m. The flood water may flow with a velocity up to 3 m/s.

The flooded water will effect 23 villages and 1 town which are located in downstream side of the Suvarnavathi dam. The population data of the flood effected villages are taken from the Internet and the number of families are calculated by considering 5 persons per family. The list of villages effected population and rehabilitation cost is tabulated in Table 8.

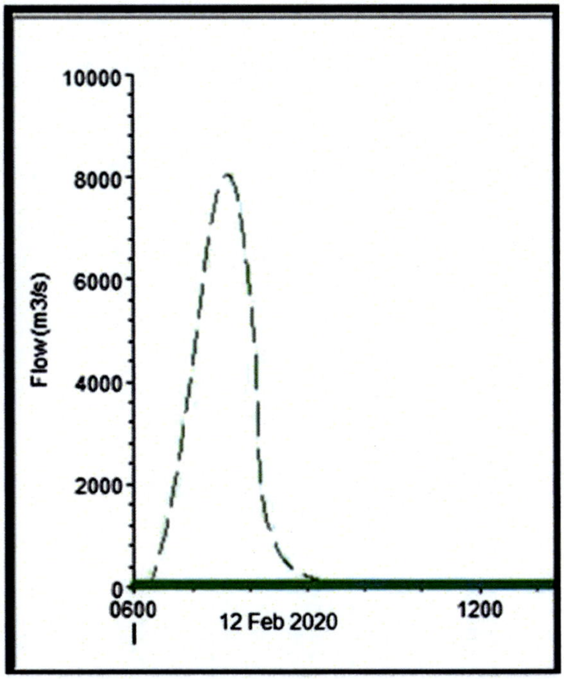

Fig. 10 Outflow flood hydrograph

Table 7 Peak flood result

S. No.	Time series	Maximum	Time at max	Volume (1000 m^3)
1	HW stage	748.28	12 Feb 2020 0609	
2	HW stage	736.57	12 Feb 2020 0739	
3	Flow	8053.91	12 Feb 2020 0738	36,216

Thus, the total cost for rehabilitation of the villages is estimated as Rs. 1802.73 crores. This has been calculated by considering 15 Lakhs per family. It includes land cost, construction cost and infrastructure cost. From RAS Mapper, it can be analyzed that, time required by flood water to reach the above listed villages are tabulated in Table 9. This information helps the authority to evacuate the people and animal life with their valuables within safe time.

The typical RAS mapper showing time of flood reach to Hongalavadi village and Yelandur city is shown in Figs. 11 and 12, respectively. The time of flood reach to various villages and city are tabulated in Table 9 by using the typical technique as shown in Figs. 11 and 12. The depth of inundation map in contour plot layer is shown in Fig. 13.

Table 8 Rehabilitation cost of villages due to flood under piping failure

S. No.	Villages	Population affected	No. of families (Approx.)	Rehabilitation cost in lakhs
1	Hongalawadi	398	80	1,194.00
2	Karadipura	70	14	210
3	Kanikere	97	19	291
4	Devarajaura	78	16	234
5	Dollipura	620	124	1,860.00
6	Byadamudlu	2355	471	7,065.00
7	Basavanapura	857	171	2,571.00
8	Hebbasur	5351	1070	16,053.00
9	Boodithittu	920	184	2,760.00
10	Kudluru	4054	811	12,162.00
11	Handarakalli	2216	443	6,648.00
12	Amble	6166	1233	18,498.00
13	Avalakandalli	1804	361	5,412.00

(continued)

Table 8 (continued)

S. No.	Villages	Population affected	No. of families (Approx.)	Rehabilitation cost in lakhs
14	D Kandahalli	672	134	2,016.00
15	**Yelandur**	8779	1756	26,337.00
16	Changarahalli	3769	754	11,307.00
17	Honganuru	6539	1308	19,617.00
18	Uppinamole	1217	243	3,651.00
19	Mellegehalli	1860	372	5,580.00
20	Chamalapura	2249	450	6,747.00
21	Boodithittu	920	184	2,760.00
22	Shivakahalli	1052	210	3,156.00
23	T Hosur	2287	457	6,861.00
24	Mulluru	5761	1152	17,283.00

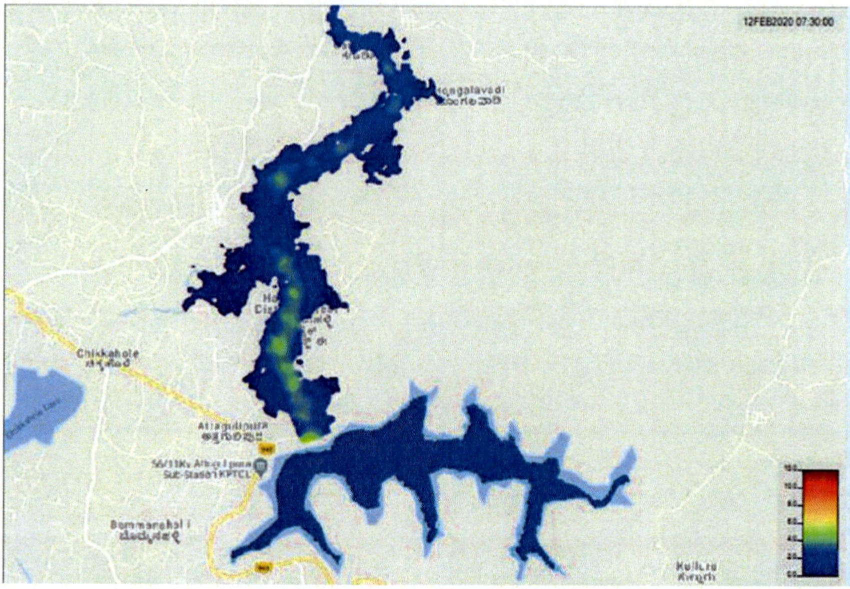

Fig. 11 Typical RAS mapper showing time of flood (1 h 30 min) to reach the village Hongalavadi

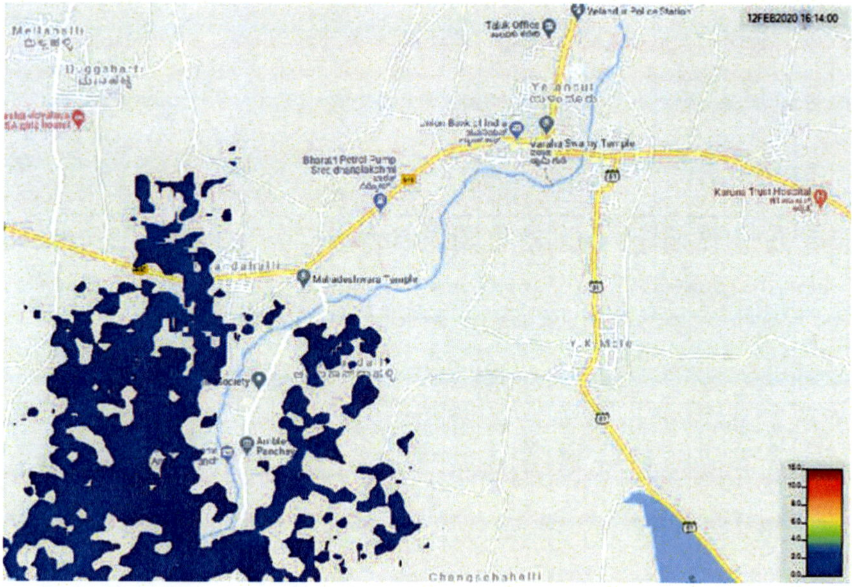

Fig. 12 Typical RAS mapper showing time of flood (10 h 14 min) to reach the Yelandur city

Table 9 Time required by flood water to reach the listed villages

S. No.	Villages	Time required to reach flood water
1	Hongalawadi	1 h 30 min
2	Karadihalla	1 h 18 min
3	Kanikere	1 h 27 min
4	Devarajaura	1 h 34 min
5	Dollipura	1 h 42 min
6	Byadamudlu	2 h 0 min
7	Basavanapura	2 h 40 min
8	Hebbasur	2 h 52 min
9	Boodithittu	3 h 18 min
10	Kudluru	4 h 07 min
11	Handarakalli	4 h 53 min
12	Amble	9 h 23 min
13	Avalakandalli	9 h 31 min
14	D Kandahalli	9 h 44 min
15	**Yelandur**	10 h 14 min
16	Changarahalli	10 h 16 min
17	Honganuru	14 h 59 min
18	Uppinamole	15 h 10 min
19	Honganuru	15 h 10 min
20	Malligehalli	20 h 16 min
21	Chamalapura	21 h 31 min
22	Boodithittu	26 h 09 min
23	Shivakahalli	29 h 11 min
24	T Hosur	41 h 24 min
25	Mulluru	55 h 04 min
26	**Cauvery River**	55 h 25 min

4 Conclusion

The Hydrological Review and Dam Break Analysis under overtopping failure and piping failure criteria are done successfully for Suvarnavathi Dam by using HEC-HMS and HEC-RAS models. From the analysis it is found that, the existing spillway was found inadequate to carry the flood discharge and hence the dam height must be raised further up to a height of 10.1 m. From RAS mapper tool, villages effected by flood are identified and the total cost of rehabilitation of villages was found to be Rs. 3440.64 crores is estimated under overtopping failure criteria. Rehabilitation cost of Rs. 1802.73 crores has been estimated under piping failure criteria. The flood water will travel 6.5 km in 47 min to reach nearest village, Hongalvadi. So 47 min is

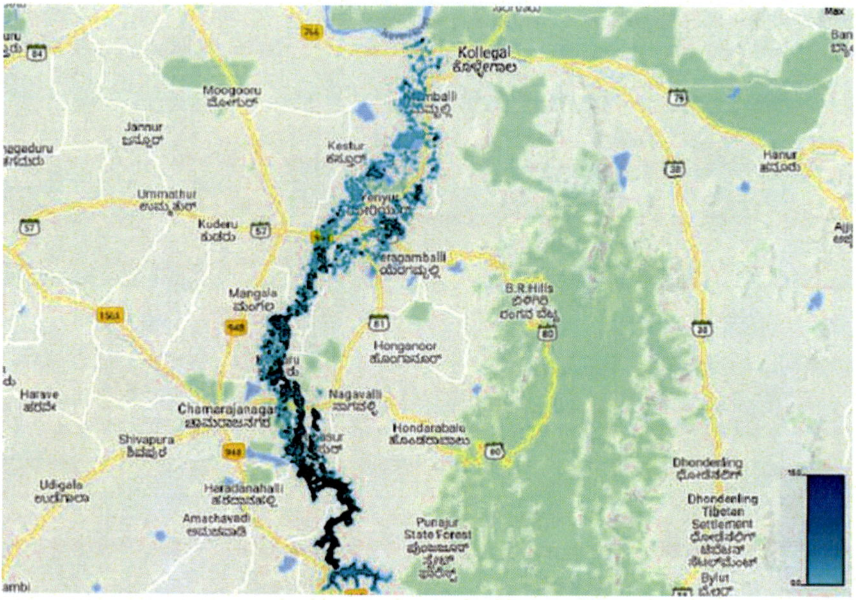

Fig. 13 RAS mapper showing depth of inundation in contour plot layer

sufficient to evacuate the peoples of Hongalavadi village. The Dam Break Analysis helps to know peak flood discharge and to prepare Inundation mapping. This project opens up to carry flood hazard mapping which in turn helps the authority to evacuate the people and animal life with their valuables within safe time.

Acknowledgements I hereby take opportunity to give my sincere thanks to Dr. K.C. Manjunath, Head of the Civil Dept, N.I.E. Mysore & Director, K.E.R.S, Mandya For their guidance and constant encouragement and support in this paper. I truly appreciate the value and their esteemed guidance and encouragement which would be remembered lifelong.

References

1. George AC, Nair BT (2015) Dam break analysis using BOSS DAMBRK. Aquat Procedia 4:853–860 (Icwrcoe)
2. Kumar M, Deputy Director (DSR), Central water commission. Dam rehabilitation and improvement project. Dam safety in India, 20 July 2017
3. Hydrological modeling system, HEC-HMS. User manual, CPD-74A, 614 Aug 2016
4. Hydrological modeling system HEC-RAS. user manual, CPD-69, 538, Feb 2016
5. Frohelich DC (2018) Dam Breach Parameters. P 1708–1719. https://doi.org/10.1061/(ASCE), Feb 2018
6. Flood estimation report for Kaveri basin subzone-3(i), Central Water Commission, Report No. CB/11/1985, Jan 1986
7. Kulkarni SR, Jagtap SA (2017) India Dam. pp. 20–23

Prediction of Seasonal Monthly Rainfall Using Back Propagation Neural Network

M. S. Karthik and M. S. Ganesh Prasad

Abstract Rainfall is a principle component in hydrological cycle and is a considerable source for both surface and ground water. Rainfall affects the maximum livelihood and all major activities that connected with it; therefore, prediction of rainfall is very important. The early prediction of rainfall is more significant to ascertain the effective use of water resources, productivity of crop, and preplanning of water structures. Due to nonlinear behavior of rainfall and complex nature of climatic system, prediction of rainfall is a challenging task. Many approaches have been practiced for the purpose of rainfall prediction. A nonlinear model such as artificial neural network (ANN) approach is considered in this work. The main objective of this study is to predict the seasonal monthly rainfall in Mysuru taluk, southern part of Karnataka state, India. Back propagation neural network (BPNN) technique has been attempted using previous 8 years datasets for climatic parameters; relative humidity, atmospheric temperature, wind speed, and rainfall. In the present study, ten different networks using BPNN were created and analyzed based on their regression results. The network result with higher regression relationship has been considered for rainfall prediction.

Keywords Rainfall prediction · Artificial neural network · Back propagation neural network · Regression · Denormalization · Simulation

1 Introduction

Rainfall prediction is a process of predicting the future rainfall values over a location with implementation of science and technology. Rainfall is a principle component in hydrological cycle and is a considerable source for both surface and ground water [1]. It affects many activities related to the field of agriculture, construction, electricity generation, etc., [2]. Agriculture is India's main practice that most of the Indian economy leans on agriculture [4, 15]. Here, rainfall plays a wide major role that

M. S. Karthik (✉) · M. S. Ganesh Prasad
Department of Civil Engineering, The National Institute of Engineering, Mysore, Karnataka 570008, India

L. Nandagiri et al. (eds.), *Sustainability Trends and Challenges in Civil Engineering*, Lecture Notes in Civil Engineering 162, https://doi.org/10.1007/978-981-16-2826-9_47

early prediction of rainfall helps farmers to manage the crop yield sufficiently. This leads to a superior growth of economy in the country [5]. Many economists agree that a strong monsoon would be a swing factor giving Gross Domestic Product (GDP) number a one-time boost [8]. Also, the prediction of rainfall is very important to some degree that it is strongly associated with adverse events such as landslides, floods, and mass movements [6]. These uncertainties in the natural cycle are mitigated and eliminated only by getting the accurate rainfall prediction technique. Due to nonlinear behavior of rainfall, accurate prediction is more complex [9, 24]. The suitable method is to be adopted to solve this nonlinear relationship.

There are many techniques of various strategies available for the purpose of prediction. Statistical method requires broad historical climatic weather data [11] which is fairly used in predicting monsoon rainfall because the method fails in building nonlinear relationship. Dynamical method is based on the system of nonlinear governing equations [7] which are responsive to the initial conditions and makes them tougher to solve equations. Hence, dynamical method gives deficient results in long range predictions. Support vector machine (SVM) technique is used in supervised learning [13]. It provides better performance and unique solution [12]. However, it is not satisfactory for rainfall prediction because it achieves poor results. The multiple linear regression (MLR) technique is used to forecast the impacts of changes from one variable to another, but this is limited to linear relationship [3, 10]. The technique is easily affected by deviation of datasets; therefore, it is not fit to use for rainfall prediction. Concern to the study, there exist a nonlinear relationship between each elements; therefore, the nonlinear model or method like artificial neural network (ANN) can be best suitable for the present study.

Artificial neural network is a computational network, which includes several processing elements which secure inputs and supply outputs based on predefined activation function [13, 36, 38]. The most attractive feature of ANN is ability to learn the exact behavior between inputs and targets. Here, the network gets trained from collected historical datasets by training which is followed by validation and testing [14]. After successful training, the network is able to perform classification, prediction, etc., and captures complex relationship between input and target [19, 20]. ANN has several network types in which the present study uses back propagation neural network to predict the seasonal monthly rainfall in Mysuru taluk.

2 Back Propagation Neural Network

Back propagation neural network (BPNN) is an important type of artificial neural network. It is a feed forward network associated with back propagation algorithm. It is present with one input layer, one output layer, and one or more hidden layer (Fig. 1). It involves two passes, one is forward pass, and another is backward pass [23, 26]. In forward pass, the synaptic weights of neurons are fixed. In backward pass, the synaptic weights of neurons are adjusted in the network in accordance with error correction rule [29]. In this network, the number of hidden layer and the number of

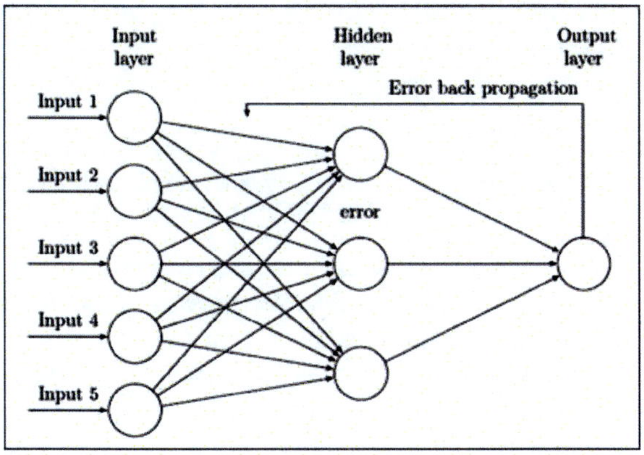

Fig. 1 Back propagation neural network

neurons in the hidden layer get selected depending upon input and target features [31, 33]. BPNN has wider applications in the field of prediction; therefore, the present study preferred this network for rainfall prediction.

3 Methodology

Following are the steps taken up in the present work:

1. Study area and data used
2. Arrangement and normalization of datasets
3. Generating rainfall prediction networks using BPNN
4. Training the created networks
5. Performance analysis
6. Conclusions.

3.1 Study Area and Data Used

The study area selected for the work is Mysuru taluk which is present in southern part of Karnataka state in India. Agriculture is the cornerstone of this area's economy, the rivers Kaveri and Kabini provide irrigation needs to it. Here, the temperature ranges from 15 °C (in winter) to 35 °C (in summer) and gets 785 mm of average annual rainfall. The prediction of rainfall in this region helps in sufficient management of crop yield and water resources.

The data used for the present work were relative humidity, temperature, wind speed, and rainfall. These were collected from Karnataka State Natural Disaster Monitoring System (KSNDMC) Web site. The Web site proactively included all the relevant information that will be easy to understand and open to the public. The data has been collected for Mysuru over an eight-year period from 2012 to 2019, and it is in Excel format.

3.2 *Arrangement and Normalization of Datasets*

The collected datasets are not in the order that required for the present work, and these should be arranged before moving it into the MATLAB workspace. At first, the datasets were grouped to input and target datasets. Parameters taken for the input datasets were relative humidity, atmospheric temperature, and wind speed where rainfall has taken for target datasets. Then, these input and target datasets were arranged in the form of matrix, i.e., [m * n]. Here, 'm' represents rows which in turn refers to parameters and 'n' represents column which in turn refers to number of collected historical data. The research considers only average seasonal month's data (June, July, August, and September) from past eight years. Thus, the total number of data used for the study is 'Thirty-two'. Then, the corresponding arranged input and target datasets were in the form of [3 * 32] and [1 * 32], respectively.

The arranged datasets should be normalized before they hit model input. Normalization is a process of rescaling one or more attributes to the value ranges from 0 to 1 [22, 28, 39]. Usually, this process is used for datasets having different SI units. In this study, parameters present in the datasets have their own SI units and are different from each other. The direct use of these datasets causes an error in the output results of network and also affects network efficiency. Thus, the datasets needs to be normalized which enhances the network by reducing total error and increasing overall efficiency. In this present study, the normalization process has been done using (Eq. 1).

$$\text{Normalized value} = \frac{(X_{\text{a}} - X_{\text{min}})}{(X_{\text{max}} - X_{\text{min}})} \tag{1}$$

where

X_{a} = actual data to be normalized in the datasets.

X_{max} = maximum value in the datasets of respected parameter.

X_{min} = minimum value in the datasets of respected parameter.

Then, the normalized values were taken into MATLAB workspace where research has been carried out using nntool by generating rainfall prediction networks using BPNN.

3.3 *Generating Rainfall Prediction Network Using BPNN*

In the present study, the rainfall prediction networks were created using back propagation neural network. It is a type of artificial neural network which is developed based on the mechanism of biological brain. Here, the construction of network has been done using neural network toolbox of MATLAB. The 'nntool' has been used because it facilitates the development of more general neural network. At first, the normalized input and target datasets were imported to MATLAB workspace. Then, nntool is opened and specified the features that it has. In this work, as shown in Fig. 2, the feed forward back propagation network type is selected with Training function 'TRAINLM', adaption learning function 'LEARNGDM', performance function 'MSE', and transfer function 'TANSIG'. All these specifications are done based on the trial work that carried using sample datasets. Here, three inputs and one target feature were used which is sufficient to use single hidden layer; therefore, the number of layers considered is two, one is hidden, and another is output layer.

Then, the hidden layer should specify the number of neurons (N). The change in the number of neurons influences the network performance and leads to achieve better results at some particular N value. Therefore, the present work performed by creating **ten** different BPN rainfall prediction networks with change in neuron value, i.e., $N = 2, 4, 6, 8, 10, 12, 14, 16, 18$, and 20. These created networks get stored in the data manager or nntool window as shown in Fig. 3.

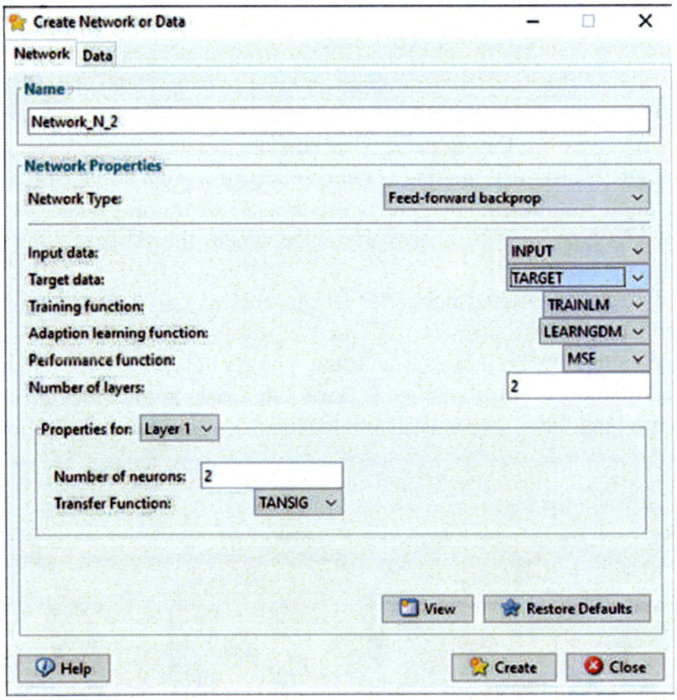

Fig. 2 Network creating window

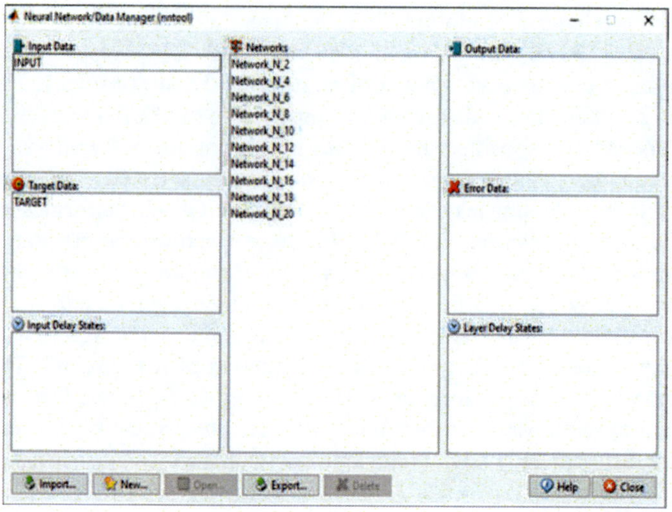

Fig. 3 Neural network window with created networks

3.4 Training the Created Networks

Training the created rainfall prediction network is very important in BPNN. The overall performance of this network depends on this training process. Training should be done effectively for this network because a well-trained network leads to achieve good results [32, 34]. Here, the process of training that carried in the initial network **'Network_N_2'** was presented where the process is same for all other created networks.

Figure 4 shows the initial network that has created using BPNN with number of neurons in the hidden layer is '2'. One can verify the created network pattern using **'view'** ribbon in the network window. Then, click on **'Train'** ribbon which opens training info as shown in Fig. 5. Here, the inputs and targets were assigned using imported input and target data from MATLAB workspace. After this, click on

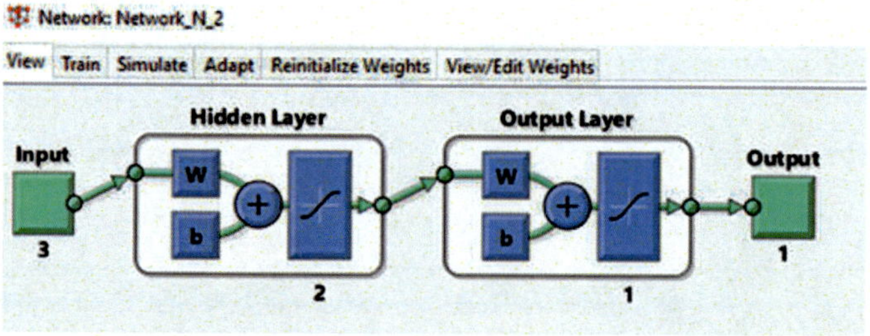

Fig. 4 Rainfall prediction network (Network_N_2)

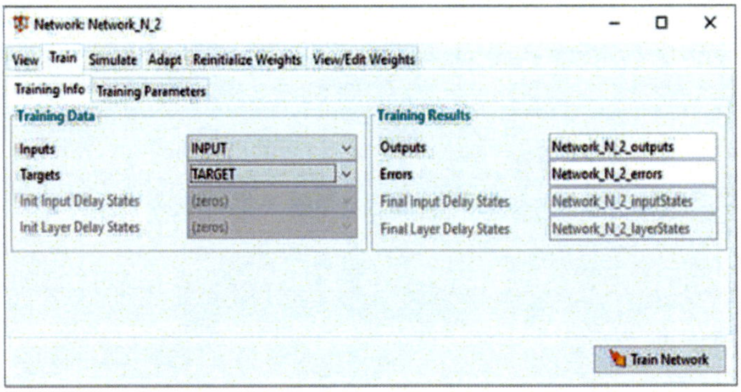

Fig. 5 Training info window

'Train Network' to train the Network_N_2 which has done more than 50 times here to achieve good training.

Then, the network should also perform the same training process by changing training parameters because network training by changing training parameters gets influence over network accuracy.

Click on training parameters. It involves show, epochs, time, goal, min_grad, max_fail, mu, etc., as shown in Fig. 6. However, present study concentrates on the change of epochs and max_fail because both of these represent iterations in the network. Iterations are very important here for network training because an increase in iterations leads to reach higher accuracy in the network results. Therefore, the present study has been performed the training for default settings of training parameters and by changing the epoch and max_fail to 100, 500, 1000, and 1500.

After the completion of training process, simulation has done for the network. Click on **'Simulate'** ribbon in the network window which opens simulation data as shown in Fig. 7. Here, the inputs and targets were specified again using input and

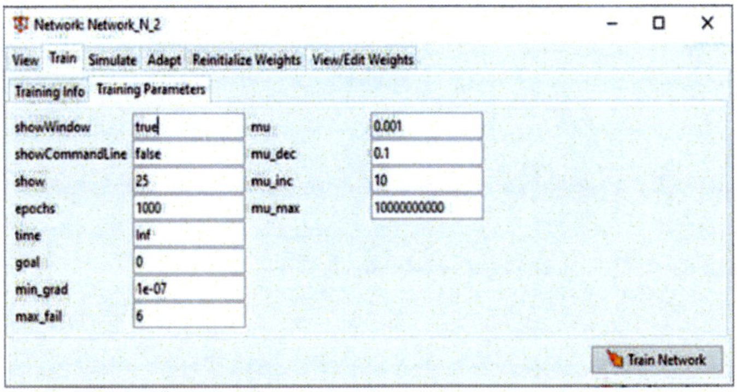

Fig. 6 Training parameters window

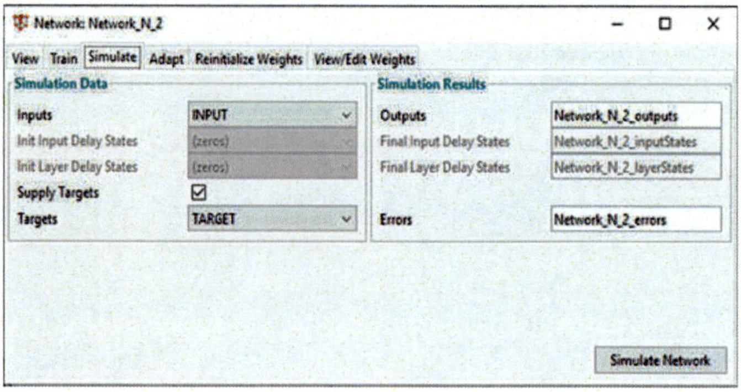

Fig. 7 Simulate window

target data. If require, the output and error name can be changed here. After this, click on **'Simulate Network'** to simulate the network outputs and corresponding errors. This simulated output and error results get stored automatically in the respective output and error data box of data manager window as shown in Fig. 8. These output and error results were exported to MATLAB workspace using **'Export'** option.

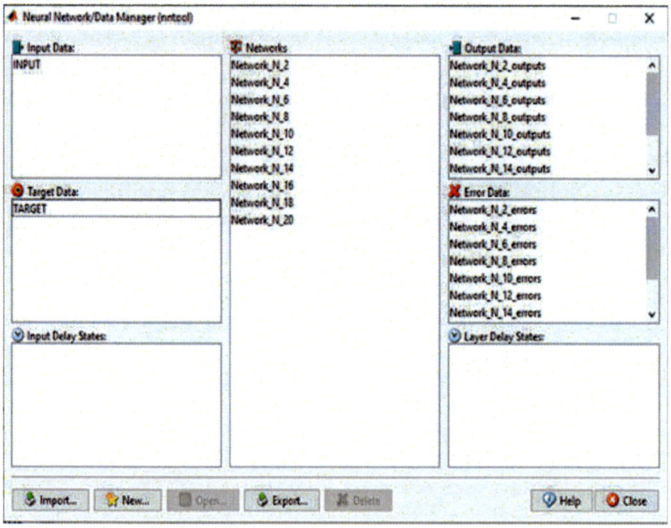

Fig. 8 Neural network window with network outputs and errors

4 Results and discussions

In this present study, the performances of ten different BPN networks were analyzed through their respective regression (*R*) value at higher iteration. *R* is defined as correlation coefficient which is used to measure the relationship between any two variables [17]. Here, correlation coefficient '*R*' is used to measure the strong relationship between output and target of the networks. In this, output represents the results obtained from the rainfall prediction network as a network output and target represents the datasets that given to the network to reach as a network target.

Based on the observation from Table 1, a good relationship can be seen in between the network $N = 10$ and $N = 16$. Here, among ten rainfall prediction BPN networks, the network at **N = 12** achieved higher correlation, i.e. **R = 0.941**. A good relationship in the network promotes good results in the prediction; therefore, output results of the network $N = 12$ were taken for the performance analysis in detail. The analysis has been done using regression results and simulated results.

4.1 Regression Results

The regression results of this network include plots of all the three stages that involved in the network which are training, validation, and testing.

1. **Training stage**

The training stage is the initial and more presiding stage where further stages are dependent on this [21]. Generally, the whole datasets that used in the work are divided into training datasets, validation datasets, and testing datasets. Usually, more than 50% of total datasets are used for training the network. The remaining datasets are equally divided for validation and testing [30].

In this study 'Nntool' has been used where the tool has default settings in datasets distribution for training, validation, and testing stage, i.e., 60% of total datasets are distributed for training, 20% for validation, and remaining 20% for testing. However, the particular study has 8 years monthly average datasets where 60% of datasets utilized the first 4 years of data for training purpose. The corresponding result of this was presented below.

Table 1 Regression value for ten different rainfall prediction networks

Network	$N = 2$	$N = 4$	$N = 6$	$N = 8$	$N = 10$
R value	0.66	0.862	0.876	0.885	0.935
Network	**N = 12**	$N = 14$	$N = 16$	$N = 18$	$N = 20$
R value	**0.941**	0.922	0.915	0.895	0.847

Fig. 9 Regression plot of training stage

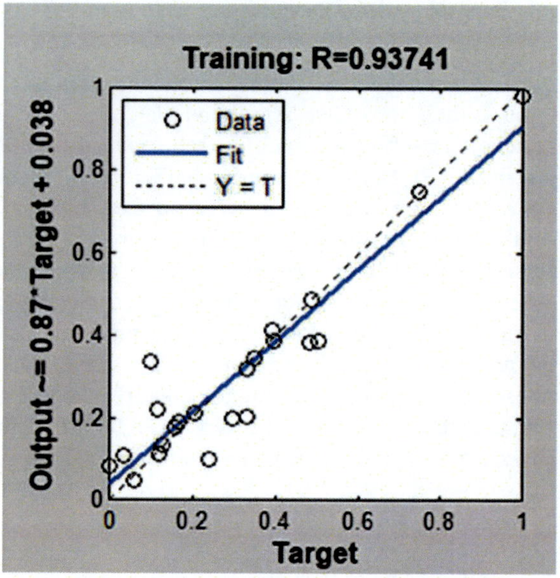

Figure 9 shows the regression plot of training stage. Here, the training process achieved correlation coefficient of 0.937 between output and target. Here, output represents the results obtained from the network at training level, and target represents the datasets that given to the network to reach as a network target. For any artificial neural network, training is the main priority. A good relationship in the training process leads to an increase of network efficiency. In this study, the network achieved a good relationship between output and target which indicates that the network is well trained in its sequence.

Then, the output values were calculated for the training stage using the output equation that is available in the regression plot of training stage. The calculated output values were in the normalized form where these values should be de-normalized.

De-normalization is the method of storing the join of superior normal form relations as base relation [35, 40]. It is very important to de-normalize the output values in this study because the prediction has to be made in the normal form of SI units. Therefore, the de-normalization has been performed in this work for the output values using (Eq. 2).

$$\text{De - normalized value} = (\text{Normalized value} \times (\text{Max}(X) - \text{Min}(X)) + \text{Min}(X)) \tag{2}$$

where

'X' refers to a specific parameter that considered for de-normalization.
Max(X) = maximum value in the datasets of specified parameter.
Min(X) = minimum value in the datasets of specified parameter.

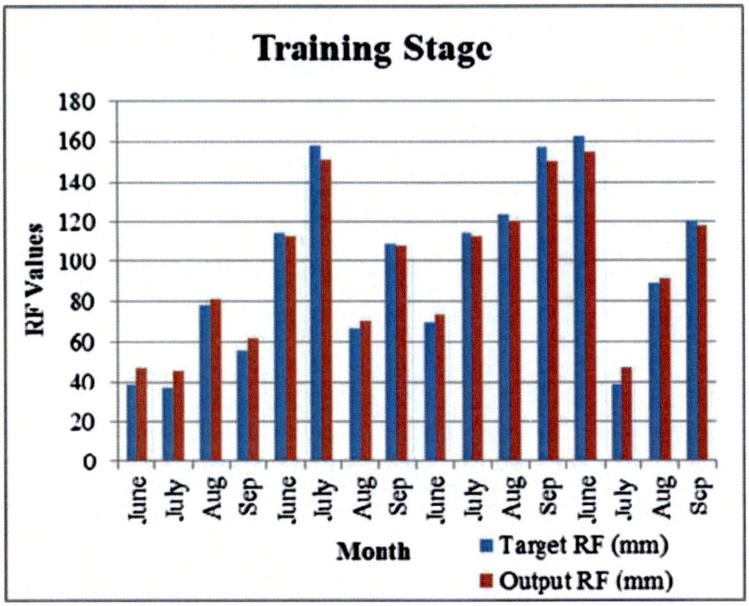

Fig. 10 Comparison graph for training stage

Then, comparison graph was plotted between the de-normalized output and target rainfall values for training stage. From the comparison graph (Fig. 10), it is observed that the output and target rainfall values are very similar to each other. A small difference is observed between them. Here, the results of regression plot and comparison graph conclude that the training stage achieved good results with better accuracy, and the network is suitable for prediction activity.

2. **Validation stage**

The training process is followed by validation process. Here, the learned model has been validated using validation datasets. The process improves the network if any variance occurs in it. In the present study, the validation process utilized next two years of data for validating the trained network. The corresponding result of this were shown below.

Figure 11 shows the regression plot of validation stage. Here, the network achieved good correlation between output and target, i.e., $R = 0.979$. Then, the output values were calculated using the output equation that available in the regression plot of validation stage.

The calculated output values were de-normalized using (Eq. 2). Then, comparison graph was plotted between the de-normalized output and target rainfall values for validation stage. From the comparison graph (Fig. 12), it is observed that the output and target rainfall values are similar to each other with small difference. Here, the results of regression plot and comparison graph conclude that the validation

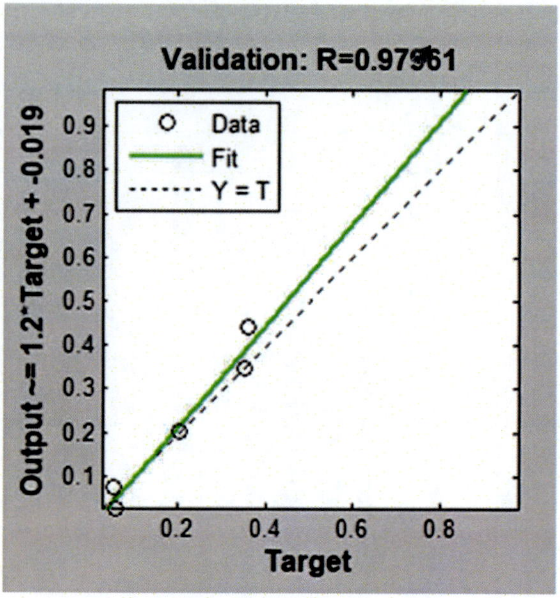

Fig. 11 Regression plot of validation stage

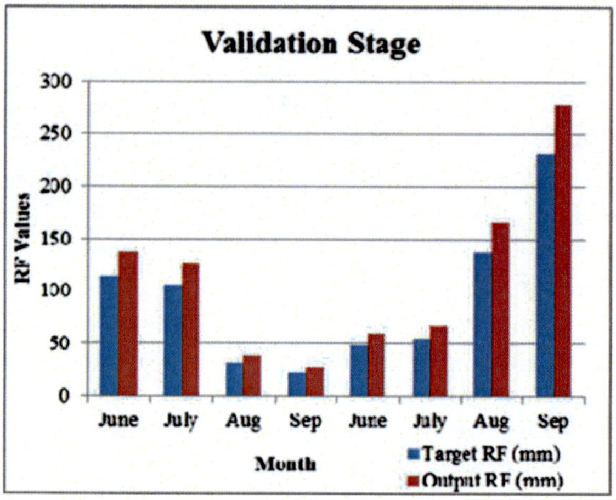

Fig. 12 Comparison graph for validation stage

stage achieved good results with better accuracy, and the network is validated in its sequence.

3. **Testing stage**

The validation process is followed by testing process. Here, the validated network has been tested using testing datasets. The testing process describes the network capability of handling the datasets [18, 37]. The present work utilized the last two

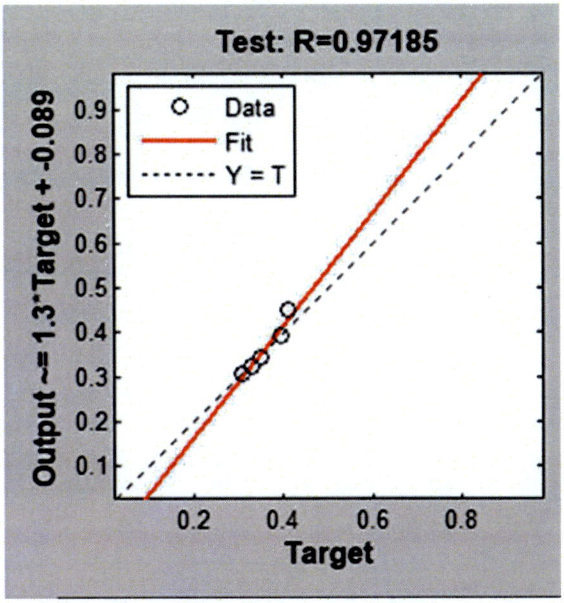

Fig. 13 Regression plot of testing stage

years of datasets to test the validated network, and the corresponding results were presented below.

Figure 13 shows the regression plot of testing stage. Here, the network achieved good correlation between output and target, i.e., $R = 0.971$. It indicates that the network is highly correlated and tested in its sequence and attained better accuracy. Then, the output values were calculated using output equation that available in the regression plot of testing stage. The calculated output values were de-normalized using (Eq. 2).

Then, comparison graph was plotted between the de-normalized output and target rainfall values for testing stage. From the comparison graph (Fig. 14), it is observed that the output and target rainfall values are not highly similar to each other which indicate that the network should be improved in this process.

Thus, from the demonstration of all the three process, the network with $N = 12$ supports to carry the action of rainfall prediction using its simulation results.

4.2 *Simulation Results*

The simulation results were obtained directly from the network. The analysis of this result has been made in the platform of Microsoft Excel. At first, simulated values were de-normalized using (Eq. 2), and then, de-normalized output and target rainfall values were compared by plotting the comparison graph. Along with the comparison graph, the linear graph was also plotted to analyze the simulated results.

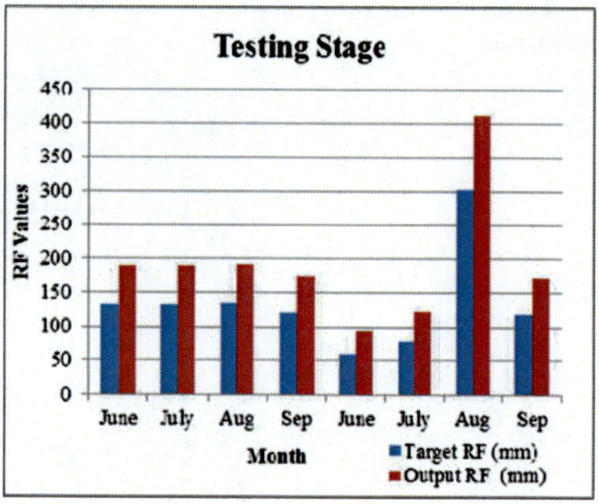

Fig. 14 Comparison graph for testing stage

Figure 15 shows the comparison graph for simulated result of rainfall prediction network $N = 12$. From comparison graph, it is observed that very similarities in the output and target rainfall values with some small difference (in between). It indicates that the network achieved good results with better performance.

From linear graph (Fig. 16), the value of R^2 and corresponding linear equation were obtained. Here, the R^2 represents the coefficient of determination. It is an output of regression analysis. R^2 is interpreted as the portion of the variance in the dependent variable that is predictable from the independent variable. Also, it is the square of correlation (R) between output and target values; thus, it ranges from 0 to 1 [16, 25,

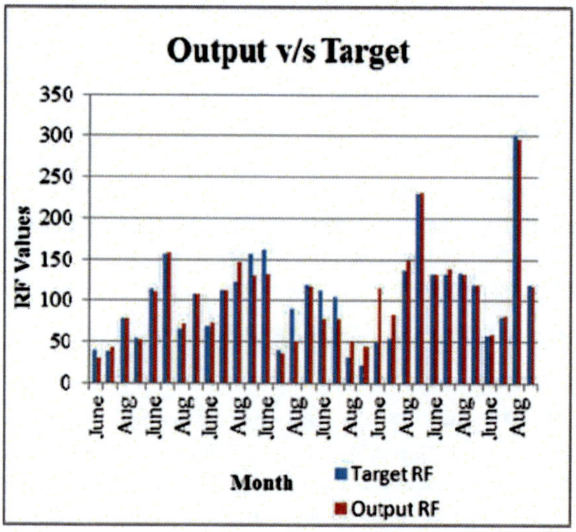

Fig. 15 Comparison graph for simulated result of $N = 12$

Fig. 16 Linear graph for simulated result of $N = 12$

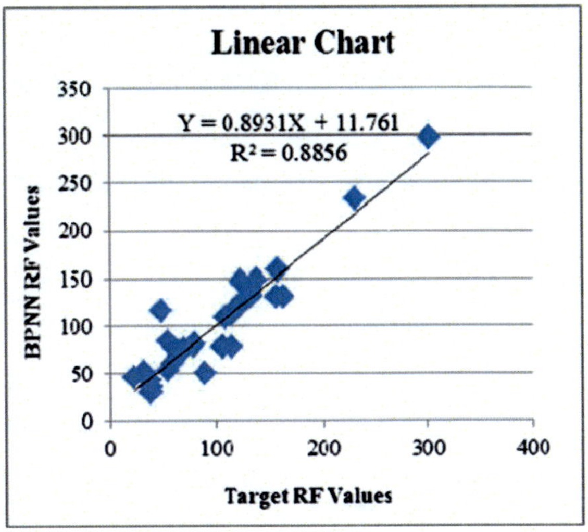

27]. Here, if R^2 is near to zero or lower value indicates that the dependent variable cannot be predicted from the independent variable. If R^2 is higher or near to one indicates that the dependent variable can be predicted from the independent variable. In this present work, the R^2 reached the value of 0.885 which is somewhat good value indicates a good relationship between output and target rainfall values in the overall network. Therefore, the linear equation obtained from this linear graph is best suitable for the prediction of future rainfall values.

4.3 Prediction Results

In this study, the back propagation neural network achieved its best result in the network $N = 12$. Therefore, the result of that network has been used for the prediction of future rainfall values. However, the prediction has been made using linear equation obtained from the linear graph (Fig. 16). The linear equation obtained was $Y = 0.8931 * X + 11.761$. Here, 'X' uses the average of monthly seasonal average rainfall values to predict the upcoming average monthly rainfall values for seasonal months. Then, the prediction of rainfall values in 'mm' has made for the Mysuru taluk and they are 92, 90, 119 and 116 (June, July, August and September) respectively.

5 Conclusions

The created rainfall prediction networks using BPNN were performed individually and achieved good results. Among ten different networks, the higher regression relationship and good fit were observed in the network '$N = 12$'. The network also achieved good results with high accuracy in the process of prediction. The predicted rainfall values in 'mm' are 92, 90, 119, and 116 (June, July, August, and September), respectively. In this study, BPNN proved as a good technique for rainfall prediction. Here, data used is not sufficient for long range prediction. The prediction accuracy can be increased further by considering larger historical datasets.

References

1. Saxena A, Verma N, Tripathi KC (2013) A review study of weather forecasting using artificial neural network approach. Int J Eng Res Technol 2(11):2029–2035
2. Gurung AR (2017) Forecasting weather system using artificial neural network (ANN): a survey paper. Int J Latest Eng Res Appl 02:42–50
3. Arya R, Pai ML (2018) Rainfall prediction using an optimised genetic-artificial neural network model. 119(10):669–678
4. Bala Sai Tarun G, Sriram JV, Sairam K, Teja Sreenivas K, Santhi MVBT (2019) Rainfall prediction using machine learning techniques. Int J Eng Adv Technol 8(6):4179–4183
5. Biju MJ, Aneesha N (2017) BPN Based rainfall forecasting: a review. Int Adv Res J Sci Eng Technol 4:50–54
6. Charaniya NA, Dudul SV (2012) Focused time delay neural network model for rainfall prediction using indian ocean dipole index. In: International conference on computational, intelligence and communication networks. 851–855
7. Chaudhari MS, Choudhari NK (2017) Study of various rainfall estimation and prediction techniques using data mining. Am J Eng Res 7:137–139
8. Nayak DR, Mahapatra A, Mishra P (2013) A survey on rainfall prediction using artificial neural network. Int J Comput Appl 72(16):32–40
9. Vamsidhar E, Varma KVSRP, Rao SP, Satapati R (2010) Prediction of rainfall using backpropagation neural network model. Int J Comput Sci Eng 2(4):1119–1121
10. Fallah-Ghalhary GA, Mousavi-Baygi M, Habibi-Nokhandan M (2009) Seasonal rainfall forecasting using artificial neural network. J Appl Sci 9(6):1098–1105
11. Chang FJ, Liang JM, Chen YC (2001) Flood forecasting using radial basis function neural networks. IEEE Trans Syst 31(4)
12. Harshani RK, Nagahamulla UR, Ratnayake AR (2011) Monsoon rainfall forecasting in sri lanka using artificial neural networks. In: International conference on industrial information systems, pp 305–309
13. Pant H, Singh Y, Pal A (2015) Rainfall Prediction techniques using artificial neural networks—a brief review. Int J Sci Res Dev 172–177
14. Sofian IM, Affandi AK, Iskhaq I, Apriani Y (2018) Monthly rainfall prediction based on artificial neural networks with backpropagation and radial basis function. Int J Adv Intell Inform 4(2):154–166
15. Ilaboya IR, Igbinedion OE (2019) Performance of multiple linear regression (MLR) and artificial neural network (ANN) for the prediction of monthly maximum rainfall in Benin city, Nigeria. Int J Eng Sci Appl 3(1):1–18
16. Jiaxing Du, Bin Zhao, Shaohui Miao (2012) An application on the immune evolutionary algorithm based on back propagation in the rainfall prediction. In: International conference on computer science and electronics engineering. vol 1, pp 313–317
17. Abbot J, Marohasy J (2017) Application of artificial neural networks to forecasting monthly rainfall one year in advance for locations within the murray darling basin, Australia. Int J Sustain Dev Plan 12(8):1282–1298

18. Joseph J, Ratheesh TK (2013) Rainfall prediction using data mining techniques. Int J Comput Appl 83(8):11–15
19. Solaimani K (2019) Rainfall-runoff prediction based on artificial neural network (A case study: Jarahi watershed). Int J Curr Microbiol App Sci 8(05):1328–1334
20. Kavitha Rani B, Srinivas K, Govardhan A (2014) Rainfall prediction with TLBO optimized ANN. J Sci Ind Res 73(10):643–647
21. Nagendra KV, Jahnavi Y, Haritha N (2007) A survey on support vector machines and artificial neural network in rainfall forecasting. Int J Future Revolution Comput Sci Commun Eng 3(11):20–24
22. Abhishek K, Kumar A, Ranjan R, Kumar S (2012) Rainfall prediction model using artificial neural network. In: IEEE control and system graduate research colloquium
23. Darji MP, Dabhi VK, Prajapati HB (2015) Rainfall forecasting using neural network: a survey. In: International conference on advances in computer engineering applications ICACEA 2015, pp 706–713
24. Porte P, Isaac RK, Mahilang KKS, Sonboier K, Minj P (2018) Groundwater level prediction using artificial neural network model. Int J Curr Microbiol Appl Sci 7(2):2947–2954
25. Prabakaran S, Naveen Kumar P, Sai Mani Tarun P (2017) Rainfall prediction using modified linear regression. 12(12):3715–3718
26. Liu Qi, Zou Y, Liu X, Linge N (2019) A survey on rainfall forecasting using artificial neural network. Int J Embedded Syst 11(2):240–249
27. Rasel HM, Imteaz MA (2016) Application of artificial neural network for seasonal rainfall forecasting: a case study for South Australia. In: Lecture notes in engineering and computer science vol 2223, p 130–134
28. Refonaa J, Lakshmi M, Allu CSK, Anantha RT (2019) Rainfall prediction using Apriori Algorithm. Int J Recent Technol Eng 8(2) Special Issue 3:593–596
29. Renuga Devi S, Venkatesh C, Agarwal P, Arulmozhivarman P (2014) Daily rainfall forecasting using artificial neural networks for early warning of landslides. In: International conference on advances in computing, communications and informatics, pp 2218–2224
30. Shaikh L, Sawlani K (2017) A rainfall prediction model using artificial neural network. Int J Sci Res Netw Secur Commun 5 (April)
31. Hudnurkar S, Rayavarapu N (2018) Performance of artificial neural network in nowcasting summer monsoon rainfall: a case study. In: 1st International conference on data science and analytics, PuneCon 2018, pp 1–5
32. Lee S, Cho S, Wong PM (1998) Rainfall prediction using artificial neural networks. J Geogr Inf Decis Anal 2(2):233–242
33. Kashiwao T, Nakayama K, Ando S, Ikeda K, Lee M, Bahadori A (2017) A neural network-based local rainfall prediction system using meteorological data on the Internet: a case study using data from the Japan Meteorological Agency. Appl Soft Comput J 56:317–330
34. Nourani V, Mehrvand M, Baghanam AH (2014) Implication of SOM-ANN based clustering for multi-station rainfall-runoff modelling. J Urban Environ Eng 8(2):198–210
35. Venkata Ramana R, Krishna B, Kumar SR, Pandey NG (2013) Monthly rainfall prediction using wavelet neural network analysis. Water Resour Manage 27(10):3697–3711
36. Chandwani V, Vyas SK, Agrawal V, Sharma G (2015) Soft computing approach for rainfall-runoff modelling: a review. In: International conference on water resources, coastal and ocean engineering, vol 4 pp 1054–1061
37. Wahyuni EG, Fauzan LMF, Abriyani F, Muchlis NF, Ulfa M (2018) Rainfall prediction with backpropagation method. J Phys: Conf Ser 983(1)
38. Wong KW, Wong PM, Gedeon TD, Fung CC (2003) Rainfall prediction model using soft computing technique. Soft Comput 7(6):434–438
39. Dash Y, Mishra SK, Panigrahi BK (2018) Rainfall prediction of a maritime state (Kerala), India using SLFN and ELM techniques. In: International conference on intelligent computing, instrumentation and control technologies. pp 1714–1718
40. Wang Zl, Sheng HH (2010) Rainfall prediction using generalized regression neural network: case study Zhengzhou. In: International conference on computational and information sciences. pp 1265–1268

Factors Influencing Productivity of Construction Labour—A Survey

R. Abhishek, K. C. Sachin, and S. R. Shashikumara

Abstract Labour productivity is a major factor which influences the timely completion of the project within the allocated budget. Factors having impact on labour productivity were recognized and rated for this study. Analysis was carried out based on the questionnaire survey consisting of three main sections having 33 various factors. First and second set of questionnaires consist factors affecting the productivity of labours. A set of data analysis was carried out for the collected responses with the help of Statistical Package for the Social Sciences (SPSS) software. To determine the internal consistency of each factor, Cronbach's Alpha was computed. Relative Importance Index (RII) analysis helped to recognize importance of factors affecting the labour productivity based on ranking, in the view of contractors and professional consultants. The monotonic correlation between the data was measured by Spearman's correlation coefficient. Third set of questionnaire consists those measures which can be instigated for improving the productivity in construction sector. Based on ratings given by participants, RII was carried out for these responses. This work would assist the construction managers to avoid the productivity loss and to achieve optimum work from labours. Experience and effective management techniques can ameliorate productivity.

Keywords Productivity · Relative Importance Index · Cronbach's Alpha · Spearman's Correlation Coefficient

1 Introduction

One of the world's most challenging and largest industries is construction, also the second largest in county like India next to agriculture [1]. Human resource is a key element in construction productivity as it uncontrollable and varying [2]. Productivity

R. Abhishek (✉) · S. R. Shashikumara
Department of Civil Engineering, JSS Academy of Technical Education, Bengaluru, India
e-mail: abhishekr@jssateb.ac.in

K. C. Sachin
RV College of Engineering, Bengaluru, India

L. Nandagiri et al. (eds.), *Sustainability Trends and Challenges in Civil Engineering*,
Lecture Notes in Civil Engineering 162, https://doi.org/10.1007/978-981-16-2826-9_48

can be defined as the quantum of tasks accomplished with number of labour involved in it [2, 3]. Low productivity problem is still being encountered by construction industry.

In construction other than resources like materials, capital and equipment, human resource is one of the major inputs as a resource which is flexible for management. So in construction, productivity can be defined as labour productivity [4]. 30–50% out of complete project cost comprises the labour cost, hence construction industry is labour-intensive industry [5]. Under many research, it has been identified that labour productivity is the key reason for delay and cost overrun of the projects. As labour productivity improvement is critical, initial step in identifying those factors which affect the productivity of labours is crucial for the success of the project [6]. The idea of development of task level productivity is neglected in construction sector compared with manufacturing sector [2].

The academicians and the practitioners have a great concern about the poor performance in the industry. It was recognized that deprived labour productivity as a key cause for delay in Indian construction projects [7]. Even though the performance of construction industry is relayed on productivity, in India, the measurement of labour productivity is too rare; due to which the losses in productivity are not accounted. There is an essential for research, to identify the factors affecting labour productivity in construction industry and also to look into the opportunities for improving the productivity. Labour productivity was interdependent and cannot be controlled completely [2]. Improving productivity in construction sector is a prime method of enhancing the profit with little or no increment in investment [8].

The objective of the project is to identify the both internal and external factors which influence the loss of productivity.

2 Methodology

Important factors affecting productivity of construction sector were shortlisted based on reviews of previous work [1–5] for further research. Questionnaire survey is most suitable method for collecting the data related to factors and identifying the relative importance by gauging the perception of every individual in construction industry [2]. Pertinent to practical working condition questionnaire was designed considering 41 factors. Based on pilot study, these factors were further modified to 33 factors.

Questionnaire was redesigned considering 33 factors under different heads are mentioned Table 1. Certain factors were merged or redesigned based on the similarities and interdependencies. Few factors were deleted as they had very low impact on productivity based on the respondent values [8]. Respondents were requested to rate the questions ranging between 1 and 5 where 1 represents strongly disagree and 5 represents strongly agree that factors affect the labour productivity [9]. Process carried out for the survey analysis is depicted in Fig. 1.

Table 1 Factors considered for the study

Internal factors	
A	**Manpower**
Q1	No proper supervision
Q2	Skill of labour
Q3	Unclear instructions to labour
Q4	Labour strikes
Q5	Difficulty in recruitment of workers
Q6	Non-payment to labour contractor
Q7	Absenteeism at worksite
B	**Management**
Q1	Resource management
Q2	Motivation
Q3	Safety issue
Q4	Lack of equipment
Q5	Disruption of power/water supplies
Q6	Stoppages because of dispute with owners/consultants
Q7	Stop work orders because of infringements of government regulations (The action of breaking the terms of law, agreement, etc.)
C	**Design factors**
Q1	Build ability (Design of a building facilitates ease of construction)
Q2	Design changes
D	**Site factors**
Q1	Site access
Q2	Congested work area
Q3	Repair and repetition works
Q4	Idling due to non-availability of materials
External factor	
Q1	Health
Q2	Climatic conditions
Q3	Force majeure (traffic/site accidents)
Productivity improvement technique	
A	**Employee-based productive improvement technique**
Q1	Quality of supervision
Q2	Skill enhancement
Q3	Time management (minimize time wasters, e.g.: inability to delegate work)

(continued)

Table 1 (continued)

Q4	Learning curve (experience and expertise)
Q5	Better living
Q6	Financial incentives (e.g.: Bonuses)
Q7	Fringe benefits (e.g.: medical insurance)
B	**Task-based productivity improvement technique**
Q1	Planning and scheduling (Right materials at the right place at right time)
Q2	Job safety design (A safe workplace)
Q3	Method engineering/work simplification (Elimination of unnecessary task to reduce time)

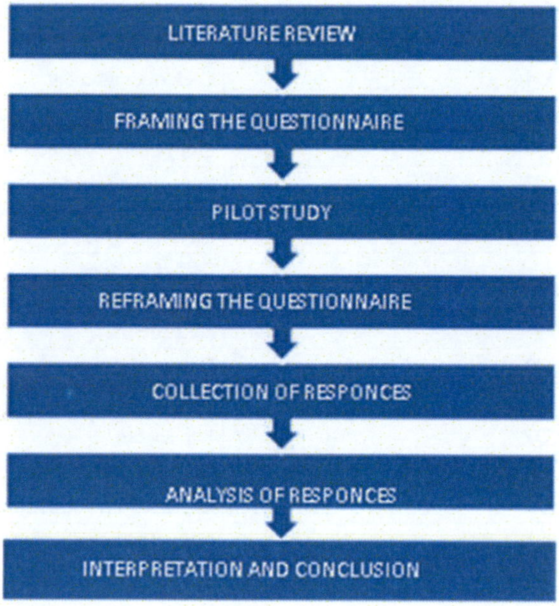

Fig. 1 Process of the analysis

The survey was carried out at Bengaluru (Bangalore) city, Karnataka, India amongst contractor, project management consultants, site engineers, senior site engineers and general manager.

The respondents were well qualified professionals with a minimum experience of 10 years in relevant fields. As this sector had variety of people from various parts of India with different experiences, this study depicts the national level response about labour productivity in construction sector.

A set of data analysis was carried out for the collected responses with the help of Statistical Package for the Social Sciences (SPSS) software. Reliability analysis was carried out to check the internal consistency of the factors. Each factors' inter-correlation was examined, and Cronbach's Alpha was computed to determine the internal consistency of each factor. The minimum advisable level was considered as 0.7 [10, 11]. The factor having computed Cronbach's Alpha greater than 0.7 was considered as highly reliable and internally consistent.

Relative Importance Index (RII) analysis was carried out for the collected data. It helps to identify importance of factors affecting the labour productivity based on ranking, from the viewpoint of contractors and professional consultants. 33 factors prominently affecting productivity of labours were classified under five headings, and the collected responses were ranked using RII. This method is best suited for respondent satisfaction rating [2].

The monotonic correlation between the data was measured by Spearman's correlation coefficient. This identifies the extent of variation that the value one variable changes to another. The value varies between −1 and +1, positive value represents the change in both variables in same direction where in negative value represents the change in in opposite direction.

3 Result and Discussion

152 responses were collected from construction industry professionals, which includes 60 responses from site engineers, 48 from quantity surveyors, 25 from contractors and 19 from project managers. Cronbach's Alpha was computed for the collected responses. The Cronbach's Alpha values for manpower, management and site factors ranged from 0.755 to 0.843, which was larger than the recommended value of 0.700. The alpha value for design factors is 0.344 less than 0.7, which is not reliable. Figures 2, 3 and 4 show the RII analysis carried for the collected responses. It shows the ranking of various factors according to their importance. The major factors affecting labour productivity under respecting heading are as follows,

- Manpower factors—No proper supervision with 73.20%.

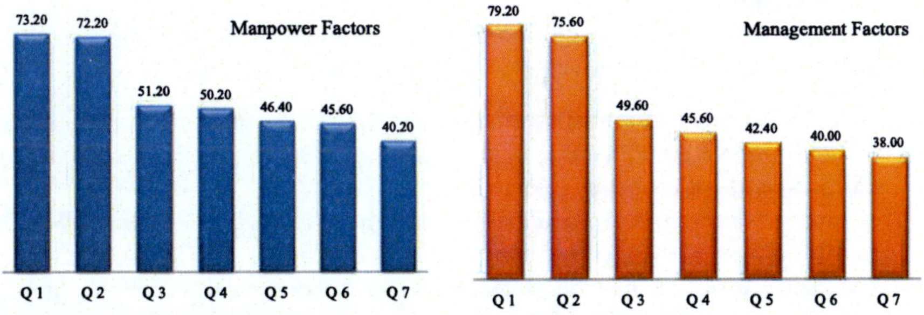

Fig. 2 RII analysis for factors related to manpower and management

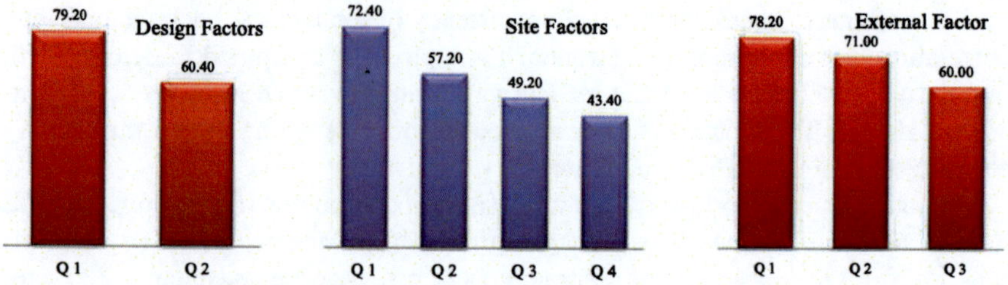

Fig. 3 RII analysis for design, site and external factors

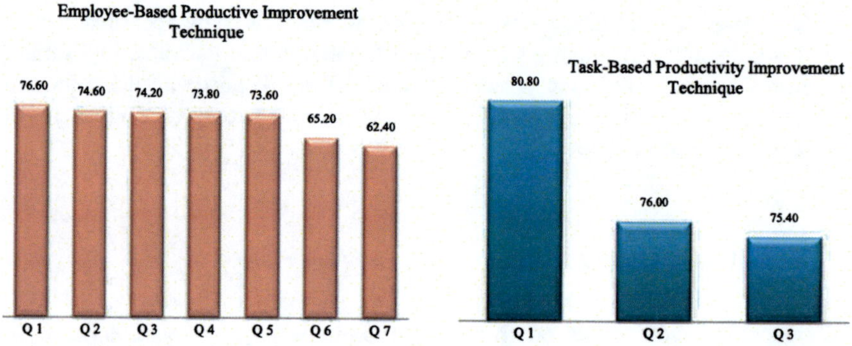

Fig. 4 RII analysis for productivity improvement technique

- Management factors—Resource management with 79.20%.
- Design factors—Build ability with 79.20%.
- Site factor—Site access with 72.40%.
- External factor—Health with 78.20%.

With RII analysis for productivity improvement techniques, it was observed that factors like quality of supervision, skill enhancement of workers and time management with proper planning and scheduling have better contribution for improving the labour productivity.

Based on Spearman's correlation coefficient, it was observed that few factors were internally correlated with each other and together affects the productivity.

- Under factors related to manpower, difficulty in recruitment and non-payment to labours is correlated with value 0.764, i.e. recruitment of proper skilled labours is difficult without proper payment.
- No proper supervision and skill of labour are correlated with value 0.745. It indicates due to poor supervision there was no improvement in skills of labours.
- With value 0.742, it can be stated that lack of communication and proper instructions will lead to labour strikes.

- Under factors related to management with a value 0.857, it was observed that infringements of government regulations may lead to dispute with owners/consultants which results in stoppage of works.
- Value 0.639 indicates that repair/repetition of work is difficult due to congested work space.
- Value 0.732 reflects that bad climatic conditions will affect the labours health.

4 Conclusions

This survey was carried to highlight the factors responsible for variation in productivity amongst human resources involved in construction industry. Many resources are essential for accomplishing a construction project out of which human resource is a vital. Human resource management has higher importance due to the challenges of handling people with various physiological, psychological and social background.

This analysis reflects that all the shortlisted 33 factors under different heads are affecting the productivity, and many factors are correlated with each other. It is difficult to improve the skills of labours without proper training and consistent monitoring. Recruitment or retainment of skilled labour is even though uneconomical, and it reduces wastage of material, saves time and hinders the chances of safety hazards. Thus, hiring skilled labours and effective management of labours are most crucial for accomplishment of every project. Management factors like poor resource management and lack of proper motivation affect duration of work completion leading to loss of productivity. Progress of the work is also suppressed due to the disputes between owners/consultants because of government regulation infringements and poor ability to analyse the design at execution level. Delay in works occurs due to congested work space and bad climatic conditions. Labours health and site conditions have its impact on the productivity.

Proper enlightenment about these factors to the execution team and periodic quality improvement trainings to the labours will help the organization to overcome the low productivity problems.

Acknowledgements Authors extend their sincere thanks to the JSS Academy of Technical Education and R V College of Engineering, Bengaluru for their infrastructure support during the conduction of this survey work. Also the supports of project managers, site engineers, consultants of various organization are highly appreciable.

References

1. Agrawal A, Halder S (2019) Identifying factors affecting construction labour productivity in India and measures to improve productivity. Asian J Civ Eng 569–579. http://doi.org/10.1007/s42107-019-00212-3

2. Karthik D, Rao CBK (2019) Identifying the significant factors affecting the masonry labour productivity in building construction projects in India. Int J Constr Manag 1–9. http://doi.org/10.1080/15623599.2019.1631978
3. Ameh OJ, Osegbo EE (2011) Study of relationship between time overrun and productivity on construction sites. Int J Constr Supply Chain Manag 1:56–67. https://doi.org/10.14424/ijcscm101011-56-67
4. Karim NA, Hassan SH, Yunus JN, Hashim MZ (2013) Factors influence labour productivity and the impacts on construction industry. Caspian J Appl Sci Res 349–354
5. Lamka AHV, Masu SM, Wanyona G, Dianga S, Gwaya AO (2014) Factors influencing effective productivity on construction sites in Nairobi County. Int J Soft Comput Eng 4:24–30
6. Dixit S (2018) Analysing enabling factors affecting the on-site productivity in Indian construction industry. Period Polytech Arch 49:185–193. https://doi.org/10.3311/PPar.12710
7. Doloi H, Sawhney A, Iyer KC, Rentala S (2011) Analysing factors affecting delays in Indian construction projects. Int J Project Manage 30:479–489. https://doi.org/10.1016/j.ijproman.2011.10.004
8. Hiyassat MA, Hiyari MA, Sweis GJ (2016) Factors affecting construction labours productivity: a case study of Jordan. Int J Constr Manag 16:1–12. https://doi.org/10.1080/15623599.2016.1142266
9. Jarkas AM, Al Balushi RA, Raveendranath PK (2015) Determinants of construction labour productivity in Oman. Int J Constr Manag 15:332–344
10. Durdyev S, Mbachu J (2017) Key constraints to labour productivity in residential building projects: evidence from Cambodia. Int J Constr Manag 18:332–344. https://doi.org/10.1080/15623599.2017.1326301
11. Bryman A, Cramer D (2011) Quantitative data analysis with IBM SPSS: a guide for social scientists. Routledge, New York

Assessment of Surface Water Quality Parameters of Panchganga River

Chidanand Patil, Sneha S. Bandekar, Sateesh Hosamane, Sanjeev Sangami, and Amrut Adavimath

Abstract This article presents physico-chemical parameters of Panchganga River, Maharashtra, India. The river receives varieties of domestic, industrial and agricultural wastes located around the river. The samples are collected from seven different locations P1 to P7 along Panchganga River. Parameters such as pH, turbidity, electrical conductivity (EC), total dissolved solids (TDS), acidity, alkalinity, chlorides, total hardness, calcium sodium and potassium were determined. Dissolved oxygen (DO) deficit assessed using Streeter-Phelps equation from March to May 2019 on monthly basis. Results showed turbidity was out of desirable limit at all sampling locations. Most of parameters showed increase in value as river flows downstream. Calcium and alkalinity showed higher values than desirable limit as per IS 10500-2012 at last four sampling points. From results, it is concluded that water quality is getting polluted moderately as river flows from Kolhapur to Ichalkaranji. And there is decrease in DO content of water as river flows downstream; critical DO ranging between 2.66 and 4.14 mg/L is observed at distance of 3396–4475 m from discharge point. Finally, the attempt has been made to study the water quality index (WQI) of the river flow and the results showed that variation in quality of water from the upstream to the downstream.

1 Introduction

Water is about 71% of the earth's surface and yet it is one of the scarcest resources. Water is abundant earth as a whole but clean and safe water for consumption is very less in ecosystem. Rivers are the most important resources of drinking water in different countries of the world. As mankind is growing, changes occurred in land

C. Patil (✉) · A. Adavimath
Department of Civil Engineering, KLE DR MSSCET, Belagavi, India

S. S. Bandekar · S. Hosamane
Department of Chemical Engineering, KLE DR MSSCET, Belagavi, India

S. Sangami
Department of Civil Engineering, Jain College of Engineering, Belagavi, India

L. Nandagiri et al. (eds.), *Sustainability Trends and Challenges in Civil Engineering*, Lecture Notes in Civil Engineering 162, https://doi.org/10.1007/978-981-16-2826-9_49

and water usage, and hydrological conditions due to coarser use of water. There is an increasing measure of water pollution throughout the world. There are mainly two types of sources of pollution namely point and non-point source pollution. In India, the rivers are getting polluted day by day. Today intense pollution is superior in many rivers such as Krishna, Bramhaputra, Tapti, Hoogly, Ganga and Narmada. As the water flows, it collects minerals, salts and silt from the soil and rock in the riverbed and riverbank. Numerous different pollutants enter stream as it flows downstream, including human waste, animal waste, agricultural runoff, urban runoff, industrial effluents, and mining waste, because of which most of the rivers are confronting pollution problem or under danger of pollution [1].

Self-purification is a process by which a stream is partially or totally restored after insertion of foreign substances in quantity and quality, by way of natural processes, to cause a distinctive change in the physical, chemical and/or biological properties of the stream. The introduction of foreign matter to the environment is usually through manmade activities [2].

The process of self-purification can be divided into four zones; "zone of degradation" is characterized by dark and turbid water and DO is as low as 40% of saturation value. Anaerobic decomposition takes place here. "Zone of active decomposition" is marked by heavy pollution. Water in this zone is gray and no fishes are found in this zone because DO decreases to lowest value. The recovery process starts within the "Zone of recovery". DO again reach 40% of saturation. Close to the end of the zone, microscopic aquatic life reappears. In "clear water zone" becomes clean and more aesthetically pleasant as DO reaches saturation value and fishes reappear. There are many factors affecting the process of self-purification including dilution, current, temperature, sunlight and rate of oxidation.

The oxygen deficit at any point in the stream is equal to the difference between saturation dissolved oxygen and actual dissolved oxygen at particular time. The resulting oxygen deficiency can be obtained by addition of the oxygenation and reaeration curves algebraically and the obtained curve is called the oxygen sag curve [3].

The WQI theory was first proposed by Horton [4] and which compares of water quality of the rivers with any statutory standards are given by various organizations. It gives different weights to the different parameters based on the limits of the standards and it is very is for us to determine the quality of water of the river [5]. Many researchers have developed WQI models successfully and applied to check the quality of water [6, 7].

The Panchaganga River basin is situated on the eastern side of the Sahyadri ranges. It is in the northern part of Kolhapur district of Maharashtra, located at 15° 43′ and 17° 17′ North latitude and 73° 40′ and 74° 42′ East longitude and is constituted of five northeast flowing streams, namely Bhogavati (83 km), Tulsi (30 km), Kasari (69 km), Kumbi (48 km) and Dhamani (41 km). The Bhogawati River is renamed Panchaganga from Prayag Chikhali. The river flows and meets Krishna River at Narsinhwadi, Tal: Shirol, Dist: Kolhapur. There are 174 villages, 2 municipal towns (Ichalkarnji and Kurundwad) and one city (Kolhapur) situated on the bank of the river [8].

Kolhapur and Ichalkaranji cities are situated on the bank of Panchganga River, which is the main water source for many purposes. Increase in industrial activities and rapid urbanization is polluting river water. The wastewater from MIDC Shiroli, Kolhapur and textile mills of Ichalkaranji is reaching the river without any treatment. Due to this man-made activities, there is problem in drinking water supply to surrounding areas. Water pollution created severe health impact on human life. Hence, it is most important to assess the quality of river Panchganga. The main objective of this work is to assess and analyze the water quality of the Panchganga River by analyzing WQI, and the attempts have been made to establish the dissolved oxygen sag curve for Panchganga River.

2 Study Area

Samples are collected from the river Panchganga at particular locations which are shown in Figs. 1 and 2. One liter of samples was collected from each location in the study area. The samples were collected in the middle of river from 7 different locations for the months of March, April and May 2019. Samples are collected in new clean air-tight plastic containers marked with numbers. Grab sampling method is used for the collection of sample (Table 1).

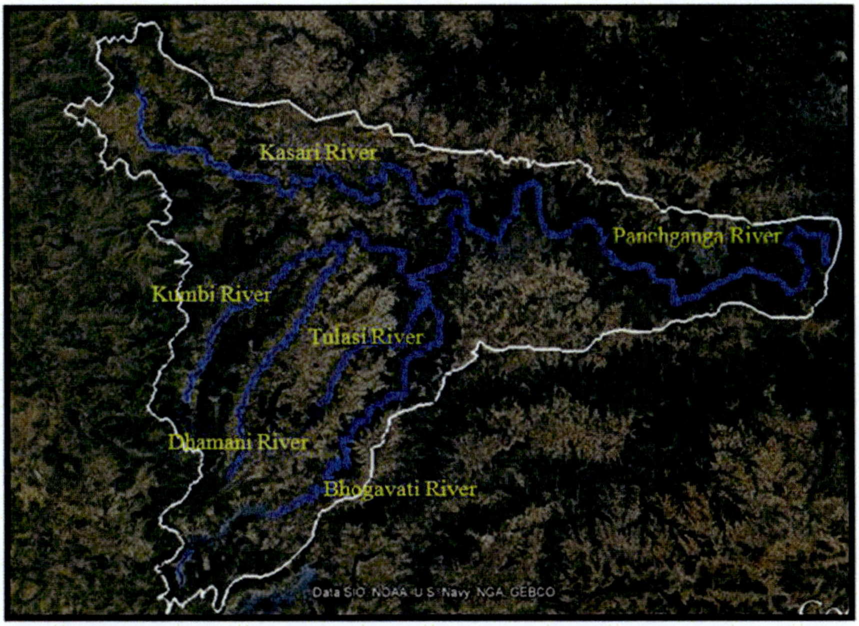

Fig. 1 Satellite image of Panchganga river basin

Fig. 2 Water sampling points

Table 1 Details of sampling points

S. No.	Site code	Distance (km)	Location		Elevation from mean sea level (m)
			Latitude	Longitude	
1	P1	0.0	16.6917568	74.1654932	543.0
2	P2	07.72	16.7615540	74.2621537	538.0
3	P3	21.88	16.7121682	74.2805517	536.9
4	P4	29.77	16.6660892	74.4757330	535.8
5	P5	03.37	16.6810911	74.5082206	534.0
6	P6	17.70	16.6818394	74.5759011	531.9
7	P7	14.80	16.6918400	74.5965658	531.0

3 Materials and Methodology

The parameters pH, turbidity, electrical conductivity (EC), total dissolved solids (TDS), acidity, alkalinity, chlorides, total hardness, calcium, sodium and potassium were analyzed as per procedure prescribed by APHA 2005.

3.1 Oxygen Sag Curve

Assume the river is completely mixed and the concentration of water and waste at $x = 0$ is [9]:

$$C_o = \frac{Q_r C_r + Q_w C_w}{Q_r + Q_w}. \tag{3.1}$$

where,

C_o initial strength at the starting point after mixing (mg/l).
Q_r discharge (m^3/s) of the stream.
C_r before mixing (mg/L).
C_w wastewater strength (mg/L).

The saturated oxygen is equal to

$$D_s = \frac{475}{33.3 + T}. \tag{3.2}$$

where,

D_s DO saturation.
T Temperature in °C.

The deoxygenation constant is obtained by the equation.

$$K_{20} = 0.3\left(\frac{H}{8}\right)^{-0.434}. \tag{3.3}$$

$$K_T = K_{20^\circ} \times 1.047^{(T-20)}. \tag{3.4}$$

where,

H average depth of river (m).
U average velocity of river flow (m/s).

The reaeration constant was determined

$$R = \frac{294(D_L U)^{0.5}}{H^{1.5}}. \tag{3.5}$$

$$D_L = 1.760 \times 10^{-4} \text{m}^2/\text{d} \times 1.037^{(T-20)}. \tag{3.6}$$

where,

R reaeration constant.
H average depth of river (m).
U average flow velocity (m/s).
D_L diffusion coefficient (m^2/day).

Several assumptions have been made for the calculation of dissolved oxygen. We assumed that the mixed one is evenly across the width of the river and is rapidly mixed with the stream water. The two necessary parameters are the rate of deoxygenation (K) and the rate of reaeration (R). And the term x indicates the distance of downstream from the waste discharge point and u in the equation indicates the velocity of the river water, respectively [9].

$$D_t = \frac{KL_0}{R-K}\left(e^{-K(x/u)} - e^{-R(x/u)}\right) + D_0 e^{-R(x/u)} \tag{3.7}$$

where,

D_t DO deficit at anytime t.
L_o BOD remaining at time $t = 0$.
K deoxygenation rate.
R reoxygenation rate.
x downstream distance from the waste discharge point.
u velocity of the river water.

4 Results and Discussion

4.1 *Physico-chemical Parameters*

The pH, turbidity, electrical conductivity (EC), total dissolved solids (TDS), acidity, alkalinity, chlorides, total hardness, calcium, sodium and potassium of the river water samples collected during March to May 2019 are explained below.

4.2 *pH*

In the present study, the pH of water is in the range of 7.09–8.29. From Table 2, it is noticed that the pH is well within the permissible range. There is no much variation of pH at different locations as shown in Fig. 3a. In majority of the cases, the water is nearer to neutral range. The minimum pH is recorded at P2 (7.09) in May and

Table 2 Variation of physico-chemical parameters at sampling locations

Location	pH	Turbidity (NTU)	EC (μS/cm)	TDS (mg/L)	Acidity (mg/L)	Alkalinity (mg/L)	Chlorides (mg/L)	Total hardness (mg/L)	Calcium (mg/L)	Sodium (mg/L)	Potassium (mg/L)
March 2019											
P1	7.45	3.2	72.5	40	5	68	16	84	28	3.19	0.8
P2	7.2	4.1	241.1	100	8	134	26	90	60	11.92	1.18
P3	7.13	4.4	167.7	30	12	98	23	88	78	7.54	1.21
P4	7.34	5.5	621.3	300	18	244	78	156	60	58.01	1.65
P5	7.56	13	832.2	400	12	252	109	232	102	84.55	1.28
P6	8.25	16.1	505	200	14	170	68	134	72	46.09	1.46
P7	8.29	4.3	599	300	8	198	64	186	112	52.33	1.26
April 2019											
P1	7.18	2.8	202.9	100	6	86	20.85	70	50	18.58	0.18
P2	7.14	4.2	486	200	10	112	22.83	94	76	25.66	0.35
P3	7.42	6.3	238.5	100	8	94	24.82	84	66	13.42	0.23
P4	7.18	4.5	617	300	12	116	36.73	90	84	55.84	0.18
P5	7.27	1.4	952.8	500	10	202	193.61	182	136	87.48	0.09
P6	7.33	4.3	836.2	400	12	196	64.53	130	120	47.89	0.41
P7	7.78	5.1	807.5	400	14	232	79.43	154	98	62.38	0.24
May 2019											
P1	7.24	2.4	215.3	100	4	78	15.88	60	32	21.72	0.1
P2	7.09	4.7	472.5	200	6	90	21.84	68	58	23.72	0.17
P3	7.55	2.9	236.8	100	8	86	25.81	52	46	23.91	0.23
P4	7.36	5.1	294.1	100	8	102	61.55	96	80	53.77	0.43

(continued)

Table 2 (continued)

Location	pH	Turbidity (NTU)	EC (µS/cm)	TDS (mg/L)	Acidity (mg/L)	Alkalinity (mg/L)	Chlorides (mg/L)	Total hardness (mg/L)	Calcium (mg/L)	Sodium (mg/L)	Potassium (mg/L)
P5	7.85	1	798.4	400	10	178	113.18	164	118	89.77	0.61
P6	7.37	3.8	854.7	400	12	182	57.58	156	106	47.45	0.18
P7	7.91	2.1	958.2	500	10	204	83.4	192	104	80.44	0.33

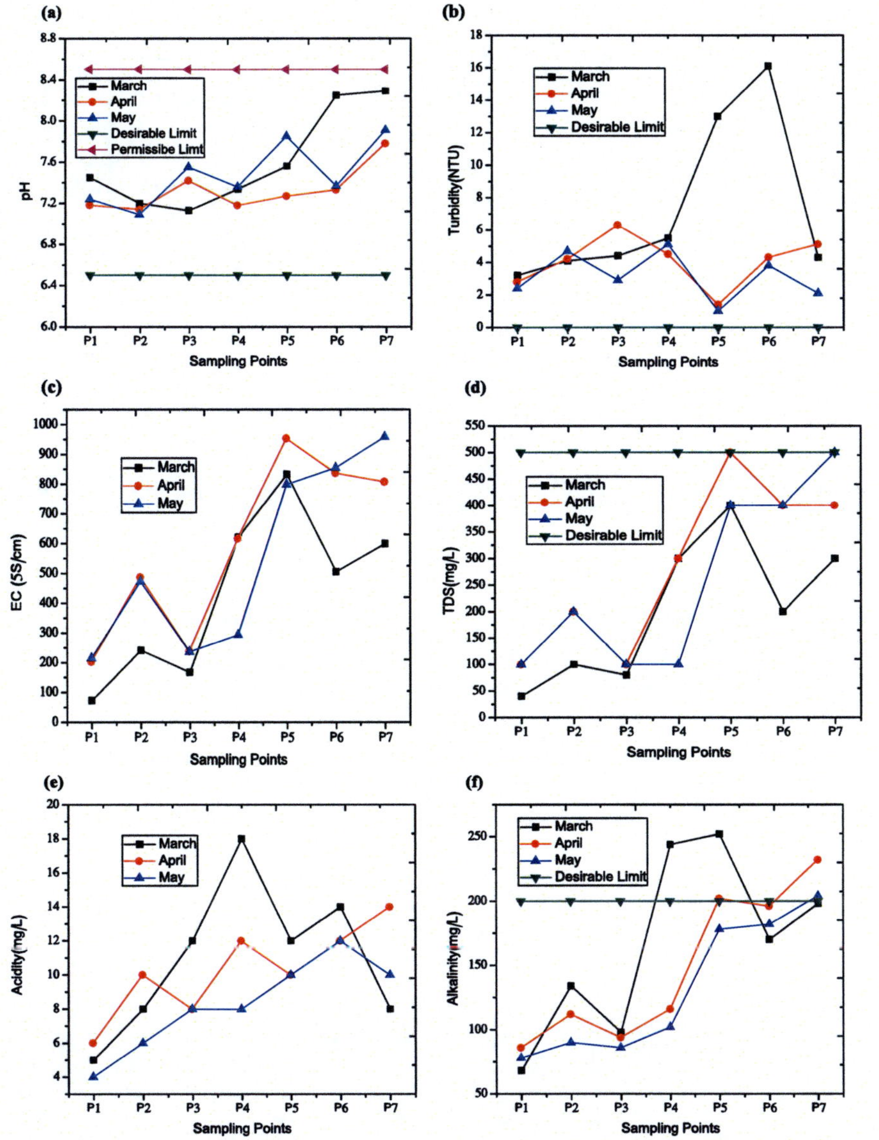

Fig. 3 Variation of physico-chemical parameters at sampling locations. **a** pH, **b** turbidity, **c** electrical conductivity (EC), **d** total dissolved solids (TDS), **e** acidity, **f** alkalinity, **g** chlorides, **h** total hardness, **i** calcium, **j** sodium, **k** potassium

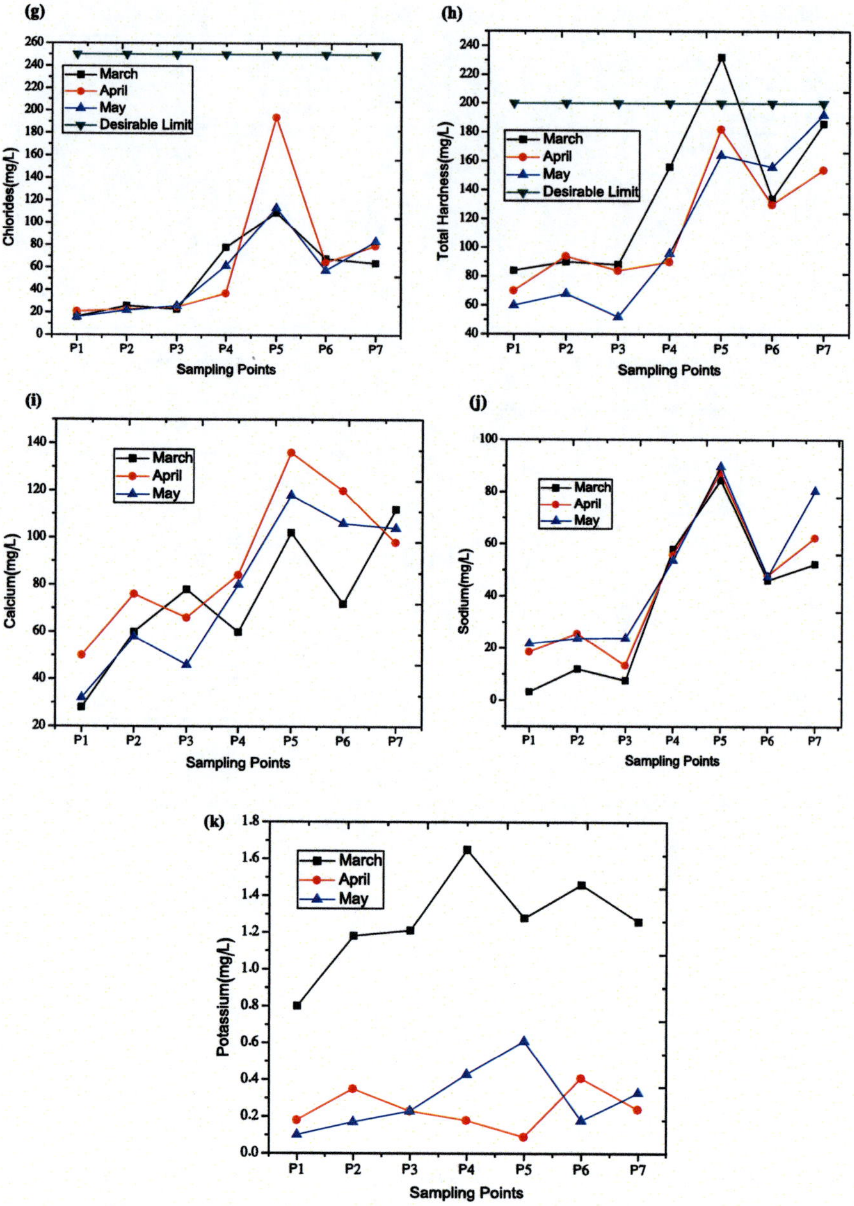

Fig. 3 (continued)

maximum pH is recorded at P7 (8.29) on 1st March. pH is slightly increased at end points due to the mixing of untreated sewage discharge from Ichalkaranji Nallah.

4.3 Turbidity

However, turbidity observed in the river waster varies from 1.0 to 16.1 NTU. The values of turbidity obtained are given in Table 2. Figure 3b represents the variation of turbidity along the Panchganga River. The turbidity is maximum at P6 (16.1 NTU) in March and sharp decline in turbidity is obtained at P5 (1.0 NTU) in May. This is attributed to the sudden deposition of relatively coarser particles derived from the upstream area and further finer sediments are carried away and remain in suspension.

4.4 Electrical Conductivity

Conductivity of water is its ability to conduct an electric current. In the present study, the Electrical Conductivity of waster varies from 72.5 to 958.2 µS/cm. Maximum was recorded at P7 in the month of May 2019. As the sampling was done in summer season, significantly high EC was recorded in the downstream. The values of EC are as given in Table 2. The variation of EC along the course of the river from Kolhapur to Ichalkaranji is shown in Fig. 3c. From the figure, it is evident that there is a considerable increase in conductivity from upstream to downstream of the study area because of discharge of domestic waste and agricultural runoff. Maximum value was obtained at P5.

4.5 Total Dissolved Solids

It is the amount ionized and non-ionized matter in water. The desirable limit of total dissolved solids in drinking water as per BIS (IS 10500-2012) is 500 mg/L. The variation of TDS along the river is shown in Fig. 3d. Evaporation and lower flow volume might be reason for higher TDS at downstream. But, the values are within the desirable limits. The recorded TDS value is low at upstream of Kolhapur (P1) in March and high at P5 and P7 in April and May, respectively.

4.6 Acidity

Acidity is caused by the presence of organic and inorganic acids and salts. Carbon dioxide which is dissolved is usually the main reason for acidity in unpolluted surface waters. The acidity of sample varies from 5 to 18 mg/L. Maximum acidity was recorded at P4 in the month of March 2019. The values of acidity are given in Table 2.

4.7 Alkalinity

The BIS (IS 10500:2012) specified desirable limit of alkalinity in drinking water is 200 mg/L (as $CaCO_3$). In the present study area, there is a wide variation in alkalinity. It varies between 68 and 252 mg/L. Table 2 shows the alkalinity variation with respect to time and space. The variation of alkalinity along the course of the river from Kolhapur to Ichalkaranji is shown in Fig. 3g. From the figure, it is evident that there is a significant increase in alkalinity at P5 is 252 mg/L in March is may be due to weathering of carbonate minerals in rocks and soil.

4.8 Chloride

Chloride irons are essential for life. However, above a concentration of 250 mg/L chloride, the water may taste salty. The BIS (IS 10500:2012) standard of chloride is 250 mg/L. The concentration of the chloride showed variation both spatially and temporally. This may be an indication of the transported soil and solute particles which reaches the river water as non-point sources. The graphical representation of chloride distribution is shown in Fig. 3h. This shows that the chloride content is within desirable limit. The higher concentrations at P5, P6 and P7 may indicate pollution by sewage and industrial waste.

4.9 Total Hardness

The concentrations of total hardness obtained during the study period are given in Table 2. The variation of total hardness in the study area is shown in Fig. 3i. It is quite evident that there is a significant increase in the total hardness in the downstream area which could be due to the nature and composition of soil particles suspended in the water. Total hardness is low at P1 (60 mg/L) in May and maximum at P5 (232 mg/L) in March.

4.10 *Calcium*

The concentration of the calcium showed variation both spatially and temporally. This is caused due to the deposition of minerals. The variation of calcium along the course of the river from Kolhapur to Ichalkaranji is shown in Fig. 3j. From the figure, it is evident that from sampling point P4 to P7 values exceeds desirable limit. This may be due to deposition of minerals due to industrial, residential and agricultural discharges.

4.11 *Sodium*

Sodium is always present in natural waters. It is also an essential dietary requirement in very small quantity. The sodium concentration along the river is shown in Table 2. It ranges from 3.19 to 89.77 mg/L. These sudden increase and decline of sodium content are referred to hydrological changes resulted from rainfall and discharge characteristics. The variation of sodium along the course of the river is shown in Fig. 3j. The peak of sodium concentration observed at P5 in March, April and May months could be due to sewage and runoff characteristic which indirectly influences the water quality parameters.

4.12 *Potassium*

Potassium is present in geological formations. In spite of high concentration of sodium in the river water, it is recorded that the potassium content is found to be very low in all months. This kind of character of ions indicates the source of sodium as natural rocks and soil type whereas potassium is contributed mainly through non-point sources, particularly from agricultural lands. High concentration is observed in March. The lowest potassium content is recorded at P5 in April and highest content is recorded at P4 in March. Leaching of artificial fertilizer is maybe the reason higher concentrations (Fig. 3k) of potassium.

4.13 *DO Sag Curve*

The DO sag curves were developed for Panchganga River where the industrial wastes discharges into the mainstream. There was discharge of 2 MLD of wastewater from MIDC Shiroli, Kolhapur (P3). The observed BOD discharge is of wastewater found to be 900, 932 and 998 mg/L for the months March, April and May 2019, respectively. And from Ichalkaranji Nallah (P5) 1.5 MLD of wastewater is discharged into the river and BOD load of 810, 865 and 912 mg/L for the months March, April and May 2019, respectively. The observed values are abnormally high. Once they joined the mainstream, the DO showed significant fall which is shown below (Fig. 4).

The DO concentrations 500 m upstream from the point of discharge of Kolhapur MIDC effluent were 6.89, 6.02 and 6.76 mg/L. At the point of discharge of waste, the DO concentrations were 6.15, 4.98 and 5.62 mg/L months March, April and May, respectively. The critical DO concentrations of 4.14, 2.66 and 3.03 mg/L were observed at critical distances 4475, 3773 and 3924 m downstream from the discharge point, respectively.

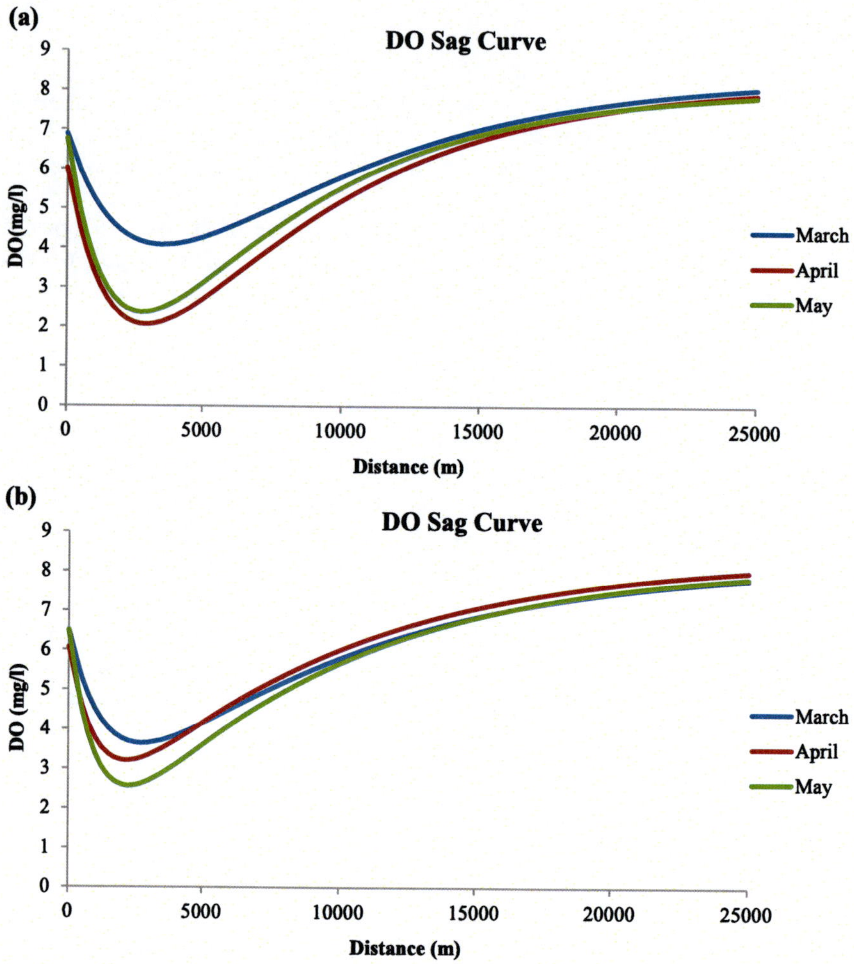

Fig. 4 DO sag curve for **a** MIDC Shiroli, Kolhapur, **b** Ichalkaranji Nallah

Table 3 Desirable limit as per IS10500-2012 [11]

Parameters	IS10500-2012 (desirable limit)
pH	6.5–8.5
Turbidity (NTU)	5
TDS (mg L^{-1})	500
Chlorides (mg L^{-1}) as Cl	250
Total hardness (mg L^{-1}) as $CaCO_3$	300
Calcium (mg L^{-1})	75
DO (mg L^{-1})	Should not be less than 4
Potassium (mg L^{-1})	

The DO concentrations 500 m upstream from the point of discharge of Ichalkaranji Nallah were 6.50, 6.07 and 6.57 mg/L. At the point of discharge of waste, the DO concentrations were 5.72, 5.18 and 5.38 mg/L months March, April and May 2019, respectively. The DO concentrations decreased to critical value of 3.94, 3.68 and 3.23 mg/L at 4019, 3122 and 3396 m which is the critical distance downstream from the discharge point. The DO concentration increases as flow takes along the distance.

4.14 Water Quality Index (WQI)

Water Quality Index (WQI) is going to decide the suitability of the water for different purpose. It is very important to specify the purpose before the analysis. In the present study, WQI is calculated based on the 8 water quality parameters namely pH, turbidity, total dissolved solids (TDS), chlorides, total hardness, calcium, DO, potassium. The weights are assigned based on the important parameter for the particular use. The desirable limit for the drinking purpose is shown in Table 3. The Bargavas [10] method is used for the calculation of water quality index as it is very simple and more easy toward analyzing water quality for various purpose. The water quality criteria for drinking purpose are shown in Table 4 and the calculated water quality Index is shown in Table 5. It is observed that water quality at P4, P5 and P6 are not acceptable with the some parameters. This may be due to the industrial discharge

Table 4 Water quality criteria based on Bhargava water quality index

Criteria	Range of values
91–100	Excellent
71–90	Good
51–70	Fair
41–50	Marginal
<40	Poor

Table 5 Water quality index

Month	Sampling points						
	P1	P2	P3	P4	P5	P6	P7
March	84	83	68	56	45	61	63
April	86	81	65	58	46	63	58
May	76	78	70	50	41	56	67

points and also during the may month the water quality is not acceptable, due to the heavy rainfall before the sampling.

5 Conclusions

This study investigated the physico-chemical and dissolved oxygen sag of Panchganga River which is polluted by various sources. The discharge of industries, agricultural chemicals and domestic sewage are the main sources of pollution.

From the study, the following conclusions were drawn:

- The water quality parameters have been observed during study period of March, April and May 2019. Turbidity, calcium and alkalinity showed very higher values at P4, P5, P6 and P7, which are more than desirable limit in all three months at downstream. Whereas total hardness exceeded desirable limit only at P5 in the month of March.
- It is noticed that there is a significant change in water characteristics as the river flows from Kolhapur to Ichalkaranji. Water pollution increases in Ichalkaranji area due to sewage and textile mill waste disposal in river without treatment.
- The DO concentrations of Panchganga River in upstream area of Kolhapur (P3) were 6.89, 6.02 and 6.76 mg/L and in upstream of Ichalkaranji (P5) were 6.50, 6.07 and 6.57 mg/L for months March, April and May 2019, respectively. These concentrations were fit for survival of aquatic life.
- The critical DO concentrations for MIDC Shiroli, Kolhapur were 4.14, 2.66 and 3.03 mg/L at critical distances of 4475, 3773 and 3924 m and for Ichalkaranji Nallah critical DO were 3.94, 3.68 and 3.23 mg/L at 4019, 3122 and 3396 m, respectively.
- Finally, the WQI confirms the zone of pollution and are P4, P5, P6, respectively.

References

1. Thorvat AR, Sonaje NP, Mujumdar MM, Swami VA (2012) A study on the physico-chemical characteristics of Panchaganga River in Kolhapur City, Maharashtra, India. Res J Chem Sci

2:76–79
2. Meenakshi P (2012) Elements of environmental science and engineering, 2nd edn. PHI Learning Private Limited, New Delhi, pp 122–132
3. Jadhav SD, Jadhav MS, Jawale RW (2013) Physico-chemical and bacteriological analysis of Indrayani River water at Alandi, Pune District (Maharashtra) India. Int J Sci Eng Res 4(11)
4. Horton RK (1965) An index number system for rating water quality. J Water Pollut Control Fed 37(3):300–306
5. Vasant W, Dipak P, Aniket M, Ranjitsinh P, Shrikant M, Nitin D et al (2016) GIS and statistical approach to assess the groundwater quality of Nanded Tehsil (MS) India. In: Proceedings of first international conference on information and communication technology for intelligent systems, vol 1. Springer, Cham, pp 409–417
6. Selvam S, Manimaran G, Sivasubramanian P, Balasubramanian N, Seshunarayana T (2014) GIS-based evaluation of water quality index of groundwater resources around Tuticorin coastal city, South India. Environ Earth Sci 71(6):2847–2867
7. Wagh VM, Panaskar DB, Muley AA, Mukate SV (2017) Groundwater suitability evaluation by CCME WQI model for Kadava river basin, Nashik, Maharashtra, India. Model Earth Syst Environ 3(2):557–565
8. Gaikwad SS (2014) Assessment of heavy metal pollution of Panchganga river with reference to diversity of molluscan fauna
9. Dunnivant FM (2004) Environmental laboratory exercises for instrumental analysis and environmental chemistry. Wiley, Hoboken
10. Bhargava DS (1983) Use of a water quality index for river classification and zoning of the Ganga river. Environ Pollut B6:51–67
11. BIS (2012) 10500: 2012. Indian standard drinking water-specification (second revision). Bureau of Indian Standards, New Delhi

A Review on Efficiency of Polypropylene Fiber-Reinforced Concrete

Aishwarya Lakshmi, Poornachandra Pandit, Yamuna Bhagwat, and Gopinatha Nayak

Abstract Sustainability of concrete is affected by cracks which are proved to be detrimental for both mechanical and durability properties as they tend to propagate under the influence of loads and result in entry of aggressive agents from surrounding environment. Polypropylene fiber in concrete acts as crack arrester along with altering the fresh and hardened properties due to improper packing and dispersion issues which has adverse effect on concrete. The effects of polypropylene fiber concrete are reviewed in the following section based on study of workability, compressive strength, and cracks. The issues are encountered to certain extent by combined usage of fiber and pozzolanic materials. This combination contributes to strength and durability along with effective utilization of industrial waste products.

Keywords Polypropylene fiber · Concrete · Pozzolanic material

1 Introduction

Construction process has experienced modification in materials, methods, and architecture from ancient times till date and it is a continuous process as new needs demand modifications in the process. Conventional concrete is combination of binder, water, and aggregates in their coarse and fine form. This combination in fresh state is of flowable consistency and as time passes gets hardened resulting in a brittle material. Concrete is a better performer on encountering compressive forces but a poor performer when tensile forces are considered. To improve tensile strength, concrete is reinforced with steel and is called as reinforced concrete. Fibers are added to concrete mix, so that when cracks start to propagate these fibers act as crack arresters. The brittle nature of concrete is compensated by adding fibers and called as fiber-reinforced concrete (FRC). For this purpose, natural, metallic, and synthetic fibers are used which are available commercially. In olden days, sun-dried

---

A. Lakshmi (✉) · P. Pandit · Y. Bhagwat · G. Nayak
Department of Civil Engineering, MIT, MAHE, Manipal, Karnataka, India

G. Nayak
e-mail: gopinathanayak@manipal.edu

L. Nandagiri et al. (eds.), *Sustainability Trends and Challenges in Civil Engineering*, Lecture Notes in Civil Engineering 162, https://doi.org/10.1007/978-981-16-2826-9_50

bricks were reinforced with straw (pueblo houses) which depicts the concept of fibers usage in construction [1]. Later in 1900 asbestos fiber usage was started in concrete. The evolution of fiber types used in construction has recorded the use of asbestos fibers. In 1950s, fibrous concrete made buzz. Later in 1960s, manmade fibers (glass, steel, basalt, textile and synthetic fibers) gained importance as asbestos fibers had cons regarding health [1]. Till date the research work on implementation of fibers in concrete is continued which demonstrated the fibers work as crack arrester, shrinkage reducer (both plastic and drying), permeability reducer, improver of strength along with improving the quality of concrete to bear impact loads, freeze thaw actions at varying temperature. Fibers are added to concrete mix, so that when cracks start to propagate these fibers act as crack arresters. The brittle nature of concrete is compensated by adding fibers and called as fiber-reinforced concrete where the matrix and fiber will be bonded in the form of physical, chemical, friction, or mechanical anchorage. For this natural, metallic and synthetic fibers are used which are available commercially. Presence of fibers in concrete will result in concrete failing in ductile manner under the loads [2].

FRC is gaining importance in construction industry but there is reluctancy in using it due to lack of application-based study of fibers on different types of structures. The pros and cons of using different fibers in concrete should be validated through studies.

The fiber usage in concrete depends on its strength and elastic modulus along with its capacity to bond with materials of the concrete. From the studies, it is clear that there exists mechanical bonding between fibers and concrete composites. Fiber-reinforced concrete is gaining its importance in construction [3] but its utilization is not done effectively as people in construction field are not readily accepting the changes based on truths revealed through research works carried out, the reason may be preset mind or economic conditions. Type of fibers, orientation, geometry (aspect ratio), density, and distribution affect the concrete property. Following paper specifically deals with usage of polypropylene fiberss (PPF) in concrete as secondary reinforcement in the form of discrete fibers. Material added to concrete should improve its behavior either in fresh or hardened state without hindering mechanical or durability properties. Still more synthetic fibers like acrylic, aramid, carbon, nylon, polyester, and polyethylene are used in concrete.

2 Polypropylene Fiber

2.1 Properties

These are thermoplastic fibers of synthetic formulation that are lightweight as shown in Fig. 1. Available in monofilament and fibrillated form, which has fine diameter with lower density, modulus of elasticity and do not absorb water (hydrophobic) [1]. They do not participate in chemical reaction with matrix (Table 1). They do not

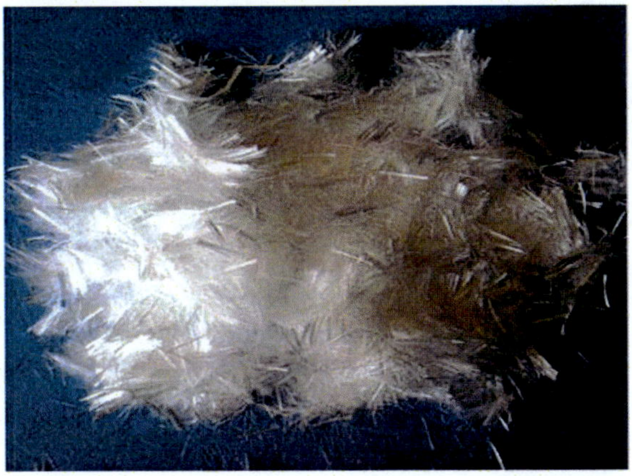

Fig. 1 Polypropylene fibers

Table 1 Properties of polypropylene fiber

Fiber	Properties			
	Specific gravity	Density (kg/m^3)	Melting point (°C)	Resistance to acid/alkali
Polypropylene	0.91	910	165	High

involve in chemical bonding with concrete composite instead they are mechanically held by binder, i.e., there exists only mechanical interaction, usually used in concrete to improve resistance to shrinkage cracks, resistance to spall, improve impact strength and energy absorption. A very small dosage is efficient to resist cracks induced due to plastic shrinkage.

2.2 *Advantages*

The following properties of polypropylene makes them suitable for use in concrete.

- High resistance to alkali/acid.
- Does not absorb water.
- Post-cracking behavior of concrete is improved.
- They induce shrinkage restraining effect.
- Small dosages do not affect strength of concrete.
- Economically feasible.

2.3 Disadvantages

These factors have negative impact on utilization of PP fiber in concrete.

- Low modulus of elasticity (1–8 GPa) [2].
- Lower tensile strength (140–690 MPa) [2].
- Susceptibility to deterioration in the presence of sunlight and oxygen.
- Evaporates at higher-temperature living pores in concrete, i.e., heat resistance is low.
- Acts as void connector in the case of improper packing within concrete.

3 Polypropylene Fiber Concrete

The behavior of concrete composite with fibers is influenced by distribution and orientation of fiber, interaction between fiber and cementitious matrix, and dense nature of it [2]. Fresh and hardened properties are influenced by the addition of fiber. Hence, it is necessary to observe the behavior of FRC. The main criteria for using PPF in concrete are its non-absorbent nature and care should be taken toward uniform dispersion of it in the mix which can be achieved by proper dry mixing of ingredients. Water absorption of concrete is a major matter of concern any modification in concrete should not severely alter the properties of concrete. In one of the results, it was seen that these fibers did not have major effect on water absorption [4]. Short fibers tend to limit the spread of cracks [5]. The compatibility and durability of PPF in Ordinary Portland cement (OPC) is found to be satisfactory and the same with pozzolanic materials in combination with OPC [5]. Fiber dosage in the range of 0.1–0.3 is called low volume and 0.4–0.8 is called high volume dosage [1]. It is evident from the research works that PPF in concrete has the calcium silicate hydrate (CSH) gel enclosing the fibers and providing bridging effect due to mechanical bonding and has the tendency to lower the permeability and capacity to block the pores [6], which becomes the major criteria for slowing down the depassivation of concrete surrounding the steel surface. PPF are resistant to alkalis due to which they are not degraded by the matrix.

3.1 Workability of Concrete with Polypropylene Fibers

Being a basic property it is observed that workability is reduced on addition of these fibers.

Ramezanianpour et al. [6] evaluated that the physical, durability, and mechanical behavior of sleepers produced with PPF was studied (dosage—0.5, 0.7, 0.9, 1.5, 2 and 4 kg/m^3). Due to the lower slump value implemented for the sleepers, the slump test showed further decrease in workability with increase in the fiber dosage.

Karimipour et al. [7] evaluated the effect of 0.1% and 0.3% PPF inclusion on workability of self-compacting concrete by conducting flow table test. Results showed that with increase in fiber content the ease with which the concrete can be handled decreased. Choosing optimal fiber dosage is necessary in order to get workable concrete.

Chen and Liu [8] determined the contribution of three types of fiber (carbon, steel, and polypropylene) addition in lightweight concrete. The workability of PPF decreased compared to other two fibers. There was up to 20.8% reduction and this was considered as the result of holding up effect of fibers, i.e., when the fibers are distributed throughout the mix and more paste is held up by fibers. There was decreased sedimentation of aggregates and bleeding simultaneously increasing the viscosity with slump of 240 mm for no fiber concrete and 190 mm for concrete with 1% PPF.

Shen et al. [9] studied the effect of length of polypropylene fibers in resisting the crack development in high-performance concrete (HPC). The lengths of fibers chosen for study were 60 mm, 54 mm, and 42 mm with dosage of 8 kg/m^3. There was slight variation in the workability for the concrete with three different sizes of fiber that ranged as 142 mm, 131 mm, and 120 mm, respectively, which shows decrease in slump with decrease in fiber size.

Alwesabi et al. [10] studied the behavior of fibrous concrete containing polypropylene fiber and mentioned that for 1% fiber dosage the workability decreased. This proves that these fibers tend to increase viscosity of the mix. Greater the number of fibers in the mix more is the contact surface which increases the viscosity and hence lowers the workability.

Yap et al. [11] investigated concrete with PPF (0.1%), steel fiber (0.9%), and the hybrid combination on mechanical and fresh properties of concrete with w/c 0.30. Higher the dosage of PPF, i.e., >0.25% poorer the dispersion in concrete which effects the workability negatively. On considering PPF only a notable range of 15–30 mm, decrease in workability is observed. large surface area of fibers requires more paste to interact which makes concrete more viscous thus decreasing the workability.

Karahan and Atiş [12] work intended to study the durability features of concrete (freeze thaw resistance and drying shrinkage) mass implemented with combination of fly ash and PPF. 15% and 30% of cement were replaced with fly ash (w/c = 0.35) and PPF were used in following volume fraction 0.05%, 0.1%, and 0.2%. The density showed decreasing trend with increase of fiber content which is the point to be noted. Ease of handling the concrete decreased making the mix harsh with the addition of fibers but fly ash showed a positive impact on the workability.

Yap et al. [13] determined the effect of polypropylene and nylon fibers with dosage of 0.25%, 0.50%, 0.75% in oil palm shell concrete. On study of workability following observation is made, i.e., smaller the length of fiber greater is the effective surface area available for interaction between surrounding material. The bond efficiency between fiber and cement paste will be higher which increases the viscosity of mix, i.e., smaller the fiber lower is the workability. 10–15 mm reduction of slump was observed.

Vairagade and Kene [14] used PPF of length (15, 20, 24 mm) that were used with dosage of 0.4% in M20 concrete. No admixtures were used for this study and influence of fibers on concrete workability was explained as having inverse relationship with length of fibers that is with increase in length of fibers the workability reduced.

Mohebi et al. [15] study showed the similar trend where workability reduced with increase in fiber dosage 0.15%, 0.30%, and 0.45%, and it is considered that there exists an inverse relation between fiber dosage and workability.

3.2 Compressive Strength

Concrete effectively encounters compressive forces. Any additions to the mix should not alter its positive contribution toward bearing compressive stresses. Addition of fibers reduces the brittleness of concrete meanwhile number of studies are conducted to ensure the effect of fibers on compressive strength of concrete due to type and dosage of fibers.

Ramezanianpour et al. [6] work showed compressive strength reduction due to incompatibility between the fibers and paste phase. Meanwhile, the tensile and flexure capacity showed an increase for 0.7 kg/m^3 of PPF which decreased as time passed. This phenomenon is the result of pores in the concrete due to fibers. Scanning electron microscope (SEM) image in Fig. 2 gave a clear picture of bridging capacity of the fibers in microstructure of 0.7 kg/m^3 of PP dosage where gel entangles the fibers and pores are more compared to control sample where the pore was more compact.

Karimipour et al. [7] work results showed that when PPF concrete is subjected to elevated temperature, i.e., in the case of fire emergency as the fibers are thermoplastic, they melted at 160 °C forming cavities post-fire exposure which acted as relief channels of pressure. At 250 °C, the reduction of strength was lower for 0.1% dosage compared to 0.3% dosage of PPF. It is noteworthy that at room temperature for former dosage concrete showed 10% increase in strength and later did not have significant effect on strength.

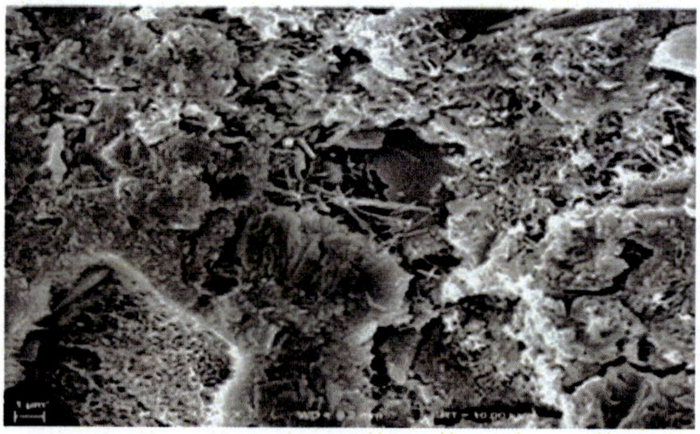

Fig. 2 Microstructure of 0.7 kg/m^3 of polypropylene fiber matrix [6]

Chen and Liu [8] study results showed that 1% fiber in concrete lowers the compressive strength but this drop was smaller compared to [4] here the compressive strength was 45 MPa which was just 0.05 MPa lower than that of control specimen. Lack of uniform dispersion of fibers due to large dosage was considered as the reason for drop in strength.

Alwesabi et al. [10] from the obtained results it is observed that for smaller dosage of (0.1%) PPF the aggregate–paste bonding improved, but compressive strength of 1% PPF 24.7 MPa and that of control specimen was 38.5 MPa. It is noted that for PPF dosage below 0.25% in combination with steel fiber improved this result.

Yap et al. [11] study depicted that irrespective of type of concrete and hybrid fiber combination, dosage of PPF beyond 0.25% negatively impacts the strength especially for single fiber concrete with 1% PPF the strength reduction was 7.4 MPa compared to control mix with strength of 36.5 MPa. The lower stiffness of PPF results in lesser impact on compressive strength but higher dosage causes dispersion problems resulting in lowered compressive strength.

Karahan and Atiş [12] study revealed that compressive strength decreased with increased dosage of both fly ash (FA) and PPF. The lowest value recorded was 42.4 MPa among all the mixes. This is attributed to the weak interaction between matrix and fibers. The highest value was 64.9 MPa for mix with 0% FA and 0.05% PP fibers.

Yuan and Jia [16] investigated the combination of glass and PPF in concrete of which PPF dosage was 0.45%, 0.90% and 1.35% of total volume. Control specimen was overall better performer compared to fiber concrete. w/c and fiber dosage both have combined effect on compressive strength. When fibrous concrete of two different water cement ratio was compared, 0.45% PPF concrete had higher compressive strength for 0.30 w/c and 0.90% PPF concrete had higher compressive strength for 0.35 w/c. Figure 3 shows with increase in dosage from 0.45 to 0.90% compressive strength drop is observed and for further increase 1.5 MPa increase is observed.

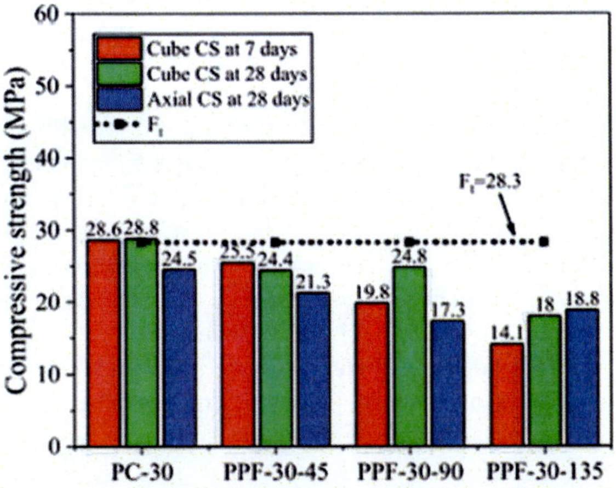

Fig. 3 Compressive strength of polypropylene-reinforced concrete [16]

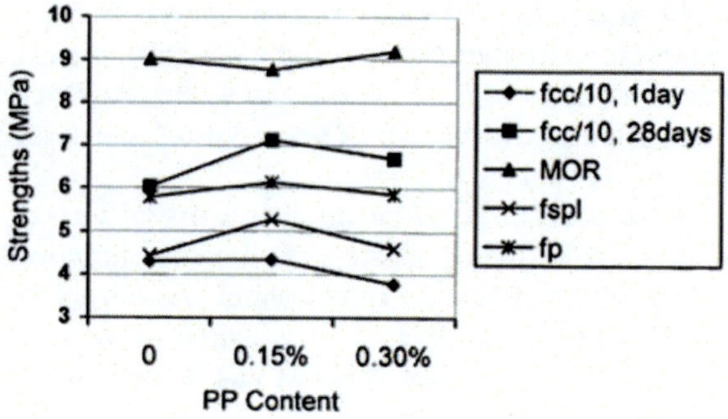

Fig. 4 Influence of polypropylene content on mechanical properties [17]

Compared to other dosages, 0.45% PPF specimen is better specimen but its strength is 3.2 MPa lower than that of control specimen.

Qian and Stroeven [17] developed hybrid fiber concrete with steel and polypropylene fibers combination. Figure 4 shows the mechanical properties like tensile strength (fspl), and compressive strength at 1 and 28 days took a rise at 0.15% PPF dosage and dropped at 0.30% and this condition was the other way for modulus of rupture (MOR); this shows with increase in fiber volume flexural strength is improved perfect packing of particles and fibers is achieved at smaller dosage of PPF which enhances the concrete properties hence 0.15% was considered as optimum dosage.

Vairagade and Kene [14] study showed a different trend, with increase in the length of both the fibers the strength factor also increased. PPF in concrete showed an increase of 11.32% in compressive strength when compared with control specimen.

3.3 *Cracks in Concrete*

Shen et al. [9] study specifically focused on the effect of length of PP fibers (42 mm, 54 mm, 60 mm) in resisting the crack development in high-performance concrete (HPC). Based on integrated criterion, crack resistance increased with respect to length of fiber as shown in Fig. 5.

Yap et al. [13] arrived at the conclusion that addition of PPF due to their flexible and ductile nature offers improved resistance to growth of cracks, crack bridging effect, and ductility to concrete best dosage of polypropylene for better crack resistance is 0.1%. Based on observations made through first crack deflection graph, it is observed that single fiber (1% PPF) showed poor performance compared to smaller dosage by showing small deflection at the time of first crack formation (0.6–0.8 mm) compared to control specimen with 1 mm deflection.

Yuan and Jia [16] study revealed that under optimum condition of packing the fibers tend to bridge the gaps which makes these fibers the crack arresters and excess fibers decreased the strength as the densification was incomplete.

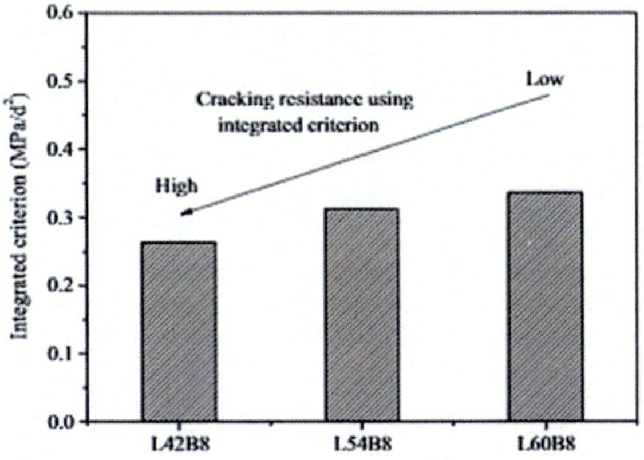

Fig. 5 Relation between cracking resistance and length of fiber [9]

Nosheen et al. [18] study revealed the effect of fiber-reinforced pretensioned concrete made with the dosage of 0.4% and 0.6% polypropylene fibers. The ductility index (displacement at mid-span/first yield displacement) of control specimen and 0.4% polypropylene was similar with increased dosage ductility experienced a drop. Crack width increased with increase in dosage of PPF. For former and latter dosage was 7.5 mm and 9.5 mm.

Zhang et al. [19] studied properties of concrete under fracture with 0.04%, 0.06%, 0.08%, and 0.1% PPF. Being lower elastic modulus flexible fibers these form network in the matrix which on contact with crack stresses induced due to thermal or drying shrinkage will encounter by the bridging effect.

Banthia and Gupta [20] investigated the effect of PPF for dosage ranging from 0.1%, 0.2%, and 0.3% on plastic shrinkage of concrete. It is observed that crack width was 1.32, 1.04, 0.89 mm, respectively, which shows with increase of dosage of fiber the crack width decreased this implies increased dosage of fibers results in crack width control.

Sadrinejad et al. [21] attempted to evaluate the effect of three types of fibers—PP, polyolefin, and steel fibers. All three fibers were used individually and two fibers in combination with PPF. The total fraction of fiber volume was limited to 1%. Meanwhile, the PPF usage was limited to 0.1% both individually and in combination. The study made an effort to emphasize on the influence of combination of these fibers on beams under sustained loads and accelerated corrosion. M40 Grade FRC was used. Experimentally, it was concluded that the bridging nature of fibers controlled the crack propagation. Study depicted that PPF tend to reinforce the concrete around the bars. On the basis of crack studies, PP was found effective in lowering the cracks. The maximum crack width with 0.1% PP was in the range between 4 and 6 mm, which was comparatively lower than other fibers and combinations. The cracking pattern varied in all the beams with single type and hybrid fiber dosages which failed to explain the trends of crack control. The best combination of fibers was steel and polypropylene. This worked well in controlling of cracks.

Sun and Xu [22]'s work focused on examining the presence of PPF in concrete on mechanical and physical properties. The interfacial transition zone between aggregate and paste was studied by SEM imaging and outcomes revealed that PP fiber reduced the crystallization of calcium hydroxide a by-product of cement hydration, which do not contribute to the strength. The fibers formed network bridging the microcracks at interfacial zone of transition. 0.9 kg/m^3 was considered to be optimum, i.e., 0.1% dosage of PPF of total concrete volume.

Carlesso et al. [23] proposed a numerical model relating increase in crack opening and mechanical performance where residual stresses were represented by considering non-cracked section. Crack opening was mentioned to be independent of fiber content and type. Statically significant influence of fiber type or content was not observed on slump and compressive strength.

3.4 Interfacial Transition Zone

The reason for strength drop in PPF concrete is explained on the basis of transition zone. Concrete is three-phase material among which transition zone is responsible for all the properties in hardened concrete. Figure 6 represents the matrix of concrete without fibers.

Liu et al. [24] developed numerical model simulated based on Fick's model with regard to diffusion of chlorides by considering interfacial transition zone. Chloride diffusion coefficient was high for concrete with polypropylene fiber and is influenced by volume and length of fiber.

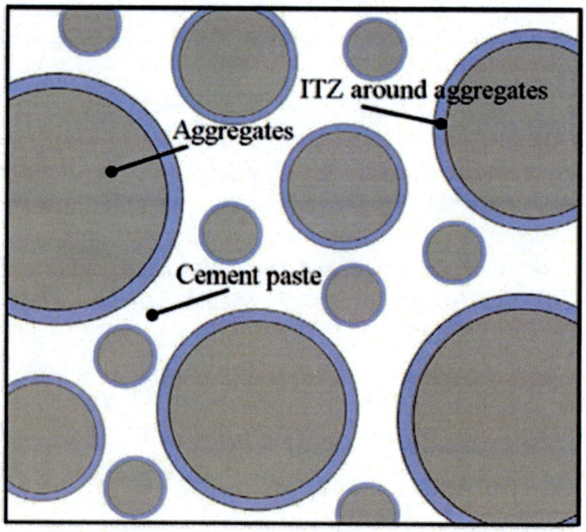

Fig. 6 Concrete matrix without fiber [24]

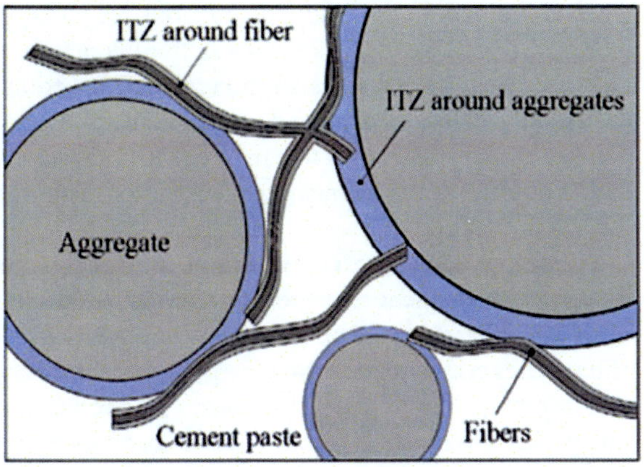

Fig. 7 Concrete matrix with fiber [24]

In FRC, the region next to fiber is also termed as transition zone. This zone decides the quality of composite. Presence of voids in this region gets connected to pores in the presence of fibers leading to interconnections which acts as connected pathway for propagation of cracks. Factors which effect these connections are volume and length of fibers used.

Since the PP are hydrophobic, thin layer of water gets accumulated around the fiber in cylindrical form represented by Fig. 7 which in later stages of hydration leads to the formation of porous layer around the fiber which is similar to that around aggregates. When fiber dosage is high, the porosity increases resulting in strength drop as observed in above results hence deciding optimum dosage to suit the requirement of concrete is vital.

4 Polypropylene Fiber with Pozzolanic Cement Concrete

It is inevitable fact that though there is enormous energy utilization in the process of cement manufacture, we cannot completely replace it with eco-friendly product. Innovations are tried by researchers to implement materials that have less impact on nature. The selection of binder depends on the necessity of the structure. It is the paste (cement + water) that holds up the aggregates by the virtue of hydration products entangling them. As the hydration proceeds, the strength of the composite increases due to improvement in gel structure. Type of concrete to be used is selected based on the requirements of the structure. OPC is commonly and widely used cement for structures whose various versions are produced to suit the requirement based on exposure conditions. Mineral-based materials are gaining their importance in construction one among them is Portland pozzolanic cement (PPC). Cement reacts with water to give CSH gel and calcium hydroxide, between these two CSH gel binds the materials together and its quality depicts the composite strength. Calcium hydroxide acts as a con factor as it does not contribute to the strength. This product when combined with pozzolanic material secondary hydration is triggered forming more CSH which enhances the strength of concrete composite.

It is evident from the observation that construction field is deliberately neglecting the use PPC due to its extended time to gain strength compared to OPC and a smaller number of researches with this regard. The total service life of PPC is found superior to that of OPC as the former performs better in aggressive environment, i.e., in long run the former serves better than latter.

Alhozaimy [25] aimed to determine the effect of polypropylene fibers ($V = 0.05$–0.3%) on mechanical strength of concrete in the presence of pozzolanic materials (25% FA, 25% slag, and 10% silica fume). ANOVA showed there was no significant effect of fibers on compressive strength and flexural strength but the impact resistance was improved due to fibers. The pozzolans reduced the impact resistance but there was positive interaction between pozzolans and fibers which ultimately contributed to impact resistance by 82% increase which implies effectiveness of fibers in the presence of pozzolans. The work was on M35 grade concrete girder reinforced with steel and fibers both used individually and in hybrid form. Durability aspects of FRC was investigated. The PP dosage was 0.40%, and 0.60% for individual mix was chosen. Water absorption was not affected by fiber reinforcement which was a topic of major concern. Compared to control mix, the PPF concrete showed lower absorption and in concrete this is attributed to the curling nature of PP fibers. On SEM analysis of PPF concrete, the fibers were attached to matrix or curled to the aggregate and steel fibers had developed rust in small amount due to exposure to open atmosphere.

Mohseni et al. [26] studied the effect of two pozzolanic materials, i.e., rice husk ash and metakaolin as partial replacement for cement and PP fiber added as additive. Compressive strength was higher than control specimen for a sample of 10% metakaolin + 20% rice husk ash + 0.3% PPF, with the influence of pozzolanic action the crack control capacity of fibers contributed to the strength of concrete and this result is opposite to that of other works which showed PP fibers have no influence on compressive strength. On testing of flexure strength, it was noticed that there was forty percent increase, and this is due to the crack bridging nature of the fibers. The test on water absorption was majorly influenced by combination of metakaolin + PPF as the presence of fibers created vacant spaces that led to moisture movement but the absorption was lower compared to control specimen. The quality of sample was determined by ultrasonic pulse velocity test (UPVT) and the specimen with only PPF and combination of PPF with pozzolanic material showed better performance than control specimen due to filler effect of pozzolanic material.

Mohod [27] used PPF with 0.5% increase in dosage ranging from 0 to 2% was used with two grades of concrete (M30 and M40) of PPC with two curing conditions one being exposure curing and other controlled curing. The concrete was stiff in fresh condition for fiber dosage greater than 1% which denotes workability loss in presence of higher volume of fibers. Based on three major mechanical properties, the optimum dosage of fibers in PPC concrete was studied. Flexural strength obtained was highest for lowest fraction of fibers for both concretes under two curing conditions. Tensile strength was maximum for M30 concrete with 0.5% fibers beyond which the tensile strength decreased with higher dosages. The curing condition is seen to have influence on compressive strength of nominal mix concrete without fibers after 7 days as in pozzolanic cement concrete secondary hydration is triggered after primary hydration, as a result of which control cured specimens performed better. 0.5% was considered as optimum based on mechanical strength observations.

5 Conclusion

- Low elastic modulus fibers cannot prevent crack formation on encountering of high stress.
- Increase in fiber dosage negatively impacts on concrete strength as optimum packing condition is not achieved.
- Significant loss of workability is observed with increase in fiber content.
- Dispersion of fibers in concrete has got influence on strength hence uniform distribution of fiber is expected.
- These fibers individually up to the dosage of 0.1% and in hybrid combination up to 0.25% improve crack control features in concrete.
- Dosage of lower volume (0.1%) is effective as secondary reinforcement in concrete as there is no effect on compressive strength.
- Presence of pozzolanic materials aids in dispersion of fibers.

References

1. Daniel JI et al (2002) Report on fiber reinforced concrete. ACI Committee 544 report, pp 4–42
2. Bentur A, Mindess S (2007) Fibre reinforced cementitious composites. Taylor & Francis, UK, pp 31–186
3. Caggiano A, Gambarelli S, Martinelli E, Nisticò N, Pepe M (2016) Experimental characterization of the post-cracking response in hybrid steel/polypropylene fiber-reinforced concrete. Constr Build Mater 125:1035–1043
4. Usman Rashid M (2020) Experimental investigation on durability characteristics of steel and polypropylene fiber reinforced concrete exposed to natural weathering action. Constr Build Mater 250
5. Furlan Jtinior S, de Hanai B (1999) Prestressed fiber reinforced concrete beams with reduced ratios of shear reinforcement. Cement Concr Compos 213–221
6. Ramezanianpour AA, Esmaeili M, Ghahari SA, Najafi MH (2013) Laboratory study on the effect of polypropylene fiber on durability, and physical and mechanical characteristic of concrete for application in sleepers. Constr Build Mater 44:411–418
7. Karimipour A, Ghalehnovi M, de Brito J, Attari M (2020) The effect of polypropylene fibres on the compressive strength, impact and heat resistance of self-compacting concrete. Structures 25:72–87
8. Chen B, Liu J (2005) Contribution of hybrid fibers on the properties of the high-strength lightweight concrete having good workability. Cem Concr Res 35(5):913–917
9. Shen D, Liu X, Zeng X, Zhao X, Jiang G (2020) Effect of polypropylene plastic fibers length on cracking resistance of high performance concrete at early age. Constr Build Mater 244

10. Alwesabi EAH, Bakar BHA, Alshaikh IMH, Akil HM (2020) Experimental investigation on mechanical properties of plain and rubberised concretes with steel–polypropylene hybrid fibre. Constr Build Mater 233
11. Yap SP, Bu CH, Alengaram UJ, Mo KH, Jumaat MZ (2014) Flexural toughness characteristics of steel-polypropylene hybrid fibre-reinforced oil palm shell concrete. Mater Des 57:652–659
12. Karahan O, Atiş CD (2011) The durability properties of polypropylene fiber reinforced fly ash concrete. Mater Des 32(2):1044–1049
13. Yap SP, Alengaram UJ, Jumaat MZ (2013) Enhancement of mechanical properties in polypropylene- and nylon-fibre reinforced oil palm shell concrete. Mater Des 49:1034–1041
14. Vairagade VS, Kene KS (2013) Strength of normal concrete using metallic and synthetic fibers. Procedia Eng 51:132–140
15. Mohebi ZH, Bahnamiri AB, Dehestani M (2019) Effect of polypropylene fibers on bond performance of reinforcing bars in high strength concrete. Constr Build Mater 215:401–409
16. Yuan Z, Jia Y (in press) Mechanical properties and microstructure of glass fiber and polypropylene fiber reinforced concrete: an experimental study. Constr Build Mater 266
17. Qian CX, Stroeven P (2000) Development of hybrid polypropylene-steel fibre-reinforced concrete. Cement Concr Res 63–69
18. Nosheen H, Qureshi LA, Tahir MF, Rashid MU (2018) An investigation on shear behavior of prestressed concrete beams cast by fiber reinforced concrete. Arab J Sci Eng 43(10):5605–5613
19. Zhang P, Liu CH, Li QF, Zhang TH (2013) Effect of polypropylene fiber on fracture properties of cement treated crushed rock. Compos B Eng 55:48–54
20. Banthia N, Gupta R (2006) Influence of polypropylene fiber geometry on plastic shrinkage cracking in concrete. Cem Concr Res 36(7):1263–1267
21. Sadrinejad I, Ranjbar MM, Madandoust R (2018) Influence of hybrid fibers on serviceability of RC beams under loading and steel corrosion. Constr Build Mater 184:502–514
22. Sun Z, Xu Q (2009) Microscopic, physical and mechanical analysis of polypropylene fiber reinforced concrete. Mater Sci Eng A 527(1–2):198–204
23. Carlesso DM, Cavalaro S, de la Fuente A (in press) Flexural fatigue of pre-cracked plastic fibre reinforced concrete: experimental study and numerical modeling. Cem Concr Compos 115
24. Liu J, Jia Y, Wang J (2019) Calculation of chloride ion diffusion in glass and polypropylene fiber-reinforced concrete. Constr Build Mater 215:875–885
25. Alhozaimy AM, Soroushiad P, Mirzac F (1996) Mechanical properties of reinforced concrete and materials polypropylene fiber the effects of pozzolanic
26. Mohseni E, Yazdi MA, Miyandehi BM, Zadshir M, Ranjbar MM (2017) Combined effects of metakaolin, rice husk ash, and polypropylene fiber on the engineering properties and microstructure of mortar. J Mater Civ Eng 29(7)
27. Mohod MV (2015) Performance of polypropylene fibre reinforced concrete. IOSR J Mech Civ Eng (IOSR-JMCE) 12(1):28–36

Corrosion of Reinforcing Bar in RCC Structures—A Review

Yamuna Bhagwat, Gopinatha Nayak, Aishwarya Lakshmi, and Poornachandra Pandit

Abstract Corrosion of the reinforcements in the concrete has been considered as the severe factor which reduces the concrete structure's durability. The structures present in marine environments are prone to the corrosion attacks. The damage of concrete is evidenced through expansion, cracking, and gradual concrete cover spalling. The non-uniformity of corrosion throughout the rebar and reduced cross section causes sudden failure of reinforced members without prior indication. With intense earnestness, the corrosion effect on normal steel and high-strength steel which is used in prestressed concrete is discussed here to understand the state of the art. The study included reasons for corrosion of bar and effect on bond strength, structural performance of normal and prestressed concrete. The paper concludes with pores distribution on the concrete, concrete–steel and an aggregate–paste interface contribute more for rebar corrosion and lowers the bond strength, structural performance of RCC structures.

Keywords Corrosion · Deterioration · Concrete structures · Prestressed concrete · Prestressed steel

1 Introduction

The corrosion of steel used in the reinforced concrete to enhance the strength of concrete is the considerable factor for the study to improve the serviceability of the structure. The corrosion is not only the sole reason in affecting the durability of the

Y. Bhagwat (✉) · G. Nayak · A. Lakshmi · P. Pandit
Department of Civil Engineering, MIT, MAHE, Manipal, Karnataka 576104, India
e-mail: yamuna.bhagwat1@gmail.com

G. Nayak
e-mail: gopinathanayak@manipal.edu

A. Lakshmi
e-mail: aishwarya.lakshmi@learner.manipal.edu

P. Pandit
e-mail: pc.pandit@manipal.edu

L. Nandagiri et al. (eds.), *Sustainability Trends and Challenges in Civil Engineering*,
Lecture Notes in Civil Engineering 162, https://doi.org/10.1007/978-981-16-2826-9_51

structure, but it has the significant contribution [1, 2]. Therefore, it has attracted the major concern while improving the durability aspects as well as to explore other dimensions that indirectly affects the corrosion.

The calcium silicate hydrated (C–S–H) gel and calcium hydroxide ($Ca(OH)_2$) are produced during hydration process when cement is mixed with water. C–S–H gel helps to determine the property of the concrete and $Ca(OH)_2$ being alkaline maintains pH near to 13 and helps to resists the corrosion of reinforcements [3]. Because of the pores present in the concrete, carbon dioxide and chlorides penetrate inside and reach up to the reinforcement and finally destroy the passive layer formed against the corrosion. Figure 1 schematically represents the flow of charges in the electrochemical process of corrosion. The result of this begins the deposition of an oxide layer on the reinforcement due to the potential difference created inside it. Generally, this deposition can increase the volume of reinforcement up to 4–6 times to its original volume and causes radial tension along the length of the reinforcement to the surrounding. This results in cracking and breaking the bonding between concrete and reinforcement as shown in Fig. 2.

The main factors for corrosion of reinforcement are

- existence of two metals or single metal at two locations at different energy levels;
- presence of an electrolyte (in the reinforcement corrosion concrete act as electrolyte);
- metallic connection between the two metals [4].

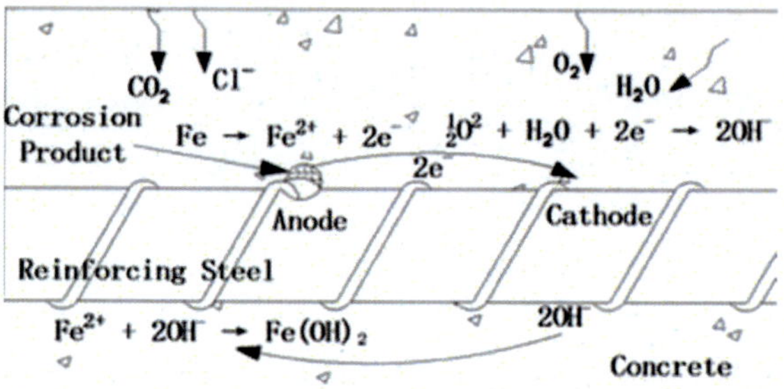

Fig. 1 Diagrammatic representation reinforcing steel corrosion in concrete [13]

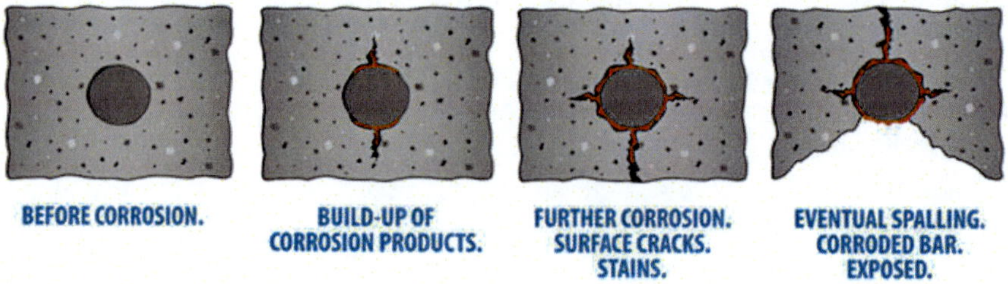

Fig. 2 Reinforcement corrosion leads to cracking and spalling [14]

Fig. 3 Stages in development of corrosion of steel [8]

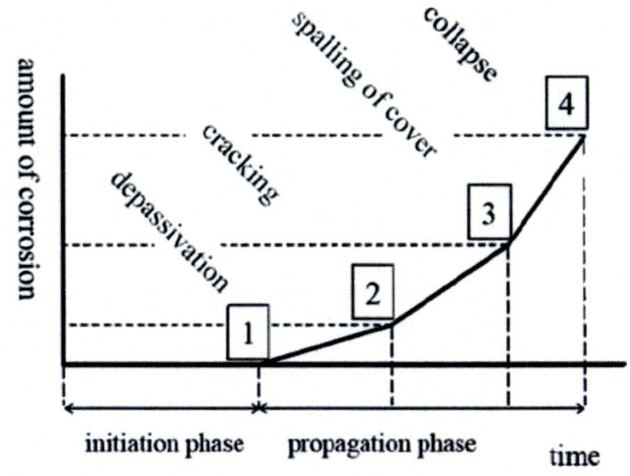

The corrosion of the steel happens mainly because of the following phenomenon—carbonation and chloride-induced corrosion. The atmospheric carbon dioxide (CO_2) ingression in the concrete causes carbonation corrosion of steel, and the chloride ions penetration through water or present in the raw material causes chloride-induced corrosion [5–7]. The chloride-induced corrosion has more wild attack than carbon-induced corrosion.

Four stages are considered in the corrosion; in the first stage, aggressive materials enter into the concrete by pores or crack present. The depassivation takes place, no color changes can be detected by formation of rust. In the second stage, the corrosion starts and depending on the moisture and oxygen access corrosion rate increases, tensile stresses develop due to products formation. In the third stage, corrosion stresses reach the cover and spall-off of cover happens because of detachment due to cracking; reinforcing bar loss is accepted in the final stage and eventually structure becomes less strengthened [8]. These all stages are called the propagation stage as shown in Fig. 3. Since the corrosion significantly affects the durability of the structure, there is a strong requirement to find out the effective solution to reduce the corrosion effect. Here, an attempt is made of collective study of the state-of-the-art papers on the corrosion of reinforcements effect.

2 Influencing Factors for Steel Corrosion in Concrete

There are several influencing factors for corrosion rate such as moisture content, temperature, pH value, permeability, carbonation and chlorides, exposure condition, thickness of cover used for reinforcement. The pH of the concrete is reduced by the ingression of the chemicals or other acidic gases such as sulfur dioxide and nitrogen dioxide. The entering of foreign substances is possible when concrete is permeable, and permeability depends on density of the concrete.

Chloride entering rate and carbonation rate depends on the permeability coefficient, and also precracking existed on the surface of the concrete member due to shrinkage, etc. [9–12].

Two types of transition zones are considered in the concrete, namely steel–concrete and aggregate–paste interface. The porous present in the aggregate paste interface has no much effect but it transports chloride ions [15]. Pores present in the concrete very adjacent to the reinforcement, i.e., steel–mortar interface, is the place where there is no immediate presence of alkalis with limited supply of oxygen [15, 16]. As a result, corrosion begins typically as localized corrosion at the interface [17]. For heedful study, defects of steel–concrete interface beneath the steel are locations which contribute less corrosion resistance [18]. Blended cements are found to be comparatively more resistant for the ingression of chloride ions [19].

Intuitively, corrosion initiation on the steel surface is reciprocal to the distance between the steel surface and the aggregate [20]. But, for randomly placed multiple aggregates with different sizes effect has to be studied thoroughly.

Corrosion rate is greatly influenced by the exposure time, temperature, and state of the environment. At low relative humidity conditions, corrosion is generally slow. Humidity conditions more than 90% with high temperature corrosion rate increases exponentially. High humid conditions form rust in a very less period and spalling of concrete cover commences within a very short period [21].

Corrosion in reinforcement also depends on the [Cl^-/OH^-] ratio, if it exceeds the threshold value then corrosion begins [22]. The threshold value or critical value of the chloride is defined as the required amount of chloride for depassivation of steel or chloride content which is associated with acceptable cracking of the reinforced concrete member [23]. Many experiments show the threshold value of [Cl^-/OH^-] between 0.3 and 3 [22]. This threshold value cannot be a constant since it depends on the many other factors like temperature, water–cement ratio, C_3A, and alkali content in cement [22, 24–26]. In America, the critical value of chloride content is taken as approximately 0.4% of weight of cement and water soluble chloride content considered as 0.1% of weight of cement [6]. The chloride threshold values as per different international standards are listed in Table 1.

3 Effects of Steel Corrosion in Concrete

3.1 Bond Strength

Other than cracking of the concrete, the tension created from the volume increases due to corrosion products, decreases the bond strength. Bond strength is the main reason for concrete to act as composite material. Using pullout test, bond strength of the concrete is studied for different levels of steel corrosion. The bond strength is responsive to corrosion level in non-confined deformed bars. Usually, bond strength in non-confined deformed bars decreases with increase in corrosion level. But, in the

Table 1 Chloride threshold value from different standards [27]

Name of standards	Details	Threshold value of chloride
British standards	For reinforced concrete structures	0.4% of cement
	Prestressed concrete structures	0.1% of cement
Australian standard	For concrete structures	0.8 kg/m^3 of concrete
Canadian standard	For RC structures	0.15–1% of cement
	Prestressed concrete structures	0.06% of cement
ACI 318-11	For reinforced concrete structures	0.15–0.3% of cement
Indian standards IS 456	Reinforced concrete	0.6 kg/m^3
	Prestressed concrete	0.4 kg/m^3
	Normal concrete	3 kg/m^3
Hong Kong standards of practice	Reinforced concrete	0.35% of concrete
	Prestressed concrete	0.1% of concrete mass
New Zealand standards	Prestressed concrete	0.5 kg/m^3 of concrete
	Reinforcement under moisture	0.8 kg/m^3 of concrete
	Reinforcement under dry	1.6 kg/m^3 of concrete

confined bars, no much effect on bond degradation was found [28]. According to the literature, bond strength improved when the corrosion level was very low, and when corrosion reached to 9%, bond strength decreased to one-third of the value of without corrosion. Smooth bars without confinement showed increase in bond strength till 2–4% increase in corrosion. But, in confined bars, up to 5% corrosion level bond strength increment has been observed [28, 29]. The corrosion loss more than 9.26% or crack formed due to corrosion effect wider than 0.67 mm decreased bond strength between concrete and corroded strands [30]. Also, the failure by bond degradation depends on the method of inducing corrosion [31]. But, when experimentally studied deterioration of members underestimated in artificially induced corrosion method when compared with natural corrosion. The presence of stirrups too affects the bond degradation, more dense stirrups decrease the bond degradation rate and surface cracking from corrosion, increase the value of pullout load [32]. Many analytical models to find out the amount of strength degradation have been developed with reinforcement corrosion considering uniform and non-uniform corrosion [33–35]. The variation of bond strength for different corrosion level from the literature is listed in Table 2.

Table 2 Bond strength variation for corrosion level from the literatures

Corrosion level	Bond strength	Literature
Up to 6.24%	Increased	Wang et al. [30]
More than 9.26%	Decreased	Wang et al. [30]
Up to 5% (rebar corrosion prior to the concrete placement)	Increased	Choi et al. [31]
Up to 1% (rebar corrosion after concrete placement)	Increased	Choi et al. [31]
Up to 5.3% (uncracked concrete)	Increased	Jiradilok et al. [37]
Up to 3% (cracked concrete)	Increased	Jiradilok et al. [37]
More than 5.3% (cracked concrete)	Decreased	Jiradilok et al. [37]
5%	Increased	Fang et al. [28]
9%	Decreased (1/3 of non-corroded specimen)	Fang et al. [28]

3.2 Structural Performance

The corrosion propagation directly affects the structural performance of reinforced concrete member. Decrease in steel c/s area greatly affects the flexural performance of the member, and concrete spalling because of splitting of the bond reduces the stability of the member. The cracks develop more intensely in a dry environment in an accelerated corrosion process than in a moist environment.

The concrete beam which was allowed to corrode in a humid environment and dry environment studied and compared for flexural performance. Since most of the corrosion products in dry concrete remained at the steel and concrete interface, it develops more pressure and leads to the development of tensile stresses and correspondingly cracks formation at this interface. Pit formation in the rebar is considered to be more in the wet-accelerated corrosion process than in the dry process. Formation of the pits on rebars reduces the load bearing capacity, for average rebar loss residual load capacity and corroded residual load capacity decreased up to 30% and 40%, respectively. For localized corrosion, corroded residual load capacity reduced up to 60% of its designed value [36].

Reduction of load bearing capacity changes the failure mode from ductile to brittle for increased corrosion of rebars [38]. Corrosion shows a very harmful effect on the mechanical behavior of reinforcement, reduced ductility observed in the corroded bars than non-corroded bars. In one of the research works, with ductile response the corroded beam failed in flexure; in non-corroded beams, brittle shear failure was observed [39].

Therefore, it is evident that decrease in c/s area of reinforcement affects the flexural capacity and also the mode of failure from shear to flexure. Presence of precrack has accelerated the performance degradation and deterioration [40] in the beam, and it increases the loss of steel bar cross section and significant effect on mechanical properties correspondingly affects the mode of failure of the beam (for each percentage loss of local c/s area of steel reduces load carrying capacity by 0.9%). The performance reduction and deterioration progress of naturally corroded reinforcement beam results have been shown in Fig. 4 [41].

The strength capacity of the bars affected by its minimum c/s area, whereas deformation capacity affected by minimum c/s area as well as distribution of c/s area throughout the bar [42]. Also, the reinforcement corrosion reduces the impact resistance and stiffness of the beam [43]. From experimental investigations, it is found that for the same level of deterioration of reinforcement due to corrosion shows different

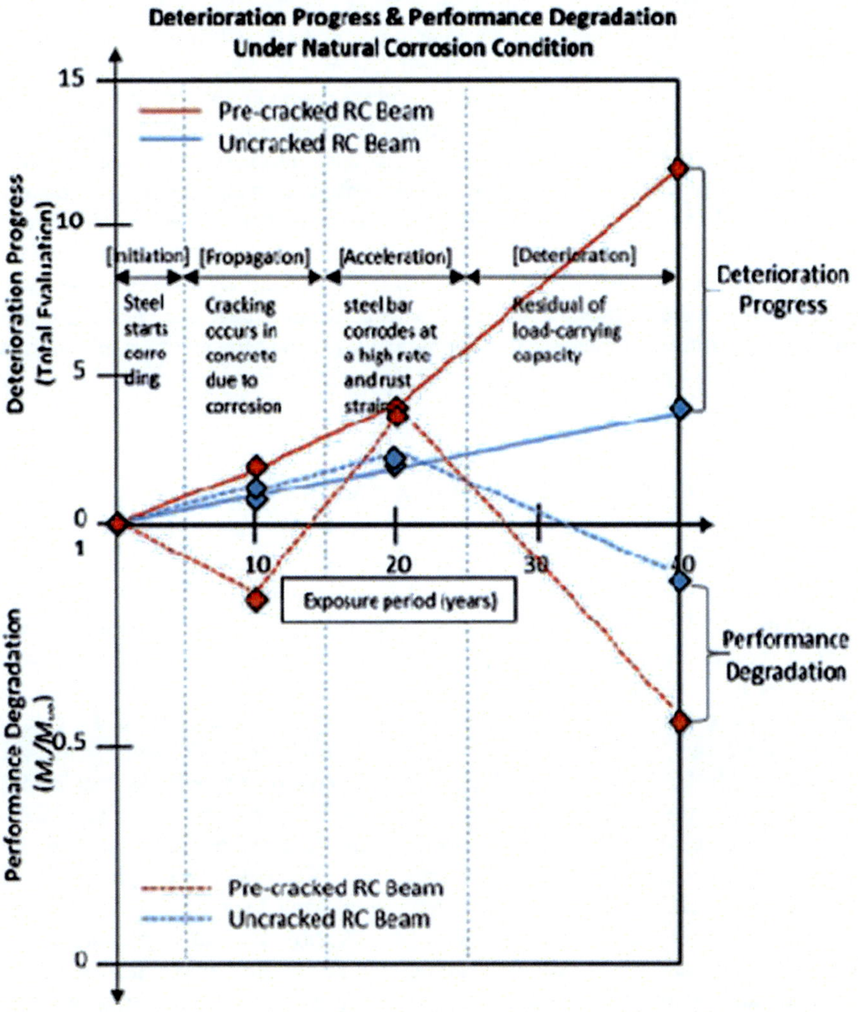

Fig. 4 Representation of deterioration progress and performance degradation [41]

fracture properties and performance of members under different action of static and impact loading.

4 Prestressed Concrete

The demand for lighter and stronger components in the construction industry has led to the development of prestressed concrete. Prestressed concrete is essentially a concrete with additional application of acceptable internal stresses of magnitude and distribution, so as to overcome a reasonable degree of the stresses arising from the external loads. High-strength steels are used to incorporate additional internal stresses in the concrete. The high tensile strength steel is formed by increasing the carbon content. The minimum characteristic strength of high-strength steel is considered as 980 MPa. Proof stress should not be less than 80% of characteristic strength and elongation at rupture should be 10% for the considered gauge length. Since steels are subjected to high stress levels in the prestressed concrete, potential threat of corrosion in prestressed concrete is much higher than the normal concrete [44–46]. The application of prestressing force increases the stiffness by 40% compared with the normal concrete and retards the crack development on the surface of the member. So, a fully prestressed member is comparatively resistant to corrosion than normal concrete [47].

4.1 High-Strength Steel Corrosion and Effect

The steel corrosion mechanism in prestressed concrete is same as that of normal concrete. Additionally, stress corrosion cracking and hydrogen embrittlement of steel is mainly considered in the prestressed steel corrosion. Steel in prestressed concrete made liable for high tensile stress and it results in development of distance between the molecules. The corrosion products deposited between this molecular gap cause stress corrosion cracking which is the reason for prestressed concrete failures in most of the cases. Hydrogen embrittlement has been observed in the steel which was placed in high alkaline medium pH about 13.2, when it is exposed to hydrogen [48].

The main causes and effect of corrosion in high-strength steel is listed in Table 3. The stress corrosion cracking develops "microcracking and microvoid" in high loaded steel, because of stress concentration in pits formed due to corrosion. The prestressing levels influence the stress corrosion cracking. Increase in the prestressing load increases the corrosion in a linear rate [49, 50]. The effect of temperature changes, chloride content, oxygen availability, and pressure waves has the significance on the start of stress corrosion cracking. This is proved in the study conducted on electrochemical behavior of high-strength steel wire with solutions of high alkaline medium similar to in concrete [51]. Increase in prestressing level increases corrosion amount as shown in Fig. 5. But, it does not influence the corrosion product

Table 3 Effect of corrosion on high-strength steel bar [53]

Effect	Causes
Stress corrosion cracking	Crack is formed reason of combined effect of stress and corrosion
Hydrogen embrittlement (hydrogen assisted cracking)	Crack formation because of the formation of brittle hydrides when exposed to hydrogen
Fretting fatigue	Due to the cyclic load, oscillating motion of the metal with the contact surface wearing happens and protective cover on surface gets destroyed and corrosion starts
Corrosion fatigue	Degradation of metal due to corrosion and cyclic loading

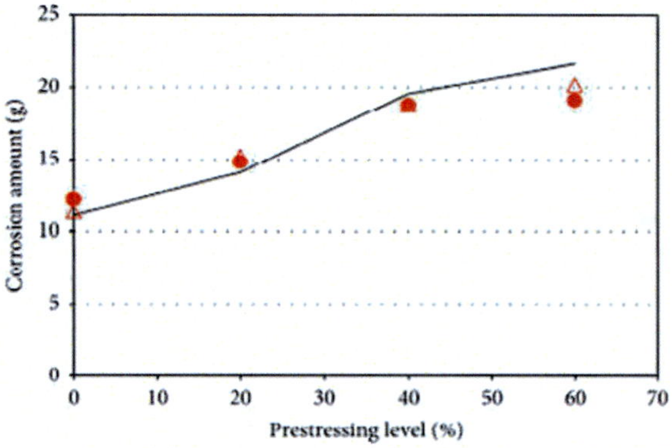

Fig. 5 Variation of corrosion amount with prestressing level [44]

composition. With an increase in corrosion amount and initial prestressing, the ultimate load reduces in the prestressing steel like in normal steel. The stress–strain curve comparison for the prestressed tendon is shown in Fig. 6 [49, 50, 52]. Also, presence of chloride forms the pitting corrosion in the prestressed steel, causing brittle failure [52]. The local defects present on the steel surface increase the possibility of cracking due to stress corrosion. Because of the localized corrosion, the steel wires do not show any reduction in the stress globally. But, there exists a loss in the tensile strength of the bar with increase in the corrosion percentage as shown in Fig. 7. Because of this reason, damage detection analysis also cannot help in most of the cases and hence without any prior indications they end up with sudden failures [49]. From the experiments, it is seen that mass loss and ductility loss are more in the stressed corroded steel wire than in unstressed corroded steel wire as shown in Figs. 8 and 9.

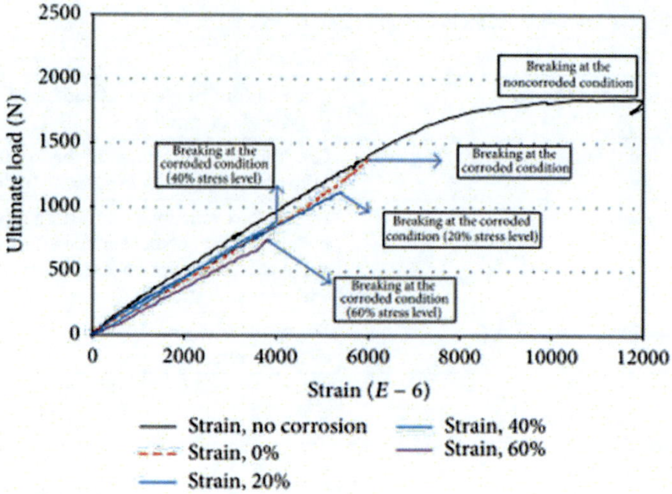

Fig. 6 Stress–strain curve for corroded tendon [50]

Fig. 7 Degradation of tensile strength of corroded steel strand [54]

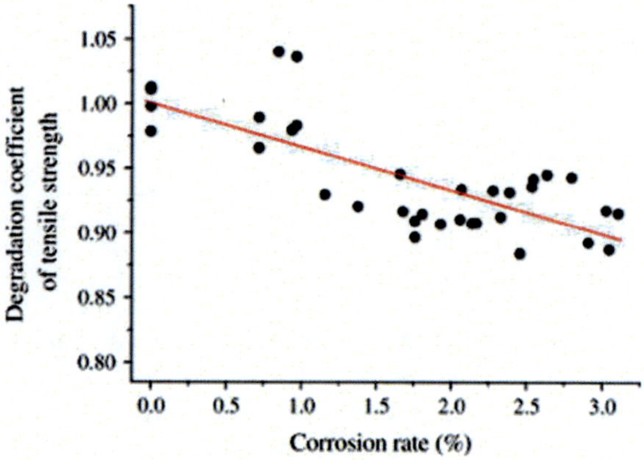

Fig. 8 Comparison of loss of mass in stressed and unstressed corroded steel wire [49]

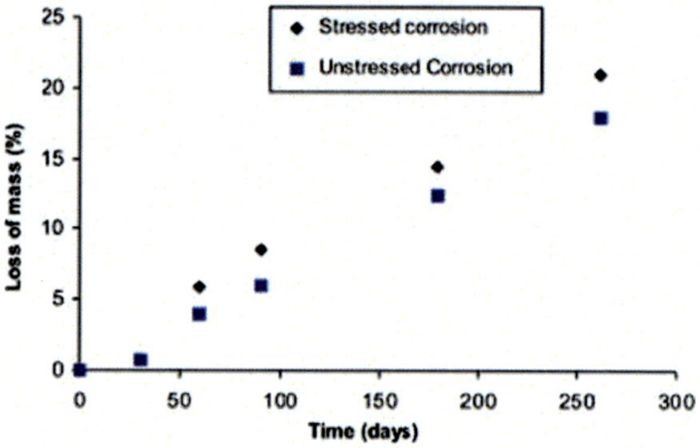

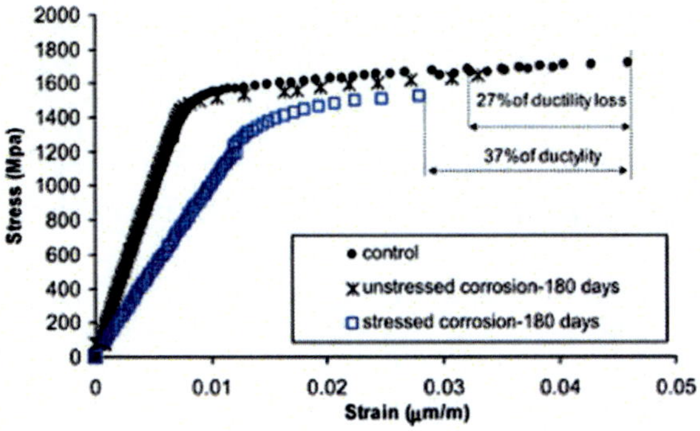

Fig. 9 Comparison of stressed, unstressed corroded steel wire and uncorroded steel wire [49]

5 Conclusions and Perspectives

Corrosion plays the major role in deterioration of the reinforced structures. After reviewing the current studies done on the effect of corrosion of reinforcement, it is clear that the main challenge in understanding the corrosion inside the reinforced concrete is heterogeneity of the reinforced concrete. The pores present in the concrete are the main reasons to carry the corrosion influential material inside the concrete. The local defects present on the steels used in the reinforcements also should be the main concern. Non-uniformity in the concrete–steel interface and aggregate–paste interfaces contributes their effect in the corrosion process initiation.

- More studies are required to understand the corrosion mechanism in the reinforced concrete considering pores distribution on the concrete and transitional zones.
- Since most of the studies assume uniform corrosion in the performance analysis of the structure, more concentration needs to be given to the non-uniform corrosion study.
- More focus should be given to reduce precracks in the member because precracks on the concrete member surface are again one of the reasons for corrosion acceleration of the embedded bars.
- Pitting corrosion is one of the main severe corrosions since it reduces the cross section. Therefore, more focus should be given in that study.
- Since most of the works were conducted with the simulation of concrete medium in prestressing steel corrosion study, rigorous study has to be conducted more effectively by considering steel in concrete. There is substantially different effect of steel corrosion in normal concrete and prestressed concrete. Precracks formation is very slow in the prestressed concrete. Therefore, it is evident that much focus should be given to know the other factors which contribute to that.
- Dense and non-porous concrete design can also be one of the solutions for these corrosion protection phenomena.
- Improvement in design method, handling of materials, and quality of concrete can be better options against the corrosion process. More focus has to be given for finding out the methods for identifying, preventing, or slowing down the

process of corrosion for reducing this method of deterioration of concrete and for a cost-effective solution.

References

1. Bertolini L (2008) Steel corrosion and service life of reinforced concrete structures. Struct Infrastruct Eng 4(2):123–137
2. Angst UM (2018) Challenges and opportunities in corrosion of steel in concrete. Mater Struct Constr 51(1):1–20
3. Shetty MS (2007) Concrete technology. S Chand Publications, pp 1–443
4. Portland Cement Association (2002) Types and causes of concrete deterioration. The Portland Cement Association—concrete information, vol PCA R&D Se, pp 1–16
5. James A et al (2019) Rebar corrosion detection, protection, and rehabilitation of reinforced concrete structures in coastal environments: a review. Constr Build Mater 224:1026–1039
6. Zhou Y, Gencturk B, Willam K, Attar A (2015) Carbonation-induced and chloride-induced corrosion in reinforced concrete structures. J Mater Civ Eng 27(9):04014245
7. Belgacem ME, Neves R, Talah A (2020) Service life design for carbonation-induced corrosion based on air-permeability requirements. Constr Build Mater 261:120507
8. Polder R, Peelen W, Klinghoffer O, Eri J, Leggedoor J (2007) Use of advanced corrosion monitoring for risk based management of concrete structures. Heron 52(4):239–250
9. Ibrahim MA, Sharkawi AEDM, El-Attar MM, Hodhod OA (2018) Assessing the corrosion performance for concrete mixtures made of blended cements. Constr Build Mater 168:21–30
10. Bertolini L, Carsana M, Gastaldi M, Lollini F, Redaelli E (2016) Corrosion of steel in concrete and its prevention in aggressive chloride-bearing environments. In: International conference on durability of concrete structure, ICDCS 2016, pp 13–25
11. Ranjith A, Balaji Rao K, Manjunath K (2016) Evaluating the effect of corrosion on service life prediction of RC structures—a parametric study. Int J Sustain Built Environ 5(2):587–603
12. Sangoju B, Gettu R, Bharatkumar BH, Neelamegam M (2011) Chloride-induced corrosion of steel in cracked OPC and PPC concretes: experimental study. J Mater Civ Eng 23(7):1057–1066
13. Zhao X, Gong P, Qiao G, Lu J, Lv X, Ou J (2011) Brillouin corrosion expansion sensors for steel reinforced concrete structures using a fiber optic coil winding method. Sensors 11(11):10798–10819
14. Ramesh MN (2020) Concrete protection coatings for reinforced concrete structures, vol 17, pp 1–17
15. Glass GK, Buenfeld NR (2000) Chloride-induced corrosion of steel, pp 448–458
16. Vonezawa T, Ashworth V, Procter RPM (1988) Pore solution composition and chloride effects on the corrosion of steel in concrete, vol 44, no 7, pp 489–499
17. Cai Y et al (2020) Characteristics of the steel-concrete interface and their effect on the corrosion of steel bars in concrete. Constr Build Mater 253:119162
18. Shi J, Ming J (2017) Influence of defects at the steel-mortar interface on the corrosion behavior of steel. Constr Build Mater 136:118–125
19. Kwon SJ, Lee HS, Karthick S, Saraswathy V, Yang HM (2017) Long-term corrosion performance of blended cement concrete in the marine environment—a real-time study. Constr Build Mater 154:349–360
20. Razmjoo A, Poursaee A (2016) Effects of coarse aggregates on the corrosion of steel bars in cement paste. J Mater Civ Eng 28(8):04016061
21. Balafas I, Burgoyne CJ (2010) Environmental effects on cover cracking due to corrosion. Cem Concr Res 40(9):1429–1440
22. Hussain SE, Rasheeduzzafar, Al-Musallam A, Al-Gahtani AS (1995) Factors affecting threshold chloride for reinforcement corrosion in concrete. Cem Concr Res 25(7):1543–1555

23. Angst U, Elsener B, Larsen CK, Vennesland Ø (2009) Critical chloride content in reinforced concrete—a review. Cem Concr Res 39(12):1122–1138
24. Mangat PS, Molloy BT (1992) Factors influencing chloride-induced corrosion of reinforcement in concrete. Mater Struct 25(7):404–411
25. Ann KY, Song HW (2007) Chloride threshold level for corrosion of steel in concrete. Corros Sci 49(11):4113–4133
26. Torbati-sarraf H, Poursaee A (2020) The influence of phase distribution and microstructure of the carbon steel on its chloride threshold value in a simulated concrete pore solution. Constr Build Mater 259:119784
27. Mahima S, Moorthi PVP, Bahurudeen A, Gopinath A (2018) Influence of chloride threshold value in service life prediction of reinforced concrete structures. Sadhana Acad Proc Eng Sci 43(7):1–19
28. Fang C, Lundgren K, Chen L, Zhu C (2004) Corrosion influence on bond in reinforced concrete. Cem Concr Res 34(11):2159–2167
29. Tondolo F (2015) Bond behaviour with reinforcement corrosion. Constr Build Mater 93:926–932
30. Wang L, Zhang X, Zhang J, Yi J, Liu Y (2017) Simplified model for corrosion-induced bond degradation between steel strand and concrete. J Mater Civ Eng 29(4):04016257
31. Choi YS, Yi ST, Kim MY, Jung WY, Yang EI (2014) Effect of corrosion method of the reinforcing bar on bond characteristics in reinforced concrete specimens. Constr Build Mater 54:180–189
32. Koulouris K, Apostolopoulos C (2020) An experimental study on effects of corrosion and stirrups spacing on bond behavior of reinforced concrete. Metals (Basel) 10(10):1–14
33. Chen H-P, Nepal J (2016) Analytical model for residual bond strength of corroded reinforcement in concrete structures. J Eng Mech 142(2):04015079
34. Zhu W, Dai JG, Poon CS (2018) Prediction of the bond strength between non-uniformly corroded steel reinforcement and deteriorated concrete. Constr Build Mater 187:1267–1276
35. Wu YZ, Lv HL, Zhou SC, Fang ZN (2016) Degradation model of bond performance between deteriorated concrete and corroded deformed steel bars. Constr Build Mater 119:89–95
36. Torres-Acosta AA, Navarro-Gutierrez S, Terán-Guillén J (2007) Residual flexure capacity of corroded reinforced concrete beams. Eng Struct 29(6):1145–1152
37. Jiradilok P, Wang Y, Nagai K, Matsumoto K (2020) Development of discrete meso-scale bond model for corrosion damage at steel-concrete interface based on tests with/without concrete damage. Constr Build Mater 236:117615
38. Pandit P, Babunarayan KS (2011) Effect of corrosion on performance of reinforced concrete structure using pushover analysis. Int J Earth Sci Eng 04(06):20410432
39. Zhu W, François R, Coronelli D, Cleland D (2013) Effect of corrosion of reinforcement on the mechanical behaviour of highly corroded RC beams. Eng Struct 56:544–554
40. Zhang W, François R, Cai Y, Charron J, Yu L (2020) Influence of artificial cracks and interfacial defects on the corrosion behavior of steel in concrete during corrosion initiation under a chloride environment. Constr Build Mater 253:119165
41. Dasar A, Hamada H, Sagawa Y, Yamamoto D (2017) Deterioration progress and performance reduction of 40-year-old reinforced concrete beams in natural corrosion environments. Constr Build Mater 149:690–704
42. Ou YC, Susanto YTT, Roh H (2016) Tensile behavior of naturally and artificially corroded steel bars. Constr Build Mater 103:93–104
43. Tamai H, Sonoda Y, Bolander JE (2020) Impact resistance of RC beams with reinforcement corrosion: experimental observations. Constr Build Mater 263:120638
44. Dai L, Wang L, Bian H, Zhang J, Zhang X, Ma Y (2019) Flexural capacity prediction of corroded prestressed concrete beams incorporating bond degradation. J Aerosp Eng 32(4):04019027
45. ACI Committee 222 (2009) 222.2R-01 corrosion of prestressing steels. Man Concr Pract 1–29
46. Treadaway KWJ (1971) Corrosion of prestressed steel wire in concrete. Br Corros J 6(2):66–72
47. Moawad M, El-Karmoty H, El Zanaty A (2018) Behavior of corroded bonded fully prestressed and conventional concrete beams. HBRC J 14(2):137–149

48. Recio FJ, Alonso MC, Gaillet L, Sánchez M (2011) Hydrogen embrittlement risk of high strength galvanized steel in contact with alkaline media. Corros Sci 53(9):2853–2860
49. Vu NA, Castel A, François R (2009) Effect of stress corrosion cracking on stress-strain response of steel wires used in prestressed concrete beams. Corros Sci 51(6):1453–1459
50. Lee BY, Koh KT, Ismail MA, Ryu HS, Kwon SJ (2017) Corrosion and strength behaviors in prestressed tendon under various tensile stress and impressed current conditions. Adv Mater Sci Eng 2017
51. Díaz B, Freire L, Nóvoa XR, Pérez MC (2009) Electrochemical behaviour of high strength steel wires in the presence of chlorides. Electrochim Acta 54(22):5190–5198
52. Li F, Yuan Y, Li CQ (2011) Corrosion propagation of prestressing steel strands in concrete subject to chloride attack. Constr Build Mater 25(10):3878–3885
53. Adrian T, Ciolko PE (2005) Corrosion and prestressed concrete bridges. In: Structures congress, pp 1–12
54. Li F, Yuan Y (2013) Effects of corrosion on bond behavior between steel strand and concrete. Constr Build Mater 38:413–422

Effect of Soil Structure Interaction on Multi-storey R.C.C Structure Under Earthquake Load

Sabyath Shetty and K. N. Jeevan Kumar

Abstract The buildings situated in areas vulnerable to earthquakes, the impact of soil structure interaction should be taken into account. The main objective of the research is to analyse the importance of the influence of the interaction of the soil structure on the multi-storey R.C.C structure. An attempt was made during the current study to compare the behaviour of R.C.C buildings with fixed and flexible support conditions subjected to seismic forces. Here, ground+15 (G+15) storey R.C.C building with a 30 m × 20 m rectangular plan, uniform across the height, is considered and analysed using ETABS 18 software for gravity and lateral loads. The functioning characteristics such as storey drift, storey displacement, base shear and natural time period are evaluated. Since storey displacements and storey drifts are less than flexible support condition, the fixed support condition structure shows good performance in resisting lateral loads.

Keywords Soil structure interaction · Earthquake force · Storey displacements · Storey drift · Natural time period · Base shear · ETABS

1 Introduction

In this era of rapid growth and urbanization along with the scarcity of land, one is forced to build structures even in relatively soft soil, which in the past has been considered inappropriate for building purpose. However, with the improvement in various ground improvement techniques, it is possible to build the structures. The designers eventually considered the interaction between soil and structure and observed the soil and structure's dynamic behaviour as a single system.

A structure's response to seismic shaking is affected by the interactions between the three interconnected systems: the structure, the base and the foundation [1]. Analysis of the soil-structure interaction tests the collective response of these systems

S. Shetty (✉) · K. N. Jeevan Kumar
Department of Civil Engineering, N.M.A.M Institute of Technology, Nitte, India
e-mail: shettysabyath@nitte.edu.in

L. Nandagiri et al. (eds.), *Sustainability Trends and Challenges in Civil Engineering*, Lecture Notes in Civil Engineering 162, https://doi.org/10.1007/978-981-16-2826-9_52

to the movement of the systems specified. The key issues involved in the occurrence of soil-structure interactions are the seismic waves transmitted through the soil during an earthquake, and at the intersection of the soil and structural foundations, a discontinuity is experienced in the wave propagation medium.

The change in material properties results in scattering, variation, refraction and reflection of seismic waves at the soil foundation interface. The mechanism in which the soil's response influences the structure's motion, and the structure's motion influences the soil's response is referred to as soil-structure interaction (SSI) [2].

Under load application, the flexibility of soil mass induces the rotation of footings and differential settlement. The relative stiffness of the soil, foundation and structure influences the structure-foundation-soil system interaction behaviour [3]. Thus, the physical property of the foundation medium is an important factor in the seismic response of structures supported on it [4–6].

The Indian Earthquake Design Code IS: 1893 (Part 1) (BIS, 2002) indicates that studies of soil structure interaction (SSI) can be ignored for rock or rock-like soil seismic design [7]. However, for the structure to rest on the stiff or soft soil, during the analysis, it is important to consider the results of the interaction.

2 Objective

In this project, G+15 storey R.C.C structure is analysed to study the effect of lateral forces such as seismic force (Zone III) for a building resting on fixed and flexible footing condition.

Type of Structure Analysed:

- RCC Structure with Fixed Base condition.
- RCC Structure resting on Hard Soil.
- RCC Structure resting on Medium Soil.
- RCC Structure resting on Soft Soil.

The software to be used for the analysis is ETABS 18. The comparison of structural behaviour is observed such as storey displacement of building, storey drift, natural time period, base shear and conclusions are made on the basis of findings, and this research finds a stronger structural system.

3 Methodology

Subsequent step-by-step procedures are followed to attain the above objective;

- Conducted a literature review to assess the goals of the project work.

- In the present investigation, a G+15 storied building is considered, having general arrangement measurement of 30 m × 20 m with a bay size of 5 m in both directions along the *X* and *Y* directions.
- Four structural systems are adopted in this work, i.e. one structure with fixed base condition and others are structures resting on hard, medium and soft Soil.
- Analyse all selected models using ETABS 18 Software by applying design loads as per IS 875.
- Evaluate the results of the analysis and check the criteria for the geometric limitations.

4 Problem Statement

Following types of structural arrangement is studied;

- Reinforced concrete multi-storey building with fixed base condition.
- Reinforced concrete multi-storey structure resting on soft, medium and hard soil.

4.1 Geometrical Data

- No of Stories: G+15
- No. of Bay in *X*-Direction: 6
- No. of Bay in *Y*-Direction: 4
- Type of Building Use: Residential
- Plan: 30 m × 20 m
- Typical Storey Height: 3.0 m
- Bottom Storey Height: 3.0 m
- Total Height: 51 m.

4.2 Materials

- Concrete Grade: M20, M25, M30
- Steel (Rebar): Fe500.

4.3 Member Properties

- Thickness of Slab: 150 mm
- Column Size: 600 mm × 600 mm
- Beam Size: 230 mm × 450 mm.

4.4 Loads Considered

- Dead Load: Auto
- Live Load: 3 kN/m^2
- Floor Finish: 1.5 kN/m^2
- Wall Load: 13 kN/m (9″ Thick)
- Other Loads: Seismic Load.

4.5 Seismic Load

In accordance with IS: 1893:2016 [7], seismic design shall be done. The structure considered is situated in Mangaluru in India which lies in earthquake zone III. The parameters to be used for analysis and design are given below.

- Zone: III
- Zone Factor: 0.16 [IS 1893 (Part 1)]
- Importance factor: 1.2
- Response Reduction: 5.0 (SMRF)
- Structure Type: RC Frame Structure.

4.6 Soil Characteristics

The structure's response is influenced by the dynamic analysis of the structure and its interaction with soil. Foundation dimensions and elastic properties of soil influence the interaction between them. The versatility of the foundation in the study is considered by replacing the foundation with statically equivalent springs (Table 1).

5 Foundation Characteristics

According to the equations given by FEMA 356, modelling of foundation soil was done by using spring constants as shown below [8] (Fig. 1).

Table 1 Soil parameters

Soil type	Hard soil	Medium soil	Soft soil
Modulus of elasticity (kN/m^2)	65,000	35,000	15,000
Poisson's ratio (μ)	0.3	0.4	0.4

Degree of Freedom	Stiffness of Foundation at Surface	Note
Translation along x-axis	$K_{x,sur} = \frac{GB}{2-\nu}\left[3.4\left(\frac{L}{B}\right)^{0.65}+1.2\right]$	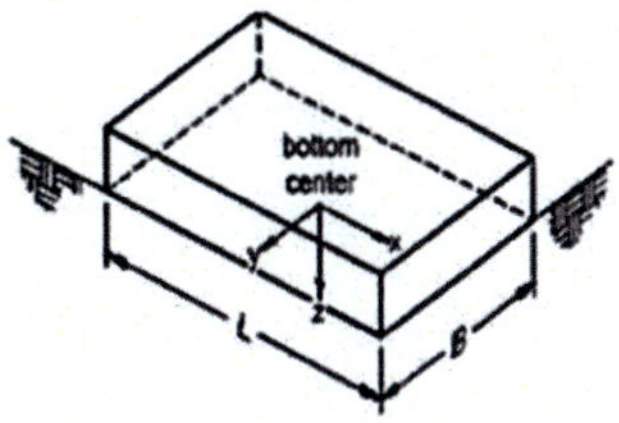
Translation along y-axis	$K_{y,sur} = \frac{GB}{2-\nu}\left[3.4\left(\frac{L}{B}\right)^{0.65}+0.4\frac{L}{B}+0.8\right]$	
Translation along z-axis	$K_{z,sur} = \frac{GB}{1-\nu}\left[1.55\left(\frac{L}{B}\right)^{0.75}+0.8\right]$	
Rocking about x-axis	$K_{xx,sur} = \frac{GB^3}{1-\nu}\left[0.4\left(\frac{L}{B}\right)+0.1\right]$	
Rocking about y-axis	$K_{yy,sur} = \frac{GB^3}{1-\nu}\left[0.47\left(\frac{L}{B}\right)^{2.4}+0.034\right]$	Orient axes such that $L \geq B$
Torsion about z-axis	$K_{zz,sur} = GB^3\left[0.53\left(\frac{L}{B}\right)^{2.45}+0.51\right]$	

Degree of Freedom	Correction Factor for Embedment	Note
Translation along x-axis	$\beta_x = \left(1+0.21\sqrt{\frac{D}{B}}\right)\cdot\left[1+1.6\left(\frac{hd(B+L)}{BL^2}\right)^{0.4}\right]$	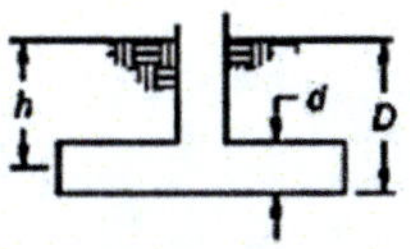
Translation along y-axis	$\beta_y = \beta_x$	
Translation along z-axis	$\beta_z = \left[1+\frac{1}{21}\frac{D}{B}\left(2+2.6\frac{B}{L}\right)\right]\cdot\left[1+0.32\left(\frac{d(B+L)}{BL}\right)^{2/3}\right]$	d = height of effective sidewall contact (may be less than total foundation height)
Rocking about x-axis	$\beta_{xx} = 1+2.5\frac{d}{B}\left[1+\frac{2d}{B}\left(\frac{d}{D}\right)^{-0.2}\sqrt{\frac{B}{L}}\right]$	h = depth to centroid of effective sidewall contact
Rocking about y-axis	$\beta_{yy} = 1+1.4\left(\frac{d}{L}\right)^{0.6}\left[1.5+3.7\left(\frac{d}{L}\right)^{1.9}\left(\frac{d}{D}\right)^{-0.6}\right]$	For each degree of freedom, calculate $K_{emb} = \beta K_{sur}$
Torsion about z-axis	$\beta_{zz} = 1+2.6\left(1+\frac{B}{L}\right)\left(\frac{d}{B}\right)^{0.9}$	

Fig. 1 Spring stiffness equations based on FEMA 356

5.1 Footing

- Length of footing **L**: 4.7 m
- Width of footing **B**: 4.7 m
- Depth of footing **d**: 0.8 m
- Depth of foundation from ground level **D**: 3 m
- Depth of centroid of effective sidewall contact **h**: 2.6 m

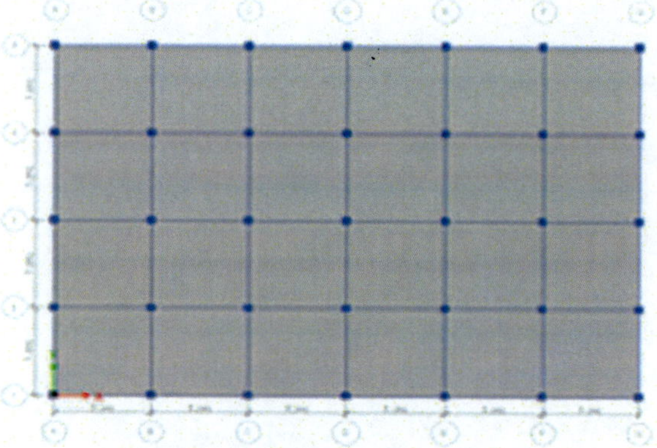

Fig. 2 Plan of the models

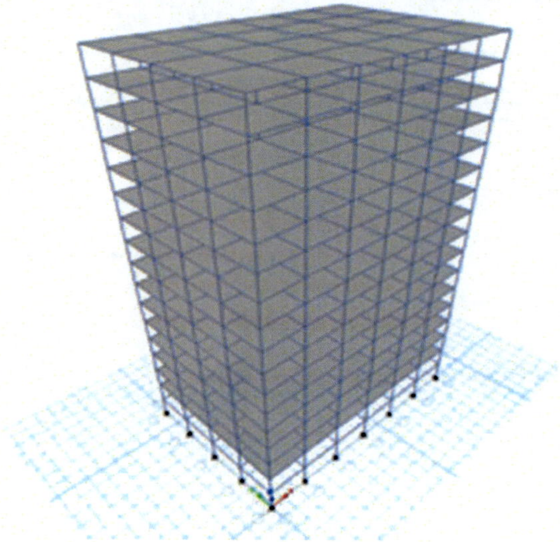

Fig. 3 Fixed base building (3D view)

6 Modelling of the Structure

See Figs. 2, 3 and 4.

7 Results and Discussion

- Response spectrum analysis and time history analysis is carried out for regular building without and with soil structure interaction.

Fig. 4 Flexible base building (3D view)

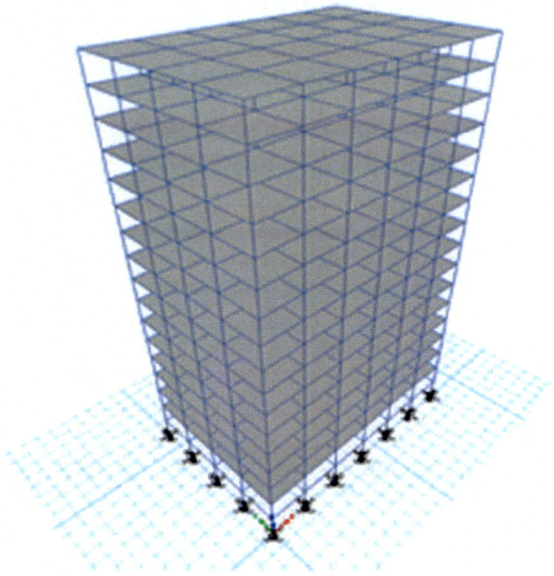

- The models are checked for storey drift, storey displacement, base shear and natural time period (Figs. 5, 6, 7 and 8).

7.1 Max. Storey Displacement

- It is total displacement of the top storey with respect to ground.

7.1.1 Fixed Base

See Table 2.

7.1.2 Building Resting on Hard Soil

See Table 3.

7.1.3 Building Resting on Medium Soil

See Table 4.

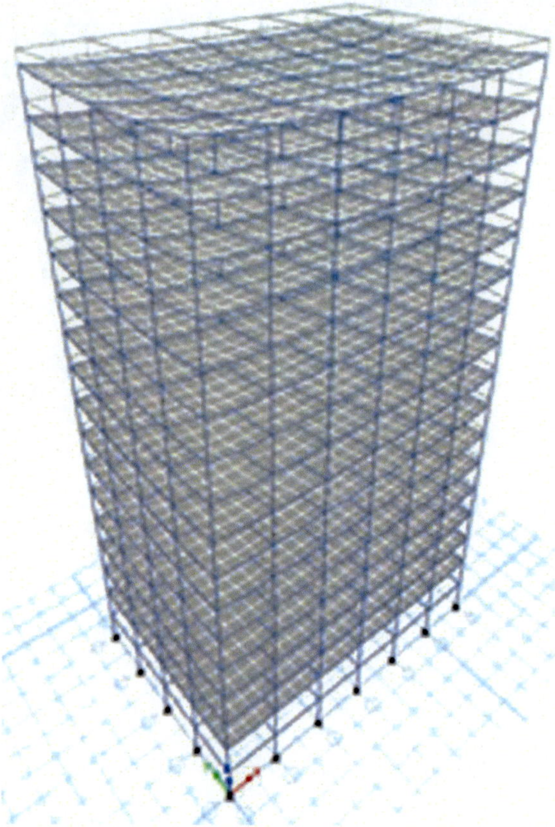

Fig. 5 Displacement (3D view)

7.1.4 Building Resting on Soft Soil

See Figs. 9 and 10; Table 5.

7.2 Comparison of Max. Storey Displacement

See Fig. 11 and Table 6.

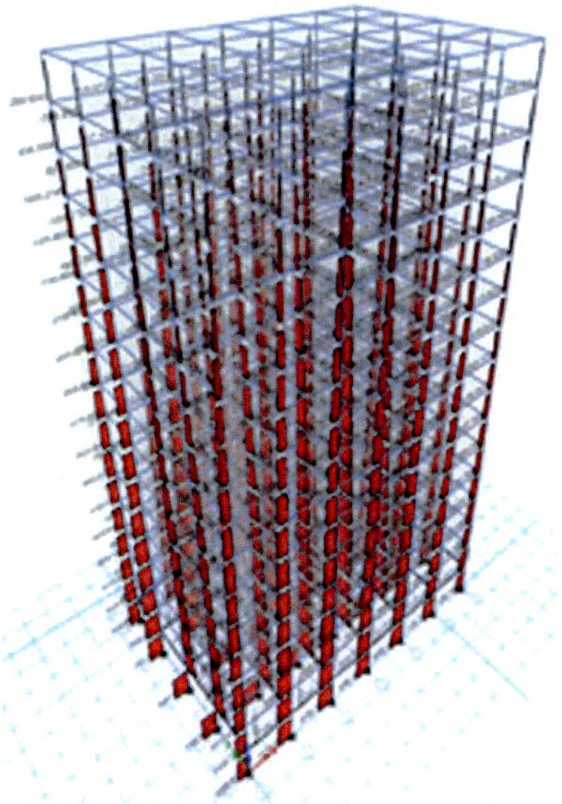

Fig. 6 Axial force diagram (3D view)

7.3 *Max. Storey Drift*

7.3.1 Fixed Base

See Table 7.

7.3.2 Building Resting on Hard Soil

See Table 8.

7.3.3 Building Resting on Medium Soil

See Table 9.

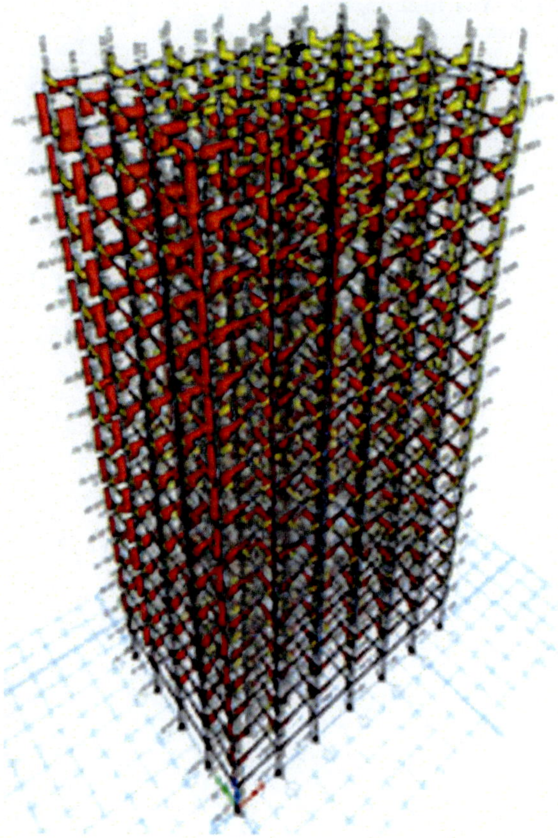

Fig. 7 Shear force (3D view)

7.3.4 Building Resting on Soft Soil

See Figs. 12 and 13; Table 10.

7.4 Comparison of Max. Storey Drift

See Fig. 14 and Table 11.

7.5 Natural Time Period

- Natural time periods are the significant factors that can influence the structure's seismic behaviour.

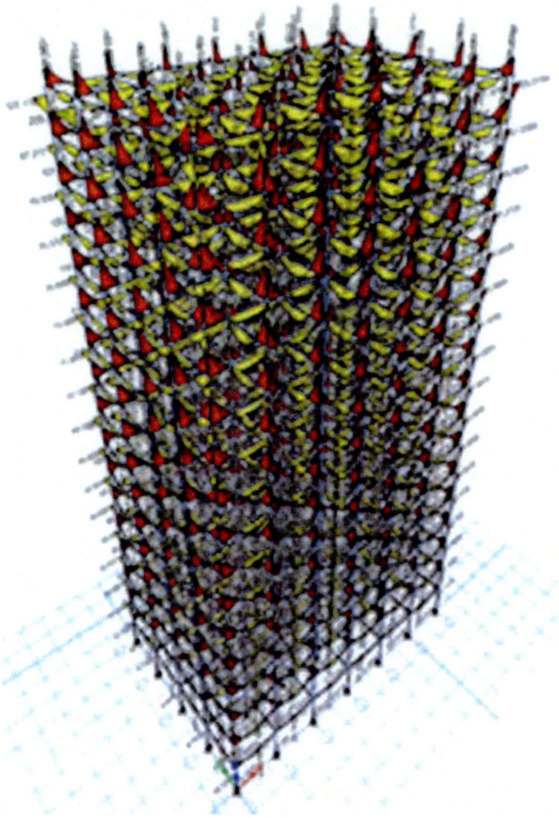

Fig. 8 Bending moment (3D view)

- Study has been made, and it shows the variation in fundamental time period for different braced structure as shown in figure (Fig. 15; Table 12).

7.6 Seismic Base Shear

Seismic base shear mirrors the seismic lateral vulnerability and is considered as one of the primary input for seismic design. The variation in base shear for structure resting different type of soil is as shown in figure (Fig. 16; Table 13).

Table 2 Max. storey displacement (mm) of fixed base building

Floor level	Fixed base building	
	X-direction	*Y*-direction
15 Floor	105.92	117.165
14 Floor	104.531	115.242
13 Floor	102.38	112.46
12 Floor	99.382	108.941
11 Floor	95.575	104.933
10 Floor	91.051	100.124
9 Floor	86.398	94.71
8 Floor	80.954	88.898
7 Floor	74.752	82.218
6 Floor	67.902	74.718
5 Floor	60.674	66.452
4 Floor	52.71	57.673
3 Floor	44.07	48.216
2 Floor	34.83	38.075
1 Floor	25.095	27.379
Ground floor	15.082	16.395
Plinth	5.545	5.973
Footing	0	0

8 Conclusion

From the results discussed with respect to the building models considered, leads to the following conclusions;

- The effect of SSI on storey displacement of G+15 storey buildings has been studied. It is observed that the displacement increase occurs in SSI models.
- Due to the impact of soil structure interaction, storey displacement, storey drift and time period values are magnified. The highest values of these response parameters are given by soft soil.
- The soil structure interaction effect contributes to the decrease of the building's base shear.
- Compared to medium and hard soils, soft soil exhibits greater reduction of base shear.
- In the case of building with a fixed base in first mode, the natural time period is 2.935 s, and in the case of hard soil, medium and soft soil conditions, it increases to 3.016 s, 3.537 s, 4.049 s.
- Compared to the traditional approach of assuming a fixed base, the response of R.C building with SSI showed a substantial increase.

Table 3 Max storey displacement (mm) of building resting on hard soil

Floor level	Fixed base building	
	X-direction	*Y*-direction
15 Floor	109.311	129.025
14 Floor	107.019	125.505
13 Floor	104.034	121.249
12 Floor	100.258	116.172
11 Floor	95.713	110.314
10 Floor	90.459	103.757
9 Floor	84.571	96.596
8 Floor	78.182	88.922
7 Floor	71.423	80.968
6 Floor	64.292	72.772
5 Floor	56.923	64.255
4 Floor	49.093	55.362
3 Floor	40.839	45.985
2 Floor	32.247	36.174
1 Floor	23.328	26.006
Ground floor	14.192	15.696
Plinth	5.527	6.023
Footing	0.405	0.409

- It is therefore inferred that fixed base structure provides lower response parameter values as compared to flexible base structure. It is therefore necessary to consider the impact of the soil structure interaction on the building resting on medium and soft soil in order to obtain the building's acceptable response.

Table 4 Max. storey displacement (mm) of building resting on medium soil

Floor level	Fixed base building	
	X-direction	*Y*-direction
15th Floor	138.275	187.906
14th Floor	134.68	181.714
13th Floor	130.308	174.684
12th Floor	125.077	166.701
11th Floor	119.041	157.786
10th Floor	112.287	148.002
9th Floor	104.902	137.435
8th Floor	96.958	126.182
7th Floor	88.872	114.344
6th Floor	80.213	102.015
5th Floor	70.975	89.623
4th Floor	61.371	76.847
3rd Floor	51.244	63.569
2nd Floor	40.564	49.924
1st Floor	29.42	35.923
Ground floor	18.037	21.782
Plinth level	7.276	8.568
Footing level	0.875	0.907

Fig. 9 Max. storey displacement of structure resting on various types of soil along *X*-direction

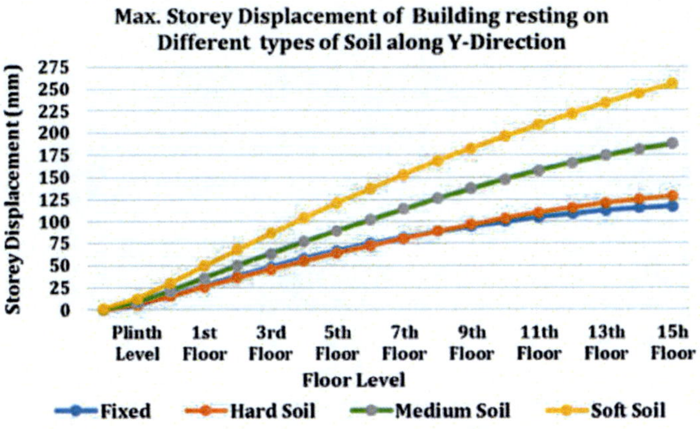

Fig. 10 Max. storey displacement of structure resting on various types of soil along *Y*-direction

Table 5 Max. storey displacement (mm) of building resting on medium soil

Floor level	Fixed base building	
	X-direction	*Y*-direction
15 Floor	168.586	256.07
14 Floor	162.904	245.551
13 Floor	156.557	234.385
12 Floor	149.437	222.463
11 Floor	141.544	209.777
10 Floor	132.916	196.716
9 Floor	123.605	182.965
8 Floor	113.695	168.513
7 Floor	103.669	153.37
6 Floor	93.068	137.548
5 Floor	81.976	121.351
4 Floor	70.813	104.443
3 Floor	59.096	86.807
2 Floor	46.869	68.49
1 Floor	34.215	49.621
Ground floor	21.432	30.589
Plinth	9.347	12.813
Footing	1.896	2.106

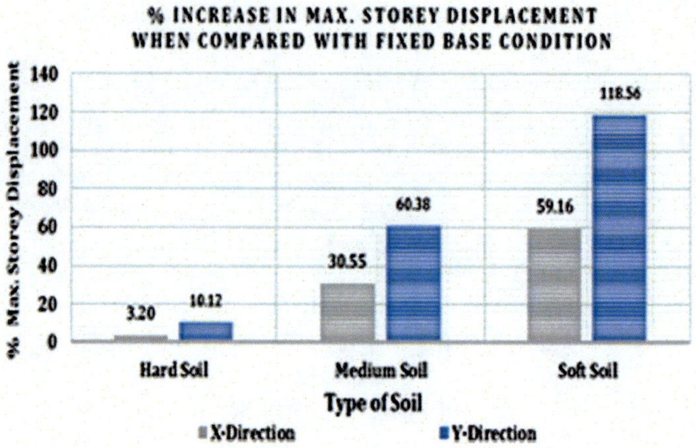

Fig. 11 % Increase in max. storey displacement

Table 6 % Increase in max. storey displacement

% Increase in max. storey displacement				
Type of footing condition	Storey displacement		% Increase in max. storey displacement	
	X-dir	*Y*-dir	*X*-dir	*Y*-dir
Fixed	105.92	117.17	–	–
Hard soil	109.31	129.03	3.20	10.12
Medium soil	138.28	187.91	30.55	60.38
Soft soil	168.59	256.1	59.16	118.56

Table 7 Max. storey drift of fixed base building

Floor level	Fixed base building	
	X-direction	*Y*-direction
15 Floor	0.000492	0.000641
14 Floor	0.000769	0.000927
13 Floor	0.001074	0.00124
12 Floor	0.001361	0.001536
11 Floor	0.001616	0.001801
10 Floor	0.001837	0.002041
9 Floor	0.002027	0.002247
8 Floor	0.002194	0.002423
7 Floor	0.002351	0.002592
6 Floor	0.00253	0.002796
5 Floor	0.002719	0.002991
4 Floor	0.0029	0.003204
3 Floor	0.00308	0.003394
2 Floor	0.003245	0.003565
1 Floor	0.003337	0.003661
Ground floor	0.003216	0.003507
Plinth level	0.001848	0.001991
Footing level	0	0

Table 8 Max. storey drift of building resting on hard soil

Floor level	Fixed base building	
	X-direction	*Y*-direction
15 Floor	0.000775	0.001173
14 Floor	0.001012	0.001419
13 Floor	0.001277	0.001692
12 Floor	0.001525	0.001953
11 Floor	0.001751	0.002186
10 Floor	0.001963	0.002387
9 Floor	0.002146	0.002558
8 Floor	0.002303	0.002704
7 Floor	0.002437	0.002845
6 Floor	0.002553	0.002964
5 Floor	0.002657	0.003064
4 Floor	0.002764	0.003158
3 Floor	0.00288	0.003277
2 Floor	0.002988	0.003389
1 Floor	0.003045	0.003438
Ground floor	0.002931	0.003261
Plinth level	0.001709	0.001881
Footing level	0	0

Table 9 Max. storey drift of building resting on medium soil

Floor level	Fixed base building	
	X-direction	*Y*-direction
15 Floor	0.001198	0.002076
14 Floor	0.00146	0.002366
13 Floor	0.001754	0.002686
12 Floor	0.002037	0.002989
11 Floor	0.002294	0.003261
10 Floor	0.00252	0.003522
9 Floor	0.002715	0.003751
8 Floor	0.002882	0.003946
7 Floor	0.003026	0.00411
6 Floor	0.003153	0.004247
5 Floor	0.003275	0.004364
4 Floor	0.003424	0.004487
3 Floor	0.003575	0.004589
2 Floor	0.003715	0.004691
1 Floor	0.003794	0.004714
Ground floor	0.003645	0.004447
Plinth level	0.002161	0.00259
Footing level	0	0

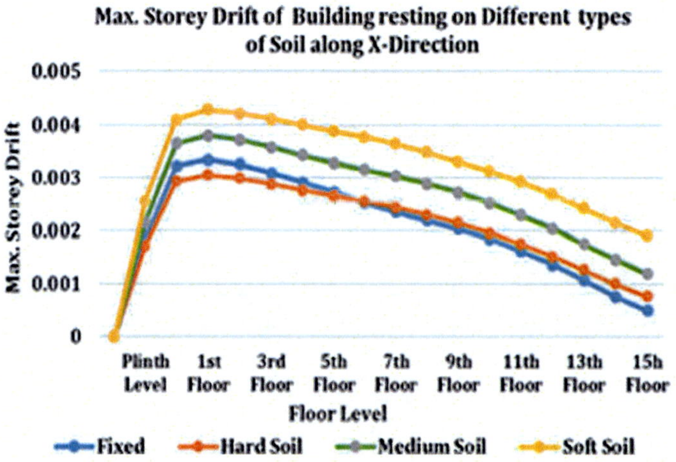

Fig. 12 Max. storey drift of building resting on different types of soil along *X*-direction

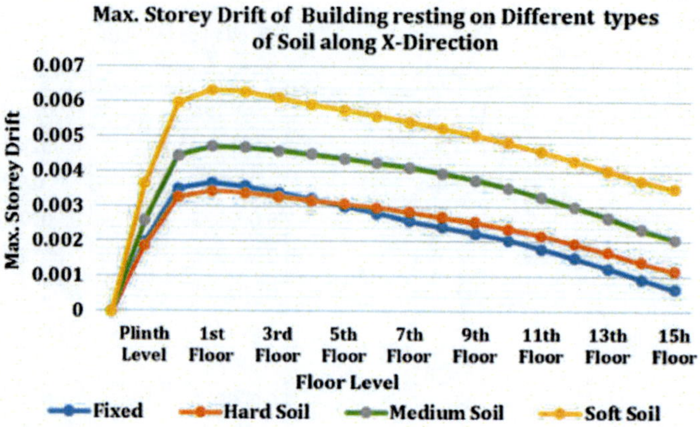

Fig. 13 Max. storey drift of structure resting on various types of soil along *Y*-direction

Table 10 Max. storey drift of building resting on medium soil

Floor level	Fixed base building	
	X-direction	*Y*-direction
15 Floor	0.001916	0.003518
14 Floor	0.002158	0.003755
13 Floor	0.002431	0.004032
12 Floor	0.002692	0.004308
11 Floor	0.002927	0.004573
10 Floor	0.003133	0.004821
9 Floor	0.003311	0.005048
8 Floor	0.003489	0.00525
7 Floor	0.003643	0.005427
6 Floor	0.003774	0.005592
5 Floor	0.003886	0.005768
4 Floor	0.004003	0.005921
3 Floor	0.004118	0.006112
2 Floor	0.004218	0.00629
1 Floor	0.004282	0.006344
Ground floor	0.0041	0.005974
Plinth level	0.00256	0.003647
Footing level	0	0

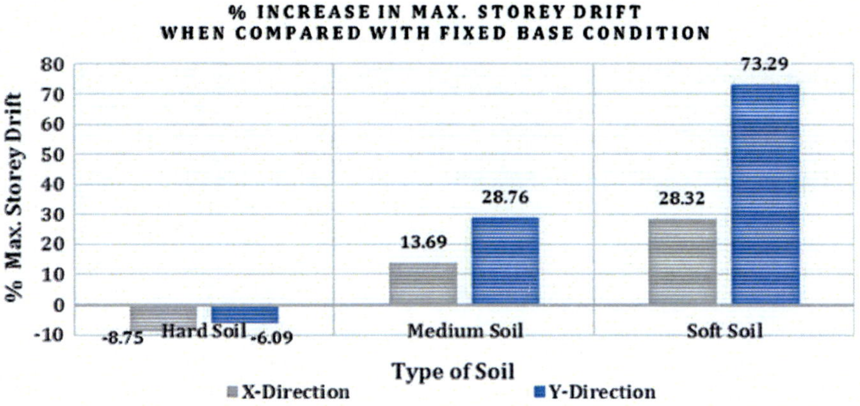

Fig. 14 % Increase in max. storey displacement

Table 11 % Increase in max. storey drift

% Increase in max. storey drift				
Type of footing condition	Storey drift		% Increase in max. storey drift	
	X-dir	*Y*-dir	*X*-dir	*Y*-dir
Fixed	0.00334	0.00366	–	–
Hard soil	0.00305	0.00344	−8.75	−6.09
Medium soil	0.00380	0.00471	13.69	28.76
Soft soil	0.00429	0.00634	28.32	73.29

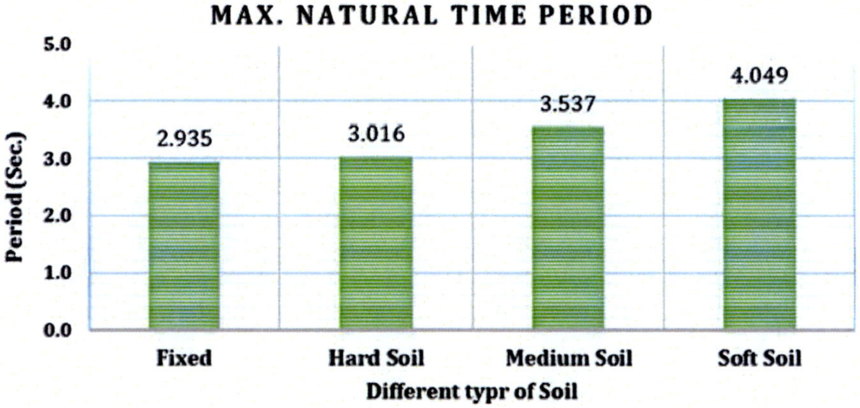

Fig. 15 Peak time period for structure resting on different types of footing

Table 12 Peak time period for building resting on different of footing

Peak time period (s)	
Types of footing condition	Time period (s)
Fixed	2.935
Hard soil	3.016
Medium soil	3.537
Soft soil	4.049

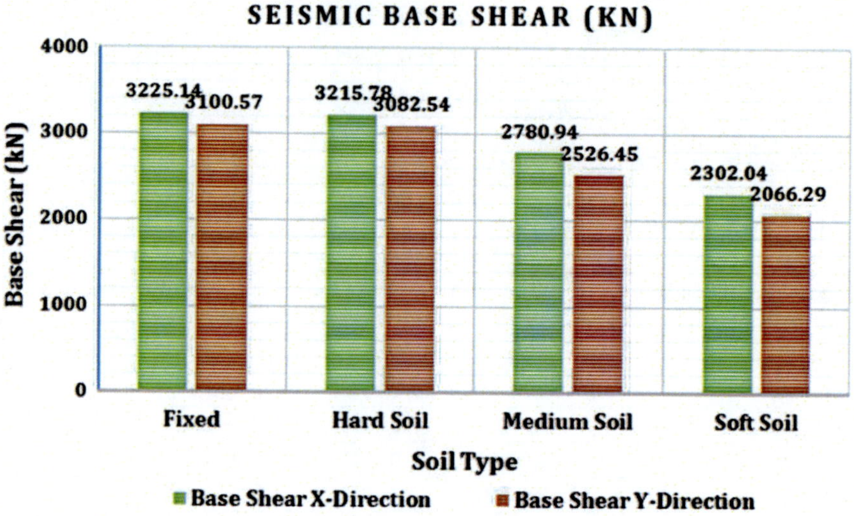

Fig. 16 Seismic base shear for structure resting on different of footing

Table 13 Seismic base shear for structure resting on different of footing

Seismic base shear		
Types of footing condition	Base shear (kN)	
	X-dir	*Y*-dir
Fixed	3225.14	3100.57
Hard soil	3215.78	3082.54
Medium soil	2780.94	2526.45
Soft soil	2302.04	2066.29

Acknowledgements It is time to acknowledge my deep sense of gratitude to all those who have extended their cooperation and helped me to complete my work.

I would like to document the appreciative original work of all the authors of the different technical papers I referred to, without which it would have been very difficult to complete this project successfully.

References

1. American Society of Civil Engineers (2017) Minimum design loads and associated criteria for buildings and other structures. ASCE/SEI 7-16
2. Prakasam ML, Siva Rama Raju V (2018) Effect of soil structure interaction on seismic behaviour of multi-storey [G+5] building with different types of slabs. IRJET 5(5)
3. Battise AS, Ghuga YM (2018) Effect of soil-structure interaction on seismic analysis of structures. IJARSE 7(4)
4. Jain M, Sanghai SS (2020) Seismic response control of unsymmetrical RCC framed building using base isolation considering soil structure interaction. In: ICRRM: international conference on reliability, risk maintenance and engineering management, pp 170–178
5. Kumar M, Mishra SS (2019) Study of seismic response characteristics of building frame models using shake table test and considering soil–structure interaction. Asian J Civ Eng 20:409–419
6. Cruz C, Miranda E (2017) Evaluation of soil-structure interaction effects on the damping ratios of buildings subjected to earthquakes. Soil Dyn Earthq Eng 100:183–195
7. IS 1893-1 (2016) Criteria for earthquake resistant design of structures, general provisions and buildings
8. Federal Emergency Management Agency (FEMA), ASCE, American Society of Civil Engineers. FEMA-356 (2000)

Groundwater Quality Analysis for South Belagavi City, India; A Case Study

Snehal D. Renake, Arjun S. Virupakshi, and Akshata Shagoti

Abstract In the current paper, groundwater status of seven fields situated in South Belagavi City, Karnataka, India is evaluated with the objective of calculating the water quality index. Water quality index categorizes the water bodies by providing a numerical value which describes the physical state or overall quality of water. In the present study, water quality is examined based upon the physicochemical analysis of eight parameters in particular pH, electrical conductivity (EC), total dissolved solids (TDSs), total hardness (TH), calcium hardness (CH), chlorides, alkalinity and acidity. The statistical regression method of analysis for the ground water from all the selected fields is carried out. It is based on the computation of correlation coefficient between varied physicochemical parameters in consideration. The outcome reveals that the water quality index in the seven selected fields fall under the category of poor to unsuitable indicating pollution in the groundwater and is not acceptable for consumption.

Keywords Groundwater · Water quality standards · Water quality index · Correlation coefficient · Regression analysis · Belagavi city

1 Introduction

Water is a natural resource and is vital for the existence of various forms of life on earth. Maintaining the water quality of the various forms of water resource is essential as it caters to the domestic needs as well as industrial and agricultural needs [1]. Access to clean water either surface or groundwater has deteriorated as a consequence of human interference in various forms like Industrialization, urbanization, agricultural practices [2–5] giving rise to associated problems [6–9]. Pollution is adding to the present state of water stress by discharging untreated sewage [10]. To

S. D. Renake (✉) · A. S. Virupakshi
Department of Civil Engineering, KLE Dr. M.S. Sheshgiri College of Engineering and Technology, Belagavi, India

A. Shagoti
Department of Architecture, M.S. Ramaiah Institute of Technology, Bangalore, India

L. Nandagiri et al. (eds.), *Sustainability Trends and Challenges in Civil Engineering*, Lecture Notes in Civil Engineering 162, https://doi.org/10.1007/978-981-16-2826-9_53

control the further deterioration of these resources regular monitoring is essential and many researchers have under taken contextual investigations in our nation and various other parts of the world [5, 7, 11–14]. Water quality is assessed by collecting the water sample for certain time duration at a fixed location along with testing of numerous parameters. Access to clean drinking water is a basic need for individuals as well as to different organisms, at the same clean water ensures good health and protection against various diseases [15]. Water should be free from organic and inorganic contaminants along with parameters like pH, electrical conductivity, TDS, dissolved oxygen, chlorides, total hardness, alkalinity, acidity, nitrates should be under permissible limits [16]. Water quality assessment in the present study is addressed with concept of water quality index. Water quality assessment in terms of water quality index evaluates quality of water for the usefulness of drinking water sources [17–19] for consumption. WQI is investigated by the use of weighted arithmetic index method [18] as it eases the comparisons between different sampling fields.

2 Experimental

2.1 *Study Area*

Belagavi is situated at 15.870 N, 74.500 E. The city is placed at an elevation of 747.47 m above mean sea level (MSL). Belagavi city is based in the north western parts of Karnataka and lies at the frontier of two states, Maharashtra as well as Goa.

Belagavi is the coldest in winter and experiences monsoon in the time of June till September. The average annual precipitation is over 1200 mm. Belagavi City Corporation has been divided into two parts depending on the functioning purposes namely North Belagavi and South Belagavi [11]. Figure 1 shows the sampling points located in South Belagavi City namely field 1. Piranwadi, field 2. Khadarwadi, field 3. Macche, field 4. Angol, field 5. Tilakwadi, field 6. Shahapur, field 7. Inox.

2.2 *Materials and Methodology*

- Collection of Samples: The sampling is carried out from December 2019 to March 2020 for a period of four months. The specimen collection is done using a bottle of one liter and 8 physico chemical parameters are analyzed.
- Water Quality Analysis: Eight physicochemical parameters namely. pH, electrical conductivity (EC), total dissolved solids (TDS), total hardness (TH), calcium hardness (CH), chloride, alkalinity, acidity are analyzed in the laboratory as per standard methods prescribed by the APHA and is used to calculate the water quality index. The pH is evaluated by Digital pH meter (Visso, MK VI). The electrical

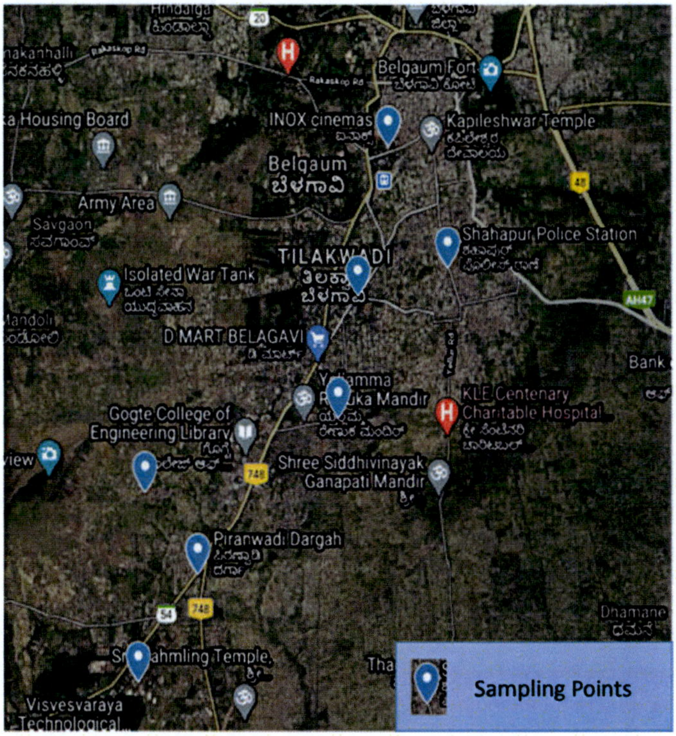

Fig. 1 Sampling points on Belagavi map

conductivity is evaluated Using TDS Meter (Systronics, 308). The instruments in the research facility are utilized in the restriction of exact precision and synthetics used are of high grade.

- Assessment of water quality index: The water quality index is determined using the weighted arithmetic index method [18]. Eight physicochemical parameters are used to compute the water quality index. Determination of water quality is done in line with the standards of drinking water quality put forward by the Bureau of Indian Standards (BIS) [16]. Table 1 shows the water quality index level and status of water quality. Further the quality rating is determined using the following expression

$$\mathrm{WQI} = \sum Q_n W_n / \sum W_n \tag{1}$$

wherein

Q_n is the quality rating of nth water quality parameter.
W_n is the unit weight of nth water quality parameter.
V_n the obtained real value of the nth parameter.

Table 1 Water quality index (WQI) and status of water quality

WQI level	Water quality Status
0–25	Excellent
26–50	Good
51–75	Poor
76–100	Very poor
>100	Unsuitable

V_i ideal value of the parameter [$V_i = 0$, except for pH ($V_i = 7$) and DO ($V_i =$ 14.6 mg/l).

V_s the standard value of the *n*th water quality parameter.

Unit weight W_n is estimated using the relation

$$W_n = k/V_s \tag{2}$$

As "*k*" is the proportionality constant and is given as

$$k = 1/\left[\sum 1/V_s = 1, 2, 3, \ldots, n\right]$$

3 Results and Discussion

The concept of water quality index is used to evaluate the water quality of groundwater in the present study [19]. Weight arithmetic WQI integrates statistics from varied water quality parameters into a mathematical formula that ranks the physical state of the water body with a numerical value [18]. It mirrors the combined influence of various parameters which is crucial for assessing and maintaining the nature of water [18]. Table 2 shows the Physicochemical parameters studied at South Belagavi City.

From the above table, we can note that the pH of the groundwater samples range from 5.5–6.76. The Bureau of Indian Standards (BIS) has fixed the acceptable limit of 6.5–8.5 [16]. The average pH values of 5 fields are slightly acidic with 2 fields exhibiting desired pH of the groundwater. The EC values of the groundwater range from 203.90 to 611.11. The average EC values at 5 fields (1–5) are within the permissible limits, field 6, 7 are exhibiting EC values beyond acceptable limit prescribed by the BIS and high values can be attributed to sewage discharge from residential areas. TDS values at all the 7 fields are under the prescribed limit as per BIS. Total hardness of the entire 7 sample is under the permissible limit as per BIS. Chloride content in all the 7 groundwater samples is under the prescribed limit. Alkalinity is under the permissible limit as per the standards of BIS.

Table 2 Physicochemical parameters studied at Belagavi city

Source month	Field no	pH	Conductivity	TDS	TH	CH	Chlorides	Alkalinity	Acidity	WQI
Dec	1	6.27	203.9	101.95	62	49	38.72	86	40	124.12
Jan	1	6.22	247.45	123.725	94	76	21.84	108	45	135.66
Feb	1	6.28	218.5	109.25	82	63	31.77	156	112	138.91
Mar	1	6.2	353.8	176.9	76	58	21.84	304	76	145.92
Dec	2	6.52	217.9	108.95	127	108	50.64	92	22	89.50
Jan	2	6.44	261.53	130.765	156	130	42.69	138	50	109.86
Feb	2	6.39	297.76	148.88	94	70	56.59	172	98	121.44
Mar	2	6.39	349.6	174.8	94	70	56.59	172	125	127.00
Dec	3	5.9	210.2	105.1	60	38	67.51	62	86	90.00
Jan	3	5.85	229.5	114.75	65	46	61.55	98	104	199.35
Feb	3	5.66	277.8	138.9	56	42	80.42	158	186	244.27
Mar	3	5.5	346.4	173.2	67	62	58.58	94	108	254.35
Dec	4	6.57	272.1	136.05	82	57	78.43	166	40	82.06
Jan	4	6.65	293.6	146.8	130	124	68.5	184	73	83.16
Feb	4	6.55	378.1	189.05	98	78	89.35	178	39	87.75
Mar	4	6.63	499.9	249.95	109	76	83.4	188	102	88.84
Dec	5	6.18	411.7	205.85	110	102	32.76	76	52	146.15
Jan	5	6.12	288.19	144.095	102	88	43.68	88	70	156.72
Feb	5	5.86	227.9	113.95	123	102	34.75	106	94	201.33
Mar	5	5.82	289.5	144.75	150	130	55.6	112	90	209.88
Dec	6	6.39	271.4	135.7	122	84	33.75	56	39	109.81
Jan	6	6.41	316.8	158.4	115	80	37.65	88	76	114.27
Feb	6	6.65	458	229	121	119	49.8	67	51	101.91
Mar	6	6.29	501.4	250.7	130	112	50.63	105	65.9	134.61
Dec	7	6.76	480.68	240.34	150	108	100.28	118	84	68.40
Jan	7	6.71	506.3	253.15	130	120	58.58	174	42	69.02
Feb	7	6.7	582.24	291.12	110	132	66.88	198.7	49	73.50
Mar	7	6.66	611.11	305.555	143	100	79.4	215.3	57	79.14

The average water quality index at field 7 is 72.515 falling under the category of poor and average value at field 4 is 85.45 and falls under the category of very poor. From Table 2 we can see the average WQI in fields 1, 2, 3, 5, 6 is (>100) and hence not suitable for drinking.

Table 3 Correlation coefficient (r) of water quality parameters

	pH	Conductivity	TDS	TH	CH	Chlorides	Alkalinity	Acidity
pH	1							
Conductivity	0.519	1						
TDS	0.519	1	1					
TH	0.519	0.422	0.422	1				
CH	0.467	0.466	0.466	**0.898**	1			
Chlorides	0.263	0.411	0.411	0.076	0.006	1		
Alkalinity	0.320	0.374	0.374	−0.016	−0.004	0.223	1	
Acidity	−0.560	−0.147	−0.147	−0.401	−0.434	0.204	0.135	1
	Strong 1	**Moderate 10**	**Weak 8**	**Negative 7**				

The value 0.898 is made bold to highlight the strongly correlated value between CH and TH
Similarly parameters which are strong, moderate, weak and negative are highlighted in bold

3.1 *Estimate of Water Quality in Terms of Correlation and Regression Analysis*

The regression analysis of the groundwater quality parameters has mild degree of significance in the correlation coefficient as indicated in Table 3. Pearson's correlation coefficient (r) measures the linear correlation and determines the correlation between parameters.

From Table 3, we can observe that CH and TH are strongly correlated ($r = 0.898$). Conductivity, TDS and TH are correlated with pH with a value of 0.519. In Contrast acidity has negative correlation with pH (−0.560), Conductivity (−0.147), TDS (−0.147), TH (−0.401), CH (0.434). These parameters are responsible to decrease acidity in the water. The correlation among few of these parameters has some level of influence on the water quality.

3.2 *Linear Correlation Coefficient and Regression Equation for Different Set of Parameters*

The regression equation is used to determine the existence of relationship between several dependent features of water quality by substituting the value for the independent parameters in a mathematical form [20, 21]. The relation between concentrations of every constituent in concern is found and a regression equation is formulated. Table 4. Summarizes the linear correlation and regression equation for different set of parameters showing significant relation.

From Table 4, the values indicate weak correlation within alkalinity and CH ($R = 0.004$), Chlorides and TH ($R = 0.076$), Chlorides and CH (0.006), Alkalinity and Chlorides ($R = 0.223$). Considerably higher correlation is observed within electrical

Table 4 Summary of linear correlation and regression equation for various pairs of parameters

Set of parameters	Regression coefficient		Regression equation	R^2	P	F
	A	B				
EC-pH	179.575	−789.438	EC = −789.483 − 179.575 pH	0.519	0.041	9.563
TDS-pH	89.787	−394.719	TDS = −394.719 − 89.787 pH	0.519	0.041	9.563
TH-EC	0.104	69.949	TH = 69.949 − 0.104 EC	0.422	0.000	5.619
TH-TDS	0.208	69.949	TH = 69.949 − 0.208 TDS	0.422	0.000	5.619
CH-EC	0.114	47.395	CH = 47.395 − 0.114 EC	0.466	0.005	7.216
CH-TDS	0.228	47.395	CH = 47.395 − 0.228 TDS	0.466	0.005	7.216
Chlorides-pH	15.791	−44.144	Chlorides = −44.144 − 15.791 pH	0.263	0.543	1.936
Chlorides-EC	0.071	31.003	Chlorides = 31.003 − 0.071 EC	0.411	0.010	5.291
Chlorides-TDS	0.142	31.003	Chlorides = 31.003 − 0.142 TDS	0.411	0.010	5.291
Chlorides-TH	0.053	49.809	Chlorides = 49.809 − 0.053 TH	0.076	0.003	0.151
Chlorides-CH	0.004	55.053	Chlorides = 55.053 − 0.004 CH	0.006	0.000	0.001
Alkalinity-CH	−0.007	134.913	Alkalinity = 134.913 + 0.007 CH	0.004	0.001	0.000
Alkalinity-Chlorides	0.616	100.129	Alkalinity = 100.129 − 0.616 Chlorides	0.223	0.004	1.359

conductivity (EC) and pH ($R = 0.519$), total dissolved solids (TDSs) and pH ($R = 0.519$), total hardness (TH) and electrical conductivity (EC) ($R = 0.422$). Correlation between total hardness (TH) and total dissolved solids (TDS) is ($R = 0.422$), calcium hardness (CH) and electrical conductivity (EC) are correlated with a value of ($R = 0.466$). Calcium hardness (CH) and total dissolved solids (TDS) are correlated with a value of ($R = 0.466$).

4 Conclusion

- The water quality analysis from the 7 selected points in South Belgaum City is carried out using weighted arithmetic WQI method. Based on the experimental analysis we can conclude that among the 7 selected areas field 7. Inox and field 4. Angol have shown comparatively low WQI ranging from 72.515 to 85.45 although they are lesser than the other areas but is still of poor quality to be recommended for human consumption.
- The overall water quality index in the selected sampling points fall under the category of poor water quality to unsuitable.
- Anthropogenic impacts likely correlate with sewage discharge from residential areas.
- The results of the correlation studies indicate strong correlation between calcium hardness and total hardness. pH is moderately influenced with Conductivity, TDS, TH,CH. Conductivity is also moderately influenced by TH,CH and Chlorides. TH, CH, Chlorides has influence on TDS.
- It can be established that from the values of water quality index, ground water requires treatment before consumption and also plan an appropriate water management facility.

Acknowledgements Snehal D. Renake thanks the Technical Assistant Joyceton Horta, Department of Civil Engineering for their co-operation during the water quality analysis. Authors are grateful to the KLE Dr. M.S.S.C.E.T for extending research facilities to carry out this study.

References

1. Yusuf KA, Oluwole SO, Abdusalam IO, Majolagbe AO (2013) Assessment of spatial variation of surface water quality in Lagos, using multivariate statistical techniques. J Environ 2:94
2. Khatri N, Tyagi S (2015) Influences of natural and anthropogenic factors on surface and groundwater quality in rural and urban areas. Front Life Sci 8(1):23–39
3. Bhatnagar A, Devi P (2016) Impact of mass bathing and religious activities on water quality index of prominent water bodies: a multilocation study in Haryana, India. Int J Environ Sci 3(2)

4. Mukate SV, Panaskar DB, Wagh VM, Baker SJ (2020) Understanding the influence of industrial and agricultural land uses on groundwater quality in semiarid region of Solapur. J Environ Dev Sustain 22(4)
5. Raju NJ, Shukla UK, Ram P (2011) Hydrogeochemistry for the assessment of groundwater quality in Varnasi: a fast-urbanizing centre in Uttarpradesh, India. J Environ Monit Assess 173:279–300
6. Singh DK, Singh AK (2002) Groundwater situation in India: problems and perspective. IJWRD 18(4):563–580
7. Pophare AM, Lamsoge B, Katpatal Y, Navale VP (2014) Impact of overexploitation on groundwater quality: a case study from WR-2 Watershed, India. J Earth Syst Sci 123(7):1541–1566
8. Shah T (2005) Groundwater and human development: challenges and opportunities in livelihood and environment. J Water Sci Technol 51(8):27–37
9. Tularam GA, Krishna M (2009) Long term consequences of groundwater pumping in Australia: a review of impacts around the globe. JASES 4(2):151–166
10. Kumar S, Meena HM, Verma K (2017) Water pollution in India: its impact on the human health: causes and remedies. IJAES 12(2):275–279
11. Mucchandi SS, Raikar RV, Virupakshi AS, Pharalad P (2017) Assessment of lake water quality using factor analysis: a case study for North Belgaum City, India. Asian J Chem 29:1
12. Naik VK, Suresh B (2015) Assessment of ground water quality outside Belgaum City, Karnataka State, India. JMESS 1(1)
13. Bora M, Goswami DC (2017) Water quality assessment in terms of water quality index (WQI): case study of the Kolong River, Assam, India. J Appl Water Sci 7:3125–3135
14. Daraigan SG, Wahdain AS, Ba-Mosa AS, Obid MH (2011) Water quality index and correlation study for the assessment of water quality and its parameters of Yercaud Taluk, Selam District, TamilNadu, India. Int J Environ Sci 1(1):139–149
15. Annan K (2005) Water for life decade. United Nations Department of Public Information, 32948, DPI/2378
16. Indian Standard Drinking Water Specification, IS: 1050-2012 (2012)
17. Tyagi S, Sharma B, Singh P, Dobhal R (2013) Water quality assessment in terms of water quality index. J Water Res 1(3):34–38
18. Brown RM, McClelland NI, Deininger RA, Tozer RG (1970) A water quality index: do we dare? J Water Sewage 117(10):339–343
19. Vasanthavigar M, Srinivasamoorthy K, Gandhi R (2010) Application of water quality index for groundwater quality assessment: Thirumanimuttar sub-basin, Tamilnadu, India. J Environ Monit Assess 171(1–4):595–609
20. Jothivenkatachalam K, Nitya A, Chandra Mohan S, Rasayan (2010) Correlation analysis of drinking water quality in and around Perur block of Coimbatore District, Tamil Nadu, India. J Chem 3(4):649–654
21. Kumar N, Sinha DK (2010) Drinking water quality management through correlation studies among various physico-chemical parameters: a case study. Int J Environ Sci 1(2)

Recent Advances in Construction of Masonry Structure by Waste Materials

Kumble Pooja and Prashanth Shreelaxmi

Abstract Masonry being one of the oldest form of construction built by joining building units with gluing agents termed as mortar. Conventional masonry is built by using adobe, bricks and cement. Use of these materials adversely affects the environment by marking large carbon foot print and huge consumption of natural resources. In the present review, use of different waste materials in the development of sustainable, alkali-activated masonry units and hence their properties in the masonry system are studied. These masonry structural based material are free of OPC. Development of masonry structural components using alkali activation of industrial wastes is expected to lead to both an economic and sustainable option in the building sector. This study is in two parts. Firstly, alkali activation of industrial wastes in the development of building blocks and in the second phase, bricks with different types of mortars are discussed. Literature review revealed that there is a potential use of industrial wastes in construction as building units with different alkali activators of different proportions and concentration to achieve the required physio-chemical and durability properties. A knowledge about the alkali-activated products are not widely commercialized. Hence more studies on alkali activation of waste materials are still required in order to develop a code of practice for a masonry system based on this technology as an alternate option to the conventional masonry which intends sustainability.

Keywords Building units · Mortar · Alkali activation · Masonry structure

K. Pooja (✉) · P. Shreelaxmi
Department of Civil Engineering, Manipal Institute of Technology, Manipal, India
e-mail: poojakumblek@gmail.com

P. Shreelaxmi
e-mail: shreelaxmi.p@manipal.edu

L. Nandagiri et al. (eds.), *Sustainability Trends and Challenges in Civil Engineering*, Lecture Notes in Civil Engineering 162, https://doi.org/10.1007/978-981-16-2826-9_54

1 Introduction

Construction sector is upraising due to the demand for infrastructure by increasing population and the resources required to meet the demand of construction is high. Conservation of natural resources is of key importance to the society. To compensate for large-scale use of natural resources which is also leading to large consumption of energy, waste materials can be used as an alternative. Wastes are the materials that have no further use. Wastes are generated during human activities such as extraction of resources, processing them and also in the consumption of final products. Different types of wastes which can be used in the construction industry are agricultural wastes, industrial wastes, construction and demolition wastes, E-wastes, medical wastes, plastic wastes, etc.

India generates 62 million tons of wastes every year of which 60% of it is collected and 15% is processed, remaining wastes end up in landfills [1]. This is a matter of concern of environmental pollution in case of efficient management. Consumption of wastes as the remedy of waste management is primarily important in the view of resource management. Efficient use by recycling or reusing of waste materials has to be planned upon for a sustainable environment. Thus, the waste materials can be incorporated in the construction sector. The manufacturing of conventional building materials like cement, clay bricks etc., produce more wastes and pollutants and also releases huge amount of CO_2 to the atmosphere, so this issue needs to be mitigated from the society by using the industrial by-products as raw materials in construction sector.

Different industries generates variety of wastes such as ash formed by burning Bagasse, Rice husk, coal, etc. and also ground granulated blast furnace slag, boiler slag, etc. in large quantities as industrial by-products. Cement is one of the major materials used widely in construction, but it has adverse effect on the environment as its manufacturing process is not a sustainable one. To overcome the problem in producing a cementitious material several industrial wastes are to be incorporated in the preparation of binder using alkali activation method. Thus, the requirement of sustainability can be met with the use of industrial wastes in the construction.

Masonry is the structural element used as load bearing/non-load bearing walls. Earlier adobes were used [2] for construction of walls, then the sun-dried clay bricks were used [3], later burnt clay bricks were used to obtain more strength and durability [4]. But the clay used is the fertile top layer of the soil whose continuous use is not sustainable. Then, the production of cement blocks was evolved with different types and shapes. But the use of cement in manufacturing building units does not meet the criteria of sustainability due to several adverse reasons. Incorporation of industrial by-products as raw materials in the preparation of building bricks/blocks will be of benefit to the society. So in order to make a walling material sustainable, present trend is toward alkali activation of materials, which tends to form cementitious substance without using OPC. Many researchers have studied on alkali activation of different waste materials as in use for building units, building mortars as well as to substitute concrete. Industrial wastes which has large amount of silica and alumina are suitable for alkali activation. Activation of chemical compounds by alkali materials results to form a gel, hardens at a particular curing condition to obtain required strength and durability [5].

2 Review on Incorporation of Waste Materials in Masonry Systems

Extensive research on efficiently using waste materials in the preparation of bricks with different mortars to form a masonry wall are discussed in the following two sections. Many researchers have examined the alkali activation of wastes with various types of activators.

2.1 Development of Bricks/ Blocks from Industrial Wastes Through Alkali Activation

This process substitutes a portion or entire amount of cement and clay with the industrial wastes for the development of bricks or blocks used for building walls. This section reports the literature of different industrial wastes and binder materials used in the manufacturing of bricks or blocks, the mechanical properties of the masonry units and the structural elements subjected to different types of loads and also the thermal properties and durability of it. The following section presents the literature pertaining to masonry units and their properties (see Table 1).

Diop and Grutzeck [6] investigated the strength and durability of Bafoundou tuff (Alumino silicate rich rock formed by the consolidation of volcanic ash) based brick. Bricks were prepared with the addition of caustic soda (sodium silicate and sodium hydroxide) and water to the basic mix of Bafoundou tuff and keeping them under the sun for curing. This resulted in increased durability without changing the process of manufacturing the brick. Different molars of alkali solutions (4, 8 and 12 Molar NaOH) were mixed with Bafoundou tuff considering curing as a function of both temperature and time. It was found that 8M NaOH was sufficient to give good strength at reduced cost. If stronger brick is required then 12 molar NaOH can be used [6].

Zhao et al. [7] developed water absorbing material with the use of coal flyash and fluxing agent (sheet glass powder) by drying the mixture at around 100 °C for 24 h and milling them to form a homogenized material. 30 wt% of NaOH solution was added to the homogenized material to alkali-activate the mix. 10% additives and 13% foaming agent were added and obtained shaped aggregates by pouring the slurry into the mold. The molds were subjected to sintering in a muffle at 1050 °C for 2 h then cooled at 1.5 °C/min. Strong and porous aggregates were produced due to the presence of silico-aluminates. Compressive strength was enhanced by potential

Table 1 Alkali-activated building units made of industrial wastes

Raw materials	Alkali activator	Properties studied	Findings	References
Bafoundou tuff	Caustic soda (sodium silicate) and sodium hydroxide	Microstructure, compressive strength, X-ray diffraction	8M NaOH was sufficient to give good strength at reduced cost.12 molar NaOH gave stronger brick	[6]
Coal flyash, fluxing agent 10% additives 13% foaming agent	30 wt% of NaOH	Apparent density, water absorption, scanning electron microscope (SEM), compressive strength, X-ray diffraction (XRD)	Strong and porous aggregates were produced due to the presence of silico-aluminate	[7]
Waste kaolin [Sandy (RGC) and clay (RFC)]	Calcium hydroxide and Sodium silicate	Compressive strength, X-ray diffraction (XRD), microstructural analysis	Pozzolonic reaction was not observed without calcination of RFC, After calcination strength was developed in both activations RGC, with and without calcination did not perform well in increasing mechanical strength	[8]
Class F fly ash and GGBFS, manufactured sand (M-sand), recycled water	Sodium hydroxide and Sodium silicate	Dry density, initial rate of absorption, water absorption, scanning electron microscope, compressive strength, flexural strength, X-ray diffraction, shear bond strength	Enhancement in mechanical strength and shear bond strength properties was found compared to traditional masonry units Strength increases with the addition of sodium silicate ratio	[9]

(continued)

Table 1 (continued)

Raw materials	Alkali activator	Properties studied	Findings	References
Coal gangue powder mixed with, urea (pore forming agent), polyvinyl	Aluminum hydroxide	Mineral phase—X-ray diffraction Porous structure—scanning electron microscopy Mechanical properties—mercury injection and compressive strength	The compressive strengths decreased with increasing mass ratio of the calcined coal gangue and $Al(OH)_3$	[10]
Cement, gypsum, lime and fly ash or sand, water and aluminum	Calcium hydroxide	Manufacturing process of AAC block	Blocks showed higher compressive strength and weighs 80% less compared to conventional clay bricks	[11]
Glass powder, palm oil flyash, lime, crusher dust, palm oil fibers, water	NaOH solution	Physical characteristics, mechanical properties, thermal studies and microscopic studies	Strength is gained by the reaction between glass powder and palm oil flyash with the presence of silica Palm fibers inclusion was beneficial to reduce the thermal conductivity of the bricks	[12]
Cohesive soil, and fibers of straw, gypsum, Elazig ferrochrome slag (EFS)	NIL	Compressive strength, water absorption, drying shrinkage, ultrasonic pulse velocity (UPV), density	Gypsum showed higher compression strength compared to EFS	[13]
Red clay brick waste RCBW, concrete waste CW and glass waste GW, ordinary portland cement (OPC)	NaOH or NaOH + Water glass	Physiochemical characterization through scanning electron microscope, X-ray diffraction, X-ray fluorescence, compressive strength	Best mechanical performance among the samples tested was observed for RCBW + 20% OPC + NaoH and water glass solution Second best performance was observed for CW mix	[14]

(continued)

Table 1 (continued)

Raw materials	Alkali activator	Properties studied	Findings	References
Coarse aggregate dust, gypsum, cement, fly ash	NIL	Compressive strength, efflorescence, water absorption	Lower particle size and larger curing time is responsible for better compression strength	[15]
Sand and recycled AAC powder (R-AAC)	NIL	Optical microscope, scanning electron microscope, XRD, density, water absorption, humidity, compressive and flexural strength	Fine sand replaced AAC wastes powder enhances strength of AAC	[17]
Fly ash, cement, eco sand plastics, hot water, silica fume, admixtures	NIL	Compressive strength, water absorption test, efflorescence	Thus 10 kg of silica fume when added with the mixture shows enhanced properties	[19]
flyash, foundry sand, bentonite	NaOH and Na_2SiO_3	Compressive strength, water absorption test, density, soundness, efflorescence, and hardness	Production cost is reduced Geo polymer bricks were tested for basic characteristics and found to have good performance	[21]
Co-fired blended ash (CBA) 80% with rice husk 20% and stone dust (SD)	Sodium silicate to sodium hydroxide	Physical, chemical, mineralogical and thermal stability properties	As the ash content increased there was reduction in density and compressive strength with increase in water absorption. Durability and thermal properties of the bricks proved better masonry	[22]
Fly ash, alccofine and M sand	Sodium silicate and sodium hydroxide	Compressive strength, water absorption, bulk density	The optimum compressive strength was noted for 30% replacement of alccofine to fly ash Compressive strength increases with decrease in water absorption and increase in bulk density	[23]

(continued)

Table 1 (continued)

Raw materials	Alkali activator	Properties studied	Findings	References
Fly ash, ground granulated blast furnace slag, natural coarse aggregate, natural fine aggregate, coarse recycled asphalt aggregate, fine recycled asphalt aggregate, sulphonated naphthalene formaldehyde	NaOH sol., Na_2SiO_3 sol.	Workability, unit weight, compressive strength abrasion resistance, water absorption	Workability reduces Unit weight and mechanical properties of the paver block increased Water absorption reduces with increase in RAP content Cost of manufacturing is also reduced	[24]

anorthite crystals and elongated grains. This water absorbing material results in exhibiting good physical and mechanical properties [7].

Gomes et al. [8] carried out a research to use different sizes of industrial wastes generated from kaolin processing. In different stages of processing, kaolin wastes are separated according to the size namely sandy (RGC) and clay (RFC). The study was to evaluate the ability of alkaline activators to activate waste kaolin. Sodium silicate activator was compared with calcium hydroxide. The wastes were calcined at 750 °C/2 h. RFC without calcination did not show pozzolonic activity, when calcined strength was gained with the use of both activators. RGC, both with and without calcination did not perform well in increasing mechanical strength. RFC's has more potential than RGC in preparing building blocks [8].

Venugopal et al. [9] developed the geo polymer masonry units by mixing the industrial by-products such as Class F Flyash and GGBS with manufactured sand and solutions of 8M NaOH and Na_2SiO_3. Prepared bocks were then subjected to air curing at room temperature. Alkaline solution (silicate and hydroxide) proportions were altered considering mortar proportion and fluid-binder ratio maintained constant as 1:1 and 0.20, respectively. Physical mechanical and microstructural studies were conducted. The results showed enhancement in flexural and shear bond properties of geo polymer blocks when compared to traditional masonry units. Thus, arrived at a conclusion that with increase in sodium silicate ratio strength of the geopolymer masonry blocks increases. Microstructural analysis revealed that the all the flyash particles were not hydrated fully by the activator due to its low molarity [9].

Wu et al. [10] carried out a study on water permeable nepheline based bricks. Calcination of Na_2CO_3 mixed coal gangue was done by treating it at 773 K for 2 h and then grinded to get coal gangue powder. Brick was manufactured by coal gangue powder mixed with aluminum hydroxide, urea (pore forming agent), polyvinyl acetate and water. Tests were conducted on the specimen by taking into consideration the preparation parameters such as (1) Material mass ratio kept constant with molding pressures of 180, 200, 220 MPa (2) Molding pressure of 200 MPa constant and material mass ratios was altered. Both the parameters of the bricks were studied to provide a balance between compressive strength and water permeability. The porous structure, mineral phase and mechanical properties were characterized. Results indicate that with the increase in molding pressure, and decrease in raw material ratio compression strength of the brick increased. Thus, the bricks meet the Chinese industry standard [10].

Wahane [11] studied the light weight aerated autoclaved concrete block made by a mix of cement, gypsum, lime and fly ash or sand. Aluminum powder of about 0.08% of total dry materials in the mix was added along, to accelerate the reaction with Calcium Hydroxide and water to release hydrogen gas with the creation of small air cells which tends to increase the volume of concrete by about three times its original volume. The AAC blocks were later subjected to high-pressure steam curing for further strengthening. The test results of these blocks showed higher compressive strength and weights 80% less compared to conventional clay bricks [11].

Raut and Gomez [12] worked on the development of thermally efficient bricks made of industrial wastes such as fine powdered glass and flyash obtained from palm oil industry. Tests to evaluate the physical characteristics, mechanical properties, thermal studies and microscopic studies were performed on the bricks. Lime was used as a binder material. Strength is gained by the reaction between glass powder and Palm oil flyash with the presence of silica. Palm fibers inclusion in the experiment was beneficial to reduce the thermal conductivity of the bricks [12].

Türkmen et al. [13], studied the effect of using Elazig Ferrochrome slag and gypsum in preparing sustainable building blocks. Mechanical and physical properties of unfired earth bricks (UEBs) stabilized with gypsum and EFS additives are determined. Four different types of unfired earth brick (UEB) samples made of soil, gypsum, Elazig Ferrochrome slag and fibers of straw. Compressive strength, drying shrinkage, water absorption, ultrasonic pulse velocity (UPV) test and density of the samples were tested and found that gypsum stabilizer showed higher compression strength compared to EFS stabilizer [13].

Robayo-Salazar et al. [14] elucidated the mechanical performances of alkali-activated cements AAC, made of recycling red clay brick waste RCBW, concrete waste CW and glass waste GW separated from construction demolition wastes (CDW). These materials were activated by the solution of NaOH or NaOH + water glass. Best mechanical performance among the samples tested was observed for RCBW + 20% OPC + NaoH and water glass solution. Second best performance was observed for CW mix. Alkali-activated cement can be obtained by adding appropriate proportion of activator and curing at room temperature [14].

Kaushik et al. [15] developed a 230 × 110 × 90 mm sized brick made of flyash with different particle size 425, 600, 825 μm cured for 7 days, 14 days and 21 days under the sun. To obtain the optimum mix for the development of bricks, materials such as fly ash (15–50%), cement (5–30%), gypsum (2%) and coarse aggregate dust (45–55%) were used and employed Taguchi method to evaluate the improvement of water absorption tightness, compressive strength and Efflorescence. The best optimum mix was found to be with the 10% fly ash, 35% cement, 3% gypsum and 52% coarse aggregate dust for particle size 600 μm at Curing Time 21 Days. Lower particle size and larger curing time are responsible for better compression strength [15].

Dahmen et al. [16] performed life cycle assessment of two types of blocks based on the impact on environment. The first block made use of compacted, stabilized engineered soil. Second type of blocks was made by adding alkali activator to clay for promoting geo polymerization reaction. These blocks showed same strength and durability as conventional concrete blocks. Blocks made of compacted stabilized engineered soil offers less embodied carbon than conventional concrete, but water consumption was high. Compacted alkali-activated blocks showed lesser embodied carbon than the compacted stabilized engineered soil blocks and conventional concrete blocks [16].

Rafiza et al. [17] carried out study on AAC block made by using recycled powdered AAC, a construction waste, as a replacement to sand in different ratios by weight. Microscopic analyses were conducted using optical microscope and scanning electron microscope and found that developed AAC block with very less recycled contents consists huge pores compared to the one with the highest recycled content. Chemical analysis was conducted using XRD and confirmed that 30% replacement of recycled contents increases the mechanical strength. Further increment of recycled

AAC contents mainly consist of calcite and quartz due to which the improvement in the mechanical strength cannot be expected. Mechanical properties were also tested and for 30% replacement results were found effective. Thus, it is concluded that with use of AAC recycling powder, compressive strength and flexural strength can be enhanced in comparison with the conventional AAC blocks and blocks made by utilizing industrial waste products [17].

Gavali et al. [18] studied the feasibility of using different industrial wastes and their activation through different alkali solutions for the production of strong, durable and sustainable bricks. Factors that influence the performance of alkali-activated bricks are discussed. Supplementary materials (waste materials) up to 20% can be mixed to reduce the dosage of alkali activator. Compaction of the bricks should be within the limit because it removes the liquid and affect the properties of the brick. Maximum of 25 N/mm^2 forming pressure was suggested. Curing temperature of 100 °C was found to be sufficient and any further increase in temperature induces cracks in the bricks. Better utilization of waste materials was recommended considering the physical properties of the bricks. Sodium based alkali activators were recommended than the other alkali activators. Life cycle study, embodied energy and carbon footprint of these bricks are to be studied in broader aspects [18].

Harshini et al. [19] studied the effect of addition of different amounts of silica fume, plastic wastes, BASF, Conplast SD110 and hot water on the bricks made of flyash, eco sand and cement mixture. Addition of 3.3% silica fume or plastic wastes in the mixture shows enhanced properties of strength in compression, water absorption and efflorescence were nil [19].

Zhou [20] elucidated the effect of solid activator activation on slag/flyash bricks by microstructural and chemical analysis. C-S-H gels are observed as products of alkali activation. After 28 days of activation, these gel crystalize to give solid structure and provides good bonding thus enhancing the parameters of strength [20].

Philip et al. [21] investigated the geo polymer bricks made of alkali-activated flyash, foundry sand and bentonite as additives for improving their properties. Geo polymer bricks were tested for basic characteristics and found to have good performance compared to clay burnt bricks. The use of industrial wastes (flyash and foundry sand) reduces the production cost of the bricks [21].

Gavali and Ralegaonkar [22] developed a brick made of co-fired blended ash (CBA) 80% with rice husk 20% and stone dust (SD) by studying the physical, chemical, mineralogical and thermal stability properties. To find the optimum mix design, tests were conducted altering the molarity of sodium hydroxide of 6M to 10M in intervals of 1M, keeping the sodium silicate and sodium hydroxide ratios at 1:1, 1.5:1 and 2:1 for different mix ratios of CBA:SD. The content of alkali activators with base material was fixed to 35% and found that 8M sodium hydroxide with sodium silicate and sodium hydroxide ratio 1:1 for all the three mixes of CBA:SD was effective. Bricks of size 230 × 100 × 80 mm were manufactured with the optimum mix and tested for density, compressive strength, flexural strength, water absorption, efflorescence, compressive bond strength, flexural bond strength, durability and thermal properties. By analyzing the test results, the brick masonry units with 2:1 CBA:SD ratio were found to be more efficient. As the ash content increased there was reduction in density and compressive strength with increase in water absorption. Masonry strength and bond strength increased, durability and thermal properties of the bricks were found to be better as compared to conventional flyash bricks [22].

Reema et al. [23] worked on geopolymer flyash bricks with the % variation of flyash with alccofine. The mix of flyash, alccofine and fine aggregates is activated through Sodium hydroxide and sodium silicate. Bulk density, compression strength and water absorption of bricks were tested and best properties were achieved by replacing 30% of flyash with alccofine [23].

Hossiney et al. [24] studied the alkali-activated paver blocks made of Recycled Asphalt Pavement (RAP) aggregates. Physical properties were investigated and results show that workability reduces, while unit weight, compressive strength and resistance of block to abrasion increased, Water absorption reduces with increase in RAP content. Cost of manufacturing is also reduced. Thus, they could be used for pedestrian paths [24].

2.2 *Properties of Masonry Constructed with Alkali-Activated Blocks and Different Mortars*

Bricks/blocks made of different types of industrial wastes glued with or without mortar, and their bond strengths, in the structure as a whole, are reviewed. This section deals with the properties of the wall made of different combinations of mortar and bricks (see Table 2).

Miranda et al. [25] studied the mechanical properties of earthen interlocking blocks made of Alkali Activation of flyash with mortar joint and without mortar joint were treated for shear strength, durability tests and compressive strength and it was found that there was improvement in the performance of masonry with the use of mortar joints compared without mortar joint. Hence it tends to be more brittle than masonry without mortar joints [25].

Raj et al. [26] carried out the studies to improve shear bond strength of an autoclaved aerated cement blocks. The plain blocks, single grooved (SGB) and double grooved (DGB) blocks glued with sand cement mortar as bonding material were subjected to experimentation. Bond strength against shear sliding and compressive strength of grooved AAC masonry showed that DGB masonry tenders high compressive load bearing capacity and shear strength compared to conventional AAC blocks, whereas PB masonry showed the least strength. Results were adopted by conducting simple analytical models and hypothesis testing [26].

Benedetti [27] studied the characterization of materials used for masonry. Fibre reinforced (FR) mortar of 15 mm layer on both the faces of the wall were applied and tested for Diagonal compression. The experiment was conducted with different combinations of mortars and units. Homogenization based on elements in series and parallel combination exhibits better performance [27].

Table 2 Bond strength properties of masonry components

Materials/methodology	Study	Purpose	Key findings	References
Interlocking compressed earth blocks (ICEB) and soft mortar (AABEM1) and stiff mortar (AABEM2) Both were produced by soil, flyash and alkali activators	Two types of mortars ere used as bed joint mortars ICEBs laid dry-stack (DSS) and both mortared joints, were tested for mechanical properties	To understand the importance of the stiffness difference between units and joints on the mechanical performance of the masonry	Improvement in the properties with the use of mortar joints compared to joints without mortar was found effective	[25]
The 220 × 120 × 140 mm^3 size plain block Single groove block Double groove block Sand-cement mortar	Bond strength against shear sliding and compressive strength of grooved AAC masonry was studied. The models were developed to simplify and compute the compressive and shear strength of the masonry	To investigate shear bond strength of AAC masonry enhancement by using grooved AAC blocks	DGB masonry tenders high compressive load bearing capacity and shear strength compared to conventional AAC blocks Whereas, PB showed the least strength	[26]
The mortars includes the]reference mortar used in composing the walls (REF), two high strength mortars (BTF, BTG) and one high adhesion mortar (BTC) The walls were reinforced with three different lime mortar compounds with layer thicknesses of 12, 15 and 30 mm, and reinforced with two different glass fibre meshes and two different arrangements of steel micro- wire strips	14 diagonal compression tests of masonry system reinforced with GFRM were built encompassing different FRM combinations were tested Finite element model was prepared and based on this a new theoretical model was defined	To study the benefit of reducing mortar thickness in the external layers and also increase mortar strength by using Fibre Reinforced Mortar (FRM)	With even less than 15 mm of strengthening layers on the two wall faces. The shear strength increased more than 100% Homogenization based on elements in series and parallel combination exhibits better performance The masonry wall made of this technique is tested experimentally and analytically in the diagonal stressing type to understand the failure pattern	[27]

(continued)

Table 2 (continued)

Materials/methodology	Study	Purpose	Key findings	References
Commercially available lightweight AAC blocks were used AAC block of size 600 × 200 × 100 mm were cut into cubes of 50, 75, 100 mm in the direction perpendicular to the rise (along the length of the block) Two types of ACC block assemblages of block size 200 × 100 × 100 mm were prepared using high strength thixotropic mortar	Investigated the physical properties, and mechanical properties Load was applied on the cubes in two directions. Compressive strength was evaluated and found that peak stress decreases, while the strain increases as the size of AAC blocks increases	Structural properties and overall behavior of the AAC blocks are studied in order to compute the models for AAC masonry-infilled framed structures	By compressive strength test peak stress decreases while the strain increases as the size of AAC blocks increases Compressive strength of specimen tested in perpendicular direction of rise was greater than specimen tested in parallel to rise Mathematical expressions of joint strength and total joint force were derived by linear and parabolic stress distribution of joints	[28]
Lightweight plasters were manufactured of 1. Hydrated lime mortar 2. Ground granulated blast furnace slag and two innovative alkali-activated slag-based mixtures Sodium metasilicate pentahydrate:potassium hydroxide:sodium carbonate = 7:3:1 The alkaline reagents equal to 20% and 24% by binder mass was used	1. To evaluate the essential characteristics of the mortars at different lightweight aggregate content and to identify the ideal mix proportioning that meets the general requirements in terms of strength and specific mass 2. Previously selected mortar were tested to obtain a more detailed characterization of its behavior, and to evaluate the compatibility between the mortar and the GFRP mesh	Development of lightweight cement-free reinforced plaster to be applied in GFRP-reinforced jacketing interventions for the energy up-grading and seismic retrofitting of poor quality masonry buildings The effectiveness of the reinforcement deriving from the use of alkali-activated slag-based plasters on three leaf masonry wall systems were studied	Portland-free plaster is able to provide a 28-day compressive strength equal to 8 MPa and a thermal conductivity of 0.35 W/m K due to density close to 700 kg/m^3 By using an alchil-alcoxisilane-based coating is possible to reduce the water absorption of mortar up to 80% Use of glass fibre reinforced mesh was best suited to alkaline environments	[29]

(continued)

Table 2 (continued)

Materials/methodology	Study	Purpose	Key findings	References
CSEB was made of 4 variants of cement and laterite mixed with or without sand and motor oil In the first phase These CSEB was tested for different properties compared to sandcrete blocks In the second phase, all the four different types of CSEB are tested against six varieties of mortars (Cement mortar, earth mortar, cement + earth (contains laterite)] mortar with and without oil)	Dry shear bond strength and wet shear bond strengths between the CSEBs and mortars were determined by an innovative loading system	Improving adhesive bond between compressed stabilized earth bricks (CSEB) and mortar	Plain CSEB absorbs less water and there was no improvement in the resistance to water absorption with the addition of motor oil and sand Increase in dry unconfined compressive strength and decrease in wet unconfined Best bond was cement + earth (contains laterite) mortar with earth for all the variants of CSEB in wet condition	[31]
Three mortar mixes 1:3, 1:4, 1:5 were prepared in varying percentage 0–30% increment of 5% styrene-butadiene rubber (SBR) latex into the mortar These walls are tested for compressive strength, stiffness, ductility, microstructural properties and the impact of lateral forces	1. Compressive strength of cement mortar 2. Compressive strength of brick masonry 3. Cyclic load test investigates the bond characteristics of infilled masonry walls built by mortar modified with latex	To understand the influence of styrene-butadiene rubber (SBR) masonry subjected to cyclic load	Optimum percentage of 20% latex modified rubber can be used for achieving better compressive strength In comparison of mortar with and without addition of latex modified rubber, there showed increase in pre yield stiffness and ductility ratio by 13% with the addition of latex	[32]

Bhosale et al. [28] investigated the physical and mechanical properties of AAC masonry infilled frame structure and mathematical modeling of the same was discussed. AAC blocks of size 600 × 200 × 100 mm were cut into cubes of 50, 75, 100 mm along the length of the block and the load was applied on the cubes in two directions. Strength of the specimens tested for compressive strength was evaluated and found that, peak stress decreases while the strain increases as the size of AAC blocks increases (from 50 to 100 mm). Specimen tested for compressive strength in perpendicular direction of rise were greater than those of specimen tested parallel to rise. In the second stage, bond and tensile strength of two types of AAC block assemblages made of 200 × 100 × 100 mm size ACC blocks and high-strength thixotropic mortar were evaluated and found that initial rate of absorption of the AAC blocks were same as clay brick, hence does not affect the bond strength of the masonry. Mathematical expressions of joint strength and total joint force were derived by linear and parabolic stress distribution of joints [28].

Coppola et al. [29] evaluated the characteristics of plaster made up of different proportions of ground granulated blast furnace slag, hydrated lime, light weight aggregates, air entraining agents and alkaline reagent (sodium metasilicate pentahydrate: potassium hydroxide: potassium carbonate in the ratio 7:3:1) by the binder mass. Density and compressive strength of the plasters were tested and found that the plaster with light weight aggregate (Expanded recycled glass aggregates) improved the performance of the plaster but water absorption rate was high. By the application of alchil-alcoxisilane-based coating water absorption decreased. Use of glass fibre reinforced mesh was best suited to alkaline environments [29].

Henrique Nalon et al. [30] investigated the effect of cement-lime mortar and grout on the mechanical properties and structural behavior of stack bonded hollow concrete block prism. Results of the failure mechanism explained the effect of prism strength and stiffness by the variation in component properties of lime cement mortar and hollow concrete blocks. Development of strength prediction models for different cases of failures modes improved the predictive ability of current design code formulation. The structural performance was better elucidated on the impact of incompatibility of component characteristics [30].

Kolawole et al. [31] aimed at enhancing the bond between the compressed stabilized earth bricks (CSEB) and mortar. CSEB was made of 4 variants of cement and laterite mixed with or without sand and motor oil. These CSEB was tested for moisture content, Atterberg limit, water absorption and compressive test in comparison with sandcrete blocks. The results showed that plain CSEB absorbs less water and there was no improvement in the resistance to water absorption with the addition of motor oil and sand, this is because of the use of finer material (laterite). Increase in dry unconfined compressive strength and decrease in wet unconfined compressive strength of CSEB with the addition of motor oil and sand were found. In the second phase, all the four different types of CSEB were tested against six varieties of mortars (cement mortar, earth mortar, cement + earth (contains laterite) mortar with and without oil) for wet and dry shear bond strength determination and found the best bond was achieved by Cement + earth (contains laterite) mortar with earth for all the variants of CSEB in wet condition [31].

Jebadurai et al. [32] conducted cyclic load tests to investigate the bond characteristics of infilled masonry walls built by mortar modified with latex. Three mortar mixes 1:3, 1:4, 1:5 were prepared in varying percentage 0–30% increment of 5% styrene-butadiene rubber latex into the mortar. These walls were tested for compressive strength, stiffness, ductility, microstructural properties and the impact of lateral forces. It was found that optimum percentage of 20% latex modified rubber can be used for achieving better compressive strength. Results on microstructural studies show that latex works in enhancing the bond characteristics by forming film around the cement particles, hence lateral forces can also be resisted. A comparison of mortar with and without addition of latex modified rubber showed increase in pre yield stiffness and ductility ratio by 13% with the addition of latex [32].

3 Discussions

Literature pertaining to various materials as base material along with different alkali activators for preparation of blocks/bricks are illustrated here. Earlier investigations on masonry blocks/bricks made of different industrial waste materials has been discussed by many authors. From the literature study, it is known that to improve the use of different industrial wastes to reduce landfill, construction industry is striving to incorporate these materials in making building materials such that it meets the environmental issues and also caters to the requirement of construction. Various properties of clay bricks, solid and hollow bricks with different mortars were discussed. Physical, mechanical and durability properties of different types of bricks/blocks and masonry prisms were discussed. First Section deals with the studies on use of flyash, bottom ash, and slag from different industries, construction demolition wastes and burnt bricks alkali activated and also for geopolymer bricks. Different alkali activators were discussed which suits the precursors. Thermal, physical and mechanical properties were tested and confirmed by microstructure and chemical analysis. Raw material ratio and molding pressure of the bricks were discussed. It was found that many researchers used NaOH and Na_2SiO_3 for the activation. Second Section deals with the properties of masonry made of bricks/blocks with different mortars. Bond characteristics of mortars with different masonry units were studied for different grades, types and mix proportion of the mortar. Shear strength, tensile strength, compressive strength of the mortar and bond strength are determined for the construction of brick/block prisms. The bond strength of the masonry was found to be directly related to the compressive strength of the mortar and also in order to avoid the accumulation of stress in the bed joints of the masonry assemblage, strength of the mortar used should be lower than the strength of blocks.

References

1. Swaminathan M (2018) How can India's waste problem see a systemic change? Econ Polit Wkly 53(16)
2. Illampas R, Ioannou I, Charmpis DC (2009) Adobe: an environmentally friendly construction material. WIT Trans Ecol Environ 120:245–256
3. Djamil B (2016) Sun-dried clay for sustainable constructions. Int J Appl Eng Res 11(6):4628–4633
4. Fiala J, Mikolas M, Krejsova K (2019) Full brick, history and future. IOP Conf Ser Earth Environ Sci 221(1)
5. Ahmari S, Zhang L (2015) The properties and durability of alkali-activated masonry units, vol 1. Woodhead Publishing Limited
6. Diop MB, Grutzeck MW (2008) Low temperature process to create bricks. Constr Build Mater 22(6):1114–1121
7. Zhao Y, Ye J, Xiaobin Lu, Liu M, Lin Y, Gong W, Ning G (2010) Preparation of sintered foam materials by alkali-activated coal fly ash. J Hazard Mater 174(1–3):108–112
8. Gomes KC, Rocha BD, Ferreira DTA , Lira EC, Torres SM, De Barros SR, Barbosa NB (2012) Activation alkaline waste kaolin for fabrication of building blocks. Key Eng Mater 517:622–627
9. Venugopal K, Radhakrishna, Sasalatti VM (2016) Ambient cured alkali activated flyash masonry units. IOP Conf Ser Mater Sci Eng 149(1)
10. Xiu-Wen Wu, Ma H-W, Nan Wu, Shi C-Q, Zheng Z-Y, Wang Y-F (2014) Nepheline based water permeable bricks from coal gauge and aluminium hydroxide. Environ Prog Sustain Energy 33(3):676–680
11. Wahane A (2017) Manufacturing process of AAC block. Int J Adv Res Sci Eng 6(2):4–11
12. Raut AN, Gomez CP (2017) Development of thermally efficient fibre-based eco-friendly brick reusing locally available waste materials. Constr Build Mater 133:275–284
13. Türkmen İ, Ekinci E, Kantarcı F, Sarıcı T (2017) The mechanical and physical properties of unfired earth bricks stabilized with gypsum and Elazığ ferrochrome slag. Int J Sustain Built Environ 6(2):565–573
14. Robayo-Salazar RA, Rivera JF, Mejía de Gutiérrez R (2017) Alkali-activated building materials made with recycled construction and demolition wastes. Constr Build Mater 149:130–138
15. Kaushik S, Raj N, Alam P, Ali S, Palariya NC (2017) Optimization of compressive strength for fly ash building bricks using Taguchi method. 8(1):394–402
16. Dahmen J, Kim J, Ouellet-Plamondon CM (2018) Life cycle assessment of emergent masonry blocks. J Clean Prod 171:1622–1637
17. Rafiza AR, Chan HY, Thongtha A, Jettipattaranat W, Lim KL (2019) An innovative autoclaved aerated concrete (AAC) with recycled AAC powder for low carbon construction. IOP Conf Ser Earth Environ Sci 268(1)
18. Gavali HR, Bras A, Faria P, Ralegaonkar RV (2019) Development of sustainable alkali-activated bricks using industrial wastes. Constr Build Mater 215:180–191
19. Harshini J, Abinaya D, Manikandan A, Srinivasan K, Natarajan N (2019) Performance of fly ash bricks with differential composition, no 1, pp 4550–4556
20. Zhou J (2019) Study on composite solid alkali-activated slag/fly ash concrete. IOP Conf Ser Earth Environ Sci 384(1)
21. Philip S, Ajin A, Shahul F, Babu L, Tina J (2019) Fabrication and performance evaluation of lightweight eco-friendly construction bricks made with fly ash and bentonite. Int J Eng Adv Technol 8(6):2090–2096
22. Gavali HR, Ralegaonkar RV (2020) Design development of sustainable alkali-activated bricks. J Build Eng 101302
23. Reema KR, Shajan SS, Soumya VS, Vinod S, Jose V. Experimental Investigation on geopolymer masonry units. In: Dasgupta K et al (eds.) Proceedings of SECON'19. Springer International Publishing, Berlin

24. Hossiney N, Sepuri HK, Mohan MK, Arjun HR, Govindaraju S, Chyne J (2020) Alkali-activated concrete paver blocks made with recycled asphalt pavement (RAP) aggregates. Case Stud Constr Mater 12:e00322
25. Miranda T, Silva RA, Oliveira DV, Leitão D, Cristelo N, Oliveira J, Soares E (2017) ICEBs stabilised with alkali-activated fly ash as a renewed approach for green building: exploitation of the masonry mechanical performance. Constr Build Mater 155:65–78
26. Raj A, Borsaikia AC, Dixit US (2019) Compressive and shear bond strengths of grooved AAC blocks and masonry. Mater Struct Constr 52(6):1–15
27. Benedetti A (2019) In plane behaviour of masonry walls reinforced with mortar coatings and fibre meshes. Int J Archit Herit 13(7):1029–1041
28. Bhosale A, Zade NP, Davis R, Sarkar P (2019) Experimental investigation of autoclaved aerated concrete masonry. J Mater Civ Eng 31(7):1–11
29. Coppola L, Co D, Crotti E, Marini A, Passoni C, Pastore T (2019) Lightweight cement-free alkali-activated slag plaster for the structural retrofit and energy upgrading of poor quality masonry walls, vol 104
30. Henrique Nalon G, Santos CFR, Pedroti LG, Ribeiro JCL, de Veríssimo GS, Ferreira FA (2020) Strength and failure mechanisms of masonry prisms under compression, flexure and shear: components' mechanical properties as design constraints. J Build Eng 28
31. Kolawole JT, Olalusi OB, Orimogunje AJ (2020) Adhesive bond potential of compressed stabilised earth brick. Structures 23:812–820
32. Jebadurai SVS, Tensing D, Pradhan PM, Hemalatha G (2019) Enhancing performance of infill masonry with latex modified mortar subjected to cyclic load. Structures 23:551–557

Comparison of Hyperspectral Atmospheric Correction Algorithms for Precise Mapping of Rice Crop

Balla Vivek, B. E. Bhojaraja, and Amba Shetty

Abstract For millions of people, rice means life, and therefore, it is harvested in many regions of the world. Two rice species are primarily cultivated in the world, namely Asian and African rice. It grows primarily in major river deltas, such as Asia and Southeast Asia. Conventional method of mapping rice crop area is tedious and time-consuming job and more often subjected to erroneous results. In this study advanced remote sensing technique is used for mapping, to map precisely hyperspectral remote sensing with different atmospheric algorithms were compared for better accuracy. Also different supervised classification techniques were compared for the accurate area mapping of rice crop. The ASD field spec Pro hand held spectroradiometer is used for reference spectra collection. And high accuracy GPS device is used to collect ground truth information. Results show that both FLAASH and HAC algorithms produce a good spectrum with respect to the rice spectra obtained from ASD handheld spectroradiometer. SAM image classification and Parallelepiped classifier were used for classification of imagery. From the accuracy assessment performed, accuracy of 88% by using SAM and 84% obtained using Parallelepiped classifier for Hooghly region and 93% using SAM and 87% using Parallelepiped for West Godavari region. From the study, it was found that the best approach for rice crop mapping in Hooghly and West Godavari is SAM classification. The study helps to map the rice crop area accurately; it can be used for yield estimation, indirectly which is helpful for policy makers and to estimate the export, import potential.

Keywords Atmospheric correction algorithms · Rice crop · Supervised classification · Rice crop · Mapping

B. Vivek · A. Shetty
Water Resourses and Ocean Engineering Department, NITK Surathkal, Surathkal, India

B. E. Bhojaraja (✉)
Civil Engineering, NMAM Institute of Technology, Udupi, India

L. Nandagiri et al. (eds.), *Sustainability Trends and Challenges in Civil Engineering*, Lecture Notes in Civil Engineering 162, https://doi.org/10.1007/978-981-16-2826-9_55

1 Introduction

For millions of people, rice means life, and therefore, it is harvested in many regions of the world. Two rice species are primarily cultivated in the world, namely Asian and African rice. It grows primarily in major river deltas, such as Asia and Southeast Asia. Oryza sativa (Asian rice) and Oryza glaberrima (African rice) [1]. In major river deltas such as Asia and Southeast Asia, it grows mainly [2]. Rice has been cultivated for over 10,000 years, mainly in Asia, but has developed in emerging economies like India. Moharana et al. [3], and it is the second largest rice producers in the world. From the last financial year analysis, India produced over 100 million tons of rice [4]. Becoming the largest rice producer, India has been one of the world's largest producers of rice. From the last financial year analysis, it is also estimated that India exported over 8 million tons of rice. UAE, Saudi Arabia, Senegal, Iran and South Africa are importing rice from India, and also it is seen that rice cultivation is the serious business module in India [4]. Among the 29 states in India more than 20 states are best in producing rice. Some of the largest rice-producing states are West Bengal and Andhra Pradesh.

About 2000 million people in Asia receive 60–70% of the calories from rice and its related products in Asia [5]. The United Nations General Assembly proclaimed 2004 the "International Year of Rice" (IYR) by identifying the essential role and significance of rice. The theme of the "International year of rice" is "Rice is life". In India rice is the staple food for above 65% of the people and it is part of every meal for Indian people and it shows that rice cultivation is how much important for people. Demand for rice is expected to increase day by day due to rapid increase in population, so cultivation and production of the rice also need to be increased. Day by day the agricultural land under rice cultivation is reducing and hence, there is a need to increase the rice productivity and rice cultivation [6].

West Bengal is called "rice granary of India", state has 50% of land which is arable is under rice cultivation. In West Bengal, Hooghly is one of the districts which comes under high productivity group (High productivity—Yield more than 2500 kg/ha) [7]. Andhra Pradesh is another leading rice cultivated state in India, which contributes 12% of the total rice producing in India [8], In Andhra Pradesh, West Godavari is known as "rice granary of Andhra Pradesh" [9].

The aim of the present study is to find the influence of atmospheric correction algorithms of rice agricultural fields in various districts in India using existing spectrum. And to compare four popularly used atmospheric correction algorithms (QUAC, ATCOR, FLAASH, 6S) in deriving rice spectra from Hyperion data in agricultural fields of West Godavari district in Andhra Pradesh and Hooghly district in West Bengal. Also to compare HAC (hybrid atmospheric correction algorithm) with the best atmospheric correction algorithm among four component algorithms and the spectra obtained from ASD handheld spectroradiometer in deriving rice spectra from Hyperion data. Finally, to map the rice crop by using the Spectral Angle Mapper classifier and Parallelepiped classifier techniques.

From the literature review, it is evident that there is no study is available for rice crop using the hyperion data and the influence of the different atmospheric correction algorithm on the paddy reflectance spectrum by using the hyperspectral remote sensing data. Despite the extensive coverage of spaceborne hyperspectral images, a few studies focus on the investigation of rice crops. Advances in technology in this sector are anticipated to attract more research and boost the use of these information.

2 Methodology

2.1 Study Areas

The study area selected for investigation is located in different parts of India. The criteria used in selection of study region is that they should be predominantly rice growing regions. Accordingly, the following districts were chosen for analysis the rice crop.

The districts are West Godavari and Hooghly, West Godavari district (16.9174° N, 81.3399° E) or paschima Godavari Jilla is one of the districts of state of Andhra Pradesh. West Godavari district situated at the coastal Andhra region of Andhra Pradesh. The district is a flat and slightly slope on the eastern side along the rivers of the West Godavari districts flow. It is bounded by the Bay of Bengal on the south and the Khammam district on the North and Godavari river separates it from the east and Tammileru River and Kolleru lake separate it from west. Rivers in the district generally flows from west to east, rivers like Godavari, Yerrakaluva and Tammileru flows through the district. It is known as "rice granary of Andhra Pradesh". West Godavari yields rice about approximately 3322 kg/ha every year. Alluvial, Black cotton and Red Ferruginous and a small belt along the coastal belt of arenaceous sandy soils are the Soils form.There are eight major types of soils in the District namely Red Sandy loams, Clay loams, Alluvial, Sandy Alluvial, Deltaic Alluvial, Black cotton. Study area is dominated by black cotton soil.

Hooghly district (22.8963° N, 88.2461° E) is district of the West Bengal in India. The district is situated on the bank of the Hooghly River; hence, it is named as the Hooghly. The land of the Hooghly is totally flat and the land elevation is not more than 200 m. The River Hooghly borders it to the east. Figure 1 shows the location map of the study areas.

3 Overall Methodology

In Fig. 2, the overall approach adopted for the analysis is shown. It shows the steps adopted for the study.

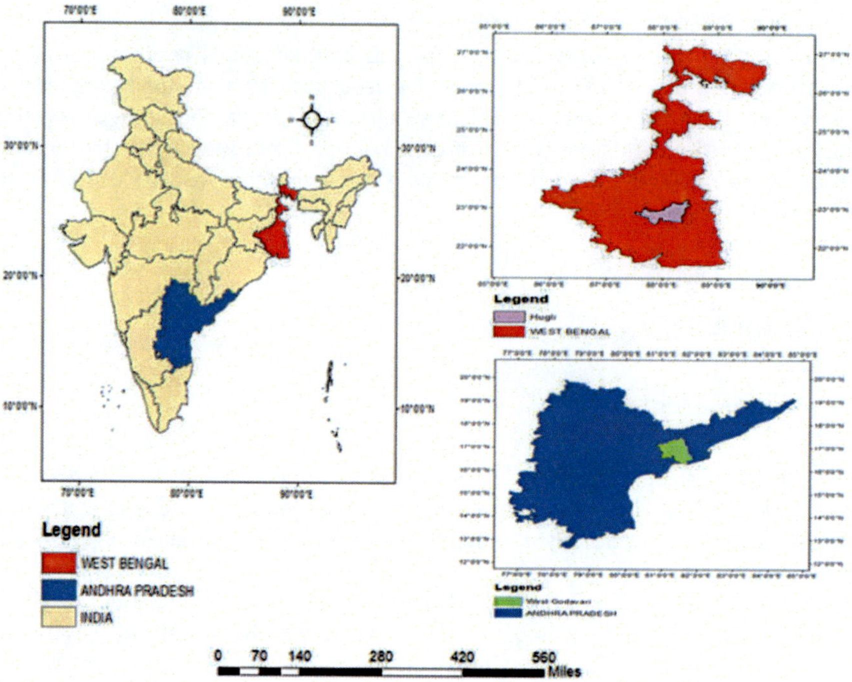

Fig. 1 Location map of the study areas

1. **Global Positioning System Data**
   GPS points of 12 locations were collected over paddy fields from Hooghly sites near to hyperion satellite data acquisition. Rice sites were geolocated using the Global Positioning System (GPS with 2 m accuracy). These locations were collected to ensure pure rice spectra over the Hyperion scene.
2. **Spectroradiometric Data**
   The spectral signatures of the paddy field were collected using ASD fieldspec Pro spectroradiometer. The collected spectral signatures were used to build a spectral library utilized for supervised classification.

Ancillary Data

The mean temperature, mean pressure and daily rainfall of the climate data of the study area are near to the satellite data acquisition (Hyperion). Obtained from the Weather Online's "global surface summary of day" product (www.weatheronline.in) and World Weather online are shown in Table 1. Climatic data from the nearest climatological station of the study were downloaded from the Web site. The nearest station of West Bengal site is Hooghly (located at 22° 39′ 32″ N, 88° 30′ 15″ E) and that of (located at16° 15′ 00″ N, 80° 55′ 00″ E). Atmosphere was clear in all case on the date of acquisition of images.

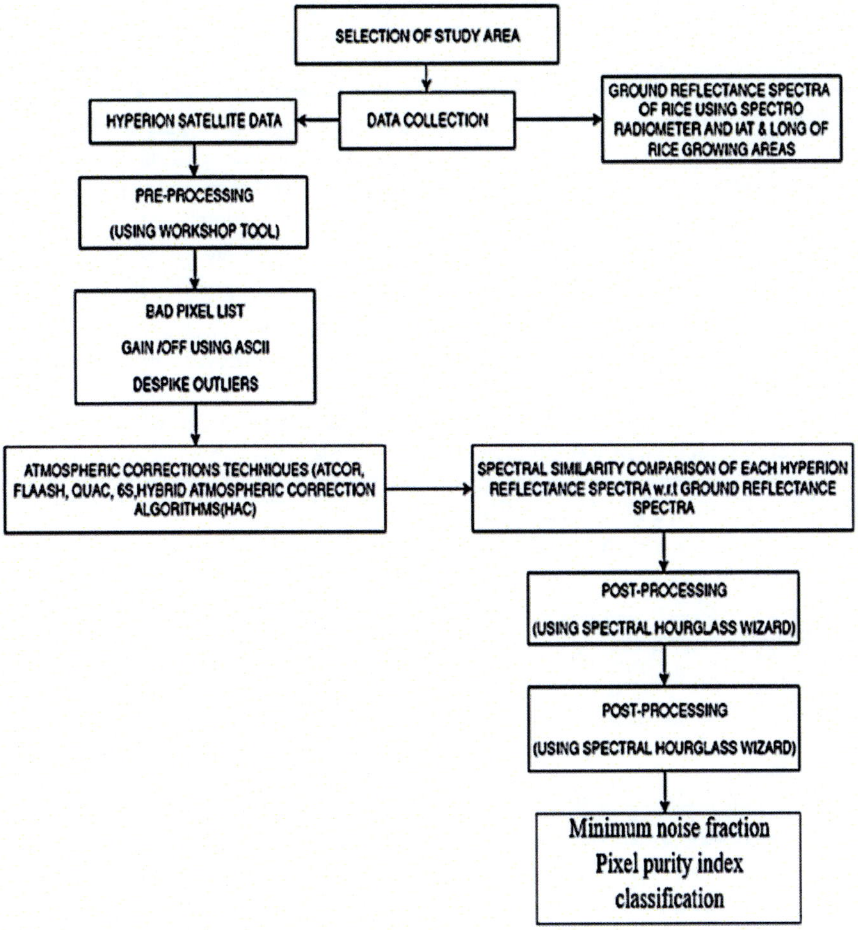

Fig. 2 Flow chart methodology

Table 1 Climate data for the acquisition dates

Place	Year-month-date	Mean temperature (°C)	Mean pressure (mbar)	Daily rainfall (mm)
Hooghly (West Bengal)	2002-07-11	33.3	999.1	0
West Godavari (Andhra Pradesh)	2008-04-18	33	1006.5	1.32

4 Pre-processing

Pre-processing of Hyperion imagery is carried to remove bad pixels and recalibration of data was obtained by application of the "Gain/offset values with ASCII files". Further processing is done by the removal of outliers and dropouts known as "blackout errors" by using "Fix out of range data" and then removal of extreme dropout and extreme outliers is done by using "Despike outliers".

The following atmospheric correction tool corrects the atmospheric errors by selecting the input reflectance image.

FLAASH is a first-principles atmospheric correction.
QUAC is a semi-empirical algorithm for aerosol retrieval and atmospheric correction.
The 6S code is the simple RT code used in the atmospheric correction algorithm for the calculation of lookup tables.
ATCOR uses lookup tables based on MODTRAN 4/5.
HAC1 is the first hybrid atmospheric model.
HAC2 is the second hybrid atmospheric model.

5 Post-processing

(a) Spectral hourglass, (b) Minimum Noise Fraction, (c) Pixel Purity Index, (d) n-D visualizer retrieving endmembers, (e) Identification of Rice crop (f) Classification techniques:

Data is classified using the SAM (Spectral Angle Mapper), and it is the physical-based spectral classification. The end members extracted from the image are used as the Spectral Angle Mapper (SAM) classification input. It uses an n-D angle to match pixel to the reference spectrum. The algorithm finds out the spectral similarity by calculating the angle between the two spectra, in a space with dimensionality treating them as vectors which is equal to the number of bands (Campbell et al. 2009). The spectra of endmember used by the SAM in the study is extract from the spectral library. Boardman et al. (1993) developed SAM algorithm. The spectral angle kept as 0.1 rad.

In this study, total two study sites are taken West Godavari (A.P) and Hooghly (W.B) and Hyperion data is used each for one study area and pre-processing and post-processing and classification is done for each study area separately.

6 Results and Discussion

6.1 Pre-processing

Standard pre-processing of the Hyperion image is carried out before atmospheric analysis. In this study, five currently available atmospheric correction methods were used to apply this transformation. They are:

QUAC, FLAASH,6S,ATCOR based on MODTRAN 4/5 and hybrid algorithms such as HAC.

Figure 3 shows reflectance spectrums obtained after application of four different atmospheric correction algorithms.

After the application of four different atmospheric correction methods spectrums were extracted from hyperion image of 6 locations sof six sof West Godavari.

After the application of four different atmospheric correction methods spectrums were extracted from hyperion image of 6 locations of Hooghly.

Figure 4 shows the reflectance spectrums obtained after application of four different atmospheric correction algorithms of Hooghly location.

Reflectance spectra of rice crop were obtained. To compare four popularly used atmospheric correction algorithms (QUAC, ATCOR, FLAASH, 6S) in deriving rice spectra from hyperion data. FLAASH corrected signals estimate rice better than ATCOR, 6S and QUAC corrected signals. The rice spectral feature between 560 and 610 nm is retained by FLAASH corrected signal, this may be because of FLAASH's narrow spectral sampling distance in that spectral region. because of high spectral resolution of hyperspectral imaging, fine atmospheric absorption feature will be detected these diagnostics features are typically of a very narrow spectral appearance.

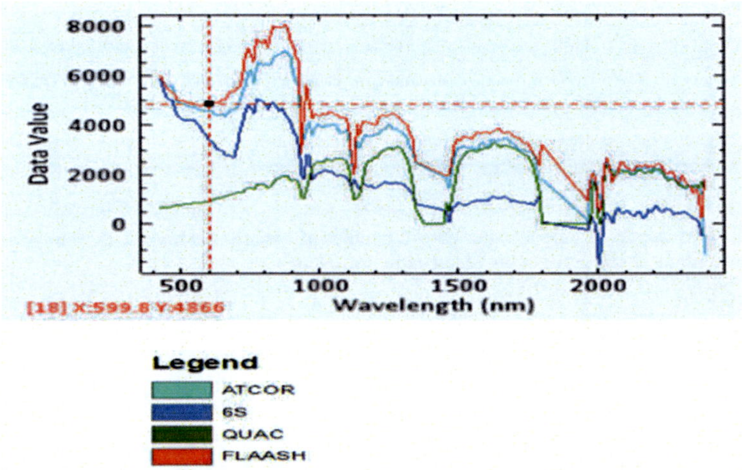

Fig. 3 Sample reflectance spectrums of location-1 of West Godavari

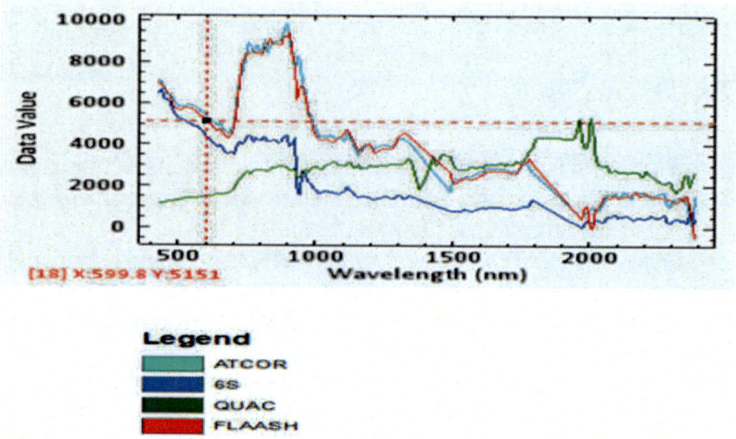

Fig. 4 Reflectance spectrums of location-1 of Hooghly

The wavelength of unique water absorption dips is extracted after the application of different atmospheric correction is shown below. Table 2 shows the wavelength unique absorption dips spectra values. Table 3 shows the wavelengths of unique water absorption dips value.

Table 2 Wavelengths of unique water absorption dips of spectra are extracted after the application of different atmospheric corrections of 6 locations of West Godavari

Rice locations	FLAASH (nm)	ATCOR (nm)	QUAC (nm)	6S (nm)
Location-1	579.25	650.15	942.05	691.23
Location-2	589.45	671.02	823.45	681.02
Location-3	579.45	589.62	701.55	691.37
Location-4	599.80	650.52	823.65	671.02
Location-5	630.32	690.37	762.60	665.02
Location-6	600.08	660.85	752.60	691.37

Table 3 Wavelengths of unique water absorption dips of spectra is extracted after the application of different atmospheric corrections of 6 locations of Hooghly

Rice locations	FLAASH (nm)	ATCOR (nm)	QUAC (nm)	6S (nm)
Location-1	640.05	691.37	701.55	695.35
Location-2	650.85	681.35	691.45	683.05
Location-3	661.02	672.08	700.27	681.02
Location-4	660.85	671.65	700.55	691.37
Location-5	671.02	691.37	695.55	694.37
Location-6	569.27	650.02	691.05	681.37

7 Comparision of Atmospheric Correction Algorithms (HAC and FLAASH)

To compare four popularly used atmospheric correction algorithms (QUAC, ATCOR, FLAASH, 6S) in deriving rice spectra from hyperion data. FLAASH corrected signals estimate rice better than ATCOR, 6S and QUAC corrected signals. The rice spectral feature between 560 and 610 nm is retained by FLAASH corrected signal, this may be because of FLAASH's narrow spectral sampling distance in that spectral region.

In this section, there will be two methods used to compare. One is a code called HAC (Hybrid Atmospheric Correction) and the other one called FLAASH (Fast Line-of-sight Atmospheric Analysis of Spectral Hypercubes).

Comparison of atmospheric correction methods of one out of six of West Godavari (Figs. 5 and 6).

Comparison of atmospheric correction methods of location one out of six of West Godavari.

Table 4 Wavelengths of unique water absorption dips of spectra are extracted after the application of atmospheric corrections FLAASH and HAC of 6 locations of West Godavari (Table 5).

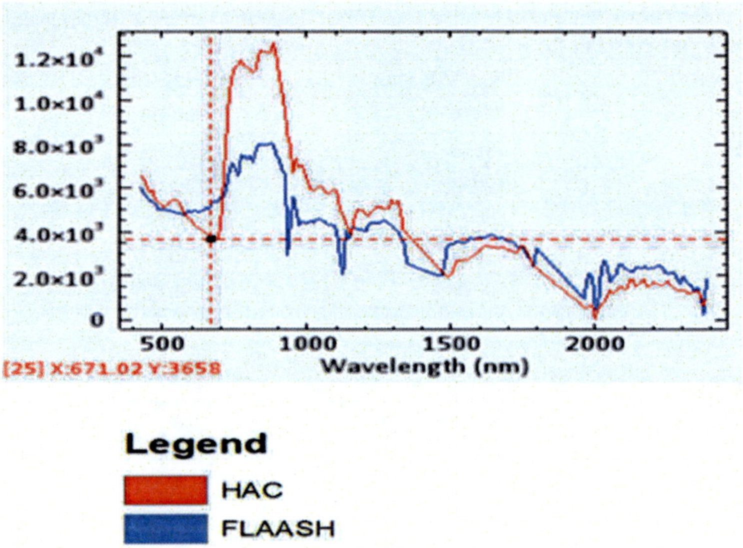

Fig. 5 Reflectance spectrums of location-1 of West Godavari

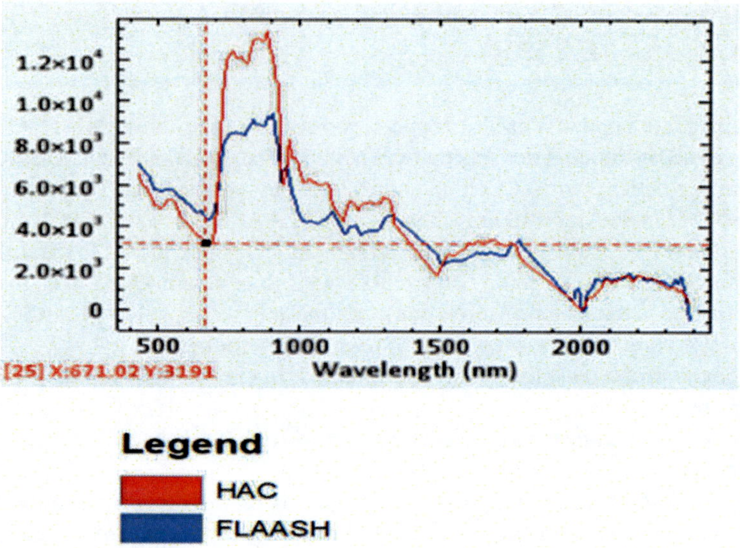

Fig. 6 Reflectance spectrums of location-1 of Hooghly

Table 4 Reflectance spectrums of location-1–6 of West Godavari extracted after the application of atmospheric corrections FLAASH and HAC

Rice locations	FLAASH (nm)	HAC (nm)
Location-1	579.25	601.09
Location-2	589.45	600.05
Location-3	579.45	582.75
Location-4	599.80	591.02
Location-5	630.32	650.25
Location-6	600.08	598.02

Table 5 wavelengths of unique water absorption dips of spectra is extracted after the application of atmospheric corrections FLAASH and HAC of 6 locations of Hooghly

Rice locations	FLAASH (nm)	HAC (nm)
Location-1	640.05	660.02
Location-2	650.85	675.85
Location-3	661.02	611.02
Location-4	660.85	667.55
Location-5	671.02	621.36
Location-6	569.27	585.07

8 Post-processing and Mapping of Rice Species

8.1 *Spectral Hourglass Wizard*

The processing in spectral hourglass uses the spectrally over-driven nature to find out the spectrally unique and pure pixels called "endmembers" in the dataset which is available and to find or mapping their sub-pixel abundance and location. The processing flow in spectral hour glass begins by using atmospherically corrected radiance or reflectance data as an input and begins with MNF (Minimum Noise Fraction) and PPI (Pixel Purity Index) techniques.

8.2 *Minimum Noise Fraction*

The Minimum Noise fraction technique is mainly used to reduce the noise in the data, to find out the inherent dimensionality of image data and to lessen the computational requirements for upcoming processing. After the processing is over it shows the MNF eigenvalues plot window. The MNF eigenvalues plot obtained after the processing shows the eigenvalue and eigenvalue number in the y-axis and the x-axis for each MNF-transformed band.

8.3 *Pixel Purity Index*

Pixel Purity Index (PPI) use the MNF (Minimum Noise fraction technique) result as an input to find the most spectrally pure (extreme) pixel in the image. After the processing PPI plot window appear. The PPI (Pixel Purity index) shows the number of pixels in y-axis and the PPI iterations in the x-axis.

8.4 *n-D Visualizer Retrieving Endmembers*

The results from PPI use as the input for the n-D visualizer and uses the n-D Visualizer with MNF data spatially subset using only the purest pixels calculated from the PPI. The n-D visualizer locates, identifies and clustering and retrieving the extreme spectral responses (endmembers) and the purest pixels in dataset in the n-dimensional space. n-D Visualizer and n-D Controls windows are appearing after processing is over used in retrieving end members.

8.5 Mapping of Rice Species Using SAM (Spectral Angle Mapper) and Parallelepiped Classifier Techniques

Data is classified using the SAM (Spectral Angle Mapper) and Parallelepiped classification techniques, SAM is the physical-based spectral classification. The end members extracted from the image are used as the Spectral Angle Mapper (SAM) classification input. The n-D angle is used to align the reference spectrum with the pixel. By evaluating the angle between the two spectra, the algorithm figures out the spectral similarity in a space with dimensionality treating them as vectors which is equal to the number of bands (Campbell 2009). The angle used for this study is 0.05 rad.

From the spectral hour glass technique, 48 end members are extracted. The spectra of the rice are extracted by using the Field spec® ASD handheld spectroradiometer in the 375–1075 nm range with 1 nm spectral resolution. After that matching the reference spectra with the 48 endmembers (spectrally unique) and finding out the pixel pure spectrum by using the spectral hourglass technique. The spectrums are calculated on the basis of the absorption dips (wavelength and intensity) and are obtained from different literature and the assumption that the dips are unique does not alter until some damage, stress and deficiency have been exposed to the species [10]. Reflectance wavelength and intensity is not matched as the seasonal information of the reference spectrum is unknown and integral area of the spectrum is also checked. The reference rice spectrum obtained from the literature is matched with the one of the spectrum from 48 spectrums. Then, these spectra are used to map the rice crop of the study area (Figs. 7, 8, 9 and 10; Table 6).

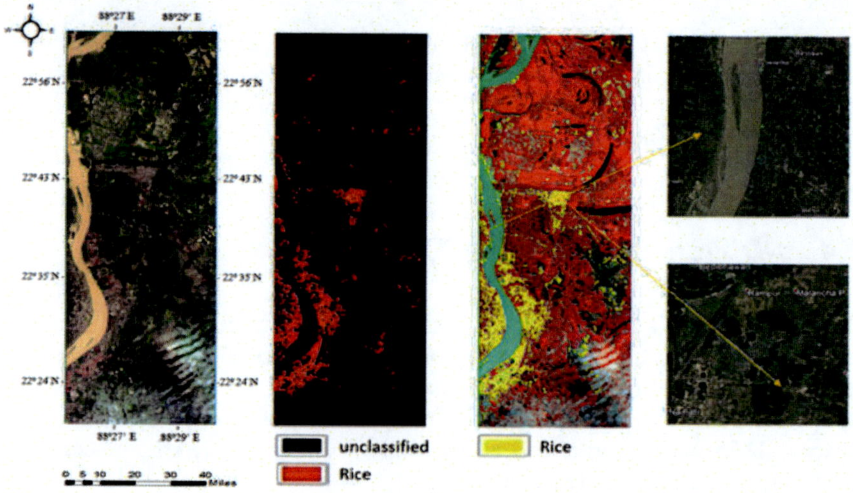

Fig. 7 Classification using SAM (Spectral Angle Mapper) classifier technique of Hooghly

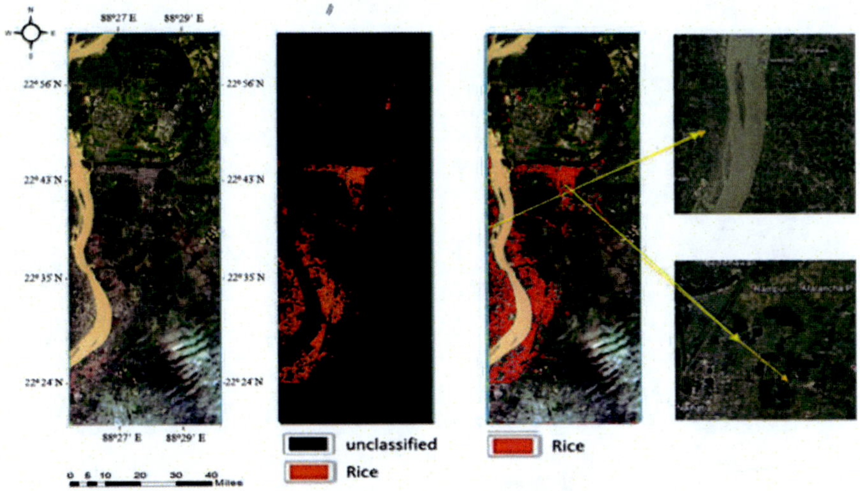

Fig. 8 Classification using parallelepiped classifier technique of Hooghly

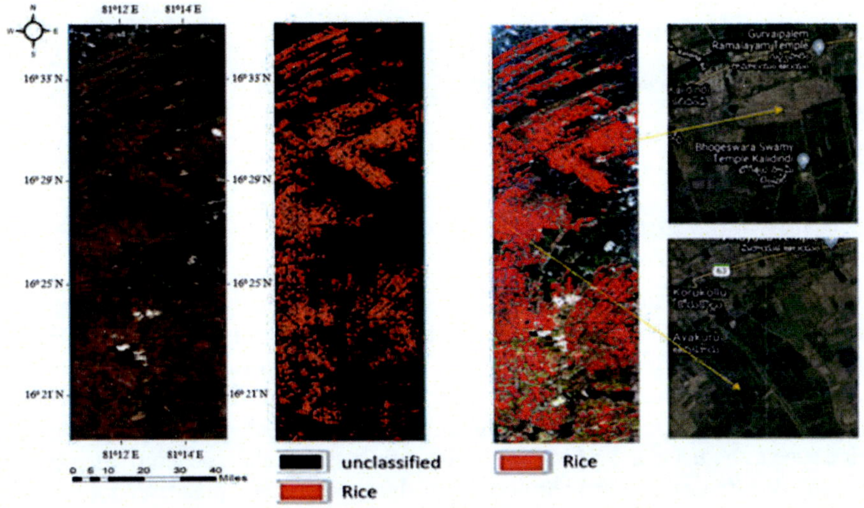

Fig. 9 Classification using SAM (Spectral Angle Mapper) classifier technique of West Godavari

The study had a classification accuracy of 88% by using SAM and 84.6% by using Parallelepiped classifier for Hooghly region and 93.9% using SAM and 87.7% using Parallelepiped for West Godavari region. The research found SAM in Hooghly and West Godavari as the best approach to rice crop mapping.

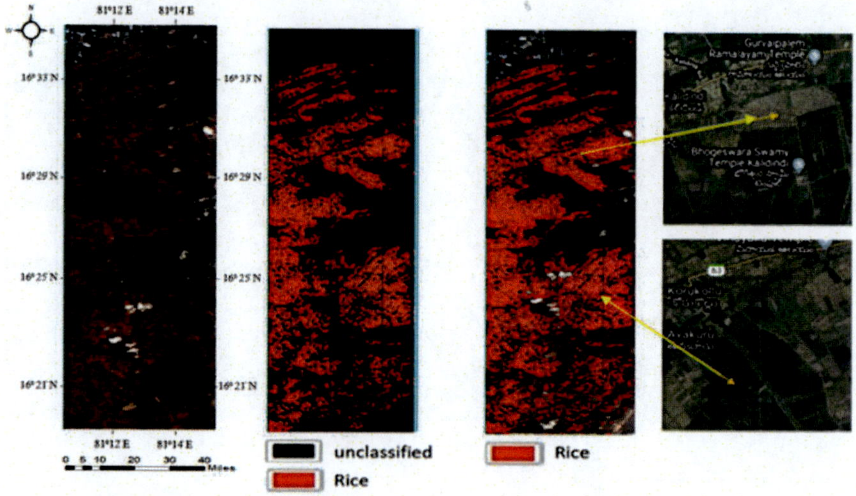

Fig. 10 Classification using parallelepiped classifier technique of West Godavari

Table 6 Overall accuracy of different classification algorithms of West Godavari and Hooghly

Classifier algorithm	West Godavari (%)	Hooghly (%)
SAM (Spectral angle Mapper)	93.93	88.04
Parallelepiped	87.72	84.67

9 Summary and Conclusions

In the present research work, initially a study to compare the reflectance spectra extracted from the five atmospheric correction algorithms are tested on EO-1 Hyperion images of 6 different locations of the West Godavari district of Andhra Pradesh and the Hooghly district of the West Bengal with the field VNIR/SWIR reflectance spectra extracted by using the spectroradiometer Field spec® and the mapping of the rice species by using the SAM (Spectral Angle Mapper) classification technique. It was to identify best atmospheric correction algorithm method to be used for satellite data analysis.

The primary objective of the research work is to assess the effect of the atmospheric correction algorithms on the spaceborne hyperspectral data. Commonly used atmospheric correction algorithms QUAC, ATCOR, FLAASH and 6S algorithms are tested on EO-1 Hyperion images, and rice reflectance spectra extracted from the corrected images of 6 different locations of the West Godavari district of Andhra Pradesh and the Hooghly district of the West Bengal, the locations of both West Godavari and Hooghly are obtained by using the GPS (Global Positioning System). To improve estimation further, HAC (Hybrid Atmospheric Correction) was compared with their component algorithms in deriving rice spectra from Hyperion data by using the reference spectra, which is extracted by using the ASD handheld spectroradiometer this research will open up a wide scope for the mapping of rice.

10 Conclusions

The research aims to compare the impact of atmospheric correction algorithms on precise mapping of rice crops in different districts in India. using reference spectrum. To compare four popularly used atmospheric correction algorithms (QUAC, ATCOR, FLAASH, 6S) in deriving rice spectra from Hyperion data in agricultural fields of West Godavari district in Andhra Pradesh and Hooghly district in West Bengal with the spectra obtained from ASD handheld spectroradiometer.

The study concentrates on overall rice characteristics and spectra of two famous regions of INDIA (West Godavari and Hooghly). Firstly, four atmospheric correction algorithms were analysed and compared with the Hyperion reflectance using spectral features (unique water absorption dips).

Secondly, for the determination of the wavelengths of unique water absorption dips, the efficiency of each atmospheric correction algorithm was analysed.

To compare four popularly used atmospheric correction algorithms (QUAC, ATCOR, FLAASH, 6S) in deriving rice spectra from Hyperion data in agricultural fields of West Godavari district in Andhra Pradesh and Hooghly district in West Bengal with the spectra obtained from ASD handheld spectroradiometer.

To compare four popularly used atmospheric correction algorithms (QUAC, ATCOR, FLAASH, 6S) in deriving rice spectra from hyperion data. FLAASH corrected signals estimate rice better than ATCOR, 6S and QUAC corrected signals. The rice spectral feature between 560 and 610 nm is retained by FLAASH corrected signal, this may be because of FLAASH's narrow spectral sampling distance in that spectral region.

In this research, FLAASH performed better than four component algorithms (QUAC, ATCOR, FLAASH, 6S), as compared to spectra corrected in terms of spectral similarities by their component algorithms.

To compare HAC (hybrid atmospheric correction algorithm) with the best atmospheric correction algorithm among four component algorithms and the spectra obtained from ASD handheld spectroradiometer in deriving rice spectra from Hyperion data.

The results show that both FLAASH and HAC algorithms produce a good spectrum with respect to the rice spectra obtained from ASD handheld spectroradiometer.

Estimation of rice from spaceborne hyperspectral imagery is influenced by:

- the type of atmospheric correction algorithm chosen.
- spatial arrangement of fields.

Mapping of the rice crop will be carried out by using the Spectral Angle Mapper classifier and Parallelepiped classifier techniques.

From the above study, it can be concluded that rice crop maps are generated for study areas. SAM image classification and Parallelepiped classifier were used in this study.

The spectra matched to the spectrum obtained from the ASD handheld spectroradiometer were used to identify and map the rice crop as an input for data classification with classifiers such as Spectral Angle Mapper (SAM), Parallelepiped. In this study, accuracy assessment was performed, and the study had a classification accuracy of 88% by using SAM and 84.6% by using Parallelepiped classifier for Hooghly region and 93.9% using SAM and 87.7% using Parallelepiped for West Godavari region. The study found SAM as the best approach for rice crop mapping in Hooghly and West Godavari.

References

1. Linares (2002) African rice (Oryza glaberrima): History and future potential. Proc Nat Acad Sci 99(25):16360–16365
2. Kuenzer C, Knauer K (2012) Remote sensing of rice crop areas. Int J Remote Sens 34(6):2101–2139
3. Moharana S, Dutta S (2014) Hyperspectral remote sensing of paddy crop using in-situ measurement and clustering technique. The international archives of the photogrammetry, vol XL-8, 2014 ISPRS technical commission VIII symposium, 09–12
4. Elena (2018) Top 10 Largest rice producing states in India. Trendrr
5. FAO (2004) Addressing marketing and processing constraints that inhibit Agri-food experts: a guide for policy analysts and planners. In: FAO Agricultural service Bulletin 160, Rome, Italy
6. Mahajan G, Chauhan BS, Jabran K (2017) Rice production worldwide. Springer International, USA
7. Bag MK, Adhikari B, Bhowmick MK, Kundu C (2011) Status paper on rice in West Bengal. Rice Research Station, West Bengal
8. Fazlani (2018) 10 Largest rice producing states in India. FEPL, India
9. Chakraborty A, Emmanuel V murray (2011), Rice production & productivity in Andhra Pradesh. National Bank for agriculture and rural development, Andhra Pradesh Regional Office, Hyderabad
10. Salghuna NN, Pillutla RCP (2017) Mapping mangrove species using hyperspectral data: a case study of Pichavaram mangrove ecosystem, Tamil Nadu. Springer International Publishing AG, Part of Springer Nature, Berlin
11. Datt B, McVicar TR, Van Niel TG, Jupp DLB, Pearlman JS (2003) Pre-processing EO-1 hyperion hyperspectral data to support the application of agricultural indexes. IEEE Trans Geosci Remote Sens 41(6)
12. FLAASH (2009) FlAASH module: user's guide. In version 4.3

Dynamic Analysis of Geodesic Dome Structure

M. Roopa, Kavitha B. Lakshmi, and H. Venugopal

Abstract In the recent years, the dome structures are the commonly used geometrical shapes to cover the large span. Specifically the "geodesic dome" structure is the most attractive structure which provides more space. Geodesic is a Latin word where "geo" means earth and "desic" means dividing. It is having spherical shape with lightweight triangular faces and struts in the frame structure. Geodesic dome has triangles in the stable network form. The study considers the response of four different geodesic dome frame structure for seismic analysis and the results are compared.

Keywords Geodesic dome · Frequency of dome · CADRE geo software · SAP2000 software

1 Introduction

Domes are the oldest structures. We can see the dome structures in temples and other monuments. The upper half of the hollow sphere simulates the architectural part of the dome. The dome structure is use to provide more space without any obstructions that is, the dome structure embeds the larger volume with the smaller quantity of material without interrupting columns. Dome structures are used in the early days as symbolical representation of the kingdoms. These structures are economical for covering larger area in a single span and they are light in weight. They are elegant structures that provide economical solutions for covering large areas with their splendid aesthetic

M. Roopa (✉) · H. Venugopal
Department of Civil Engineering, Sri Siddhartha Institute of Technology (Sri Siddhartha Academy of Higher Education), Tumkur, Karnataka, India
e-mail: roopam@ssit.edu.in

H. Venugopal
e-mail: Venugopalh@ssit.edu.in

K. B. Lakshmi
CAD Structures, Department of Civil Engineering, Sri Siddhartha Institute of Technology (Sri Siddhartha Academy of Higher Education), Tumkur, Karnataka, India

L. Nandagiri et al. (eds.), *Sustainability Trends and Challenges in Civil Engineering*, Lecture Notes in Civil Engineering 162, https://doi.org/10.1007/978-981-16-2826-9_56

appearance. They consist of [1] one or more layers of elements that are arched. These types of structures are used in temples, sports stadium, assembly halls, exhibition centres, swimming pools, shopping centres and industrial buildings. Dome structures are classified into various types, namely beehive dome, cloister vault dome, crossed-arch dome, geodesic dome, monolithic dome, inflated dome, etc. In this commonly used dome structure is geodesic dome.

2 Geodesic Dome Structure

2.1 Geodesic Dome

The structure with hemispherical shape and thin shell-like structure is known as geodesic dome. The dome structure that is formed by lightweight straight elements will form the interlocking polygons. Geodesic dome can be constructed easily and quickly. Walther Bauersfeld was the first person who designed the geodesic dome after World War I. Later after twenty years, Buckminster Fuller, an American architect, designed many geodesic dome structures. For better understanding, the geodesic structure an example of soccer ball can be considered. The geodesic dome has the straight lines rather than curved faces in the geodesic playdome and soccer ball as shown in Fig. 1. Half of the ball resembles the half sphere.

Important parts of the geodesic dome is shown in Fig. 2.

Real geodesic dome structure can be seen in many countries, like full sphere geodesic dome structure is in Spaceship Earth Florida in USA, Calico Dome at Ahmedabad in Gujarat India, and in Karnataka we can see geodesic dome structure at Infosys Mysore, Sri Siddhartha Institute of Technology Tumkur. The pictures of the geodesic dome structure in Karnataka are shown in Fig. 3.

Geodesic Playdome

Soccer Ball

Fig. 1 Geodesic playdome and soccer ball

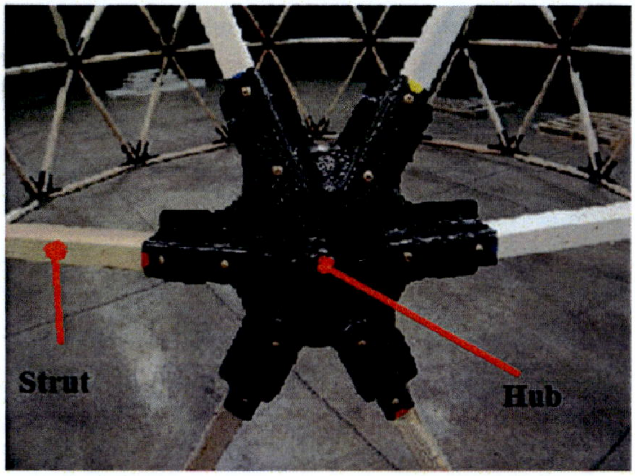

Fig. 2 Parts of geodesic dome

Fig. 3 Real geodesic dome structures in Karnataka

2.2 *Classification of Geodesic Dome*

Classification of the dome structure is made on different basis such as based on the polyhedron where we can find three types of polyhedron, namely tetrahedron with four sides and it is three-sided pyramid all three faces having same dimensions, octahedron with eight sides and it is two four-sided pyramids and icosahedron with twenty equilateral triangles, five triangles forms the pentagon around each pole, as shown in the Fig. 4.

Each triangle can be subdivided that is faces can be expanded. The degree to which each face is subdivided is known as frequency where v represents the frequency of the dome structure. Based on frequency, the dome structure is classified as 1v, 2v, 3v, 4v, 5v and 6v structure as shown in Figs. 5 and 6. These 1v to 6v geodesic dome structures are the types in modelling whereas in construction we have various methods like wooden domes having wooden struts and stainless steel hub called hub-and-strut dome, panelled domes here steel hubs are not required, steel frame domes and aluminium frame domes, whereas the cladding may be glass or plastic

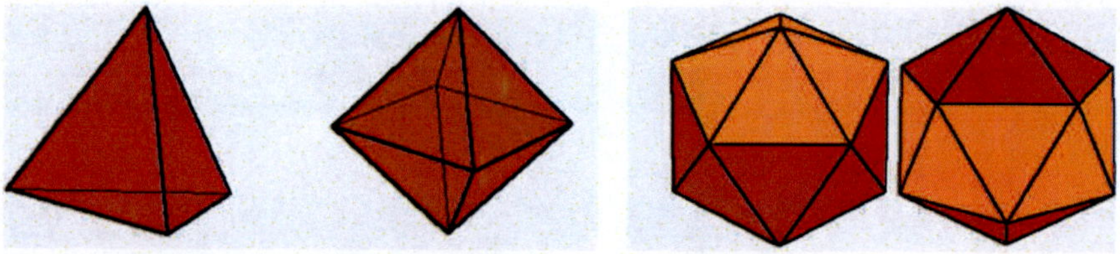

Fig. 4 Polyhedron shapes—tetrahedron, octahedron and icosahedron

Fig. 5 Frequency representation for geodesic domes

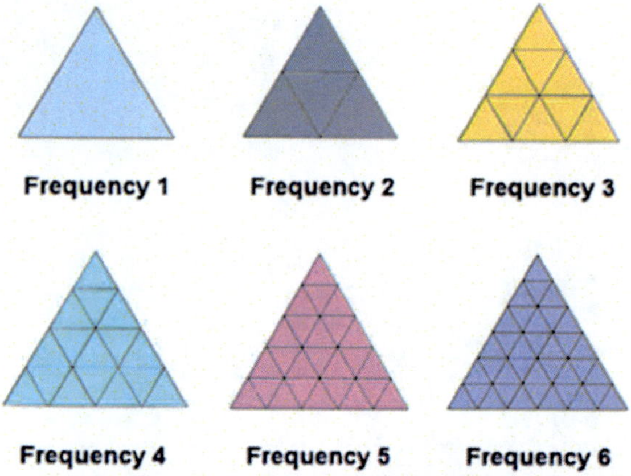

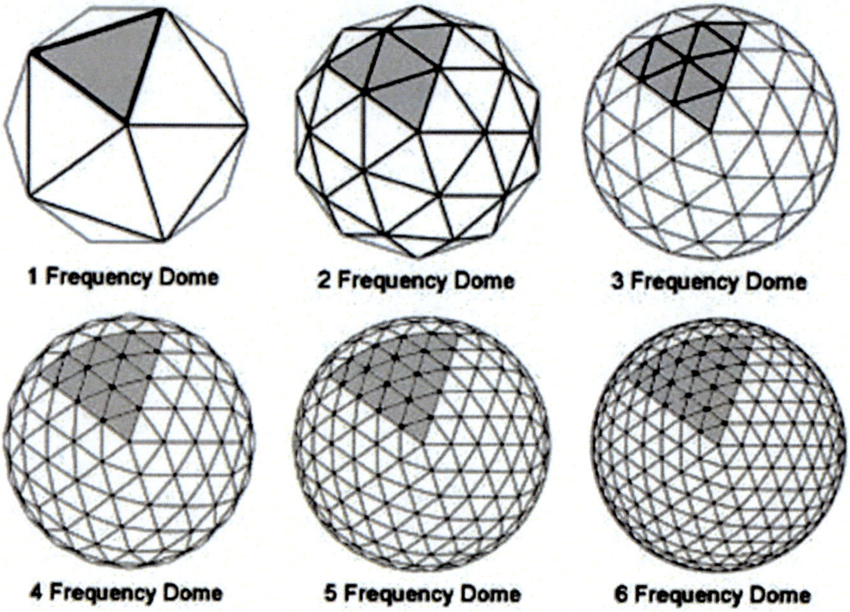

Fig. 6 .

sheets having PVC coping and in some domes steel frame with concrete cladding is used with chicken wire meshing. Here in this research only the frame structure of the geodesic dome is considered that is cladding is not considered.

3 Earthquake Analysis of Geodesic Dome Structure

Earthquake is such a phenomena that will occur suddenly, so the structure which is existing in that area gets affected by this. In order to avoid the damage in the structure due to earthquake, the structure should be well designed to take that earthquake load, thus the earthquake analysis of structure plays an important role. In case of the special structures like geodesic dome structure, this earthquake analysis place an important role as they have no intermediate columns to take the load. There are two major types of earthquake analysis namely static analysis and dynamic analysis. In dynamic analysis, there are various types namely linear analysis in that response spectrum analysis and time-history analysis, dynamic analysis in that we have pushover analysis. Here in this paper, the response spectrum analysis which is a linear dynamic earthquake analysis is done using SAP2000 version 20 software.

3.1 Literature Survey

Nie and Liu [2], in their work, experimental tests to check the seismic behaviour [3] of the single-layer reticulated dome was carried [4]. The test structure was a single-layer reticulated dome with a span of 2.7 m and a height of 0.9 m. Two tests with the same models with and without isolated supports were conducted using the shaking table. They found that the structure with the isolated support had a promising reduction effect on the acceleration response and stress of the member.

Niel et al. [5], in their research [5], in order to have a better understanding of the mechanical performance and to verify the failure pattern under seismic motion, they conducted shaking table test of a single-layer reticulated dome. The mechanical performance of the model under weak and strong seismic motions was investigated experimentally by using shaking table test. They measured material properties of the members, [5] yield strength, member [5] section, member length and its Young's modulus and the geometric position of the structural nodes. They have given compassion between shake table test results and the results of the model in finite element software ABAQUS in their work. **Jayminkumar and Vahora** [6], they studied the behaviour of flexible dome, where the dome under external load was nonlinear, and this necessitates them to carry out nonlinear analysis in order to know the realistic response of the dome. The steel dome structure was analysed using STADD Pro—Software with different diameters and rise of dome as parameters. Patel [1] different types of domes are explained and the analysis of Schwedler dome for different diameter and height has been done to find the best H/D ratio that was carried. They found

Fig. 7 Buddha temple of Thailand

that for Schwedler dome the best-fitted *h*/*d* ratio from parameters to resist the wind load is 0.4 for all diameters and *H*/*D* ratio of 0.2 is least fitted. **Anwar et al.** [7], they have carried the analysis and design of single-layer reticulated dome structure [8] which is inverted monk bowl (Buddha temple of Thailand as shown in Fig. 7) in shape. The Buddha hall dome which is located in Saraburi, Thailand, having base dia is 65 m, middle dia is 86 m and the height of the dome is 40 m with gravity load case was evaluated based on the nonlinear staged construction analysis including P-Delta effect. In their work, wind load is considered by performing the wind tunnel test and the earthquake load is considered from Thai earthquake calculation guidelines. The model of single-layer steel dome was developed using SAP2000 V14.2.4 to examine the design of the structure which is having inverted monk bowl shape. After they amend the structural details of some components, it is found that the global and local structural response was improved to resist both wind and earthquake.

Zhu and Ye [4], they conducted a vulnerability analysis [4] on three shake table test models of a single-layer reticulated dome built at a scale of 1:10 to find out its vulnerable failure scenarios [9]. In model 1, all members are made up of steel tubes of same sections [9]. In model 2, all the members are made up of steel tubes of various sections and six artificial weak links. In model 3, all the members are made up of different steel tube sections but without weak links. They conducted the shake table test for the three models. Their study shows that the shake table test method proposed [4] can identify the weaknesses of a structure under load and accurately predict the damage locations and failure modes of the structure.

From the literature survey, it is found that the analysis of the dome structure can be carried using software like STAAD.Pro, SAP, ABACUS.

3.2 Adapted Software

Modelling of the geodesic dome:

CADRE Geo is a modelling software where we can use it for modelling of different dome structures.

- CADRE Geo is a software used for modelling of the domes.
- Different types of domes can be modelled using this software.
- Reticulated dome, geodesic dome, full/half/level-based dome can be modelled.
- Here in this project, CADRE Geo Version 7 is used for modelling.
- AutoCAD Software is used to scale dome the model for required dimensions.

Analysis of Geodesic Dome:

- For analysis of the dome structure here SAP2000 software is used.
- SAP2000 version 20 is a CSI Software for structural analysis.

3.3 Response Spectrum Analysis

For dynamic analysis, two different methods can be used, response spectrum and time-history analysis. In response spectrum method, the peak response of a structure during an earthquake is obtained directly from the earthquake response (or design) spectrum. This method is also known as modal method or mode superposition method. The method is applicable to those structures where modes other than the fundamental one significantly affect the response of the structure.

Response spectrum analysis of the geodesic dome structure with different frequencies is done and the results are compared.

3.4 Structural Details of Geodesic Dome

In the present study, the seismic analysis of 3V, 4V, 5V and 6V geodesic dome structure with constant diameter is discussed. The diameter of the geodesic dome selected = 31 m

- Configuration of the dome = Geodesic spheroid icosahedron.
- The dome model is scaled done to the scale of 1:100.
- And the material for the frame structure is assumed as Fe250-5 mm round rod.

4 Analysis

Analysis is carried using SAP2000 20 software.

Load case: Self-weight of the dome structure is considered. Along with the earthquake load EQX and EQY, for response spectrum analysis, the load case of RSX and RSY is taken. Analysis was carried for individual load pattern. It is done as per IS1893:2016 [10] that is the zone factor, importance factor, base shear and so on.

4.1 Results

Results like base shear, modal period, member forces are obtained and the same are enlisted in Tables 1, 2, 3, 4, 5 and 6 respectively. The comparison between the various frequency of the dome structure is made.

Table 1 Base reactions for 3V geodesic dome

Load case	FX	FY	FZ
	N	N	N
Dead	−3.823E−15	−2.449E−15	19.4
EQ-X	−1.58	−4.177E−15	1.932E−15
EQ-Y	−3.483E−15	−1.58	−1.187E−15
RS-X Max	0.21	5.432E−09	2.452E−08
RS-Y Max	1.756E−09	0.21	7.479E−08

Table 2 Base reactions for 4V geodesic dome

Load case	FX	FY	FZ
	N	N	N
Dead	−2.085E−15	−6.994E−15	25.81
EQ-X	−2.14	−3.053E−15	3.386E−15
EQ-Y	−1.705E−15	−2.14	2.599E−15
RS-X Max	0.29	6.132E−08	6.285E−08
RS-Y Max	6.212E−08	0.29	0.000000273

Table 3 Base reactions for 5V geodesic dome

Load case	FX	FY	FZ
	N	N	N
Dead	−8.105E−15	6.106E−15	32.19
EQ-X	−2.7	2.193E−15	−1.094E−14
EQ-Y	3.844E−15	−2.7	3.803E−15
RS-X Max	0.36	4.574E−07	1.417E−08
RS-Y Max	0.000000448	0.36	7.511E−08

Table 4 Base reactions for 6V geodesic dome

Load case	FX	FY	FZ
	N	N	N
Dead	−3.331E−16	3.469E−15	38.56
EQ-X	−3.27	3.712E−15	−4.569E−15
EQ-Y	1.249E−15	−3.27	−3.532E−15
RS-X Max	0.43	7.339E−07	3.875E−08
RS-Y Max	7.315E−07	0.43	6.31E−08

Table 5 Modal period of the geodesic dome

S. No.	Frequency of dome	Mode	Period (s)
1	3V	1	0.001102
2	4V	1	0.001083
3	5V	1	0.001071
4	6V	1	0.001061

Table 6 Member forces of geodesic

Frequency of dome	Max (in N)	Min (in N)
3V	1.39	−1.36
4V	1.38	−1.34
5V	1.26	−1.30
6V	1.34	−1.26

4.2 *Figures*

The obtained results as schematic with respect to the geodesic dome in 3D are shown in the Figs. 1, 2, 3, 4, 5, 6, 7, 8, 9, 10, 11, 12, 13, 14, 15, 16, 17, 18, 19, 20, 21, 22, 23, 24, 25, 26 and 27.

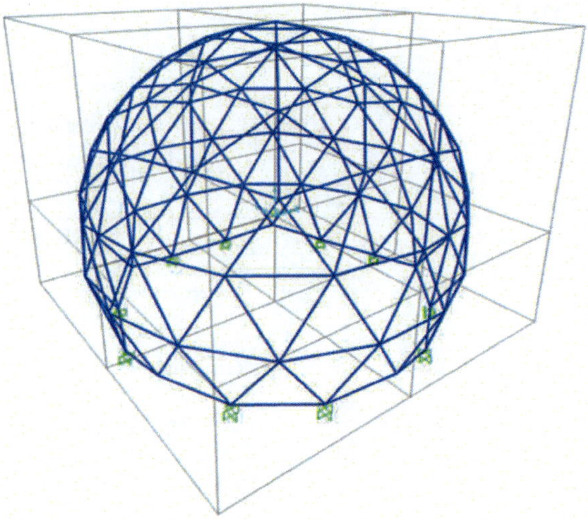

Fig. 8 3V geodesic dome in SAP2000

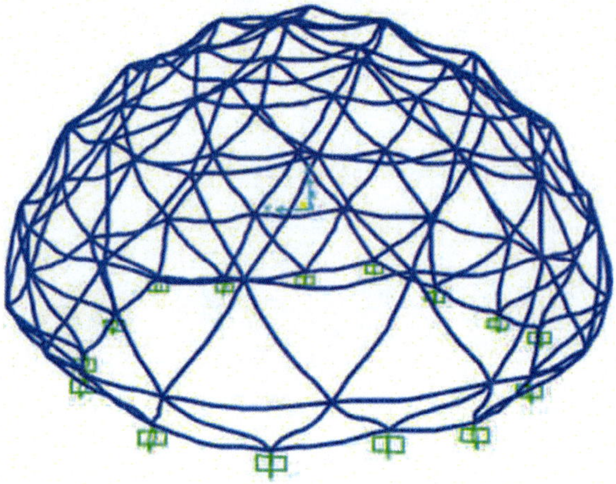

Fig. 9 Deformed shape of 3V geodesic dome

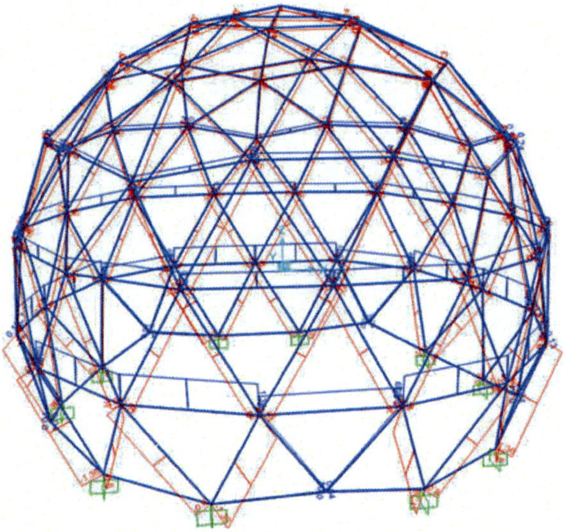

Fig. 10 Axial force for 3V geodesic dome

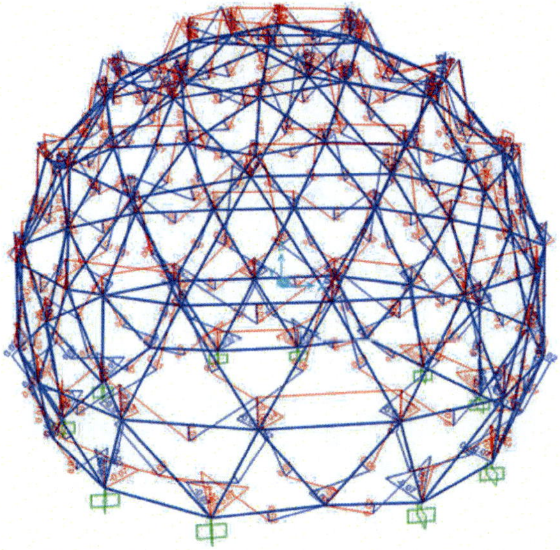

Fig. 11 Moments for 3V geodesic dome

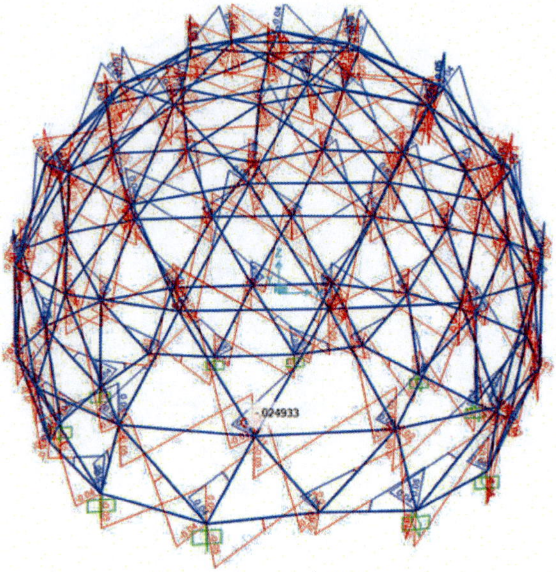

Fig. 12 Shear force for 3V geodesic dome

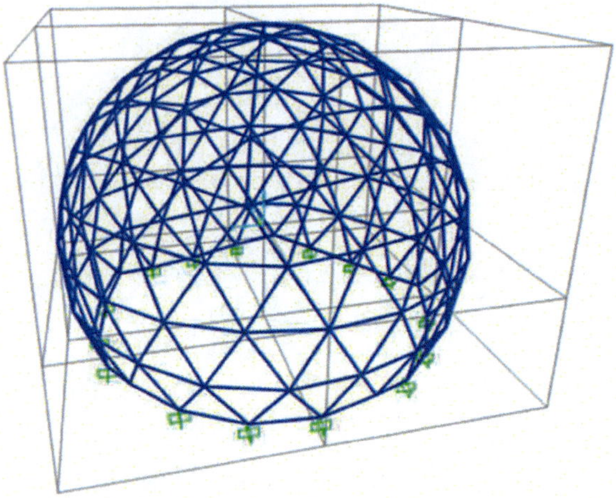

Fig. 13 4V geodesic dome in SAP2000

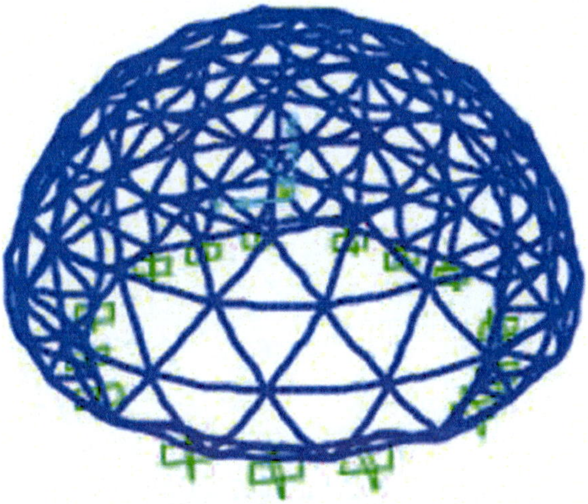

Fig. 14 Deformed shape of 4V geodesic dome

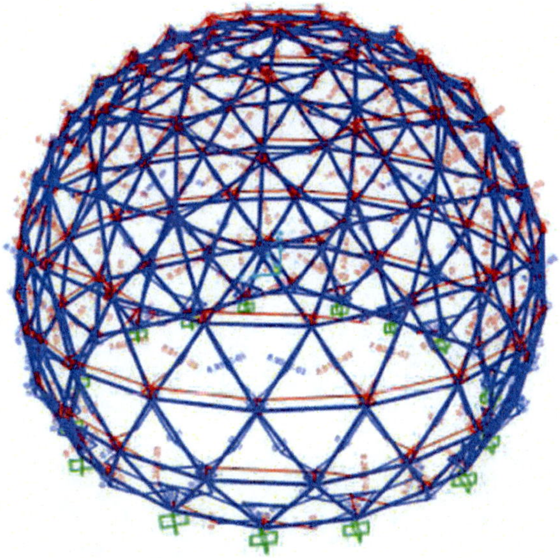

Fig. 15 Moment for 4V geodesic dome

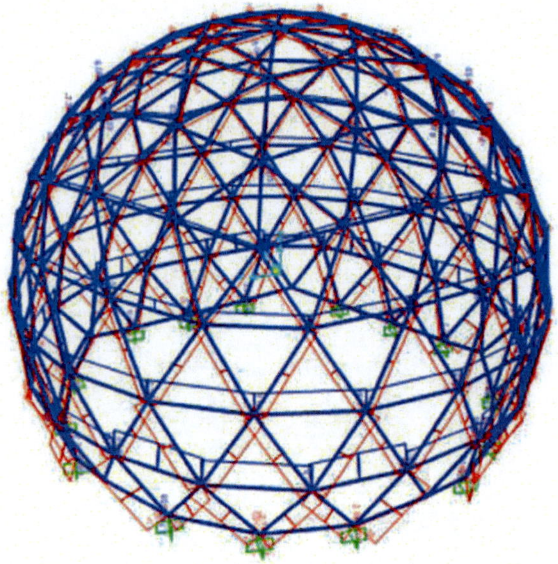

Fig. 16 Axial force for 4V geodesic dome

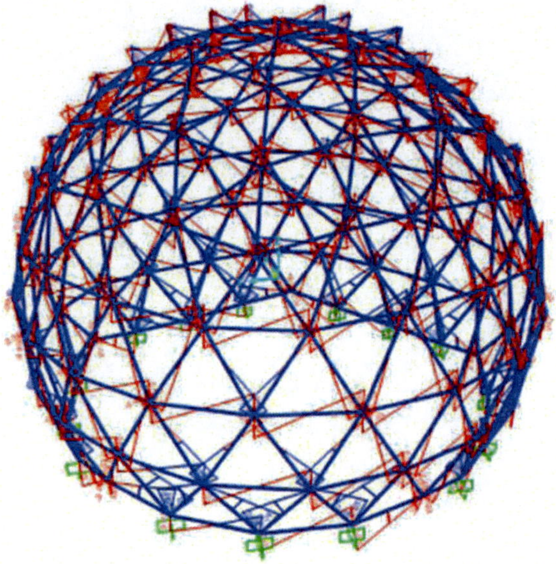

Fig. 17 Shear force for 4V geodesic dome

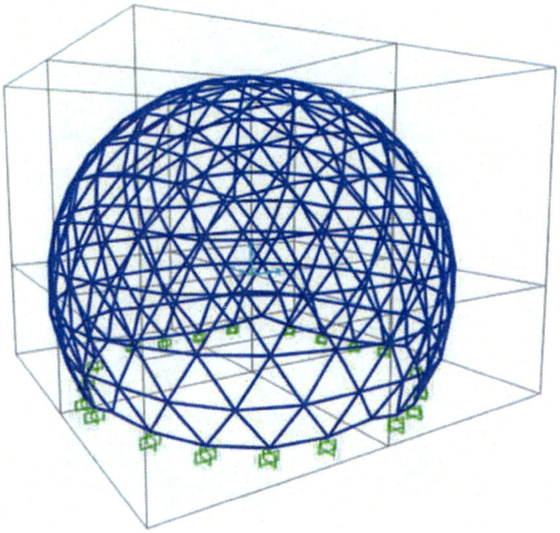

Fig. 18 5V Geodesic dome in SAP2000

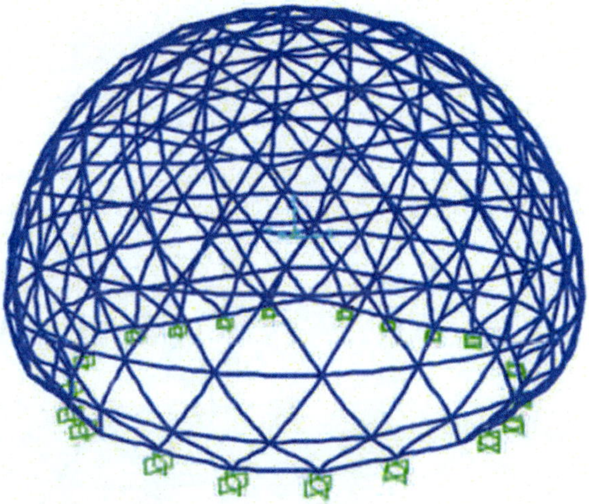

Fig. 19 Deformed shape of 5V geodesic dome

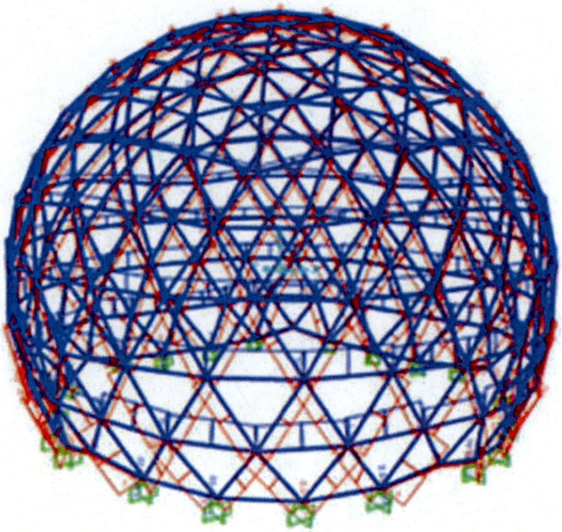

Fig. 20 Axial force for 5V geodesic dome

Fig. 21 Moment for 5V geodesic dome

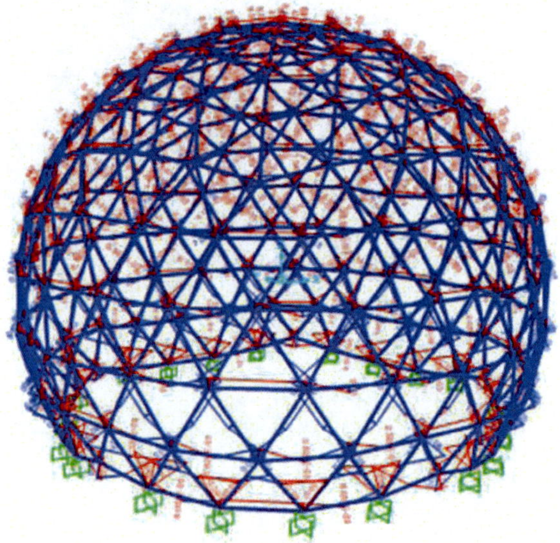

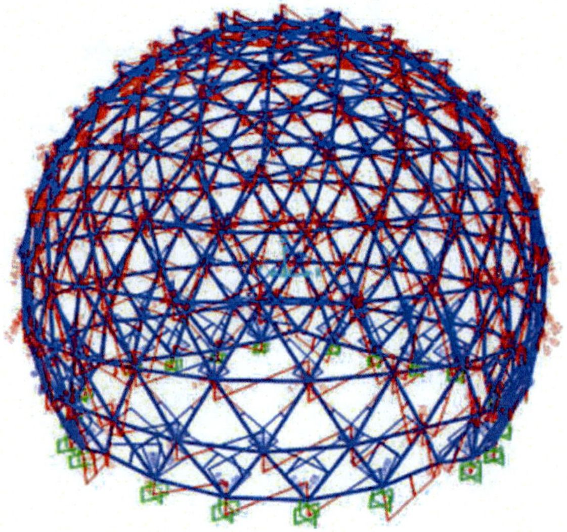

Fig. 22 Shear force for 5V geodesic dome

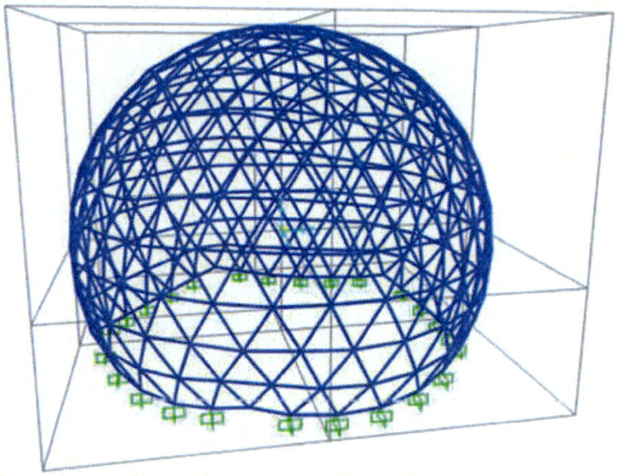

Fig. 23 6V Geodesic dome in SAP2000

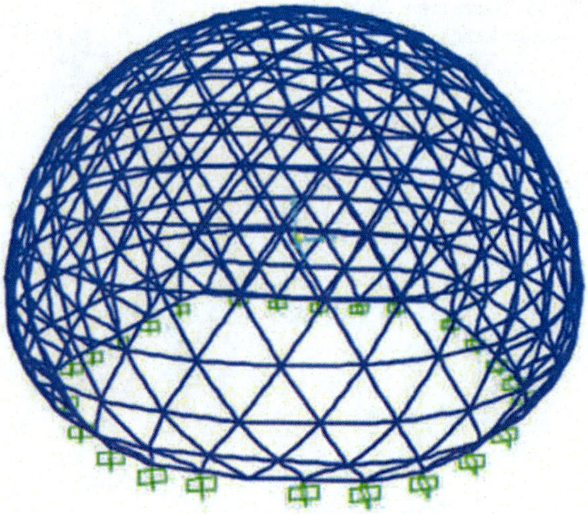

Fig. 24 Deformed shape for 6V geodesic dome

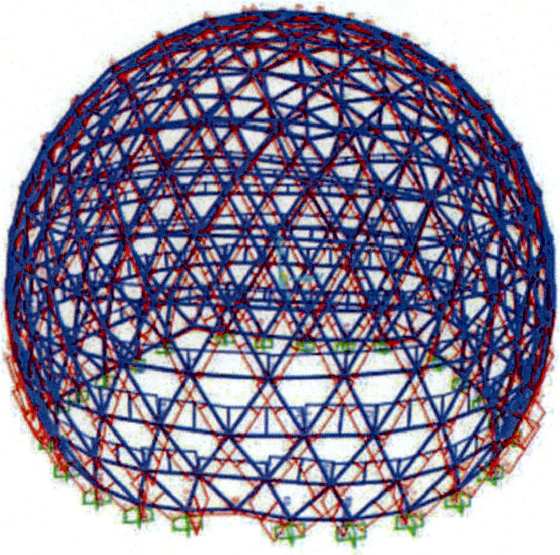

Fig. 25 Axial force for 6V geodesic dome

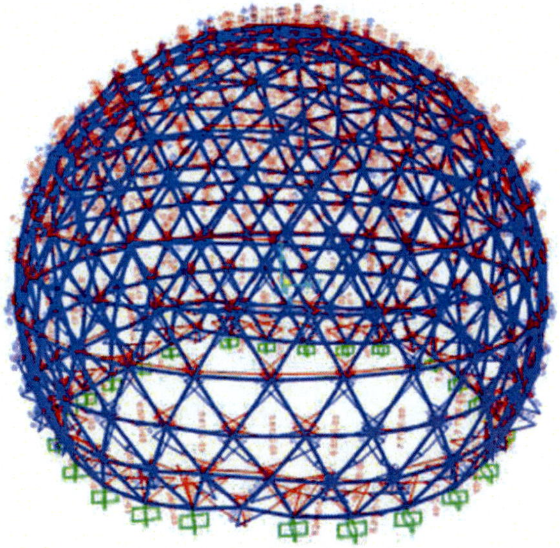

Fig. 26 Moment for 6V geodesic dome

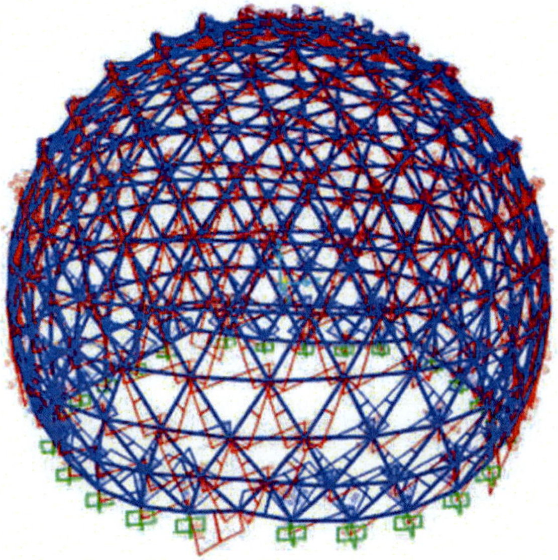

Fig. 27 Shear force for 6V geodesic dome

4.3 Comparison

1. Base shear.

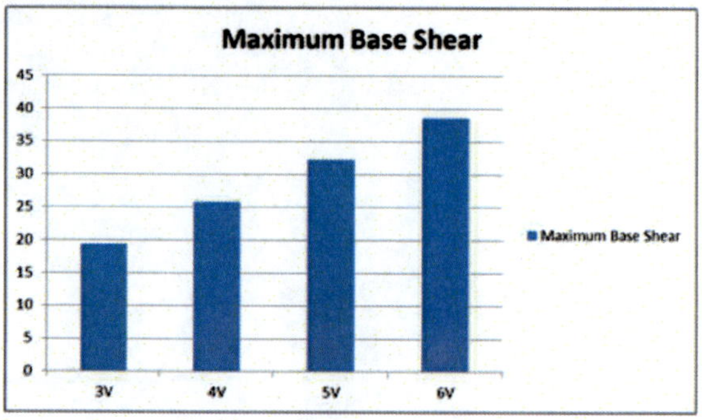

2. Elemental force.

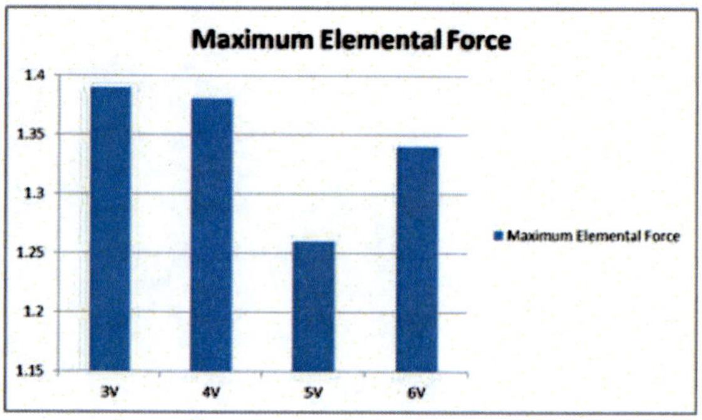

4.4 Conclusion

The study is carried out on geodesic dome structure for zone II according to IS1893:2016 [10]. Different frequency geodesic domes with same radius and materials are analysed. The obtained results show that base reaction will be more in 6v dome compared to the low-frequency dome structure. The axial force is greater in 3v dome compared to other frequency domes, whereas for 5v geodesic dome the axial force is the least. Finally, the conclusion can be drawn that the dome structures considered for the study are earthquake resistant and design are also safe.

5 Future Scope

- Scaled model of different frequency with same dimensions can be considered and shake table test can be conducted.
- Diameter of the geodesic dome can be varied and the test results can be compared.
- Height-to-diameter ratio can be considered and the model for fixed H/D ratio can be made and test results of various domes can be compared.
- Experimental model of the geodesic dome can be computed, and it can be tested in shake table.
- The experimental and the analysis results can be compared for further study.

Acknowledgements This project is conducted in Sir Siddhartha Institute of Technology (A Constituent College of Sir Siddhartha Academy of Higher Education) Maralur, Tumkur. Thankful to the College and Department of Civil Engineering for their support and encouragement. And also we are thankful to KSCST Karnataka State Council for Science & Technology Indian Institute of Science Campus, Bengaluru-560012 Student Projects Programme (SPP) 43rd series 2019–2020.

References

1. Patel KR (2016) Earthquake analysis of different configuration of steel domes using computer programme. Int J Res Eng Technol
2. Nie G, Liu K (2018) Experimental studies of single-layer reticulated domes with isolated supports. American Society for Civil Engineering
3. Nie G-B, Zhi X-D, Fan F, Dai J-W (2014) Failure pattern of singe-layer reticulated dome under severe earthquake and its shaking table test. Int J Steel Struct
4. Zhu N, Ye J (2014) Structural vulnerability of a single-layer dome based on its form. American Society for Civil Engineering
5. Niel G, Zhu X, Zhi X, Wang F, Dai J (2017) Study on dynamic behavior of single-layer reticulated dome by shaking table test. Int J Steel Struct
6. Jayminkumar SY, Vahora F (2016) A parametric study on steel dome structures. Int J Technol Res Eng
7. Anwar N, Norachan P, Aung TH (2015) Case study: challenges of a single-layer reticulated dome. In: IABSE conference—structural engineering: providing solutions to global challenges. J Constructional Steel Res
8. Guscetti G, Lelli L (2015) The new UEFA building in Nyon: prestressed light slabs and other structural aspects. In: IABSE symposium report
9. Jihong Y, Nian Q (2017) Progressive collapse simulation based on DEM for single-layer reticulated domes. J Constr Steel Res
10. IS 1893-1 (2016) Criteria for earthquake resistant design of structures. Part 1: general provisions and buildings
11. Anuj C (2014) Analysis and design of steel dome using software. Int J Res Eng Technol 3(3):2321–7308
12. Zhou H, Zhang Y, Fu F, Wu J (2018) Progressive collapse analysis of reticulated shell structure under severe earthquake loading considering the damage accumulation effect. American Society of Civil Engineering

Studies on Dry Lean Concrete with New Mix Design Approach

Somanath Khot, B. M. Mithun, Archana N. Shagoti, and Nitendra Palanakar

Abstract The present study investigates the properties of dry lean concrete (DLC) produced using new mix design approach. DLC layer is considered as the sub-base course layer, and the minimum required compressive strength is 10 MPa after 7 days of water curing. As per the IRC:SP:49-2014 guidelines, the current rational practice of mix design includes the calculation of quantities of ingredients based on the fixed cement to aggregate ratios without taking into consideration of the volume calculations. The earlier approach leads to volume irregularities and hence leads to problems related to quantity calculations. The current study is focused to prepare DLC mixes by taking into account the volume occupied by each of the ingredients calculated based on specific gravities of the materials. In the present study, the DLC mixes with cement contents of 155, 162, and 171 kg/m^3 have been prepared for different water contents, and the optimal mix is identified. The prepared DLC mixes are evaluated for their engineering properties, namely compressive strength and flexure strength, water absorption, volume of permeable voids (VPV), and ultrasonic pulse velocity. From the results, it was observed that the mix containing 162 kg/m^3 cement with moisture content of 6.02% attained highest compressive strength. The optimal water content was observed in the range of 6.02–6.52%. The new method of DLC mix design can be suitably implemented to design DLC mixes by eliminating the drawbacks of the rational method adopted by the IRC: SP: 49-2014.

Keywords Dry lean concrete · Mechanical properties · Durability · New mix design approach

S. Khot · A. N. Shagoti · N. Palanakar (✉)
Department of Civil Engineering, KLS Gogte Institute of Technology, Belagavi, India
e-mail: nnpalankar@git.edu

S. Khot
e-mail: smkhot@git.edu

A. N. Shagoti
e-mail: anshagoti@git.edu

B. M. Mithun
Department of Civil Egineering, NMAM Institute of Technology, Karkala, Udupi, India
e-mail: mithunbm@nitte.edu.in

L. Nandagiri et al. (eds.), *Sustainability Trends and Challenges in Civil Engineering*, Lecture Notes in Civil Engineering 162, https://doi.org/10.1007/978-981-16-2826-9_57

1 Introduction

With a total road network of about 5.89 million km in length, India is having one of the largest road networks in the world. Out of the total road network, around 3.5% is made with the concrete and of about 58.8% is made up with bituminous material. With an average growth of commercial traffic at the rate of about 7% yearly and with increase in the economy, the government of India is encouraging the construction of expressways and national highways with cement concrete roads through National Highway Development Programme (NHDP) [1]. The government is even developing the rural road network with cement concrete roads across the country through various rural road development projects. The versatility of cement concrete along with some added advantages like less maintenance cost and longer service life makes it one of the most demanding construction material across the world. Generally, rigid pavements consist of pavement quality concrete as wearing course, dry lean concrete as base course or sub-base course followed by prepared subgrade [2]. The structural and functional performance of the rigid pavement is greatly influenced by the underlying type base course. By and large, in India, dry lean concrete is constructed as base course layer to provide uniform support to the pavement quality concrete. Dry lean concrete is most common sub-base/base course in the rigid pavement components [3]. DLC is no slump concrete with high ratio of aggregate to cement in comparison with conventional cement concrete. The minimum required cement quantity for manufacturing of DLC is 140 kg/m^3 [2]. The high ratio of aggregate to cement in DLC makes it a perfect material for base course layer which also acts as drainage layer. A suitably designed and constructed DLC not only provides uniform support to the pavement quality concrete (PQC), but also brings some added advantages like strong underlying support to distribute loads, minimizing the mud pumping, confrontation to deformation, easy movement of construction equipment, firm support at joints, ease of form work fixing for PQC, proper placement of dowel bars, and most importantly designer can decrease the thickness of PQC slab from the point view of axle load distribution [3]. DLC is commonly manufactured with ordinary Portland cement (OPC) as per IRC: SP: 49-2014 specification. This specification (IRC: SP: 49-2014) advocates the use of other cement such as Portland Pozzolana Cement (PPC), Portland slag cement (PSC) also in the manufacture of DLC. The minimum required compressive strength is 10 MPa at 7-days curing and minimum ordinary Portland cement (OPC) content of 140 kg/m^3 [2]. Nevertheless, there is no clear instructions such as cement content, aggregate to cement ratio, moisture content, etc., for DLC manufactured with PPC and PSC. The actual thickness of the DLC will be governed by the design requirements; usually, 150 mm thickness is provided for National highways and State highways. If the DLC is used as sub-base layer other than national highways and state highways, then 100 mm thickness is sufficient. It is recommended to extend the DLC base/sub-base layer beyond the pavement edges up to 500 mm so as to support the movement of construction equipment and extended width of DLC would improve the flexural strength characteristics of PQC [3]. IRC: SP: 49-2014 advocates use of industrial by products as mineral admixtures like fly ash and ground granulated blast furnace slag at a 15–30% and 25–30% by the total weight of cementitious material in the DLC [2].

Although the DLC is such an important component in the rigid pavement, scanty studies have been done on DLC. Apart from industrial by products mentioned in the code, attempts have been made to produce dry lean concrete using different materials like Reclaimed Asphalt Pavement (RAP), copper slag, etc., RAP extracted from two different pavements with service life of 20 and 2.5 years was used in production of DLC as a base course, and it was found that RAP obtained from 20 years old pavements have the potential of 75% replacements to coarse aggregates and up to 50% replacements to fine aggregates [4]. Linz-Donawitz slag was used as a replacement for coarse aggregates in preparation of the dry lean concrete with 25, 50, and 100% replacements, and fine aggregates were replaced with loamy soil, red soil, and black cotton soil and concluded that LD slag has the potential to replace the natural coarse aggregates in DLC with some selected combination for fine aggregates replacements [5]. Dry lean concrete was prepared with a blend of stone dust and copper slag as a replacement for fine aggregates with 20, 40, 60, and 80% proportions. It was observed that maximum of 40% replacement for copper slag as fine aggregates yielded the minimum required compressive strength [6].

DLC laid as a base course layer for PQC with extended width analyzed for vertical stresses and flexural stresses using finite element method. Results indicated in reduction of flexural stresses developed in PQC, and because of reduced stresses, thickness of the PQC can be reduced [3].

DLC layer considered as the sub-base course layer, and the minimum required compressive strength is 10 MPa after 7-Days water curing. As per the IRC: SP: 49-2014 guidelines, the current rational practice of mix design includes the calculation of quantities of ingredients based on the fixed cement to aggregate ratios without taking into consideration the volume calculations. In IRC: SP: 49-2014 guidelines, the suggested water contents are 5, 5.5, 6, 6.5, and 7% for OPC of minimum 140 kg/m^3 and maximum cement to aggregate ratio of 1:14. This approach leads to volume irregularities and hence leads to problems related to quantity calculations. In the current study, it is focused to prepare the DLC mixes by taking into account the volume occupied by each of the ingredients calculated based on specific gravity of the materials. The DLC mixes with cement contents of 155, 162, and 171 kg/m^3 have been prepared for different water contents, and the optimal mix is identified.

2 Experimental Investigation

2.1 Materials

The dry lean concrete mixes prepared using 43 grade ordinary Portland cement confirming to IS 8112-2013 [7]. The compressive strength of cement assessed for

Table 1 Physical properties of coarse aggregates and fine aggregates

Properties	Coarse aggregates	Fine aggregates
Specific gravity	2.62	2.67
Aggregate crushing strength (%)	27.82	–
Los Angles abrasion value (%)	18.60	–
Aggregate impact value (%)	10.28	–
Water absorption (%)	0.39	1.10
Bulk density (kg/m^3)		
(a) Dry loose	1495	1475
(b) Dry compact	1653	1548
Silt content (%)	–	2

3 days and 7 days curing and found to be 24 MPa and 32 MPa, respectively. The specific gravity of the OPC was 3.15, and Blaine's fineness found to be 325 m^2/kg. 32% reported as normal consistency, and initial and final setting time found to be 95 min and 328 min, respectively. The aggregates used in present study are procured from local stone crushing industry and were assessed as per specification advocated in IS: 383-2016 and IS: 2386-1963 (Reaffirmed 1997) to check their suitability [8]. The physical properties of coarse aggregates and fine aggregates are evaluated and presented in Table 1.

The gradation of coarse aggregate and fine aggregate was performed, and the gradation was as confirming to the IRC: SP-49:2014 guidelines. The maximum size of aggregates used was 26.5 mm. The fine aggregates gradation found to be in the Zone II. The combined gradation of coarse aggregates and fine aggregates was evaluated, and they fulfill the required specifications [2].

The potable water available in the laboratory is used for mixing of raw mixes of different dry lean concrete mixes, and the same is used for curing of the specimens.

2.2 Mix Design and Metodology

In the present investigation, it was intended to examine the performance of DLC mixes using new mix design approach.

1. Need for the New Mix Design
   The mix design was done by fixing the water contents. The IRC: SP: 49-2014 code provides flexibility to the designers to arrive at the mix design by varying the cement contents, aggregate to cement ratio and water content. As per the guidelines, the minimum cement content should be greater than 140 kg/m^3 and maximum aggregate ratio to be 1:14 for ordinary Portland cement. As per the guidelines, the moisture content is fixed based on the dry weight of the ingredients. However, IRC: SP: 49-2014 guidelines does not provide any

methodology or guidelines to consider the specific gravities of the ingredients to calculate the volume during the mix design. Hence, there is a possibility that the mix design may lead to volume irregularities. As an example, let us consider a cement content of 180 kg/m^3 and aggregate cement ratio of 1:14 with a moisture content of 6%. A mix design made with these values will lead to theoretical density of 2862 kg/m^3 which is falling under the category of high density concrete and hence accommodating higher volume than 1 m^3. Further, the density will keep increasing with the increase in cement contents. The main drawback of such a mix design methodology is due to the fact that the volume calculations are not done based on the specific gravities of the ingredients.

2. New Mix Design Approach
   In the present study, the mix design is carried out by calculating the volume of the each of the ingredients by keeping the ratios of the materials within the limits as given by the IRC: SP: 49-2014 guidelines. Initially, the water content is fixed in order to arrive at moisture contents of 5.49, 6.02, 6.52, and 7.02 for various cement contents of 155, 162, and 171 kg/m^3. Based on the volume calculations, the quantities of aggregates are adjusted and the aggregate to cement ratios are calculated to be within the limit of 1:14 as mentioned by the IRC: SP: 49-2014.

In the present investigation, the cement content is varied from 155, 162, and 171 kg/m^3 based on the mix design calculations. The DLC mixes were prepared in the tilted drum mixer as per standard procedures. The cubes of size 15 cm × 15 cm × 15 cm were cast and compacted using Vibratory Jack Hammer [9]. The specimen were cured for 7-days of water curing. The details of the materials used for the preparation of DLC mixes are given in Table 2. The performance of the DLC mixes was assessed for various mechanical properties such as compressive strength and flexure strength. Durability properties such as water absorption, volume of permeable voids, and ultrasonic pulse velocity were analyzed. The consistency of the fresh DLC mixes examined via Vee-Bee consistometer test. The compressive strength and flexure strength tests were performed for 7 days curing [10]. Ultrasonic pulse velocity (UPV) test was performed on cubes as per the guidelines provided in ASTM C 597-09.

3 Results and Discussion

The prepared DLC mixes evaluated for their engineering properties, namely compressive strength and flexure strength, water absorption, volume of permeable voids (VPV), and ultrasonic pulse velocity test.

Table 2 Mix proportions for different cement contents

Mix No	M 1	M 2	M 3	M 4
Cement content of 155 kg/m^3				
Cement content (kg)	155	155	155	155
Cement/aggregate ratio	1:14	1:13.79	1:13.61	1:13.42
Moisture content (%)	5.49	6.02	6.52	7.03
Water content (kg)	135	147	158	169
Fresh density (kg/m^3)	2413	2433	2464	2458
Mix No	M 5	M 6	M 7	M 8
Cement content of 162 kg/m^3				
Cement content, kg	162	162	162	162
Cement/aggregate ratio	1:13.36	1:13.16	1:12.98	1:12.80
Moisture content, %	5.49	6.02	6.52	7.03
Water content, kg	135	147	158	169
Fresh density, kg/m^3	2432	2447	2487	2479
Mix No	M 9	M 10	M 11	M 12
Cement content of 171 kg/m^3				
Cement content, kg	171	171	171	171
Cement/aggregate ratio	1:12.61	1:12.42	1:12.25	1:12.08
Moisture content, %	5.49	6.02	6.52	7.02
Water content, kg	135	147	158	169
Fresh density, kg/m^3	2433	2452	2498	2490

3.1 Compressive Strength

The compressive strength of DLC mixes for different cement and varying moisture contents cured at 7 days presented in Fig. 1. From the results, it may be noticed that the compressive strength for all the mixes was in the range of 8.5–14.21 MPa. The mix M6 with fixed cement content of 162 kg/m^3 and water content of 6.02% recorded highest compressive strength. From the results, it was observed that for the mixes with the increase in the binder content increase in the compressive strength was observed up to 6.52% water content. However, further at 7.02% of water content, a slight decrease in the strength was noticed. The increasing trend in the compressive strength may be accounted for the increased mobility of the mix with increasing water content as the mix for 5.49% water content found to be very dry. However, the DLC mixes with the water contents of 6.02 and 6.52% attained good consistency for mixing and hence leading to better compaction and resulting in good compressive strength. Furthermore, the mixes with water content beyond 6.52% attained much better consistency but, however, more water content lead to the decrease in the compressive strength. This trend was common for all binder contents. From the

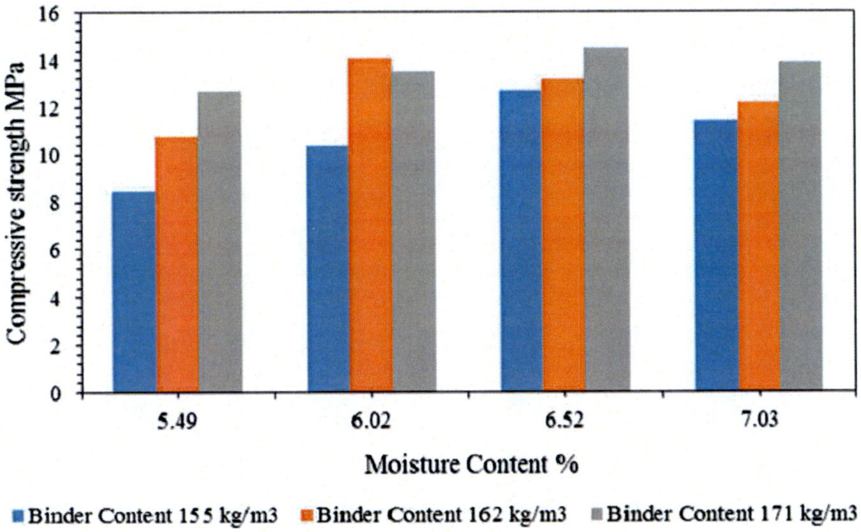

Fig. 1 Compressive strength of DLC mixes for different cement and moisture contents

overall scenario, it can be understood that the optimal water content for which better compressive strength was obtained is in the range of 6% to 6.5% irrespective of the binder content. The mixes with higher cement content achieved higher compressive strengths. However, from the point of application as the sub-base, the target strength was 10 MPa. Hence, considering that into account, the optimal mix out of all DLC mixes considered to be as 162 kg/m^3, and mixes with high strengths cannot be considered as optimal mixes [2].

3.2 Flexural Strength

The flexure strength of DLC specimen was assessed for two point loading conditions [10] and the results depicted Fig. 2. From Fig. 2, it may be noticed that flexural strength of mixes increased as the binder content increased for the water content up to 6.52%. For the mixes having water content 5.49%, found to be very stiff and resulted in the poor flexure strength. However, the mixes having water contents of 6.02% and 6.52% attained good consistency and proper compaction resulted in considerable increase in the strength. Furthermore, as the water content increased beyond 6.52%, reduction in flexure strength was noticed. This may be due to the increased mobility of the mixes as the binder content is increased.

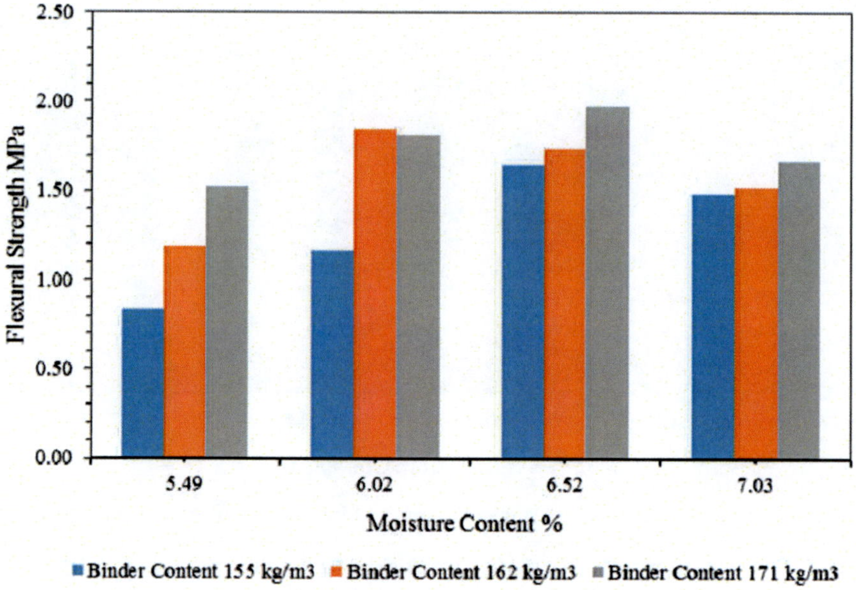

Fig. 2 Flexural strength of DLC mixes for different Cement and Moisture contents

3.3 *Water Absorpion and Voume of Permeable Voids*

The volume of permeable voids (VPV) and water absorption of DLC mixes cured at 7 days of water curing were investigated as per ASTM C 642-06 guidelines and given in Table 3. The mixes M1 to M4 containing 155 kg/m^3 of binder content

Table 3 Engineering properties of DLC mixes cured for 7 days

Mix No	Water absorption (%)	VPV (%)	UPV (km/s)	Vee-Bee consistency (s)
M1	6.98	18.44	3.75	19.42
M2	5.88	17.37	3.9	18.44
M3	4.96	16.78	4.15	16.37
M4	5.57	16.28	4	15.54
M5	5.42	15.68	3.81	18.21
M6	5.26	15.33	4.41	16.1
M7	4.89	15.14	4.2	14.25
M8	4.45	14.33	4.1	13.6
M9	5.33	13.83	4	10.14
M10	5.14	13.2	4.22	8.34
M11	4.34	13.00	4.3	7.37
M12	4.13	12.72	4.14	5.91

displayed higher water absorption and VPV as compared to mixes M5–M8 containing 162 kg/m^3 binder content and M9–M12 containing 171 kg/m^3 binder content.

It may be observed that the mixes containing lower binder contents displayed higher water absorption and volume of permeable voids as compared to those mixes with higher binder contents. This may be attributed to the higher cement to aggregate ratio in mixes containing lower binder content.

3.4 Ultra-Sonic Pulse Velocity Test

In the present study, UPV test is carried out on the DLC mixes cured for 7 days of water curing, and the results were given in Table 3. From the results, it may be noted that, the mixes with the lower binder content, i.e., 155 kg/m^3 and high aggregate ratio recorded lower UPV values. This may be related to the higher VPV values in such mixes. However, DLC mixes having binder contents of 162 kg/m^3 and 171 kg/m^3 have noted higher UPV values. This could be because of the increased binder content in DLC mixes leads to good consistency and better mobility of the mixes leading to proper compaction. All the mixes attained UPV values in the range of 3.75–4.41 km/s indicating the quality of concrete to be ranked well.

3.5 Vee-Bee Consistometer Test

In the present study, the consistency of the DLC mixes was measured in terms of the Vee-Bee seconds, and the results were given in Table 3. From the results, it was noticed that the DLC mixes were harsh mix in nature [5]. The results depict that, for mixes with binder content of 155 kg/m^3 displayed higher Vee-Bee time as compared to those mixes with higher binder contents which recorded lower Vee-Bee times. The mixes containing higher binder content had lower cement to aggregate ratios which indicates higher paste content, thus leading to better consistency and compaction. For mixes with binder content of 155 kg/m^3, the mix M1 recorded 19.42 s, while mixes M2, M3, and M4 recorded 18.44 s, 16.37 s, and 15.54 s, respectively. It was also observed that with increase in water contents irrespective of the binder content, the Vee-Bee time displayed lower values with increase in water contents. Similar trends were observed with mixes containing binder contents of 162 and 171 kg/m^3.

4 Conclusions

The conclusions drawn from the present investigation are summarized as below.

1. From the results, it was observed that, as the cement content increases the consistency and mobility of the DLC mixes increased. The optimal water content was found to be in the range of 6.02–6.52% irrespective of the binder content. The cement content of 162 kg/m^3 has recorded the targeted compressive strength as well as the flexural strength.
2. Water absorption and volume of permeable voids decreased as there is an increment in the binder content. Mixes with lower binder contents displayed higher water absorption and VPV values.
3. The mixes with higher binder contents displayed better UPV values as compared to low cement content mixes. All the mixes attained UPV values in the range of 3.75–4.41 km/s indicating the quality of concrete to be ranked good.
4. Irrespective of the binder content, increase in water contents in the mixes led to lower Vee-Bee time in the DLC mixes.

It may be understood that the new method of mix design can be suitably implemented to design DLC mixes by eliminating the drawbacks of the rational method adopted by the IRC: SP: 49-2014.

References

1. MORTH (2016–17) Basic road statics of India
2. IRC (2014) Guidelines for the use of dry lean concrete as sub-base for rigid pavement, New Delhi
3. Suryateja S (2019) Effect of extended dry lean concrete base on flexural stresses in concrete pavements. J IRC 1:1–476
4. Singh S, Ransinchung GD, Monu K (2019) Sustainable lean concrete mixes containing wastes originating from roads and industries. Constr Build Mater 209:619–630
5. Chandrashekhar SY, Madhu KA, Anil Kumar MS, Rajakumara HN (2019) Production of dry lean concrete by sustainable materials to use in road sub-base layer for rigid pavement. Int J Recent Technol Eng 8(3):750–757
6. Kumar B (2013) Properties of pavement quality concrete and dry lean concrete with copper slag as fine aggregate. Int J Pavement Eng 14(8):746–751
7. Bureau of Indian Standard (BIS) (2013) IS: 8112-1989, Specification for 43 grade ordinary portland cement. Bureau of Indian Standard, Delhi
8. Bureau of I. S. (BIS) (1970) IS 383: 1970 Specification for coarse and fine aggregates from natural sources for concrete. Indian Standard, pp 1–24
9. Kumar R (2016) A comparative study on dry lean concrete manufactured with OPC vis-a-vis PPC to be used for the construction of concrete roads. Indian Concr J 70–76
10. IS 516:2014 (2004) Method of tests for strength of concrete, IS 516–1959 (Reaffirmed 2004), New Delhi, India

Properties Enhancement Strategy for Fibre Reinforced Standard Concrete Using Foundry Sand (FS) and Crushed Concrete Waste (CCW)

H. M. Yajnodbhavi, C. M. Ravi Kumar, T. Maruthi, and S. R. Pruthviraj

Abstract Nowadays, many research works are carried for all grades of concrete to make the concrete most economical and durable by adding the supplementary cementitious materials and alternative replacement aggregates. In this research work the experimental investigation of mechanical properties of the M30 and M50 grade concrete by replacing the fine and coarse aggregate by foundry sand and crushed concrete waste, respectively. The type of mix design is used in this paper as per IRC44:2017 Guidelines and recommendations. Proper dosage of super plasticizer (SP) was maintained in the concrete to make it better performed. In this present investigation, a poly-propylene fibre (PPF) of 0.3% by weight of the cement is used. Mechanical properties were determined by preparing the respective mould sizes for specific test and are cured for 7, 14 and 28 days and result obtained for respective days were tabulated and discussed.

Keywords M30 and M50 grade concrete · Poly-propylene fibres · IRC guidelines · Mechanical properties

1 Introduction

Concrete is the maximum substantially used production material in the world, next to water. Increasing rate of industrialization and urbanization has led to over exploitation of natural resource such as gravel and river sand, which is giving rise to sustainability issues. The High Strength and High Performance Concrete (HSC and HPC) are becoming important concrete material for conventional concrete. The High Strength and High Performance Concrete (HSC and HPC) are obtained often, by decreasing

H. M. Yajnodbhavi (✉)
Department of Civil Engineering, PES Institute of Technology and Management, Shivamogga, Karnataka 577204, India
e-mail: yagna.hm@pestrust.edu.in

C. M. Ravi Kumar · T. Maruthi · S. R. Pruthviraj
Department of Studies in Civil Engineering, University B.D.T College of Engineering (A Constituent College of Visvesvaraya Technological University), Davangere, Karnataka 577004, India

L. Nandagiri et al. (eds.), *Sustainability Trends and Challenges in Civil Engineering*, Lecture Notes in Civil Engineering 162, https://doi.org/10.1007/978-981-16-2826-9_58

the amount of W/C ratio, with the use of special admixture that also improves the workability [Admixtures such as water-reducing agents, plasticizers (P), super plasticizers (SP), hyper plasticizers (HP) and hyper-hyper plasticizers (HHP)]. Proper Mix design for concrete plays a major role in getting the desirable concrete in construction field. Proper maintenance of W/C ratio makes the concrete more workable and Durable. By restricting the W/C ratio to extremely lesser percentage can achieve more strength in terms by using plasticizers (P) and super plasticizers (SP). Also, Proper gradation of the fine aggregates (FA) and coarse aggregates (CA) as per standard specifications and guidelines also becomes the factors for achieving strength of concrete.

1.1 IRC Recommendations

- Salient features of IRC44:2017 [1]
  The main objectives of IRC 44-2017 [1] (Third revision) are to design the mixes to arrive at the very economical and practical combinations and the proportions of different ingredients to produce the concrete that will meet the performance requirements of the desirable grade of concrete under specified environmental condition. The main Scope of the IRC 44-2017 is to design the concrete mix as per the requirements.
- Salient features of IRC: 58-1988 [2] (Guidelines for the design of rigid pavements for highways).

  The basic design parameters and assessments of their design values are as follows.

(i) Traffic parameters
(ii) Environmental parameters
(iii) Concrete characteristics
(iv) Design of slab thickness
(v) Design of joint
(vi) Design of reinforcement

- Salient features of IRC: SP: 62-2014 [3]
  The code IRC: SP: 62-2014 [3] deals with the design and construction of cement concrete pavements (CCP) for low volume roads.

This specific code covers the design principles of rigid pavements of the low volume roads of about 3.75 m wide concrete. Also, in this code the factors governing the design of concrete with respect to the traffic conditions are discussed.

- Salient features of IRC: 15-2011.

Scope of IRC: 15-2011:

This specific code intended to consider the good quality practices for the construction of cement concrete pavements including the components of rigid pavements such as subgrade and sub-base.

- Salient features of IRC: SP: 46-2013 [4] (Guidelines for design and Construction of Fibre Reinforced Concrete Pavements-First Revision)
  Scope: IRC: SP: 46 discuss the use of fibre reinforced concrete in concrete pavements and the aspects are differentiated with conventional concrete. Discussed the steel and polymeric fibres only with short length-60 mm discontinuous fibres in the concrete pavements. Discussed the use of fibres in a cement-based matrix such as concrete, micro-concrete and in use of concrete pavement and related repair structures.

2 Literature Review

Sriram Aaleti et al. [5] investigated the usage of a thin layer of ultra-HPC overlaying a normal strength concrete (NC) deck. They analysed the behaviour of the HPC by considering the performance and the durability. Also examined the UHPC and normal concrete composite deck for strength, roughness and curing conditions on the shear transferring behaviour of the concrete such as ultra-HPC and conventional concrete were determined.

Wadekar et al. [6] carried out the Experimental study to assess mechanical properties of high strength fibre reinforced concrete of grade M80 Mpa. They used the material for investigation such as silica fume, fly Ash and three types of fibres such as hooked end steel, flat steel fibre and crimped steel fibre. Replacement was done for silica fume as 50% and for fly ash it is 20%. For this investigation, maintained W/C ratio about 0.25. Each type of fibre of volume fraction is taken from 0.5 to 4.0% with an increment of 0.5% in difference of each by weight of cement.

Jayachandra et al. [7] conducted an experimental study on the possessions of concrete by using foundry sand as replacement of fine aggregate in high strength concrete (HSC). In this review, it was found that replacement of sand by replacement of foundry sand which results in the marginal increase in the Hardened properties of the concrete, and the foundry sand gives up to the optimum percentage of 25%.

Puneeth HC et al. [8] studied the strength properties of Recycled Construction and Demolition Waste in concrete is investigated with five different replacement ratios including 20, 40, 60, 80, and 100%. The replacement of recycled materials was replaced for CA in M30 and M40 grade of concrete. From this study, it is found that an effective replacement of 45.75 and 23.35% of RCA for M30 and M40 grade, respectively.

Milind V. Mohod et al. [9] studied the effects of addition of various proportions of PPF on the houses of High strength concrete (M30 and M40 mixes). The main aim of the investigation programme is to study the effect of PPF mix by varying content such as 0, 0.5, 1, 1.5 and 2% and finding the optimum PPF content. It was concluded that the increasing percentage volume of fibre brought into the concrete would lead the workability decreased.

Table 1 Properties of cement used

SL	Properties	Test results	IS: IS: 8112-1989 requirements
1	Specific gravity	3.1	No standard valve
2	Standard consistency (%)	30	
3	Fineness (m^2/kg)	300	Not less than 225
4	*Setting time (mm)*		
	(a) Initial Setting time	57	Not less than 30
	(b) Final Setting time	370	Not more than 600
5	*Compression strength (Mpa)*		
	For 43 grade: 3 days	24.5	23
	7 days	37.8	33
	28 days	44.2	43

Chandrasekar R et al. [10] carried out the Experimental results carried out for utilization of waste foundry sand (WFS) in High strength concrete. The waste foundry sand was replaced in the place of normal sand with four different percentages (10, 20, 30 and 40%). In this test results, there is sudden decrease in compressive strength due to the increase in percentage of foundry sand from 30%.

3 Materials

From the above listed materials, priority of the materials is selected based on the many factors and the role in consideration. The following materials are selected for the preparation of concrete, they are.

3.1 Cement

In this experimental work, OPC-43 and 53 grades were used. The property of cement that was used conforms to IS: 8112-1989 as shown in Table 1.

3.2 Fine Aggregate (FA)

3.2.1 River Sand and Foundry Sand (RS and FS)

RS and FS materials can be taken as surface Clean and Dry sand is used as locally available material. The various properties of RS and FS are listed in Table 2 [11].

Table 2 Properties of fine aggregate

SL	Properties	Test results	IS: 383 requirements
I	*Foundry sand*		
1	Specific gravity	2.46	2.2–2.7
2	Bulking of sand	4%	–
3	Silt content	Nil	Less than 8%
4	Abrasion test	0.235	
5	Particle size distribution	2.7	Fine:FM:2.2–2.6 Medium: F.M:2.6–2.9 Coarse: F.M:2.9–3.2
II	*River sand*		
1	Specific gravity	2.45	2.2–2.7
2	Bulking of sand	6%	–
3	Silt content	4	Less than 8%
4	Particle size distribution	3.2	Fine:FM:2.2–2.6 Medium: F.M:2.6–2.9 Coarse: F.M:2.9–3.2

3.3 *Coarse Aggregate (CA)*

3.3.1 Natural Aggregate and Crushed Concrete Waste (NA and CCW)

NA and CCW aggregates are taken passing through 20 mm sieve and retained on 12.5 mm sieve and as given in IS: 383-1970 [11] is used for all the specimens. The various properties of NA and CCW are listed in Table 3.

3.4 *Water*

Casting and curing of concrete specimens were done with the potable water. Water used in the preparation of concrete should be free from dirt and organic matters.

3.5 *Super Plasticizer (SP)*

To achieve the workability for concrete, super plasticizers are used. In this, poly carboxylic ether is used. Conplast SP430: Conplast SP430 should complies with BIS: 9103-1999 and BS: 5075-part3 and ASTM C494. Super plasticizer molecules and cement grains are oppositely charged and hence repel each other and the properties of super plasticizes can be listed in Table 4.

Table 3 Properties of coarse aggregate

SL	Properties	Test results	IS: 383 requirements
I	*Natural aggregate*		
1	Specific gravity	2.46	2.2–2.7
2	Crushing strength (%)	26.60	Not less than 30%
3	Flakiness (%)	5.36	Less than 15%
4	Elongation (%)	14.76	Less than 15%
5	Impact value	18.82	10–20% strong
6	Water absorption (%)	2.37	–
II	*Crushed concrete waste*		
1	Specific gravity	2.6	2.2–2.7
2	Crushing strength (%)	16.39	Not less than 30%
3	Flakiness (%)	11.4	Less than 15%
4	Elongation (%)	14.8	Less than 15%
5	Impact value	4.79	< 10% exceptionally strong
6	Water absorption (%)	2.75	–

Table 4 Properties of super plasticiser

SL	Properties	Test results	BIS: 9103 = 1999
Conplast SP430			
1	Specific gravity	1.18	1.0–2.0
2	Colour	Light brown	
3	Plasticizer type	High performance super plasticizer	
4	Air entrainment (%)	1%	Less than 2%

3.6 Poly-Propylene Fibres (PPF)

The raw material of poly-propylene is derived from monomeric C_3H_6 which is purely hydrocarbon. Its mode of polymerization, its high molecular weight and the way it is processed into fibres combine to give poly-propylene fibres very useful properties as explained in Table 5.

Table 5 Properties of poly-propylene fibres

SL	Properties	Test results
1	Colour	Natural white
2	Length	40 mm
3	Cross section	Rectangular (1 × 0.5 mm)
4	Density	0.91 kg/m^3
5	Specific surface area	250 m^2/kg
6	Water absorption (%)	Nil
7	Melt point	160 °C
8	Ignition point	365 °C
9	Thermal and electrical conductivity	Low
10	Acid resistance	High
11	Alkali resistance	100%

4 Experimental Procedure as Per IRC

4.1 Mix Design, Means, Modes and Methods

In this experiment conducted, the grades of concrete M-30 and M-50. The mix design was carried out as per IRC 44-2017 [12]. The trials have been prepared on these different grades.

4.2 IRC Method of Concrete Mix Design

The IRC method of mix design calculated for different grades such as M30 and M50 grade of concrete. The mix proportions of various grades can be listed in Table 6.

4.3 Mixing of Samples

The mixing of ingredients is done with proper care and all materials were weighted properly and mixed in the laboratory concrete mixer. In addition, water super plasticizes (SP) was used in this experiment. To find the optimum dosage of super plasticizer (SP) we conducted Marsh Cone test.

- Marsh Cone Test (Flowability test)
  Marsh cone testing method is used for finding the saturation dosage. Observation was made by taking the Super plasticizer (SP) dosage of 0, 0.25, 0.5, 0.75, 1.0,

Table 6 Mix proportion as per IRC 44-2017 [12]

Name of mixture	Proportions in kg/m^3							
	Designation	Cement	Fine aggregate		Coarse aggregate		Admixture	Water
			NS	FS	NA	CCW		
M30 Grade								
M1	CC	360	650	0	1198	0	4.5	150
M2	FS15		552	98	1018	180		
M3	FS30		455	195	838	360		
M4	FS45		357	293	658	540		
M5	CCW10		585	65	1078	120		
M6	CCW20		520	130	958	240		
M7	CCW30		455	195	838	360		
M50 Grade								
M1	CC	420	467	0	1405	0	5.2	126
M2	FS10		420	47	1265	140		
M3	FS20		374	93	1124	281		
M4	FS30		327	140	983	422		
M5	CCW5		444	23	1335	70		
M6	CCW10		420	47	1265	140		
M7	CCW15		397	70	1195	210		

1.25, 1.5, 1.75 and 2.0% to the weight of cement into Consideration. Active performance of the cement with Super plasticizers (SP) is discussed in Table 7.

By observing the graph in Fig. 1, the optimum dosage of super plasticizer (SP) is taken at 1.25%.

Table 7 Marsh cone test results

SL	Dosage in %	Time in s
1	0	37.70
2	0.25	27.50
3	0.50	25.60
4	0.75	26.20
5	1.0	24.50
6	**1.25**	**23.10**
7	1.5	22.90
8	1.75	22.90
9	2.0	23.00

Note: By observing the above Table, 1.25% is the Optimum dosage of super plasticizer (SP) is taken

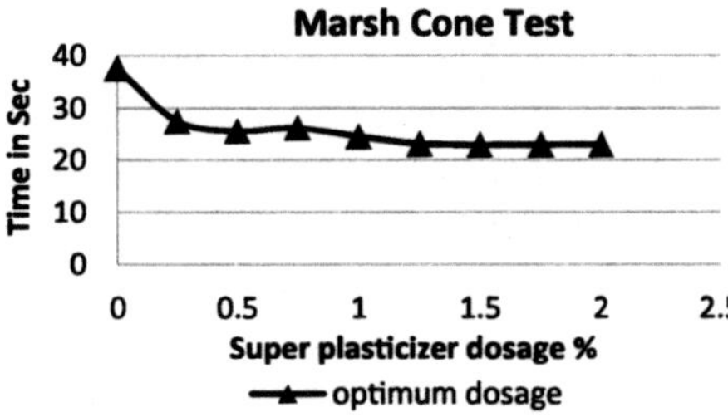

Fig. 1 Marsh cone test result

4.4 Casting of Specimens

After all the materials are collected and mixed with proper manner. The cubes were filled of size 15 × 15 × 15 cm by partially replacement of FS and C&D waste and also specimens casted for the test on Hardened concrete-like cubes in size 15 × 15 × 15 cm, cylinders in size 10 × 50 cm and beams in size 10 × 10 × 50 cm. Compacted by using table vibrating machine Tamping rod.

4.5 Curing of Specimens

After casting the specimens like cubes, beams and cylinders. All the specimens were kept for 24 h till the testing days up to 7, 14, and 28 days.

4.6 Testing of Specimens

After completion of the curing, the testing samples are dry about half an hour. After completely surface drying is done then testing was done by universal testing machine of each percentage of three cubes. The prepared specimens are tested Mechanical properties of concrete for 7, 14, and 28 days of curing.

5 Experimental Observation

5.1 Tests for Fresh Concrete

- Slump test
  Slump test is one of the easy tests to carry out the workability of the concrete both in field and laboratory. This type of test is done when the concrete is more workable and the slump value of concrete depends on the *W/C* ratio.

- Compaction factor test
  Compaction factor test is carried in the laboratory only to determine the compaction factor of the concrete. The concrete of very hard concrete can also determine using this test. Very low workable concrete can also be determined through this test.

5.2 *Test for Hardened Concrete*

- Crushing strength (CS)
  Crushing strength test is carried to determine the amount of the compression load taken by the specific dimension of the concrete cube. The cube may be prepared with a standard size of 150 mm × 150 mm × 150 mm and are cured and tested through compression test arrangement. The crushing strength of the cube can be calculated as compression load acting (P) on the cube to the area (A) of the specimen. The crushing strength of concrete can be expressed in terms of Mpa.
- Split Tensile Strength (STS)
  Split tensile strength test is carried out to determine the tensile capacity of the concrete sample in its hardened state. The cube prepared for split tensile strength of concrete having cylindrical in shape dimensioned 150 mm in diameter and 300 mm length. The formula used based on IS: 5816=1970 guidelines.
- Flexural strength test (FST)
  Flexural strength test can be carried out through one point or two points loading excluding supports. The mould of size 100 mm × 100 mm × 500 mm was prepared and cured for respective curing periods. To analyse the flexural behaviour of the concrete in post-cracking stages can be done through this test and can also determine the toughness characteristics of the concrete.
- Shear Strength Test (SST)
  Shear strength is the most extreme load required to remove an example such that the subsequent pieces or totally certain of each other. This is experienced lust before a material crack. The formula used based on IS: 516-1959 guidelines.
- Impact test (Dropping Weight test) (IT)
  The impact test or dropping weight test is one of the methods used to assess the impact properties of polymer. Computation of the impact strength was as follows

$$\text{Impact strength of sample} = (W * h * n) \text{ N-m}$$

where

W = Weight of hammer = 4.5 kg = 45 N.
H = Height of hammer −0.457 m.
N = Number of blows.

6 Experimental Results

6.1 Tests on Fresh Concrete

- Slump test results as per IS 1199:1959
  The slump values for various compositions are reported in Table 8, and graphical representation of those values is indicated in Fig. 2.

In Fig. 2, it can be observed that the various concrete mixes such as M1–M7 gives the respective slump valves. In mix M1 gives the more slump valve about 48 mm

Table 8 Slump values of the concrete

SL	Concrete type	M30 grade slump in mm		M50 grade slump in mm	
		0% PPF	0.3% PPF	0% PPF	0.3% PPF
1	M1	48	40	20	17
2	M2	36	28	15	13
3	M3	30	25	15	12
4	M4	22	20	12	10
5	M5	20	20	14	12
6	M6	18	16	12	10
7	M7	18	15	12	8

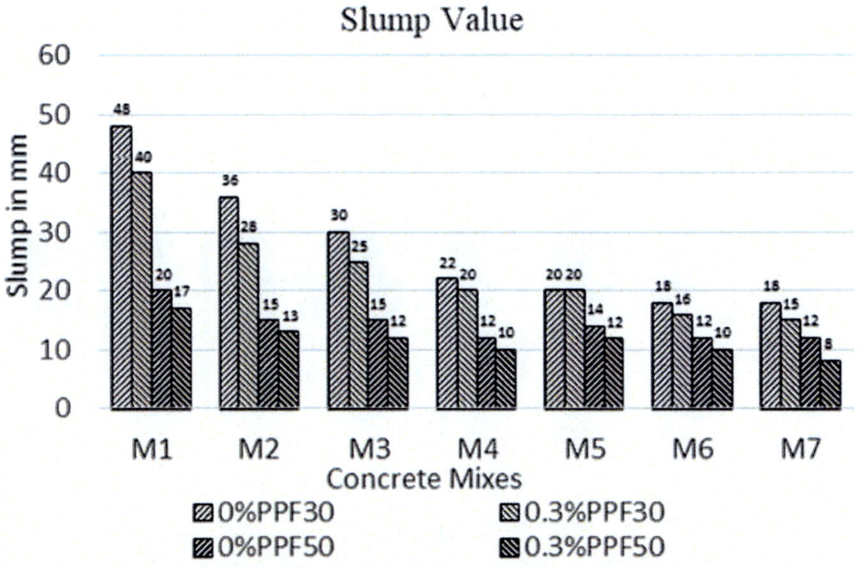

Fig. 2 Graphical representation of slump test results

for without PPF and 40 mm for with PPF and in the same manner for FS and CCW replaced materials are also calculated the workability will decrease with the grade of concrete.

- Compaction factor test as per 1199:1959
  The workability in terms of compaction factor for various compositions is reported in Table 9 and graphical representation of those values is indicated in Fig. 3.

In Fig. 3, it can be observed that the various concrete mixes such as M1–M7 gives the respective compaction factor values from these values it can be observed that for M30 grade of concrete it varies from 0.90 to 0.95. M50 grade of concrete it varies 0.78–0.88 and also after adding the PPF to the same mixes the compaction factor will decrease.

Table 9 Compaction factor values of the concrete

SL	Concrete type	M30 grade compaction factor		M50 grade compaction factor	
		0% PPF	0.3% PPF	0% PPF	0.3% PPF
1	M1	0.95	0.94	0.88	0.86
2	M2	0.94	0.92	0.85	0.84
3	M3	0.94	0.93	0.84	0.81
4	M4	0.92	0.91	0.83	0.81
5	M5	0.91	0.90	0.80	0.78
6	M6	0.90	0.88	0.80	0.77
7	M7	0.90	0.86	0.78	0.77

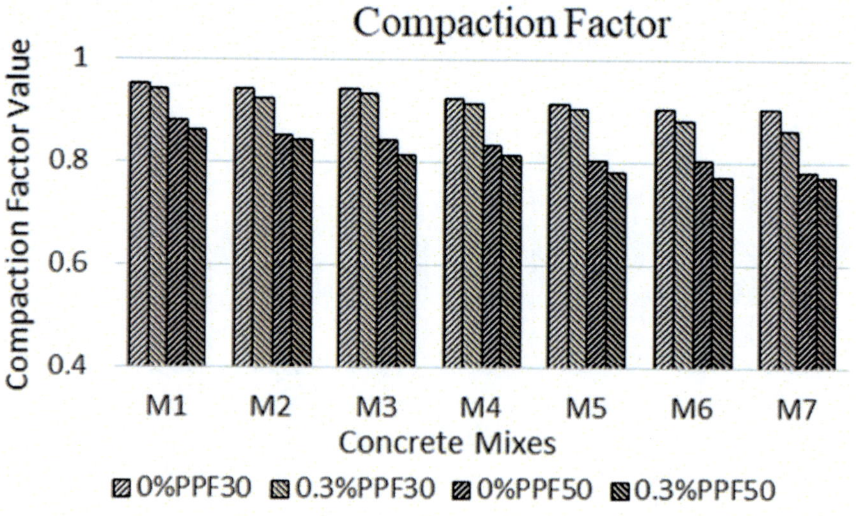

Fig. 3 Compaction factor test results

6.2 Tests on Hardened Concrete to Find Optimum Replacements of FS and CCW

- Crushing strength test results for partially replacement FS and CCW.
  The crushing strength results for FS and CCW of various percentage replacements for 7, 14 and 28 days are detailed in Table 10 and the graphical representation of each test results are shown separately in Figs. 4, 5 and 6.

In Fig. 4, it can be observed that the Crushing strength (CS) of various mixes from M1–M7 for 7 days of curing. In which the Crushing strength (CS) as compared to the both with and without PPF. It shows that with fibres shows maximum Crushing strength (CS) as compared to the without fibres.

In Fig. 5, it can be observed that the Crushing strength (CS) of various mixes from M1–M7 for 14 days of curing. In which the Crushing strength (CS) as compared to the both with & without PPF. It shows that with fibres shows maximum Crushing strength (CS) as compared to the without fibre.

In Fig. 6, it can be observed that the Crushing strength (CS) of various mixes from M1–M7 for 28 days of curing. In which the Crushing strength (CS) as compared to the both with and without PPF. It shows that with fibres shows maximum Crushing strength (CS) as compared to the without fibres.

- Water absorption test for FS and CCW
  The Water absorption test results for FS and CCW of various percentage replacements for 28 days are detailed in Table 11.

i. Tests on Hardened Concrete results for optimum replaced materials
   The test on Hardened concrete test results for Optimum replaced materials FS and CCW are detailed in Table 12 and the graphical representation of each test results are shown separately in Figs. 7, 8, 9 and 10.

In Fig. 7, it can be observed that the Crushing strength (CS) of various grades of concrete such as M30 and M50 for 7 and 28 days of curing, respectively. By comparing these optimum mixes of M3, M5 and M6 to the conventional mix M1 mix. The Crushing strength (CS) of M3 and M6 has shown increased strength than the M1 mix for M30 grade, the Crushing strength (CS) of M3 and M6 has shown increased strength than the M1 mix for M50 grade for both 7 and 28 days of curing.

In Fig. 8, it can be observed that the Flexural strength test (FST) of various grades of concrete such as M30 and M50 for 7 and 28 days of curing, respectively. By comparing these optimum mixes of M3, M5 and M6 to the conventional mix M1 mix. The Flexural strength test (FST) of M3 shown increased strength and M6 shown decreases strength than the M1 mix for M30 grade, the Flexural strength test (FST) of M3 and M6 has shown increased strength than the M1 mix for M50 grade for both 7 and 28 days of curing.

In Fig. 9, it can be observed that the Shear strength test (SST) of various grades of concrete such as M30 and M50 for 7 and 28 days of curing, respectively. By comparing these optimum mixes of M3, M5 and M6 to the conventional mix M1

Table 10 Crushing strength result of FS and CCW for 7, 14 and 28 days of curing

Grade	FS							CCW						
	Mix	0% PPF			0.3% PPF			Mix	0% PPF			0.3% PPF		
		7 days	14 days	28 days	7 days	14 days	28 days		7 days	14 days	28 days	7 days	14 days	28 days
M30	M1	26.07	36.14	40.88	27.99	37.47	48.29	M1	26.07	36.14	40.88	27.99	37.47	48.29
	M2	25.32	32.14	34.81	30.84	38.36	47.99	M5	22.66	32.88	36.29	23.55	24.59	40.88
	M3	29.92	36.44	41.66	34.66	39.85	53.47	M6	26.22	38.51	44.88	29.62	39.69	51.10
	M4	23.84	27.55	31.99	30.07	36.14	38.81	M7	21.18	23.70	38.69	20.29	34.07	39.99
M50	M1	39.70	56.59	61.48	44.29	58.22	65.18	M1	39.70	56.59	61.48	44.29	58.22	65.18
	M2	38.96	55.40	62.37	44.55	58.66	66.07	M5	41.77	57.48	62.66	45.33	58.51	64.29
	M3	43.25	57.77	64.59	47.55	60.29	68.88	M6	42.37	58.37	64.29	46.37	60.44	66.07
	M4	42.22	56.44	62.51	45.33	58.96	65.77	M7	41.18	56.44	61.03	42.96	57.33	64.29

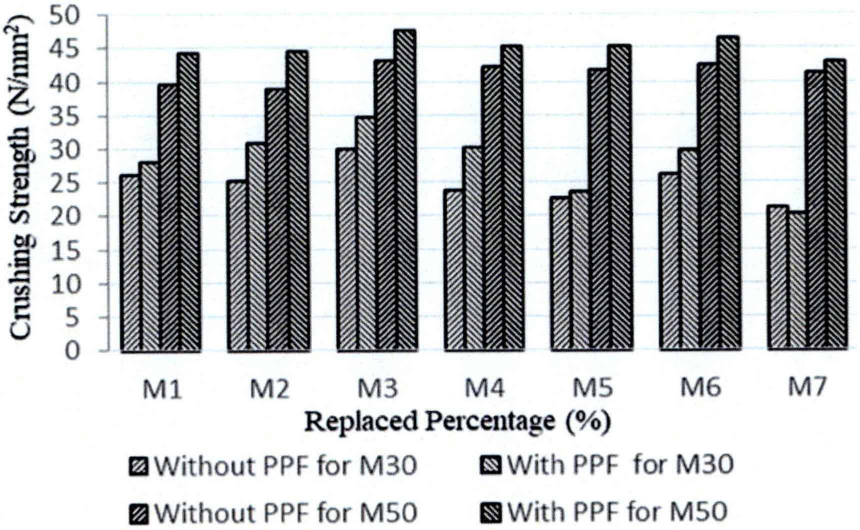

Fig. 4 Graph showing crushing strength of FS and CCW for 7 days curing

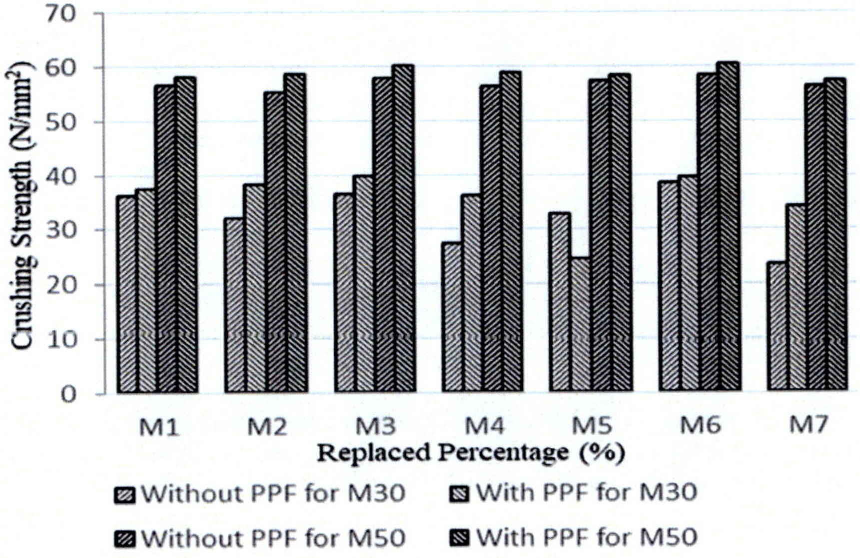

Fig. 5 Graph showing crushing strength of FS and CCW for 14 days curing

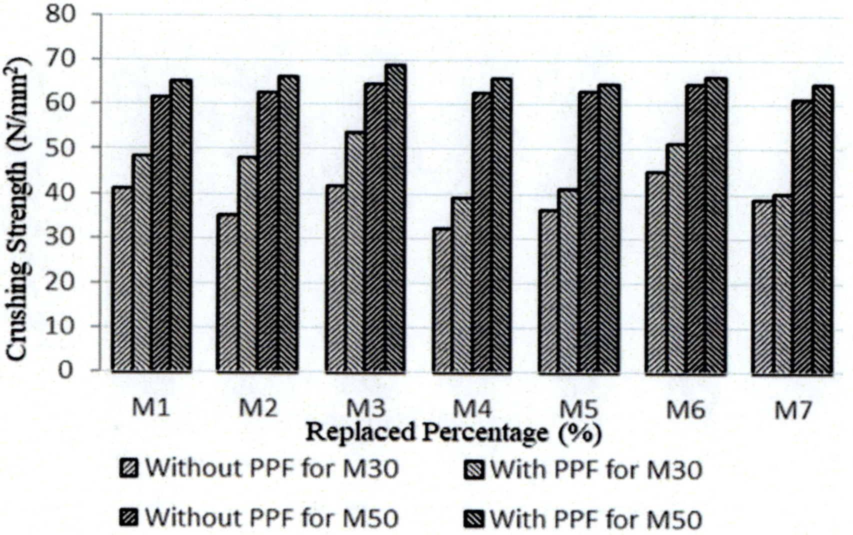

Fig. 6 Graph showing crushing strength of FS and CCW for 28 days curing

mix. The Shear strength test (SST) of M3 and M6 shown increased strength than the M1 mix for M30 grade, the Shear strength test (SST) of M3 and M6 has shown increased strength than the M1 mix for M50 grade for both 7 and 28 days of curing.

In Fig. 10, it can be observed that the Split tensile strength (STS) of various grades of concrete such as M30 and M50 for 7 and 28 days of curing, respectively. By comparing these optimum mixes of M3, M5 and M6 to the conventional mix M1 mix. The Split tensile strength (STS) of M3 and M6 shown increased strength than the M1 mix for M30 grade, the Split tensile strength (STS) of M3 and M6 has shown increased strength than the M1 mix for M50 grade for both 7 and 28 days of curing.

In Fig. 11, it can be observed that the Impact test (IT) of various grades of concrete such as M30 and M50 for 7 and 28 days of curing, respectively. By comparing these optimum mixes of M3, M5 and M6 to the conventional mix M1 mix. The Impact test (IT) of M3 shown increased strength M6 shown decreases strength than the M1 mix for M30 grade, the Impact test (IT) of M3 has shown increased strength M6 shown decreases strength than the M1 mix for M50 grade for both 7 and 28 days of curing.

6.3 Observations

Following observation are made based on the experimentation conducted for determining the mechanical properties of concrete with replacement of foundry sand and crushed concrete waste in concrete with and without adding the poly-propylene fibres are below

Table 11 Water absorption test result of FS and CCW

Grade	FS					CCW				
	Mix	0% PP		0.3% PP		Mix	0% PP		0.3% PP	
		surface dry	Oven dry	Surface dry	Oven dry		Surface dry	Oven dry	Surface dry	Oven dry
M30	M1	8250	8060	8190	7090	M1	8250	8060	8190	7090
	M2	8260	8150	8200	8030	M5	8230	7980	8200	8070
	M3	8250	8130	8195	7990	M6	8200	7990	8170	7950
	M4	8300	8050	8210	8000	M7	8220	8110	8210	8000
M50	M1	8550	8360	8430	8220	M1	8550	8360	8430	8220
	M2	8620	8270	8590	8370	M5	8530	8310	8410	8200
	M3	8600	8310	8570	8350	M6	8540	8300	8440	8210
	M4	8680	8380	8650	8440	M7	8555	8340	8430	8190

Table 12 Hardened concrete test results of FS and CCW for 7 and 28 days of curing

SL	Tests	Days	M30 grade with 0.3%PP			M50 grade with 0.3%PP		
			M1	M3	M6	M1	M3	M6
1	Compressive strength test	7	27.99	34.66	29.62	44.29	47.55	46.37
		28	48.29	53.47	54.10	65.18	68.88	66.07
2	Flexural strength test	7	3.14	3.82	3.02	3.38	4.75	3.88
		28	3.73	4.24	3.77	4.02	6.41	5.77
3	Shear strength test	7	5.92	7.14	6.41	10.74	13.70	11.14
		28	11.11	14.92	12.44	15.55	20.00	17.74
4	Split tensile strength test	7	1.98	2.59	2.74	2.45	3.25	2.89
		28	2.78	3.63	2.93	3.77	4.33	4.08
5	Impact test	7	16.55	19.89	16.06	22.66	23.55	21.18
		28	21.52	23.55	20.93	41.77	42.37	40.88

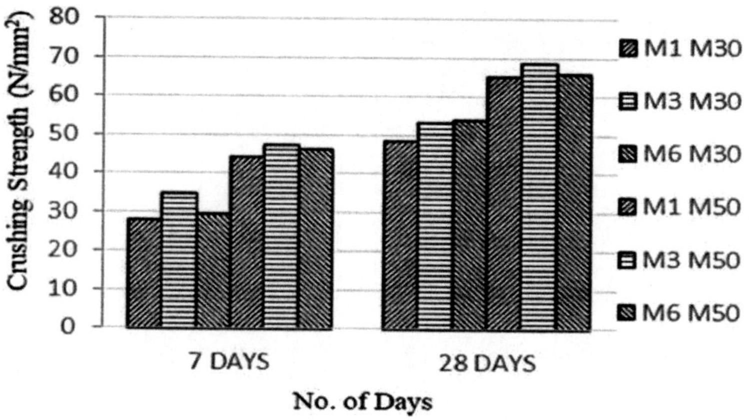

Fig. 7 Graph showing Crushing strength of FS and CCW for 7- and 28-days curing

1. It is observed that materials which procured are functioned well and all the materials are satisfied the basic tests and strength requirements.
2. HPC with replacements are measured from slump and compaction factor and observation made that, the degree of workability goes on decreasing as the percentage of replacements goes on increasing. This is due to the fineness of the foundry sand, its surface area and uniform in size and in coarse aggregate, due to the presence of pores in the crushed concrete waste, the absorption of water increases. With the addition of poly-propylene fibre into the concrete, still decrease in the value of the slump thereby decreasing the workability of the concrete.
3. Crushing strength of crushed concrete waste goes on decreased with increase in the percentage from 0 to 40% (for every 10% intervals) but the same concrete

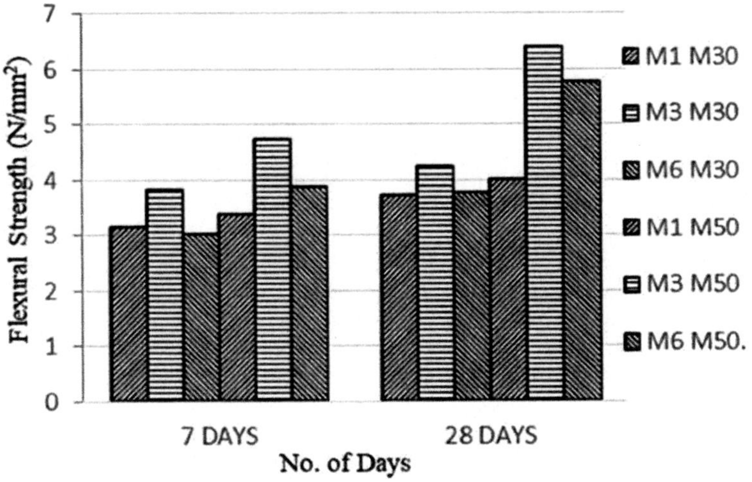

Fig. 8 Graph showing flexural strength of FS and CCW for 7- and 28-days curing

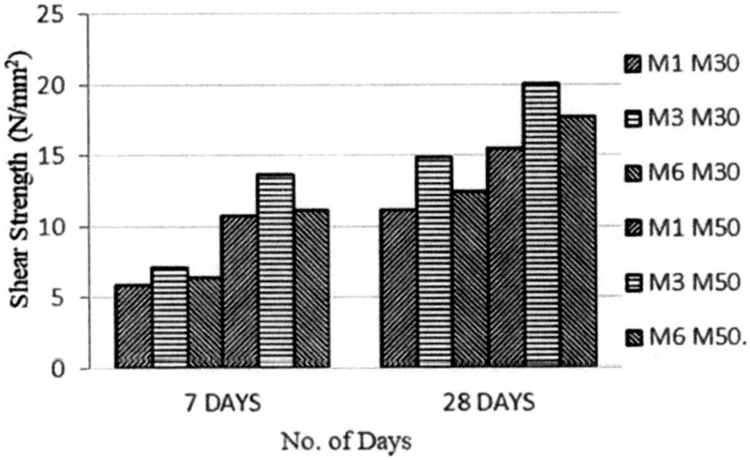

Fig. 9 Graph showing shear strength of FS and CCW for 7- and 28-days curing

with poly-propylene fibre gives 5–10% increase in the crushing strength thereby getting the optimum value of concrete with crushed concrete waste as 10% with the addition of dosage of fibre as 0.3% and for 28 days of curing.

It is observed that the crushing strength of foundry sand goes on increased up to 15% replacement with and without adding the fibre into the concrete. But there is a 5–10% increase in the value of crushing strength of the concrete with the use of fibres. This is due to the internal bondage between the fibre and the concrete, initially concrete fails by compression thereby transformation of load

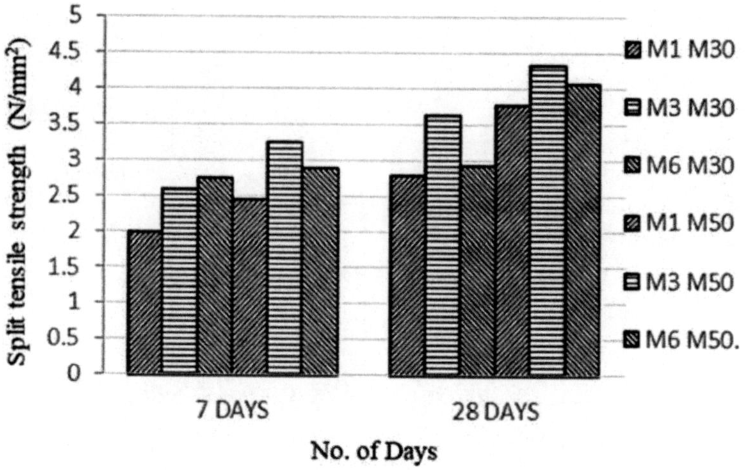

Fig. 10 Graph showing split tensile strength results of FS and CCW for 7- and 28-days curing

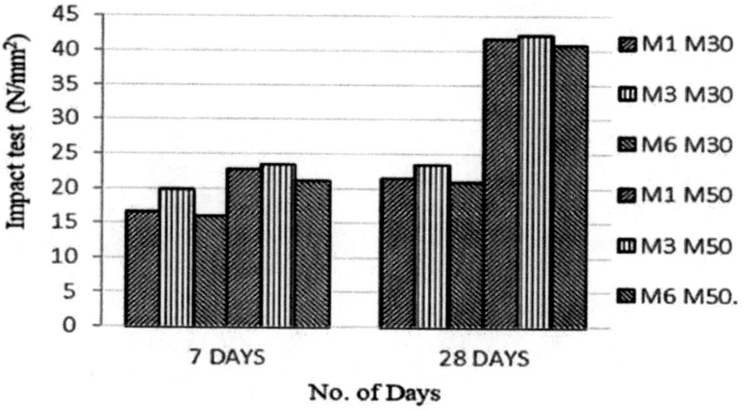

Fig. 11 Graph showing impact test results of FS and CCW for 7- and 28-days curing

from concrete to fibre and since, the proper presence of gradation, aspect ratio of the fibre, tensile taking capacity of fibre, there is an increase in the crushing of concrete.

4. Flexural strength results obtained for HPC with the addition of fibres improved but with the addition of replacement aggregates to the concrete, there is a decrease in the Flexural strength. Observation says that the bondage between the fibre and concrete may not be conditioned for bending test.
5. There is a minor variation in the split tensile strength of the concrete by observing the result obtained from split tensile strength. But the function of fibres must be appreciable.

6. Impact strength value of HPC will be doubled by adding the poly-propylene fibre which is observed from the test results. In this present investigation, 229% increase in the impact value of the fibre reinforced HPC compared to conventional concrete.
7. There is a remarkable increasing in the shear strength of the concrete with the addition of the foundry sand to 10%. Due to the presence of fibres in the concrete, there is an increase in the shear value of the concrete for 7 and 28 days of curing.

7 Conclusions

- Workability property of the FRHPC with replacement decreases the slump value and same condition reflects in the fibred concrete also. But for the case compaction factor maintained less than 1 (Compaction factor < 1). Shear slump obtained for all type of HPC in this present investigation.
- Optimum value of crushing strength of Standard Concrete (SC), High Strength Concrete (HSC) and High Performance Concrete (HPC) compared with and without adding the poly-propylene fibres (PPF) for 7, 14 and 28 days of curing. Results obtained for concrete increase in its crushing strength for an optimum replacement of foundry sand from 15% up to 30% and Crushed Concrete Waste 10–20% for above mentioned three types of concrete.
- Optimum replacements obtained from the crushing strength after 280 days of curing were selected and for those concrete mixes, the Flexural strength test of the concrete was conducted.
- There is a substantial increase in the Flexural strength of the HPC by providing the poly-propylene fibres of dosage about 0.3% by volume of cementitious material for both conventional and replaced concrete.
- Impact strength value of HPC will be doubled by adding the poly-propylene fibre which is observed from the test results. In this present investigation, 229% increase in the impact value of the FRHPC compared to conventional concrete.
- Split tensile strength of the fibre reinforced HPC for 10% CCW and 15% FS was noted worth readings such as 6.08 Mpa and 6.50 Mpa, respectively.
- Shear strength of fibre reinforced HPC increases its value for 15% replacement of foundry sand and decrease with 10% replacement of CCW compared with conventional M80 grade concrete. The value of strength of 15% FS and 10% CCW is 54.45 Mpa and 45.56 Mpa, respectively.

Acknowledgements We are grateful to our beloved Principal Dr. Mallikarjun S. Holi, who always been thriving for enhancement and excellence for collage, department facilities and student welfare. We deeply indebted to Head of the Dept. Dr. Nagarajappa D. P. and PG Coordinator Dr. H. R. Prabakara for his moral support and encouragement.

References

1. IRC 44-2017 Guidelines for cement concrete mix design for pavements
2. IRC: 58-1988 Guidelines for the design of rigid pavements for highways
3. IRC: SP: 62-2014 deals with the design and construction of cement concrete pavements (CCP) for low volume roads
4. IRC: SP46-2013 Guidelines for design and construction of fiber reinforced concrete pavements-First Revision
5. Aaleti S Quantifying bonding characteristics between UHPC and normal-strength concrete for bridge deck. Application Iowa State University, sri@iastate.edu, Follow this and additional works at: https://lib.dr.iastate.edu/ccee_pubs
6. Wadekar, Maqtedar MA, Moid, et al (2016) Study of high strength fiber reinforced concrete for M80 grade by using different types of steel fibre article August 2016. IJSRD-Int J Sci Res Develop 4(06) | ISSN (online):2321-0613
7. Jayachandra, Shashi Kumar A (2015) Strength behaviour of foundry sand on modified high strength concrete. Int J Res Eng Technol (IJRET) 04(05). eISSN: 2319-1163 | pISSN: 2321-7308, May-2015, Available http://www.ijret.org
8. Puneeth HC, Mahendra SP (2018) Replacement of recycled construction and demolition waste coarse aggregates in pavement quality concrete. Int J Appl Eng Res 13(7):139–145. ISSN 0973-4562. Research India Publications. http://www.ripublication.com
9. Mohod MV (2015) Performance of polypropylene fibre reinforced concrete. IOSR J Mech Civil Eng (IOSR-JMCE) 12(1):28–36, e-ISSN:22781684, p-ISSN:2320-334X, Ver.I (Jan-Feb.2015). https://www.iosrjournals.org
10. Chandrasekar R, Chilabarasan T (2017) Development of high strength concrete using waste foundry sand. J Chem Pharm Sci JCPS 10(1), January-March 2017. https://www.jchps.com
11. IS:383:1970 (Reaffirmed 1997) Specification for coarse and fine aggregates from natural sources for concrete. Bureau of Indian Standards (IS), New Delhi
12. IRC 44-2017 Concrete mix proportioning Indian Road Congress (IRC), New Delhi

Response Analysis of Berthing Structure with Soil–Structure Interaction

Sushmitha Shettigar, B. R. Jayalekshmi, and Katta Venkataramana

Abstract The berthing structures including piles and diaphragm walls are supported on soft marine soils. The soft soils under severe loading are likely to undergo vertical and lateral movement. The anchored diaphragm wall is provided to support the open berth structure against backfill. In this paper, finite element analysis of berthing structure has been carried out using a finite element program ANSYS APDL. The soil strata is modelled as 3D continuum. The response analysis of diaphragm wall for different pretension forces in anchor rod has been carried out. The variation in displacement, shear force and bending moment along the depth of wall is plotted. The result is compared with the case without considering soil–structure interaction. The optimum value of pretension force is obtained as 1050 kN which effectively reduced the deflection of diaphragm wall. The percentage increase in maximum lateral displacement, shear force and bending moment of wall without considering soil–structure interaction effect was found to be 25.265%, 52.523% and 892.944%, respectively.

Keywords Berthing structure · Diaphragm wall · Anchor rod · Soil–structure interaction · Displacement · Shear force · Bending moment

1 Introduction

The ports and harbours consists of a major platform called as berth which provides facilities for berthing and mooring of ships, embarking and disembarking of passengers and loading, unloading and storage of cargo. The berth mainly includes structures like deck slab, beams, piles and diaphragm wall. The berthing structures are liable to both vertical loads and lateral loads. The vertical loads are exerted by self-weight and live load acting on the berth. Lateral loads are imparted by approaching vessels striking the berth, due to pulling action of mooring lines, motion of wind and waves,

S. Shettigar (✉) · B. R. Jayalekshmi · K. Venkataramana
Department of Civil Engineering, National Institute of Technology Karnataka, Surathkal, India

B. R. Jayalekshmi
e-mail: brjaya@nitk.edu.in

L. Nandagiri et al. (eds.), *Sustainability Trends and Challenges in Civil Engineering*, Lecture Notes in Civil Engineering 162, https://doi.org/10.1007/978-981-16-2826-9_59

differential water pressure and earth pressure [1]. The studies show that mooring load is more critical than other lateral loads [2]. Diaphragm wall and tieback elements are provided to provide a lateral support to open berth structure [3]. Diaphragm wall is a reinforced concrete retaining wall which holds the lateral loads exerted by the backfill soil. It also experiences lateral forces from soil during dredging process [4, 5]. Tieback elements such as anchor rods are installed as reinforcement into the soil with its one end secured to the diaphragm wall and another end secured to a stable concrete deadman or into soil with sufficient resistance. Tie-rod anchors can increase the strength of the structure, resist the lateral loads and largely reduce the deflection [6]. The tieback-deadman structure resists the force that cause the wall to lean. It is adopted in berthing structures as it reduces the lateral displacement and bending moment, thus resulting in an economical construction of structure with the reduction in the required quantity of reinforcement [3]. Open face type berth structure is shown in Fig. 1. The anchored diaphragm wall is as shown in Fig. 2.

Soil–structure interaction (SSI) is a mutual relation between structure and underlying soil which influences the motion and behaviour of each other. SSI is important in the areas of structural dynamics, soil dynamics and earthquake engineering. The offshore structures like berthing structures are built on soft marine soils which undergo vertical and lateral displacement under severe loading. The process of dredging affects the lateral movement of soil bed. The amount of displacement of soil depends on factors such as properties of soil, properties of structure and sequence of dredging which in turn affects the diaphragm wall deflection [5]. The lateral loads can be resisted by the soil–pile interaction effect [7]. The adhesion of pile with the surrounding soil and the type of interface between the structure and soil will be responsible for the ultimate lateral soil resistance [8]. The piles in group experience significant interaction effects which are different for drained soil and undrained soil condition [9].

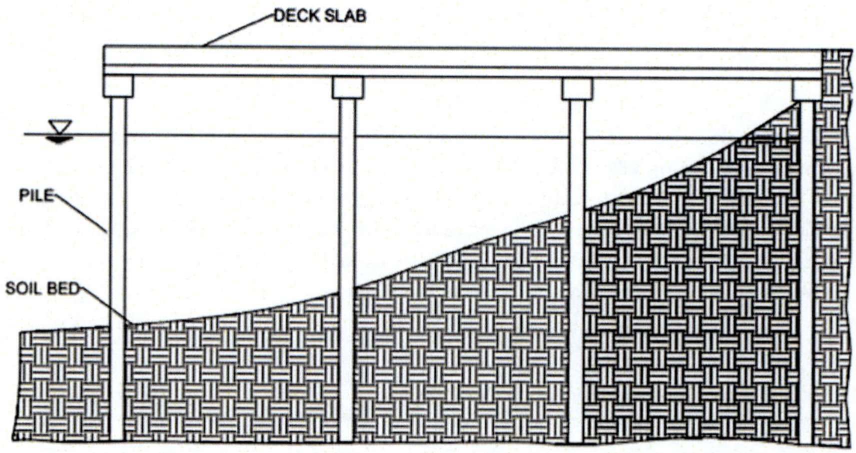

Fig. 1 Open face type berth structure

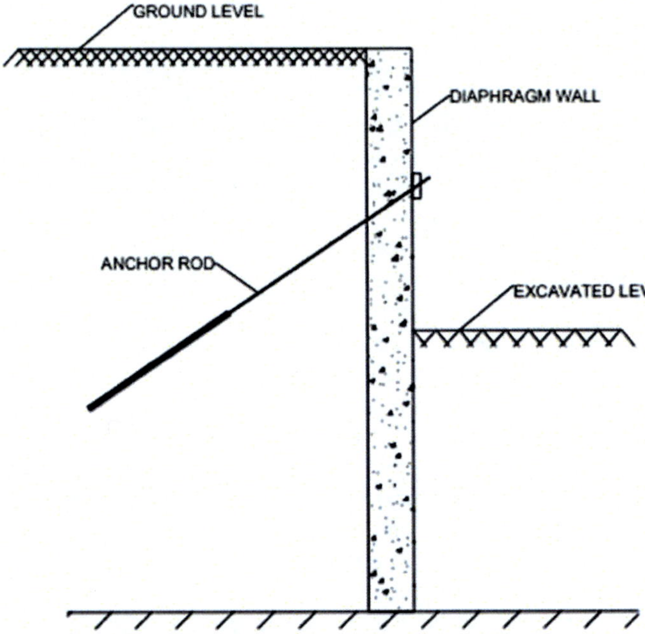

Fig. 2 Cross section of anchored diaphragm wall

The investigations of SSI can be carried out experimentally or by analysis research [10]. The tremendous progress and advancement in the computer technology has given rise to numerical simulation method. Finite element method (FEM) is a powerful numerical simulation method which is mostly applicable for complex problems. It discretizes the large system to a group of small elements of certain size, achieved through meshing. It can deal with complex geometries, higher loads and nonlinear behaviour of objects. ANSYS, PLAXIS, ABAQUS, etc., are some of the FEM software being used by structural engineers and geotechnical engineers. FEM provides certain approximations to predict true solutions. Researchers have shown that adopting Mohr–Coulomb and Drucker–Prager models in numerical modelling are satisfactory [11]. FEM has become an important tool in geotechnical engineering in studying the behaviour of large complex structures [12]. ANSYS APDL is one such finite element program. APDL is an abbreviation for ANSYS parametric design language, a powerful scripting language that allows one to parameterize the model and automate common tasks. It is the foundation for all sophisticated features, offering many conveniences such as parameterization, branching and looping, and complex math operations. ANSYS has been used by structural engineers for design of various complex structures such as skyscrapers, bridges, and dams. The designers and engineers are able to analyse the behaviour, strength and safety of structures by experimenting the various designs in a virtual environment.

2 Literature Review

The finite element analysis of diaphragm wall installation showed that the three-dimensional effect resulted in less lateral soil movement when compared with the plain strain case of diaphragm wall installation. Also, the panel length of wall affects the lateral stress reduction near the diaphragm wall [13]. The studies have shown that the diaphragm wall undergoes cantilever-like deflection under lateral loading with maximum horizontal displacement at top which reduces along the depth [14]. In a study of ground movement control by anchored diaphragm wall, it was found that the lateral earth system supports the wall and hence limits the deflection, and the installation of tieback elements through over stressed soil layers leads to ground surface settlement [15]. Some investigators have successfully modelled and analysed berthing structure using finite element program PLAXIS and concluded that the results obtained from numerical method agrees well with full-scale field test results [4]. The dredging process being carried out in ports and harbours disturbs the marine soil bed. Researchers have found that the dredging increases the lateral deflection of the berthing structure [4, 16]. The bed slope also affects the berthing structure. The increase in bed slope caused by dredging process increases the lateral deflection of pile and diaphragm wall [4]. Kavitha et al. have studied the soil–structure interaction effect on the behaviour of berthing structure supported on sloping ground. It was found that the increasing modulus of soil increases the lateral load intensity on the berthing structure. Increasing the pile diameter increases bending moment of berthing structure. Also, the response of structure depends on whether the change in the slope of bed is caused by deposition or by erosion of soil [17]. Some researchers have carried out the study on the influence of tie rods on the behaviour of diaphragm wall. Premalatha et al. observed that the length of tie rod influences the diaphragm wall deflection. Deflection of the wall decreased as the length of tie rod was increased from 6 to 18 m. Beyond 18 m, there was no significant reduction in the deflection of wall [6]. Tie rod increases the stability of diaphragm wall [14, 16]. Yajnheswaran et al. studied the influence of tie rod by analysing berthing structure using finite element software PLAXIS 3D. The analysis was carried out for absence of tie rod and for different positions of tie rod. It was revealed that the location of tie rod affects the deflection, shear force and bending moment of the diaphragm wall [3]. Extensive literature review showed that the lateral response of berthing structures depends on structural material, dimensions of structure, soil properties and slope of soil bed. Increasing the stiffness of anchor rod largely reduces the top lateral displacement of the diaphragm wall [18].

The studies have shown that nonlinear stress–strain characteristics of soil has important role in the computation of soil–structure interaction [19]. The examination of effect of soil layers on lateral loading behaviour of piles has been carried out. The presence of different soil layers affects the lateral resistance offered by the each layer. The lower layer of soil has significant effect on the lateral resistance of the upper layer and vice versa [20]. The active soil wedge also has significant effect on the lateral capacity of piles [21]. The presence of adjacent surcharge increases the lateral soil movement and increases the stress on the structure in the vicinity [22]. Li et al. carried out three-dimensional finite element analysis using ANSYS software on soil–structure interaction system. It was observed that soil–structure interaction effect has considerable effect on displacement peak value at bottom part of the structure. The effect of SSI on displacement of the structure is larger with decrease of shear modulus of soil and larger with increase of rigidity of the structure [10]. The modelling of pile and surrounding soil can be done using either finite element method or boundary element method since both the methods can accurately solve the problem under both static and dynamic loadings [23].

In this paper, effect of SSI on lateral displacement of diaphragm wall is studied. Soil is modelled as three-dimensional continuum, and its effect on the behaviour of diaphragm wall is analysed using a general purpose finite element program ANSYS APDL. The diaphragm wall is pre-stressed with different pretension force and arrived at the optimal value of pretension force which results in minimum lateral displacement. The variation of displacement, shear force and bending moment of diaphragm wall are plotted against depth. The result is compared with the structure without considering SSI effect by removing the soil strata and providing rigid base.

3 Details of Berthing Structure

3.1 Details of the Berth

For the purpose of analysis, the deep draft multipurpose Berth 18 of New Mangalore Port Trust is considered. It is located opposite to Kudremukh iron ore berth. The berth is of length 325 m and width 25 m. The entire length of 325 m is divided into five panels of length 65 m each. In this study, one panel of dimension 65 m × 25 m is considered. The plan of a typical single panel is shown in Fig. 3. The entire berthing structure is supported by 55 rows of piles in transverse direction having 4 piles in each row. Out of these, 22 rows from seaward end of berth are of 1.3 m diameter and the remaining 33 rows are of 1.2 m diameter. The piles rest on weathered rock at a depth of 30 m. Piles are provided at a spacing of 6 m centre to centre along transverse direction. Each row of piles is named as A, B, C and D. The dredge depth is 17 m for group D piles, 15 m for group C piles, 13 m for group B piles and 10.5 m for group A piles. The diaphragm wall is located at a distance of 25 m from the pile near the sea side. The thickness of diaphragm wall is 1.1 m, and its depth is 22 m. The steel tie rod anchors of diameter 0.05 m are provided at a spacing of 2.5 m centre to centre at the top of diaphragm wall and are anchored into soil at an inclined angle of 45°. The deck slab of dimension 65 m × 25 m and thickness of 0.45 m is provided over piles. The slab is supported on cross beams of dimension 1.1 m × 1.1 m in the lateral direction and secondary beams of dimension 0.8 m × 1.1 m in the longitudinal direction. Crane beams of dimension 1.2 m × 1.1 m are provided in the longitudinal

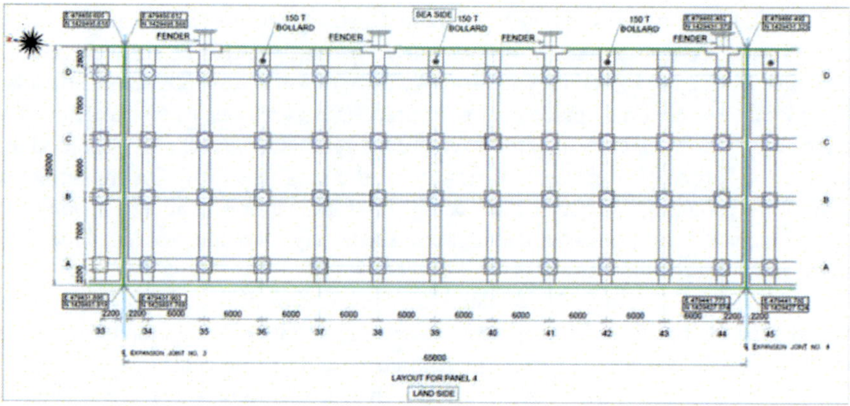

Fig. 3 Plan of a typical panel of the berthing structure

direction. Along the slab edge on the sea side, fender beams of dimension 1 m × 4 m are provided.

3.2 *Details of Sub Soil*

From borehole investigation, soil types and soil depth details are obtained by New Mangalore Port Trust. The cohesion, angle of internal friction and unit weight of soil are obtained for different soils along the depth. The soil profile is shown in Fig. 4. The soil properties of different layers are as given in Table 1.

4 Description of Approach

In this study, a single deck panel of dimension 65 m × 25 m is modelled using the finite element software ANSYS APDL. The deck slab is supported over piles of 1.2 m diameter. The diaphragm wall of thickness 1.1 m is provided for a length of 25 m. The various loads acting on berthing structure are calculated as per IS: 4651(Part III)-1974 [24]. Dead load (DL) is applied as gravitational acceleration in the software. A uniform live load (LL) of 55 kN/m^2 is applied on the deck slab. The berthing force (BL) imparted to fender system by berthing ships is calculated as 1682 kN and applied as a point load on the face of fender beam. The mooring force (ML) due to pulling action of mooring lines is applied at the bollard position. The mooring force is calculated as 1500 kN. The pressure developed due to difference in water levels (DWP) at the fill side and the water side is also taken into account. It is applied as a pressure on the diaphragm wall surface. Active earth pressure (AEP)

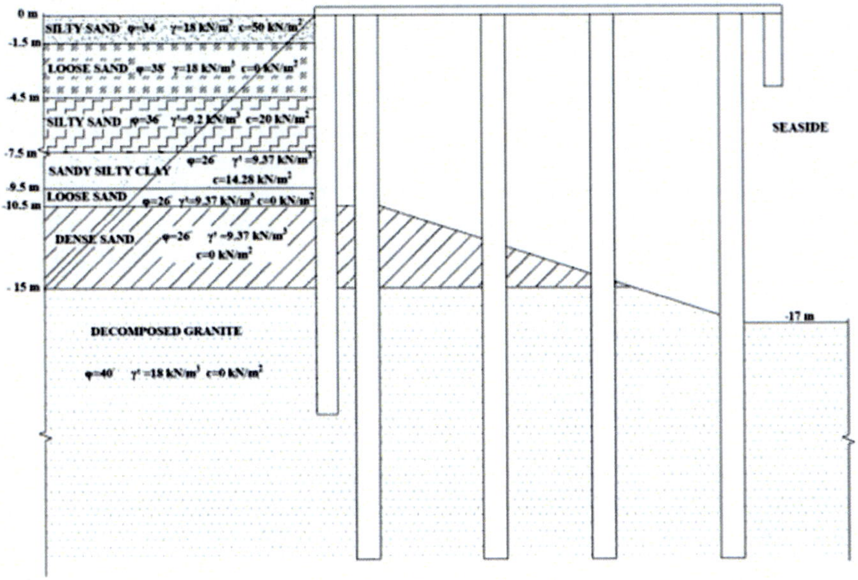

Fig. 4 Soil profile

Table 1 Properties of different soil layers

Soil layer level		Properties			
Top (m)	Bottom (m)	Soil type	Angle of internal friction (°)	Unit weight (kN/m^3)	Cohesion (kN/m^2)
4.662	3.162	Silty sand	34	18	50
3.162	0.162	Loose sand	38	18	0
0.162	− 2.838	Silty sand	26	9.2	20
− 2.838	− 4.838	Sandy silty clay	26	9.37	14.28
− 4.838	− 5.838	Loose sand	26	9.37	0
− 5.838	− 10.338	Dense sand	26	9.37	0
− 10.338	− 27	Decom-posed granite	40	18	0

exerted by soils on one side of the diaphragm wall is calculated as per IS: 9527(Part III)-1983 [25]. The analysis has been carried out for critical load combination (1.5DL + 1.5LL + 1.5ML + DWP + AEP).

The piles and beams are modelled as line element of type Beam 188. The diaphragm wall and slab are modelled as shell element of type Shell 281. Tie rod is designed as link element of type Link 180.

4.1 Assumptions Made in the Analysis

All the piles are fixed at the bottom. In the case of presence of soil strata, the diaphragm wall is supported by soil, whereas in the absence of soil strata, the wall is assumed as fixed at the bottom. The bonded contact has been considered between soil and piles interface. Two lateral faces of soil at a distance of 10 m on one side and 6 m on another side are restrained against lateral movements in x and z directions. Mohr–Coulomb constitutive model has been adopted for both reinforced concrete structure and soil.

4.2 Modelling of Berthing Structure with SSI

The cross section of structure modelled in ANSYS APDL is shown in Fig. 5. Soil is modelled as a 3D structure in order to introduce the effect of SSI. Different soil layers are modelled, and properties are assigned in terms of modulus of elasticity, poisons ratio and density [26]. The structural members and soil are discretised to an optimum element size where the solution has approximately converged.

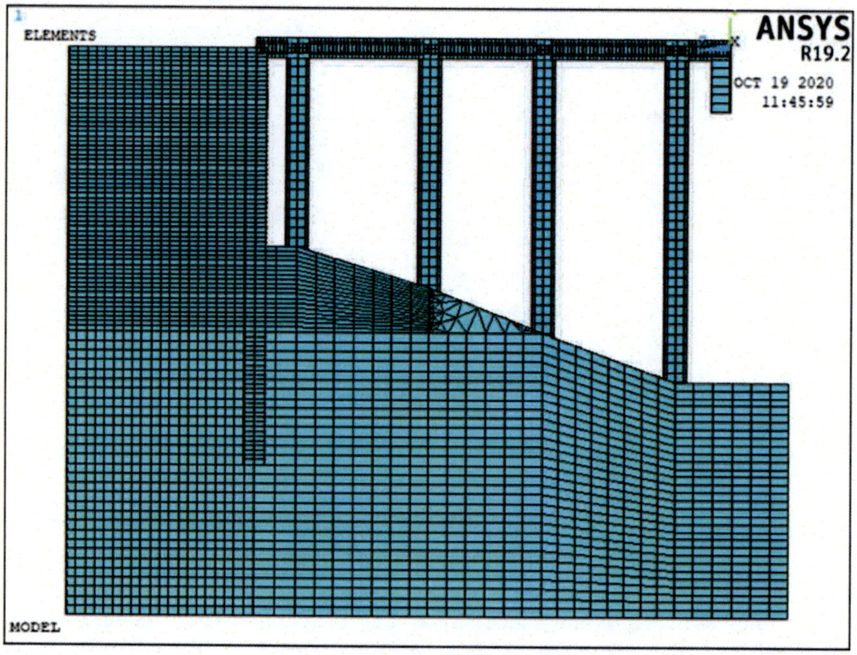

Fig. 5 Cross section of modelled berthing structure with 3D soil strata

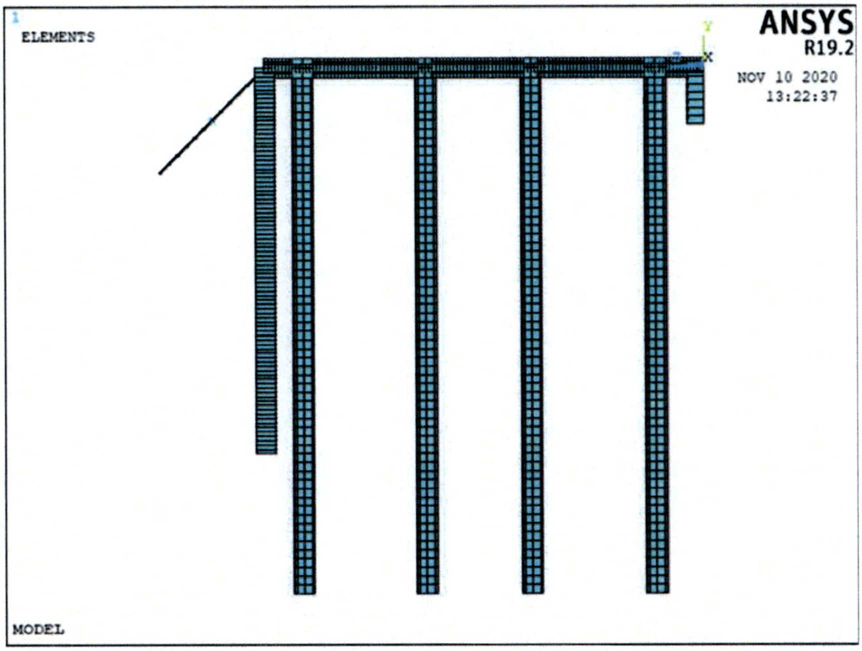

Fig. 6 Cross section of modelled berthing structure without 3D soil strata

4.3 *Modelling of Berthing Structure Without SSI*

The cross section of berthing structure without soil strata is shown in Fig. 6. Piles and diaphragm wall are fixed at the bottom. Tie rods are pinned at the bottom. All vertical and lateral forces are applied as in the case with soil strata.

The analysis has been carried out for the structure with soil strata and without soil strata for the following cases:

Case 1: Response of diaphragm wall for different pretension forces in anchor rod with SSI.

Case 2: Response of diaphragm wall for different pretension forces in anchor rod without SSI.

5 Results

5.1 *Case 1: Analysis of Berthing Structure with SSI*

The anchor rod has been pre-stressed to different forces. The force was gradually increased from 200 kN. The minimum lateral displacement was obtained when anchor rod was pre-stressed to 1050 kN with zero displacement at top and bottom. The maximum diaphragm wall displacement for 200, 600 and 1050 kN pretensioning was found to be 0.01451 m, 0.007671 m and 0.001078 m, respectively. Increasing the pretension force beyond 1050 kN resulted in increasing lateral displacement of wall towards the soil. Figure 7 shows the variation in lateral displacement of wall for different pretension forces.

The variation in shear force of diaphragm wall with SSI for different pretension forces in shown in Fig. 8. The considerable variation in shear force is due to change of soil properties at various depths. Since dredging has been carried out up to 10.5 m depth, active earth pressure alone acts on the wall till 10.5 m depth. Beyond 10.5 m, forces due to passive earth pressure also start acting on the diaphragm wall. The maximum shear force at 1050 kN pretension force is found to be 37.848 kN at 15 m depth.

Figure 9 shows the variation in bending moment along the depth of diaphragm wall for various pretension forces. The bending moment is found to be maximum for

Fig. 7 Lateral displacement of diaphragm wall for different pretension forces with SSI

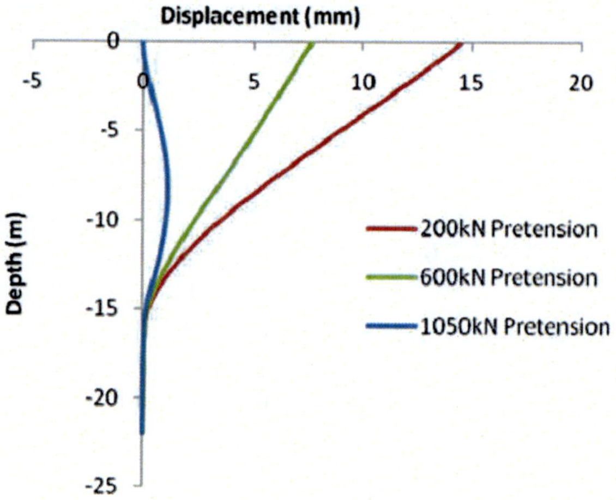

Fig. 8 Variation in shear force of diaphragm wall for different pretension forces with SSI

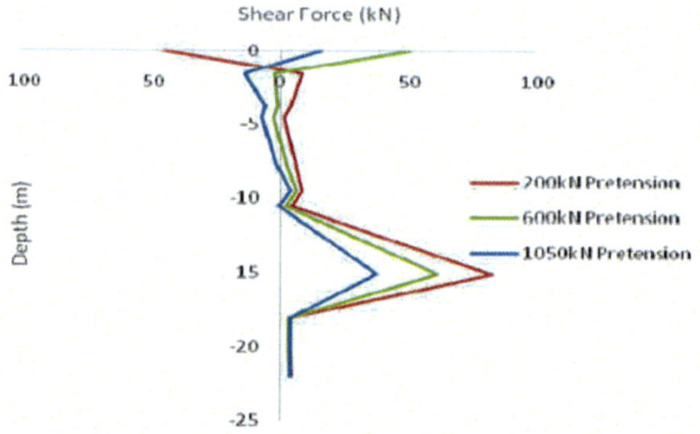

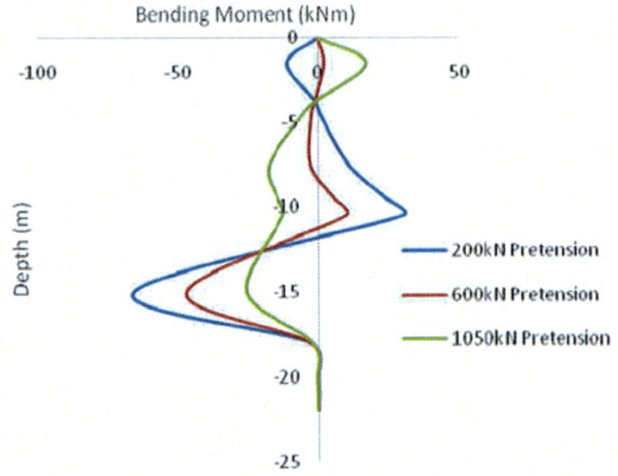

Fig. 9 Variation in bending moment of diaphragm wall for different pretension forces with SSI

200 kN pretension forces. A considerable decrease in bending moment is obtained for 1050 kN pretensioning.

5.2 *Case 2: Analysis of Berthing Structure Without SSI*

The lateral displacement of diaphragm wall for different forces of pretensioning anchor rods without considering soil strata is shown in Fig. 10. The maximum

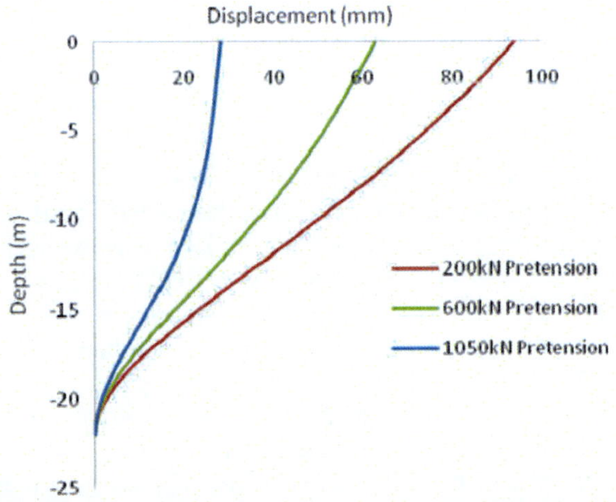

Fig. 10 Lateral displacement of diaphragm wall for different pretension forces without SSI

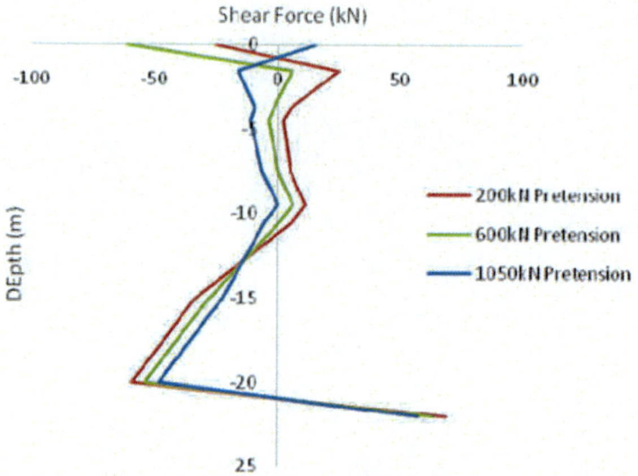

Fig. 11 Variation in shear force of diaphragm wall for different pretension forces without SSI

displacement of wall for 200, 600 and 1050 kN pretensioning is obtained as 0.09378 m, 0.06297 m and 0.02831 m, respectively. A considerable increase in the lateral displacement of wall has been observed when compared with the results of structure with soil strata.

The variation in shear force along the depth of diaphragm wall without modelling soil strata is shown in Fig. 11. The maximum shear force for 1050 kN pretension force is found to be 57.227 kN at the bottom of the diaphragm wall.

The bending moment variation in diaphragm wall without SSI effect is as shown in Fig. 12. For 1050 kN pretension force, the maximum bending moment is found to be 256.1 kN m at the bottom of the diaphragm wall. The bending moment value for all pretension forces without soil strata was found to be more than that of with soil strata.

6 Conclusions

From the response analysis of diaphragm wall with anchored rod, it can be concluded that anchor rod plays a major role in controlling the lateral displacement of diaphragm wall. In the considered model, pretension force of 1050 kN can be taken as optimal value which could successfully hold the diaphragm wall at the top against active earth pressure causing zero displacement at top and the minimum displacement at the middle portion. Tie rod increases the stability of berthing structure. The comparison of results of analysis of behaviour of diaphragm wall with and without soil strata resulted in considerable variation in the result. The percentage increase in maximum displacement, maximum shear force and maximum bending moment for the case without considering SSI are 25.265%, 52.523% and 892.944%, respectively. In the presence of soil strata, the resistance imposed by the surrounding soil for the lateral

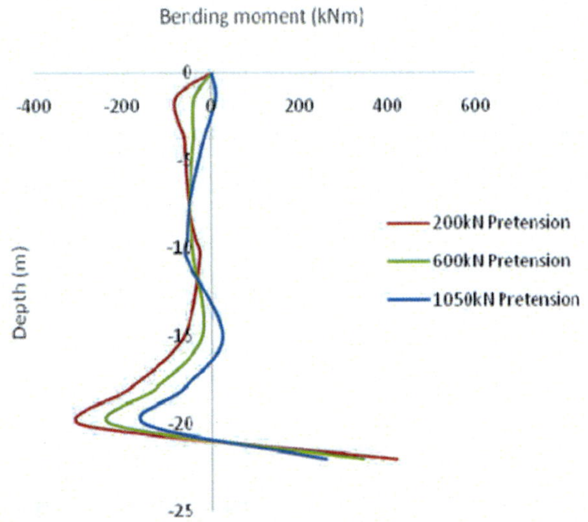

Fig. 12 Variation in bending moment of diaphragm wall for different pretension forces without SSI

movement of the structure resulted in lower lateral displacement and forces in the wall. The interaction occurring between soil and wall surfaces in contact plays a role in reducing the lateral deflection of the structure. Hence, it can be concluded that SSI effect should be considered in the analysis of berthing structure to arrive at the accurate and real field behaviour. It results in reduction in the quantity of reinforcement required for the construction of berthing structure and hence results in an economic and cost-effective construction.

Acknowledgements The authors are grateful to the Dean of National Institute of Technology Karnataka, Surathkal and Head of the Department of Civil Engineering for providing the opportunity to conduct the research work and for the constant guidance and support.

References

1. Chopra HB, Patal PG (2015) Application of forces acting on jetty structure. Int J Sci Technol Eng 1(11)
2. Premalatha PV, Muthukkumaran K, Jayabalan P (2011) Effect of dredging and tie rod anchor on the behaviour of berthing structure. Int J Eng Sci Technol 3(6):5099–5115
3. Yajneswaran RHS, Rao S (2015) Analysis of the effect of anchor rod on the behaviour of diaphragm wall using Plaxis 3D. Aquatic Procedia 4:240–247
4. Muthukkumaran K, Sundaravadivelu R (2007) Numerical modeling of dredging effect on berthing structure. Acta Geotech 2:249–259
5. Muthukkumaran K, Sundaravadivelu R, Gandhi SR (2007) Effect of dredging and axial load on a berthing structure. Int J Geoeng Case Hist 1(2):73–88

6. Premalatha PV (2009) Analyzing the optimum length of tie rod anchors for a berthing structure. Indian Geotech Soc 5:61–67
7. Rollins KM, Sparks A (2002) Lateral resistance of full scale pile cap with gravel backfill. J Geotech Geoenviron Eng 128(9):711–723
8. Chen L, Poulos HG (1993) Analysis of pile-soil interaction under lateral loading using infinite and finite elements. Comput Geotech 15(4):189–220
9. Chen CY, Martin GR (2002) Soil–structure interaction for landslide stabilizing piles. Comput Geotech 29(5):363–386
10. Li P, Lu X, Chen B, Chen Y (2004) Computer simulation on dynamic soil structure interaction system. In: 13th world conference on earthquake engineering, Paper No. 3233
11. Tschuchnigg F (2010) Optimization of a deep foundation with diaphragm wall panels employing 3D FE analysis. Geotechnical Challenges in Megacities, Moscow, pp 471–478
12. Cakir T (2015) Finite element based investigation of backfill effects on seismic behaviour of a cantilever wall. Procedia Earth Planet Sci 15:231–236
13. Gourvenec SM, Powrie W (1999) Three-dimensional finite-element analysis of diaphragm wall installation. Goetechnique 49(6):801–823
14. Lewandowska AS (2005) The displacements of anchored diaphragm walls. Studia Geotechnica et Mechanica 27(1–2):155–168
15. Dimitros CK, Whittle AJ, Scharner B (2004) Control of ground movements for a multi-level-anchored diaphragm wall during excavation. In: Proceedings, fifth international conference on case histories in geotechnical engineering, New York, USE, pp 1–7
16. Subha IP (2012) Behaviour of an open type berthing structure under earthquake condition—a numerical approach. Int J Eng Res Appl 6(12):46–56
17. Kavitha PE, Beena KS, Narayanan KP (2017) Influence of soil structure interaction in the structural behaviour of a berthing structure in sloping ground. J Inst Eng India Ser A 98(4):469–476
18. Ibrahim KMHI, Ibrahim TE (2013) Effect of historical earthquakes on prestressed anchor tie back diaphragm wall and on near-by building. HBRC J 9(1):60–67
19. Jardine RJ, Potts DM, Fourie AB, Burland JB (1986) Studies of the influence of non-linear stress-strain characteristics in soil-structure interaction. Geotechnique 36(3):377–396
20. Yang Z, Jeremic B (2005) Study of soil layering effects on lateral loading behavior of piles. J Geotech Geo-environ Eng 131(6):762–770
21. Kavitha PE, Muthu VM, Sundaravadivelu R (2015) Soil structure interaction analysis of dry dock. In: Aquatic procedia, international conference on water reseources, coastal and ocean engineering, vol 4, pp 287–294
22. Zhao MH, Liu DP, Zhang L, Jiang C (2008) 3D finite element analysis on pile soil interaction of passive pile group. J Central South Univ Technol 15(1):75
23. Mardfekri M, Gardoni P, Roesset JM (2013) Modeling laterally loaded single piles accounting for nonlinear soil-pile interactions. J Eng
24. IS: 4651-Part III (1974) (Reaffirmed 2012) Code of practice for planning and design of ports and harbours-Part III: Loading
25. IS: 9527-Part III (1983) (Reaffirmed 2012) Code of practice for design and construction of ports and harbour structures–Part III: Sheet pile walls
26. Bowles JE (1997) Foundation analysis and design. The McGraw-Hill Companies Inc., Singapore

Seismic Response of Multi-storey Building with Different Plan Configuration Using X-Bracing

Shaik Shaista Farheen and B. Rohini

Abstract In recent times, it is seen that RCC buildings which appear strong in appearance may collapse in the blink of eye during earthquakes. This raises the question of its capability to withstand the strong motion. In order to study the performance of buildings subjected to seismic loads, evaluation is carried out in both linear and nonlinear static methods. Also the bracing systems which have good structural importance when it comes to the RCC building are also considered in the present study. To study the objective, a (G + 5) building structure is considered, with and without X-bracing for Rectangle, L and T shape plan configurations and the analysis is carried out in ETABS. The parameters considered for the comparison are storey displacement, overturning moment, base shear and storey drift.

Keywords Building · Earthquake · Bracing · Base shear · Storey displacement · Storey drift · Overturning moment

S. S. Farheen (✉) · B. Rohini
Department of Civil Engineering, RGMCET, Nandyal, Kurnool, Andhra Pradesh 518501, India

L. Nandagiri et al. (eds.), *Sustainability Trends and Challenges in Civil Engineering*, Lecture Notes in Civil Engineering 162, https://doi.org/10.1007/978-981-16-2826-9_60

1 Introduction

Structures undergo ground motion when subjected to seismic waves caused by the earthquake effect and the consequences are huge destruction of property and loss of life. While it is of major concern for the structural designers to provide structural safety with good serviceability, adequate stability and strength to the building under severe earthquake. Hence, one needs to understand a building under seismic loading requires the keen knowledge of the structural performance under large inelastic deformations.

In a recent development, the nonlinear static analysis called pushover analysis has been followed. Nonlinear static analysis is carried out by applying a lateral load in increasing the level of a structure up to ultimate strength to approximately know the strength of a structure beyond its elastic limit up to its ultimate strength in a post-elastic range. Equivalent Static Analysis is linear analysis where the loads are factored to give design force in a single shot. Therefore, to handle this issue, procedures in analytical mode should be developed which ensures the structure to withstand the occurrence of minor earthquakes and indicate caution when prone to major earthquake events.

Mani Deep et al. [1] studied a G + 9 Building in all seismic zones and analysed it by pushover analysis in SAP 2000 Software. From the results, it was observed that when zone varies base shear, displacement, and time period were increased indicating the severity of seismic activity. It is concluded that damage in the building is less and columns in bottom storeys can be retrofitted.

Mohod et al. [2] carried out work on a 12 storied building considering 9 different irregular plan configurations the effect of irregularity on the structure was analysed in STAAD PRO Software. From the study, the irregular plan configurations such as L, C, H, T and E shapes have shown higher deflection than regular geometries. Hence, it is concluded that simple geometry attracts less force and must be adopted as they perform well during the effect of earthquake.

Haamidh et al. [3] analysed RCC frame structure incorporated with various bracings. The structural response of bracings like X, V and inverted V is analysed for 8, 12 & 16 storey buildings by pushover analysis using ETABS Software. Therefore, it can be concluded that the X-bracing frame performs good with less displacement and the base shear of bracings is satisfactory and for development of plastic hinges which hold high importance compared to other frames.

Balappa and Malagavelli [4], investigated a (G + 10) multi-storey building for 4 models with and without bracings by nonlinear static analysis using SAP-2000 software. A parametric study of different parameters has been carried out. Therefore, it is seen that storey displacement and storey drift decreased in case of models with bracings especially model 2 have shown effective reduction. Time period obtained was less for building with bracings compare to unbraced model. Also base shear was found to be minimum with provision of bracing as they have high lateral stiffness. Hence, RC Frame without bracing got (CP) level. Model 2 performed better compared to other models.

G. Sai Prasanna Kumar Reddy and Dr. V. Ranga Rao [5] carried out analysis of a (G + 10) multi-storey building for four models using SAP-2000 Software. The parameters studied are base shear, displacement, pushover curve and location of hinges. Hence, it was concluded that the time period obtained was less for building with bracings compared to unbraced model. Storey displacement and storey drift was reduced in case of models with bracings especially model 2 has shown effective reduction. While base shear was found to be minimum with presence of bracing as they have high lateral stiffness. Hence, RC Frame without bracing got collapse prevention (CP) level. Model 2 performed well compared to other models.

Teruna [6] studied the response of a six-storey building of three-bay RC Frame with different mass introduced at second storey (M2), fourth storey (M4) and sixth storey (M6) and stiffness irregularity at first floor (Mo) by pushover analysis and time history analysis (THA) using SAP 2000 software. The response is studied and development of plastic hinges is discussed. The pushover analysis generated damage control performance level, while THA exhibited limited safety range performance level for the whole frame and after evaluating frames in plastic hinge status it was indicated that frame M6 varies from Immediate occupancy to the collapse prevention level and in both methods it has shown poor performance level. Hence, it can be concluded that the results obtained from pushover analysis are lesser than the results of time history analysis, while time history analysis provides us with accurate results.

Rofooei et al. [7] analysed five different special moment-resisting steel frames, namely two, five, ten, fifteen and twenty storey buildings with different load patterns and new approach by static and dynamic pushover analysis using Drain-2dx nonlinear analysis programme. Therefore, it was concluded that the dynamic pushover with a trilinear approximation of pushover curve and the newly defined effective modal load pattern (SRM) possessed the least error in resembling the target displacements, specifically in both 15 and 20 stories structural models, while nonlinear static pushover analysis was satisfactory, no precise inclination is observed in using static procedure as per code.

The main aim of the study is to perform analysis of a (G + 5) multi-storey building subjected to seismic load by equivalent static method and pushover method considering Rectangular building, L shape building and T shape building in ZONE V, with and without bracings by providing the bracings at the outer side of the plan configuration using ETABS Software. And also to study and compare the response in the terms of storey displacement, storey overturning moment, base shear and storey drift. Finally, the shape desired for construction in seismic zones is found from the analysis.

2 Methodology and Description of Building

Structure type	SMRF (G + 5) building
Plan area	(30 × 20) m^2
Each floor height	3 m
Bottom floor height	4 m
Type of concrete	M25
Type of steel	Fe415
Beam Dimensions	300 mm × 450 mm
	400 mm × 500 mm
	400 mm × 550 mm
	450 mm × 600 mm
Column Dimensions	500 mm × 500 mm
	600 mm × 600 mm
	700 mm × 700 mm
	800 mm × 800 mm
Slab details	150 mm (Shell-Thin)
	M25 Grade
Live load (on slab)	5 kN/m^2
Diaphragm	Rigid diaphragm
Bracing considered	ISWB (Pinned)
	X-Bracing
Masonry wall thickness	230 mm
Poisson's ratio of concrete	0.2
Seismic Zone	Zone V
Importance Factor	1
Parapet wall dimensions	Height—0.90 m
	Thickness—115 mm
Site Type	Type II
Density of concrete	25 kN/m^3
Response Reduction Factor	5
Damping Ratio	5%
RCC Design Code	IS 456:2000
Earthquake Design Code	IS 1893:2002

DEAD LOAD (On frame):

2.1 Wall Load Calculations

(1) Wall load on other floors:

Unit weight of brick × thickness of masonry wall × height of the floor
$= 20\ \text{kN/m}^3 \times 0.230\ \text{m} \times 3\ \text{m} = 13.8\ \text{kN/m}.$

(2) Wall load (Top floor—on parapet wall):

Unit weight of brick × thickness of masonry wall × height of the floor
$= 20\ \text{kN/m}^3 \times 0.115\ \text{m} \times 0.90\ \text{m} = 2.07\ \text{kN/m}.$

2.2 Building Models in ETABS Software

Figure 1 represents rectangular plan configuration in ETABS. Figures 2 and 3 represent 3D view of building with rectangular plan without and with X-bracings respectively. Figure 4 represents L-shape plan configuration in ETABS. Figures 5 and 6 represent 3D view of building with L-shape plan without and with X-bracings respectively. Figure 7 represents T shape plan configuration in ETABS. Figures 8 and 9 represent 3D view of building with T-shape plan without and with X-bracings respectively.

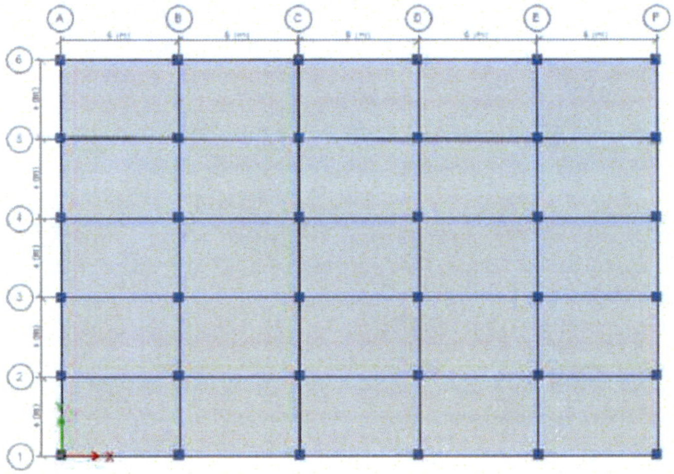

Fig. 1 Rectangle plan shape

Fig. 2 3D view of rectangle plan shape

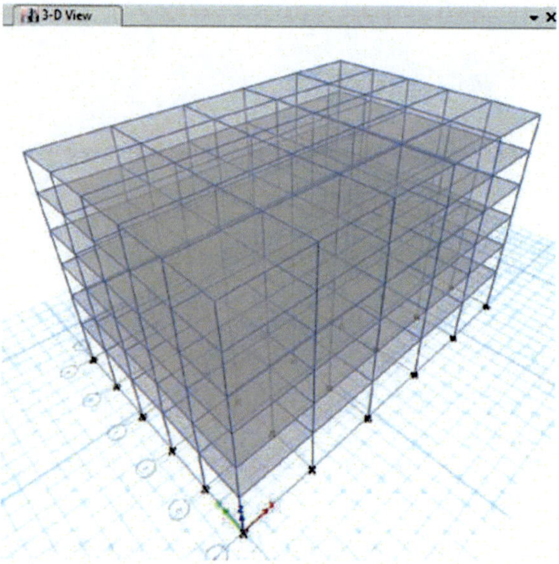

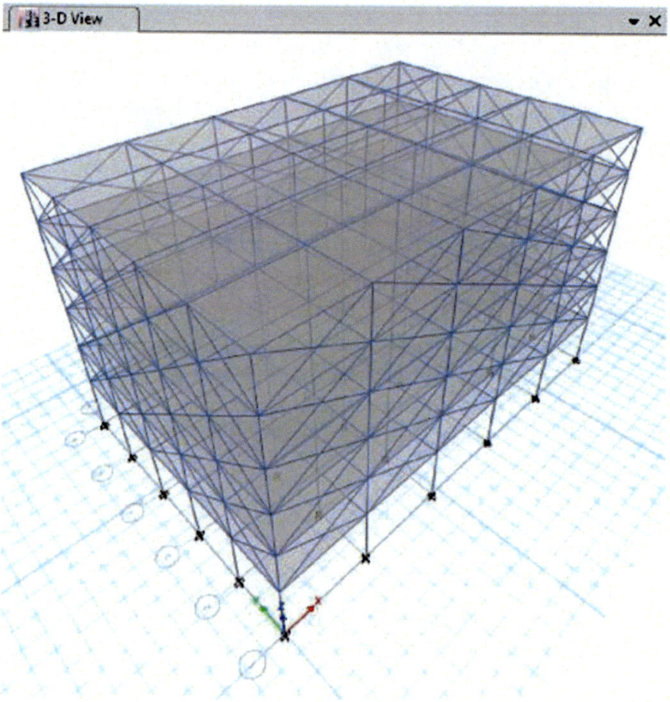

Fig. 3 3D view of rectangle plan shape with X-bracing

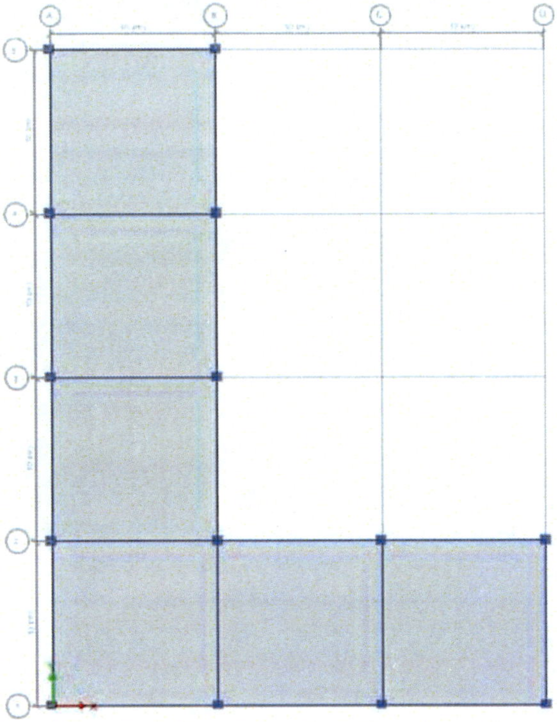

Fig. 4 L-Plan Shape

3 Results and Discussions

3.1 Comparison of Storey Displacement in Equivalent Static Analysis

In Fig. 10, the variation of the storey displacement for different plan configuration along with storey number is plotted. It is studied that in the absence of bracing for the three plan configurations the storey displacement correspondingly increases with the storey number. For rectangle plan configuration, there is a decrease of 75% of storey displacement when X-bracing is used. While for L shape plan configuration with bracing (w/b), the storey displacement reduces by 64.28% similarly there is a decrease of 56.5% displacement for *T* shape plan configuration. According to IS 456-2000 and IS 1893(Part-1) 2002, maximum storey displacement is limited to *H*/500, where *H* indicates total height of the building. Hence, it is concluded that storey displacement did not exceed the limits and is recorded highest for L shape plan configuration with (*w*/*b*) and without (*w*/ob) X-bracing.

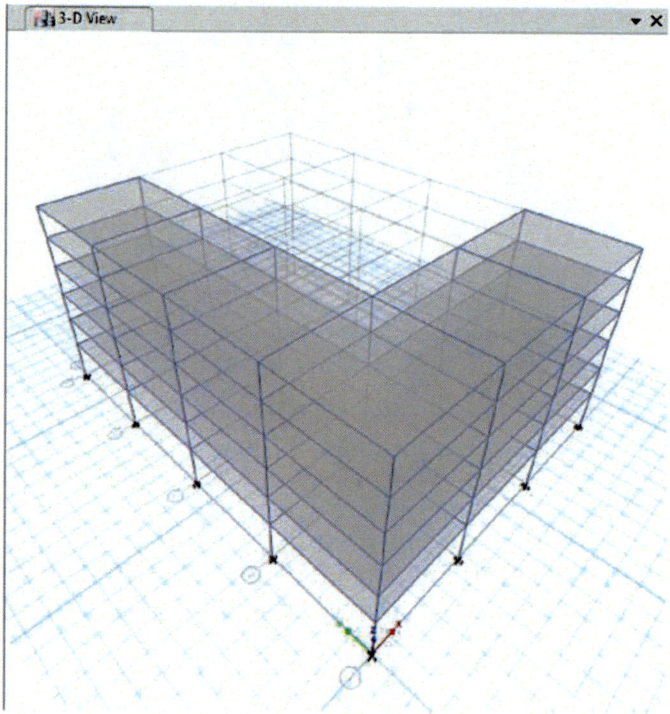

Fig. 5 3D view of L-plan shape

3.2 *Comparison of Storey Overturning Moment in Equivalent Static Analysis*

The storey overturning moment variation for different plan configuration in the presence of bracing is shown in Fig. 11. It can be observed that there is marginal change in the presence of bracing. *T* shape plan configuration with bracing (*w/b*) records highest value and the least value is found in rectangular buildings without bracings (*w*/ob) which indicates that less moment is required to overturn the storey.

3.3 *Comparison of Storey Drift in Equivalent Static Analysis*

From Fig 12, it can be seen that the storey drift varies in parabolic manner with storey number. It can be observed that maximum drift is at storey 2 thereafter it goes on reducing up to storey 6. In the presence of X-bracing storey drift is found to be maximum at storey number 1 for every plan configuration considered. As per the code IS 1893(Part-1)2002 the maximum drift limit is 0.004*h*, where *h* indicates

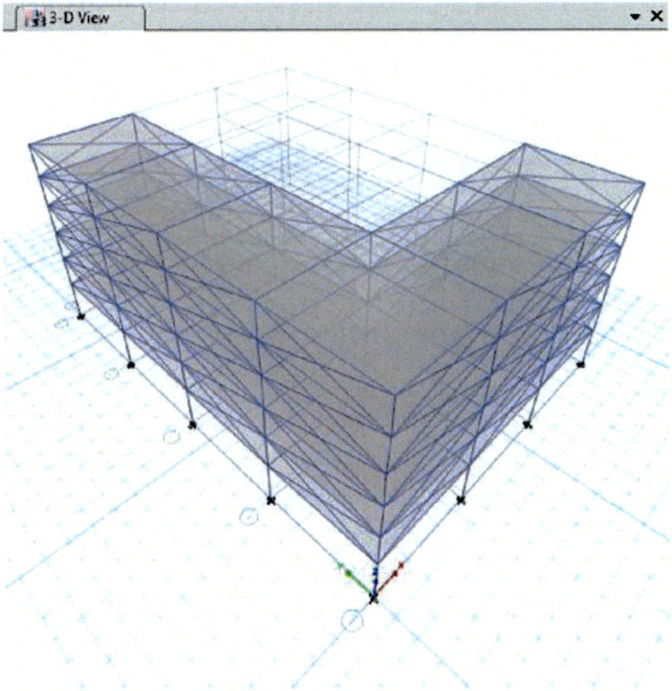

Fig. 6 3D view of L-plan shape with X-bracing

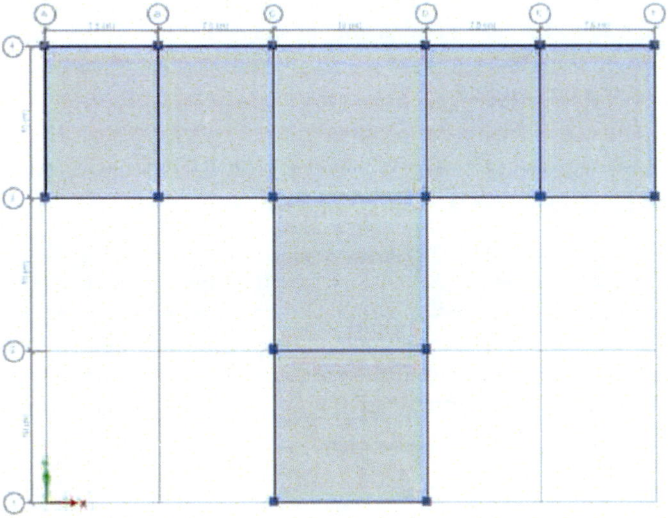

Fig. 7 T-PLAN Shape

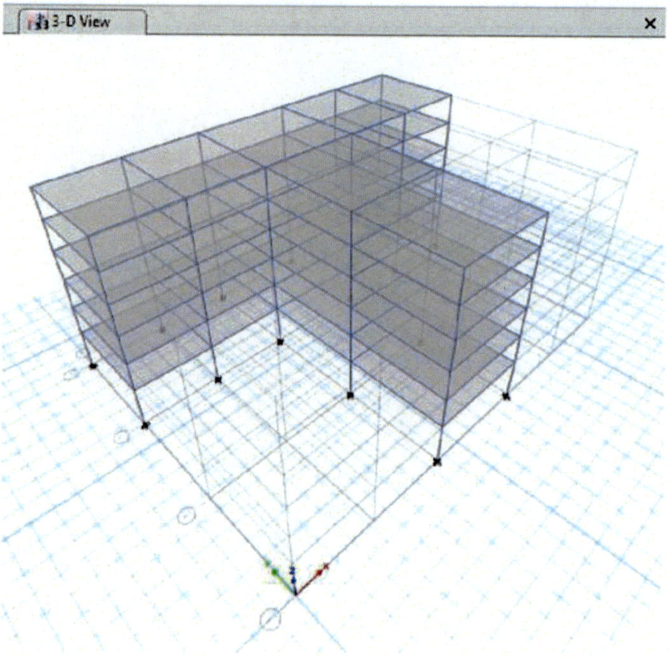

Fig. 8 3D view of T-plan shape

height of the storey. Storey drift here did not exceed the limits and is found to be highest in *L* shape plan configuration with bracings (*w*/b).

There is a percentage decrease of 14.28, 36 and 15% for rectangle, *L* shape and *T* shape plan configurations.

3.4 Comparison of Storey Displacement in Pushover Analysis

From Fig. 13, it is clear that in Push X load case the storey displacement is less in the case of *T* shape plan configuration without bracings (*w*/ob) while the highest value is seen in rectangular shape plan configuration with bracings (*w*/*b*). Hence, it can be concluded that the storey displacement here did not exceed the limits as per the code IS 456-2000 and its values correspondingly increase with increase in the height of structure.

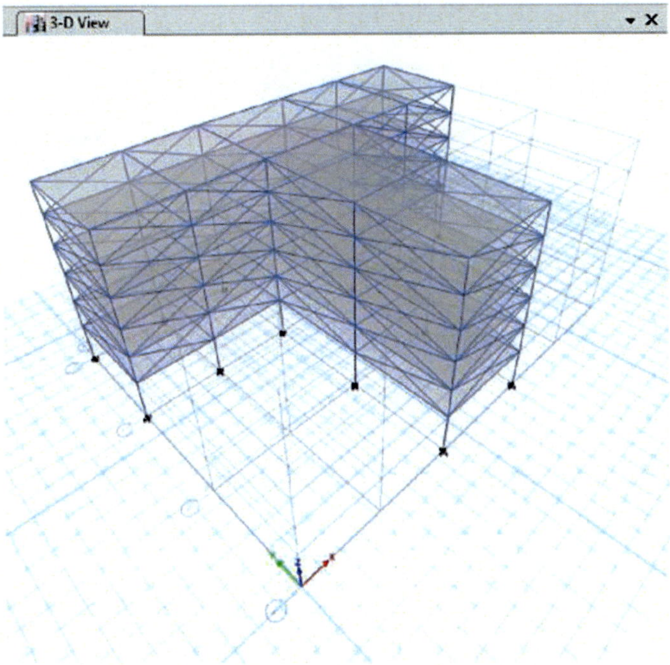

Fig. 9 3D view of T-plan shape with X-bracing

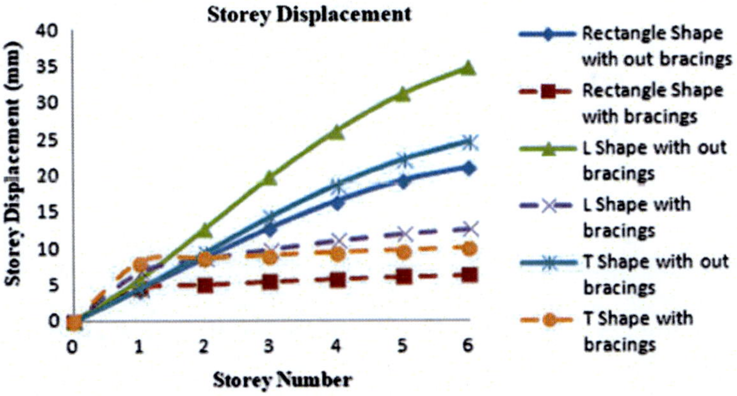

Fig. 10 Storey displacement

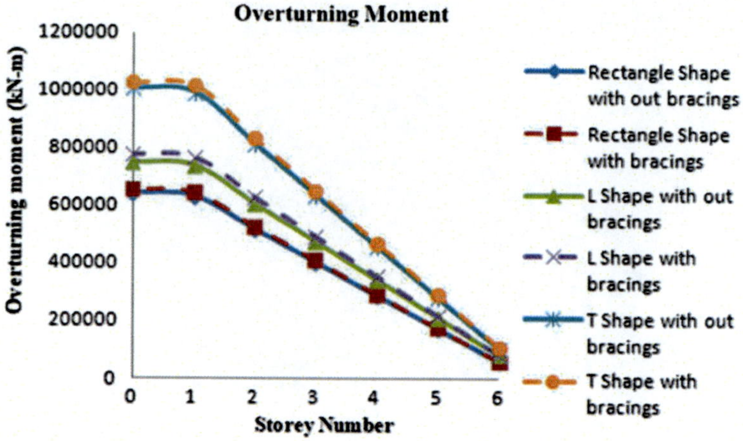

Fig. 11 Overturning moment

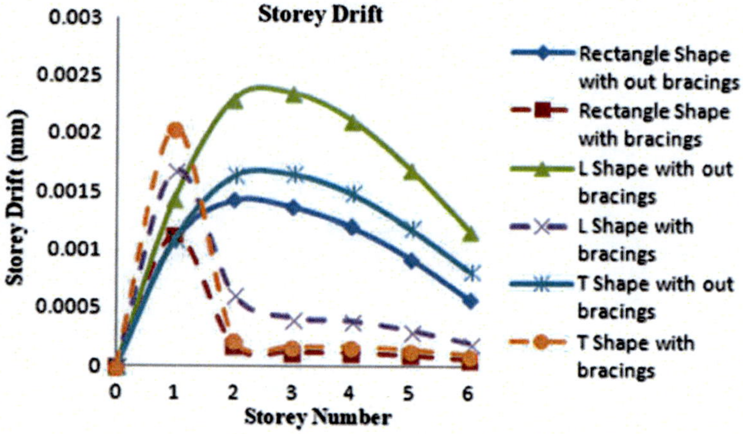

Fig. 12 Storey drift

3.5 Comparison of Storey Overturning Moment in Pushover Analysis

The storey overturning moment variation case for different plan configuration for Push X load case in the presence of bracing is illustrated in Fig. 14. And also storey overturning moment records the highest value in *T* shape plan configuration with bracings (*w*/*b*) and least in rectangular plan configuration without bracings (*w*/ob) which indicates that less moment is required to overturn the storey.

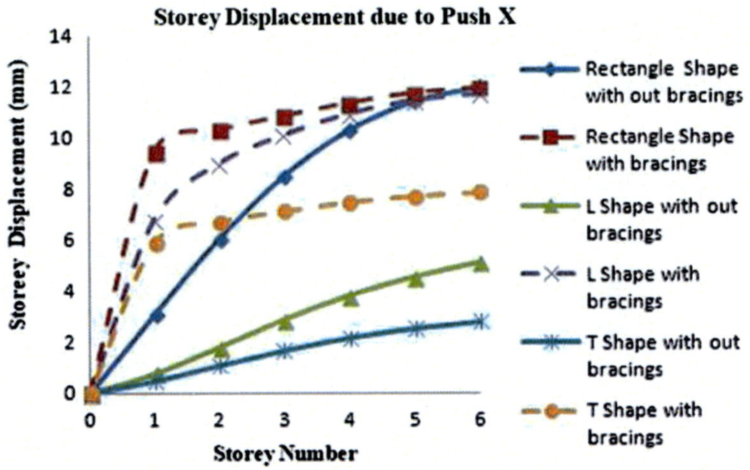

Fig. 13 Storey displacement due to Push X

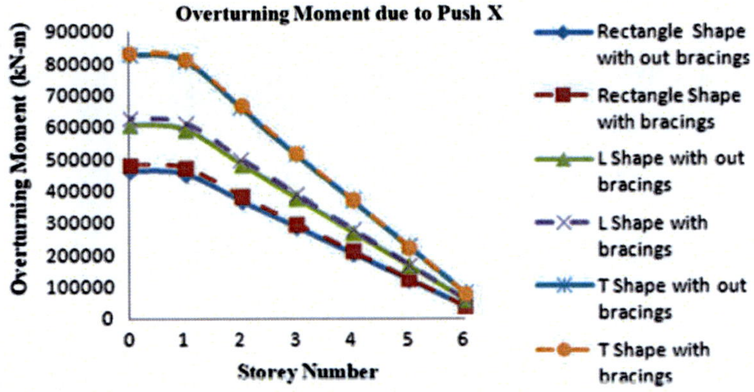

Fig. 14 Overturning moment due to Push X

3.6 Comparison of Storey Drift in Pushover Analysis

From Fig. 15, it can be observed that the maximum drift for Push X load case is maximum in rectangle shape with bracings (*w*/*b*) followed by *L* shape plan configuration with bracings (*w*/*b*). *T* shape plan configuration without bracing (*w*/ob) has shown least value. As per the code IS 1893(Part-1)2002, the maximum drift limit is given by 0.004*h* where *h* indicates height of the storey. Therefore, the drift of storey here did not exceed the limits in the Push X load case.

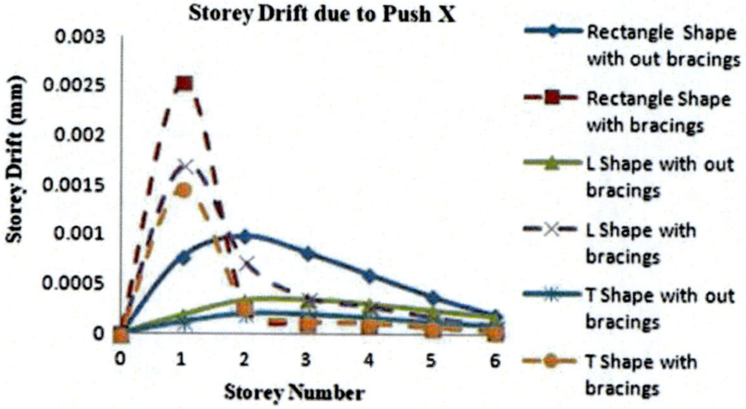

Fig. 15 Storey drift due to Push X

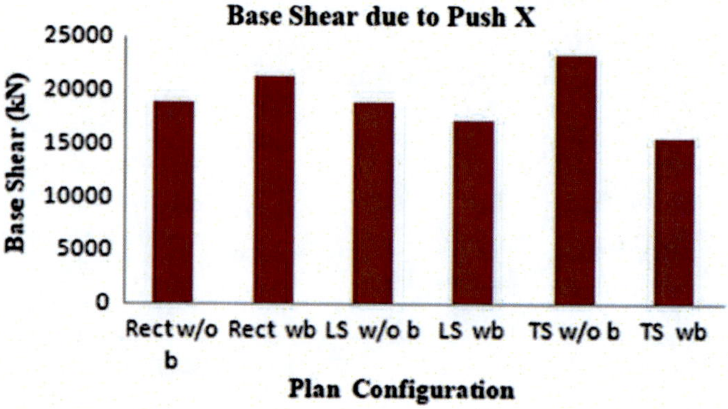

Fig. 16 Base shear due to Push X

3.7 Comparison of Base Shear Due to Push X Load Case

Base shear due to Push X load case shown in Fig. 16. It is found to be higher in T shape plan configuration without bracings (*w*/ob) due to rise of lateral displacement in the building and least in T shape plan configuration with bracings (*w/b*). This is because of increasing formation of plastic hinges in the structure.

3.8 Comparison of Base Shear Due to Push Y Load Case

From Fig. 17, base shear due to Push Y load case is found to be higher in the L shape plan configuration with bracings (*w/b*) due to increase of lateral displacement

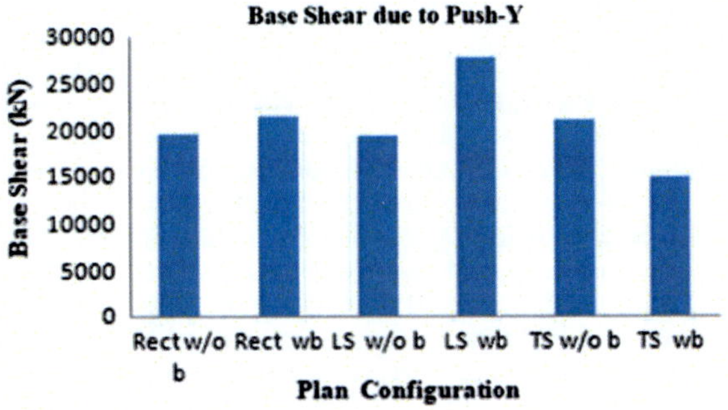

Fig. 17 Base shear due to Push Y

in the building and least in the *T* shape plan configuration with bracings (*w/b*). This is because of increasing formation of plastic hinges the structure.

4 Conclusions

From the observations and results, the following conclusions drawn are:

1. The storey displacement is found to be within limits and is least in rectangle shape building with bracing and without bracing in Equivalent static analysis.
2. The values of storey drift satisfy the permissible limits, and it was recorded highest in L shape building while the least in rectangular building in Equivalent static analysis while the storey drift values are almost equal in pushover analysis and equivalent static analysis.
3. Overturning moment got its highest values for *T* shape building with bracings and without bracings in both pushover and equivalent static load cases.
4. The values of base shear are found to be highest in *L* shape and *T* shape building while the least values were recorded in *T* shape model with bracings for both Push X and Push Y load cases.

From the results obtained it is concluded that the displacement, drift, overturning moment have drastically decreased in the case of models with bracings due to increase in stiffness and adequate ductility compared to the models without bracings. The base shear is higher due to irregularity in plan configurations and the structures of simple regular geometry with bracings are best sought after for construction in seismic zones.

References

1. Mani Deep V, Polu Raju P (2017) Pushover analysis of RC building: comparative study on seismic zones of India. Int J Civ Eng Technol (IJCIET) 8(7). ISSN Print 0976-6308
2. Mohod MM (2015) Effect of shape and plan configuration on seismic response of structure. Int J Sci Technol Res 4(09). ISSN 2277-8616
3. Haamidh A et al (2020) Seismic response assessment of stiffened RCC frames located in Gujarat. Int J Recent Technol Eng (IJRTE) 8(5). ISSN 2277-3878
4. Balappa DL, Malagavelli V (2018) Pushover analysis of high rise buildings with and without bracings. Int J Civ Eng Technol (IJCIET) 9(9). ISSN Print 0976-6308
5. Sai Prasanna Kumar Reddy G, Ranga Rao V (2018) Effect of bracings in controlling the structural response under seismic force. Int J Civ Eng Technol (IJCIET) 9(7). ISSN Print 0976-6308
6. Teruna DR (2017) Comparison of seismic responses for reinforced concrete buildings with mass and stiffness irregularities using pushover and nonlinear time history analysis. In: IOP conference series: materials science and engineering vol 180, pp 012145
7. Rofooei FR, Attari NK, Rasekh A, Shodja AH (2006) Comparison of static and dynamic pushover analysis in assessment of the target displacement. Int J Civ Eng (IJCE) 4(3)

Studies on High-Performance Concrete Containing Aluminium Dross

B. M. Mithun, Shriram Marathe, and Gururaj Acharya

Abstract This paper focuses on the studies on high-performance concrete (HPC) incorporating Aluminium dross which is a by-product material obtained from smelting of Aluminium as a Shrinkage reducing admixture. Aluminiums dross liberates vapours of hydrogen peroxide when it comes in contact with water-which leads to a swelling behaviour. Due to this swelling behaviour when comes in contact with water, it could be effectively used as an admixture instead of Shrinkage reducing admixture in high-performance concrete. In the present work, the comparative study was also been presented with normal HPC mixes for strength and shrinkage. By concept, high-performance concrete should be workable, durable, low permeability, high-dimensional stability and long lasting. Whilst in potential application, the name HPC is sometimes used for the representation of concrete, this sometimes does not have any of shown above characteristics. In specific, the use of extremely active pozzolans, including the use of silica fume, with low water-binding concrete mixes has been found to be considerably moisture absorption, especially without sufficient curing. This behaviour can mostly be due to an improvement in autogenous drying reduction and perhaps a decline in creep for improved shrinking cracking in these kinds of mixes. The research findings on use of Aluminium dross in HPC—resulting in an appreciable reduction in the drying shrinkage at 1%. In order to consider the hydration of concrete with the use of Aluminium dross, micro-structural observations have also been conducted out from hardened concrete extracted samples

Keywords Aluminium dross · High-performance concrete · Shrinkage cracking · Shrinkage reducing admixture · Micro-structure

1 Introduction

Development in the field of construction has been known for many years both as materials as well as structural development. Construction materials cost is raising day-by-day which has to be controlled. Construction industry has to support a world

B. M. Mithun (✉) · S. Marathe · G. Acharya
Department of Civil Engineering, NMAM, Institute of Technology, Nitte, India

L. Nandagiri et al. (eds.), *Sustainability Trends and Challenges in Civil Engineering*, Lecture Notes in Civil Engineering 162, https://doi.org/10.1007/978-981-16-2826-9_61

of continuing population growth and economical growth. Concrete is India's most used building tool and uses over 100 million cubic metres annually. Experts agree that traditional, compressive strength-based concrete may not sufficiently act in accordance with many practical specifications, like those of low permeability, freeze–thaw resistance, thermal cracking. Therefore, as result of the efficacy of the rise in alternative building materials and construction techniques are being sought. Aluminium dross is a wastes from the manufacture of aluminium from Aluminium smelting. It contains copper, salts as well as other non-metallic materials. Duping of Aluminium dross is problematic as it affects the ecosystem since it comprises fluorides and other salts [1]. Through this research, the Aluminium dross was used as a admixture and its features and Efficiency in concrete. High-performance concrete (HPC) is a specialty concrete sequence designed to offer certain advantages in constructing concrete buildings, which cannot be regularly accomplished using traditional materials, standard mixing and curing processes. Better mixes with low-water binding ratios, which contain extremely active pozzolanic materials such as silica fume, were observed to experience substantial shrinkage, particularly when adequate cure is not given. To address problem associated with dry shrinkage compensating cement is used which consist of Shrinkage reducing admixture (SRA) which significantly reduces the shrinkage but is not cost-effective.

Many researchers have contributed to the field utilization of Industrial wastes in producing quality concrete. Haneef and Harish [2] indicated that utilizing aluminium dross upto 5% in conventional concrete could improve prolonged concrete/mortar stength. Pereira et al. [3] shown that utilization of industrial by-product materials like Aluminium dross is a must, which otherwise causes environmental degradation since it lead to a toxic and hazardous waste product. Dai [4] have indicated that Aluminium dross can be successfully used as an engineering material in the concrete industry. If it is not utilized, it would create environmental issues. Ewais et al. [5] shown that consumption of aluminium dross in generating calcium aluminate cement based concrete has shown a promising response-which will lead to improved performance of concrete. Mailar et al. [6] from their experimental results Noted that the initial set-up period of the recycled Aluminium dross concrete extended to 20% by around 30 min. This property of recycled Aluminium dross concrete makes it ideal for hot climate conditions. Based on the finding, the substitution of cement with 20% of the Al dross produces better mechanical strength and durability.

In this work, a study has been carried out to develop a shrinkage compensating concrete by studying the properties and performance of aluminium dross with concrete. Aluminium is found in the dross mainly as metal Aluminium—31.5%, Aluminium nitride—9.5% and Aluminium oxide—49%. Aluminium dross reacts with water to liberate Methane, Ammonia and Hydrogen. It contains some Aluminium, but consists principally of byproducts obtained during the refinement of Aluminium. Cubes of size 10 cm × 10 cm × 10 cm were cast and compressive strength tests are taken on cast cubes of concrete an average of 3 cubes for single result of test. The tests are taken after 7 days and 28 days to check compressive strength of Aluminium dross mixed concrete. 6 beams of size 10 cm × 10 cm × 50 cm used for flexural strength and shrinkage measurement. Flexural strength measured at 28 days and shrinkage measured using length comparator for 120 days.

2 Experimental Methodology

2.1 Materials

In the current study, the cement used for research is ordinary portland cement (OPC) 43 grade manufactured by Ultratech cement, obtained from local supplier. The specific gravity of which was about 3.12. Silica fumes used for the present study is obtained from A N metals Belgaum was used. The specific gravity of the Silica fume used is 2.24 and its colour is greyish black. Aluminium dross is a byproduct obtained during smelting of aluminium. For this particular project, the waste dross was collected from AA metals Belgaum. It has a specific gravity of 2.85 and its colour is dark grey. A fine powder passing 90 μm seive is used.

The superplasticiser employed was Conplast SP430G8 complying with IS:9103:1999. Conplast SP430G8 is developed on sulfonated napthalene polymers and is readily diffusible in water as a brown liquid. Conplast SP430G8 was primarily developed for high-water reductions by up to 25% without loss of workability to manufacture quality concrete with decreased permeability. This was a specific gravity of 1.23, Zero content of chloride. And air entrainment were less than 1%. A portable water was used in the investigation in compliance with IS 456 [7].

Locally available crushed stone aggregate of 20 mm maximum size was used as coarse aggregate. The coarse aggregate passing through 20 mm and retaining 4.75 mm was used for experimental work. This was collected from a quarry at Parapady. The material had specific gravity of 2.74 and water absorption was 1%. The locally available normal river sand, passing through 4.75 mm and retained on 0.15 mm and Zone 3 sand as per IS:383 [8] was used in this experimental work which was collected from Gurupura river bed. The material had specific gravity of 2.65 and water absorption was 1.5%.

2.2 Mix Design

The mix design has been developed based on Indian Standard method of mix design IS:10262–2009 [9] for a strength of 50 MPa. The various ingredients used to produce concrete with and without the usage of Aluminium dross is indicated in Table 1. All the quantities are in kg per cubic metre of concrete. The 0% mix indicates the initial basic mix which can be used for the comparison. Further, the variation of mix with 0.0, 0.5, 1.0, 2.5, 5.0 and 10.0% dosage of Aluminium dross was made.

Table 1 Mix proportion details in kilogram per cubic metre

% of aluminium dross	0	0.5	1.0	2.5	5	10
Mix ID	C-1	C-2	C-3	C-3	C-5	C-6
OPC kg/m^3	407	405	402	396	363	396
Silica fume kg/m^3	33	33	33	33	33	33
Aluminium dross kg/m^3	0	2.2	4.4	11	22	44
Fine aggregate kg/m^3	645				636	630
Coarse aggregate kg/m^3	1175				1160	1150
Water kg/m^3	174.4					
SP kg/m^3	2.64					

2.3 *Preparation of Specimens*

Using Table 1, dry mixing of Cement, silica fumes, Aluminium dross and aggregates are done in mixer. Water and superplastisizer are added uniformly to the mix and wet mixing of concrete is done. Workability of concrete is found-out using slump cone test and Cubes of size 10 cm × 10 cm × 10 cm were cast and compressive strength tests are taken on cast cubes of concrete an average of 3 cubes for single result of test. The tests are taken after 7 and 28 days to check compressive strength of Aluminium dross mixed concrete. Prisms (7.5 cm × 7.5 cm × 28.5 cm), conforming to ASTM C 157, were cast to find out free shrinkage. After 24 h, the prisms were demolded and stored thereafter at 20 °C and 50% RH. Dimension change measurements were made on a weekly basis. 3 beams of size 100 mm × 100 mm × 500 mm used for flexural strength test. Flexural strength measured at 28 days and for carrying out shrinkage test, beams of size 10 cm x 10 cm x 50 cm were cast and kept in a dry condition in absence of moisture. Shrinkage measured using length comparator for 120 days. For carrying out shrinkage test, beams of size 10 cm × 10 cm × 50 cm were cast and kept in a dry condition in absence of moisture. Cast cubes are shown in Fig. 1 from which a measurable amount of swell can be visually noticed.

After mixing concrete is poured in the moulds and vibrated. After 24 h demoulded, the test specimens are put in water bath for curing. The top surface of these specimen should be made even and smooth. The cube specimens were tested for their compressive strength using standard compression testing machine after 7 and 28 days of water curing. The procedure suggested by Indian Standard method of testing suggested in IS:516 was adopted [10].

SEM is performed for microstructure studies. In QA of concrete, SEM provides vital evidence about degree of hydration of cement, development and dispersal of hydration products, bond to aggregates and homogeneity of cement paste. The SEM was preferred as the observing apparatus for the reason that of its separate advantages to straight observe the development and/or propagation of micro cracks. These micro cracks will exists in concrete even prior to the application of load in the form of shrinkage micro cracks (initial bond micro cracks). As the compressive load

Fig. 1 Cube specimens with aluminium dross (noticable swelling at 10% can be observed)

increases, these micro cracks widen and propagate until failure occurs [11]. In the present work, the hardened concrete specimens subjected to SEM analysis at the facility available at NITK Surathkal, and the results were analyzed.

3 Results and Discussions

The workability of fresh concrete was done using slump test, which was conducted as per Indian Standards [12]. The test results have shown in Table 2. The results indicates that as the slump value decreases as the percentage of Aluminium dross dosage. Initial design mix was having slump value of 70 mm, which further reduced to about 45 mm for 2.5% addition of dross again it was increased to 105 mm due to air liberation started during the slump test, which lead to the shear slump.

Compressive strength results on the concrete cube specimens were also indicated in Table 2. From the results, it can be seen that the use of Aluminium dross has negative effect on compressive strength. The 28 day compressive strength reduced to about 30.3 MPa due for 10% Aluminium dross. Marginal reduction observed upto 1% aluminium dross addition when compared with conventional high-performance concrete with silica fume.

Table 2 Compressive strength test results of various mixes

Mix ID	Slump (mm)	Compressive strength (MPa)	
		7 days	28 days
C-1	70	40.9	62.1
C-2	60	38.6	59.3
C-3	55	36.1	56.4
C-4	45	33.9	50.7
C-5	60	25.3	37.9
C-6	95	19.6	30.3

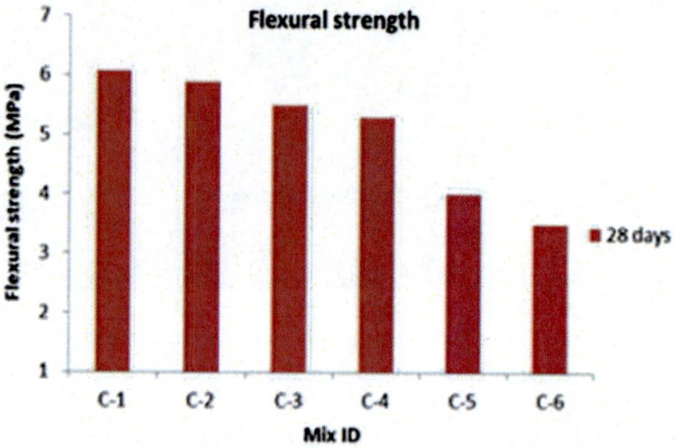

Fig. 2 Flexural strength test at 28 day

The results of flexural strength tests presented in Fig. 2 on various mixes, also showed similar trend as that of compressive strength. Increase in Aluminium dross decreased the flexural strength which further reduced greater extent for 5 and 10% dross. This is because more the quantity of dross more and more gas liberated leading to increased air entrainment at higher amount and also more micro cracks appeared because of the further development in expansion after hardening of concrete. Which can be evidenced from the SEM images and from the Figs. 3 and 4.

Drying Shrinkage and expansion test are conducted using length comparator instrument and change in length of specimen is obtained from which percentage change in dimension is calculated. The test results were depicted in Figs. 3 and 4. The results show that as the percentage of Aluminium dross increases upto 1%, there is a decrease in shrinkage percentage. It was also observed that 1.0% aluminium dross performed better for shrinkage reduction, the air entrainment due to reaction with dross increased the volume of concrete which compensated the shrinkage. Which could make 1.0% as the optimum dosage of the Aluminium dross for the optimum performance of concrete. Beyond 1.0% aluminium dross made the concrete very much expand to around 2.2% for 10% dross.

From Figs. 5, 6, 7, 8, 9, 10, 11 and 12 Aluminium dross increment shows more number of air entrainments reflected as voids when compared to 0.5 and 1%

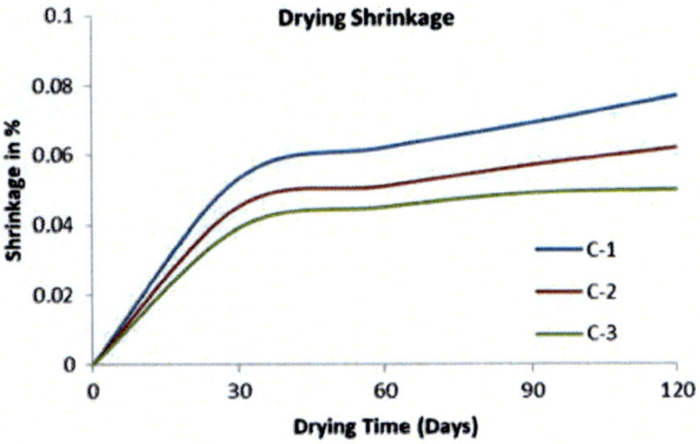

Fig. 3 Drying shrinkage of HPC with/without alumimium dross (ASTM C 157, 1-day water cured)

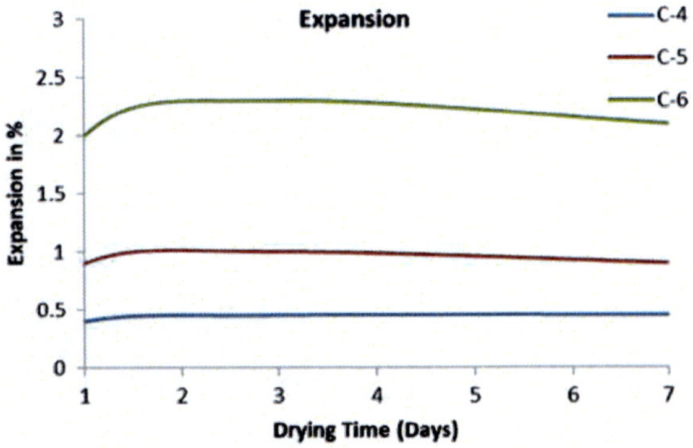

Fig. 4 Expansion of HPC with dosage of 2.5%, 5% and 10% alumimium dross content

Fig. 5 SEM for 0.5% replacement of aluminium dross specimen (C-2)

Fig. 6 SEM for 0.5% replacement of aluminium dross specimen showing nearly spherical air voids formation (C-2)

Fig. 7 SEM for 1% replacement of aluminium dross specimen (C-3)

Fig. 8 SEM for 1% replacement of Aluminium dross specimen with clear air voids formation (C-3)

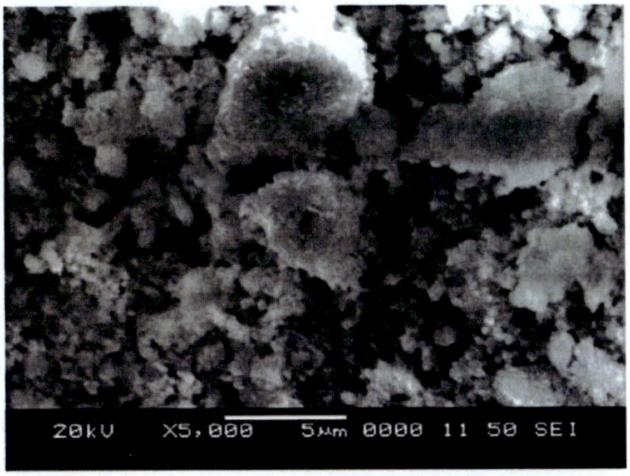

Fig. 9 SEM for 2.5% air entrainment or air voids for the aluminium dross specimen (C-4)

Fig. 10 SEM for 5% replacement of aluminium dross specimen showing micro cracks formed due to expansion (C-5)

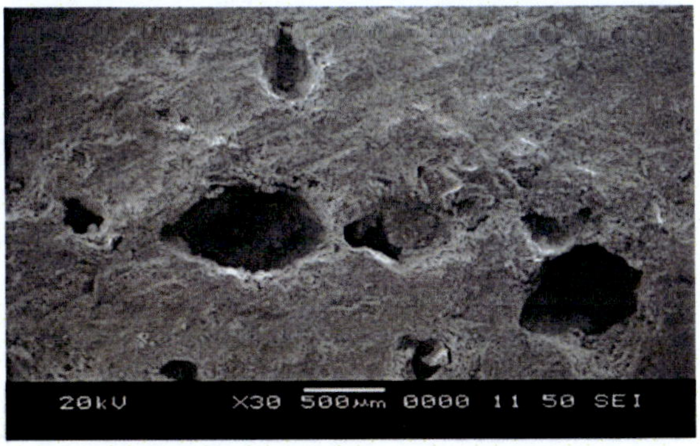

Fig. 11 SEM voids for 5% replacement of aluminium dross specimen showing increased air voids size (C-5)

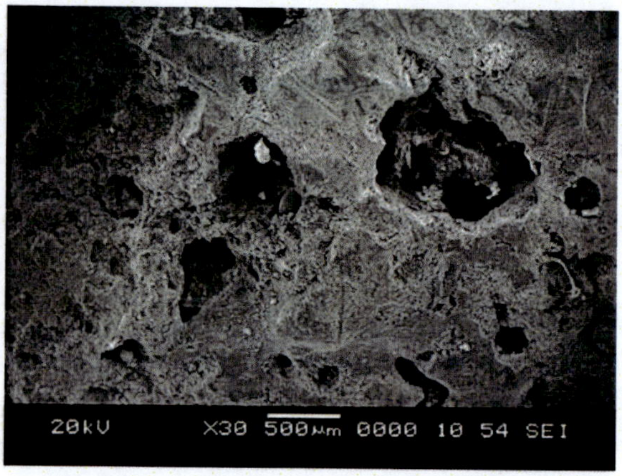

Fig. 12 SEM for 10% replacement of aluminium dross specimen (C-6)

Aluminium dross replaced concrete specimen. Fig. 10 shows micro cracks development as the dosage of aluminium dross increases. The white patches around the dark products represents unhydrated products and dark spots represents hydrated products, nearly round ball like particles are air entrainment.

Further, the cost of high-performance concrete remains the same, whereas the cost of Shrinkage reducing admixture can be saved by using Aluminium dross as the latter is an industrial waste.

4 Conclusions and Future Scope

From the experimental investigations and the analysis of the results, following key conclusions can be drawn:

- The workability of aluminium dross concrete decreased up to 2.5% later show marginally incremental changes as compared to 0% dross. This is due to more immediate air release of dross admixed concrete at high dosages. But these changes are very much marginal. Hence, one can say that the workability were almost similar to the standard concrete
- The addition of aluminium dross to concrete in smaller dosage upto 1% increases the shrinkage resistance of concrete.
- By increasing the dosage of dross more than 2.5% in concrete showed severe expansion, swelling and concrete becomes more porous and compressive strength decreases.
- The microstructure study showed increase in air entrainment and micro cracks as the dosage of aluminium dross is increased.

References

1. Satish Reddy M, Neeraja D (2018) Aluminium residue waste for possible utilisation as a material: a review. Sadhana Acad Proc Eng Sci 43(8):1–8
2. Haneef SM, Harish P (2016) An experimental investigation on use of secondary aluminium dross in cement concrete 2(6):204–207
3. Pereira DA, De Aguiar B, Castro F, Almeida MF, Labrincha JA (2000) Mechanical behaviour of Portland cement mortars with incorporation of Al-containing salt slags. Cem Concr Res 30(7):1131–1138
4. Dai C (2012) Development of aluminium dross-based material for engineering applications by submitted to the Faculty of the Degree of Master of Science in Material Science and Engineering
5. Ewais EMM, Khalil NM, Amin MS, Ahmed YMZ, Barakat MA (2009) Utilization of aluminium sludge and aluminium slag (dross) for the manufacture of calcium aluminate cement. Ceram Int 35(8):3381–3388
6. Mailar G, Sujay Raghavendra N, Sreedhara BM, Manu DS, Hiremath P, Jayakesh K (2016) Investigation of concrete produced using recycled aluminium dross for hot weather concreting conditions. Resour Technol 2(2):68–80
7. IS 456-2000: Plain and Reinforced Concrete, Bureau of Indian Standards
8. IS 383-2016: Specification for coarse and fine aggregates from natural sources for concrete, Bureau of Indian Standards
9. IS 10262-2019: Concrete Mix Proportioning Guidelines, Bureau of Indian Standards
10. IS 516-2017: Indian standard methods of tests-for strength of concrete, Bureau of Indian Standards
11. Derucher KN (1978) Application of the scanning electron microscope to fracture studies of concrete 13:135–141
12. IS 1199-2016 Methods of sampling and analysis of concrete, Bureau of Indian Standards
13. ASTM C157/C157M–17 Standard test method for length change of hardened hydraulic-cement mortar and concrete

Morphological Characteristics of the Kanakapura Watershed, Arkavathi River Basin, Karnataka, India—Using GIS and DEM

H. C. Hema, S. Govindaiah, K. J. Suresha, and Arunkumar Yadav

Abstract This research article is attempted on the study of morphological characteristics in Kanakapura watershed, located in Karnataka state, India. Kanakapura watershed is demarcated at the southern end of the Arkavathi River Basin. Morphological characteristics like flow accumulation, flow direction, stream network, and stream link have been extracted based on the digital elevation module (DEM), and results were interpreted and analyzed. Geographic Information System (GIS)-based approach with the use of DEM facilitates the understanding of different morphological characteristics that were represented in thematic raster data maps, and the correlation between the flow accumulations, flow direction, upstream, downstream pattern, stream order, and stream network was explored well. The study represents the morphological features extracted by using DEM in Arc Hydrology tools effective in understanding the spatial distribution of the stream network and identifying the groundwater potential locations.

Keywords Cartosat-1 DEM · Flow accumulation · Flow direction · Stream network · Stream link

1 Introduction

The drainage basin study is vital in hydrological examinations. Understanding the hydrological establishment of rivers is key elementary fact for decision making in several areas typically environmental planning and the utilization of water resources in the basin of the river. The essential geographic boundaries of watersheds, subwatersheds, and stream network support in evaluating data for the activities of watershed management. Morphological plays a major role in understanding the controls on

H. C. Hema (✉) · A. Yadav
Department of Civil Engineering, CMR Institute of Technology, Bengaluru, India

S. Govindaiah
Department of Civil Engineering, DSCE, Bengaluru, India

K. J. Suresha
Department of Civil Engineering, ATMECE, Mysuru, India

L. Nandagiri et al. (eds.), *Sustainability Trends and Challenges in Civil Engineering*, Lecture Notes in Civil Engineering 162, https://doi.org/10.1007/978-981-16-2826-9_62

the variability of the river process. Assessing the morphological parameters includes upstream and downstream, stream order, flow direction, flow accumulation, stream link, and stream network can well be interpreted using digital elevation model (DEM) [1, 2]. In the earlier days, topographical maps were the key sources of information for the origin of the catchment characteristics in hydrological models; currently, DEM is highly significant in hydrological studies, topography characterization as it delivers fast, economic, and consistent information.

The automatic extraction data from DEM has replaced the manual demarcation of stream networks and watershed boundary from topographic maps widely [3]. A few research works suggested that stream networks that are present in the toposheets and also in the satellite images are matching with delineated stream network DEM data [4]. To analyze the spatial morphometric parameters, it is essential to delineate the stream network [5]. DEM derivatives have a wide range of applications in the field of engineering aspects like elevation extraction, cut and fill analysis, land reclamation, and slope determination to understand the geomorphology of the area [6]. Remote sensing (RS) and GIS delivered essential results in delineating the geography and landforms [7–9]. In recent days, RS and GIS are commanding tools with an advantage over digital analysis with no laborious and provide the required output in less time [10, 11]. No research work has been carried out in the Kanakapura area in this aspect by using the DEM data. Hence, the study is expected to be useful for future researchers in the aspects of water resources, geology, and geomorphology, etc.

2 Material and Methodology

Kanakapura town is situated at Kanakapura taluk of Ramanagara district, Karnataka State, India. Kanakapura watershed covers the geographically 815.5 km^2 area, and it is located between the coordinates 12° 16′ 0″N to 12° 34′ 0″N and 77° 17′ 30″E to 77° 38′ 0″E. The location map (3D perspective representation) of the present study is represented in Fig. 1.

Extracting the morphological aspects by using geospatial technology is the key objective of this study. Morphological analysis interpreted using Carosat-1 DEM as the data source in the Arc GIS tool Cartosat-1 DEM data is downloaded from the Bhuvan ISRO Web site, which is having a pixel 30 m * 30 m grid cell size. The spatial reference of the study area is the World Geographic Coordinate System of 1984 and UTM 43N. Using Arc Hydrology tools, basic parameters for the watershed are derived. About the Arc GIS manual of the "Arc Hydrology environment," the methodology flowchart is prepared. The results derived using Cartosat DEM are flow direction, flow accumulation, and stream network using GIS software. Figure 2 flowchart depicts the methodology used in this work. Watershed is named as Kanakapura watershed since Kanakapura town is a headquarters of Kanakapura taluk. Subwatershed boundaries are delineated using the DEM and named to the prominent village in that particular zone deriving the hydraulic functions based on the DEM involves the following steps of works. The pour point algorithm eight direction flow

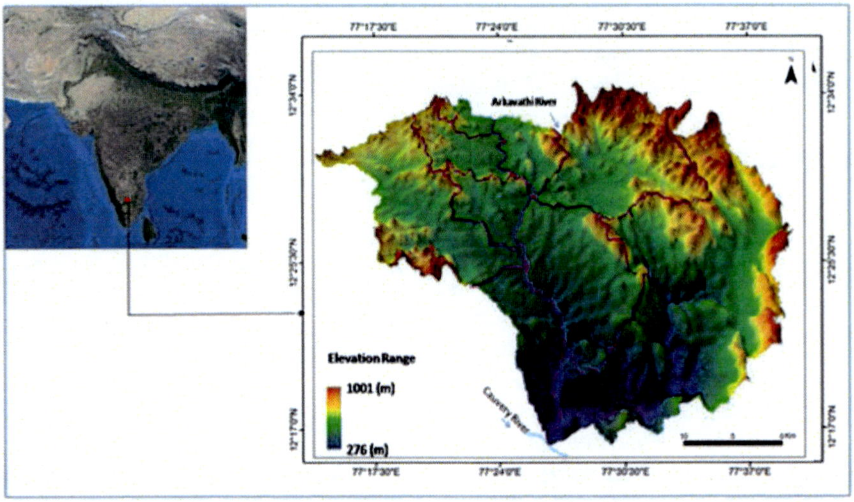

Fig. 1 Location map

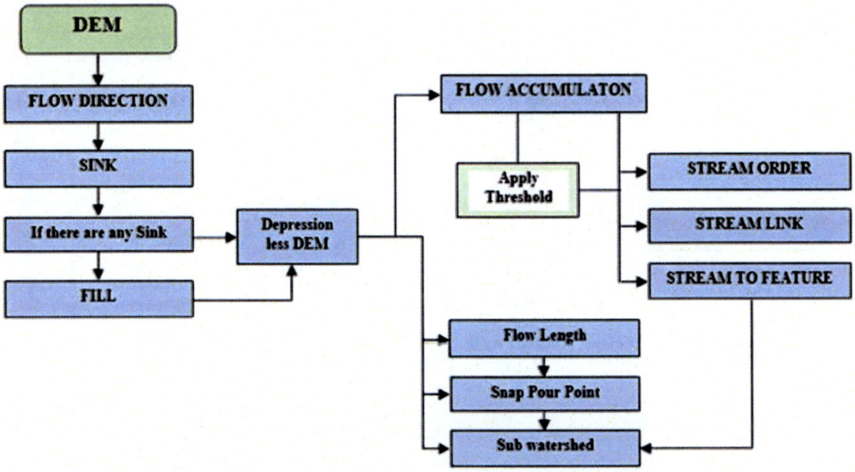

Fig. 2 Flowchart methodology

code (D-8) which assigns, for each, the cells in flow direction represents the steepest downhill slope as demarcated based on the input of DEM.

Eight neighbor cells flow toward the water flows. If the cells have the terrain depression, it is not possible to derive flow direction, so that DEM has to fill in to arrive in the flow direction. Toward the trend of the steepest, descent water flows to one of its nearby grid. The flow route takes one out of eight probable values. The defined threshold value for all the grid cells forms a part of the stream network. The two connective successive junctions segments of a stream channel, a junction, and an outlet present the stream links.

3 Result and Discussion

Morphological infer of the present study is as follows:

A. *Upstream and Downstream*: These are vital morphological parameters used for understanding the behavior of the streams in various directions on the surface. Upstream and downstream help interpret the groundwater recharge, groundwater infiltration, and flood directions [2]. The river morphology aspects are indicated in upstream and downstream maps. Upstream and downstream are the key factors in controlling groundwater potentiality [12, 13]. High upstream (Fig. 3) and low downstream (Fig. 4) values indicate that northern and eastern parts were good for potential groundwater zone. The center and southern portions of the Arkavathi River near the Sangam region indicate the high upstream values, the slope is toward the south, and drainage flows from the

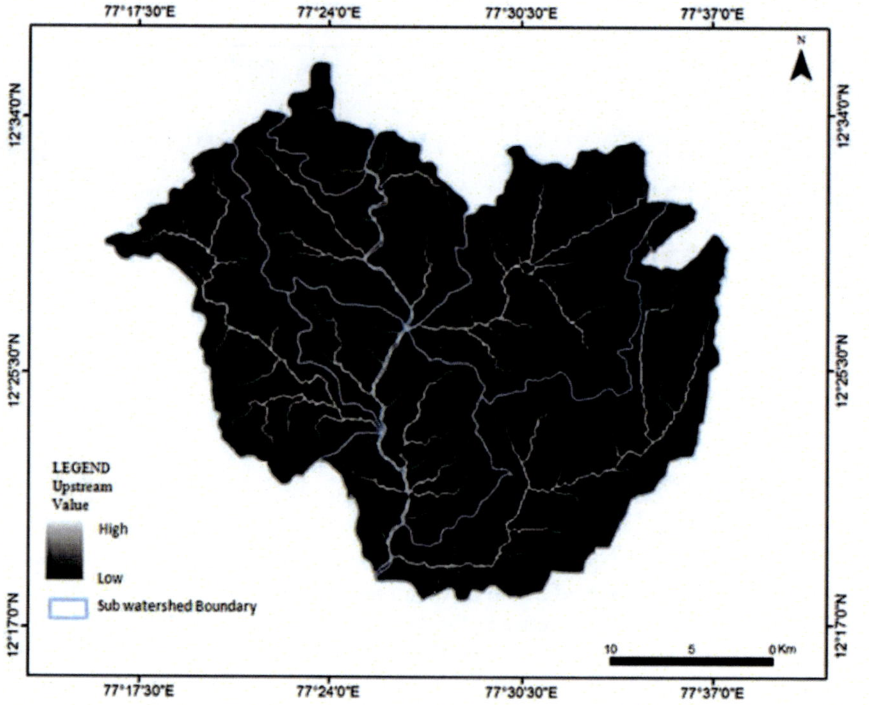

Fig. 3 Upstream of Kanakapura watershed

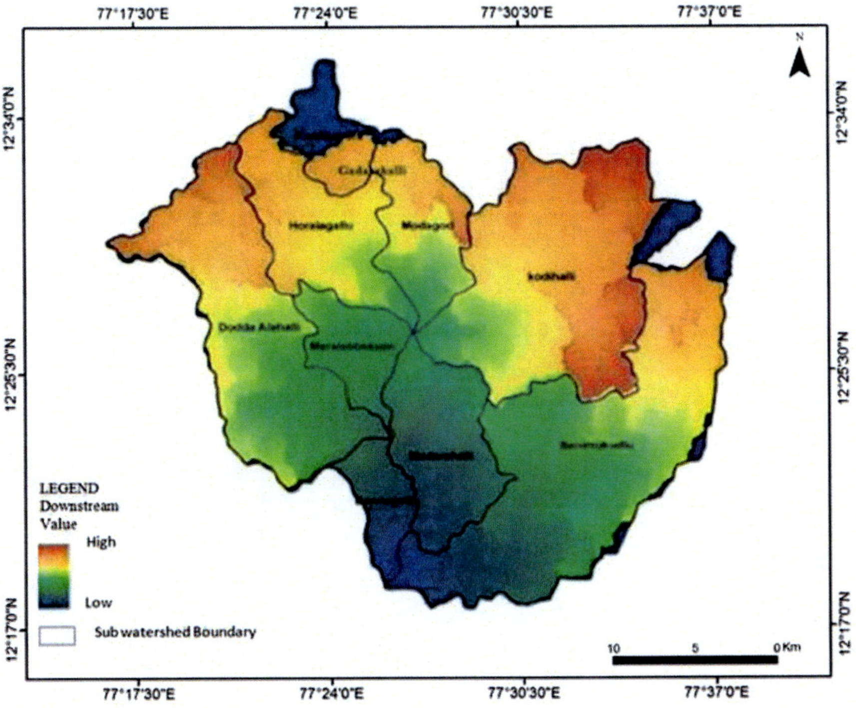

Fig. 4 Downstream of Kanakapura watershed

north to the south direction which is shown in the downstream direction map. Deep observation indicates that the flow trend of the groundwater is in NE, NW to the south, and joins river Cauvery at Sangam.

B. *Flow direction and Flow accumulation*: In a hydrological assessment is to compute the flow direction factor, it is essential to arrive at the flow direction of streams from every grid. The maximum flow accumulation drop of values from each cell and the direction of steepest descent indicate the direction of streamflow. The eight directions output, relating to the eight neighboring cells where the flow might move, is called as an D8-eight direction flow model [10]. The accumulated flow of the accumulated weight of all neighboring cells in the raster, which are rolling into respective downslope cells, generates a final raster of flow accumulation. It represents the valley fills which are the depositional landforms and acts as good aquifers for tapping groundwater at Horalagallu sub-watershed.

The quantity of upslope cells that move into each downslope cell is known as flow accumulation. Stream networks are delineated by providing the threshold values of 100 to the flow accumulation. Flow accumulation depends on the flow direction, which in turn is used to determine the stream network by giving a certain threshold.

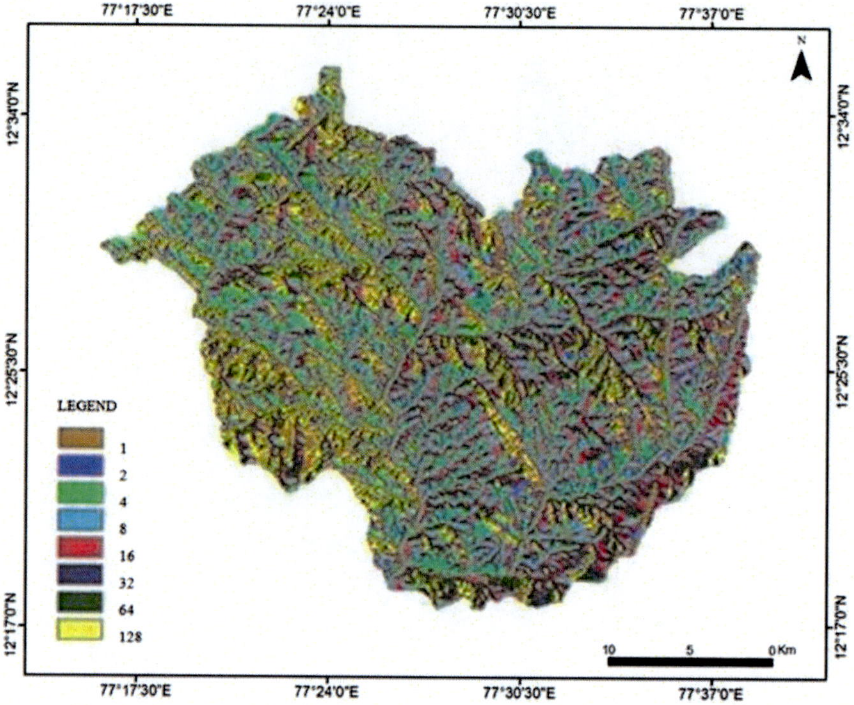

Fig. 5 Flow direction of Kanakapura watershed

Flow direction and flow accumulation are effective tools of GIS in giving the output of morphological characters of the stream network.

The flow direction in this study was derived using the sinks free DEM. The greater rates flow direction of the stream was observed in the Sangam region on the southern side, represented in Figs. 5 and 6 Flow direction and accumulation represent the spatial pattern in the watershed. Another important step is to pour point placement in the delineation of the watershed boundary. A pour point must present inside an area of greater flow accumulation to compute the total contributing water flow to that given point. Sub-watershed boundaries are delineated and named about the prominent village of that particular zone, which is represented in Fig. 7. Sub-watershed boundaries were delineated using GIS software where in every grid the point in a DEM is assigned with input flow direction and upstream points count [14, 15].

The flow direction generated the lowest cell value in the neighborhood, and the corresponding output grid receives the flow direction value. Further, this creates the network between streams with the cell centroids. High values specify confluences of streams, whereas values of zero represent the delineation of watershed boundaries. Based on this flow of methodology, it further created the stream link and stream

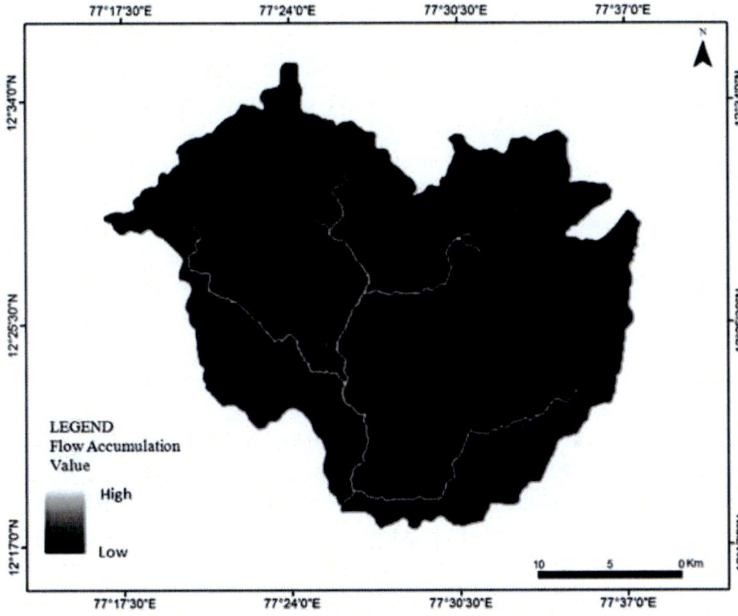

Fig. 6 Flow accumulation of Kanakapura watershed

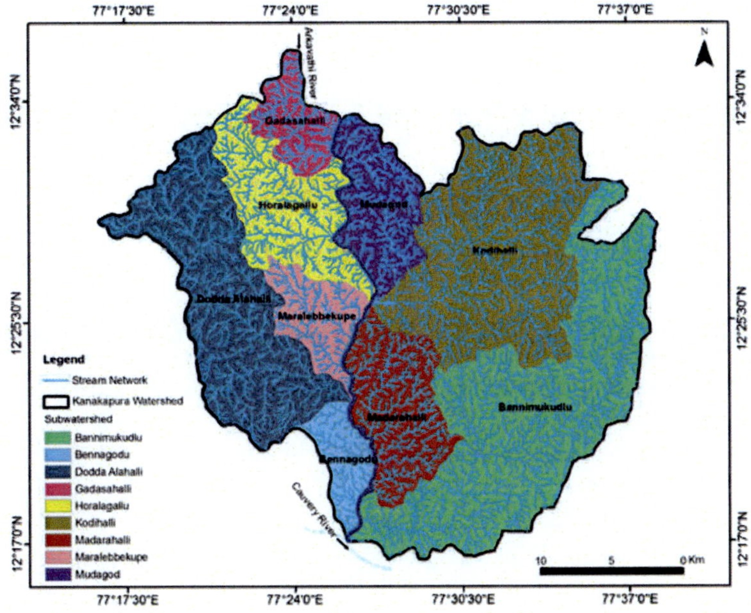

Fig. 7 Sub-watershed boundary and stream network of Kanakapura watershed

network. The characteristics of longer slopes and finer textures of the area were represented by relatively smaller values of streams [4, 13, 16].

C. *Stream Network and Stream Link*: These hydrological parameters revealed the characteristics of surface runoff.
Stream network was derived from the output of flow accumulation function by applying a threshold. Stream links are the successive drainage junctions where two stream channels connect. Longer values represent flatter gradients, and variation from its normal performance specifies that the surface is having dissimilarity in rock types and also in surface topography. The network of streams is formed by connecting all individual streams as shown in Fig. [17]. The linear morphometric parameters are derived as represented in Table 1.

D. *Drainage Lineament Density*: Lineaments indicate linear structures which are important in detecting appropriate locations for groundwater recharge [11, 17]. According to [18], lineaments are linear structures that are formed due to the tectonic action which replicates overall external appearance of underground fractures. Lineament density is calculated on the basis of lineaments cumulative length of per unit area [15], as represented in Fig. 8. It is noted in the field observation that in the Maralabekkupe sub-watershed, the Horobele dam is constructed in the high lineament density region.

4 Conclusions

The study indicates that the Cartosat-1 DEM data GIS-based application useful in analyzing the morphological characteristics, which facilitate the understanding of different morphological characteristics and the relationship between flow accumulation, flow direction, stream order, stream network, upstream and downstream arrangement, are explored well. The study specifies that efficient analysis of morphological features by the integration of GIS and DEM used to know the spatial distribution of stream network characteristics and which are used in deriving details of linear morphometric parameters indicated the maximum stream order of 6 having the length of 13.7 km. Manual calculation of these features from the topographic maps or 2D satellite images is a time-consuming process. Lineament density map which is derived using the DEM indicates the high groundwater recharge locations in the Kanakapura watershed.

Table 1 Linear morphometric parameters

SW No.	SW name	Stream order						Stream length (Km)						Stream Length Ratio					
		I	II	III	IV	V	VI	I	II	III	IV	V	VI	II/I	III/II	IV/III	V/IV	IV/V	II/I
1	Bannimukudlu	1056	218	46	11	2	1	316.2	142.2	70.5	35.4	21.91	13.7	0.45	0.50	0.50	0.62	0.63	0.45
2	Bennagodu	130	30	6	3			39.84	18.9	7.7	4.3			0.47	0.41	0.56	0.00		0.47
3	Dodda Alahalli	795	157	26	9	1		209.8	98.35	50.0	24.61	24.50		0.47	0.51	0.49	1.00	0.00	0.47
4	Gadasahalli	172	36	11	2			49.34	28.14	10.8	5.4			0.57	0.38	0.50	0.00		0.57
5	Horalagallu	403	86	18	7	2		121.05	47.5	22.10	11.19	10.73		0.39	0.47	0.51	0.96	0.00	0.39
6	Kodihalli	951	178	44	11	2	1	253.93	132.41	58.6	33.9	13.09	10.53	0.52	0.44	0.58	0.39	0.80	0.52
7	Madarahalli	388	71	16	3	1		101.4	41.98	29.2	15.9	0.44		0.41	0.70	0.54	0.03	0.00	0.41
8	Maralebbekupe	184	36	7	2			53.01	22.60	13.06	7.88			0.43	0.58	0.60	0.00		0.43
9	Mudagod	305	61	9	3			87.2	33.15	20.17	7.9			0.38	0.61	0.39	0.00		0.38

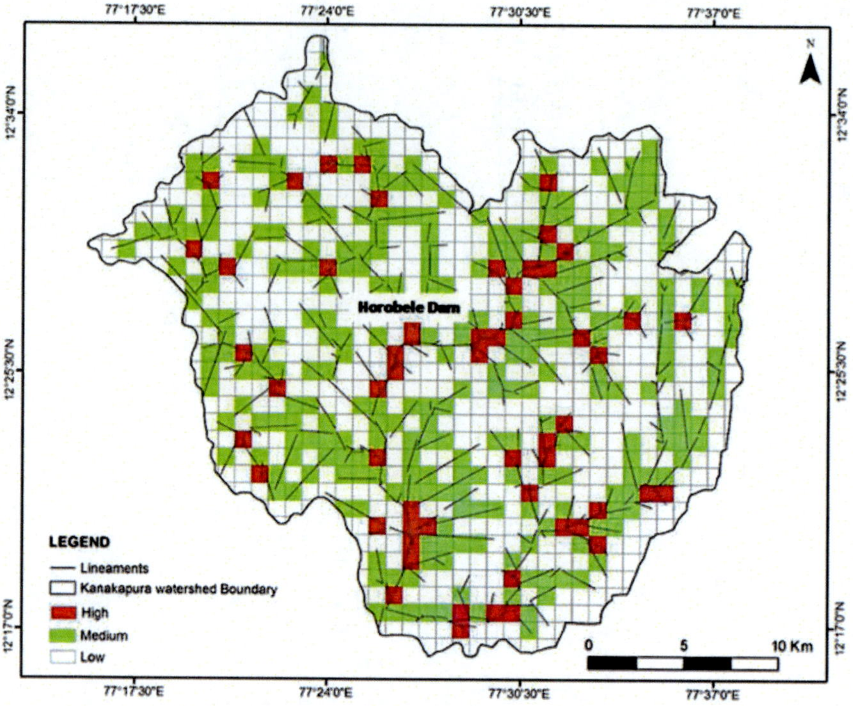

Fig. 8 Lineament density map of Kanakapura watershed

References

1. Mahtab A, Narender B (2003) Satellite derived digital elevation model and terrain parameters—generation, accuracy assessment and validation. J Ind Soc Remote Sens 31:19–24. https://doi.org/10.1007/BF03030748
2. Venkatramanan S, Ramkumar T, Anithamary I, Ganesh P (2011) Morphological characteristics of Tirumalairajan river, East Coast of India—a GIS approach. Arab J Geosci 6:1871–1881. https://doi.org/10.1007/s12517-011-0492-3
3. Visharolia US, Shrimali NJ, Prakash I (2017) Watershed delineation of Purna River using geographical information system (GIS). Int J Adv Eng Res Develop 4:690–695
4. Bhagwat TN, Shetty A, Hegde VS (2011) Spatial variation in drainage characteristics and geomorphic instantaneous unit hydrograph (GIUH); implications for watershed management—a case study of the Varada River basin, Northern Karnataka. Catena 87:52–59. https://doi.org/10.1016/j.catena.2011.05.007
5. Qadir A, Yasir M, Abir IA, Akhtar N, San LH (2020) Quantitative morphometric analysis using remote sensing and GIS techniques for Mandakini River Basin. IOP Conf Ser Earth Environ Sci 540:1–10
6. Arunkumar Y, Dodamani BM, Dwarakish GS (2018) Shoreline analysis using Landsat-8 satellite image. ISH J Hydraulic Eng 1–9. https://doi.org/10.1080/09715010.2018.1556569
7. Arbind Verma K, Madan Jha K (2017) Extraction of watershed characteristics using gis and digital elevation model. Int J Eng Sci Invention 6:1–6

8. Vittala SS, Govindaiah S, Gowda HH (2004) Morphometric analysis of sub-watersheds in the Pavagada area of Tumkur district, South India using remote sensing and GIS techniques. J Ind Soc Remote Sens 32:351–362. https://doi.org/10.1007/BF03030860
9. Arunkumar Y, Kappadi PK, Hafeezunnisa (2019) Crop pattern change and crop water requirement judgment using remote sensing and GIS techniques: a research on Tungabhadra Dam Right Canal, vol 8, pp 652–657. https://doi.org/10.35940/ijrte.B1120.0782S319
10. Zhao G, Gao J, Tian P, Tian K (2009) Comparison of two different methods for determining flow direction in catchment hydrological modeling. Water Sci Eng 2(4):1–15. https://doi.org/10.3882/j.issn.16742370.2009.04.001
11. Suresha KJ, Jayashree TL (2019) Demarcation of groundwater potential zones in Devalapura Sub Watershed Mysuru District using remote sensing and GIS. J Remote Sens GIS 8:267–273
12. Agarwal CS (1998) Study of drainage pattern through aerial data in Naugarh area of Varanasi district UP. J Indian Soc Remote Sens 26:169–175. https://doi.org/10.1007/BF02990795
13. Oguchi T, Saito K, Kadomura H, Grossman M (2001) Fluvial geomorphology and paleohydrology in Japan. Geomorphology 39:3–19
14. Ajay Kumar V, Mondal NC, Ahmed S (2020) Identification of groundwater potential zones using RS, GIS and AHP techniques: a case study in a part of Deccan volcanic province (DVP), Maharashtra, India. J Ind Soc Remote Sens 48:1–15. https://doi.org/10.1007/s12524-019-01086-3
15. Rai PK, Chandel RS, Mishra VN, Singh P (2018) Hydrological inferences through morphometric analysis of lower Kosi river basin of India for water resource management based on remote sensing data. Appl Water Sci 8:1–16
16. Pradhan B (2009) Groundwater potential zonation for basaltic watersheds using satellite remote sensing data and GIS techniques. Central Eur J Geosci 1:120–129
17. Narendra K, Rao KN, Latha PS (2013) Integrating remote sensing and GIS for identification of groundwater prospective zones in the Narava basin, Visakhapatnam region, Andhra Pradesh. J Geol Soc India 81:248–260
18. Pérez-Peña JV, Azañón JM, Azor A, Delgado J, González-Lodeiro F (2009) Spatial analysis of stream power using GIS: SLk anomaly maps. Earth Surf Proc Land 34:16–25. https://doi.org/10.1002/esp.1684

Automatizing the Khasra Maps Generation Process Using Open Source Software: QGIS and Python Coding Language

Rohit Sharma, M. K. Beg, B. E. Bhojaraja, and U. Pruthviraj

Abstract Humans are trying to acquire a piece of land from the time they have come into existence. In modern era, the management of land and its ownership is taken up by the Land and Revenue Department of the State. In order to do that, they need maps with specific objectives, so that even a laymen can understand and use it. The process explained in this paper automate the process of map making after getting the digitized shapefile of the khasra (property identification number), as a single village is divided into numerous grids and it is a tedious work and can have lots of errors while doing it manually. So in order to do the process in swift manner and without having any errors, the process was developed using the Quantum Geographic Information System (QGIS) and Python. The proposed method involves making the use of models built in QGIS along with the Python console. It helps to run the whole process on its own with taking the required input parameters and storing the outputs in a specific folder designed for them. The requirement of the project was to do the same operations on a village file and to get the final khasra map from the village polygon file. Depending upon the village area and its dimensions, the numbers of grids for a particular village is decided and the same GIS tools need to be run on each grid files which make this process a tedious work and more prone to errors. By making use of the method suggested in the paper, all the work can be done error proof with the use of Python. The use of Python code helps to do work in just couple of seconds which would have taken days to complete.

Keywords PyQGIS · Mapping · Python · Open source software · Khasra · Property identification number

R. Sharma (✉) · U. Pruthviraj
WREM, NITK, Surathkal, Mangalore, India
e-mail: rs.192rs016@nitk.edu.in

M. K. Beg
Chhattisgarh Council of Science and Technology (CGCOST), Raipur, India

B. E. Bhojaraja
Civil Engineering Department NITTE, Udupi, Karnataka, India

L. Nandagiri et al. (eds.), *Sustainability Trends and Challenges in Civil Engineering*, Lecture Notes in Civil Engineering 162, https://doi.org/10.1007/978-981-16-2826-9_63

1 Introduction

The process of making the khasra maps is completely manual even in this digital era. The khasra maps are taken from the British period, and then, it is traced using a tracing paper, and afterward, it is used by the Land and Revenue Department officers for further economic activities.

The motive of the project is to make the khasra/property identification number maps for a whole state. The projects start from the digitizing the scanned copies of khasra maps. This digital format is then used to make the new property identification number/ khasra maps for the Land and Revenue Department.

Process of making khasra maps before was completely done by tracing the previously available maps, and it was a very tedious job and took lots of concentration and human effort for making a single map. After digitizing the available khasra maps, main challenge was to make a usable map out of it using the software as previously human were trained to trace the map in a specific manner, so that it can be used on field. But training the same process to the software was a challenge.

The process of making the data useful for map is done by making use of model builder in QGIS and also automating the models by making use of Python-embedded environment in QGIS known as PyQGIS.

The recommended process makes the process to be done at the blink of eyes with the desired results. The process can be further automated to make the pdf or jpeg format outputs instead of shapefile output in this case.

2 Literature Review

The process of making the khasra maps is completely manual even in this digital era. The khasra maps are taken from the British period, and then, it is traced using a tracing paper, and afterward, it is used by the Land and Revenue Department officers for further economic activities. The process of tracing a khasra maps for an entire village takes days to finish, and it is full of human errors also called as operator error. The use of technology helps to work fast and is error-free giving much better results.

(1) *Maps*: A map is a two-dimensional representation of the features on the Earth's surface. In this paper, the objective is to map the land boundaries for property identification number purpose. Maps help to consize a large area on a sheet of paper and from that it can help in planning purposes.

(2) *QGIS*: QGIS stands for Quantum Geographic Information System. It is one of the leading open source software available to do the GIS analysis. Apart from the general tools available for GIS analysis, this platform also provides various interface which helps in doing advance level analysis such as:

- Model builder: By using models in QGIS various tools in GIS can be placed in series or parallel in graphical notation, and it also runs the tools depending on the way it is placed in the graphical builder. The one of the most important feature of the models is that it can be converted into Python scripts, so that it can be directly used in Python codes.

- PyQGIS: It is the Python interface of the QGIS. In this window, various GIS tools can be altered according to the need and models can also be used in the form of Python syntax. So it gives the freedom for the end user to change the inputs, attributes, and storing the result. Using Python in QGIS helps to execute long process immediately and with no errors.

3 Objectives

- To make property identification number (PIN)/khasra maps of private property for the entire state.
- Maps are to be used by Patwari's for the purpose of plot cutting, identification and selling.
- Map scale should be 1:500 and to be printed on an A0 (841 * 1189 mm) size paper.
- None of the private land should be left out and should also not get repeat.
- Maps should be easy to read and understandable, so that even an uneducated person can easily understand it.
- The adjacent maps should be overlapping, so that it can be placed one over the other if necessary.

4 Software's Used

Quantum Geographic Information System (QGIS): This software was used to perform all the basic GIS operations along with model making and using the Python console, so that process can run automatically. The model making and PyQGIS are the backbones of this project. The version used is QGIS Desktop 3.10.8.

Atom: This is an interactive text editor compatible with various coding language. It was used to write the PyQGIS codes, so that the codes are easily readable and understandable.

5 Dataset Resource

The dataset used for this study was given by the Space Application Center of Chhattisgarh Council of Science and Technology. The data is in the form of the vector shapefile format, and it contains the geometry and attribute values like khasra number, village name, tehsil name, and all other geometric details. But in this analysis, there is no such use of the attribute of the data. The geometry of the khasra number is the key on which the whole analysis is done.

6 Methodology

The process of getting the desired maps from a vector shape file of a village involves using of many GIS operations in a systematic way, and the same process has to be repeated again and again to get the maps of whole state. In order to do this process efficiently and to remove human errors while executing these, it is better to make models of the GIS operations and then use the models in Python code to execute the sequentially on its own. The process can be describe as follows:

- Making models of GIS operations.
- Getting the Python script of models.
- Writing the Python code using the models Python script.
- Executing the Python code.

The whole process can be understood using the flowchart (Fig. 1).

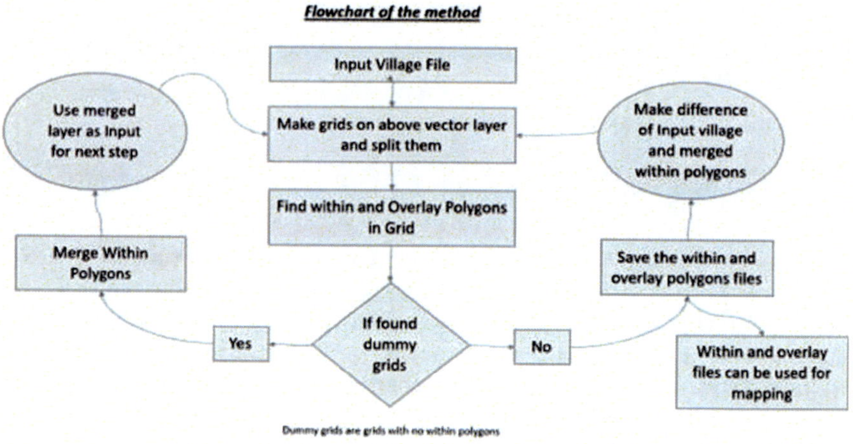

Fig. 1 Flowchart of method

7 Process

7.1 Making Parent Grids

The village file is taken as an input, and grids were made on them with the desired height and width. The height and width of a grid were calculated by doing the back calculation as the paper size and scale were already defined, so the size of grid is 408 * 490 m. Now as the size of a grid is known by using the village extend and buffering it over a distance of 500 m, an array of grid is made. Then by using the 'create grid' tool, the grids were created on the buffered extend. It creates a grid on the whole extend but that is not required as grid is clipped using the "Extract by location," tool which helps to remove the grids which are not intersecting the village layer. By doing the "Extract by location" tool, the default indexing of the grids is altered, and in order to rectify that, "Field Calculator" is used to give the proper indexing to the grids. The series of grid made by using the above tools were named as A Series as there will be several other series in the further process, so it would be easy to identify the correct grid with its series name and number. Finally, the array of indexed grids in A series was split, and several individual grids were formed. All these grids were stored in separate folder, so that it would be easy to identify the files and map them afterward.

The model which was made to carry the whole process which was explained earlier is shown below and also the result (Fig. 2) obtained after executing that model.

The village khasra polygon had been dissolved to view only the boundary of the village.

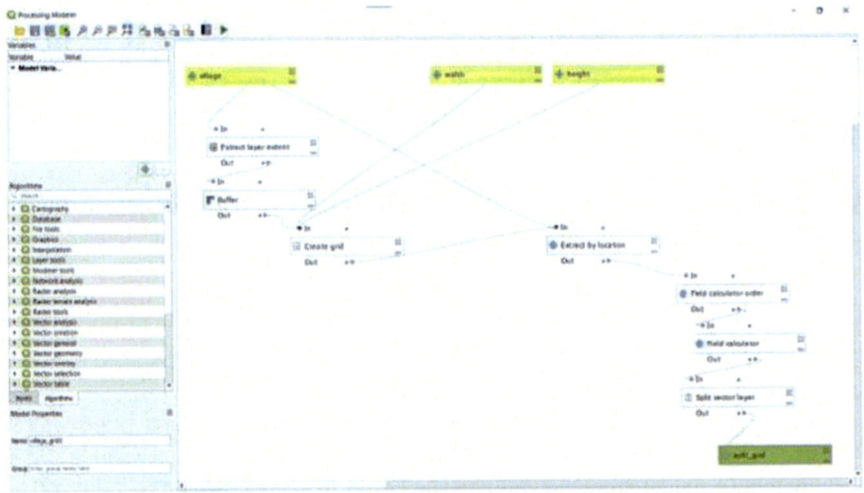

Fig. 2 Result after merging the outputs of model for making parent grid

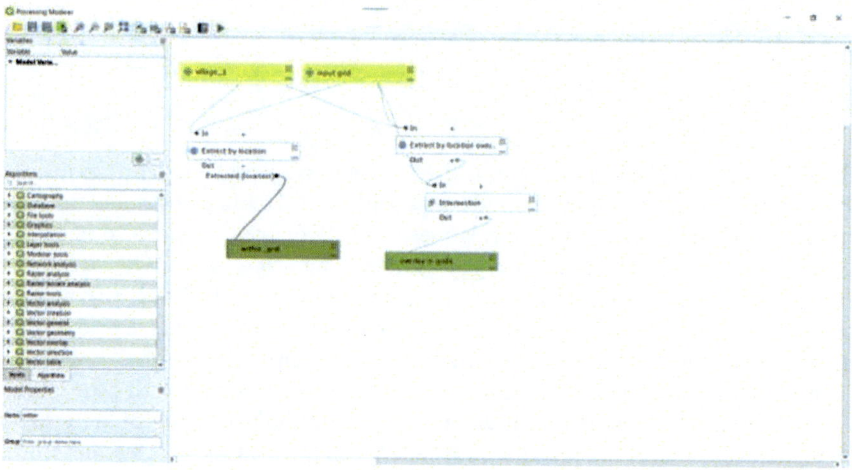

Fig. 3 Model for getting within and overlay polygons in grid

7.2 *Getting the Polygons Inside the Grids*

As lots of A series grids were made in the previous model, and now, the polygons which are completely inside as well as which are not completely inside need to me identify. In order to do so, a model was build up which takes up grid and village layer as input and give the polygons completely inside and intersecting with boundary as separate outputs. To find the inside polygons, "Extract by location: Within" tool was used, and a separate "Extract by Location: Overlay" was used to find the polygons which were intersecting boundary of the grid.After that, the result was intersected with the grid to remove the unwanted polygons lines which were outside the grid.

The same process was repeated for all the grids files which were obtained in the making parent grids part. The model (Fig. 3) build for this process and the output (Fig. 4) of a single grid file are shown below.

In the above figure, the solid lines represent the polygons that are inside the Grid A18 and the dashed line represents the polygons which were not completely inside the grid.

7.3 *Merging All the Inside Polygons*

Before making the next array of grids, it is required to club all the polygons which are already been mapped or came completely inside the grids. In order to achieve this, the Python syntax of the merging of the vector files is run with list of the vector files which need to be merged as the input. By doing this step, it has two benefits, firstly it helps to identify the grids which does not hold any grids inside them and

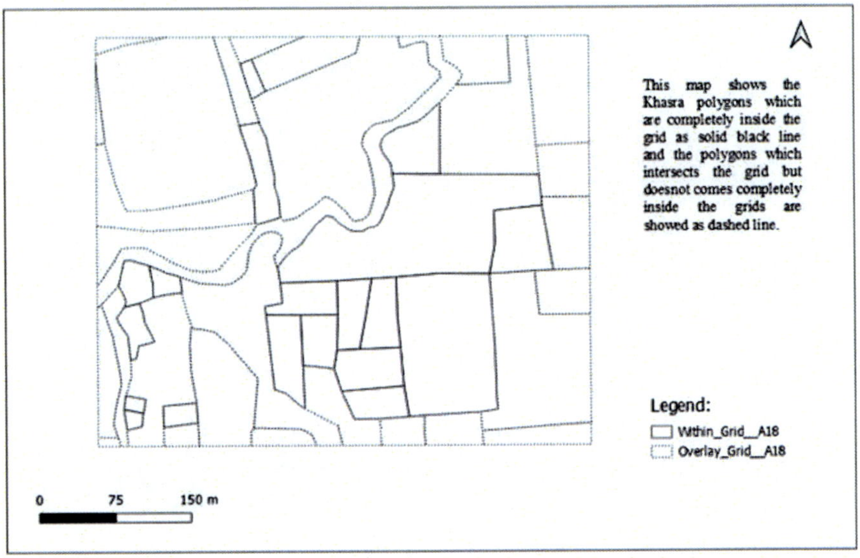

Fig. 4 Result of executing within and overlay model for a single grid

secondly the inverse of this layer with village khasra polygon will be used as an input for making next series of grids.

The output (Fig. 5) of merging all within polygon by dissolving it is shown below:

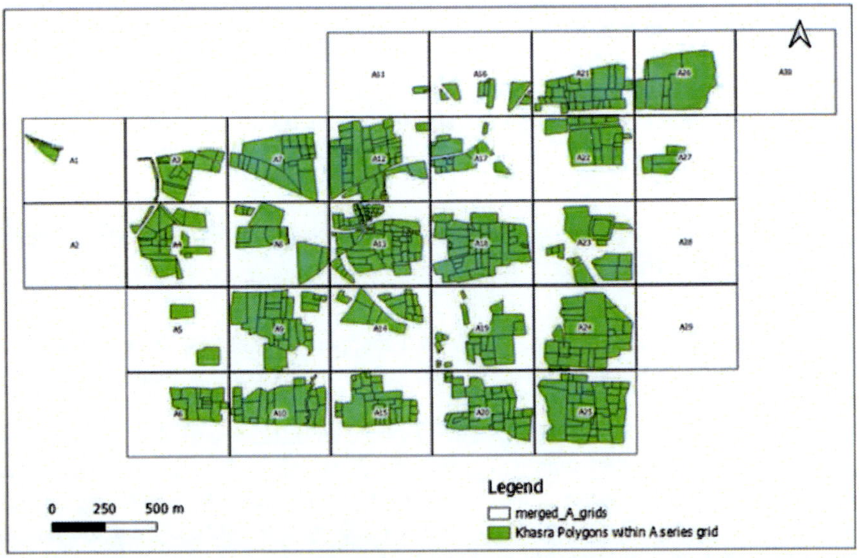

Fig. 5 Dissolved within A series polygon

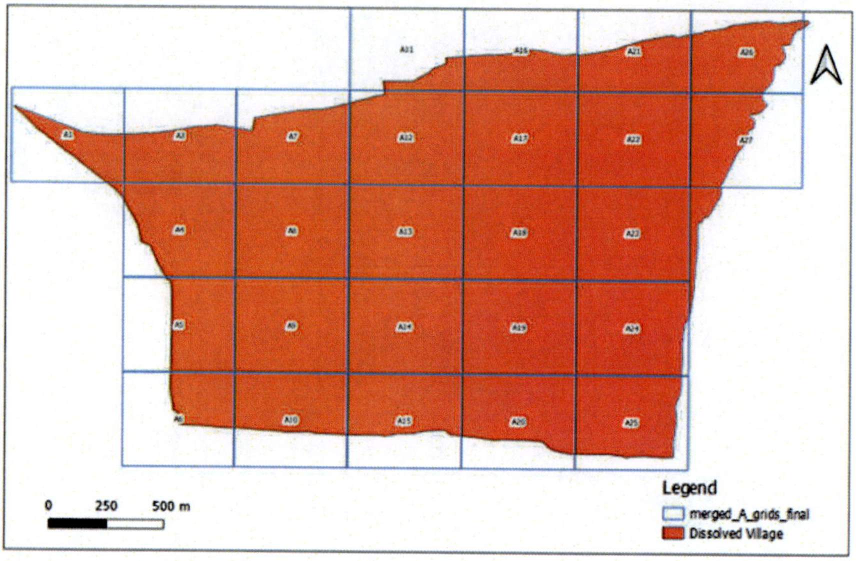

Fig. 6 Within A series polygon after removing useless grids

It is clearly visible from the above figure that Grid A2, Grid A28, Grid A29, and Grid A30 do not hold any polygons inside them, so it is not required to map them. So to omit those grids and also not altering the index of the series, the process which was explained earlier "making parent grids" and "getting the polygons inside the grid" needs to be performed again but taking the merged polygons file which was obtained from previous process instead of the village file. The result (Fig. 6) of doing the process is shown below:

As it is shown in above figure, the Grids A2, A28, A29, and A30 which were there earlier in Fig. 5 have now been removed, and only the required number of grids are made

7.4 Making Affinegrids

As there are polygons which have not been mapped in A series, the grids need to adjust a bit in order to accumulate the polygons which were not able to come in A series.

The new B series of grids are made with a little adjustment with the A series. The grids array is translated in such a way that the origin of the B series grid coincides with the centroid of A series grid. So the grids are translated by a distance of half the side in X-direction (i.e., 490/2 = 245 m) and by a distance of half the length in Y-direction (i.e., 408/2 = 204 m). The B series are created in such a way that it

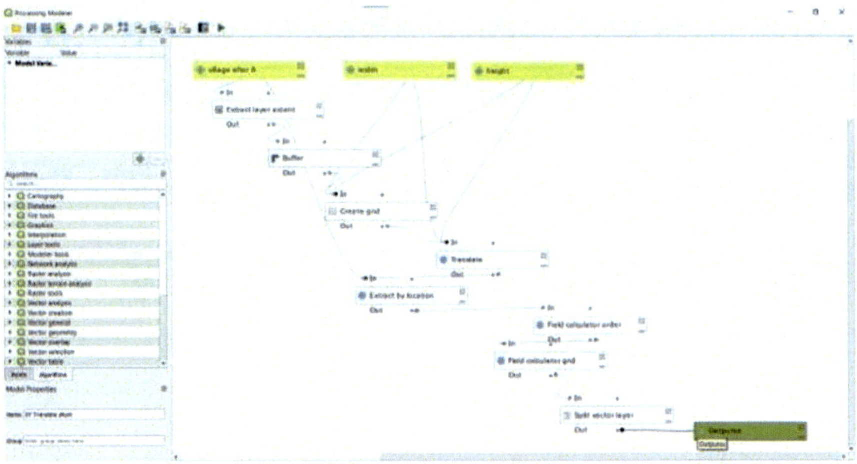

Fig. 7 Model for translating the parent grid

overlaps 25% of the area with the parent A grid. The model (Fig. 7) and the result (Fig. 8) of executing that model are shown below:

The above figure shows the placement of B series grids w.r.t the A series.

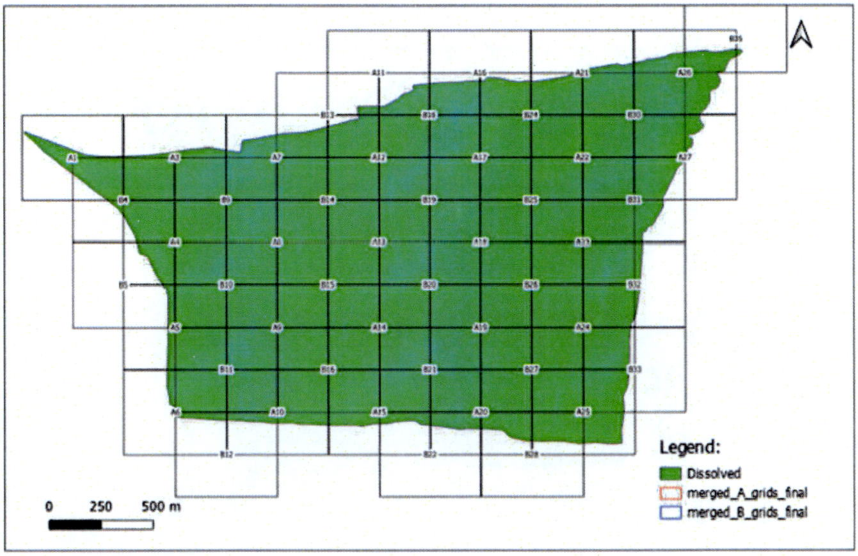

Fig. 8 A and B series relative position

7.5 Getting the Within Polygons in B Series

By using the same model which was used earlier to get within and overlay polygons in A series, the polygons inside B series are determined just giving input as village layer which is obtained after taking the difference of original village layer and merged A within polygons layer, so that the polygons which are mapped earlier are not mapped again.

Now, the within polygons are merged and then used as an input layer for creating grids and then getting final within polygons as it was done earlier in the A series.

7.6 Getting the C and D Series

In order to accumulate most possible polygons without repeating anyone, two another series were created, i.e., C and D series. In C series, the parent A series were translated to only half of the distance in *X*-direction (i.e., 490/2 = 245m), and in D series, the parent A series were translate to only half of the distance in *Y*-direction (i.e., 408/2 = 204 m) such that each C and D series grids overlap 50% with the A series. The C (Fig. 9) and D (Fig. 10) series are shown.

Once the grids for C and D series are obtained, the within model was executed with the input layer village as:

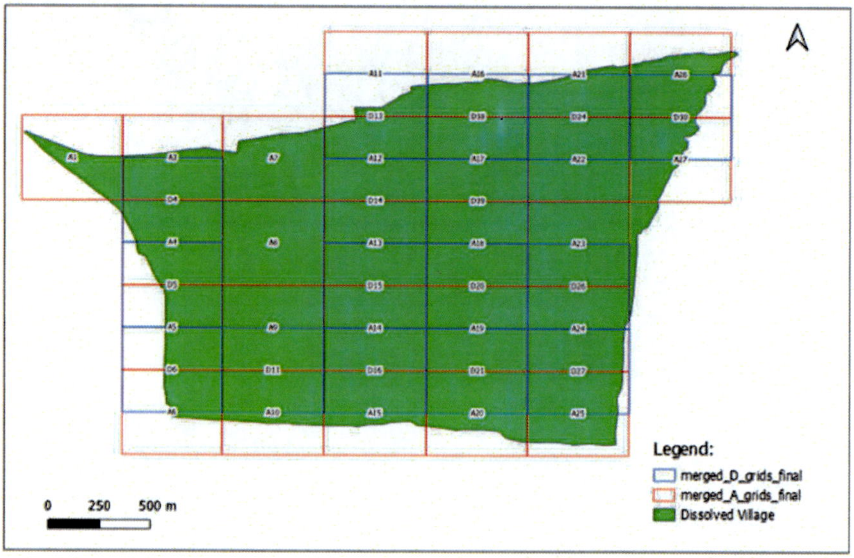

Fig. 9 A and D series relative position

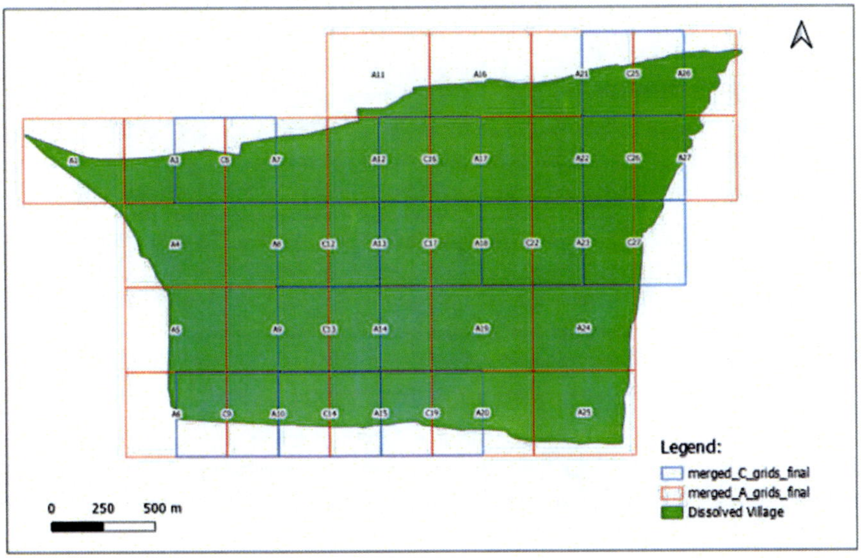

Fig. 10 A and C relative position

For C series: (Original Village)—(Merged Polygons within A series)—(Merged polygon within B Series)

For D series: (Original Village)—(Merged Polygons within A series)—(Merged polygon within B Series)—(Merged polygon within C series) to not form any duplication of any polygons.

Then, to obtain the final within and overlay polygons in C and D series, the same procedure was repeated as earlier for A series but keeping the input file as mentioned above.

8 Running Python Console

All the above-mentioned procedure is run with the help of Python code. The code contains all the syntax for making separate folders to store respective files in them. It also has syntax for the models in the order to run the models on their own and taking the required input values and files and storing the outputs in the respective folder.

By making use of the Python console, all the whole process can be completed in a very short duration of time and the operational errors can be brought down to zero as running the models one by one for many number of times can lead in errors and it is also a very tedious job. By making use of the code, the duration of completing the process of making the files required for mapping is brought down from a one day job for a village to some couple of seconds with no errors.

9 Result

After running the Python code, the village khasra vector file is divided into numerous grids with the required separate files having the polygons inside the boundary of grid and the polygons cutting the boundary of the grid. This operations let the creation of several vector, files and the name and purpose of each file are explained below:

- *Grid files*: It is a simple vector file having rectangles of desired grid dimensions on the village boundary.
- *Within grid files*: It contains all the polygon which are completely inside the grid and are to be marked as the continuous lines in the map.
- *Overlay grid files*: It contains all the polygon which are not completely inside the grid and has to be displayed as dashed lines in map.

The shape files obtained after executing the Python script were stored in a specific folders, and it is necessary as the files are to be used later for making maps, and to make a single map, it involves the involvement of many shapefiles which are related to it. The OS module was used to manage the files and store them properly. The directory (Fig. 11) looks like following after executing the code:

This "amlidih" folder (Fig. 12) was created by the code itself under the directory which was provided and it had following folders inside it:

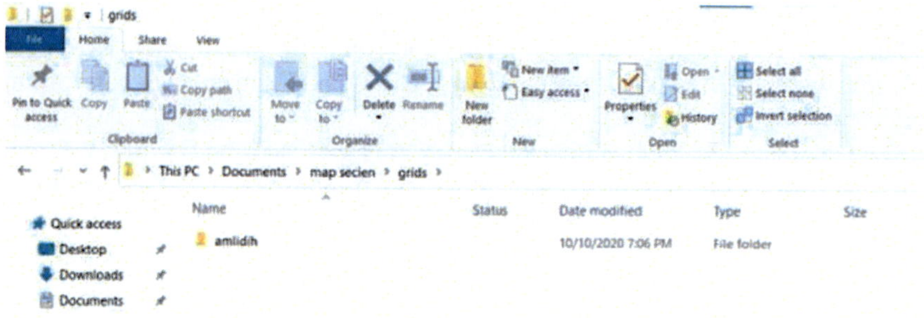

Fig. 11 Folder with village name created after running Python code

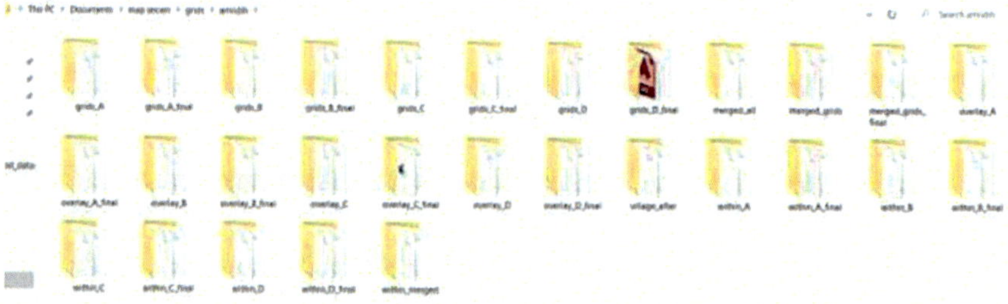

Fig. 12 Folder containing various shapefile for further mapping

The folders suffix with "final" are the folders which contain the grids, within, and overlay files which had to be mapped. The grids series A, B, C, and D are shown in the figure below.

From this, the positioning of all grids series with respect to other grid series can be examined easily. The final image gives the position of each grid with respect to another grids, and this image (Fig. 13) will be used as a reference image in the map.

In order to map the khasra (polygons) inside the (let us say Grid A18), then all the Grid A18 files inside the Grids_A_final, overlay_A_final, and within_A_final folders need to put on the map canvas and the required geometry can is obtained. The main data frame of map will look like this for the Grid A18 (Fig. 14).

Here, the polygons in solid line represents the khasra polygons inside the grid A18 and the dashed one are those whose are not completely inside the Grid A18. Similarly for other series of grid:

For grid B5 (Fig. 15):

Here, it is clear that only three khasra polygons are within grid, so those polygons are counted here. Similarly, it goes for the C and D series.

From Fig. 16, it is showed that the polygons in dark red color are the polygon which were within the grids of A series and similarly gray, red, and green for the within B, within C, and within D series, respectively. The polygons which are in pink color are the polygons which are not in any of the grid series, these were left out as the dimensions of the polygons were big, mostly these are government-owned lands, and hence, it was not an objective to cover these polygons. So there is no drawback in leaving these polygons.

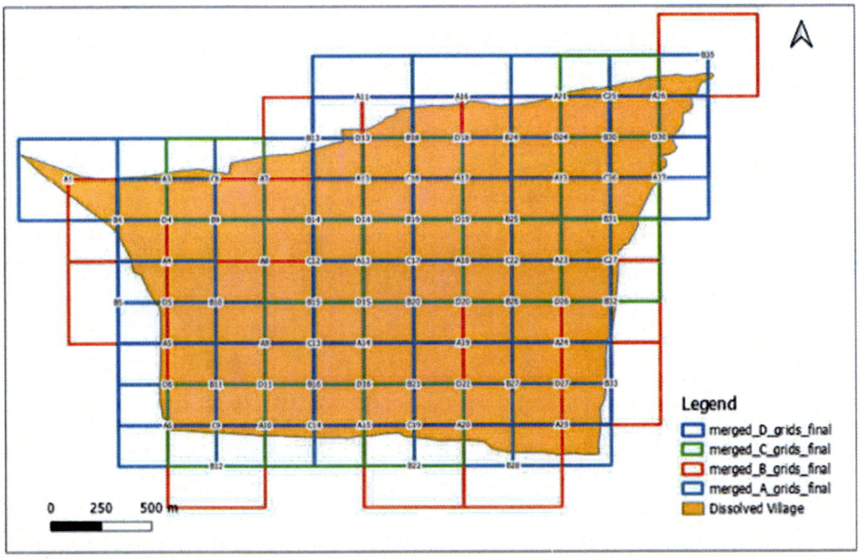

Fig. 13 A, B, C, and D series combined on village layer

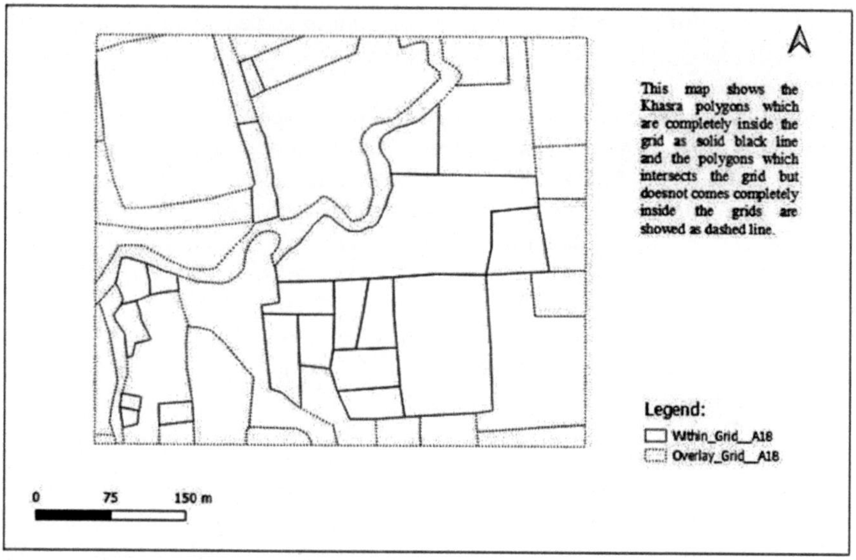

Fig. 14 Within and overlay polygons of Grid A18

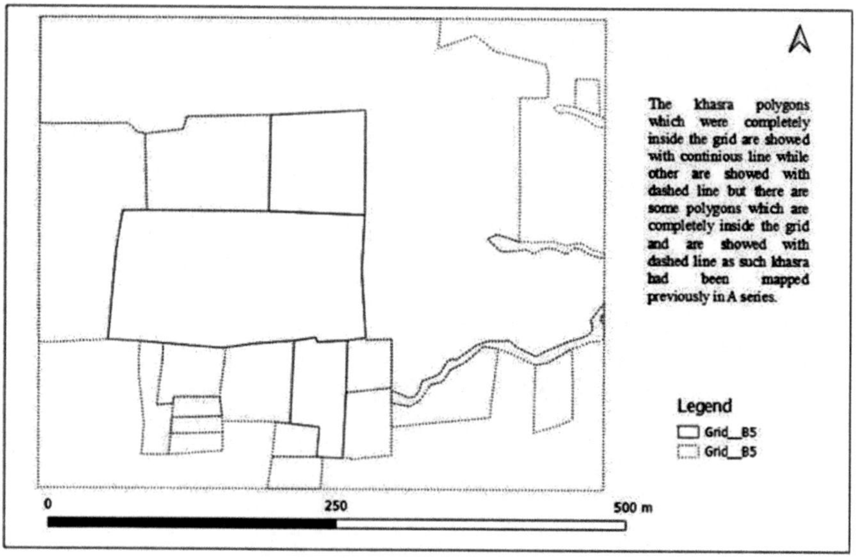

Fig. 15 Within and overlay polygons of grid B5

Fig. 16 All polygons which were covered in A, B, C, and D series

The village Amlidih is a total of 3.85577574 km^2 in area. And to cover the whole village according to the methodology proposed, a total of 79 grids files were built in which 26, 25, 12, and 16 belongs to A, B, C, and D series, respectively. The size of the grid was 408 * 490 m (Fig. 17).

10 Application

The codes developed for this projcct hclp to do thc projcct in quick span of time and with no error. So the same model can be used in other states of India to create the khasra maps of other villages.

The process can be further automated by making the maps also using the Python libraries to generate the PDF/JPEG copies of maps.

11 Conclusion

The use of Python in geospatial technologies leads to the advancement of technology in domain as well makes it easier and effective to do the analysis. The use of open source software's makes the analysis globally accepted, so that everyone around world can perform at no cost and also various developer's forums and documentations help in understanding of the software's and working of the tools.

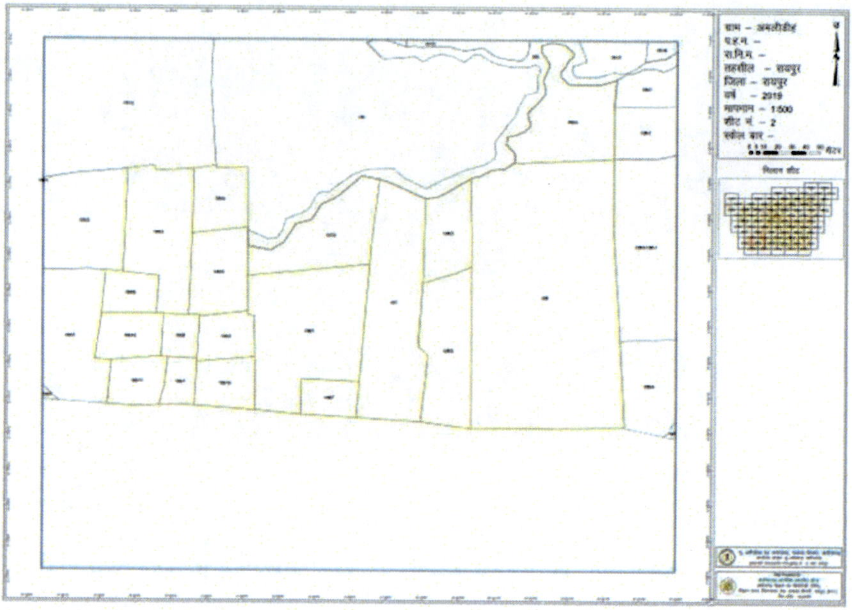

Fig. 17 Final khasra map

Acknowledgements "I would like to thank all those who had contributed in one way or another to complete my project. Words are inadequate to express my gratitude to all of them."

References

1. PyQGIS developer cookbook, Release 3.4
2. QGIS PYTHON (PYQGIS) API

Shoreline Change Model: A Review

Sheetal Mutagi, Arunkumar Yadav, and Chandrashekarayya G. Hiremath

Abstract Coastal regions are complex and underrated environments with vast variability of processes, source causing erosion, accretion, and shoreline evolution every year. To reduce the future changes in shoreline and coastal environment models are developed by different approaches, suited timescales, and incorporation of parameters responsible for the change. This current paper gives an overview of different available models used alongshore, cross-shores, sandy coasts, estuaries for both long-term and short-terms periods. Models are upgraded with the incorporation of new modules, omitted parameters, updated equations, and a combination of 1D/2D/3D systems to increase the potential to predict and prevent future loss.

Keywords Shoreline change · Modeling · Sediment transport morphology · Erosion

1 Introduction

The shoreline is unique features and complex clearly defined as a line/interface where water and land interconnect. Waves, winds, climate change, currents, rise in sea level, and other factors are the causes of change in the shoreline. Shorelines vary with location depending on the surroundings of shoreline and sand supply. Shoreline change may cause either erosion or accretion or even both in the same area. Accretion is a gradual increase in land by deposition of alluvial material and Erosion is the opposite described as the removal of material by action of waves or tides that replace the old dry area with water. Using the present and past information of a particular area the shoreline change data helps in the prediction of future changes and reducing the loss of life and property. This needs a multidisciplined approach of finding and the changes by incorporating varying coastal processes and

S. Mutagi (✉) · C. G. Hiremath
Department of Water and Land Management, Visvesvaraya Technological University, Belagavi, Karnataka, India

A. Yadav
Department of Civil Engineering, CMR Institute of Technology, Bangalore, Karnataka, India

L. Nandagiri et al. (eds.), *Sustainability Trends and Challenges in Civil Engineering*,
Lecture Notes in Civil Engineering 162, https://doi.org/10.1007/978-981-16-2826-9_64

human and natural factors causing the change into a shoreline model which is uniform throughout and also with the location change. Models are the representation of reality and are formed to solve serious problems with a set of equations or simulations to predict future changes with available past data. Due to the dynamic environment and varying factors and processes causing the change, it becomes difficult to model, and knowledge to understand and model with vast variability factors remains still finite [1]. Due to the long-term shoreline movement includes both long-term and short-term process interactions modeling of shoreline at different timescales is difficult [2, 3]. Modeling of coastal areas can be done for different categories of shoreline movements like a rotation of beach, cross-shore, long shore migration based on long-term, short-term or mid-term time scales, and different approaches to evaluating various factors like change in climate, sea-level rise, currents, and other. In the past, the potential of models was confined as information about the shoreline positions and the absence of tools required for analysis to predict the future trend gave the results with less accuracy and reliability [4]. Modern geographic information system (GIS) technology is useful in accessing the information about the shoreline position change, rate and helps to visualize the erosion and accretion trends. GIS helps in various initial stages of model formation by the collection of data, analysis, prediction, and evaluation. Shoreline change modeling has evolved from simple physical models, numerical models to one-line shoreline models, multi-line shoreline models, conceptual or equilibrium type models (2D/ 3D process-based beach profile models), statistical model approaches, reduced complexity models, scale aggregated semi-empirical models, behavior-oriented models, statistical approaches were used to calculate the change rates of shorelines to predict the changing shoreline positions. With advance of the geographic information system (GIS), DSAS known as digital shoreline analysis system tool was developed in the 1990s that has many statistic methods like endpoint rate, linear regression rate, net shoreline movement which is the best conventional method in terms of both coast and time and helps in the future management of the coastal area [5]. Average of rate (AOR), end point rate (EPR), and linear regression rate (LRR) models are used to predict the changes and rate of shoreline movement using historical data and finding the change in shoreline positions. Results obtained from this model help in the comparison between the rate of changes occurring every year and forecast future trends [6]. Any randomly selected parameter from the representation of the model will not give the exact prediction of shoreline change rate due to varying points and dynamic shoreline environment. To predict the exact shoreline change or position, the model should maintain continuity [4]. Framework in modeling should also address processes that are physical such as shoreline recovery or constraints accordingly and should also ensure consistency among the different coastal processes and drivers and generalization of modeling approaches when advanced from event to multidecadal and centennial timescales [7]. The simplified approach of models helps to solve any long-term evolution of shoreline by simulation and each model focuses only on limited problems and only individual problems can be studied [8]. The combination of different models helps to address most of the problems and improve predictive capacity with fewer errors.

This current paper gives an overview of different available models used alongshore, cross-shores, sandy coasts, estuaries for both long-term, and short-terms periods.

2 Models

Different models for prediction of shoreline change using different approaches and equations like numerical models, analytical models, profile models, one-line models, equilibrium models, process-based models involving 1D, 2D, and 3D are listed below. Uncertainty may occur due to process or source inputs carrying to the whole process. Models are updated to include the omitted module or any parameter to eliminate the restrictions on using the model to solve only particular related problems and build up a good predictive capacity to known about future scenarios of change. Table 1 shows the model name, overview, and outcome resulted from the type of model.

3 Conclusions

We observe that with the evolution of time, changes in landforms and also the evolution of coastal area is seen and prediction and calibration of model for shoreline are complex because of its dynamic nature and less knowledge to understand the variability of source and process and this paper presents the review of many different models based on a different variation of time scales. Models stated in this paper have their different approaches or combination of the idea of approaches/processes with certain timescales in the end to stimulate and predict the evolution of beaches in short-, long-term periods including cross-shores, long shores, sandy coast, tidal basins. Most main important parameters other than physical were omitted, but incorporated in other models to upgrade the simulation of processes causing evolution, predict future trends, and prevent shoreline change.

Table 1 Detail of all models

Model name	Description	Outcome
Brunn Rule and Modified Brunn rule	• Relationship between sea-level rise, indicating shore face being in equilibrium changes due to sea-level rise [9, 10] • Modified Brunn rule came into existence to inspect the cross-shore totally from landward to seaward depending on current storm conditions and omission of some important factors like sediment transport in landward due to over wash and Aeolin process [11] • Equilibrium profiles are used here to known about the profile responses by sea level rise for combination from full landward to full seaward sediment transport [11] • Equations provided are in same form for both cases and profile change and volumetric transport change depending on sea level rise without over wash is illustrated with 3 cases as seaward (S), landward (L),both (S&L)	• Modified Brunn rule also predicts landward transport accounting sources, gradients • Brunn rule was tested in fields to determine the validation of the rule for all types [12–16] • With sea level rise, future long-term shoreline responses cannot be quantified for long-term landward transport rates • Main problem in Brunn rule is it completely assumes that only waves create a sand movement and have restrictive assumptions with incorrect relationships
Delft3D	• 2D /3D model investigates hydrodynamics, simulates wave propagation, sediment transport, currents, morphology, estuarine, and coastal environments [10] • Presence of wave-current interaction and online visualization • Delft3D-RAM is used for Rapid assessment of morphology to simulate bottom changes [19].	• Provides a neat picture of near shore conditions anywhere in the world • Delft3D provides wave, current and sediment depictions[18]

(continued)

Table 1 (continued)

Model name	Description	Outcome
ONE-LINE	• 2-Dimensional transport models alongshore to predict the evolution of shoreline [20] • Simulation of changes by a series of contour lines with 3D-beach change numerical model • New model NLINE 3D beach evolution carried out at Gasparilla island used fewer calculations and helps to calculate the movement of the contour line due to sediment [21] • Cross-shore profile in NLINE model is represented as coupled horizontal layers, uses the left hand coordinate system and mathematical equation given for slope function and different parts interrelated to sediment transport rate in cross-shore direction	• Used in extensive beach changes as well as in local modeling • During the storm events NLINE model observes the mode of erosion dominant that is in cross shore transport [21] • NLINE model came into existence to be applied even for change near inlets, calculate open beach responses and long-term changes in shoreline due to cross-shore [22]

(continued)

Table 1 (continued)

Model name	Description	Outcome
GENESIS Generalized Model For simulating shoreline change	• Found by Peldnard- Considere [23, 24] • One-line behavior-oriented model and needs boundary conditions at lateral ends like moving, pinned or gated boundaries assuming shoreline positions to be constant or moving at a constant rate • Simulate long-term shoreline change from long shore transport of difference in sediments • Suited and best for shoreline that has a systematic trend in long-term shoreline position change and formulations are not included for non-uniform gradations of sand [26] • Used in Coastal engineering project design and analysis of cost or benefit by the behavior of beach predictions	• Applicable for a period of 1–10's years, but not for a change in shoreline induced by storms of the short-term period • Doesn't provide statistical answers and one should depend on, quality of inputs and individual knowledge to reevaluate and find a solution [25] • Includes only inputs adjacent to inlets not within the inlets as each of them requires a new domain in the model • Model is deterministic and highly complex with less availability and non-quantifiable data perform any function
SBEACH Strom-induced Beach Change	• Developed by the US Army Corps of Engineers • Empirical coefficients are derived numerically from process - based models to simulate long-term processes simulation of storm-induced beach change along cross shore [27] • SBEACH is updated with the application of the energy flux and dissipation theory • NSBEACH model is compared with Collaroy-Narrabeen data of beach from NSW and SUPER TANK experiment data [28]	• SBEACH model simulated only for cross-shore and erosion events of the short-term period • NSBEACH results showed model can simulate both short-term and longer-term time scales for both accretion and erosion events [28] • Sediment interactions and fluid in the swash zone act as boundary conditions and varying boundary conditions which help to predict erosion and accretion

(continued)

Table 1 (continued)

Model name	Description	Outcome
CASCADE Coupled Regional Coastal Ocean Model	• Simulate the long-term evolution of shoreline and long shore and cross shore sediment transport • Ability to act simultaneously with different temporal and spatial scales falling from regional to local with complex regional shoreline change Trend • Model is applied in mouth shore of Long Island, New York [29]	• Used at a large scale for (up to 100+ km) and from regional to local scale information is passed and extends to other scales like large or small • Regional changes in the coastal area and sediment transport are found and predict erosion and accreted volumes of up drift and downdrift inlets [29]
ASMITA Aggregated scale morphologic interaction between tidal basins and the adjacent coast	• Introduced by Stieve in 1998 and used because of its simplicity [30] • Mainly to also simulate ecologically important Coastal areas with estuaries and lagoons	• Solve problems of macroscale morphology changes which is not the only function of change of sedimentation but also dependent on mean sea level • This aggregated model was able to assess the response of tidal basins and edd- tidal deltas at the macroscale [30]
GEOMBEST Geomorphic model for barrier, estuaries, and shore face translation	• Morphologic-behavior model used for simulation of coastal morphology evolution and stratigraphy resulted from change in sea level and volume of sediment within shore face, estuaries, and barriers • Tested on outer banks of North Carolina and steeper, Washington [31]	• GEOMBEST 3D model stimulates alongshore, cross-shore, and vertical [31] • Forms the main link between the concept of coastal process and geological data • Coastal tract divided as shore face, back barriers, and estuary treating as individual function

(continued)

Table 1 (continued)

Model name	Description	Outcome
SCAPE Soft cliff and erosion platform	• Introduced by Walkden and Hall and the first modeled site was the Naze peninsula in Essex *for* shores and eroding soft rocks. [32] • Model boundaries are located in net flux sediment expected out of the model and at boundaries transport of sediment is estimated expect for tidal coasts and estuaries [33] • The influx of erosion of cliffs and outflux through longshore transport is used to determine beach volume • From particle size distribution surveys beach forming proportion of material within cliffs and shore is determined empirically	• Models are used for long and shorter periods • Predicts the behavior of coast and process which causes erosion neglecting offshore losses and hence can be regulated • Simulation of erosion and regional or local cost and benefit analysis of coastal management can be evaluated [33] • iCOASST project uses SCAPE + to stimulate the behavior of coupled cliff and barrier and long-term development of the shore profile

(continued)

Table 1 (continued)

Model name	Description	Outcome
MIKE21FM	• Applied for simulation of hydrodynamics and hydraulics related phenomenon in estuaries, coastal water, seas, short waves, sediment transport, wave dynamics, harbor • MIKE21 FM is flexible and can be applied in curved for curved coastlines • MIKE21 FM tries to bridge between 2DH and 1D model similar to 2DH with constrained morphology like to prescribe coastal profiles using modified one-line equations for shoreline movement for each strip • Model is set up for 3 different options targeted sand placement, submerged control structure and artificial headlands using the entire morphologic and shoreline model to quantify initial evolution and alongshore redistribution, respectively [36]	• MIKE21 spectral wave model helps in a simulation of decay growth, the transformation of waves generated by wind [34] • Uses fewer assumptions and has a framework to uncertainty caused by the variability of predicted impacts which is dependent on pre-described coastal profiles • The model should produce the same results as observed variations and will be easier in persistent erosion/accretion patterns in particular locations [36] • MIKE21 has different modules such as MIKE21 WA, MIKE 21 SW helpful for engineers, coastal engineers [35]

(continued)

Table 1 (continued)

Model name	Description	Outcome
CEM Coastal evolution model	• Predicts evolution in the sandy, wave-dominated coastline • Represents the underlying geology of coastline and shore face • Snell's law in combination with energy flux conservation, wave propagation is computed shows good results for straight, but when applied to complicated coastline geometries they are unreliable • 1D model does not simulate the sea-level rise and wave climate change, CEM2D was developed which gives cause-effect relationship [37] • Similar to CEM, CEM2D model shoreline is located by boundary between wet and dry	• Used for time scales year to millennia, 1–100 km spatial scales • CEM2D predicts morphodynamic coastline evolution, depositional feature in Mesoscale, and cause of sediment transport due to water level change [37] • In CEM2D model, in order to prevent loss of material and conserve sediment there is No-flow conditions • Model domain is controlled by periodic boundary conditions [37]
LX-Shore	• Designed by Robinet mainly for sandy coasts and cliffs beaches • The incoming and outgoing fraction of sediment caused by cross-shore and long shore transport, each step sediment fraction is updated to each shoreline cell [38] • In shoreline cells if sediment fraction gets lower then 0 or specified value adjustments are made with neighboring cells, if excess or deficit is removed and taken from specific land cells	• Applied to handle complex shoreline geometries and non-erodible areas like headlands and coastal defenses [38] • One-line model with 2D plan view cellular-based where cross-shore due to wave conditions, long shore transport and also non erodible areas also can be specified • Simulates changes in shoreline in long shore and cross shore due to incident wave energy variability

(continued)

Table 1 (continued)

Model name	Description	Outcome
COCOONED Coupled Cross-shore, long shore, for Dune evolution	• The model was applied in NBSC of Washington for 35 years • Long and cross-shore models are integrated by differential equations based on different process-driven coastal models even with wide timescales ranging from short to long • Jacobian Free Newton Krylov method used to solve equations and reduce inputs and computational efforts in accurate manner • FDE model used the values obtained by manual extraction to avoid uncertainty obtained from values derived geometrically and eroding fordune TWL exceeds fordune toe elevation • The hybrid shoreline change-fore dune erosion model gives response for longer periods from dense to large regions undergoing coastal floods and erosion [39]	• Prediction of future climate change is in the Probabilistic framework • Results from this model are nearer to measured data and the obtained result was similar to the observed trend of shoreline change and rotation [39] • Absence in consideration of wave conditions with local variations and omits the other natural conditions causing shoreline change
EPR Model End Point Rate	• Modeling and prediction of shoreline [40] • Prediction of shoreline rate of change using the oldest point (p_1,q_1) and newest (p_n, q_n) shoreline position dataset • Future shoreline position is given by equation as ($q = m_{EPR}$ *(p-p_n) + q_n) [40]	• For long-term shoreline position [42] • Uses historical data and best for prediction of future trend • Main limitation of this method is that it considers only two points (old and new) and erosion /accretion results may be inaccurate [43]
LRR Model Linear Regression Rate	• Prediction of shoreline rate based on all shoreline data collected	• Short-term change f shoreline position [42]

References

1. Stive MJF, Aarnikhof SGJ, Hamn L, Hanson H, Larson M, Wijnberg KM, Nicholls RJ, Capobianco M (2002) Variability of shore and shoreline evolution. Coastal Eng 47(2):211–235
2. Ranasinghe R (2016) Assessing climate change impacts on open sandy coasts: a review. Earth-Sci Rev 160:320–332
3. Toimil A, Losada IJ, Camus P, Díaz-Simal P (2017) Managing coastal erosion under climate change at the regional scale. Coast Eng 128:106–122
4. Srivastava A, Niu X, Di K, Li R (2005) Shoreline modeling and erosion prediction, pp 7–11, Mar 2005
5. Thieler ER, Himmelstoss EA, Zichichi JL, Ergul A (2009) The digital shoreline analysis system (DSAS) version 4.0—an ArcGIS extension for calculating shoreline change. U.S. Geological Survey Open-File Report 2008-1278
6. Mondal I, Thakur S, Juliev M, Bandyopadhyay J, De TK (2020)Spatio-temporal modeling of shoreline migration in Sagar Island, West Bengal, India. J Coast Conserv 24:1–20
7. Toimil A et al (2019) A review of existing modeling frameworks to assess climate change-driven shoreline changes the title of the deliverable
8. Hanson H et al (2003) Modelling of coastal evolution on yearly to decadal time scales. J Coast Res 19(4):790–811
9. Bruun P (1954) Coast erosion and the development of beach profiles. Tech Memorandum 44:82
10. Bruun P (1962) Sea-level rise as a cause of shore erosion. Am Soc Civil Eng J Waterways Harbours Division 88:117–130
11. Rosati JD, Dean RG, Walton TL (2013) The modified Bruun Rule extended for landward transport. Mar Geol 340:71–81
12. Larson M, Donnelly C, Jimenez JA, Hanson H (2009) Analytical model of beach erosion and overwash during storms. Proc ICE Marit Eng 162(3):115–125
13. Dean RG (1990) Beach response to sea level change. In: Le Mehaute B, Hanes DM (eds) Ocean engineering science, v. 9 of the Sea. Wiley, New York, pp 869–887
14. Kaplin PA, Selivanov AO (1995) Recent coastal evolution of the Caspian Sea as a natural model for coastal responses to the possible acceleration of global sea-level rise. Mar Geol 124:161–175
15. List JH, Sallenger AH, Hansen ME, Jaffee BE (1997) Accelerated relative sea-level rise and rapid coastal erosion: testing a causal relationship for the Louisiana barrier islands. Mar Geol 140:347–363
16. Leatherman SP, Zhang K, Douglas BC (2000) Sea level rise shown to drive coastal erosion. Eos Trans Am Geophys Union 81(6):55–57
17. Cooper JAG, Pilkey OH (2004) Sea-level rise and shoreline retreat: time to abandon the Bruun Rule. Global Planetary Change 43(3–4):157–171
18. Dykes JD, Hsu YL, Kaihatu JM (2003) Application of Delft3D in the nearshore zone: In: Proceedings of the AMS coastal conference, pp 57–61
19. Roelvink JA, Jeuken MCJL, Van Holland G, Aarninkhof SGJ, Stam JMT (2001) Long-term, process-based modelling of complex areas. Coast Dyn 01:383–392
20. Gravois U, Callaghan D, Baldock T, Smith K, Martin B (2016) Review of beach profile and shoreline models applicable to the statistical modeling of beach erosion and the impacts of storm clustering. Bushfire and Natural Hazards CRC
21. Dabees M, Kamphuis J (2000) ONELINE: efficient modeling of 3-D beach change. In: Proceedings of the 27th international conference on coastal engineering. ASCE, pp 2700–2713
22. Hanson H, Larson M (2001) Simulating coastal evolution using a new type of N-line model. Coast Eng 2000:2808–2821
23. Peldnard-Considere R (1956) Essai de theorie de l'evolution des formes de rivage en plages de sable et de galets. In: 4th Journees de l'Hydraulique, Les Energies de al Mer. Question III. Rapport No. I, pp 289–298

24. Hanson H, Kraus NC (1989) GENESIS: generalized model for simulating shoreline change. Report 1: Technical Reference. U.S. Army Engineering Waterways Experiment Station, Coastal Engineering Research Center, Vicksburg
25. Young R, Pilkey O, Bush DM, Robert Thieler E (1995) A discussion of the generalized model for simulating shoreline change (GENESIS). J Coast Res 11(3):875–886
26. Thomas RC, Frey AE (2013) Shoreline change modeling using one-line models: General model comparison and literature review. No. Erdc/chl-chetn-ii-55. Engineer research and development center vicksburg ms coastal and hydraulics lab
27. Roelvink DJA, Brøker I (1993) Cross-shore profile models. Coast Eng 21:163–191
28. Yuan F, Cox R (2013) Modelling coastal process for long-term shoreline change
29. Larson M, Kraus N, Hanson H (2002) Simulation of regional longshore sediment transport and coastal evolution—The "Cascade Model"
30. Stive MJF, Wang Z (2003) Morphodynamic modeling of tidal basins and coastal inlets. Adv Coast Model 67:367–392
31. Stolper D, List J, Thieler E (2005) Simulating the evolution of coastal morphology and stratigraphy with a new morphological-behavior model (GEOMBEST). Marine Geol 218:17–36
32. Walkden MJ, Hall JW (2005) A predictive Mesoscale model of the erosion and profile development of soft rock shores. Coast Eng 52:535–563
33. Walkden M (2011) A mesoscale predictive model of the evolution and management of a soft-rock. Coast J Coast Res 27:529–543
34. Parvathy KG, Gopinath DI, Noujas V, Thomas KV (2014) Wave Transformation along Southwest Coast of India Using MIKE 21. Int J Ocean Climate Syst 5(1):23–34
35. Warren IR, Bach HK (1992) MIKE 21: a modeling system for estuaries, coastal waters, and seas. Environ Softw 7:229–240
36. Kaergaard K, Mortensen SB, Kristensen SE, Deigaard R, Teasdale R, Hunt S (2014) Hybrid shoreline modelling of shoreline protection schemes, Palm Beach, Queensland, Australia. Coastal Eng Proc 1(34):23
37. Leach C, Coulthard T, Barkwith A, Parsons D, Manson S (2019) The coastline evolution model 2D (CEM2D) V1.1. Geosci Model Develop Discussions 1–32
38. Robinet A, Idier D, Castelle B, Marieu V (2018) A reduced complexity shoreline change model combining longshore and cross-shore processes: the LX-Shore model. Environ Model Software
39. Anderson D, Ruggiero P, Kaminsky GM (2019) Predicting climate-driven coastlines with a simple and efficient multiscale mode. J Geophys Res Earth Surf 124:1596–1624
40. Chenthamil Selvan S, Kankara RS, Rajan B (2014) Assessment of shoreline changes along Karnataka coast, India using GIS & remote sensing techniques. India. J Geo Mar Sci 43:1–7
41. Li R, Liu J-K, Felus Y (2001) Spatial modeling and analysis for shoreline change detection and coastal erosion monitoring. Mar Geodesy 24(1):1–12
42. Mondal I, Thakur S, Juliev M, Bandyopadhyay J, De TK (2020) Spatio-temporal modeling of shoreline migration in Sagar Island, West Bengal, India. J Coastal Conserv 24(4):1–20
43. Genz AS, Fletcher CH, Dunn RA, Neil Frazer L, Rooney JJ (2007) The predictive accuracy of shoreline change rate methods and alongshore beach variation on Maui, Hawaii. J Coastal Res 23(1):87–105

Assessment of Groundwater Quality Using WQI and GIS in Nacharam and Mallapur Industrial Development Areas, Hyderabad, India

Durgasrilakshmi Hari, V. Navya, and V. Sai Nikhil

Abstract Groundwater is the key source which satisfies the major water demands. Groundwater is under great stress concerning quality and quantity due to uncontrolled withdrawal and improper management, especially in urban agglomerations. Groundwater quality in Nacharam and Mallapur Industrial Development Areas (IDAs) which comes under Medchal-Malkajgiri District of Telangana State, India has special importance and requires significant care of all concerned. To conserve groundwater resources for domestic, drinking and irrigation uses, therefore it is important to determine and monitor the quality of groundwater. The current study is undertaken to analyse the key parameters of groundwater quality such as Colour, Turbidity, Hydrogen ion concentration (pH), Electrical conductivity (EC), Total Dissolved Solids (TDS), Total Hardness (TH), Total Alkalinity (TA), Dissolved Oxygen (DO), Chloride (Cl^-), Phosphate (PO_4^{3-}), Nitrate (NO^{3-}), Fluoride (F^-), Sulphate (SO_4^{2-}) and Lead (Pb) for 45 bore well locations in the study area. Water Quality Index (WQI) for each sample was calculated using the parameters obtained from the physicochemical analysis. Water Quality Index values are very high in and surrounding areas of the study area range from 81.6 to 671.57. Using the Inverse Distance Weighted (IDW) interpolation method in ArcGIS, the results of the physicochemical parameters and Water Quality Index of all sampling points are spatially interpolated. The spatial distribution map of Water Quality Index indicated that groundwater quality in and surrounded 1 km area of Nacharam and Mallapur Industrial Development Areas is very poor and unfit for drinking and other purposes as groundwater are highly polluted.

Keywords Groundwater · Water Quality Index (WQI) · Assessment · GIS

1 Introduction

Groundwater is a vital key component of life-supporting ecosystem system hence it should be carefully used and conserved. Groundwater used for various purposes

D. Hari (✉) · V. Navya · V. S. Nikhil
Department of Civil Engineering, Vardhaman College of Engineering, Hyderabad, India

L. Nandagiri et al. (eds.), *Sustainability Trends and Challenges in Civil Engineering*, Lecture Notes in Civil Engineering 162, https://doi.org/10.1007/978-981-16-2826-9_65

like for drinking, domestic, agriculture, industrial, geothermal generation and many more. Pollutants may enter through natural processes and human activities due to which groundwater aquifers get contaminated. The groundwater used for various purposes should be free from physical, chemical and biological pollutants. And it may have potential health hazards using it without proper treatment. Therefore, continuous assessment and monitoring of groundwater quality are essential to the management of groundwater resources. Geographical Information System (GIS) is an advanced technology used in various real-world applications. GIS techniques are widely applied in surface and groundwater quality monitoring and spatial assessment of pollution spread. The study aims to explore the groundwater quality status through water quality spatial analysis. This study will also use geostatistical methods to derive the Water Quality Index. To prepare the groundwater Quality parameters and WQI spatial distribution maps, the ArcGIS Inverse Distance Weighting (IDW) interpolation tool is used. The Groundwater Quality Index map is useful to visualize and understand the relationship among the measured points and areal extent of Groundwater contamination in the study area.

2 Study Area Portrayal

The present study area considered is Nacharam and Mallapur Industrial Areas located in Medchal-Malkajgiri district of the state Telangana in India. Study area is situated at 17° 25′ 24.34″N latitude and 78° 32′ 47.18″E longitude and at the mean sea level of 494 m. It covers an area of 3.72 Sq. Km. Average annual rainfall is about 763.2 cm and annual average temperature of 26.7 °C. The study area is identified as a Multi-Industrial Cluster. The Study Area map is shown in Fig. 1.

2.1 Statement of the Problem

The study area Nacharam and Mallapur Industrial Area surrounded by small scale manufacturing industrial units of clothing exports, fabrication, rubber material, tyres, and furniture. It is famous for its shoe factories. The industrial area comprises of various types of industries like pharmaceutical industries, food industries, electronic workshops, chemical and plastic industries, etc. Over 10,000 people live within a 2–3 km radius of the IDAs. Mallaiah Colony, Vasavi Nagar colony, Gayathri Hills, Jyothi Nagar Colony and Tirmala Meadows Colony residents are the most affected. Nacharam, Mallapur industrial development areas residents have complained an alarm over the foul smell and groundwater pollution originating from adjoining industries, which they claim has aggravated during monsoon. Residents are complaining of health issues such as asthma and bronchitis. Residents nearby IDAs raised concern about disposing of untreated industrial wastes through the resident's drainage system. This may seriously affect the nearby water bodies as well as the groundwater. The

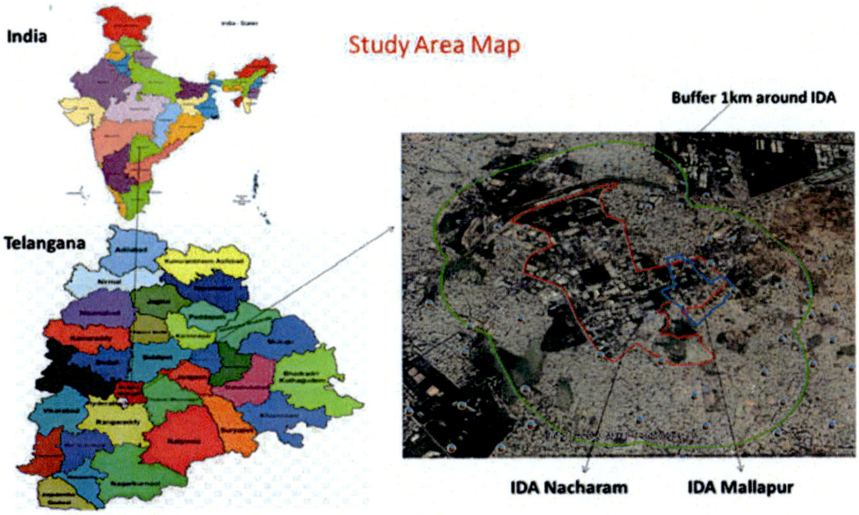

Fig. 1 Study area map

bore well depth in the study area ranges from 100 to 500 ft. The colour of water in some regions of the study area was found reddish-brown to yellow. In addition to this industrial effluent from IDA Nacharam and Mallapur contain appreciable amounts of inorganic and organic chemicals as their by-products also contaminating groundwater in the study area. Drinking water is a primary concern and is delivered on alternative days or once in three days in the study area, Even the quantity of water supplied is meagre and is insufficient for daily domestic uses. Therefore, people living in nearby colonies in study area are hardest-hit with the problem of unsafe groundwater. As per 2019 pollution index scores out of 75 industries in Nacharam area 55 in Red, 14 in Orange and 6 in green category. And out of 94 industries in Mallapur 66 in Red, 19 in orange and 9 in green category. Many of the resident colonies near industrial development areas are on the edge of pollution and have raised issues related to water and air quality. In the study area, the majority of individuals depend on and use groundwater for drinking and domestic purposes. Therefore, to protect human health and safeguard groundwater resources, periodic water quality assessment in the study area is necessary.

2.2 *Scope of the Study*

- The scope of the research is to assess the quality of the groundwater in and around the Nacharam and Mallapur industrial areas and to evaluate the feasibility of the groundwater for specific applications.

- This study creates public awareness of the study area's groundwater quality status and assists administrators and policy makers in the preparation of groundwater management plans.

2.3 *Objectives of the Study*

- To analyse the intensity of the groundwater contamination by assessing its physicochemical characteristics and Water Quality Index.
- To prepare spatial maps on distribution of groundwater quality index and all other parameters in study area using GIS techniques.
- To distinguish the boundaries of the affected areas and suitability of groundwater quality status in and around Nacharam and Mallapur Industrial areas.

3 Methodology

To achieve the study objectives, the following methodological steps are taken. The methodology considered for the study is shown in Fig. 2.

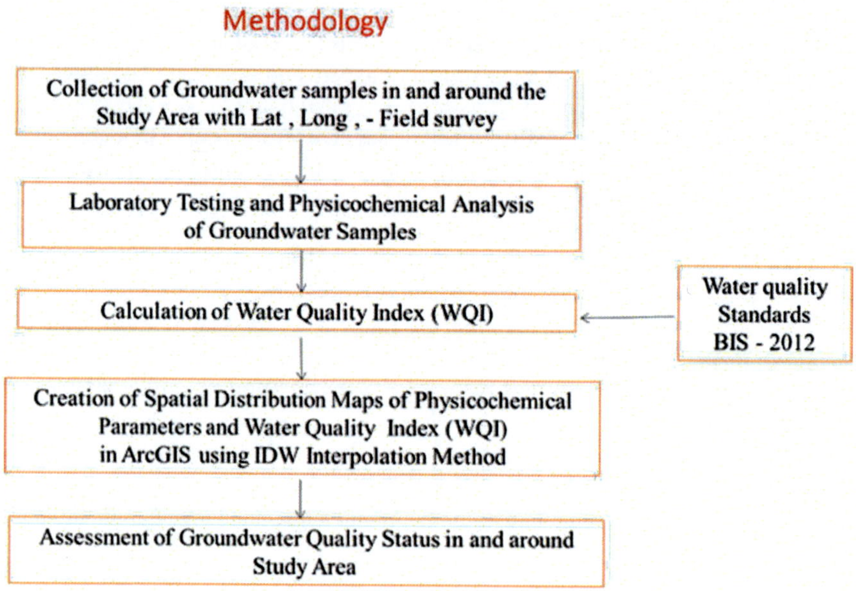

Fig. 2 Methodology flow chart

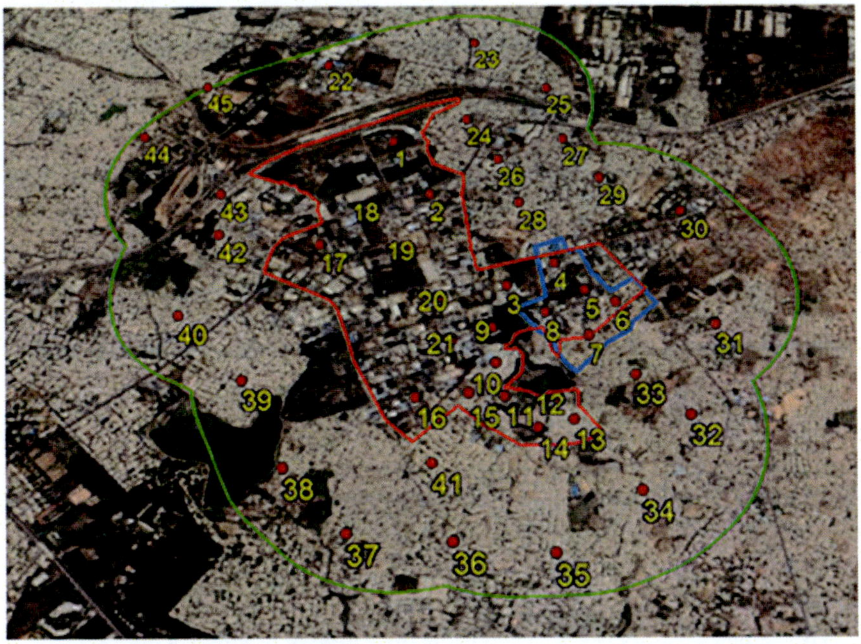

Fig. 3 Location of groundwater samples

3.1 *Data Collection*

The sample locations for groundwater quality were chosen so that they reflect the prevailing influence of the study area's water quality. Throughout the study area, collected about 45 groundwater samples from various bore wells. The bore wells sample locations are plotted on the map as given below in Fig. 3.

3.2 *Laboratory Testing Procedure for Physicochemical Analysis*

The 45 samples of groundwater acquired were carried to the laboratory and 14 different physicochemical parameters were analysed. Physicochemical analysis performed following standard testing methods and procedures.

3.3 Calculation of Water Quality Index (WQI)

WQI is a ranking reflecting the cumulative effect on the total quality of water of various water quality parameters. Water quality information can be shown in a single value using WQI. WQI is generally used for identification and assessment of water pollution. In this study the Water Quality Index was calculated for parameters viz.: Colour, pH, Conductivity, Turbidity, Total dissolved solids, Total hardness, Alkalinity, Dissolved oxygen, Chlorides, Phosphates, Nitrates, Sulphates, Fluorides and Lead.

The estimation of WQI was carried out by Horton's method in this work. Using the following expression, the WQI is calculated.

$$\mathrm{WQI} = \sum q_n W_n \Big/ \sum W_n \quad (1)$$

where

q_n nth water quality parameter quality rating
W_n nth water quality parameter unit weight.

Quality rating (qn).
The rating of quality is calculated using the expression provided by

$$q_n = (V_n - V_{io})/(S_n - V_{io}) \times 100 \quad (2)$$

V_{io} = ideal value (for pH = 7 and for other values it is zero and for DO it is 14.6).
S_n = permissible limit as given in BIS 2012 standards.
Where V_n = estimated chemical value of sample
Unit Weight (W_n)

- In the first step, weights are assigned from 1 to 5 for the designated water parameters (pH, turbidity, hardness, conductivity, DO, nitrates, etc.), depending on parameters relative value in terms of overall water quality for drinking purposes.

In the next step, calculate the Unit weight (Wn) for each chemical parameter using the below equation

$$W_n = w_i / \sum w_i \quad (3)$$

where W_n is unit weight of parameter

The following Table 1 shows the Weight and Unit Weight of physicochemical Parameters considered for this study.

Table 1 Weight and unit weight of physicochemical parameters

S.No.	Parameter	BIS standards (IS 10500: 2012) permissible limits	Weight (wi)	Unit Weight $W_n = w_i/\Sigma w_i$
1	Colour (pt/co)	5	2	0.04
2	pH	8.5	5	0.1
3	Conductivity (μs/cm)	3000	3	0.06
4	Turbidity (N T U)	5	2	0.04
5	TDS (ppm)	2000	5	0.1
6	Hardness(ppm)	600	4	0.08
7	Alkalinity(ppm)	600	3	0.06
8	Dissolved oxygen (ppm)	No Relaxation	5	0.1
9	Chloride (ppm)	1000	4	0.08
10	Phosphates (ppm)	0.1	2	0.04
11	Nitrate (ppm)	No Relaxation	4	0.08
12	Fluoride(ppm)	1.5	4	0.08
13	Sulphate (ppm)	400	3	0.06
14	Lead (ppm)	No Relaxation	4	0.08
			$\Sigma w_i = 50$	$\Sigma W_n = 1$

3.4 Creation of Spatial Map on Distribution of Water Quality Index (WQI) in ArcGIS Using IDW Interpolation Method

Geographic Information System (GIS) deals with the analysis and modelling of geographic data and information. GIS is an advanced technology which provides spatial decision making for water quality studies. GIS allows the interpolation to create unknown location water quality from known sample point values to get total area water quality based on different water quality parameters.

In the present study, Water Quality Index is calculated from the results of the physicochemical analysis of 45 groundwater samples. Spatial distribution maps of groundwater quality index are created using Inverse Distance Weighted (IDW) interpolation method in ArcGIS.

4 Result and Discussion

4.1 Physicochemical Analysis of Groundwater Samples

The collected 45 water samples of the study area Nacharam and Mallapur Industrial Development Areas were analysed in the laboratory for 14 water quality parameters such as Colour, pH, Conductivity, Turbidity, Total dissolved solids, Total hardness, Alkalinity, Dissolved oxygen, Chlorides, Phosphates, Nitrates, Sulphates, Fluorides and Lead using various specific analytical instruments and standard procedures. The physicochemical analysis results are presented in Table 2.

The results obtained are compared with the BIS 2012 standards and the study area ranges are noted as below in Table. 3

The analysed physicochemical parameters were compared with the BIS 10500 limits (2012). The results indicating that all 14 parameters analysed are not within the permissible limits of drinking water. The water quality assessment at all the sampling points in and around Nacharam and Mallapur Industrial Development Areas showed that groundwater quality is very poor. The values of TDS, Total Hardness, Alkalinity, Chlorides, Phosphates, Nitrates, Fluorides, Sulphates, and Lead are very high. This may lead to potential health hazards in and around study area.

4.2 Water Quality Index (WQI) Calculation

The drinking water quality requirements specified by the Bureau of Indian Standards—2012 were considered in this analysis for the calculation of the Water Quality Index. Weight and Unit Weight considered for Water Quality Index calculation of all the physicochemical Parameters are shown in Table.1. The weightages are assigned and quality rating for parameters is calculated by using estimated values of parameters and their permissible limits using Vn and Sn value and the qn is calculated from Eq. 2. Then the Water Quality Index is calculated using formula mentioned in Eq. 1 for all the parameters.

4.3 Water Quality Index (WQI) and Status Groundwater Quality

WQI-based classification of groundwater quality status of Nacharam and Mallapur Industrial Development Areas is presented in Table 4.

Water Quality Index values are very high in and surrounding areas of study area which ranges from 81.6 to 671.57. According to the groundwater quality status, Classification of Nacharam and Mallapur Industrial Development Areas based on WQI no sample is there in excellent category. Around 2.2% of groundwater samples

Table 2 Physicochemical analysis results of selected groundwater samples in Nacharam and Mallapur Industrial Development Areas

S. No.	Colour (HU)	Turbidity (N T U)	pH	EC (µs/cm)	TDS (ppm)	TH (ppm)	TA (ppm)	DO (ppm)	Cl^- (ppm)	$PO_4{}^{3-}$ (ppm)	$NO_3{}^-$ (ppm)	F^-(ppm)	$SO_4{}^2$ (ppm)	Pb (ppm)
1	56	3.8	6.19	4040	2180	938	200	8.20	967.2	0.3	408	0.96	20	1.40
2	83	5.3	6.24	1419	734	916	220	8.10	277.41	9.9	52	0.6	136	0.60
3	150	5.2	6.41	3130	1690	1672	650	5.60	774.75	0.08	4	1.07	40	1.10
4	290	91	5.61	7990	5000	5164	130	2.30	1382.07	1.6	20	0.42	384	0.10
5	71	4.9	7.02	458	267	416	180	7.80	59.981	9.9	3	0.56	36	0.10
6	89	5.3	6.44	2140	1090	1508	250	2.50	322.4	0.4	408	0.64	12	0.10
7	66	3.9	6.48	2090	1140	1760	210	2.70	302.406	0.52	28	0.77	50	0.30
8	215	49.9	7.44	2040	1120	772	340	8.10	267.41	34.5	53	1	116	0.50
9	55	3.2	6.44	2660	1450	1936	380	7.50	422.37	10.1	5	1.08	24	0.60
10	81	3.3	6.86	2440	1360	1324	220	7.70	399.876	0.03	57	1.26	12	0.10
11	34	3.2	6.48	2510	1300	1420	300	8.00	349.892	0.52	106	0.75	356	0.10
12	133	5.8	6.57	890	505	500	160	6.90	124.96	0.7	9	0.91	52	0.10
13	43	4	6.44	2090	1110	1060	330	5.70	282.41	12	9	1.61	105	0.10
14	14	3.2	6.44	2490	1280	1408	420	7.50	474.85	6.4	4	0.97	172	0.40
15	64	3.8	6.52	2080	1110	1480	250	2.70	357.38	0.6	37	0.79	34	0.30
16	53	3	6.39	1932	1050	2412	260	4.50	487.34	12.2	176	1.41	55	0.10
17	67	3.7	6.49	1938	1021	904	200	6.50	249.92	8.4	104	0.77	272	0.70
18	165	3.3	6.41	2220	1140	1384	400	3.30	354.88	6.5	13	0.99	97	0.20
19	112	3.4	6.16	3000	1710	2296	310	7.30	299.9	4.1	10	1.1	7	0.10
20	95	3.9	6.24	4210	2340	3564	300	2.50	537.3	1.4	20	1.2	102	0.30

(continued)

Table 2 (continued)

S. No.	Colour (HU)	Turbidity (N T U)	pH	EC (µs/cm)	TDS (ppm)	TH (ppm)	TA (ppm)	DO (ppm)	Cl^- (ppm)	$PO_4{}^{3-}$ (ppm)	$NO_3{}^-$ (ppm)	F^-(ppm)	$\underline{SO_4{}^2}$ (ppm)	Pb (ppm)
21	63	3.5	6.5	2600	1530	2016	280	2.70	504.84	1.8	11	0.98	284	0.20
22	64	3.7	6.55	1354	775	1056	200	3.90	172.446	2.2	8	1.29	95	0.20
23	62	3.8	6.4	1482	807	1076	450	7.70	149.95	7.3	200	0.88	145	0.50
24	76	7.1	7.71	2270	1230	552	330	3.10	649.79	4.2	27	1.13	188	0.20
25	95	3.1	6.35	1444	789	1084	6020	5.40	147.45	0.12	60	1.12	140	0.60
26	60	5.3	6.29	1484	804	1292	560	2.50	112.46	0.76	2	1.11	80	0.30
27	54	2.6	6.82	1329	741	728	370	2.50	150	0.4	3	1.01	81	0.10
28	65	3.3	6.76	1511	806	1248	320	6.40	172.44	0.08	4	1.03	92	0.40
29	45	4.3	6.31	1582	814	784	550	6.80	307.4	2.4	106	1.48	122	0.10
30	83	3	6.88	3060	1480	1236	340	4.90	524.83	6.9	3	0.88	95	0.10
31	58	3.6	7.04	557	311	468	480	3.30	117.46	4.6	5	0.17	40	0.00
32	53	4.8	6.67	1404	787	1244	470	5.80	114.96	2.9	21	0.95	92	0.20
33	127	3.3	6.79	1705	888	956	450	6.80	357.38	20.8	10	1.08	150	0.10
34	59	4.6	6.52	1790	925	1164	340	2.40	182.443	1.8	110	0.86	166	0.40
35	53	5.9	6.61	1173	682	932	290	5.60	149.95	7.1	50	1.48	120	0.10
36	79	4.9	6.47	1491	832	1224	440	2.40	154.95	9	4	0.75	150	0.30
37	89	5.2	6.77	1235	681	984	290	2.80	157.45	8.8	10	1.57	165	0.00
38	74	3.7	6.55	1498	825	720	420	7.10	184.94	4	4	1.09	77	0.10
39	61	3.4	6.54	1452	757	572	290	8.10	487.34	8.3	22	1.12	110	0.10
40	83	4.5	6.62	1385	737	652	330	7.50	202.43	0.8	106	0.98	91	0.20

(continued)

Table 2 (continued)

S. No.	Colour (HU)	Turbidity (N T U)	pH	EC (µs/cm)	TDS (ppm)	TH (ppm)	TA (ppm)	DO (ppm)	Cl^- (ppm)	$PO_4{}^{3-}$ (ppm)	$NO_3{}^-$ (ppm)	F^-(ppm)	$\underline{SO_4}^2$ (ppm)	Pb (ppm)
41	68	3.4	6.73	880	673	632	340	3.30	109.96	4	5	1.03	69	0.20
42	77	3.8	6.73	1664	951	1288	590	3.00	212.34	0.4	55	1.22	53	0.20
43	106	2.8	6.37	1994	1003	1664	290	7.60	262.4	0.5	10	1.19	20	0.20
44	105	4.5	6.56	1317	761	892	250	3.10	147.45	0.9	3	0.26	42	0.10
45	52	2.9	6.23	1034	1150	672	200	3.10	122.16	2.7	8	1.41	65	0.20

Table 3 Comparative statistical summary of the physicochemical parameters analysed with BIS—2012 standards

S. No.	Water quality parameter	BIS standards (IS 10500: 2012) Desirable–Permissible	Study area ranges Min–Max
1	Colour (HU)	5–15	34–215
2	pH	6.5–8.5—No relaxation	6.19–7.71
3	Electrical Conductivity (μs/cm)	700–3000	557–4210
4	Turbidity (N T U)	1–5	2.9–49.9
5	TDS (ppm)	500–2000	267–5000
6	Total Hardness (ppm)	200–600	468–3564
7	Alkalinity (ppm)	200–600	130–6020
8	Dissolved Oxygen (ppm)	5 (ICMR)—Not defined	2.30–8.30
9	Chloride (ppm)	250–1000	112–1384
10	Phosphates (ppm)	0.08–0.1	0.3–34.5
11	Nitrate (ppm)	45—No relaxation	3–408
12	Fluoride (ppm)	1–1.5	0.17–1.61
13	Sulphate (ppm)	200–400	7–384
14	Lead (ppm)	0.01—Not defined	0–1.40

could be useful for domestic, industrial and irrigation uses under the Good Water Quality Status. The 20% of samples are in poor state of water quality and can be used for irrigation but not for other applications. As 37.7% of samples are in very poor water quality status and these are restricted to use for irrigation also. 40% of samples are Unfit for drinking and cannot be used for either for irrigation or industrial purpose appropriate treatment is needed prior to use.

4.4 *Spatial Distribution of Groundwater Quality*

Geostatistical methods and GIS for Groundwater studies provide a detailed synthesis of water quality and quantity-related issues. Spatial interpolation methods are generally used to calculate the values of physicochemical parameters and WQI of groundwater in unknown locations wherever they are not calculated. A total of 45 samples of 14 physicochemical parameters were analysed in this analysis and the analysed parameters were compared to the BIS specifications. The physicochemical parameters results and Water Quality Index of all groundwater samples are used as input for spatial interpolation. The spatial distribution of all the analysed samples for the 14 water quality parameters and Water Quality Index were mapped using ArcGIS, based on the Inverse Distance Weighted (IDW) process and the results are presented in Figs. 4, 5, 6, 7, 8, 9, 10, 11, 12, 13, 14, 15, 16, 17 and 18.

Table 4 WQI-based classification of groundwater quality status of Nacharam and Mallapur Industrial Development Areas

S. No.	Lat	Long	Location	WQI	Groundwater quality status
1	17.444282	78.56149	Old meerpet	570.0627	Unfit for drinking
2	17.439428	78.564345	Narsimha nagar	430.23	Unfit for drinking
3	17.431718	78.569699	Nacharam	508.61	Unfit for drinking
4	17.433503	78.572554	Industrial area Mallapur	524.70	Unfit for drinking
5	17.431362	78.574482	Industrial area Mallapur	164.59	Poor
6	17.430434	78.576338	Ambedkar Nagar	260.56	Very poor
7	17.427935	78.574767	Ambedkar Nagar	354.60	Unfit for drinking
8	17.429648	78.572126	Industrial area Mallapur	545.81	Unfit for drinking
9	17.428506	78.568985	Tirumala colony	585.32	Unfit for drinking
10	17.425936	78.569342	Tirumala colony	204.82	Very poor
11	17.423581	78.569985	Hema Nagar	174.74	Poor
12	17.423295	78.573126	Boddupal	218.765	Very poor
13	17.42201	78.573982	Mallaiah Nagar	167.32	Poor
14	17.421439	78.571983	Ambedkar Nagar	386.31	Unfit for drinking
15	17.423795	78.567914	Tirumala Nagar	351.20	Unfit for drinking
16	17.423509	78.564916	Hema nagar	214.58	Very poor
17	17.435073	78.557849	Ravindra Nagar	671.57	Unfit for drinking
18	17.438143	78.560919	Nacharam Industrial Estate	349.99	Unfit for drinking
19	17.434788	78.563489	Industrial area Nacharam	234.62	Very poor
20	17.43079	78.565844	Industrial area Nacharam	415.62	Unfit for drinking
21	17.427578	78.56663	Hema Nagar	283.19	Very poor
22	17.452134	78.555993	Musi nagar	258.82	Very poor
23	17.45449	78.566201	HB Colony	518.88	Unfit for drinking
24	17.446424	78.566201	Shanthi nagar Colony	285.67	Very poor
25	17.449636	78.571555	Krishna Nagar	665	Unfit for drinking
26	17.442569	78.568557	Annapurna Colony	341.58	Unfit for drinking
27	17.444496	78.57284	Berappa hills	168	Poor
28	17.438643	78.570127	K.L Reddy Nagar	419.51	Unfit for drinking
29	17.440856	78.575338	Venkatramana colony	174.79	Poor
30	17.437857	78.580621	Industrial area	205.43	Very poor

(continued)

Table 4 (continued)

S. No.	Lat	Long	Location	WQI	Groundwater quality status
31	17.42872	78.582334	RNS Colony	81.6	Good
32	17.422296	78.580621	Veera reddy Nagar	254.04	Very poor
33	17.42508	78.57748	Annapurna Colony	232.06	Very poor
34	17.417485	78.577746	Sai Bhavani Nagar	434.20	Unfit for drinking
35	17.413773	78.573177	East Balaji Hill Colony	175.57	Poor
36	17.41444	78.567752	Anand Nagar Colony	356.24	Unfit for drinking
37	17.414915	78.562136	Kalyanpuri	125.39	Poor
38	17.418913	78.558139	North Kalyanpuri	179.25	Poor
39	17.424719	78.554808	Raghavendra Nagar	169.34	Poor
40	17.429478	78.550049	Veera reddy Nagar	277.6	Very poor
41	17.419199	78.566229	Hema Nagar	255.53	Very poor
42	17.43595	78.551286	Baba Nagar	285.62	Very poor
43	17.439472	78.550715	Durganagar	296.25	Very poor
44	17.444802	78.544433	Siripuri Colony	204.29	Very poor
45	17.44985	78.54776	Sanjay Gandhi Nagar	243.45	Very poor

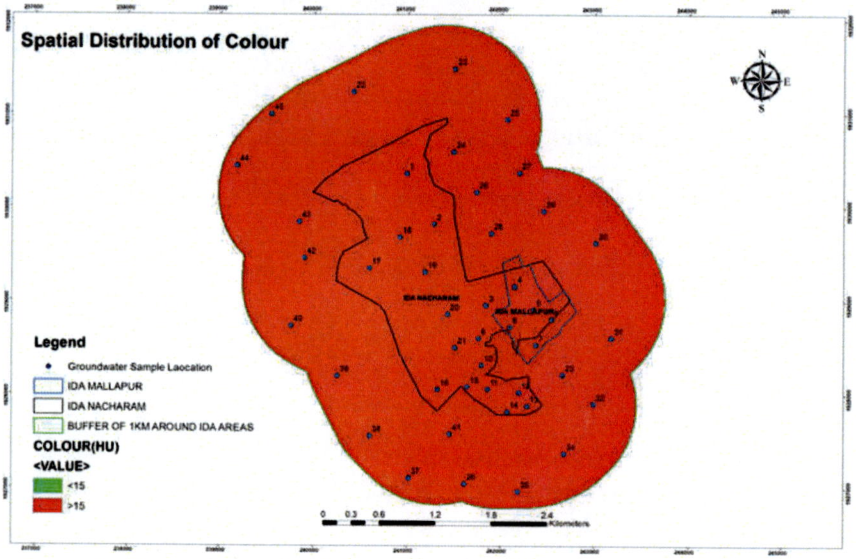

Fig. 4 Spatial distribution map of colour

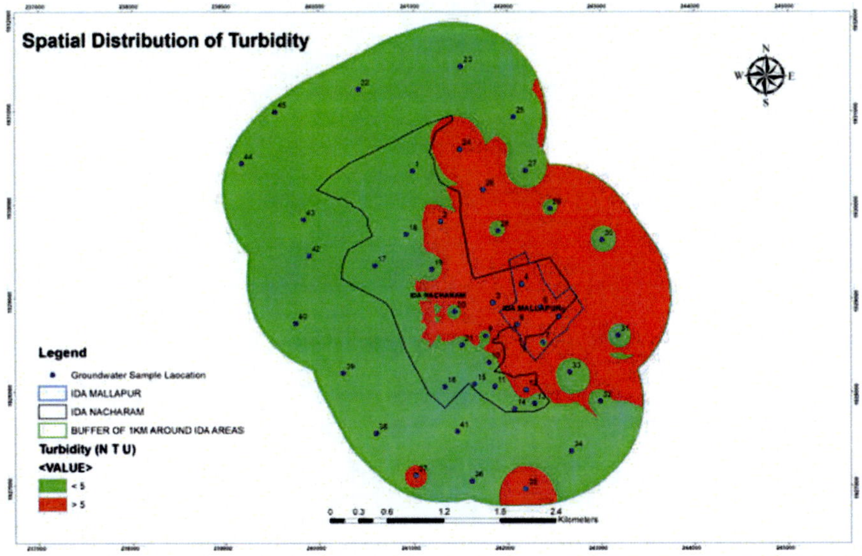

Fig. 5 Spatial distribution map of turbidity

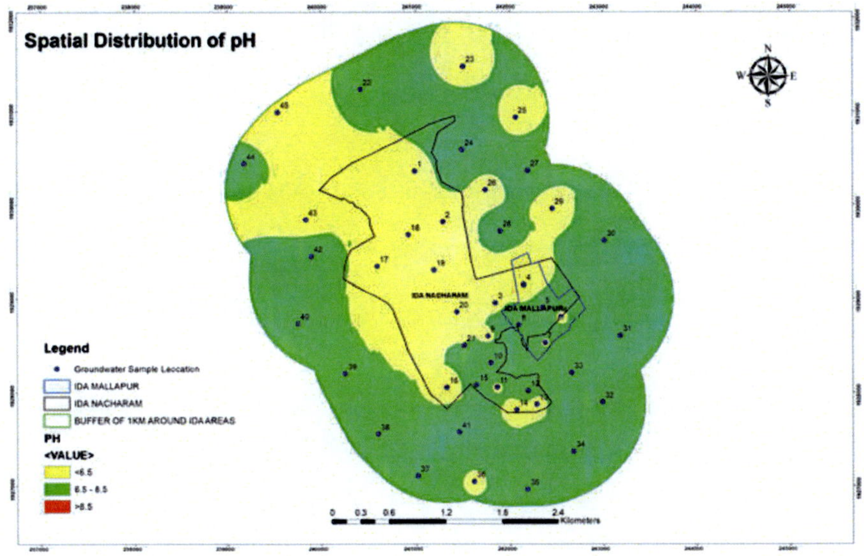

Fig. 6 Spatial distribution map of pH

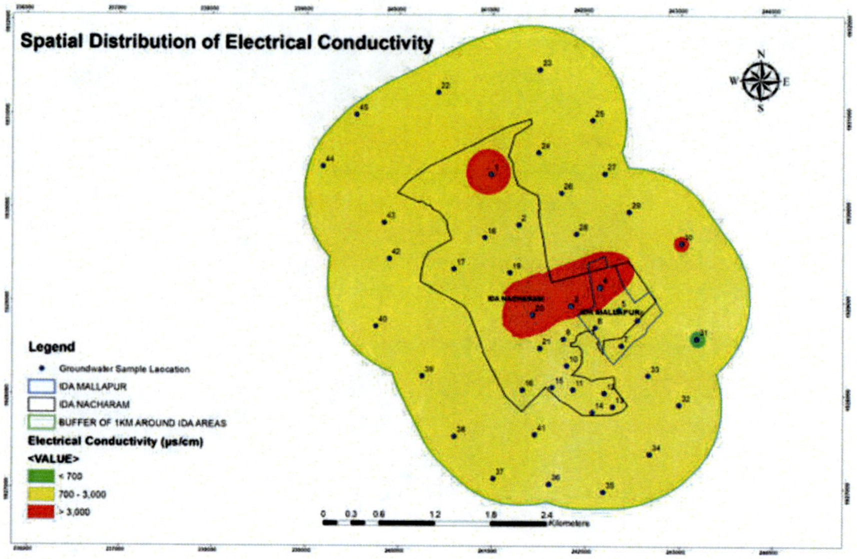

Fig.7 Spatial distribution map of EC

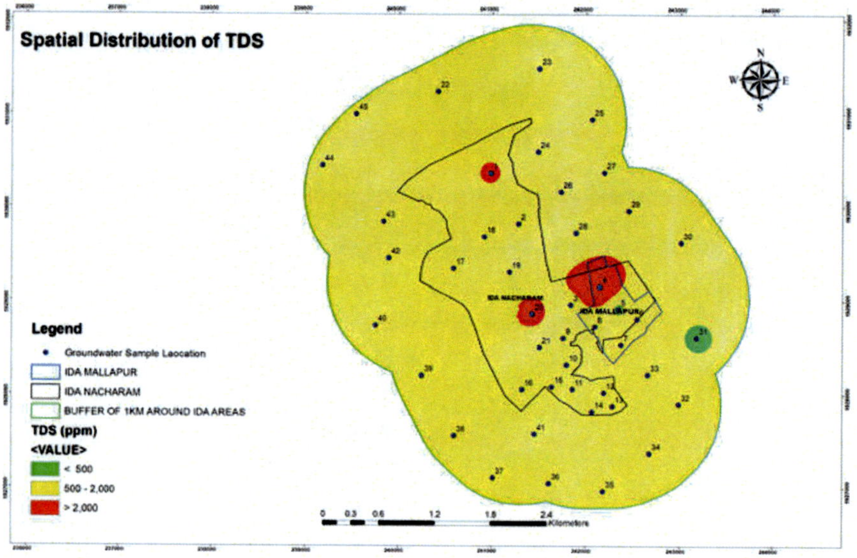

Fig. 8 Spatial distribution map of TDS

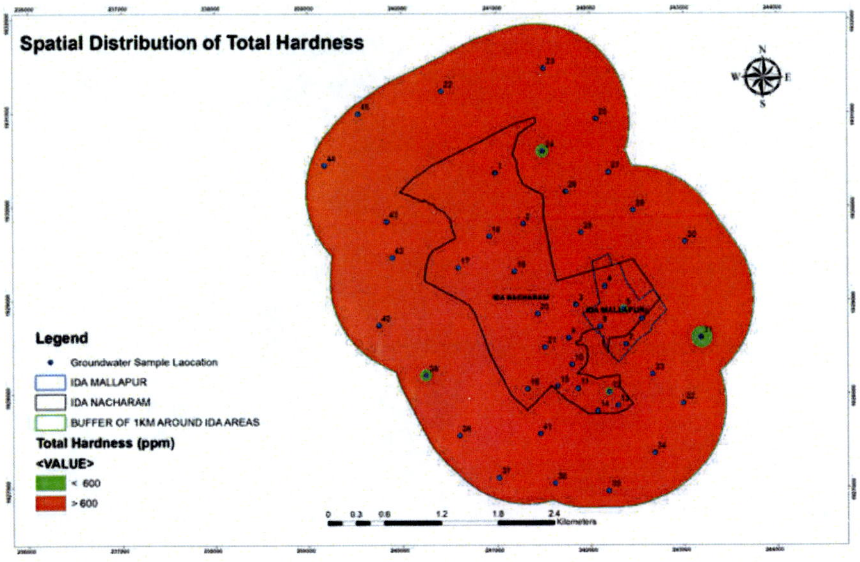

Fig. 9 Spatial distribution map of total hardness

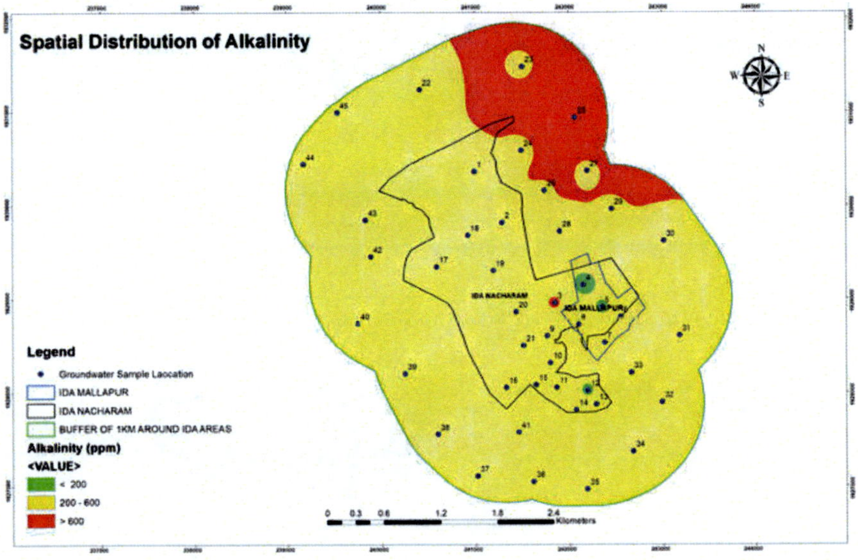

Fig. 10 Spatial distribution map of alkalinity

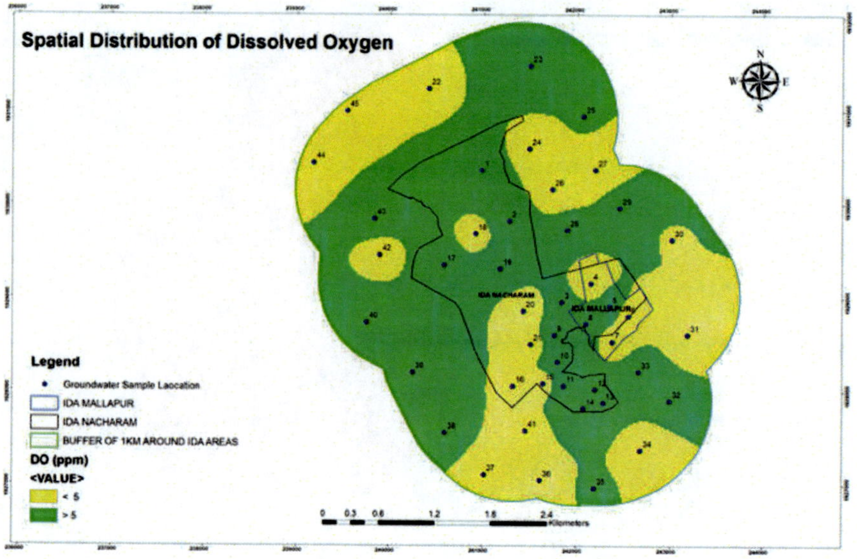

Fig. 11 Spatial distribution map of DO

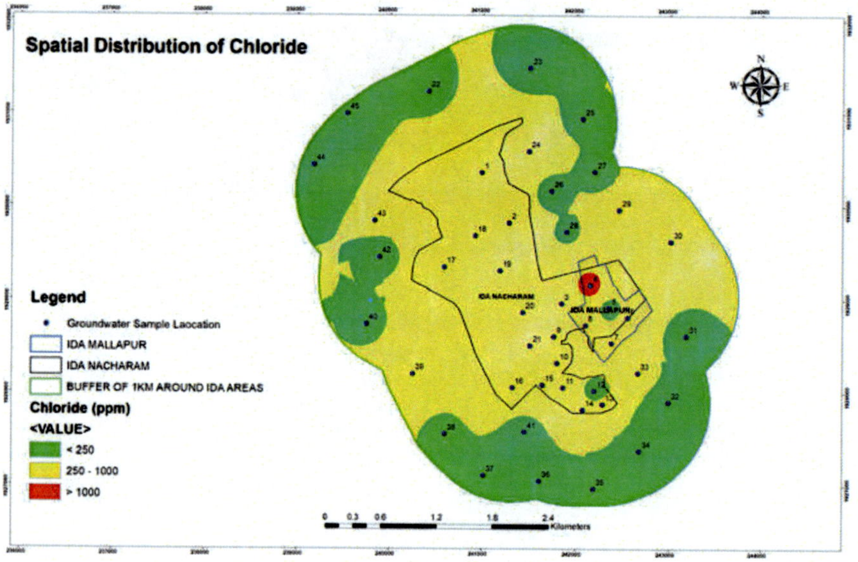

Fig. 12 Spatial distribution map of chloride

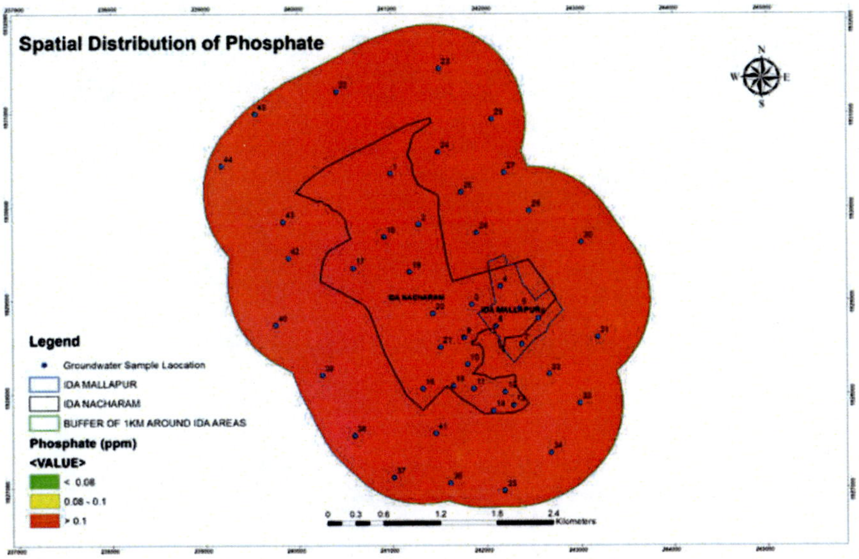

Fig. 13 Spatial distribution map of phosphate

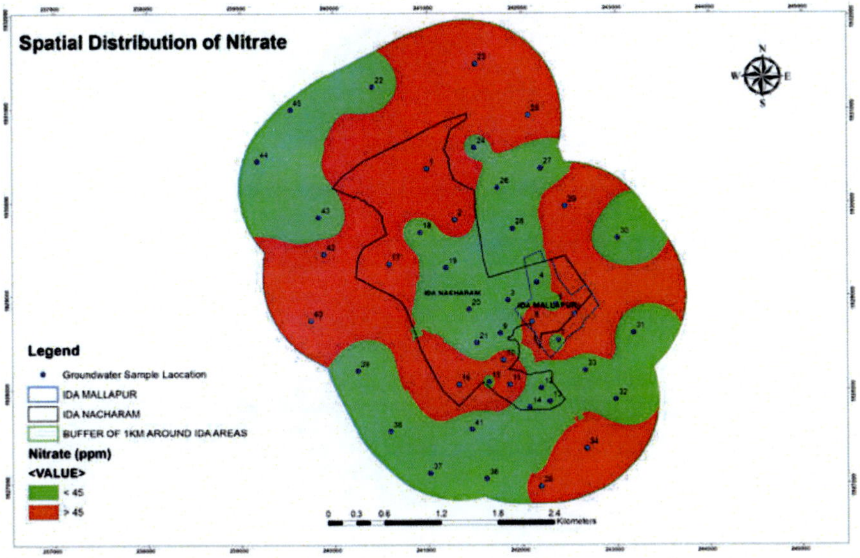

Fig. 14 Spatial distribution map of nitrate

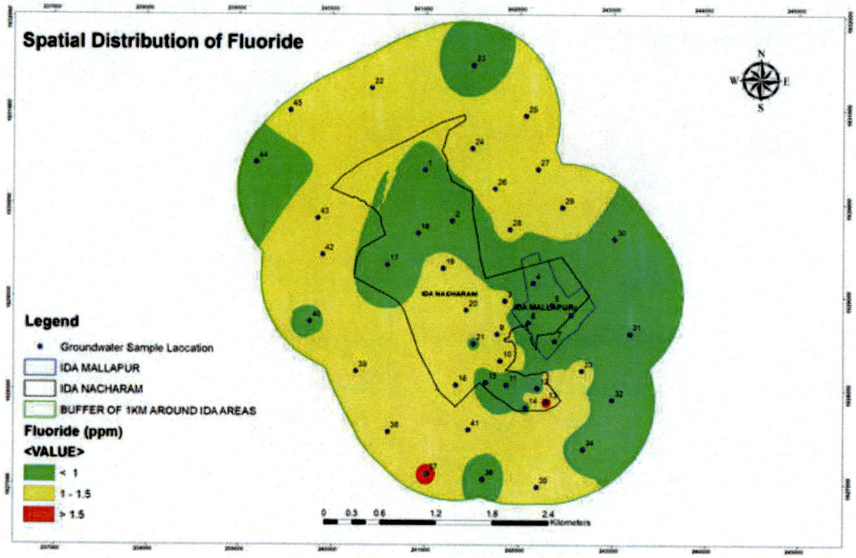

Fig. 15 Spatial distribution map of fluoride

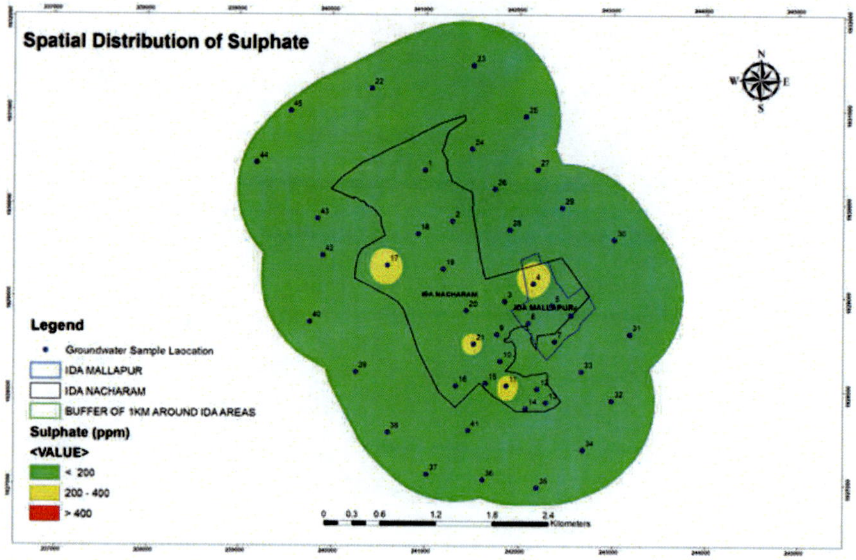

Fig.16 Spatial distribution map of sulphate

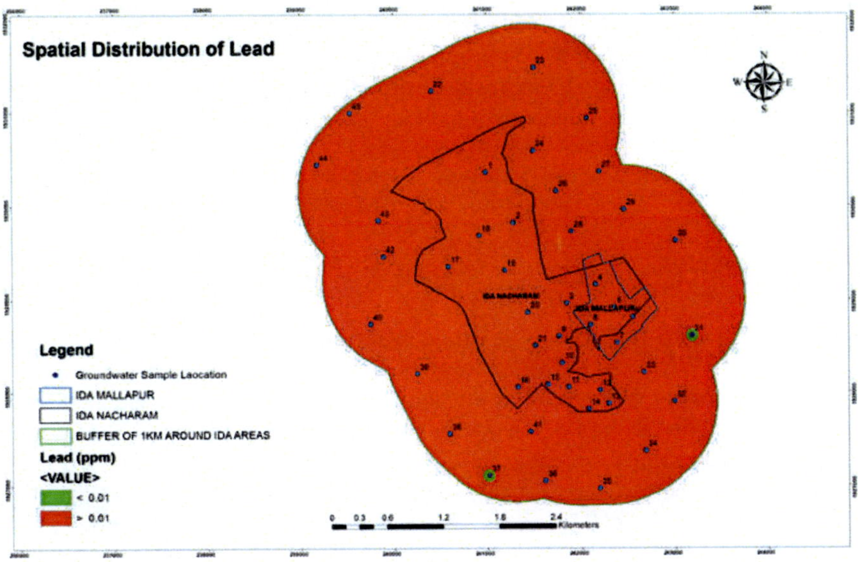

Fig.17 Spatial distribution map of lead

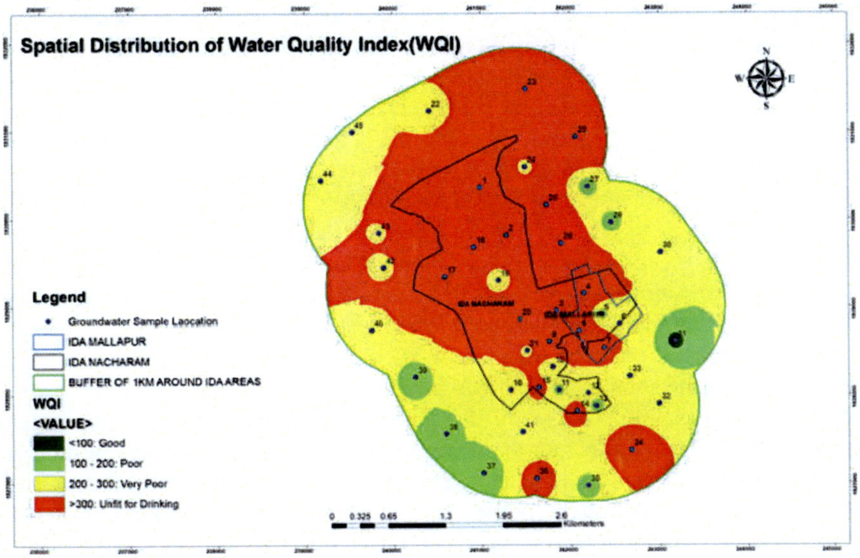

Fig.18 Spatial distribution of Water Quality Index and status of groundwater quality in and around Nacharam and Mallapur Industrial Development Areas

WQI map was created showing the spatial variability of groundwater quality Status of the Nacharam and Mallapur Industrial Development Areas.

The groundwater quality index spatial distribution assessment around 1 km buffer area and in study area showed that groundwater quality is mostly very poor and unfit for drinking.

5 Conclusions

The current study is carried out for Groundwater Quality Assessment Using Water Quality Index (WQI) and GIS Technologies in Nacharam and Mallapur Industrial Development Areas. GIS and WQI methods were applied in this research to visualize the spatial distribution of the quality of groundwater in the study area. A total of 45 bore well samples are collected in the study area and physicochemical analysis was performed for 14 parameters such as Colour, pH, Conductivity, Turbidity, Total dissolved solids, Total hardness, Alkalinity, Dissolved oxygen, Chlorides, Phosphates, Nitrates, Sulphates, Fluorides and Lead. The physicochemical parameters analysed were compared to the limits of BIS 10500 (2012). It is inferred from the findings that all the 14 parameters analysed are not within the permissible drinking water standards. The concentrations of TDS, Total Hardness, Alkalinity, Chlorides, Phosphates, Nitrates, Fluorides, Sulphates, and Lead are very high. This may lead to potential health hazards in and around the study area. Water Quality Index values are very high in and surrounding areas of the study area range from 81.6 to 671.57. According to the groundwater quality status, Classification of Nacharam and Mallapur Industrial Development Areas based on the WQI sample categories as follows 1—Good water quality status. The other categories include 9—poor, 17—very poor, and 18—Unfit for drinking and no sample is there in Excellent category. The analysis indicated that most of the groundwater in and surrounding buffer of 1 km of the study area is Unfit for drinking, domestic purposes and cannot be used for either for irrigation or industrial Application. The groundwater is highly contaminated in the study area and suitable treatment is required prior to use. This research shows that Water Quality Index and GIS methods combined can provide useful insights for the assessment of groundwater quality. This study illustrated that industrial effluents and human impacts are adversely affecting the groundwater of the study area and require an effective plan to manage further degradation and contamination of the groundwater. The conclusions drawn in this study made it possible to bring awareness of the crisis of degraded quality of groundwater in the study area among the citizens, municipal authorities and the government. The state and central Governments should take necessary actions to control the degradation of groundwater quality in the study area through frequent monitoring of groundwater quality and indorsing appropriate laws and implementing them strictly.

References

1. APHA (1998) Standard methods for the examination of water and wastewater, 20th edn. American Public Health Association, Washington D. C, USA (1998) https://www.mwa.co.th/download/file_upload/SMWW_1000-3000.pdf
2. BIS (IS: 10500) (2012) Bureau of Indian standards specification for drinking water (2012). http://cgwb.gov.in/Documents/WQ-standards.pdf
3. Latha S et al (2010) Assessment and spatial distribution of quality of groundwater in Zone II and III, Greater Visakhapatnam, India using water quality index (WQI) and GIS. Int J Environ Sci 1(2):198–212
4. Ketata M, Gueddari M, Bouhlila R (2011) Use of geographical information system and water quality index to assess groundwater quality in El Khairat deep aquifer (Enfidha, Central East Tunisia). Arab J Geosci
5. Vinod J et al (2013) Assessment of water quality index of industrial area surface water samples. Int J ChemTech Res 5(1):278–283
6. Venkatesan G et al (2013) Water quality assessment in Tiruchirappalli India. Asian J Water Environ Pollution 10(4):33–42
7. Sridhar N et al (2014) Spatial analysis of groundwater quality for Tarangabadi taluk, Nagappatinam district, Tamilnadu using GIS. Int J Emerg Technol Adv Eng 4(9):251–258
8. Tiwari AK et al (2014) GIS-based evaluation of water quality index of ground water resources in West Bokaro Coalfield, India. Curr World Environ 9(3). https://doi.org/10.12944/CWE.9.3.35
9. Sharma P et al (2016) Groundwater quality assessment using water quality index and GIS in rajasthan, India. Int J Adv Remote Sens GIS and Geogr 4(2):12–26
10. Sudha RP et al (2016) Assessment of groundwater quality using WQI method around Vellalore Municipal Solidwaste Disposal Site in Coimbatore, Tamilnadu, India. Int J Chem Sci 14(1):229–243
11. Bodrud-Doza Md et al (2016) Characterization of groundwater quality using water evaluation indices, multivariate statistics and geostatistics in central Bangladesh. Water Sci
12. Muralitharan J et al (2018) GIS based water quality index method for groundwater quality assessment—using hydro geochemical data: Karur District, Tamil Nadu, India. IJSRST 4(5):833–842
13. Chourasia LP et al (2018) Assessment of groundwater quality using water quality index in and around Korba City, Chhattisgarh, India. Am J Software Eng Appl 7(1):15–21
14. Khan A, Qureshi FR (2018) Groundwater quality assessment through water quality index (WQI) in New Karachi Town, Karachi, Pakistan. Asian J Water Environ Pollution
15. Verma P, Singh PK, Sinha RR et al (2020) Assessment of groundwater quality status by using water quality index (WQI) and geographic information system (GIS) approaches: a case study of the Bokaro district, India. Appl Water Sci 10:27. https://doi.org/10.1007/s13201-019-1088-4

A Case Study on Estimation and Composition of Construction and Demolition Waste in Bengaluru

M. Abhishek and Ashwin M. Joshi

Abstract Quantification of construction and demolition waste (CDW) is vital for building up an effective waste management system as it helps in developing practical approaches, establishing waste collection centres, arranging labour and transport facilities. The present work focuses on the estimation of CDW generated during the time of demolition, construction, renovation of small-scale residential structures and identifying the disposal site for dumping of CDW. Data regarding total CDW disposed at the dumpsite and information related to the challenges faced by demolition contractors were collected. Based on the collected data, correlation analysis, regression analysis, forecasting and energy calculations were performed. The total CDW generated at the generation sites, dumpsites and their Waste Generation Index (WGI) was calculated. About 84% of CDW was contributed by demolition works with WGI of 57.55 kg/m^2, where brick masonry and concrete are the major contributors of CDW with 31% and 21%, respectively. There was not much variation between the observed and predicted CDW generation in case of construction works, but small variation was witnessed in demolition works which was negligible. On forecasting the CDW to be collected at a dumpsite (Rampura), it was estimated to collect about 51,680 tonnes for a year. The maximum energy consumption in transporting CDW from site to dump yard was contributed by dump trucks compared to tractors. The average distance between the CDW generation site and the disposal dumpsite per trip was about 21 km.

Keywords Construction and demolition waste · Dumping · Estimation · Waste generation index · Forecasting

1 Introduction

Construction and demolition wastes (CDW) are produced during any construction, repair and/or demolition work on structures like flyovers, subways, renovation works,

M. Abhishek · A. M. Joshi (✉)
Infrastructure Construction and Management, RASTA—Centre for Road Technology, Bengaluru, Karnataka, India

L. Nandagiri et al. (eds.), *Sustainability Trends and Challenges in Civil Engineering*, Lecture Notes in Civil Engineering 162, https://doi.org/10.1007/978-981-16-2826-9_66

etc. It generally comprises of inert and non-biodegradable materials, for example, concrete, mortar, metal, wood, blocks, and so forth. These wastes consist of generally substantial mass and volume, and thus, consume space. Due to the unavailability of disposal sites and stringent disposal mechanisms, they are either disposed near roads, empty and barren sites, water bodies, etc. [1] as shown in Figs. 1 and 2.

It was not exceptional to notice massive loads of CDW dumped across streets particularly near large-scale construction and demolition sites, causing traffic jam and interruption. In India, the CDW generated is generally poorly managed via illegal disposal in open space, roadsides and water bodies due to unavailability of the recycling plant for processing [2].

Fig. 1 CDW dumped across road sides

Fig. 2 Construction and demolition waste disposed in an abandoned site

It is assessed that the CDW generation in India was around 10–12 million tonnes of waste yearly [1]. Nearly 55,000 million m^3 of aggregates are required for the housing sectors and an extra 750 million m^3 aggregates for road sector projects. Reusing of materials from CDW possibly will lessen the burden of supply of natural aggregates for these sectors [3].

Materials, viz. blocks, wood, metal and tiles can be reused. Concrete and masonry waste along with excavated soil, representing over 50% of the CDW and the most portion of which remain un-utilised in India [4, 5], whereas countries like the United Kingdom, the United States of America, France, Denmark, Germany and Japan are already well equipped with the process of reusing waste materials [6]. According to an estimation conducted by Technology Information, Forecasting and Assessment Council (TIFAC) in 2000, 70% of the construction industry is not aware of recycling measures [3]. A reliable estimate of CDW generation is not available currently in India [7].

There is enormous potential to utilise the CDW and other marginal materials like slag in construction materials. Researchers have made efforts to utilise these materials in the production of construction materials including masonry blocks, viz. stabilised adobe blocks [5, 6], concrete blocks [5], stabilised mud-concrete blocks [8], controlled low strength materials for various applications [9] and also utilised in self-compacting concrete [10]. However, there is enough scope to further pursue the interest in collecting data on CDW and thereby adopt best practices to dispose of these inert materials efficiently by causing minimal impact on ecology and environment.

The current study focuses on the significance of reusing CDW, creates awareness about the issue of managing such wastes and estimates the quantity of CDW generation in localities of Bengaluru.

1.1 Generation of CDW

Thc construction industry has witnessed tremendous development over the past decade, and the pace of increase of CDW generation was critical. In general, there are two hotspots for generating CDW materials, to be specific, mass or bulk generators and retail or small generators. The sectors which are the main sources for CDW generation are (1) infrastructure development sector and (2) real estate sectors. Construction and repairing of streets, spans, flyovers, etc. are grouped under the infrastructure development sector. The land part of the real estate sector comprises lodging, industrial and commercial building development, destruction of unapproved structures and so on. Small business ventures and individual house construction groups are considered as retail or small generators. The sources of CDW in a project are shown in Fig. 3.

It has been recognised that appropriate evaluation of CDW is vital for building up an effective management system. Quantification refers to the value of CDW generation in a specific project. It can help the project managers to alter the material purchase plan, organizing the stockpiling on-site and to decide the potential

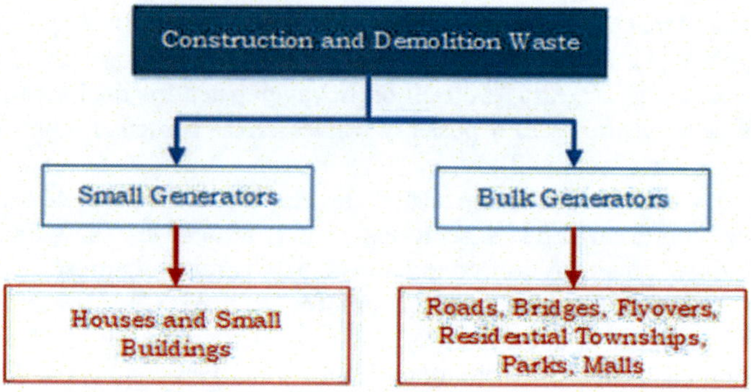

Fig. 3 Sources of CDW generators

Table 1 Major and minor components of CDW

Major components	Minor components
Brick masonry	Conduits
Concrete	Wood
Metal	Pipes
Bitumen	Electrical fixtures
Excavated soil	Panels

waste recycling advantage and dumping cost. The data of waste generation can help construction professionals in adapting a suitable methodology, facilitate establishing waste collection centres, arranging labour and transport facilities [11].

The types of CDW that are generated during the construction, demolition, and/or renovation of different types of structures are indicated in Table 1.

1.2 Objective and Scope of the Study

The scope of the current study was limited to collecting information related to the number of dump sites located in and around Bengaluru and also estimate the total CDW collected at the dump sites per day. Secure information regarding the waste generated at construction sites and demolition sites. Conduct data analysis using appropriate data analysis tools.

2 Methodology

Questionnaires were framed based on literature survey and the requirements of the current study. It has been circulated among different sections like demolition contractors, construction and demolition sites, etc. The different basic data collection methods adopted were site visits, face-to-face interviews, telephonic interviews, and online platforms such as WhatsApp messenger and emails.

The following data were collated,

- The location and number of CDW disposal sites in Bengaluru (marking their location through appropriate tools such as Google Maps / Google Earth).
- Identify the coordinates of the location marked; information related to the number of tractors/trucks visiting the disposal sites for dumping.
- Identify the per vehicle load capacity, estimate the waste collected/disposed of at the dump yard per day.
- Information about the demolition and construction site such as number of storeys, Gross Floor Area (GFA), identify the location of the waste generation site and estimate the quantity of waste generated per square metre.
- Information such as waste generated per work, type of transport used, practices followed during demolition, awareness of CDW disposal, etc.

The methodology adapted is indicated in Fig. 4. The images of dump sites is shown in Fig. 5..

Using appropriate data analysis tools, the above information was analysed through the following:

- **Correlation analysis**: Correlation test was performed to uncover the effect between the factors/variables. Higher the correlation, the higher is the effect and vice-versa. This analysis is done between those factors which are responsible for the generation of CDW. In statistics, the correlation strength of variables can be

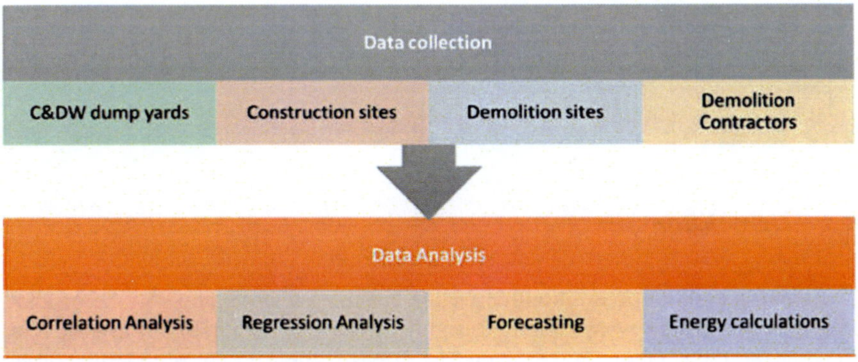

Fig. 4 Methodology followed during research

Fig. 5 Disposing of construction and demolition wastes in dump site and GPS images of Rampura and Kogilu

estimated using correlation coefficient (r). The following scale can be considered. If $r = \pm 0.5$, it is large, $\pm$ 0.3 is medium, and $\pm$ 0.1 is small.

- **Regression analysis**: Post the correlation test, regression analysis was carried out to those factors which show correlation value beyond the above-mentioned scale. It helps to predict the actual value and compare it with the obtained/ observed values.
- **Forecasting**: The data obtained using the above data collection methods is then forecasted and estimated for a set of days.
- **Energy calculations**: The energy consumed during the process of transporting the CDW from the demolition site to the dump yards was also calculated.

3 Case Study

3.1 Demolition Site

The identification of demolition/renovation sites was challenging. A total of about nine demolition sites were identified, and field visits were carried out to each identified site to collect the required data. The different types of CDW identified at the site and their nomenclature are shown in Table 2.

Table 2 Nomenclature for the material composition

M	Masonry
W	Wood
BT	Broken tiles
S	Steel waste
C	Concrete waste
ES	Excavated soil

Table 3 Demolition waste generated per work and Waste Generation Index

Site No.	Area, m^2 (GFA)	Total CDW generated per work, tonnes	WGI, kg/m^2	Types of CDW collected
1	672	25	37.2	M, W, S
2	929.03	30	73.39	M, BT
3	743.22	30	53.82	M, BT
4	557.42	35	37.67	S, M
5	734.31	30	40.36	C, W
6	445.93	45	61.28	M, C
7	408.77	75	134.55	M, BT, W
8	557.42	30	67.28	S, M, W
9	– *	180	–	M, W, C, ES

*Still undergoing demolition

The total CDW generated per work in tonnes and the Waste Generation Index in kg/m^2 were calculated as shown in Table 3. The Waste Generation Index is calculated using Eq. (1),

$$\text{Waste Generation Index} = \frac{\text{Total C\&D waste generated per work in kg}}{\text{Gross Floor Area}} \tag{1}$$

Tables 4, 5 and 6 show the correlation analysis between the factors that are involved in calculating the total CDW generated during demolition and construction. The *R*-values (correlation coefficient) shown in the table depict the correlation strength between the given factors.

3.2 *Construction Site*

Similar to identifying demolition sites, it was also difficult to identify suitable construction sites. A target of five construction sites were set, and field visits were carried out to each site and the required data was collected.

Table 4 Correlation analysis (demolition waste)

	GFA	Total trips per work	Per trip capacity	Total CDW generated per work	WGI
GFA	1				
Total trips per work	0.70914	1			
Per trip capacity	−0.5851	−0.8321	1		
Total CDW generated per work	−0.0243	0.0456	0.4932	1	
Waste generation index	−0.4927	−0.2835	0.70003	0.87549	1

Table 5 Construction waste generated per work and waste generation index

Site No.	Area, m^2 (GFA)	Total CDW generated per work, tonnes	WGI, kg/m^2	Types of CDW collected
1	5574	3	1.08	C. ES
2	468.23	2	4.27	C, M, ES
3	278.7	2	7.18	M, ES
4	222.96	4	8.97	C, M
5	499.16	2	6.01	ES

The Waste Generation Index was calculated using the same formula as mentioned in the previous section. The result and the waste materials identified at the sites are also shown in Table 5.

Table 6 Correlation analysis (construction waste)

	GFA	Total trips per work	Per trip capacity	Total CDW generated per work	WGI
GFA	1				
Total trips per work	−0.2843	1			
Per trip capacity	0.99871	−0.25	1		
Total C&D waste generated per work	0.21508	0.875	0.25	1	
WGI	-0.8464	0.64478	-0.8223	0.23364	1

3.3 CDW Dump Yard

Bengaluru had about eight C&D waste disposal sites in 2016. Out of which, only three were active, i.e. Mallasandra, Anjanapura, Srinivaspura and Kogilu [12]. However, interaction with Bruhat Bengaluru Mahanagara Palike (BBMP), indicated that the CDW dumping have been stopped at these sites and used only for dumping other solid wastes (temporarily). It was also witnessed during site visits that there were a number of private lands which were used for disposing CDW and one among them was identified and visited at Rampura. Required data was collected this site for four days.

(1) *Transportation energy*

Transportation energy is defined as *"the energy required for transporting CDW from the demolition or construction site to the dump yard or disposal site"*. Various steps involved in computing the transportation energy were,

- Estimating the per load capacity of the vehicle used to transport.
- Number of trips done for the transportation of demolition waste to the dumpsite (Table 7).
- Based on the site visits, for the sites undergoing/undergone demolition and the designated CDW disposal sites, the average transportation distance was found to be about 21 km per round trip, and it was considered for the calculation of transportation energy. Tractors of capacity 5 tonnes and dumpers of capacity 15 tonnes were used for transporting the demolition waste to the dumpsite. The fuel efficiency of the vehicles used viz., the tractor and the dumper were fixed based on the information provided by operators at 6 kmpl and 3 kmpl respectively.
- Fuel consumed per trip and total fuel consumed for transporting the demolition waste to disposal site were calculated.
- Transportation energy was computed by multiplying the total volume of fuel consumed with the energy density of diesel, i.e. 38.7 MJ/litre [2] (Table 8). The details of per day CDW collected at Rampura site is presented in Table 9.

3.4 Demolition Contractors

As per the survey, the following observation was noted:

- Depending on the site conditions, approximate 15–40 tonnes of waste were generated per work.
- An average of eight trips were carried out to transport the CDW per site.
- Maximum distance covered between the loading and unloading point varies from 10 to 20 km.
- Tractors, 6 wheeler truck and 10 wheeler trucks were used. The type of vehicle used to collect the waste was dependent on the site conditions; the 10 wheeler truck was commonly used.

Table 7 Information collected at dump site

Location	Area, acres	Land ownership	Type of vehicle	Total trips per day	Per trip capacity, tonnes	Total CDW generated per day, tonnes	Used for	Remarks
Kogilu (co-ordinates = 13.1043270, 77.6457292)	15	Unauthorised site	Tractor	50	5	250	Land filling	Data collected was only collected for one day
			Dump truck	Nil	Nil	Nil		No dump trucks were allowed
Rampura (co-ordinates = 13.0488580, 77.6786429)	190	Private site	Tractor	96	5	478	Land filling	Average of 4 days data collected
			Dump truck	6	15	90		6 trips were done only once as per the data collected

Table 8 Summary of total transportation energy of seven building wastes transported till the dump yard

Description	Notation	tractor				Dump truck		
		B1	B2	B3	B4	B5	B6	B7
Total energy consumed for transportation of demolition debris to CDW dump yard	TE	1354.5	1896.3	1354.5	1625.4	2515.5	1006.2	6037.2
TE in GJ		1.35	1.90	1.35	1.63	2.52	1.01	6.04

- 28.6% of respondents claimed that they segregate the waste and 57.1% claimed that they segregated the waste at the demolition site. Due to the lack of equipment

Table 9 Per day CDW collected for four days at Rampura

Days	Trips per day (tractor)	CDW per day in tonnes	Date of visit
1	100	500	29/2/2020
2	90	450	28/2/2020
3	95	475	27/2/2020
4	97	485	26/2/2020

for segregation, the only waste material that was separated was steel. Post demolition, the broken steel reinforcements and other metals were separated from the debris and sent to the scrap buyers.

- 42.9% of respondents indicated that they do not record the waste generated during demolition and 71.4% said they do not have any previous year record of waste collected. It was because, during the time of demolition, the data was not recorded and maintained.
- Only 28.6% of the demolition contractors claimed that they have received training on demolition waste management. Hence, few demolition contracts follow a waste management strategy.
- Most of the demolition contractors were aware of the utilization of CDW in concrete and other construction material, but only 28.6% were aware of the methodology of processing and reusing.
- Land clearance, machine maintenance, etc. were amongst the major challenges faced.

4 Result

4.1 Total CDW Generated Per Work and Waste Generation Index

Table 10 provides the contribution of various activities contributing to CDW generation. Demolition contributes nearly 84% of the total CDW generated from the data collected as indicated in Fig. 6. Construction and renovation activities contribute 10% and 6%, respectively. The average WGI for demolition activities was found to

Table 10 Total waste generated per work and waste generation index

Type	Number of sites	Total waste generated per work in tonnes	Average WGI, kg/m^2
Demolition	9	480	57.55
Construction	4	150	5.72
Renovation	2	4	2.99

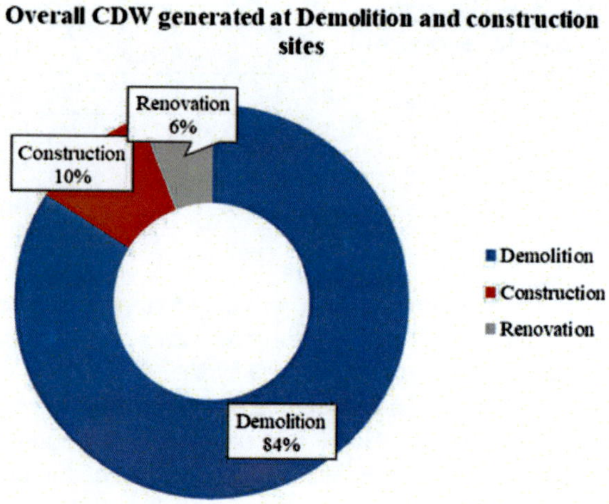

Fig. 6 Overall CDW generated at demolition and construction sites

be 57.55 kg/m^2. And the average WGI for construction and renovation activities are 5.72 kg/m^2 and 2.99 kg/m^2, respectively.

Figure 7 shows a mere 8% of the waste is recycled and the material sent for recycling was reinforcing steel, window bars, and steel gates because of its scrap value. The extracted steel was sent to the scrap vendors, and the remaining CDW materials were directly disposed of at the dumpsites.

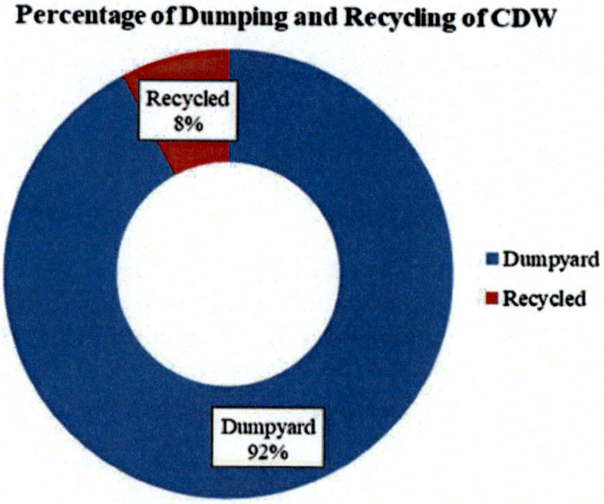

Fig. 7 Percentage of dumping and recycling of CDW

4.2 Demolition Works

Table 11 and Fig. 8 show the observed and predicted values of demolition waste generated per work. The predicted values are generated by performing regression analysis between total CDW generated per work and total trips per work because the correlation coefficient value between these two factors is more than the mentioned scale.

Table 11 Observed and predicted demolition waste generated per work

Observed demolition waste generated per work, tonnes	Predicted demolition waste generated per work, tonnes	Total trips per work
25	36.84	5
35	35.79	7
30	36.32	6
30	36.32	6
45	37.89	3
30	38.42	2
30	38.42	2
75	36.84	5
30	33.16	12

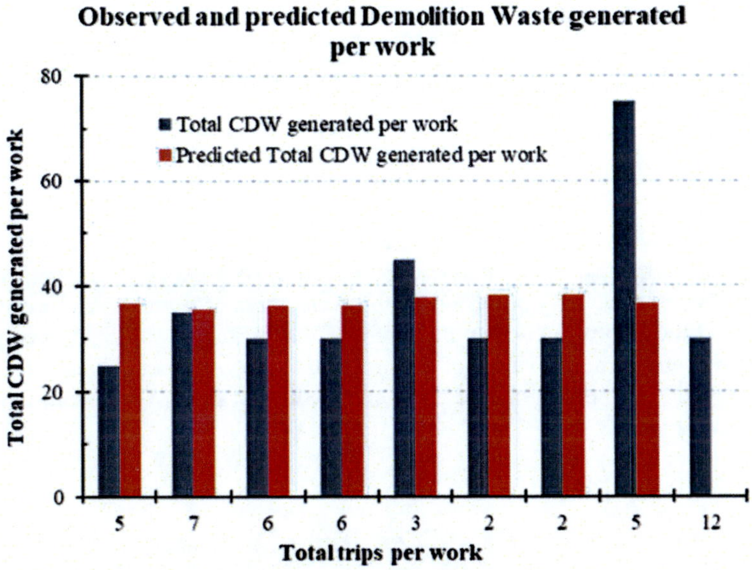

Fig. 8 Observed and predicted demolition waste generated per work

Table 12 Observed and predicted construction waste generation per work

Observed construction waste generated per work, tonnes	Predicted construction waste generated per work, tonnes	Total trips per work
3	2.25	1
2	2.25	1
2	2.25	1
4	4.00	2
2	2.25	1

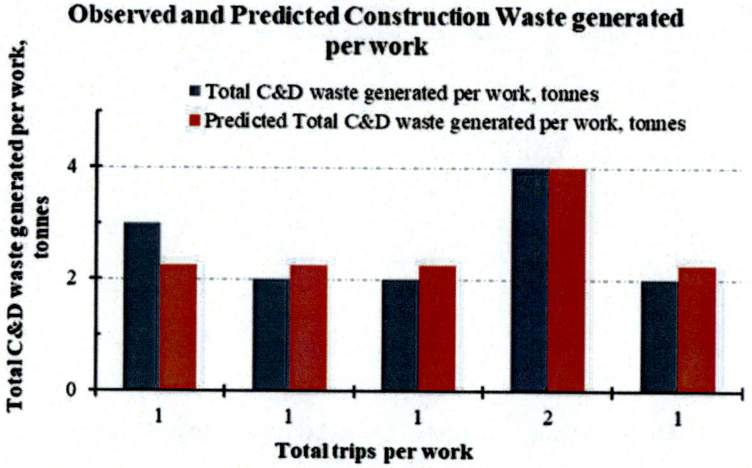

Fig. 9 Observed and predicted construction waste generated per work

4.3 Construction Work

Similar to demolition waste data, the observed and predicted values were generated by carrying out the regression analysis for the total CDW generated per work and the total number of trips per work to verify the correctness of observed and estimated values, and the results are shown in Table 12. The estimated values showed good correlation with the actual values of construction waste which was generated.

Figure 9, shows the bar graph of actual and predicted values of construction waste generated per work.

4.4 Construction and Demolition Waste Composition

Figure 10, indicates the average composition of CDW (%) observed in various construction and demolition sites. Demolished brick masonry constituted about 31%

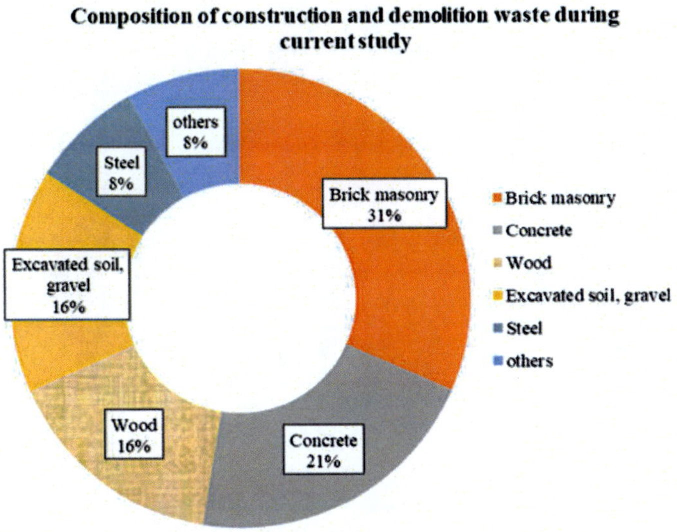

Fig. 10 Composition of construction and demolition waste during current study

of the overall composition followed by concrete debris and excavated soil. Joshi et al [4–6] and Sequeria et al [9] also indicated a similar trend in the composition of CDW.

Figure 11 shows the comparison of the typical composition of CDW obtained from different studies and the present study. Figure 12 shows the typical composition of

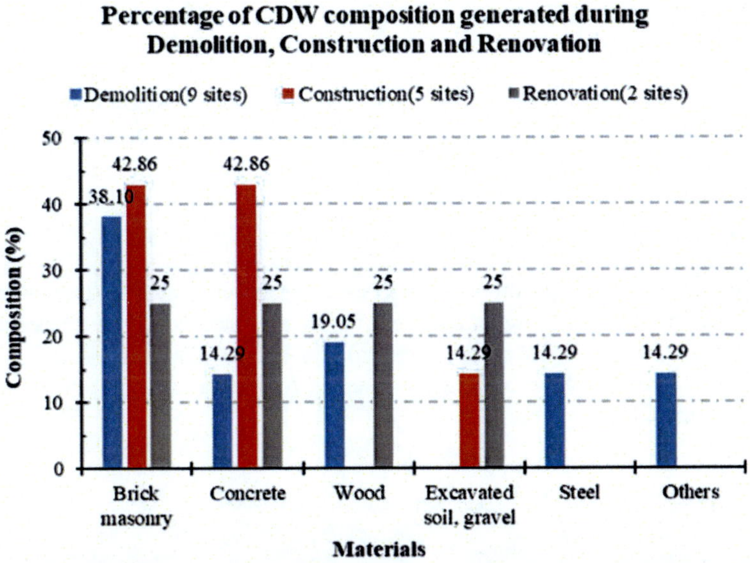

Fig. 11 Percentage of CDW composition generated during demolition, construction and renovation

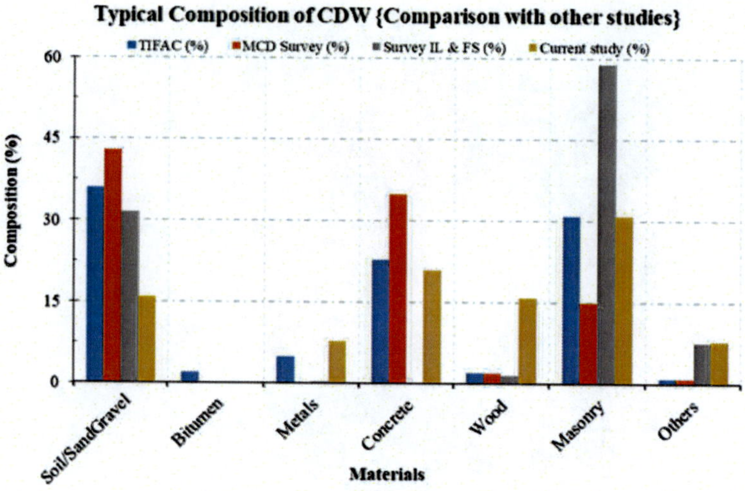

Fig. 12 Typical composition of CDW comparison with other studies

CDW as indicated by various studies along with the values observed in the current study. The soil/sand and gravel proportions are high as compared to the current study and the percentage of masonry is not available with other surveys.

4.5 CDW Disposal Site/Dump Site

(1) Data forecasting

Figure 13 shows the data forecasting on CDW collected at Rampura for a duration of up to 320 days. The value stands at about 9230 tonnes for 20 days and 51,680 tonnes for about 320 days, which can be considered as the approximate volume of waste disposed over a year after deducting for days with lean supply.

(2) Energy consumption

Figure 14, indicates that amongst the seven cases, B6 consumes the highest energy of about 6.04 GJ due to a higher number of trips as compared to other cases. When the number of trips increases, the total volume of fuel consumed also increases. Figure 15, shows the share of energy consumed by tractor and dump truck used to dispose of the CDW.

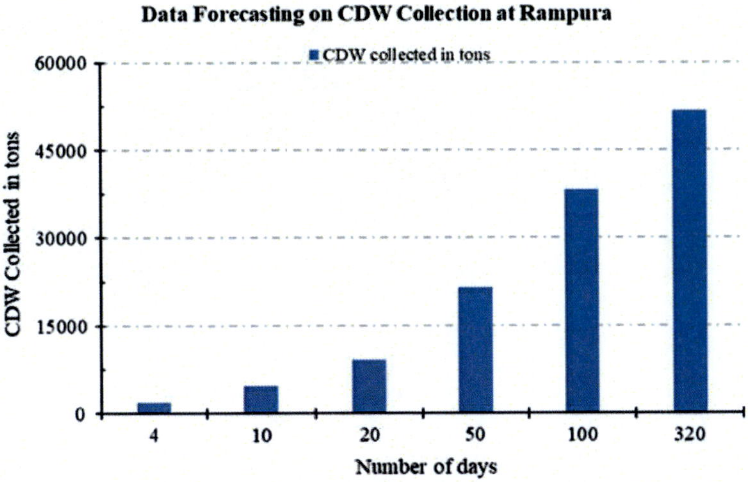

Fig. 13 Data forecasting on CDW collected at Rampura for 10, 20, 50, 100 and 320 days

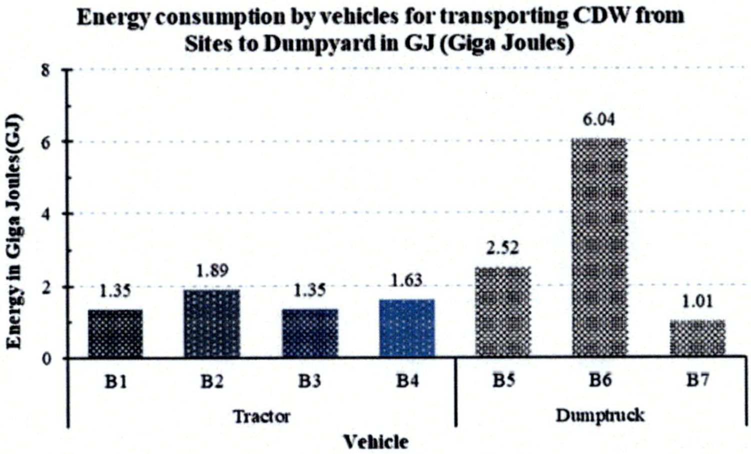

Fig. 14 Energy consumption by vehicles for transporting CDW from seven building sites to dump yard in GJ (Giga Joules)

4.6 Demolition Contractor

From the results of the regression analysis, the graph is plotted between the total trips and maximum distance and showing the comparison between observed and predicted total trips per work as shown in Fig. 16. The variation between one of actual and predicted values were attributed to the error in data secured from the demolition contractor. All other values indicate less deviation.

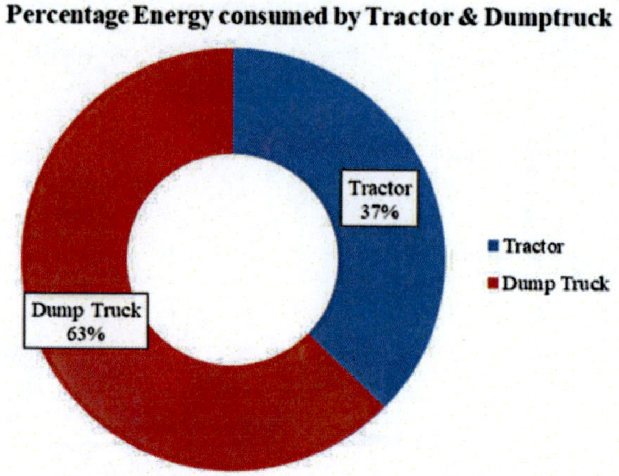

Fig. 15 Percentage energy consumed by tractor and dump truck

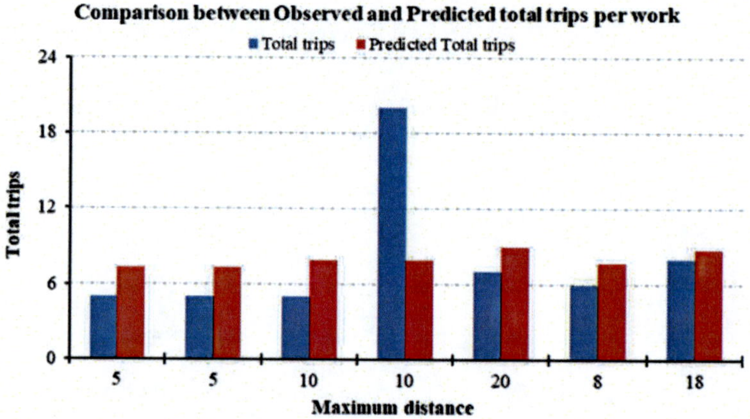

Fig. 16 Comparison between observed and predicted total trips per work

5 Conclusion

From the results obtained, conclusions are summerised as follows,

- The absence of precise knowledge of CDW has led to ineffective policies and poor management in handling CDW across many Indian cities. Thus, there is a need to adopt an effective methodology to estimate the CDW generation. The current study is one small step in this direction.
- The average WGI for demolition, construction and renovation debris was found to be 57.55 kg/m^2, 5.72 kg/m^2 and 2.99 kg/m^2, respectively.
- As per the survey of demolition contractors, demolition debris of about 15–40 tonnes were generated per site based on size of the structure.

- CDW disposal sites need to be designated by the government agencies, as it is difficult to locate CDW collection centre in Bengaluru. With the absence of a designated location, most of the CDW was disposed of near the vicinity of roads, water bodies, abandoned sites, etc. As per the results obtained from forecasting data at Rampura dumpsite alone, the CDW debris was expected to reach 51,680 tonnes/year.
- Masonry debris, concrete debris and excavated natural soil consisted of a major proportion of waste collected at dumpsite. The trend was similar in other major cities of India.
- In the absence of an established market for consumption and recycling facilities, these materials were dumped in landfills or in abandoned sites. There is an immediate requirement to set up CDW crushing or recycling facilities in every zones of Bengaluru and thereby reduce the impact on the environment to promote sustainability.
- The energy consumed in transporting CDW from the demolition site to the disposal site was substantial and was led by dump trucks. Effective recycling facilities in designated locations of Bengaluru may also help in the reduction of energy and fuel consumption for CDW disposal.
- Currently barring steel, wood and similar materials with residual scrap value, other materials, viz. masonry and excavated soil debris remains unused or recycled. This can be a valuable resource which can be effectively utilized in the preparation of construction materials, viz. masonry blocks, mortars, low strength concrete and controlled low strength materials [4–6, 8–10].

Acknowledgements Authors would like to express sincere gratitude and regards to the parent institution, RASTA-Centre for Road Technology in bringing out this work. Authors would also acknowledge and offer sincere regards to Prof. K S Jagadish, Retired Professor of Civil Engineering, Indian Institute of Science (IISc), Bengaluru for motivation and valuable guidance in bringing out this work.

References

1. Ministry of Urban Development (2016) Manual on municipal solid waste management. [Online]. Available: http://cpheeo.gov.in/upload/uploadfiles/files/Part2.pdf
2. Rakesh S, Keshava M (2019) A study on Embodied energy of recycled aggregates obtained from processed demolition waste. In: National conference on recent trends in architecture & civil engineering towards energy efficient and sustainable development, Tiruchirapalli
3. ICI–Guidelines (2015) Recycling, use and management of C&D WASTES. Indian Concrete Institute, Chennai
4. Joshi AM, Basutkar SM, Keshava M, Raghunath S, Jagadish K (2018) Performance of Masonry units prepared using construction and demolition waste as fine aggregates. In: 10th International Masonry conference (IMC), Milan. https://doi.org/10.13140/RG.2.2.22838.55364
5. Joshi AM, Basutkar SM, Ahmed MI, Keshava M, Raghunath S, Jagadish K (2019) Performance of stabilized adobe blocks prepared using construction and demolition waste. J Build Pathol Rehabil 4(13). https://doi.org/10.1007/s41024-019-0052-x

6. Joshi AM, Raghunath S, Keshava M, Jagadish K (2018) Stabilized adobe using demolished brick masonry waste. In: National seminar on emerging building materials and construction technologies, New Delhi. https://doi.org/10.13140/RG.2.2.24096.84481
7. Ram VG, Kalidindi SN (2017) Estimation of construction and demolition waste using waste generation rates in Chennai, India. Waste Manage Res 35(6):1–8
8. Joshi AM, Basutkar SM, Jagadish K (2020) Stabilized mud concrete for sustainable construction. In: Sustainable materials in building construction, vol 11. Springer, Cham. https://doi.org/10.1007/978-3-030-46800-2_6
9. Sequeira VL, Joshi AM, Kerekoppa MD, Bharadwaj N (2020) Physical and microstructural properties of construction and demolition waste based masonry units. In: 3rd international conference on innovative technologies for clean and sustainable development (ITCSD 2020), RILEM Bookseries, vol 29. Springer, Cham. https://doi.org/10.1007/978-3-030-51485-3_27
10. Joshi AM, Nagaraj A, Shamanna G, Bharadwaj N (2020) Experimental study on developing SCC using IS 10262: 2019 by utilizing C&D waste for sustainable construction. In: 3rd international conference on innovative technologies for clean and sustainable development (ITCSD-2020), Chandigarh
11. Wu Z (2014) Quantifying construction and demolition waste: an analytical review. Waste Manage 1–9. https://doi.org/10.1016/j.wasman.2014.05.010
12. Vunnam V, Ali MS, Singh A, Asundi J (2016) Construction and demolition waste (CDW) utilization for recycled products in Bengaluru: challenged and prospects. Center for Study of Science, Technology & Policy (CSTEP), Bengaluru. https://doi.org/10.13140/RG.2.2.10058.77765
13. Ponnada MR, Kameswari P (2015) Construction and demolition waste management—a review. Int J Adv Sci Technol 19–46. https://doi.org/10.14257/ijast.2015.84.03
14. Kolaventi SS, Tezeswi TP, Kumar MVNS (2018) A modelling approach to construction waste management. In: ASCE India conference 2017. https://doi.org/10.1061/9780784482032.002
15. Jain S, Singhal S, Jain NK (2018) Construction and demolition waste in India: generation rate and implications of C&DW recycling. Int J Construct Manage 1–10. https://doi.org/10.1080/15623599.2018.1523300
16. Gagan, Arora S (2015) Recycled aggregates: a sustainable solution of construction and demolished waste. IOSR J Mech Civil Eng (IOSR-JMCE) 58–63. Available: https://www.iosrjournals.org/iosr-jmce/papers/AETM'15_CE/2/20-CE-137.pdf

Effect of Buckling Due to Wind Load on Analysis of Natural Draught Cooling Tower

C. L. Mahesh Kumar, K. G. Shwetha, B. C. Shanthappa, and K. Manjunatha

Abstract This paper presents the study of buckling effect on natural draught cooling towers. Natural draft cooling tower(NDCT) is the describing milestones of intensity stations and are utilized as heat exchangers in Thermal power plants. Two cooling towers i.e. Cooling Tower 1 (100.0 m), Cooling Tower 2 (200.0 m) with variation in elevation and thickness of shell are considered for the analysis. Using ANSYS Software, which is a FEA-based, the Eigen value buckling analysis is carried out for two cooling towers. Natural draught cooling towers have been modelled considering top end to be free and bottom end to be fixed. The study has been done in detail about the finite shell element. Here three various cases have been considered for the study of the effect of buckling on analysis of the natural draught cooling tower. Case 1—Height is kept constant and thickness of shell is varied from 400 to 800 mm. Case 2—Height is varied and thickness of shell is kept constant. Case 3—Both height and cooling tower thickness of shell is varied. The behavioural changes of considered cooling towers are examined using ANSYS, with Variation of its height and thickness for initial 6 buckling modes. Maximum Deformation, Maximum Principal Stresses are obtained and compared.

Keywords Cooling tower · Maximum principal stress · Deformation · Thickness · Height

C. L. M. Kumar (✉)
Department of Civil Engineering-VTU RRC, Jnana Sangama,VTU, Belagavi 590018, India
e-mail: Maheshkumar.cl@nmit.ac.in

C. L. M. Kumar · K. G. Shwetha
Department of Civil Engineering, Nitte Meenakshi Institute of Technology, Yelahanka, Bengaluru 560064, India
e-mail: Shwetha.kg@nmit.ac.in

B. C. Shanthappa
SJM Institute of Technology, NH-4 Bypass, PB No 73, Chitradurga 577501, India

K. Manjunatha
Department of Studies in Civil Engineering, U B D T College of Engineering, Hadadi Road, Davangere 577004, India

L. Nandagiri et al. (eds.), *Sustainability Trends and Challenges in Civil Engineering*, Lecture Notes in Civil Engineering 162, https://doi.org/10.1007/978-981-16-2826-9_67

1 Introduction

The most effective and economical way for the cooling of the thermal power plant by using very less water and by reducing the thermal pollution of the natural water resources, the natural draught hyperbolic cooling tower is the common measure for the same. By doing this they balance the investment and cost of operation and environmental factors and energy supply. This natural draught cooling tower contributes to the energy output and the careful balance of environment. Due to the good strength and stability the shape of the hyperbolic tower is considered [1]. Due to the high demand for the modern construction technology the design and the construction practices and technology towers with minimum thickness is preferred [2].

The present-day Natural draft cooling towers are exceptional structures in view of their sheer size and complexities. It takes a shot at the rule of temperature contrast between the air inside and outside the peak. Hyperbolic state of cooling tower is generally favoured because of its quality and soundness and bigger accessible territory at the base. Hyperbolic cooling towers are viably utilized for cooling enormous amounts of water in thermal power stations, treatment facilities, nuclear power plants, steel plants, cooling and other mechanical plants. Hyperbolic cooling tower also called as Natural draft cooling tower (NDCT) [1] is the describing milestones of intensity stations and are utilized as heat exchangers in Thermal power plants. They contribute both to a good yield and to a cautious offset to site condition. These shell structures are exposed to environmental loads, for example, Seismic and warm slopes that are stochastic in nature.

A thorough literature survey study was done on the Analytical work carried out on buckling analysis of cooling towers. In the analytical study conducted by Busch et al. [3], A Discussion is done on the requirement on the necessity of the extended durability of the tower, and the explanation of the durability design concept has been done in detailed, and illustration has been carried out for the same. Sabouri-Ghomi et al. [4] has discussed the importance of efficient and economical reinforced concrete cooling tower which has led the engineers to design the tall and thin structure with high slenderness aspect ratio. Stiffening rings has been added to increase the structural buckling stability, parameters such as number and dimension, and location of the stiffening rings has been considered while designing the cooling tower. By using all the parameters, the structural buckling stability has been increased of the cooling tower. Kulkarni and Kulkarni [5] has considered the different height and different thickness of the cooling tower for the study of the effect of wind on cooling tower, the cooling tower chosen for study was Bellary thermal power station. Staad pro was used for the analysis of the cooling tower where bottom end is fixed and top end is free [6]. Studies have been done on the variation of the maximum and minimum principal stresses and the variation of displacement vs thickness has also been considered for the study. The material properties of the cooling tower are density of RCC is 25 kN/m^3, Poisson ratio is 0.15 and young's modulus is 2.1 Mpa. Wind load has been applied in the form of the wind pressure according to the codes IS 875 part-3 (1987). Lang and Strau [7] has done the work on the comparison of the cooling towers and has considered various aspects in the comparison. The two natural draught cooling towers located in the city of lunen were used for comparison with both having different heights and different number of columns in the cooling tower. Comparison was done between the structural aspects of the cooling tower, like the design and analysis and the structural systems and constructive requirements. Construction stages of natural draught cooling tower like the erection of columns and lower ring beam were discussed. Design and analysis method have also been conceded for the study.

Due to the high demand for the natural draught cooling towers which are economic and efficient, this has led the engineers to design the towers which are tall, thin and high slenderness aspect ratio. By designing the economical and thin and tall cooling towers the cost, time and the materials can also be reduced [1].

2 Modelling

This paper gives the discussion on the effect of buckling due to wind load on analysis of the natural draught cooling tower. The prime reason for the study is to determine the changes in the stresses and the deformation which have been calculated from the eigen buckling analysis. The height and shell thickness of cooling tower is varied considered for the analysis. To study the effect of buckling due to the wind load on analysis of cooling tower, the eigen value buckling analysis is carried out for below mentioned three cases with varying different parameters [2].

The shell thickness is varied to 400, 600 and 800 mm.

- Case 1—cooling tower height is fixed and the thickness of the shell is changed from 400 to 600 mm.
- Case 2—cooling tower height is changed and shell thickness is fixed.
- Case 3—both cooling tower height and shell thickness of cooling tower is changed.

2.1 Selection of Finite Element Tool

ANSYS software is used for the finite shell elements convergence study and analysis to validate the shell element properly and accurately. By Using the linear static analysis of natural draught cooling tower, The Maximum principal stress, deformation are obtained from the analysis, the optimum mesh size and finite shell element were found by successful trials.

2.2 Introduction to ANSYS

Manually producing a great finite element component is a long-time consuming process with regarding money, the preparation of the huge quantity of the finite component can be challenge with regard to the computer output. ANSYS workbench is a graphically based software package, model is formulated throughout the software compared and imported to ANSYS workbench. ANSYS workbench is a computer-aided engineering tool that generates the finite element model (pre-processor) along with the connected results (post-processing). Using the united automatic technologies ANSYS workbench allows the design engineers to publish the change solid products involving computer parts along with the forecasts its behaviour with design optimization. The pre-processor software helps to predict the one for to edit model series and it made possible for import geometric model through the solid modelling [8].

2.3 Analysis of Buckling Behaviour

Linear buckling analysis (LBA) is typically preferred to find the critical buckling load acting on cooling tower structure. This basic type of analysis is considered for the obtaining of the results related to the eigen value problem and its related eigen value associated vectors. These analyses produces the set of the load factor and the buckling shape and mode, whereas the least value of these can be considered for the buckling failure.

2.4 Material Properties

The different Material properties used for modelling of natural draught cooling tower for buckling analysis are (Table 1).

Table 1 Material properties of cooling tower

Material properties	Values
Young's modulus of concrete	34,000 N/mm^2
Poisson's ratio	0.2
Compressive strength	35 N/mm^2
Tension Strength	2.67 N/mm^2
Young's modulus of steel	2.1×10^5 MN/m^2

2.5 Models of Cooling Tower

The representation models of cooling tower are mentioned here of various height and thickness of shell of the tower are shown here. 100 m cooling tower with 400 to 800 mm and 200 m cooling tower with 400 to 800 mm of shell thickness (Figs. 1 and 2).

The present cooling tower considered for the effect of wind pressure on the cooling tower, the wind pressure calculations are according to the IS 875 Part 3 [9].

Basic Design Wind (V_B) speed is 39 m/s, considering the Mangalore region. Considering the Category 1 class A (Exposed open terrain with no obstructions).

So, we have $V_Z = V_B * K1 * K2 * K3$
Where, $K1 = 1.06$, from page no 11 of IS 875 Part-3 (1987)
$K2 = 1.26$, from page no 8 of IS 875 Part-3 (1987)
$K3 = 1$, for tower under consideration
$V_Z = V_B * K1 * K2 * K3 = 39 * 1.06 * 1.26 * 1$
$V_Z = 52.0884$ m/s
P_Z = design wind pressure
$P_Z = 0.6\,(V_z)^2 \ldots \text{N/m}^2$
$P_Z = 0.6\,(52.0884)^2$

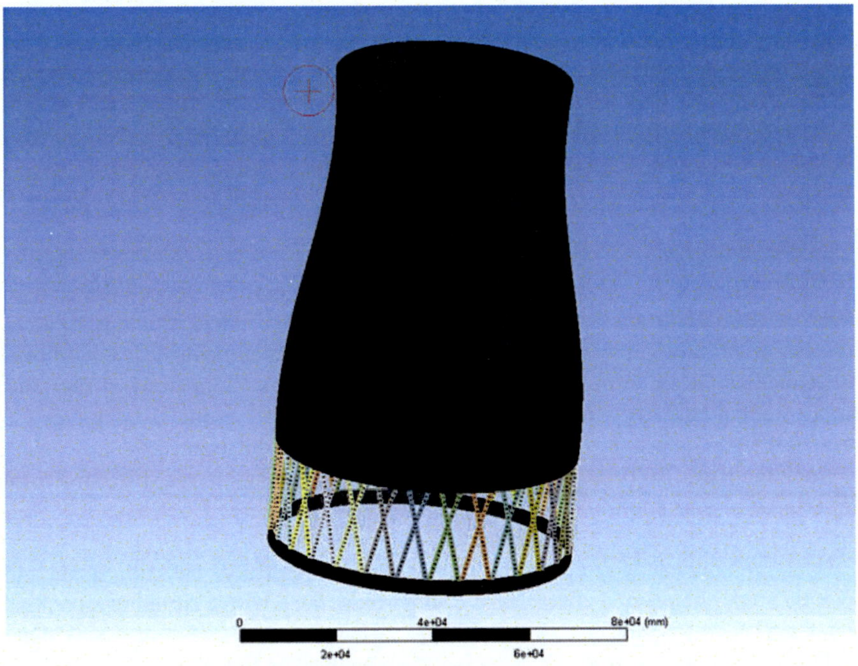

Fig. 1 Meshing details of 100 m cooling tower

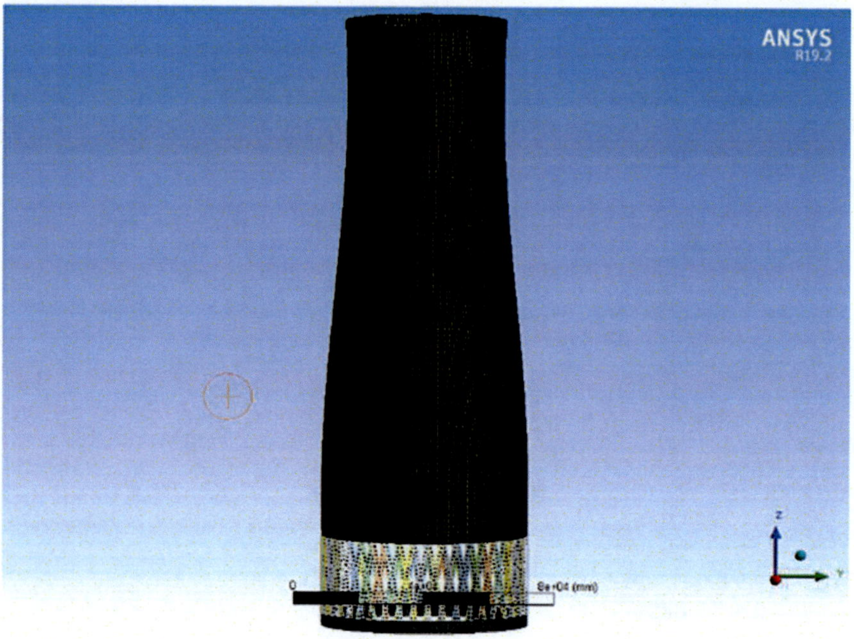

Fig. 2 Meshing details of 200 m cooling tower

$P_Z = 1627.9$ N/m^2.

Therefore, P_Z is 0.001628 N/mm^2 (Figs. 3 and 4).

3 Results and Conclusions

Eigen value buckling analysis gives the strength of the elastic ideal structure. It calculates the structural eigen values of the given system loading and boundary conditions. It is called as the Euler buckling analysis. Eigen value buckling analysis Explains the theoretical buckling strength of a Good ideal elastic structure. It calculates the structural eigen values for the given system loading and constraints [2].

Buckling loads for certain designs are available. However, in real-life, structural imperfections and nonlinearities prevent most live structures from reaching their eigen-value predicted buckling strength.

Buckling analysis is a technique used to determine critical buckling loads and buckled mode shapes (the characteristic shape associated with a structure's buckled response) [2].

The flow procedure of eigen-value buckling analysis is as follows.

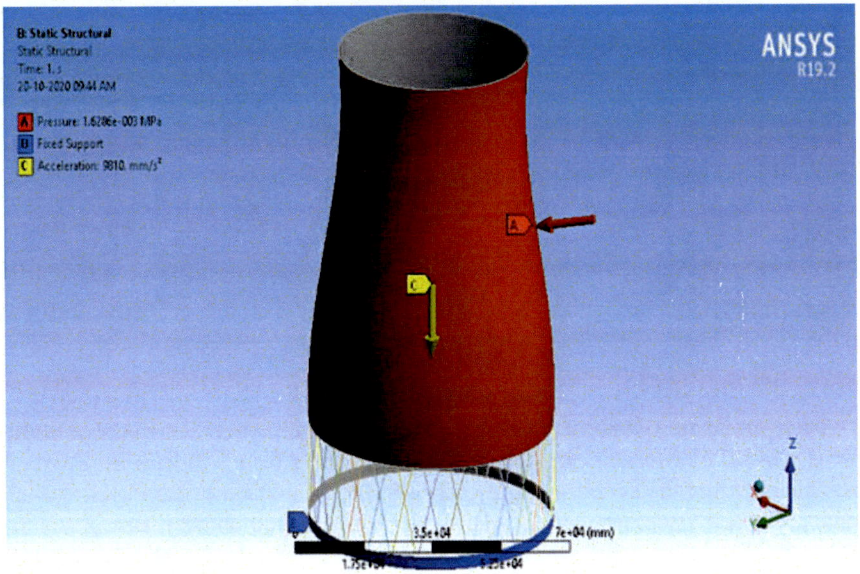

Fig. 3 Representation of the loading details of the cooling tower of 100 m height

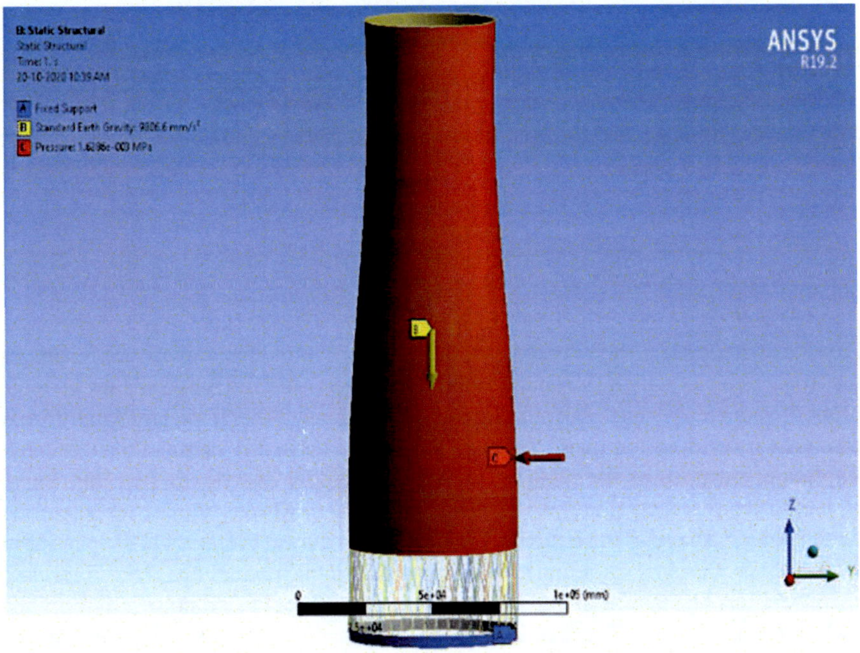

Fig. 4 Representation of the loading details of the cooling tower of 200 m height

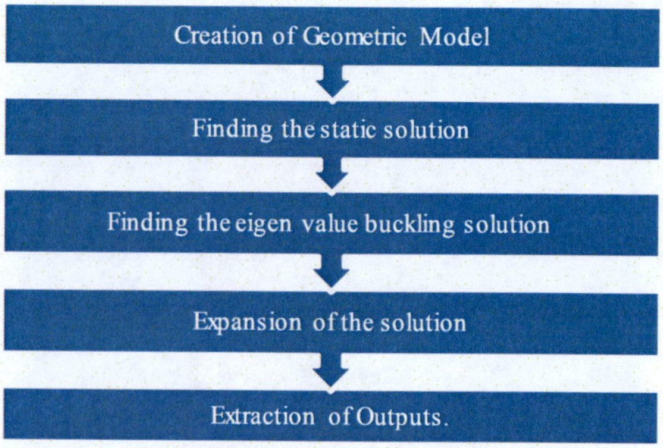

There are some techniques which are available in the ANSYS are Ansys Multiphysics, ANSYS Mechanical, ANSYS Structural and ANSYS functional programs for calculation of the buckling mode shape of a structure. Non-linear buckling analysis, and linear buckling analysis can be performed using above methods. As the above methods can yield different results. Here The analysis used for study is Eigen value buckling analysis. The concrete shell thickness is generally governed by self-weight and wind loads.

3.1 Static Structural Results

The total deformation and the maximum principal stress results have been discussed here (Figs. 5, 6, 7 and 8).

3.2 Modal Analysis

Physical structures always tend to vibrate, the modal analysis gives the results of the frequencies at which the natural vibration occurs, and the mode shapes give the properties of the system. The structures always resonate therefore any small forces acting on it result in the major deformation and can possibly damage the structure. The major purpose of the natural frequency or modal analysis is that the design of the structures such that the resonance is reduced. Modes shapes represent the dimensionless shape of the structure that undergoes during vibration at the particular frequency i.e. the corresponding natural frequency of the mode shape. The vibrational characteristics i.e. natural frequencies and mode shapes of the structure is determined by the modal analysis. This can lead to the continuation of the starting point of the another or more detailed dynamic analysis. For the dynamic loading condition, the natural frequencies and the mode shapes are important parameters in the design of the structures (Figs. 9, 10, 11 and 12, 13 and 14; Tables 2, 3 and 4).

Fig. 5 Representation of the total deformation details of the cooling tower of 100 m height and thickness of 400 mm

4 Conculsions

The following conclusions can be drawn from the work

- The total deformation in the static structural analysis increases gradually as the height and the thickness of the cooling tower increases. The values for the same have been observed increase in height 100 and 200 m and thickness is from 400 to 800 mm.
- The maximum principal stress in the static structural analysis remains approximately the same for the height 100 and 200 m. and the slight variation of the stresses in observed in the thickness variation in cooling tower.
- The total deformation in the eigen value buckling analysis has the slight variation from 1.1 to 1.8 mm considering all the cases with different height and the different thickness of shell including the 6 modes of all cases as shown in the deformation contour plots.
- The maximum principal stress in the eigen value buckling analysis has different values for the different height and the shell thickness, in height of 100 and 200 m cooling tower the stress observed is reducing as the thickness is increasing as shown in the stress contour plots.

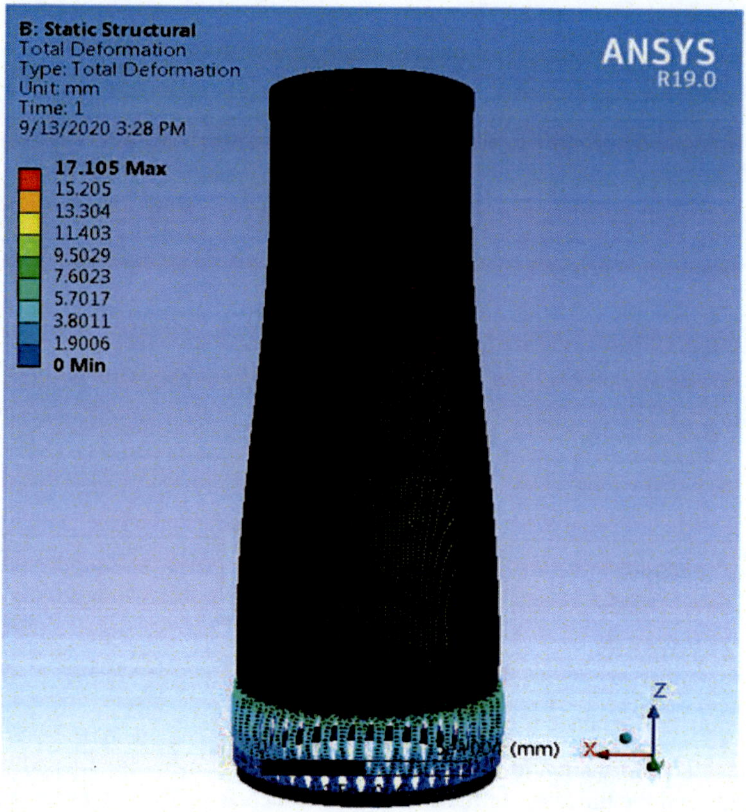

Fig. 6 Representation of the total deformation details of the cooling tower of 200 m height and thickness of 400 mm

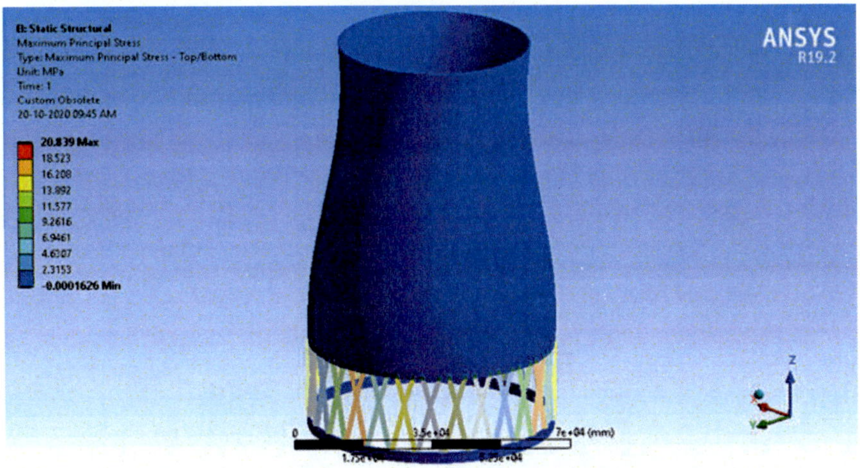

Fig. 7 Representation of the maximum principal stress details of the cooling tower of 100 m height and thickness of 400 mm

Fig. 8 Representation of the maximum principal stress details of the cooling tower of 200 m height and thickness of 400 mm

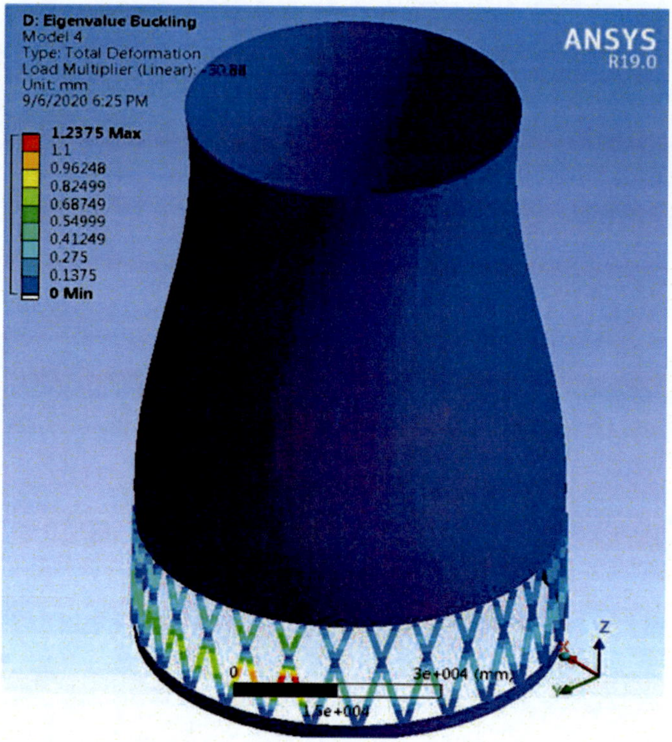

Fig. 9 Representation of the eigen value buckling of total deformation details of cooling tower of 100 m height and thickness of 800 mm

Fig. 10 Representation of the eigen value buckling of total deformation details of cooling tower of 200 m height and thickness of 800 mm

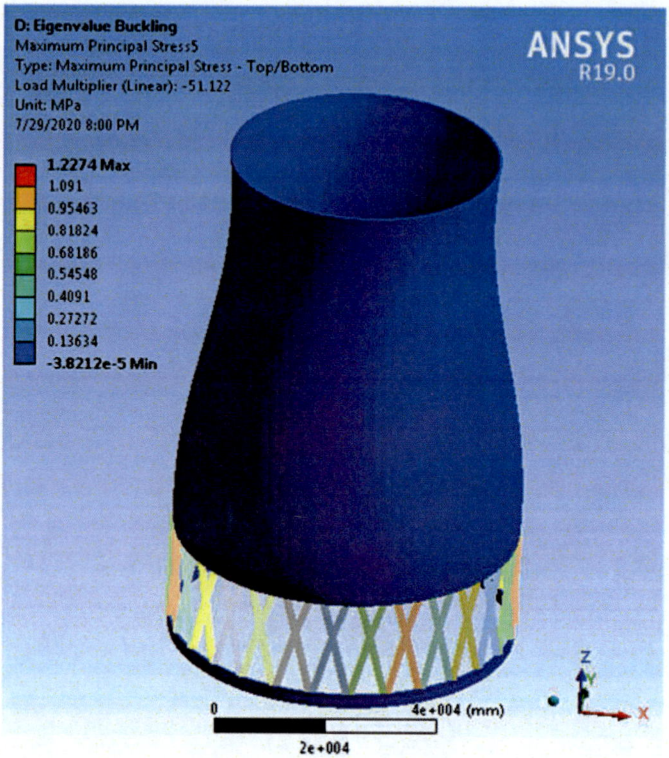

Fig. 11 Representation of the eigen value buckling of maximum principal stress of cooling tower of 100 m height and thickness of 400 mm

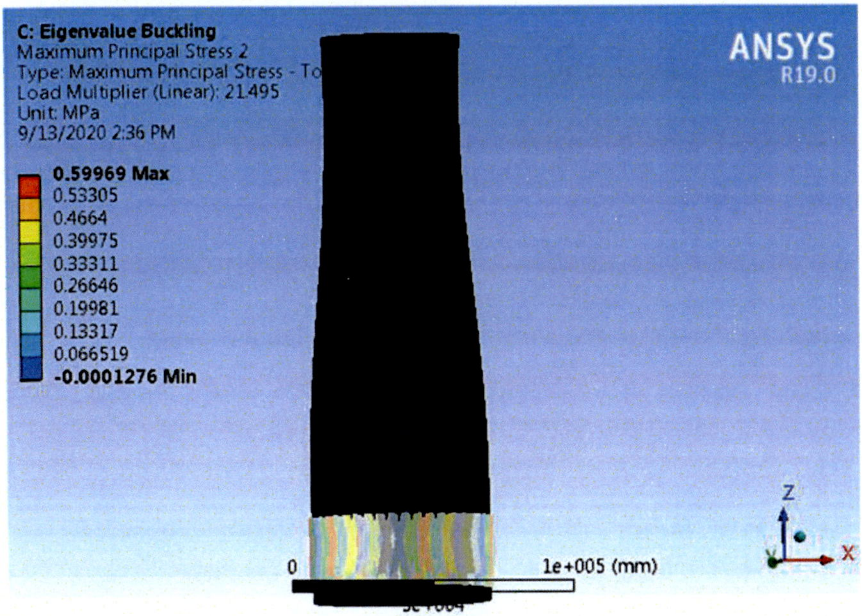

Fig. 12 Representation of the eigen value buckling of maximum principal stress of cooling tower of 200 m height and thickness of 400 mm

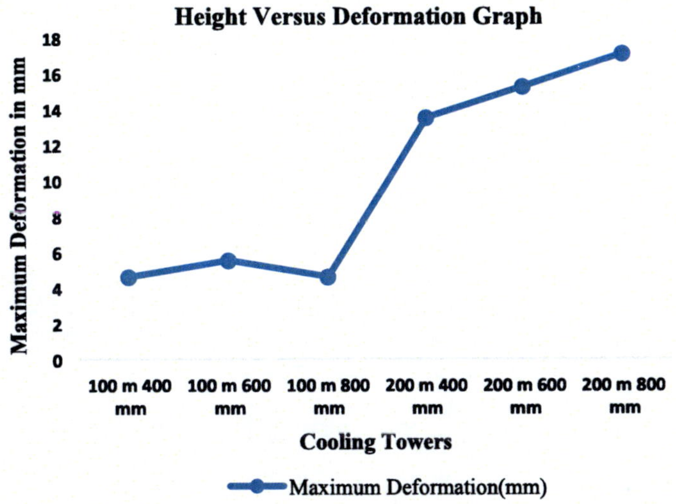

Fig. 13 Graph of cooling tower height versus maximum deformation values

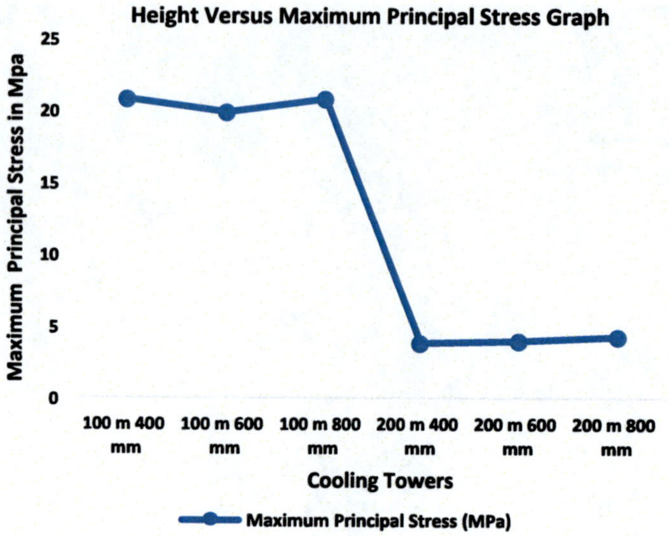

Fig. 14 Graph of cooling tower height versus maximum principal stress values

Table 2 Maximum deformation and maximum principal stress values for 100 m height 400 mm thickness cooling tower

Buckling modes	Maximum deformation (mm)	Maximum principal stress (MPa)
Mode 1	1.1297	1.0555
Mode 2	1.1377	0.71947
Mode 3	1.0087	0.71947
Mode 4	1.0216	0.71947
Mode 5	1.0614	0.71947
Mode 6	1.0212	0.71947

Table 3 Maximum deformation and maximum principal stress values for 200 m height 400 mm thickness cooling tower

Buckling modes	Maximum deformation (mm)	Maximum principal stress (MPa)
Mode 1	1.1541	0.14785
Mode 2	1.1458	0.17705
Mode 3	1.0208	0.16944
Mode 4	1.0885	0.097674
Mode 5	1.3358	0.20127
Mode 6	1.1292	0.12068

Table 4 Maximum deformation and maximum principal stress values for 100 m height and 200 m height cooling tower

Cooling towers	Maximum deformation (mm)	Maximum principal stress (MPa)
100 m 400 mm	4.6489	20.839
100 m 600 mm	5.6156	19.937
100 m 800 mm	4.6489	20.839
200 m 400 mm	13.521	3.9327
200 m 600 mm	15.306	4.0476
200 m 800 mm	17.105	4.2996

References

1. Mahesh Kumar CL, Shanthappa BC, Manjunatha K (2021) A study on dynamic response of hyperbolic cooling tower for fixed base condition. In: Sitharam TG, Pallepati RR, Kolathayar S (eds) Seismic design and performance. Lecture notes in civil engineering, vol 120, pp 269–275, Springer, Singapore. ISBN 978-981-334-4004-6. https://doi.org/10.1007/978-981-33-4005-3_22
2. Kulkarni S, Kulkarni AV (2016) Effect of buckling on analysis of hyperbolic cooling towers. Int J Eng Res Technol (IJERT). ISSN: 2278-0181
3. Busch D, Harte R, Kratzig WB, Montag U (2002) New natural draft cooling tower of 200 m of height. Eng Struct 24(12):1509–1521
4. Sabouri-Ghomi S, Kharrazi MHK, Asghari A, Javidan P (2006) Effect of stiffening rings on buckling stability of RC hyperbolic cooling towers. Thin Walled Struct 44:152–158
5. Kulkarni P, Kulkarni SK (2015) Wind effect on hyperbolic RCC cooling tower. Int J Curr Eng Technol 5(6):3513–3517. E-ISSN 2277-4106
6. Jacob M, Mohanan A (2017) Study the effect of wind load and dead load on RC hyperbolic cooling tower by the provision of stiffeners. Int J Innov Sci Eng Technol 4(4)
7. Lang C, Strau J (2018) Natural draft cooling tower design and construction in Germany—past (since 1965), present and future
8. ANSYS mechanical APDL structural analysis guide & ANSYS reference guide
9. Goyal A, Shankar AN, Sethy SK (2017) Parametric analysis of hyperbolic cooling tower under seismic loads, wind loads and dead load through Staad. Pro. Int J Eng Res Sci 3(8). ISSN: 2395-6992